B. J. Merkel • B. Planer-Friedrich • C. Wolkersdorfer (Eds.)

Uranium in the Aquatic Environment

W0107048

Springer-Verlag Berlin Heidelberg GmbH

B. J. Merkel
B. Planer-Friedrich
C. Wolkersdorfer
(Editors)

Uranium in the Aquatic Environment

Proceedings of the
International Conference Uranium Mining
and Hydrogeology III and the
International Mine Water Association Symposium
Freiberg, Germany, 15-21 September 2002

With 453 Figures, 151 Tables and a CD-ROM

 Springer

Prof. Dr. Broder J. Merkel
Dipl-Geol. Britta Planer-Friedrich
Dr. Christian Wolkersdorfer

Technical University Bergakademie Freiberg
Department of Geology
Gustav-Zeuner-Str. 12
09596 Freiberg, Germany

http://www.geo.tu-freiberg.de/umh/

Cataloging-in-Publication Data Applied for

Die Deutsche Bibliothek - CIP-Einheitsaufnahme
Uranium in the aquatic environment: proceedings of the International Conference
Uranium Mining and Hydrogeology III and the International Mine Water Association
Symposium Freiberg, Germany, 15 - 21 September 2002; with CD-ROM / ed.: Broder Merkel
.... - Berlin; Heidelberg; New York; Barcelona; Hongkong; London; Mailand; Paris; Tokio:
Springer 2002 ISBN 978-3-642-62877-1   ISBN 978-3-642-55668-5 (eBook)
            DOI 10.1007/978-3-642-55668-5

This work is subject to copyright. All rights are reserved, whether the whole or part of the
material is concerned, specifically of translation, reprinting, reuse of illustrations, recitati-
on, broadcasting, reproduction on microfilm or in any other way, and storage in data
banks. Duplication of this publication or parts thereof is permitted only under the provisions of the German Copyright Law of September 9, 1965, in its current version, and permission for use must always be obtained from Springer-Verlag. Violations are liable for
prosecution under the German Copyright Law.

http://www.springer.de

© Springer-Verlag Berlin Heidelberg 2002
Originally published by Springer-Verlag Berlin Heidelberg New York in 2002
Softcover reprint of the hardcover 1st edition 2002
The use of general descriptive names, registered names, trademarks, etc. in this publicati-
on does not imply, even in the absence of a specific statement, that such names are exempt
from relevant protective laws and regulations and therefore free for general use.

Production: PRO EDIT GmbH, 69126 Heidelberg, Germany
Cover Design: Erich Kirchner, Heidelberg, Germany
Typesetting: Camera-ready by the editors

Printed on acid-free paper   SPIN: 10866911   30/3130So-5 4 3 2 1 0 -

Contents

Preface

Uranium is a radioactive element and a heavy metal which is naturally occurring in ground and surface water. Although uranium is enriched in granites and gneiss ground water from these host rocks often shows low to intermediate uranium concentrations, while some ground waters from sandstone and carbonate aquifers show elevated uranium concentrations up to several hundred mg/l without man made impact. On the other side, surface water contains increased anthropogenic uranium concentrations due to the intensive use of phosphate fertilizers and in mining areas due to mining and milling activities. Saxony and Thuringia both being states of the reunified Germany are probably an area where uranium mining activities have impacted the environment more severely than in any other part of the world. Thus, the federal government of Germany allocated huge amounts of money for the rehabilitation work, a unique proceeding without precedent in mining history.

In October 1995 the first international conference on Uranium Mining and Hydrogeology (UMM I) was held in Freiberg being organized by the Department of Geology at the technical University Freiberg by the support of the Saxon State Ministry of Geology and Environment. Due to the large scientific interest in the topic of uranium a second conference (UMH II) took place in Freiberg in September 1998. Then, in 2002 scientists working on the topic of uranium mining and hydrogeology from all over the world gathered at Freiberg to attend the third consecutive conference (UMH III) which was jointly held together with the International Mine Water Association (IMWA) Symposium 2002. The conference addressed scientists and engineers in the field of uranium mining and milling sites, clean up measures, emissions of nuclear power plants as well as low and high level radioactive waste disposal. Besides these hydrogeologists, water work managers and decision makers are interested in uranium since its chemical toxicity is exceeding the radioactive toxicity by far which is expressed in the 2 ppb standard recommended by the World Health Organization (WHO).

Freiberg, September 2002

Broder J. Merkel   Britta Planer-Friedrich   Christian Wolkersdorfer
Technical University Bergakademie Freiberg

Foreword Saxon State Minister of Environment and Agriculture

The immense interest in the Third International Conference on Uranium Mining and Hydrogeology shows the still existing significance of an international exchange of scientists in uranium mining remediation issues. About 300 scientists from more than three dozens countries are presenting the results of their work or discussing new ways to optimize the environmental situation during and after mining activities. Since the first conference in 1995 an international network of experts came into being. Common international research projects have been initiated.

Natural processes controlling radionuclide emission from former or present mining sites will be a central topic of this conference. The knowledge about these processes has been growing strongly. It is essential for the development of cost effective passive treatment methods to minimize the emissions not only of radionuclides but also of arsenic and heavy metals into the rivers. It will help to make future mining more environmentally benign.

Saxony, especially the traditional Technical University of Freiberg, provides manifold experience and knowledge in the field of mining and mining remediation. This is one good reason to meet here in Central Europe. Another reason is the presence of traces of ancient mining as well as from yesterday's work in the ore mines. And last but not least the beautiful landscape of the nearby ore mountains invite for a stay.

All participants of the Third International Conference on Uranium Mining and Hydrogeology I want to give a cordial welcome in Saxony. Most of them have been participating at the last two conferences and helped to make them a success. I hope that this year's conference will be equally successful and wish all participants a good time in Saxony.

Dresden, September 2002

Steffen Flath
Saxon State Minister of Environment and Agriculture

Foreword IMWA president

Since its inception in 1978 the International Mine Water Association (IMWA) has organised International congresses and symposia in many countries namely Australia, Austria, Brazil, England, Hungary, Poland, Portugal, Slovenia, Republic of South Africa, Spain, Ukraine, United States of America, and Zambia. We will thus take an important step with the symposium in Freiberg, Saxony, as many of the mining technologies that were used in the past – and still are used – emerged from Saxonian mining engineers and scientists

Co-organising a meeting in conjunction with the 3rd Conference on Uranium Mining and Hydrogeology is an outstanding opportunity to gather engineers and scientists from two fields that otherwise don't meet regularly. Therefore, I believe that the intended topics are very appropriate to the country and the symposium in Freiberg.

Once again, the publications arising from this symposium will add to the already existing comprehensive collection of information associated with mine water and the environment as well as uranium mining These publications and the interdisciplinary constitution of the symposium's delegates clearly show, that IMWA is both a European and international platform for the exchange of knowledge on mine water and its environmental impacts.

On behalf of all the IMWA members and visitors to the symposium in Freiberg we would like to thank the organisers, co-sponsors, and the Institute for Geology, Technical University Bergakademie Freiberg/Germany, in anticipation for the symposium. This meeting will once more add great value in the fields of engineering, science of mine water and associated problems and foster great international friendship which will always be an important consideration with all of us at IMWA.

Peet Nel
IMWA President

Dose Limits and Maximum Concentration Limits (MCL´s) for Radionuclides – Implication on Remediation of Uranium Mining and Milling Facilities in Saxony, Germany

Michael Kinze

Saxon State Agency of Environment and Geology, Zur Wetterwarte 11, 01109 Dresden

Abstract. In 2000 and 2001 the EU framework for Community action in the field of water policy and simultaneously a new radiation protection regulation came into force in Germany. Both affect the remediation of uranium mining and milling activities in Saxony. Radionuclide emissions from sources like waste rock piles and flooding waters of mines contribute to increased levels of radionuclides in the Mulde River and in the groundwater. Drinking water supplies along the Mulde River have to guarantee a good potable water quality. The dose limit of the drinking water regulation has no real impact on the toxic importance of uranium. For the chemical toxicity of uranium no MCL exists. In Saxony the radiation protection and the water authorities are well aware of these circumstances. It is an urgent task for the German and the European authorities to set up a MCL for uranium in water.

Introduction

Saxony has a long tradition of ore mining. It is known, that it took place for at least 800 years. The region with the highest ore contents is the traditional mining area Erzgebirge, the German term for "ore mountains".

From the early fifties until the late eighties of the past century uranium mining was going on intensively. Eastern Germany produced about 200.000 tons of uranium and therefore was worldwide number three in the field of uranium production. About 50% of it came from Saxony ore.

As a consequence of extensive mining radionuclides were mobilized by precipitation water from tailings and waste heaps as well as by flooding water from the underground mines.

This is the background why in Saxony radionuclides are of special interest as radiologically and chemically toxic elements.

Natural background in Saxon waters and man made contaminations

The natural background of radionuclide concentrations in ground and surface waters in Saxony depends on the geological situation on the one hand, on the other hand on the hydrogeochemical situation.

Rock formations with higher radionuclide contents

They are found in the Erzgebirge (hydrothermal vein deposits) in the southern part of Saxony. In the Permian Döhlen Basin (hard coal) west of Dresden some coal seams show high uranium contents. In some places of the lower Saxon Cretaceous we find increased uranium contents (Königstein sandstone deposit). In all these formations uranium was mined. In some tertiary lignite seams in north-western and south-eastern Saxony, which partly also show increased uranium concentrations, it was prospected but not mined.

Hydrogeochemical conditions controlling radionuclide migration

Different radionuclides have different chemical properties. Uranium for example is very well soluble under oxidising conditions, while radium can be found in increased concentrations under more reducing conditions. Both can be used as indicators for the presence of other natural radionuclides. If uranium and radium show very low concentrations, normally no other natural radionuclides can be found in waters.

If natural organic matter (NOM) is present uranium mainly is transported in organic complexes. In this case it is much more mobile. On the other hand if there is immobilized NOM in the water, uranium is much more immobile.

Rock surfaces as well can be very important for radionuclide mobility. Iron or manganese hydroxides on rock surfaces as well as carbonates can restrain radionuclide mobility in two ways: no radionuclides can escape from the rocks and the water soluble radionuclides can be bound at the large surfaces of the hydroxides.

From the above one can conclude, that areas
- with increased uranium concentrations in the rocks,
- where the uranium bearing rock is close to the surface (high redox)

- with increased contents of NOM

are prone to have naturally increased uranium concentrations in the ground and surface water.

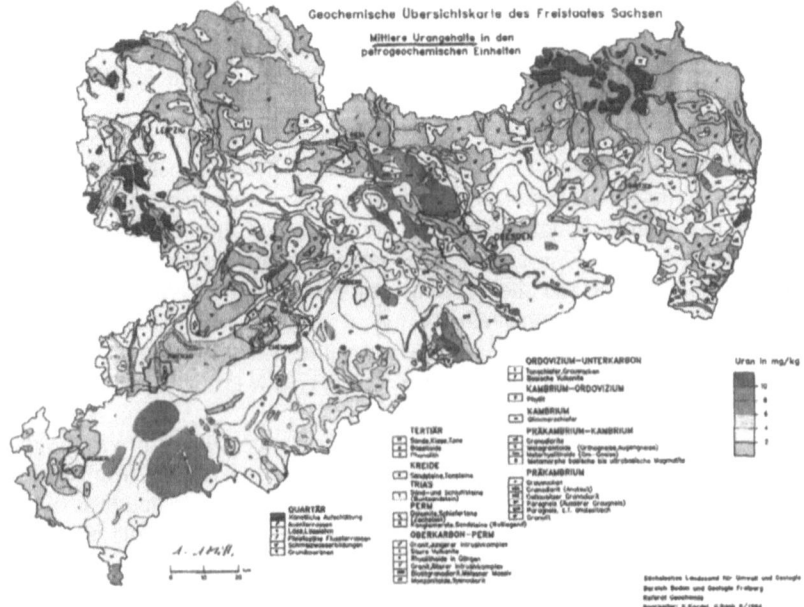

Fig. 1. Geochemical Map of Saxony. Mean uranium contents in Saxon rocks. The spots with the high uranium concentrations in the south west are part of the Erzgebirge.

However, areas
- with increased uranium concentrations in the rocks
- where the uranium bearing rock is deep enough under the ground water table (low redox)

are prone to have naturally increased radium concentrations in the ground water.

In areas where uranium mining or mining of other ores with increased uranium concentrations took place the water naturally shows more increased radionuclide concentrations. The reason is an enlargement of the rock surface by the mining activities (tailings, waste rock) creating better conditions for the leaching agent (rain water) and alteration of the hydrogeochemical situation (e.g. flooding of underground mines to a lower level than the natural one).

The natural background concentrations of uranium in water of the three Saxon areas of former uranium mining are as follows:

a) In the Erzgebirge natural background samples could only be taken where no remnants from uranium mining exist. 6 samples from creekwater showed a mean uranium concentration of 0,17 µg/l. Groundwater samples showed concentrations between 1 and 2 µg/l.

b) Upstream samples of tailings ponds from the Döhlen Basin showed uranium concentrations between 0,7 and 0,8 µg/l.

c) Close to the ore body of the Königstein deposit uranium concentrations are generally close to or below the detection limit.

These generally low natural uranium concentrations in water from areas of ore deposits are not surprising. If the hydrogeochemical conditions had mobilized uranium, no ore deposits would exist.

Anyway, natural radium concentrations can be very high in the closer surroundings of uranium ore deposits.

In the tertiary lignite areas the situation is much more differentiated in regard to uranium concentrations. The concentrations vary between below detection limit and a maximum concentration of about 300 µg/L in a ground water well south of Leipzig. However, it is not clear if this very high concentration has been induced by a local lowering of the groundwater table. In this case, oxygen is transported into the groundwater and the solubility of uranium is increased. Most samples show uranium concentrations below 1 µg/l.

For other areas in Saxony only random samples exist. Surface waters show uranium concentrations of 0.11 to 4.5 µg/l. The very few ground water samples show uranium concentrations below 1 µg/l.

In contrast to these concentrations, uranium concentration is very high at all places where uranium mining was going on. In seepage waters from waste piles, in ground waters of the surroundings of tailings ponds and in flooding water from underground mines uranium concentrations of several mg/l are the rule. It can be observed, that during the flooding of mines uranium concentrations decrease and radium concentrations increase. On the other hand mine waters from Saxon locations flooded several decades ago show low uranium and low radium concentrations. But there are too few examples to prove that they are representative for the general geochemical development of mine waters.

Most former uranium mining objects in Saxony are located in the catchment area of the Mulde River. Today in the Mulde River concentrations between about 10 and about 80 µg uranium/L can be measured. Some creeks in the catchment area of the Mulde River show concentrations up to 600 µg uranium/L.

The catchment area of the Mulde river covers nearly a quarter of the area of Saxony. As long as uranium mining remediation activities are going on the Mulde water will show increased radionuclide concentrations.

German and international regulations and standards

§ 7 of the German Drinking Water Regulation of May 21, 2001 says that "…in water for human consumption indicator parameters of the standards in attachment 3 have to be followed". Attachment 3, number 20 says, that a total individual dose of 0,1 mSv/a is allowed. This regulation includes all radionuclides except of ^{3}H, ^{40}K, Rn and Rn daughters.

With this paragraph a corresponding regulation of the European Council Directive 98/83/EC of 3 November 1998 on the quality of water intended for human consumption came into force for Germany.

The radiation protection regulation of July 20, 2001 gives standards for the ingestion of radionuclides with drinking water. Looking at the standards for infants from 0 to 1 year age one can calculate concentrations for different natural uranium isotopes from 1.59 to 1.73 Bq/L (128 to 130 µg/L).

In all these regulations uranium is only considered as a radionuclide, not as a heavy metal. Uranium as a heavy metal causes kidney diseases and is – if ingested - much more toxic than as a radionuclide (ATSDR 1999, Gilman et al. 1998a, Jacob et al. 1997 and Mc Donald-Taylor et al. 1992).

The World Health organisation (WHO) has established a Tolerable Daily Intake (TDI) for uranium of 0.6 µg/kg body weight and day. This is based on adverse affects observed by Gilman et al. (1998b) with kidneys of rats at uptakes of 60 µg U per kg and day. The WHO provisional guideline for drinking water quality recommends 2 µg uranium per litre as a standard.

Zamorra (1998) was the first to publish a study on the effects of chronic ingestion of uranium with drinking water on humans. He found that kidney function is affected by uranium uptakes considered to be safe in the publications based on animal studies.

International standards and recommendations are taking the chemical toxicity of uranium into consideration:

- Health Canada recommends an interim maximum acceptable concentration (IMAC) of uranium in drinking water of 20 µg per litre.
- The U.S. Environmental Protection Agency (EPA) Rule on Radionuclides in Drinking Water prescribes a maximum contaminant level for naturally occurring uranium of 30 µg per litre. EPA determines a safe level of 20 µg per litre assuming that an adult with a body mass of 70 kg drinks 2 litres of water per day and that 80% of exposure to uranium is from water. For cost considerations, however, EPA established a standard of 30 µg per litre rather than 20.
- The U.S. EPA Preliminary Remediation Goals for Superfund objects recommend 2.22 µg per litre for uranium in tap water while the U.S. EPA Groundwater Standards for Remedial Actions at Inactive Uranium Processing Sites recommend 30 pCi/L for U-234 and U-238 (44 µg per litre where secular equilibrium obtains).
- The U.S. Nuclear Regulatory Commission (NRC) made occupational annual limits on intakes (ALI) for oral ingestion of 10 µCi (=14.8 g).
- The Australian Drinking Water Guidelines say that the concentration of drinking water should not exceed 0.02 mg/L (= 20 µg/L). (http://www.antenna.nl /wise/uranium/utox.html)

As mentioned above in federal German regulations no standards for uranium as a chemical toxic exist. But e.g. the environmental department of the state Hessen in 1998 laid down a standard of 2 µg/L for uranium in drinking water.

Due to increased natural concentrations in drinking waters in northern Bavaria the Bavarian water authorities are presently discussing to recommend a standard of 5 µg/L for drinking water.

Also the German State Commission on Water Issues (Länderarbeitsgemein-schaft Wasser) presently is discussing a level of insignificance (Geringfügig-keitsschwellenwert) of 1 µg/L for natural uranium in surface and ground water.

It is to state, that the standards for the radiation doses and for the chemical toxicity are not comparable. Additionally the residual risk from chemical toxicity regarded acceptable usually is orders of magnitude lower than from radiation. Unfortunately there are still not sufficient data available for the chemical toxicity of uranium. These may be the reasons for the relatively wide spreading values in the different international regulations.

Proposal for an EC regulation

A standard for uranium as a heavy metal, i. e. as a chemical toxic in drinking water is not only needed for Germany. The European Commission should take care for a common regulation for Europe.

The toxicity of uranium is comparable to that of Arsenic. The EC standard for arsenic is 10 µg/L. For that reason it would be plausible to discuss a standard of 10 µg/L for uranium in drinking water.

Taking into consideration the facts shown above by the responsible persons in the EC it should be only a question of time until a standard for uranium is part of a European drinking water directive.

We do not have data of uranium concentrations in waters of other European countries. But regarding the situation in Germany one can guess, that more than 90% of all drinking water supplies in Europe have waters with uranium contents below 1 µg uranium/L. Drinking water supplies with uranium concentrations higher than 10 µg/L in their waters will be far below 1%. They should – in case of the establishment of a standard - get a special support by the water authorities to minimize economic disadvantages.

Impact of a regulation on uranium mining remediation

Drinking water regulations definitely can have a substantial influence on the demands of the regulator concerning the way and quality of the remediation method. In the case of a precautionary uranium standard for example, water authorities have to demand water treatment methods that keep back higher contents of uranium. This can lead to higher expenses for the deposition of the solid remnants of the treatment. Additionally it can lead to longer water treatment times.

Therefore it is very important to work out optimized remediation concepts from the beginning.

In the first place such concepts should take into account that the hydrogeochemical development of radionuclide migration strongly depends on the specific on-site conditions. Especially if prognoses let expect long treatment times one should develop and apply passive treatment methods at the earliest possible time.

However, drinking water standards are no remediation standards. In Saxony remediation decisions are made on the base of the respective present utilization of waters. For this reason in most cases a standard for uranium as proposed above would have no direct consequences on remediation decisions.

References

ATSDR – U. S. Agency for Toxic Substances and Disease Registry (1999): Toxicological Profile for Uranium

Council Directive 98/83/EC of 3 November 1998 on the quality of water intended for human consumption

Directive 2000/60/EC of the European Parliament and of the Council of 23 October 2000 establishing a framework for Community action in the field of water policy

Gilman A P, Villeneuve D C, Secours V E (1998a): Uranyl nitrate: 91-day toxicity studies in the New Zealand white rabbit. Toxicological Sciences Vol 41 No. 1: 129-137

Gilman A P, Villeneuve D C, Secours V E (1998b): Uranyl nitrate: 28-day and 91-day toxicity studies in the Sprague-Dawley rat. Toxicological Sciences Vol 41 No. 1: 117-128

Jacob P, Pröhl G, Schneider K, Voß J-U (1997): Machbarkeitsstudie zur Verknüpfung der Bewertung radiologischer und chemisch-toxischer Wirkungen von Altlasten, Umweltbundesamt Texte 43/97 Berlin: 145 p

Mc Donald-Taylor C K, Bhatnagar M K, Gilman A (1992): Uranyl nitrate-induced glomerular basement membrane alterations in rabbits: A quantitative analysis. Bulletin of Environmental Contamination and Toxicology Vol 42 No. 6: 367-373

Verordnung über den Schutz vor Schäden durch ionisierende Strahlen (Strahlenschutzverordnung – StrlSchV) vom 20. Juli 2001 (BgBl, I S. 1714) BGBl. III 751-1-8.

Verordnung zur Novellierung der Trinkwasserverordnung vom 21. Mai 2001, Bundesgesetzblatt, Jahrgang 2001, Teil I, Nr. 24, ausgegeben zu Bonn am 28. Mai 2001, S. 959-980

WHO - World Health Organisation (1998): Guidelines for drinking water quality. Second edition, Addendum to Vol. 2: Health criteria and other supporting information. WHO/EOS/98.1, Geneva, 283 p.

Zamorra M L, Tracy B L, Zielinski J M, Meyerhof D P, Moss M A (1998): Chronic ingestion of uranium in drinking water: A study on kidney bioeffects in humans. Toxicological Sciences Vol 43 No. 1: 68-77

Remediation oriented use of conceptual site models at WISMUT GmbH: Rehabilitation of the Trünzig tailings management area

Alexander T. Jakubick, René Kahnt

WISMUT GmbH, Chemnitz

Abstract. Sites with complex environmental liabilities, such as present at the former U (surface) mining and processing sites of Seelingstädt, comprising multiple sources of environmental contamination require a phased modeling approach to support the site rehabilitation activities. The rehabilitation planing is best developed on the basis of a conceptual site model that structures the sub-models approximating the remedial objects, site conditions, the contaminant transporting media (such as the receiving streams and the aquifer on the site) and combines these with the process models calculating contaminant release, transport and geochemical interaction. The remedial problem determines the selection of the model code and the model set up process. The need for additional remedial investigation is measured by the requirements of the remediation oriented conceptual site model, thus directing the development of the database. The methodology is outlined on the case of the Trünzig tailings management site. By performing model simulations and feasibility assessment of the potential rehabilitation measures in terms of environmental and economic gains it was possible to derive verifiable rehabilitation decisions for the site.

Specification of the remedial problem for modeling

The former U (surface) mining and processing units of Culmitzsch and Trünzig at the Seelingstädt site comprise several types of contaminants sources, tailings ponds/deposits and waste rock piles (partly serving as tailings pond dams). The dominant constituents of contamination are uranium, radium, nickel and sulfate. The impact of the multiple sources is overlapping. The contaminants transporting media on the site are the receiving streams and partly the ground water. The different contaminant sources and the receiving streams are shown in the left part of figure 1. The geological boundaries of the Culmitzsch Sandstone are indicated by a dotted line in the right part of figure 1. The main discharge route from the site is through the streams Finkenbach (East) and Culmitzsch (North). The site is located in Eastern Thuringia near the city of Gera. A cross sections of the main geological and hydrogeological formations is presented in figure 2:

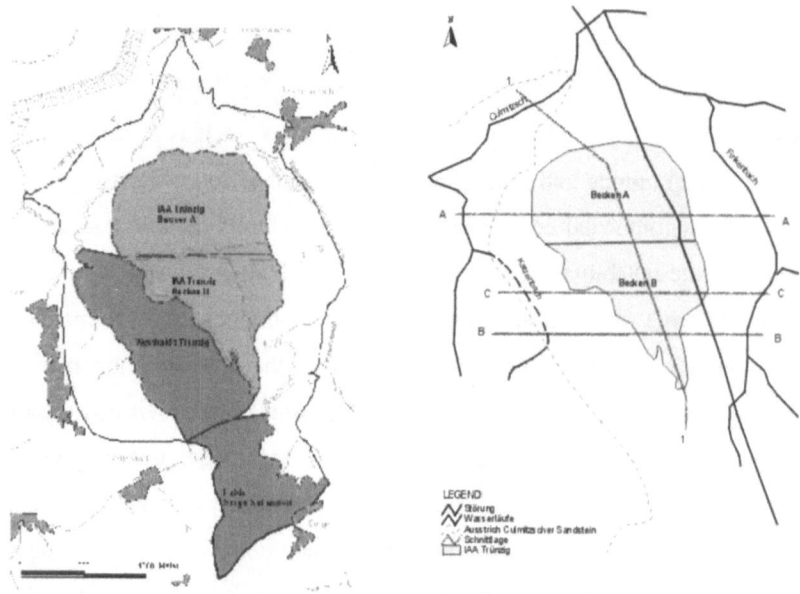

Fig. 1. Illustration of the different contaminant sources at the Trünzig site and the receiving streams. The geology along the cross section 1 is shown in figure.2

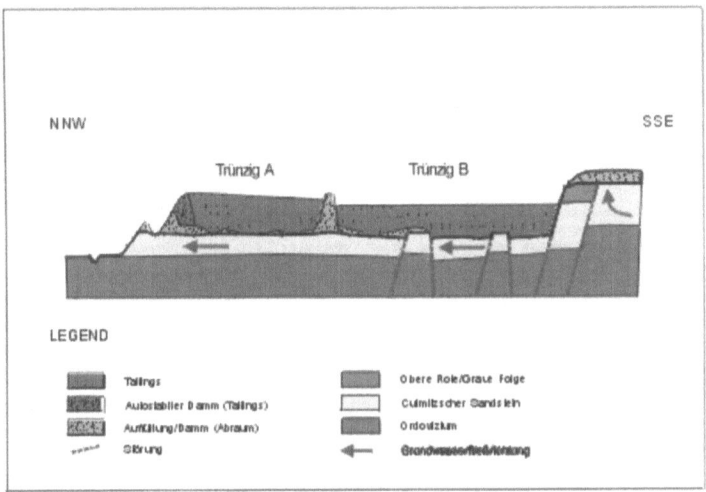

Fig. 2. Geology of the Trünzig site along the NNW- SSE cross section (a-a').

The hydrogeological system is dominated by the Culmitzscher Sandstone aquifer having a thickness of approximately 20 m and more. The Sandstone is partly eroded and fractured by the Culmitzsch-fault system which caused relative vertical shifts of up to 10 m. The faults system partly acts as hydraulic barrier. The weathered upper 10 m of the Ordovician bedrock is a smaller aquifer of minor importance. The main groundwater flow follows the basis of the Culmitzscher Sandstone. The global discharge of the groundwater is into the streams Finkenbach in the East and into the Culmitzsch in the North. The groundwater table is few meters below the surface in the discharge area and some 10 meters below the piles. The water table in the region of the tailings ponds is below the bottom of Trünzig pond "A" and inside the tailings in pond "B".

The modeling was needed for justification of the remediation and for evaluation of the feasibility/optimization of the planned remedial measures. The effect of various remedial measures was to be assessed from a long term perspective (200 years). The short term problems, such as radon exhalation and dust development have been handled at the beginning of the remediation. The release of contaminants into the surface and ground water was the main long term concern.

Conceptual site model

In view of the limited site specific data and need to progress with the remediation it was considered best to approach model development in phases and, as additional data become available, adjust the model in course of the project. The approach and the role of the conceptual site model are illustrated in figure 3.

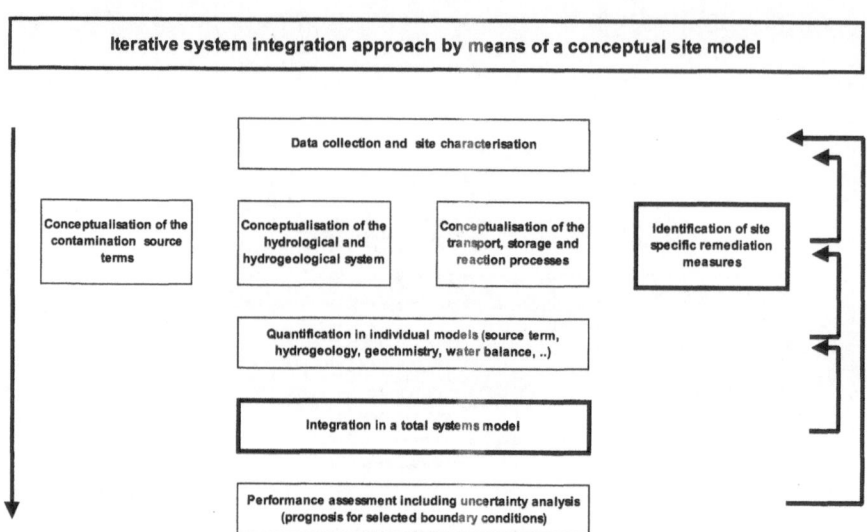

Fig. 3. Illustration of the role of the conceptual site model in the process of preparation of remediation measures.

The assessment of the long term environmental impacts and effectiveness of remedial measures at a complex site like Trünzig requires the use of a number of highly specialized models and the subsequent integral evaluation of the modeling results to estimate the overall site impact and obtain the best remedial strategy. To achieve this we proceeded by combining the required (sub)models into structured assembly which constitutes what we term the Conceptual Site Model[1].

Pending on the possible degree of simplification of the remedial problem and required dimensionality for the site description, either a numerical site model or a multi-compartmental model could be chosen.

In a numerical model the site (along with the remedial objects acting as contaminant sources including driving processes) is described in detail in a regional flow model, within which the contaminant transport is simulated as well.

Compartmental models have a heuristic character, i.e. use a very simplified approximation of the real site conditions. A multi-compartment model, however, can serve to combine and interrelate the output of other more detailed, more sophisticated models. The multi-compartment model can than be used to calculate/predict the integral response of the site in terms of contaminant loads and mass balances in ground water and receiving streams.

Due to the fact that the model simulation involves very different time scales for surface/ground water flow and contaminants release/transport, it is advantageous to select a multi-compartment model for the particular problem. In the multi-compartmental model the site is subdivided into representative, well defined units (representing the remedial objects, receiving stream and ground water) which

---

[1] The term conceptual model is used in an extended way (i.e. beyond the development of conceptual understanding) including model selection, model set up and conversion into a simulation model.

are/can be connected with detailed process models handling geomechanical, hydrogeological and geochemical questions. For correctly setting up of the compartments an understanding of the global hydrological and ground water flow conditions on the site must be acquired first.

The compartments used in the case of Trünzig are smart compartments, not common black boxes[2]. Each of the compartments, if required, can be resolved in greater detail; this refers to both spatial resolution and/or consideration of specific processes.

The compartmentalization into sub-models receiving as input the results of flow and contaminant transport simulation from autarkic models, rather than simulation of all processes in one numerical site flow model, creates the possibility of answering highly specialized questions at any stage of model and (project) completion. At the same time the global perspective can be continuously maintained by a Top- Down-Approach.

The implementation of the conceptual site model was realized under the Goldsim code (a development of Golder Ass., Seattle).

Model assumptions, considered processes and selection of codes

Selection of model input contaminants/substances

Before modeling started, the substances were identified which were to be quantified in terms of loads and concentrations. From an environmental perspective, the particularly relevant substances at the site are uranium and nickel. In addition, chloride and sulfate were selected for consideration because of high sulfate content in the tailings pore water and because of the tracer-like behavior of chloride. Heavy metals and radium were looked at as well but their contribution to the overall site impact turned out to be of little significance.

Ground water recharge/percolation

The model domain includes, beyond the area of immediate relevance, also the area which may become affected by contaminant transport in the future. The boundary of the domain was selected such that no ground water inflow takes place across the boundary. This allows the calculation of the underground discharge from the model domain as the ground water recharge and storage are known. A distinction is made between the natural catchment area, waste rock piles and specific segments of the tailings pond.

---

[2] The compartments used are not simple black box - every box compartment may have a complex internal structure

Natural catchment area

The ground water recharge for the site was calculated with the GEOFEM Light '98 program system. A recharge rate of approximately 70 mm/a resulted for all natural catchment areas.

Waste rock piles

Following to the type of vegetation, the waste rock pile surfaces were subdivided into different hydrotopes (grass and forest). The type of hydrotops determined the code selected for simulation. While grass covered areas the HELP code was used, simulation forested areas is more reliable to employ the soil code. The sum of underground and hypodermic was estimated to be approximately 160 mm/a for grass covered surfaces and approximately 120 mm/a for forested areas.

Water balance of the tailings ponds

For the tailings ponds, several approaches are used to estimate percolation.

Global water balance

Due to the basin like structure of the tailings ponds there is no surface runoff from the tailings. The average percolation rate for basin A, which had been capped with an interim cover for several decades, could, therefore, be established when the evapotranspiration and the pumped off water volumes were accounted for. This approach could not be used for basin B, which was used for storage purposes till 1996.

Water balance modeling of specific tailings pond zones using the HELP code

Water balance calculations analogous to those for the waste rock piles were carried out for the (interim) covered tailings pond A using the meteorological records. The tailings soil parameters were taken into consideration to differentiate between the sandy and slimy parts of the tailings deposit.

Percolation history in the slimes zone based on the consolidation state

From the measurements of depth dependent distribution of pore ratio and hydraulic conductivity and using finite strain distribution code, the tailings discharge and consolidation history can be reconstructed and the representative material parameters back-calculated. By using the pore ratio and taking into consideration the hydraulic conductivity distribution, the maximum percolation rate can be derived.

Percolation history based on concentration profiles

Assuming that percolation through tailings is mainly vertical, the presently observed depth dependent distribution of chloride concentration can be used to derive the average historical percolation. For the calculation the assumption is made that initial pore water concentration both in the sandy tailings and in the slimes was equal to the concentration presently observed in the bottom layer of the slimes zone. The chloride profiles are shown in figure 6.

Percolation rates used in the model

After calibration of the two simulation approaches the percolation rates history was derived indicating for the sandy tailings zone percolation rates of approximately 140 to 180 mm/a and for the slimes zone 80 to 120 mm/a.

Infiltration rates for different cover options

One of the requested applications of the conceptual site model was the evaluation of the impact of different tailings covers on the water and contaminant balances. The long term effectiveness of the various covers was calculated using the HELP and the SOIL (COUPMODEL) codes. The use of SOIL became necessary because HELP is inappropriate for forested areas. In addition, coupled to the SOIL code, the Hydrus2d code (a development of Jansen) was used to calculate saturation in the individual layers. The long term performance (200a) of three cover designs was measured by the decrease of the percolation rate in addition to the effect of the existing interim cover.

Table 1. List of key parameters for the considered cover options

Cover option	Infiltration rate (mm/a)	effective oxygen diffusion constant (m²/s)
Interim cover	180 (beach area)	$1 \cdot 10^{-7}$
P1 – single layer cover	82	$2.6 \cdot 10^{-9}$
P2 – two layer cover (moisture storage layer + loess loam layer)	43	$5.4 \cdot 10^{-10}$
P3 – two layer cover (moisture storage layer + clay layer)	6	$4.8 \cdot 10^{-10}$

Oxygen diffusion

As indicated by the geochemical condition, it can be expected that pyrite oxidation could have a significant effect on contaminants mobilization in the tailings, particularly in the Trünzig A pond. The oxygen penetration into the tailings was analyzed to evaluate this effect for both the present and the rehabilitated condition. For the assessment of the present state, a back-calculation of the pyrite oxidation history in the tailings had to be performed.

For the cover options the unsaturated flow field was calculated based on historic meteorological data by SOIL. This code developed by Jansen works similar to HELP. It gives more reliable results for forest vegetation. The results of SOIL were fed into Hydrus2d, to simulate the transient saturation states and calculate the effective diffusion coefficient for the partially saturated conditions for each layer.

Based upon saturation history of tailings, an effective oxygen diffusion coefficient of 10^{-7} m²/s was derived for the present interim tailings cover. The calculated effective diffusion coefficients for the final cover options are given in Table 1.

Ground water flow

As the main purpose of the modeling was to balance water and contaminant flows, a simplified approximation of the ground water flow was acceptable. Nevertheless, the subdivision of the model domain into compartments had to be based on a predictive numerical ground water flow model (MODFLOW). The simplified approximation is, however, fully acceptable for the assessment of the cover options because covering does not effect the main percolation and ground water flow directions or the areas of discharge.

The ground water flow field and the global flow directions on the site are shown in figure 4.

The subdivision of the model domain into compartments in accordance with the main geological features and flow conditions is shown in figure 5 along with the illustration of the major water flows and contaminant fluxes at the site.

The right part of figure 5 shows the Goldsim containers representing the geologic elements.

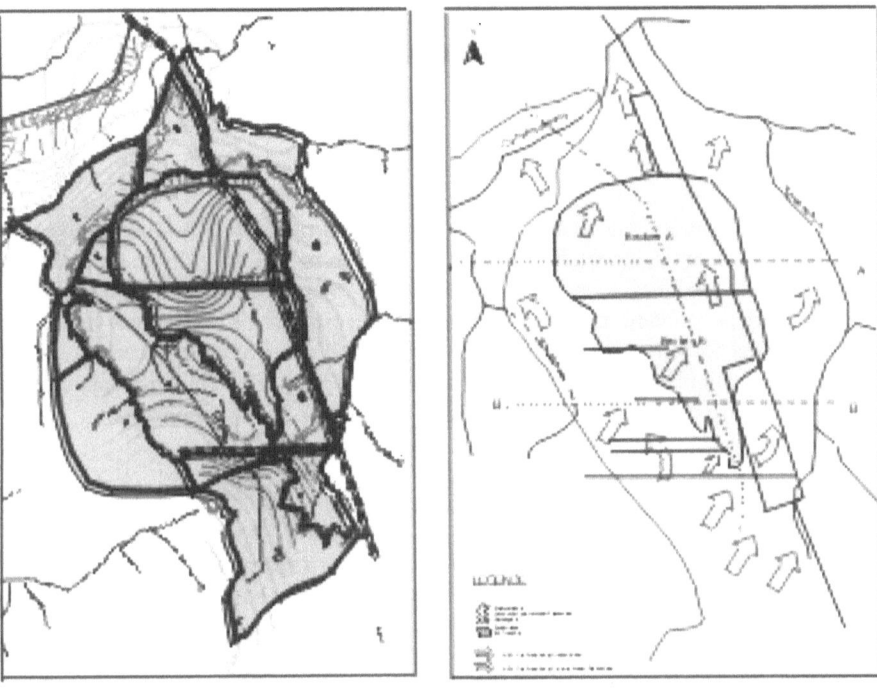

Fig. 4. Contour lines of the present state ground water conditions and resulting flow directions in the two main aquifers, the Culmitzsch sandstone and the weathered part of the Ordovician formation.

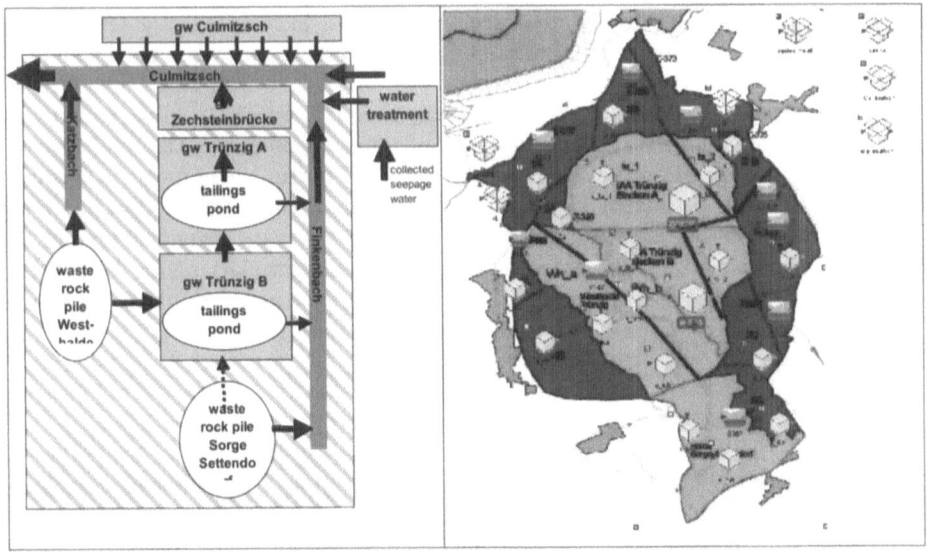

Plenary

Fig. 5. Subdivision of the area into ground water compartments and schematic picture of the dominating water flows and contaminant fluxes.

Contaminant transport in ground water

In the site model, the contaminant transport was calculated by simple mixing cells. The approach is acceptable because in the site model only the mass balancing of the contaminants is essential. To this end, average concentrations of all measured substances relevant to the problem and representative of the current state were established for all compartments. These average concentrations were used to calibrate the loads released from the remedial objects. The contaminants concentration history could not be estimated because of lack of measurements.

Water balance in the receiving streams

In the absence of flow records for the receiving streams located within the model domain, the rate of flow had to be calculated on the basis of the catchment areas. For this end, topographical catchment areas were delineated. The receiving streams were the represented by mixing cell type of compartments.

Mass balance in surface waters

Ground water flowing leaving the area under consideration fully discharges into the receiving streams which are part of the model domain. For this reason, contaminant concentrations in the receiving streams become exactly reproducible (under consideration of the water throughput in the stream), when all contaminant sources are implemented and the release quantified. However, this is only the case for the Finkenbach and Katzbach streams, while concentrations in the Culmitzsch stream are dominated by the tailings ponds of the Culmitzsch site unit. Accordingly, only the additional load coming from the Trünzig site unit can be specified and only the associated increase of concentration calculated for this receiving stream. Similar to the use of the mean concentrations in the individual ground water compartments, the average concentration (in particular) of the Finkenbach stream was used as a measure for water and contaminant mass balancing (figure 8).

Contaminant release from waste rock piles

As the site model is to be used primarily for evaluation of the tailings covering options, a simplified approximation of the contaminant release from waste rock piles was justified. The release rate was considered constant after being derived from and calibrated to the present measurements of the infiltration rate and concentrations observed in ground and surface waters. Once more detailed data become available, these may be included in the site model. The model upgrading procedure here is similar as in case of Trünzig basin A. However, the additional data (and model upgrade) can only affect the long term releases since the current state was used for calibration.

Consolidation of tailings

The settlement of fine tailings (slimes) and the associated squeezing out of pore water is of key relevance for remediation. For this process to be properly simulated, a special modeling and data acquisition concept was developed at Wismut. The focus of the concept is to reconstruct the history of the tailings discharge/deposition in order to back-calculate the geomechanical parameters which led to the present state of consolidation. The computer codes employed for the task are Plaxis, FS-Consol and the advanced Consol2d code (a joint WISMUT/TU Chemnitz development). The main use of the consolidation codes is in conjunction with the transient release of pore water and contaminates from the tailings ponds. The reactive contaminant transport in the tailings is simulated by the TEN3d code which can interface with the consolidation code and is capable of handling the time dependent outputs, such as change of porosity and layer thickness. Since the first simulation assignment was the evaluation of the final cover options, a full consolidation of the tailings could be assumed and evaluations of transient states were not required.

Contaminants release from the tailings deposits

Because of the limited data base, the evaluation had to be restricted to the Trünzig A basin. Two different simulation approaches were used: a simplified approximation of the essential processes within the Site Model (Goldsim code), i.e. directly in the overall systems model and an approximation with the reactive contaminant transport model TEN3D which is tailor-made for the tailings problem.

Simplified modeling using the Site Model

In this approach, the sandy and slimy parts of the tailings pond were subdivided into zones for which the pore water development was back-calculated taking into consideration the ground water recharge/percolation, oxygen diffusion, solids contents and solubilities. The simulation results were calibrated on the basis of the currently observed pore water quality. For purposes of illustration the back-calculated concentrations for chloride and for sulfate are presented in figures 6 and 7. For the calculations it was assumed that the dominating water movement through the tailings is vertical and the present pore water quality in the bottom layer of the fine tailings is equal to the initial pore water quality.

Because of lack of data on depth distribution of the relevant substances a single mixing cell approach was selected for the basin B. The calibration was carried out by using the concentrations observed below and within the tailings and at ground water measuring points. In addition, it turned out necessary to consider the supernatant tailings pond water which was present till 1996, in particular the volume of pond water which percolated through the beach area.

Contaminant release modeling with consideration of reactive processes

In order to model contaminant release from tailings ponds, HPC (a Gera based company) under conceptual guidance of the authors was commissioned to develop a specialized code (TEN3d) for the prognosis of contaminant release from tailings. The code was based on a tested code (TENSIC) used by WISMUT for contaminants release from waste rock piles. The TEN3d code was employed to produce model prognoses for the Trünzig A basin. The code is capable of handling all relevant transient processes.

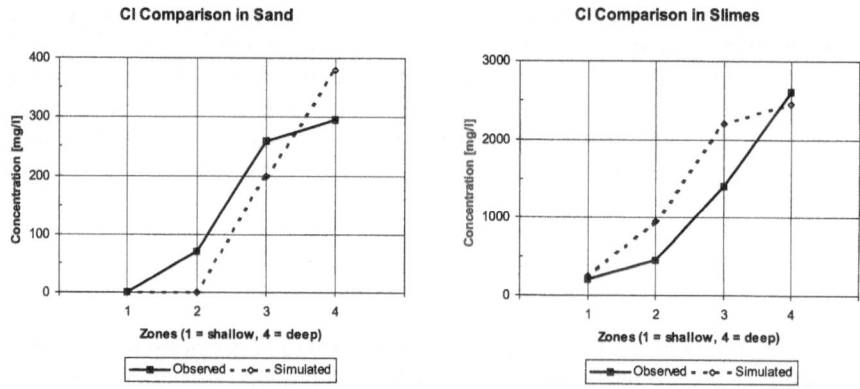

Fig. 6. Vertical chloride profiles from back-calculation compared with current observations in the sandy and slimes zones of the Trünzig A tailings pond. Based on this chloride profile the probable range of historical infiltration rates was established.

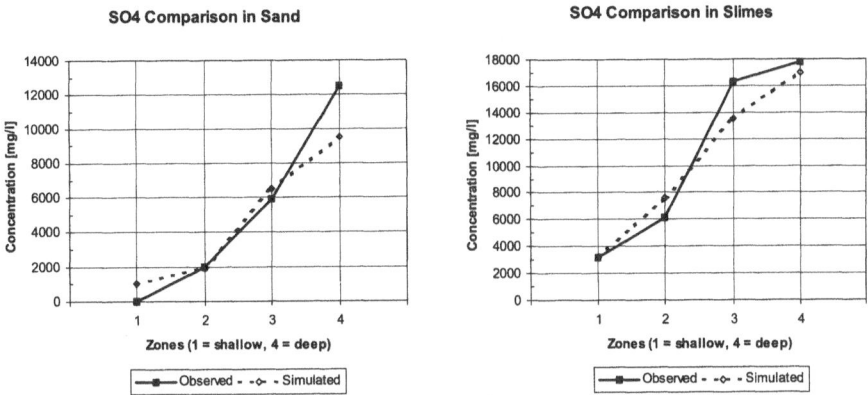

Fig. 7. Vertical sulfate profiles back-calculated from historical records compared with currently measured profiles in the sand and slime zones of the Trünzig A tailings pond. The present concentrations could only be arrived at if oxygen diffusion and pyrite oxidation was considered.

The results of the model simulation are the water volumes and contaminant loads released from the bottom tailings layer as a function of time. The calculations were carried out for the sandy and the slimes zones and the results fed in form of time series into the site model (Goldsim code). As the present modeling task focused on the long term development of the release, it is admissible to neglect the transient effects of consolidation.

Implementation of the site model and model setup

Implementation of the site model proceeded in the following sequence:
- Implementation of the ground water system including geometry, infiltration and hydraulic conductivities
- Calibration of source terms on the basis of relevant seepage and ground water concentrations
- Development of a detail model for the Trünzig A tailings pond and reproduction of the observed concentration profiles by back-calculation of the historical development on the basis of the pore water composition in the bottom layers of the fine tailings (slimes)
- The calibration of the chloride concentration profiles (having the advantage of a narrow range of variation of ground water recharge) was followed by estimation of the production rates and effective solubilities for the considered substances and derivation of the relationship to oxygen diffusion.
- Implementation of the simplified release model for basin B. The contaminants release was derived on the basis of the water quality measurements in the wells while consideration was given to the changing extent

of the coverage of tailings by pond water, to the composition of the supernatant pond water and of the pore water.

- Calibration of the water movement on the basis of observed concentration distributions in ground- and percolation water.
- Implementation of the simulation results from the TEN3D code.

Because the interaction of the environs of the tailings pond with the tailings deposit is very limited, the processes occurring within the tailings could be analyzed and modeled independently of the overall hydrological and hydrogeological flow field. With this reasoning, it was possible to dispense with the development of a combined flow and mass transport model. Independent models suitable for detailed analysis of the objects were developed instead. The degree of details of these models was in accordance with the available data for the specific task. The fundamental mass balances obtained were integrated within the conceptual site model. Although in the site model the balance analysis was restricted to 12 substances, the detailed geochemical models (of the tailings pond A) considered all other substance affecting the geochemical conditions. Of particular importance from the environmental point of view were uranium and nickel. Chloride and Sulfate were used for model calibration.

The objective of the entire modeling effort was the integration of all relevant information and data which were very heterogeneous into a consistent picture. Consistent model simulations of the impact of the various rehabilitation measures on the overall site performance required the quantification (calibration) of the relationship between system parameters and observed quantities and concentrations of the relevant substances in the pore water and seepage.

Model simulation results

Based on model simulations, the following statements can be made:
- Infiltration rate in the sandy beach zone of basin A amounted historically to approximately 140 - 180 mm/a;
- Infiltration in the slimes zone of basin A was historically approximately 100 - 120 mm/a;
- Pore water composition in sandy beach zone of basin B and hence the runoff towards Finkenbach stream was dominated by composition of the supernatant pond water;
- Water balance of basin B is dominated by percolation in the beach zone increased by the alternating inundation by pond water;
- Basin A drains to more than 90 % via the Zechstein bridge to the north;
- The beach of basin B drains partly via an old drainage system towards Finkenbach, the slimes drain to approximately 50 % towards Finkenbach and to 50 % via the Zechsteinbrücke;
- The state of basin A can be very well described already on the basis of simplified approaches (leaching, pyrite dissolution).

Figure 8 below illustrates for one of the cover options how currently observed average concentrations (broken line) of uranium and nickel at the lower course of the Finkenbach stream can be reproduced by back-calculation of the 40 years history.

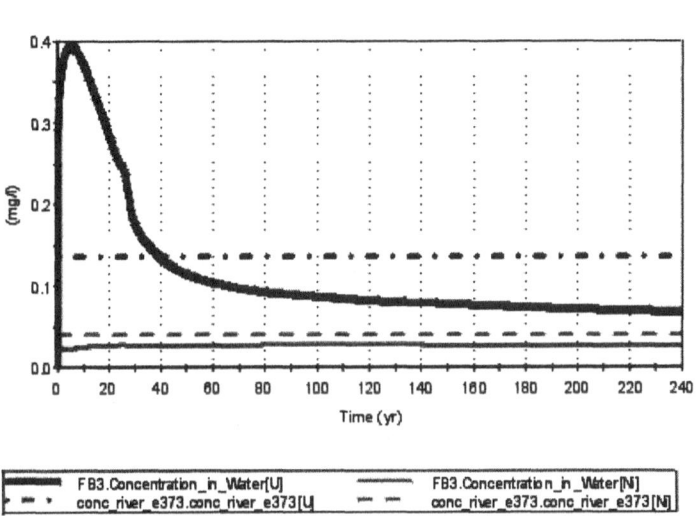

Fig. 8. Plotted is the average concentration of uranium and nickel at the lower course of the Finkenbach stream. The broken line depicts currently observed concentrations (i.e. year 40 of the tailings pond).

With the input as specified previously, the site model successfully simulated the past and future global water and contaminant mass balances for the Trünzig site unit.

Conclusions

The description of the heterogeneous remedial objects (waste rock piles and different zones of two tailings deposits) and media conducting the contaminants (aquifers and receiving streams) in a multi-compartment site model proved to be a very useful approximation for the Trünzig site. By integrating the objects in a site model and establishing the necessary assumptions (unavoidably connected with model simplifications), it was possible to arrive at plausible and consistent water and contaminant mass balances. However, for this it was necessary to assume a substantial retardation of the heavy metals in the downstream ground water.

Model simulation results showed that the main contaminant release from the tailings deposit is driven by percolation through the sandy beach zones. Geochemically, the release is dominated by pyrite oxidation which commenced at the time prior to remediation and proceeds to a limited degree even after construction of the interim cover. Based on the conceptual site model the remaining uncertainties could be identified, such as the uncertainty of duration of contaminants release.

Most importantly, the present application of the conceptual site model illustrates that meaningful results can be achieved with a multi-compartment approximation using limited data even for a complex remedial problem. The conceptualization and quantification of the remedial problem accompanied by a consistent and plausible mass balancing over the whole problem domain allows the project manager already at the early stages of the project (when only limited data are available) to focus investigations to factors which genuinely determine the system. As the site-specific data increase and the need for specific answers in the project develops, it becomes necessary to add more complex sub-models to the site model. For this reason the architecture of the conceptual model was such to allow an easy extension and combination of compartmental and process models. In order to maintain the remedial problem orientation of the Conceptual Site Model, in spite of increasing complexity, the overall model is structured in a Top-Down manner. The approach is illustrated in figure 3.

Finally, it should be pointed out that the philosophy of the presented modeling approach is laid out to follow the path of highest problem solving efficiency: While detailed analysis of aspects, which have already turned out to be irrelevant for decision-making is abandoned, aspects of particular relevance are systematically followed up till satisfactory solutions have been reached.

Acknowledgements

The authors want to thank for the numerous contributions and discussions, in particular the IAA Project management staff and the Engineering Department of WISMUT GmbH. Particular thanks are due to contributors of various partial aspects and especially to Mr. Schöpe and Mr. Hähne of C&E Chemnitz for providing background/initial data and Mr. Eckart and the staff of HPC Gera for the the geochemical modeling with TEN3d and the development of the code.

Deep mine hydrogeology after closure: insights from the UK

Paul L Younger

Hydrogeochemical Engineering Research & Outreach (HERO), Department of Civil Engineering, University of Newcastle NE1 7RU, UK
(p.l.younger@ncl.ac.uk)

Abstract. Although the UK has had an extremely rich history of deep mining, at the start of the 21st Century most mines are already abandoned and in various stages of flooding. Lessons of generic value have been obtained in the UK over the last decade, particularly in relation to the reactivation of mine subsidence processes during flooding of old workings, and the generation and release of polluted drainage. As researchers and practitioners have attempted to characterise and (in some cases) predictively model such processes, new analytical techniques have been developed. These range from simple 'rules of thumb' for pollution intensity prediction to sophisticated 3-D models of flow through networks of mine roadways routed through variably-saturated porous media.

'A Tainted Legacy'

The former prodigious productivity of the UK's deep mining industry not only fuelled the industrial revolution, yielding riches which continue to circulate in global financial markets, but also left a significant heritage of historical mine sites. For instance, it is nowadays possible for tourists in the UK to marvel in safety in ancient underground mine voids dating from the Neolithic Age and the Bronze Age, as well as in several well-preserved examples of modern deep coal mines. However, less attractive legacies of millennia of deep mining have become all too apparent as most major UK coalfields (and all but a few mines working for industrial minerals and metals) have finally closed over the last decade. Once mines are abandoned and dewatering systems are withdrawn, the workings flood, in many cases giving rise to a range of environmental problems. Most prominent among these problems is the generation and release of polluted drainage, but hazards due to renewed mine subsidence and / or accelerated emissions of dangerous mine gases (i.e. explosive or asphyxiative) are also locally exacerbated by the process of deep mine flooding. Political decisions in the early 1990s have resulted in the UK abandoning entire coalfields long before this has taken place in most other European countries. For this reason alone, mining and environmental professionals in the UK have learned a number of lessons over the last decade which may well be applicable to many other deep mine settings elsewhere in the world. This paper aims to summarise some of these lessons for the benefit of the wider international mine water management community. As such, this paper emphasises recent UK literature, but it should be noted that references to the wider international literature can be found in the works cited here.

Hydrogeological lessons from UK deep mine closures

Some of the key lessons learned from the experiences of coalfield / mine closure in the UK in the decade 1992 - 2002 have concerned the following topics:
- reactivation / exacerbation of land subsidence by processes associated with flooding of strata and old workings by rising mine waters
- the mineralogical and geochemical aspects of water quality deterioration during the flooding of mine voids
- the functional relationship between the duration of mine flooding and the temporal changes in pollutant loadings after decant of the flooded workings to the surface and / or adjoining aquifers
- scale-appropriate modelling of mine water 'rebound' processes (i.e. flooding of extensive networks of underground voids)

We will now consider each of these in more detail, each of them in turn in the sections which follow.

Rising mine waters and reactivation of subsidence

The prediction, control and (where necessary) mitigation of the effects of surface subsidence due to deep mining has now reached the level of a fine art in many advanced mining countries. In the UK, the pinnacle of achievement in developing a predictive capacity for mining subsidence was attained in 1975, with the publication of the second edition of the national "Subsidence Engineers' Handbook" (SEH) a compendious practical manual which has never been superseded (NCB 1975). The SEH was developed from an enormous database of observed subsidence and associated strain measurements, and it remains extremely useful and generally reliable for predicting both surface subsidence (especially that due to longwall mining) and the subsurface strains associated with the process of settlement above caving workings in Carboniferous coal-bearing strata. The ability to predict maximum tensile strain at a specified stratigraphic position above active longwall workings is particularly useful in the design of safe undersea and sub-aquifer workings (e.g. Orchard 1975; Dumpleton 2002). One of the principal drivers behind the development of the SEH was the need for officers of the national-ised coal industry to be able to fairly adjudicate legal claims for damage due to mining subsidence which were continuously received from surface property own-ers in coalfield areas. From the database which underpins the SEH, it is clear that all significant surface movements caused by the goafing of a given longwall panel occur within a few years (usually 2 - 3, rarely as long as 5) after the completion of panel extraction. Indeed, this observation came to be enshrined in the legal codes governing subsidence compensation arrangements.

Given this background, subsidence engineers can be forgiven for concluding that once the collieries they used to work in had been closed for a few years, they would at least be spared from adjudicating further claims for compensation for subsidence. After all, flooded workings benefit from the additional buoyant sup-port offered by the water they contain, which helps to minimise long-term settle-ment of voids. However, before mines flood to equilibrium levels, certain proc-esses which occur while water levels are still rising can temporarily exacerbate subsidence. Significant subsidence damage and associated geomorphological changes have recently been recorded in various UK coalfields subject to ongoing rebound, giving rise to compensation claims which the present custodians of the historic liabilities of the coal industry (i.e. the subsidence section of the govern-mental Coal Authority) did not expect to be receiving. From the scattered evi-dence gathered to date, it would appear that there are at least three mechanisms by which mine water rebound can reactivate surface subsidence:

(i)   the weakening of the floors (and to a lesser extent the roofs) of mine voids by the reaction of certain types of strata to sudden submergence, and

(ii)   direct erosion of mine voids, shaft caps and adit plugs etc by the pressure and kinetic energy of rapidly-flowing mine water

(iii)   reactivation of previously-dormant faults which are intersected by old mine workings subject to recent flooding for the first time.

Moisture weakening void roof and / or floor strata

It is a common characteristic of many coal-bearing sequences that the workable coal seams are immediately underlain by soft, relatively plastic mudstone horizons often termed "seat-earths" (Murchison and Westoll 1968). These seat-earths are essentially buried soils, which often contain the fossilised rootlets of the swamp vegetation which flourished in the coal-forming swamps. When dry, these seat-earths can retain sufficient strength for most mining purposes (e.g. support for pil-lars of intact coal, and firm bedding for mine infrastructure, such as conveyor lines, rails etc). However, when they are wetted by the impingement of mine wa-ters following the cessation of dewatering after mine closure, seat-earths can lose much of their strength. No longer able to bear the stresses passed on to them by loaded support pillars, the seat-earths can deform rapidly, leading to pillar collapse and surface subsidence (e.g. Smith and Colls 1996), often in the form of crown holes. In some cases, crown holes of this type can intersect surface water bodies, leading to a dramatic increase in water ingress to the flooding workings below.

Such an example occurred in November 2000 adjacent to Newcastle International Airport, when a large collapse crater developed above actively-flooding workings led to drainage of an entire pond of surface flood waters (estimated volume ~ 5 Ml) draining into the underground voids in the space of a few hours.

A class of subsidence features which are more subtle in appearance than crown holes, but which affect far larger surface areas (up to several km^2), are thought to be associated with flooding of abandoned longwall panels in some areas, notably in eastern Northumberland (Jackson N., pers. comm. 2001). The formation of these large-scale reactivated subsidence troughs presumably involves a combination of mechanisms, such as rib pillar failure and / or weakening of shaley clasts in goaf by wetting, leading to greater settlement under the same imposed load as the clasts deform and porosity is reduced. This latter mechanism is well known from recently-flooded opencast backfill (see Younger *et al.* 2002), which is geotechnically very similar to longwall goaf. At the time of writing, the characterisation of these large-scale subsidence troughs is under investigation.

Finally, it should be noted that flooding of deep coal mine strata can in some cases lead to 'upsidence' (i.e. raising of the ground surface), presumably due to wetting of expansive clays in the sequence. Although this process has been documented from the Netherlands (Bekendam and Pottgens 1995), but has not yet been reported in the UK.

Direct erosion by pressure / kinetic energy of flowing mine water

Where mine water is free to flow through open voids, especially above the water table, it can erode the walls and floor of the voids sufficiently that collapse occurs. In recent years, such erosive behaviour has been the cause of several major UK mine water discharges spontaneously shifting their points of surface emergence. Table 1 summarises some of these cases, highlighting incidents in which this process led to bypassing of mine water treatment facilities, flooding of residential areas / public roads and other nuisances.

Where rising mine water is impounded behind piles of roof-fall debris and / or engineered stoppings (e.g. shaft caps or adit portal plugs, which are generally installed for public safety reasons when the mine is abandoned) hydraulic head can build up to such an extent that the strength of the impounding feature (or of its interface with the surrounding strata) is overwhelmed, resulting in a sudden outburst of mine water. An infamous example of this genre led to the major outburst of ochreous water from the Wheal Jane mine in Cornwall in January 1992 (see Younger and LaPierre 2000). A variant of this process occurred in June 2000 at the former Low Lands Colliery, Co Durham (UK). The brick-lined shaft at this site had been capped with a 0.3m concrete slab when the shaft was abandoned in the 1960s. Exceptionally high heads were recorded throughout the local coalfield area in early 2000, following the record wet winter of 1999-2000. Not only did the outburst at this site destroy the shaft cap, but the water which emerged from the shaft was so rich in ferrous iron that the entire downstream course of the adjoining river was stained with ochre as the Fe^{2+} oxidised in the channel. Shaft-cap destruction by pressurised mine water has occurred more than once in this district: in

1979, a factory in the nearby town of St Helen Auckland had to be demolished after pressurised mine water destroyed a shaft cap which was present in the foundations of the building (Younger *in press*).

Table 1. Summary of some recent cases of erosion of mine voids by flowing water in abandoned UK deep mines, and some consequences of this phenomenon.

Site (*MINE TYPE*)	Year	Event	Consequences	Source
Randolph Pit (Fife, Scotland) (*COAL*)	1985	Decant of deep mine water to adjoining Michael Colliery by erosion of pathway through goaf	Large increases in water make and turbidity at Michael Colliery	Younger and LaPierre (2000)
Gripps Level (Leadhills, Scotland) (*LEAD*)	1991	Adit roof collapse due to erosion by water led to impoundment of a large volume of mine water	Tension cracks formed on hillside and began to discharge water, threatening village below with severe flooding.	Schmolke (1998)
Nangiles Adit, Wheal Jane (Cornwall, England) (*TIN / ZINC*)	1992	Unanticipated accumulation of ~ 50 Ml of acidic water behind a pile of roof-fall debris, which suddenly gave way, releasing water.	Spectacular outburst of acid water caused major ochreous plume in Fal Estuary, causing public outcry which led to major treatment scheme.	Younger and LaPierre (2000) and references therein
Spittal (Berwick, England) (*COAL*)	1998	Erosion of adit blockage by impounded water	Flooding of 19 houses and a hotel with ochreous water	Younger (2001a)
Gwenffrwd (South Wales) (*COAL*)	1999	Erosion of roof-fall impoundment in lower adit led to down-draining of water hitherto decanting via upper adit.	New emergence point was distant from existing RAPS-based treatment plant; local river became polluted once more	Younger and LaPierre (2000)
Tailrace Level (Co Durham) (*FLUORITE*)	2002	Erosion of adit by mine water leads to collapse, impounding mine water which then decants at a new point 250m upstream of the existing treatment plant site	Increased the length of the Rookhope Burn which is polluted and made long-term treatment more difficult to implement.	J Teall, Environment Agency (pers. comm.)
Sheephouse (West Yorks) (*COAL*)	2002	Erosion of adit by water led to relocation of mine water discharge	Temporary flooding of the main Sheffield - Manchester road with ochreous water: messy and a hazard to traffic	A. England, IMC Ltd (pers. comm.)

Fault reactivation

Faults play an ambiguous role in mining hydrogeology (Younger and Adams 1999), functioning variously as conduits for, or barriers to, groundwater flow. Which role a given fault will play depends on local geological conditions and the state of the fault plane itself (e.g. is it clean, or filled with fault-gouge). In major UK coalfields, the largest faults (which have throws of as much as 200m) displace the strata to such an extent that mine workings were usually discontinued on either side of them. After mine abandonment, these large faults tend to function as barriers to mine water flow (Younger 1998). On the other hand, some of the lesser faults (i.e. those with throws of up to a few tens of metres) caused significant nuisance in some sub-aquifer workings, as they reduced the effective cover to water-bearing rocks, thus allowing ordinary longwall faces to induce extraordinarily voluminous feeders (Saul 1970). The interaction of mine water and fault planes appears not to be limited to the period of longwall extraction. For instance, since the abandonment of the last deep coastal collieries of County Durham (northern England) in 1994, faults which had previously been utterly dormant have begun to move again (Young and Culshaw 2001). The displacements, though relatively modest and almost aseismic in comparison with fault movements in the world's tectonically active regions, are sufficient to cause severe damage to buildings, and cracking of public road surfaces. The timing of the renewed fault movement appears to coincide with the flooding of horizons of extensive abandoned coal workings which penetrate the faults. Similar examples of fault reactivation coinciding with flooding of deep workings are also recorded from the coalfields of the English Midlands and South Wales (Donnelly 2000).

Deterioration of water quality during mine flooding

By far the most widespread and environmentally damaging consequence of deep mine abandonment is the generation and surface discharge of poor quality mine water. Being the major problem associated with mine abandonment, it has also received the greatest coverage in print, and will therefore be dealt with only briefly here, by outlining the significant advances in the understanding and characterisation of the problem which have emerged in the UK in recent years.

Dissolution of 'acid-generating salts'

Most classic descriptions of the phenomenon of 'acid mine drainage' focus on the direct oxidation of pyrite to form dissolved iron, sulphate and proton acidity, typically described by a summary reaction such as:

$$FeS_2 + \frac{7}{2}O_2 + H_2O \rightarrow Fe^{2+} + 2SO_4^{2-} + 2H^+ \qquad [1]$$

While this may be an adequate model for surface mines or waste rock piles in which frequent rainfall washes most surfaces, it is not very appropriate for working deep mines. In a deep mine, the water make normally enters via a few major feeders, and is led by a limited number of drainage pathways to dewatering sumps. Thus most of the mine is not in contact with flowing water, albeit limited moisture is ubiquitous in the form of the humidity of the mine atmosphere. Under these conditions, pyrite oxidation does not proceed all the way to dissolved products in one reaction sequence, but rather results in the pyrite being replaced by secondary minerals. This occurs by reactions such as:

$$3\,FeS_2 + 11\,O_2 + 16\,H_2O \rightarrow Fe^{(2+)}Fe_2^{(3+)}(SO_4)_4.14\,H_2O + 2\,SO_4^{2-} + 4\,H^+ \quad [2]$$

in which pyrite (FeS_2) is weathered to form a ferrous-ferric hydroxysulphate mineral, in this case römerite. Römerite is only one of a number of similar minerals formed by wetathering of pyrite in water-sparse environemnts; other such minerals include melanterite, coquimbite, szomolnokite, copiapite, aluminocopiapite and various types of jarosite (Bayless and Olyphant 1993). Besides the acidic metals these minerals contain, they also tend to sorb protons (such as the four H^+ ions shown on the RHS of the preceding reaction). These minerals have been collectively termed "acid generating salts" (AGS) (see Bayless and Olyphant 1993; Younger et al. 2002). AGS tend to form conspicuous white / yellow efflorescent crusts on pyritic beds within mines. When the beds hosting these efflorescent crusts are finally flooded after mine closure, the AGS dissolve congruently and (with the possible exception of jarosite; Saaltink et al. 2002) extremely rapidly. In doing so, they impart to the water a low pH (typically around 2.5) and extremely high dissolved concentrations of problematic metals (Fe, Al, Zn, Cu, Cd etc). It should be noted that acidic pH values can even develop where the orebody is limestone-hosted, so fast are the dissolution kinetics of the AGS compared to that of calcite (e.g. Johnson and Younger 2002). Younger (1998) termed this abrupt deterioration in mine water quality a 'geochemical trauma', and since it has often caught unsuspecting mine owners off their guard, it can lead in turn to a management trauma. In essence, therefore, the formation of AGS during the working of a deep mine results in large-scale storage of soluble acidity in the unsaturated zone, but this acidity is rapidly released to solution upon flooding of the mine voids after dewatering ceases.

'Stone dust' and other emollients

Fortunately AGS are not the only minerals contributing to the solute loadings of waters rising through abandoned mine workings. Other minerals with relatively rapid dissolution kinetics serve to counteract the acidification of mine water. Most notable is calcite, but dolomite can also be important locally. However, neither of these two minerals are particularly abundant in UK Carboniferous coal-bearing sequences. Far more common in that geological setting are various iron-rich carbonate minerals, such as siderite ($FeCO_3$) and ankerite (ferroan dolomite). While these will locally react with acidic waters to neutralise them, the beneficial impact on pH persists only as long as the water remains anoxic, for upon aeration the dissolved ferrous iron imparted by these minerals is rapidly oxidised to the ferric form, each mole of which will hydrolyse to release three moles of proton acidity. For this reason, the iron carbonate minerals have no net neutralisation potential (see Younger et al. 2002, for further discussion / literature review).

In terms of naturally-present carbonate minerals, therefore, the UK Coal Measures are relatively poor in carbonate-based neutralisation potential. However, an ubiquitous mine-safety practice inadvertently serves to counteract this paucity of natural carbonates: in order to minimise the risk of coal powder explosions, mining regulations in the UK (as in many other countries) require the liberal spreading of finely-comminuted limestone powder in all major roadways. This limestone powder (referred to as 'stone dust' in the industry) is so fine-grained as to be somewhat hydrophobic, but once suspended in the water column its fine-grained nature ensures it is highly chemically reactive, serving to rapidly neutralise dissolved acidity. The dissolution of stone dust was inferred by Wood *et al.* (1999) to explain the coincidence of peak Fe, SO_4, Ca and HCO_3^- concentrations during the 'first flush' period of several abandoned underground mines in Scotland.

Even where stone dust was never used (as in metal mines) or where it has been totally dissolved, neutralisation reactions clearly continue to influence mine water quality (e.g. Banwart and Malmström 2001), even in the absence of *any* naturally-occurring carbonate minerals (e.g. Strömberg and Banwart 1994). It is now clear that such neutralisation reflects the incongruent dissolution of aluminosilicate minerals, which have very sluggish dissolution kinetics when compared with carbonates (Banwart *et al.* 2002). Aluminosilicate minerals are, of course, abundantly present in virtually all mine water settings, and they can serve to significantly affect the proton balance of a given water provided this has sufficient hydraulic residence time in the strata (e.g. Younger *et al.* 2002). Issues of 'hydraulic residence time' are only one manifestation of an important (but frequently overlooked) facet of the geochemical evolution of mine waters: the mediating role of hydrological processes, which control both the availability of key reactants and the transport of dissolved products from the reaction sites. The following section considers recent findings on this topic in a little more detail.

The longevity of mine water pollution

It has been recognised since at least the 1970s (Cairney and Frost 1975) that the deterioration in mine water quality which occurs during the flooding of previously dry workings is a temporary phenomenon, so that water quality gradually improves again once a steady outflow from the workings (by natural decant or pumping) is re-established. The fact that mine water quality is subject to natural amelioration processes over time has considerable engineering significance, for it provides a conceptual framework within which holistic, long-term remedial strategies for polluted waters can be developed and sustained (e.g. Younger 1997, 2001). For this reason, various UK authors have studied this phenomenon in some detail, most notably Frost (1979), Glover (1983) and Younger (1997, 1998, 2000a). The current state-of-the-art in predicting the duration of various episodes of contamination in a given mine water discharge is set out in detail by Younger (2000a), and is thus only summarised below.

The peak concentrations of contaminants in water flowing from a flooded deep mine typically occur shortly after the time at which water first begins to decant from the workings. Thereafter, the contaminant concentrations tend to decline exponentially (Fig. 1) during a period of water quality improvement which Younger (1997) termed "the first flush". After completion of the first flush, a long-term asymptotic water quality is established which represents the dynamic balance between ongoing pollutant release (e.g. by seasonal formation and dissolution of AGS within the zone of water table fluctuation) and pollutant attenuation (dissolution of carbonates, aluminosilicates etc). Of key interest to design engineers are the following aspects of this time-variant behaviour:

- the peak contaminant concentration
- the duration of the first flush
- the long-term 'asymptotic' contaminant concentration

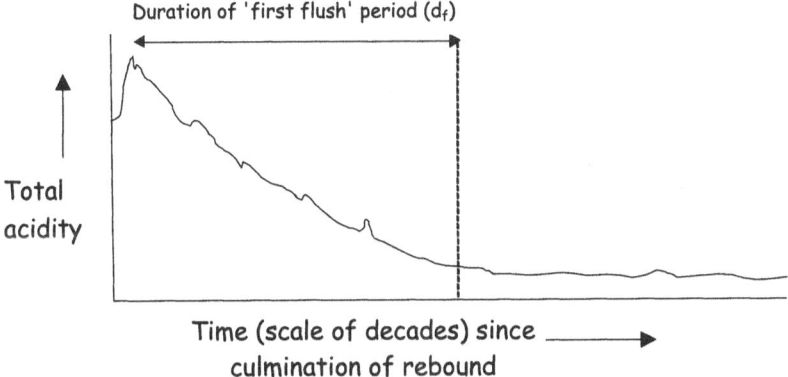

Fig. 1. Typical exponential decline in contaminant concentrations observed over time after water begins to discharge from newly-flooded deep mine workings.

Peak contaminant concentrations

Scrutiny of all readily-available records from UK mines has supported the development of predictive criteria relating peak contaminant concentrations (especially Fe concentrations) to the pyrite content of the strata. Where direct assays of pyrite content are not available (as is often the case), the more widely-available data for total sulphur content can be used instead. On this basis, the following broad classes of peak iron concentration are estimated (Table 2):

Table 2. Estimating peak Fe concentrations in water flowing from newly-flooded deep mines as a function of the total sulphur content of the mined horizons.

Total sulphur content (wt %)	Typical observed range of values for peak Fe concentration (mg/l)
< 1	0.01 - 0.5
1 - 2	0.5 - 100
2 - 3	100 - 350
3 - 4	350 - 1200
4 - 5	1200 - 1500

In many cases, a multiple-seam sequence will have been worked in which each seam has a different mean sulphur content. Predictive use of Table 2 for such cases demands the calculation of a weighted average total sulphur content, for which the weights are provided by the relative extent of workings in each of the seams.

Where no total sulphur data are available at all, the only recourse is to estimate likely pollutant generation potential on geological grounds (Younger 2000a, 2001), such as proximity of the mined bed to a 'marine band' (i.e. a bed which clearly underwent deposition / diagenesis by marine waters, and which is therefore likely to contain reduced inorganic sulphur compounds such as pyrite).

Duration of the 'first flush'

The first flush phenomenon is essentially a process of displacement of the initial flooded volume of mined ground by less-contaminated water which recharged the mined ground after completion of rebound. As such, the duration of the first flush is principally a function of the void volume of the flooded mined system and the rate of recharge. While it is possible to estimate the void volume of the flooded mined system from mine survey records, considerable uncertainties are associated with such estimates for at least two reasons:

(i)   the volume of voids includes not only those represented by open mine voids, but also those for areas of 'goaf' (collapsed roof strata), for which direct measurement is usually impractical, and reliable estimation techniques are few (e.g. Rogoż 1994)

(ii)   There is usually insufficient information available to allow accurate estimation of the volume of unmined strata around the mine voids which have been dewatered by mine drainage (e.g. Adams and Younger 2002).

If the volume of voids is difficult to estimate accurately, the situation is even worse in relation to recharge rate, which is notoriously difficult to characterise in virtually all hydrogeological systems (Lerner *et al.* 1990).

Fortunately the very process of mine water rebound sheds some light on the magnitude of both parameters, for rebound rate is also a function of void volume and recharge rate. Logically, therefore, a functional relationship ought to exist between the duration of the first flush period (d_f - see Fig. 1) and the time which it took for the mine to flood completely following the cessation of pumping (d_r). From analysis of available records for UK deep mines, Younger (2000a) found that:

$$d_f \approx 4 \cdot d_r \quad [3]$$

Beyond the first flush: long term prognoses

Once the limit time d_f has finally been exceeded, an 'asymptotic water quality' will persist until exhaustion of the various pollutant source / sink minerals in the zones of ongoing oxidation. To a first approximation (Younger 2000a), it is normally found that the asymptotic concentrations of key solutes after the first flush are typically an eighth to a tenth of the initial peak concentrations (e.g. divide the Fe concentrations in Table 2 by some factor in the range 8 - 10). However, as reliable data upon which this approximation is based available relate only to time-scales of up to 100 years after mine abandonment (e.g. Wood *et al.* 1999), it is not possible to extend predictions to the long term in this simple manner without site-specific geochemical assessments. For instance, where sulphides are likely to be depleted before carbonates, for instance, alkaline conditions will persist in the long term. However, where the carbonates are likely to be depleted before the sulphides, re-acidification of mine waters may eventually occur. (Examples of the latter phenomenon are now beginning to emerge at certain sites in northern Europe; e.g. Younger 2000b; Strömberg and Banwart 1994). Where accurate site-specific data exist for discharge flow rates and chemistry, and for the mineralogical composition of the mined strata, it is possible to use mass-balance based geochemical modelling techniques to obtain more credible estimates of the rates of depletion of the key minerals, and therefore to calculate the likely duration of a given level of acidity (e.g. Banwart and Malmström 2001; Banwart *et al.* 2002).

A key point to bear in mind in such exercises is that, in flooded mines, sulphide oxidation is effectively limited to the unsaturated and seasonally-saturated zones only, and simply cannot occur at any great depth below the water table. On the other hand, dissolution of buffering carbonate and aluminosilicate minerals can occur both above and below the water table. Hence a simple mineralogical balance of sulphides versus carbonates (such as is yielded by conventional 'acid-base accounting' techniques) does <u>not</u> tell use whether a system will stay alkaline or eventually become acidic; rather, such data must be interpreted within a holistic framework which takes present and future hydrogeological conditions into account (e.g. Hattingh *et al.* 2002; Younger *et al.* 2002).

Rebound modelling

For effective planning of remedial interventions to intercept polluted mine waters before they cause damage to the surface environment, it is necessary to be able to predict d_r (equation [3]). A number of techniques have been derived for this purpose, which have been extensively reviewed by Younger and Adams (1999). Of the various predictive methods developed to date, two categories have been found to be particularly useful. These are briefly described below. Whichever modelling technique is eventually selected for a particular application, parameterisation of the models is a major undertaking for all but the simplest of mined systems (Sherwood 1997; Adams and Younger 2001, 2002). To assist the process, the use of 3-D visualisation software has recently been shown to provide a cost-effective means for manipulating the vast quantities of spatially-distributed data upon which conceptual hydrogeological modelling of large systems of inter-connected mine workings must be based (Dumpleton *et al.* 2001).

Semi-distributed lumped-parameter models

Also known as 'pond models' (e.g. Younger and Sherwood 1993; Sherwood 1997; Younger and Adams 1999; Banks 2001; Whitworth 2002) and 'box models' (e.g. Rogoż 1994; Gatzweiler *et al.* 1997), these models reduce the complexity of both mined void volumes and recharge processes by 'lumping together' large volumes of extensively-interconnected mine voids as single hydrological units ("ponds"). While such models can be implemented using purpose-written codes (e.g. the GRAM model; Younger and Sherwood 1993; Sherwood 1997) they are often sufficiently simple that they can be implemented within a spreadsheet environment (e.g. Banks 2001; Whitworth 2002). Within any one pond, the water table will be virtually flat, and water can migrate between neighbouring ponds only via a limited number of 'decant features' (such as old mine roadways, and / or areas of coalesced goaf; see Sherwood 1997 for real examples of such features). In practice, the identification of decant features can be greatly aided by means of appropriately-designed mine water tracing experiments (for guidance, see Wolkersdorfer 2002).

In applying semi-distributed 'pond' models to real systems, it is important to determine whether the rate of water inflow to the flooding working during rebound will be independent of, or dependent upon, the current hydraulic head within the workings (Younger and Adams 1999). In some cases, inflows during rebound have proven to be head-independent, and the models which assume a constant recharge rate to the workings (e.g. the GRAM code; Younger and Sherwood 1993; Sherwood 1997) have been successfully applied (Younger and Adams 1999; Burke and Younger 2000). In other cases, much of the inflow to the mine voids proves to be head-dependent, so that a modified version of the GRAM algorithm must be used, such as that described by Banks (2001) and applied to an abandoned base-metal mine in Bolivia by Banks *et al.* (2002).

In general, pond-based rebound models are best applied to relatively large-scale mine systems (e.g. underlying areas of hundreds to thousands of km^2). The principal applications of the GRAM model to date have been for systems of several hundred to more than 2000 km^2 in aerial extent. Where a smaller system of voids must be analysed in greater detail, it is probably best to use the more physically-realistic models described in the following section.

Physically-based, distributed models

In the most generic conceptualisation, deep mines undergoing flooding are most reasonably regarded as systems of conduits (representing mine roadway networks / stopes, in which flow may well be turbulent) routed through heterogeneous, variably-saturated porous media (representing the enclosing rockmass, both intact strata and rocks which have fractured in response to mining of voids nearby). Models based on this type of conceptualisation are best based on orthodox physical equations, and demand spatial discretisation and distribution of parameter values. For this reason, they are termed "physically-based, distributed models", and they are invariably solved using numerical methods such as finite differences (in distinction to the largely analytical methods used to solve 'pond' models).

There are a number of ways in which a physically-based model suitable for mined voids may be represented mathematically. For instance, a multiple-fracture system based on Navier-Stokes theory is one option. In many natural aquifers, in which the "conduits" are irregular, planar fractures, this approach is probably most appropriate. In most mined systems, however, the major conduits are tube-like roadways which are better represented as pipes rather than planar fractures (Younger and Adams 1999). A number of different representations of turbulent flow in a pipe network are commonly used in practice (e.g. the Darcy-Weisbach formula and the Hazen-Williams formula). Detailed analyses of the performance of the most well-used formulae for the range of hydraulic conditions likely to be encountered in most mine water rebound problems found no great difference in the results produced between one formula and another (Sherwood 1997; Younger and Adams 1999). For this reason, the most computationally-inexpensive formulation is the logical choice for a physically-based rebound modelling code. The VSS-NET is a purpose-written code in which a 3-D pipe network formulation (based on the Darcy-Weisbach formula and the Gradient Algorithm network solver) is routed through a 3-D, column-oriented (as opposed to layer-oriented), block-centred finite difference grid which is configured to solve for saturated-unsaturated flows (Adams and Younger 2001).

The gains in realism associated with using complex physically-distributed models such as VSS-NET are offset by the relatively high costs of parameterisation (either in terms of gathering the spatially-distributed data needed to support parameter selection, or in the sheer-scale of data manipulation that may be needed before runs can be commenced). For this reason, VSS-NET and similar models are probably best applied to systems comprising just one or two adjoining mines, underlying areas of $< 100 \text{ km}^2$.

Conclusions

Lessons learned from the closure of British deep mines (principally coal mines, but to a lesser extent also gold and base metal mines) over the last decade have yielded a range of practical conceptual modelling approaches and software tools which are potentially applicable to the prediction and management of deep mine closure in many other geological settings. Some of the key advances relate to the following phenomena / techniques:

- exacerbation of land subsidence by processes including slaking of seat-earths and fault reactivation by rising mine waters
- the role of AGS in the deterioration of mine water quality during flooding, and the roles of carbonates, aluminosilicates and 'stone dust' in buffering
- the duration of the first flush (typically four times the time it took for the mined system to flood) and peak pollutant concentrations (typically ten times greater than long-term values)

-   scale-appropriate modelling of mine water 'rebound' processes (i.e. flooding of extensive networks of underground voids) using both semi-distributed 'pond' models and physically-based, hybrid pipe network / porous media modelling codes such as VSS-NET

Application of generic characterisation / modelling techniques which have now been widely-tested for UK systems provides the essential background for long-term planning of sustainable management of abandoned underground mines (Younger 2000a), not least the rational design of active and passive water treatment systems (e.g. Younger *et al.* 2002; Wolkersdorfer and Younger 2002).

References

Adams R, Younger P L (2001) A strategy for modeling ground water rebound in abandoned deep mine systems. Ground Water 39: 249-261

Adams R, Younger P L (2002) A physically based model of rebound in South Crofty tin mine, Cornwall. In Younger P L, Robins N S (editors) Mine water hydrogeology and geochemistry. Geological Society, London, Special Publication. (in press).

Banks D (2001) A variable-volume, head-dependent mine water filling model. Ground Water 39: 362–365

Banks D, Holden W, Aguilar E, Méndez C, Koller D, Andia Z, Rodriguez J, Saether O M, Torrico A, Veneros R, Flores J (2002) Contaminant source characterisation of the San José Mine, Oruro, Bolivia. In Younger P L, Robins N S (editors) op. cit.

Banwart S A, Malmström M E (2001) Hydrochemical modelling for preliminary assessment of minewater pollution. J. Geochem. Exploration, 74: 73 - 97

Banwart, S.A., Evans, K., Croxford, S (2002) Predicting mineral weathering rates at field scale for mine water risk assessment. In Younger P L, Robins N S (editors) op. cit.

Bayless E R, Olyphant G A (1993) Acid-generating salts and their relationship to the chemistry of groundwater and storm runoff at an abandoned mine site in southwestern Indiana, USA. J. Contam. Hydrol. 12: 313-328

Bekendam R F, Pottgens J J (1995) Ground movements over the coal mines of southern Limburg, the Netherlands, and their relation to rising mine waters. Int. Assoc. Hydrol. Sci. Publ. 234: 3-12

Burke S P, Younger P L (2000) Groundwater rebound in the South Yorkshire Coalfield: a first approximation using the GRAM Model. Q. J. Eng. Geol. Hydrogeol. 33: 149–160

Cairney T, Frost R C (1975) A case study of mine water quality deterioration, Mainsforth Colliery, County Durham. J. Hydrol. 25: 275-293

Donnelly L J (2000) The reactivation of geological faults during mining subsidence from 1859 to 2000 and beyond. Trans. Inst. Mining & Metall. (Sect A: Mining Technol.) 109: A179-A190

Dumpleton S (2002) Effects of longwall mining in the Selby Coalfield on the piezometry and aquifer properties of the overlying Sherwood Sandstone. In Younger P L, Robins N S (editors) op. cit.

Dumpleton S, Robins N S, Walker J A, Merrin P D (2000) Mine water rebound in South Nottinghamshire: risk evaluation using 3-D visualization and predictive modelling. Q. J. Eng. Geol. & Hydrogeol. 34: 307-319

Frost R C (1979) Evaluation of the rate of decrease in the iron content of water pumped from a flooded shaft mine in County Durham, England. J. Hydrol. 40: 101-111

Gatzweiler R, Hähne R, Eckart M, Meyer J, Snagovsky S (1997) Prognosis of the flooding of uranium mining sites in east Germany with the help of numerical box-modeling. Proc 6th Intl. Mine Water Assoc. Congress (Bled, Slovenia, 8 - 12th September 1997). Vol. 1: 57-64

Glover H G (1983) Mine water pollution - an overview of problems and control strategies in the United Kingdom. Water Sci. Technol. 15: 59-70

Hattingh R, Pulles W, Krantz R, Pretorius C, Swart S (2002) Assessment, prediction and management of long-term post-closure water quality: A case study - Hlobane Colliery, South Africa. In Younger P L, Robins N S (editors) op. cit.

Johnson K L, Younger, P L (2002) Hydrogeological and geochemical consequences of the abandonment of Frazer's Grove carbonate hosted Pb/Zn fluorspar mine, North Pennines, UK. In Younger P L, Robins N S (editors) op. cit.

Lerner D N, Issar A S, Simmers I (1990) Groundwater recharge. A guide to understanding and estimating natural recharge. Intl. Contribs. to Hydrogeology, Volume 8. Intl. Assoc. Hydrogeologists. Verlag Heinz Heise, Hannover. 345pp.

Murchison D G, Westoll T S (editors) (1968) Coal and coal-bearing strata. Oliver and Boyd, Edinburgh. 418pp.

NCB (1975) The subsidence engineers' handbook. National Coal Board, Mining Department, London. 111pp.

Orchard R J (1975) Working under bodies of water. Mining Engineer 170: 261-270

Parkin G, Adams R (1998) Using catchment models for groundwater problems: evaluating the impacts of mine dewatering and groundwater abstraction. In Wheater H, Kirby C (editors) Hydrology in a Changing Environment. Volume II. Wiley, Chichester. 269–279.

Rogoż M (1994) Computer simulation of the process of flooding up a group of mines. Proc. 5th Int. Mine Water Congress (Nottingham, UK). Vol. 1: 369–377

Saaltink M W, Domènech C, Ayora C, Carrera J (2002) Modelling the oxidation of sulphides in an unsaturated soil. In Younger P L, Robins N S (editors) op. cit.

Saul H (1970) Current mine drainage problems. Trans. Inst. Mining & Metall. (Sect. A), 79: A63 - A80

Schmolke C M R (1998) Leadhills, Scotland – flood and pollution threat from disused lead workings. In Fox H R, Moore H M, McIntosh AD (editors) Land Reclamation: Achieving Sustainable Benefits. Balkema, Rotterdam. pp. 517 – 523

Sherwood J M (1997) Modelling minewater flow and quality changes after coalfield closure. Unpublished PhD Thesis, Dept Civil Eng., University of Newcastle, UK. 241pp.

Smith J A, Colls J J (1996) Groundwater rebound in the Leicestershire Coalfield. J. Chartered Inst. Water & Env. Mgt. 10: 280-289

Strömberg B, Banwart S A (1994) Kinetic modelling of geochemical processes at the Aitik mining waste rock site in northern Sweden. Appl. Geochem. 9: 583-595

Whitworth K R (2002) The monitoring and modelling of mine water recovery in UK coalfields. In Younger P L, Robins N S (editors) op. cit.

Wolkersdorfer C (2002) Mine water tracing. In Younger P L, Robins N S (editors) op. cit.

Wolkersdorfer C, Younger P L (2002) Passive Grubenwassereinigung als Alternative zu aktiven Systemen. Grundwasser 7: 67-77

Wood S C, Younger P L, Robins N S (1999) Long-term changes in the quality of polluted minewater discharges from abandoned underground coal workings in Scotland. Q. J. Eng. Geol. 32: 69–79

Young B, Culshaw M G (2001) Fissuring and related ground movements in the Magnesian Limestone and Coal Measures of the Houghton-le-Spring area, City of Sunderland. British Geological Survey, Technical Report WA/01/04

Younger P L (1997) The longevity of minewater pollution: a basis for decision-making. Sci. Total Env. 194/195: 457-466

Younger P L (1998) Coalfield abandonment: geochemical processes and hydrochemical products. In Nicholson K (editor) Energy and the Environment. Geochemistry of Fossil, Nuclear and Renewable Resources. (Society for Environmental Geochemistry and Health). McGregor Science, Aberdeen. pp 1 - 29

Younger P L (2000a) Predicting temporal changes in total iron concentrations in groundwaters flowing from abandoned deep mines: a first approximation. J. Contam. Hydrol. 44: 47-69

Younger P L (2000b) Holistic remedial strategies for short- and long-term water pollution from abandoned mines. Trans. Inst. Mining and Metall. (Sect. A: Mining Technol.) 109: A210 - A218

Younger P L (2001) Mine water pollution in Scotland: nature, extent and preventative strategies. Sci. Total Env. 265: 309-326

Younger P L (in press) Coalfield closure and the water environment in the EU: Perspectives from the Fifth Framework research community. Trans. Inst. Mining and Metall.

Younger P L, Adams R (1999) Predicting mine water rebound. Environment Agency R&D Technical Report W179. Bristol, UK. 108pp.

Younger P L, LaPierre A B (2000) "Uisge Mèinne": Mine water hydrogeology in the Celtic lands, from Kernow (Cornwall) to Ceapp Breattain (Cape Breton, Nova Scotia). In Robins N S, Misstear B D R (editors) Groundwater in the Celtic regions: Studies in Hard-Rock and Quaternary Hydrogeology. Geological Society, London, Spec. Publ. 182: 35-52

Younger P L, Robins N S (editors) (2002) Mine water hydrogeology and geochemistry. Geological Society, London, Special Publication. (in press).

Younger P L, Sherwood J M (1993) The Cost of Decommissioning a Coalfield: Potential Environmental Problems in County Durham. Mineral Planning 57: 26-29

Younger P L, Banwart S A, Hedin R S (2002) Mine Water: Hydrology, Pollution, Remediation. Kluwer Academic Publishers, Dordrecht. (ISBN 1-4020-0137-1). 464pp.

Strategy Concept Elbe

Heinrich Reincke[1], Stephanie Hurst[2], Petra Schneider[3]

[1] ARGE ELBE - Arbeitsgemeinschaft für die Reinhaltung der Elbe, Wassergüte-stelle Hamburg, Neßdeich120-121, D-21129 Hamburg, wge@arge-elbe.de
[2] Sächsisches Landesamt für Umwelt und Geologie, Zur Wetterwarte 11, D-01109 Dresden, Germany. e-mail: stephanie.hurst@lfugdd.smu.sachsen.de
[3] Hydroisotop-Piewak GmbH, Oberfrohnaer Str. 84, D-09117 Chemnitz, hydroiso-top-piewak@t-online.de

Abstract. The pollution of the Elbe River and especially the catchment area of the tributary Mulde with rising groundwater-level in the mining areas and tailings of the old mines in the Ore Mountains is one of the great environmental problems of this catchment. In 1998 the strategy concept was installed to improve the Elbe water quality, reducing the impacts of uranium mining on the Elbe River. One main focus in the strategy concept was set on the use of passive water treatment methods: use of reactive materials and wetlands.

Introduction

The Elbe River is one of the major rivers in the Western Europe. From its spring in the Giant Mountains (Czech Republic) to its mouth at the North Sea near Cuxhaven (Germany) it covers a distance of 1091 kilometers and a catchment area of 148 268 km^2 - one third of it located in the Czech Republic and two thirds in the Federal Republic of Germany (figure 1). Along its way the catchment drains some of north and central Europe's major cities including Prague, Dresden, Berlin and Hamburg.

The Elbe River arises in the Giant Mountains and flows through the Bohemia Chalk Basin, the Mid-Bohemia Highlands and the Elbe Sandstone Mountains before it reaches the middle course downstream the Castle Hirschstein (between the cities Meißen and Risa), the Middle and North German Lowland. Downstream from the city of Lauenburg there is the lower course of the river Elbe, comprises the stretch from the weir at Geesthacht to Cuxhaven and further on the North Sea. This is the tidal part, that means, that the flow is controlled by the tide (see figure 1). The water quality in the catchment area of the Elbe River has highly improved in the last twelve years (ARGE ELBE, 2000).

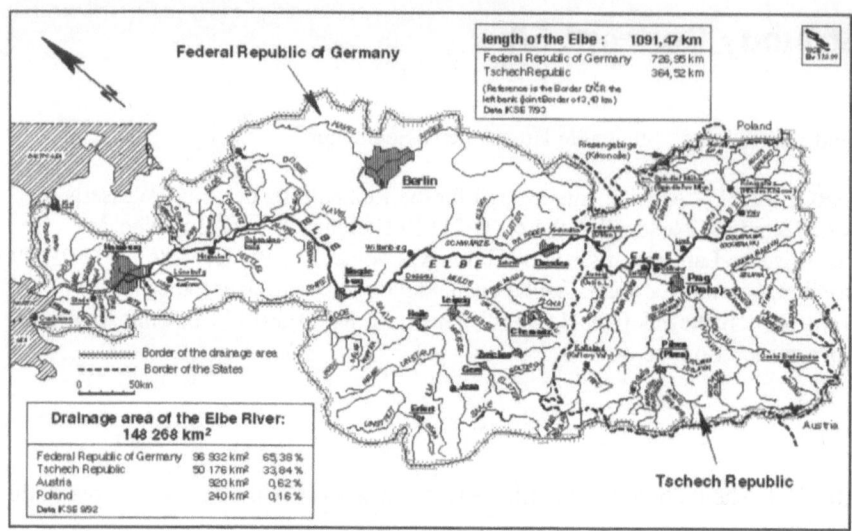

Drainage area of the Elbe River

Fig. 1. Drainage area of the Elbe river.

239 great municipal treatment plants were built since 1990, with a capacity of 25.5 million inhabitant equivalents. You can find 61 in the Czech Republic, 177 in Germany and 1 plant in Austria. All communities with more than 20 000 inhabitants in the catchment area of the Elbe River have modern treatment plants now. Due to technology variations in the industrial and chemical plants and a better handling of the industrial wastewater the share pollutants from industrial areas has decreased.

More success has been watched concerning the number of fish-species comparing the time of German Reunification and now. Now we have 94 fish-species in the whole catchment area, 36 of it in the Czech area. Salmons are expected in the tributaries of the River Elbe of "Swiss Bohemia" as soon as possible. Never the less there are some "sorrows of tomorrow", which have to be solved in the next years, to get a good ecological condition of the tributaries in the old catchment area of Elbe River according to the European Water Framework Directive (2000/60/EG). The pollution of the river Elbe and especially the catchment area of the tributary Mulde with rising groundwater-level in the mining areas and tailings of the old mining in the Ore Mountains ist one of the great sorrows.

In July 1998 the conference of the Elbe River responsible ministers decided to develop a strategy concept to improve the Elbe water quality, reducing the impacts of uranium mining on the Elbe River. Beyond the existing research on geochemical engineering methods to treat water with increased concentrations of heavy metals from ore mining has to be pointed out. Strategies to avoid or to increase heavy metal concentrations in the Elbe catchment area have to be developed and possibilities to finance research have to be found.

Feasibility studies of Passive Water Treatment Methods

The effective fixation of heavy metals on the surface above the water table is not simple to realize. For very big and diffuse emissions of reservoirs with often more than 10 – 100 years residence time, especially from mining (mine-buildings, surface mining, deposits, tailings) costly technical solutions are not tenable of economical reasons. In many cases there must be realized a combination of monitoring, based on a fixing of the sources (isolation of the contaminant species, pH-rising, multi barrier system) and a handling afterwards in similar-to-nature systems (wetlands).

With respect to the costs of mining remediation passive water treatment systems are the only possible methods for a long term treatment of waters from mine sites. The passive treatment methods should be applicable with a minimum of energy, manpower and without the need of permanent renewal of chemicals. For the treatment of surface waters internationally mainly constructed wetlands are in practice. For the treatment of groundwater contamination there are only a few sites supplied with reactive permeable walls consisting of zero valent iron. Experiences from hydrogeochemical and biogeochemical research and from conventional water treatment methods are not much tested yet and there are still not sufficient investigations to optimize existing methods (Hurst, 2001).

One of the most important aspects of uranium mining remediation is the long-term durability of remediation methods. In this context the development of mine and seepage water quality is of special interest. As another item of research the combination of biological and chemical methods is of interest. In the strategy concept Elbe the focus was put on three topics concerning passive water treatment methods:

- constructed wetlands
- reactive materials for in situ mine water treatment
- infiltration and injection methods

In the following the concept will be introduced in details and will be reported about the update for preparing project sketch and the following realization of a concrete project for application reactive covering systems and geotechnical handling concepts of the old mining sites. First research results are already presented. It will be expected that the presented research topics will be realized with the support of the Federal Ministry of Research and the countries of Germany, especially the countries Saxony and Thuringia to realize the demands of a good ecological condition according to EC Water Framework Directive also in the catchment areas with old mining (metallogenic catchment type, Schneider et. al. 2002).

Feasibility study project: Reactive Covering Systems

The long-term mitigation of pore waters of acid waste rock dumps formed during uranium mining requires new remediation approaches. A pilot study was performed to evaluate the feasibility of reactive covering systems (RCS) as part of an alternative mineral covering system for uranium mining dumps (Schneider et al 2002). This kind of technology is a combination of geotechnical and geochemical methods. Some of the effluent waters of the rock dumps are characterized by pH values as low as 3 due to residues of acid from ore processing and pyrite oxidation.

Due to the high costs of classical pump-and-treat technologies, reactive barriers have been used increasingly in the last decade as an alternative strategy for remediation of water (U.S. Dept of Energy 1996). Reactive barriers are zones of high geochemical reactivity, where contaminants are immobilized in-situ by redox processes, co-precipitation, adsorption or biological processes. Usually they are classified as naturally formed or man-made (artificial) geochemical barriers. The development of reactive barrier systems for removal of radionuclides and heavy metals from percolating waters requires an improved understanding of the elementary processes that control the interactions between dissolved contaminants and barrier material (Schneider et al 2002). The reactive covering system was conceived especially or metal contaminated dumps. The high permeability of the dumped material provides the migration paths for the distribution of the reactive solution.

Principles of a reactive covering system

A special type of a reactive barrier is the reactive covering system (RCS), where a layer of the reactive material is located under the mineral soil of an alternative covering system. The reactive surface barrier will only be activated if there is a hydraulic breakdown of the mineral soil cover (Schneider et al 2002). When the covering system has lost its functionality, precipitation will percolate through the mineral soil cover and chemical reactions with the barrier will be initiated (see figure 2).

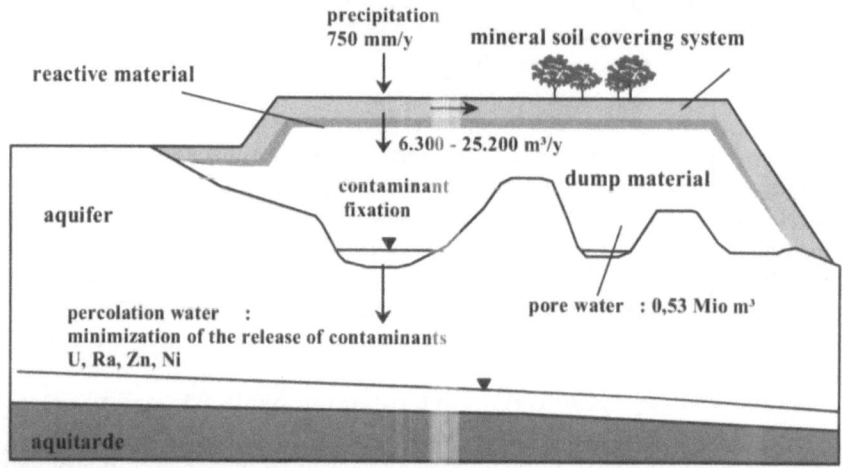

Fig. 2. Schematic cross section of a reactive covering system (Schneider et al 2002). The infiltration rate assumes the hydraulic breakdown of the soil cover.

After leaching, the dissolved reactive substances will be transported into the dump material and react there with the contaminated pore waters. The aim of this study is to evaluate the feasibility of several reactive materials with suitable chemical properties for efficient mitigation of uranium and radium-226 in an acid milieu. The results of the study should be applicable to many other uranium dumps with similar geochemical characteristics (Schneider et al 2002).

Suitable reactive materials

One topic of the investigation is to evaluate suitable reactive materials for the mitigation of radionuclides and heavy metals in an acid milieu. In a theoretical pilot study PHREEQC geochemical modeling were included equilibrium and mixing calculations to evaluate the chemical interactions between dump waters and reactive materials (Schneider et al 2002). The engineering feasibility of RCS was evaluated calculating a mass balance considering different dump water hydraulics, layer thickness and pore water concentrations. The feasibility of using several RCS-suitable reactive materials for the mitigation of radionuclides and heavy metals was evaluated for the acid mine dump Schüsselgrund (Saxony, Germany) on theoretical scale. The main data on hydrogeology, hydrology and geochemistry of the Schüsselgrund site are given in Schneider et al. 1999 and Schneider et al. 2001. The main contaminants of the pore waters are uranium (20-30 mg/l) and radium-226 (about 1 Bq/l). In addition, contaminants such as zinc (50-150 mg/l), nickel (2-4 mg/l), and sulfate (2-4 g/l) are present in the pore water.

The main findings are that a RSB of zero-valence iron (Fe^0) causes a long-term mitigation of uranium and zinc. Alkaline hydroxides ($Ca(OH)_2$, $Ba(OH)_2$) cause the mitigation of radium-226. In the case of nickel, mitigation by Fe^0 and $Ca(OH)_2$ only occurred when the dump water constituted less than 30 % of the mixing solution. Fe^0 may be the most suitable reactive material to mitigate uranium and zinc in an acid milieu. The changes in geochemical milieu by oxidation of Fe^0 cause a mitigation of uranium. In addition to redox changes, uranium-sorptive iron hydroxides will be formed after transformation of Fe^0 to Fe^{2+}. According to its chemical properties (dissolution rate), a lifetime of 2150-8500 years has been calculated for the reactive barrier (Schneider et al 2002).

Alkaline hydroxides have been identified to be the only suitable reactive material for the mitigation of radium-226. In contrast to Fe^0 and alkaline hydroxides, PO_4-compounds have no redox-effective properties. The reactivity of these materials is characterized by the formation of insoluble uranium-phosphate-complexes. The feasibility of PO_4-compounds as a RCS for uranium mitigation was not definitively determined. The theoretical study strongly suggests that the use of RSB can provide a sustainable mitigation concept for radionuclides and heavy metals in an acid milieu (Schneider et al 2002).

Taking into consideration the prognostic character of the theoretical modeling of the pilot study, the next step will be lab and field measurements. Based on the results of this feasibility study, laboratory experiments have been initiated. According to the different reactivities of the investigated barrier materials, a mixture of different reactive materials has to be considered as a combined mitigation concept. Our experiments will investigate if mixed reactive materials remain reactive for the mitigation of radionuclides and heavy metals when barrier material interactions are taken into account (Schneider et al 2002)..

Feasibility study project: Injection Methods

Another type of geotechnical methods for the mitigation of uranium mining sites are injection methods. This kind of mitigation concept will be investigated for contaminated uranium sites with a very low permeability, e.g. tailings. The technological principles of this remediation concept base on the minimization of the permeability of the contaminated source. In special drillings reaching the contaminated source will be injected clay minerals to decrease the pore volume of the tailings. The technical conception and optimization is topic of this feasibility study project. Otherwise it will be tested to combine injection methods and reactive materials in order to inject reactive solutions to cause an in-situ mitigation of contaminants.

Conclusions

In Saxony, the draining water from nearly all of the uranium mining sites flow into the Mulde River, which is a tributary to the Elbe River. The aim of the Saxonian regulatory authorities is to minimize the radionuclide concentrations in the Mulde, its sediments, and its meadowlands. Therefore, an alternative treatment method must be developed in the catchment area of the Mulde, at least until the mines have been flushed 6 to 8 times (Hurst 2002). If the mine water or seepage water is needed as drinking water, or if the runoff flows into a fishing area, measures have to be found to remove the radium and, sometimes also the arsenic, from the water. However, conventional methods of doing this have high long term costs and produces wastes that has to be disposed of. Passive treatment methods are being assessed. For passive water treatment to be effective, the hydrogeochemical system and the water chemistry have to be known very well. The potential benefits of natural attenuation processes also have to be considered. The use of passive water treatment methods will minimise the catchment management costs especially due to the demands of the EC Water Framework Directive.

References

ARGE ELBE (2000) Die Wassergüte der Elbe im Jahr 2000, published by Wassergütestelle Elbe.

Hurst S, Staupe J, Leder F (1998) German regulations referring to long-term surveillance and maintenance stewardship – in theory and in practice, U.S.DOE: Long-Term Stewardship Workshop, June 2-3, 1998

Hurst, S.; Schneider, P.: (2001) Improvement of long term effectiveness of passive water treatment systems, Proceedings on Side meeting at ICEM`01 on „Natural Attenuation in Uranium Mine and Mill Remediation" Thursday, October 4, 2001 Brügge, Belgium

Hurst S (2002) Remediation 700 years of Mining in Saxony: A Heritage from Ore Mining. J Mine Water and the Environment, Special Issue: Large Scale Modeling of the Environment 21: 3 - 6.

Schneider P, Voerkelius S, Nindel K, Forster M, Schreyer J (2001a) Release of Contaminants from Uranium Mine Waste - Laboratory and Field Experiments. In: Mine Water and the Environment. Journal of the International Mine Water Association (20): 30 - 38.

Schneider P, Neitzel PL, Schaffrath M, Schlumprecht H (2002): Leitbildorientierte physikalisch-chemische Gewässerbewertung - Referenzbedingungen und Qualitätsziele. in: Abschlussbericht zum F&E-Vorhaben des Umweltbundesamtes (FKZ 200 24 226), Chemnitz, 2002, 149 S.

Schneider P, Neitzel PL, Osenbrück K, Noubactep C, Merkel B, Hurst S (2001b) In-situ Treatment of Radioactive Mine Water Using Reactive Materials - Results of Laboratory and Field Experiments in Uranium Ore Mines in Germany. Acta Hydrochim. Hydrobiol. 29: 129 - 136

Schneider P, Osenbrück K, Neitzel PL, Nindel K (2002) In-situ mitigation of effluents from acid waste rock dumps using reactive barriers - A feasibility study, J Mine Water and the Environment, Special Issue: Large Scale Modeling of the Environment 21: 36 - 44.

U.S. Dept of Energy (1996) Research and application of permeable reactive barriers, DOE-K0002000 (Subcontract No. DE-AC 13-96DJ87335)

Environmental hydrogeology of in situ leach uranium mining in Australia

Gavin M. Mudd

Formerly Research Fellow, Dept. Civil Eng., University of Queensland, St Lucia, QLD Australia now P.O. Box 6405, St Lucia, QLD 4067 angelb@netspace.net.au

Abstract. The use of the 'in situ leach' technique of uranium mining is a new development in Australia's expanding uranium industry. To date there have been three sites of pilot leach mines at Beverley (1998) and Honeymoon (1982, 1998-2000) in South Australia and at Manyingee (1985) in Western Australia. The Beverley and Honeymoon projects gained regulatory approvals in early 1999 and late 2001, respectively. The principal concerns with solution mining relate to impacts on groundwater and whether this naturally attenuates or requires active restoration. These environmental hydrogeological issues are reviewed in detail.

Overview of In Situ Leach Uranium Mining in Australia

Until recently, Australia had a short and mostly experimental history with in situ leach uranium mining, also known as solution mining (Mudd 2001a). Pilot scale testing has been undertaken at three deposits: Honeymoon (1982, 1998-2000), Beverley (1998) and Manyingee (1985; locations shown in Fig. 1). The recent approvals of Beverley and Honeymoon as commercial mines has allowed the technqiue to become a part of Australia's expanding uranium industry.

The Beverley mine began commercial operations in late 2000 with production of 219 and 327 t U_3O_8 in the June and December halves of 2001. The Honeymoon project, which now also includes the adjacent East Kalkaroo deposit, obtained regulatory approvals in late 2001 and is currently planning and moving towards construction. Commercial operation is expected by early 2003. The Manyingee deposit, however, is still being re-assessed (slowly) with no firm plans.

The various pilot mines operated at all three sites were developed to investigate both commercial and engineering requirements as well as the groundwater impacts and necessary management strategies, such as pH control to minimise mineral precipitation. No pilot site was required to restore impacted groundwater. The specific hydrogeological and environmental impacts of each site is now reviewed.

Fig. 1. Location of potential in situ leach uranium mines in Australia (Mudd 2001a)

Beverley, South Australia

The Beverley project, owned by General Atomics of the USA, has succeeded in becoming Australia's first commercial acid-based in situ leach uranium mine. For a broader history of Australian in situ leach mines (Cu, Au) see Mudd (1998, 2001a) and for the Beverley project's development history HR (1998a, b) and Mudd (1998, 2001a).

Geology and Hydrogeology

The geology and hydrogeology of the Beverley uranium deposit is described by Haynes (1975), SAUC (1982), HR (1998a, b) and Mudd (1998, 2001a).

Located in the western part of the Frome Embayment, the deposit is overlain by about 100 m of alluvial fans comprising lenses of gravels, sands, silts and clays. The uranium mineralisation occurs within aquifer sands that resemble a concealed fluvial system or palaeochannel. The deposit contains three ore zones, Northern, Central and Southern, each with increasingly higher salinity, respectively. Beneath the ore zone aquifer is a thick mudstone sequence and the Cadna Owie sandstone of the Great Artesian Basin at 300 m depth. The hydrogeology is complicated by structural deformation and faulting, which may provide vertical interconnection between the deeper aquifers, while possibly truncating aquifers in the shallower sediments (Hancock 1986). A regional hydrogeological cross-section is shown in Fig. 2. A plan of the ore zones is shown in Fig. 3. A compilation of groundwater quality is given in Table 1.

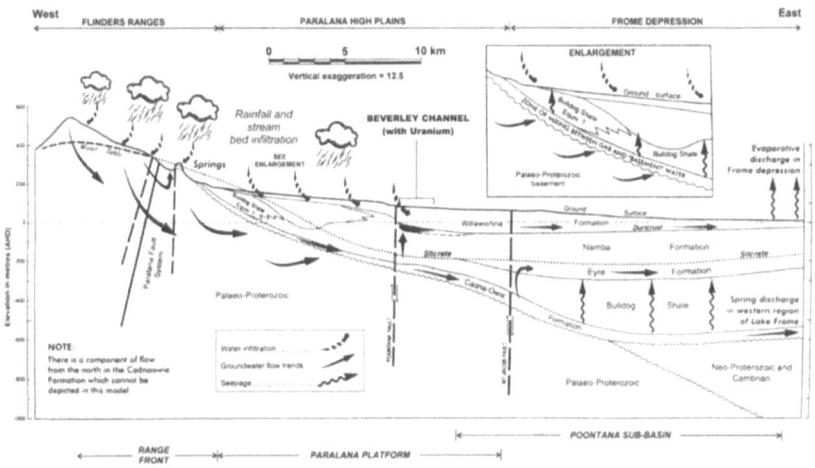

Fig. 2. Regional hydrogeological cross-section showing the Beverley deposit (HR 1998a)

Table 1. Groundwater and mining solution quality at Beverley: Northen, Central and Southern ore zones, Northern field leach trial data (Injection and Extraction averages March to July 1998) and Retention Pond (July 1998; adapted from Mudd 2001a)

units	pH -	TDS g/L	SO$_4$ g/L	Cl g/L	F mg/L	Na g/L	K mg/L	Ca mg/L	Mg mg/L
Northern	7.3	3-6	1.6	2	0.85	1.2	42	380	198
Central	7	6-10	2.1	ng	ng	ng	ng	610	ng
Southern	6.8	11-13	2.6	ng	ng	ng	ng	850	ng
Injection	1.93	11.5	4.79	2.0	7.67	1.43	59	610	337
Extraction	1.97	11.7	4.84	2.0	7.33	1.43	59	600	337
Ret. Pond	2.10	62.1	29.5	6.1	5.50	15.1	105	460	369

(mg/L)	Al	Fe	Mn	Si	SiO$_2$	U	^{226}Ra [a]	^{222}Rn [a]
Northern	0.2	0.7	0.2	48	ng	0.076	22-967	500-2,000
Central	ng	ng	ng	ng	ng	1.91	1.2-3,100	5-32,140
Southern	ng	ng	ng	ng	ng	0.70	13-111	20-585
Injection	91	109	0.7	138	294	2.9	8414	ng
Extraction	91	105	0.8	133	283	162	9881	ng
Ret. Pond	39	39	0.9	99	211	272	1713	ng

(µg/L)	B	Ba	Cd	Co	Cr	Cu	Ni	Pb	Se	V
Northern	1,600	53	0.2	100	20	30	4	40	1	1
Injection	1,000	37	117	20,000	100	200	8,470	160	410	1,100
Ext'n	1,100	39	116	20,000	580	200	8,330	790	410	1,130
R Pond	3,400	76	49	6,600	260	180	2,480	70	310	780

[a] ^{226}Ra and ^{222}Rn in Bq/L. Note: ng - not given; no redox data available.

The impact of unsealed exploration bores (when Beverley was planned as an open pit mine in the early 1970s), which increases the risk of excursions, has been recognised (Hancock 1986, 1988) although downplayed in more recent times.

Impact on groundwater from radionuclide emission

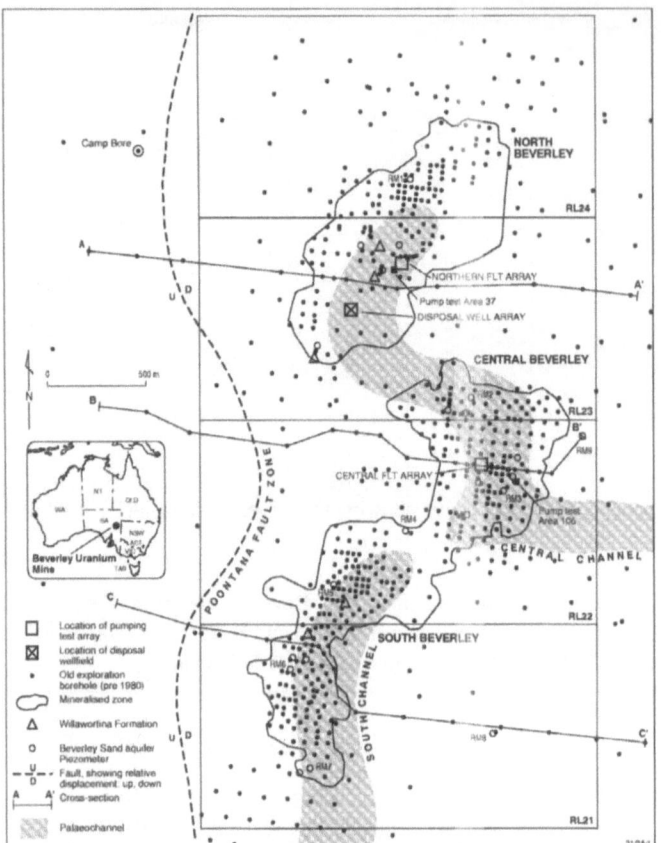

Fig. 3. Plan view of the ore zones and exploration bores at Beverley (Habermehl 1999).

Hydrogeological Impacts

After operating a trial acid leach mine during 1998 (producing 33.27 t U_3O_8) and releasing their environmental impact statement, further reviews were conducted to assess the hydrogeological impacts, centred on water quality and the 'semi-isolated' nature of the aquifers. Regulatory approvals were given in April 1999, making Beverley the western world's first acid leach uranium mine (OECD 2000).

In contrast to Beverley, commercial leach mines in the USA all use alkaline chemistry, generally dispose of liquid wastes through evaporation (or very deep groundwater injection) and are required to restore all impacted groundwaters to their pre-mining quality and state (Mudd, 1998, 2001a). As allowed, Beverley uses acid chemistry, disposes of liquid wastes by injection to the ore-zone and is not required to restore impacted groundwater. The basis for the approvals was that following mining the pH, metals and radionuclides will return to pre-mining conditions given several years, although no mechanisms or data were provided.

Impact on groundwater from radionuclide emission

The Beverley aquifers were believed to be effectively isolated, posing minimal risk to surrounding groundwater if the above argument proved wrong.

The ore zone contains low sulfide (0.13%), organic carbon (0.05%), carbonate (0.06%), Fe, Mn and clay content (Hancock 1986, 1988; HR 1998a). Hancock (1986, 1988) argued that the exchangeable and soluble calcium and carbonate in the clays and sands surrounding the ore zones would be sufficient to neutralise the residual acid from migrating mining solutions and therefore precipitate gypsum. Due to the minimal degree of exploration data beyond the confines of the ore zones, however, this remains an untested hypothesis and no data has been released to demonstrate this mechanism could perform satisfactorily at Beverley. There is no redox data in public documents, thus precluding an accurate geochemical assessment of possible attenuation rates or reactions (Mudd 2001a).

The high Ca and SO_4 levels of the Beverley ore zones, especially the Central and Southern ore zones, create the potential for gypsum precipitation (see Table 1). In the USA, gypsum formation was shown to be related to elevated salinity and radium in post-restoration groundwater (Mudd 2001a).

Honeymoon, South Australia

The Honeymoon project operated the first pilot solution mine in Australia during 1982 but failed to be developed after the withdrawal of government support in 1983. After a hiatus until the late 1990's, new owners Southern Cross Resources of Canada operated the old pilot mine again from 1998-2000 and received approvals for a commercial project by November 2001. Construction is set to start during 2002 with commercial operation expected by early 2003. The development history is given by Minad (1980, 1981), Mudd (1998, 2001a) and SCR (2000a, b).

Geology and Hydrogeology

The geology and hydrogeology of the Honeymoon and East Kalkaroo uranium deposits is described by Brunt (1978), Minad (1980, 1981), Curtis et al. (1990), SCR (2000a, b) and Mudd (1998, 2001a).

The Honeymoon and East Kalkaroo uranium deposits are located within the Yarramba palaeochannel in the southern Frome Embayment. The channel sands consist of three distinct layers which form the Basal, Middle and Upper aquifers. The Upper aquifer is occasionally used by pastoralists in the region while the Basal sand contains the Honeymoon deposit. Traces of uranium exist in all three sands, with the Yarramba deposit to the north located in the Middle sand. The hydraulic head is identical for all three sands, suggesting a high degree of vertical interconnection. Hydrogeological cross-sections are shown in Fig. 4, with a geological map shown in Fig. 5. Groundwater quality data is given in Table 2. The Honeymoon/East Kalkaroo deposits have several unique features related to the use of ISL, including pyrite at 5-15% compared to less than 2% in most USA deposits; higher salinity; low organic content (0.3%); and direct hydraulic connections between the palaeochannel aquifers due to gaps in the clay layers (Mudd 2001a).

Impact on groundwater from radionuclide emission

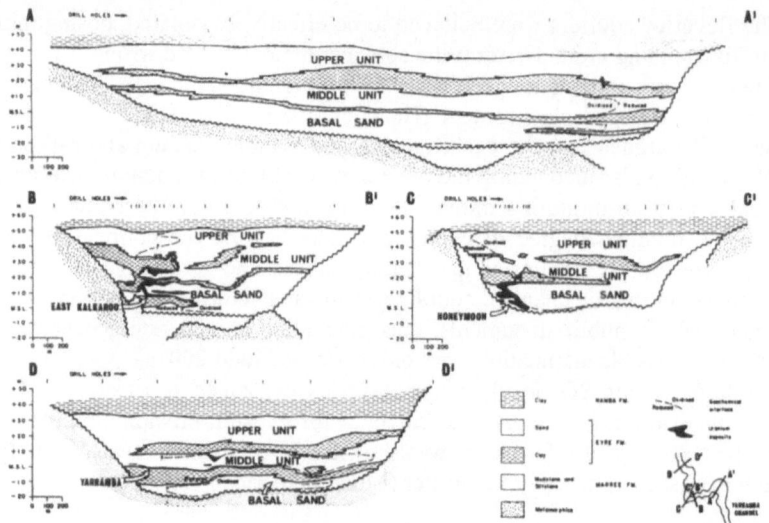

Fig. 4. Hydrogeological cross-sections of the Yarramba palaeochannel showing the Honeymoon, East Kalkaroo and Yarramba uranium deposits (Brunt 1978)

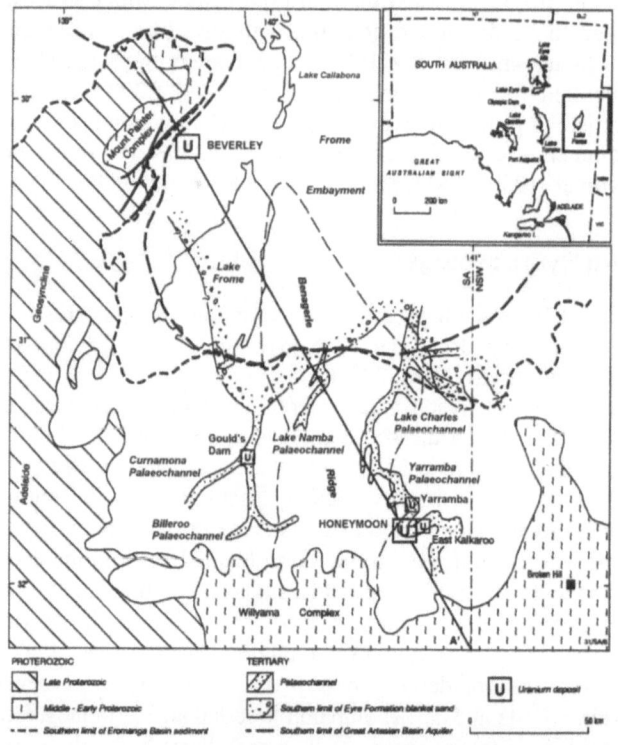

Fig. 5. Regional geological map of the Frome Embayment (adapted from Habermehl 2000)

Impact on groundwater from radionuclide emission

Table 2. Average groundwater and mining solution quality at Honeymoon: Upper, Middle and Basal sands, mining solutions and liquid wastes (adapted from Mudd 2001a)

	pH	TDS	SO_4	Cl	F	Na	K	Ca	Mg
Units	-	g/L	g/L	g/L	mg/L	g/L	mg/L	mg/L	mg/L
Upper	7.5	10.3	1.45	4.80	0.8	2.81	ng	478	260
Middle	7.0	11.4	1.54	5.37	0.5	3.39	ng	560	270
Basal	6.9	16.1	1.75	7.85	0.5	4.31	20.7	906	390
Leaching	2.2	16.43	5.30	8.47	0.6	6.17	ng	940	210
Wastes	2.3	19.8	6.11	8.02	1.9	5.60	27.5	1,000	430

(mg/L)	Org. C [a]	HCO_3	Al	Fe	SiO_2	U_3O_8	^{226}Ra [b]	^{222}Rn [b]
Upper	ng	187	<1.0	1.0	6.5	0.022	3.1	23
Middle	ng	160	<1.0	1.0	6.4	0.018	7	7
Basal	1.2	145	<1.0	1.0	7.6	1.2	205	5,000
Leaching	2	<5	15	260	ng	75	830	12,700
Wastes	ng	<5	28.3	200	101	2.3	405	ng

(μg/L)	Co	Cr	Cu	Mo	Ni	^{210}Pb [b]	Se	V	Zn
Upper	ng	ng	8	9	ng	3.6	14	<20	200
Middle	ng	ng	30	10	ng	6.6	19	<20	200
Basal	60	20	20	13	78	0.8	30	<20	190
Leaching	Ng	ng	7,000	3	ng	ng	55	4,000	110,000
Wastes	2,200	100	1,800	7.4	3,530	560	79	1,100	56,300

[a] Org. C - organic carbon. Note: ng - not given. [b] ^{226}Ra, ^{222}Rn and ^{210}Pb in Bq/L.

The 1982 trial encountered significant operational difficulties due to the precipitation of jarosite (Mudd 1998, 2001a). The 1998-2000 trial had an excursion in late 1999 where leach solutions were thought to have migrated through the lower clay confining unit (SCR 2001), despite assurances of minimal risk of excursions (eg. SCR 2000a, b). The excursion was controlled through remedial action, (ie. additional pumping), although it highlighted inadequacies with monitoring bores.

The only redox data available on the public record is from an analysis of mining solutions using different oxidising agents during the 1998-2000 trial (SCR, 2001). The use of oxygen, hydrogen peroxide, ferric sulfate and sodium chlorate gave redox potential values in leaching solutions of 415, 650, 684 and 970 mV, respectively, with the pregnant (recovered mining) solution being about 415 mV.

The approvals for Beverley set important precedents for acid leach mining in Australia that have major implications for the Honeymoon project (Mudd 2001a): 1) the project proposes to re-inject all liquid wastes into the Lower palaeochannel aquifer which is known to be hydraulically connected to the Upper aquifer occasionally used by pastoralists; 2) the potential for 'natural attenuation' is uncertain, although this depends on the reactivity of pyrite (or other reducing agents) remaining after mining; and 3) the Yarramba palaeochannel is the only groundwater resource in the region (the velocity is about 18 m/year; MINAD 1980).

Geochemical modelling of the interactions of Honeymoon mining solutions and liquid wastes with groundwater quality was presented by Pirlo (2000, 2001). This was based on samples of groundwater from the Honeymoon and East Kalkaroo ore zones plus a sample of wastewater from the Honeymoon trial mine.

Impact on groundwater from radionuclide emission

No field measurement of redox potential was undertaken, being calculated by Zn^{2+}/ZnS. Although mixing of the various solutions suggested that precipitation effects in the aquifer would be minimal and that heavy metals would not remain mobile after sufficient dilution, Pirlo (2000) acknowledged that kinetic effects are not incorporated in this mixing approach. As with SCR (2000a, b), there is no data or analysis presented to justify the high dilution ratio of 10:1 (groundwater:mining solutions) used in mixing and geochemical modelling.

No published field evidence from the pilot leach trial corroborates the analysis by Pirlo (2000, 2001), especially concerning the redox state in the aquifers, nor does it demonstrate that natural attenuation has or will work at Honeymoon.

Manyingee, Western Australia

The Manyingee uranium deposit, discovered in 1974, was the site of pilot-scale alkaline leach mining in 1985. No more work has been undertaken at the site and it is currently owned by Australian explorer Paladin Resources Ltd.

The geology and hydrogeology of Manyingee is described by Valsardieu et al. (1981). The palaeochannel is buried beneath approximately 70 m of Cainozoic and Cretaceous sediments. Uranium mineralisation is generally found within the lower part of the Lower Cretaceous Birdrong sandstone, ranging from 70 to 110 m in depth, controlled by redox state and geologic structure. The sandstone units often contain abundant carbonaceous matter, including lignitic and wood fragments, as well as pyrite. The groundwater is of moderately low salinity of 3.4 g/L, being mainly Na and Cl with minor Mg, HCO_3 and SO_4.

The environmental and hydrogeological data from the 40.5 ML alkaline trial mine, as with Honeymoon and Beverley, has not been publicly reported (Mudd, 1998). The site did apparently undertake some groundwater restoration activities, although the extent or success of this work is unknown.

Discussion & Conclusions

The key for driving natural attenuation is for active reducing agents to be present, primarily organic carbon or sulfide (Buma 1979). In contrast, Riding et al. (1979) state that many roll-front uranium deposits in the Colorado Plateau of the USA showed poor correlation between reducing agents and uranium ores.

Both Honeymoon and Beverley contain low organic matter, with Honeymoon containing abundant pyrite. It would seem reasonable that the organic matter may have been consumed during formation of the respective roll-fronts and uranium precipitation. It is the organic matter remaining after acid leach mining, however, and the impacts of acidic, oxidised liquid wastes on aquifer sediments which will mainly determine if reducing conditions will re-establish after mining.

Morris (1984) stated that "reliance on this process [natural attenuation] has never been tested". The time period and the rates at which natural processes could attenuate such levels of pollution are yet to be firmly established (Rojas 1987).

Impact on groundwater from radionuclide emission

During operations, there is potential for excursions due to unsealed exploration bores, as well as excursions due to well casing failures (Marlowe 1984). Curiously, approvals for Beverley included provisions that liquid waste reinjection only occur in the Northern zone – the area of least exploration drilling and best quality groundwater (TDS ~3-6 g/L). This salinity is similar to regional pastoral use (excluding radionuclides), although often mines in Western Australia use saline groundwater up to 240 g/L (Sparrow & Woodcock 1993).

The use of acid in the USA was considered problematic due to restoration difficulties and higher salinity and some radionuclides in post-restoration groundwaters (related to gypsum formation during mining; Mudd 2001a). Pilot mines are used as the public basis for assessing commercial mines, and as such, acid leach mines have never been approved in the USA (Mudd 1998, 2001a).

Given the complexity of the geochemistry of in situ leach mines, it should be expected that detailed hydrogeological and geochemical studies be done for each proposed project. In Australia, the results from all leach mine trials at Beverley, Honeymoon and Manyingee have never been fully published and thus information is limited on their short and long-term impacts. Critically, the issues of geochemical conditions in the groundwater following mining have not been satisfactorily addressed, with no clear field evidence of natural attenuation at any site.

The standards of the Beverley and Honeymoon projects - acid leaching with no restoration of polluted groundwater - is more akin to practices in Eastern Europe and the Former Soviet Union, where the available evidence suggests that natural attenuation fails to reduce the impacts from such mines (Mudd, 2001a, b). This suggests that natural attenuation appears spurious at worst, ineffective at best.

Australia has the lowest continental rainfall with water a limiting and highly valuable environmental resource (Smith, 1998). The standards applied at the Australian acid leach uranium mine sites are not considered an acceptable approach for arid regions that are almost entirely dependent on groundwater.

References

Brunt DA (1978) Uranium in tertiary stream channels, Lake Frome area, South Australia. AusIMM Proceedings, 266: 79-90

Buma G (1979) Geochemical arguments for natural stabilization following in-place leaching of uranium. In: In Situ Uranium Mining and Ground Water Restoration, New Orleans Symp., SME of AIME, Feb. 19, 1979, 113-124

Curtis JL, Brunt DA & Binks PJ (1990), Tertiary Palaeochannel Uranium Deposits of South Australia. In: Geology of the Mineral Deposits of Australia & Papua New Guinea, Ed. FE Hughes, AusIMM, Monograph 14, 2: 1631-1636

Habermehl MA (1999) Assessment of the Beverley uranium mine proposal - Beverley uranium mine environmental impact statement by Heathgate Resources P/L. Prepared by Bureau of Rural Sciences for Environment Australia, Canberra, ACT, March 1999

Habermehl MA (2000) Professional opinion: Honeymoon uranium project - Comparison of the hydrogeology of the Beverley and Honeymoon uranium projects. Prepared by Bureau of Rural Sciences for Environment Australia, Canberra, ACT, Dec. 2000

Hancock S (1986) Environmental aspects of proposed in-situ leach mining at Beverley - South Australia. In: 8th Australian Geological Convention, Feb. 16-21, 1986, 94-95

Hancock S (1988) In-situ leach mining: the next quantum leap ? In: 3rd International Mine Water Congress, IMWA and AusIMM, Melbourne, VIC, Oct. 1988, 883-889

Haynes RW (1975) Beverley Sedimentary Uranium Orebody, Frome Embayment, South Australia. In: Economic Geology of Australia & Papua New Guinea, Ed. CL Knight, AusIMM, Monograph 5, 808-813

HR (1998a, b) Beverley uranium mine - draft / response document and supplement to the environmental impact statement. Heathgate Resources Pty Ltd, June 1998 / Oct. 1998

Marlowe JI (1984) An environmental overview of unconventional extraction of uranium.U.S. Environmental Protection Agency, EPA-600/7-84-006, Jan. 1984, 130 p

MINAD (1980, 1981) Honeymoon project: draft / final environmental impact statement. Mines Administration Pty Ltd (MINAD), Nov. 1980 / March 1981

Morris LJ (1984) Solution Mining. In: 8th Australian Groundwater School, Vol. 2, Chap. 14, AMF, Adelaide, SA, Aug. 27-Sep. 7, 1984, 131 p

Mudd GM (1998) An environment critique of in-situ leach mining: The case against uranium solution mining. Research report, Melbourne, VIC, July 1998, 154 p

Mudd GM (2001a) Critical review of acidic in-situ leach uranium mining: 1 USA and Australia. Env Geology 41: 390-403

Mudd GM (2001b) Critical review of acidic in-situ leach uranium mining: 2 Soviet block and Asia. Env Geology 41: 404-416

OECD (2000) Uranium 1999: resources, production and demand. OECD-NEA & IAEA

Pirlo MC (2000) Applications of geochemical modelling to groundwater management at the Honeymoon uranium project. In: 25th Environmental Workshop, Minerals Council of Australia, Perth, WA, Oct. 29-Nov. 2, 2000, 479-497

Pirlo MC (2001) Honeymoon uranium project - Terms of reference for additional evaluation of aquifer. Prepared for Environment Australia, Canberra, ACT, July 2001

Riding JR, Rosswog FJ, Buma G & Tweeton DR (1979) Ground water restoration for in-situ solution mining of uranium. In: In Situ Uranium Mining and Ground Water Restoration, New Orleans Symp., SME of AIME, Feb. 19, 1979, 67-85

Rojas JL (1987) Introduction to in situ leaching of uranium. In: In Situ Leaching of Uranium: Technical, Environmental and Economic Aspects, Technical Committee Meeting, Vienna, Austria, November 3-6, 1987, IAEA TECDOC-492, 7-20

SAUC (1982) Beverley project - draft environmental impact statement. South Australian Uranium Corporation (SAUC), Frewville, SA, July 1982, 329 p

SCR (2000a, b) Honeymoon uranium project - draft environmental impact statement / response supplement. Southern Cross Resources Australia Pty Ltd, Toowong, QLD, May 2000, 372 p / Nov. 2000, 114 p

SCR (2001) Honeymoon uranium project - field leach trial April 1998-August 2000. Southern Cross Resources Australia Pty Ltd, Toowong, QLD, Sept. 2001, 39 p

Smith DI (1998) Water in Australia - resources and management. Oxford Uni Press, 406 p

Sparrow GJ & Woodcock JT (1993) Gold ore treatment at medium-sized Australian gold plants. In: The Sir Maurice Mawby Memorial Volume 2nd Edition, Ed's JT Woodcock & JK Hamilton, AusIMM Mono. 19, Vol. 2: 1035-1059

Valsardieu CA, Harrop DW & Morabito J (1981) Discovery of uranium mineralization in the Manyingee channel, Onslow region of Western Australia. AusIMM Proceedings, 279: 5-17

Reliance on existing wetlands for pollution control around the Witwatersrand gold/uranium mines of South Africa - Are they sufficient?

Henk Coetzee[1], Peter Wade[2] & Frank Winde[3]

[1]Council for Geoscience, Private Bag X112, Pretoria, 0001, South Africa, henkc@geoscience.org.za
[2]Dynamic Strategies, Suite 210, Postnet X4, Menlo Park, 0102, South Africa, pwade@dynamicstrategies.co.za
[3]Department of Geography, Potchefstroom University for CHE, PO Box 19436, Noordbrug, Potchefstroom, 2522, frank.winde@gmx.de

Abstract. The gold deposits of the Witwatersrand contain appreciable concentrations of uranium, which have at times been exploited commercially. Conditions downstream of mines have led to the development of extensive wetland systems, which have been shown to concentrate heavy metals including uranium and uranium series radionuclides from water discharged by mining operations and leachates derived from mine wastes. While these wetlands act as pollution sinks, geochemical simulations and models indicate that metals may be remobilised and pose a hazard to downstream water users.

Introduction

Since the commencement of mining activities in the Witwatersrand Goldfields in the late 19th Century, liquid effluents have been entering the natural surface and ground water systems. These effluents are a cocktail of the direct discharges of both process and fissure water from mines, run-off and infiltration from mine waste materials and contaminated areas and other industrial and domestic wastewaters from the cities and towns which have developed around the mining industry.

The salt and nutrient content of these waters has resulted in the development and maintenance of large wetland systems, which, while active and submerged tend to scavenge metals as well as organic contaminants from river water. Airborne radiometric survey data from the Witwatersrand confirm the concentration of uranium-series radionuclides in wetlands downstream of mines, while chemical analysis of sediment samples collected downstream of mines show concentration of uranium downstream of mining activities.

These observations raise three important questions regarding the efficacy of these wetlands in the mitigation of water pollution:

1. Are they currently effectively reducing pollutant levels to acceptable levels before the water reaches potential downstream water users under normal flow, drought and flood conditions?
2. Do the contaminants bound to wetland sediments remain in place, or is there a slow downstream migration of a contamination front?

If the water and salt inputs due to mining were to be stopped, and nutrient inputs from human settlements better managed, would the wetlands remain in place, and what would happen to the accumulated pollution due to more than a century of mining?

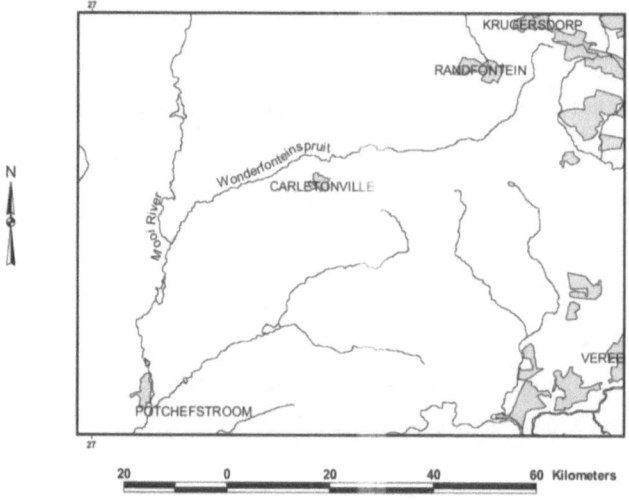

Fig. 1. Location of the Wonderfonteinspruit and Mooi Rivers, flowing towards Potchefstroom. The greatest concentrations of mining activities are in the Krugersdorp-Randfontein and Carletonville areas.

This study investigates the role of wetland sediments in pollution attenuation and possible remobilisation in the Wonderfonteinspruit Catchment. This river system (See Fig. 1) drains the mining areas of Krugersdorp, Randfontein and Carletonville. A large portion of the catchment is underlain by Proterozoic dolomites of the Malmani Group. The dolomitic environmental waters and the underlying rock play an important role in buffering the pH of the system, in contrast to other portions of the Witwatersrand basin where acid drainage is more common.

Current reduction of pollution levels

Adsorption and precipitation processes in the environment as well as dilution with environmental waters reduce pollution levels in streams in mining areas. A study conducted by the South African Department of Water Affairs and Forestry on the Mooi River System, during 1997-98 concluded that while radionuclide levels are elevated in streams in mining areas, water quality appears to improve downstream, with radionuclides posing a minimal hazard (IWQS 1999) to users in the town of Potchefstroom, downstream of the mines (See Fig. 1). In this study it was concluded that adsorption/precipitation processes in the environment reduced the radionuclide content of the water. It should be noted that a large portion of this area is underlain by dolomite, and the resulting neutralisation of acid minewaters will play a significant role in pollution mitigation. Other Witwatersrand mining areas, where the surface geology is not dolomitic appear to have significantly lower ability to attenuate mine-related water pollution.

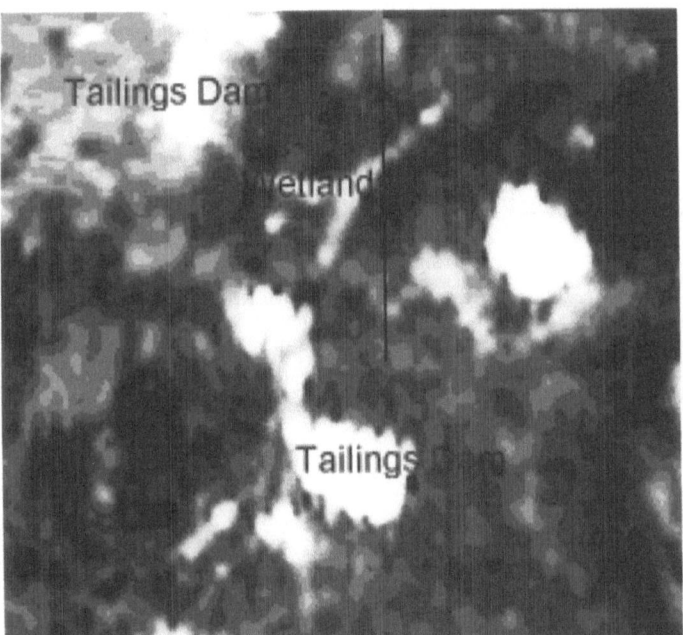

Fig. 2. Airborne radiometric image of a portion of the Wonderfonteinspruit Catchment, with lighter colours indicating higher gamma ray emission rates. Note the contamination of wetlands downstream of mining activities.

Ground and airborne radiometric surveys, as well as sediment sampling has shown uranium and other radionuclide levels in river sediments to be elevated above natural background levels several kilometers downstream of mining activities. Uranium concentrations above $10kBq \cdot kg^{-1}$ have been measured in farm dams and close to informal settlements. Whether or not these pose an immediate risk to the inhabitants of these areas or to people consuming agricultural produce from these farms has not yet been determined, but these levels of radioactivity exceed the legislated level for regulatory control of $200Bq \cdot kg^{-1}$ (Wade et al. 2001).

Mobility of adsorbed metals in wetlands

To date, it has been believed that once adsorbed in wetlands, radionuclides and other metals will remain fixed in place. Recent studies have shown that the adsorbed phases of uranium in this system occupy a relatively narrow area in Eh-pH space. Sequential chemical extractions (Ure et al. 1993, Wade et al 2001) performed on sediment samples (See Fig. 3) have shown that changes in both pH and redox conditions within environmentally plausible ranges can release uranium and radium. Furthermore, these experimental studies indicate that uranium is bound to multiple sites in approximately equal quantities, so that any change in environmental conditions could release pollutants.

Sequential Chemical Extraction

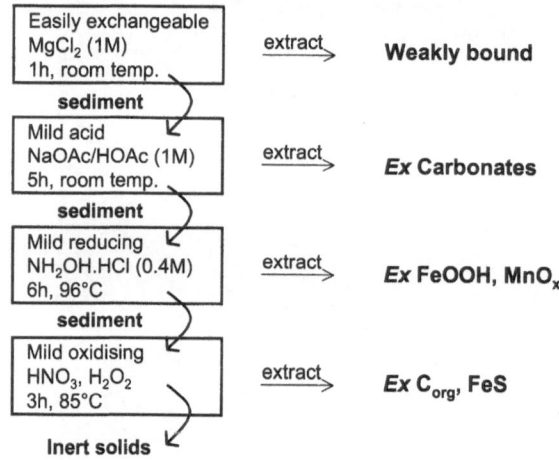

Fig. 3. Sequential extraction method used to determine the speciation of metals in sediment samples. The different extraction stages are used to simulate environmentally plausible conditions, which could impact on contaminated sediments

Sequential extractions as described in Fig. 3 have been performed on a suite of thirteen samples collected from a dam downstream of the Carletonville mining area. The samples were found to have a mean uranium concentration of 275ppm, using ICP-MS. The distribution of uranium between the different phases in the sediment is indicated on Fig. 4.

Impact on groundwater from radionuclide emission

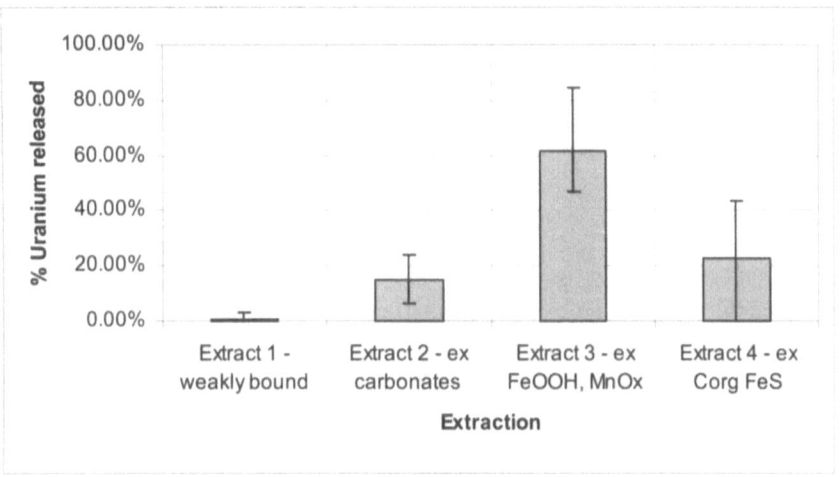

Fig. 4. Percentage uranium released from thirteen samples in a sequential extraction, as described on Fig. 3. Vertical bars indicate the mean percentage of the total uranium per sample released, while error bars show the range of percentages released.

These results indicate that most of the uranium is bound to FeOOH and MnO_x complexes, but that a significant proportion is bound to carbonates and in reduced phases on organic carbon or pyrite. Scanning electron microscope studies have shown that the pyrite within the sample is authigenic, indicating that the transport of uranium is in a dissolved phase, rather than as particulate derived from mining and milling of ores.

The long-term future of polluted wetlands

Many wetlands downstream of current mining areas are fed by water pumped from operating mines. Large quantities of nutrients also enter these systems from informal settlements, overloaded or poorly maintained sewage treatment works and industries. As formal housing becomes available, old sewage works are upgraded, higher environmental standards are imposed upon industry and mines discontinue pumping, it is possible that the water and nutrient inputs into wetlands may be reduced. Previously non-perennial rivers where flow has been maintained by mine dewatering could once again dry up during the dry seasons, exposing the wetlands and allowing oxidation to release adsorbed contaminants.

Mine closure poses an additional long-term environmental hazard. Contaminated areas outside of mine properties may not be subjected to the same type of regulation or restriction as those on mines. This could lead to the use of old contaminated sites for agricultural or other purposes in the future, after the financial responsibility for clean-up has been transferred to the State. The areas in question,

Impact on groundwater from radionuclide emission

as well as the volumes of material and nature of contamination could place a significant burden onto future generations.

Conclusions and recommendations

Wetlands have proved an excellent means to attenuate pollution from Witwatersrand gold/uranium mines. The uncontrolled manner in which this has been allowed to happen is however of extreme concern. Firstly, it has allowed to build-up of contamination on private and public land off mining properties and secondly it has happened in a haphazard and uncontrolled fashion. Artificial wetlands may offer a partial solution to problems related to discharge of water on these mines, but it should be borne in mind that the cost of maintenance and monitoring should be borne by the polluters and that the eventual fate of the contaminants should remain the responsibility of the mines involved.

The sequential extraction study undertaken shows that while wetland sediments have a significant capacity to adsorb and immobilise pollutants, plausible chemical conditions may also lead to the remobilisation of these pollutants. The reservoirs of contaminants stored downstream of mining activities could therefore pose a long-term risk to downstream water users.

The processes that concentrate radionuclides in sediments also concentrate gold. Consequently, decontamination of contaminated sites may be able to generate sufficient revenue to partially fund rehabilitation. The current trend is however to limit rehabilitation to those areas where gold concentrations make rehabilitation profitable, neglecting less gold-rich areas unrehabilitated. An holistic approach is needed, where environmental priority areas are also rehabilitated.

References

IWQS (1999) *Report on the Radioactivity Monitoring Programme in the Mooi River (Wonderfonteinspruit) Catchment.* Institute for Water Quality Studies. Department of Water Affairs and Forestry. Report No.: N/C200/00//RPQ/2399.

Ure A M, Quevauviller P H, Muntau H and Griepink B (1993) Speciation of Heavy Metals in Soils and Sediments. An Account of the Improvement and Harmonization of Extraction Techniques Undertaken under the Auspices of the BCR of the Commission of the European Communities. *Intern. J. Environ. Anal. Chem.*, 51, 135.

Wade P W, Woodborne S, Morris M W, Vos P and Jarvis N V (2001) Tier 1 risk assessment of selected radionuclides in sediments of the Mooi River Catchment, Water Research Commission, Pretoria, 107pp.

Radon, water and abandoned metalliferous mines in the UK: Environmental and Human Health implications

Gavin K. Gillmore[1], John Grattan[2], F. Brian Pyatt[3], Paul S. Phillips[4], Gillian Pearce[5].

[1] Department of Environmental Science, University of Bradford, West Yorkshire, BD7 1DP, UK.
[2] Institute of Geography and Earth Sciences, University of Wales, Aberytswyth, SY23 3DB, UK.
[3] Department of Life Sciences, Nottingham Trent University, Clifton Lane, Nottingham, NG11 8NS, UK.
[4] University College Northampton, Boughton Green Road, Northampton, NN2 7AL, UK.
[5] Devon and Cornwall Prospecting Society, 44 Lancaster Drive, Paignton, Devon, TQ4 7RR, UK.

Abstract. A study has been undertaken of radon gas levels in the atmosphere in abandoned metalliferous mines that were exploited primarily for tin in South-West England, UK, and compared to levels in an old lead mine in northern England, UK. Measurements have been taken since 1992 in the South-West of radon in the air using a variety of measuring techniques. Extremely high radon gas levels have been noted in a number of these mines, one of the highest levels recorded in Europe was recorded at 3,932,920 Bq m^{-3} in a shallow adit of an ex-uranium mine. The health implications for casual users/explorers of such mines are shown to be considerable. Even outside such mines, in adit entranceways, very high atmospheric radon levels were recorded of, for example, 200,000 Bq m^{-3}. The heavy metal content of stream-water that flows from such mine adits together with dissolved radon content has also been measured and assessed in terms of potential health effects. A combination of heavy metal pollution and radiation can have a considerable impact on health and this research recommends that further research should be undertaken in such environments.

Radon and safety

Radon (^{222}Rn) is a naturally occurring gas which is formed in the decay series of ^{238}U. It is a colourless and odourless gas, which is very dense and chemically unreactive. Radon decays to highly radioactive heavy metallic daughter products (e.g. ^{218}Po) known as progeny. Some of the progeny decay by alpha particle emission. These daughter products, if retained in the lung through inhalation, will irradiate lung tissue by alpha particle emission (alpha particles are very energetic, travel short distances and cause localised damage).

There is a substantial body of evidence to show that raised radon levels in some mines can cause lung cancer in miners (Muirhead et al. 1993). Muirhead et al. (1993) have shown that, in studies of over 50,000 miners 686 lung cancer deaths might be expected, but 2,299 were noted. The ICRP (1993) have concluded that excessive radon levels are hazardous to health. Studies by Lubin and Boice (1997) and Darby et al. (1998) have shown that raised radon levels in the built environment (particularly in regions designated as radon Affected Areas) are a health hazard. There is a clear link between radon and lung cancer (Lorenz 1944; Field et al., 2000; Gillmore et al. 2000a; Kendall 2000), and it has been suggested that radon in homes causes some 2,500 lung cancers deaths each year in the UK (Thompson et al. 1998). The US Environmental Protection Agency (EPA) has estimated that 20,000 lung cancer deaths each year in the US can be linked to radon.

Radon has also been implicated in cancers other than those of the lung/respiratory tract. Henshaw et al. (1990) has suggested that there is a link between radon levels and myeloid leukaeimia, cancer of the kidney and melanoma.

The ICRP (1981) has concluded that a dose of 10 mSv is received from an exposure to radon and its daughters of 1 Working Level Month (WLM). It is not unreasonable in the built environment and to a certain extent in enclosed spaces such as caves and mines, to assume an equilibrium factor (F) of 0.5 (Sperrin et al. 2000). This permits the derivation of the relation that 1 mSv of dose is received by exposure to 126,000 Bq m^{-3} of radon gas (Gillmore et al. 2000b). This has been used by a number of authors to estimate dose received from radon to occupants in built overground environments (Denman and Parkinson 1996; Denman and Phillips 1998; Phillips et al. 2000).

The level of radon exposure that the population experiences in the built environment involves a complex inter-relationship between geology, meteorology, micro-meteorology, social habit and building type. There have therefore been extensive studies on radon levels in homes, hospitals, schools and other workplaces in areas of increased risk. In the UK various studies led the NRPB in 1990 to suggest an Action Level of 200 Bq m^{-3} in homes (O'Riordan 1990; Miles et al. 1992) and 400 Bq m^{-3} in the workplace. Above these levels action should be undertaken to remediate. Under the Ionising Radiation Regulations (IRR 1999) employers are required to remediate if radon levels are greater than 400 Bq m^{-3}.

Gillmore et al. (2000a) have shown that elevated radon levels in caves in the UK can increase radon risk for cave explorers. Gillmore et al. (2001a, b) have further shown that exposure to high levels in abandoned mine workings is also a significant risk. This work further highlights this, pointing out some record radon gas levels in the old metallic ore workings in Devon and Cornwall. A number of mines were surveyed for radon levels in 1992-1994 by the Devon and Cornwall Prospecting Society (DCPS). These have been resurveyed by the authors in 2000 for comparative purposes and the results presented here.

There is a not insignificant community of dedicated explorers that visit abandoned mines in Cornwall and Devon. The DCPS estimate that there are around 4000 regular visitors to such mines (in South-West England) as the ones examined in this study (Pearce, pers comm.). Therefore, there is a small section of the population that may be putting themselves into areas of considerable radon risk.

Geological background

Tin had been produced from mines in the South-West of England since at least the 12th Century (Hoover and Hoover 1986). The vast quantities of copper produced in Cornwall have come from high-temperature veins with which were associated cassiterite, wolframite, arsenopyrite, sphalerite, pyrite and small amounts of pitchblende (U_3O_8), cobalite, argentite, stibnite and galena (Dunham et al. 1978; Jackson et al. 1989). The uranium minerals pitchblende and coffinite often occur in north-south structures with polymetallic mineralization, that is, nickel, cobalt, bismuth and silver (Darnley et al. 1965). In the Land's End area pitchblende occurs with chalcopyrite in tin lodes (e.g. Geevor Mine). Small amounts were produced at several mines (e.g. Wheal Owles, East Pool, St. Austell, New Crow Hill and Trenwith: Burt et al. 1987), but only South Terras mine (national grid reference SW 935 524; see Figure 2), south-west of the St. Austell granite pluton, was at one time worked solely for uranium, according to Dunham et al. (1978) and Smale (1993). Both Smale (1993) and Burt et al. (1987) suggest that tin and iron were produced from this mine in the late 19th Century. It is interesting to note that South Terras was owned by the Societe Industrielle du Radium Ltd. in 1913, although Burt et al. (1987) also noted that it was not worked in that year. Uranium Mines (national grid reference also SW 935 524) is recorded by Burt et al. (1987) as a producer of uranium as well as lead (plumbago) and arsenic pyrite (no detailed production returns are available on the latter two). Smale (1993) suggests that the active life of S. Terras mine was from 1870-1930, and uranium and radium were produced.

The highest production year for uranium in the late 19th to early 20th Centuries is 1905 when 103 tons of ore was extracted from Uranium Mines worth approximately £10,000 (Smale 1993). The fact that the Uranium Mines and S. Terras mines are given the same national grid reference by Burt et al. (1987) suggests that either these mines are right next to each other, or this is in fact the same mine sett. Indeed, according to Smale (1993), Uranium Mines Ltd. was formed in 1889 to acquire mining rights and work certain deposits in St. Stephen-in-Brannel parish. Uranium Mines were noted by A.J. Leese, the Secretary of Uranium Mines Ltd. (see Smale 1993), as being formerly known as South Terras.

Impact on groundwater from radionuclide emission

In 1978 Dunham et al. estimated that around 2,000 tonnes of uranium ore had been produced from the Devon and Cornwall region. Dines (1956), Dunham et al. (1978), and Jackson et al. (1989) suggested that 750 metric tons was produced from the South Terras mine from a 60m deep vein. South Terras is 0.5 km SW of St. Stephen-in-Brannel, Cornwall around Tolgarrick Mill. The workings have been flooded since 1928 (Darnley et al. 1965). In the area of this mine the uraniferous mineralization is of pitchblende in siderite. There is also pitchblende-coffinite in quartz. Smale (1993) suggests that the primary ore worked at S. Terras was uraninite and pitchblende, with a secondary zone of enrichment in the upper levels of the mine in the form of torbernite and autunite (hydrated phosphates of copper and uranium and calcium and uranium). The uranium ore lode at South Terras was over 450 metres long with almost unbroken uranium mineralisation (Smale 1993). Some parts of the lode assayed 31 percent uranium. Very high radon levels were recorded in South Terras.

The country rock consists of grey and brown Lower Devonian slate of the Meadfoot Group with greenstone and elvan dyke intrusions (Smale 1993).

Kings Wood mine (SX 713 665) is 2.5 km west of Buckfastleigh, South Devon, on the south-eastern side of Dartmoor. This mine was explored for copper (Harris 1992) but also contained argentiferous galena, sphalerite, cobalt, nickel, fluorspar and barytes together with pitchblende-coffinite (Darnley et al. 1965; Beer and Scrivener 1982; Harris 1992). The mineral vein in this mine is in Devonian slate and was probably derived from waning hydrothermal emissions from the granite according to Beer and Scrivener (1982). These mineralised cross-courses are younger than the east-west sulphide veins seen in mines in the Dartmoor and Gunnislake areas (Beer and Scrivener 1982; see also Gillmore et al. 2001a). The vein containing uranium ore in Kings Wood is only a centimetre to tens of centimetres in thickness and sometimes splits into two or three branches (Darnley et al. 1965). Uranium concentration is highest where the mineral lode is particularly brecciated. Pitchblende fills the fractures in the vein quartz with coffinite forming much of the matrix. Darnley et al. (1965) noted that mineral concentrates from these veins had intense point sources of alpha activity attributable to radium.

Methods

The mines chosen for this study were ones that were known by the Devon and Cornwall Prospecting Society for their uranium ore content. They have also been frequently visited in the past, although one cave (South Terras) was gated in 2000 to prevent cattle wandering into the adit entrance. While South Terras workings were once quite extensive a number of levels have collapsed, so it was only possible to visit the adit entrance section. Kings Wood is a much smaller mine being a simple adit cut into the hillside by a stream.

Radon gas levels were measured during the months of January to December in 1992 to 1994 and in July 2000 for comparative purposes, using a variety of measuring methods. Mostly passive alpha track etch detectors were employed, following the method laid out by Green et al. (1992), from an NRPB approved source. In addition to the time-averaged track etch and activated carbon detectors electronic real time devices were used, such as a Pylon WLx and Radhome P. A Rad7 was also used but radon levels proved to be outside of the measuring range of that device.

Due to the high levels of radon gas in the S. Terras mine detectors were placed and the authors retreated outside the mine entranceway to minimise exposure times.

Results

Results of this analysis are shown in Tables 1 and 2.

Table 1. Radon levels measured at South Terras mine, Cornwall, UK.

Date	Position	Detector	Radon level
Last week 5/92	70m from entrance	Alpha track	>41,667 Bq m-3 (saturated)
1st week 6/92	70m from entrance	Activated carbon	1,300,000 Bq m-3
Last week 8/92	2.4m from entrance	Alpha track	194,000 Bq m-3
Last week 8/92	17m from entrance	Alpha track	748,000 Bq m-3
Last week 8/92	52m from entrance	Alpha track	1,490,000 Bq m-3
Last week 8/92	52m from entrance	Pico rad	3,000,000 Bq m-3
Last week 8/92	70m from entrance	Alpha track	1,080,000 Bq m-3
Last week 8/92	70m from entrance	Pico rad	1,800,000 Bq m-3
08/10/92	Inaccessible inner workings	Alpha track	>1,900,000 Bq m-3 (saturated)
08/10/92	70m from entrance – head height	Alpha track	>3,390,000 Bq m-3 (saturated)
08/10/92	70m from entrance – ground level	Alpha track	>3,400,000 Bq m-3 (saturated)
Late 12/92	52m from entrance	Alpha track	200,000 Bq m-3
10/04/93	52m from entrance	TN-IR-31	0.37 WL
03/06/93	52m from entrance	Alpha track	379,000 Bq m-3
16/07/93	52m from entrance	Pylon WLx	29.9 WL
31/07/94	52m from entrance	Alpha track	3,200,000 Bq m-3
22/07/00	Outside mine entrance, 2m away	Radhome P	7,600 Bq m-3
22/07/00	52m from entrance	Radhome P	2,983,600 Bq m-3
22/07/00	52m from entrance – ground level	Alpha track	3,932,920 Bq m-3
22/07/00	52m from entrance – 1 metre from ground level	Alpha track	2,154,560 Bq m-3
22/07/00	One mine visit (1 hour)	Volalpha Personal Dosemeter	18 mSv

Radon levels in South Terras mine can be extremely high in the mines inaccessible inner workings. The furthest point into the S. Terras adit that could be easily

Impact on groundwater from radionuclide emission

reached was 70m in. Measurements were taken at both head height and at floor level in order to see if there was any significant difference in radon levels. As can be seen from Table 1 there was no significant difference when measured in October 1992, both being over 3 MBq m^{-3}. However, the 2000 year measurements demonstrate that at ground level 3,932,920 Bq m^{-3} of gas was measured, while 1m above ground level the radon gas was measured as 2,154,560 Bq m^{-3}. The lowest radon gas level measured was in May while the highest level at the same point was measured in October of the same year. In other words, levels were lowest in the spring and highest during the winter months. The radon gas levels in this mine are all consistently extremely high.

Table 2. Radon levels measured at Kings Wood mine, Devon, UK.

Date	Position	Detector	Radon level
25/01/92	Ore chamber	Alpha track	25,400 Bq m^{-3}
Early 6/92	Ore chamber	Activated carbon	30,000 Bq m^{-3}
31/07/93	Furthest point in	Pylon WLx	2.354 WL
31/07/93	Ore chamber	Pylon WLx	2.66 WL
31/07/93	Ore chamber	Pylon WLx	2.71 WL
22/07/00	Outside mine entrance, 2m away	Radhome P	13,400 Bq m^{-3}
22/07/00	40m from entrance	Radhome P	37,000 Bq m^{-3}
22/07/00	40m from entrance – ground level	Alpha track	32,257 Bq m^{-3}

It is also clear from Table 1 that there is a distinct gradation of radon gas concentration in S. Terras from the entranceway at 194,000 Bq m^{-3} to 17m in (748,000 Bq m^{-3}) to 52m in (3,000,000 Bq m^{-3}). It is interesting to note that radon levels then fell to 1,080,000 Bq m^{-3} at the 70m point. This is probably because it is at 52m that the pitchblende ore is exposed in the adit.

It is interesting to note that even standing 2 metres outside the mine entrance at South Terras will still expose an observer to radon gas levels of 7,600 Bq m^{-3} (Table 1). A very short distance into the mine, in the entranceway, levels were measured of 194,000 Bq m^{-3}.

Working Levels were also measured. In April 1993 52m into the mine a TN-IR-31 working level meter gave a reading of 0.37 WL. By July 1992 this level had risen to 29.9 WL.

The Volalpha personal dosemeter was left down the S. Terras mine for an hour and then retrieved. This gave a dose level of 18 mSV. It is recommended by the IRR (1999) that a member of the publics dose should not be greater than 1 mSv in a year, while a radiation workers maximum yearly dose is 6 mSv and a Classified Workers limit being 10mSV. Thus one hours visit would be equivalent to 18 years dose for a member of the public.

Radioactivity in Water

According to Durrance (1986) the natural radioactivity of water is highly variable. Generally, groundwater sources for drinking water have a higher level of radioactivity than water derived from surface flow (Durrance 1986). Obviously the range of radioactivity in groundwater will depend heavily on the geology and groundwater residence time.

Lewis (2001) examined radon levels in groundwaters in Pennsylvania, USA to assess the variability of levels. Average levels varied in springs from 100 to 2,000 pCi l[-1] (3,700-74,000 Bq m[-3]) over a number of years. However, the average for 28 groundwater sources was 465 pCi l[-1] (17,205 Bq m[-3]). It is interesting to note that even low radon in groundwater can lead to raised indoor air levels if there is a large enough volume of water involved (Lewis 2001).

Allen-Price (1960) undertook a study on the distribution of cancer in west Devon, UK and suggested a link with radioactivity in drinking water in the region. The activity of drinking water in the study area was up to 500 Bq l[-1] which was primarily attributed to [222]Rn (Abbot et al. 1960). The result of this study was that the most active water supplies were not exploited. The US Environmental Protection Agency (EPA) estimates that 168 cancer deaths per year are caused by radon in drinking water; 89% is lung cancer as a result of de-gassed radon and 11% stomach cancer (Jasensky 2001).

Turner et al. (1961) demonstrated that there were particularly high [222]Rn concentrations in drinking water from Cornwall. At St. Ives the level was 400 Bq l[-1]. Durrance (1986) has pointed out that [226]Ra and other [238]U daughter products also contribute to radioactivity in groundwaters. Kenny et al. (1966) suggests that activities in areas underlain by granite can be up to 800 Bq l[-1], although activity in the order of 1,000 Bq l[-1] has been noted in one spring (Durrance 1986).

In Devon, on the Dartmoor granite, a substantial public water supply was established extracting groundwater from an alluvial deposit. A survey of [222]Rn showed activities of around 550 Bq l[-1]. A degassing plant was erected to remove [222]Rn and CO_2. The plant was constructed underground so some care had to be taken to avoid high [222]Rn levels in the plant (Durrance 1986).

According to Durrance (1986) the effect of Rn in drinking water is regarded by many authors as unclear. The effect however consists of two aspects – ingestion and inhalation of radioactivity. Anderson and Nilson (1964) suggested that the half life of [222]Rn in the body following consumption via water is only 30 minutes, based on an activity of 2,500 Bq l[-1]. However, if there is an activity of 3,300 Bq l[-1] and the tap water intake is around 2.2 l per day the dose equivalent per year is 5,000 mSv (Durrance 1986). Some authors have pointed out that the half-life is so short the radiation dose would not be significant. It is important to recognise that most drinking water comes from surface water in granite areas in the UK where activity is much less (Durrance 1986).

Lindel (1968) suggested that a maximum permitted activity in tap water (assuming an intake of 1 l) should be 3,700 Bq l[-1]. The UK's National Radiological Protection Board suggests a maximum permitted activity of around 30,000 Bq l[-1], based on an annual oral intake for the general public of 2×10^8 Bq with a daily intake of 2 1 of tap water. If we assume that the whole body is the critical organ an average dose equivalent would be approximately 50,000 mSv per year (Durrance 1986). According to Lewis (2001) and Jasenksy (2001) the EPA has yet to publish a final standard for radon in drinking water. The current proposed standards are 300 pCi l^{-1} MCL (Maximum Contaminate Level) or 4,000 pCi l^{-1} AMCL (Alternate Maximum Contaminate Level) (11,100-148,000 Bq m^{-3}).

The source for much of the radioactivity in the South Terras mine region is the uranium ore, pitchblende. It is interesting to note that elements such as Mo, Rb, Sr, Cd, Ag, Pd, Sb, Sn, Cs, I, Ba may be removed from pitchblende by solution, depending on the chemistry of that solution (Durrance 1986).

Dissolved radon was measured in a number of mines in the study area. The levels of dissolved radon were in the order of 10,000 Bq l^{-1}. It should be stressed here that it was dissolved radon that was measured rather than radon as gas bubbles in the water, which yielded much higher results. The method employed here was that developed by Alan Worley of Track Analysis Systems Ltd and field tested by the DCPS.

Gamma spectrometry was also undertaken for mine water from South Terras mine and the adjacent Tolgarrick mine. Gamma spectrometry only enables certain sections of the uranium and thorium decay chains to be measured, these were notably of ^{226}Ra (plus daughters) and ^{228}Acc (plus daughters). For South Terras one water sample yielded a result of 300 Bq kg^{-1} of ^{226}Ra, while 5 Bq kg^{-1} of ^{228}Acc was measured. This analysis was undertaken by Harwell Radiation Spectrometry Unit, AEA Environment and Energy in July 1992. Harwell also analysed the uranium levels in mine water from S. Terras. Their results indicated that there was 360 +/- 20 Bq l^{-1} of ^{238}U, with 14 +/- 2 Bq l^{-1} ^{235}U and 370 +/- 20 Bq l^{-1} ^{234}U.

Equilibrium Factors and Dose

The ration of radon to progeny (the Equilibrium factor, F) is relatively constant in homes. So radon concentration is often used to determine dose (Phillips et al. 2000). If we assume that the radon levels and their progeny did not waver significantly in July from year to year, then the Equilibrium Factor (F) in S. Terras may be in the order of 0.2 to 0.5. Snihs and Ehdwall (1976) measured F in working Swedish mines. In most of these Swedish mines, F was 0.4 to 1, with an average F (from 37 mines) of 0.7. This suggests that our assumption of F of 0.5 is not unreasonable.

We can, using F of 0.5, calculate dose rates following Denman and Parkinson (1996) and Gillmore et al. (2000b).

$$\text{Effective Dose (mSv)} = \frac{(\text{Radon Concentration, Bq m}^{-3}) \times (\text{duration, hours})}{126,000}$$

Assuming that F is 0.5 in such mines as these, and each visit lasted approximately 2 hours (following Gillmore et al. 2000b), exposure to the levels in South Terras mine (at a maximum of 3,932,920 Bq m^{-3}) would give an the effective dose of approximately 62 mSv per visit. This is considerably higher than the 1 mSv per year recommended as a maximum dose for a member of the public.

Acid Mine Water Drainage

Water that drains from mine workings may be termed acidic as a result of natural oxidation of sulphide mineral when waste is exposed to air and water (Bell 1998). Iron pyrites (FeS_2) often occurs as a gangue mineral in the mines in South-West England in this study. The pH of mine water from these mines was generally between 4 and 5, However, Kingswood mine water was measured with a pH of 3. Such acid mine drainage is responsible for water pollution in metal mining areas around the world. Acid generation can lead to elevated levels of heavy metals such as copper and zinc together with sulphate in water. The acid Kingswood mine water sample had raised levels of copper. However, the rate and nature of acid mine drainage is essentially controlled by chemical and biological reactions (Bell 1998). The correct conditions for the generation of such waters depends on the combination of mineralogy, as well as pH value, temperature, oxygen, surface area of exposed metal sulphide etcetera. The presence of the acidophilic autotrophic bacteria in the form of *Thiobacillus ferrooxidans* may accelerate oxidation of sulphides of minerals such as arsenic, cadmium, copper, lead and zinc. When mine water discharge reaches watercourses it is rapidly oxidised, and the majority of dissolved metals may be precipitated onto the bed sediment (Hill 1999). This precipitate may give an orange or yellow ochre colour to the bed of the watercourse. There are very distinctive biological effects of such discharge. These might include; a depletion of the numbers of sensitive, and diversity of all, aquatic organisms; a loss of spawning gravel for fish; direct fish mortalities; and lastly a variety of other sub-lethal effects (Hill 1999).
sonable.

Hill (1999) noted that, using the Bureau Communite de Reference method (BCR), results from heavy metal analysis of a number of stream beds in the Yorkshire region showed that Cononley Becks contained high levels of zinc, lead and cadmium. Hill (1999) recorded a zinc BCR mean in the less than 63 micron fraction bed sediment in 1998 of around 2,000 mg kg^{-1} dry weight. A lead BCR mean of 5,000 mg kg^{-1} dry weight was also noted, together with a cadmium BCR mean of over 17 mg kg^{-1} dry weight. Hill (1999) suggests that at Cononley there has been co-mineralisation of cadmium ore with galena and sphalerite. Levels of nickel, iron, manganese, chromium, calcium, copper (copper BCR mean of approximately 42 mg kg^{-1} dry weight) and aluminium (BCR mean of 1250 mg kg^{-1} dry weight) were also noted by Hill (1999) in Cononley Beck, but were generally less significant in comparison to the other stream beds examined.

Hill (1999) suggests that a cocktail of metals present in the sediment act antagonistically upon benthic macroinvertebrates, more than one metal producing toxic effects either working individually or in combination.

Heavy metal analysis (Ni, Pb, Cd, Zn, Ba, Cu) of water from South Terras suggests that many of these metals occur in the water, sometimes in high concentrations. Zinc was found in all samples from both mines in this study, ranging from 0.123 to 0.217 ppm in S. Terras. This was above the normal range for freshwater (5-50 ppb). It is very interesting to note that similar analyses of mine water from Kingswood mine showed significantly raised levels of nickel (1.69 ppm) in winze water where the pitchblende and arsenopyrite ores are exposed. Nickel is toxic to aquatic life at 0.15 ppm. Copper (0.606 ppm) was also higher in this water than water collected from elsewhere in Kingswood mine. The ecotoxicity of copper to aquatic life in freshwater is 0.01 ppm. Lead was higher here than elsewhere in the mines in this study at 0.346 ppm.

Analysis of the sediment samples collected shows that high levels of heavy metals exist in some of the fine sediments in streams from the mine entranceways. In samples from Brookwood mine, close to Kingswood, a maximum of 6,700 mg kg^{-1} of copper was noted. A level that would be toxic to life. In the samples from Kingswood mine copper in the sediment was much lower at less than 300 mg kg^{-1}, but levels of nickel were raised. Heavy metals in South Terras mine sediments were generally not of environmental concern.

Conclusions

The analyses undertaken for the abandoned mines in this study suggest that raised radon levels within the mines poses a significant health risk. The amount of dissolved radon in the mine water is also significant and may pose a real risk to health if ingested for any length of time (e.g. via private spring-fed water supplies). The levels of some heavy metals in the water were significant where ore veins were exposed in the water courses. Analysis of heavy metals within the sediments seems to indicate that metals transported by mine water and deposited into streamways should be investigated further.

References

Abbot JD, Lakey JRA, and Mathias DJ, (1960) Natural radioactivity in west Devon water supplies. Lancet 2: 1272.

Allen-Price ED (1960) Uneven distribution of cancer in west Devon with particular reference to the diverse water supplies. Lancet 1: 1235.

Anderson ID, Nilson I (1964) Exposure following ingestion of water containing radon-222. Proceedings of the IAEA Symposium on Assessment of Radioactive Body Burdens in Man II. Vienna, International Atomic Energy Agency.

Beer KE, Scrivener RC (1982) Metalliferous Mineralisation. In: Durrance, E.M. and Laming DJC (Eds.), The Geology of Devon, University of Exeter Press, UK: 346pp.

Bell FG (1998) Environmental Geology. Principles and Practice. Blackwell.

Burt R, Waite P and Burnley R (1987) Cornish Mines. Metalliferous and Associated Minerals 1845-1913. The University of Exeter, UK: 562pp.

Darby S, Whitley E, Silcocks P, Thakrar B, Green M, Lomas P, Miles JCH, Reeves G, Fearn T and Doll R (1998) Risk of lung cancer associated with residential radon exposure in south-west England: a case-control study. Br J Cancer 78(3): 394-408.

Darnley AG, English TH, Sprake O, Preece TH and Avery D (1965) Ages of uraninite and coffinite from south-west England. Mineralogical Magazine 34: 159-176.

Denman AR, Parkinson S (1996) Estimates of radiation dose to National Health Service workers in Northamptonshire from raised radon levels. Short communication. The British Journal of Radiology 69: 72-75.

Denman AR, Phillips PS (1998) A review of the cost effectiveness of radon mitigation in domestic properties in Northamptonshire. J Radiol Prot 18(2): 119-124.

Dines HG (1956) The metalliferous mining region of south-west England. Mem geol Surv Gt Br 2 vols: 795pp.

Dunham K, Beer KE, Ellis RA, Gallagher MJ, Nutt MJC, Webb BC (1978) United Kingdom. In: Bowie SHU, Kvalheim A, Haslam HW (Eds.) Mineral deposits of Europe Volume 1: Northwest Europe, The Institution of Mining and Metallurgy, The Mineralogical Society, London: 263-317.

Durrance EM (1986) Radioactivity in Geology. Principles and Applications. Ellis Horwood, Chichester, UK.

Field RW, Steck DJ, Smith BJ, Brus CP, Neuberger JS, Fisher EF, Platz CE, Robinson RA, Woolson RF, Lynch CF (2000) Residential Radon Gas Exposure and Lung Cancer: The Iowa Radon Lung Cancer Study. American Journal of Epidemiology 151(11): 1091-1102.

Gillmore GK, Sperrin M, Phillips P, Denman A (2000a) Radon Hazards, Geology, and Exposure of Cave Users: A Case Study and Some Theoretical Perspectives. Ecotoxicology and Environmental Safety 46(3): 279-288.

Gillmore GK, Sperrin M, Phillips P, Denman A (2000b) Radon-prone geological formations and implications for cave users. Technology 7(6): 645-655.

Gillmore GK, Sperrin M, Pearce G (2001a) Radon in a disused mine in Cornwall, UK. Journal of Environmental Management and Health 12(5): 500-509.

Gillmore GK, Phillips P, Denman A, Sperrin M, Pearce G. (2001b) Radon levels in abandoned Metalliferous Mines, Devon, South-West England. Ecotoxicology and Environmental Safety 49(1): 281-292.

Green BMR, Lomas PR, O'Riordan M.C. (1992) Radon in dwellings in England. National Radiological Protection Board Report R254: 1-72.

Harris H (1992) The Industrial Archaeology of Dartmoor. Peninsular Press, UK.

Henshaw DL, Eatough JP, Richardson RB (1990) Radon as a causative factor in induction of myeloid leukaemia and other cancers. Lancet 335: 1008-1012.

Hill M (1999) Relative impact of bed sediment metal pollutants and water column organic pollutants on benthic macroinvertebrates of 19 watercourses in Yorkshire. Unpublished PhD thesis. University of Bradford.

Hoover HC, Hoover LH (translators) (1986) Georgius Agricola. De Re Metallica, 1556. Dover Publications republication of 1912 addition: 638pp.

Ionising Radiations Regulations (1999) Health and Safety Executive, Statutory Instrument 3232, HMSO.

International Commission on Radiological Protection (1981) Limits for inhalation of radon daughters by workers. ICRP Publication 32. Ann ICRP 6(1).

Impact on groundwater from radionuclide emission

International Commission on Radiological Protection (1993) Protection against radon-222 at home and at work. ICRP Publication 65. Ann ICRP 23(2).

Jackson NJ, Willis-Richards J, Manning DAC, Sams MS (1989) Evolution of the Cornubian Ore Fields, Southwest England: Part II. Mineral Deposits and Ore-Forming Processes. Economic Geology 84: 1101-1133.

Jasensky J (2001) Residential waterbourne radon removal study using granular activated carbon (GAC) filtration systems. AARST 2001 International Radon Symposium, October 21-26, Daytona Beach, USA: 120-127.

Kendall G (2000) Doses from radon to organs other than lung. Environmental Radon Newsletter 23: 4.

Kenny AW, Crooks RN, Kerr JRW (1966) Radium, radon and daughter products in certain drinking waters in Great Britain. J Inst Water Eng Sci 20: 123.

Lewis RK (2001) An investigation of radon occurrence in Pennsylvania fish and boat commission fish culture station. AARST 2001 International Radon Symposium, October 21026, Daytona Beach, USA: 52-67.

Lindell B (1968) Ingested radon as a source of human radiation exposure. Proceedings of International Congress on Radiation Protection, Rome.

Lorenz E (1944) Radioactivity and lung cancer: A critical review of lung cancer in the miners of Schneeberg and Joachimsthal. J Natl Cancer Inst 5: 1-15.

Lubin JH, Boice JD (1997) Lung cancer risk from residential radon: Meta-analsis of eight epodiomiological studies. J Natl Cancer Inst 89(1): 49-57.

Miles JCH, Green BMR, Lomas PR (1992) Radon affected areas: Derbyshire, Northamptonshire, and Somerset. Doc Natl Radiol Prot Board 3(4): 19-28.

Muirhead CR, Cox R, Sather JW, MacGibbon BH, Edwards AA, Haylock, RGE (1993) Estimates of late radiation risks in the UK population. Doc Natl Radiol Prot Board 4(4): 15-157.

O'Riordan MC (1990) Recommendations on Radon in Homes. Doc Natl Radiol Prot Board 1(1): 17-32.

Phillips P, Denman A, Gillmore GK (2000) Radon, Schools and Health: Implications for policy and practice of a comparative study of programmes in Poland and the UK. Fresenius Environmental Bulletin 9: 711-718.

Smale CV (1993) South Terras. Cornwall's Premier Uranium and Radium Mine. Journal of the Royal Institution of Cornwall. New Series 1(3): 304-321.

Snihs JO, Ehdwall H (1976) Supervision of radon daughter exposure in mines in Sweden. In: Personal monitoring and area monitoring suitable for radon and its daughter products, Proceedings of the Nuclear Energy Agency Specialist Meeting, October: 191-197.

Sperrin M, Denman T, Phillips, PS (2000) Estimating dose from radon to recreational cave users in the Mendips, UK. Journal of Environmental Radioactivity 49: 235-240.

Thompson A, Hine PD, Poole JS, Greig JR (1998) Environmental geology in land use planning: A guide to good practice. Report to the Department of the Environment, Transport and the Regions (DETR), Symonds Travers Morgan, East Grinstead, UK.

Turner RC, Radley JM, Mayneord WV (1961) Naturally occurring alpha-activity of drinking waters. Nature 189: 348.

Physico-chemical surface water conditions of catchments with metallogenic origin: A contribution to the establishment of the EC Water Framework Directive 2000/60/EG in Germany

Peter L. Neitzel[1], Petra Schneider[1], Helmut Schlumprecht[2]

[1]Hydroisotop-Piewak GmbH, Oberfrohnaer Str. 84, D-09117 Chemnitz
[2]Büro für Ökologische Studien, Alexanderstr. 5, D-95444 Bayreuth

Abstract: For the realisation of the EC Framework Water Directive in Germany, the reference conditions of metals in surface waters have to be determined taking into consideration the so called surface water landscape. In catchments of metallogenic origin metals and radionuclides like uranium influence the quality of the surface waters. In this study, the quality status of metallogenic (Ore Mountains, Harz, Rhine Slate Mountains) and other surface water landscapes were compared in order to differentiate factors influencing the natural background in similar geological formations.

Introduction

In the year 2000 the EC Water Framework Directive (2000/60/EG) passed the parliament of the European Community (EC), which then became legal law in all countries of the EC. The main objective of the EC-directive is to achieve a good quality of all surface, ground-, estuary and coastal waters. This includes an assessment of the chemical status of the surface waters (environmental quality standards for nearly 30 priority compounds and substances) and a five-stage ecological classification of waters. According to the regulations three groups of characteristics are important for the European surface waters, namely: hydrobiology (priority), hydromorphology (supporting hydrobiology) and physico-chemical conditions (supporting hydrobiology). "High water quality" (class I) hence reflects natural conditions totally, while class II refers to "good water quality". Additionally the maximum ecological potential is taken as reference for heavily modified and artificial water bodies (Irmer 2000).

Within an investigation project, financed by the Environmental Federal Agency (UBA) of Germany, all available analytical data of low polluted or close to natural surface waters in 30 different water landscapes of Germany were collected and analysed by statistical methods (Schneider et al. 2002, Neitzel et al. 2002). The data base included summary and effect indices, major ions, salts and nutrients, heavy metals (Cd, Cr, Ni, Hg, Pb, U, Zn), as well as As and Cu. The aim of the study was the elaboration of a physico-chemical data base for the assessment of the natural background of surface water bodies in Germany (Neitzel et al. 2002). This data base was used to define the reference status according to their hydro-morphological and geological setting. Regarding the quality of surface waters, the EC Framework Water Directive requires the level 'good ecological and chemical quality' to be reached in the year 2016. In this study, the quality status of metal-logenic (Ore Mountains, Harz, Rhine Slate Mountains) and other surface water landscapes was compared in order to differentiate factors influencing the natural background in similar geological formations.

Physico-chemical surface water assessment of catchments with metallogenic origin

Assessment of physico-chemical data of surface waters

In order to meet the requirements of the EC Framework Water Directive in Germany, reference concentrations (= natural background concentrations (Fergusson 1990)) of metals have to be determined taking into consideration the so called surface water landscape. In catchments of metallogenic origin, heavy metals and radionuclides like uranium can influence the quality of the surface waters. As investigations showed, the following geological formations were of metallogenic origin: granite, slate, gneiss, diabase, and Grauwacke. The data were compared to reference concentrations of metals and uranium in surface waters with low anthropogenic influence.

Data pool

LAWA (Länderarbeitsgemeinschaft Wasser/German Water Association) defined the reference status as natural background concentrations of a non-anthropogenic influenced catchments in 1998. The investigated data base of 560 measuring locations of 30 different surface water body types contains 3.500 analytical data sets. All data were analysed using statistical methods. Mean, 10-, 50-, and 90-percentils as well as minimum and maximum values were calculated. The representative data base obtained the assessment of the sustainability of the environmental quality standards as defined by LAWA.

Impact on groundwater from radionuclide emission

Surface water landscapes with metallogenic origin

Part of this study was the investigation of the quality status of metallogenic surface water landscapes. As the results of the physico-chemical data assessment show, the Ore Mountains, the Harz Mountains and the Rhine Slate Mountains differ significantly in their metal content from the other catchment types. Hence the metallogenic surface water landscapes have to be given a special status.

Results

Main results of the investigation of all surface water landscapes

Physico-chemical data of 30 different types of surface water landscapes were investigated according to the surface water landscape classification of Briem (2001). The results of the statistical analysis show, that the surface water bodies in Germany can be summarised in six catchment types (Fig. 1):
- bogs and bog riverside meadows
- salinic catchment type
- carbonatic-dolomitic catchment type
- sandy-clayey catchment type
- silicatic catchment type and
- metallogenic catchment type

Heavy metals bound to particular matter (based on 50 percentils)

For all catchment types, with the exception of the metallogenic landscapes, the reference status of lead, chromium and mercury met the quality standards of class I (high water quality standard) of LAWA, while cadmium, copper, nickel and zinc values were found to be increased in all catchment types. A reference status for these parameters cannot be defined in the particular matter of metallogenic landscapes with the exception of mercury.

Heavy metals in the total liquid phase (based on 50 percentils)

A reference status referring to high water quality standard (class I) was evaluated for following heavy metals:
- chromium: all catchment types
- copper and lead: carbonatic-dolomitic catchment type
- nickel: all types excluding sandy-clayey catchment type and salinic type
- zinc: carbonatic-dolomitic type, silicatic type, sandy-clayey catchment type

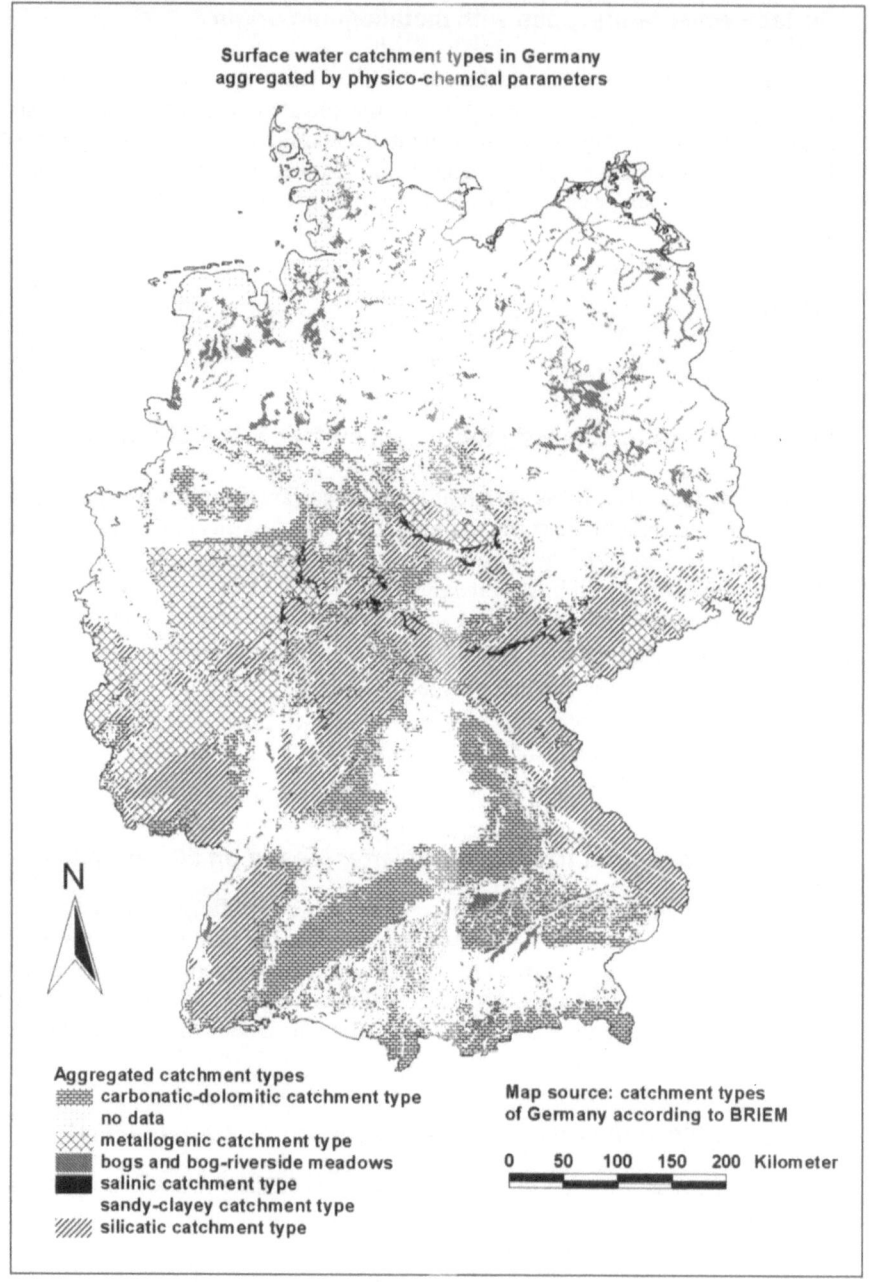

Fig. 1. Aggregated surface water landscapes according to the results of the physico-chemical assessment.

Impact on groundwater from radionuclide emission

No reference status was found in whole Germany for cadmium and mercury considering NOEC-standards (No Observed Effect Concentration) with respect to the most sensitive water organism. There are different causes for this situation: various diffuse and historical pools as atmospheric deposition, use of fertilizers and solution effects of acid rain in the soils.

Metals/metalloides in the liquid phase (based on 50 percentils and means)

No LAWA quality aims exist for aluminium, arsenic, barium, iron, cobalt, manganese and uranium. The results of the statistical analysis provide a data base that can be used do define a reference status for these parameters. These results also correspond to data found in literature.

Main results of the investigation of the metallogenic catchment types

The results of the investigation of the metallogenic catchment types in comparison to the quality class I (the natural background) of LAWA (1998) and the other surface water catchment types are summarized in the tables 1 and 2.

Table 1. Metals in the suspended phase - comparison to the quality class I of LAWA (1998) (based on 50 percentiles)

metal (mg/kg)	quality aim class I (LAWA)	bogs and bog riverside meadows	carbonatic-dolomitic type	silicatic catchment type	sandy-clayey type	salinic type	**metallogenic type**
lead	≤ 25	18	22	30	24	26	**175**
cadmium	≤ 0,3	1,0	0,4	0,5	0,3	1,0	**17,5**
copper	≤ 20	7,0	39	37	21	9,0	**81**
nickel	≤ 30	11	56	47	29	22	**120**
mercury	≤ 0,2	no data	0,1	0,2	0,1	no data	**0,2**
zinc	≤ 100	36	127	140	92	62	**1100**

Table 2. Metals in the liquid phase - comparison to the natural background of LAWA (1998) (based on 50 percentiles)

metal	natural background LAWA	bogs and bog riverside meadows	carbonatic-dolomitic type	silicatic catchment type	sandy-clayey type	salinic type	**metallogenic type**
lead (µg/L)	0,4 - 1,7	1,0	1,4	1,0	0,9	1,0	**2,1**
cadmium (ng/L)	9 - 36	250	100	150	130	150	**339**
chromium (µg/L)	1,3 - 5	0,5	0,5	0,5	0,6	0,9	**2,5**
copper (µg/L)	0,5 - 2	4,8	1,4	1,3	0,9	4,0	**2,4**
nickel (µg/L)	0,6 – 2,2	2,0	1,3	1,8	2,0	4,7	**5,6**
mercury (ng/L)	5 - 20	50	100	500	50	20	**188**
zinc (µg/L)	1,8 - 7	10	5,0	13	5,0	12	**28**

Impact on groundwater from radionuclide emission

An increased metal status in the liquid phase in comparison to the natural background concentrations of the other surface water landscapes has to be considered for:

- Ore Mountains: cadmium, copper, zinc, lead,
- Harz Mountains: nickel, zinc,
- Rhine Slate Mountains: nickel, zinc, lead, copper.

The most increased metal status of all investigated geological metal rich formations (granite, slate, gneiss, diabase, Grauwacke) was found in slate. To evaluate the reference status of surface waters of mixing catchment types and catchments with anthropogenic influence a runoff-concentration-analysis was calculated using the Hellmann-method. There was no usable result due to the variability and occurrence probability of the catchment influences.

Conclusions

The results of the physico-chemical surface water assessment study provide the reference status of the metallogenic catchment types as base for the implementation of the EC Water Framework Directive in Germany. The results of the investigation of the reference status of surface waters show, that a special physico-chemical surface water status for catchments with metallogenic origin has to be considered. All reference data of metallogenic catchments where measured in areas with no mining activities. There is no suitable way to use the high LAWA quality aims for the implementation of the EC Water Framework Directive in metallogenic catchments. Especially the natural occurring metals cadmium, nickel, copper, zinc and lead cause a high potential of metals in the liquid and the suspended phase of the rivers draining metallogenic catchments.

The authors would like to thank the Environmental Federal Agency (UBA) of Germany for financial support, given to this R&D-project.

References

Briem E (1998): Die Fließgewässerlandschaften der Bundesrepublik Deutschland - Teil 1: Fließgewässerformen-, -strukturen und -typologie. Ein kleines Handbuch der abiotischen Fließgewässerkunde (Bericht), Dörrenbach, 90 S.

European Commission (2000): Richtlinie 2000/60/EG des Europäischen Parlaments und des Rates vom 23.10.2000 zur Schaffung eines Ordnungsrahmens für Maßnahmen der Gemeinschaft im Bereich der Wasserpolitik. in: Amtsblatt der Europäischen Gemeinschaften, L 327/1, Luxemburg

Irmer U (2000): The New EC Framework Water Directive: Assessment of the Chemical and Ecological Status of Surface Waters. Acta hydrochim. hydrobiol. 28, 7-14

Fergusson JE (1990): The Heavy Metals - Chemistry, Environmental Impact and Health Effects. Pergamon Press, Oxford, p. 275

Impact on groundwater from radionuclide emission

LAWA (1998): Beurteilung der Wasserbeschaffenheit von Fließgewässern in der Bundes-republik Deutschland - chemische Gewässergüteklassifikation, LAWA (Hrsg.), Kul-turbuchverlag Berlin GmbH, Berlin, S. 1-35

Neitzel PL, Schneider P, Schlumprecht H (2002): Umsetzung der EG-Wasserrahmenrichtlinie: Leitbildorientierte physikalisch-chemische Gewässerbewer-tung - Referenzbedingungen und Qualitätsziele am Beispiel deutscher oberirdischer Binnengewässer. in: Lecture-Proceedings Jahrestagung der Wasserchemischen Gesell-schaft in der GDCh, Eichstätt, S. 148-151 (ISBN 3-936028-05-2)

Schneider P, Neitzel P L, Schaffrath M, Schlumprecht H (2002) Leitbildoreintierte physika-lisch-chemische Gewässerbewertung - Referenzbedingungen und Qualitätsziele. in: Abschlußbericht zum Forschungsvorhaben des Umweltbundesamtes (Förderkennzei-chen 200 24 226), Chemnitz.

Impact on groundwater from radionuclide emission

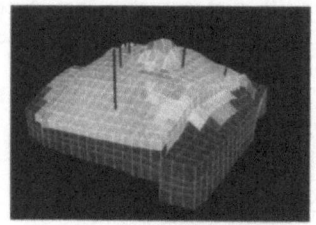

Geocomputing – Visualization and Simulation in Geoengineering

Geocomputing links computer science with the classical engineering sciences, particularly with geotechnical engineering, mining, gas and oil exploration, and groundwater management. Computer Sciences and simulations are increasingly used for process management, and for planning and modelling in geoengineering.

1st – 4th term (graduate studies)
* Mathematics, Computer science, Geosciences
* Choice of two out of three modules:
  - Geotechnical Engineering
  - Reservoir Engineering, Ground Water Hydraulics, Oil and Natural Gas Production, and Storage Technology
  - Mining, Special civil Engineering
* Internship (minimum 6 weeks)
* oral examination after 4th term
* Master thesis in the 4th term

Admission
* Bachelor of Science in „Network Computing"
* Bachelor/Master of Science in „Computer Sciences" or „Mathematics"
* equivalent access entitlement

Duration/Degree: 4 semesters/Master of Science in Engineering (Mag. Geocomputing)
Start: winter term

Job opportunities
Application of visualisation and simulation of processes in Geoengineering and Earth Sciences; National and international positions in resource-related industries, environmental and engineering consulting, banks, software development, and all levels of state institutions; Particular option in internet-based development

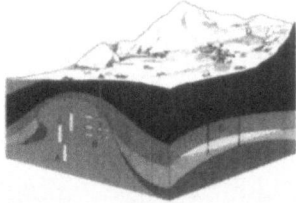

Radionuclides in underground water in an area contaminated with uranium mill waste

Polona Vreček[1], Ljudmila Benedik[1], Urška Repinc[1], Peter Stegnar[1], Ivan Gantar[2]

[1]Jožef Stefan Institute, Jamova 39, 1000 Ljubljana, Slovenia
[2]Uranium mine Žirovski vrh, Todraž 1, 4224 Gorenja vas

Abstract. The uranium mine and mill at Žirovski vrh (RŽV), Slovenia stopped mining and milling in 1990. Mill tailings containing about 20 % moisture were transported to a disposal site called Boršt. The objective of this work was to study the migration of the radionuclides U-238, Ra-226, Pb-210 and Po-210 from the Boršt waste disposal site. For determination of radionuclides, gamma and alpha spectrometry, beta counting, neutron activation analysis and liquid scintillation techniques were used.

Introduction

The uranium mine and mill at Žirovski vrh (RŽV) produced its first yellow cake in late 1984 and stopped mining and milling in 1990. About 600,000 tonnes of ore were processed in this period. The facilities are located 35 km west-northwest of Ljubljana in a valley drained by the Brebovščica stream, a tributary of the river Poljanska Sora. The facilities consist of an underground mine, a mill and several deposits of mining and hydrometallurgical wastes. Fig. 1 shows a 3D view of the surroundings of the uranium mine and the waste disposal sites.

The mill tailings and mine waste pile are deposited on Carnian clastites and Noric dolomites of the Triassic strata. The Carnian clastites consist of brightly coloured siltstones, sandstones, tuffites, and more limestone developed into lens bodies. During the Alpine Orogeny the entire sequences of the region were folded, heavily sheared, cut by north-south (N-S) and northwest-south east (NW-SE) faults and overthrusted with permocarboniferous strata. Fractured porosity developed in the clastites. We have observed caverns and karst caves inside the limestone at the Boršt and Jazbec sites. Fracture porosity is also typical of Upper Triassic dolomite. Only tuffites represent a barrier to water because they are impermeable, even in fractured zones. Under the mill tailings tuffites divide the ground water table into two parts. Water above the tuffite lens is polluted and represents a potential landslide problem. Water under the tuffite lens is unpolluted.

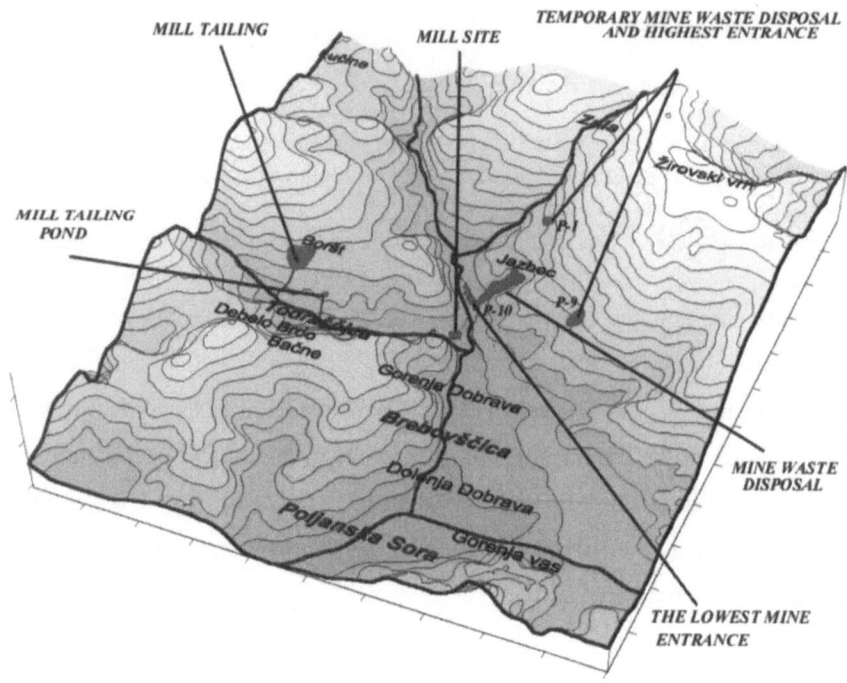

Fig. 1. 3D view of the mine region (by T. Beguš).

The RŽV mill design capacity was 160000 tonnes of ore processing per year at an average grade of about 0.1 % U_3O_8. The ore was crushed and leached with sulphuric acid. The mill tailings with about 20 % moisture were transported to the disposal site at Boršt, nearly 3 km from the processing plant. These tailings (altogether 700000 tonnes) are deposited on a sloping area. The tailings consist of sands and slimes less than 28 mesh and contain on average 8630 Bq kg^{-1} ^{226}Ra, 3930 Bq kg^{-1} ^{230}Th and 1000 Bq kg^{-1} ^{238}U. The tailings dump incorporates a system of capillary or wick drainage to capture and collect maximum amounts of moisture from the tailings deposit and discharge the water via a channel to the river. Due to heavy rains in November, 1990, a substantial part of the tailings deposit with nearly 2.5 million cubic metres of the underlying natural bedding rock began to slide. Survey results indicated an average daily movement of 1.2 mm at the beginning of the slide. Investigations by experts indicated that land movement was due to the heavy inflow of ground water. A stability study showed that lowering of the water table by about 20 m would stop the slide. Therefore a drainage tunnel was constructed under the sliding area to collect the seeping ground water. A series of drainage water wells to lower the water table are envisaged. The water from the collection ditch flows out to the local surface stream (Todraščica). A

Impact on groundwater from radionuclide emission

sketch map of the area around the mill tailing site and the general direction of the underground water flow are shown in Fig. 2.

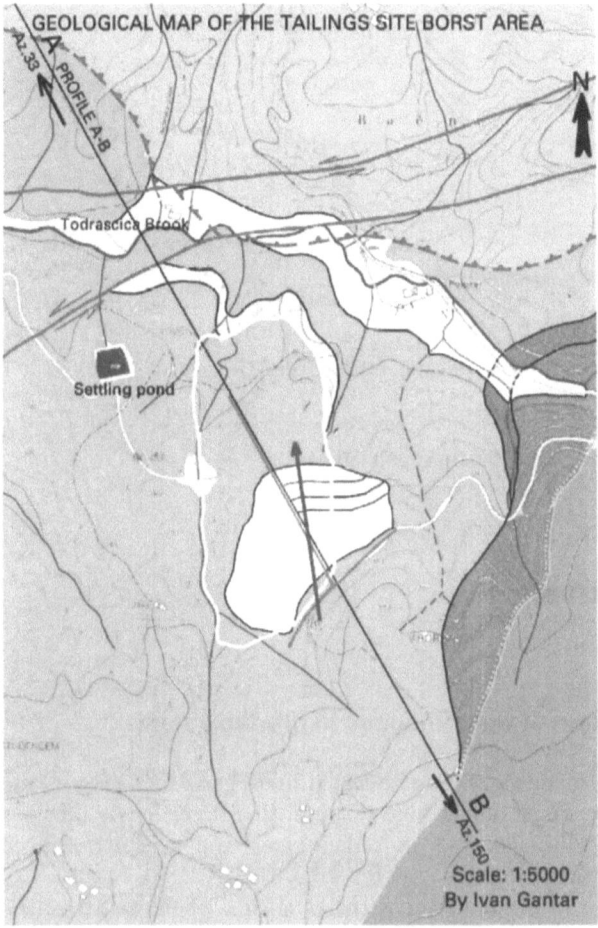

Fig. 2. Geological sketch map of the area around the mill tailings site, and general direction of the underground water flow.

The water contains dissolved uranium (1340 - 1570 mg m^3), radium (5430 - 6340 Bq m^{-3}) and high concentrations of ammonium, calcium, sulphate and chloride ions. The surface is covered temporary by thin soil layer. The effluent flows are underground water beneath the mill tailings (sampling point SDIJ), surface out-flow via the mill tailings pond Boršt and to a minor degree the underground drain-age tunnel. The flow of the underground water flow below the mill tailings (SDIJ) decreased considerably after the construction of the drainage tunnel. The concen-trations of contaminants rose five times due to lack of dilution by underground water. Fig. 3. shows a cross section of the Boršt mill tailings.

Impact on groundwater from radionuclide emission

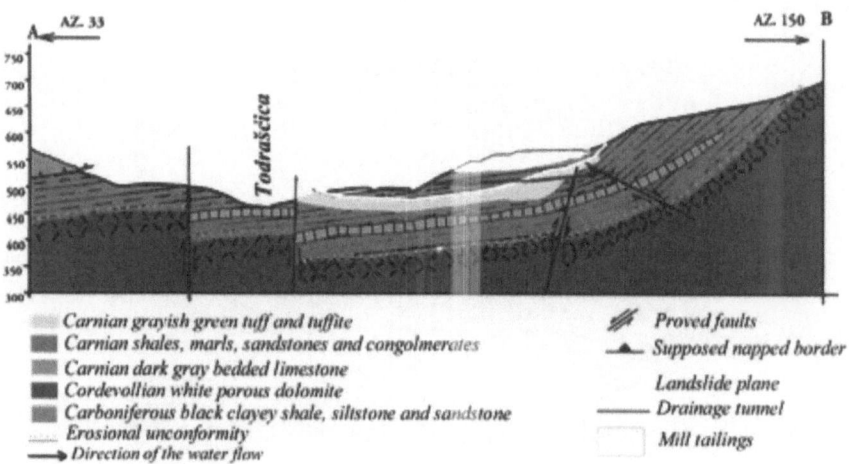

Fig. 3. Cross-section of the Boršt mill tailings.

Experimental

Determination of radionuclides in groundwater

Samples of groundwater were collected in the base rock (Fig. 4) and in the body (Fig. 5) of the Boršt mill tailings disposal site. Immediately after sampling, water were filtered through 0.45 μm filter paper and acidified with HNO_3 (1 mL HNO_3/L of water).

According to the radioactive characteristics of the radionuclides of uranium and their decay products, we determined them using radiochemical neutron activation analysis, alpha spectrometry, beta counting and liquid scintillation counting.

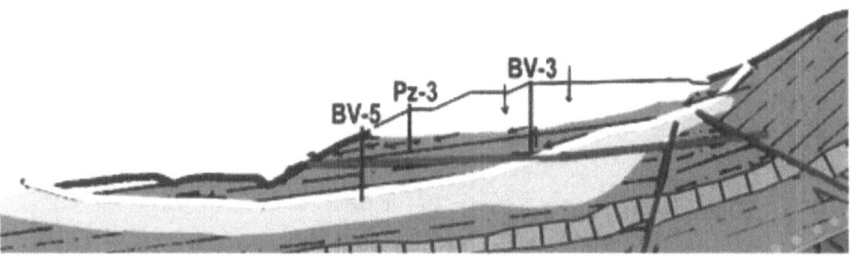

Fig. 4. Sampling points of underground water in the base rock of the Boršt mill tailings disposal site.

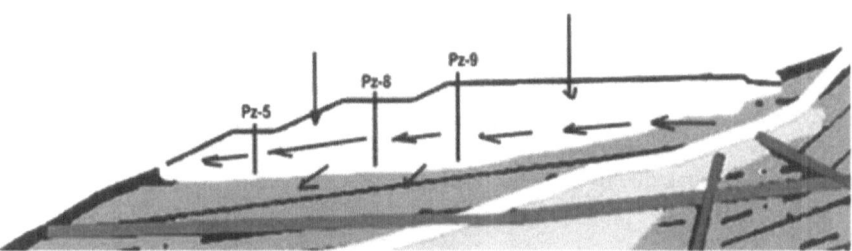

Fig. 5. Sampling points of underground water in the body of the Boršt mill tailings disposal site.

Determination of uranium by radiochemical neutron activation analysis:

Water samples were weighed and sealed into clean polyethylene ampoules. The sample weights ranged from 2.5g to 3g. The sealed ampoules were further encapsulated in polythene foil to avoid superficial contamination during irradiation and handling. Samples were irradiated in the Institute's TRIGA MK II reactor in the pneumatic tube (rabbit system) at a neutron fluence rate of 4×10^{12} ncm^{-2} s^{-1} for up to 5 min with a uranium standard ($\sim$100 ng). When ^{238}U is irradiated in a reactor the following capture reaction takes place:

$$^{238}U(n,\gamma)^{239}U(t_{1/2} = 23.5 \text{ min}) \rightarrow {}^{239}Np \ (t_{1/2} = 2.35 \text{ d}) \rightarrow$$

Based on earlier work (Byrne and Benedik 1988, 1995) a very sensitive method for uranium determination using selective extraction was used. The short-lived ^{239}U nuclide was extracted with 50 % TBP in toluene from 5M nitric acid following wet-ashing in the presence of uranium carrier and conversion of uranium to the U^{6+} form with addition of perchloric acid. The organic phase was cleaned up with two washes of 5M nitric acid containing 0.2% HF. The gamma peak of ^{239}U at 74.7 keV of the sample and standard solution was measured for determination of ^{238}U content. Use of a relatively large amount of uranium carrier (50mg) allows the chemical yield to be evaluated in the gamma spectrum of the isolated uranium fraction from the 185.7 keV γ peak of ^{235}U. Gamma ray spectrometry was performed in an HP Ge well-type detector connected to a Canberra MCA by Genie-2000 Software.

Determination of ^{226}Ra by a liquid scintillation technique

The radiochemical separation procedure for determination of radium was adopted from Burnett (Burnett and Tai 1992). 1L of water was transferred to a glass beaker. After addition of ^{133}Ba tracer together with Ba-carrier, the sample was heated and stirred for approximately 30 min. The procedure was based on co-precipitation of radium with barium sulphate after addition of sulphuric acid, conversion to carbonate, nitrate and finally dissolution of nitrate in water. The chemical yield was determined after separation by measurement of the ^{133}Ba gamma peak at 355.86 keV. After the separation 10 mL of Packard ULTIMA GOLD AB scintillation solution was added. Before direct measurement, the scintillation solutions were pre-count delayed (10 min of adaptation in the dark and thermostatted in the counting chamber), and then counted immediately after separation, when only the photopeak due to ^{226}Ra (4.78 MeV) was observed.

Determination of ^{210}Pb and ^{210}Po in water

For determination of ^{210}Pb and ^{210}Po in water 25 mg Pb^{2+} carrier and ^{208}Po (~ 0.3 Bq) tracer were added to 5 L of water and then the radionuclides co-precipitated with MnO_2 (Minczewski et al 1982). The precipitate was then dissolved with a mixture of H_2O, HCl and H_2O_2, evaporated to dryness and the residue dissolved in 30-50 mL 2M HCl and loaded on a Sr resin column (Eichrom Industries Inc.). The analytical method is based on selective separation of lead and polonium by extraction chromatography with bis-4,4'(5')-t-butyl-cyclohexano-18,6-crown ether (Vajda et al 1997). The non-retained ions were washed from the column with 100 mL 2M HCl. Polonium was stripped with 6M HNO_3 while lead was removed with 6M HCl solution. A polonium source was prepared by spontaneous deposition of polonium radioisotopes onto a copper disk at 50°C and pH 1 (Benedik and Vreček 2001). Polonium radioisotopes were then measured by alpha spectrometry. Lead was precipitated as lead sulphate and the beta activity of its daughter ^{210}Bi measured at equilibrium on a beta proportional counter.

Results

The activity concentrations of dissolved long-lived radionuclides of ^{238}U, ^{226}Ra, ^{210}Pb and ^{210}Po in water in Bq m^{-3} are shown in Table 1.

Table 1. Activities of dissolved long-lived radionuclides in water in Bq m^{-3}.

Sampling point	U-238	Ra-226	Pb-210	Po-210
PZ-5	12800 ± 500	1480 ± 100	241 ± 19	184 ± 23
PZ-8	38500 ± 1200	3270 ± 130	954 ± 74	857 ± 93
PZ-9	35900 ± 1100	1620 ± 180	205 ± 16	102 ± 12
PZ-3	54 ± 6	165 ± 27	30 ± 2	17 ± 6
BV-3	13 ± 2	10 ± 18	135 ± 11	9 ± 3
BV-5	10700 ± 250	520 ± 50	431 ± 34	40 ± 6
SDIJ	9000 ± 300	500 ± 60	80 ± 8	81 ± 15

Conclusions

The aim of this study was to determine the activities of uranium, ^{226}Ra, ^{210}Pb and ^{210}Po in underground water in the base rock and in the body of the Boršt mill disposal site. The concentration of radionuclides in mill tailings were determined by direct gamma spectrometry. The mill tailings deposited on the Boršt disposal site contain on average 850 Bq kg^{-1} of uranium, 7700 Bq kg^{-1} of radium and 6500 Bq kg^{-1} of Pb-210. The results of our measurements of dissolved radionuclides show that during seepage through the pile water leached these radionuclides resulting in contamination of underground waters. The underground water in the body of the Boršt mill tailings (sampling points PZ-5, PZ-8, PZ-9) is much more contaminated than the underground water in the base rock of the disposal site (sampling points BV-3, PZ-3 and BV-5).

Acknowledgement

This work was financially supported by Ministry of Education, Science and Sport of Slovenia (Project group P0-0532-0106). We acknowledge the assistance of the Uranium mine Žirovski vrh in sampling.

References

Benedik L, Vreček P (2001) Determination of ^{210}Pb and ^{210}Po in environmental samples, Acta Chem. Slov., 48: 99-213

Burnett WC, Tai WC (1992) Determination of Radium in Natural Waters by α Liquid Scintillation, Anal. Chem., 64: 1691-1697

Byrne AR, Benedik L (1988) Determination of uranium at trace levels by radiochemical neutron-activation analysis employing radioisotopic yield evaluation, Talanta, 35: 161-166

Byrne AR, Benedik L (1995) Simultaneous determination of trace uranium and thorium by radiochemical neutron activation analysis, J. Radioanal. Nucl. Chem., 189: 325-331

Minczewski J, Chwastowska J, Dybczynski R (1982) Separation and preconcentration methods in inorganic trace analysis, John Wiley & Sons Ltd. England

Vajda N, La Rosa J, Zeisler R, Danesi P, Kis-Benedek G (1997) A novel Technique for the Simultaneous Determination of ^{210}Pb and ^{210}Po Using a Crown Ether, J.Environ. Radioactivity, 37: 355-372

Impact of Uranium Mining and Processing on the Environment of Mountainous areas of Kyrgyzstan

I.A. Torgoev[1], U.G. Aleshyn[1], H. B. Havenit[2]

[1] Research Center "Geopribor", Bishkek, Kyrgyzstan
[2] University of Liege, Belgium

Abstract. In this report the results of analysis of modern geo-ecological situation in areas of uranium mining and milling in the territory of Kyrgyzstan are presented. Major threats for the mountain environment and citizens which are connected to mining and processing of the uranium are due to the contamination of the hydrosphere by radio nuclides, and due to the stimulation of dangerous natural and technological processes (landslides, mudflows, destruction of the tailings and dumps) with unfavorable ecological consequences.

Introduction

During the last 100 years the territory of Kyrgyzstan served for the Czar Russia at the beginning and later the USSR as one of the major mineral and raw materials base (radium, uranium and rear earth elements). First findings of radioactive uranium–radium minerals in the mountains (Ferghan valley) occurred at the same time as in Kuri couple at the end of XIX century (Aleshin and others, 2000). Similar minerals were found in Teo – Moyunsky mine which is located 30 km on the South – West from Osh city (Fig.1)

From 1907 to 1913 820 000 kg of ores was mined from this uranium–radium deposit, 655 000 kg was brought to St. Petersberg. The Teo–Moyunsky mine was an underground mine, at the time of it's closing the depth was more than 220 m.

In the 40th of the 20 century with the practical implementation of atomic energy techniques mostly for military purposes a booming development of uranium industry happened. At that time 10 uranium deposits in Shekaftar, Kyzyl–Jar, Mayly Suu, Rishtan and other place (Fig.1) were started out in Kyrgyzstan in the Northern bench of Ferghana valley. Industrial exploitation of the large uranium deposit Mayly Suu started in 1945. The Lenynabad (Ferghansky) chemical plant was build in 1947 in a very short period of time; Mayly Suu and TeoMounsky mines were part of this complex, being the first soviet uranium production for military purposes. Enterprises at this site including the hydro steel plant in Mayly Suu were processing ores from Eastern Germany (Erzgebirge), Chekhoslovakia (Yakhimov), Bulgaria (Bukhovo) and Tadjikistan (Taboshar, Adrasman) utill the middle of 1950. About 75% of the uranium was brought from the former German Democratic Republic ("Wismut").

At the beginning of 1950 uranium mining started in Myn Kush and Kadjy-Say. In 1951 the Karabaltynsky rock plant in the north of Kyrgyzstan was the biggest plant of the USSR. In our days, uranium which is mined by in situ leaching in Kazakhstan is processed in the plant. Starting in the middle of 1950, Kyrgyzstan became one of the major uranium producers in the former USSR. It should be mentioned that in Kyrgyzstan uranium was mined by all known technologies including recovery from the lake Issyk–Kul.

As result of this activities there are 30 tailings with a total area of 6 500 000 m^2 and 50 000 000 m^3 of tailings in different areas of Kyrgzstan (Fig.1). As a result of mining 25 dumps with a total area of 230 000 m^2 and 4 000 000 $м^3$ of rocks and raw uranium were created. Analysis of modern geo-ecological situation in the uranium mining and processing areas in Kyrgyzstan show that negative impact of mines and wastes on the environment has 2 major forms:

- Stimulation of dangerous natural and geotechnical processes (landslides, mudflows, accidents on the waste storage) with unfavorable ecological consequences with regional and global character
- Contamination of the environment and especially hydrosphere by radionuclides and toxic components.

Dangerous geo-ecological processes in the area of storage of mine waste

The number of uranium mines, total amount of uranium waste and sizes of the tailings are relatively small in comparison to USA, Canada, and East Germany. But due to natural phenomena of mountainous landscape, character and scale of the impact of uranium industry products on the environment is much higher in Kyrgyzstan.

Among unfavorable factors connected to the mining specifics are the following: complex mountainous landscape (3-dimensional); high seismic and tectonic activity; instability of slopes; variety and intensity of destructive natural processes in the mountains (erosion, denudation, landslides, mudflows, extreme sediments and change of the climate); high destruction of mountainous regions by geotechnical factors, especially during mining activities (Torgoev & Aleshyn 2001).

Long time history of mining, milling and processing in Kyrgyzstan including uranium shows that combination of above mentioned factors and processes was often the reason of the accidents in regional and global scale. The high probability of such cascade catastrophes is connected to the fact that the majority of mines were built on unstable mountain slopes. And radioactive dumps were located adjacent to riverbeds due to lack of suitable places.

Impact on groundwater from radionuclide emission

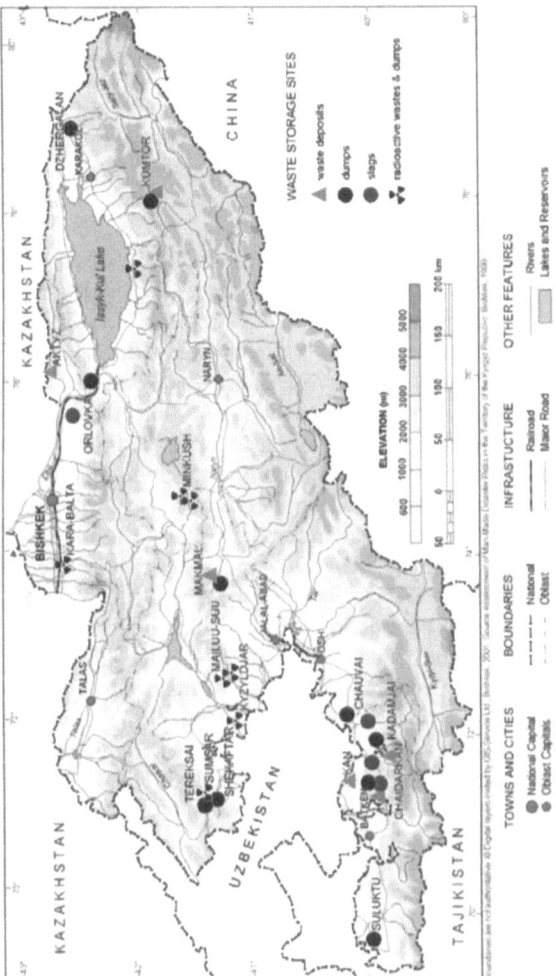

Fig. 1. Waste Storage Sites in Kyrgyzstan

In April, 1958 after an earthquake and heavy rains the weak dam of the tailing #7 in Mayly Suu was destroyed, being constructed only 30 m away from the river Mayly Suu(Fig.2). 600 000 m^3 of the tailings (about 50 % of the total volume) were spilled into the river. The radioactive mudflow destroyed a lot of houses in the town, people were killed and the tailings were spread over 40 km down by the river, contaminating flood plains. Cleaning up agricultural fields was going on for many years. Such accidents also took place on other tailings in Kyrgyzstan (Torgoev, Aleshyn 2001).

Evaluation of geo-ecological and radiological risk in the areas of mining and ore processing of radioactive minerals is showing the high risk of destruction of radioactive waste storages during earthquakes, mudflows and abnormal precipita-

Impact on groundwater from radionuclide emission

tion in Myn Kush, Kadjy Say, Ak Tuz and especially when landslides in Mayly Suu (Fig.1) happened.

Landslide processes in Mayly Suu region started to develop in 1953 to 54, when underground works, building of mining constructions and infrastructure measures were intensified. Around mines most of the landslides were caused by underground voids with depth up to 30 - 40 m, and by dumps constructed on unstable mountain surfaces. More than 50 different active landslide events within the town's area. were reported (Torgoev, 2001).

In Mayly Suu landslides developed in loess-like soils of the red color sandstone clay deposits of the Massagetsky stage (oligocenic-miocen). The largest landslides with a volume of more than 1 million m^3 (Tectonic, Koy-Tash and others) happened on slopes that are related to sandstone-clay deposits of the Cretaceous-Paleogenic with anticlinale folds in the upper slopes and sinclinale in the lower part. Due to high level of ground waters (1.5 - 5.0m), the activation of landslide processes is happening every year in relation to heavy rainfall events.

The situation in Mayly Suu is even worse due to the fact that some large landslides are dangerous not only for living areas but also to tailings. Destruction of the tailings (3,5,7,8,9) with spills of radioactive waste to the river can cause ecological catastrophes in the far field with possible radioactive contamination of heavy populated territories in Kyrgyzstan and Uzbekistan.

Calculations show that destruction of tailings might deposit 1,2 million m^3 in to the river with total amount of radio nuclides 15 000 Ku. The overall contaminated area will be 300 km^2. In order to prevent such catastrophic scenarios a project for re-burial of waste was started in 1996. At the same time an automatic monitoring of the landslide processes started and is going on. It includes a sub-system of emergency alarm for landslides (Torgoev et. al. 20001).

Contamination of the hydrosphere

Besides the narrow location of the radioactive waste sites to lakes and rivers there is also the poor geotechnical constructions of the tailings, absence of hydro-isolation and degradation of protective constructions to mention. Especially old tailings from the time after the II World War tailings which were build at the early stages of uranium mining form 1946 to 55 are of special concern. This period of time is characterized by serious under estimation of ecological danger caused by uranium mining and processing. Today it is clear that serious mistakes were made (Torgoev and Aleshyn 2001).

The beginning stage of uranium producing in Mayly Suu can be considered as an example. In the valley of Maylyu Suu river, about 30 km from the border to Uzbekistan radioactive waste was stored in 23 tailings and 13 dumps (Fig.2). The total volume of radioactive waste is 2 million m^3 and the total activity is 50 000 Ku. Investigations show that some tailings (3,5,7,8,9,10,18) and dumps in the riverbed of Myly Suu are the reason of systematic radioactive contamination of the river due to poor dam construction, high water content of the tailings and not reliable hydro-isolation.

Uranium concentrations in the water shown higher value than background with $4,4*10^{-6}$ g/l above the tailing#3 and $1,7*10^{-2}$ g/l below tailing#3, that means several thousands times above background concentration. 30 km from the tailing in the village of Madanyat, the uranium concentration in the river water is still 10 - 15 times higher than the background with $1,8*10^{-5}$ g/l (Aleshyn and others 2000). In many cases seepage water from the tailings infiltrates into the ground water.

This can be shown on the example of uranium tailing in Kara Balta. (Torgoev and Aleshyn 2001). In spite of a mighty aeration zone (up to 85 m) the ground water is contaminated with high level of sulfates, nitrates and heavy metals. Within contaminated areas the sulfate and nitrate concentrations are 5-8 fold the natural ones. An intensive contamination zone is spread on 15 km from the tailing and being 100 - 120 m in depth. A change of the intensive contamination zone was determined in 1999 based on samples from ground waters.

In the drill holes upstream the tailings the concentration of uranium in the ground waters was found to be 0.008 mg/l, in the drills downstream the town uranium concentrations reach up to 0.30 mg/l (Torgoev, Aleshyn 2001).

There are some major reasons of radioactive and toxic contamination of the soil and underground water in the area of Kara Balty tailing. One of them is a loss of the quality of the tailing's hydro-isolation. Leaching rates from the tailing were calculated by balance method for 1996 to 1998 to be 70 to 100 m^3 /year.

In the area around Kadjy Say in the period of 1948 to 1966 uranium bearing coal was mined for leaching of uranium oxide ($U_3 O_8$). During operation of the plant and after it's liquidation 400 000 m^3 of waste were buried in the tailing and an ash dump. The tailing is located 1.5 km far from the coast of the mountainous lake Issyk Kul. area. This is the Djety Oguz radon spa and beach by the village Jenish with radioactive monazite sand. Gamma radiation from cosmic rays is 0.466 mSv/a on the lake altitude (1608 m). Thus, in Issyk Kul province average amount of radiation from the natural sources is 1.726 mSv/a, which is much smaller than the background value 5 mSv/a (Mylius 1997).

Migration of seepage water from the tailing towards the lake is likely to occur. However, no monitoring wells are available. It should be mentioned that the natural elevated uranium concentration in the lake's water are affected by radioactivity and hydro chemical characteristics of the surface drainage water and subsurface radon springs. On average, Issyk Kul's lake water contains $3.0*10^6$ % of uranium which is one order of magnitude higher than Ocean water (10^{-7} %).

In 1997 specialists from the German Federal Institute for Geosciences from Hannover (BGR) jointly conducted measurements of radiation level in 680 places around lake Issyk Kul with specialists from Kyrgyzstan.

This measurements show that average earth gamma radiation is about 1.26 mSv/a, 66% of measured data is lower than that, 28% of measurements is 1.26 - 1.77 mSv/a. High numbers (6%) were measured in the Kadjy Say uranium mining district.

Impact on groundwater from radionuclide emission

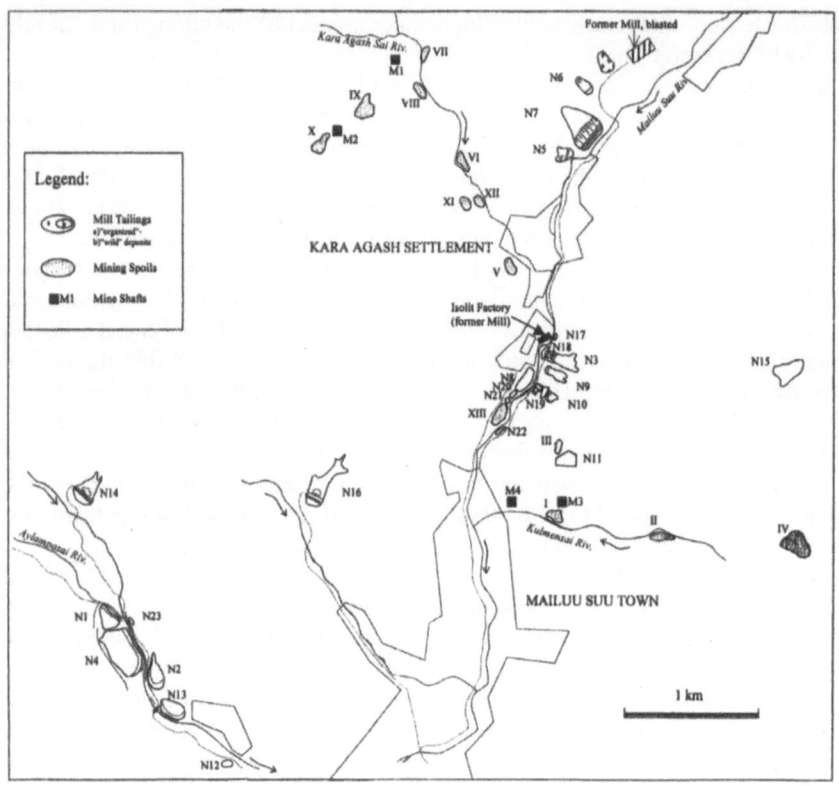

Fig. 2. Location of Mill Tailings and Waste Rock Dumps in the Mailuu Suu Area

References

Aleshin U. G. Torgoev I. A. Losev V. A. 2000, Radiation ecology of Mayly Suu. Bishkek "Ilim",96 p.(Rus)

Torgoev I. A. Aleshin U. G. 2001, Ecology of the mountain industry complex in Kyrgyzstan. Bishkek "Ilim". 181 p.(Rus)

Torgoev I. A. Losev V. A. H. B. Havenith 2001, Monitoring of sliding processes in populated territories of Kyrgyzstan. Materials of Intern. Symposium ''End G'eol City – 2001''. Ekaterinburg, Vol 1: pp 385-393

Mylius H.- G. (1997) Bestimmund des Strahlengrund niveaus im Issyk-Kyl-Gebiet. BGR: http: //www.bgr.de

Impact on groundwater from radionuclide emission

Uranium content in ground water in Stara Planina Triassic sediments

Zoran Nikić[1], Jovan Kovačević[2], Branislav Radošević[3]

[1]Forestry Faculty, Kneza Višeslava 1, 11030 Belgrade, Yugoslavia.
[2]Geoinstitut, Rovinjska 12, 11000 Belgrade, Yugoslavia.
[3]"Advanced Systems", Vojislava Ilića 18, 11000 Belgrade, Yugoslavia.

Abstract. Uranium investigations in the Stara Planiana Triassic sediments were carried out for several decades. Some of these geological and hydrogeological results that are related to an established uranium deposit in the Dojkinci-Brlog-Jelovica-Ponor area are presented in this paper. Hydrogeological investigations were performed to investigate the chemical properties of ground water (e.g. anion-cation composition, microelement content, radioactive element content, pH value, Eh value). On the basis of these data a hydrogeological model was established for ground waters with increased uranium content.

Introduction

The Stara Planina massive, within the Carpathian-Balkan arch, is partly situated in western part of Bulgaria and partly in the easternmost part of Serbia.

In the southeast part of Stara Planina, belonging to Serbia, a uranium deposit is found, as well as several occurrencers. During several years of exploration numerous data were collected and many geological and hydrogeological problems were solved, helping to iterpret more precisely the increased uranium contens in ground waters of the Dojkinci area (N.N. *et al* 1985). The performed investigations covered an area of about 55 km^2.

Most of this area is uninhabited. The only settlements are Dojkinci, Jelovica, Rosomača and Senokos, which are connected by asphalt road to Pirot, the industrial and cultural centre of the area. The main occupation of the inhabitants is cattle breeding and agriculture to a lesser extent.

The main features of the investigated area

The area is featured by prominent mountain relief, ranging from 800 m to over 2000 m ASL (above sea level). The highest peak of this part of Stara Planina is Midžor (2169 m) with several others over 1700 m ASL (e.g. Crni Vrh: 1831 m, Kopren: 1935 m, Bratkova Strana: 1934 m, Tri Čuke: 1937 m).

The whole area has typical mountain climate, with sharp winters and a short, relatively warm, summer period. The snow often holds until mid May. The average annual rainfall for the period 1960-1990 ranges from 712.2 mm (Kamenica) to 981.1 mm (Dojkinci). Uneven distribution of rainfall is prominent, where much higher rainfall is represented in mountainous parts than in the valley parts of the area.

The significant streams in the area are the Kamenička, Rosomačka, Jelovička, Dojkinačka, and Toplodolska rivers. During their flow from the mountain heights these streams encounter a zone of karstified limestones with numerous sinks in the riverbeds. During the recession period, with low proticaj, all drained water sinks, so that down stream of the karstified limestones Kamenička, Rosomačka and Do-jkinačka rivers are dry for several weeks during the year.

Geological composition and tectonic fabric of the investigated area

Stara Planina is a part of the Carpathian-Balkan geotectonic unit, with a strike of NW-SE through eastern Serbia, continuing to the north in Romania and to Bulgaria in the southeast (N.N. *et al* 1977).

The south-eastern part of Stara Planina, which is the subject of this work, is composed of quartz-albite-muscovite-chlorite schists of Riphean-Cambrian (R, Cm) age (R, Cm) with intrusions of granodiorite ($\delta\gamma$) and granodiorite-porphyrite ($\delta\gamma\eta$). This schistose complex composes a wide belt on the crest of Stara Planina, with a supposed thickness of over 600 m (Fig. 1).

Transgressively over the crystalline schists lies the clastic complex of Lower Triassic (T_1), where sandstones and conglomerates are predominant rocks. Lower Triassic sediments are developed in form of a continuing narrow belt which strikes from Rudinja village in the northwest, and over the Kopren mountain towards the Yugoslav-Bulgarian border. Within this formation, in the area of Dojikinci and Jelovica, the lower member "Kopren" and the upper member "Dojkinci", composed of varicoloured sandstones, are identified (Fig. 1).

The "Kopren" member is mostly composed of coarse grained quartzitic sandstones and red conglomerates, showing distinct cross-bedding. The thickness of this member is 200—250 m. Above the sediments of the "Kopren" member, conformably lies the varicoloured sandstones of the "Dojkinci" member. The varicoloured sandstone series is composed of dark reddish sandstones, greenish sandstones, siltstones and intra-formational conglomerates. The thickness of this member is 120—200 m.

Impact on groundwater from radionuclide emission

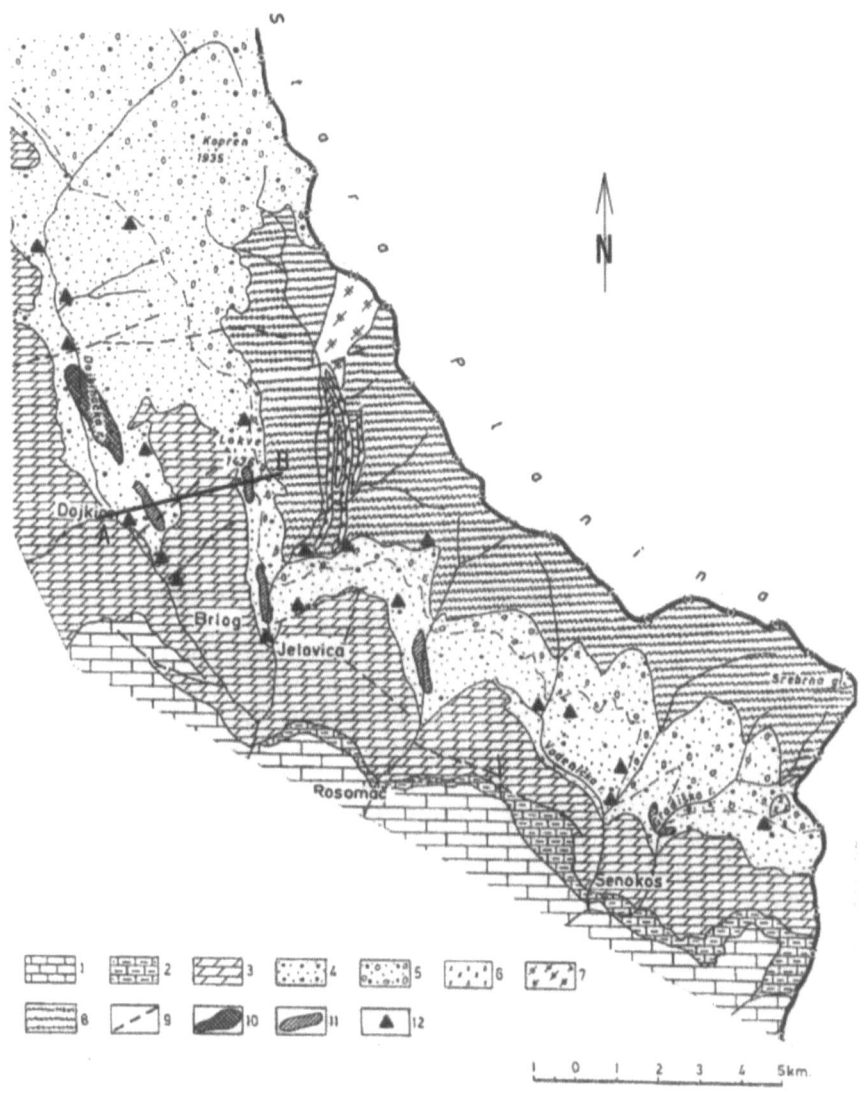

Fig. 1. Geological map of the investigated area. 1. Jurassic sediments, 2. Limestones sandstones and claystones (T,J), 3. Limestones, dolomitic limestones and dolomites, 4. Varicoloured sandstones (T₁), 5. Quartz sandstones and conglomerates (T₁), 6. Granodiorite- porphyrites, 7. Gabbro, 8. Quartz-albite-muscovite-chlorite schists (R, Cm), 9. Fault, 10. Uranium deposit, 11. Uranium occurrence, 12. Minor uranium mineralisation .

In the lower level of the varicoloured sandstones a geochemical barrier is identified with a strike NW-SE, up to 200 m wide and 5—25 m thick, representing the uranium bearing horizon (Kovačević 1997). The main feature of the uranium hori-

Impact on groundwater from radionuclide emission

zon is intensive alternation of sediments, reflected in changes of lithological composition, granulation, porosity, colour and presence of organic matter, sulphides and uranium mineralisation.

Over the "Dojkinci" member, lies a limestone and dolomite complex of Middle Triassic (T_2) age. The feature of this complex is alternation of thick beds of pure and karstified limestones with dolomitic limestones, dolomites and knotty limestones, somewhat less karstified. The thickness of this complex is about 250 m.

Outside the investigated area, upwards in the geological column of the Stara Planina anticline, there are flysch-like sediments of Jurassic and Lower Cretaceous age.

All the mentioned series have a mild dip 5—20°, rarely higher, and are generally inclined to the southwest.

Hydrogeological features of the investigated area

Important for this work is the existence of porous and permeable rock complexes, which are at the base limited by impermeable rocks.

The complex of limestones and dolomites of Middle Triassic (T_2) is identified as a hydrogeological area with aquifers of dissolutional and dissolutional-fracture porosity. The degree of their karstification is indicated by numerous typical karst occurrences and forms (e.g. sinks, caves, springs, caverns, sinkholes). In the period of high waters, complete underground rivers flow through caverns, with water yields of over 3 m^3/s for some springs (Veliko Jelovičko spring).

The sediments of Lower Triassic (T_1) age belong to the hydrogeological category of aquifers with fracture porosity. The intensity of their fracturing and existence of hydraulically connected fracture systems is indicated by the large number of permanent springs, with minimal water yields of 1—3 l/s.

The facies of greenschists of Riphean-Cambrian (R, Cm) age belongs to the hydrogeological category of more or less waterless rocks. This rock complex has a significant function of base impermeability and a lateral hydrogeological barrier for ground water in fractured Lower Triassic sediments.

Ground waters in the Dojkinci and Jelovica areas are according to their chemical composition hydrocarbonate, rarely hydrocarbonate-chloride and hydrocarbonate-sulphate, and very rarely hydrocarbonate-sulphate-chloride, sulphate-hydrocarbonate-choride and purely sulphate. By their cation composition they are mostly calcium, calcium-magnesium and rarely magnesium-calcium. The hydrocarbonate types are represented in the whole of the investigated area, hydrocarbonate-chloride types are found in the northern part in chlorite schists, while hydrocarbonate-sulphate and purely sulphate types are found in quartz conglomerates and conglomeratic sandstones north of Jelovica.

The content of radioactive elements (U, Ra, Rn) in ground waters of the Dojkinci-Jelovica area are found in the following ranges: uranium 0.1—40.9 µg/l, radium, 0.036—0.294 Bq/l and radon 0.52—63.2 Bq/l.

The mean content of uranium in ground water of the investigated area is 0.68 μg/l. The analysis of data, from 100 water samples, on normal dispersion and uranium and radium content range, in ground water of the investigated area, show that the increased uranium content is found in about 38%, and radium in about 34% of the water samples. The anomaly coefficient for uranium varies from 2.01 to 25.71, while for radium it varies from 1.0 to 8.2. Increased uranium contents were found in the area of Dojkinci-Paleški Vrh, in the area of Brlog-Jelovica and north of Ponor.

In the Dojkinci-Peleški Vrh, the increased content of uranium is found in ground water genetically connected with the varicoloured sandstones (0.3—6.8 μg/l). The anomaly coefficient varies from 1.85 to 15.22. The redox potential (Eh) of these waters ranges from +130 to 135 mV, indicating the transitional oxide-reduction environment with a pH value of 7.1—7.5, representing a neutral to mildly alkaline environment. According to the chemical composition, these waters belong to the hydrocarbonate-calcium type.

In the Brlog-Jelovica area several springs with an increased uranium content (1.9—40.9 μg/l) were found, situated in conglomerates, varicoloured sandstones and Granodiorite-porphyrites. The anomaly coefficient of uranium in these waters ranges from 2.01 to 25.71. Their redox potential varies from +125 to +135mV, indicating the transitional oxidational-reductional environment, while the pH value from 6.9 to 7.5 points to neutral to mildly alkaline waters. According to the chemical composition, these waters mostly belong to the hydrocarbonate-calcium type, with occurrences of hydrocarbonate-magnesium-calcium types. In the vicinity of Jelovica, the ground waters from the varicoloured sandstones and quartz conglomeratic sandstones are, according to their chemical composition, of mixed type, where hydrocarbonate-sulphate-calcium and purely calcium types dominate.

North of Ponor, in some ground waters within chlorite schists, increased contents of uranium were found, ranging from 1.5 to 20.4 μg/l, while their coefficients vary from 2.01 to 3.21. The Eh values of these waters range from 5.7 to 6.3, meaning that these waters are mildly acidic, indicating transitional oxidation-reduction environment. According to the chemical composition these waters are of the hydrocarbonate-chloride-calcium and sulphate-hydrocarbonate-chloride-magnesium type.

The increased content of radium in the investigated ground waters partly corresponds to the waters where increased contents of uranium were found. This mostly applies to the Dojkinci-Peleški Vrh area, and partly Jelovica area, while in the area north of Ponor increased contents of radium were not identified.

The SiO_2 content in the waters of the investigated area is generally low, ranging from 2 to 24 mg/l, with a mean value of 10 mg/l. This fact indicates that most of the ground water in the investigated area is in the zone of filtration, meaning that they are of surface origin and there is a fast percolation of water (Nikić 2001).

Impact on groundwater from radionuclide emission

Hydrogeological model of ground water formation with increased uranium content in Dojkinci-Jelovica area

The general characteristics and some particularities presented earlier represent a base for understanding and interpreting the role of hydrogeological conditions in the Dojkinci area and formation of ground water with an increased content of uranium (Fig. 2).

Due to favourable climate, hydrogeological and morphological conditions of charge of the karst aquifer in the Middle Triassic limestones (T_2) is done by infiltration of rain and snow water and sinking of surface water courses, flowing from surrounding non-karstic areas. Drainage of such an aquifer is done partly through springs and partly by sinking downwards. In the waters of karst springs an increased uranium content was not identified.

Charging of the fracture aquifer, formed in Lower Triassic sediments (T_1), is done by infiltration of rain and snow water, filtration of ground water from the limestones in the base, and surface water courses which flow from the surrounding impermeable areas. In the surface, the outcrops of Lower Triassic sediments, represent the areas of direct charge to the aquifer, while the indirect charge is from the karst aquifer which is in the base.

By filtration along hydraulically connected fissure systems, the water comes in the contact with the uranium bearing horizon. By contact of water with uranium ore, new uranium compounds are formed, contaminating the ground water transported into favourable geochemical conditions. Migrativity of uranium in ground water of Dojkinci-Jelovica area, varies and depends on chemical properties of ground waters, controlled by geological composition. The chemical composition of waters with a high HCO_3 content is mildly alkaline and mostly in an oxidation environment, indicating that these ground waters represent an aggressive environment collecting and maintaining uranium in form of complex alkaline uranilocarbonate. The coefficient of uranium migration in the waters of hydrocarboneta types varies from 0.012 to 0.448, with a mean value of 0.145. In hydrocarbonate-magnesium calcium waters the migration coefficient ranges from 0.013 to 0.101, with a mean value of 0.061. The hydrocarbonate-sulphate-calcium waters have a uranium migration coefficient ranging from 0.096 to 1.363, with a mean value of 0.330. The hydrocarbonate-sulphate-calcium waters and hydrocarbonate-chloride-calcium-magnesium waters have an increased uranium migration coefficient ranging from 0.112 to 1.482, with a mean value of 0.769.

Due to the higher position of ground water within the Lower Triassic aquifer, in relation to the local erosional base, represented by the Dojkinačka Reka river bed, favourable dip of hydrogeological base and the aquifer in relation to the erosional base, plentiful quantities of infiltrational waters and the existence of hydraulically connected fissure systems, the process of gravitational discharge of ground water, enriched by uranium, is done dosed through some springs. The locations of these springs are determined by the distribution of fault systems, fractures, and fissures that are found on the surface. For the motion of ground water within the aquifer and their spontaneous discharge on the surface, the influence of distribution of water energy in space (mechanical energy) in the form of potential energy is of importantance.

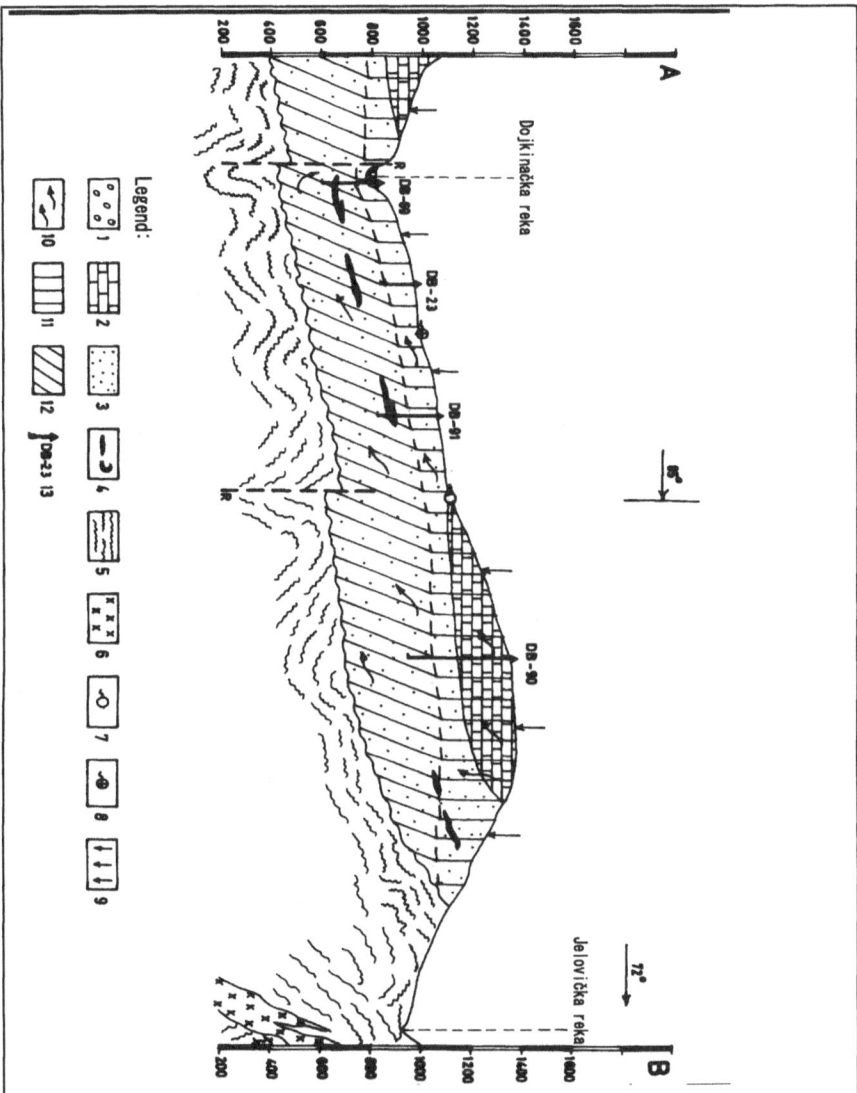

Fig. 2. Thematic hydrogeological model of formation of ground water with increased uranium content in Dojkinci-Jelovica area. 1. Alluvial deposits, 2. Limestones and dolomites, 3. Sandstones and conglomerates, 4. Uranium mineralization, 5. Schists, 6. Granodiorites, 7. Karst spring, 8. Spring. Water with increased uranium content, 9. Atmospheric precipitation, 10. General direction of ground water flow, 11. Zone of faster filtration (water exchange), 12. Zone of slower filtration, 13. Exploratory drill hole.

Impact on groundwater from radionuclide emission

Discharge of water from the Lower Triassic sediments is done mainly through fracture systems of largest dimensions, where the spring yield is the highest, while at the end of the recession period, during the lowering of the ground water level, the water is discharged from even smallest fissure systems, with distinguished participation of water from the lower part of the aquifer. In that way, together with ground water discharged along fracture systems, the uranium comes to the surface in form of solution.

Conclusion

The variety of hydrochemical features of ground waters indicates their genetic bond with particular rocks with which they were in contact during their flow along connected fracture systems. Increased content of radioactive elements in ground waters from some drill holes and springs, points to a conclusion that it is a consequence of their direct contact with uraniferous horizons formed in Lower Triassic sediments. The factors controlling the occurrence of increased uranium contents in ground waters of Stara Planina, in the vicinity of Dojkinci-Brlog-Jelovica-Ponor, are of geological, hydrogeological, tectonic, climate, and geomorphological type and at the same time in simultaneous interaction.

References

N.N. et al (1977) Geological map and booklet for shets Pirot and Breznik, scale 1:100 000. Federal Geological Survey, Belgrade.

N.N. et al (1985) Uranium investigation in the area of Permian-Triassic sediments of Stara Planina. Geoinstitute reports library. Belgrade.

Kovačević J (1997) Metallogeny of uranium in Permian-Triassic sediments of Stara Planina. Masters thesis. University of Belgrade, Faculty of Mines and Geology. Belgrade.

Nikić Z (2001) The role of hydrogeological conditions for formation of small waters, medium and small watercourses in Serbia. Doctor's dissertation. University of Belgrade, Faculty of Mines and Geology. Belgrade.

Contaminant transport from proposed Jabiluka mine uranium tailings paste repositories – 10 000 year dispersion and dilution analysis

C.R. Dudgeon [1], F.R.P.Kalf [2]

[1] Water Research Laboratory, School of Civil & Environmental Engineering, University of New South Wales, Manly Vale 2093, Australia
[2] Kalf and Associates Pty Ltd, 52 York Terrace, Bilgola 2107, NSW, Australia

Abstract. This paper describes the extension of a hybrid numerical/analytical modelling investigation of the possible movement of potential contaminants from tailings repositories towards wetlands adjoining the Jabiluka uranium mining lease in northern Australia. Predictions extending to 10 000 years were requested by a UNESCO scientific committee investigating claims that mining would endanger Kakadu National Park, a World Heritage Area that surrounds the lease. Because of the uncertainty in the model parameters, a Monte Carlo approach was used to generate a large number of possible concentration profiles for radium. An additional regional model was employed to predict the ultimate fate of uranium.

Introduction

It is proposed to mine uranium at the Jabiluka mine site about 230km east of Darwin and 20km north of the existing Ranger Uranium Mine in the Northern Territory of Australia. The proposed mine is situated within a 73km^2 mineral lease near the edge of, but surrounded by, Kakadu National Park which covers an area of approximately 19 800km^2. The mine will be underground and accessed by a decline from the surface. The area to be occupied by the mine's surface facilities is 27ha (0.27km^2). The decline has already been constructed and the first drive into the orebody commenced in 1998. Work has been suspended pending a final decision by the current mining company to proceed. The mining proposal involves milling the ore at Jabiluka. This option would involve storage of mill tailings in deep subsurface repositories to meet current regulatory requirements. The repositories would include the mine void and about 180 specially constructed vertical silos with their tops about 100m below ground level. The current proposal is to dispose of the tailings as a paste after partial dewatering and the addition of cement.

The uranium tailings paste proposal has been discussed by Dudgeon & Waite (1999) whilst two papers by Dudgeon and Kalf (2001) and Kalf and Dudgeon (2001) present a conceptual description and model studies for an investigation of the possible extent of movement of potential contaminants from the repositories.

Topography, Drainage and Climate

The Jabiluka ore body proposed for mining lies in hilly terrain adjacent to the Magela floodplain (Fig. 1 & 2). Ground levels range from several metres above mean sea level over the floodplain to 160m in the hills. (Mine RL's shown in figures are plus 1000m). The hilly terrain forms a broad north-south ridge which is intersected in an east-west direction by numerous drainage gullies situated on both sides of the topographic divide.

Surface water drainage at the mine site is towards the Magela floodplain, which lies about 1.5km to the west of the topographic divide that separates the drainage valley from the Swift Creek catchment. On the eastern side of the divide, surface water flows eastward towards Swift Creek which joins the Magela floodplain several kilometres further north.

The region is subject to consistent annual wet and dry seasons. It has an average yearly rainfall of about 1500mm. Monthly average rainfall increases steadily from near zero in September to a peak in January/February (approximately 370mm in each month) and decreases rapidly to near zero again in June.

Geology

The orebody is contained within the Cahill Formation which is mostly schist but includes some carbonate. To the west, the Cahill formation underlies the Magela floodplain and forms the bedrock that dips east and south beneath the overlying Kombolgie sandstone Formation.

The Kombolgie Formation is mainly quartz sandstone with a little siltstone. It forms the broad north-south topographic ridge and the elevated plateau further east. Most of the sandstone is better described as quartzite because of the deposition of secondary silica, although some relatively friable layers do occur. The intergranular porosity is very low and the groundwater flow at the mine site is restricted mainly to the joint and fracture system. Along the Magela floodplain, clays (generally dark organic clay overlying grey clay of marine origin), silts and sandy alluvial sediments overlie the Cahill Formation. The bedrock in contact with the sediments is weathered.

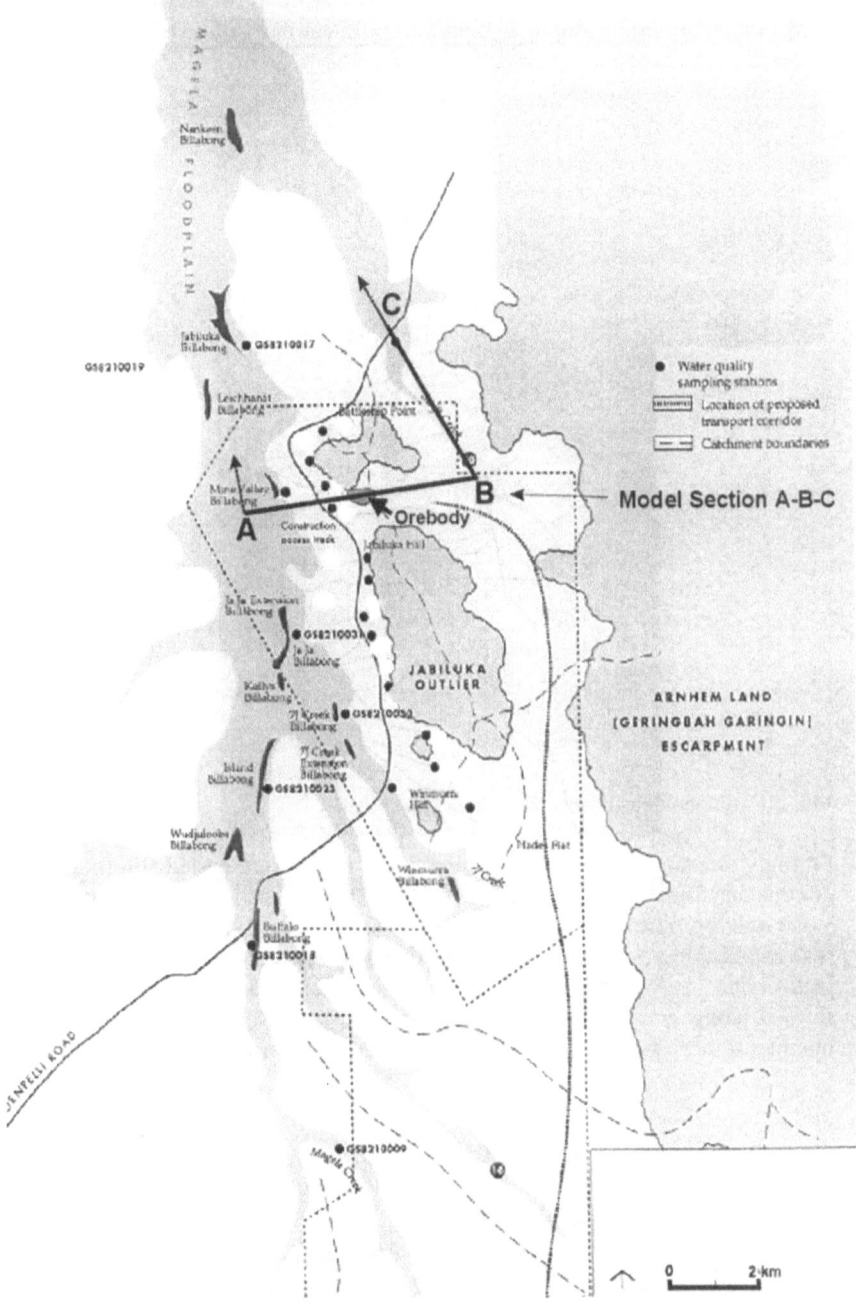

Fig. 1. Plan of Jabiluka site showing key drainage features and sections

Long term storage of radioactive waste

Immediately east and west of the topographic divide the weathered bedrock in the lower drainage valley slopes is overlain by sands and silts. Drilling has revealed that weathered bedrock can occur up to 50m below ground surface.

A conceptual hydrogeological section, A-B-C in Fig. 1, is given in Fig. 2.

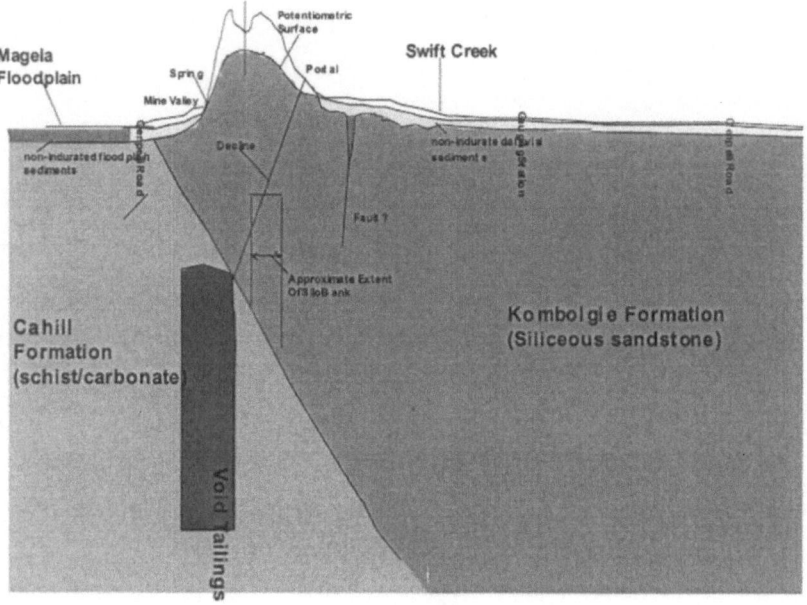

Fig. 2. Hydrogeological conceptual model plane A-B-C

Strongly developed lineaments comprising joint/fracture systems in the sandstone are evident from aerial photographs of the Jabiluka outlier and the elevated sand-stone outcrop north of orebody No 2. These structures strike at 60 to 80 degrees with another less dominant set at 350 degrees. The structural lineaments are less well defined in Mine Valley. However, it is possible that Mine Valley may have formed along zones of rock weakness created initially in the past by one or a number of these structures that have now been filled in with weathered material.

Groundwater Flow Directions

Recharge of the groundwater system is by direct infiltration of rainfall into the shallow fractured aquifers and aquitards. Water table levels in boreholes show that groundwater flows to the east and west away from the Mine Valley topographic ridge. On the western side of the ridge water flows towards the Magela floodplain and then north along the flood plain. In the east it flows towards Swift Creek.

Leaching and Transport Model

Groundwater flowing through the tailings will leach contaminants from the paste. The contaminants will then be carried in the direction of the hydraulic gradient imposed on the aquifer by water table differences caused by the annual recharge to the Kombolgie sandstone aquifer. Dispersion (both lateral and longitudinal) and adsorption onto aquifer/aquitard material will reduce concentrations.

Concentrations of various contaminants leaving the repositories will depend on the adsorption/desorption characteristics of the contaminants in the paste, geometry of the repositories, the local hydraulic gradient and the permeabilities of the aquifer and the paste. Groundwater flow velocities, dispersion characteristics of the aquifer, adsorption characteristics of the contaminants in the aquifer and dilution will control concentrations in the groundwater further downstream.

With tailings paste permeability lower than that of the rock surrounding the repositories, groundwater will preferentially flow around them. The lower the permeability that can be achieved for the set paste mass, the lower will be the rate of flow through and leaching from it.

The main potential contaminants of concern included uranium and radium 226. Absolute paste pore water concentration values based on data obtained from the nearby Ranger uranium mine and the literature were: sulfate 20.000mg/L; magnesium 5 000mg/L; manganese 500mg/L; uranium 15Bq/L and radium 226 15Bq/L.

The leaching and transport of contaminants by the groundwater flow, and beneficial or detrimental effects such as those described, were investigated by a predominantly modelling approach. A hybrid model was used to quantify both concentrations and time of travel of potential contaminants emanating from the repositories towards the Magela wetlands.

The models used were:
1. A regional scale two dimensional (2D) section finite element section model to determine head distributions, flow directions and the range of Darcy velocities along Section A-B-C (Fig. 1) parallel to the groundwater flow lines.

2. A local scale three-dimensional (3D) finite difference solute transport model applied to a 1m thick horizontal layer of horizontal flow through and around a repository to determine the concentrations of contaminants leached from the tailings.

3. An analytical contaminant transport model to determine concentrations along the flow paths represented by the finite element flow model. The effects of advection, dispersion in three co-ordinate directions and retardation are accounted for. The model uses as input the flow velocities determined from model 1 above and source concentrations determined from model 2 above. This model was combined with Monte Carlo simulations to determine concentration profiles for a large number of parameter values within selected ranges.

Details of the contaminant and aquifer characteristics, models and model parameters and results are given in Kalf and Dudgeon (2001). In summary 1 000 year simulations conducted for radionuclides uranium and radium 226 indicate that the bulk of these contaminants is restricted in movement to within several hundred metres from the repositories. Provided that adequately low permeability can be achieved in the tailings paste, the concentrations will remain at background levels within the wetlands. This situation would also apply to manganese.

Sulfate was found to be the most mobile contaminant but concentrations emanating from the tailings paste would be low if the tailings paste permeability were 10^{-4}m/day or lower. Sulfate concentrations in the wetlands currently occur at high levels due to naturally occurring processes in the acid sulfate soils of the floodplain. Some sulfate from the tailings will reach the floodplain to the west, but concentrations are predicted to be less than those that occur naturally in this area. It is significant that pre-mining sulfate levels currently in the Magela floodplain are substantially reduced by dilution and flushing during the annual wet seasons.

Ten Thousand Year Dispersion and Dilution Analysis

For radium 226, a Monte Carlo simulation was conducted to determine the concentration profiles based on 255 runs (realizations) with a uniform distribution assumed for all aquifer parameters. The cumulative probability that the 50% normalized concentration will lie within a given distance from the source is shown in Fig. 3. Thus, for example, there would be an 88% probability that the 50% normalized concentration level will lie within 200m from the source.

50 Percentile Cumulative Probabiltiy - Radium 226 -10,000 years -Western Area

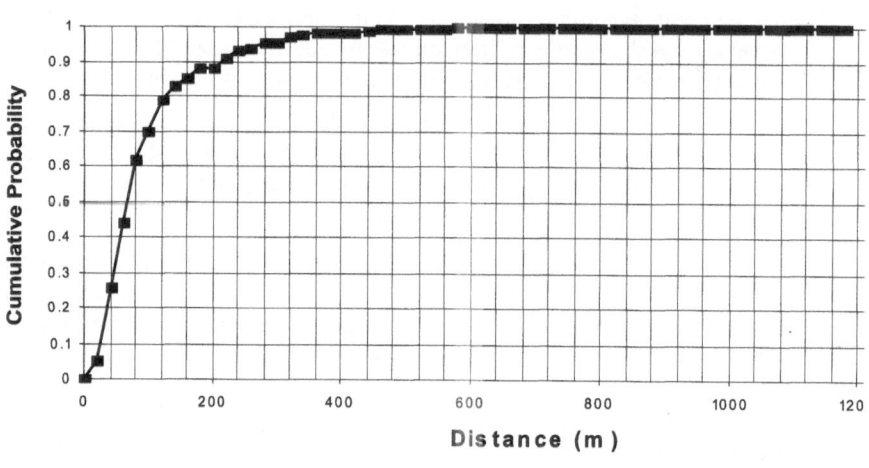

Fig. 3. Cumulative probability of travel distance of 50% concentration

Long term storage of radioactive waste

For uranium, the type of analysis applied to radium was not considered suitable because of the greater dispersion in the far field. The leaching model was therefore run over 10 000 years and the mass lost from the silos was determined. The concentration profile through two typical silos with a paste permeability of 10^{-4} m/day is shown in Fig. 4 with groundwater flowing from left to right. Concentrations in Fig. 4 are shown as a percentage of the source concentration versus distance. Because of the high levels of dispersion and dilution over this time, the bulk concentration levels were subsequently determined in the Magela flood plain.

For Fig. 4, the most conservative scenario, about 40% of the 3800 tonnes of uranium is leached during 10 000 years. Assumption of a 10% upward flow flux yields total solute uranium flux of about 15 kg/year. By comparison, data provided by the Supervising Scientist, Environment Australia indicate that there is some 800 kg/year of uranium recycled from leaf litter and grasses. Surface water inflow carries an additional 100 kg/year of uranium. Wasson (1992) has also estimated that 5 000 tonnes per year of sediment with a uranium concentration of 6mg/kg is deposited on the flood plain; i.e. 30kg of uranium per year. The conclusion is that the dispersed uranium mass created by leaching is small compared to that coming from other sources. For the assumed upward groundwater flux of 10% the uranium concentration increase is calculated to be about 0.014 µg/L.

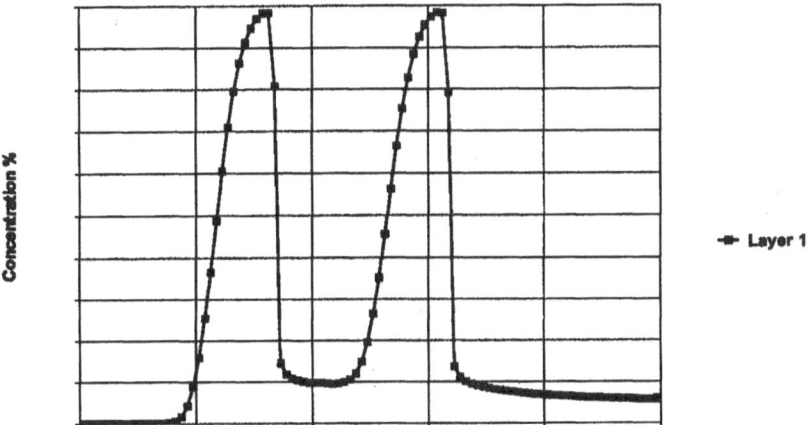

Fig. 4. 10 000 year concentration profile through two typical silos

To determine more precisely the vertical flux of groundwater into the Magela Creek drainage system during dry periods, an additional section model extending north along the Magela floodplain was developed as shown in Fig. 5.

The results indicate that the upward flux component of the groundwater down valley flow would amount to 0.1% if there is no ponded water at the surface. With wet season ponding of water at the surface, the upward flow would be reversed. Hence the uranium flux would be much less than that calculated on the basis of 10% upward flow. In addition high adsorption of uranium could be expected within the clay layers that overlie the thinner sandy layer shown in Fig. 5.

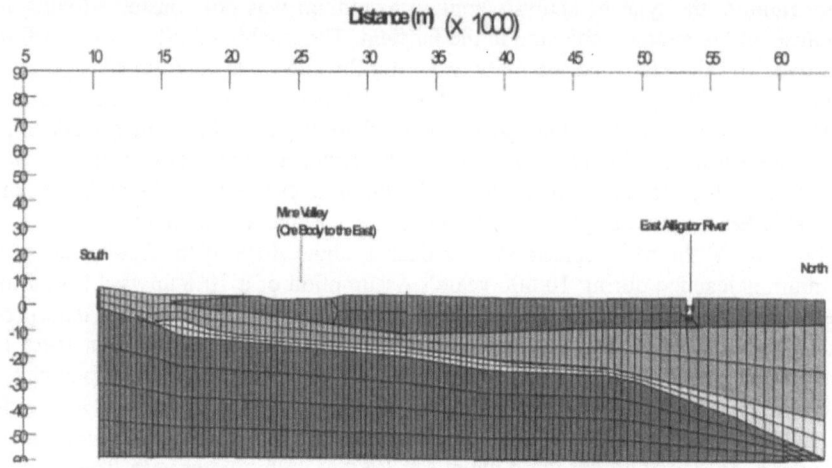

Fig. 5. Finite element model to determine upward flux component.

Acknowledgements

The study described in this paper was funded by the Supervising Scientist, Environment Australia. The authors express their appreciation of the assistance of staff of the Supervising Scientist and Energy Resources of Australia Ltd.

References

Dudgeon C.R. and Kalf F.R.P. (2001) Contaminant transport from proposed Jabiluka mine uranium tailings paste repositories – conceptual description. IMWA 7th International Congress, Poland, Proc 103-113.

Kalf F.R.P. and Dudgeon C.R. (2001) Contaminant transport from proposed Jabiluka mine uranium tailings paste repositories – model analyses. IMWA 7th International Congress, Poland, Proc 115-125.

Waite J.D., Dudgeon C.R. and Fell R. (1998) Review of Jabiluka Mine Alternative Tailings Management Proposal. Report 35051, Unisearch Ltd, Kensington, NSW.

Wasson, R. J. (1992) Modern sedimentation and late quaternary evolution of the Magela Creek Plain. Supervising Scientist for the Alligator Rivers Region, Research Report 6, Australian Government Publishing Service, Canberra.

Natural Uranium Fluxes and their Use in Repository Safety Assessment

David Read[1], Karl-Heinz Hellmuth [2], Juha Kaija[3], Lasse Ahonen[3]

[1]Enterpris, University of Reading, PO Box 227, Whiteknights, Reading, RG6 6AB
[2]STUK, Laippatie 4, PO Box 14, FIN-00881 Helsinki, Finland
[3]GTK, Betonimiehenkuja 4, PO Box 96, FIN-02150, Espoo, Finland

Abstract. Studies of uranium series geochemistry have made relatively little impact on performance assessment where simplistic advection-dispersion models are still being used to derive dose and risk estimates. Supplementing the models with a scientifically sound representation of radionuclide migration would greatly enhance the rigour and transparency of the safety case. This paper outlines a methodology for employing natural geochemical fluxes as alternative indicators of repository safety by reference to case studies in Finland. The approach may be applied to any situation where a comparison between ambient exposure rates and increments received due to industrial activities is required.

Introduction

The regulations used to judge the acceptability of an underground repository for nuclear waste are invariably based on dose and/or risk (*e.g.* Vieno and Nordman 1999; Nirex 2000). The most commonly used measure is radiological risk, where conventionally a value of less than 10^{-6} per annum is deemed acceptable. In Finland, where the spent fuel disposal programme is advanced, an effective annual dose of 0.1mSv to the most exposed individual is taken as the limit. Similar constraints are now being employed when devising remediation targets for radioactively-contaminated land or former uranium mining areas. The choice of these measures, though routine with operational nuclear plant, causes many problems in an environmental context, since:

- There is great controversy surrounding the health effects of prolonged exposure to low doses of ionising radiation.
- Local factors are not easily taken into account, for example natural background varies substantially within and between countries.
- Assumptions need to be made concerning site evolution and future human behaviour that are almost impossible to justify.
- Neither dose nor risk is a concept easily understood by non-technical audiences.
- There are a number of shortcomings with the models employed.

For these reasons, the IAEA, among others, has initiated research into alternative indicators of safety, which may be used initially as an adjunct to dose and risk (Miller 2000). The current paper describes part of this work, focussing on the application of natural uranium series concentrations and fluxes in Finland.

Natural safety indicators

Analogues and performance assessment

Similarities between natural uranium occurrences and repositories for spent nuclear fuel have been recognised for many years, accounting for the large number of 'natural analogue' investigations undertaken world-wide (*e.g.* Chapman et al. 1991; Read et al. 1993; Duerden et al. 1994; Blomqvist et al. 2000). Surprisingly, and despite the resources expended by waste disposal agencies in carrying out the work, these studies have made relatively little impact on performance assessment (PA) procedures to date. This may be due to a number of reasons, including *inter alia*:

- The data derived are often qualitative and do not conform to the input requirements of existing models.
- The geochemical concepts are complex and have not always been well articulated by specialist investigators.
- Inertia on the part of workers involved in safety assessment calculations who are reluctant to abandon 'tried and trusted' methods, even where shortcomings are evident.

In addition, many of the analogue investigations, particularly the earlier studies, were self-contained and were not carried out with performance assessment in mind. Nevertheless, and whatever the cause, the direct consequence of this lack of interaction is that overly simplistic, physical dispersion models are still being used to derive desired dose and risk estimates for repositories many thousands of years after closure. Problems with such an approach are well documented. They may fail to satisfy either scientific reviewers, owing to omission of geochemical reactions, or the general public, due to the lack of transparency in the final results.

The main concern with performance assessment is that the modelling tools take no account of interactions between geochemical and hydrological processes. Further, they fundamentally misrepresent the known geochemical behaviour of the uranium series elements (Read and Hooker 1992; Burns and Finch 1999; Hellmuth 2001). The calculations are biased towards dispersion and dilution of activity in the far field where, normally, dose rates are estimated for a drinking well scenario (*e.g.* Vieno and Nordman 1999). Consequently, if there were to be an accumulation of activity, as is often the case, sinking a well into groundwater above the repository would give rise to higher doses than those calculated.

A number of biosphere receptors (*e.g.* peat) also have the ability to concentrate radioactive species (Armands 1967; Halbach et al. 1980; Read et al. 1993; Porcelli et al. 1997) and, thus, neither the geosphere nor the biosphere can be regarded purely as a dilution medium. Indeed, Hellmuth (2001) states that "(the assumption of dispersion and dilution)... of ions released from the waste, transported through the geosphere by groundwater and finally reaching the biosphere is not supported by any observation from natural anomalies."

The complexity of the natural environment necessitates approximation but, in the past, the simplification has been taken to extremes. All geochemical processes affecting radionuclide release are reduced to two 'equivalent' parameters (limiting solubility and Kd) and, therefore, the mineralogical and geochemical controls are obscured. Of the various reasons advanced for the simplification process, or 'disaggregation' as it is often called, few stand up to scrutiny. The need for statistical sampling of parameters need not apply as deterministic treatment may well be adequate and limited computational power is far less of a constraint than hitherto.

The issue then is one of making natural system studies more relevant in an assessment framework while, at the same time, devising more justifiable radionuclide migration models that can be verified against observations from natural systems. The two aims are not incompatible.

The advantage of safety indicators

The purpose of PA is to help convince observers that disposal of radioactive waste in an underground repository is 'safe'. Therefore, the process of quantifying safety cannot be divorced from the need to communicate the results effectively. With operational plant, radiological dose can be calculated readily but, in a long-term performance assessment, the estimates are heavily reliant on assumptions made of barrier evolution, climate change and future human behaviour.

The situation is slowly changing and several countries have now embraced the concept of using natural geochemical concentrations or fluxes directly as indicators of repository safety (Miller 2000; Hellmuth 2001 Traber 2002). Such an approach offers the facility to compare the results of safety calculations directly with natural sources of exposure to ionising radiation; a concept more accessible to non-technical audiences than probabilistic dose or risk estimates.

This confers a number of advantages:

- abundant data are already available, in many cases for systems very similar to those envisaged for disposal.
- the values are measurable directly and readily understood; they can be compared directly to drinking water or other safety standards.
- reasoning by analogy is a powerful argument in convincing sceptical observers.
- the processes leading to contaminant migration and retention from an ore body should be similar to those that would operate in a repository, at least for simple waste matrices such as spent fuel.

Long term storage of radioactive waste

- estimation of dose requires prior calculation of spatial concentrations as a function of time and, therefore, no additional modelling capability is required. Comparison of model output with natural concentration ranges is an obvious means of assessing parametric uncertainty. It provides a check on the realistic bounds of the simulations and this is where emphasis has been placed to date. However, conceptual uncertainty is at least as important and is perhaps, the main area where indicators offer potential for enhancing performance assessment exercises. No amount of statistical sampling will compensate if the parameters being sampled do not reflect the processes actually occurring. For this reason, both parametric and conceptual uncertainties are addressed in the succeeding sections.

Methodology

For natural indicators to form part of the basis for a safety case there needs to be consensus on:
- The aspects of a natural system that may be compared to a waste disposal situation
- The means by which parameter values can be calculated.
- Methods used to compare them to repository-derived equivalents.
- The constraints on the approach.

Each of the above is addressed briefly to examine how such a consensus could be reached.

Parameter selection

Any one of three measures, either individually or in combination, may be used to place a repository-derived impact in the context of natural systems. These are respectively, concentration, flux and accumulated mass.

Concentration, expressed in mass terms or as activity, is the simplest measure to convey. It can be demonstrated, for example, that the uranium concentration in natural groundwaters falls within a given range and, even above ore deposits, does not exceed a certain threshold value owing to mineralisation reactions in the subsurface. Current PA modelling tools do not include these reactions and so such a demonstration is not possible.

The number of factors affecting absolute radioelement concentrations in groundwaters is very large. However, source mineralogy, aqueous speciation and the solubility of derived phases constitute the major controls. Where two or more components can be shown to derive from the same host, relative measures of concentration provide a better guide to mobilisation and redistribution patterns. This is evident from the enormous body of research dealing with fractionation along the homologous rare earth (REE) series. The approach can also be applied to uranium and thorium, where conventionally 'immobile' Th is taken to be invariant in mass balance calculations (*e.g.* Braun et al. 1993). Thorium satisfies most of the requirements of an immobile element, notably very low aqueous solubility and occurrence in phases resistant to weathering. However, caution is needed when interpreting elemental as opposed to isotopic patterns since the mineralogical associations of ^{232}Th and ^{238}U do not necessarily coincide.

Nuclide flux (the rate of mass transfer per unit area normal to the direction of mass flow) can be measured directly in some situations, for example at springs or in spas but, more commonly, has to be inferred from other evidence. In the case of surface processes, including rock/soil weathering, surface flows, estuarine mixing and lake or sea deposition, there is an abundance of data from geochemical surveys and investigations in model catchments (*e.g.* Porcelli et al. 2001; Tarvainen et al. 2002.

Investigations in areas of high natural radioactivity demonstrate clearly that few trace elements are dispersed following alteration of primary ores. These include uranium (*e.g.* Miller et al. 1994 and references therein; Blomqvist et al. 2000); thorium (*e.g.* Braun et al. 1993; Read et al. 2002) and the rare earths (*e.g.* Cuney and Mathieu 2000). The secondary deposits so formed may reflect very significant mass accumulation and, in some cases, constitute more important sources to groundwater than the primary deposit (*e.g.* Blomqvist et al. 2000). Again this feature is not accounted for in the existing PA methodologies.

The uranium-thorium series nuclides with their wide range of daughter half-lives offer an additional advantage in that they provide the means of assigning time frames to contaminant migration events on a scale commensurate with the needs of a safety assessment. One example of the approach is the Broubster study in northern Scotland where source rock U depletion was matched with accumulation in surface peat deposits. The peat was dated radiometrically and this information, used in conjunction with mass balance calculations, allowed accumulation and historical flux to be inferred (Read et al. 1993). In a similar study on the Finnish Peräjävuoma peat, Porcelli et al. (1997) obtained enrichment factors relative to local waters of 10^5 and 6×10^5 for U and Th, respectively.

Comparison of parameter values

In a recent investigation, Miller (2000) calculated elemental fluxes for four diverse geochemical environments in the UK; the Carnmenellis Granite, the Chalk aquifer of the London Basin, the thermal springs at Bath and that part of the Oxford Clay formation surrounding the upper reaches of the Thames. He then compared gross-averaged concentration and flux data for uranium and other elements with the results of earlier PA calculations (Nirex 2000). He also alluded to the large total mass and activity contained in these systems when viewed against the inventory of a spent fuel repository in Scandinavia.

This type of exercise, though useful for illustrating the potential of natural safety indicators, needs considerable refinement before application to a repository safety case. In particular, it has to be ensured that:

    a)   There is commonality of scale between the natural sources cited and a repository. For example, the volume of the Carnmenellis

Granite is estimated to be 800km³. With large areas, the number of potential discharge points is increased and the effects of intrusion into the body correspondingly reduced. As a consequence the impact on individual receptors will be markedly different.

b) The mineralogy is comparable to the waste form being considered and would generate a similar source term (*e.g.* uraninite and spent fuel, monazite and synthetic phosphates).

c) The models used in the performance calculations accurately represent the processes occurring at the locations considered.

The most complete and relevant data set from which to develop an improved methodology applicable to a safety case for spent fuel disposal in Finland is Palmottu (Blomqvist et al. 2000). This site forms the focus of the case study described in the next section.

Scenario development

A crucial element of PA modelling is the definition of future states. At its simplest level, the process merely involves specifying the states *a priori* (*e.g.* Vieno and Nordman 1999). The other extreme is represented by the so-called 'dynamic simulation' approach in which an attempt is made to model system evolution explicitly (Thompson and Sagar 1993). Most commonly, however, both waste disposers and those regulating the industry favour the scenario development approach (*e.g.* Bonano et al. 1988; Eng et al. 1994).

Each scenario (*i.e.* a postulated sequence of future events) is derived by defining the separate components of the system of interest and their respective interactions. Various methods have been devised for representing the interactions and for formalising the judgement process. The weighting assigned to interactions is then used to indicate the level of treatment needed in the performance assessment calculations.

Over the past few years, considerable attention has been placed on 'FEP' (Features, Events, Processes) lists for scenario development, notably in Sweden (*e.g.* Skagius et al. 1995). Recognition that *processes* (chemical, physical, biological) act on *features* of the system (barriers, rock mass, groundwater) to produce *events* is the first stage in constructing defensible scenarios that may be justified by recourse to nature.

At the very least, observations on natural systems provide a check on the realism of scenarios derived by elicitation or other methods. At best, they constitute the basis of a viable alternative to current approaches.

Palmottu case study

Site characteristics and model conceptualisation

The Palmottu U-Th mineralisation in SW Finland (Fig. 1) was discovered by airborne radiometric surveys in the late 1970's and was a target of intensive uranium exploration from 1979 to 1984. The ore body consists of narrow pegmatite veins (2-10m) cutting granite and mica gneiss. The mineralisation extends to depths of at least 400m and total resources are estimated at around 1,000 tonnes (Kaija et al. 2002). The deposit proved to be too small for mining purposes, but provided an excellent opportunity for studying radionuclide transport along well-identified groundwater pathways in the fractured crystalline rock of the Fennoscandian Shield. Analogue studies have been in progress at Palmottu since 1987 (Blomqvist et al. 2000), aimed at characterising the geology, hydrology and hydrochemical setting of the uranium mineralisation.

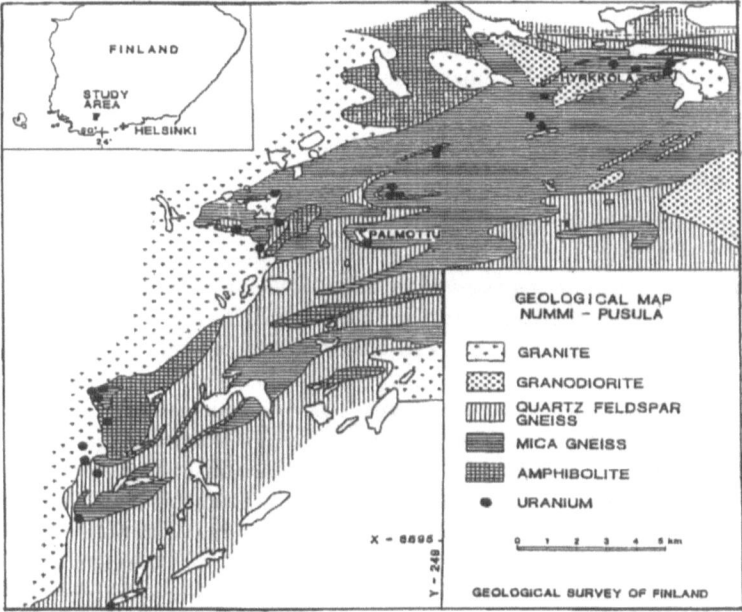

Fig. 1. Geology of the Palmottu study site

The present groundwater flow system dates back to the last deglaciation and the subsequent emergence of the landmass some 10,000 years ago. The uranium deposits extend from near the bedrock surface through a redox transition zone into a strongly reducing groundwater environment. This has allowed the full cycle of uranium redox chemistry under natural conditions to be studied.

Glacial tills or glaciofluvial formations largely cover the bedrock. In local depressions, post-glacial clays were deposited during different stages of Baltic Sea ingress and retreat. Soon after deglaciation, peat started to form in these depressions and has now reached a thickness of 3-4m. Near the exposed U mineralisation it contains several hundreds of ppm of uranium in ash.

In many respects, the geological history of the Palmottu deposit resembles the postulated evolution of the repository proposed for spent U fuel at Olkiluoto (Crawford and Wilmot 1998). These include the inventory, host rock and geographical setting. The deposit is limited in size, contains uranium mainly in the form of UO_2 and has been subjected to a series of perturbations since its emplacement, including an extensive drilling campaign as also envisaged at Olkiluoto. The site experienced continental ice margin conditions and a unique feature of the Palmottu exercise is that the influence of palaeoclimatic changes on groundwater geochemistry formed a central part of the investigation throughout. However, the modelling approaches used at the analogue site (Blomqvist et al. 2000) are very different to those employed in the most recent Finnish safety case, TILA-99 (Vieno and Nordman 1999). Here, a number of discrete canister failure scenarios (small/large hole, canister disappearance after 10,000 years) were combined with hydrodynamic dispersion in the far field. Thus, rather than consider the repository as a geochemically anomalous 'anthropogenic ore body', the safety case assumes that the system becomes homogenised and indistinguishable from its surroundings. The effects of this fundamental difference in model conceptualisation are now explored.

Concentration and flux estimates

The TILA-99 performance assessment was designed to be transparent and is readily amenable to comparison with alternative approaches. Dose calculations relate to a well for drinking water that is assumed to be located in the vicinity of the repository or in the groundwater discharge zone. This is the only exposure pathway considered. The model employed assumes that the annual releases from the repository into the biosphere are diluted in 100,000 m^3 of water and that an individual drinks 500 dm^3 of this water per year (Vieno and Nordman 1999).

The activity entering the biosphere essentially depends on the values chosen for solubility in the near field and equilibrium distribution coefficients (Kd) for barrier components and the host rock. In the case of uranium, solubility ranges from $3x10^{-7}$ to 10^{-4} mol dm^{-3} for reducing and oxidising conditions, respectively. Such values do not seem unreasonable (*e.g.* Burns and Finch 1999) but when combined with sorption, diffusion and dispersion terms the net effect is marked.

No ^{238}U releases are predicted for the reference damaged canister scenarios. Even for cases where the canister is assumed to 'disappear', the maximum release rate from the geosphere is given as only 1.2 Bq y^{-1} (and then after 1 million years). This equates to a concentration of $4x10^{-15}$ mol dm^{-3}. In comparison, uranium concentrations measured at Palmottu range from around $4x10^{-9}$ mol dm^{-3} for the most reducing waters to $>10^{-6}$ mol dm^{-3} nearer the surface (Kaija et al. 2002). The safety case considered release from only one canister, however scaling up to match the total inventory at Palmottu (~1000 t), is insufficient to account for such a large discrepancy. If all the canisters were assumed to disappear there would still be 10^3-10^6 times less uranium in the simulated well water than is actually observed in the geologically similar Palmottu system (Fig. 2). The difference is entirely due to assumptions made regarding the geochemical behaviour of uranium. This is not only a feature of TILA-99 (Vieno and Nordman 1999) but of all performance assessment calculations (*e.g.* SKB 1999; Nirex 2000). Indeed, the Finnish approach is rare in its clarity of procedure, allowing such checks for realism to be made.

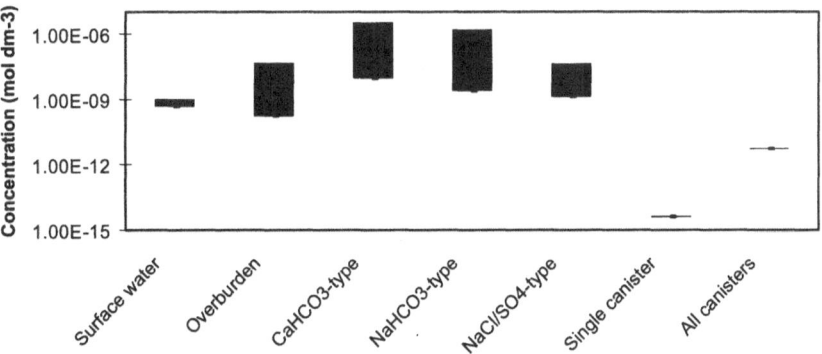

Fig. 2. Uranium concentrations in Palmottu surface and groundwaters compared to PA predictions for a drinking water well.

It is important to note that these calculations do not suggest that the disposal concept is 'unsafe'; the highest concentration of U found in Palmottu groundwaters would produce a dose of less than $0.25\mu Sv\ y^{-1}$ constrained by the solubility of secondary β-uranophane, which is abundant at shallow depths. The issue is one of increasing confidence in the predictions.

The approach is easily extended to alternative exposure pathways or other inventory components. For example, the maximum ^{226}Ra flux to the biosphere is estimated at around $2MBq\ y^{-1}$ in the reference scenarios and in the (absolute) worst case scenario as $70MBq\ y^{-1}$. Bearing in mind that the disposal concept involves emplacing a significant radioactive heat source at depth, it is not unreasonable to compare these values to those for natural thermal spas. It can be seen from Table 1 that the annual ^{226}Ra release, even for the worst case scenario and regarded as highly pessimistic by the authors (Vieno and Nordman 1999), is substantially less than that found, for example at Bath Spa (Kellaway, 1991). The latter is by no means exceptional in European terms (Carlé 1975; Traber 2002).

Table 1. Annual releaes of ^{226}Ra calculated from the recent PA exercise and Bath Spa.

	Reference	High flow	Worst case	Bath Spa
^{226}Ra (MBq/y)	6.7×10^{-6}	1.9	70	200

Long term storage of radioactive waste

As a drinking water well was the only exposure route considered in the recent PA exercise, no quantitative data on surface accumulation of activity are available. Nevertheless, the authors comment on the long term evolution of the candidate sites and the mix of freshwater, forest, peatland etc. environments that could exist. Data for a number of surface media are available from Palmottu, including stream and lake water, overburden, sediments, till and peat (Kaija et al. 2002). Since the last glaciation, the lake sediments are estimated to have accumulated about 130kg uranium with an additional 100kg in two local peat bogs. The Palmottu peats are not particularly radiogenic and the values may represent a lower bound. More than 1t U was found in peat associated with the much smaller ore body at Broubster (Read et al. 1993) whereas several percent U has been found in other Scandinavian and Russian peat bogs.

Similar verification calculations to those described above may be carried out for concentrations, fluxes and accumulated masses of other naturally occurring elements. Nor are artificial isotopes necessarily precluded as discussed below.

The potential role of natural indicators

Several radiologically important elements do not occur at significant concentrations in nature. Miller (2000) highlights this as a severe limitation of natural safety indicators and on the use of analogues in general. Though true to an extent (chemical analogues exist in the majority of cases and artificial isotopes are now widespread in the environment) it applies equally to any alternative measure of performance currently available. Importantly, there are no *additional* problems when compared to the use of dose or risk constraints but there are important advantages, including the transparency of the method and the ease of communicating the results to regulators, scientific reviewers and the Public. At the very least, the use of natural safety indicators, by focussing on the behaviour of geochemical systems, provides a firm basis for extrapolation. Thus, though the use of natural safety indicators is not problem-free, most limitations are procedural and may be overcome by enhancements to the current methodology. For example, new research into the weathering of DU munitions is already furnishing useful data on the likely fate of trace impurities.

Now that arguments in favour of natural safety indicators have become more widely accepted, it is worthwhile assessing the extent to which they could replace dose/risk estimates or whether, as suggested by Miller (2000), they should be viewed as complementary.

A common feature of all PA modelling carried out to date is the hierarchical sequencing of computer programs whereby detailed process-specific models are progressively simplified until one reaches a stylised representation of the entire 'disposal system'. Only at the last stage are parameters sampled and thus, the often-substantial uncertainties associated with the more fundamental aspects of modelling tend to be hidden. This provides much of the rationale for natural indicators, *i.e.* a more transparent safety case. Criticism of PA calculations is at least partly justified, therefore, as in actuality the models do not describe 'the system' but are merely simplistic calculations of groundwater-mediated radionuclide transport. Calibration against more detailed models (where carried out) rather than provide surety often tends to introduce additional errors.

It is possible that, in future, the additional layer of 'PA modelling' could be excluded from safety assessments in favour of an explicit treatment of geochemical transport. This would provide a means for direct comparison of output with natural fluxes and related indicators. If required, conversion of nuclide concentration profiles to dose or risk values could be performed easily by post-processing. In the short term, however, it is likely that both approaches will be applied in tandem.

One consequence of adopting natural indicators as performance measures is the greater visibility of the models used to simulate radionuclide migration. Chemical transport models have to be employed and shortcomings with respect to geochemical site data, thermodynamic/kinetic constants, hydrogeological characterisation etc. are more readily apparent. This is not necessarily a disadvantage if it stimulates the research needed to remedy the deficiencies.

References

Armands G (1967) Geochemical prospecting of a uraniferous bog at Masugnysbyn, N. Sweden. A Kvalheim Ed Geochemical Prospecting in Fennoscandia. Interscience: 127-154

Blomqvist R, Ruskeeniemi T, Kaija J, Ahonen L, Paananen M, Smellie J, Grundfelt B, Pedersen K, Bruno J, Pérez del Villar L, Cera E, Rasilainen K, Pitkänen P, Suksi J, Casanova J, Read D, Frape S (2000) The Palmottu natural analogue project - Phase II: Transport of radionuclides in a natural flow system at Palmottu. CEC Report EUR 19611

Bonano EJ, Davis PA, Cranwell R (1988) Demonstration of a performance assessment methodology for high level waste disposal in basalt formations. Sandia Report 86-2325

Braun J-J, Pagel M, Herbillon A, Rosin C (1993) Mobilisation and redistribution of REEs and thorium in a syenitic lateritic profile: A mass balance study. Geochim Cosmochim Acta 57: 4419-4434

Burns PC, Finch RJ Eds (1999) Uranium: Mineralogy, geochemistry and the environment. Min Soc Am, Reviews in Mineralogy 38:679p

Carlé W (1975) Die Mineral und Thermalwässer von Mitteleuropa. Geologie, Chemismus, Genese. Bucher der Zeitschrift Naturwissenschaftliche Rundschau. Wissenschaftliche Verlagsgesellschaft MbH, Stuttgart

Chapman NA, McKinley IG, Shea ME, Smellie J (1991) The Poços de Caldas Project: Summary and implications for radioactive waste management. SKB Report 90-24

Crawford MB, Wilmot RD (1998) Normal evolution of a spent fuel repository at the candidate sites in Finland. Posiva Report 98-15

Cuney M, Mathieu R (2000) Extreme light rare earth element mobilisation by diagenetic fluids in the geological environment of the Oklo natural reactor zones, Franceville basin, Gabon. Geology 28: 743-746

Duerden P, Lever DA, Sverjensky DA, Townley LR (1994) Summary of findings: Alligator Rivers Analogue Project. UKDOE Report DOE/HMIP/RR/92.072

Eng T, Hudson J, Stephansson O, Skagius K, Wibourg M (1994) Scenario development methodologies. SKB Technical Report TR-94-28

Halbach P, van Borstel D, Gundermann KD (1980) The uptake of uranium by organic substances in a peat bog environment on a granitic bedrock. Chem Geol 29: 117-138

Hellmuth K-H, Backman B, Blomqvist R, Kaija J, Salminen R, Savolainen H, Tarvainen T (2001) Natural geochemical concentrations and fluxes on the Baltic Shield of Finland as indicators of nuclear waste repository safety. STUK Progress Report, IAEA CR.

Kaija J, Rasilainen K, Blomqvist R (2002) The use of selected safety indicators (concentrations, fluxes) in the assessment of radioactive waste disposal. Site-specific natural geochemical concentrations and fluxes at the Palmottu U-Th mineralisation (Finland) for use as indicators of nuclear waste repository safety. STUK Report (draft), IAEA CRP

Kellaway GA Ed (1991). Hot Springs of Bath. Bath City Council

Miller W (2000) Potential natural safety indicators and their application to radioactive waste disposal in the UK. Nirex Report 6297A-1, Version 2

Miller W, Alexander R, Chapman N, McKinley I, Smellie J (1994) Natural analogue studies in the geological disposal of radioactive wastes. Studies in Environmental Science 57 Elsevier Amsterdam 395p

Nirex (2000) Generic performance assessment. Nirex Report S/REP/331665

Porcelli D, Andersson PS, Wasserburg GJ, Ingri J, Baskaran, M (1997) The importance of colloids and mires for the transport of uranium isotopes through the Kalix River watershed and the Baltic Sea. Geochim Cosmochim Acta 61: 4095-4113

Porcelli D, Andersson PS, Baskaran M, Wasserburg GJ (2001) Transport of U and Th series nuclides in a Baltic Shield watershed and the Baltic Sea. Geochim Cosmochim Acta 65: 2439-2459

Read D, Hooker PJ (1992) Using data from natural geochemical environments to improve models of uranium speciation in groundwaters. J Geochem Explor 46: 63-81

Read D, Bennett DG, Hooker PJ, Ivanovich M, Longworth, G, Milodowski A, Noy D (1993) The migration of uranium into peat-rich soils at Broubster, Caithness, Scotland. J Contam Hydrol 13: 291-308

Skagius K, Ström A, Wiborgh M (1995) The use of interaction matrices for identification, structuring and ranking of FEPs in a repository system. SKB Report TR-95-22

SKB (1999) Deep repository for spent nuclear fuel: SR97 – post closure safety. SKB Report TR-99-06

Tarvainen T, Backmann B, Hellmuth K-H, Hatakka T, Savolainen H (2002) The use of selected safety indicators (concentrations, fluxes) in the assessment of radioactive waste disposal. Erosion rates on the Baltic Shield of Finland for use as indicators of nuclear waste repository safety. STUK Report (draft)

Thompson BGJ, Sagar B (1993) The development and application of integrated procedures for post-closure assessment based upon Monte Carlo simulation: The probabilistic systems assessment (PSA) approach. J Rel Eng Sys Safety 42: 125-160

Traber D (2002) The use of selected safety indicators (concentrations, fluxes) in the assessment of radioactive waste disposal. Geohydraulic and geochemical basis of geochemical fluxes from deep groundwaters observed at the Earth's surface. STUK Report (draft)

Vieno T, Nordman H (1999) Safety assessment of spent fuel disposal in Hästholmen, Kivetty, Olkiluoto and Romuvaara, TILA-99. Posiva Report 99-07

Long Term Safety of the Asse Salt Mine

Guido Bracke, Bernd Förster, Herbert Meyer, Gerd Hensel, Götz-Wolfram Thauer, Thomas Taylor

GSF – Forschungszentrum für Umwelt und Gesundheit, Forschungsbergwerk Asse (Research Mine Asse), Am Walde 2, D - 38 319 Remlingen

Abstract. The strategy for performance assessment of the long term safety of a salt mine with radioactive waste is complex. It must include various aspects of geochemistry, fluid flows, geomechanics, geology, hydrology and radiation protection. The concept of geochemical engineering to ensure low radionuclide solubilities and enhance sorption as well as program tools to model fluid flows in the galleries and shafts with and without technical barriers have to be developed and fine tuned for selecting technical measures to enhance the long term safety. These results have to be combined with the hydrogeological model and the geologic long term prediction. The radiological exposition of man has to comply with the limits given by the federal laws for radiation protection to prove the long term safety. The current status of assessment of long term safety of the Research Mine Asse is presented here.

Introduction

The Asse salt mine operated from 1909 to 1964. In this period chambers were excavated with a total volume of more than 4 million m^3. The salt mine was acquired by the GSF as a research and development site for nuclear waste disposal in 1965 in contract with the federal government. The mine was used subsequently as a test site for low-level nuclear waste disposal from 1967 to 1978. The salt mine was used thereafter as a research and development site for high level waste disposal. Since termination of research in 1995 the mine is backfilled and prepared for closure. A prerequisite for the mine closure in 2013 is an assessment of the long term safety in order to comply with federal laws.

Important boundary conditions for the long term safety of the Asse salt mine are:

1. A site-specific property of the Asse Salt Mine is large excavated volumes and a network of galleries and shafts.
2. Brine (saturated with halite) is leaking into the Asse Salt Mine for several years now. The origin of the brine is not exactly known. There are indications of a supply with groundwater.

Brine will corrode the disposed waste containers. Radionuclide from the waste will disperse in the salt solution and will be transported to the biosphere finally. In addition the brine alters carnallitite rock leading to geomechanical instabilities.

Therefore, a series of technical measures are considered to ensure as low as reasonably possible potential radiological consequences in future times. This is called the Asse closure concept.

Calculations did show the necessity of technical measures to achieve a lower level of radiological exposition.

Multi Barrier System and Technical Measures

The multi-barrier system for an underground waste repository is illustrated in Fig. 1. Neither can the disposed radioactive material, its cemented or bituminized waste matrix and waste container be strengthened as a barrier nor the host rock, geosphere and biosphere. The remaining technical barriers include emplacement chamber, disposal level and the mine. They can be enhanced by technical measures.

Radioactive Material (RM)	
Technical Barrier (TB)	Waste Matrix (WM)
	Waste Container
	Waste Chamber
	Disposal Level
	Mine Building
Host Rock (HR)	
Geosphere (G)	
Biosphere (B)	
Man (M)	

Fig. 1. Multi Barrier System.

The closure concept aims to lower the release of radionuclides and to limit and hinder their transportation to the geo- and biosphere. Technical measures are possibly applied to the emplacement rooms, disposal level (near field), the mine and shafts. The various aspects of technical measures are:

Waste chamber

1. Minimizing the fluid volume in the chamber (backfilling)

2. Hindering fluid inflow into the chamber (barriers, plugs)
3. Lowering the solubility of radionuclides (geochemical buffer)

Disposal level

1. Hindering fluid flows in the mine (barriers)
2. Hindering fluid flow through the chambers (barriers, plugs)
3. Avoidance accumulation of gas in the chambers (buffer, transport)
4. Avoidance of brines of different densities

Mine

1. Minimizing the convergence rate (backfilling)
2. Avoidance of brines with different densities (backfilling)
3. Stabilizing the mine and minimizing host rock deformations (backfilling)
4. Minimizing carnallitite alteration (backfilling)

Shafts

1. Sealing the shafts

Closure Concept

1. The present closure concept includes the option to lower the solubility of radionuclides using a geochemical buffer.
2. Barriers and plugs are build at various places to hinder fluid flows in chambers and galleries (disposal level).
3. Backfilling of the salt mine with crushed salt provides stability of the mine.
4. Backfilling of the remaining pore volume with $MgCl_2$-rich brine lowers the convergence rate significantly and reduces alteration of carnallitite significantly.

Long Term Safety

The performance assessment of the long term safety comprises several steps:

Scenario Development

The basis for an assessment of long term safety of the Asse salt mine is a site-specific scenario analysis, which identifies a number of potential evolutions –

Long term storage of radioactive waste

called scenarios. They should be assessed for their potential consequences. The scenario analysis includes the Multi Barrier System along with site-specific screening of catalogues of features, events and processes (FEP). This, for the Asse salt mine, led to two main scenarios, which are fluid mediated:

1. "Asse Closure Scenario" implying technical measures as outlined in the closure concept
   and
2. "Asse Natural Filling Scenario" implying natural filling of the mine with saturated NaCl-brine.

Transport of radionuclides in the mine

The inventory of the disposed radioactive waste is documented. Using the inventory, a source term is calculated applying a geochemical speciation code (EQ3/6). This assumes a quasi-closed system in the emplacement chamber as depicted in Fig. 2 (Schuessler et al. 2001a, b). After equilibration the fluid may be squeezed out (small arrows) by the convergence of the surrounding host rock (thick arrows).

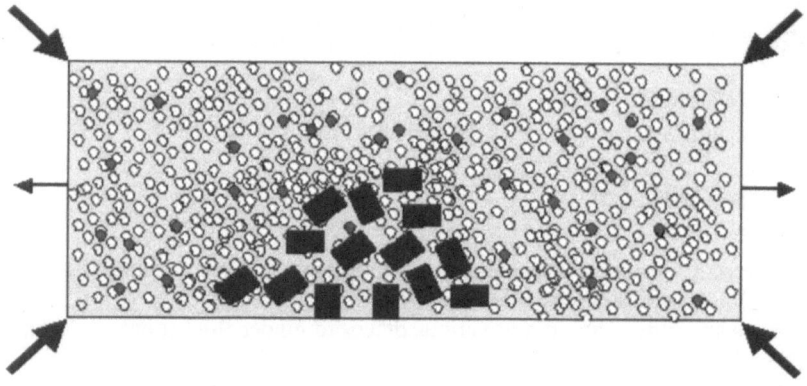

Fig. 2. Quasi-closed system and convergence.

Applying the source term to a model of the mine the pathways and dilution processes of the nuclides are calculated. The main driving force for squeezing out the fluid is the convergence. Minor driving forces are gas accumulation, temperature and density changes of the brine, which can cause additional fluid flows in the mine.

A sketch of the mine and its first model representation is given in Fig. 3. This simplified model showed that technical measures at the disposal level are sufficient to hinder fluid flows in the mine. A more detailed version of this model will be used to calculate the distribution, dilution and transport of radionuclides in the mine.

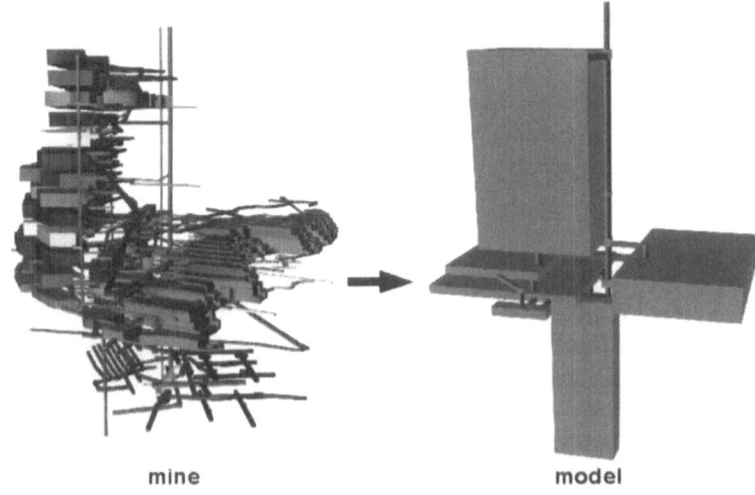

mine   model

Fig. 3. Derivation of a simplified model of the mine

Transport of radionuclides in the overlying rock

According to analytical data (groundwater and brine chemistry, isotopic data) it is likely that the brine entering the mine contains meteoric water. The precise location of the brine leakage into the Permian salt formation is still unknown despite hydrogeologic and geophysical research (seismic, tracer experiments). A possible source and pathway of the brine entering the mine can be described as follows:

1. Groundwater penetrates through the overlying rocks of the Asse salt ridge
2. The underground passage leads to an increase of salinity of groundwater due to rock / water interaction
3. The brine is saturated (halite) by passing outer salt rock areas
4. Deformation processes resulting from the mining activities and long-term convergence caused cracks and fractures in the host rock of the salt ridge thus opening pathways for brines
5. The brine enters the mine at the weakest Permian salt rock zones.

The present leakage of brine into the mine is assumed to be the leakage for the contaminated fluid squeezed out to the geosphere. Using a hydrogeological model, the pathway of contaminated fluid into the biosphere is modelled through the overlying rock. Dilution, sorption and precipitation may reduce the amount of radionuclide reaching the biosphere.

Transport of radionuclides into the biosphere

Transport into the biosphere occurs by direct uptake of contaminated groundwater via plants and animals. Transfer coefficients are applied.

Long term storage of radioactive waste

Potential Exposition

Combining all models yields the potential radiological exposition of man by the disposed radioactive waste for the given scenarios.

Summary and Conclusion

Two fluid mediated scenarios are the basis for performance assessment of the long term safety of the Asse salt mine.

The radionuclide inventory and its composition are documented.

Using this data gas generation is calculated.

The source term with application of a buffer is modeled using a geochemical code. The economic efficiency of its application still has to be proved.

A simple model of the mine showed that technical measures at the disposal level are sufficient to hinder fluid flows in the mine. Nevertheless a more detailed model is necessary to simulate the transport of radionuclides in the mine to the geosphere.

The discussion of the hydrogeological model of the brine leakage into the mine rsp. transport of brine in the geosphere is still in progress since various pathways are possible.

Up to date, only a normalized radiological exposition of man via the biosphere is available.

Combining all parts of the performance assessment of long term safety of the Asse salt mine is scheduled for 2004.

The outlined methodology has been proven to be successful at other sites to assess the consequences. The final performance assessment has to comply with the federal law for radiation protection in order to prove the long term safety and allow closure of the Asse salt mine.

References

www.gsf.de/asse

Schuessler W, Kienzler B, Meyer H, Foerster B (2001a) Geochemical based source term assessment for the ASSE salt mine. 8th Internat. Conf. on Chemistry and Migration Behaviour of Actinides and Fission Products in the Geosphere (Migration '01), Bregenz, A, September 16-21, Book of Abstracts, p 196

Schuessler W, Kienzler B, Meyer H, Foerster B, Stippler R (2001b) Geochemical based source term assessment for the ASSE salt mine. Gesellschaft Deutscher Chemiker Jahrestagung, Würzburg, Kurzreferatband, Nuklearchemie p 52

Wilhelm S, Schwarz R, Klubertanz G, Siegel P, Foerster B (2002) Influence of chemical reactions on the flow system and contaminant transport in a former salt mine. Uranium Mining and Hydrology, Abstracts

Risk Assessment for Underground Disposal of Radioactive Wastes in Clay Strata

Regina Dashko, Alexey Korotkov

Saint Petersburg State Mining Institute (Technical University), 2, 21 Line, St. Petersburg, Russia

Abstract. Radioactive wastes (RAW) storage in clay strata followed by long-term radiation effect on clays results in a noticeable decrease of shear strength and deformation property, with intensification of microbiota activity being a negative factor. Clay properties change provokes the disturbance of hydrodynamic condition and enhances a contamination hazard of water-bearing horizons. Fundamental hydrogeological and geoenvironmental problems require implementation of special engineering treatments for protection of water-bearing horizon used for water supply.

Introduction

Accumulation of considerable amount of radioactive wastes (RAW) as a result of various manufactures activity in the Leningrad region, should stimulate the solution of the problem of its safe storage and / or processing in the short time. RAW have been buried in surface storages of the specialized industrial complex "Radon" and in burials of the Leningrad atomic power plant (LAPP) for a long period of time. At present due to absence of RAW storage the problems of disarrangement of nuclear pipelines which service garantees have been expired at LAPP and nuclear fuel removal can't be cardinally solved.

Some experts in nuclear physics recommend underground disposal of RAW of the Leningrad region to be embedded within Vendian Upper Cotlin clays and Lower Cambrian blue clays , while the blue clays are preferred ones due to its higher plasticity, content of clay particles as well as to its surface occurrence over the pre-glintal territory, where the wastes are supposed to be buried.

As the thickness of blue clays does not exceed 130 m in the region observed, RAW must be buried in special underground workings.

General features of Lower Cambrian blue clays.

In the design of RAW underground disposal in clay massif is based on the assumption that such deposits are fine-porous non-cracking medium with the low filtration capacity, low diffusive permeability and high sorption capacity. Besides, development engineers declare that clays are resistant to the action of radioactive irradiation, high temperatures; at the same time they are inert to corrosive mediums.

The studies have shown that Lower Cambrian blue clay strata should be considered as a unitized medium with a deep zone structure. Each zone is characterized by the intensity of different genesis of fissuring: tectonic, glacio-tectonic, litho-genetic, weathering, reconsolidation and others. Tectonic and litho-genetic fissuring form an unitized structure of blue clay massif: a bloc size increases regularly from the top downward. The depth of a reconsolidation zone usually does not exceed 25 m and is characterized by higher values of moisture content and lower values of density. Each zone distinguished according to its depth has a definite variability of parameters of physical and mechanical properties depending on fissuring intensity and a degree of clay reconsolidation.

Radiation stability of rocks and soils.

At present there are few researches done concerning the influence of radioactive irradiation on rocks and soils, including clay deposits and rock-forming minerals. Radiation stability of minerals usually reduces through quartz to alumosilicates: quartz – microcline – kaolin – hydromica – montmorillonite. According to the statements of the theory of radiation damages in a solid body, crystalline structure of minerals is changed owing to interaction among radiating particles and mineral constituent elements, which are disturbed from equilibrium condition at the points of the lattice and may cause the displacement of other elements. Such processes lead to crystalline structure disordering, vacancies making and forming defects on the micro-level, finally all these phenomena cause the change of mineral properties owing to their amorphisation. Special investigations have shown that amorphisation of quartz after being irradiated is accompanied by an increase in its volume to more than 17%. First of all, the irradiation influences on structural bonds existing between minerals and its aggregates. For example, granite samples are desintegrated to pieces, the size of grains being 1-5 mm at 2.8×10^{20} neutron/cm^2 irradiation.

The exposure of different types of concrete to radioactive irradiation results in the decrease of its strength from 20 to 90% depending on the conditions of irradiation, its action duration, cement composition and filling materials.

It is necessary to pay a special attention to radioactive stability of clay minerals, which interlayer bonds usually results from unlike charged surfaces, or cations which are between the same charged layers. Similar bonds are easily broken down when exposed to radioactive irradiation simultaneously with the structural bonds existing between mineral particles. Change in valent bond vibrations of Si - O and Al - O inside a layer up to its fracture is observed on long-term radioactive exposure. Besides there is dehydratation of clay minerals with consecutive isolation of crystallization water, and structural water in the form of OH⁻ – groups from the crystal lattice. The removal of crystallization water has an impact on chemical, physico-chemical and physical and mechanical properties of clays, and the removal of structural water leads to a complete fracture of clay mineral.

Being part of clay minerals of Lower Cambrian blue clays are represented by high-alkaline minerals, interlayer links destruction will take place at low enough radiation doses. The main structural defects in clay minerals are developed on radiation exposure, which is 10^5-10^6 Grey. Thus the process of destruction of silica-alumina nucleus of clay minerals is becoming more active in a particular sequence of cation outlet of crystal lattice: $Fe^{3+} > Ca^{2+} > Mg^{2+} > Na^+ > K^+ > Si^{4+}$. Except for the destruction of a solid phase of clays, there is radiolysis of pore water, which results in the formation of free radicals and the molecular components causing the change of acid-alkaline and oxidation-reduction conditions in a clay strata. In blue clays organics and sulphides may be oxidized actively while forming the process of radiolysis of pore water at absorbed doses up to 10^2-10^3 Grey. Besides, a rise in temperature of a soil massif with radioactive decay generally causes activation of the process of water radiolysis. For example, when the chain process of radiation oxidation of substances is in progress, a rise in temperature from 20 up to 80 °C results in approximately a twice rate increase. Complex action of radiation and temperature factors also promotes a deeper degradation of clay minerals.

Transformation of blue clays under the influence of radioactive irradiation.

Studies of Lower Cambrian clays, which have been exposed to radioactive irradiation for 50 years at the bottom of one of RAW storages, have shown considerable transformations of these clays. It is necessary to note that low-active wastes (LAW) with the α – radiation level that did not exceed 10^5 Bk/kg and β – radiation level, less than 10^6 Bk were disposed in the considered storages. Under the action of long-term irradiation the amount of clay minerals, which are identified by means of X-defraction analysis has reduced; in d < 0,002 mm fraction composition only illite and a muscovite were found, while unirradiated clays contained chlorite, hydromuscovite, hydrobiotite, glauconite, less often montmorillonite. Amorphisation of clay minerals is confirmed by the results of analysis for adsorption capacity of blue clays. In blue clays the value of adsorption capacity is 12 mg-eqv/100 gr of soil, in irradiated ones the value of this index has practically double increased - 20-22 mg-eqv/100 gr of soil. Amorphisation of clay minerals has also led to a rise of hydrophylity of blue clays and it has invoked the increase of plasticity index and capacity of these soils to swelling. The value of free swelling of irradiated blue clays has considerably increased in comparison with unirradiated soils. Besides the long-term radiation has essentially lowered parame-

soils. Besides the long-term radiation has essentially lowered parameters, such as shear strength and deformation capacity of blue clays (Table 1).

Table 1. Blue clays parameters modification on exposure to radioactive influence.

Region of re-search	Depth, m	Moisture content	Plasticity index	Value of swelling, %	Shear strength parameters		The module of deforma-tion,
					C, MPa	$\varphi 0$	E, MPa
Storage of wastes (Sil-lamyaye)	5.0-0.0	0.19-0.23	0.31-0.41	14.0-20.0	0.019-0.075	0	2.9-7.8
Narva region	5.0-0.0	0.16-0.28	0.14-0.22	4.5-6.0	0.10-0.14	3-4	10.0-28.0
Nikolskoye town, the Leningrad region	3.0-8.0	0.20-0.23	0.19-0.27	5.0-8.0	0.075-0.117	3-8	19.0-24.0

As shear strength and deformation capacity determines the stability of rocks and soils in underground workings, it is necessary to take into consideration a decrease of such parameters on exposure to radiation while designing both underground storages of the radioactive wastes storages in blue clays and underground atomic power stations. Even visual evaluation of Lower Cambrian blue clays that suffered the long-term radioactive irradiation, reveals an intensive degree of their desinte-gration and fissuring.

Intensification of microbiological activity under the influence of low-active wastes.

While arranging and using LAW storages special attention should be paid to a possible change in micribiological activity in enclosing soil massif. From the point of view of survival rate of microorganisms on exposure to radioactive irradiation, first of all, it is necessary to distinguish the microorganisms in an active state. The former which are continuously capable of elliminating radioactive damages and slumbering ones, in which there is no metabolism and reproduction. Radiation in-tensity is more important for «active» microorganisms than the rate of the proc-esses of damages recovery. The integral dose, obtained in "a rest" state, is much more important for "slumbering" microorganisms.

It should be noted that only recendly researchers have started to investigate the problem of microbiological activities at RAW burials, whereas a few studies paid attention to the role of microorganisms at deep RAW disposal. As a rule the field of microbiological activity studies should include: a) survival of microbes in irridiation conditions; b) possibility of formation for products of their metabolism and finally biochemical gas; c) role of corrosion of container's walls, caused by microbes; d) microbiological degradation of waste products and materials.

The microbiological degradation of waste products draws a special concern at burial of low-active wastes as they frequently contain a significant amount of organic materials, which are subjected to biochemical degradation. It is necessary to note that according to the data of some specialists the microbiological activity in the zones of burial of low-active wastes exists everywhere and has a significant impact on the geoenvironment. The studies have been conducted in Great Britain in the zone of the underground burial of low-active wastes for 42 years, have shown that microbiological disintegration and corrosion of materials have caused a considerable deterioration of the natural ecological situation. The general trend of the processes was bound up with the formation of reduction conditions and methane generation. The latter could lead to its self-ignition and was especially dangerous.

In our case the first stage of study of the microbiological activity at the bottom of the tailing dump of the low-active wastes storage Sillamyaye was held to estimate bacterial mass at the blue clays bottom section. The value of bacterial mass was estimated by Bredford's biochemical method, which allows to obtain the content of microbial protein in blue clays (Table 2).

Table 2. Microbial protein (MP) content in blue clays

Depth from the roof of blue clays	1,8-3,4	3,5-11,0	12,0-19,0
MP, microgr/gr	$\dfrac{204\text{-}465*}{284/9}$	$\dfrac{150\text{-}365}{234/10}$	$\dfrac{104\text{-}275}{172/10}$

* In numerator: minimum - maximum values of an index; in denominator: average value and number of analyses.

Precise regularity of MP values decreasing with the depth (Table. 2) may be explained by the existence of the surface radiation source. As a rule MP does not exceed 10-30 microgr/gr for the considered clays outside irradiation zones and other technogenic effects, where the depth influences insignificantly.

The presence of microbiota, the cells of which are sorbed on dispersed particles, operates as a negative factor, promote additional strength loss, development of plastic properties and generating biocorrosive attack to building and defensive constructions. All these factors should be taken into consideration while designing RAW storages and atomic power stations.

Hazard of water-bearing horison contamination.

It should be noted that Lower Cambrian blue clays serve as the upper confining layer occuring at the top of the Lomonosov water-bearing horizon, which is used for water supply in the western parts of the Leningrad Region (the towns of Ivangorod, Kinguiseppe, Slantsy) and in Estonia (Narva).

At present it is planned to bury RAW of the "Radon" plant and Leningrad atomic plant within the Koporye area, which is 10 km from Finland Bay, where the Lomonosov water-bearing horizon is discharged. Therefore, its contamination has a regional significance resulting not only in contamination of water-bearing horison, used for water suply, but also in the waters of Finland Bay. The considered area of burial is located within the limits of two transcontinental lineaments covering, one of which – Lapland- Nilsk with kinematics of shift and contraction has the meridional direction, and the second one - a northwest direction of strike (extent), - Narva-Amudarya is characterized by the existence of gapping deformations. The territory is crossed by a system of regional and transregional fractures of sublatitude submeridional and northwest directions. At present in the clays of the sedimentary blanket of Pre-quarternary deposits the intensification of fractures takes place when while the intensity of fissuring of water-bearing and waterproof soils increases accordingly.

Desintegration increase of waterproof blue clays strata on exposure to radiation and temperature causes the growth of clay massif permeability, where RAW are to be buried. Besides, it should be noted that the active use of the Lomonosov water-bearing horizon for the purpose of water supply will predetermine the development of a depression, which will cover the Koporye area. Deterioration of waterproof properties of blue clays, the change of the hydro-dynamic situation caused by intensive use of the Lomonosov water-bearing horizon will result in the flowing from upper underground waters. Descending water flow passing through the zone influence of radioactive wastes storage will be enriched by radionuclides and transfer them to the water-bearing horizon used.

Such serious hydrogeological and geotechnical problems require the development of special engineering means of water-bearing horison protection with regard to long-lasting negative changes in waterproof clay strata.

Long term storage of radioactive waste

Uranium Migration in Argillaceous Sediments as Analogue for Transport Processes in the Far Field of Repositories (Heselbach Site, Germany)

Thomas Brasser[1], Ulrich Noseck[1], Dagmar Schönwiese[2]

[1]Gesellschaft für Anlagen- und Reaktorsicherheit (GRS) mbH,
38122 Braunschweig, Germany
[2]Technical University, 38106 Braunschweig, Germany

Abstract. Essential goal of investigations at Heselbach site has been, so far, to identify an U-accumulation at the eastern rim of Wackersdorf lignite basin - known from literature - and to check its suitability as natural analogue for transport processes in the overburden of a final repository for radioactive wastes. On the basis of at first four exploration drillings, located along a profile vertically to the basin rim, this task has been completed and a follow-on project has started.

Introduction

Long-term safety of underground repositories for radioactive wastes is based on a multi-barrier concept in which clay materials often play an important role as geological barriers. For performance assessment (PA) of those repositories it is essential to consider the potential mobility of critical radionuclides through relevant rock material under representative long-term conditions. Information on these time scales can only be obtained by studying natural occurrences of radioisotopes and trace elements.

The Heselbach site has been chosen, since its similar geological but different geochemical conditions (near surface, more oxidizing environment) are able to complete the results obtained at the Natural Analogue Site Ruprechtov, Czech Republic. The aim of the Natural Analogue Study in general is to understand and to quantify relevant geochemical and hydraulic processes at the site and to identify the mechanisms leading to transport and immobilisation of uranium, thorium and radium. After detailed characterization of the site, flow and transport models will be applied to describe the migration of these three elements at Heselbach site. Therefore this study intends to contribute to data acquisition and model testing for performance assessment (PA).

Geology

Heselbach site is located at the NE rim of Wackersdorf Tertiary lignite basin (30 km N of Regensburg, Bavaria), within a small E-oriented side-bay of former mining district 'Wackersdorf Nord' (Fig.1) (Brasser et al. 1998).

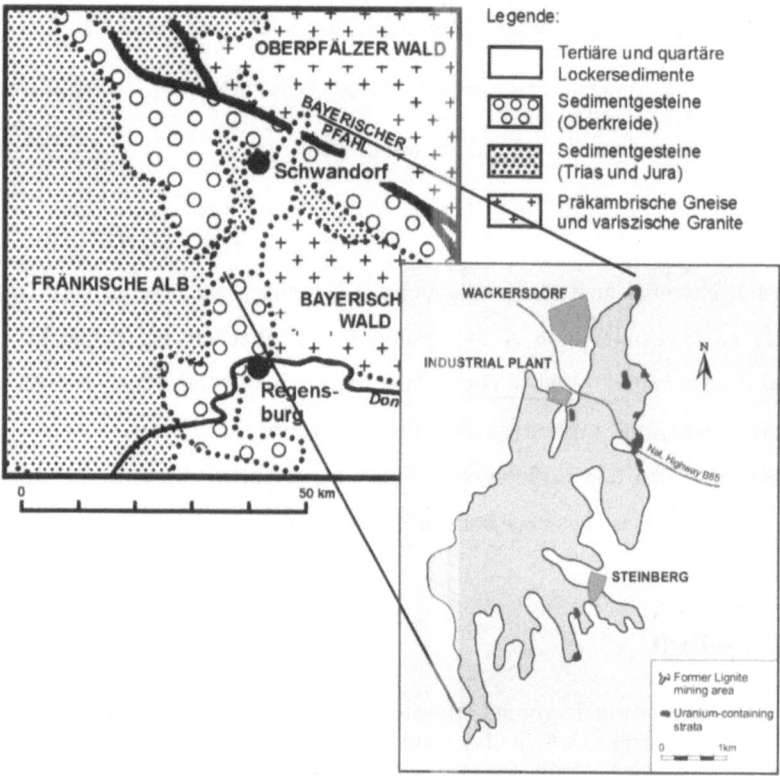

Fig. 1. Outline of geological situation of Oberpfalz region (left) and Wackersdorf lignite basin (right); project area is located at E-rim of basin, directly N of federal highway B85.

The actual research area comprises mainly three essential geological units which are also represented by core material received from drilling of four wells (Fig.2):

- Burgsandstein (middle Keuper - ‚Keuper Sandstone') as underlying strata, resp. eastern surrounding of lignite basin,
- Remaining relicts of Tertiary lignite seams after termination of open-pit coal mining as well as
- Inner dump of former mining district 'Wackersdorf Nord' with raised material of different origin up to former level of land surface.

Long term storage of radioactive waste

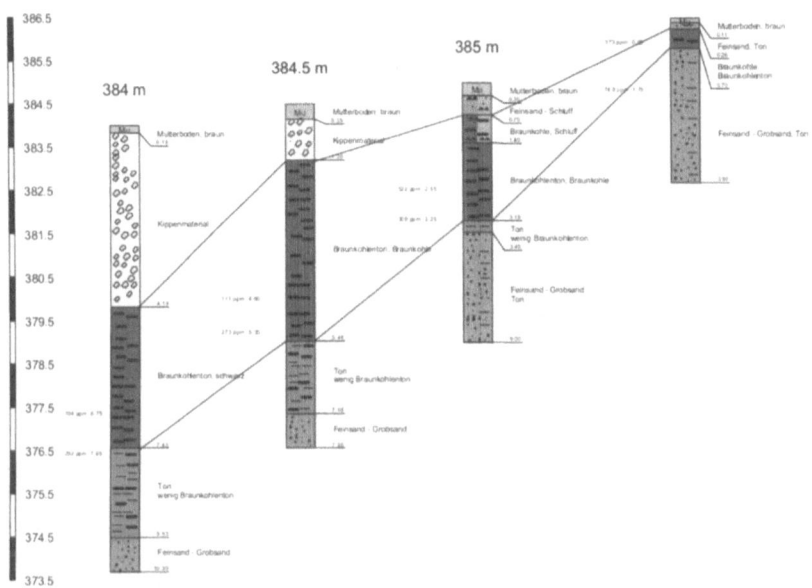

Fig. 2. Geological correlation of exploration drillings at Heselbach site (B3 - B2 - B1 - B4, from left to right); assumed groundwater flow is from right to left.

Hydrogeology

Investigations of groundwater chemistry for characterization of hydrochemical milieu comprised 'normal' lab-analytics, on-site measurements of T, EC, pH and Eh as well as on-site spectrophotometric determination of Fe^{2+}, Fe_{tot}, SO_4^{2-} and S^{2-}.

Results from these on-site measurements already illustrate the different hydrochemical milieu with respect to Ruprechtov site. Along with comparable total mineralization at both sites, groundwaters from Heselbach site clearly show lower pH- and positive Eh- values.

Altogether, groundwaters from Heselbach site are from Ca-SO_4-Cl-type. TIC-concentrations correspond with a partial pressure of CO_2 which is just slightly above partial pressure of CO_2 in atmosphere. At pH-values around 5, the predominant percentage is represented by H_2CO_3, as well as solved CO_2. Main components are plotted in a (modified) Piper-diagram (Fig.3).

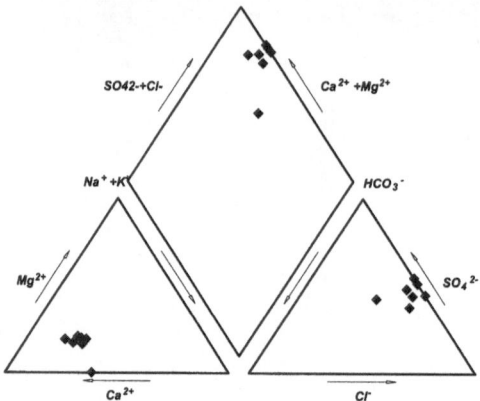

Fig. 3. Plot of groundwater-analyses (B1 - B4; main components) in a modified Piper-diagram

Radiometry

All 4 exploration drillings are characterized by two horizons of U-accumulation, the upper one coming across the lignite seam itself, the lower one across the transition zone between lignite and underlying clays. Differentiation between both horizons is decreasing with increasing depth (Fig.4).

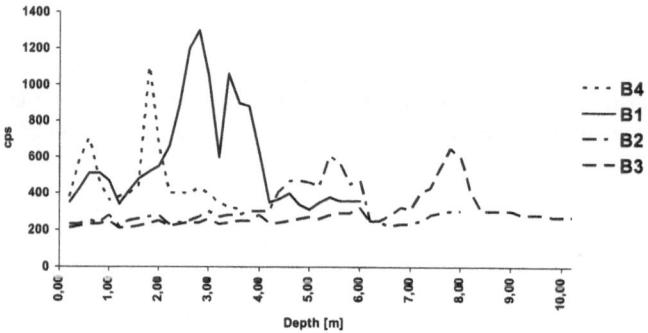

Fig. 4. Measurement of gamma-radiation along sediments of exploration drillings B1 - B4 with help of a portable differential-spectrometer.

Two samples per borehole with high gamma-activity have also been analysed for alpha-activity. Results of both, gamma- and alpha-measurements are compared in Fig.5 and indicate a rather good correlation.

In Fig.6, activities of relevant isotopes of the U-decay chain are plotted. Fig.7 shows activity ratios of $^{230}Th/^{238}U$ versus $^{234}U/^{238}U$. Except one sample from lower horizon of B1 all samples are characterized by $^{230}Th/^{238}U$-ratios which are <1.

Long term storage of radioactive waste

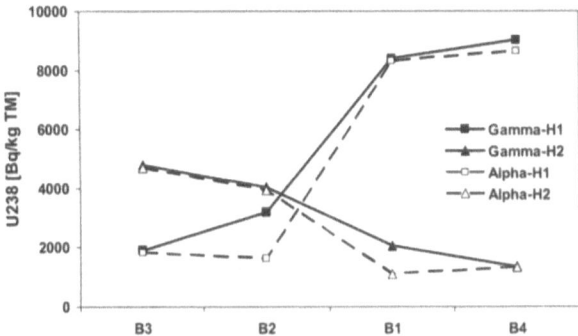

Fig. 5. Distribution of ^{235}U-activity along profile B3 - B2 - B1 - B4 (H1 = upper horizon, H2 = lower horizon).

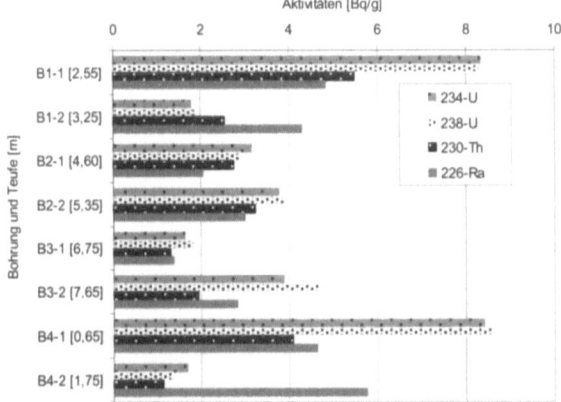

Fig. 6. Activities of isotopes of U-decay chain (sediment samples from B1 - B4).

In Fig.8 selected results, only from lignite samples (upper horizon), are shown. It is assumed that groundwater discharge takes place in the lignite horizon. Ratios of ^{234}U/^{238}U-activity show minor variations around ‚1'. This corresponds to ratios measured in groundwater samples from B1 - B4. The ratio of ^{230}Th/^{238}U-activity is lowest in borehole B4 (0,47), generally increasing with flow direction. Since data are significantly minor 1, an U-deposition has to be assumed for all lignitic sediments. Increase of ^{230}Th/^{238}U-activity ratio with groundwater flow direction might be attributed to a decrease of the U-precipitation rate. This assumption is supported by generally high U-concentrations in boreholes B4 und B1 and a decrease of U-concentrations in groundwater with flow direction. At present, no sufficient explanation for strong variance of ratios of ^{230}Th/^{238}U-activity in underlying clay layers can be given. This open question will hopefully be solved with help of additional drillings and samples in the current follow-on project.

Long term storage of radioactive waste

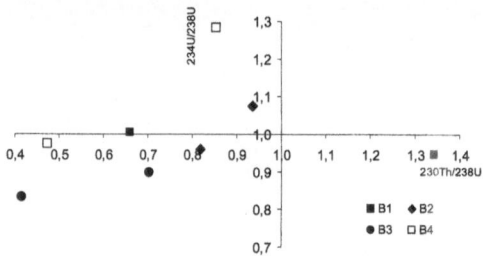

Fig. 7. Ratios of $^{230}Th/^{238}U$- versus $^{234}U/^{238}U$-activities (all analysed samples from boreholes B1 - B4).

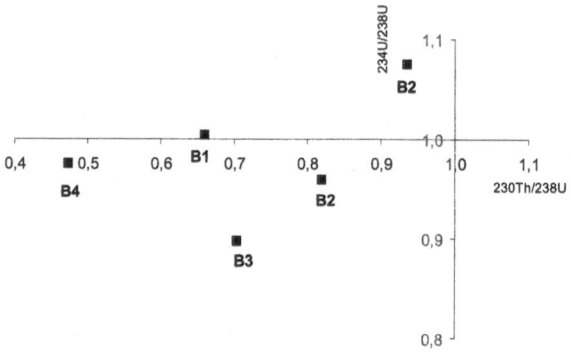

Fig. 8. Ratios of $^{230}Th/^{238}U$- versus $^{234}U/^{238}U$-activities (samples from lignite horizons only).

The ratios of isotope activities from U-decay chain measured in groundwater samples are plotted in Fig.9. Ratios of $^{234}U/^{238}U$-activity in groundwater samples from boreholes B2 and B3 are close to ‚1' (with regard to margins of error). This might indicate that chemical and transportation processes run off relatively fast and, in consequence, no influence of α-recoil effects on distribution of isotopes in groundwater can be detected. Since ratios of $^{234}U/^{238}U$-activity are close to ‚1' also in sediments a reversible exchange of mobile and non-mobile uranium might occur. Explanation of low ratios of measured $^{234}U/^{238}U$-activity needs expansion of investigation area and number of samples which is one of the targets of the follow-on project. Low ratios of $^{230}Th/^{234}U$-activity are likely be caused by low solubility of Th.

Geochemical modelling

First geochemical modelling has been performed with PHREEQC-code and Wateq 4.0 database, not taking into consideration the uranylphosphate-complexes

$UO_2(HPO_4)_2^{2-}$ and UO_2HPO_4 respectively With regard to present geochemical conditions uranium should exist as hexavalent ion. Measured redox-values vary between 300 and 550 mV, pH-values between 4,7 and 5,4 in most groundwaters.

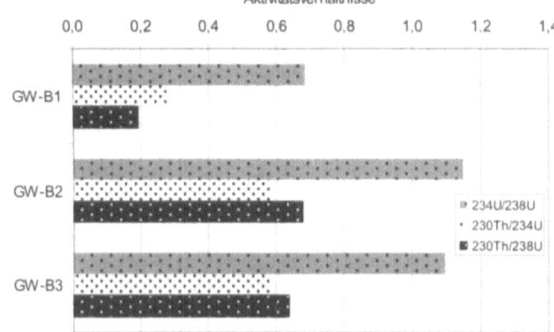

Fig. 9. Ratios of isotope activities (different isotopes from U-decay chain) in water samples (GW = groundwater)

Calculations of U-speciation have been carried out for a pH-range from 4,5 - 6. Additionally phosphate- und TIC-concentrations have been varied from 0,01 - 0,2 mg P/l and 0,1 - 2,5 mg C/l respectively (Noseck et al. 2000, 2002). First results showed that the mobile U-phase is defined mostly by free UO_2^{2+} - cation and UO_2HPO_4-complex, the last one becoming more important at higher pH-values.

Calculations with U-concentrations in range observed at the site indicate that - even at highest values (4 µg/l) - no saturation of U(VI)-minerals is to be expected. Similarly, a variation of redox-values up to 100 mV doesn't result in saturation of U-minerals. Again, this can be taken as indication for U-retention mainly originated by sorption.

Conclusions

So far, results from geochemical, mineralogical and radiometric investigations at Heselbach-site yield in following conclusions:

Within two distinct horizons (upper lignitic and underlying clay layer) U-accumulations up to 640 ppm could be detected. Activity-ratios in lignite clearly indicate U-enrichment processes. U-concentrations in groundwater amount to appr. $3 \cdot 10^{-9}$ mol/l. Decreasing U-concentrations (sediment as well as groundwater samples) with groundwater flow direction imply U-retention mainly by sorption processes. Similarly, geochemical modelling indicates U-minerals to be far from saturation. In-situ - distribution-coefficients (calculated from ratio of non-mobile to mobile U-concentrations) amount to more than 100 m^3/kg and are explicitly higher than typical values for U-sorption on clay minerals or iron oxides.

It is still an open question, how or to what extent geochemical conditions at Heselbach site have been influenced by long lasting open-pit mining. Possibly groundwater lowering has caused oxidation of sulfidic minerals in deeper horizons. The recovery of groundwater level after termination of mining, in consequence, could have originated lower pH-values. Groundwater chemistry might also be influenced by dumping material used for recultivation of mining district.

Based on current knowledge a first hypothetical model for U-transport at Heselbach site has been set up which is sketched in Fig.10. Due to overall geological and hydrogeological situation only Keuper sandstone (Burgsandstein) has to be considered as U-source.

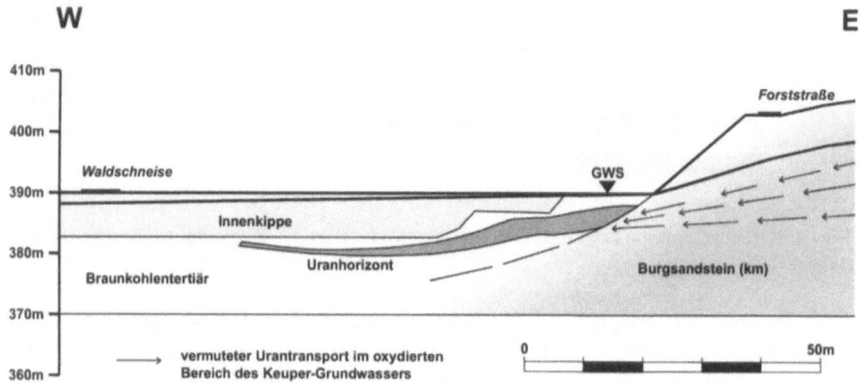

Fig. 10. Sketch of geologic-hydrogeological situation at Heselbach site; arrows indicate groundwater flow and direction of U-transport.

Potential retention processes for uranium are represented by sorption on iron oxides, organic matter and clay material as well as U-co-precipitation processes. An extended follow-on project has started with 10 new boreholes, so far, and new samples for a more detailed investigation of the site.

Acknowledgement

This work has been funded by German Bundesministerium fuer Wirtschaft und Technologie (BMWi) under contract-No. 02 E 9128.

References

Brasser Th et al. (1998) Untersuchung der Uran-Thorium-Mobilisation als natürliches Analogon für den Radionuklidtransport im Deckgebirge eines Endlagers für radioaktive Abfälle. GRS-A-2652: XII + 228 S.
Noseck U et al. (2000) Uranium Migration in Tertiary Sediments. International Conference on Radioactive Waste Disposal, DISTEC: 514-519
Noseck U et al. (2002) Tertiäre Sedimente als Barriere für die U-/Th-Migration im Fernfeld von Endlagern. GRS-176: 284 pp

Translating natural concentrations and fluxes into safety indicators for radioactive waste repositories

Francisco Lemos[1], Terry Sullivan[2], Kurt Friese[3], Timothy Ross[4], Maria Barbosa[5]

[1]Comissão Nacional de Energia Nuclear, Brazil.
[2]Brookhaven National Laboratory, USA.
[3]UFZ-Centre for Environmental Research, Department for Inland Water Research, Brückstr. 3a, D-39114 Magdeburg / GERMANY
[4]Dept. of Civil Engineering, The University of New Mexico, NM 87131, USA
[5]UFOP- Universidade Federal de Ouro Preto, Departamento de Geologia da Escola de Minas, Campus do Morro do Cruzeiro, Ouro Preto, Brazil, CEP - 35400-000

Abstract. Natural deposits of radioactive elements can provide important and very useful information for safety evaluation of radioactive waste repositories. The information is especially helpful for public communication. However, a simple comparison between the calculated fluxes and concentrations in a proposed repository and those natural ones can be ambiguous. This ambiguity arises from high complexity, imprecise knowledge, different data formats and units. This work suggests the use of a methodology, based on fuzzy logic tools, to handle this ambiguous information. A case study is presented as an example.

Introduction

There are many sources of uncertainties in the traditional dose and risk calculations used in Performance Assessment (PA). A main source of uncertainties is the need for making assumptions regarding human future habits, and the repository performance during its useful life. (IAEA1999; Kozak 1997)

In an attempt to improve confidence and public acceptance of the results of the performance assessment calculations, the use of complementary safety indicators has been suggested. (Miller et al. 2000) One of these complementary methodologies would be to compare predictions of repository releases with natural fluxes and concentrations of chemical species. This methodology would have an additional advantage which is the use of a natural context for safety demonstration making it easier for the public to understand.

However, the use of natural fluxes and concentrations requires the use of a huge amount of data that may not be readily available. The lack of data together with spatial variations, are important sources of uncertainties and ambiguousness in data analysis.

Some solutions to this problem, such as a global average flux value (for example, the global average activity flux due to groundwater discharge) masks considerable variation in the fluxes which occur at different sites, and in different geological and climatic environments.

This shows that the use of these complementary safety indicators will not reduce uncertainties, rather they have the advantage of placing the calculations in a framework that can be compared to natural processes. While these safety indicators do enhance confidence, there still exist ambiguousness in the results due to the uncertainties.

It is within this context that this work suggests the use of a methodology, based on fuzzy logic tools, which is designed to handle ambiguous data and allows the use of natural language terms for the comparisons between repository system and natural environments.

Fuzzy relations

Fuzzy relations are calculated through logic compositions. The mapping of elements of one universe Y to other universe X is made through a cartesian product of the two universes. The strength of the relation is measured with a membership function. Methods to accomplish this are described in (Ross 1995). One of these methods is the max-min. It can be imagined as the links of a chain. The strength of a chain is equal to the strength of the weakest link. In case of two parallel chains, the strongest one will determine the strength of the two of them. This can be shown by the equations:

$T = R \cdot S$, where : R is a fuzzy relation on the cartesian space $X \times Y$ and S is a fuzzy relation on the cartesian $Y \times Z$ space, and T is a fuzzy relation on the $X \times Z$ space. Then:

$$\chi_T(x,z) = \bigvee_{y \in Y} (\chi_R(x,y) \wedge \chi_S(y,z)) \qquad (1)$$

Where χ_T is the characteristic function of T in the interval [0,1]. This function measures the strength of the relation, i.e., a value of 1 means full relation and 0 no relation.(Ross 1995)

$\wedge$ is minimum and, $\vee$ is a maximum value.

Fuzzy pattern recognition

Site parameters can be defined as fuzzy sets. In fuzzy sets, the known patterns typically are represented as class structures, where each class structure is described by a number of features. A typical problem in pattern recognition is to collect data from a physical process and classify them into known patterns or rank them according to a pre-determined criteria.

Suppose we have patterns represented as fuzzy sets A_i on $X(i=1,2....m)$ and a new piece of data, perhaps consisting of a group of observations, is represented by a fuzzy set B on X. The task now is to find which A_i the sample B most closely matches.

According to (Ross 1995) if we define two fuzzy vectors, say $\mathbf{A}$ and $\mathbf{B}$, then if the vectors are identical (same length and same elements) their inner product $\mathbf{A} \bullet \mathbf{B}^T$ reaches a maximum value as their outer product, $\mathbf{A} \oplus \mathbf{B}^T$ reaches a minimum value. These two norms can be used simultaneously in pattern recognition studies because they measure closeness or similarity.

$$(A,B)_1 = (A \bullet B) \wedge \overline{(A \oplus B)} \tag{2}$$

$$(A,B)_2 = \frac{1}{2}\left[(A \bullet B) + \overline{(A \oplus B)}\right] \tag{3}$$

In particular, when either of the values of (A,B) above approaches 1, then the two fuzzy sets $\mathbf{A}$ and $\mathbf{B}$ are more closely similar. When either of the values are close to zero they are more far apart or dissimilar. As some of the features may be more important than others, weights can be introduced, ω_j, where:

$$\sum_{j=1}^{m} \omega_j = 1 \tag{4}$$

Therefore, equations 2 and 3 are then modified for each known pattern $(i=1,2.......c)$:

$$(B, A_i) = \sum_{j=1}^{m} \omega_j (B_j, A_{ij}) \tag{5}$$

Sample $\mathbf{B}$ is closest to pattern A_j when,

$$(B, A_j) = \max_{1 \le i \le c} \{(B, A_i)\} \tag{6}$$

Where $\mathbf{B}$ is a collection of fuzzy sets, $\mathbf{B} = \{B_1, B_2,.........B_n\}$, and when $\mathbf{B}$ is a collection of crisp singletons, i.e., $\mathbf{B} = \{x_1, x_2,. ...x_n\}$ then equation (3) reduces to

$$\mu_{Ai}(x) = \sum_{j=1}^{m} \omega_j \cdot \mu_{Aij}(x_j) \tag{7}$$

in the maximum approach degree, sample x is closest to pattern A_j when equation (5) reduces to

$$\mu_A(x) = \max_{1 \le i \le c} \{\mu_{Ai}(x)\} \tag{8}$$

The use of elemental flux as a natural safety indicator

Calculated fluxes of naturally occurring materials are result of a series of processes (or features) in the surface and subsurface environments (Miller et al. 2000). In order to keep this example simple, a few of the most important features will be considered in the analysis. This list can be changed upon experts agreement.

Typically, it is very difficult to obtain a consistent database of natural geochemical and process rate data, and therefore accurate determinations of average values. A number of assumptions have to be made when quantifying natural concentrations and fluxes to be compared against the repository source term.

Processes driving natural fluxes have considerable variation in their rates. However, these variations are not always due to differences in the inherent properties of the geological materials (such as hydraulic conductivities). They can also be a consequence of external factors such as climate which alters the impact of certain processes (e. g., erosion) and indeed, climate is indicated to be one of the largest causes of erosion of a granitic pluton (Miller et al.2000).

Elemental fluxes may be calculated for specific processes and so a range of mass fluxes corresponding to different processes can be generated for the same element (e.g. flux due to groundwater discharge, erosion, river flow, etc.). In this manner the most significant mass transport mechanism can be readily identified. In terms of providing direct comparisons with repository releases, it is anticipated that fluxes associated with processes which drive the transfer of materials from the groundwater discharge (solute transfer) and erosion (solid transfer) will dominate.

Repository system

The calculated fluxes inside a repository will depend on a number of parameters and processes. For example, intrinsic factors of the waste: the radionuclide's waste stream, waste form, and container control release from the waste (Sullivan, 1993). These factors together with radionuclide specific parameters such as half life, solubility limits, transport parameters (Kd), water flow and moisture contents, initial conditions and boundary conditions will fully describe the problem.

However, due to the large number of different container types and waste forms, it is not always possible to have precise values for all the parameters in order to model the release processes. Rather, analysts professional experience is used to find a model representative of the system.

Let's say containers fall into 3 types, A, B, and C. Type A has an expected lifetime of 1 – 150 years. Type B from 30 – 500 years, and Type C from 300 – 1000 years. A deterministic, conservative (early failure or worst case) model would assign lifetimes of 1, 30 and 300 years to each categories. A probabilistic approach would sample along the ranges and values combined randomly. However, as can been seen, the categories blend in each others intervals, therefore it would not be clear during calculations how a container which expected life time of 100 years would be classified in the category A or B.

A fuzzy set approach would address the problem by using language terms to define the containers conditions such as category A (short life), B (medium) and C (long life). Now a container with life time of 100 years would be placed in both categories A and B, however with different degrees of memberships. The same rationale would be used for determining the release mechanisms in order to describe classes of waste forms.

Upon analysts agreement, it is possible to determine a group of features or parameters (fuzzy sets), to compose vectors for comparison between the repository and the sites features. An example of this approach will be given in the next section.

Example calculations

Suppose one wishes to demonstrate how differently a disposal unit would affect candidate sites' environments. The pattern recognition technique can be used in two ways. First, according to the degree of similarity, the approach can be used to provide a measure of the similarity between each site and the repository. Second, it can determine an ordering of similarity between each site and the repository features. In other words, what site would the repository most closely match.

In a traditional procedure a list of 20 or more sites would be screened for use as a repository site. Simple screening criteria would be evaluated to narrow the list to 5 potential repository sites. These five sites would be presented to decision makers for further consideration. For these sites, it is required to know the effect repository construction and performance would have on each of the near field environment of these sites. Assuming that the sites which are least impacted by the repository should receive further consideration, two sites can be selected.

At the point where there is a list of 5 sites, it would be necessary to conduct a more detailed analysis, with a more detailed data collection and more complex performance analysis. However, making a complete site characterization for 5 sites would be extremely expensive. Even for well studied sites, such as Poços de Caldas in Brazil, the lack of data, force analysts to use natural language (ambiguous) to describe site conditions. (Lemos et al. 2001)

A question remains on how to enhance confidence that one meets the objectives of the site selection, i.e., a list of 5 sites with acceptable degrees of safety and how the repository will affect the selected site.

In this example, a list of 5 candidate sites will be studied and two that have the closest match to the repository will be selected for further characterization, just for demonstration purposes. Then the influence of the repository on the sites will be assessed.

Upon experts agreement the features to be used as comparison factors in this example are:
A- Inventory concentration
B- Redox fronts
C- Sorption

D- Dispersion/diffusion
E- Water flow rate
F- pH
G- Speciation
H- Colloid concentrations

Table I shows a set of features, for each site and repository, after a study of their respective characteristic functions. Some of these features may be typically very different inside the repository and in the environment or between two different sites and this does not necessarily mean that one site has better performance features. How can we then compare the features at both sites? For example, how can one make a comparison between the repository and the site if pH inside the repository is between 9 and 11, and in the environment it is between 6-8 If either range of pH has very little influence on the calculated fluxes for their respective context how should they be compared? Conversely, if pH has a large impact on predicted flux what is the basis for comparison?

An answer would be the characteristic function χ which is defined in the interval [0,1]. This function measures the strength of the link on a relation. In this example, the link is a measure of the impact of a parameter on contaminant flux to the environment. For the first parameter, pH, if Repository pH (between 9-11 due to cementitious materials used to construct the repository) has a very weak link to flux (where $T = pH \cdot Flux$) then $\chi(pH)$ will be "low", the same is valid for site pH 6-8. If this range of pH, for any reason, has a weak link to flux it will also generate a "low" characteristic function. Now the characteristic functions can be compared and in this example they would be similarly "low".

This reasoning can be applied to other features such as colloid concentrations, dissolution limits and others. Cs 137, for example, has a high sorption capacity and therefore a high water flow rate may not have a high impact on the calculated flux. However, the transport can be facilitated by the presence of colloids from package corrosion. So, instead of simply comparing water flow rate, it would be more effective to compare between the characteristic function of its link to the flux, depending on each context.

Table I: Example of characteristic functions for each site and repository to be compared

Mode (process)	$\chi(pH)$ $\omega_1=0.05$	$\chi(Redox)$ $\omega_2=0.1$	$\chi(Gwt$ flow rate) $\omega_3=0.1$	$\chi(Inven-$ tory com.) $\omega_4=0.3$	$\chi(sorp-$ tion) $\omega_5=0.05$	$\chi(colloid$ conc.) $\omega_6=0.2$	$\chi(disper-$ sion/diff.) $\omega_7=0.2$
Site 1	Medium	Medium	High	Low	High	Medium	High
Site 2	Low	High	Very high	Very low	High	High	Low
Site 3	Medium	Low	Medium	High	Medium	Very high	Low
Site 4	Very low	Low	Medium	Low	Medium	Medium	High
Site 5	Low	Medium	Low	High	High	Low	Medium
Repository	Low	Medium	Medium	Medium	High	Low	Medium

Fig. 1 shows a representation of fuzzy sets low and medium $\chi(pH)$
for repository and site 1 respectively. Applying equation (1) to find the degree of compatibility between site 1 and repository for the comparison factor pH gives :

$\mu($ Repository χ (pH) $\bullet$ site 1 χ (pH)) $=$ max([(0∧1),(0∧0.75), (0.4∧0.4), 0.25∧0.75), (1∧0)]= 0.4

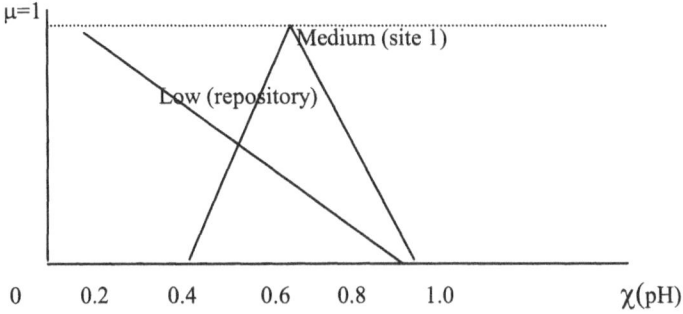

Fig. 1. Example of a comparison between fuzzy sets describing influence of pH on flux, χ(pH), between repository and site 1.

In the above expression, the values for the membership function are evaluated over the domain of the characteristic function at several points. At each point, the degree of compatibility between the two fuzzy sets (pH characteristics in the repository and site 1) is taken as the minimum at each point. The maximum value from this set of minimums is the degree of compatibility, Equation 1.

The final degree of compatibility will be the sum of each of the features degrees of approaching along with respective weights, Equation 6. For site 1 the analysis found the following::

(Repository, site 1) = 0.4·0.05 + 0.3·0.1 + 0.5·0.1 + 0.3·0.3 + 0.8·0.05 + 0.1·0.2 + 0.5·0.2 = 0.02 + 0.03 + 0.05 + 0.09 + 0.24 + 0.02 + 0.1 = 0.55

This same calculation is repeated for all pairs (site, repository), and the following results were obtained, Table II.

Table II: Comparison of compatibility of the repository with each site.

Repository/site	Degree of compatibility or approaching
Site 1	0.55
Site 2	0.50
Site 3	0.3
Site 4	0.6
Site 5	0.2

This analysis indicates that the proposed repository will be more closely similar to sites 4 and 1 regarding the selected features. After a more detailed analysis, the values of 0.6 or 0.55 could also lead to the conclusion that, as it is close to 1, it would not have a strong impact on the site's environment, while a value of 1 would suggest no impact. It is important to recognize that the fuzzy set approach has taken the ambiguous data in Table II and permitted a ranking among the sites. This clearly could not be done by inspection of Table I.

Conclusions

A fuzzy logic based approach has been developed to examine site information which are usually given in ambiguous expressions, so they can be treated in a mathematical basis and yet keep its natural language characteristics.

The major advantages of the approach are :
- a-   It translates language expressions into mathematical values, or fuzzy sets.
- b-   The use of natural language makes it easier for the public and decision maker to be more familiar with the meaning of the results.

A simple example that examined the compatibility of five hypothetical repository sites with the proposed repository conditions was performed and it was found that the approach successfully met its objective to give support for a site selection decision that would best match natural conditions with those envisaged for the repository.

This calculation has another advantage of being easier to communicate to the public as it uses natural language expressions which are familiar to public and decision makers. In addition, the approach is flexible and readily permits incorporation of new information into the analysis as it becomes available.

References

IAEA (1999)-International Atomic Energy Agency: Safety Assessment for Near Surface Disposal of Radioactive Waste. IAEA Safety Standards Series, Safety Guide No. WS-G-1.1. Vienna.

Kozak M(1997), Sensitivity, Uncertainty, And Importance Analyses , 19th Annual DOE LLW Conference, Salt Lake City, May).

Ross T(1995) Fuzzy Logic With Engineering Applications , McGraw-Hill.

Lemos L, Sullivan T, Silvia M, Friese K (2001), "Using Fuzzy Logic to Assist in Performance Assessment of a Nuclear Waste Repository", Waste Management'01, Tucson, USA.

Miller B, Lind A, Savage D, Philip M, Robinson P (2000) Natural Elemental Comcentration and Fluxes: Their use as indicators of repository safety, Quantisci, QSL-6180-GEN/2, UK.

Sullivan T (1993) Disposal Unit Source Term (DUST) Data Input Guide. Brookhaven National Laboratory, BNL-NUREG-52375, USA.

Evaluation of the long-term durability of engineered dry covers for mining wastes, and consideration of associated design constraints

Stephan Kistinger, Guido Deissmann, Martin Goldsworthy, Ralf H. Stollenwerk

Brenk Systemplanung GmbH, Heider-Hof-Weg 23, D-52080 Aachen, Germany

Abstract. To reduce pollution of water resources it is often necessary to cover uranium mining waste heaps and tailings. In this paper the failure mechanisms which can be significant for a mineral sealing layer are set out and discussed. On the basis of this, the requirements for the construction of a cover are defined. The necessity for post-closure care is discussed and, finally, the question is addressed of when it is sensible to include a mineral sealing layer in a cover system.

Introduction

An important issue at many mining sites is the reduction of detrimental effects caused by the discharge of contaminated seepage from mine wastes. A frequently used method for the mitigation of environmental impacts from mining wastes is the construction of engineered soil covers on waste dumps or tailings. The purpose of such rehabilitation measures is to reduce the environmental impacts to acceptable levels as well as to facilitate future use of the former mining sites, e.g. for recreational purposes and/or to create wildlife habitats.

Although construction of a suitable cover does not reduce the contamination potential of the mine waste, it can delay the contaminant release and thus reduce the contaminant loads in the seepage per unit of time, i.e. reduce the rate of contaminant release, so that lower contaminant loadings are discharged over a longer time span. Ideally, the cover should reduce the contaminant concentration in the seepage to levels which are so low that even without seepage treatment the contamination of the receiving water bodies does not exceed acceptable levels. However, even if some water treatment is still necessary, the long-term costs will be lower than for a collect-and-treat option due to the financial discounting of future expenditures and the potential for improvement in water treatment technologies (e.g. passive or microbiological treatment).

The design and the construction of a cover system has to integrate carefully all site-specific aspects of the ecosystem, land-use and climate, with the characteristics of the waste and the cover materials, in order to minimise the short and long-term rehabilitation costs for the mine (i.e. costs for cover construction, water treatment (if necessary), cover maintenance and monitoring), as well as the future environmental risks. This paper discusses essential design and construction requirements to ensure the long-term durability of dry cover systems for mining wastes, focussing on cover systems that include a compacted and water-saturated clay sealing layer, to reduce the fluxes of oxygen and water into the waste materials.

Cover design

Dry cover systems for acid-generating or potentially acid-generating sulphide–bearing mine wastes often require a particular constructive component to minimise the fluxes of oxygen and water into the waste. The objective is to reduce contaminant mobilisation resulting from sulphide oxidation processes and to minimise the amount of contaminated seepage. This can be achieved by covers containing compacted soil/clay layers at high water saturation or industrially modified mineral materials or by synthetic liners (e.g. polyethylene geomebranes). Capillary barriers can reduce the water flux into a waste dump, but will not present a restriction for oxygen ingress.

For cost reasons, only the first mentioned type of cover was generally considered for the remediation of the waste dumps at the uranium mining sites in the eastern part of Germany. An example of a synthetic liner combined with a bentonite-gravel layer is that used in the cover system for uranium tailings at Ellweiler (in Rhineland-Palatinate, in the western part of Germany).

Experience from landfill sites indicates that a drainage layer on top of a clay layer increases the risks of desiccation and cracking of the clay barrier. A drainage layer improves the geomechanical stability of the system and acts as a capillary barrier to reduce the moisture flow from the clay into the upper layers of the cover system in dry periods. The disadvantage of a drainage layer is that water is removed in wet periods and the conditions in a subsequent dry period become more extreme. There is an ongoing discussion concerning the benefits and risks of a drainage layer in a cover system.

In the following discussions of design and construction requirements for engineered covers on mining wastes, only soil covers are considered, focussing on cover systems which include a compacted, water-saturated soil/clay layer (oxygen barrier) overlain by a substrate layer as required for the sustainable development of vegetation.

Long term storage of radioactive waste

Potential failure mechanisms

Mechanisms which may threaten the long-term integrity of a compacted clay/soil barrier in a dry cover system on mine wastes include:

- cracking due to differential settlement of the mine waste,
- erosion,
- human intrusion,
- cryoturbation, i.e. disturbance due to frost,
- bioturbation, i.e. disturbance by animals
- root penetration,
- desiccation and cracking,
- slope failure.

A suitable, sufficiently thick substrate layer is required in these cover systems in order to ensure the long-term integrity of the clay layer and to facilitate the development of a sustainable cover of vegetation. Disturbance of the substrate layer, which can subsequently result in failure of the clay barrier, may result from:

- erosion,
- intrusion by humans (e.g. mountain biking, motocross) or animals (rodents, wild boars, etc.),
- uprooting of trees,
- slope failure.

In the following, some of the processes which may impair the performance of a cover system are discussed in more detail.

Differential settlement

Cracking due to differential settlement of the mine waste can be avoided by

- compaction of the waste material - possibly during the construction of the waste heap, and/or
- choosing an appropriate combination of clay material and moisture content to ensure deformation occurs without cracking under the anticipated range of movements.

Erosion and human intrusion

For the remediation of former uranium mines a minimum primary period of 200 years is considered when selecting and designing the protective measures. (During the selection of the measures a secondary 1000 year period is considered qualitatively.) For such periods protection against erosion becomes of central importance.

As there is generally no building on the remediated areas suitable vegetation can provide erosion protection. For this, the greatest risk is in the early stages, when the vegetation has not yet developed a sufficient protective effect. In this phase erosion protection must be provided through an appropriately intensive after-care.

Longer term, and especially under extreme weather conditions, the surface water drainage system becomes an erosion weak point, as the energy of the water becomes concentrated within it.

In the densely populated parts of Germany damage to vegetation and the cover resulting from inappropriate human activities (e.g. motocross) is a significant factor.

Frost / Animals / Roots

In the regions of Thuringia and Saxony affected by uranium mining the maximum frost depth is about 1 m, although locally greater depths cannot be excluded. In order to protect the sealing layer from burrowing animals (mice, rabbits, foxes, badgers) a minimum 0.3 m frost-free zone must be available in the vegetation substrate layer. This reduces the incentive for the animals to burrow into the less accessible, compacted sealing layer. From this a minimum thickness of 1.3 m can be derived (locally however more) for the vegetation substrate.

The rooting depth of plants depends essentially on the soil and the water conditions. According to the literature the rooting depth of native trees in Germany is up to 3 m, and 4 m under conditions of water shortage. Some meadow plants root deeper than trees. Saturated and dense (>1.8 g/cm^3) soil conditions restrict the depth of roots. When there is sufficient water and nutrients plants do not root deeper than 1.5 m in such conditions.

Plants require a certain degree of mechanical support, in order to resist loadings from storms and to a lesser degree from snow and from growing inclined due to uneven light availability. In natural conditions the soil is often denser at depth and there may be a rock weathering zone. A cover system is, in contrast, looser and has potential slip surfaces at layer boundaries (in the case considered here, between the sealing layer and the vegetation substrate). Trees growing on such a cover system are in considerably more danger of being uprooted than trees in natural locations. According to assessments obtained from forestry experts, with a vegetation substrate depth of 1.5 m the risk of uprooting becomes acceptable.

Desiccation / Cracking

The most significant failure mechanism for a sealing layer is the development of cracks. This occurs when a critical suction level, exceeding the tensile strength, is reached in the sealing layer. With further drying or in subsequent dry periods cracks can grow until finally the complete depth of the sealing layer is penetrated.

Laboratory and field experiments made during the BMBF research programme on the further development of waste deposit sealing systems show that on rewetting the cracks in the sealing layer close again, but that they remain as water pathways. This means that a long-term increase in the permeability to gases can also not be excluded.

For a typical clay material we calculated the critical suction value for different construction methods. The results ranged between 100 and 200 kPa for compaction with a machine of medium weight at a moisture content higher than the Proctor optimum, and from 500 to 800 kPa for compaction with a heavy machine at a low moisture content. At a low moisture content means here at the optimum for the specific compaction equipment, but because the compaction energy is higher than in the standard Proctor test the moisture content would be below the standard Proctor optimum. The results show that the critical suction value can be significantly increased by compacting thoroughly at the lowest practical moisture content. The attainable critical suction value is however still far below that which can be generated by plants (the permanent wilting point is pF=4.2, which corresponds to 1500 kPa).

The vegetation substrate layer must be such that the critical suction is not exceeded at the top of the clay seal. It must be thick enough for the plants to find sufficient water and nutrients without generating high suctions at the boundary to the sealing layer. Cracks resulting from such suctions become accessible to roots, and can then be widened as further water is extracted. From this is derived the requirement that the vegetation substrate layer should have the highest possible storage capacity for water which is available to plants. In the literature, a minimum of 20 cm of water is given.

The necessary thickness of the vegetation substrate can be calculated as a coarse approximation with computer programs such as HELP. Calculations which we made with Version 3.05 for a specific uranium mining waste heap in Thuringia showed that for drying to the permanent wilting point not to occur the following layer thicknesses had to be assumed:

- plateau and north facing slope 1.5 m;
- east and west facing slopes 2 m;
- south facing slope 3 m.
- Use of the program for this purpose was however found to be problematic, as:
- it required daily weather data;
- the capillary rise of water was not considered;
- the behaviour of plants in dry periods was not described correctly, in that the amount extracted by them was independent of the soil water content.

We consider that the calculated thicknesses are too great.

For a vegetation substrate above a sealing layer the minimum thickness given in the literature is 1.5 m. For trees, a thickness of 2 m or more is typically given. It is recommended that the minimum thickness required is determined for the specific site.

Long term storage of radioactive waste

Slope stability

For the system analysed here the sealing layer is directly overlain by the vegetation substrate. In the context of the stability of the cover system doing without a drainage layer has a number of weaknesses;

- there is a critical slip plane between the sealing layer and the vegetation substrate;
- in wet periods a significant amount of water can build up on top of the sealing layer, reducing the available shear resistance.

In demonstrating that the system is sufficiently stable the wind loads from trees should be considered. As the roots should not penetrate into the sealing layer they cannot be considered to have a stabilizing effect.

Cover requirements

Sealing layer

The sealing layer should have the lowest possible field conductivity. In our opinion it is more appropriate to define the required field conductivity than the hydraulic permeability of the material. This, along with other requirements, can be derived from the desired field conductivity. The pore air volume should also be as low as possible, in order to minimize the passage of oxygen (and radon).

To obtain the desired properties in the sealing layer the following requirements must be met during its construction:

- The material should contain a wide range of particle sizes (for low residual permeability and low tendency to cracking);
- compaction should be with the heaviest possible equipment with the objective of reaching a high density (e.g. $\geq 100\%$ of the Proctor maximum density);
- compaction should be at a water content which is optimized for the equipment in use and is therefore lower than the standard Proctor optimum water content;
- at a minimum in the joint zones between adjacent construction panels, compaction should be in two layers in order to avoid local water pathways there.

The necessary logistic and technical measures must be taken to avoid material being delivered and compacted in too moist a condition.

Vegetation substrate (water storage layer)

From the discussion of failure mechanisms it can be seen that to protect the sealing layer the thickness of the vegetation substrate should ideally be 3 to 4 m, but at least 1.5 m. This thickness does not provide absolute security against a long-term degradation of the functioning of the sealing layer, but, on the other hand, there are no reported observations indicating that a loss of effectiveness is to be expected.

If a thin vegetation substrate layer is aimed for, in order to minimize remediation costs, this must be considered in planning the after-care, as the expenditure required for the long-term monitoring and maintenance rises as the thickness of the vegetation substrate layer is decreased.

In order to obtain the desired properties the following requirements should be met during the construction of the vegetation substrate layer:

- the thickness should be at least 1.5m
- the available field capacity, i.e. the water storage capacity between field capacity (pF=1.8) and permanent wilting point (pF=4.2), should be at least 15 vol.% or 22.5 cm of water;
- in its upper zone (to a depth of 1.5 m) the layer should be constructed with the minimum possible compaction to achieve a density of less than 85% of the Proctor maximum density;
- any part of the layer below a depth of 1.5 m should be heavily compacted (in order to increase the resistance to penetration by animals and roots).

Plants and animals

The following points suggest that woodland vegetation should be aimed for on waste piles with this type of cover:

- sowing grass, and establishing a woodland in parallel to this, hinders the spread of plants with particularly aggressive roots (e.g. dock, thistle);
- woodland restricts unwanted human activity on the cover (e.g. motocross);
- a mixed woodland provides the best long-term erosion protection;
- in temperate climatic zones the natural long-term vegetation is woodland, and by planting desired species those with aggressive roots can be restricted.

To protect the sealing layer against mice and rabbits the presence of their natural predators can be encouraged, e.g. by providing perches for birds of prey. To the extent that the intended use of the area permits this, the populations of wild boar and of badgers should be kept as low as possible by hunting.

Slope stability

In order to ensure the necessary long-term stability of the slopes we recommend the following:

- in the determination of the allowable slope angle and of the necessary shear strength parameters of the sealing layer material it must be taken into account that under extreme weather conditions full saturation of the vegetation substrate layer can occur (for thickness up to about 2 m) in longer slopes (some 10's of metres long);

- if a cohesion value is to be included in the calculations of slope stability this should be determined in careful triaxial tests carried out under drained conditions at the appropriate low stress level;
- for a vegetation substrate layer thickness of 1.5 m the slope gradient should not exceed 1:3, but flatter would be better.

Construction quality

To minimize the risk that the cover system will fail as a result of material deficiencies or construction errors, reliable quality assurance must achieved during construction. Appropriate recommendations (e.g. in Germany the "GDA-Empfehlungen") should be applied.

After-care

The institutional controls and after-care must be so arranged that the cover functions as intended for the time period which the specialists consider to be necessary. For residues from uranium mining the 200 year primary time period which was mentioned above provides some indication of the timescale. After-care generally includes:
- maintenance of the cover itself (repair of erosion damage);
- maintenance of the surface water drainage system to ensure undiminished effectiveness;
- maintenance of the vegetation to protect against erosion and against the spread of species with aggressive roots.
- If this is necessary, collecting and treating contaminated water is also part of the after-care, but this is not the subject of this Paper.
- Repairs to the cover should not first be made when the sealing layer is already damaged, but rather when damage is to be expected if an observed process is allowed to continue. The reasons for this are:
- damage to the environment should be avoided if possible;
- a repaired sealing layer is unlikely to have as high a quality of the original, as smaller equipment will be used and there will be more joint zones;
- erosion damage can extend very quickly, so early repair can be the cheapest option.

Concluding remarks

As a conclusion we would like to address the question of whether it is sensible to construct a cover with a mineral sealing layer when taking into consideration the expenditure in construction and after-care and the risks of failure.

Long term storage of radioactive waste

In our opinion the answer must depend on the individual circumstances, but the following principles can be established:

- If the degradation of all the affected environmental media by substances from the source <u>remains and will remain</u> at a tolerable level, even without sealing, then a sealing layer (with the objective of reducing the outflow of contaminants) should not be constructed.
- If all significant seepage from the source is collected (i.e. the remaining outflows can be tolerated) <u>and</u> long-term disposal of this without causing any damage is possible <u>and</u> this is economically viable, then a sealing layer need not be constructed.
- In our opinion the pre-requisite for this is that those responsible for the rehabilitation commit themselves to long-term water treatment or that a fund for this purpose (or an equivalent arrangement) is set up. (For operating facilities it is therefore advisable to reach a binding agreement on the time, the extent and the responsibility for after-care before establishing a new potential source of contamination.)
- In cases when the contaminant inventory makes it imperative that the long-term sealing effect of the cover is ensured, and the consequences of degradation cannot be accepted as a residual risk, then it is our view that an appropriate geomembrane should be included in the cover system.

Long term storage of radioactive waste

Consultants GmbH

We offer you services in the following sectors:

- **IT consulting,**
- **databases,**
- **geographical information systems,**
- **cartography,**
- **geology and natural resources,**
- **hydrogeology,**
- **handling of contaminated sites,**
- **soil protection, and**
- **landscape planning.**

MULTIPLE SKILLS AND DISCIPLINES and the application of advanced computer aided technologies are the cornerstones of our company philosophy. Geologists, geochemists, hydrogeologists, ecologists, biologists, computer scientists, mathematicians, cartographers, surveyors, and process engineers work hand in hand on the solution to all of your problems.

During the realization of our services we also utilize, in addition to our own resources, the INTERNATIONAL KNOW-HOW and the experience of the other members of the BEAK group of companies that operate worldwide.

Beak Consultants GmbH
Am St. Niclas Schacht 13
09599 Freiberg
Germany
Phone: (49) 3731 781 350
Fax: (49) 3731 781 352
Mail: postmaster@beak.de

www.beak.de

Hydrology of Engineered Covers of Waste Rock Dumps towards Heavy Rainfall Phenomena: A Case Study in Uranium Mill Tailings Mitigation

Petra Schneider[1]; Schaffrath, M.[1]; Nindel, K.[2]

[1]Hydroisotop-Piewak GmbH, Oberfrohnaer Straße 84, D-09117 Chemnitz,
[2]WISMUT GmbH, Jagdschänkenstraße 29, D-09117 Chemnitz

Abstract. The in-situ remediation of uranium mine dumps by engineered and planted mineral soil covers plays an important role in long-term mitigation of uranium mining sites in Eastern Germany. This case study discusses results of soil mechanical and hydrological investigations of a mine dump with a 10 years old engineered mineral soil cover. A heavy rainfall event was simulated using irrigation equipment including a precipitation measurement gauge assuming 120 mm precipitation within 24 hours ($N_{120/24}$). Soil moisture and water content were measured by a frequency domain response (FDR) sensor.

Introduction

The site under study is the Schüsselgrund Mine Dump, a uranium mine tailing constructed in 1967 as a part of the Königstein mine (see figure 1). After the end of the mining activities in 1990, Schüsselgrund Mine Dump is part of the remediation activities involving the Königstein mine. Schüsselgrund mine dump currently is made up from mining waste containing uranium and radium. General information on hydrogeology, hydrology and geochemistry of the Schüsselgrund site are given in Schneider et al. 1999 and Schneider et al. 2001.

The Schüsselgrund Mine Dump covers 24.2 ha (including 8.0 ha slopes) with a total volume of 3.9 mio m³. The dump has a high priority in the complete remediation activities due to the chemical, mineralogical and radiological composition of the materials disposed at this site. Currently, a continuous release of radioactive matter occurs via air and water paths. The investigations reported in the following have been motivated by the need for estimates concerning the long-term stability of mineral soil covers. As a basis for this study the existing engineered soil covers have been investigated in order to obtain hydrological and soil mechanical data of the engineered covers of the slope during periods of heavy rain events. The long-term remediation strategy includes a reforestation of the dump plateau. For the dump plateau, a double layer mineral cover has been projected. The prospected utilisation of the site is forestry.

Site Characterisation

Location and structure of the tailings

The testground is located south-southeast to Dresden City, the capital of German state of Saxony. The site is situated in the environmental protection area ‚Säch-sische Schweiz' in a distance of as close as 1 km to the river Elbe and about 2.5 km from the ‚Sächsische Schweiz' national park (see Fig. 1).

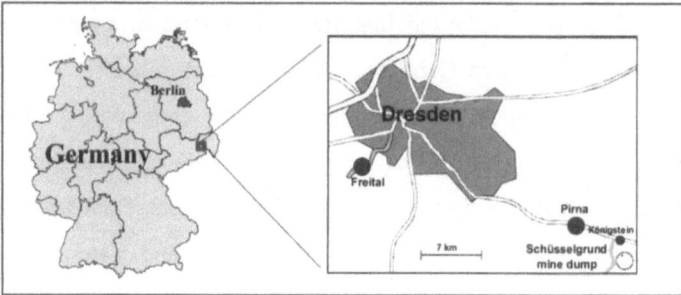

Fig. 1. Location of the study site

The area surrounding the site (except the western part) is under commission of the state forest. To the west facilities of the Königstein mining plant are located. Further beyond the facilities, areas are used agricultural. Height of the dump plateau is about +300 m NN, the base being at +250 m NN. The slope at the northern and eastern side of the dump has been covered with mineral soils of about 1 m thickness between years 1991 and 1998. The mineral soil cover is constructed on an engineered basis of mining waste with a height of about 10 m. On a width of 60 m towards the center of the dump, tailings with increased radioactivity have been dumped.

Soil mechanical characterisation

The slopes are constructed with tips with a height of 10 m separated by almost horizontal sections with a depth of about 5 m. The slope of the visible side has an average of 1:2.5. The soil of the existing cover is made up from about 25 % silt and 75 % sand (result of a particle size analysis). At the time of construction, densities (dry) in the range 1.6 g cm^{-3} to 1.8 g cm^{-3} and permeabilities in the range 5.6 $\cdot$ 10^{-10} m s^{-1} to 9.6 $\cdot$ 10^{-10} m s^{-1} (kf; according to DIN 18130) have been obtained for the engineered layers. As a result of chemical analysis a pH 5.7, low amounts of humics (nearly 0.2 %), leachable phosphorus (nearly 18 ppm) and potassium (50 mg – 70 mg) have been found. Leachable magnesium (50 ppm – 70 ppm) was in the average while leachable calcium (about 0.5 %) is above average.

Starting in 1988, tips and horizontal sections of the slopes have been successivley cultivated according to a planting schedule developed by Tharandt University of Forestry. A section of the tailings basement has been planted by coniferous and deciduous trees (pines, spruces, alders, beeches). The slopes have been covered with both spreading and grouping grass species together with brushwood fascicles for additional stabilisation. The covers on the slopes are drained by trenches and ditches to ease the run-off of surface water. In 1996, the engineered covers have been investigated at four locations to understand the effects of rootage. The measured permeabilities (from infiltration measurements according to DIN 18162) were in the range $4.4 \cdot 10^{-5}$ m s^{-1} and $6.6 \cdot 10^{-5}$ m s^{-1}. At the time of that study, the engineered covers have been in operation for about 5 years with dry densities of 1.21 g cm^{-3} and 1.7 g cm^{-3}.

Climate and general hydrologic setting

The site has a moderate climate. Average annual precipitation is about 750 mm. Since the beginning of quantitative observations, German Meteorological Office (DWD) recorded an absolute annual high of 1172 mm and an absolute annual low of 484 mm. Average annual temperature is about 7.6 °C. DWD gives recurrence intervals for heavy rainfall events listed in Table 1.

Table 1. Reoccurrence intervals for heavy rainfall at Schüsselgrund Mine Dump.

Repetition period in years	6	10	20	50	100
mm precipitation	70	80	93	109	122

Field experiments

The stability of the engineered covers of the slopes has been investigated under unsaturated (natural soil moisture) and saturated conditions in former geomechanical investigations. According to geomechanical calculations the stability of the engineered covers decreases with increasing penetration of the saturated zone. At a depth of 0.5 m, the stability limits will be reached. Such a penetration depth of the saturated zone may be possible only after a heavy rainfall event. In the course of a long-term performance assessment for the stability of the engineered covers, these modeling results have motivated irrigation experiments at four selected sections of the slopes, each covering an area of 20 m^2. As a scenario, a 100 year reoccurrence event was simulated (see Table 1). During the irrigation experiments, soil moisture was recorded continuously in a depth of 0.5 m. In addition, field measurements have been supplemented by laboratory experiments.

Irrigation studies

A 100 year reoccurrence rainfall event was simulated using irrigation equipment including a precipitation measurement gauge assuming 120 mm precipitation within 24 h ($N_{120/24}$). The water was provided in a 4 m^3 mobile tank above the irrigation fields. Irrigation was regulated by an irrigation computer. The amount of precipitation was divided into 10 min intervals using a digital rain recorder with 0.2 mm resolution. Soil moisture was measured by a frequency domain response (FDR) sensor in a depth of 50 – 60 cm. Selection of the irrigation fields intended to represent the variability in vegetation conditions of the slopes. The locations of the four selected sites, VF1 to VF4, are given in Figure 2.

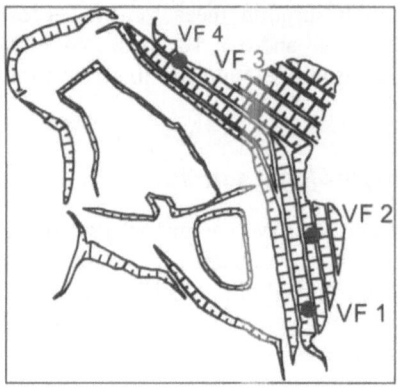

Fig. 2. Locations of the irrigation fields VF1 to VF4.

The water content (in vol-%) was recorded by the FDR sensor with a resolution of 0.1 vol-%. The resolution allowed to record minor variations in the water content during the irrigation period. Soil samples of the irrigated fields have been collected before and after the irrigation to obtain soil physical parameters of

- shear resistivity according to DIN 4096,
- density according to DIN 18125 – F 62,
- shear parameters (friction angle and cohesion) according to DIN 18137 Part 3 (E),
- grain size distribution according to DIN 18123,
- grain density according to DIN 18124,
- water content according to DIN 18121 T 1.

Density and water content of the engineered cover materials have been determined from two soil samples from each of the four irrigation fields taken from 30 – 35 cm, 50 – 55 cm and 50 – 60 cm, respectively. Shear resistivity, considered as a measure for stability of the engineered cover, has been obtained from one sample for each section taken in 50 – 60 cm depth. Because of a heavy natural rainfall immediately before starting the investigations, irrigation was reduced to 70 mm at section VF1.

Results and Discussion

Irrigation experiments

The soil in the irrigated sections consitst of silty fine grained sand to sandy silt. The clayey components important for the water binding capacity, has a relative contribution of 5 to 9 percent. In section VF 3 a higher content in gravel was found. Soils are classified as stiff to medium solidified. Furthermore, a decrease in soil moisture of 0.15 was found in the engineered cover going from the upper to the downer part of the slopes. A hydraulic gradient directing to the center of the mining waste has to be assumed because the dump material has a higher permeability compared to the engineered cover.

Evolution of soil moisture changes during irrigation

Results of continuous soil moisture measurements during irrigation are given in Figure 3 for the four irrigation sections. Soil moisture maxima have been observed for sections VF3 and VF4 after about 10 hours irrigation while soil moisture developed almost continuously for the other both sections. After irrigation, soil samples showed an increase in soil moisture between 1.7 vol-% (minimum; VF2) and 5.4 vol-% (maximum; VF3).

Table 2. Soil mechanic characteristics in 50 – 60 cm depth following irrigation.

irrigation field	VF 1		VF 2		VF 3		VF 4	
sample number	1	2	1	2	1	2	1	Nr 2
specific weight γ in kN/m³	20.4	20.2	20.1	20.7	18.8	19.3	20.1	20.7
angle of friction φ' in °	34.2		34.6		29.7		32.3	
cohesion c´in kN/m²	25.55		35.09		15.27		24.54	

The soil was classified as stiff. A certain undersaturation of the soils was indicated by comparing calculated densities and measured bulk densities for 50 cm – 60 cm depth, while no significant changes could be observed for cohesion and density.

Long term storage of radioactive waste

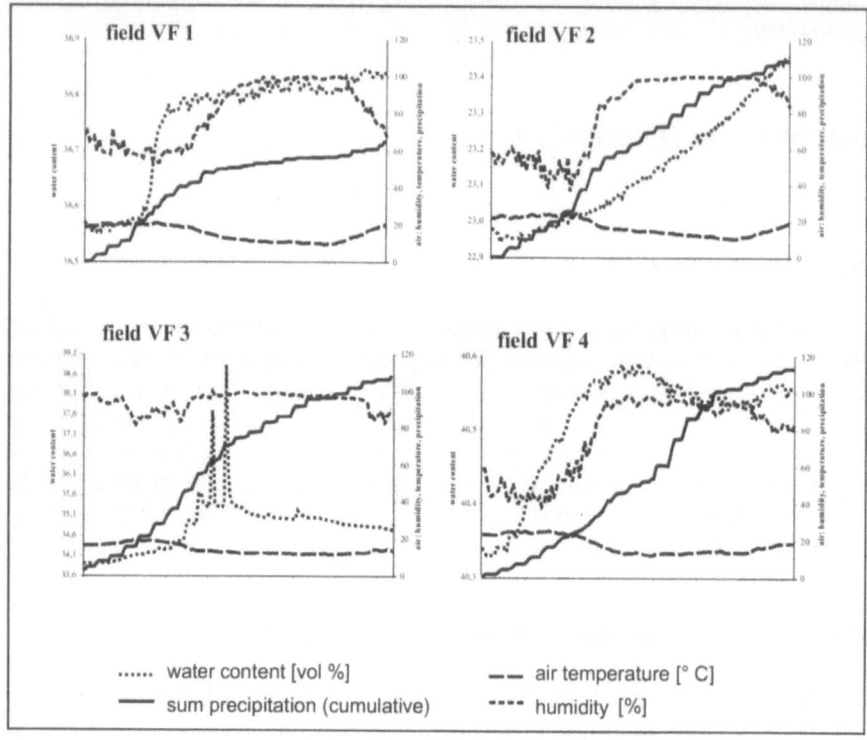

Fig. 3. Results of continuous soil moisture measurements during irrigation.

In the sloped sections (30 to 35 cm) a slight under saturation of the pore volume was found on basis of measurement results given in Table 3. After irrigation, a local saturation could be observed in some upper parts of the slopes causing surface runoff.

Table 3. Soil mechanic characteristics in 30 – 35 cm depth.

30 – 35 cm depth	VF 1	VF 2	VF 3	VF 4
dry density [g/cm³] before irrigation	1.488	1.8705	1.566	1.722
% water content before irrigation	24.95	12,7	24,55	17,7
pore volume	0.4404	0.2968	0.411	0.3761
calculated density at saturation g/cm³	1.93	2.17	1.98	2.10
measured density g/cm³	1.86	2.11	1.92	2.03

Due to surface run-off, the saturation zone of the slopes can be found in a depth between 0 cm and 35 cm. It is furthermore possible that in sandy locations observation of a saturation zone may fail as a consequence of local variability in material composition, that is indicative of a constant hydraulic gradient.

Long term storage of radioactive waste

Results of hydraulic modeling

Hydrological modeling using HYDRUS-2D was prepared to verify the hydraulic and hydrologic processes in the mineral soil cover for mean climatic conditions and for the heavy rainfall phenomena. The computer program HYDRUS-2D was developed by the Soil Physics group (Simunek, J, Sejna, M. and M. Th. Van Genuchten 1999) of the U.S. Salinity Laboratory, USADA-ARS, Riverside, CA. It is a software package for simulating water, heat and solute movement in two-dimensional variably saturated media. The model numerically solves the Richards` equation for saturated-unsaturated water flow and the convection-dispersion equation for heat and solute transport. The flow equation incorporates a sink term to account for water uptake by plant roots. Model calibration and validation was made with the measured soil mechanical and hydrologogic data. The results are shown in Figure 4.

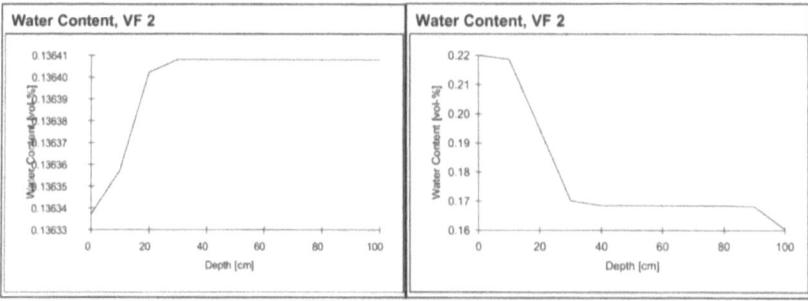

Fig. 4. Results of hydrologic modeling of heavy rainfall events at irrigation field 2. Diagram 1 before irrigation, diagram 2 after irrigation.

The results of the modeling confirm the measured data. Evaluation of experimental data indicates a stability of the covering layer even in case of a heavy rainfall event with reoccurrence of one in hundred years. During the investigations it has been found impossible to reach a complete saturation of the covering mineral soil layers.

Conclusions

Based on field and laboratory experiments several conclusions concerning the engineered slopes of Schüsselgrund Mine Dump can be pointed out:
- the results of the irrigation experiments simulating a worst-case scenario with regards to rainfall and soil moisture indicate stability of the covers even for an heavy rainfall event with reoccurrence of 100 years,
- locally, in the slopy sections of the engineered cover, erosion channels and partial land slides of limited extention may form during continuous precipitation periods,

Long term storage of radioactive waste

- in these sections, additional drainage and bioengineering have to warrant a sufficient protection against erosion of the engineered covers.

References

Schneider P, Osenbrück K, Nindel K, Voerkelius S, Forster M, Schreyer J (1999) Current and future impact of a uranium mine waste disposal site on groundwater. Proc Symp of the International Atomic Energy Agency on Isotope Techniques in Water Resources Development and Management, Vienna, 1999. IAEA-SM-361-20. Published on CD-rom S4_Schneider.pdf.
Schneider P, Voerkelius S, Nindel K, Forster M, Schreyer J (2001) Release of Contaminants from Uranium Mine Waste – Laboratory and Field Experiments. Journal of Mine Water and the Environment, 20(1): 30-38.

Hard pan formation on mining residuals

Dieter Rammlmair

BGR, Stilleweg 2, D-30655 Hannover, Germany E-mail: rammlmair@bgr.de

Abstract. Hard pan formation is a widespread natural phenomenon which is able to reduce erosion by wind and water, prevent water infiltration and air circulation due to particle agglutination and pore filling, and to accumulate high amounts of potentially toxic elements in secondary mineral and gel phases. The process is driven by capillary forces, occurs with different climates from the polar circle to the equator, and is depending on material composition, climate, and deposition history. The size of matrix flow cells and the development of focused rock drainage systems controls the effectiveness of this self-sealing process.

Introduction

Hardpan formation is a widespread natural phenomenon. It is known from semi-arid to arid climates such as calcrete, silcrete, ferricrete, but has been observed also in humid and tropical areas. These crusts are in general characterized by reduction of pore volume and permeability, relative or absolute increase or depletion of certain elements as well as by the formation of characteristic mineral phases stable under these environmental conditions (Thornber et al. 1987).

This natural phenomenon can be observed and characterized macro- and microscopically, chemically and geophysically already a few years after deposition on anthropogenic depositories like slagheaps, tailings, waste rock, etc. (Blowes et al. 1991; Chermak and Runnells 1996; Dold and Fontboté 2001; Rammlmair 1996; Schuiling & van Gaans, 1997, Ettner and Braastad 1999; Hammarstrom et al. 1999; McGregor et al. 1998; Jung et al. 2000; Grissemann et al. 2000; Rammlmair and Grissemann 2000; Rammlmair et al. 2000; Niederleithinger et al. 2000).

The processes of hardpan formation on anthropogenic depositories is of interest since the generation of hardpans will positively influence the natural attenuation processes in multiple ways. The process may last several to hundreds of years showing all stages from onset to final. The chemical and mineralogical inventory will be controlled by the precursor material, but will show individual patterns even in homogeneous matter.

The aim of this paper is to highlight the process of hardpan formation and to focus on some of the aspects supporting natural attenuation.

Field observations

A number of depositories – slagheaps, tailings, low grade ore etc.- have been investigated from the polar circle via humid, semiarid to arid climates to elucidate their potential in forming hardpans. Hardpans could be documented in all climatic zones and all materials ranging in size from the sub-mm- to meter- scale either at the surface or under a cover of a few cm of non-aggregated particles (Fig.1). The hardpans show an extreme variety in density ranging from incipient edge agglutination to almost 100% pore filling.

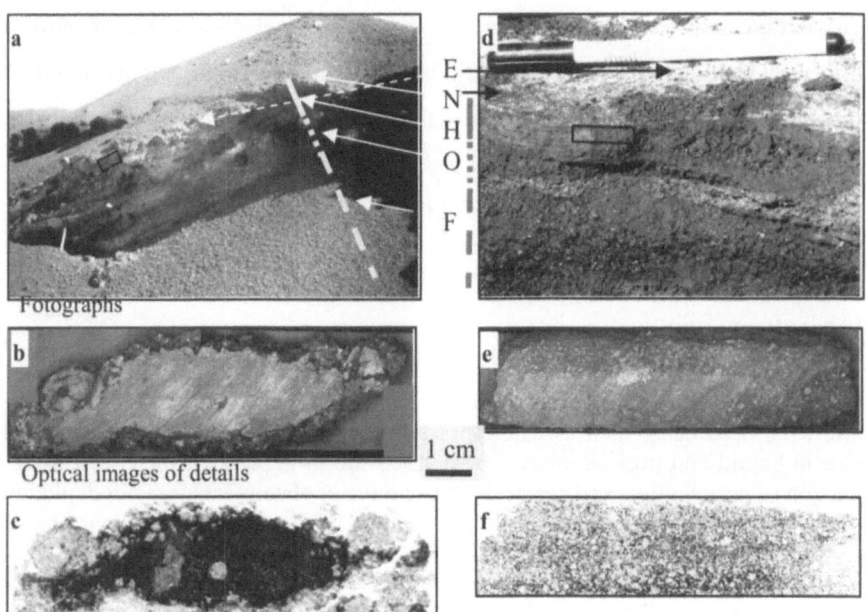

EDXRF-microscopy element distribution: Mo-Tube, 45 kV, 30 mA, 100µm step, 0.5 sec

Fig. 1. Example of hardpan profile on a slagheap (Selebi Phikwe, Botswana) showing efflorescence (E), not agglutinated particles (N), hardpan (H), an highly oxidized layer (O) and primary unoxydized material (F). At the flank within 10 yrs a 40 cm dense hardpan has evolved (a). On the lorry road the broken spheres are compacted, and a 3 cm thick dense hardpan was formed within one year (d). The optical image of detailed sections of the hardpans show coarse and fine grained slag particles from the flank (b, c) and the road (d, e), respectively, agglutinated by secondary Ni-Cu-phases or -gels. The inverted EDXRF-elemental scans show high Ni-Cu-concentration in dark, particles in grey (c,e).

Long term storage of radioactive waste

At the moment the scattered data do not allow to quantify the hardpan formation potential of a certain material in a certain climatic zone, but qualitative and process information can be already obtained at this stage of research.

Slag material often appears to be highly reactive depending on glass content and phase crystallinity and composition. This material might form already within one to ten years very compact hardpans.

On the other hand tailings derived from modern treatment plants often show very small amounts of reactive material only. The potential for hardpan formation might be limited or is far beyond effectiveness.

Alteration pattern

After deposition material will be exposed to weathering and diagenesis. The available and accessible surface area of the particles will interact with fluids, gases, bacteria, fungi and plants. Particles are composed of phases some of which are highly reactive, whereas others will not react in a visible way. Just after deposition dust particles and finest grains attached to coarser particles will provide elevated surface areas. The instantaneous accessibility of the micro-particles will cause expulsion in form of fluids highly enriched in soluble matter. Depending on the climatic premises and reaction time span, the generated solutions might be able to contribute to rock drainage.

With propagating alteration the accessibility and availability will become reduced by a number of factors such as loss of finest particles due to dissolution or particle transport in macro pore systems, loss of accessible highly reactive phases by selective leaching (Fig.2), shrinking cores in homogeneous matter, protective coatings and rims due to precipitation, reactions and bleaching. Thus, spontaneous dissolution will become reduced and diffusion will be the dominant control.

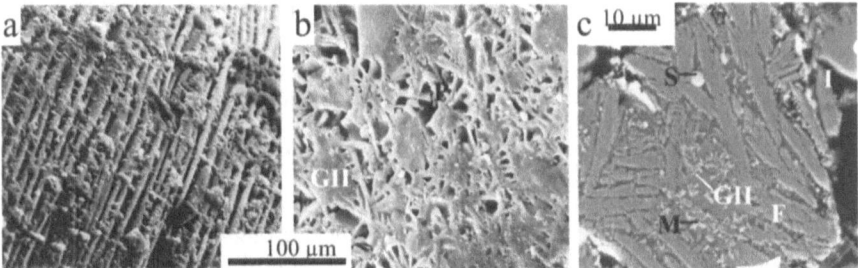

Fig. 2. Example of selective leaching of slag material from Selebi Phikwe, Botswana. (Leaching conditions 1minute in 30% H_2O_2 steam, 25°). SEM-SE images of fayalite strings with micro- channels of leached glass II from the slag sphere surface (a), and of plagioclase lath with pores from the core (b), and SEM-BSE image of polished section of a fragmentafter 150 days SOXHLET leaching at pH7 with fayalite (F), magnetite (M) sulfide spheres (S) and partially leached alkali-rich glass II.

Long term storage of radioactive waste

Since dumping of material occurs over a mine's lifetime, above dissolution steps will intermingle over a long time span.

The alteration of particles has an important side effect. Selective dissolution of phases or particles (Fig.2) might enhance the surface area up to a factor of 10,000. The retention capacity of water will increase substantially.

The evolution of micro-porosity and micro-channels due to selective leaching is further responsible for a strong increase of capillary force-driven transport, which again will speed up precipitation of gels, and secondary efflorescent phases at the wind and temperature controlled oscillating capillary rim.

Since reaction affects a number of phases the acidity and neutralization potential will compete in pH-micro-domains eventually causing spontaneous collapse of dissolved and colloidal matter in the form of gel-coating of particle surfaces or colloform-textured pore fillings.

These processes of dissolution, transport, reaction, and precipitation have to be regarded as highly dynamic and complex systems. They are concentrated in certain portions of a heap and may speed up certain processes and able to limit other ones.

The oscillating capillary rim will cause precipitation of dissolved matter over a wide zone in loose material at the flanks, where the wind pressure might be highly variable and within a narrow zone at the more compacted top of the heap where temperature variations might dominantly control evaporation.

Even most of the early precipitates are highly soluble and as efflorescent material often instantaneously available during later rainfalls hardpans still will evolve progressively with time. It is important to note that even highly soluble matter such as micro-crystallites of KCl or NaCl will survive rainfalls, due to the limited accessibility, generated by the precipitated matter in the evolving hardpans.

In the water saturated portion of a slagheap reducing conditions prevail, whereas in the under-saturated zones oxidation is the controlling feature. In this portion dissolution of a number of phases is easier to achieve than in the saturated part. Since the boundary zone in general is oscillating according to the daily and seasonal changes, availability of dissolvable matter is provided for an extended time period.

Fluid pathways

Depending upon their depositional history depositories might show significant inhomogeneities such as layering, crossbedding, changing degrees of sorting, compaction, consolidation, composition etc. These primary inhomogeneities will be diagenetically superimposed and are responsible for the development of micro-, meso- and macro-pore domains and therefore for highly variable hydraulic conditions.

A number of transport phenomena can be differentiated. The micro-pore or matrix flow takes place within the fine intragranular pore space, and appears to be relatively homogeneous, slow and achieves semi-equilibrium conditions due to the long reaction times. Besides the slow seepage an upward capillary transport in the inter-and intra-granular areas as well as along grain surfaces takes place.

On the other hand macro-pore transport occurs along pores, which are substantially larger, then the micro-pores. In depositories these macro-pore systems are provided by primary inhomogeneities during deposition and consolidation cracks or induced by earthquakes or landslides (Blowes et al. 1991; McGregor et al. 1998; Walder et al. 2000).

Selker et al. (1999) have shown that with a given gradient (gravity) the transport velocity in a macro-pore channel will increase with the square of the diameter, whilst the amount of the transported matter will increase by 10^4 relative to the surrounding matrix. But not all of the dissolved material will leave a depository via macro-pore channels as acid rock drainage (ARD). Part of it will feed the unsaturated matrix via capillary transport. In humid climate the matrix might be saturated and exchange might therefore be limited.

Macro-pore channels are additionally characterized by particle transport. Finest particles might form thin layers at the channel walls thus limiting the exchange with the confining matrix (Pivetz and Steenhuis 1995). This channel sealing process will be supported by evaporation from the matrix towards the macro-pores and precipitation of dissolvable and un-dissolvable matter as small hardpans speeding up the transport process, but limiting the reaction time.

Another aspect, which might appear to be of prime importance, is the development of so-called finger zones in tailings or other fine-grained wastes. Hill and Parlange (1976) showed that water spilled on dry and unstructured soil would follow distinct and isolated fingerlike zones. The diameter of these finger zones is indirectly proportional to the characteristic grain size distribution. In silt ($\emptyset$ approx. 0.01 mm) and in coarse sand ($\emptyset$ app. 1 mm) finger diameters of 100 cm and 2 cm, respectively will be achieved. This amount can be doubled under extreme water supply condition (Parlange et al. 1990).

Such a finger zone will remain active until the soil will become absolutely dry again, or saturated (Glass et al. 1998, and might last hundreds of years. These fingers are substantially harder then the un-weathered matrix (Selker et al. 1999). Such long-living drainage zones should - according to Walder et al. (2000) - be next to the sporadically apparent macro-pore channels basically responsible for ARD.

Onset and end of ARD is dependent on a number of factors, as Pactunc and Davé (2000, 2001) have demonstrated in long term column experiments. Long reaction times in the matrix contrast with short ones in the channels. The physico-chemical properties of fluids and material will therefore change permanently (Oelkers, 1996).

These heterogeneities can be observed in the field as well as in the laboratory, and can be described mineralogically, chemically and geophysically (Morris et al. 1995; Rammlmair 1996; Campell et al. 1999; Vaughn et al. 1999; Jung et al. 2000, 2001; Grissemann et al. 2000; Rammlmair and Grissemann 2000; Rammlmair et al. 2000, 2001; Niederleithinger et al. 2000).

Long term storage of radioactive waste

Structural reorganization

Heaped material, even if relatively homogeneous compared to the material before deposition, will evolve local in-homogeneities during the process of dumping, and undergo significant re-structuring during the diagenetic overprint. These sedimentologically and diagenetically superimposed structures have major influences on the fluid path dissolution, precipitation, and the water retention capacity. Interfering micro-, meso-, and macro-pore zones are responsible for the development of inaccessible cells in heaps. The dimension of such a cell might range from centimeters to several meters in diameter. Cells are bounded by more or less accessible portions of the heap. Based on this observation, several major zones which can develop vertically as well as laterally appear to be relevant for the structural development of a heap (Fig.3).

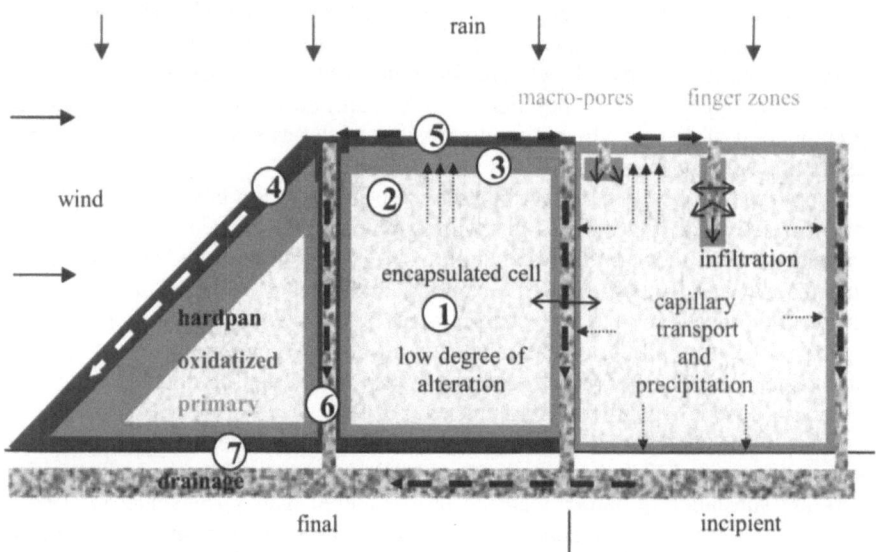

Fig. 3. Section through a slag heap showing different diagenetic stages. Zones of drainage, infiltration and capillary transport are shown in conjunction to the evolution of an oxidation rim and subsequent hardpan formation and encapsulation of cells (1-7; see text).

1. A relatively homogeneous core cell, where slow homogeneous matrix downflow predominates. This zone is controlled by long-term reaction and slow downward migration competed by upward capillary force driven transport of an almost equilibrated fluid.
2. A transition zone where cementation due to interacting fluids might cause precipitation due to sudden changes of physicochemical conditions e.g. changes of pH, fO_2. Precipitation of gels or secondary minerals might take place. If the conditions are stable over a certain time period layer like zones might develop.

Long term storage of radioactive waste

3. In a zone exposed to air where dissolution and capillary transport are the dominant features, local super-saturation might be achieved, according to the actual climatic conditions. Coating of fragments, edge and face agglutination and protected (coated) micro-channels might develop.

4. At the capillary rim precipitation is the dominant feature, resulting in coatings, and agglutination of particles, grain coarsening, and finally in a reduction in pore size and volume providing relatively dense crusts/ hard pans. Two types of hard pans can be distinguished for the same time interval on a heap. (A) A dense crust on the flat top of the heap, which is characterized by accumulation of all transported matter in a very narrow zone (Fig. 1d-f). (B) On the other hand a relatively thick, but porous crust might develop at the flanks of a heap (Fig. 1a-c). This crust is basically characterized by edge agglutinated fragments / grains. The crust on the flanks might reach up to 30 times the thickness of the dense top crust. The same amount of transported matter is dispersed over a 30 times higher volume than in the flat portion. This might be due to the changing wind pressure which controls the oscillating evaporation level at the flanks

5. The uppermost portion of the heap eventually covering an impermeable crust is characterized by intensive re-dissolution, removed fines due to erosion (wind, water), accumulation of nutrients for vegetation. The re-dissolved secondary phases, depending on the immediate availability of water, redistribute the solute into the heap or might feed the macro pore channel system.

6. A network of variable spacing characterized by leaching and fast down flow (finger zones and macro pore system) focusing into a drainage system, which penetrates the heap.

7. A bottom zone of mega pores, where total physical removal of fine-grained material has occurred, due to rapidly flowing water. As there is little time for reaction, basically Fe-hydroxide coating can be observed. The development of this zone is much dependent on the grain size distribution in a heap and of the morphological feature of the bedrock.

Conclusions

Hardpan formation appears not to be confined to one climatic zone. It has been observed from the polar circle to the equator. Even in a relatively humid climate heaped depositories provide semi-arid conditions due to their morphological features. Heaps are significantly more exposed to wind, sun and rain as their natural surroundings. Vegetation in this environment is in general limited due to missing soil or even prevented by toxicity of the environment. Endemic flora might evolve.

The depositional history is an important feature for the development of focused drainage via macro-pore channels and for the development of self-isolating zones of reduced exchange, which favor the capillary transport towards the channels and the heap surfaces.

Long term storage of radioactive waste

At the same heap different types of hardpans might evolve. At the flanks an oscillating zone up to several meters, depending on grain size, pore geometry and distribution, and climatic factors such as wind, sun, and rain exposure might evolve. Precipitation of secondary phases due to forced evaporation will occur over the wide zone of interaction. Starting with edge agglutination the filling of pores is a continuous process.

In contrast, at the heap top all the matter transported via capillaries focuses in a very narrow zone of a few centimeters only. The result after a few years might be a dense, almost waterproof hardpan.

The kinetics of hardpan formation is much dependent on the climatic condition, the sedimentological, mineralogical, and microbiological parameters. Almost waterproof, dense, one cm thin hard pans might develop within one year in highly reactive compacted fine grained slag. In contrast slimes with very small pyrite contents might show only first steps of hardpan generation even after 30 years. On the other hand blocky, low-grade ore with interstitial finer material might develop 15 cm thick highly resistive hardpans within some 500 years.

The formation of effective hardpans might be inhibited due to man-made disturbances, episodic rainstorms and wind erosion able to destroy the incipient crust, limited amounts of reactive material and fluids.

Acid rock drainage is an expression of the instantaneous availability of soluble primary matter and efflorescent secondary phases associated with the hardpan and accessible along the macro pore drainage channels. Acid rock drainage will persist till drainage channels are sealed or washed out and hardpans are dense enough to limit water introduction and therefore evaporation and precipitation.

Controls of the status quo of hardpan formation and effectiveness might be obtained by thermal imaging and various tomographic geophysical long-term monitoring techniques. It appears to be crucial to obtain information across a water infiltration cycle to be able to highlight zones of impermeability and of focused drainage.

Taking all observations into account a number of steps could be taken into consideration to speed up hardpan formation and therefore to reduce erosion, ARD and gas emissions. Surface compaction, grain size reduction or agglomeration, enhancement of reactive surfaces, bacterial disintegration, ARD recycling, and material mixing could be applied to one or the other depository.

References

Blowes DW, Reardon EJ, Jambor JL, Cherry JA (1991) The formation and potential importance of cemented layers in inactive sulfide mine tailings. Geochim. Cosmochim. Acta, 55, 965-978

Chermak JA, Runnells DD (1996) Self-sealing hardpan barriers to minimize infiltration of water into sulfide-bearing overburden, ore, and tailings piles. - In: Tailings and mine waste '96. Proceedings of the International Conference on Tailings and Mine Waste. 3: 265-273

Long term storage of radioactive waste

Campell DL, Horton RJ, Bisdorf RJ, Fey DL (1999) Some geophysical methods for tailings / mine waste work. In: Tailings and mine waste '99. Proceedings of the International Conference on Tailings and Mine Waste. 6: 35-43

Dold B, Fontboté L(2001) Element cycling and secondary mineralogy in porphyry copper tailings as a function of climate, primary mineralogy, and mineral processing. J. Geochem. Explor. 74: 3-55

Ettner DC, Braastad G (1999) Induced hardpan formation in a historic tailings impoundment, Roros, Norway. In: Tailings and mine waste '99. Proceedings of the International Conference on Tailings and Mine Waste. 6: 457-464

Glass RJ, Stenhuis TS, & Parlange JY (1989) Mechanism for finger persistence in homogeneous, unsaturated, porous media: theory and verification. Soil Sci. 148, 1: 60-70

Grissemann Ch, Rammlmair D, Siegwart C, Fouillet N (2000) Spectral induced polarisation linked to image analyses: A new approach. - In: D. Rammlmair, J. Mederer, Th. Oberthür, R.B. Heimann, and H. Pentinghaus (eds.), Applied Mineralogy in Research, Economy, Technology, Ecology and Culture, Proceedings of the 6th International Congress on Applied Mineralogy, ICAM 2000, Göttingen, Germany, 17-19 July 2000, Vol. 2: 561-564, A.A.Balkema / Rotterdam

Hammarstrom JM, Meier AL, Seal RRII (1999) Geochemistry and mineralogy of ochres associated with acid mine drainage at abandoned massive sulfide mine. In: Geological Society of America, 1999 annual meeting. Abstracts with Programs – Geol. Soc. Am. 31, 7: 224

Jung HG, Rammlmair D, Tacke KD, Tufar W (2001) Crust formation processes on Tailings Impoundments simulated by Column Experiments. In: Securing the Future, Proceedings of the International Conference on Mining and the Environment, Skellefteå, June 25- July 1 2001, Vol. 1: 327-335

Jung H, Tufar W, Rammlmair D, Bosecker K (2000) Self-organization of a mining dump in the Freiberg mining district, Germany. – In: D. Rammlmair, J. Mederer, Th. Oberthür, R.B. Heimann, and H. Pentinghaus (eds.), Applied Mineralogy in Research, Economy, Technology, Ecology and Culture, Proceedings of the 6th International Congress on Applied Mineralogy, ICAM 2000, Göttingen, Germany, 17-19 July 2000, Vol. 2: 579-582, A.A.Balkema / Rotterdam / Brookfield

McGregor RG, Blowes DW, Jambor JL, Robertson WD (1998) The solid-phase controls on the mobility of heavy metals at the Copper Cliff tailings area, Sudbury, Ontario, Canada. J Contam Hydrology 33, 3-4: 247-271

Morris WA, Hearst RB, Clarke MA (1995) Aplication of IP/resistivity and magnetic surveys for defining the acid generating potential of mine tailings. In: Sudbury'95, Conference Mining and the Environment, Sudbury Ontario May 28th-June 1, Proceedings:1027-1034

Niederleithinger E, Grissemann Ch, Rammlmair D (2000) SIP geophysical measurements on slag heaps) A new way to get information about subsurface structures and petrophysical parameters. In: D. Rammlmair, J. Mederer, Th. Oberthür, R.B. Heimann, and H. Pentinghaus (eds.), Applied Mineralogy in Research, Economy, Technology, Ecology and Culture, Proceedings of the 6th International Congress on Applied Mineralogy, ICAM 2000, Göttingen, Germany, 17-19 July 2000, Vol. 2: 607-610, A.A.Balkema / Rotterdam

Oelkers EH (1996) Physical and Chemical Properties of Rocks and Fluids for Chemical Mass Transport Calculations. - In: P.C.Lichtner, C.I.Steefel and Oelkers E.H. (eds.), Reactive Transport in Porous Media. Reviews in Mineralogy. 34, Ch. 3: 131-191

Pactunc AD, Davé NK (2000) Mineralogy of pyritic waste rock leached by column experiments and prediction of acid mine drainage. In: D. Rammlmair, J. Mederer, Th. Oberthür, R.B. Heimann, and H. Pentinghaus (eds.), Applied Mineralogy in Research, Economy, Technology, Ecology and Culture, Proceedings of the 6th International Congress on Applied Mineralogy, ICAM 2000, Göttingen, Germany, 17-19 July 2000, Vol. 2: 621-623, A.A.Balkema / Rotterdam

Paktunc AD, Davé NK (2001) A comparative Evaluation of Effluent Chemistry and Mineralogical Data from Column Leaching Experiments Using Reactive Tailings. 9p. In: Proceedings of the International Conference on Mining and the Environment) Securing the Future, Skellefteå June 25 – July 1 2001: 634-643

Parlange J-Y (1976) Capillary hysteresis and the relationship between drying and wetting curves. Water Resour. Res., 12, 224-228

Parlange J-Y, Glass RJ, Steenhuis TS (1990) Application of scaling to the analysis of unstable flow phenomena. In: Scailing in Soil Physics, Soil Science Soc. Am., Spec. Publ. 25, Madison, WI, 53-58

Pivetz BE, Steenhuis TS (1995) Soil matrix and macropore biodegradation of 2,4-D. J. Environ. Qual. 24, 564-570

Rammlmair D, Tacke KD, Jung H (2001) Application of new XRF-scanning techniques to monitor crust formation in column experiments. In: Securing the Future, Proceedings of the International Conference on Mining and the Environment, Skellefteå, June 25-July 1 2001, Volume 2, 683-692

Rammlmair D, Grissemann Ch (2000) Natural Attenuation in slag heaps versus remediation. – In: D. Rammlmair, J. Mederer, Th. Oberthür, R.B. Heimann, and H. Pentinghaus (eds.), Applied Mineralogy in Research, Economy, Technology, Ecology and Culture, Proceedings of the 6th International Congress on Applied Mineralogy, ICAM 2000, Göttingen, Germany, 17-19 July 2000, Volume 2:645-648, A.A.Balkema / Rotterdam

Rammlmair D (1996) The role of gels in self organisation of slagheaps from the arsenic production site, Muldenhütten at Freiberg, Saxony, FRG. In: A. Niedbalska, A. Szyman´ski and A. Wiewióra (eds.), Proceedings ICAM'96, Warsaw, Poland, 2-5 June 1996, ext. abstract: 378-382, SUBDAN-DRUK, Warsaw

Schuiling RD, van Gaans PFM (1997) The waste sulfuric acid lake of the TiO2-plant at Armyansk, Crimea, Ukraine. Part I. Self-sealing as an environmental protection mechanism. Appl. Geochem. 12: 181-186

Selker JS, Keller CK, McCord JT (1999) Vadose Zone Processes, Lewis Publishers, London, pp.339

Thornber MR, Bettenay E, Russel WGR (1987) A mechanism of alumosilicate cementation to form a hardpan. Geochim. Cosmochim. Acta, 51: 2303-2310

Vaughn RB, Stanton MR, Horton RJ (1999) A year in the life of a mine dump : A diachronic case study. In: Tailings and mine waste '99. Proceedings of the International Conference on Tailings and Mine Waste. 6: 475-484

Walder IF, Schuster PP, Lichtner PC (2000) Remediation and immobilization) Reactive transport modeling of mine wastes and the usefulness for cover design. In: D. Rammlmair, J. Mederer, Th. Oberthür, R.B. Heimann, and H. Pentinghaus (eds.), Applied Mineralogy in Research, Economy, Technology, Ecology and Culture, Proceedings of the 6th International Congress on Applied Mineralogy, ICAM 2000, Göttingen, Germany, 17-19 July, Vol. 2: 83-86, A.A.Balkema / Rotterdam

Long term storage of radioactive waste

Supercritical Carbon Dioxide Fluid Leaching (SFL) of Uranium from Solid Wastes Using HNO$_3$-tributylphosphate (TBP) Complex as a Reactant

Zenko Yoshida[1], Yoshihiro Meguro[1], Shuichi Iso[1], Jin Ougiyanagi[1,2], Osamu Tomioka[3], Youichi Enokida[3], Ichiro Yamamoto[4]

[1]Advanced Science Research Center, Japan Atomic Energy Research Institute, Tokai, Ibaraki 319-1195, Japan
[2]Department of Environmental Sciences, Faculty of Science, Ibaraki University, Mito, Ibaraki 310-8521, Japan,
[3]Research Center for Nuclear Materials Recycle, Nagoya University,
[4]Department of Nuclear Engineering, Nagoya University, Furo-cho, Chikusa-ku, Nagoya 464-8603, Japan

Abstract. Supercritical carbon dioxide (CO$_2$) leaching method (SFL), which is based on the efficient and selective dissolution of UO$_2$ and U$_3$O$_8$ with supercritical CO$_2$ containing HNO$_3$-tributylphosphate (TBP) complex at 333 K and 15 – 20 MPa, has been developed for the removal and recovery of uranium from the solid waste contaminated by uranium oxides. The decontamination factor of UO$_2$ or U$_3$O$_8$ of higher than 500 were attained by the recommended procedure, which was demonstrated using synthetic solid waste samples of a mixture of the uranium oxides and sea sand.

Introduction

Removal of radioactive contaminants from solid wastes such as various metallic, ceramic or organic materials and sludge or ash products from the nuclear waste treatment processes as well as mine tailings leads to safe and economical storage and disposal of the wastes. The decontamination treatment of the solid wastes, however, is not commonly conducted because of methodological limitations as follows; methods applicable to a large-scale treatment of the wastes are very limited and wet-chemical decontamination processes such as acid leaching process often generate large amount of the secondary wastes.

Recently, much attention has been paid to a novel technology for the separation of metals using the supercritical CO_2 fluid as a medium (Iso et al. 1995; Meguro et al. 1996; Lin et al. 1995; Toews et al. 1996; Erkey 2000). Supercritical CO_2 fluid extraction (SFE) with the supercritical CO_2 instead of an organic solvent has several advantages as follows. The extraction efficiency and rate are expected to be enhanced due to rapid mass transfer in the supercritical fluid phase. The rapid and complete recovery of the extracted substances from CO_2 is attained by gasification of CO_2. The extraction efficiency or selectivity can be optimized by changing optionally the properties of the CO_2 medium by tuning pressure and temperature (Meguro et al. 1998; Lin et al. 1995). The application of the SFE technology in the nuclear field exhibits particular significance for minimizing the amount of the radioactive solvent waste. The most attractive method based on the SFE is "direct extraction (leaching)" of the metals from the solid samples. Neither organic solvent nor acid solution for the pretreatment of the samples are necessary to be used in this method, which leads to a total minimization of the secondary wastes from the separation process (Tomioka et al. 2000, 2001a; Murzin et al. 1998).

The present paper aims at presenting the recent results of the supercritical CO_2 fluid leaching (SFL) of uranium from the solid wastes containing uranium oxides. The HNO_3-TBP complex was employed as a reactant to dissolve the uranium oxides efficiently and selectively into supercritical CO_2 phase.

Principle of the SFL of UO_2 and U_3O_8

The SFL method for removal and recovery of uranium from solid samples consists of two elementary processes, one of which is the dissolution of the uranium oxide with the HNO_3-TBP complex in the supercritical CO_2 and the other is the recovery of supercritical CO_2-soluble substances, which are $UO_2(NO_3)_2(TBP)_2$ formed and the HNO_3-TBP complex unreacted in the collector. The solutes, $UO_2(NO_3)_2(TBP)_2$ and the HNO_3-TBP complex, are transported with the supercritical CO_2 flow to the collector kept at an atmospheric pressure where the solutes are deposited completely by gasification of CO_2.

The oxides, UO_2 and U_3O_8, react with the HNO_3-TBP complex and the overall dissolution reactions are expressed by the following equations:

$$UO_2 + (8/3)HNO_3 + 2TBP \rightarrow UO_2(NO_3)_2(TBP)_2 + (2/3)NO + (4/3)H_2O$$
$$UO_3 + 2HNO_3 + 2TBP \rightarrow UO_2(NO_3)_2(TBP)_2 + H_2O$$

Tetravalent uranium U(IV) once dissolved in the supercritical CO_2 containing the HNO_3-TBP complex is liable to be oxidized to U(VI). It was found (Tomioka et al. 2001b) that dissolution rate increased with increasing the HNO_3/TBP ratio and the concentration of the reactant in the supercritical CO_2. The dissolution rate was practically independent on the temperature in the range of 313 to 333 K, and decreased slightly with an increase of the pressure in the range of 12 to 25 MPa. The HNO_3-TBP complex has multi-functions in promoting such elementary reaction steps involved in the dissolution reaction as protonation of the oxides, oxidation of U(IV) to U(VI) and solubilization of U(VI) in the supercritical CO_2 through a formation of $UO_2(NO_3)_2(TBP)_2$.

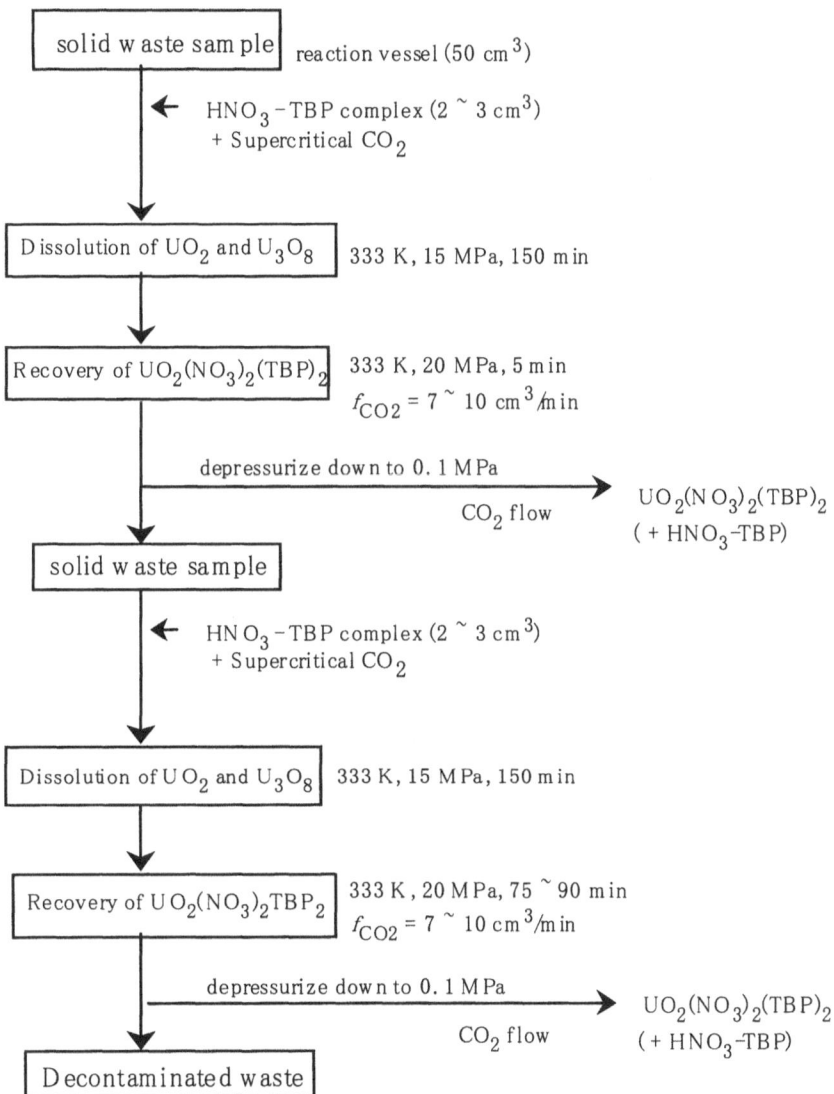

Fig. 1. SFL procedure for the decontamination of UO_2 and U_3O_8 from solid wastes.

Long term storage of radioactive waste

Recommended procedure

The flowsheet of the recommended procedure for the SFL of UO_2 and U_3O_8 from the solid waste sample which consists of two cycles of dissolution-recovery processes is shown in Fig. 1. The highest dissolution efficiency of UO_2 and U_3O_8 was observed at the pressure in the range of 12.5 to 17.5 MPa at 333 K, and the highest recovery efficiency was observed at 17.5 to 20 MPa (Meguro et al. 2002). The pressure in the recommended procedure was thus chosen to be 15 MPa for the dissolution process and 20 MPa for the recovery process.

In this flowsheet the flow rate of CO_2, f_{CO2}, is expressed as that inside the reaction vessel at 333 K and 20 MPa.

Experimental

Apparatus for the SFL was essentially identical to that reported previously (Tomioka et al. 2002). The main part of the apparatus consisted of a stainless steel reaction vessel (50 ml), which was installed in a thermostat oven (333 K), a reactant mixing vessel and a collector. Stainless steel balls of 3.2 mm in diameter were packed in the collector which was made of Pyrex glass cylinder of 300 mm in height and 10 mm in diameter with water jacket to keep temperature at 333 K. A syringe pump was used to flow CO_2.

The HNO_3-TBP complex was prepared by mixing vigorously 100 ml of 70% HNO_3 with 100 ml of TBP in a conventional extraction tube for 30 min. The HNO_3-TBP complex liquid thus obtained contained HNO_3 and TBP at mole ratio of 1.4 ~ 1.5 : 1 as the mixture of $(HNO_3)_2(TBP)$ and $HNO_3(TBP)$ complexes. The water content of the HNO_3-TBP complex was approximately 1.3 mol/l (Tomioka et al. 2001b).

Synthetic solid waste samples were mixture of ca. 100 ~ 200 mg of UO_2 or U_3O_8 powers and 20 ~ 50 g of matrix. The matrices employed were the standard sea sand (20 ~ 30 mesh), ash, and polyvinyl chloride (PVC) materials of various types such as powder, film and plate.

Results

Decontamination factors of uranium from the synthetic waste samples were determined according to the recommended procedure.

The decontamination factors obtained for the sample of the mixture of 100 mg UO_2 or U_3O_8 powders and 50 g sea sand were enough high and approximately 500. Most of uranium, i. e., 95 to 99%, contained in the sample was recovered in the collector.

The mixture of 25 g of PVC powders (Wako Pure Chemical Ind. Ltd, reagent grade PVC of about 0.2 mm in diameter) and 120 to 135 mg UO_2 or U_3O_8 powders was treated by the SFL. In this experiment, single cycle dissolution-recovery process consisting of the dissolution at 333 K and 15 MPa for 150 min and the recovery at 333 K and 20 MPa with CO_2 flow at f_{CO2} = 8 cm^3/min for 75 min was employed. The amount of uranium, which was present as a form of oxide and $UO_2(NO_3)_2(TBP)_2$ complex, remaining in the reaction vessel after the SFL was determined distinguishing the chemical form. It was found that 25 - 27% and 3 – 5% of uranium of the total uranium taken initially remained in the reaction vessel with the PVC powders as a form of uranium oxide and $UO_2(NO_3)_2(TBP)_2$ complex, respectively. The inhibition of the dissolution of the uranium oxides by the presence of the PVC was observed, which is attributable to the lowering of the oxidation potential or the proton activity arising from the HNO_3-TBP complex when the PVC coexists in the supercritical CO_2 phase. Further investigation is required to optimize the SFL condition for the decontamination of solid wastes of the PVC matrix.

The mixture of ash and uranium oxide powders was tested. In this experiment the synthetic ash sample was treated by the SFL after packed in an inner container made of porous alumina cylinder to prevent discharge of the fine ash particles from the reaction vessel to the down stream of the SFL system. Only 10 – 20% of uranium of the total uranium initially taken was removed from the ash sample. An appropriate procedure to enhance the mass transport between inside and outside of the alumina container during the dissolution and recovery processes is necessary to be developed by adopting such techniques as a "pressure-swing" method, which is now in progress.

References

Iso S, Meguro Y, Yoshida Z (1995) Chem. Lett. 1995, 365.

Meguro Y, Iso S, Takeishi H, Yoshida Z (1996) Radiochim. Acta 75, 185.

Lin Y, Smart N G, Wai C M (1995) Trends Anal. Chem. 14, 123.

Toews K L, Smart N G, Wai C M (1996) Radiochim. Acta 75, 179.

Erkey C (2000) J. Supercrit. Fluids 17, 259.

Meguro Y, Iso S, Yoshida Z (1998) Anal. Chem. 70, 1262.

Tomioka O, Enokida Y, Yamamoto I, Takahashi T (2000) Prog. Nucl. Energ. 37, 417.

Tomioka O, Meguro Y, Iso S, Yoshida Z, Enokida Y, Yamamoto I (2001a) J. Nucl. Sci. Technol. 38, 461.

Murzin A A, Babain V A, Shadrin A Yu, Smirnov I V, Romanovskii V N, Muradymov M Z (1998) Radiochemistry 40, 47.

Tomioka O, Meguro Y, Enokida Y, Yamamoto I, Yoshida Z (2001b) J. Nucl. Sci. Technol. 38, 1097.

Tomioka O, Meguro Y, Iso S, Yoshida Z, Enokida Y, Yamamoto I (2002) Proc. Internat. Solv. Extr. Conf. ISEC 2002, 1143.

Meguro Y, Iso S, Yoshida Z, Tomioka O, Enokida Y, Yamamoto I (2002) J. Supercrit. Fluids, submitted

Long term storage of radioactive waste

Release behaviour of radionuclides from contaminated concrete materials

Guido Deissmann[1], Stefan Thierfeldt[1], Adrian Bath[2], Stephan Jefferis[3]

[1] Brenk Systemplanung GmbH, Aachen, Germany
[2] Intellisci Ltd., Loughborough, UK
[3] Dept. of Civil Engineering, University of Surrey, Guildford, UK

Abstract. During the decommissioning and dismantling of nuclear facilities large quantities of radioactively contaminated concrete materials arise, which must be managed safely and cost-effectively. This paper provides an overview of a research project dealing with the development of source terms for the mobilisation of radionuclides from contaminated concrete. Key parameters to be taken into account comprise contamination sources/pathways, the physical, chemical, and mineralogical properties of the concrete and their alteration with time, as well as the environmental conditions depending on the chosen disposal/reuse option.

Introduction

During the service life of nuclear installations, concrete structures and building materials can become activated or radioactively contaminated by a variety of radionuclides through different mechanisms. As a result, during decommissioning and dismantling of the facilities, a considerable quantity of concrete material (e.g. concrete structures, rubble) arises, which is at least potentially contaminated with radionuclides and which must be managed safely and cost-effectively. The total amount of building materials arising from the decommissioning and dismantling of existing nuclear facilities in the European Community was estimated to be approximately 25×10^6 Mg up to the year 2150 (Deckert et al. 1999). Depending on the contamination of the material, different decommissioning options exists such as the direct reuse of building structures (with or without prior decontamination), the reuse of recycled building materials, the disposal of rubble in conventional landfills, or the disposal in special repositories for radioactive waste, which is by far the most costly disposal option.

In this context, the German Federal Ministry of Education and Research awarded a research project (BMBF Grant-No. 02S7900) that aims to improve knowledge concerning the release behaviour of radionuclides from contaminated concrete materials and the recycling/reuse of contaminated building materials from nuclear installations. One of the key issues addressed in this project is the development of source terms for the mobilisation of relevant radionuclides from contaminated concrete materials. These source terms might then be used for the modelling of radiological consequences (i.e. dose assessments) of disposal options for building materials from nuclear installations.

Properties of concrete

The behaviour of radionuclides in contaminated building materials depends on the nature and chemical behaviour of the contaminant, the source/pathway of the contamination, and the physical, chemical, and mineralogical properties of the material itself. Concrete is a structural material which consists generally of a binder (cement), water, sand and aggregate. The binder is typically a Portland cement (OPC) which comprises the four principal clinker phases tricalcium silicate (Ca_3SiO_5), dicalcium silicate (Ca_2SiO_4), tricalcium aluminate ($Ca_3Al_2O_6$), and calcium aluminoferrite ($Ca_4Al_2Fe_2O_{10}$), plus various other minor additives including gypsum, that react rapidly with the added mix water (hydration reaction). Aggregate forms the main component of concrete by volume/mass. Aggregates used in concrete fabrication comprise natural materials like gravel or crushed rocks (e.g. granite, basalt, sandstone) or artificial materials like blast-furnace slag. Heavyweight concrete used in parts of nuclear installations comprises additional high-density aggregates, especially barite and iron oxides (magnetite, hematite).

The chemistry of OPC hydration has been studied intensively over many years, due to the importance of cement and concrete materials to civil engineering. The chemical reactions during cement hydration are complex, because of the polymineralic nature of cement clinker, complex reaction kinetics, and the formation of hydration products with ill-defined stoichiometry. The calcium silicates react with water to form a non-stoichiometeric calcium silicate hydrogel (C-S-H), which is initially amorphous and develops some quasi-crystalline structures over time, along with some portlandite ($Ca(OH)_2$). The aluminate phases react with water and sulphate to form (iron-bearing) ettringite ($Ca_6Al_2[(OH)_4SO_4]_3 \cdot 26H_2O$) and analogous hydrated tricalcium aluminate ferrite tri phases (AF_t) and a hydrated dicalcium aluminate monosulphate phase (AF_m) Other minor solid phases present in cement paste gel are siliceous hydrogarnet ($Ca_3(Al, Fe)_2SiO_4(OH)_8$) and brucite ($Mg(OH)_2$). In addition, traces of unhydrated cement clinker phases can remain because of heterogeneity of the mix and the tendency of clinker particles to form coatings that inhibit further hydration, plus traces of gypsum and additives.

The excess of the aqueous phase that is not required for cement hydration forms the pore water of the fresh concrete and contributes to the development of submicron sized pores in the hardened concrete. Major aqueous components in cement pore water comprise the alkalis Na^+ and K^+, Ca^{2+} and the anions OH^- and SO_4^{2-} (Glasser 1997). The pH of the pore solution is at first controlled by small amounts of alkali hydroxides (NaOH, KOH) to very high values between 13 and 14. After leaching of the alkalis, the pH is buffered by $Ca(OH)_2$ and C-S-H around pH 12.5 (e.g. Taylor 1997). Cement pore waters are commonly slightly oxidising (Eh +100 to +200 mV), however pore waters in slag-rich concrete can be reducing (cf. Albinsson et al. 1996, Glasser 1997).

The main transport mechanism in cement and concrete materials is diffusion, due to the low permeability of the porous matrix. The permeability of good construction concrete is commonly in the order of 10^{-12} to 10^{-11} m/s, depending on the porosity and the amount of shrinkage cracks developed during hydration and curing. However transport properties may change with time due to mechanical and chemical degradation of concrete material. Mechanical degradation of concrete can occur due to natural processes (e.g. by freeze/thaw cycles, expansion at the cement/aggregate interface) or can be technically induced in-service or during demolition and handling (e.g. crushing, thermal treatment etc.).

Chemical degradation processes which can reduce the strength of concrete structures and alter the porosity/permeability and the internal chemical state comprise (i) leaching of alkalis and calcium hydroxide, (ii) formation of expansive minerals such as ettringite, thaumasite, or brucite due to sulphate and magnesium attack, and (iii) carbonation, i.e. replacement of portlandite by calcite, associated with a decrease in pH. The nature of potential chemical degradation processes depends on the (geo)chemical environments of concrete structures and of the subsequent disposal sites for concrete materials.

In particular, chemical degradation processes may considerably alter the mobility and transport properties of radionuclides in concrete e.g. by the formation of cracks, changes in porosity (due to leaching or sealing of pore space), the decrease in pH and ionic strength of pore fluids, the dissolution of matrix minerals, the incorporation of contaminants in the crystal structure of newly formed precipitates, or the increase/decrease in internal surface area and potential sorption sites due to mineral precipitation/dissolution.

Radioactive contamination of concrete

Radioactive contamination of concrete from nuclear installations can have a number of sources and comprise a variety of radionuclides. Air-borne contamination can occur, e.g. due to leakage of steam in nuclear power reactors, dispersion of contaminated particulates, accidental release of UF_6 at enrichment or fuel fabrication plants, fall-out/wash-out from nuclear power plant emissions to the atmosphere, or can be present as daughter products of gaseous ^{222}Rn, e.g. at ore processing sites.

Water-borne contamination, either as dissolved solutes or as colloids or other suspended particles, can be introduced into concrete materials either from the design-use of the concrete, e.g. as tank linings, pipework, etc. and especially by spillage of contaminated water or process solutions onto general concrete structural elements and floors. This type of contamination can involve nuclides in virtual any combination, depending on the type of nuclear facility.

In-situ production of radionuclides in concrete (activation) can occur when it is exposed to neutron flux, for example in concrete shielding of reactors (biological shield) and other neutron sources as well as accelerators. Several elemental (trace) components of concrete and the steel reinforcement have significant cross-sections to fluxes of high energy and, especially, thermal neutrons. Thus some of these components may absorb the neutrons and undergo nuclear transformations to form radioactive nuclides such as ^{41}Ca, ^{54}Mn , ^{55}Fe, or ^{60}Co by activation of their stable nuclides in concrete materials.

Independently of the primary contamination pathway, radionuclides that have been deposited onto a concrete surface as dry particles or in solution or that have formed in situ, may then be remobilised in aqueous or other solutions to some degree. Thus surface contamination of concrete can be redistributed over time into the concrete matrix by diffusion through continuously water-filled pores or especially in cracks.

The spectrum of relevant radionuclides present in contaminated concrete can vary greatly depending on the type of nuclear facility. Important nuclides present in contaminated concrete from nuclear power plants include ^{60}Co, ^{90}Sr, ^{129}I, ^{241}Am but may vary depending on the type of reactor (e.g. BWR, PWR, gas cooled reactors, etc.). Fission products such as Cs, Sr, Ni, I, Eu and actinide elements (e.g. Pu, Np, Cm) are likely to be more prevalent where spent nuclear fuel is stored or processed. U, Th, and Ra are naturally-occurring radioelements that occur as contamination in uranium ore processing facilities, fuel fabrication facilities and also at spent fuel handling facilities. Radionuclides addressed in detail during the first stage of the project and their approximate half lifes are summarised in table 1.

The solid speciation of the radioactive contaminants in the concrete is variable depending on the chemical behaviour of the radioelements at high pH. Some radionuclides may be precipitated under alkaline conditions as hydroxides such as $Eu(OH)_3$, $Th(OH)_{4(am)}$ (e.g. Brookins 1989). Uranium can form calcium uranates $(CaUO_4$ or $CaU_2O_7)$ or calcium uranium silicates in concrete, depending on the availability of silica (Glasser 1997). Ni may be bound in cementitious systems by the formation of Ni- and Al-containing hydrotalcite-like layered double hydroxides (Scheidegger et al. 2000). The alkaline earth elements Sr and Ra can substitute for Ca in C-S-H or can be incorporated in calcite formed during carbonation of the concrete. Sorption especially to C-S-H is another important mechanism for the fixation of radioelements such as Pu, Am, Np, Ni, Cs, I and others (e.g. Heath et al. 1996, Aggarwal et al. 2000). The sorption potential of C-S-H depends on the Ca/Si ratio. At high Ca/Si-ratios, C-S-H has a positive surface charge favouring the sorption of anionic species, whereas at low Ca/Si-ratios the surface charge becomes negative improving the sorption of cationic species. Due to the high pH of cement/concrete pore waters, the dominant aqueous species of radioelements in the pore water are significantly different from the species expected in "normal" ground or surface waters. Table 2 provides an overview on important aqueous species of the relevant radioelements in cement pore waters.

Table 1. Nuclides of the elements of interest that may occur as contaminants in concrete from nuclear facilities.

Nuclide	approximate half-life
^{3}H	12.3 a
^{60}Co	5.3 a
^{59}Ni, ^{63}Ni	7.6 x 10^4 a, 96 a
^{90}Sr	29.1 a
^{129}I	1.6 x 10^7 a
^{137}Cs, ^{134}Cs, ^{135}Cs	30.2 a, 2.1 a, 3.0 x 10^6 a
^{154}Eu, ^{155}Eu, ^{152}Eu	8.6 a, 4.8 a, 13.6 a
^{226}Ra, ^{228}Ra	1.6 x 10^3 a, 5.7 a
^{227}Ac	21.7 a
^{232}Th, ^{230}Th, ^{228}Th	1.4 x 10^{10} a, 7.7 x 10^4 a , 1.9 a
^{231}Pa	3.3 x 10^4 a
^{238}U, ^{235}U, ^{234}U	4.5 x 10^9 a, 7.0 x 10^8 a, 6.5 x 10^3 a
^{237}Np	2.1 x 10^6 a
^{239}Pu, ^{240}Pu, ^{241}Pu, ^{238}Pu	2.4 x 10^4 a, 6.5 x 10^3 a, 14 a, 88 a
^{241}Am, ^{243}Am	430 a, 7.4 x 10^3 a
^{242}Cm, ^{243}Cm, ^{244}Cm	0.4 a, 29 a, 18.1 a

Table 2. Important aqueous species of radioelements of interest at high pH.

Element	important species at high pH
H	OH^-, H_2O
Co	$Co(OH)_2^0$, $Co(OH)_4^{2-}$
Ni	$Ni(OH)_2^0$, $Ni(OH)_3^{-1}$, $Ni(OH)_4^{2-}$
Sr	Sr^{2+}
I	I^-, IO_3^-
Cs	Cs^+
Eu	Eu^{3+}, $Eu(OH)_3^0$, $EuCO_3^+$
Ra	Ra^{2+}, $Ra(OH)^+$
Ac	$Ac(OH)_3^0$, $Ac(OH)_4^-$
Th	$Th(OH)_4^0$
Pa	$Pa(OH)_5^0$, $Pa(OH)_6^-$
U	$UO_2(OH)_3^-$, $UO_2(OH)_4^{2-}$, $UO_2(CO_3)_3^{4-}$
Np	$Np(OH)_4^0$, NpO_2OH^0, $NpO_2(CO_3)_2^{3-}$, $Np(OH)_4CO_3^{2-}$
Pu	$PuO_2(OH)_3^-$, $Pu(OH)_4^0$, $PuO_2CO_3^-$, $PuO_2(CO_3)_2^-$, $PuO_2(CO_3)_3^{4-}$, $PuCO_3^+$
Am	$Am(OH)_2^+$, $Am(OH)_3^0$, $Am(CO_3)_2^-$, $Am(CO_3)_3^{3-}$
Cm	$Cm(OH)_2^+$, $Cm(OH)_3^{0-}$, $Cm(CO_3)_2^-$, $Cm(CO_3)_3^3$

Radionuclide mobilisation

The process by which a particular radionuclide might be released from contaminated concrete depends on a number of factors such as the original source of the nuclide and especially its form on/in the concrete, .which might change with time due to subsequent alteration of the concrete, resulting e.g. in the creation of new host phases or the mobilisation of nuclides into the pore solution. The rate of nuclide release can be affected by mechanical degradation of concrete surfaces because dissolution or leaching will be accelerated as the degradation enhances the ratio of surface area to volume.

The release of radionuclides from contaminated concrete (or other matrices) involves a number of processes including (i) dissolution of a radionuclide-containing solid phase, (ii) desorption of a radionuclide from a surface, or (iii) leaching of radionuclides from a solid matrix without disrupting the structure of the solid host. The presence of water is a requirement for all these processes, whose rate depends (inter alia) on the chemical conditions in solution, the continuity of the aqueous phase (e.g. in the pore space), and the flux of water. The suitability of a number of conceptual models that can be adopted to represent the processes going on during the release of radionuclides is presently being investigated.

The release of radionuclides by dissolution of particles refers e.g. to dust grains on the concrete surface that contain the radionuclide, or newly-formed minerals in which a radionuclide has been (co-)precipitated. Dissolution involves the transfer of a solid host phase into solution, and is thus controlled by the solubility and other chemical properties of the host solid rather than those of the radionuclide itself. When the dissolution rate is controlled by diffusion of solutes away from the solution-solid interface, the rate of dissolution of a small spherical particle is of the form:

$$\frac{\partial C}{\partial t} = \frac{A\,D\,(C_s - C)}{\ell}$$

where t is time, A the surface area, D the diffusion coefficient of solute in the solution boundary layer, C the concentration of solute in general solution, C_s the concentration of solute at the dissolution interface (i.e. saturation concentration), and ℓ the thickness of the boundary layer at the solution-solid interface. If the rate of dissolution is controlled by mass transfer from solid phase to liquid phase, and the two phases are close to equilibrium, the dissolution rate can be described by:

$$\frac{\partial C}{\partial t} = kA\left(1 - (C/C_s)^p\right)^q$$

where k is the reaction rate and p and q are fitted parameters.

Radionuclides that are sorbed on the outer or inner surface of concrete will be desorbed into contacting water by a number of mechanisms. The concentration of a desorbed radioelement in a contacting solution that is equilibrated is:

$$C_l = C_s \cdot K_d$$

where C_l is the concentration in solution, C_s the concentration sorbed on the solid phase, and K_d the distribution coefficient for the radionuclide between solid and liquid phase under given conditions. In reality, desorption is more complex because equilibration between sorbed radionuclides and solution may not be instantaneous but kinetically controlled and sorption may not be fully reversible.

Leaching describes a process by which the contaminant is removed into solution, e.g. by dissolution or desorption of a contaminant from the internal matrix of the solid and out-diffusion leaving an intact matrix. An appropriate model to describe this process is the "shrinking core" model (e.g. Baker and Bishop 1997). The conceptual basis of this model is the production of an altered outer layer of the solid matrix, through which solutes must diffuse from the residual unreacted inner core.

The diffusive transport of radionuclides, which have been released into the pore fluid, through the porous solid (e.g. concrete) towards the external surface can be described by Fick's Law:

$$\frac{\partial C}{\partial t} = D_a \frac{\partial^2 C}{\partial x^2}$$

where D_a is the apparent diffusion coefficient and x the distance. The apparent diffusion coefficient D_a incorporates the effects of porosity, pore tortuosity, and the contaminant retardation (e.g. by sorption), etc:

$$D_a = D_f \cdot \phi \cdot \frac{\varepsilon}{\varepsilon + \rho K_d}$$

where D_f is the free water diffusion coefficient, ϕ is a "geometric factor" including allowance for constrictivity and tortuosity, ε is porosity, ρ is bulk density and K_d the distribution coefficient.

Concluding remarks

The aim of the project presented here is the improvement of knowledge regarding the release of radionuclides from contaminated concrete from nuclear installations, taking into account especially the chemical behaviour and speciation of the radioactive contaminants within the concrete, the chemical environment and the hydraulic regime within the concrete, and concrete degradation processes and their influence on contaminant mobilisation. The detailed analysis of source terms and the modelling of long-term radionuclide mobilisation from contaminated concrete will be employed especially for

- the re-evaluation of generic clearance levels for contaminated building materials from nuclear facilities,
- the modelling of radiological consequences (i.e. dose assessments) of disposal options for building materials from nuclear installations and the optimisation of the disposal process (i.e. selection of cost-effective and reasonable disposal options), and
- the assessment of recycling/reuse options of slightly contaminated materials in order to reduce the amount of waste for disposal.

In addition, the results of the project may be useful for the evaluation of the long-term behaviour of cementitious waste forms (especially of LLW) and the performance assessment of cement solidification/stabilisation for non-radioactive wastes or naturally occurring radioactive materials (NORM), but the transferability of the derived source terms to other applications has to be evaluated carefully in the future.

References

Aggarwal S, Angus MJ, Ketchen J (2000) Sorption of radionuclides onto specific mineral phases present in repository cements. NIREX Safety Studies Report NSS/R312, UK Nirex, Harwell.

Albinsson Y, Andersson K, Börjesson S, Allard, B (1996) Diffusion of radionuclides in concrete and concrete-bentonite systems. J Contaminant Hydrology 21: 189-200.

Baker PG, Bishop PL (1997) Prediction of metal leaching rates from solidified/stabilized wastes using the shrinking unreacted core leaching procedure. J Hazardous Materials 52: 311-333.

Brookins DG (1989) Aqueous geochemistry of rare earth elements. Reviews in Mineralogy 21: 201-225.

Deckert A, Thierfeldt S, Kugeler E, Neuhaus I (1999) Definition of clearance levels for the release of radioactively contaminated buildings and building rubble. Final report Contract No C1/ETU/970040.

Glasser FP (1997) Fundamental aspects of cement solidification and stabilisation. J Hazardous Materials 52: 151-170.

Heath T, Ilett DJ, Tweed CJ (1995) Thermodynamic modelling of the sorption of radioelements onto cementitious materials. Scientific Basis for Nuclear Waste Management XIX: 443-449.

Scheidegger AM, Wieland E, Scheinost AC, Dähn R, Spieler P. (2000) Spectroscopic evidence for the formation of layered Ni-Al double hydroxides in cement. Environ Sci Technol 34: 4545-4548.

Taylor HFW (1997) Cement chemistry, 2nd edition. Thomas Telford Publishing, London.

Solubility of an uranium(IV) amorphous phase under geochemical conditions representative for the direct disposal of spent nuclear fuel in Boom Clay

Grégory Delécaut[1], Pierre De Cannière, Lian Wang and Norbert Maes

[1] SCK•CEN, Waste and Disposal, Boeretang 200, B-2400 Mol, Belgium

Abstract. The solubility of hydrous UO_2 was studied in Boom Clay porewater under in situ conditions (reducing conditions, $pH \sim 8$, $\log f\,CO_2$ -2.4). Dithionite, iron powder and sulphide were used to ensure reducing conditions and to minimise the risk of uranium(IV) oxidation during the experiments. The average total uranium concentration was 3.9×10^{-6} mol·l^{-1}. The determined solubility is generally higher than the values reported in the literature but the effect of the natural organic matter present in the Boom Clay interstitial water is not known.

Introduction

During more than 20 years, the Belgian policy concerning the management of the spent nuclear fuel has been favourable to reprocessing. However, these last ten years, the alternative of the direct disposal of spent fuel is envisaged as well. At present, the disposal in deep geological formations is considered at the international level as the most promising option for the management of high level and long-living nuclear waste. In Belgium, research is concentrated on argillaceous formations and the Boom Clay is the candidate host rock presently studied. The Boom Formation is characterised by its high plasticity, a low hydraulic conductivity, strongly reducing (pyrite, natural organic matter) and slightly alkaline ($pH \sim 8.2$) conditions, a high specific surface and a high cation exchange capacity. Due to these properties, the migration of radionuclides in Boom Clay is very slow and mainly controlled by diffusion (advection plays only a negligible role).

Performance assessment studies on the direct disposal of spent fuel have indicated that uranium isotopes and their daughters would be the main contributors to the total dose rate at very long-term (after ~ 1 million years). Therefore, it is particularly important to correctly understand the uranium migration in the clay formation, the main barrier of the whole disposal system. The migration behaviour of radionuclides in the Boom Clay is determined by their solubility, their interactions with inorganic and organic solid surfaces and their complexation with naturally occurring dissolved inorganic (OH^-, CO_3^{2-}) and organic ligands. In addition, uranium is a redox sensitive element and its oxidation state controls all its geochemistry. Therefore, the reducing conditions prevailing in the Boom Formation limit on its solubility. According to our geochemical calculations, uranium should be mainly present in its tetravalent state under the reducing conditions measured *in situ*, E_h between –250 and –400 mV *vs* SHE (about –300 mV if pyrite is the E_h controlling phase). As a consequence, the uranium concentration in Boom Clay porewater would be solubility controlled and the neutral species $U(OH)_4(aq)$ would be the dominant aqueous species of uranium (Fig.1).

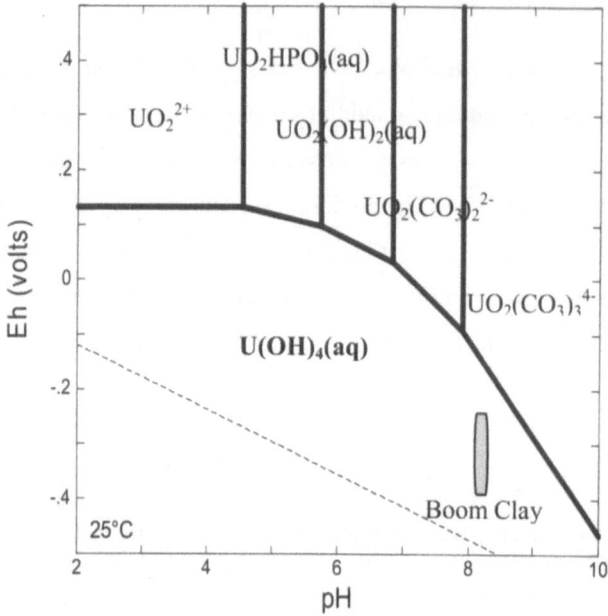

Fig. 1. E_h-pH diagram of uranium aqueous species for the Boom Clay porewater system. Code: *The Geochemist's Workbench*-3.2.2 (Bethke, 2001); NEA database (*Grenthe et al.*, 1992).

Under the influence of α-radiolysis, a partial oxidation and dissolution of the UO_2 matrix, the main component of spent fuel, and the subsequent uranium(VI) mobilisation will occur. Once mobile uranium will reach the undisturbed clay where strongly reducing conditions are prevailing, it would reprecipitate in a poorly crystallised form. So, the solubility study of the uranium(IV) amorphous phase is important to determine the maximum uranium concentration in the interstitial Boom Clay water in direct contact with the spent fuels.

This work presents the results of a first series solubility experiments with uranium(IV) amorphous precipitate in Boom Clay water in absence or in presence of different reducing agents.

Experimental

The solubility tests were carried out in duplicate at 20 °C in a glove box with a controlled Ar - 0.4 % CO_2 atmosphere to simulate the *in situ* partial pressure of CO_2 (log $f CO_2$ = -2.42). The solubility was approached by the undersaturation direction since the amorphous uranium(IV) precipitates were suspended in Boom Clay porewater sampled from the EG/BS piezometer in the HADES underground research laboratory at Mol. These suspensions were equilibrated for one week. The composition of the interstitial water is given in the Table 1. The porewater was equilibrated with the CO_2 partial pressure of the glove box by bubbling with the working gas (Ar – 0.4 % CO_2) overnight. The experiments were conducted in absence or in presence of reducing agents: iron powder, sulphide (Na_2S) and dithionite ($Na_2S_2O_4$). All solutions were freshly prepared under anaerobic conditions from degassed Milli-Q water (18 MΩ·cm) and further degassed by bubbling with Ar just before use.

Table 1. The Boom Clay porewater composition used for solubility experiments and for the modelling.

Cations	mg·l⁻¹	mol·l⁻¹	Anions	mg·l⁻¹	mol·l⁻¹
Na^+	400	1.7×10^{-2}	HCO_3^-	730	1.2×10^{-2}
K^+	8.6	2.2×10^{-4}	Cl^-	25.4	7.2×10^{-4}
Mg^{2+}	2.6	1.0×10^{-4}	F^-	2.8	1.5×10^{-4}
Ca^{2+}	3.1	7.7×10^{-5}	HPO_4^{2-}	2.4	2.5×10^{-5}
Fe	0.8	1.4×10^{-5}	Br^-	0.55	6.9×10^{-6}
B	7.9	7.3×10^{-4}	SO_4^{2-}	< 0.25	$< 2.6 \times 10^{-6}$
Si	2.2	7.8×10^{-5}	NO_3^-	< 0.25	4.0×10^{-6}
Al	< 0.2	$< 7.4 \times 10^{-6}$	TOC	103	—

To perform the precipitation of a suitable uranium solid phase, uranium(IV) is prepared by electrochemical reduction of natural uranium(VI) in air at a constant cathodic potential. This technique presents the main advantage that the introduction of undesirable chemical impurities in the solution is avoided. The electroreduction is performed in 1M HCl background electrolyte since the uranium(VI) reduction is promoted in acidic medium as described by the reaction

$$UO_2^{2+} + 4H^+ + 2e^- \rightarrow U^{4+} + 2H_2O$$

Moreover, unlike HNO_3, H_2SO_4 and $HClO_4$, HCl is not an oxidised acid. The optimal potential, derived from polarisation curves, is –200 mV vs Ag/AgCl reference electrode. At lower potentials the reduction of water starts and $H_2(g)$ is formed at the cathode. The potential at the working electrode is continuously measured with respect to the reference electrode and adjusted to the desired value by the potentiostat. Platinum electrodes are used as working electrode (cathode) and counter electrode (anode) and the reference electrode is an Ag/AgCl electrode with $E_0 = +210$ mV vs SHE (25°C).

The uranium(VI) solution was prepared by weighing 0.136 g of $UO_2(NO_3)_2 \cdot 6H_2O$, removing nitrate by heating the salt in concentrated HCl and dissolving the residue in 27 ml 1M HCl to obtain a 10^{-2} M uranium(VI) solution. If this salt was directly dissolved in 1M HCl, the presence of nitrate (2×10^{-2} M) could introduce interferences since it can be reduced itself. To avoid any possible complications, nitrate had to be removed before electroreduction. So, uranyl was first dissolved in 3 ml 38 % HCl and the solution was evaporated under an infra-red-lamp. The process was repeated three times and then the solid residue was finally dissolved in 1M HCl. This solution was electroreduced for 25 hours at -200 mV vs Ag/AgCl reference electrode under constant stirring to enhance the transport of the reactive species towards the electrodes. The oxidation state of uranium was controlled by UV-Vis spectrophotometry. The UV-Vis spectra show the disappearance of the characteristic absorption band of uranium(VI) at 414 nm and show the typical absorption bands of uranium(IV) at 428, 495, 549 and 648 nm (Fig.2).

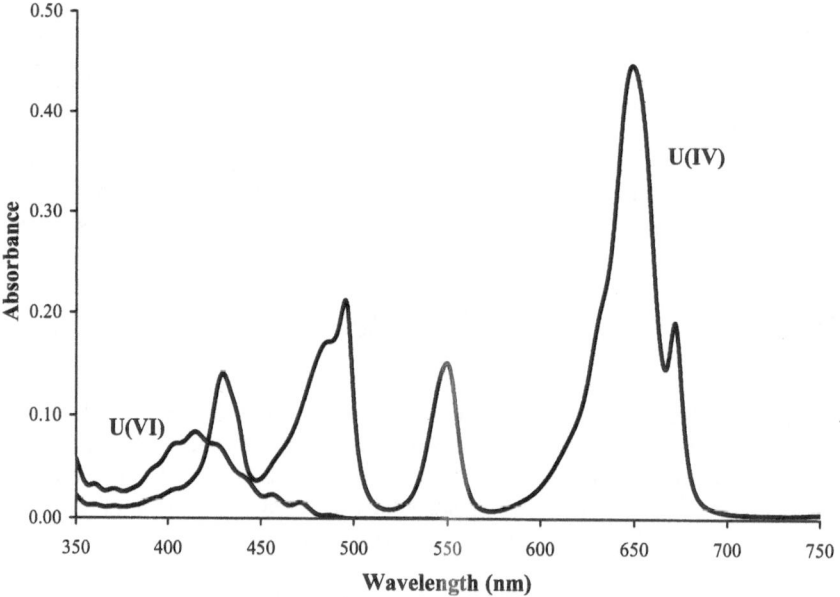

Fig. 2. UV-Vis spectra of 10^{-2} M natural uranium solution in 1M HCl before and after 25 hours of electroreduction at -200 mV vs Ag/AgCl electrode.

Prior to its introduction inside the anaerobic glove box, the uranium(IV) acid solution was degassed by bubbling with Ar during 1 hour. *Nalgene* centrifuge tubes (16 ml) were filled with 3 ml of the 10^{-2} M uranium(IV) solution which was afterwards titrated by an equivalent amount of degassed 1M NaOH (prepared inside a N_2 glove box to avoid carbonatation). A greenish solid appeared in all tubes. To ensure reducing conditions during the precipitate ageing, 4 ml degassed 5×10^{-3} M $S_2O_4^{2-}$ solution (pH 12) were added to the 6 ml of the neutralised solution. The amorphous phase was allowed to age overnight. The suspensions were centrifuged for 15 min at 8 500 g and the resulting solid phase was washed two times with 10 ml degassed 5×10^{-3} M $S_2O_4^{2-}$ solution at pH 12 inside the glove box. To avoid oxidation during the centrifugation step performed outside the glove box, the 16 ml centrifuge tubes were inserted inside a 50 ml centrifuge tube (*Falcon*) and the interstice was filled with degassed Milli-Q water.

Several sets of experiments were conducted with these washed precipitates and Boom Clay porewater (Table 2):
- without addition of reducing agents,
- with dithionite (5×10^{-3} M $S_2O_4^{2-}$),
- with sulphide (10 mg·l^{-1} S^{2-}),
- with iron powder (1 g·l^{-1} Fe).

Each 16 ml centrifuge tube containing the uranium(IV) amorphous phase was filled with 15 ml of Boom Clay porewater. Dithionite and iron were added by weighing the 13 mg of $Na_2S_2O_4$ and 15 mg of Fe powder respectively. The Fe powder was washed before use with degassed 0.1M HCl during 10 minutes to eliminate any trace of oxy-hydroxide of Fe^{3+} and rinsed 5 times with degassed Milli-Q water. A 75 μl aliquot of degassed 2 g·l^{-1} S^{2-} stock solution was added to the tubes for the tests conducted in presence of 10 mg·l^{-1} sulphide. The tubes were sealed and the suspensions were equilibrated for one week.

Table 2. The different sets of solubility experiments performed.

Tests No.	10^{-2} M uranium(IV) (ml)	1M NaOH (ml)	Reductants added
S 1-1	3.0	3.0	no
S 1-2	3.0	3.0	no
S 1-3	3.0	3.0	dithionite
S 1-4	3.0	3.0	dithionite
S 1-5	3.0	3.0	sulphide
S 1-6	3.0	3.0	sulphide
S 1-7	3.0	3.0	Fe powder
S 1-8	1.7	3.0	Fe powder

At the end of this equilibration period, the redox potential was measured with a combined platinum-Ag/AgCl electrode (*Metrohm*) calibrated against quinhydrone buffers at *p*H 4, 7 and 9. The *p*H was measured with a combined glass electrode (*Metrohm*) calibrated against *p*H buffers 7 and 9. To avoid the possible removal of uranium(IV)-organic matter complexes by ultracentrifugation, micro-filtration through a 0.45 μm filter was preferred. A 5 ml aliquot was sampled in each tube, of which 1 ml was used for conditioning the filter, *i.e.* saturation of possible adsorption sites on the filter. This first 1 ml filtrate was discarded while the next filtered 4 ml were analysed for TOC and uranium. The uranium concentration in the filtrates was measured by Inductively Coupled Plasma-Mass Spectrometry (ICP-MS).

Results and discussion

The results of these first solubility experiments are presented in Fig. 3. The average solubility of the uranium(IV) precipitate in Boom Clay porewater is 3.9×10^{-6} mol·l^{-1}. The minimum and the maximum uranium concentration obtained are 4.4×10^{-7} moles·l^{-1} and 6.6×10^{-6} mol·l^{-1}, respectively. Except two points (S 1-6 and S 1-7) which present lower uranium concentration (4.4 and 9.2×10^{-7} mol·l^{-1}), the results are well reproducible and give an average uranium concentration of 4.9×10^{-6} mol·l^{-1}.

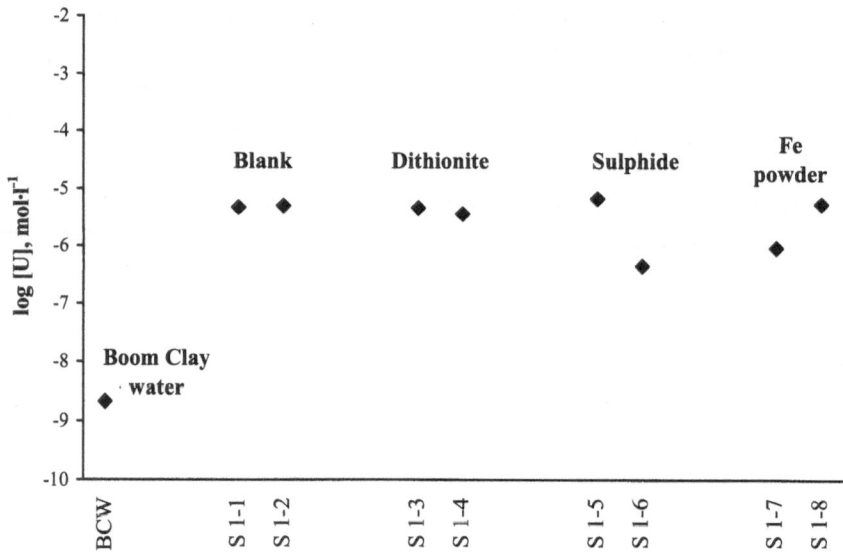

Fig. 3. Measured solubilities of amorphous uranium(IV) oxide in Boom Clay porewater in absence and presence of reductants (experiments in duplicate).

The concentration of natural uranium in Boom Clay porewater is 2.1×10^{-9} mol·l^{-1} which is much lower than the solubility determined in this work so that the background uranium concentration does not interfere on the measured values. However, this could indicate that the natural uranium concentration present in Boom Clay interstitial water is controlled either by well crystallised uranium(IV) phase or by additional phenomena occurring besides solubility. These processes could be sorption, complexation by immobile natural organic matter or co-precipitation.

As mentioned in the introduction, under *in situ* Boom Clay conditions, geochemical calculations predict that the dominant aqueous species of uranium is U(OH)$_4$(aq) even if the boundary of the domain where the highly soluble UO$_2$(CO$_3$)$_3^{4-}$ species dominates is close (Fig.1). The concentration of U(OH)$_4$(aq) is solubility limited by uranium minerals. In order to identify the phases which could be precipitated in our experiments, the saturation index (SI) is calculated for the higher and lower uranium concentrations measured. The geochemical conditions of our system implemented in the code *The Geochemist's Workbench*®-3.2.2 running with the NEA database are Boom Clay porewater composition (Table 1), pH of 8.2, E_h of -275 mV and log f CO$_2$ = -2.4. In our experiments, the Boom Clay interstitial water was oversaturated with respect to the following uranium phases: UO$_2$(c), USiO$_4$(c), UO$_{2.25}$(c), UO$_{2.25}$(beta) and UO$_{2.3333}$(beta). In a second time, the solubility of the identified phases is calculated for Boom Clay conditions (Table 3). The uranium solubility calculated varies from 4.0×10^{-10} to 3.5×10^{-7} moles·l^{-1} depending on the controlling phase. However, these solubilities are in all cases lower than the experimental values. This could be due to either the amorphous character of the precipitates or to the presence of uranium(VI) species. It is also important to stress that these geochemical calculations do not take into account the complexes of uranium with natural organic matter.

Table 3. Solubility of uranium phases in the Boom Clay porewater system at pH 8.2 and E_h -275 mV. Code: *The Geochemist's Workbench*-3.2.2 (Bethke, 2001); NEA database (*Grenthe et al.*, 1992).

Solubility controlling phases	Solubility (mol·l^{-1})
UO$_2$(c) (uraninite)	4.0×10^{-10}
USiO$_4$(c) (coffinite)	1.4×10^{-09}
UO$_{2.25}$(c)	4.8×10^{-08}
UO$_{2.25}$(beta)	4.9×10^{-08}
UO$_{2.3333}$(beta)	3.5×10^{-07}

In comparison with literature data concerning the solubility of amorphous uranium(IV) oxide, our results are in the range of the reported values which vary over more than five orders of magnitude. Bruno et al. (1987) measured in 0.5 M NaClO$_4$ at 25 °C in the pH range 6-10, a solubility for amorphous UO$_2$ of 4×10^{-5} mol·l^{-1} while Rai et al. (1990) determined in dilute solutions uranium solubilities of about 10^{-8} mol·l^{-1} in the same pH interval. Yajima et al. (1995) obtained in dilute NaCl solutions 2×10^{-9} mol·l^{-1} for the solubility of UO$_2$(s) at $pH > 2$. All the data of the above mentioned studies were interpreted by assuming that the only dissolved uranium species under alkaline conditions is $U(OH)_4(aq)$. Although these results are obtained in carbonate free system, they can be compared with our data since the expected dominant aqueous species of uranium in Boom Clay interstitial water is $U(OH)_4(aq)$ as well. Moreover, Rai et al. (1998) have demonstrated that the uranium(IV) carbonate complexation does not influence the $UO_2(am)$ solubility at bicarbonate concentration value typical for the Boom Clay porewater (HCO_3^- 1.2×10^{-2} $mol \cdot l^{-1}$). The uranium concentrations of our study are about two orders of magnitude higher than the solubility reported by Rai et al. (1990) and Yajima et al. (1995). This difference could be explained either by a contamination of our system by uranium(VI) despite the use of reductants, or by the presence of natural organic matter, an important component of the Boom Clay porewater which was not present in the others studies. Indeed, we presently do not know the effect of natural organic matter on the $UO_2(am)$ solubility.

The pH value of Boom Clay porewater was 8.4 after equilibration with the working gas of the glove box. In general, the pH values at the end of the tests were lower than the initial value (Table 4). The decrease of pH could result from the fact that the uranium(IV) amorphous phase was precipitated at acidic pH. Indeed, the uranium(IV) 1M HCl solution was titrated by an equivalent amount of 1M NaOH resulting in a roughly neutral pH. However, the hydrolysis of the uranium(IV) cation produces 4 protons decreasing the pH during the solid formation:

$$U^{4+} + x\ H_2O \rightarrow UO_2 \cdot (x\text{-}2)H_2O + 4\ H^+$$

Therefore, the dithionite added to ensure reducing conditions during the precipitate formation could be unstable in this acidic solution giving rise the production of sulphate through the reaction:

$$2\ S_2O_4^{2-} \rightarrow 2\ S + 2\ SO_4^{2-}$$

The question arising from this observation is the possible precipitation of other phase than amorphous uranium(IV) oxide, $e.g.$ phases containing sulphate. According to Rai et al. (1997) $UO_2(am)$ is the solubility controlling phase even in solutions ranging in concentrations up to 0.1M sulphate. Moreover, geochemical calculations show that the uranium(IV) sulphate phases are very soluble at the sulphate concentration expected in our work. Therefore, hydrous $UO_2(am)$ is considered as the solubility controlling phase in our experiments. The higher pH measured in the S 1-8 test is explained by the fact that 1M NaOH solution was added in excess to the uranium(IV) 1 M HCl solution (Table 2). From these observations, it appears that to avoid pH fluctuations during the equilibration period, the pH at the end of the precipitation step has to be measured and adjusted to the pH value of Boom Clay porewater. The lowest values (pH 7.3) were measured in the samples containing dithionite. This observation is well known and explained by the dithionite oxidation which acidifies the system:

$$S_2O_4^{2-} + 2\ H_2O \rightarrow 2\ SO_3^{2-} + 4\ H^+ + 2e^-$$

The TOC variation is very low taking into account the 10 % of error on the measurements (Table 4). This observations indicate that the filtration at 0.45 μm do not remove the organic matter in our experimental conditions.

The E_h values show that strongly reducing conditions were still present at the end of the tests (Table 4). Nevertheless, the redox measurements are only given as an indication because of the large uncertainty affecting these data.

Table 4. Total uranium concentration, Total Organic Carbon (TOC), pH and E_h at the end of the tests.

Tests No.	[U] mol·l^{-1}	Log [U] mol·l^{-1}	TOC mg C·l^{-1}	pH	E_h mV
BC water	2.1×10^{-09}	-8.68	103	8.4	—
S 1-1	4.6×10^{-06}	-5.34	85	7.8	-227
S 1-2	5.0×10^{-06}	-5.30	84	7.9	-353
S 1-3	4.5×10^{-06}	-5.35	92	7.3	-421
S 1-4	3.5×10^{-06}	-5.45	91	7.4	-428
S 1-5	6.6×10^{-06}	-5.18	86	7.9	-465
S 1-6	4.4×10^{-07}	-6.36	88	7.9	-478
S 1-7	9.2×10^{-07}	-6.04	88	8.1	-451
S 1-8	5.3×10^{-06}	-5.27	102	8.5	-353

BC: Boom Clay

Conclusion

The solubility of hydrous UO_2(am) in Boom Clay porewater (pH ~ 8, HCO_3^- ~ 1.2×10^{-2} mol·l^{-1}, TOC 103 mg C·l^{-1}) was determined under geochemical conditions representative for Boom Clay; *i.e.* reducing conditions and log f CO$_2$ -2.4. The average total uranium concentration measured in the tests is 3.9×10^{-6} mol·l^{-1}. The expected dominant aqueous uranium species is $U(OH)_4$(aq), carbonate complexes do not increase the uranium(IV) solubility at HCO_3^- concentration of Boom Clay porewater. There is not visible effect of the added reducing agents.

Our solubility value is one order of magnitude lower than the values reported by Bruno et al. (1987) but ~2-3 orders of magnitude higher than data from Rai et al. (1990) and Yajima et al. (1995). This difference could arise for an uranium(VI) contamination despite the precautions taken during the tests. Another possible explanation is the organic matter complexes which could increase the hydrous UO_2(am) but this hypothesis needs to be confirmed by further experiments. Since the formation of the precipitate occurred at low pH due to hydrolysis of U^{4+}, the dithionite added to maintain reducing conditions during the precipitate ageing could have produced sulphate. However, $UO_2(am)$ is assumed to be the uranium controlling solubility phase as uranium(IV) sulphate solids are very soluble.

The results presented in this paper are from a first solubility experiment and need to be confirmed by further tests. Next experiments are planned to be performed with improved procedure. The following parameters will be studied: equilibration time and effect of organic matter on the uranium solubility. To assess the impact of natural organic matter, solubility experiments will be repeated again in Boom Clay porewater but also in synthetic clay water, both in presence and absence of organic matter.

Acknowledgements

This work is financed by the Belgian National Organisation for Radioactive Waste and Fissile Materials (NIRAS/ONDRAF). The European Commission (EC) is also acknowledged for its financial support in the frame of the TRANCOM-II Project (contract No.FIKW-CT-2000-00008). The authors also express their thanks to the R&D Geological Disposal colleagues for their collaboration and especially to Christelle Cachoir for reviewing this paper.

References

Bruno J, Casas I, Lagerman B, Munoz M (1987) The determination of the solubility of amorphous $UO_{2(s)}$ and the mononuclear hydrolysis constants of uranium (IV) at 25°C. Mat. Res. Soc. Symp. Proc. 84. 153-160

Rai D, Felmy A R, Ryan J L (1990) Uranium(IV) hydrolysis constants and solubility product of $UO_2.xH_2O(am)$. Inorg. Chem. 29. 260-264

Grenthe I, Fuger J, Konings R J M, Lemire R J, Muller A B, Nguyen-Trung C, Wanner H (1992) Chemical thermodynamics of uranium. Edited by Wanner H. and Forest I., Elsevier Science Publishers B. V., Amsterdam

Yajima T, Kawamura Y, Ueta S (1995) Uranium(IV) solubility and hydrolysis constants under reduced conditions. Mat. Res. Soc. Symp. Proc. 353. 1137-1142

Rai D, Felmy A R, Sterner S M, Moore D A, Mason M J, Novak C F (1997) The solubility of Th(IV) and U(IV) hydrous oxides in concentrated NaCl and $MgCl_2$ solutions. Radiochimica Acta 79. 239-247

Rai D, Felmy A R, Hess N J, Moore D A (1998) A Thermodynamic model for the solubility of UO_2 (am) in the aqueous K^+-Na^+-HCO_3^--CO_3^{2-}-OH^--H_2O system. Radiochimica Acta 82. 17-25

Bethke C M (2001) The Geochemist's Workbench. A User's Guide to Rxn, Act2, Tact, React and Gtplot. Release 3.2. University of Illinois

Uranium migration in argillaceous sediments as analogue for transport processes in the far field of repositories (Ruprechtov site, Czech Republic)

Ulrich Noseck[1], Thomas Brasser[1], Ales Laciok[2], Mirek Hercik[2], Frantisek Woller[3]

[1]Gesellschaft für Anlagen- und Reaktorsicherheit (GRS), Theodor-Heuss-Str. 4, 38122 Braunschweig, Germany
[2]Nuclear Research Institute, plc (NRI), 20568 Rez, Czech Republic
[3]Radioactive Waste Repository Authority (RAWRA), Gorazdova 24, 12000 Prague 2, Czech Republic

Abstract. For long-term performance assessment of radioactive waste repositories it is essential to consider the potential mobility of critical radionuclides through the relevant rock material under representative long-term conditions. The aim of this natural analogue study is to understand and quantify the relevant long-term geochemical and hydraulic processes at the Ruprechtov site in Czech Republic and to identify the mechanisms leading to mobilisation/immobilisation of uranium, thorium and radium. The actual results from the bilateral project of NRI and GRS will be presented.

Introduction

Underground repositories are based on a multi-barrier concept in which clay materials often play an important role as geological barriers. Detailed investigations of suitable geological analogues may lead to a better understanding of the complex interrelations between transport and sorption of radionuclides in argillaceous media under natural conditions and especially on very long time scales relevant for performance assessment (PA).

The Ruprechtov site has been chosen because its geological and geochemical conditions are similar to sediments which in many cases cover host rocks for underground waste repositories. The Ruprechtov site is situated in the western part of the Czech Republic where Tertiary basins with clay and organic material (coal, lignite) commonly occur, in some cases with high concentrations of uranium. Within a bilateral project this site has been investigated by GRS, Germany and

NRI, Czech Republic in order to identify the main mobilisation/immobilisation processes for the PA-relevant elements uranium, thorium, and radium. The previous investigation program consists of three reconnaissance drillings (NA1-NA3) and two additional drillings for more detailed investigations (NA4, NA5). Geophysical logging performed in the latter two boreholes indicated two horizons of higher permeability within a nearly impermeable sequence. Consequently, these two boreholes were lined as groundwater measuring points at around 35 m (NA4) and 20 m depth (NA5). Uranium enrichments have been observed in organic rich layers as well as in the underlying kaoline in the depth of 33 - 38 m.

Geology of the site

The Ruprechtov site is situated in the Sokolov basin in the north-western part of the Czech Republic. This basin with its Tertiary sediments of Oligocene and Miocene age is surrounded or underlain by Carboniferous granitic rocks which have intensively been kaolinized in wide areas during Mesozoic and Cenozoic Age (Svoboda et al. 1966). The thin basal Staré-Sedlo formation fills flat depressions in the basement and is overlaid by the so-called volcano-detritic formation which is composed mainly by volcanogenic series and lignite seams. Former exploration activities for uranium in this region gave hints that most of the uranium accumulation is of sedimentary (hydrological) origin and strongly correlated with organic matter, i.e. lignite seams or coaly sediments. A morphological profile of the site is shown in Fig. 1.

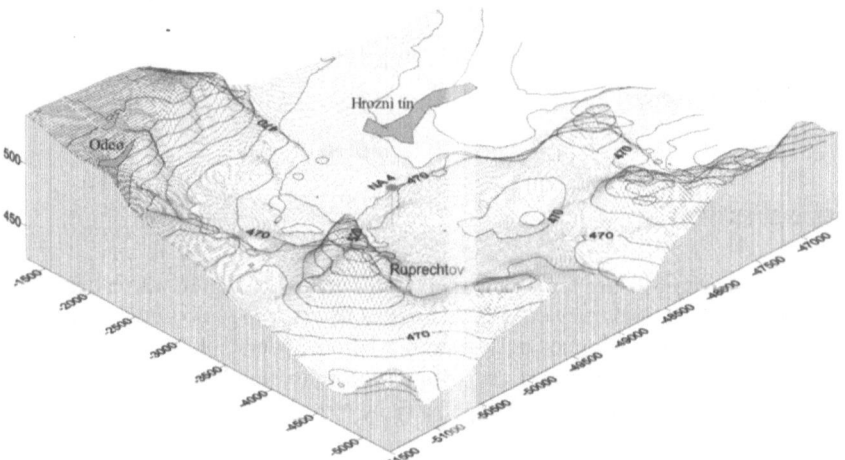

Fig. 1. Morphology at the Ruprechtov site. View from SW to NE. Location of the drillings indicated by dot NA 4.

Results

Up to now altogether five boreholes down to 40 m deep have been drilled covering only a small area of 200 m². A number of geochemical, mineralogical, hydrogeological and radiometrical investigations of drill core samples have been performed. Additionally groundwater from the horizons at 20 m and 35 m depth as well as porewater samples have been analyzed in detail. Selected results are presented.

Hydrology

The findings from well logging, i.e the existence of two water bearing horizons in 19-20 m and 34-36.5 m depth, have been confirmed by measurement of hydraulic conductivities. From pumping tests hydraulic conductivities $kf = 1,8 \cdot 10^{-6}$ m/s for the deeper aquifer and $kf = 1,5 \cdot 10^{-7}$ m/s for the upper aquifer have been determined. The thickness of both aquifers is estimated from core analyses to about 1-2m. The aquifers are divided by very low permeable clay layers with hydraulic conductivities in the range of 10^{-11} m/s.

Furthermore, there is some information about the hydrogeology of the site available in literature. Due to this information the hydraulic situation of the site is characterized by two flow regimes, one in the tertiary sediments and another one in the underlying granite. Both systems are divided by the very low permeable kaolin layer of some ten m thickness. Based on data about groundwater levels from previous drilling wells, which are no more accessible today, the general flow direction is expected to be from south-west to north-east. With hydraulic gradients of about 10 m / 500 m in the model area flow velocities in the lower aquifer are estimated to be about 1-2 m/y.

Mineralogy

Analytical results show that the sediments are mainly composed of quartz, kaolinite, muscovite/illite, montmorillonite and anatase. In many samples siderite and pyrite have been detected as trace minerals. Due to the existence of lignite horizons in the depths between 34 and 36.5 m a number of samples contain a very high amount of organic carbon, up to 50 weight %. A characteristic feature of the site is the heterogeneity of sediments at small scale.

Unevenly distributed uranium accumulations have been identified mainly in the lignite/clay-sand horizon at depths between 33 and 37 m. Furthermore, rather high uranium concentrations have been detected in the upper part of the kaoline in 37-39 m depth. Figure 2 illustrates that these high uranium concentrations occur in very narrow sections extending only over a few decimeters in the surrounding of the detected aquifer. The highest concentrations, measured in borehole NA5, are about 300 mg/kg. By scanning electron microscopy the existence of ningyoite

(CaU[PO4]₂) and traces of uraninite minerals in the zones of highest U-enrichment in the lignite/clay-sand horizon has been demonstrated.

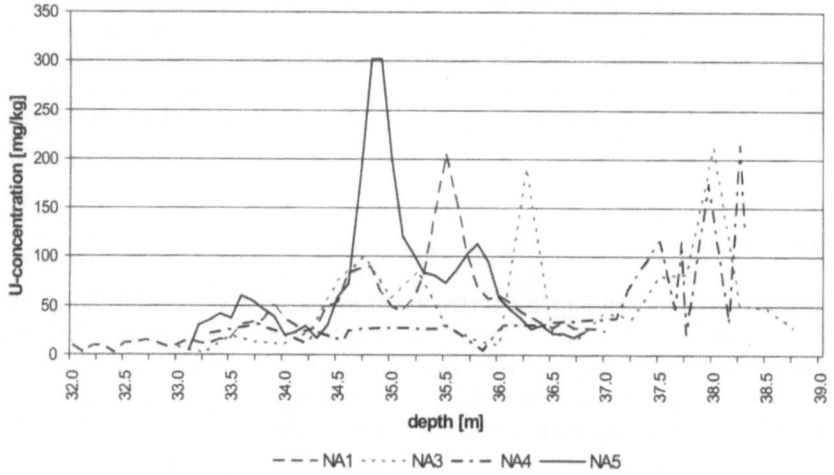

Fig. 2. Uranium enrichment in different boreholes between 33 and 39 m depth.

Geochemistry

Analysis of groundwater samples characterize them as Ca-Na-HCO3 - type waters with clear distinctions between upper and lower horizon. The geochemical conditions of water from the lower horizon, which is in the focus of interest, are summarized in Table 1. The pH-values are about 6.9. Eh-values measured after pumping in an isolated flow-through-cell and corrected to standard hydrogen electrode are about $0 - 50$ mV. However, these values are probably to high, since the existence of pyrite and siderite minerals is observed in the surrounding of the aquifer. This fact together with S^{2-}/SO_4^{2-}-ratios in the range of $0.5 - 5$ % (from on-site analysis) indicate Eh-values below -150 mV. In-situ measurements of Eh and S^{2-} concentration, will be done to clarify this important parameter.

Table 1. Geochemical conditions of the groundwater from the horizon in 35 m depth (NA4)

pH	6.9	Ca [mg/l]	52	Si [mg/l]	29
Eh	0-50 mV	Mg [mg/l]	25	P [mg/l]	0.05
T [°C]	9	S [mg/l]	25	Al [mg/l]	0.3
Na [mg/l]	24	HCO₃⁻[mg/l]	350	Fe [mg/l]	2.0

At Eh-values below -150 mV U(IV) is the thermodynamically stable redox state of uranium. However, in this range the system is very sensitive against changes of Eh and pH. A slight increase of Eh and/or pH will increase the U(VI) fraction significantly. Since at now, only ningyoite and traces of uraninite minerals have been detected, geochemical calculations with PHREEQC and database Wateq 4.0 have been performed to check if these minerals are stable under the conditions at Ruprechtov site. Calculated saturation indices for the reference case (Eh = -165 mV) and variations of Eh and phosphate concentration are shown in table 2.

Table 2. Saturation indices calculated with PHREEQC for reference groundwater and few parameter variations. Element concentrations are taken from Table 1. $U = 1 \cdot 10^{-4}$ mg/l

Eh-value [mV]	uraninite	ningyoite	Phosphate [mg/l]	ningyoite
-165	0.04	-0.81	0.05	-0.81
(reference case)			(reference case)	
-230	0.05	-0.74	0.1[a]	-0.21
-130	-0.12	-0.97	0.125	-0.01

The calculation results for the reference groundwater show uraninite saturation. If the phosphate concentration is increased only about a factor of 2.5 compared to the reference water, also ningyoite will reach saturation. This is a strong indication that both minerals could have been formed under Ruprechtov conditions. Due to the recommendation of NEA (Grenthe 1992) and (Duerden 1992) the phosphate complexes $UO_2(HPO_4)_{aq}$ and $UO_2(HPO_4)_2^{2-}$ have not been considered in this calculation. Additionally no uranium complexation with organic ligands and especially colloids like humics or fulvics have been taken into account. First laboratory measurements indicate relatively low colloid concentrations in the deeper aquifer (Noseck et al. 2002).

Radiometry

The alpha and gamma-ray measurements of single isotopes of the uranium decay chain show significant non equilibrium states in core sections with elevated uranium concentrations from 33 – 38 m depth. In Figure 3 the ratio of $^{234}U / ^{238}U$ versus $^{230}Th / ^{238}U$ is plotted for samples from four boreholes. The values for most of the samples are accumulated in the same quadrant of the diagram with $^{230}Th^{238}U$-activity ratios and $^{234}U^{238}U$-activity ratios significantly lower than one.

At the same time high $^{234}U/^{238}U$-activity ratios in groundwater and porewater samples in the range between 2.5 and 3.5 have been detected (s. Table 3). This is an indication for slow moving uranium. The observed activity-ratios of water and sediments are rather typical for uranium enrichments due to the Roll Front model (Osmond et al. 1983), which is shortly discussed in the following chapter.

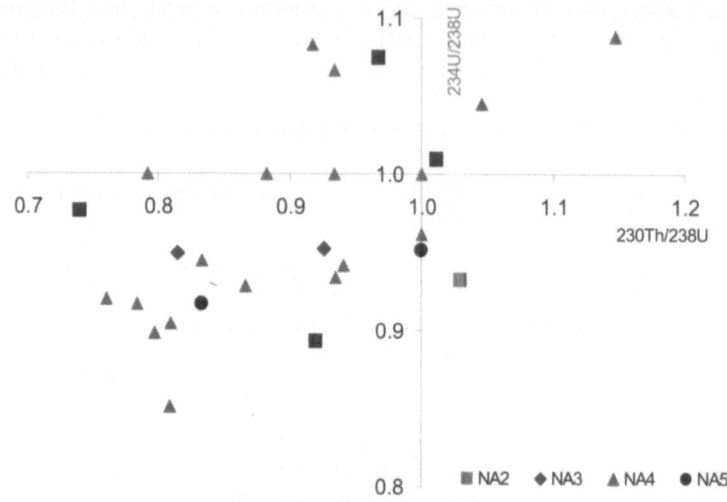

Fig. 3. Activity ratios for isotopes of the uranium decay chain. Data from different drill cores at Ruprechtov site

Table 3. Activity ratios in groundwater and porewater samples

	$^{234}U/^{238}U$	$^{230}Th/^{238}U$
groundwater	3.5 ± 0,52	n.d.
porewater, 1 fraction	2.48 ± 0,4	0.18
porewater 2. fraction	2.93 ± 0,45	<0.07

Discussion

One frequently observed natural uranium deposit is the so called "roll front orebody". This name describes the behavior of the uranium deposit, which "rolls" downgradient as the upgradient edge dissolves and is redeposited near the downgradient edge. Such a uranium body is build under distinct geochemical conditions: Uranium is usually dissolved from granite or rhyolite rocks in oxidizing and/or carbonate rich waters and is transported along an aquifer into horizons, where reducing substances prevail. In this area reduction of uranium(VI) and precipitation of uranium(IV)-minerals occur. Further inflow of oxidizing water leads to oxidation of the reducing substances and causes a movement of the redox front in flow direction. Due to the different mobility of the three radionuclides of the uranium decay chain ^{238}U, ^{234}U and ^{230}Th a characteristic concentration profile in the sediment is formed, as schematically shown in Fig. 4. A detailed description of the model is given in (Osmond et al. 1983).

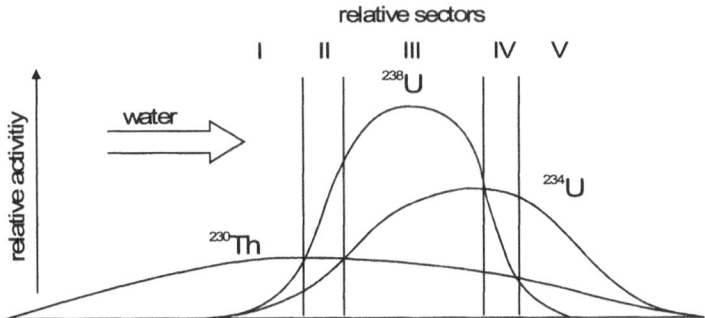

Fig. 4. Relative activity distribution of ^{238}U, ^{234}U and ^{230}Th in sediments of roll front systems (Osmond et al. 1983)

The major enrichment zone is characterized by ^{234}U/^{238}U- and ^{230}Th/^{238}U-activity ratios lower 1 (s. also Fig. 5). In the area of the major enrichment zone ^{234}U/^{238}U-ratios in the groundwater are rather high (about 2-5).

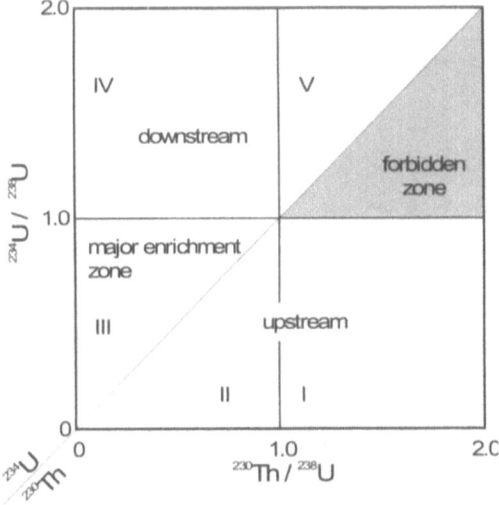

Fig. 5. Characterization of uranium enrichments in roll front systems by activity ratios of the isotopes ^{238}U, ^{234}U and ^{230}Th (Osmond et al. 1983)

These are the conditions observed in most of the sediment samples in the horizon in 33 – 38 m depth and the corresponding groundwater from this horizon at Ruprechtov site. Furthermore other typical features of roll front enrichments, as reducing conditions, relatively low uranium concentrations in the water, and an aquifer of limited thickness are found at Ruprechtov site.

Based on the radiometric data and all other findings the conceptual model of uranium transport has been developed so far and is schematically presented in Fig. 6. According to literature it is very likely that the surrounding "Erzgebirge granite" was/is the uranium source, because the concentration of U in this granite type is high (20 - 40 ppm) and the U/Th ratio is >1. Minor supply due to *in situ* dissolution of U-bearing minerals or from discrete tectonic zones in underlying granite could also contribute to the accumulation of U. It is supposed that uranium has been mobilized under oxidizing conditions and transported through the aquifers of 1-2 m thickness. Especially, in the vicinity of lignite seams or lenses, reducing conditions have led to reduction and subsequent precipitation and / or sorption of uranium. Ningyoite and traces of uraninite minerals have already been detected in samples with high uranium content in the surrounding of the aquifer.

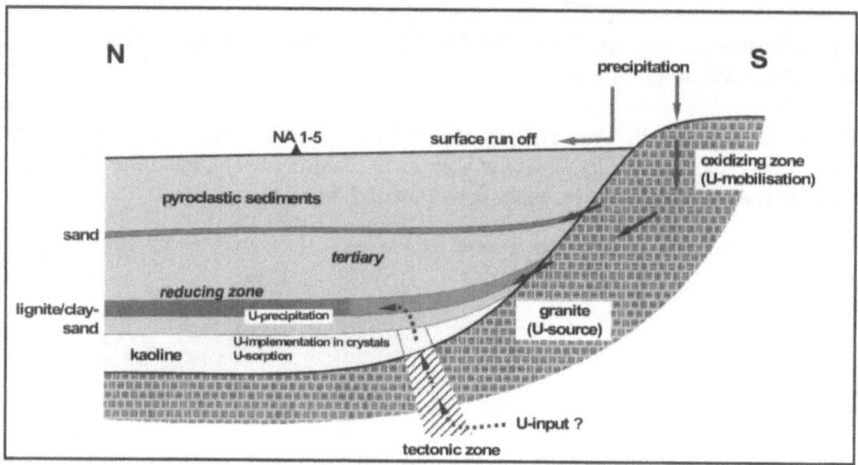

Fig. 6. Conceptual model with simplified geometry for uranium transport at Ruprechtov site.

Up to now only a restricted area has been investigated. Future investigations will be focused on site-scale characterization, e.g. the connectivity of permeable horizons. Further boreholes will be drilled, especially upstream in direction of the outcropping granite, where the main uranium source is assumed. Comprehensive analyses of sediments and groundwater will be performed to confirm or modify the conceptual model.

Based on measurements of water level and natural isotopes a more sophisticated groundwater flow model will be developed. The overall aim of the study is the application of PA transport models to describe the migration of uranium, thorium and radium at the Ruprechtov site.

Acknowledgment

This work has been funded by the German Federal Ministry of Economics and Technology (BMWi) under the contract numbers 02 E 9128 and 02 E 9551 and by the Czech Radioactive Waste Repository Authority (RAWRA) under the contract no. 1999/31/NACH.

References

Duerden P, Lever, DA, Sverjensky, DA, Townley LR (1992) Alligator Rivers Analogue Project.- Final Report. Volume 1 (1992).

Grenthe I, Fuger J, Konings R, Lemire RJ, Muller AB, Nguyen-Trung C, Wanner H (1992) Chemical Thermodynamics of Uranium.- Nuclear Energy Agency/OECD. Elsevier Science Publ., Amsterdam

Langmuir D (1997) Aqueous Environmental Geochemistry.- Prentice Hall, New Jersey

Noseck U, Brasser T, Woller F, Laciok A, Hercik M (2000) Uranium migration in tertiary sediments - natural analogue study for repository systems at Ruprechtov Site.- in: "Proccedings of the International Conference on Radioactive Disposal, DISTEC 00", 4.-6. September 2000, Berlin

Noseck U, Brasser T., Pohl W. (2002) Tertiäre Sedimente als Barriere für die U-/Th- Migration im Fernfeld von Endlagern. GRS-176, Gesellschaft für Anlagen und Reaktorsicherheit, Braunschweig

Osmond JK, Cowart JB, Ivanovich M (1983) Uranium Isotopic Disequilibrium in Ground Water as an Indicator of Anomalies.- Int. J. Appl. Radiat. Isot. 34/1 : 283-308

Svoboda, J et al. (1966) Regional Geology of Czechoslovakia, Part I: The Bohemian Massif.- Geol. Survey of Czechoslovakia, Prague

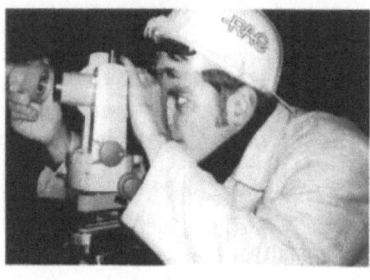

Geodesy and Mine-Surveying

1st – 4th term (undergraduate studies)
* Mathematics, Physics, Mechanics
* Computer Science
* Geosciences
* Economics, Law
* Geodesy
* Internship (60 shifts) in industries
* oral examination after 4th term

5th – 9th term (graduate studies)
* Surveying (surveying techniques, surface and mining damages)
* Geodesy (engineering, integrated, astronomic, and satellite geodesy)
* Photogrammetry and Remote Sensing
* Geodesy
* Surveying, Landscape planning
* Digital image processing and Geo-information systems
* Geotechnics (Soil and Rock Mechanics)
* Geomodelling and Geokinematics
* Mining Economics
* Mining Law
* Mining (underground and open-pit)
* Safety practices
* Internship (60 shifts) in industries
* Masters thesis in the 9th term

Admission
* „Abitur"/Baccalaureate (A-level) or
* Specialized college degree or
* equivalent access entitlement

Duration/Degree: 9 semesters/Master of Science in Engineering (Dipl.-Ing.)
Start: winter term

Job opportunities
Surveying in exploration and mining; Survey departments of industries and forestry; Federal institutions like regional mining offices and departments for environmental protection; Surface and underground mining; Traffic- and hydro-engineering; Private surveying and engineering offices; Research institutes of mining and civil engineering

Groundwater flow systems and the export of uranium from nuclear waste repositories to the biosphere

Klaus-Peter Seiler, Wolfgang Graf, Hans Lang, Reinhard Zapata

GSF-Institute of Hydrology, Neuherberg, Germany

Abstract. Numerical flow simulations with common hydraulic conductivity/depth distributions result in an active and passive turn-over zone, both occurring world wide. Geochemical modeling of uranium species distribution under oxidizing conditions in carbonate and sulfate waters mostly yields negatively charged or neutral UO_2-carbonate complexes, which do not sorb on negatively charged mineral surfaces (pH 5 to 8), which is in good agreement with batch experiments. Contrary, in crystalline aquifers Uranium species are mostly positively charged, favoring sorption. Dilution in the passive turn-over zone is dominated by transverse hydrodynamic dispersion and gets enhanced by orders of magnitudes in bi-porous sediments through matrix diffusion. At the transition from the passive into the active turn-over zone dilution gets further enhanced.

Introduction

Uranium primarily occurs in granites and in sedimentary rocks from weathered granites. In central Europe these granites are of Paleozoic age (e.g. Massive Central, Black forest, Bohemian block) and sediments of Cretaceous or Tertiary age (e.g. Saxony, Thuringia, Bavaria). Some of these resources have been mined in the past; today they are either abandoned or are prepared to get closed.

Under natural conditions uranium occurs in granites as uraninit, carnotit, torberneit and uranium-glimmer, which all

- behave immobile in their natural mineralogical phases,
- however, reach a high mobility if an oxidizing environment is introduced to the uranium resource through exploitation and

- by open-air storage of ores and ore residuals.

Nowadays, we are also faced with radioactive wastes, stored in host rocks with very low groundwater flow in its vicinity, which could endanger the biosphere on a long run. Since physical and chemical processes will always act in the same way along the flow path of uranium, the objectives of the following discussion deals with these physical, chemical and hydrodynamic boundary conditions influencing the migration of uranium in subsurface waters.

Contrary to uranium mining, in the surroundings of a deep nuclear waste repository, physical, chemical and hydrodynamic conditions are not significantly changed as compared to natural. This is considered as an important difference in assessing the uranium migration in the stage of exploitation as compared to the stage of storage.

Since uranium is not only radioactive but also toxic, the risk of any release from a waste repository has to be assessed under both of these aspects over a run of more than 100,000 years. Within this time span the climatic and geologic boundary conditions as well as the exploration and exploitation of groundwater resources may change the present groundwater flow pattern.

Types of groundwater and the chemical behavior of uranium in aquifers

Groundwater chemistry reflects the mineralogy of the most soluble rocks along the flow path of the groundwater. Since silicate, carbonate and salt rocks are most frequently occurring, in hydrogeology it is generally distinguished between crystalline, carbonate and sulfate dominated waters; a generalized chemical composition of these groundwater is shown in table 1.

These chemical compositions in mind, a geochemical calculation was performed with 10^{-6} Mol/L uranium added to the respective groundwater type, in order to describe quantitatively the respective abundance of species of uranium and its electric charge using the program Hydroql (Papelis et al. 1988).

Results are:

- Under reducing conditions U^{4+} dominates; it chemically precipitates and is strongly sorbing on sediments and rocks.
- In groundwater with redox >-400 mV U^{6+} uranyl prevails and no chemical precipitation occurs. Under these conditions species distribution (Table 2) of uranium in carbonate and sulfate water is dominated by neutral and negatively charged uranyl carbonate complexes,
- in crystalline waters UO_2^{2+} and UO_2OH^{1+} (20-65 %) are the dominant species and responsible for significant sorption.

Table 1. Generalized chemical composition of groundwaters in crystalline, carbonate and sulfate rocks

		Groundwater in		
	Dimension	Crystalline rocks	Carbonate rocks	Sulfate rocks
pH		5.4	7.3	6.7
Ion strength	mMol/L	12.55	12.55	51.6
Ca^{2+}	mg/l	10.0	115	380.6
Mg^{2+}	mg/l	4.5	25	95.7
Na^+	mg/l	3.5	5.5	26.5
K^+	mg/l	2.5	2.5	10.6
HCO_3^-	mg/l	25.0	410	490
SO_4^{2-}	mg/l	15.0	47.5	938
Cl^-	mg/l	6.0	12.5	23.2

Table 2. Species distribution and electric charges of species of 10^{-6} Mol/L uranium in sulfate, carbonate and crystalline waters. pH ranges has been selected according to table 1

	pH	UO_2	UO_2 SO_4	UO_2 CO_3	UO_2 $(CO_3)_2$	UO_2 $(CO_3)_3$	Ca_2UO_2 $(CO_3)_3$	UO_2 OH	UO_2 $(OH)_2$
Electric Charge		2+	0	0	2-	4-	0	1+	0
	7.0	-	-	-	24.3	16.1	59.2		
Sulfate	6.5	-	-	1.1	27.2	7.2	64.5	-	-
Water	6.0	-	-	13.3	58.1	2.8	25.3	-	-
	8.0	-	-	-	23.5	68.7	7.6	-	-
Car-	7.5	-	-	-	33.5	38.4	27.8	-	-
bonate water	7.0	-	-	-	23.5	68.7	7.6	-	-
Crys-	6.0	3.0		62.1	10.4	-	-	15.2	7.3
talline wa-	5.5	17.1	2.7	46.1	1	-	-	27.3	4.1
ter	5.0	40.8	7.0	14.9	-	-	-	25.2	4.1

As a rule, mineral surfaces at pH 5 to 8 are negatively charged. The results of geochemical modeling are that for carbonate and sulfate waters most ionic and complexed U-species are either neutral or negatively charged and consequently do not sorb substantially; this was also stated by batch experiments yielding sorption coefficients in between 1 and 13 cm^3/g (Wolfrum et al. 1988). Only U-species in groundwater out of crystalline rocks undergo sorption due to the occurrence of positively charged uranyl and uranium hydroxide complexes.

Results of geochemical modeling and laboratory experiments show that chemical precipitation as well as sorption does not play a dominant role for U^{6+}, except in groundwater of the crystalline type. This also holds for NaCl-waters and changes to significant sorption if the chemical environment turns into reducing conditions (U^{4+}). As a consequence, under oxidizing conditions dilution plays an important role in assessing the propagation of uranium.

Depth related groundwater flow and dilution

The majority of the groundwater belongs to a long-term reserve (Freeze and Witherspoon 1967; Seiler and Lindner 1995), because groundwater recharge does not discharge in equal distribution through the entire thickness of aquifer systems. Instead, it predominantly flows through the aquifer systems close to the ground surface; this results from numerical flow simulations with generally known hydraulic conductivity/depth distributions (Fig. 1) and has been documented by respective field research (Seiler and Lindner 1995). The calculated amounts of turn-over in the individual aquifers (Fig. 1) lead to the conclusion, that more than 85% of it occurs in near-surface and less than 15 % of the groundwater recharge reaches also deep aquifers. Related to this, groundwater in near-surface aquifers are young (< 50 years) and significantly age with growing depth (Fig. 3A): lots of groundwater recharge flows in an active, only few in a passive, deep groundwater turn-over zone (Fig. 2). The thickness of the active turn-over zone depends upon effective recharge, the storage and drainage of the aquifer system in which it occurs and ranges mostly between meters to about 100 m; the passive groundwater turn-over zone itself achieves a thickness of many 100 m and is underlain (Fig. 2) by connate or formation waters (v. Engelhardt 1960), which did not return into the biosphere since being entrapped during sedimentation.

With respect to pollution near-surface aquifer systems offer easy access and quick transport of contaminants to the biosphere. Contrary, in deep groundwater
- Long mean residence times occur, which favor
  - quantitatively physical and chemical reactions with even slow kinetics and
  - radioactive disintegration;
- Hydrodynamic dispersion yields some dilution without and strong dilution with matrix diffusion and
- Additional mixing occurs, when deep groundwater crosses near surface groundwater to join the biosphere.

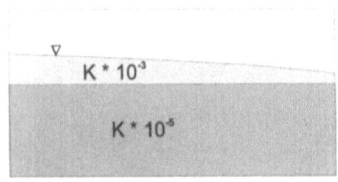
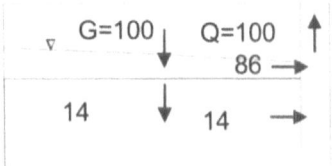

Permeabilities (k)
in m/s

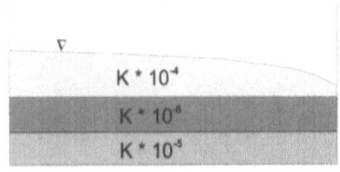
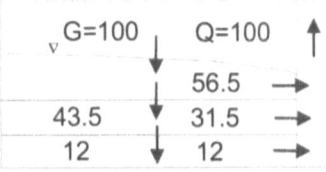

Recharge G = 100 ‰

Discharge Q = 100 ‰

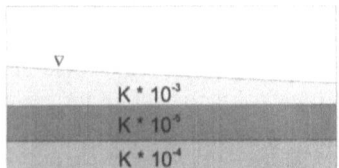
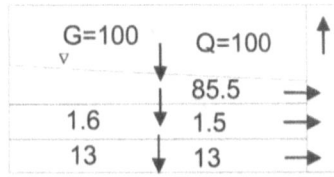

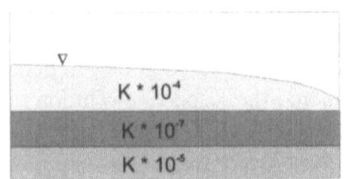
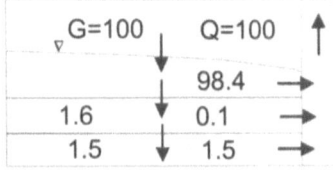

Fig. 1. Selected examples of changing hydraulic conductivities with depth, as frequently occurring in nature (left column) and the distribution of the groundwater discharge (in percent of recharge) upon the individual layers (right columns). G = recharge, Q = discharge.

Long term storage of radioactive waste

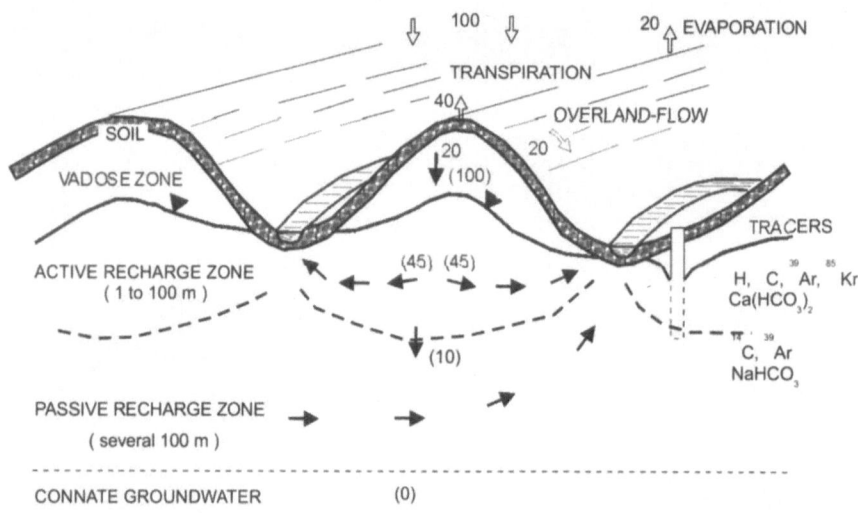

Fig. 2. Subdivision of the aquifer systems in an active and passive groundwater turn-over and connate groundwater zone; out of scale. 100 = 100% of precipitation, (100) = 100% of groundwater recharge.

It proved practical to define the boundary between these two zones by ^{3}H, changing from measurable to non detectable concentrations (Seiler and Lindner 1995). ^{3}H, a radioactive environmental tracer with a half-life of 12.34 years, is introduced to groundwater in considerable concentrations only through the water cycle and yields water ages of less than 50 years, which is typical for the active turn-over zone.

Uranium, continuously released from waste repositories gets mixed according to the contribution from the source area to natural groundwater flow. All further mixing in deep groundwater is dominated in rocks without matrix porosity only by transverse dispersion. Since in central Europe most Mesozoic rocks dispose of a matrix porosity (sandstones, chalk, reef carbonates, dolomites, compacted clays), hydrodynamic dilution gets enhanced by orders of magnitudes through matrix diffusion (Sudicky and Frind 1981). As a worth case, Fig. 4 presents an example of mean turn-over times and dilution from a hypothetical inflow area at 600 m depth to the deep groundwater into a mono-porous media and the respective dilution till reaching the biosphere at a distance of only 2 km.

At the transition to the biosphere, waters from the passive turn-over zone have to join and mix with waters from the active turn-over zone. According to the distribution of groundwater recharge following Fig. 1, this mixing leads to a further dilution of uranium of a factor minimum of 6, maximum of 50.

Long term storage of radioactive waste

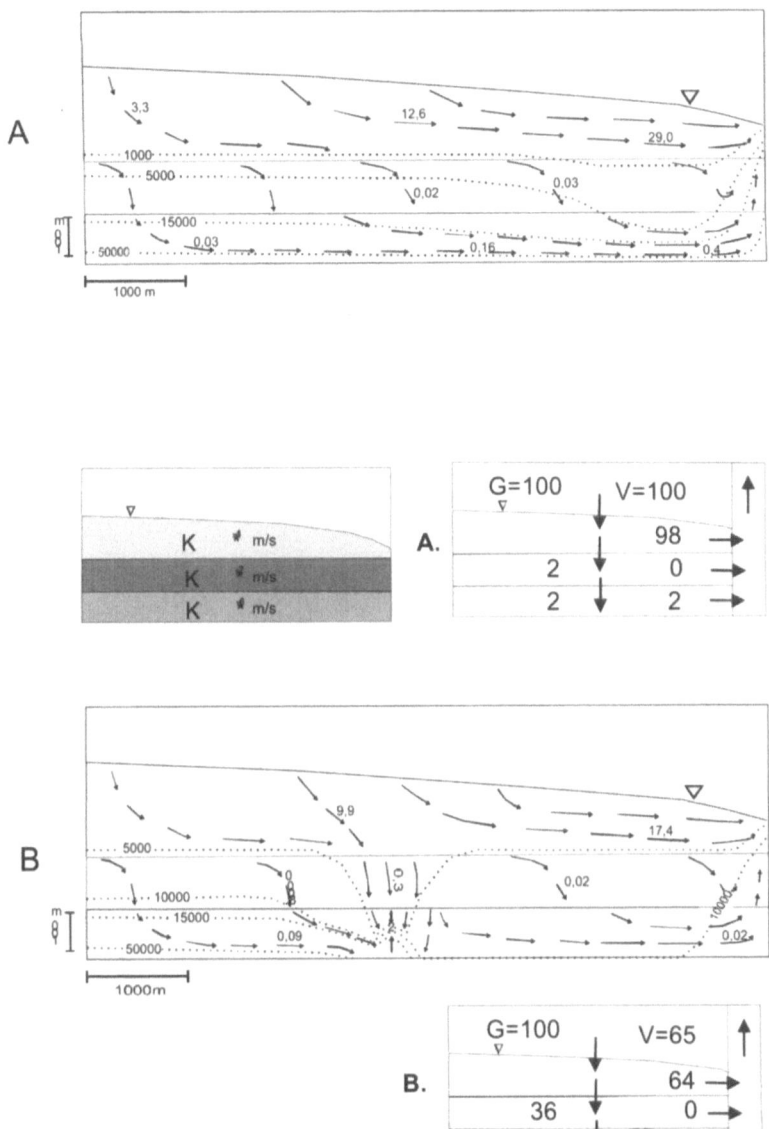

Chap. III 1.1 Fig. 5

Fig. 3. Flow lines (arrows), distance velocities of the groundwater (m/d) and age distribution (years) in the groundwater (dotted lines) at a certain hydraulic conductivity distribution (m/s) in the aquifer systems. A without, B with groundwater abstraction from great depth.

The two mixing processes, hydrodynamic dispersion and mixing of waters at the transition from deep to near surface groundwater, can clearly be recognized from Fig. 4; from this hypothetical worst case a dilution of 4 to 5 orders of magnitude on a distance of only two kilometers results: it would significantly increase if matrix diffusion was also considered.

Although, under natural conditions the biosphere will not be significantly impacted from uranium releases (Fig. 4) out of a hypothetical repository, it should be kept in mind that any groundwater exploration or exploitation activity in the vicinity of an uranium flow path could lead to serious short- or long-term impacts to the biosphere (Fig. 3B; Seiler and Lindner 1995). Consequently, any release of uranium from a repository will mostly lead to some restrictions in future groundwater exploration and exploitation (Ghergut et al. 2001).

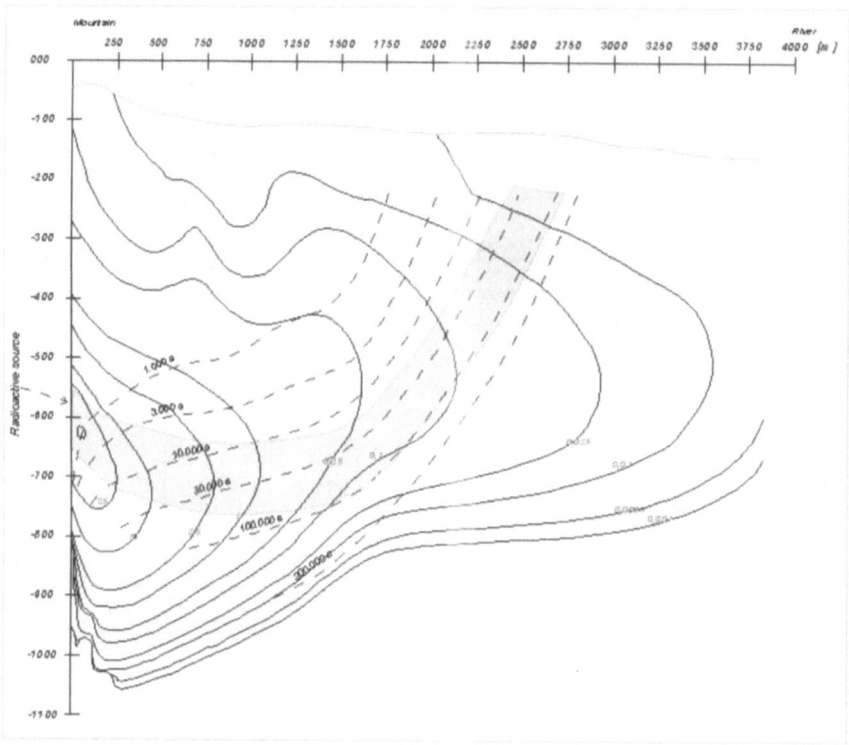

Fig. 4. Dilution of uranium released at about 600 m below ground surface into a monoporous media and propagating (gray channel) on a distance of 2 km to the biosphere. Included are isolines of concentration starting with 100 units at the release site and the isochrones reaching from 1,000 to 300,000 years till reaching the biosphere.

Long term storage of radioactive waste

Conclusions

The storage of radioactive wastes takes place deep below the active turn-over zone of groundwater; in these depths turn-over times are very long, independently from the apparent hydraulic conductivity, because groundwater flow is only linked to a small percentage to the actual water cycle.

Undoubtedly, main emphasis for radioactive waste storage must be given to safely enclose uranium in the storage facility. A rest risk by releases to the hydrosphere, however, cannot definitely be ruled out.

Under natural conditions such potential releases undergo

- strong sorption only in low mineralized and in reducing groundwater,
- negligible sorption in carbonate and sulfate waters with some oxidizing conditions and
- sorption is not significantly influenced by the presence of NaCl;
- significant dilution in the passive groundwater turn-over zone only through transverse hydrodynamic dispersion, which gets enhanced by several orders of magnitude through matrix diffusion in rocks with matrix porosity, and
- further mixing occurs along the flow path of deep through shallow groundwater.

All these hydrodynamic processes reduce the concentration from the release point to the biosphere by at least many orders of magnitude.

Contrary to these conditions, it should be realized that any groundwater exploration or exploitation activity close to the propagation path of uranium should be avoided because it would create artificial short cuts to the biosphere and because of dilution of uranium in deep groundwater, which was mostly not efficient enough on short distances to protect health and live. These short cuts are produced under long-term transient conditions (Seiler and Lindner 1995).

References

Wolfrum C, Klotz D, Bode W (1988) Bestimmung des Sorptions-Desorptionserhaltens ausgewählter Radionuklide an Sedimentproben des Asse-Deckgebirges. GSF-Bericht 25/88

Engelhardt v. W (1960) Der Porenraum der Sedimente. 207 p, (Springer) Berlin

Freeze R A, Witherspoon P A (1967) Theoretical analysis of regional groundwater flow: 2. Effect of water table configuration and subsurface permeability variations. Water Resources Res. 3, 623-634

Gherguth J, Maloszewski P, Seiler K-P (2001) Age, environmental isotope, and contaminant transport uncoupling in heterogeneous aquifers following hydraulic stress. In: Seiler K.-P. and Wohnlich S. (Eds.): New Approaches Characterizing Groundwater Flow, Vol. 1, 327-334

Papelis C, Hayes K F, Leckie J O (1988) Hydraql: A Program for the computation of chemical equilibrium composition of aqueous batch systems including surface complexation modelling of ion adsorption at the oxid/solution interface. Stanford University California, Technical report No. 306, Sept.1988

Seiler K-P, Lindner W (1995) Near surface and deep groundwater. J. Hydrol. 165, 33-44; Amsterdam

Sudicky E A, Frind E O 1981 Carbon-14 dating of groundwater in confined aquifers: Implication of aquitard diffusion. Water Res. Research 17, 1060-1064

Long term storage of radioactive waste

Natural U occurrences as a palaeo-hydrogeological indicator – observations from the Palmottu natural analogue site, Finland

Urho J. Suksi[1], Kari Rasilainen[2], Timo Ruskeeniemi[3], Nuria Marcos[4], Karl-Heinz Hellmuth[5]

[1]University of Helsinki, Laboratory of Radiochemistry, 00014 HY, Finland
[2]VTT Processes, 02150 Espoo, Finland
[3]Geological Survey of Finland, 02151 Espoo, Finland
[4]Helsinki University of Technology, 02015 TKK, Finland
[5]Radiation and Nuclear Safety Authority Finland, 00881 Helsinki, Finland

Abstract. Use of natural U in studying recharge of oxygenated waters in crystalline bedrock is discussed. Uranium is a redox-sensitive element and therefore easily mobilised when conditions change from reducing to oxidising. It has two independent decay chains and occurs easily accessible to groundwater in the bedrock, thus forming a sensitive network of "probes". Glacial melt waters rich in dissolved oxygen may penetrate the bedrock, producing signals in the U network along and around the flow routes. The signals record U behaviour as U-series disequilibria that are detectable long after the forming of the signal. Relative roles of modern groundwater flow and glacial melt waters as forming these signals are considered.

Introduction

Present climate models indicate that the next ice age in northern Europe will take place after some tens of thousands of years. The processes most likely to affect subsurface conditions are permafrost and recharge of oxygenated melt waters (*e.g.* Boulton et al. 2001). The possible effects of ice ages on disposal safety must be considered in performance assessments of geological disposal of nuclear waste in northern countries, like Finland.

Traces of glacial melt water recharge are being studied in site investigations for possible nuclear waste repository sites. Despite clear oxygen isotope evidence of deep penetration of cold climate recharge (glacial melt water; *e.g.* Blomqvist et al. 1989; Blomqvist et al. 2000), convincing geochemical evidence of the influence of dissolved oxygen still remains to be found. Slight impact on mineral phases may be expected as a result of oxidation but the interpretation often lacks age constraints. Fracture fillings have been considered as key samples since they represent specific sampling points in the groundwater flow channels, but all past changes in the conditions in the fracture have immediately affected the chemical and isotopic composition of these samples. Rock matrix adjacent to water-carrying fracture appears to represent a better protected but still accessible sample material. All changes in the groundwater in the fracture are conveyed into the rock matrix via diffusion, and one can assume that the most recent responses are closest to the fracture while the older ones are further in the matrix. In the present paper we discuss the prospects of using natural U in rock matrix for palaeo-hydrogeological investigations applying observations from the Palmottu natural analogue study site, in southern Finland.

U-series disequilibria record uranium mobility

Natural U in the form of mobile secondary compounds (labile U; Stuckless and Ferreira 1976) appears useful target material for geochemical research. The known mobility of U and immobility of Th in groundwater conditions (Langmuir 1978 and Langmuir & Herman 1980) offer a means to consider geologically recent U mobilisation, providing the basis for coupling U mobilisation with U-series disequilibria (Ivanovich and Harmon 1992).

Uranium forms two independent radioactive decay chains (Fig. 1). Due to long half-lives of several radionuclides (greater than a thousand years) the decay chains are particularly useful, offering possibility to date geological deposits or determine migration time-scales in case of radionuclide mobilisation. In a sealed rock mass with no mass transfer decay chains attain radioactive equilibrium when radionuclides have equal radioactivity, i.e. their radioactivity ratio equals one. The equilibration time is determined by the longest-lived daughter and is up to 5 times of its half-life period. For a specific radionuclide pair the shorter half-live determines equilibration time; e.g. for the $^{230}Th/^{234}U$ pair equilibration time ~ 350 ky is determined by ^{230}Th (75.2 ky). Qualitatively it is clear that the most recent and intensive mass flow results in the largest deviation from equilibrium. The larger the deviation is the easier it can be measured. In practice, analytical technique determines how accurately equilibration can be determined. Using e.g. α-spectrometry practical equilibration time is ~ 300 ky.

If radioactive disequilibrium is found, *i.e.* ratios are found not to equal unity, then the rock matrix has been affected, probably by groundwater transport of radionuclides away from or into the rock. The value of the radioactive disequilibrium depends on the intensity of U mass flow and the time since the mass flow occurred. If $^{230}Th/^{234}U$ disequilibrium is found it directly indicates U mobilisation within 300 ky. The activity ratio above unity indicates U release, and below unity indicates U accumulation. Based on U chemistry the former can be said to indicate a change from anoxic to oxic conditions and the latter a converse change, from oxic to anoxic conditions.

Long term storage of radioactive waste

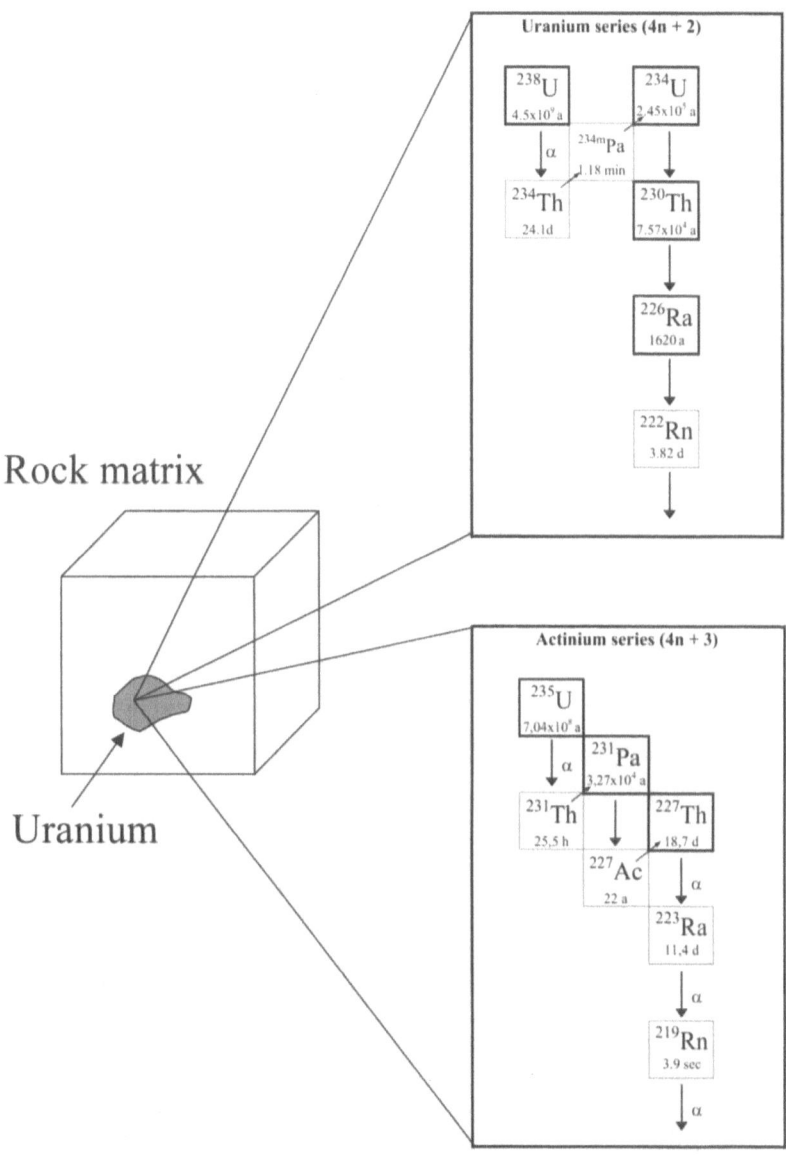

Fig. 1. The U decay chains. The longest-lived radionuclides useful for migration research and geological dating are indicated (darkened). Whether U has been transported away from or into the rock matrix during water-rock interaction can be judged from radioactivity ratios of mobile U and its immobile daughters Th and Pa.

Long term storage of radioactive waste

Did glacial melt water penetration cause the U release observed at Palmottu?

The EU-funded Palmottu natural analogue project reported several examples of U mobilisation, which may be related to the late stage of the glacial cycle (Blomqvist et al. 2000). Rasilainen et al. (2002) studied the possible role of glacial melt water pulses in U release observed in rock matrix adjacent to water-carrying fractures in a well-characterised groundwater flow path (Fig. 2). The working hypothesis was that oxygen dissolved in the flowing groundwater diffused into the surrounding rock matrix, oxidised the rock matrix and triggered U release (cf. U concentration and U series disequilibrium profiles in Fig. 3). The aim of that study was limited to test a novel approach for interpretation, however. In the current paper we discuss the subject in a broader context based on what we know of the site, glaciations and other palaeo-hydrogeological evidence.

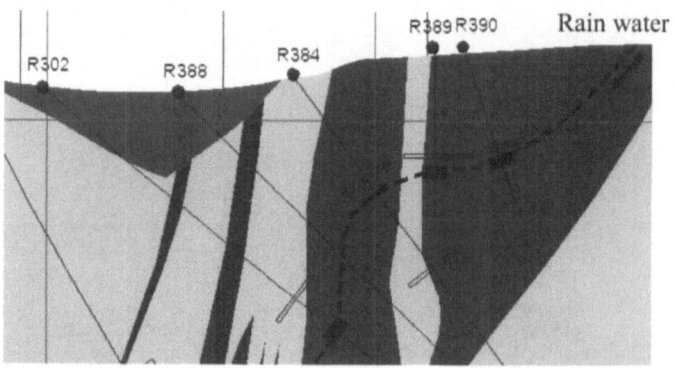

Fig. 2. Sampling locations of the drill cores along the defined groundwater flow path (Blomqvist et al. 1998).

The technical approach of interpretation in Rasilainen et al. (2002) was based on combining USD simulations with standard mass balance calculations. The USD simulations were done to study how rapidly the U that was concluded "missing" in Fig. 3 (sample R384 as an example) was released. In order to generate the observed $^{230}Th/^{234}U$ disequilibria (>>1) the U release must have been considerably faster than what the chain decay can "repair". The results in Fig. 4 showed that the U release had to have taken place during a time period of less than 300 ka, otherwise the resulting disequilibria are too low.

Mass balance calculations were done to assess the time needed to oxidise the rock matrix to the observed distance of 5 cm from the fracture. Mass balance studies were also done to estimate the time required for the released U inventory to be transported away via diffusion. The sum of these time periods represent the minimum time required for the U to have been released and transported away. The extent of the oxidised zone in the rock matrix was estimated as follows (Fig 5).

Long term storage of radioactive waste

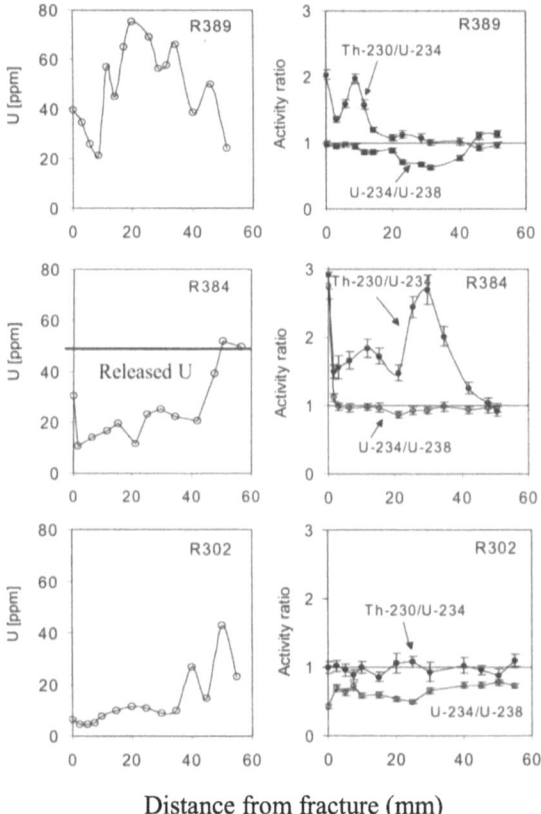

Distance from fracture (mm)

Fig. 3. U-series profiles in the drill core samples taken along the well-defined groundwater flow route outlined in Fig. 2. Recent (<200 ky) U release is evident in the two uppermost samples ($^{230}Th/^{234}U \gg 1$); oxic conditions (redox-front) have spread from the fracture in the rock matrix, triggering U out-diffusion. The inventory of released U was estimated as an integrated area between the straight line and U concentration profile in sample R384. The development of $^{230}Th/^{234}U$ activity ratio as a function of time of U release is simulated in Fig. 4. The propagation of oxic conditions from fracture into the rock matrix is further studied in Fig. 5.

First the mass flow of dissolved oxygen was calculated by Fick's law. This mass flow was subsequently multiplied by the stoichiometric factor of the redox reactions and set equal to the consumption rate of stationary Fe(II) in the rock matrix caused by the propagation rate of the oxidising front. The original redox state of the rock matrix was assumed to be anoxic; this will be discussed in a moment. It turned out that the time required for the released U to be transported away by diffusion is around 30 ka. Thus the release must have lasted much longer than the

Long term storage of radioactive waste

whole postglacial period (10 ka). The input of dissolved oxygen into bedrock is believed to have been stopped by the ice cover and it follows that the last U release can have started only 10 ka ago. This in turn means that the U release has most probably taken place in at least two episodes, i.e. during at least the last two deglaciations 100 ka apart. The simulations appeared useful. Certain scenarios could be rejected because generated Th-230/U-234 activity ratios were far from those measured.

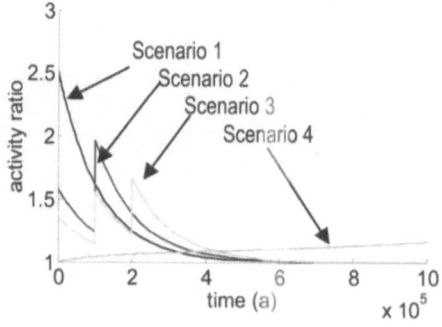

Fig. 4. The ^{230}Th/^{234}U activity ratio simulations in assumed U release histories. Scenario 1 represents instantaneous release in one go, scenario 2 release in two episodes 100 ka apart, scenario 3 release in 3 episodes 100 ka apart and scenario 4 continuous constant rate release. In the multistage release scenarios 2 and 3 the individual release batches are defined so that equal fractions of the existing inventory are released each time. In the continuous release rate scenarios the rate of release was obtained by dividing the released amount by the time period.

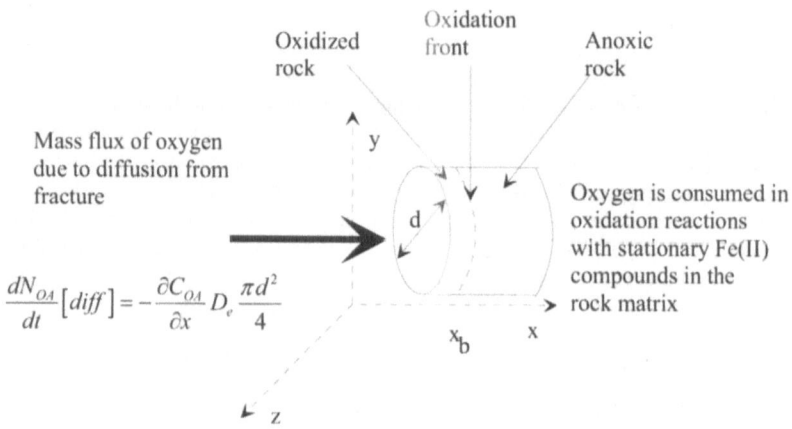

Fig. 5. Conceptual model of the propagation of redox front in rock matrix due to diffusion of dissolved oxygen.

Long term storage of radioactive waste

Discussion of the previous conclusions

For the time being there are clear geochemical observations in groundwater at Palmottu that can be explained by glacial melt water penetration (Rasilainen 2002). Mineralogical indications can be sought, for instance on fracture coating material, which would indicate reactions with oxygen, but this data does not have any direct time constraints and the reactions with oxygen may even have been as old as hydrothermal. Rare earth elements (REE) have been studied in drill core samples at Palmottu. They have many features in common with the USD observations, but like the mineralogical indications, time constraints are difficult to obtain. Against this general background the USD method with its absolute dating potential appears a useful supplement for the palaeo-hydrogeological toolbox.

The USD based conclusions about the role of glacial melt water in the U release from rock matrix were based on a hypothesis of dissolved oxygen spreading from fracture into rock matrix, oxidising the rock matrix and triggering U release. The fundamental assumption of anoxic rock matrix conditions before melt water pulse appears reasonable. The huge amount of Fe(II) containing minerals that are dissolved gradually thus releasing a steady Fe(II) flux in the rock matrix can rapidly consume all oxygen trapped in the bedrock after oxygen input is stopped, see Rasilainen et al. (2002) for details. This conclusion is supported by the observations at Palmottu that rock matrix appears anoxic even if the water-carrying fracture would be clearly oxic (Eh>0) (Blomqvist et al. 2000).

The respective roles of modern groundwater and glacial melt water in contributing to the oxidation of rock matrix can be studied by sensitivity studies. This is based on the fact that the extent of the oxidised zone in rock matrix is a function of dissolved oxygen concentration and of time. Therefore a difference can be made if the two parameters can be estimated for the two water type candidates. For modern groundwater the oxygen concentration is obviously much smaller and is quicker consumed than from melt waters (see *e.g.* Guimerà et al. 1999). If this is true then modern groundwaters cannot have important role in the oxidation of the rock matrix.

The researchers also discussed the possibility of local advection in rock matrix as a means to convey dissolved oxygen into rock matrix. This was based on the fact that, according to Darcy's law pressure differences cause advection, even if the hydraulic conductivity were small. Pressure differences could, in principle, appear in the local input of over pressurised glacial melt water, see Rasilainen et al. (2002) for details. It is worth noting that diffusion is by far the most probable mechanism, advection can occur only in special locations of hydraulic discontinuity.

Earlier USD observations of fracture coating material at Palmottu clearly support periodic mobilisation of U (Fig. 6) but their dating poorly coincides with the known details of glacial cycles (Suksi et al. 2000). It is believed that this is because the fracture coatings have experienced several successive mobilisation events with respective USD responses on top of each other's. This stack of responses was lost in the sampling procedure that produced a weighted average that is representative for the whole U mass in the sample but not representative for any individual responses.

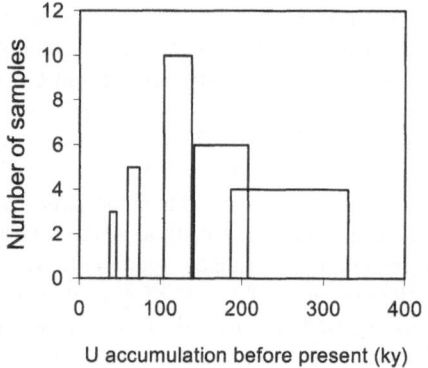

Fig. 6. Model age clusters of fracture coating samples (mainly calcites). Several U accumulation events within the time period of 110 ky before present are indicated. It should be noted here that the whole coating layer was removed for analysis. The sampling may thus represent more than one U accumulation event, providing a weighted average and, consequently, an upper limit for the age.

The above dating problem of fracture coating material exemplifies the uncertainties of the USD method and, indeed, all palaeo-hydrogeological methods. The possibility that one's sampling material has recorded many successive responses is unavoidable in nature. In terms of fracture-controlled responses it must be further stated that the average distance between adjacent fractures at Palmottu is around 30 cm. This means that the further away from fracture into rock matrix one's sampling point is, the higher is the probability that the sampling point has recorded responses controlled also by adjacent fractures. In the case of the distributions in Fig. 3 the probability of responses by other fractures is low because the actual distance covered by the sampling (5 cm) is quite short as compared to the average fracture distance.

Concluding remarks

The signs of glacial melt water penetration into bedrock observed at Palmottu were for shallow depths (around 30-40 m). As such melt water penetration to shallow depths is not surprising. The question of how deep the dissolved oxygen can penetrate into a flow system in bedrock after the next glaciation is, however, of interest to performance assessments of geological disposal of nuclear waste in Nordic countries.

Experimentally this question can be approached by the USD method by taking samples beyond current redox boundaries both along the flow route and within the

Long term storage of radioactive waste

surrounding rock matrix. The possibility is based on the fact that datable U-series disequilibria caused by past dissolved oxygen pulses will remain in the system for around 300 ka also in the case temporary oxic conditions would return to the original reducing. Thus it is possible to track the range of glacial melt water pulses provided the pulses have been long enough to release U from rock matrix. A concentrated effort with site investigation routines and other palaeo-hydrogeological methods would of course be necessary, as all palaeo-hydrogeological indications must match in a correct explanation.

References

Blomqvist, R., Ahonen, L. and Hakkarainen, V., 1989. Alustava hydrogeologinen tulkinta Outokummun Sukkulansalon alueesta. Abstract: Preliminary hydrogeological interpretation of deep groundwater in the Outokumpu area. Geological Survey of Finland, Nuclear Waste Disposal Research. Report YST-67. 47 p. + 15 app.

Blomqvist, R., Kaija, J., Lampinen, P., Paananen, M., Ruskeeniemi, T., Korkealaakso, J., Pitkänen, P., Ludvigson, J-E., Smellie, J. T. A., Koskinen, L., Floría, E., Turrero, M. J., Galarza, G., Jakobsson, K., Laaksoharju, M., Casanova, J., Grundfelt, B. and Hernan, P., 1998. The Palmottu natural Analogue Project – Phase 1: Hydrogeological evolution of the site. European Commission, Report EUR 18202, 95 p. + 1 Appendix.

Blomqvist, R., Ruskeeniemi, T., Kaija, J., Ahonen, L., Paananen, M., Smellie, J., Grundfelt, B., Pedersen, K., Bruno, J., Pérez del Villar, L., Cera, E., Rasilainen, K., Pitkänen, P., Suksi, J., Casanova, J., Read, D. & Frape, S. 2000. The Palmottu natural analogue project – Phase II: Transport of radionuclides in a natural flow system at Palmottu. Luxembourg: European Commission. 174 s. (Nuclear Science and Technology Series EUR 19611 EN.)

Boulton, G.S., Kautsky, U. and Wallroth, T., 2001. Impact of long-term climate change on deep geological repository for spent nuclear fuel. SKB Technical Report TR-99-05. Swedis Nuclear Fuel and Waste Management Co., 117 p.

Ivanovich, M. and Harmon, R.S., 1992. Uranium-series disequilibrium, applications to earth, marine, and environmental sciences, Clarendon Press, Oxford (2nd ed.), 910 p.

Langmuir, D., 1978. Uranium solution-mineral equilibria at low temperatures with applications to sedimentary ore deposit. Geochim. Cosmochim. Acta 42, 547-569.

Langmuir, D. and Herman, J.S., 1980. The mobility of thorium in natural waters at low temperatures. Geochim. Cosmochim. Acta 44, 1753-1766.

Guimerà, J., Duro, L., Jordana, S. & Bruno, J. 1999. Effects of ice melting and redox front migration in fractured rocks of low permeability, Stockholm: Swedish Nuclear Fuel and Waste Management Co. 86 p. (Technical Report TR-99-19.)

Rasilainen, K., Suksi, J., Ruskeeniemi, T., Pitkänen, P., and Poteri, A., 2002. Release of U from rock matrix – A record of glacial melt water intrusions? J. Contamin. Hydrol. (in print).

Rasilainen, K. (ed.) 2002. Nuclear waste management in Finland - Final report of Public Sector's Research Programme JYT2001 (1997-2001). 258 p. Ministry of Trade and Industry Studies and Reports 15/2002, 258 p.

Long term storage of radioactive waste

Suksi, J., Rasilainen, K., Casanova, J., Ruskeeniemi, T., Blomqvist, R. and Smellie, J.A.T., 2000. U-series disequilibria in a groundwater flow route as an indicator of uranium migration processes. J. Contamin. Hydrol. **47**/2-4, 187-196.

Stuckless, J.S. and Pires Ferreira, C., 1976. Labile uranium in granitic rocks. Exploration for uranium ore deposits. Proceedings of a symposium on exploration of uranium ore deposit, 29 March - 2 April 1976, Vienna. IAEA-SM-208/17, 717-729.

Long term storage of radioactive waste

Calculation of radioactive field in nuclear accident contaminated area from data acquired in aerial survey

Xueshi Lu[1], Yanxin Wang[2], Pingan Luo[2], Xiaobin Li[1] and Xiaoqi Zhang[2]

[1]Zhuzhou Institute of Technology, Wenhua Road, Zhuzhou, Hunan Province, 412008, China P.R.
[2]Faculty of Engineering, China University of Geosciences, Wuhan, Hubei Province, 430074, China P.R.

Abstract. By using the convolution theorem and the deconvolution theorem published recently, this paper solved the problem of calculating radiative field distribution generated from any plane radioactive deposit from the data acquired in aerial gamma-ray survey. In nuclear accident contaminated area, aerial survey can provide radiative field data at certain height. The distribution on the ground can be calculated from these measured data by using the deconvolution theorem. By using the convolution theorem, the distribution at any height can be calculated from the ground distribution in turn. Because of the use of FFT arithmetic, the computing speed is high enough to do real-time calculating.

Introduction

Since its discovery in the early 20th century, nuclear energy has not only been widely employed in military applications making weapons such as nuclear bombs, H-bombs and neutron bombs, but also been used in civil applications, such as nuclear power plants, scientific research, medical and industrial fields. The application of nuclear energy not only brings us huge economic benefits, but also potential threatening. Explosion of nuclear weapons and accidents of nuclear reactors can all cause disasters (for example, the explosion of nuclear bombs dropped in Hiroshima and Nagasaki in the WWII, the accident at Pennsylvania's Three Mile Island nuclear power plant in 1979, and the Chernobyl nuclear accident in 1986 (Deping Li et al. 1990)). The resulting residue radioactive field has a long lasting effect in a large area.

In a large area that possibly has been contaminated by the residue radiative field (typically 10 to 20 kilometer wide, tens or even hundreds kilometer long), conventional ground survey is not enough. In recent years, various computer simulation programs have been developed, as well as satellite survey system (only used for early stage monitoring). Currently, aerial survey systems have attracted a lot of interest.

The radiative field resulting from nuclear accidents or explosions is a complex field consisting of hundreds of radioactive elements. Its distribution strongly depends on the factors such as the way the accident or explosion happened, the geography and weather conditions. Before the radioactive dust completely settles, the field distribution changes continuously. Even after the deposition finishes, the field distribution still changes all the time because of the decay of the radiative elements. Even so, the real time survey is still of great importance. In order to develop efficient nuclear accident emergency plans, it is important to get real time field distribution by aerial survey.

Because the contaminated area is a surface source, a simple point source model can not describe it well. An altitude correcting coefficient based on a point source model can not be used to get the real distribution, because the coefficient can be different for different surface source at the same spot and same height, it can be different even for the same surface source at the same spot if it's at a different height. Calculating the surface distribution of the radiative field on the ground from the distribution at a height is mathematically to solve a two dimensional or three dimensional integral function. This itself is a difficult problem in mathematics. That is why the problem has remained unsolved till now.

This problem can be solved by using the Deconvolution Theorem (Pingan Luo and Chang Miu 1999).

Physics model

For simplification, let's assume
1. the surface density distribution of radioactivity of the radioactive materials on the ground is A(x, y, 0), {-∞<x<+∞, -∞<y<+∞, z=0}, unit is Ci.
2. no contribution from the radioactive materials suspended in air.
3. air does not absorb or diffract the γ ray.
4. no contribution from α, n, β rays.
5. one decay emits one γ ray.

For a source of radioactivity of A Ci (one decay emits m photons of different energy), at a distance r meters, the illumination can be described as (Xinghong Li et al. 1982):

$$\dot{X} = \frac{A\Gamma}{r^2} \qquad (1)$$

Based on the assumption in our model, for an A(x, y, 0) surface source, a differential element is as shown in Fig. 1.

Long term storage of radioactive waste

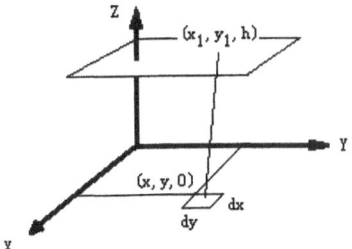

Fig. 1. Illumination at the height of h meters generated by a differential element of a surface source.

In a plane at z=h, the illumination at the spot (x1, y1, h) is

$$d\dot{X}(x_1, y_1, h) = \frac{\Gamma A(x, y, 0)dxdy}{(x - x_1)^2 + (y - y_1)^2 + (0 - h)^2} \qquad (2)$$

hence

$$\dot{X}(x_1, y_1, h) = \int_{-\infty}^{+\infty} dy \int_{-\infty}^{+\infty} \frac{\Gamma A(x, y, 0)}{(x - x_1)^2 + (y - y_1)^2 + h^2} dx \qquad (3)$$

Let

$$H(x_1, y_1, h) = \frac{\Gamma}{x_1^2 + y_1^2 + h^2} \qquad (4)$$

From the definition of two dimensional convolution, we have

$$\dot{X}(x_1, y_1, h) = A(x_1, y_1, 0) * H(x_1, y_1, h)$$

Here * is the operator of convolution operation.

For the sake of traditional convention, x1 and y1 can be replaced by x and y.

$$\begin{cases} \dot{X}(x, y, h) = A(x, y, 0) * H(x, y, h) \\ H(x, y, h) = \frac{\Gamma}{x^2 + y^2 + h^2} \end{cases} \qquad (5)$$

Long term storage of radioactive waste

For a complex γ radiation field, the calculation of the illumination coefficient Γ is very complicated. From the reference (Xinghong Li et al. 1982) we can get

$$\Gamma = 2.361 \times 10^{-3} \sum_{i=1}^{m} n_i \left(\frac{\mu_{en}}{\rho}\right)_i E_{\gamma,i} \qquad (6)$$

Since the spectra distribution in the contaminated area is too complicated and hard to measure, for simplification, it can be considered as illumination coefficient of one radioactive element. ^{60}Co is normally used in instrument calibration, we use the Γ coefficient of ^{60}Co as the illumination coefficient in the calculation. From the reference (Xinghong Li et al. 1982) we have

$$\Gamma = 1.32 \quad (m^2 Ci^{-1} R/h)$$

Now, illumination at any spot at z=h can be calculated from equation (5) as long as surface source distribution A(x, y, 0) is known. But in reality, $\dot{X}(x, y, h)$ at any spot at z=h can be measured by survey. For example, the distribution at the height of 1 meter can be obtained by ground survey, the distribution at the height of h meters can be obtained by aerial survey. The problem turns to calculating A(x, y, 0) from $\dot{X}(x, y, h)$ and H(x, y, 0).

This is just a problem that deconvolution will solve. The deconvolution theorem is described as follows (Pingan Luo and Chang Miu 1999).

If (1) y(t), h(t), x(t) meet Dirichlet condition over an arbitrary range [a, b], and $\int_{-\infty}^{+\infty}|y(t)|dt$, $\int_{-\infty}^{+\infty}|h(t)|dt$, $\int_{-\infty}^{+\infty}|x(t)|dt$ converge, ie. these integrals are absolutely integratable.

(2) The convolution y(t)=h(t)*x(t) exists.

(3) X(ω) only has definite zero points on an arbitrary range.

(4) $\dfrac{Y(\omega)}{X(\omega)}$ is absolutely integratable over $(-\infty, +\infty)$.

here Y(ω), H(ω) and X(ω) are Fourier transformations of y(t), h(t) and x(t) respectively, then there must be a value α_n exists so that

$$h(t) = \lim_{n \to \infty} F^{-1}\left[\frac{Y(\omega)}{X(\omega) + \alpha_n(\omega)}\right]$$

here $|\alpha_n(\omega)| \le a_n$ n=1, 2, ..., $\lim_{n \to \infty} a_n = 0$

and $|X(\omega) + \alpha_n(\omega)| > 0$, $|X(\omega) + \alpha_n(\omega)| \ge \dfrac{1}{k} \max\{|X(\omega)|, |\alpha_n(\omega)|\}$

(n=1, 2, ...,   k is a positive constant)

a_n is relevant to n, irrelevant to ω.

Long term storage of radioactive waste

For example: let $a_n = \dfrac{2}{n}$, $\alpha_n(\omega) = \begin{cases} 0, & |X(\omega)| \geq \dfrac{1}{n} \\ \dfrac{2}{n}, & |X(\omega)| < \dfrac{1}{n} \end{cases}$

It can be proved that $\alpha_n(\omega)$ meets the above conditions. The deconvolution theorem is also true in multi-dimensional space.

In this article, $\dot{X}(x,y,h)$, $H(x,y,h)$ and $A(x,y,0)$ correspond to y(t), h(t) and x(t) respectively. An example of using a software package based on the deconvolution theorem to solve the problem is as follows.

Supposing in a 10.8km×23.6km plane area, there is a uniform surface source of 7.55Ci/m². By using equation (3), the calculated distribution at the heights of 1m, 50m, 80m, 100m, and 200m are as shown in fig.2, fig3, fig4, fig.5, fig.6, and Fig. 7. (Note: sampling space is 100m×100m). In aerial survey, suppose the airplane flying at the height of 100m, the measured distribution should be as shown in Fig. 5. By using the deconvolution theorem, the distribution on the ground (at the height of 0m) can be calculated. An example for a_n=0.00001 is shown in Fig. 8.

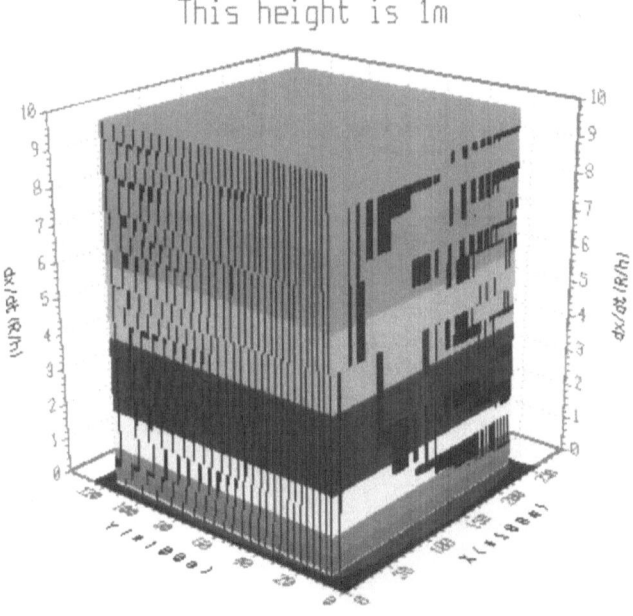

Fig. 2. The distribution at z=1m.

Long term storage of radioactive waste

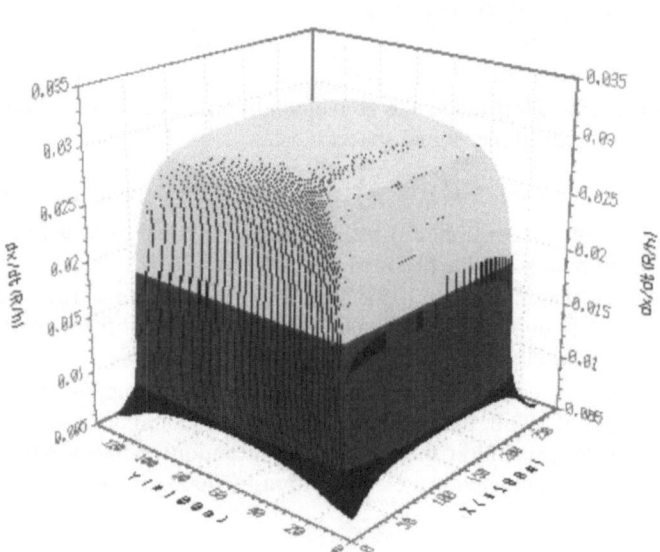

Fig. 3. The distribution at z=50m.

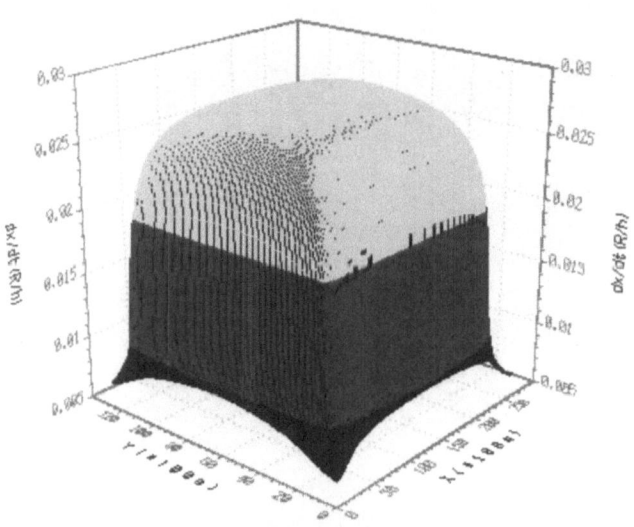

Fig. 4. The distribution at z=80m.

Long term storage of radioactive waste

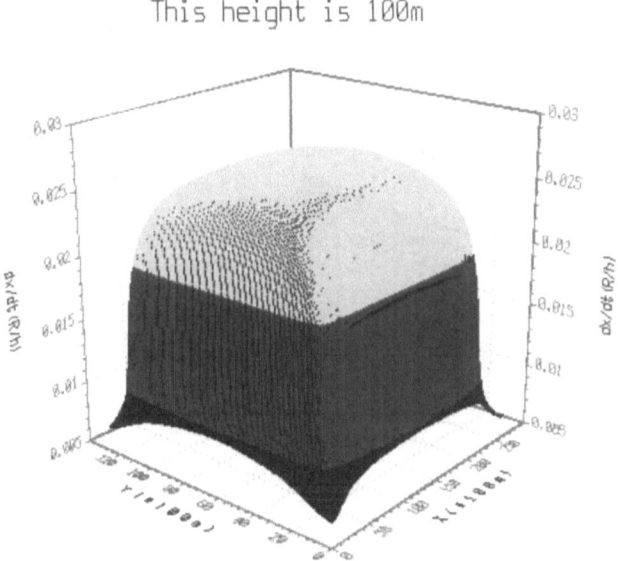

Fig. 5. The distribution at z=100m.

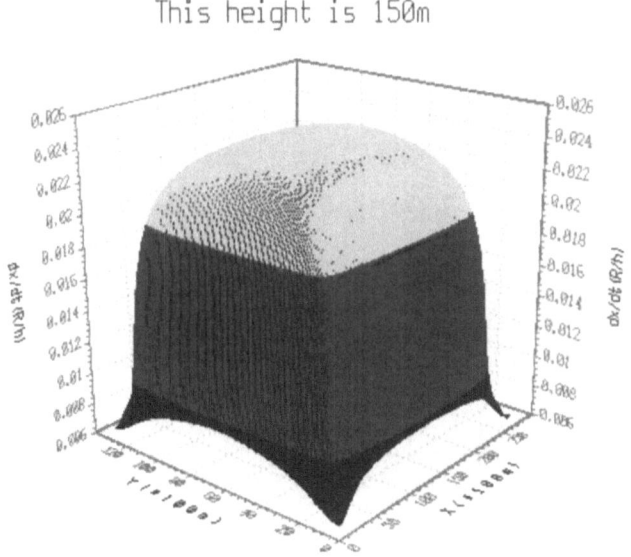

Fig. 6. The distribution at z=150m.

Long term storage of radioactive waste

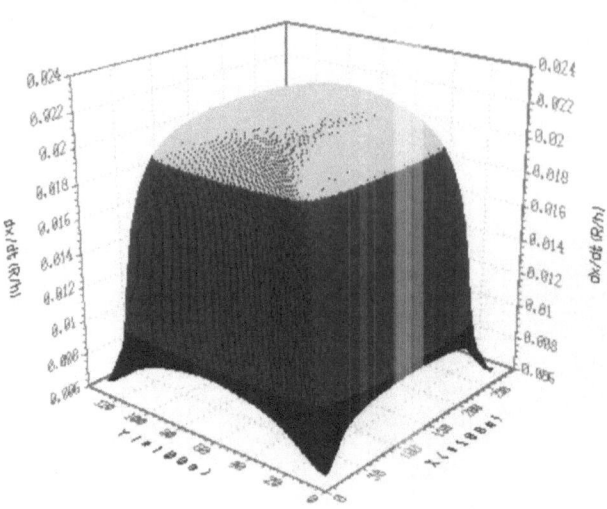

Fig. 7. The distribution at z=200m.

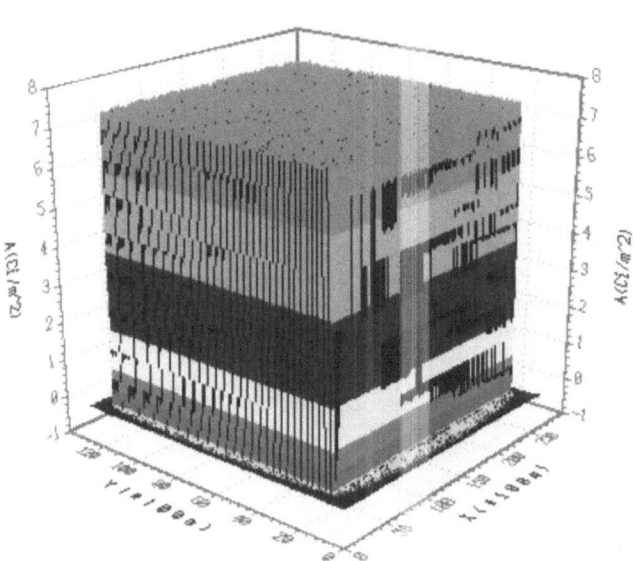

Fig. 8. Radioactivity distribution at z=0m calculated from the illumination distribution at z=100m by using deconvolution theorem.

Long term storage of radioactive waste

Discussion and conclusion

1. After getting a plane distribution at certain height through measurement, the surface distribution on the ground can be calculated by using the deconvolution theorem. The distribution at any height can be calculated from the found distribution in turn by convolution theorem.
2. At the spot (12km, 6km), the illumination at the height 1m, 50m, 80m, 100m, 150m, and 200m are shown in fig.9.
3. Table 1 shows the comparison between surface source and point source models.

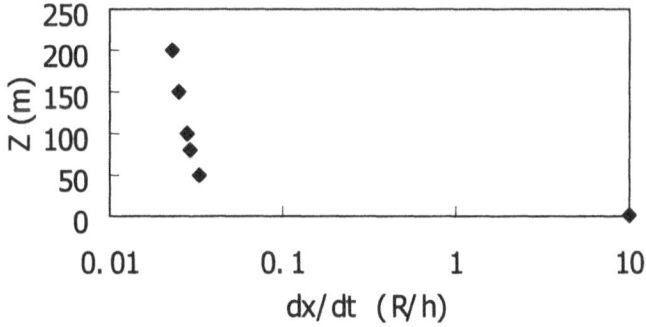

Fig.9. Illumination versus height in the uniform plane source model. (dx/dt=10R/h at z=1m)

Table 1. Comparison between surface source and point source

Z (m)	dx/dt (R/h)	Surface Source (Ci/m^2)	Point Source (Ci)	Point/Surface
1	10	7. 55	7. 58	1. 00
50	0. 033	7. 55	62. 5	8. 28
80	0. 029	7. 55	140. 61	18. 62
100	0. 028	7. 55	212. 12	28. 10
150	0. 025	7. 55	426. 14	56. 44
200	0. 023	7. 55	696. 97	92. 31

Long term storage of radioactive waste

4. From the simulation result, we can see that the simple altitude correcting coefficient can not be used for a surface source. For example, the illumination at 1m is about 300 times of the value at the height of 50m, 340 times, 360 times, 400 times and 430 times of the values at the height of 80m, 100m, 150m, and 200m respectively.
5. Fig.8 shows that the calculated value fits the theoretical value $7.55Ci/m^2$ very well. This means the calculation based on the deconvolution theorem is reliable. The oscillation on the edge is caused by FFT, which can be reduced by choosing appropriate window function.
6. The calculation employed fast Fourier transformation, for 256×128 spots, sampling space 100m×100m, on a computer with 266MHz CPU, the calculation took a few seconds. This calculation method can be used for fast data processing.

Acknowledgement

The authors thank professor Yuanhe Huang of department of chemistry at Beijing Normal University for help.

References

Deping Li, Ziqiang Pan et al. (1990) Manual of radiation protection III – radiation safety: 721-732

Xinghong Li et al. (1982) Fundamental of radiation protection: 110-120

Pingan Luo, Chang Miu (1999) Deconvolution Theorem. Nuclear electronics and Detection technology 6: 110-120

Experimental large diameter well in the clays for radioactive waste Repository

Konstantin Yefimov, V.N. Chernonozhkin[1], Sergey Dmitriev, Lev Prozorov, Aleksey Tkachenko[2]

[1] Institute of Ecotechnologies
[2] MosNPO "Radon"

Abstract. Since 1997 MosNPO "Radon" has been testing large diameter wells (LDW) as low and intermediate level waste (LILW) repositories. If waste is deposited below the frost zone damage to engineered barriers due to climatic factors is practically impossible. Construction of the repository by means of drilling prevents large disturbances of the host rocks, as it might happen during excavation work. When LDW are used, the usage factor which is the ratio of repository surface area (natural barrier) to total repository surface area is about 1, as almost all the surface of the repository is covered with host rocks.

Background

Application of radioactive substances in industry, medicine and scientific research leads to occurrence of radioactive wastes. The problem to improve reliability of isolating radioactive waste is becoming more and more acute, and requirements for ecological safety are getting more and more stringent.

Near surface repositories are considered to be acceptable for storage of waste with low and intermediate levels of activity (LILW), which decay to safe levels in some hundreds of years. However, the experience with LILW isolation in the near surface facilities in Russia has shown that some operational and natural factors may have an impact on the engineered barriers and may cause failure of the isolation. In addition, the exploitation of old repositories and construction of new ones require more space.

Significant increase in LILW isolation reliability and space saving may be achieved by wastes disposal/storage in large diameter wells (LDW), drilled in clayey sediments. Thus MosNPO "Radon" specialists developed the LDW construction technology with the aim to use such wells for LILW isolation in moraine clays. The diameter of a LDW type repository may range from 1 to 5 or 6 m depending on drilling rig capabilities and performance parameters of the repository in whole. The depth of the wells depends on geological and hydro-geological conditions of the site.

A project for constructing a demonstration unit for LILW storage in large diameter wells at the MosNPO "Radon" disposal site near Sergiev Posad was developed. The aim of the project is further development of the technology and pilot operation of the new repository for 25-30 years.

The project provides for construction of 9 wells with diameters from 1.9 to 4 m and with depths from 40 to 50 m. The wells will be drilled in a 7.2 x 7.2 m grid. For radiation and geological monitoring 16 wells will be constructed.

The LDW construction is based on multi barrier criteria. The engineered barriers provided in LDW have optimal shape and perform at stable conditions:

- The barriers have a closed ring configuration, which provides the ability to resist more load in comparison to open box shapes;
- The barriers have minimal contact with other barriers, decreasing negative impacts of one barrier on another.

However, the reliability of the engineered barriers does not reduce the role of the natural barrier, the hosting deposits. The post-closure long-term safety assessment and evaluation of likely radionuclide release to the environment in the case of engineered barrier failure credits the surrounding rocks with providing a medium that is able to constrain the contamination to remain within the site boundaries.

Geology of the test site

The uppermost geologic deposits in the area are Quaternary moraine clays between 0.0 and 62 m in depth. Below this layer a water-bearing Cretaceous horizon was found. The aquifer is composed of sands and clays (in the top) and various grained sands (in the bottom). Ground water in the aquifer is artesian, and the water table is at the 41.8 m. depth. Between the depths of 45.7 and 48.2 m, clays have locally lenses of water-saturated sands. Due to the hydro-geological conditions, the LDW depth was limited to 40 m.

During construction of the first wells it was discovered that the clayey deposits of the Moscow moraine, where the wells are situated, include local water-containing lenses and thin layers of sands and loamy sands, sometimes with gravel. Thickness of several sandy layers is 1-2 cm, seldom exceeding 5-10 cm. According to hydro-geological pump-down results, the hydraulic connection between the layers is absent or very slow. The hydraulic conductivity (Kf) of:

- thin sand layers is 0.08-0.1 m/day,
- gravel loamy sands - 1.0 m/day,
- moraine deposits in average - 0.008 m/day.

Besides the ground water in the Moscow moraine stratum there is a water horizon in the clayey sands of intermoraine fluvioglacial bed at the depth of 44.8 m. The clayey sand thickness is 1.6 m. The aquifer has a water head of 33 m (piezometric surface is at 11.4 m depth). The water has a bicarbonate-magnesium-calcium content and 0.4 g/l salinity. Hydraulic conductivity of the aquifer is 0.94 m/day.

Average sorption and filtration features of the Moscow moraine interbeds and the intermoraine fluvioglacial clayey sands are given in Table 1.

Table 1. Average sorption and filtration features of interbeds.

	Soils			
	loam prQII	loam gQIIms	Loam gQIIdn	loamy sand flgQII dn-ms
Average thickness, m	3.1	41.6	8,0	1.6
Plasticity index, %	16	9	11	-
Fraction content <0.005 mm, %	26	14	16	-
Colloid activity	0.61	0.68	0.70	-
90Sr Kd	78	27	26	24
137Cs Kd	3850	2300	2300	2260
Kf, m/day	0.0094	0.008	0.0013	0.94

As may be seen from Table 1, sorption properties of the water-containing intermo-raine clayey sands are close to the properties of the loams. Therefore, in the case of hypothetical radionuclide release to this aquifer, the contamination should not move too far because of the high sorption capacity and low permeability and hy-draulic gradient (~0.001) of the horizon. Because of low water yield, the aquifer is not used for local water supply; nevertheless, it should be observed for possible radionuclide migration during the time of LDW site operation.

Pilot LDW Site

In accordance with the project plan, two LDW repositories have been constructed at the MosNPO "Radon" Site. These are wells with 1.5m internal diameter and 40 m depth (Fig.1a).

The wells were drilled with the A-50 drilling rig by means of rotary drilling a 1.9 m diameter borehole, using clay mud as drilling fluid. At completion, the well was cased with a steel pipe of 1.5 m diameter with a 16-20 mm wall thickness. Af-ter the casing string had been set, the drill string borehole annulus was filled with bentonite-cement mortar for providing isolation of the casing from surrounding rocks. When the bentonite-cement mortar solidified, the clay drilling mud was pumped from the well.

Long term storage of radioactive waste

In the dry well, containers of solidified radioactive wastes were put with a special installing column. Metal 200-liter drums 0.6 m in diameter were used as containers.

It is worth noting that the wells may be used for both storage and disposal. In the first case the LILW containers are stored in a dry well and may be recovered at any time. In the second case, the void space between the containers and casing pipe is filled with bentonite-cement mortar and the wastes are not subject to recovery (Fig.1b).

Control of near field rocks and leak proofness of the repository is provided by a monitoring system. The system consists of wells equipped with a set of high resolution seismic gages and radiometric equipment for monitoring any possible radionuclide release out of the repository boundaries. A layout of the wells and special channels for monitoring of LDW repositories are shown in Fig. 2.

Geological monitoring wells are located between adjacent repositories. In geological monitoring wells, rock conditions during LDW drilling are observed. The results of stress deformation field observations will serve as a base for determination of a safe distance between LDW repositories from the point of view of rock stability. Radiation is controlled in the working zone of the repository, in the casing pipe annulus and in hosting rocks. The radiation monitoring results provide the basis for conclusions about isolation of LILW containers and the repository as a whole.

In the working zone radiation monitoring is performed by means of periodic gamma-logging in a work string. Using the same monitoring channel, possible water penetration into the repository is detected and controlled.

To control the isolation of a LDW repository, three metal perforated pipes as channels are set at the total depth of the repository in the casing string annulus. Within these channels gamma-logging is performed periodically and the hydrodynamic mode of underground water is controlled.

Long term storage of radioactive waste

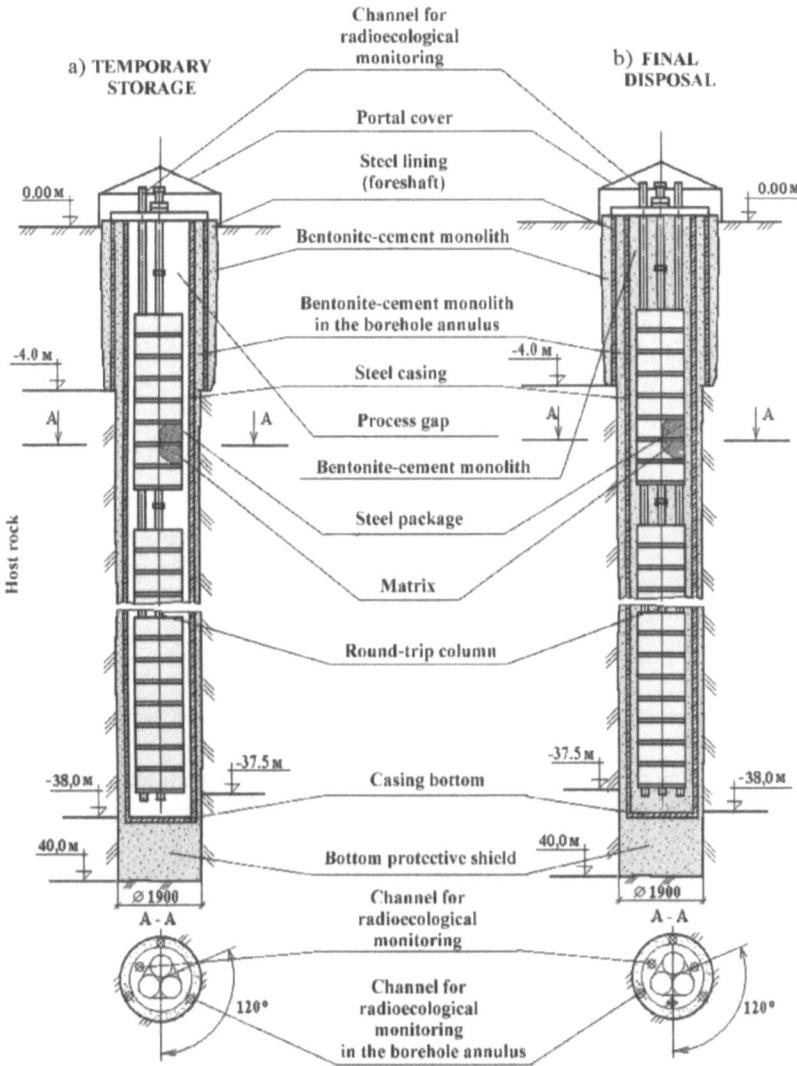

Fig. 1. Large diameter wells applied as a LILW repository (a) and as a LILW disposal (b).

Under certain conditions at the Sergiev Posad site, control of possible radionuclide releases from a repository in the host rocks can be performed using a well, located 2 m from the construction wall along the ground water flow. This well taps an aquifer, located at about 47 m depth. Control of soil background radiation, hydrodynamic mode of the aquifer, and radionuclide content of underground water are determined using gamma-logging and water sampling methods

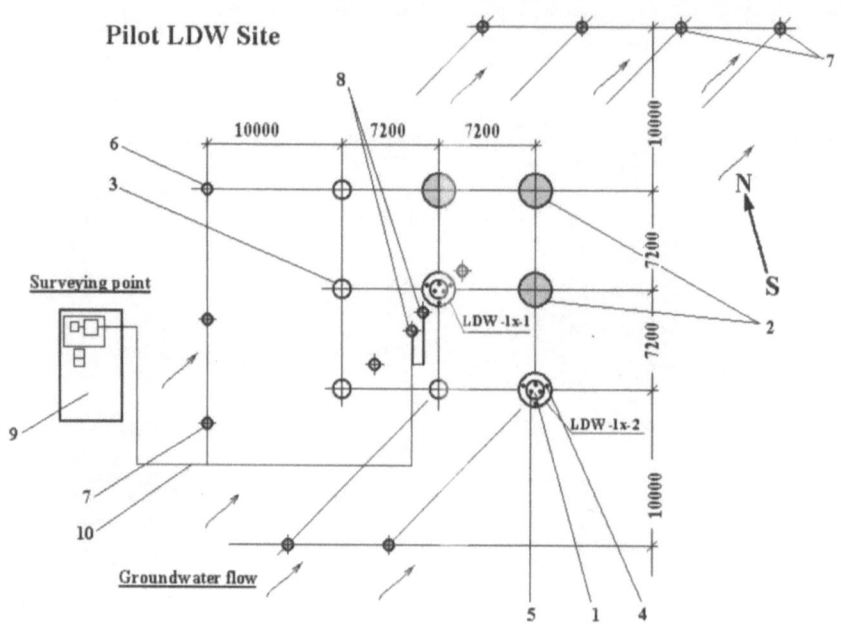

Fig. 2. Pilot LDW site layout.

The developed monitoring system allows controlling construction impermeability and surrounding rock stability with high level of reliability.

LDW Advantages

Nowadays typical near surface repositories are widely used for long-term storage of conditioned low and intermediate level radioactive wastes (LILW) at "Radon" facility sites. But their essential disadvantage has been shown as a result of long operating experience and field monitoring at the sites (Prozorov et al. 1998). This disadvantage is a decrease in LILW isolation reliability due to negative artificial and natural factors impact on the engineered protective barriers of the shallow ground repositories. In some cases there has been radionuclide migration into the near field, which leads to radioactive contamination of the ground around the storage facility. Decontamination of such ground results in generation of secondary wastes and in growth of storage costs.

Construction of new shallow ground repositories at the operating sites within the boundaries of the lease area is problematic because of the shortage of free area after 30-40 years of exploitation or for ecological safety reasons. Creation of new sites for LILW storage is difficult because of high costs of engineering surveys and construction works, long times required for obtaining regulatory authorization and license, and there is a strong negative public opinion about new repository construction for radioactive waste storage in Russian regions. These problems may be solved by increasing LILW repository reliability and by more effective use of existing sites.

In the case of large diameter wells (LDW) for LILW storage, the waste is removed from the scope of human activity to a stable geological strata. Waste is placed below the frost zone where engineered barrier damage because of climatic factors is practically impossible. Construction of the repository by means of drilling prevents large disturbances of the host rocks, as happens during excavation work. When LDW are used, the usage factor (ratio of repository surface area, protected with natural barrier, to total repository surface area) of the geological barrier (host rocks) is about 1. The construction of a LDW repository employs multi-barrier protection of the environment from radioactive waste. There are 6 barriers blocking likely radionuclide release, as follows:

1) the cement matrix;
2) the steel drum;
3) space between drums and the casing filled with bentonite-cement mixture in the case of disposal;
4) steel casing;
5) bentonite-cement stone around the casing column;
6) surrounding clayey soils. Shape and properties of the bentonite-cement stone protective barrier in a drilled borehole annulus depend on drilling technology and geology of surrounding strata.

The engineered barriers are made during the well construction. They have optimal shape and operate in stable conditions. At the expense of some increase in capital costs per m^3 of radioactive wastes (Prozorov et al. 2000), casing of a well with metal pipe with walls 10-22 mm in thickness, and grouting the bentonite-cement mortar into the space outside of the casing wall, the reliability of the LDW repositories is greatly increased.

Thus, the key differences between the LDW repositories and near surface ones are:

- there are a metal casing and bentonite-cement stone in the drill annulus, as additional protective barriers;
- the zone of hosting soils, broken during the drilling, becomes stronger after bentonite-cement mortar grouting;
- LDW depth may be up to 100 m, when depth of near surface repository bottoms is generally 3-4 m, not more than 10 m. Thus, the hosting soils play a more effective role in providing safety than in the case of typical near surface repositories.
- due to the depth, impact of climatic factors on LDW construction and waste leaching is minimal.

These factors help lead to the conclusion that, even in the case of total destruction of all engineered barriers, which is very unlikely, radionuclides will be constrained in a small volume of the hosting rocks and their release to the human activity sphere is virtually impossible.

Long term storage of radioactive waste

Thus, LILW storage in large diameter wells may be used to satisfy current requirements for long term radioactive waste storage. This type of repository appears to be promising for application at existing operational "Radon" sites.

Conclusion

Technologies used during LDW construction help make engineered barriers with desired parameters, which meet geological conditions of a given site. Such barriers will provide reliable isolation of radioactive wastes.

Monitoring systems will provide the possibility to control conditions during waste storage both inside the repository and outside, and allow remedial maintenance in case of an emergency. Hoisting mechanisms and charge-discharge systems allow removing waste packages at the end of the storage period or in the case of a need to retrieve the containers.

In spite of rather high storage costs, the LDW repositories guarantee more reliable radioactive waste isolation and, consequently, higher levels of safety for the public and the environment. The implementation of LDW repositories at operating sites for LILW storage, which are located in populous East-European regions of Russia, instead of typical near surface ones, will allow conservation of existing waste repository areas and help solve the problem of negative public opinion about new disposal sites construction. Therefore, consideration of the matter from every side, taking into account all social and economic issues, will allow acceptance of the decision to establish radioactive waste placement sites.

References

Prozorov L.B., Veselov E.I., Rybakov A.I. 1998. State of Subsurface Radioactive Waste Repositories. Proc.19th Low-Level Radioactive Waste. Management Conference. Salt Lake City, 1998.

Prozorov L.B., Litinsky Y.V, Tkachenko A.V., Titkov V.I., Tarasov V.L 2000. Economic Aspects of Large Diameter Well Construction and Operating. Proc.of WM2K, CD, Tucson, USA, 2000.

Redox processes in uranium deposits – Lessons from Natural Analogue Studies

Jordi Bruno, Lara Duro, Esther Cera and Mireia Grivé

EnvirosQuantiSci (Spain), Parc Tecnològic del Vallés
Avda, Universitat Autònoma 3, Cerdanyola 08290 (Spain)

Abstract. The measurement and understanding of the processes that control the redox intensities (Eh) and capacities (RDC) of uranium deposits are critical to assess the long-term stability of nuclear waste repositories, the migration of toxic metals from uranium mine tailings and other contaminated sites. We present the redox modelling work performed in connection with the geochemical characterisation of five uranium deposits used as Natural Analogue systems: Poços de Caldas (Brazil), Cigar Lake (Canada), El Berrocal (Spain), Oklo (Gabon) and Palmottu (Finland). The data from these five sites indicate that $Fe(III)/Fe(II)$ equilibrium controls the redox state of surface and intermediate groundwaters. However, the redox state of undisturbed groundwaters in contact with U mineralisations appears to be controlled by $U(IV)/U(IV)$ equilibria in the solid phase.

Introduction

The geochemical stability of nuclear waste repositories and uranium mine tailings largely depends on the redox state of the system. This is because U solid phases are the main components of these systems and U alteration and mobility is strongly linked to the oxidation of $U(IV)$ to $U(VI)$.

Geochemical stable conditions are partially defined by the various intensive parameters, eg. pH and Eh. However, the capability of the disposal system to react against chemical disturbances is not completely described by the intensive parameters since their persistence depends on the associated extensive parameters: alkalinity and redox buffer capacities.

The definition of redox conditions has been largely hampered by the fact that it is difficult to establish to which extent a Natural Water system is in redox equilibrium and which is the redox couple that controls the system. This has been the result of both the difficulties on measuring reliable redox potentials and/or the relatively slow kinetics of some redox processes. This could be the case of multielectron homogenous redox processes with a major structural rearrangement like the U(VI)/U(IV) redox reactions (see Bruno, 1997 for a more detailed discussion).

The present nuclear waste disposal programs have had a profound impact on the development and improvement of studies of the redox conditions of deep groundwater, as well as in the establishment of precise redox measurements (Wikberg, 1987 and Ahonen, 1994).

Natural Analogue studies have been a dynamic element in the development of the redox measurement technologies and the understanding of the associated geochemical processes.

We present in this paper the main outcome from the investigations of redox processes in five uranium deposits that have been investigated in the connection with Natural Analogue studies. These are Poços de Caldas (Brazil), Cigar Lake (Canada), El Berrocal (Spain), Oklo (Gabon) and Palmottu (Finland).

Brief description of the sites investigated

A detailed description of the investigations performed in these sites may be found in Miller et al (2000). What follows is a brief indication of the main geochemical attributes of the investigated systems.

1. Poços de Caldas (Brazil), a Mesozoic volcanic ring-structure hosting a U-Th mineralisation (Chapman et al., 1993).
2. Cigar Lake (Canada), a 1.3 billion year old uranium deposit located in a water saturated sandstone where the ore is surrounded by a clay-rich halo (Cramer and Smellie, 1994).
3. El Berrocal (Spain), a granitic Hercinian massif intercepted by a quartz vein with associated primary uranium mineralisations (Enresa, 1996).
4. Oklo (Gabon), a fossil natural nuclear reactor systems located in a Precambrian sedimentary basin (Louvat et al., 2000).
5. Palmottu (Finland), a uranium deposit located within Precambrian metamorphosed supracrustal and sedimentary rocks in SW Finland (Blomqvist et al., 2000).

Cross-sections of the investigated sites are shown in Figure 1. The diagrams include information on the hydrogeology of the systems, together with the measured pH and Eh.

A common trend in all five sites is the presence of more or less concentrated uranium mineralisations, placed at various depths in the host rock.

Description of the geochemistry of the sites investigated

The different geochemical environments considered can be classified into three groups:

Geochemical and reactive transport modeling

- Granitic media: El Berrocal (Spain) and Palmottu (Finland)
- Sandstone-Clay media: Oklo (Gabon) and Cigar Lake (Canada)
- Alkaline volcanic media: Poços de Caldas (Brazil)

The main characteristics of these sites according to the mineralogy and to the composition of fracture filling material are summarised below. Also some indications on the hydrochemistry are given.

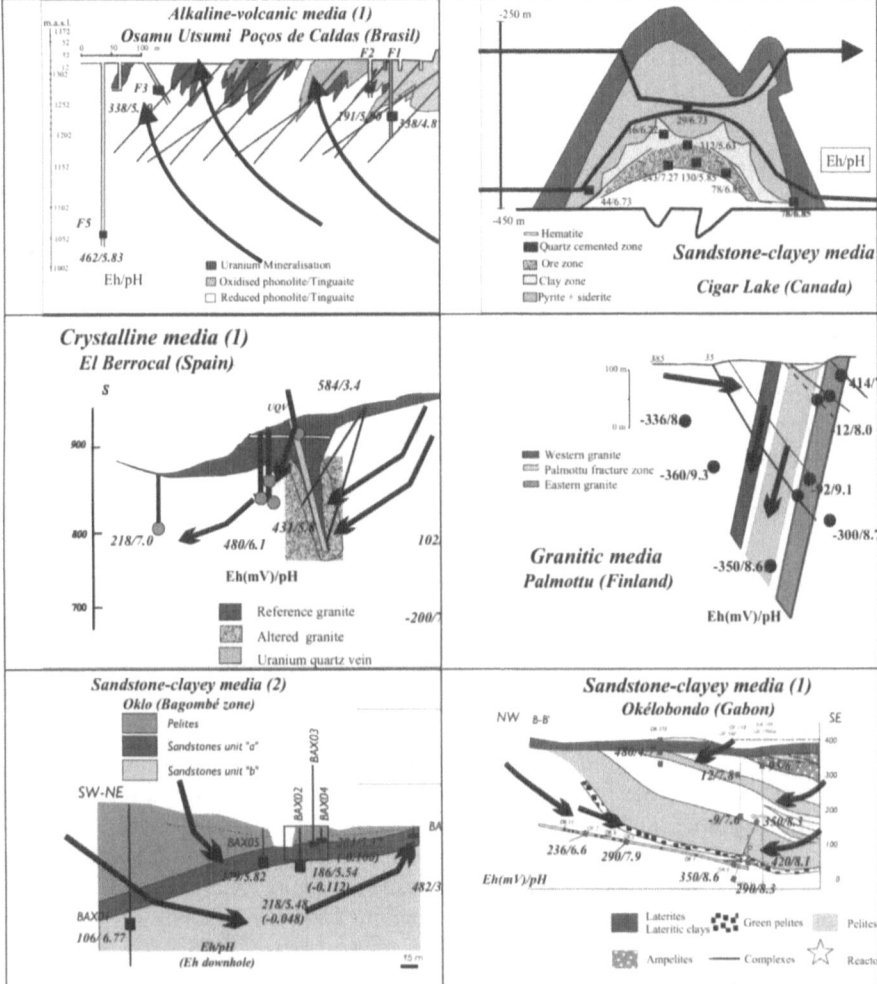

Fig. 1. Cross section diagrams of the various sites presented in this study. The big arrows indicate the flow direction. Eh and pH data are plotted in the approximated location of the sampling boreholes.

Geochemical and reactive transport modeling

El Berrocal (Spain)

According to the hydrochemical investigations conducted at the site (Gómez et al., 1996) the groundwater sampled can be classified into two main groups. The Ca-SO_4 type waters in the vicinity of the Uranium Quartz Vein (UQV), due to pyrite oxidation, and Ca-HCO_3 type waters in the boreholes located farther from the UQV, that have greater alkalinity. The pH of the samples ranged from acidic (pH: 3-5) in the vicinity of the UQV to values around 7 in the deepest samples of boreholes S14 and S11. The redox potential is oxidising in the more shallow waters, achieving reducing values in the deepest samples. A complete description of the El Berrocal project can be found in ENRESA (1996).

Palmottu (Finland)

According to the groundwater geochemical modelling in the Palmottu site performed by Gimeno and Peña (1999), the composition of the water sampled in the site changes from Ca-HCO_3-type near the surface to Na-HCO_3-type at intermediate depths. At greater depths, the waters are of Na-SO_4-Cl type with the Na-SO_4 groundwater apparently confined to the mineralised area. In general, an increase in total dissolved solids and a decrease in the redox potential are observed with depth. The pH values of the groundwater samples in Palmottu range from 7.5 to 9.3. The measured Eh values are normally reducing, in the range −12 to −360 mV, except for the shallowest sample, which gave a redox potential of 414 mV. A complete description of the Palmottu Natural Analogue study can be found in Blomqvist et al. (2000).

Oklo (Gabon)

In the Oklo region two sites were investigated, Okélo and Bagombé.

According to the geochemical modelling reported in Peña et al. (2000), the groundwater samples can be classified into 4 different groups:. (1) Samples of the west gallery (Okélo C), that present low mineralisation, slightly acidic to neutral pH values and oxidising conditions. (2) Samples from boreholes drilled from the surface, sampled in the Complexes Formation, with neutral pH and reducing Eh values; (3)Samples close to the reactor (Okélo E), corresponding to groundwaters which have interacted with pelites, with alkaline pH and oxidising Eh values; (4) Surface waters, with acidic pH values and oxidising redox conditions.

In the Bagombé area quartz, chlorite goethite and clay minerals mainly compose the host rock, while the fracture fillings contain quartz, carbonates, sulphides (pyrite), chlorite, Fe oxy-hydroxides and illite. Near the reactor quartz, chlorite, illite, kaolinite, pyrite, Fe oxy-hydroxides, and organic matter mainly compose the fracture fillings. All the samples are of Ca-HCO_3 type. These groundwaters are moderately acidic to neutral (pH values from 4 to 7). The redox potentials are discussed in Louvat et al. (2000), they range between 100 and 400 mV.

A complete description of the Oklo Natural Analogue study can be found in Louvat et al. (2000).

Cigar Lake (Canada)

The groundwaters at depth are weakly mineralised and highly reducing, with an ionic strength in the range 10^{-4} to $5 \cdot 10^{-3}$ and low Eh (Cramer et al., 1994). From the study of the groundwater evolution at the site (Cramer and Nesbitt, 1994), it appears that the bulk composition of the groundwater is controlled by interactions with clay minerals, mainly illite-kaolinite, and that the redox geochemistry is controlled by the iron and sulphur redox couples.

A complete description of the Cigar Lake Analogue study can be found in Cramer and Smellie (1994).

Poços de Caldas (Brazil)

Two sites were investigated in Poços de Caldas, the Osamu Utsumi uranium mine and the thorium deposit of Morro de Ferro.

The groundwaters at the Osamu Utsumi mine are of K-Fe-SO_4 type, with F much more important than Cl. This groundwater composition is unusual and is due to the weathering of the hydrothermally altered complex of intrusive phonolites. K is in higher concentration than Na and other common cations in most groundwaters. For example, Ca and Mg are commonly below 1 mg/dm^3, comparable in concentration to Sr and Ba. The pH values are slightly acidic, between 5 and 6.5, with high concentrations of dissolved silica.

A complete description of the Poços de Caldas Natural Analogue study can be found in Chapman et al. (1993).

Results of the geochemical modelling

Once reliable redox and general hydrochemical data have been obtained for the five sites , we attempted to evaluate which were the main processes responsible for the redox data obtained in the field. This was done by using the geochemical modelling codes PHREEQC (Parkhust and Appelo, 1999) and Medusa (Puigdomènech, 1998). The details of the modelling have been published elsewhere.

In all the systems studied most of the measured pH and Eh values fall into the $Fe(OH)_3(s)$ /$Fe(II)$ redox equilibrium line. This confirms the previous findings by Grenthe et al (1992), concerning the active role of this couple in controlling the redox state of slightly oxidised and transitional groundwaters. This can be the result of the interaction between oxidised surface waters with the deeper groundwaters, either by the disturbance through the sampling or as a result of the groundwater mixing.

This universal behaviour can be observed in Figure 2, where all the pH, Eh measured data have been plotted in a common diagram. The calculated lines have been obtained by assuming a $Fe(OH)_3(s)$ of varying cristallinity and consequently a different log Kso spanning three orders of magnitude.

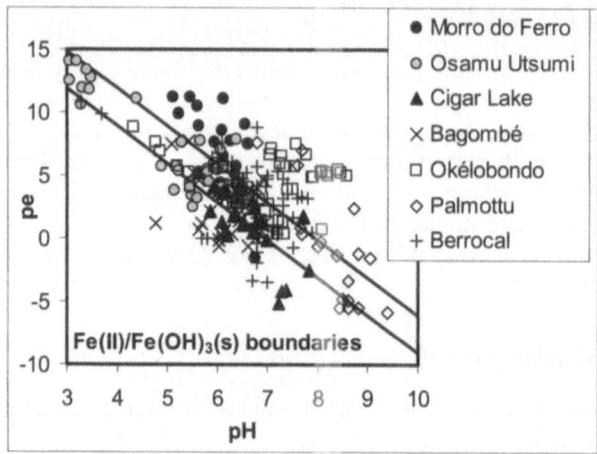

Fig. 2. Measured pH and Eh for the redox data collected in the investigated sites. The calculated lines have been obtained by assuming a log Kso for $Fe(OH)_3(s)$ which spans over the reported variability of stability as a function of cristallinity (Langmuir and Whitthemore, 1971).

However, some of the groundwaters that have had the opportunity to contact with the uranium mineralisations present at these sites and have had sufficient time to reach equilibrium, their redox potential appears to be controlled by the U(VI)/U(IV) couple.

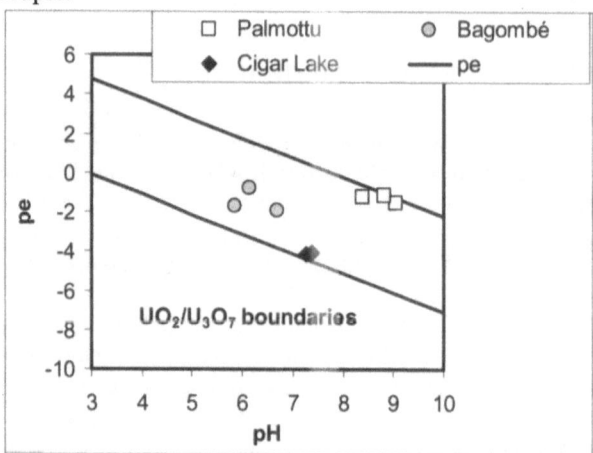

Fig. 3. Measured Eh, pH data for the sections of the investigated sites where contact with uranium mineralisation was indicated by the mineralogical analyses. The theoretical lines are calculated assuming the redox equilibrium between slightly oxidised $U_3O_7(s)$ and $UO_2(s)$.

Geochemical and reactive transport modeling

This is the case of some of the groundwaters in Palmottu that are positively in contact with the uranium mineralisation, groundwaters sampled in the reactor zone of Bagombé (Oklo), and the groundwaters that were measured in the ore body of Cigar Lake (Figure 3). In this case the two main conditions to reach equilibrium for a multielectron transfer are satisfied. This is long residence time of the groundwaters (>10.000 years) and contact with active uranium mineral surfaces which are known to promote fast electron transfer.

Summary conclusions

The redox characterisation work performed in connection with Natural Analogue studies indicates that a proper quantification of the redox processes responsible for the redox state of undisturbed groundwaters is possible.

This is particularly true in sites like Palmottu, where a thorough redox measurement work was done as an integral part of the overall Natural Analogue project.

The main processes which control the redox state of superficial groundwaters are the $Fe(OH)_3(s)/Fe(II)$ equilibrium, as already pointed out in Grenthe et al. (1992). This can be the result of mixing with oxic groundwaters, either through the sampling and measurement procedure or as a consequence of natural mixing processes.

In addition, in the vicinity of the U mineralisations, redox transformations involving U(IV) and U(VI) appear to be quite relevant in poising the redox state. This is particularly relevant for groundwaters with long residence times and in close contact with active mineral surfaces which are a prerequisite for equilibrium in multielectron transfer reactions.

References

Ahonen, L.;Ervanne, H.;Jaakkola, T.;Blomqvist, R. (1994). Redox Chemistry in Uranium-rich Groundwater of Palmottu Uranium Deposit, Finland. Radiochimica Acta, 66/67, 115-121

Almen, K., Andersson, O., Fridh, B., Johansson, B.-E., Sehlstedt, M., Gustavsson, E., Hansson, K., Axelsen, K. and Wikberg, P. (1986). Site investigation equipment for geological, geophysical, hydrogeological and hydrochemical characterization. SKB (Swed. State Power Board), Stockhol, Tech. Rep. 86-16, 141 pp.

Blomqvist R., Ruskeeniemi T., Kaija J., Ahonen L., Paananen M., Smellie J., Grundfelt B., Pedersen K., Bruno J., Pérez del Villar L., Cera E., Rasilainen K., Pitkänen P., Suksi J., Casanova J., Read D. and Frape S. (2000) The Palmottu natural analogue project. Phase II: Transport of radionuclides in a natural flow system at Palmottu. EUR report 19611.

Bruno, J (1997). Trace Element Modelling.In: Modelling in aquatic Chemistry. Ed: Grenthe, I. and Puigdomènech, I.. OECD Nuclear Energy Agency.

Cramer J. and Nesbitt W. (1994) Groundwater evolution and redox geochemistry. Section 3.5.4 in Cramer J. and Smellie J. eds. (1994) Final report of the AECL/SKB Cigar Lake Analog Study. SKB Technical Report 94-04.

Cramer J. and Smellie J. eds. (1994) Final report of the AECL/SKB Cigar Lake Analog Study. SKB Technical Report 94-04.

Cramer J. Vilks P. Miller H. and Bachinski D. (1994) Water sampling and analysis. Section 3.5.2 in Cramer J. and Smellie J. eds. (1994) Final report of the AECL/SKB Cigar Lake Analog Study. SKB Technical Report 94-04

Chapman N.A., McKinley I.G., Shea M.E. and Smellie J.A.T. eds. (1993) The Poços de Caldas Project: natural analogues of processes in a radioactive waste repository. Elsevier Amsterdam.

Enresa (1996) El Berrocal project. Characterization and Validation of natural radionuclide migration processes under real conditions on the fissured granitic environment. ECC F12W/CT91/0080 Topical Reports.

Gimeno M.J. and Peña J. (1999) Geochemical modelling of groundwater evolution in the Palmottu natural system. The Palmottu Natural Analogue Project. Technical Report 99-04.

Gómez P., Turrero M.J., Martínez B., Melón A., Gimeno M.J., Peña J., Mingarro M., Rodríguez V., Gordienko F., Hernández A., Crespo M.T., Ivanovich M., Reyes E., Caballero E., Plata A. and Fernández J.M. (1996) Hydrochemical and Isotopic Characterization of the groundwater from the El Berrocal site, Spain. Topical Report 4. In El Berrocal project. Characterization and Validation of natural radionuclide migration processes under real conditions on the fissured granitic environment. ECC F12W/CT91/0080 Topical Reports Volume I. ENRESA (1996

Grenthe, I., Fuger, J., Konings, R.J.M., Lemire, R.J., Muller, A.B., Nguyen-Trung, C.and Wanner, H. (1992): Chemical Thermodynamic 1; Chemical Thermodynamics of Uranium.

Langmuir, D. And Whittemore, D.O. (1971). Variations in the stability of precipitated ferric oxihydroxides. In: R.F. Gould (Editor), Nonequilibrium Systems in Natural Water Chemistry. Adv. Chem. Ser., 106: 209-234

Louvat D. Michaud V. and von Maravic H. eds. (2000) Oklo Working group. Proceedings of the second joint EC-CEA Workshop on OKLO- Phase II. Helsinki, 16 to 18 June 1998. EUR report 19116 EN.

Parkhurst, D. And Appelo, C.A.J. (1999). User's Guide to phreeqc (version 2). A computer program for speciation , batch-reaction, one-dimanesional transport and inverse geochemical calculations. Water-resources investigations report 99-4259. U.S. Department of the interior U.S. Geological Survey. Denver, Colorado.

Peña J., Turrero M.J., Gómez P., Garralón A., Duro L., Arcos D. and Bruno J. (2000) Water-rock interaction processes in Okélobondo. In (Louvat, Michaud and von Maravic eds.) Oklo Working groups. Proceedings of the second joint EC-CEA Workshop on OKLO- Phase II. Helsinki, 16 to 18 June 1998. EUR report 19116 EN.

Puidomènech, I. (1998). MEDUSA (Make Equilibrium Diagrams Using Sophisticated Algorithms). Vers. 9 Oct. 1998.

Wikberg, P. (1987). The chemistry of Deep Groundwaters in Crystalline Rocks. Thesis, Department of Inorganic Chemistry, The Royal Institute of Technology, Stockholm.

Flooding of the Königstein uranium mine - Aquifer reactivity versus dilution

Broder Merkel

Technical University Bergakademie Freiberg, Department of Geology, 09599 Freiberg

Abstract. The Königstein underground and in situ leaching mine was operated from 1967 to the end of 1991. After ten years of tests, research and pre-works the flooding was started in January 2001. Mine water will be pumped and treated at least until the first flush of contaminants has passed. The question is when active treatment will be stopped and passive treatment take over. One option is diverting the mine water without treatment into the Elbe river. Due to an average flow of about 300 m^3/s compared to 0.028 m^3/s of mine water there will be only a minor impact on the river water quality. Another option is using mine voids or the downstream aquifer as reactive treatment zones. Geochemical simulation were performed by means of PHREEQC´s one dimensional reactive transport tool taking into account dilution, double porosity, surface complexation on iron hydroxides and uranium reduction due to degradation of organic matter.

Introduction

The Uranium Königstein mine is situated about 30 km southeast of the city of Dresden close to the Elbe river and the nature resort "Sächsische Schweiz". Underground operations started at the Königstein mine in 1967 by classical deep mining in the Cretaceous sandstone ore body. Since 1984 a special in situ leaching technique was applied using sulfuric acid as leaching agent. Uranium mining was stopped at the end of 1991 after reunification of Germany (Zimmermann & Schreyer 1995).

The multilayer aquifer is build up from 4 double porosity aquifers and siltstones in between. Aquifers no 1 und 2 are unconfined and only of local interest, aquifer 3 is unconfined as well and used downstream of the mine for extracting potable water for the city of Pirna. Aquifer 4 is confined and the lowermost with the uranium ore body in it. After a ten years period of research, tests and pre-works flooding of the mine was started in January 2001. During the flooding procedure mine water is pumped from the mine and treated on site at least until the first flush has been passed. Pumping is done by means of a control adit which was build at the downstream edge of the mine; ground water and mine water is withdrawn by means of hundreds of small diameter boreholes which have been drilled through the roof of the control adits into the aquifer no 4 forming a dense drainage fan. However, the question is when the active treatment will be stopped and what strategy will then be followed up. As long as the control adit is accessible and all mine water is pumped to the surface for treatment this requires huge quantities of water to be pumped and treated. The amount of water to be pumped and treated in the long run is depended on the permeability of the sandstone of the aquifer no. 4. If the control adit is flooded and used as a horizontal well the cone of depression is much smaller and the amount of water to be pumped and treated is getting much smaller. One option is diverting the mine water without treatment into the Elbe river. Alternatively some adits and mine voids could be filled with permeable and reactive material and finally the downstream aquifer could be used as natural reactor.

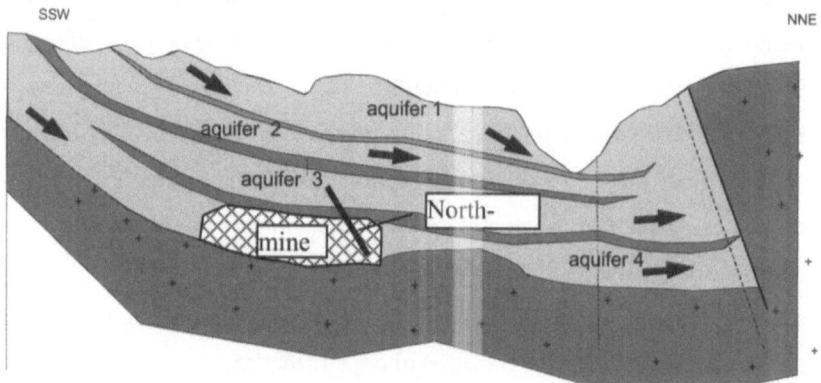

Fig. 1. cross section of the Königstein mine

From two medium scales flooding experiments in the Königstein mine and recent measurements from the ongoing flooding the mine water quality is known considerably well. Average values are given in table 1 for a selected number of elements and parameters.

Table 1. Average concentrations of selected elements in mg/L and key parameters in the flood water of the Königstein mine (first flush)

pH	Temp °C	eH mV	EC mS/cm	Ca	Mg	Na	K	SO_4^{2-}	Cl⁻
2	14	650	6.3	130	20	40	10	2700	35
F	U	Cu	Ni	Zn	Sr	Ba	Pb	As	Cd
1.5	200	0.5	4.0	55	0.9	0.03	0.65	1.65	1.4

Dilution

From flooding of other mines it is known that the concentrations (as shown in Table 1) are peak concentrations which will decrease by at least one order of magnitude within a few years. Assuming an amount of 100 m³/hour (0.028 m³/s) flowing through the mine and taken into account the average flow of the Elbe with about 300 m³/s the dilution factor is $9.3 \cdot 10^{-5}$. Thus the uranium increase for the Elbe would be 1.8 µg/L assuming an uranium concentration of 20 mg/l in the mine water respectively 0.18 µg/L, if the mine water contains only 2 mg/L. Compared with average uranium concentrations of 1 to 4 µg/L in the Elbe river an additional increase of less than 1 µg/L sounds negligible. However, taken into account the WHO Uranium MCL of 2 µg/L for drinking water, the Elbe is already at critical quality and thus sensitive for additional loads. Assuming that the concentrations of other elements (e.g. As, Cd, sulfate) are as well decreasing by 1 to 2 orders of magnitude in the mine water, dilution would make the impact of the mine water extremely small for these elements. However, it is in question whether the German regulators will accept dilution as natural attenuation measure since the yearly load would be 1.8 tons of uranium assuming 2 mg/l respectively 18 ton (20 mg/l) of uranium in the mine water diverted to the Elbe.

Backfill of reactive material

Use of zero-valent iron to remove Uranium is a common technique (Farrell et al 1999, Schneider et al. 2001) and lab and in situ field experiments of the Wismut GmbH have shown promising results (personal communication). However several questions are still to be addressed: First of all it is not yet known, whether the removal of uranium is due to sorption or reduction of U(VI) to V(IV) and thus precipitation of uranium oxide (Noubactep et al. 2001). Sorption, as a removal mechanism, is not preferred because uranium will remain in its more soluble oxidation state (i.e., U^{6+}), thereby lending itself to be easily transported by the colloidal iron corrosion products. Another concern is the potential release of soluble uranium back into the groundwater.

Reduction of U^{6+} to U^{4+} is the preferred removal mechanism since the resulting U^{4+} species is less soluble and less mobile in groundwater, assuming that the U^{4+} species is not colloidal. Thermodynamical reduction to U^{4+} is slightly favorable in strongly acidic media as indicated by the modest positive value for the standard cell potential:

$$Fe^0 + 1.5\ UO_2^{2+} + 6\ H^+ \longleftrightarrow Fe^{3+} + 1.5 U^{4+} + 3\ H_2O \qquad E^0 = +0.17\ V$$

However, results by Wersin et al. (1994) indicate that reduction to the less soluble U^{4+} (e.g., $UO_2 \cdot xH_2O$) is controlled kinetically. Similarly, reduction of U^{6+} by the ferrous ion (Fe^{2+}) has been reported to be kinetically slow except in the presence of strong acid (Baes 1953).

Geochemical and reactive transport modeling

Then a backfilling technique for getting the iron-sand mixtures into the adits and tunnels has to be developed, since only distinct areas of the underground mine are nowadays accessible. As long as adits are open the preferential flow is through the adits, however, after filling them with a sand-iron mixture the permeability will decrease by several orders of magnitude and the preferential flow might by through the mined parts (conventional and in situ leached blocks). The only adit which can be backfilled with conventional techniques comparatively simply is the control adit. This option has furthermore the advantage that most ground water will pass through as long as the permeability of the backfill is much better than the sandstone. However, the capacity for taking reactive material is limited and an exchange of the reactive material is not possible without development of new techniques.

Aquifer as reactive zone

Another alternative is using the downstream aquifers as reactive treatment zone. Since the aquiclude between the Aquifer 3 und 4 is existing in the entire area the contamination from the mine will flow downstream mainly within the aquifer no. 4 as long as the fault zone between aquifer 4 and 3 in the mine area can be controlled. A certain contamination of the aquifer no. 4 might be acceptable by the German regulators and stakeholders since this aquifer is not used due to natural elevated concentrations of radium in the ground water. What are the controlling processes? Dilution will take place and this will certainly increase the pH, thus distinct minerals might get over saturated and might precipitate. Ion exchange might take place which is strongly dependent on the pH, since uranium species are changing readily with pH. However, solution of minerals (pyrite, calcite) might occur as well. An important reaction might be the decay of organic carbon which is present in the sandstones and was actually triggering the formation of the Königstein uranium ore body at a redox barrier a couple of million years ago. However the organic matter still present in the sandstones seems to have nowadays a considerably low reactivity, since the redox potential of the ground water in aquifer no. 4 is indicating oxidizing conditions.

Distinct potential reactions were modeled by means of PHREEQC using its 1d reactive transport tool. One principle problem linked with 1d transport modeling is that dilution by the surrounding not contaminated ground water is not taken into account. However, within PHREEQC 1d reactive transport tool this can be incorporated. Unfortunately the amount of pyrite and calcite in the Cretaceous sandstone are not known very well. Thus 1% of calcite and pyrite respectively and 2% organic matter were assumed. Kinetic rate constants for calcite and pyrite solution and decay rates of organic matter were taken from the PHREEQC.dat data set (Parkhurst, Apello 1999) and partly modified . Since laboratory tests of permeability show k_f-values of some 10^{-7} m/s and pumping test data on contrary comes up with some 10^{-5} m/s it is proven that the Cretaceous sandstone is a double porosity aquifer. The effective porosity of the sandstones is between 0.1 and 0.2 (Flesch

2000), however, the porosity of fractures is much less and approximately between 0.01 and 0.05. Thus in the PHREEQC 1d transport simulation a primary (fracture) porosity of 0.01 being coupled with 0.15 secondary sandstone porosity were assumed, with diffusion being the major process of exchange between fracture and pore volume. Assuming a k_f of $1 \cdot 10^{-5}$ m/s, a flow gradient of 0.02 and a fracture volume of 0.03 the distance velocity is approximately 210 m/year. A one dimensional array with 20 cells each 40 m in length and a dispersion of 5 m was defined.

Reactive geochemical transport models showed that cation and anion exchange as well pyrite dissolution have only minor impact on concentrations in the contamination plume. On contrary it is clearly shown that calcite dilution, precipitation of iron hydroxides and consequently surface complexation of e.g. uranium and arsenic but as well dilution by uncontaminated ground water are important factors. Fig. 2 shows the breakthrough curves 8000 m downstream the mine assuming that the mine water has the condition described in table 1, which is certainly too pessimistic. Nearly all iron is precipitated as $Fe(OH)_3$ in the first cell and uranium is partly sorbed on the surfaces and thus reducing its concentration roughly by 50%. However, after 10 years the calcite embedded in the sandstone is completely dissolved in the first cell, and thus the pH is decreasing. Due to the decrease of the pH the protons are displacing uranium from the $Fe(OH)_3$-surfaces causing a sharp peak with uranium concentrations exceeding the uranium concentrations of the mine water (200mg/l) by far. Then the system stabilizes again due to still remaining calcite buffer capacities in the next cells of the model. It is important to point out the fact, that precipitation and sorption takes

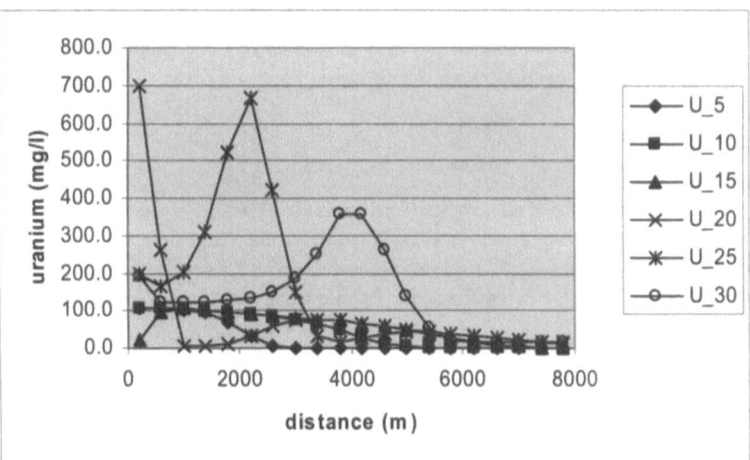

Fig. 2. Breakthrough curve for Uranium for time intervals of 5 30 years assuming concentrations of table 1. Due to iron hydroxide precipitation in the first cell of the model and consequent surface complexation of uranium on the $Fe(OH)_3$ uranium concentration is decreased from 200 to 105 mg/l in the first cell. Further decrease of uranium is due to dilution of the plume. Peaks are due to complete consumption of calcite in the first cell.

place throughout the total model time just in the first three cells of the model which is only 120 m of the aquifer downstream the mine. Besides, this is as well the only cell where gypsum precipitation occurs.

Fig. 3 shows a similar run, but assuming that the peak flush has been pumped off the mine, been treated and the major contaminants are decreased by one order of magnitude. In this case the assumed amount of 1% calcite is sufficient to buffer the mine water to a pH between 6.7 and 7.3. Precipitation of $Fe(OH)_3$ occurs only in the first cell of the model, thus clogging effects might be a problem in reality. However precipitation of $Fe(OH)_3$ was modeled thermodynamically and not kinetically since no kinetic constants were available. Gypsum saturation was not achieved in this run. Assuming double porosity condition the decrease of concentrations in the plume is even more drastic than by dilution and is controlled by the ratio of mobile to immobile pore volume and the pore geometry.

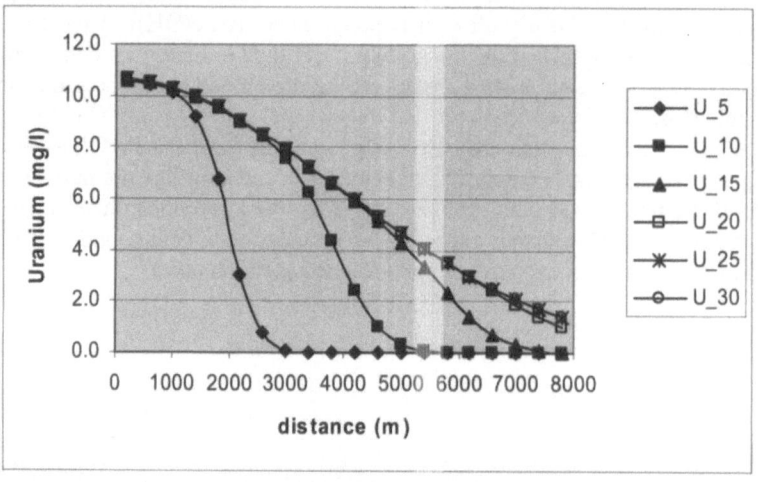

Fig. 3. Breakthrough curve for Uranium for time intervals of 5 ... 30 years assuming concentrations 1/10 of table 2. Due to iron hydroxide precipitation in the first cell of the model and consequent surface complexation of uranium on the Fe(OH)3 uranium concentration is decreased from 20 to 10.5 mg/l in the first cell. Further decrease of uranium is due to dilution of the plume.

Fig. 4, 5, and 6 show results of simulations assuming that easily degradable organic matter is available besides calcite and pyrite. In the very beginning pyrite solution lead to an increase of the sulfate concentrations. However, after about 800 days the pe has dropped from 17.6 to 4 and reduction of sulfate to H_2S starts (fig. 4). With decreasing pe redox sensitive elements like iron, arsenic, cupper, and uranium are reduced consequently as well. Fig 5 shows that iron(II) increase and iron(III) decrease start after roughly 200 days. After 490 days more or less no iron(III) is left in the system. A few days later uranium is changing more sharply from uranium(IV) to uranium(VI).

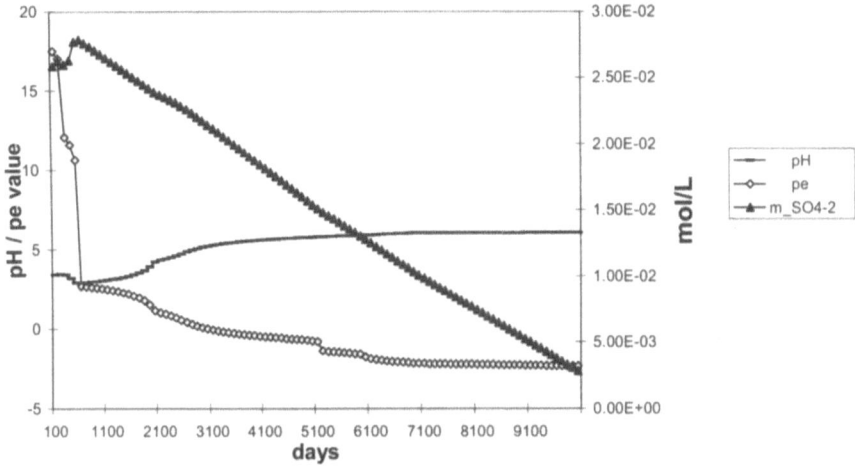

Fig. 4. Redoxpotential (pe) is dropping dramatically during some hundred days at the very beginning. The wrinkle in the pe-line at about 800 days is correlated with the sulfate peak. Decreasing sulfate concentrations is due to further microbial degradation of organic matter.

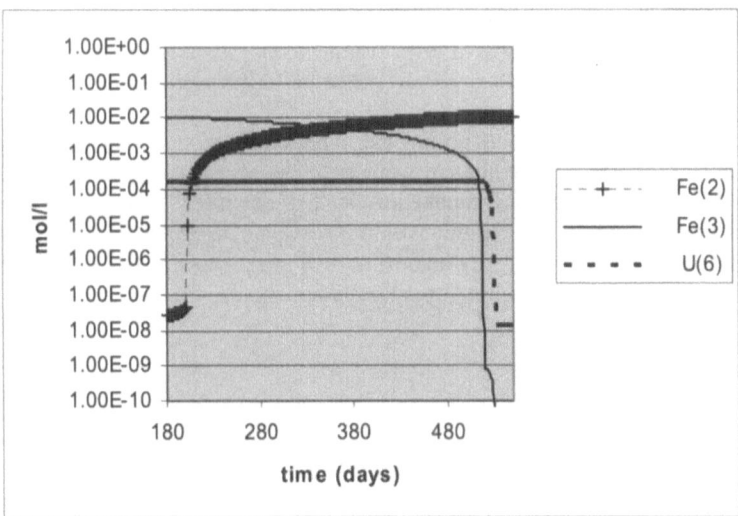

Fig. 5. Speciation change from Fe(II) to Fe(III) and reduction of U(VI) to U(IV) which is precipitated as amorphous UO_2 and thus does not shown up as aqueous species

Fig 6 is enlarging the time period from fig. 5 and displays as well the development of arsenic species. The arsenic(V) decrease line is very close to the iron(III) decrease line and the increase of arsenic(III) is negatively correlated to arsenic(V). Roughly ten days later uranium is reduced form uranium(VI) to uranium(IV) which is readily precipitated as amorphous UO_2. If the thermodynamical data used are reliable and the reactions Fe(III) to Fe(II) respectively U(VI) to U(IV) are not kinetically controlled this is a proof that sorption of uranium on $Fe(OH)_3$ and simultaneous precipitation of UO_2 or Uraninite is not likely to occur.

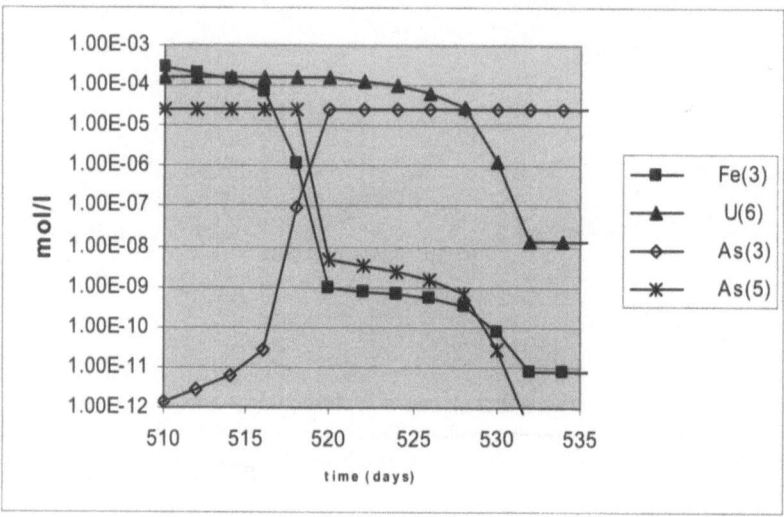

Fig. 6. Distribution of redox sensitive species of iron, arsenic and uranium versus time respectively development of pe due to kinetically controlled degradation of organic matter

However, this simulation was assuming the availability of easily degradable organic matter and there is no evidence from field or lab experiments that the organic matter in the Cretaceous sandstones is that reactive. Therefore reducing conditions in the contamination plume would have to be triggered by adding easily degradable organic matter (e.g. methanol) to the mine water. Theoretically this could be done by means of injecting organic matter into the control adit.

Conclusion

The German government has already spent several hundred millions Euro during the last ten years for rehabilitation work in the Königstein mine and will spent some hundred millions Euro for the Königstein mine in the future. Pump and treat is a suitable option to catch the first flush with high contamination potential. However at the earliest time pump and treat has to be replaced by an alternative technique. One dimensional reactive transport has proven as a handsome tool to evaluate different scenarios and boundary conditions. Dilution and double porosity aquifers can be simulated by means of PHREEQC's one dimensional transport tool, however, it is not possible to take into account both in one model. Thermodynamically and kinetically reactions can be incorporated as well as limited mineral phases (Merkel, Planer Friedrich 2002). A major problem, as often in modeling business, comes with the lack of reliable data. Gathering these missing data is thus the most important aim.

References

Baes, C. F., Jr. The Reduction of Uranium (VI) by Ferrous Iron in Phosphoric Acid Solution: The Formal Electrode Potential of the U(IV)/U(VI) Couple; Report ORNL-1581; Oak Ridge National Laboratory: Oak Ridge, TN, 1953.

Farrell J., Bostick W.D, Jarabeck R.J., and Fiedor J.N. (1999): Uranium removal from ground water using zero valent iron media. Ground water 34, 618-624.

Flesch (2000) Hydrogeologisches Modell der Sächsischen Kreide. Wiss. Mitt. Inst. Für Geologie, 14, ISSN 1433-1284

Merkel B, Planer-Friedrich B (2002) Grundwasserchemie - Praxisorientierter Leitfaden zur Modellierung von Beschaffenheit, Kontamination und Sanierung aquatischer Systeme, Springer Verlag, 220 S. 74 Abb., 56 Tab.

Noubactep C, Meinrath G, Volke P, Peter H.J, Dietrich P, Merkel B (2001): Understanding the Mechanism of the Uranium Mitigation by Zero Valent Iron in Effluents. Wiss. Mitt Inst.für Geol. TU BAF, Vol 18, 36-44, ISSN 1433-1284

Parkhurst D, Apello C (1999) Users´Guide to PHREEQC (Version 2) — A computer program for speciation, batch-reaction, one-dimensional transport, and inverse geochemical calculations. Water-Resources Investigations Report 99-4259

Schneider P., Neitzel P.L., Osenbrück K., Noubacteb, C., Merkel B. Hurst S. (2001): In-situ Treatment of Radioavtive Mine Water using Reactive Materials - Results of Laboratoty and Field Experiments in Uranium Ore Mines in Germany. Acta hydrochim. hydrobiol. 29, 2-3, 129-138

Wersin, P.; Hochella, M. F., Jr.; Persson, P.; Redden, G.; Leckie, J. O.; Harris, D. W. (1994) Geochim. Cosmochim. Acta 58 (13), 2829-2843

Zimmermann U, Schreyer J. (1995) Erstellung experimenteller Daten mittels eines Flutungsexperimentes zur Vorbereitung der Flutung der Lagerstätte Königstein. Uranium Mining and Hydrogeology I, Sven von Loga Verlag: 575-583

C&E
Consulting & Engineering GmbH

As the successor company of the in 1990 privatized engineering, research, geological and project planning branch of the former mining company SDAG Wismut, C&E Consulting und Engineering GmbH looks back on nearly 50 years of engineering experience with a regional focus on East Germany.

Environmental Engineering Services:

- Investigation of sites and contaminated areas
- Environmental consultation and assessment
- Remediation concepts and engineering
- Landfill and hydraulic engineering
- Mining consultation and engineering
- Geotechnical investigation and evaluation

These engineering and consulting services comprise a large spectrum from data evaluation and assessment, concept development, expert reports and studies to complete planning, construction supervision and project management.

C&E Consulting und Engineering GmbH is an independent saxonian company. International tasks are performed in close co-operation with national, international and local partners. This guarantees the full benefit of a high-level scientific-technical know-how pool and the necessary regional experiences.

Contact :
Jagdschänkenstrasse 52
D-09117 Chemnitz /Germany
Phone: 0049-371-881-4228
Fax: 0049-371-881-4311
e-mail: info@cue-chemnitz.de
Internet: www.cue-chemnitz.de

References:

- Rehabilitation of the tailings pond facilities Trünzig, Culmitzsch, Helmsdorf, Dänkritz; Lengenfeld, Gittersee, Borbach; *Concepts, Planning, Supervision*
- Contouring and covering of uranium waste rock dumps of the WISMUT GmbH; *Concepts, Planning, Supervision, Geotechnical Services*
- Flooding of the U-mining sites Ronneburg, Aue, Königstein;
- Prognosis, Planning, Supervision
- Rehabilitation of U-mining liabilities in Central and East European Countries (AL,BG,CZ,EST,HU,PL,RO,SK,SLO); *Concepts, Technical Planning*
- Closure of U-mining sites Buhovo, Eleshnitsa, Dospat / Bulgaria; *Supervision*

Numerical simulation of uranium transport in flooded underground mines

Sabine M. Spiessl[1], Henning Prommer[2], Martin Sauter[1], Chunmiao Zheng[3]

[1]Institute of Geoscience, Friedrich-Schiller-University Jena, Germany.
[2]Centre for Applied Geoscience, University of Tübingen, Germany and Faculty of Civil Engineering and Geosciences, Delft University of Technology, The Netherlands.
[3]Department of Geological Science, University of Alabama, USA.

Abstract. Former mining areas generally suffer numerous environmental problems such as groundwater contamination with heavy metals and acidification. A reactive hybrid transport model has been developed to allow the quantification of uranium transport in flooded underground mines. This hybrid model specifically takes into account two different transport time scales within an underground mine, which result from the existence of a network of highly conductive shafts, drifts or ventilation raises within the considerably less permeable ore material in the underground. This paper introduces the model concept and addresses potential model applications.

Introduction

The occurrence of acidic, metal-rich groundwater is often observed in mining areas. At sites of uranium mining, exposure of receptors to radioactivity might result from the migration of radionuclides within the subsurface. In many cases the oxidation of naturally abundant pyrite has shown to be the key process that causes the acidic geochemical environment (Wunderly et al. 1996). Thus, in the last few years, underground or subsurface mines have undergone controlled flooding to prevent access of oxygen and thereby the further lowering of pH due to pyrite oxidation (Bain et al. 2001). Subsurface mines are typically composed of networks of highly conductive shafts, drifts or ventilation raises within a considerably less permeable ore material. Therefore, the simulation of contaminant transport in such domains requires consideration of two distinct time scales in the transport model. As a result of the discrete nature of a pipe network, and because the quantification of contaminant flux from the solid mineral phase into the mobile water phase requires a discrete surface area, double continuum models cannot be employed for modelling of reactive transport in underground mines. Therefore, a "hybrid" transport model has been developed and subsequently coupled with a geochemical model. The underlying concepts of the resulting numerical model RUMT3D (i.e., three-dimensional reactive underground mine transport model) are the focus of the present paper as well as a demonstration of its potential use.

Hybrid flow model (CAVE)

To compute the flow field in the combined pipe-network/matrix system, the CAVE model (Clemens et al. 1996), a hybrid model coupling flow in a discrete conduit network to a continuum is employed. CAVE was originally developed for modelling the genesis of karst aquifers. However, the simulation of groundwater flow in underground mines, i.e. a system with both, a highly conductive conduit network and a considerably less permeable, porous matrix can be treated in an analogous way as karstified carbonate aquifers.

The three-dimensional continuum equation for flow in the ore material (matrix), including a further source/sink term γ that couples the continuum to the pipe-flow model, is described as follows:

$$\frac{\partial}{\partial x_m}\left(K_{m,xx}\frac{\partial h_m}{\partial x_m}\right) + \frac{\partial}{\partial y_m}\left(K_{m,yy}\frac{\partial h_m}{\partial y_m}\right) + \frac{\partial}{\partial z_m}\left(K_{m,zz}\frac{\partial h_m}{\partial z_m}\right) - W_m - \gamma = S\frac{\partial h_m}{\partial t_m} \quad (1)$$

with x_m, y_m, z_m [L] the distance along the respective Cartesian coordinate axis in the ore material, t_m [T] represents the time, h_m [L] is the hydraulic head in the ore material, $K_{m,xx}$, $K_{m,yy}$, $K_{m,zz}$ [L T^{-1}] depict the values of hydraulic conductivity along the coordinate axes in the ore material, S [L^{-1}] is the specific storage coefficient, W_m [T^{-1}] is a volumetric flux term per unit volume from a sink/source into the ore material e.g. groundwater recharge, and γ [T^{-1}] denotes the volumetric rate of fluid transfer between the ore material and the conduit system per unit volume.

A conduit system is defined in the model as a pipe network consisting of cylindrical tubes. Conduit nodes are introduced between the connecting tubes to allow for exchange of flow between the different tubes from different faces of a cell and between a conduit node and the continuum (matrix, ore material) at different locations in the model domain. Only one conduit node can be placed in an ore material cell. There are 6 potential faces of a cell so of a conduit node for tube connection in 3-dimensional model domains, i.e., top, bottom, front, back, left and right. A conduit tube may extend over one or more ore material (matrix) cells depending on the respective geometry of the mineshafts or adits, and the locations of the sinks and sources, e.g. direct recharge and fixed heads. Conduit orientations can be freely designed, i.e. they do not necessarily need to be vertical or horizontal along the same continuum layer, so that the model can easily match the actual spatial coordinates of real mine networks. It is assumed that the conduit system is fully saturated, an assumption that will hold for most flooded underground mines. Flow between the ore material and the conduit nodes is described by a linear relationship between the two systems (Barenblatt et al. 1960):

$$\Gamma_i = \alpha_i \left(h_i - h_{i,m} \right) \tag{2}$$

where Γ_i [$L^3\ T^{-1}$] represents the exchange flow rate between conduit node i and the ore material cell, α_i [$L^2\ T^{-1}$] stands for the exchange coefficient between node i and the ore material, h_i [L] is the hydraulic head at conduit node i and $h_{i,m}$ [L] is the hydraulic head in the ore material cell where conduit node i is located. The magnitude of exchange coefficient α_i depends on the hydraulic conductivity of the ore material and geometrical factors, determined by the discretization of the adjacent continuum cell.

Laminar flow in each tube, i.e., laminar flow from one conduit node to another can be calculated from an expression for discharge which can be found by substituting Hagen-Poiseuille's into Darcy-Weisbach equation. Turbulent flow is solved according to the implicit Colebrook-White law. Conservation of flow volume at any conduit node i can be determined by using Kirchhoff's law, which states that the total inflow and outflow must balance at each node. A more detailed description of the pipeflow model can be found in the work by Clemens et al. (1996).

Hybrid transport model (UMT3D)

Transport of mass within the ore material (matrix) is also simulated with a continuum approach, i.e. with the standardized three-dimensional multi-species transport model MT3DMS (Zheng and Wang 1999). In contrast, mass transport within the network of tubes is modelled discretely with a one-dimensional advective transport model. Coupling between the continuum transport model and the transport in the conduits is achieved using a sequential operator-splitting procedure (Walter et al. 1994, Steefel and MacQuarrie 1996). The transport of solutes in the continuum is solved first, followed by a second step during which the solute concentrations in the conduit system are updated and the solute mass is transported. The coupled hybrid transport model is UMT3D (three-dimensional underground mine transport model).

The general transport equation (Zheng and Wang 1999) has been extended to include a further sink/source term, i.e. ξ to couple MT3DMS to the conduit transport model. Thus, the three-dimensional transport equation in the continuum is:

$$\frac{\partial(\theta_m C_m^k)}{\partial t_m} = \frac{\partial}{\partial x_{m,i}}\left(\theta_m D_{m,ij} \frac{\partial C_m^k}{\partial x_{m,j}} \right) - \frac{\partial}{\partial x_{m,i}}\left(\theta_m v_{m,i} C_m^k \right) + q_{m,s}\ C_{m,s}^k + \xi^k + \sum_{k=1}^{N} RXN_{m,k} \tag{3}$$

with C_m^k [M L^{-3}] is the aqueous concentration of component k in the ore material, $D_{m,ij}$ [L^2 T^{-1}] depicts the hydrodynamic dispersion coefficient tensor in the ore material, $v_{m,i}$ [L T^{-1}] is the linear pore water velocity in the ore material, $q_{m,s}$ [T^{-1}] denotes the volumetric flux of water per unit volume of ore material representing sources (positive) and sinks (negative), θ_m [-] is the porosity of the ore material, $C_{m,s}^k$ [M L^{-3}] is the concentration of component k of the sources or sinks to the ore material, ξ^k [M L^{-3} T^{-1}] denotes the volumetric rate of fluid transfer between the ore material and the conduit system per unit volume multiplied by the respective concentration values of the conduit sink and source terms of component k and $RXN_{m,k}$ [M L^{-3} T^{-1}] is the chemical reaction term of the ore material with respect to component k.

For the calculation of solute transport from the continuum cell to a conduit node i, the mass transfer rate is determined by multiplying the cell concentration of the pore water in the ore material of component k, $C_{m,i}^k$, at node i, with the respective exchange flow rate Γ_i and dividing the mass flux by the volume of the respective cell, $V_{m,i}$, i.e.,

$$\frac{\Gamma_i C_{m,i}^k}{V_{m,i}} \tag{4}$$

Alternatively, if solute mass is transported from conduit node i into the matrix cell, then C_m^k is replaced by the nodal concentration, C_i^k. Fig.1 illustrates this exchange for both cases. Such mass transfer rates are then treated as mass sink/source terms in MT3DMS since conduit nodes act similarly to other sinks/sources (e.g., wells) in MT3DMS. The difference between wells as implemented in MT3DMS and conduit nodes is that the mass removed by means of a well is not returned to the ore material, while the mass removed with an entry conduit node may be returned completely through the exit conduit nodes. Amount, location and required time for these returns mainly depend on (i) the transport velocity in the different conduits, (ii) the magnitude of the exchange coefficients between the exit conduit nodes and the ore material, (iii) the magnitude of conduit sink terms, and (iv) the length of the different conduit tubes.

A one-dimensional transport equation that solely considers advection is applied to each tube, i.e.,

$$\frac{\partial C_j^k}{\partial t} = -q_j \frac{\partial C_j^k}{\partial z_j} \tag{5}$$

where C_j^k [M L^{-3}] represents the aqueous concentration of component k in conduit tube j, q_j [L T^{-1}] is the flux of water in tube j, t [T] is the time and z_j [L] refers to the distance along a respective Cartesian coordinate axis in the respective tube j. Note, there are no sink/source terms in this transport equation. The mass exchange rates from the ore material, from the six potentially connecting tubes and from conduit sink/source terms such as direct recharge, fixed head and fixed concentration as applied to the different conduit nodes are considered in terms of initial or boundary concentration values to the transport equation (5). The resulting concentration values are obtained by a weighted arithmetic mean of the single flow and transport components for each transport time step. Such an approach is common in mixing cell models (Bajracharya and Barry 1993). Mathematically, a weighted arithmetic mean of the concentration value of component k at conduit node i can be expressed as follows:

$$
C_i^k = \frac{\sum\limits_f^6 Q_{i,j}^{f+} C_{i,j,l}^{k,f} + \Gamma_i^+ C_{i,m}^k + \sum\limits_s Q_{i,s}^+ C_{i,s}^k}{\sum\limits_f^6 Q_{i,j}^{f+} + \Gamma_i^+ + \sum\limits_s Q_{i,s}^+}
\tag{6}
$$

where $Q_{i,j}^{f+}$ [$L^3\,T^{-1}$] represents the discharge of tube j connected to face f of conduit node i into the respective node i, $Q_{i,s}^+$ [$L^3\,T^{-1}$] refers to a volumetric flow rate of a conduit source term to node i, and the subscript l refers to the first or the maximum number of tube sections or segments in the different tubes depending on the flow direction. The tubes or conduits can further be divided into a user-defined number of segments or sections to decrease the effect of numerical dispersion and improve numerical stability. Also note that tubes within a network can be considerably longer than cell widths, lengths or thickness.

Eq. (5) is solved with a mass-conservative semi-Lagrangian scheme (EMCNOT). This scheme was developed by Liu et al. (2001) for modelling advection-dominated mass transport problems and is an explicit mass conservative scheme without time step limitation. With the EMCNOT method, a transport time step size of up to the minimum residence time value of a respective pipe in the conduit system multiplied by a user defined Courant number, Cr can theoretically be used for both models. The residence time of groundwater in a specific pipe transported by advection under steady state flow conditions is determined by dividing the length of a pipe, L_j by the flow velocity in a respective pipe. In mathematical form, the maximal transport time step size, Δt_{max} can be expressed as:

$$
\Delta t_{max} = Cr \left(\frac{L_j}{q_j} \right)_{min}
\tag{7}
$$

Such a time criterion may vary with each flow time step since the flow rate in each tube may change with each flow time step. Spiessl et al. (2002) demonstrated that applying the EMCNOT scheme to solve advective transport in the conduit system significantly reduces numerical dispersion compared to the standard finite difference (FD) method.

Mass balance in the conduit system is determined in a similar way as in the ore material. To check performance of both transport models, mass balance calculations are also carried out independently for the different conduit nodes in MT3DMS as sink/source terms. In contrast to the conduit transport model by Birk (2001) the above-described model uses a global approach to calculate mass balance and thus improves computational efficiency. Moreover, the global variable arrays and subroutines in the modified conduit transport model are fully compatible with those used in MT3DMS.

Reactive hybrid transport model (RUMT3D)

The comprehensive geochemical model PHREEQC-2 (Parkhurst and Appelo 1999) was coupled with the UMT3D model as a solver for the reaction term within Eq. (3), i.e., $RXN_{m,k}$ using a sequential operator splitting technique approach (Walter et al. 1994, Steefel and MacQuarrie 1996). The transport equation (3) does not have to be solved for every individual chemical species, but only for total aqueous component concentrations (Yeh and Tripathi 1989, Engesgaard and Kipp 1992), defined as:

$$C_u = c_u + \sum_{k=1,n_s} Y_k^s s_k \tag{8}$$

where C_u is the total aqueous component concentration of the uth component, c_u is the molar concentration of the uth (uncomplexed) aqueous component, n_s is the number of dissolved species that form complexes with the uth aqueous component, Y_k^s is the stoichiometric coefficient of the aqueous component in the kth complexed species and s_k is the molar concentration of the kth complexed species. The (local) redox-state, pe, is at present modeled by transporting chemicals/components in different redox states separately, while the pH is calculated based on a (local) charge balance. The transport model UMT3D needs to solve transport for n_{tot} entities, with

$$n_{tot} = n_{e,nre} + \sum_{k=1,n_{e,re}} n_{rs,k} \tag{9}$$

where $n_{e,nre}$ is the number of (mobile) chemical elements occurring in only one redox state, $n_{e,re}$ is the number of elements occurring in multiple redox states and $n_{rs,k}$ is the appropriate number of different possible redox states of the kth element. The resulting model (RUMT3D) can handle a wide range of chemically reactive processes including aqueous complexation, mineral dissolution/precipitation, and ion-exchange. Reactions might be assumed to occur as equilibrium reactions and/or kinetically controlled. More details on the incorporation of PHREEQC-2 and MT3DMS can be found in Prommer et al. (2002). Chemical reactions in the conduit system are at present assumed to have a negligible effect on the composition of the groundwater because of the typical short residence times of the solutes in the conduit system.

Scenarios of reactive transport in a coupled conduit-continuum system

In order to demonstrate the capabilities of the RUMT3D model two simplified, schematic scenarios, one of which includes the hydraulic effect of a single U-shaped conduit system have been selected. Solute fluxes (discharge of uranium species) to a river in the proximity of a subsurface mine were compared for a model setup (i) natural groundwater flow conditions and (ii) flow conditions affected by the presence of a pipe network. For both scenarios, steady state groundwater flow and reactive transport in a vertical cross-section was simulated with a continuum (matrix, ore material) domain of 1200 m by 480 m. Heads at the river were fixed at 390 m, imposing a hydraulic gradient between ore material and the discharge point. A fixed constant-head boundary of 470 m was set at the "influent" boundary of the model. A uniform groundwater recharge rate of 6×10^{-9} m s^{-1} was assumed at the surface. The ore matrix has a vertical and a horizontal hydraulic conductivity of 1.0×10^{-6} m s^{-1} and an effective porosity of 0.05.

Table 1. Chemical composition of the background groundwater, the uranium/pyrite ore body, the influent boundary and the recharge water.

Component	Background groundwater, uranium/pyrite ore body and influent boundary (mol/l)	Recharge water (mol/l)	Component	Background groundwater, uranium/pyrite ore body and influent boundary (mol/l)	Recharge water (mol/l)
pH	7.53[a]/7.53[b]/7.24[c]	7.66	S(6)	8.010×10^{-3}	1.000×10^{-3}
pe	-3.20[a]/-3.07[b]/-3.11[c]	12.99	S(-2)	1.474×10^{-12}	-
C(4)	1.220×10^{-3}	1.447×10^{-3}	U(3)	-	-
Ca	5.225×10^{-3}	1.947×10^{-3}	U(4)	-	-
Cl	1.100×10^{-3}	1.000×10^{-3}	U(5)	-	-
Fe(2)	1.001×10^{-3}	6.030×10^{-16}	U(6)	-	-
Fe(3)	5.010×10^{-10}	1.000×10^{-5}	Calcite	$1.000^{a,b}$	-
Na	5.788×10^{-3}	5.000×10^{-4}	Pyrite	$(1.000 \times 10^{-3})^{b}$	-
O(0)	-	7.000×10^{-4}	UO$_2$(am)	$(1.000 \times 10^{-3})^{b}$	-

[a] for background groundwater only
[b] for uranium/pyrite ore body only
[c] for influent boundary only.

Geochemical and reactive transport modeling

A uranium-pyrite rich ore body was placed in the centre of the flow domain as depicted in Fig. 2. The remaining continuum cells are assumed not to contain any amorphous uranium oxide or pyrite. An anaerobic homogeneous water composition was assumed at the beginning of the simulation. The reaction network consists of 15 aqueous components (see Table 1) and three minerals (amorphous uranium oxide, pyrite, calcite). In contrast to uranium oxide and pyrite, calcite was assumed to be uniformly distributed within the aquifer. Table 1 also lists the chemical composition of the background groundwater, the uranium oxide/pyrite ore body, of the influent water at the constant-head boundary and of the recharge water.

In one of the two simulations a conduit system was placed into the model domain just below the water table (see Fig.2) to investigate its influence on the uranium mobilization process. The U-shaped conduit system consists of two vertical shafts and one horizontal drift modelled with 49 conduit nodes and 48 tubes. The tubes have a diameter of 0.5 m. The horizontal drift was located in the centre of the uranium/pyrite ore body where mining activities are likely. The flow simulations show that approximately half of the conduit nodes in the pipe network are entry nodes (all on the left side of the conduit system) while the other half are exit nodes. Figs.3a and 3b show the head profiles of the flow set-up without and with the conduit system, respectively. The equipotential 450 m in Fig.3b separates entry from exit nodes. For the (assumed) exchange coefficient of 1.0×10^{-5} m^2 s^{-1}, flow conditions remain laminar throughout the conduit system.

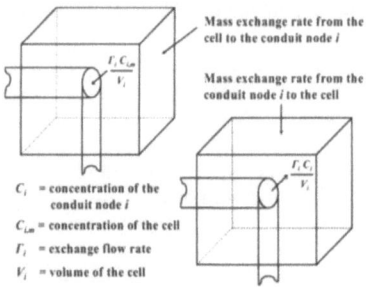

Fig. 1. Illustration of ore material cell-conduit exchange.

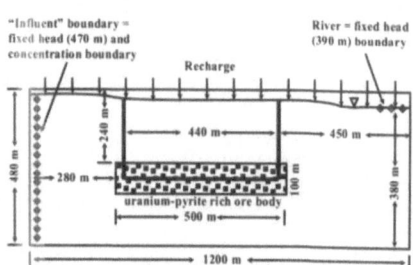

Fig. 2. Configuration of the simplified, schematic mine system.

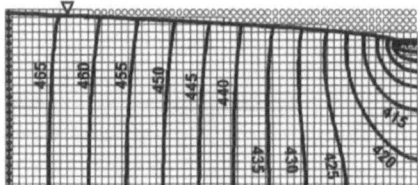

Fig. 3a. Head distribution profile for the mine system without conduit system.

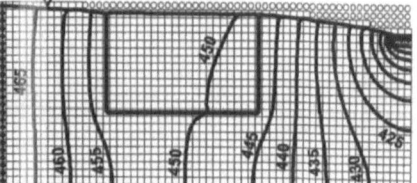

Fig. 3b. Head distribution profile for the mine system with conduits.

Initial results

Preliminary simulation results of RUMT3D indicate that the conduit system plays a major role for the release and the transport of uranium species to the discharge point. As apparent from Fig.4b, U(VI) (e.g., UO_2^{2+}) is mobilised in the area surrounding the conduit system and rapidly discharged into the river. With no conduit system present the U(VI) species remain stable within the ore body (Fig.4a), i.e. only negligible concentration/mass of uranium can be observed downgradient from the ore body.

Fig. 4a. U(VI) concentration profile for the mine system without conduit system.   **Fig. 4b.** U(VI) concentration profile for the mine system with conduits.

Summary and future work

This paper presents the reactive hybrid transport model RUMT3D developed for three-dimensional reactive underground mine transport problems. RUMT3D can handle two different transport time scales, found in underground mine systems. It is planned to further integrate surface controlled release of uranium into the model by implementing kinetically controlled reaction rate equations into the PHREEQC-2 database. Furthermore, sorption to charged surfaces (surface complexation) e.g. to iron(III) oxides will certainly play an important role for the mobility of uranium and will thus need consideration in future work. The assumption of a negligible effect of chemical reactions in the conduit system due to much shorter residence times of the components in the conduit system than in the ore material may also require further investigation.

Acknowledgements

The funding for this study was mainly provided by the European Commission (PIRAMID project, Contract EVK1-CT-1999-000021). Thanks are also given to Gaisheng Liu who prepared the numerical method EMCNOT to use for the solution of advection in the conduit transport model.

References

Bajracharya K., Barry D.A. (1993). Mixing cell models for nonlinear equilibrium single species adsorption and transport. J. Contam Hydrology 12: 227-243.

Barenblatt G.I., Zheltov I.P., Kochina I.N. (1960). Basic concepts in the theory of seepage of homogeneous liquids in fissured rocks. J Appl Math and Mech 24: 1286-1303.

Bain J.G., Mayer K.U., Blowes D.W., Frind E.O., Molson J.W.H., Kahnt R., Jenk, U. (2001). Modelling the closure-related geochemical evolution of groundwater at a former uranium mine. J Contam Hydrology 52: 109-135.

Birk S. (2001): Characterisation of karst systems by simulating aquifer genesis and spring responses: model development and application to gypsum karst. Geowissenschaftliche Arbeiten, C60, Tübingen, Germany.

Clemens T., Hückinghaus D., Sauter M., Liedl R., Teutsch G. (1996). A combined continuum and discrete network reactive transport model for the simulation of karst development. Proc ModelCARE 1996, Golden, Colorado, IAHS Publ. 237: 309-318.

Engesgaard P., Kipp K.L. (1992). A geochemical transport model for redox-controlled movement of mineral fronts in groundwater flow systems: A case of nitrate removal by oxidation of pyrite. Water Resour Res 28: 2829-2843.

Liu G., Wang P., Zheng C. (2001). An explicit and mass-conservative scheme without time-step limit for modeling advection-dominated contaminant transport, 2001 MODFLOW international conference in Golden, Colorado.

Parkhurst D.L., Appelo C.A.J. (1999). User's guide to PHREEQC (version 2) – a computer program for speciation, batch-reaction, one-dimensional transport, and inverse geochemical calculations. Water-Resources Investigations Report 99-4259, U.S. Dept. of the Interior, U.S. Geological Survey.

Prommer H., Bajyrry D.A., Zheng C. (2002). MODFLOW/MT3DMS based reactive multicomponent transport modelling. Accepted for publication in Ground Water.

Spiessl S.M., Sauter M., Zheng C., Liu G. (2002). Simulation of contaminant transport in flooded underground mines using a coupled continuum-conduit transport model: Comparison of two numerical methods for advection in the pipe network. Submitted to Proc ModelCARE 2002, Prag, 17-20 June 2002.

Steefel C.I., MacQuarrie K.T.B. (1996). Approaches to modeling of reactive transport in porous media. In: Reactive Transport in Porous Media (ed. by P. C. Lichtner P.C., Steefel C.I., Oelkers E.H.) Min. Soc. Of America, 1996, 83-129.

Walter A.L., Frind E.O., Blowes D.W., Ptacek C.J., Molson J.W. (1994). Modeling of multicomponent reactive transport in groundwater. 1. Model development and evaluation. Water Resour Res 30: 3137-3148.

Wunderly M.D., Blowes D.W., Frind E.O., Ptacek C.J. (1996). Sulfide mineral oxidation and subsequent reactive transport of oxidation products in mine tailings impoundments: A numerical model. Water Resour Res 32: 3173-3187.

Yeh G.T., Tripathi V.S. (1989). A critical evaluation of recent developments in hydrogeochemical transport models of reactive multichemical components, Water Resour Res 25: 93-108.

Zheng C., Wang P.P. (1999). MT3DMS: A modular three-dimensional multispecies model for simulation of advection, dispersion and chemical reactions of contaminants in groundwater systems; Documentation and User's Guide, Contract Report SERDP-99-1, U.S. Army Engineer Research and Development Center, Vicksburg, MS.

Stream pollution by adjacent tailing deposits and fluvial transport of dissolved uranium - dynamics and mechanisms investigated in mining areas of Germany, Southern Africa and Australia

Frank Winde

Potchefstroom University, Department for Geography and Environmental Studies (South Africa), frank.winde@gmx.de

Abstract. Although the transport of solute uranium (U) from tailings deposits in adjacent streams is one of the fastest ways of distributing U throughout the biosphere it cannot yet be sufficiently modeled. By comparing U-mining sites in different physical-geographical environments this study discusses factors controlling the migration of solute U along the aqueous pathway and within streams. It focuses on three major aspects of the transport, illustrated by examples from study sites:

- Chemical transport and immobilisation of U along the aqueous pathway
- Hydraulic mechanisms controlling U-migration via groundwater into streams
- Hydrochemical fluctuations controlling U-mobility in streams.

Introduction

The migration of highly contaminated seepage from tailings deposits of U and gold mining activities into underlying aquifers and adjacent streams is a major pathway for the transport of dissolved U and other heavy metals into the environment. The subsequent transport in streams constitutes an extremely fast way of distributing contaminants over long distances throughout the biosphere. In contrast to low-energy systems, like groundwater or porewater in tailings, U-transport in streams cannot yet be satisfactory modeled. This is mainly due to significantly higher dynamics of transport-processes in the high-energy system stream and increased numbers and complexity of influencing factors. Combined groundwater-stream transport mainly differs from U-migration in groundwater in the following

aspects:
- it includes a higher number of sediment-water systems through which U travels and where it might be retained or re-mobilised from;.
- the hydraulic movement of contaminated water is more complex and faster, including e.g. interactions at the groundwater-stream interface, and
- biological and atmospherical effects on hydrochemical parameters controlling U-mobility are more pronounced.

In this paper examples for each aspect are presented and implications for the solute transport of U discussed. By comparing sites with distinctly different natural conditions, impacts of regional peculiarities, like annual water balance, seasonal effects, geology etc. on waterborne U-transport will be identified.

Study sites and methods

The following sites and streams were studied: the "Lerchenbach" (Wismut region, East-Thuringia, Germany), affected by seepage from an adjacent U-tailings ponds ("IAA Culmitzsch"); the "Koekemoer Spruit" (NW-Province, South Africa), affected by U-polluted seepage from nearby "slimes dams" of the Buffelsfontein goldmine (these "slime dams" consist of tailings hydraulically deposited as a water-tailings mixture, "slime", from which surplus water evaporates during daytime, leaving dry tailings deposits behind); the "Magela Creek" (Northern Territory, Australia) affected by seepage from tailings of the Ranger Uranium mine and the Rössing Uranium mine near Swakopmund (Namibia), as a site with extremely arid conditions.

At all sites water and sediment from various points of the aqueous pathway were sampled and analysed for U_{nat} and other heavy metals using OES (after leaching sediments with aqua regio) and laser-spectroscopy (for Unat in water). The mobility of U in sediment-water systems was determined by distribution coefficients (Kd) calculated as the quotient of U-concentration in solids and water.

For tracking hydrodynamical processes real-time in-situ measurements of gauging heights and electrical conductivity (EC) in the stream-groundwater system were taken by a computerised data-logger, recording data in ten minute-intervals. Simultaneously pH, EC, temperature and redox-potential (Eh) of stream water as well as climatic parameters (precipitation, air temperature and relative air humidity) were recorded at the same time-interval. In the Magela Creek a submersible data-logger was used, measuring in 30min-intervals.

U-transport and immobilisation along the water-path

At all three sites elevated U-concentrations in sediment and water samples outside the fenced mining property were found. With 200ppm in wetland sediments near the Lerchenbach and 1200ppm in salt crust on topsoil near the Koekemoer Spruit these levels exceeded by far the U-concentration in tailings as actual source of contamination (100ppm and 130ppm respectively). Increased U-levels in seepage sampled at the Rössing Uranium mine (2-7ppm U) suggest that these tailings pose a potential risk to the aquifer of the nearby Khan River, which is dry for most of the year. U-concentrations in sediments of receiving watercourses range from 20-40ppm in the Lerchenbach (max. 74ppm, n=64), 40-50ppm in the Koekemoer Spruit (max. 54ppm, n=6) and 4-8ppm in the Magela Creek (n=3). The latter is surprisingly low considering the high U-levels in the adjacent tailings (400ppm). Despite increasing distance from the source of pollution no decrease of U in sediments was found in the remaining 10km-long reach of the Lerchenbach, downstream the point of impact.

Since the tailings deposits at all sites are well maintained and surrounding areas are not polluted by eroded tailings particles, the observed off-site contamination can reasonably be attributed to water-borne transport of solute U. Leached from tailings particles (which usually still contain about 10% of the original U-content of the ore) into porewater or introduced by contaminated process water, U travels along with seepage through the deposits into underlying aquifers and further into adjacent streams. Thereby it passes through several sediment-water systems in which it is partly removed from the water-phase and retained as a solid (the transition from [mobile] solute phases into [immobile] solid phases of U is called "immobilisation". It is mainly triggered by changing hydrochemical conditions that alter the speciation of U and decrease its solubility). Typical sediment-water systems of the aqueous pathway are shown in Fig.1 (featuring slimes dams as source of contamination).

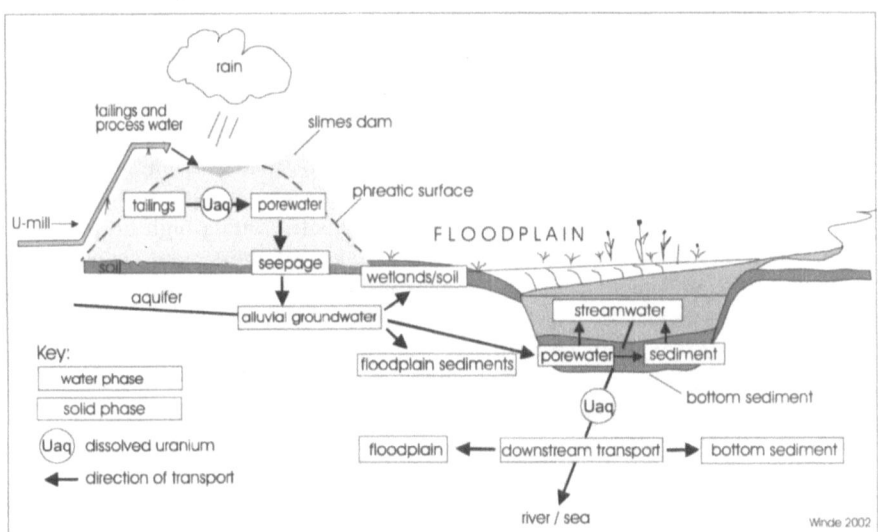

Fig. 1. Sediment-water systems of the aqueous pathway affecting U-transport

Due to high evaporation rates in South Africa and elevated groundwater levels in the vicinity of slimes dams, groundwater frequently ascends to the surfaces by capillary fringe effects and subsequently evaporates from the topsoil. The high concentrations of U and sulphate in remaining crusts (gypsum, epsomite or gosslarite) point to the formation of uranyl-sulphate-complexes in the groundwater, commonly found in sulphate-dominated waters (Markos and Bush 1982). Thee incorporation of U in such neutral complexes prevents the cation largely from adsorption onto negatively charged surfaces of clay minerals and organic matter in the floodplain sediments, which therefore show only low U-concentrations (19-28ppm) compared to wetland-sediments in the Wismut area (200ppm). Consequently, solute U-concentration in the alluvial groundwater remains high (0.5ppm) despite migrating for several hundred meters through highly sorptive sediments. Low Kd-values of 38-56 reflect the unexpectedly high mobility of U in the wetland-groundwater system.

Surprisingly, much higher Kd-values (i.e. lower U-mobility) were found in flowing water, like the Koekemoer Spruit (Kd=960) and a concrete channel in which dolomitic groundwater was discharged into the stream (Kd=771). The paradox that U is less mobile in fast-moving water with short contact-time to almost inert sediments (e.g. quartzite pebbles) than in slow-moving groundwater with long contact-time to highly sorptive sediments, is – apart from the formation of neutral complexes – mainly due to two different types of precipitation, which specifically occur in surface water and effectively remove U from the water phase.

One of it, the precipitation of calcite ($CaCO_3$) from dolomitic groundwater, was found to be dominating in the concrete channel, where contaminated sediments (better termed as scale) contain >74wt% of $CaCO_3$ and up to 54ppm U. High correlation between U and $CaCO_3$ (R=0.82; n=8) suggests that precipitation of calcite ("de-calcification") simultaneously removes U from the water. Hellmann (1999) found for the river Rhine (Germany) that de-calcification significantly reduces the concentration of dissolved heavy metals in the stream. Due to the saturation with Ca^{2+} and CO_3^{2-} this process is particular strong in dolomitic water.

In the Koekemoer Spruit, however, where contaminated groundwater seeps into the stream channel, redox-controlled precipitation dominates. While groundwater is almost oxygen-free, constituting a reducing environment with low Eh-values, the turbulently flowing streamwater is well oxygenated with a high Eh-potential. When both waters mix inside the bottom sediment of the stream channel, dissolved Fe(II) and Mn(II) from the groundwater immediately oxidise and precipitate as hydrous oxides or hydroxides ($FeOOH$, $Fe(OH)_3$), settling as amorphous gels they also cover sediment particles with thin brownish coatings (Since small particles – assumed to be of spherical form – have a larger surface-area compared to their volume than bigger particles, the total surface in fine-grained sediments covered with U-bearing coatings is higher than in coarse sediments. This is the main reason for increasing U-concentrations – and other heavy metals – in fine fractions of contaminated stream sediments). A strong correlation between U and Fe (R=0.97, n=12) and U and Mn (R=0.91), which further increases when Fe and Mn are combined (R=0.98), suggests that redox-initiated precipitation is a major immobilisation mechanism in streams.

The extraction of U from the water phase either happens by co-precipitation or subsequent adsorption onto the large surfaces of freshly precipitated gels. Since other heavy metals are accumulated as well (except for Cr/Ni all heavy metals correlate with U at R>0.90; n=12) unspecific co-precipitation seems more important than subsequent adsorption, which is more selective in terms of valency and size of ions. Redox-initiated precipitation preferably occurs at the chemo-cline between reducing and oxidising conditions inside the bottom sediments. Being a comparatively fast process, a significant amount of dissolved heavy metals can be removed from the groundwater while it filters through the sediment. Brownish layers of flocculated iron-hydroxide, floating just above the bottom sediment, were frequently observed in the critical reach of Lerchenbach. While these layers can be easily removed by flowing water, this is not the case with gels precipitated in the bottom sediment. They are largely protected from erosion and gradually crystallise into water-free solids. As a groundwater-stream interface, bottom sediments therefore act as both a geochemical barrier and a sink for heavy metals.

Hydraulic mechanisms controlling solute U-transport

In regions where negative annual water balances allow for dry deposition of tailings it was generally found that an elevated phreatic surface develops inside the depositions and drives seepage towards receiving watercourses. This occurred even under extremely arid conditions in the Namib Desert where tailings contain 10-50vol% water (Kehrberg 2000). With an average of some 20vol% water, this is comparable to old, decommissioned slimes dams of the Witwatersrand, where rainfall is more than 20 times higher.

Apart from the concentrations of U in the migrating water, stream pollution depends also on the rate at which contaminated water moves from the tailings into the stream. This movement comprises two sub-processes: a) the migration from the tailings across the floodplain towards the stream, and b) hydraulic interactions between groundwater and stream at the stream channel. While both processes were investigated, the focus here is on the latter, illustrated by an example from the Lerchenbach. For observing the processes in real time and in-situ, several datalogger controlled EC-probes were installed at the groundwater-stream interface at the "critical reach" of the Lerchenbach, where artesian groundwater seeps into the stream channel (Fig. 2). Pronounced EC-differences allows for clear discrimination between streamwater and the highly contaminated groundwater.

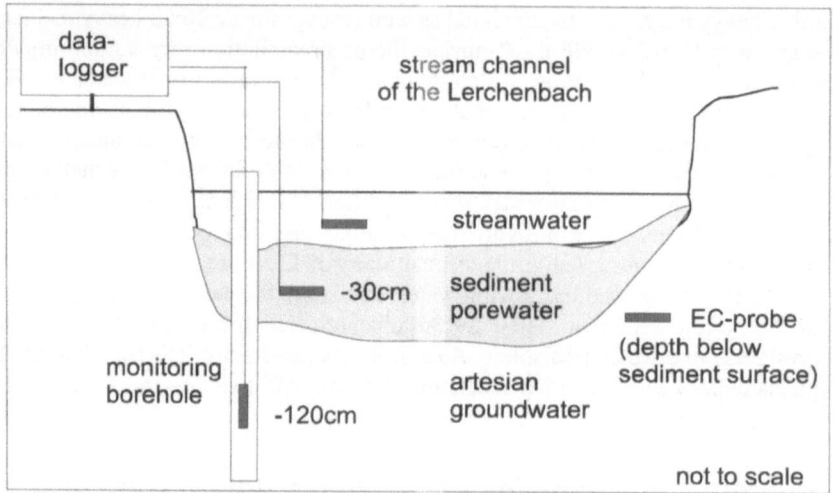

Fig. 2. Location of EC-probes in the critical reach of the Lerchenbach

As an example of stream-groundwater interactions and impacts of flood events an
EC-time-series of the groundwater, sediment-porewater and streamwater for a
three-week period (1-24/9/98) is displayed in Fig. 3.

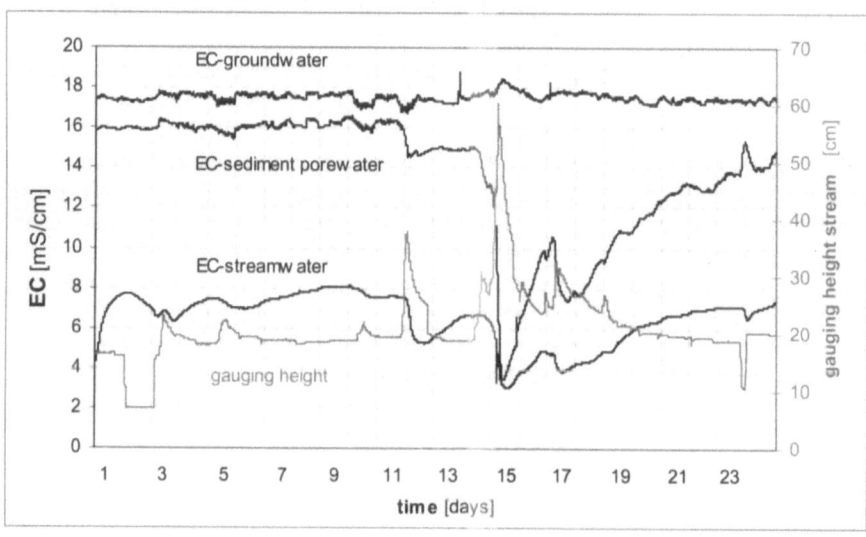

Fig. 3. EC and gauging heights indicating exchange of porewater (Lerchenbach, 1-24/9/98)

Geochemical and reactive transport modeling

Hydrochemical fluctuations and U-mobility in streams

After U has entered the stream its fate and the distance over which it will be transported downstream are mainly controlled by chemical conditions, namely pH and Eh of the streamwater. Both parameters determine to what extent U can be immobilised along the way (the term "immobilised" is, strictly spoken, only valid in a chemical sense and for groundwater systems where U is not transported any further after turning into the solid phase. This, however, is different in streams where such phases [e.g. U adsorbed onto suspended solids] are still being transported by flowing water). Real-time measurements in all three streams suggest that both parameters fluctuate significantly in daily and seasonal rhythms as well as in response to events like rainfall, waste-water spills etc. An example of diurnal fluc-

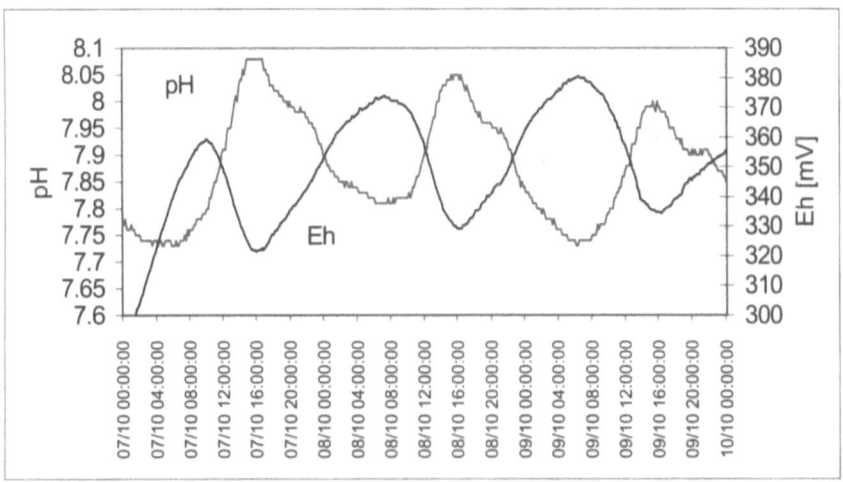

tuations of both parameters in the Koekemoer Spruit is displayed in Fig. 4.

Fig. 4. pH-Eh-fluctuations in the Koekemoer Spruit (7-10/10/1999)

Fig. 4 shows that both parameters fluctuate with inverse rhythms, indicating their close relationship. While Eh stands for the concentration of (negatively charged) electrons the pH represents the concentration of (positively charged) H^+-ions. Assuming that the net-charge in a system is neutral, rising concentrations of H^+-ions will result in rising concentrations of electrons (Eh). Since pH is an inverse indicator of the H+-concentration, the relationship between pH and Eh is inverse too.

According to Schwörbel (1987) pH-oscillations in streams are caused by photosynthesis of algae, phytoplankton and aquatic plants using dissolved CO_2 for assimilation and O_2-production, as soon as sunlight is available. Apart from biological activity the diurnal pH-cycle is also affected by daily oscillations of water-temperature controlling the solubility of CO_2 in water. While temperature oscillations are also caused by the insolation cycle and usually amplify photosynthesis-triggered pH-fluctuations, this is not the case in the Koekemoer Spruit where groundwater discharges disturb the relation (Winde 2001).

Although diurnal pH-cycles were observed in all three streams, temporal occurrence during the day, amplitude and level differs remarkably. The highest daily amplitudes were observed during spring in the Lerchenbach. Triggered by increasing temperature oscillations and growing algae-populations (cladophera) the pH fluctuated by up to two logarithmic units per day comprising changes of the H^+-concentration by a factor of 100. Due to the low buffer capacity of the non-dolomitic streamwater the daily maximum of the pH is not only higher than in the Koekemoer Spruit (9.0) but occurs earlier (shortly after noon). However, amplitudes and temporal dynamic of the diurnal pH-fluctuations vary significantly throughout the year and show the lowest intensity during winter. Distinct seasonal changes of diurnal pH-cycles do also occur in the tropical Magela Creek (Australia). Amplitude and level of the pH-cycle change drastically with the beginning of the wet season, when the system receives huge volumes of acidic rainwater. While the pH during the dry season fluctuates from 5.8 to 6.9 the pH-level drops to 4.5 in the wet season and daily amplitudes decrease to 0.2 pH-units/d (Fig. 5) (ERISS 2000).

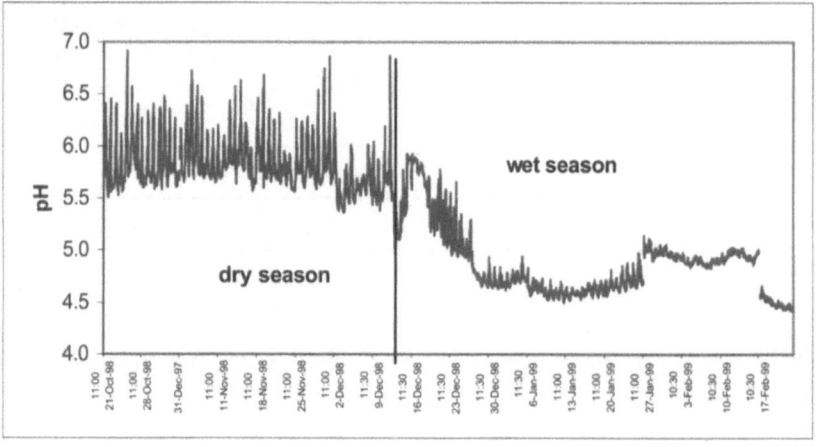

Fig. 5. Seasonal fluctuations of the pH in the Magela Creek (21/10/98 – 11/2/99)

While the temporal occurrence during the day is similar to Koekemoer Spruit (pH-maxima in the late afternoon around 4 p.m.) the pH-level is significant lower than in the other two streams, showing acidic conditions throughout the year.

In general effects of diurnal Eh-changes on the U-mobility in the stream are rather small compared to pH. The pH not only controls the precipitation of calcite that removes dissolved U from stream water (as the finding from the Koekemoer Spruit suggests) but also strongly affects the precipitation of iron-hydroxides, as the other important immobilisation process. The rate of Fe-hydroxide precipitation increases a 100 times if the pH - and thus the $(OH)^-$-concentration - rise by one logarithmic unit). With daily changes of 2 pH-units in the Lerchenbach the precipitation rate increases by up to 10,000 times (!) during the day. In addition to this the precipitation of inorganic Fe(II) as Fe(OH)$_3$ - without bacteria - only is possible in neutral-alkaline environments (pH>7). A pH>8.5 is required for precipitation of Mn as hydrous oxide or oxide (MnO$_2$) (Sigg and Stumm 1991).

Besides the co-precipitation of dissolved U with these compounds, the subsequent adsorption of U onto precipitated gels is also controlled by the pH-value. The charge of the large surface of hydrous iron oxides only turns negative when the pH rises above 8 thereby making it impossible for cations like U and other heavy metals to adsorb to the gels before this pH-level is reached. Since many of the pH-thresholds are only exceeded at certain times of the day, it is to be expected that U-concentration in streamwater varies profoundly during the day. Lower pH values in general favour higher U-mobility at night, while biological de-calcification and exponentially increased precipitation of iron-hydroxides reduce U-mobility during the day. Due to different kinetics of fast precipitation (happens within minutes) and slow re-solution of solid phases (e.g. coatings) it is unlikely that low pH-values during the night will lead to re-solution of U from sediments. This, however, might be different in case of event-related pH-changes, which often last longer and cover a wider range (Winde 2001).

Summary and conclusions

The study has established that waterborne transport of U from tailings deposits is a major mechanism of distributing the radionuclid at all four sites, even occurring under extreme arid conditions.

U migrates together with other dissolved heavy metals in seepage from tailings into groundwater and further into adjacent streams. En route to the stream, U passes through several sediment-water systems where different mechanisms partly remove the dissolved metal from the water column by turning it into solid phases (immobilisation). While retaining U in sediments helps to reduce water pollution, it simultaneously creates off-site accumulations, which sometimes exceed U-levels in the source of contamination and act as secondary sources of contamination (e.g. sulphate crusts). Within streams, precipitation of calcite and iron-manganese hydroxides was identified as the most important mechanisms for immobilising U. The latter preferably occurs inside stream-channel sediments, which act as a interface between reducing groundwater and well-oxygenated stream water and constitute a sink and geochemical barrier for solute transported metals.

Real-time in-situ measurements in this barrier suggest that the water composition in stream-sediments is highly dynamic and can drastically change during flood-events when hydraulic balances between stream and groundwater are disturbed. In case of the Lerchenbach this leads to short-term pulses of dissolved U in the early phase of the event while later on suppressing artesian baseflow significantly reduces stream pollution. The infiltration of polluted streamwater into sediments downstream of the critical reach may also explain why those sediments are contaminated, although they are not in touch with polluted groundwater.

Geochemical and reactive transport modeling

The transport of U in streams is also controlled by diurnal and seasonal fluctuations of pH and Eh in streams which affect chemical mobility of U. Triggered by photosynthesis and influenced by regional peculiarities like geology, aquatic vegetation, rainwater chemistry, etc., these fluctuations vary in amplitude and level. In general, they favour higher U-mobility during the night, while in daytime biological de-calcification and increased rates of precipitation partly immobilise U. Persistent acidic conditions in the tropical Magela Creek, however, largely prevent U-immobilisation. Together with flush effects and dilution during extreme floods in the wet season, this might explain why stream sediments are only weakly contaminated, despite much U-concentration in the tailings than at all other sites.

Acknowledgement

The study was funded by the German Academy for Natural Scientists, Leopoldina (BMBF-LPD 9801-17). I also wish to thank all involved colleagues from the Environmental Research Institute in Jabiru, the Department of Water Affairs and Forestry in Pretoria and Boskop Dam and all studied mines for their kind support and the provision of data. I thank Dr. Luke Sandham for reviewing the manuscript.

References

ERISS, Environmental Institute of the Supervising Scientists (2000): Data-logger records from the Magela Creek near Jabiluka (30min interval, raw data) Jabiru (Australia)
Kehrberg S (2000): personal communication
Markos G, Bush KJ (1982): Geochemical processes in uranium Tailings and their relationship to contamination. In: IAEA (Ed.): Management of waste from uranium mining and milling. Wien. 231-246
Hellmann H (1999): Qualitative Hydrologie. Berlin-Stuttgart
Schwörbel J (1987): Einführung in die Limnologie. Gustav Fischer Verlag. Jena
Sigg L, Stumm W (1991): Aquatische Chemie. Zürich, Stuttgart
Winde F (2001): Slimes dams as source of uranium contamination of streams – the Koekemoer Spruit (Klerksdorp gold field) as a case study. Conference on Environmentally Responsible Mining in Southern Africa, 25-28 September 2001, Muldersdrift (South Africa), Papers – Volume 1, 2c1-2c10.

Mobilities of uranium species in sedimentary interbed from the Snake River Plain

Robert A. Fjeld, John T. Coates, Alan W. Elzerman, and James D. Navratil

Department of Environmental Engineering and Science, Clemson University, Clemson, SC 29634, USA

Abstract. Laboratory column tests were conducted to characterize the mobility of uranium in a composite of sedimentary interbed samples from the Snake River Plain. The objective of the studies was to investigate the influence on uranium mobility of carbonate, fluoride, sulfate, and hydroxide complexes at pH 8, which is representative of ground water from the region. It was determined that carbonate species were the most dominant followed by hydroxide species. The observations are consistent with the results of speciation modeling. The carbonate species, which had retardation factors between 5 and 10, were considerably more mobile than the hydroxide species, which had retardation factors in excess of 300.

Introduction

Uranium contamination is found at most U. S. Department of Energy facilities. At some of these facilities, uranium is an important contaminant of concern because of the human health risk posed by exposure through the ground water pathway. Predictive models for the subsurface transport of uranium are thus needed to evaluate the need for, and effectiveness of, remediation measures. The subsurface transport models, in turn, require accurate estimates of uranium mobilities in the subsurface system. Presented in this paper are results of laboratory column tests to determine the mobility of uranium species in sedimentary interbed from the Snake River Plain, site of the Idaho National Engineering and Environmental Laboratory (INEEL). It is estimated that almost 330 tonnes of uranium, most of it depleted, are buried at this facility.

Although uranium can be found in both the tetravalent and hexavalent oxidation states in the natural environment, under Eh-pH conditions typical of the INEEL (Brookins, 1988), hexavalent uranium is expected to be the predominant

specie. Pentavalent uranium is predicted thermodynamically, but disproportiona-
tion causes the tetra- and hexavalent species to prevail (Hobart, 1990).

Under alkaline conditions such as those that exist at the INEEL, uranyl complexation with carbonate and hydroxide is expected to dominate the aqueous
chemistry. Even in waters containing humic acid, where the trivalent actinides are
influenced by organic complexation, the hexavalent actinides such as uranium are
present as carbonate or hydroxo complexes (Moulin and Moulin, 1995). Major
aqueous species for typical natural waters include $UO_2CO_3^0$ (aq), $UO_2(CO_3)_2^{2-}$,
and $UO_2(CO_3)_3^{4-}$ (Scanlan, 1977; Ciavatta, et al., 1979; Maya, 1981,1982; Grenthe
et al., 1984; Grenthe and Lagerman, 1991). Elless and Lee (1998) studied two
carbonate-rich, uranium contaminated soils from the Fernald site and reported that
the uranyl-carbonate complexes were the most stable and abundant complexes
present. Further, in an extensive review, Meinrath (1996) found that the principal
complex was $UO_2(CO_3)_3^{4-}$ and noted the lack of evidence for the other carbonate
complexes. As the carbonate concentration and pH decrease, the hydroxo complexes become increasingly important. For example, in the pH range of 4.5 to 5.5,
Kramer-Schnable et al., (1992) found the hydroxo complexes $UO_2)_2(OH)_2^{2+}$ and
$(UO_2)_3(OH)_5^{1+}$ to dominate. They also found these species to have very low solubilities. Thermodynamic modeling of ground water simulants with low carbonate
concentrations at pH 8 suggests the presence of the mononuclear di- and tri-
hydroxo complexes. At high carbonate concentrations and low fluoride concentrations at pH 6, uranyl carbonate, hydroxo-carbonate and hydroxo complexes are
predicted (Myers and Navratil, 1998).

In chemically complex waters, (e.g. waters containing ligands such as carbonate, fluoride, and sulfate) uranium speciation may be complex. Under acidic conditions, both fluoride and sulfate complexation are important. Strong mononuclear complexes with fluoride are formed, with $UO_2F_4^{2-}$ representing the upper
limit (Ahrland and Larsson, 1954; Ahrland et al., 1956; Ishigiro et al. 1977; Aas et
al., 1998). Serebrennikov and Dorofeeva, (1980) noted the presence of uranyl-
sulfate complexes, including UO_2SO_4 and $UO_2SO_4^{2-}$ in subsurface leaching zones
of epigenetic uranium ore deposits. Venkataramani and Gupta, (1990) discussed
the formation and existence of uranyl hydroxo-sulfate complexes in the laboratory. Abdel (et al. 1989) evaluated the stability constants of uranyl complexes
with typical ground water ligands and noted that the complexing power of fluoride
was greater than sulfate. Patil and Ramakrishna (1976) had noted earlier that as
fluoride concentrations increased, uranyl fluoride complexes were increasingly
important uranyl species.

The far-field subsurface environment at the INEEL is characterized by high pH
(7.0 to 9.1) and variable carbonate concentrations (50 to 500 mg/L) (Wood and
Low, 1986). The predominant uranium species in those areas where the carbonate
concentration is high will be carbonates. However, in regions where the carbonate
concentration is lower or in the vicinity of waste sites where the pH may be lower,
other uranium species are possible. These species may have mobilities that differ
from those of the carbonates. This study focused on the INEEL far-field environment and investigated the influence of selected ligands (carbonate, fluoride, sulfate, and hydroxide) on uranium mobility in sedimentary interbeds from the site.

Geochemical and reactive transport modeling

Experimental

Experiments were performed with a total of seven ground water simulants. The compositions (Table 1) of all but one of these (Simulant 7) were based on the composition of the Snake River Plain aquifer (Wood and Lowe, 1986) and unpublished measurements of perched water from the vadose zone. Simulants 1 – 6 were developed to determine the effect of carbonate, fluoride, and sulfate complexes on uranium mobility. Simulant 7 was developed to verify the role of hydroxide complexes. Simulant 1 was our approximation of a ground water with elevated concentrations of fluoride, sulfate, and carbonates. Fluoride was removed from Simulant 2, sulfate was removed from Simulant 3, carbonate was removed from Simulant 4, carbonate and fluoride were removed from Simulant 5, and carbonate and sulfate were removed from Simulant 6. Simulant 7 was distilled water adjusted to pH 8 and thus contained only hydroxide. Each of the simulants was prepared using reagent grade chemicals and distilled water. Dilute hydrochloric acid and sodium hydroxide were used to adjust pH.

Table 1. Compositons of ground water simulants. Concentrations are in mg/L unless noted otherwise.

Species	Simulant					
	1	2	3	4	5	6
Ca^{2+}	10.5	10.5	10.5	10.5	10.5	10.4
K^+	10	10	10	0	0	0
Mg^{2+}	17.5	17.5	2.3	17.5	17.5	2.3
Na^+	570	566	450	316	174	291
Cl^-	220	218	218	218	217	217
SO_4^{2-}	350	356	-	355	-	355
HCO_3^-	750	747	749	trace	trace	trace
F^-	20	-	20	20	20	-
Si^{2+}	10	10	10	10	10	10
pH	8.2±0.2	8.2±0.2	8.2±0.2	8.2±0.2	8.2±0.2	8.2±0.2
I (meq/L)	0.034	0.031	0.02	0.02	0.019	0.0085

Sedimentary interbed representative of INEEL was collected from two boreholes at five different depths and combined to form a composite. The composite sample was classified as a sandy loam. It was dry sieved, and the fraction less than 0.25 mm was retained. The dominant minerals in the sample were montmorillonite, muscovite, biotite, and vermiculite. In addition, there were small amounts of exchangeable iron (1.77 % Fe_2O_3) and manganese (1.9 ppm Mn). The cation exchange capacity was 1800 meq/g.

Spiking solutions were typically prepared by adding a known activity of uranium-233 to the appropriate simulant followed by a pH adjustment to 8.2 ± 0.2 using dilute HCl or NaOH. The concentrations ranged from 80 to 500 Bq/mL (1 x 10^{-6} to 6 x 10^{-6} M).

A diagram of the flow through column apparatus used in this study is given in Fig. 1. The apparatus consisted of reservoirs for the unspiked and spiked simulants, a peristaltic pump (Masterflex, Cole-Palmer Instrument Co.), 1.5 cm x 8 cm Plexiglas® column, and fraction collector (Universal Fraction Collector, Eldex Laboratories, Inc.). The components were interconnected with Masterflex® tubing.

Each column was packed by adding approximately 1 cm of the soil to the column and lightly tapping the column on a bench top to compress the soil until the column was completely filled. Before each successive addition of soil, the soil surface was agitated using a wooden dowel to avoid creating distinct layers of packed soil. When packed, each column had approximately 30 grams of soil corresponding to a dry bulk density of 1.7 to 1.9 g/cm^3 and a pore volume of about 6 ml.

The spiked simulant was introduced into the dry column as a finite step with a width of 1 pore volume and subsequently eluted for up to 1000 pore volumes by the unspiked simulant. Effluent fractions were collected and analyzed by liquid scintillation counting. These data were reduced to yield elution profiles, i.e., normalized concentration (effluent concentration/spike concentration) vs. reduced time (expressed as displaced pore volumes, DPV). Hydrogen-3 was used as a non-reactive tracer to monitor for channeling.

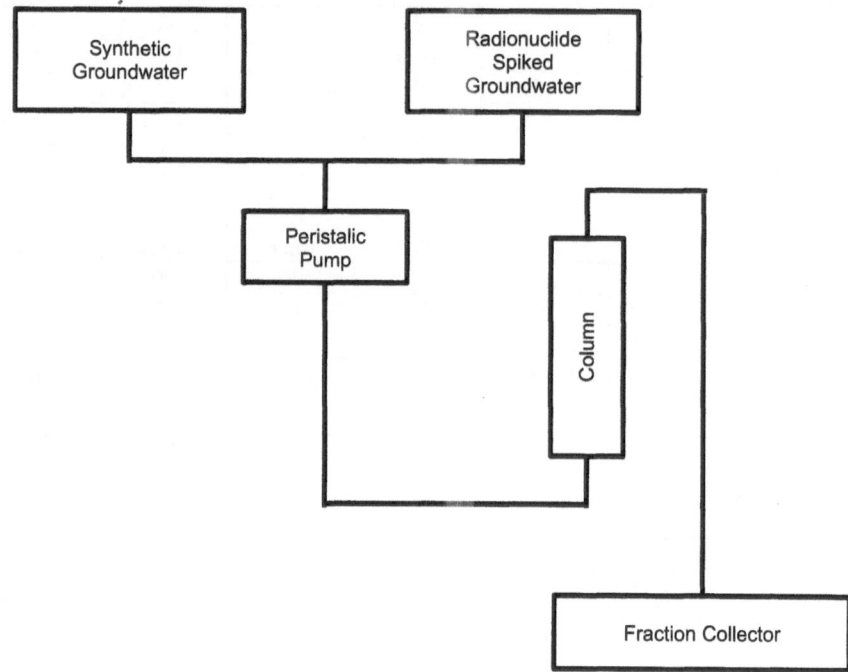

Fig. 1. Experimental apparatus.

Uranium mobilities were characterized through the retardation factor, which is the ratio of the mean contaminant travel time in the column to the mean travel time of water. Mean contaminant travel time is the first time moment of the elution profile (Fjeld et al. 2001) and the mean travel time for water is 1 DPV.

The equilibrium speciation code MINTEQA2 (Allison et al., 1991) was used to predict the uranium species in the simulants. The equilibrium constants used in the chemical equilibrium model were taken from the Data0.com.V8.R6 file of the EQ3/6 thermodynamic database (Worley 1992a, Worley 1992b).

Results

Presented in Fig. 2 are uranium elution profiles for Simulants 1, 2, 3, and 4. The curve for Simulant 1, which approximates the actual ground water at the site, is characterized by a single peak containing, within the bounds of experimental error, all of the uranium in the spike. Uranium appeared in the effluent almost immediately, peaked between 2 and 3 DPV, and was completely eluted within approximately 100 DPV. This behavior was not significantly affected by the removal of either F^- (Simulant 2) or SO_4^{2-} (Simulant 3). However, the behavior was greatly affected by the removal of CO_3^{2-} (Simulant 4), which resulted in almost the complete disappearance of the peak at 2 DPV and the appearance of a new peak that emerged around 30 DPV, reached a maximum between 200 and 300 DPV, and was not completely until eluted within 1000 DPV. These results indicate that the large peak at 2 DPV for Simulants 1, 2, and 3 was due to one or more carbonate complexes. The new peak that appeared between 100 and 1000 DPV upon removal of CO_3^{2-} remained when F^- (Simulant 5) and SO_4^{2-} (Simulant 6) were also removed (Fig. 3).

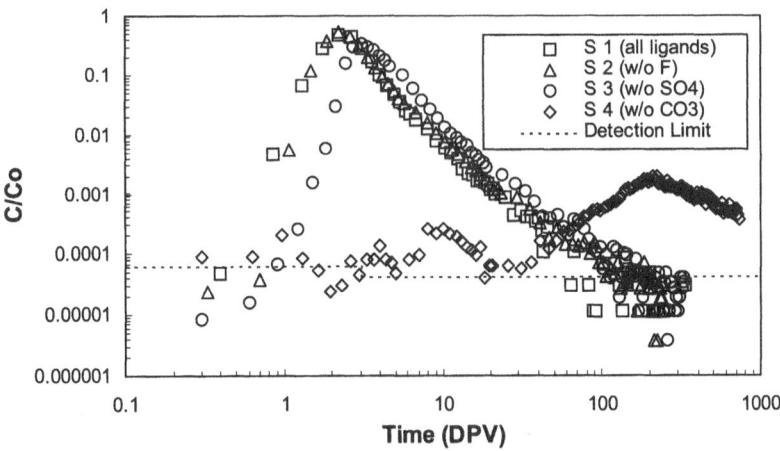

Fig. 2. Uranium elution profiles for Simulants 1, 2, 3, and 4.

Geochemical and reactive transport modeling

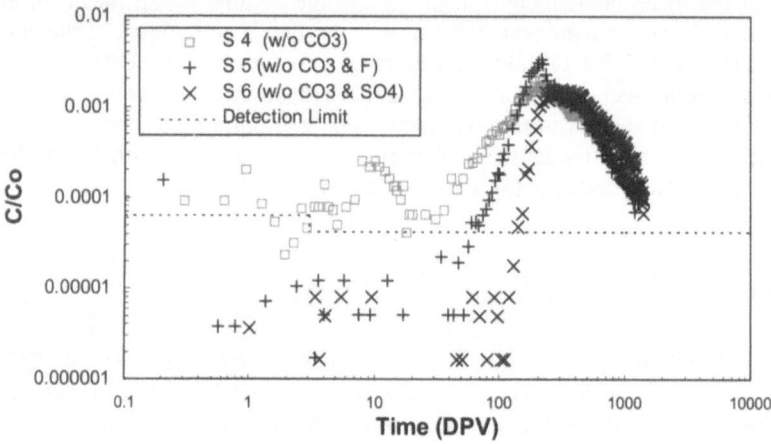

Fig. 3. Uranium elution profiles for Simulants 5, and 6.   The profile for Simulant 4 is included for comparison.

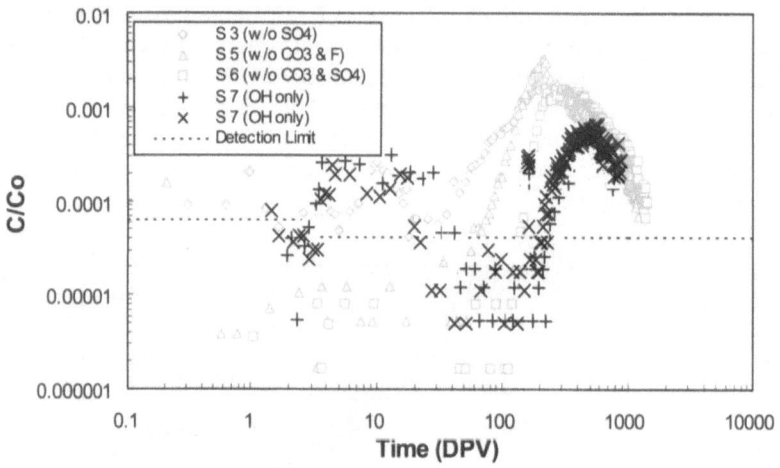

Fig. 4. Uranium elution profiles for Simulant 7.   The profiles for Simulants 3, 5, and 6 are included for comparison.

This suggests that the peak was due to a complex with another ligand, probably hydroxide. This was confirmed in a subsequent experiment in which hydroxide was the only ligand present (Simulant 7), and the elution profile contained the new peak (Fig. 4).

Geochemical and reactive transport modeling

There are a couple of interesting subtleties in the elution profiles. The first is that two of the profiles "without" carbonate (4 and 7) have a small peak between one and 10 DPV, and the other two (5 and 6) do not. Since this peak suggests a carbonate species, our hypothesis is that it is due to small amounts of carbonate in these simulants, possibly due to incomplete isolation from the atmosphere. The other is that the peaks between 100 and 1000 DPV for the simulants without carbonate may not be identical. For example, for Simulant 4, the peak began to appear around 30 DPV, for Simulant 5 it began to appear around 60 DPV, for Simulant 6 it began to appear around 100 DPV, and for Simulant 7 it began to appear around 300 DPV. Further study would be needed to determine if these differences are real or if they are within the bounds of experimental variability.

Results of speciation modeling are presented in Table 2. The modeling results and the experimental results are consistent. The model predicts $UO_2(CO_3)_3^{4-}$ and $UO_2(CO_3)_2^{2-}$ as the predominant species in the simulants with high carbonate concentrations (1, 2, and 3), and it predicts $UO_2(OH)_2$ and $UO_2(OH)_3^-$ as dominating in the simulants with low carbonate concentrations (4, 5, 6, and 7).

Table 2. Results of speciation modeling of the ground water simulants.

Simulant	Distribution (%)			
	$UO_2(CO_3)_2^{2-}$	$UO_2(CO_3)_3^{4-}$	$UO_2(OH)_2$	$UO_2(OH)_3^-$
1	3.9	96.1	-	-
2	3.9	96.1	-	-
3	4.5	95.4	-	-
4	-	-	75.6	24.0
5	-	-	74.8	23.8
6	-	-	75.0	23.4
7	-	-	78.3	21.3

Presented in Table 3 is a summary of the analyses of the elution profiles. The fractional recovery and retardation factor are based on the specified DPV range. Recoveries were complete within the bounds of experimental error for Simulants 1 through 3. For the others, the tests were terminated before the uranium concentration had returned to zero and the fractional recoveries were considerably less than one. Consequently, the retardation factors for the simulants represent lower bounds. Based on these results, we infer that the carbonate complexes, which we observed to have retardation factors in the 5 - 10 range, are much more mobile than the hydroxide species, which we observed to have retardation factors in excess of 300. This is consistent with Mahal et al., (1982) who attributed an observed increase in uranium sorption at pH values above 9 to the formation of hydroxide species.

Table 3. Summary of results from the uranium effluent profiles.

Simulant	DPV Range	Likely Complex	Fractional Recovery	Retardation Factor
S1 (all)	0 - 100	Carbonate	0.93	4.3
S1 (all) replicate	0 - 100	Carbonate	0.97	4.1
S2 (w/o F⁻)	0 – 238	Carbonate	0.93	5.1
S3 (w/o SO_4^{2-})	0 – 330	Carbonate	0.95	8.1
S4 (w/o CO_3^{2-})	0 - 20	Carbonate	0.002	12
	10 - 800	Hydroxide	0.45	340
S5 (w/o CO_3^{2-} & F⁻)	20 – 1220	Hydroxide	0.77	700
S6 (w/o CO_3^{2-} & CO_4^{2-})	100 – 1415	Hydroxide	0.74	560
S7 (OH⁻ only)	0.1 – 20	Carbonate	0.001	6
	20 - 830	Hydroxide	0.12	480
S7 (OH⁻ only) replicate	0.1 - 20	Carbonate	0.0007	5
	20 – 900	Hydroxide	0.26	560

Conclusions

This research was concerned with the mobilities of uranium species in sedimentary interbed from the Snake River Plain. The experiments were conducted at pH 8.2, which is in the middle of the range (7 to 9.1) expected in the far-field at the INEEL (Wood and Low, 1986). We confirmed that uranyl carbonate species dominate at elevated levels of carbonate and found that they have a moderately high mobility (retardation factors on the order of 10 or less). In the absence of carbonate, we confirmed that hydroxide species dominate and found that they have much lower mobilities (in excess of 300). Although neither fluoride nor sulfate appeared to have a major influence on uranium mobility in our experiments, they could be important in the vicinity of waste sites where lower pH's are possible

Acknowledgements

This research was supported in large part by the Idaho National Engineering and Environmental Laboratory under contracts K97-560265, K98-564517, and K99-181044.

References

Aas W, Moukhamet-Galeev A, Grenthe I (1998) Complex formation in the ternary U(VI)-F-L system (L = carbonate, oxalate, and picolinate). Radiochem Acta 82: 77-82

Allison J D, Brown D S, Gradac K J (1991) MINTEQA2/PRODEFA2, a geochemical assessment model for environmental systems: version 3.0. EPA/600/3-91/021

Abdel Razik A, Ali F A, Abu Attia F (1989) Evaluation of the stability constants of uranyl association complexes with chloride, fluoride, bromide, and sulfate anions in solutions of constant ionic strength. Microchem J 39(3): 258-264

Ahrland S, Larsson R (1954) The complexity of uranyl fluoride. Acta Chem Scand 8: 354-355

Ahrland S, Larsson R, Rosengren K (1956) Complex chemistry of the uranyl ion. VIII. complexity of uranyl fluoride. Acta Chem Scand 10: 705-718

Brookins D G (1988) Eh-pH diagrams for geochemistry. New York: Springer-Verlag

Ciavatta L, Ferri D, Grimaldi M, Palombari R, Salvatore F (1979) Dioxouranium (VI) carbonate complexes in acid solution. J Inorg Nucl Chem 41:1175-1182

Elless M, Lee S Y (1998) Uranium solubility of carbonate-rich uranium-contaminated soils. Water, Air, Soil Pollut 107(1-4): 147-162

Grenthe I, Lagerman B (1991) Studies on metal carbonate equilibria. 22. A coulometric stury of the uranium(VI)-carbonate system, the composition of the mixed hydroxide carbonate soecies. . Acta Chem Scand 45(2): 122-128

Grenthe I, Spahiu K, Olofsson G (1984) Studies on metal carbonate equilibria. 9. Calorimetric determination of the enthalpy and entropy changes for the formation of uranium(IV and VI) carbonate complexes at 250 C in a 3 M sodium perchlorate-perchloric acid ionic medium. Inorg Chim Acta 95(2): 79-84

Hobart D E (1990) Actinides in the environment. Proceedings of the Robert A. Welch Foundation conference on chemical research XXXIV Houston, Texas: 379-434

Ishiguro S, Kao C-F, Kakihana H (1977) Formation constants of hydrogen fluoride ion $(HFm(1-m)+)$ and uranyl fluoride ion $(UO2Fn(2-n)+)$ complexes in 1 Mol Dm-3 (sodium chloride) medium. Denki Kagaku Oyobi Butsuri Kagaku 45(10): 651-653

Kramer-Schnabel U, Bischoff H, Xi R H, Marx G (1992) Solubility products and complex formation equilibria in the systems uranyl hydroxide and uranyl carbonate at 250C. and I = 0.1 M. Radiochim Acta 56(4): 183-188

Mahai H S, Venkataramani B, Venkateswariu K S (1982) Effect of anions on the sorption of uranium (VI) on hydrous oxides. Proc Indian Acad Sci 91(14): 321-327

Maya L (1981) Dioxouranium (VI) carbonate complexation in uranium recovery by reactive ion exchange. J Inorg Nucl Chem 43(9): 2133-2137

Maya L (1982) Hydrolysis and carbonate complexation of dioxouranium (VI) in neutral-pH range at 250C. Inorg Chem 21(&): 2895-2898

Meinrath G (1996) Coordination of uranyl(VI) carbonate species in aqueous solutions. J Radioanal Nucl Chem 211(2): 349-362

Moulin V, Moulin C (1995) Fate of actinides in the presence of humic substances under conditions relevant to nuclear waste disposal. Appl Geochem 10(5): 573-580

Myers J, Navratil J D (1998) Actinide mobility at the soil consolidation unit. Project Review Mntg Bolder Co

Patil S K, Ramakrishna V V (1976) Sulfate and fluoride complexation of U(VI), Np(VI) and Pu(VI). J Inorg Nucl Chem 38: 1075-1078

Scanlan J P (1977) Equilibria in uranyl carbonate systems: II. The overall stability constant of UO2(CO3)22- and the third formation constant of UO2(CO3)34-. J Inorg Nucl Chem 39: 635-639

Geochemical and reactive transport modeling

Serebrennikov V S, Dorofeeva V A (1980) Forms of occurrence of uranium in the production solutions in areas of underground leaching of epigenetic uranium deposits. Geokhimiya 9: 1391-1395

Venkataramani B, Gupta A R (1990) Sulfate complexes of hydrolyzed uranyl species in solution: formation constant of uranyl hydroxide sulfate ((UO2)2(OH)2SO4). Indian J Chem, Sect A 29A(4): 373-375

Wood W W, Low W H (1986) Aqueous geochemistry and diagenesis in the Eastern Snake River plain aquifer system. Geol Soc Am Bull 97: 1456-1466

Wolery T J (1992a) EQ3/6, A software package for geochemical modeling of aqueous systems: Package overview and installation guide, version 7.0. Lawrence Livermore National Laboratory, Livermore, CA.

Wolery T J (1992b) EQ3NR, A computer program for geochemical modeling of aqueous-speciation-solubility calculations: Theoretical manual, user's guide, and related documentation, version 7.0. Lawrence Livermore National Laboratory, Livermore, CA.

Geochemical and reactive transport modeling

A 3-D numerical modeling of reactive transport of U(VI) in shallow groundwaters: a case study at a uranium mill tailing site in southern China

Yanxin Wang, Teng Ma, Lechang Xu

Department of Hydrogeology and Environmental Engineering, China University of Geosciences, 430074 Wuhan, P. R. China; Email: yx.wang@cug.edu.cn

Abstract. Surface complexation model DLM was coupled with groundwater flow transport model to simulate the transport of U(VI) in shallow groundwaters around a uranium mill tailing site in southern China. The sorption of U(VI) onto clayey soils has a non-linear relationship with pH, reaching maximum around neutral pH and sharply decreasing under acidic and alkaline conditions. According to our modeling results, after 1000a of decommissioning, the vertical transport of U(VI) could be down to 19m, and the direction of transport turn horizontally; on the plane, the transport could extend 800m to west, 1500m to the south and 2000m to the north, with a total contaminated area up to about 9 km^2.

Introduction

Uranium mill tailings not only occupy large areas of land, but constitute a potential source of contamination since the contaminants such as heavy metals and radionuclides in the tailings may easily be released into the environment due to leaching of the tailings by meteoric water (Dubrovsky et al. 1984; Lieser et al. 1992). In China, there have been more than 70 uranium mines and milling plants in operation since the 1960's. Large amounts of mining wastes and tailings have been produced. It was roughly estimated that there are about 280 million tons of uranium mining wastes and 300 million tons of uranium mill tailings (Li 1996). Most of the wastes and tailings are dumped in open fields. More and more monitoring data have shown that many tailings have caused contamination of groundwaters.

Sorption is an important factor in solute transport modeling. For cases when the surface charge of sorbents shows close relationship with pH, surface complexation model proves to be useful in describing sorption behaviors (Davis and Kent 1990;

Payne et al. 1994; Turner 1995; Roberto and David 1997; Langmuir 1997; Edward, 1998; Prikryl et al. 2001). In this study, groundwater flow transport model was coupled with surface complexation model to simulate the transport of uranium(VI) in shallow groundwater system around a uranium mill tailing site in southern China.

Site Hydrogeology

The uranium mill tailing site was constructed in 1963 and decommissioned in 1994. The total area of the reservoir is $1.47km^2$, with 2×10^7 tons of tailings.

There are three terraces along the east bank of the river flowing around the western border of the site (Fig.1). The I terrace is composed of Holocene sediments, the II terrace upper Pleistocene sediments, and the III terrace alluvial-diluvial sediments. The tailings are at the top of the III terrace.

The site constitutes relatively independent groundwater system: the river on the west and the water divide along the Gulongmiao-Wanggutang line on the east are natural boundaries (Fig.1). And the northern and southern boundaries are the groundwater flow lines parallel to the two ends of the Ω-shaped river. In the range delineated by these boundaries develop Quaternary porous aquifers and Tertiary fissured aquifers. The two types of aquifers are hydraulically interconnected.

Methods

At the bottom of the tailings, there is a 4-5m thick clay. Previous studies show that the retardation effect on the transport of radionuclides by this layer is quite strong: in the past 30 years: U(VI) has been retarded within 0.1-0.5m inside the clay. Since the aquifers from the study area contain very little clay and show very limited sorption capacity for U(VI), only sorption of U(VI) onto the clay layer is therefore considered in our study.

Bath experiments were done to study the U(VI) sorption behavior of the clay. XRD analysis of the clay samples revealed that the major mineral is montmorillonite. Standard uranium solutions containing 1g/L-100mg/L uranium were diluted with 0.01M $NaNO_3$ to prepare stock solutions containing respectively $4.3\times10^{-4}M$ and $4.3\times10^{-5}M$ uranium. Eight groups of 100mL stock solutions with different pH values for each uranium concentration were then prepared by adding NaOH and HNO_3. The resultant 100 mL stock solutions have pH from 2 through 9 with a unit interval. The sorption experiments were conducted by reacting 1g, 0.1g, and 0.01g samples of clay in 100 mL of the $4.3\times10^{-4}M$ and the $4.3\times10^{-5}M$ uranium stock solutions. The samples were reacted in high density polyethylene (HDPE) plastic vials for 72 h in a constant temperature shaker maintained at 25±0.5 °C. The samples after reaction were centrifuged for uranium analysis of the solution.

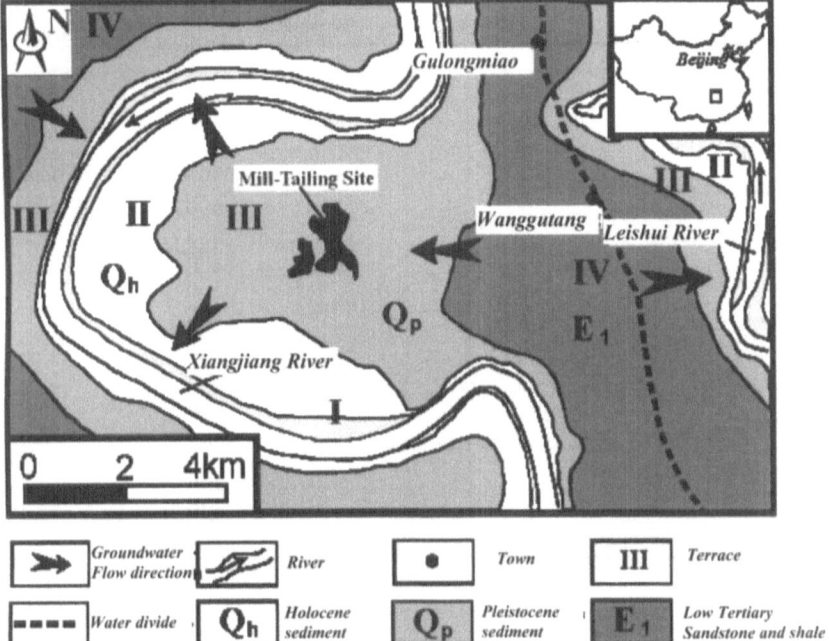

Fig 1. Hydrology map of the mill-tailing site

There are different types of surface complexation models, CCM (Constant Capacitance Model), DLM (Diffusion Double Model) and TLM (Triple Layer Model) being the most commonly used. The CCM usually applies to cases with low surface charge (ϕ < 25mV)and high ionic strength. The TLM can simulate experimental data quite well, but needs a lot of parameters. The DLM can often simulate experiment results very well, with less parameters as compared with the TLM. The DLM was used in this study. The modeling was done using the geochemical modeling code MINTEQA2/PRODEFA2 (Allison et al. 1991).

The MINTEQA2calculation results about the relation of aqueous species of U(VI) with pH at 25℃ and $P_{CO2}=10^{-3.5}$ is shown in Fig. 2. The predominant species between pH 4 and pH 7 are UO_2^{2+}, UO_2OH^+ and $(UO_2)_3(OH)_5^+$. Zachara and Smith (1994) considered two surface comlexes $\equiv X\text{-}UO_2^+$ and $\equiv X\text{-}(UO_2)_3(OH)_5^0$ for their TLM models. The following four surface complexation reactions were considered in our modeling using DLM:

$$\equiv SiOH^0 + UO_2^{2+} = \ \equiv SiO\text{-}UO_2^+ + H^+ \qquad (1)$$
$$\equiv SiO^0 + 3UO_2^{2+} + 5H_2O = \ \equiv SiO\text{-}(UO_2)_3(OH)_5^0 + 6H^+ \qquad (2)$$
$$\equiv AlOH^0 + UO_2^{2+} = \ \equiv AlO\text{-}UO_2^+ + H^+ \qquad (3)$$
$$\equiv AlOH^0 + 3UO_2^{2+} + 5H_2O = \ \equiv AlO\text{-}(UO_2)_3(OH)_5^0 + 6H^+ \qquad (4)$$

The DLM parameters used are listed in Table 1.

Geochemical and reactive transport modeling

Table 1. The DLM parameters for simulating sorption of U(VI) onto clays

Solid / solution ratio (M/V)	1g/l
Site density	$2.3/nm^2$
Surface area	$9.7m^2/g$
Total site concentration	$\equiv AlOH: 1.68 \times 10^{-5} mol/l$
	$\equiv SiOH: 2.02 \times 10^{-5} mol/l$
Ionic strength	0.1M $NaNO_3$
Concentration of $^{233}U(VI)$	$4.3 \times 10^{-4} M$
P_{CO2}	$10^{-3.5} atm$
Edge-site reactions	LogK
$\equiv AlOH^0 + H^+ \Leftrightarrow \equiv AlOH_2^+$	8.33
$\equiv AlOH^0 \Leftrightarrow \equiv AlO^- + H^+$	-9.73
$\equiv AlOH^0 + UO_2^{2+} \Leftrightarrow \equiv AlO\text{-}UO_2^+ + H^+$	2.7
$\equiv AlOH^0 + 3UO_2^{2+} + 5H_2O \Leftrightarrow \equiv AlO\text{-}(UO_2)_3(OH)_5^0 + 6H^+$	-15.00
$\equiv SiOH^0 \Leftrightarrow \equiv SiO^- + H^+$	-7.2
$\equiv SiOH^0 + UO_2^{2+} \Leftrightarrow \equiv SiO\text{-}UO_2^+ + H^+$	2.65
$\equiv SiO^0 + 3UO_2^{2+} + 5H_2O \Leftrightarrow \equiv SiO\text{-}(UO_2)_3(OH)_5^0 + 6H^+$	-15.24

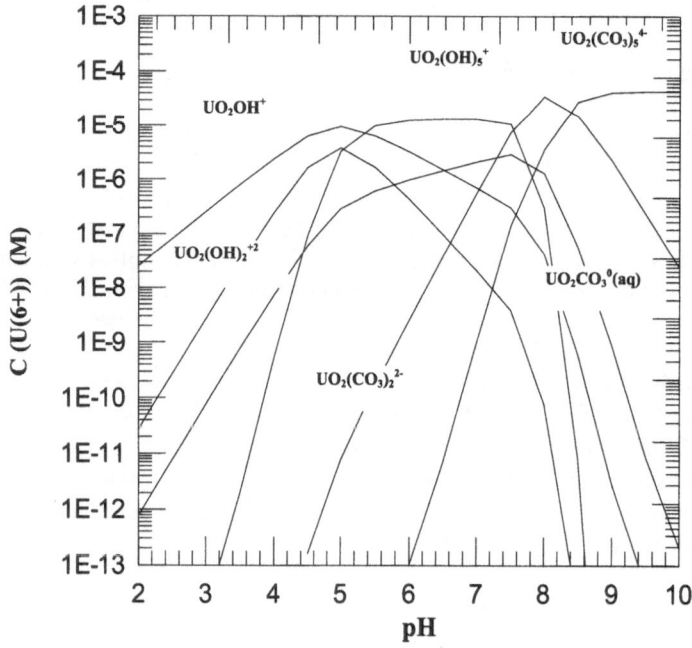

Fig. 2. The relationship plot between the aqueous speciation of U(6+) and pH at 25 °C, calculated by MINTEQA2

Geochemical and reactive transport modeling

The following equation is used to describe the transport of a contaminant in groundwater system:

$$\frac{\partial C}{\partial t} = \frac{\partial}{\partial x_i}(\frac{D_{ij}}{R_d} \cdot \frac{\partial C}{\partial x_j}) - \frac{\partial}{\partial x_i}(C\frac{u_i}{R_d}) \quad (i,j=1,2,3) \qquad (5)$$

where C is the contaminant concentration; D_{ij} the hydrodynamic dispersion coefficient, $R_d=1+\frac{\rho_b}{n}K_d$, the retardation factor; K_d the distribution coefficient, n and ρ_b are respectively porosity and matrix density; u_i is the average flow velocity.

Modeling Results

Our experimental results show that sorption of U(VI) onto clays from the study area has a non-linear relationship with pH, reaching a maximum around a pH of 7 and sharply decreasing under acidic and alkaline conditions. Traditional isothermal sorption models are unsuitable for describing the observed U(VI) sorption behavior. DLM was therefore used in this study. MINTEQA2 was run to simulate the speciation of U(VI) in aqueous phase under different pH conditions for a uranium concentration of $4.3 \cdot 10^{-4}M$ (I=0.1NaNO$_3$). The results of calculation using DLM match quite well with experimental data (Fig.3)

Rain acidification has been increasingly serious in the study area. Based on the monitoring data of rain pH in 1983-2000, a pH of 4.1 was chosen for the DLM calculation. For U(VI) concentrations of 0.705 mg/L (tailings in operation) and 0.509 mg/L(tailings decommissioned), pH=4.1, M/V=0.58, the distribution coefficients (K_d) were calculated using DLM to be 0.0025L/g and 0.0017L/g, respectively.

Visual MODFLOW provided by Waterloo Hydrogeologic was used for modeling U(VI) transport in the shallow groundwater system. The result of modeling U(VI) transport after 1000a of decommissioning is shown in Fig. 4.

It can be seen from Fig.4 that the maximum vertical transport distance will be 19m, and the direction of transport tend to turn horizontally. In a plane view, the U(VI) plume extended 800m to the west, 1500m to the south, and 2000m to the north, with a total contaminated area of 9 km^2.

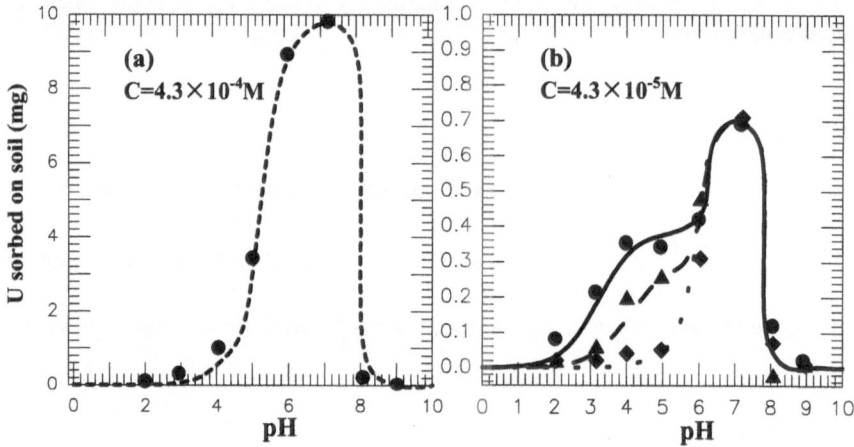

Fig. 3. Comparison of experimental data with DLM calculation results for: (a) one group of experiment with $c=4.3\times10^{-4}$mol/L (the solid circle and dash line represent respectively experimental data and calculated curve respectively); (b) three groups of experiments with $c=4.3\times10^{-5}$mol/L (the solid circle, triangle and diamond and lines respectively represent experimental data and calculated curve)

Conclusions

1.   Coupling surface complexation models with flow transport models is an effective approach for describing and predicting the transport of trace contaminants that is controlled by sorption processes in groundwater system. The coupled modeling provides information about the characteristics of contaminant transport in groundwaters, as well as the sorption behavior of the contaminants underground. Especially important is that surface complexation models can calculate distribution coefficients under various complex conditions.

2.   Sorption of U(VI) onto clays from the study area shows a non-linear relationship with pH, reaching maximum at a pH of around 7 and sharply decreasing under acidic and alkaline conditions.

3.   After 1000a of decommissioning, the maximum vertical transport distance is predicted to be 19m, and the direction of transport tends to turn horizontally. In a plane view, the U(VI)-containing plume extends 800m to the west, 1500m to the south, and 2000m to the north, with a total contaminated area of 9 km^2. Careful monitoring and reinforcement of the decommissioning systems are therefore necessary to prevent from radionuclide contamination of the shallow groundwaters.

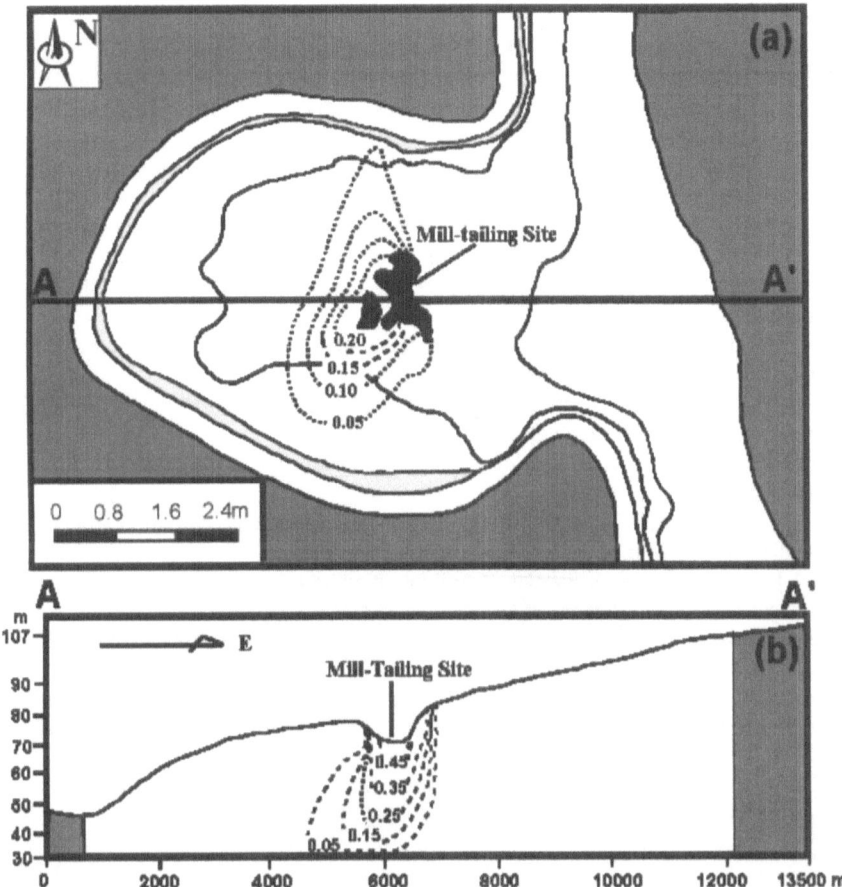

Fig. 4. Results of predicting the distribution of U() in shallow groundwaters after 1000a of decommission of the mill tailing Site

Acknowledgements

The research work was funded by National Natural Science Foundation of China (NSFC-49832005) and the Teaching and Research Award Program for Outstanding Young Teachers in Higher Education Institutions of Ministry of Education, P. R. China.

References

Allison J, Brown D, Novo-Gradac K (1991) MINTEQA2/PRODEFA2: A geochemical assessment data base and test case for environmental system. Report EPA/600/3-91/-21.Athens, U.S.EPA

Davis J, Kent D (1990) Surface complexation modeling in aqueous geochemistry, In: Mineral-Water Interface Geochemistry, Hochella M, White A (eds.), 178-249

Dubrovsky N, Morin K, Cherry J (1984) Uranium tailings acidification and subsurface contaminant migration in a sand aquifer. Water Pollut Res J Canada, 19: 55-89

Edward H (1998), Surface complexation modeling of metal removal by recycled iron sorbent. J Environ Eng 9: 13-20

Lieser K, Quandt-Klenk S, Thybusch B (1992) Sorption of uranyl ions hydrous silicon dioxide. Radiochim Acta 57: 45-50

Li X (1996) Remediation of uranium mining-related contamination. Beijing: China Environmental Science Press (in Chinese)

Langmuir D (1997) Aqueous environmental geochemistry. Upper Saddle River: Prentice-Hall Inc.

Payne T, Davis J, Waite T (1994) Uranium retention by weathered schists: the role of iron mineral. Radiochim Acta 66/67: 297-303

Prikryl J, Jain A, Turner D, Pabalan R (2001) Uranium(VI) sorption behavior on silicate mineral mixtures. J Contam Hydrol 47: 241-253

Roberto T, David R (1997) Uranium (6+) Sorption on montmorillonite: experimental and surface complexation modeling study. Aquatic Geochem 2: 203-226

Turner D (1995) A uniform approach to surface complexation modeling of radionuclide sorption, Report CN-WRA 95-001. San Antonio, Center for Nuclear Waste Regulatory Anal, TX

Zachara M, Smith S (1994) Edge complexation reactions of cadmium on specimen and soil-derived smectite. Soil Sci Soc Amer J, 58: 762−769

Flooding of the deep uranium mine Hamr I by free water from tailings pond[*]

Jan Novák, Jiří Mužák

Technical University in Liberec, Dept. of Modeling of Processes, Hálkova 6, 461 17 Liberec, Czech Republic

Abstract. The contribution is devoted to the description and the application of the mathematical model of groundwater flow and reactive-transport on the problem of flooding of the deep uranium mine by water from tailings pond. The currently performed flooding of the Hamr I mine significantly affects the hydrogeological situation in the area of Straz block. The problem of flooding the mine goes together with liquidation of the chemical treatment plant tailings pond. The flooding by alkalized tailings pond free water will avoid future acidification of mine water as a result of decomposition of oxidized pyrite minerals.

Situation in the Stráž deposit

The uranium deposit Stráž was exploited by underground acidic leaching between 1968 and 1996. In the cenomanian aquifer we can define two different areas: the area of the leaching fields and the area of the solution excursion towards the deep mine. There is more then 190 millions m^3 of contaminated water in the whole cenomanian aquifer.

High concentrations and large vertical extents of contamination can be found in the area of older leaching fields where extremely high dosages of sulfuric acid were used. Similar solutions are also in the more permeable parts of the profile, in the neighborhood of the leaching fields.

In the area of the solution escape the thickness of the contaminated horizon is decreasing from the hydraulic barrier to the drainage system of the deep mine. The purpose of this drainage system is to hold back acid water contaminated from the leaching process. The salinity of solutions varies from 1 to 10 g/l.

The co-existence of both the classical and chemical mining techniques at the same place naturally induced a complicated hydrogeological situation in the re-

---

[*] This work was supported by grant No. GACR 205/01/P047

gion. The classical mining was stopped in 1993 and the mines have been filled with concrete. The leaching has finished just now and this technique will be replaced by the remediation which will take approximately 30 years. Nowadays, the flooding of the mines will be started so that the escape of the pollutants from the leaching fields would be minimized. It will be controlled using the results of the reactive-transport modeling.

Mathematical description of unsteady unsaturated flow model

Unsteady flow problem is solved on fixed domain Ω, which includes both saturated and unsaturated zones. Let us call phreatic surface the area where both zones meet. The phreatic surface separates saturated zone $p \geq 0$ from unsaturated zone $p < 0$, where p is a pressure head. At the beginning the contact area is not a priori known. Unsteady unsaturated porous media flow is governed by the following continuity equation

$$\frac{\partial \theta}{\partial t} + \nabla \cdot \mathbf{u} = q \qquad \text{in } \Omega,$$

where θ is called the water (or moisture) content, $\nabla = \left(\frac{\partial}{\partial x}, \frac{\partial}{\partial y}, \frac{\partial}{\partial z} \right)$, $\mathbf{u}$ denotes filtration velocity and function q describes solution source density. The first fraction describes accumulation capacity of rock. The effective water content $\theta_e = \frac{\theta - \theta_0}{\varepsilon - \theta_0}$ (where θ_0 is a rest moisture and ε is a porosity) depending on the negative pressure head (capillary head) by the van Genuchten formula

$$\theta_e(p) = \frac{1}{\left(1 + \alpha|p|^m\right)^{\frac{m-1}{m}}} ,$$

where α and m are geometrical characteristics of rock (effective pore diameter). The Darcy's law will be considered in the following form

$$\mathbf{u} = -k_r(\theta)\mathbf{K}(\nabla p + \nabla z) \qquad \text{in } \Omega$$

where $k_r(\theta) \approx \theta_e^3$ is a relative hydraulic conductivity and tensor $\mathbf{K}$ represents environment hydraulic conductivity. The piezometric head equals to sum of $p+z$.

The mixed boundary conditions will be prescribed on the boundary of the domain Ω. An aquifer base is considered impermeable. Permeability of the domain cover depends on pressure head gradient and transfer coefficient σ. Thus following general Newton boundary condition is prescribed on this part of the boundary

$$\mathbf{u} \cdot \mathbf{n} - \sigma(p - p_D) = q_N$$

This condition allows even modelling of the impermeable part of the boundary letting $\sigma = q_N = 0$ respectively setting defined inflow $\sigma = 0$. On the free water surface Γ_D the non-homogenous Dirichlet boundary condition $p = p_D$ is prescribed.

Mathematical description of reactive transport model

Complex task of chemical substance transport is defined by balance equations for each substance in solution and its form for the l-th substance is:

$$\frac{\partial c^{\ell}}{\partial t} + \nabla \cdot \left(\mathbf{u} \ c^{\ell} \right) - \nabla \cdot \left(\mathbf{D} \ \nabla c^{\ell} \right) + c^{\ell} q^{-} + r^{\ell,-} \left(c^{\ell}, c^{i}, .. \right) = c^{\ell *} q^{+} + r^{\ell,+} \left(c^{k}, c^{i}, .. \right)$$

where D is tensor of diffusivity and dispersivity, which depends on molecular diffusivity of l-th substance, longitudal and transversal dispersivity and filtration velocity according to following formula:

$$\mathbf{D} = D_m \delta_{ij} + \alpha_T |\mathbf{u}| \ \delta_{ij} + \left(\alpha_L - \alpha_T \right) \frac{u_i u_j}{|\mathbf{u}|},$$

where D_m is molecular diffusion coefficient, u is filtration velocity, α_L, α_T are longitudal and transversal dispersivity coefficients. Homogenous boundary conditions are prescribed. Let us denote $\partial \Omega^{+}$ the sum of inflow border faces. If we assume uncontaminated water flows into the model area through inflow border faces, then homogenous Dirichlet boundary condition

$$c^{\ell} = 0$$

is assigned. If the type and chemical composition is known, then non-homogenous condition is assigned. Let us denote $\partial \Omega^{-}$ the sum of outflow border faces. Homogenous Neumann's condition

$$\mathbf{D} \ \nabla c^{\ell} \cdot \mathbf{n} = 0$$

characterising the case in which chemical components are brought out of model area only by convection (impact of diffusion is neglected due to low concentrations) is applied on outflow border faces.

Time decomposition of transport operator

Dominant influence on transport progress in substance migration in tailings pond surrounds problem has convection, given by following equation:

$$\frac{\partial c^{\ell}}{\partial t} + \mathbf{u} \cdot \nabla c^{\ell} + c^{\ell} q^{-} = c^{\ell *} q^{+}$$

In this equation q- denotes abstraction output and q+ injection output of solution with concentration $c^{\ell*}$ of ℓ th substance. This equation is explicitly discretized in time and solved in each time step.

Chemical changes have another important influence on transport progress. There are two types of chemical changes.

Chemical changes evoked by new balance of substances in solution. Following system of non-linear equations for separate solution components mathematically describes these changes:

$$G^m\left(c^\ell, c^i, c^k, \ldots\right) = 0 \quad \text{, pro} \quad \ell, i, k \in L_s,$$

where Ls denotes index aggregate of solution components. If there are major chemical changes within the time step, multiple recalculation of the system is necessary due to the non-linearity in chemical balance appointment within the time interval.

Chemical changes evoked by dissolve and precipitation reactions. These are much slower compared to reactions in solution and thus calculated according to kinetic model and described by following differential equations:

$$\frac{\partial c^\ell}{\partial t} + r^{\ell,-}\left(c^\ell, c^i, ..\right) = r^{\ell,+}\left(c^k, c^i, ..\right)$$

Function $r^{\ell,-}$ defines changes in concentration of ℓ th substance evoked by precipitation reactions. Function $r^{\ell,+}$ defines donation in ℓ th substance evoked by dissolve of substances from rock. Both functions include velocity coefficients set according to immediate chemical conditions. Influence of diffusion and dispersion can be also taken into account in the model, mathematically described by following parabolic type partial differential equation:

$$\frac{\partial c^\ell}{\partial t} - \nabla \cdot \left(\mathbf{D}\, \nabla c^\ell\right) = 0 \quad , \quad \ell \in L_s$$

As the influence of diffusion and dispersion is equal to the numerical diffusion of model and thus already included in the

Application of the model

An application of the model to modelling of flooding deep uranium mine after finishing the mining activities is considered. The flooding is done by decreasing the activity of the water barrier situated between the two mining techniques. On fig. 1 we can see location of the model mesh within the Straz block area. Area of interest covers approximately 120 km². The bases of the elements in the area of mining blocks are equilateral triangles with side of approximately 80 meters. Largest elements are around south and eastern edge, where the length of triangle side is approximately 800 meters. Vertically the area is divided into 12 layers with thickness varying from 5 to 20 meters. The mesh contains 21 966 spatial elements.

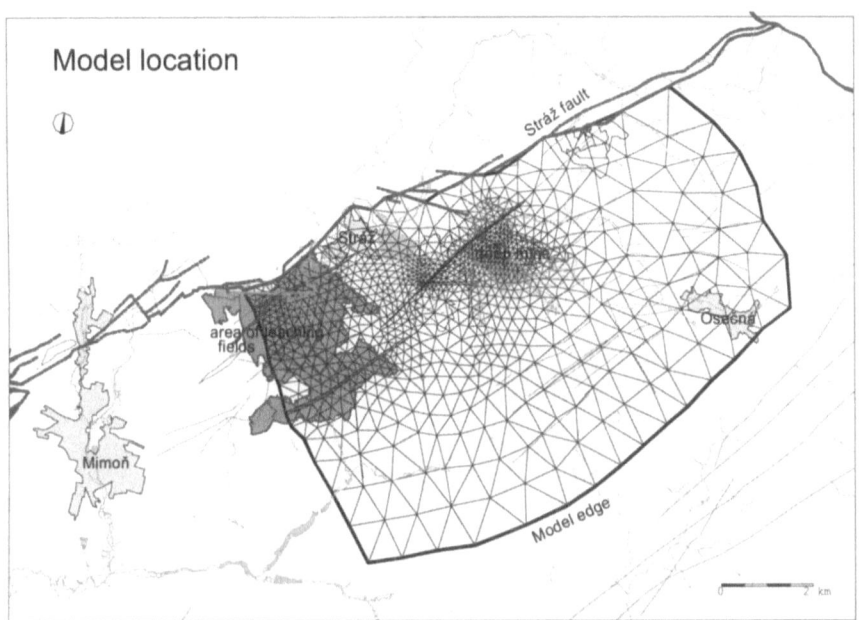

Fig. 1. The shape and location of finite element model mesh

For the flooding process the water from the tailings pond is used. It has been found that the tailings pond free water has suitable chemical properties and does not react chemically with the basic rock formation of the mine as well as with the other fresh water sources from the Lužice mountain area. The water level of the groundwater will rise by approximately 90 meters during the whole flooding process.

Concentration of TDS [g/l]

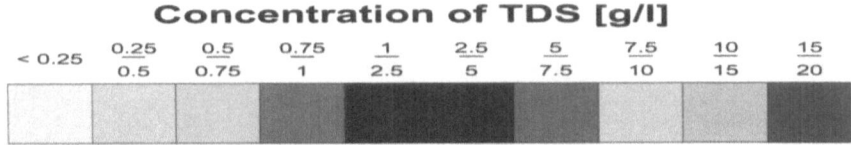

Fig. 2. The legend for figures 4 an 6

On the fig. 3 we can see the character of flow field in the modeled area 30 days since the beginning of injection of free tailings pond water into deep mine. The fig. 4 shows the distribution of TDS in corresponding time 30 days since the beginning of injection. The fig. 5 and 6 shows the situation after 2 years of injection.

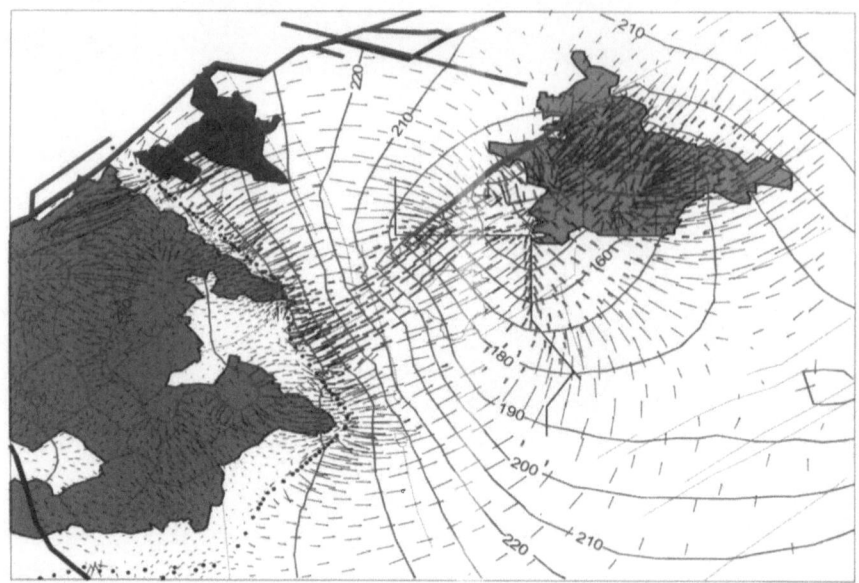

Fig. 3. The character of flow in the area 30 days since the beginning of injection

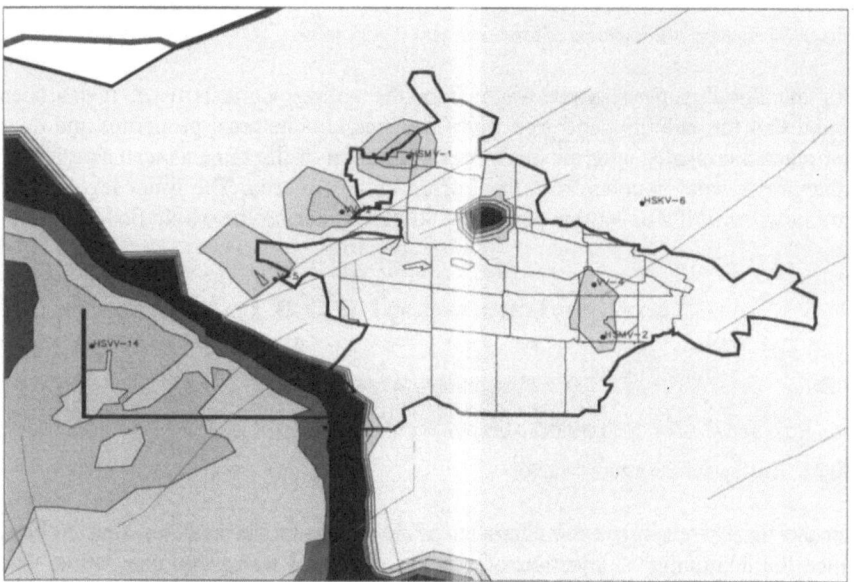

Fig. 4. The distribution of TDS after 30 days of injection

Geochemical and reactive transport modeling

Fig. 5. The character of flow in the area 2 years since the beginning of injection

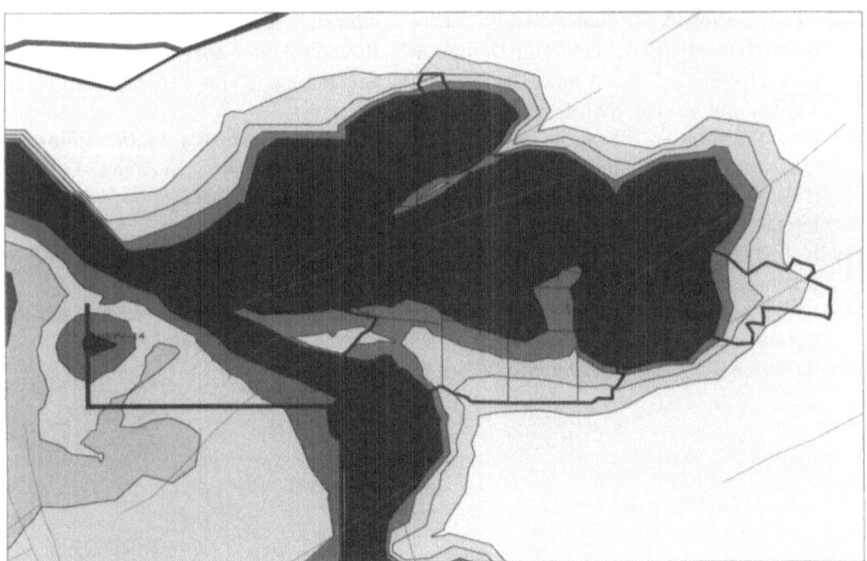

Fig. 6. The distribution of TDS after 2 years of injection

Geochemical and reactive transport modeling

References

Bear J, Verruit A (1990) Modeling Groundwater Flow and Pollution. D. Reidel Publishing Company, Dordrecht, Holland

Maryška J, Mužák J (1995) Mathematical modeling of the transport of chemical species in the contaminated underground water. In Bourgeat A P, Carraso C, Luckhaus S & Mikelič A Editors, Proceedings of the Conference Mathematical modeling of Flow Through Porous Media, pages 450-459, World Scientific

Maryška J, Rozložník M, Tůma M (1995) : Mixed-hybrid finite element approximation of the potential fluid flow, J. Comput. Appl. Math., 63:383-392

Beneš M, Stýblo M, Maryška J, Mužák J (1996) : The application of mathematical models of the transport of chemical substances in the remediation of consequences of the uranium mining, Proceedings of 3rd Workshop on Modeling of Chemical Reaction Systems, Heidelberg

Maryška J, Frydrych D (1997) : Mixed-hybrid model of unsaturated porous media fluid flow, Proceedings of ALGORITMY'97, 161-166, Slovak Technical University, Bratislava, Slovakia

Maryška J, Frydrych D, Mužák J (1997) : Mixed-hybrid model of unsteady free boundary porous media fluid flow, Proceedings of ALGORITMY'97, 149-155, Slovak Technical University, Bratislava, Slovakia

Novák J et al.(1997) : Zatápění dolu HAMR I, VZ 465/97, archiv s.p.DIAMO, Stráž p.R.

Maryška J, Rozložník M, Tůma M (1998) Schur complement reduction in the mixed-hybrid finite element approximation of Darcy's law: Rounding error analysis. Technical Report TR-98-06, Swiss Center for Scientific Computing, Swiss Federal Institute of Technology, Zurich, Switzerland

Novák J, Smetana R, Šlosar J (1998) Remediation of a sandstone's aquifer following chemical mining of uranium in the Stráž deposit, Czech Republic. In Arehart G B and Hulston J R Editors, Proceedings of the 9th International Symposium on Water-Rock Interaction, pages 989-992, Taupo, New Zealand

Frydrych D (1999) The Mathematical Modeling of the deep mine flooding process. In Proceedings of the Conference ECMS 99, Liberec

Mužák J, Novák J, Pisková E (1999) The Mathematical Modeling of the natural and the technological processes on the uranium deposit Stráž the models and the application. In Proceedings of the Conference ECMS 99, Liberec

Application of numerical simulation system on Turonian aquifer remediation control[*]

Hana Čermáková, Jan Novák, Jiří Mužák

Technical University in Liberec, Dept. of Modeling of Processes, Hálkova 6, 461 17 Liberec, Czech Republic

Abstract. This contribution presents the simulation system SATURN_2001, developed for the purpose of optimized management of remediation extraction from Turonian aquifer in Straz pod Ralskem region. The numerical model including the effect of dual porosity is used for modeling of reactive transport of contaminants. The model has been calibrated by utilizing data from VP-9C area. Application of the system is presented by one of the solved remediation variant.

Introduction

The drinking water sources for Northbohemian region lie in Turonian aquifer, part of the Czech Cretaceous Table. Due to chemical mining of uranium in Straz block the underlying Cenomanian aquifer was polluted. In spite of a 60 m thick aquitard between Cenomanian and Turonian aquifers 0.5 % of contamination migrated from the Cenomanian to the Turonian aquifer. This represents approximately 30 000 tons of dissolved solids. The contamination is characterized by isolated plums with concentration of TDS from 5 to 30 g/l.

Remediation is performed in the form of abstraction of the contamination from the ground and the polluted water is cleaned by membrane technologies, neutralized or directly discharged to rivers (only very slightly contaminated water).

To evaluate the underground situation numerical models describing groundwater flow and contaminant transport taking into account dual porosity were developed. The models were calibrated using the data from monitoring. The optimizing software tool SATURN was developed to model the processes controlling the remediation.

---

[*] This work was supported by grant No. GACR 205/01/P047

Groundwater flow and contaminant transport model with dual porosity

By "dual porosity" we express that rock matrix is divided into two pore volumes, active and inactive. The inactive pores are filled with solution but the velocity of flow in this pores is neglect able compared to the velocity in active pores.
Let us denote n_a – active porosity, n_s – inactive porosity and n – total porosity ($n = n_a + n_s$).

The transport of the contaminants is driven by the following aspects: advection and dispersion in active pores and diffusion exchange between active and inactive pores (fig. 1).

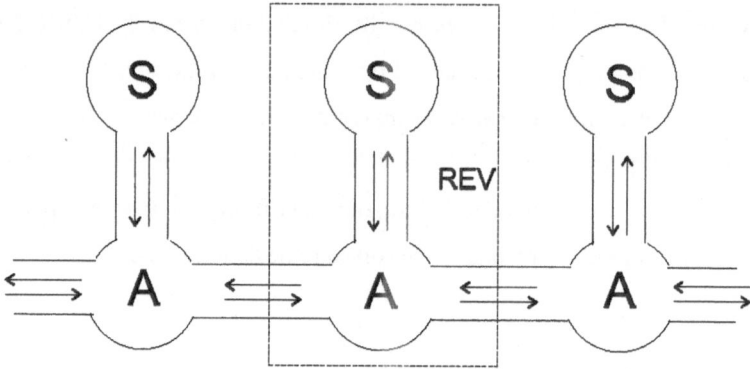

Fig. 1. Model of structures of active and inactive pores

Let us consider that diffusion flow between the volumes V_A and V_S is linearly depended on the difference of the concentrations in these volumes with proportion coefficient k

$$Q = k\,(c_A^l - c_S^l),$$

where Q is mass of the chemical component divided by time unit. The direct application of this formula in the model is not suitable. Numerical solution of parabolic equation in channel makes condition on time step similarly as in advection case. For diffusion the following equation must be valid

$$\Delta t < \frac{(\Delta x)^2}{2b}$$

where b is diffusion rate. Such approach would make it necessary to work with very small time steps slowing down the program performance. The alternative is to use analytical solution of the local exchange problem and its application for porous media. For time dependence of concentration of l-th substance in active and inactive pore volume the following exponential equations

$$c_A^l(t) = (c_A^l(0) - \bar{c}^l(0))\,e^{k\frac{V_A+V_S}{V_AV_S}t} + \bar{c}^l(0)$$

$$c_S^l(t) = (c_S^l(0) - \bar{c}^l(0))\,e^{k\frac{V_A+V_S}{V_AV_S}t} + \bar{c}^l(0)$$

are valid for any time $t \in (0, +inf)$, where $c_A^l(0)$ is concentration of the l-th substance in time $t=0$ in active pore volume, $c_S^l(0)$ is concentration of the l-th substance in time $t=0$ in inactive pore volume and $\bar{c}^l$ is average concentration of the l-th substance in time $t=0$. Computing by means of exponential functions is fast and robust. The coefficient k in the exponent includes as diffusion properties of the chemical components as geometrical properties of the soil (length and cross-section of the pores). Expressing these properties we get

$$k\frac{V_A+V_S}{V_AV_S} = D^l\frac{S}{\lambda}\frac{1}{\lambda S}\frac{n}{ns} = \frac{D^l}{\lambda^2}\frac{n}{ns} = \alpha^l$$

where D^l denotes diffusivity of l-th substance, S is the cross-section of the pore (the area between active and inactive pore volumes), λ is the distance between both volumes. We defined notation α^l for coefficient of substance mass exchange between the volumes. It can be written in the form $\alpha^l = D^l . K_r$, where D^l depends only on chemical substance and K_r is depended on the properties of the soil.

In practice it is problematic to determine local material characteristics that in addition do not directly express the rate of exchange. For calibration of the model it is suitable to define the following value

$$T_{\frac{1}{2}}^l = \frac{ln2}{\alpha^l}$$

which we call half-time of exchange. During this time the concentration changes from its initial value to value

$$\frac{c_A^l(0) + c^{-l}(0)}{2}$$

In the model we use the function

$$\frac{t}{T_{\frac{1}{2}}\ln 2}$$

and as the diffusion rate coefficient it is set the half-time of exchange $T_{1/2}$.

Groundwater flow and transport model calibration

The model was calibrated using data from monitoring of VP-9C site for the period from March 2000 to September 2001. The calculation was performed as a series of 19 steady state constant boundary conditions and abstraction scenarios in each month. The transport of the contaminants was calculated using a time step of three days.

The first phase of calibration was the setting and correction of initial conditions obtained by logging. Then the comparison of global balances in dependence on the active and total porosity and diffusion rate was performed. For each trial calculation the deviations of mass in each month and their squares were computed. The sums of deviations and their squares for the whole calculation period are two target functions that we tried to minimize by changing of material parameters.

Table 1. Resulting intervals of material parameters.

	min	max
Active porosity n_a	0,25	0,27
Total porosity n	0,07	0,08
Diffusion rate $T_{1/2}$ (days)	120	150

Optimizing system SATURN

The system serves for optimal choice of wells for remediation abstraction. In each time step the needed volume and quality of abstracted solution for surface cleaning technologies has to be assured. The system works with following technologies: neutralization, membrane technologies and direct discharge.

SATURN is built as a sequence of graphical forms that are filed by the user. The form serves as for typing of model inputs as for writing of model results.

In localization of abstraction (fig. 2) user defines requirements for spatial distribution of intensity and quality of abstracted solution within remediation area. To each leaching field we assign appropriate technology, intensity and quality of needed solution.

The form No. 5 is called pre-choice of abstraction places (fig. 3). For each technology the model can choose only the elements of finite element mesh that satisfy the required condition for quality of abstracted solution.

The form No. 7 (fig. 4) shows the optimal solution of the task. In the upper part there are results of the optimizing task: fulfilling of limits and list of chosen elements. In the bottom there are type activity of leaching fields, surface technologies character and technological effect of remediation.

The simulation is performed in half year time steps. The whole variant of remediation is composed of the results obtained in each time step. Finally the economical evaluation is done. It serves for comparison of computed variants.

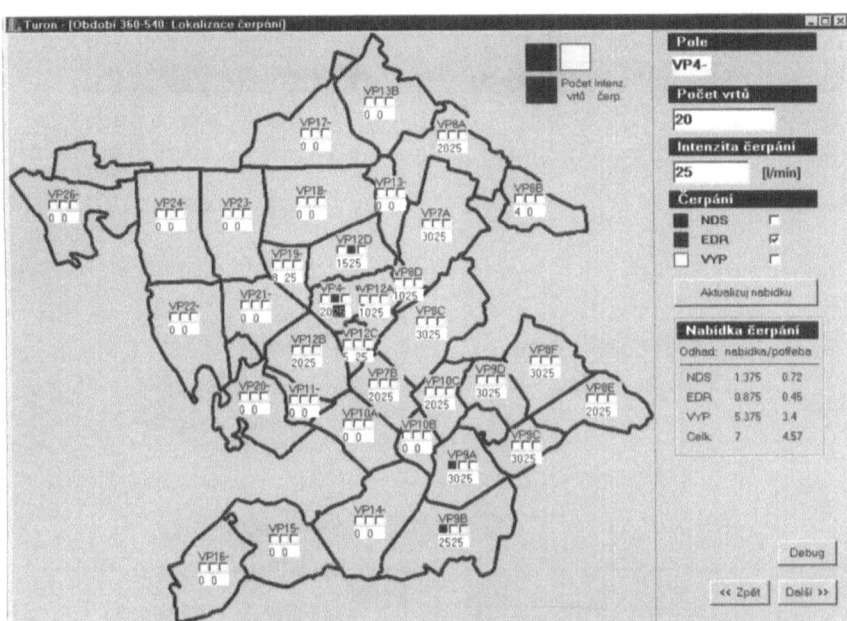

Fig. 2. The localization of abstraction

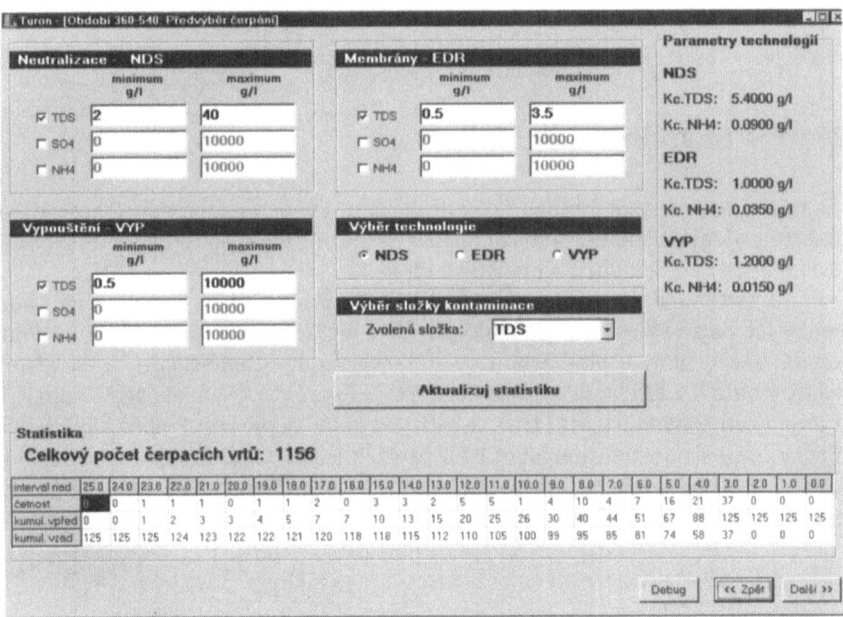

Fig. 3. The pre-choice of abstraction

Geochemical and reactive transport modeling

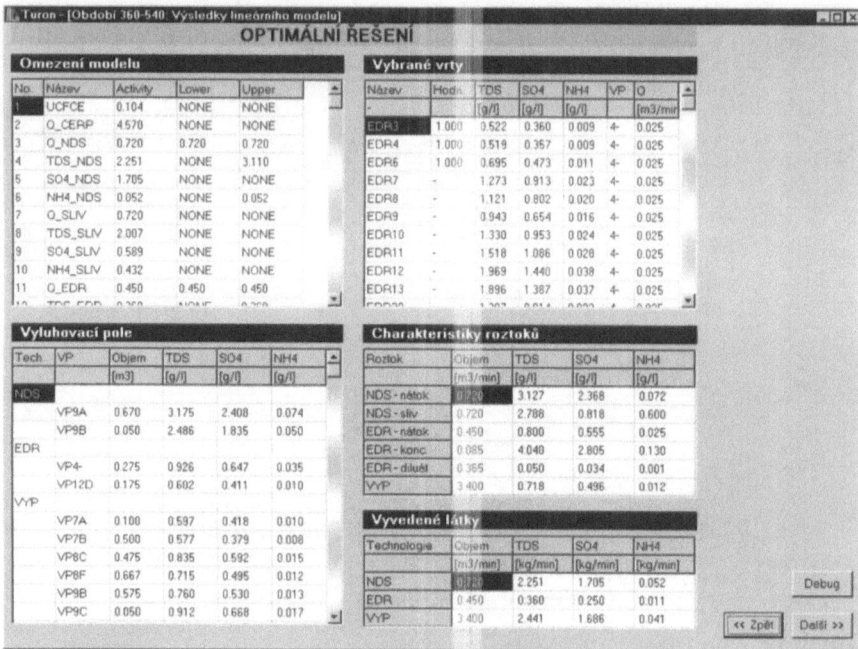

Fig. 4. The results of linear optimizing model

Results of variant calculation

The results are presented in the form of maps of contamination development and total technological and economical characteristics of the whole process. The time development of contamination is shown on fig. 5.

Fig. 6 shows the time development of concentration of TDS in abstracted solution by the cleaning surface technologies. The highest concentration has the solution for treatment by neutralization. In 2005 the strong solutions will be liquidated and then only the discharge will be performed. Fig. 7 the graph shows the amount of abstracted contaminants (TDS) in half-year time steps. Fig. 8 shows the graph of cumulated abstracted amount of NH_4 by each technology. Most of the contaminants have to be liquidated by neutralization or membranes. The limits from the regulators for direct discharge are very strict.

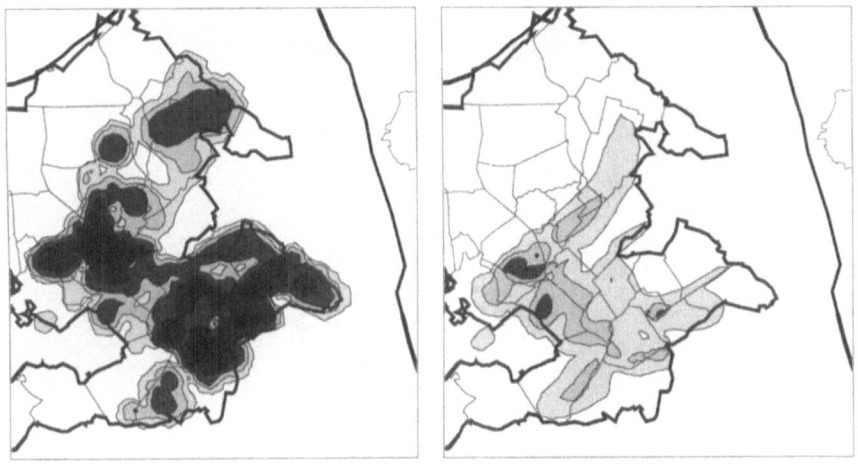

Fig. 5. The development of distribution of NH$_4$ concentration – initial state and state after remediation.

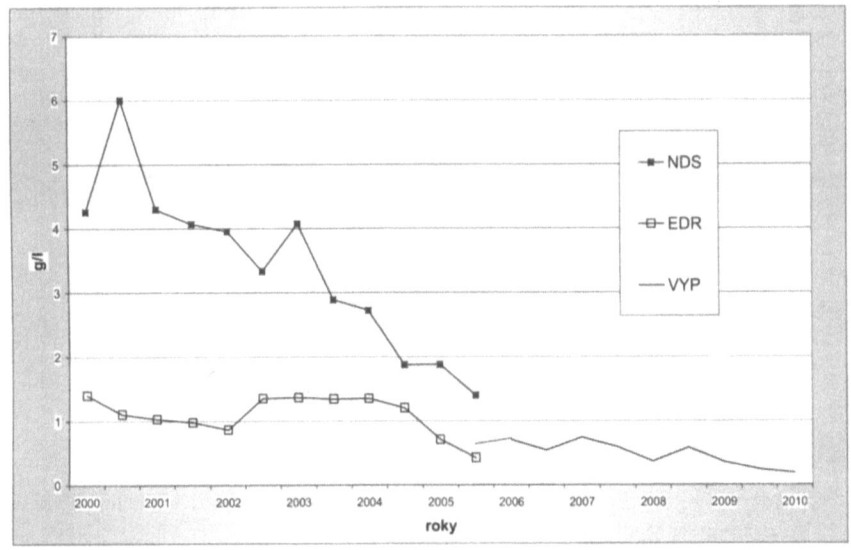

Fig. 6. The time development of concentration of TDS in abstracted solution

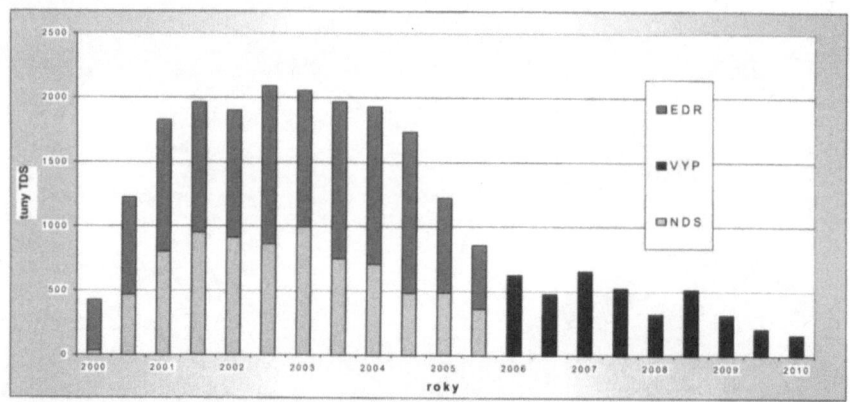

Fig. 7. The time development of concentration of TDS in abstracted solution

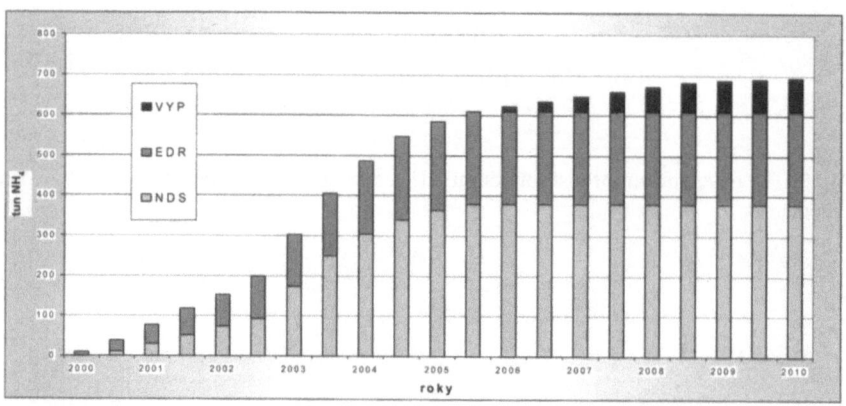

Fig. 8. The time development of concentration of TDS in abstracted solution

References

Bear J, Verruit A (1990) Modeling Groundwater Flow and Pollution. D. Reidel Publishing Company, Dordrecht, Holland

Maryška J, Mužák J (1995) Mathematical modeling of the transport of chemical species in the contaminated underground water. In Bourgeat A P, Carraso C, Luckhaus S & Mikelič A Editors, Proceedings of the Conference Mathematical modeling of Flow Through Porous Media, pages 450-459, World Scientific

Maryška J, Rozložník M, Tůma M (1995) : Mixed-hybrid finite element approximation of the potential fluid flow, J. Comput. Appl. Math., 63:383-392

Mužák J, Novák J, Pisková E (1999) The Mathematical Modeling of the natural and the technological processes on the uranium deposit Stráž the models and the application. In Proceedings of the Conference ECMS 99, Liberec

Geochemical and reactive transport modeling

Factors controlling migration and concentration of natural and technogenic radionuclides in bottom sediments of the Eastern Gulf of Finland, Baltic Sea

Andre Grigoriev[1], Alexei Marchenko[2]

[1]All-Russia Scientific Research Geological Institute (VSEGEI), 74, Sredny prospect, St.-Petersburg 199106, Russia
[2]St.-Petersburg State Mining Institute (Technical University), 2, 21st line, St.-Petersburg 199106, Russia; march@am8089.spb.edu

Abstract. Several natural and technogenic factors have been identified to control migration and concentration of radionuclides in the submarine environments. The principle natural factors are lithology of sediments, bottom relief, occurrence and type of Fe-Mn concretions, and seawater dynamics. The main technogenic impact corresponds to the Chernobyl trace (^{137}Cs) and the supply of natural radionuclides (mostly U and [226]Ra) from the southern coast. The joint trend-cluster model allows subdivision of the area into 5 regions that differ by relative distribution of radionuclides caused by diverse transport and accumulation processes.

Introduction

Characterization of radioactive parameters of landscape media is one of the most important procedures in the complex estimation of environmental conditions of territories. Radiometric and geochemical surveying and monitoring of the Baltic Sea and its Gulf of Finland is necessary due to many reasons such as their geographic position in the highly populated region of Europe, geologic setting with the presence of rocks containing high concentrations of natural radionuclides, significant technogenic impact on the Baltic Sea and surrounding area, etc.

This paper describes the principle results of the study aimed at determination of activities of natural and artificial radionuclides in the bottom sediments of the Eastern Gulf of Finland and distinguishing between different natural and technogenic factors responsible for migration and concentration of radionuclides in the submarine environments.

Geologic setting and environmental characterization

The Eastern Gulf of Finland covers the southern margin of the Precambrian Fennoscandian Shield and northern margin of the Russian Plate. The territory was glaciated in Pleistocene, and many geomorphologic features of the gulf bottom and shores allow considering this sub-aquatic area as a typical glacial shelf. Several geologic particularities are supposed to be important for the distribution patterns of natural radioactive elements and artificial radionuclides in the submarine environments:

1) Precambrian intrusive acid rocks often enriched in uranium, thorium, and potassium are widely distributed in the southern part of the Fennoscandian Shield and exposed on the northern coast of the gulf in the Karelian Isthmus, on numerous islands, and somewhere on its bottom (Koistinen 1996). The most radioactive are rapakivi granites of the Vyborg massif. Background contents of uranium in rapakivi granites are usually from 7 to 15 ppm, thorium from 15 to 40 ppm and potassium from 3 to 5 % that results in the high effective radioactivity of these rocks. The gamma radiation on the surface of rapakivi granites (2π-geometry) is 20-30 μR/h and more.

2) South of the Gulf of Finland, the Paleozoic sedimentary rocks compose the northern part of the Russian Plate. In the stratigraphic sequence there are units enriched in organic components and natural radioactive elements, mainly uranium and radium, such as combustible shales, dictionemic shales, phosphatic shales and phosphorites. Uranium content in dictionemic shales usually varies from $n\times10$ to $n\times100$ ppm and in some points up to 0.1-0.2% (Kiselev et al. 1997) that produces their effective radioactivity about 1500-3000 Bq/kg and more. Some bodies with high uranium concentrations are considered as uranium deposits of economic scale – for example, the Slavyanovskoye deposit located in the southern suburb of St.-Petersburg but not supposed to be involved in mining development because of high ecological risk. Phosphorites typically contain from 20 to 120 ppm of uranium, and their effective radioactivity is about 400-800 Bq/kg.

3) On the lands north and south of the Gulf of Finland, many radon-containing water springs have been known. The same springs probably exist somewhere on the gulf bottom yielding gaseous Rn into the sea waters, so the Rn decay solid products can be deposited in the marine sediments.

4) The lithologic composition of marine sediments varies from coarse-grained terrestrial sediments to aleuritic and pelitic ones. These types of sediments differ in their ways of sedimentation and adsorption capabilities that can make differences in their radiogeochemical properties. Additional radiogeochemical differentiation may be caused by variable forms of submarine relief and peculiarities of the marine hydrodynamics.

5) On the bottom of the Gulf of Finland, the iron-manganese concretions and crusts, often enriched in platinum group elements, has been revealed by now (Zhamoida and Grigoriev 2000). Actually, areas with increased concentrations of Fe-Mn nodules are estimated as possible economic deposits. The Fe-Mn formations predominantly lay in the north-western part of the Eastern Gulf of Finland and may include trace elements and radionuclides due to adsorption processes. It has been known that an anthropogenic influence on the Eastern Gulf of Finland is significant because of the water and sediment pollution by the solid and liquid components moving into the gulf from the St.-Petersburg city and surrounding areas. Pollutants can also be transported by air and brought to the gulf waters and sediments by atmospheric precipitations. Among the principle sources of technogenic pollution, the sites of active mining operations in the drainage basin of the Eastern Gulf of Finland should be mentioned. Of the most importance are those situated south of the gulf where deposits of combustible shales and phosphorites have been under exploitation for many years. These mining activities have led to degradation of soils on large areas in the south-western part of the Leningrad region, and involved large masses of elements that previously had been fixed in rocks into the geochemical migration, including transportation of the both stable elements (P, S, etc.) and radionuclides (U as a high-content component of shales and phosphorites, and products of uranium disintegration) by runoff from the land to the marine environment.

Besides of this mixed natural and technogenic impact of the environment (natural from the point of view onto the primary sources of radionuclides, and technogenic in the mechanisms of engaging radionuclides in active migration), the direct technogenic radioactive pollution of the territory has been known. Many areas in the western part of the Leningrad region demonstrate radioactivity of ^{137}Cs on the surface 0.25 to 1.5 Cu/km^2 and somewhere more (Davydov et al. 2000). This is so called Chernobyl trace formed in May 1986 by the radioactive precipitation from the dispersal cloud that had been produced by the Chernobyl reactor accident on the April 26, 1986. The Chernobyl's radionuclides precipitated not only on the land but on the Gulf of Finland as well.

Methods of study

This study is based upon data obtained in the case of geological mapping of the Gulf of Finland at the 1:200,000 scale carried out during last years by the team of the All-Russia Geological Institute (VSEGEI). The team of marine geologists and geochemists included Spiridonov M, Grigoryev A, Zhamoida V and others. More than 300 samples were taken from the top layer of bottom sediments and uplifted to the ship. After drying of samples, the activities of uranium decay daughter products (mostly ^{226}Ra), ^{232}Th, ^{40}K, and ^{137}Cs were determined by gamma-spectrometry. The detection limits for the radioactivities of these radionuclides were 8.2 Bq/kg Ra, 5.6 Bq/kg Th, 37 Bq/kg K, that corresponds to the equivalent contents 0.7 ppm U, 1.4 ppm Th, 0.1% K, and 2 Bq/kg ^{137}Cs. The relative error of determinations was not greater than 20%. Some of the samples of sediments were also analyzed by gamma-spectrometry for ^{60}Co (the detection limit 4 Bq/kg), and by other methods (emission spectrometry, atomic absorption spectrometry, etc.) for determination of non-radioactive elements.

Methods of data processing included those of univariate and multivariate statistics. The univariate statistical methods were descriptive statistical procedures, estimation of parameters of statistical distribution of data, tests to determine differences between means and variances, etc. A sophisticated study of radiogeochemical characteristics consisted of the multivariate statistical data modelling and analysis with the use of trend, cluster, and principal component models. Thus, the use of a joint trend-cluster model required the following sequence of operations: computing smoothed values of regional nature that correspond to the trend surface - hierarchical clustering – associative classification of these regionalized data, taking into account spatial relationships of points, into k groups where k is chosen in accordance with the results of hierarchical clustering – subdivision of the territory into k regions and mapping boundaries of these regions – estimation of statistical parameters of radionuclide activities in the regions distinguished – interpretation of results of the regionalization. Besides, deviations of radioactivity values from the trend surfaces were also studied to reveal and interpret local radiogeochemical anomalies. For data interpretation, the available additional information concerning the geologic and submarine landscape conditions, as well as human activities in the surrounding lands, was widely involved.

Results

As to background amounts of radionuclides, the activities of ^{226}Ra and ^{137}Cs show obviously an increase from coarse-grained terrestrial sediments to pelitic ones consisting of clay and mud (Table 1). On the other hand, activities of ^{232}Th and ^{40}K do not demonstrate such a strait trend. The largest potassium amounts occur in pebble sediments mostly originated from acid and/or alkaline plutonic rocks.

Table 1. Average background activities of radionuclides (Bq/kg) in bottom sediments of the Eastern Gulf of Finland

Type of sediments	^{226}Ra	^{232}Th	^{40}K	^{137}Cs
Pebble	28	35	1005	21
Sand	27	33	968	27
Mixed sand and pelites	49	30	678	61
Pelites	52	75	805	84

It should be noted that ^{137}Cs show dramatic increase in pelite sediments in the areas of the Chernobyl trace where its contents are about 560 ± 460 Bq/kg, e.g. averagely 7 times higher than outside of the polluted areas. High concentration of radionuclides in the pelite sediments is not unexpected because the facts of enrichment of clayey deposits of the Eastern Gulf of Finland in many toxic heavy metals have been known (Rybalko et al. 1998). This can be explained by the large adsorption capacity of fine-grained sediments. Results of the joint trend-cluster analysis allow to subdivide the sub-aquatic area into 5 main regions (Fig.1).

Fig. 1. Radiogeochemical regionalization of the Eastern Gulf of Finland based on contents of radionuclides in bottom sediments

These generalized regions differ by relative distribution of radionuclides (Table 2) because of diverse transport and accumulation processes.

Table 2. Average activities of radionuclides (Bq/kg) in bottom sediments of the different regions of the Eastern Gulf of Finland

Region	^{226}Ra	^{232}Th	^{40}K	^{137}Cs
1	19	35	737	86
2	41	43	537	507
3	95	47	389	813
4	35	58	285	263
5	55	50	470 and 1012[a]	134

[a] Average values for two subsets of samples due to bimodal frequency distribution

Region 1 is situated in the Neva inlet area next to St.-Petersburg. The highest ^{40}K refers to the prevailing sandy sediments in this part of the gulf with a lot of potash feldspars. The ^{137}Cs is low that responds to the global background pollution only.

Regions 2 and 3 show the largest amounts of ^{137}Cs and ^{226}Ra. The first marks a position of the Chernobyl trace, on the one hand, and the extra accumulation of this technogenic radionuclide in pelite sediments and, in part, in Fe-Mn concretions that prevail in the region 3 which has relatively high depths of the sea.

Geochemical and reactive transport modeling

Sources of ^{226}Ra are deemed to be situated on the southern coast of the gulf. The main sorces are phosphorite and combustible shale quarries, mine wastes and waters enriched in uranium and its disintegration daughter products. Even so these deposits of phosphorites and combustible shales have not been developed for uranium extraction, their impact on the environment can be compared with that of uranium mines themselves due to high contents of U and Ra in the ore bodies. Geochemical migration of U and Ra from the land into the gulf is active enough to influence the marine environment. Re-mobilization, supplemental transportation and re-deposition of U, Ra, and Cs depend on seawater hydrodynamics and bottom relief. The behaviour of ^{226}Ra in the gulf is similar to that of ^{137}Cs, but it should be noted that ^{137}Cs mostly tends to concentrate in clayey sediments while ^{226}Ra is more inclined to be accumulated in Fe-Mn concretions. The average activities of ^{226}Ra in different types of Fe-Mn concretions and crusts vary from 230 to 420 Bq/kg (much higher sediment background). Different morphologic types of Fe-Mn species significantly differ from each other on radiogeochemical parameters – thus, the largest ^{226}Ra activity is determined in spherical Fe-Mn nodules.

Region 4 shows relatively high ^{232}Th activities that can be explained by presence of terrestrial material originated from Precambrian granitoids with Th-containing minerals resistant against weathering. The characteristics of region 5 correspond to diverse lithologic composition of sediments and occurrence of Fe-Mn concretion fields. The ^{60}Co activities greater than detection limit were revealed only in single samples from the North-West of the area studied.

The principal component model shows 3 main factors responsible for distribution of radionuclides in bottom sediments. The first factor corresponds to lithology of sediments and presence of Fe-Mn concretions. The second factor responses to forms of radionuclide migration: mostly mechanical (Th and K) or dissolved with consequent adsorption (Ra and Cs). The third factor reflects similar behaviour of K and Cs in some of sub-regions due to chemical similarities of these elements while both migrate in solution with further deposition in sediments.

References

Davydov A, Igumnov S, Talalay A et al. (2000) Radioecology. Ekaterinburg, UGGGA : 351 p. (In Russian)

Kiselev I, Proskuryakov V, Savanin V (1997) Geology and mineral deposits of the Leningrad Region. St.Petersburg : 196 p. (In Russian)

Koistinen T (Editor) (1996) Explanaition to the map of Precambrian basement of the Gulf of Finland and surrounding area 1 : 1 mill. Geol Surv Finland, Spec Paper 21: 141 p.

Rybalko A, Spiridonov M, Fedorova N et al. (1998) Geoecological zonation of the eastern part of the Gulf of Finland, and specific features of migration and accumulation of chemical elements. Geoecology of the North-West of the Russian Federation, St.-Petersburg, NWRGC: 60-87. (In Russian)

Zhamoida V, Grigoriev A (2000) Distribution of platinum group elements and gold in the Fe-Mn concretions of the Gulf of Finland, Baltic Sea. 6th Marine Geological Conference, The Baltic. Hirshals, Denmark : 92.

Geochemical and reactive transport modeling

Geochemical processes related to uranium mobilisation on groundwaters in a restored uranium mine

P. Gómez, A. Garralón, B. Buil and L. Sánchez

CIEMAT. Av. Complutense, 22, 28040 Madrid, Spain

Abstract. This paper describes the uranium content analysed in the groundwaters around a restored uranium mine and their impact on the discharge area. A conceptual model for uranium mobilisation in the groundwaters has been proposed: the percolation of oxidised waters, through the fractured granite, produces the oxidation of pyrite and arsenopyrite and the precipitation of iron oxyhydroxides. The dissolution of primary pitchblende takes place and, subsequently, a release of U(VI) species to the solution is produced. These U(VI) species are fixed by silicate species, and they are also retained by iron hydroxides. Secondary uranium species are eventually formed, as reducing conditions are re-established due to water-rock interactions.

Introduction

As part of the tasks of the Spanish Nuclear Waste Management Company (ENRESA), a uranium mine restoration programme has been carried out in South-Western Spain. The Ratones mine was the most important uranium mine in the area and was mined between 1955 and 1975. It is located within the Albalá Granitic Pluton in the Iberian Massif (Spain). This mine was an underground pit and the tailings were accumulated on the surface. Stabilisation of wastes and control of effluents were adopted during restoration (1997—1999). At the same time, many scientific studies were developed to evaluate the environmental impact of the mine. Geological, structural, hydrological and hydrogeochemical studies around the mine were carried out.

Weathering of rock materials, tailings and mine dumps produce acidic and metal enriched waters that can contaminate surface and groundwaters (Abdelouas et al 1998; Al et al 1997). In some mining districts, natural attenuation is produced by the ability of the host rock to buffer acid conditions (Strömberg and Banwart 1994; Blowes and Patcek 1994; Berger at al 2000; Al et al 2000). As pH increases, metal ions tend to precipitate as hydroxide, oxy-hydroxide or sulphate phases (Jambor 1994). The purpose of this work is to understand the processes responsible for the metal attenuation, especially uranium, in Los Ratones mine and to develop a model that describes the mobilisation of uranium in surface and groundwaters around the mine.

Geological and hydrogeological background

The main geological feature identified in the area studied is the Northern fault (NF) located at the Northern edge of the site, and which comprises a 5 to 20 m wide brittle fracture. Two NNE-SSW sub-vertical dykes (27 and 27') which intersects the NF at right angles, were mined. These dikes present quartz, pitchblende, coffinite and black Uoxides (uraninites and parapitchblende). The uranium mineralisation is associated to quartz and sulphides veins. Pitchblende is usually replaced by sulphides, especially pyrite and melnicovite. SEM observations have shown evidences of formation of secondary uranium minerals: autunite (Fig 1), saleeite, phosphouranilite, coffinite and uranotile (Buil, 2002). Towards the Southern edge of the area studied, an oblique and broad fault zone, named Southern fault (SF) is observed, wich splits into three branches towards the Northeast (Pérez-Estaún, 1999).

Fig. 1. SEM micrograph of a fracture filling. Autunite lines different inherited minerals from the granite(1). Plagioclase (2) and K-Feldpars (3) is also observed.

Five boreholes were drilled around the mine and comprehensive hydrogeological and hydrogeochemical studies were carried out. The deepest borehole is 500m deep. The mine recharge area is located to the North and the flow lines follow the NE-SW fractures in a S-SW direction. The groundwater discharge takes place through dykes 27 and 27′ flowing to the Southern fault (Fig 2). A good connection between the mine and the Southern fault across the dykes has been identified (Ortuño et al. 1999).

Fig. 2. Main geological features NF (Northern Fault), SF (Southern Fault) and 27, 27′uranium dykes in the area of the Ratones mine. Solid flow lines describe the underground recharge area and dashed flow lines show the discharge area. Boreholes (SR1 to SR5) are also represented.

Uranium in groundwaters

Based on the groundwaters chemical analysis (Gómez et al 2000), four different groups are identified: <u>Group A</u>: SO_4^{2-} - Mg^{2+} type waters, that flow through the uraniferous mined dyke, with neutral pH values and tritium content similar to rainwater; <u>Group B</u>: discharge waters of the mine (SF). They are HCO_3^-–Mg^{2+}-Ca^{2+} type waters with redox conditions; <u>Group C</u>: young HCO_3^--Na^+-Ca^{2+}-meteoric-type waters, that flow through the Northern Fault (NF). They also present redox conditions; and <u>Group D</u>: Deep granitic waters (500m depth) isolated from the mine environment. They are HCO_3^--Na^+ type waters, and their estimated residence time, based on ^{14}C (corrected with ^{13}C), and noble gases, is approximately 16,000 years.

The mean uranium concentration in the four types of waters ranged from less than 1 µg/l to a maximum of 104 µg/l. The highest concentration is found close to the uraniferous dykes. The results are summarised in Table 1. A large number of chemical parameters were analysed and most of them are below the detection limit, except As, whose concentration range from 0.04 ppm (Group D) to 2.6 ppm (Group B).

Table 1. Mean chemical composition of the groups of waters in the surroundings of the Ratones Mine

	pH	Eh (mV)	HCO_3^- meq/l	SO_4^{2-} meq/l	Cl^- meq/l	Ca^{2+} meq/l	Mg^{2+} meq/l	Na^+ meq/l	U meq/l
Group A	7.2±0.3	-192±57	2.06±0.57	3.89±0.87	0.27±0.08	1.86±0.27	2.59±0.30	1.57±0.22	$1.3E^{-4}$±$1.2E^{-5}$
Group B	6.9±0.3	-113±45	2.68±0.78	1.69±1.23	0.22±0.05	1.66±0.39	1.76±0.96	1.02±0.64	$1.0E^{-4}$±$8.1E^{-5}$
Group C	7.5±0.3	-131±52	2.34±0.82	0.17±0.09	0.24±0.06	0.60±0.27	0.67±0.16	1.35±0.37	$5.7E^{-5}$±$8.0E^{-6}$
Group D	7.9±0.2	-253±54	4.78±0.47	0.09±0.06	0.45±0.13	0.41±0.08	0.50±0.09	4.06±0.60	$1.5E^{-5}$±$9.7E^{-6}$

Results and Discussion

Acid drainage and large amounts of metal ions in the effluents from mines are the main negative environmental impact of mining and milling activities. In Los Ratones mine, neutral pH values and bicarbonate-type waters suggest that neutralisation is the most important geochemical process and the key to understand the absence of environmental impact. Sulphide oxidation takes place in the granitic alteration zone, but H^+ are rapidly neutralised by carbonates (ankerite) present in fracture fillings. Furthermore, the concentrations of ligands in waters are very low. Phosphates and organic matter are below detection limit. Carbonate complexes could be the dominant species but, due to the redox conditions of the waters (Table 1), U(IV) species are predominate instead. A good agreement between the identified minerals and the uranium speciation in waters is found. The speciation calculations were made using the geochemical program PHREEQC (Parkhurst and Appelo 1999). The thermodynamic database used for the calculations was the EQ3/6 GEMBOCHS database (Delany 1991).

Saturation indexes of the main uranium minerals show that $UO_{2.25}$ is the solid phase nearest to equilibrium (Fig 3). In general, partially oxidised uraninites are the only minerals that may indicate equilibrium. Other minerals like autunite, saleeite, schoepite, or uranophane are undersaturated in the waters studied. Therefore, uraninite and coffinite are in saturated conditions.

According to previous studies (Duro et al 1997; Bruno et al 1998) the assumption that pure solid phases controlled the solubility of uranium in natural systems, caused an overestimation of the concentration analysed in groundwaters. The results of uranium concentration in equilibrium with several pure phases are shown in Figure 3.

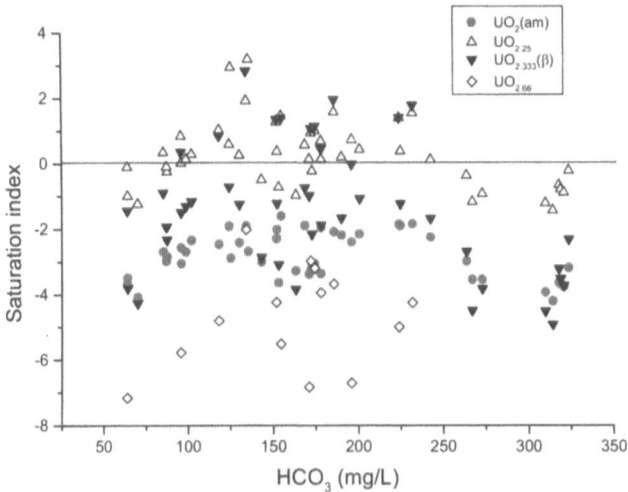

Fig. 3. Saturation indexes of uranium oxides in the waters of the Ratones mine.

As can be observed, the solubility with $UO_{2,25}$ and $UO_{2,33}$ predicts very well the determinations in all the waters except in samples from Group C and D, located far from the mine and with an uranium content lower than waters from Group A and B. On the other hand, equilibrium with autunite, schoepite, ningyoite, $UO_2(am)$ overestimate the uranium concentrations in all the samples. Coffinite and uraninite produce too low concentration with respect the uranium determined. SEM observations, μ-PIXE analysis and chemical determinations of Fe-oxyhydroxides, present in fractures fillings, reveal that uranium is sorbed on these minerals with a molar fraction of 1×10^{-4}. A hypothesis that could explain this phenomenon is to consider that U is forced to precipitate, provided a precipitation of iron hydroxide is assumed:

$$Fe^{3+} + nU^{4+} (3+2n)H_2O \rightarrow Fe(OH)_3\text{-}nUO_2 + (3+4n)H^+$$

Where n indicates the molar fraction of uranium coprecipitated with Fe-oxyhydroxides.

The uranium concentration from groups C and D is fairly well predicted by calculating the solubility for the $UO_2(am)$ based on the coprecipitation with $Fe(OH)_3$.

To verify previous hypotheses and to simulate the main geochemical process related to the chemical composition of waters, a direct model was performed. A sample from Group B was selected (Fig 4) since it represents the discharge water from the mine.

Geochemical and reactive transport modeling

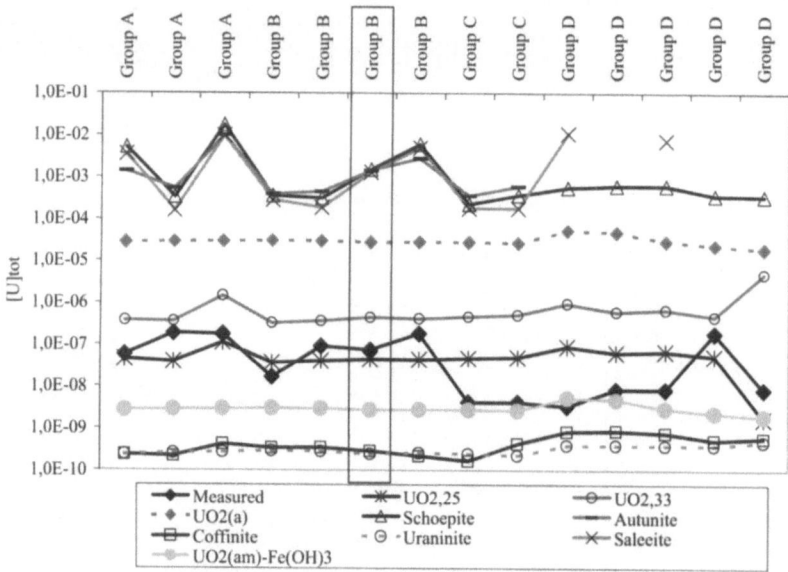

Fig. 4. Comparison between measured uranium concentration and values obtained after the simulation of equilibrium with different uranium minerals. The sample in brackets has been selected to simulate the chemical evolution of waters.

The authors considered the following reactions to simulate the waters from the mine:

1) Surface waters, containing approximately 6ppm of dissolved oxygen, react with the altered minerals in the first 6-8 meters of depth and the oxidation of sulphides, like arsenopirite, takes place. Release of arsenic, as well as dissolution of abundant minerals on the site like fluorapatite, occurs.

2) The water obtained after the first step, is equilibrated with $UO_{2,33}$ at a constant CO_2 pressure of -2.3 atm., in order to simulate the increase of CO_2 in the saturation zone.

3) When this step is finished, the water is equilibrated with albite, chalcedony, $Fe(OH)_3$; the precipitation of siderite, goethite, coffinite is forced, and dissolution of calcite is reached. Finally, the water reacts with the smectite exchange complex, causing Na release and retention of Mg and Ca. The results obtained are represented in Figure 5. As can be seen, the model fits the chemical composition of the water considered fairly well, though a small difference in the sulphate content is observed.

Results of this study show that the chemical evolution of the waters that circulate through the mineralised dykes can be explained by the combined effects of:
(a) Sulphide oxidation; (b) dissolution of carbonates; (c) precipitation of Fe oxyhydroxides and siderite; and (d) equilibrium with feldpars and partially oxidised uraninites.

These combined reactions simulate pH and pe values in agreement with pH and pe measured in situ. (Fig 6).

The model predicts the uranium concentration based on equilibrium with partially oxidised uraninites and precipitation of coffinite. Both solid phases have been detected in the uranium mineralisation and their features (eg. dissolution pits in the crystal surfaces) are in agreement with the proposed chemical processes.

The physico-chemical conditions of the waters (neutral pH and redox conditions), the geochemical control of the host rock (neutralisation of acidic drainage) and the hydrogeological features of this mine (discharge area) avoid the environmental impact of the mine in its surroundings.

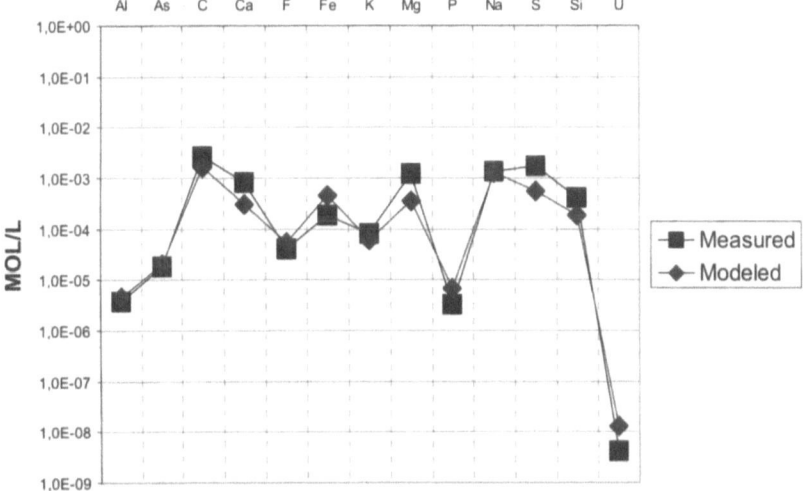

Fig. 5. Comparison between the values obtained in the direct modelling and the chemical analysis of discharge water from the mine.

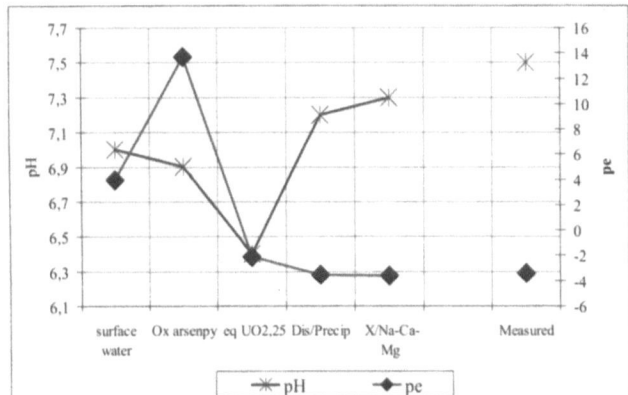

Fig. 6. Evolution of pH and pe values with the PHREEQC simulation steps. Measured values in sample selected (Fig 3) were added to compare with modelled results.

Geochemical and reactive transport modeling

Conclusions

The main conclusion obtained in this work is that uranium dissolved in waters from Los Ratones uranium mine is controlled by the equilibration between partially oxidised uraninites and the precipitation of coffinite. Geochemical modelling is used to demonstrate the importance of dissolution of carbonates, sulphide oxidation, and Fe oxy-hydroxide precipitation in waters circulating through the mineralised uranium dykes.

The coprecipitation approach of U with Fe oxyhydroxides has been used to verify that the uranium concentration in waters far from the mine is controlled by the solid phase UO_2-$Fe(OH)_3$.

Acknowledgements

This work has been financially supported by ENRESA and CIEMAT.

References

Abdelouas A, Lutze W, and Nuttall E (1998) Chemical reactions of uranium in ground water at a mill tailings site. Journal of Contaminant Hydrology 34, 343-361.

Al TA, Blowes DW, Martin CJ, Cabri LJ, and Jambor JL (1997) Aqueous geochemistry and analysis of pyrite surfaces in sulfide-rich mine tailings. Geochimica et Cosmochimica Acta 61, 2353-2366.

Al TA, Martin CJ, and Blowes DW (2000) Carbonate-mineral/water interactions in sulfide-rich tailings. Geochimica et Cosmochimica Acta 64, 23, 3933-3948.

Berger AC, Berthe CM, and Krumshansl JL (2000) A process model of natural attenuation in drainage from a historic mining district. Applied Geochemistry, 15, 655-666.

Blowes DW, and Patcek CJ (1994) Acid-neutralization mechanism in inactive mine tailings. In Environment Geochemistry of Sulfide Mine-Wastes (eds. J.L. Jambor and D.W. Blowes), Mineralogical Association of Canada Short Course, 22, 271-292.

Bruno J, Duro L, de Pablo J, Casas I, Ayora C, Delgado J, Gimeno MJ, Peña J, Linklater C, Pérez del Villar L, and Gómez P (1998) Estimation of the concentrations of trace metals in natural systems: The application of codissolution and coprecipitation approaches to El Berrocal (Spain) and Poços de Caldas (Brazil). Chemical Geology 151, 277-291.

Buil B (2002) Caracterización petrológica, mineralógica y geoquímica y evaluación del comportamiento geoquímico de las REE en la fase sólida (granitoides y rellenos fisurales) del sistema de interacción agua-roca del entorno de la Mina Ratones. Informe interno CIEMAT/DIAE/54440/1/02. RAT-CIE-IF-04.144p.

Delany J M, and Lundeen SR (1991) The LLNL Thermochemical Data Base- Revised Data and File Format for the EQ3/6 Package: UCID-21658, Lawrence Livermore National Laboratory, Livermore, California

Duro L, Bruno J, Gómez P, Gimeno MJ, and Wersin P (1997) Modelling of the migration of trace elements along groundwater flowpaths by using a steady state approach application to the site at El Berrocal (Spain). Journal of Contaminant Hydrology, 26, 35-43.

Gómez P, Garralón A, Turrero MJ, Sánchez L, Melón A, y Ruiz B (2000) Estudio del efecto de la restauración de la Mina Ratones en las aguas subterráneas. CIEMAT/DIAE/54440/1/00.

Jambor JL (1994) Mineralogy of sulfide-rich tailings and their oxidation products. In Jambor J.L.; Blowes D.W. (Eds). Short Course Handbook on Environmental Geochemistry of Sulfide Mine-Wastes. Mineralogical Association of Canada, 59-102.

Ortuño F, Floría E, Carretero G, y Suso J (1999) Caracterización hidráulica de Mina Ratones. AITEMIN. 10-AIT-IA-04.

Parkhust DL, and Appelo CAJ (1999) User's guide to PHREEQC (Version 2)- A computer program for speciation, reaction-path, 1D-Transport and inverse geochemical calculations. U.S. Geological Survey. Water Resources Investigations Report 99-4259.

Pérez Estaún A (1999) Estudios Geológico-estructurales y geofísicos en Mina Ratones. 10-CJA-IF-03. Tomo VI

Strömberg B, and Banwart S (1994) Kinetic modeling of geochemical processes at the Aitik mining waste rock in northern Sweden. Applied Gepchemistry 9, 583-594.

Study courses at the Technical University Freiberg
Akademiestr. 6   09599 Freiberg   Germany
http:/www.tu-freiberg.de   dekan.F3@fggb.tu-freiberg.de

Geoecology (Earth-System-Science)

1st – 4th term (undergraduate studies)
* Mathematics and Computer Science
* Physics and Chemistry
* Biology
* Earth Sciences and Earth System Science
* Field courses, Excursions, Mapping exercises (20 days)
* oral examination after 4th term

5th – 9th term (graduate studies)
* Compulsory: Environmental Management, Environmental Law, Geocomputing
* Choice of three out of nine modules:
  – Atmospheric Sciences (Air pollution and climate research)
  – Ecology
  – Environmental Analysis/Environmental Geochemistry
  – Environmental Geotechnical Engineering
  – Environmental Management/Environmental Law
  – Environmental Microbiology/Biotechnology
  – Hydrogeology
  – Hydrology
  – Pedology
* Field training, excursions; Laboratory training courses
* Internship (minimum 2 months)
* Masters thesis in the 9th term

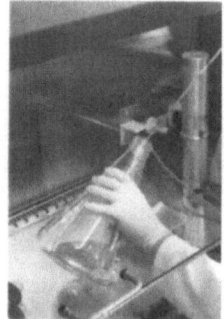

Admission
* „Abitur"/Baccalaureate (A-level) or
* Specialized college degree or
* equivalent access entitlement

Duration/Degree: 6 semesters/Bachelor of Science
9 semesters/Master of Science in Geoecology (Dipl.-Geoökol.)
Start: winter term

Job opportunities
Atmospheric emission and deposition control; Chemical and ecotoxicological environmental analysis; Ecosystems research and management; Environmental consultancy; Environmental impact and risk assessment; Land development and resource planning; Science journalism; Soil science and management; Waste management; Water science and management, including limnology; Public services on local, regional, federal and international level; Insurance companies ...

Uncertainties in Chemical Calculations

Christian Ekberg

Dept. Nuclear Chemistry, Chalmers University of Technology

Abstract. Computer simulations of chemical systems have become increasingly popular in the last decades due to the availability of fast computers. However, computer programs need input data to work and in the case of modeling chemical systems these input data may consist of, e.g. stability constants and element composition in the water. All these input data are encumbered by uncertainties. This paper discusses an approach for handling and quantifying some of these uncertainties.

Introduction

For many years chemical speciation has been a problem for chemists in many fields of research. In the older days, doing speciation calculations by hand, was a tedious work and in many cases almost impossible for realistic cases involving natural waters. However with the aid of faster and more generally available computers these kinds of calculations may be easily done in a matter of seconds. Unfortunately there are still problems remaining and one of these is the case of uncertainties. With computer simulations there are two major kinds of uncertainties (given that the code is correctly written): uncertainties in input data and conceptual uncertainties in selecting the underlying theoretical model upon which the computer program is built. In the simplest case of just a speciation calculation there are not much of conceptual uncertainties why the uncertainties in thermodynamic data are by far the most important. Speciation calculations are important since many other factors such as sorption and diffusion depend strongly on which species dominate or exist in solution. The fact that speciations are most uncertain has often been neglected in the literature where it is common to see speciation diagrams showing, e.g. hydrolysis species domination as a function of the pH in solution. If such calculations are to be meaningful it is important to include the uncertainties associated with the stability constants determining the speciation. This will be discussed in more detail below.

The next step in complexity is the task of calculating the solubility of a solid phase in a given water. This is an exercise that is rather common in the lower grades of chemical studies at a university. However, as will be demonstrated below, this is not an easy task. Although the thermodynamic theory is rather straight forward the resulting solubility depend on the stability constants used in the calculation and their values may be most uncertain. Thus the solubility will not be a single value but rather a interval with an associated distribution function.

The problem gets further complicated when reality knocks on the door and the modeler is supposed to calculate a solubility of a solid phase under ground water conditions. This task includes all the uncertainties discussed above with the addition of the conceptual uncertainty, i.e. how does the modeler think the interactions with the surrounding rock occurs. This has been studied internationally and will be further discussed here.

Speciation calculations

The "simple" speciation calculation is more important that what is normally deemed. It is not uncommon that when making some sort of decision a researcher finds a paper in the literature discussing the system he want to investigate and looks at the speciation diagram. In most cases such diagrams are rather simple to interpret with single lines describing which species dominates at, e.g different pH. An example of such a diagram is shown below in Fig 1 where the data are taken from (Ekberg et al. 2000).

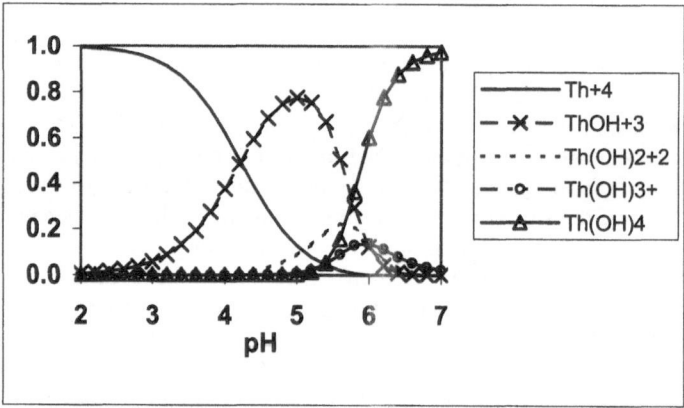

Fig. 1. Species distribution diagram for the thorium hydroxide system.

By looking at Fig 1 in the range of ph 5 to pH 6.5 there is a dominance of $ThOH^{3+}$ and $Th(OH)_4$. In this case it is clear that if one wants to investigate the effect of $ThOH^{3+}$ and $Th(OH)_4$ only on something the pH selected should be close to 6. On the other hand, if uncertainties in stability constants are included in the speciation diagram the situation is drastically changed, as can be seen in Fig. 2.

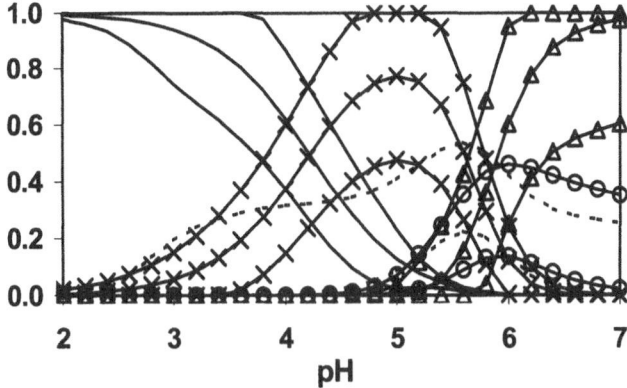

Fig. 2. Species distribution diagram for the thorium hydroxide system with appropriate uncertainty intervals based on the data from (Ekberg et al. 2000)..

Now it is no longer clear which species dominates at a pH around 6. Thus the predictive power of for example sorption based on the domination of $ThOH^{3+}$ and $Th(OH)_4$ has changed considerably. This may seem as a drawback but in reality I would argue that it is much less misleading for the decision makers. The speciation calculation including uncertainty bands were made using the LJUNGSKILE program package (Ekberg et al. 2002).

Solubility calculations

The uncertainties associated with ordinary solubility calculations may be divided into two groups:
1. Uncertainties in thermodynamic data such as stability constants or enthalpies of reaction.
2. Uncertainties in determining the composition of the water used.

The common feature of these uncertainties is that they may be analyzed by common statistical sampling tools. For this purpose the program packages SENVAR (Ekberg and Emrén 1996) and UNCCON (Ekberg and Börjesson et al. 2000) has been developed. SENVAR is handling uncertainties in thermodynamic data and UNCCON is handling how uncertainties in water composition will affect the calculated solubility. The common feature is that both programs use sampling techniques for the uncertainty analysis, e.g. simple Monte Carlo sampling or Latin Hypercube sampling (McKey et al. 1977).

Thermodynamic data

In this paper we will make a SENVAR calculation on the solubility of the solid phase $Th(OH)_4(am)$ using the same thermodynamic input data that was used for the speciation uncertainty calculation. The water used is according to Table 1.

Table 1. Composition of reference water from the Äspö site, borehole KAS 02, level 530-535m (Smellie and Laaksoharju 1992 , Nilsson 1992). Concentrations in mM.

Ca	47.2	Cl	181	U[a]	5.45E-07
Mg	1.73	C_{tot}	0.164	Sr	0.399
Na	91.3	S_{tot}	5.83	Li	0.144
K	0.207	F	7.90E-02	N_{tot}	3.52E-03
Fe[a]	4.37E-03	Br	0.501	pH	8.1
Mn	5.28E-03	P	1.61E-04	pe[b]	-4.37
Al[a]	1.00E-03	Si	0.146	Temp (°C)	15

[a] The analysis did not contain any value for this species. The concentration was estimated on the basis of other Äspö ground water samples.

[b] The pe value was adjusted from the measured value (-5.42) with regard to additions of Al, Fe, U, and the equilibrium between SO_4^{2-}/SO_3^{2-} in the solution .

The result from the SENVAR calculation is both an uncertainty analysis and a sensitivity analysis. the former giving a confidence interval and a distribution for the calculated solubility for the given criteria while the sensitivity analysis gives the species which stability constant affects the solubility the most.

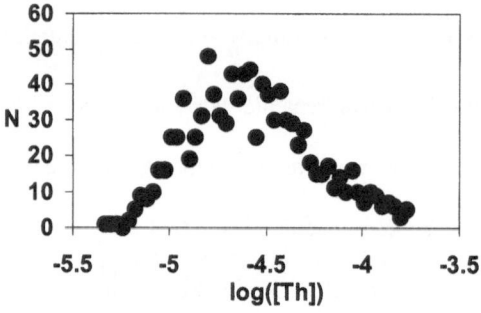

Fig. 3. Plot of the solubility distribution function resulting from a SENVAR calculation. Mean solubility is 2.77E-5 M and the standard deviation is +1.6E-6 and −1.4E-5. The skewed standard deviation in the linear scale is due to a symmetric interval in the logarithmic scale.

Clearly from Fig 3 the solubility spread is not vast but still uncertainties in thermodynamic data contribute to close to two orders of magnitude for the uncertainty in $Th(OH)_4(am)$ solubility in the given water. It has been shown elsewhere (Ekberg and Emrén 1998) that the uncertainties in calculated solubilities originating from uncertainties in enthalpies of reaction are about one order of magnitude at 80 deg C why such a calculation example is omitted in this paper.

The sensitivity analysis show that, by far, the most important species are the solid phase and $Th(OH)_4$. Thus, the main conclusion to draw from these results is that in order to improve the predictive strength of a solubility calculation of $Th(OH)_4(am)$ the stability constant for $Th(OH)_4$ and the solubility product of $Th(OH)_4(am)$ has to be determined more accurately.

Water composition

The water composition is a relatively easy thing to determine in a lab where you start with water and the chemicals you want to use but in nature it is more complicated. There, several problems arise. The first and most obvious one is uncertainties in concentration determination that may originate from different sources. Another, less obvious factor is that the water properties may change during either sampling or during transport. In addition, if calculations are to be made to simulate, e.g. ground water properties there is a problem is spatial uncertainties and costs of sampling. Therefore, in some cases, computer programs are used to calculate the composition of a water in contact with a certain rock. Naturally, in such a case uncertainties in rock composition and conceptual model used plays an important role. This has been shown elsewhere (Ekberg and Börjesson et al. 2000) and will not be discussed further here.. In the case presented here the uncertainties used are shown in Table 2.

Table 2. Uncertainties in the water composition used for the UNCCON calculations (Samuelsson 1996).

Element	Tot. uncertainty, %	Element	Tot. uncertainty, %
Na	10.3	Br	34.0
K	19.0	Si	12.5
Ca	18.5	Fe	24.4
Mg	8.9	Mn	7.8
C	24.9	Li	19.0
Cl	10.0	Sr	18.3
SO4	32.2		

In this case there is also contribution to the overall uncertainty in the solubility calculation but much smaller than for the case of thermodynamic data. This is, naturally, due to the relatively small input uncertainties in the water composition since only statistical and measurement uncertainties have been taken into account. The distribution however is in this case not normal but more of a uniform character, as seen in Fig 4.

Geochemical and reactive transport modeling

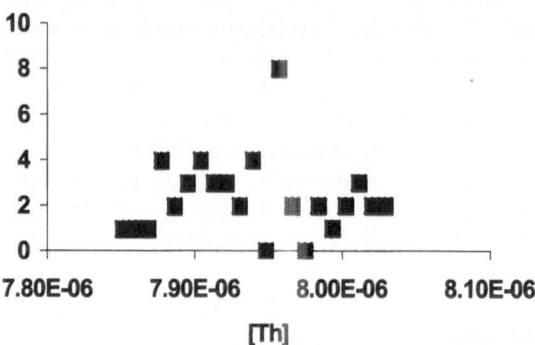

Fig. 4. Plot of the solubility distribution function resulting from an UNCCON calculation. Mean solubility is 7.9E-6 M.

The sensitivity analysis show that the most important element concentration is F. This is probably due to the relatively strong complexation of thorium with phosphate.

Conceptual uncertainties

There are of course a number of conceptual uncertainties that may arise when discussing a problem. This is illustrated in a recent study (Emrén et al. 1999), in which four modelers were asked to calculate the solubility of $Pu(OH)_4$ in a specified groundwater, which was surrounded by a rock of a specified composition with respect to mineralogy. The modelers were requested to use the same equilibrium program (PHREEQE) and the same thermodynamical data. Exchange of results was not allowed until the results were ready. The range of calculated solubilities covered six orders of magnitude. The subsequent analysis could not rule out any of the results as unreasonable. Rather, each modeler had to make assumptions on unknown parameters, and their number was large enough to result in such a spread.

Here, we will focus on a subset of the possible methods for solubility calculations under groundwater conditions, and in particular on four cases that are of interest for uncertainty analysis (Ekberg and Emrén 2000).

1) Isolated dissolution
In the first method, the water composition is supposed to be known, measured or calculated, within uncertainty limits, and the rock - groundwater interactions during dissolution/precipitation reactions are neglected. This method is analogous to laboratory measurements in which water is sampled from a borehole and then equilibrated with the desired solid phase.

2) One mineral
This method is similar to the first, with the difference that the dissolution/precipitation reactions are assumed to take place in the presence of one randomly chosen mineral. In old fractures, large areas of the fracture walls are covered with only one mineral. In such a part of the fracture, the solubility of a substance will be determined by the presence of this mineral.

3) Simulated water pumping
In the third method, the solubility is calculated in the presence of different minerals at different locations across a fracture surface. In this case, the water properties are allowed to vary locally while the water propagates along the fracture. The resulting waters are then mixed to form a simulated sample. During this simulated sampling process, no dissolution/precipitation reactions are supposed to take place.

4) Simulated random sampling
This method is similar to the simulated pumping method except that no pumping is simulated. Instead the full range of solubilities in the fracture is investigated. This method is a complement to method 2 in the sense that the solubilities are given by one mineral at a time, but the water is allowed to vary to a larger extent. Solubility values are sampled from some 3000 locations along a simulated fracture.

The results from the calculations on conceptual uncertainties are shown in Table 3 and there it is noteworthy that the case with the isolated dissolution is the same as the simulated water composition in the case described above.

Table 3. Resulting solubility interval for the different conceptual cases

Solid phase	Isolated dissolution	One mineral	Simulated water pumping	Simulated random sampling
	min ; max	min ; max	min ; max	min ; max
$Th(OH)_4$	9.02E-6 ; 3.23E-05	4.09E-06 ; 2.39E-05	1.23E-07; 5.83E-05	9.72E-08; 9.75E-05

The uncertainty interval varies significantly between the different methods and so do the distribution of the solubility. From our point of view it is not obvious which method to select and thus the modeller may influence the result more than is desired.

Conclusion

It has been shown that even rather simple chemical calculations such as speciations and solubility calculations are heavily encumbered by uncertainties of different kinds. The effect of these uncertainties vary, but in some cases the calculated solubility may vary several orders of magnitude. Thus, results of chemical calculations ought to be given as confidence intervals rather than a fixed deterministic value. By using this method many unwanted and difficult discussions of the validity of different databases could be avoided.

Acknowledgement

Arvid Ödegarrd-Jensen is acknowledged for his aid with computer runs when the time was pressing. This work was funded by the Swedish Nuclear Power Inspectorate (SKI).

References

Emrén, A. T., Arthur, R., Glynn, P. D., McMurry, J., (1999), The modellers influence on calculated solubilities for performance assessments at the Äspö hard-rock laboratory., Mat. Res. Soc. Symp. Proc., 465, 559-566.

Ekberg, C., Albinsson, Y., Comarmond, M., J and Brown, P., (2000), Studies on the Complexation Behaviour of Thorium(IV). 1. Hydrolysis equilibria., J. Solution Chem., Vol 29, No. 1, 63-85.

Ekberg, C., Börjesson, S., Emrén, A. T., and Samuelsson, A., (2000), MINVAR and UNCCON, Computer Programs for Uncertainty Analysis of Solubility Calculations in Geological Systems, Computers and Geoscienses, Vol. 26, No. 2, 121-246.

McKay, M. D., Beckman, R. J., and Conover, W. J., (1979), A Comparison of Three Methods for Selecting Values of Input Variables in the Analysis of Output from a Computer Code, Technometrics, 21, 239-245.

Smellie, J., Laaksoharju, M., (1992), The Äspö Hard Rock Laboratory: final evaluation of the hydrochemical pre-investigations in relation to existing geologic and hydraulic conditions, Swedish Nuclear Fuel and Waste Management co. (SKB), TR 92-31 Stockholm.

Nilsson, A-C., (1992), Groundwater chemistry monitoring at Äspö during 1991, Swedish Nuclear Fuel and Waste Management Co. (SKB), progress report.

Samuelsson, A., (1996), Diploma thesis, Dept. Nuclear Chem., Chalmers University of Technology, S-412 96, Gothenburg, Sweden.

Ekberg, C., Emrén, A. T., (2000), Conceptual uncertainties in solubility calculations in groundwater systems: A calculation exercise., Radioactive Waste Management and Environmental Restoration, V. 22, No. 3, 163-173.

Ekberg, C., Emrén, A. T., (1998), Uncertainties in solubility calculations aimed at a repository for spent nuclear fuel., Proceedings of the PSAM 4 conference, 13-18 September, New York City, vol. 4, 2223-2228.

Ekberg, C., Emrén, A. T., (1996), SENVAR: A code for handling chemical uncertainties in solubility calculations, Computers & Geosci., 22, 867.

The Role of Metrology and Statistics in Making Geochemical Modeling Defensible

G. Meinrath[1], B. Merkel[2] , P. Schneider[3], B. Delakowitz[4]

[1] Schießstattweg 3A, 94032 Passau
[2] TU Bergakademie Freiberg, Institute for Geology, 09599 Freiberg
[3] Hydropiewak GmbH, Oberfrohnaer Str. 84 D-09117 Chemnitz
[4] Hochschule Zittau/Gölitz, Postfach 261, 02755 Zittau

Abstract. Geochemical transport modeling of contaminants in the aquatic environment is an important tool supporting decision-making on both the technical and political level. There is a specific meaning for modeling emissions of radioactive contaminants from uranium mining and processing sites as well as from nuclear installations and waste repositories due to the longevity and toxicity of some radionuclides. The complexity of issues studied by applying reactive transport modeling is steadily growing with the development of hard- and software. Reliability of model output is of crucial importance. Reliability of a model output is intimately related to the quality of the model's input data. Thus, a quantitative measure for data quality and quality assurance becomes essential in addition to transparency and documentation.

Introduction

Geochemical modeling is a genuinely interdisciplinary field. Hydrogeology, chemistry, geochemistry, water chemistry, rheology, information science, numerical mathematics and others more contribute to its development. All contributing factors cause uncertainties in the total process. In the present discussion, concentration will be on the contributions from chemistry. The past two decades of impetuous growth in CPU clock speed and digital storage capacity have raised expectations to model and to predict complex scenarios on large time scales. Modeling codes of considerable complexity have been created raising the expectation that predictions become the more reliable the more data and details can be included into the model (Voss 1998; Prado et al. 1999; Ekberg 2000). It is meanwhile understood that this expectation has ignored several limitations in the accuracy and precision of both the input data to a model and the modeling algorithms themselves as well as our limitations in the perception of the complex processes, features and events that in total make up the real world surrounding us. Hence, information is necessary to quantify the limitations of our understanding and insight into natural processes and the rules governing them.

Thermodynamic data as results of analytical measurements

Prediction of future processes on basis of quantitative information requires a causal relationship between causes and their effects. The relationship not necessarily needs to be strictly deterministic. To take an example, metabolic processes are quite difficult to take into account. Geochemical reactive transport modeling is commonly based on thermodynamic data of those reactions considered to be relevant. During the past century a larger amount of thermodynamic data of metal ions in a wide range of aqueous solutions has been forwarded. These data are commonly collated into collections, either in printed form, on-line data base or code-specific thermodynamic data base for speciation codes (Hefter 1979; May and Murray 1991; Parkhurst 1995).

The fundamental relationship Eq. 1 links chemical reactions with thermodynamics:

$$\ln K_r' = -\frac{\Delta G_r'}{RT} \tag{1}$$

K_r' :   formation constant of a reaction r
$\Delta G_r'$:   Gibbs' free energy involved in reaction r
R :   gas constant
T :   absolute temperature

Thus, formation constant K_r' may be understood as a constant of nature under given conditions specified by the superscript ('). There are numerous ways of evaluating either $\Delta G_r'$ or K_r'. Thus, the formation constant K_r' represents an energy difference. This energy difference manifests itself in often quite sensitive equilibria of different chemical forms of the constituents of interest. The evaluation of the energy almost always must rely on analytical chemical data of the concentrations of these different chemical forms. Almost always a certain amount of the chemical forms cannot be analyzed by direct measurement but has to be estimated on basis of mass-balance considerations and other general assumptions.

Since thermodynamic data are a basis of predictive reactive transport modeling and obtained by analytical measurements, these data are affected by measurement uncertainty. Measurement results without statements of reliability (i.e. uncertainty) should not be taken seriously as soon as important decisions are based on these data. The major reason is that such data cannot be trusted because the analyst forwarding these data has not considered whether his approach and procedure is itself trustworthy (de Bièvre 1998).

Measurements in chemistry, responsibility and metrology

Chemistry has performed and interpreted analytical measurements for some centuries. Great chemists have moved chemical science by personal experience, hard work, talent and a considerable portion of try-and-error into the 21st century. The science of analytical chemistry has developed over this period and provided remarkable successes. It is well possible to enjoy the results of these successes as interesting and entertaining, as a kind of social activity. However, the successes of analytical chemistry include an important factor: responsibility. In forwarding a result of analytical measurement, the analyst takes responsibility for the conclusions drawn on basis of this measurement result. Chemical measurements of various kinds are playing a rapidly expanding role in modern society and increasingly form the basis of important decisions. Chemical measurements form a basis for agreement in situations of discordant interests and even distrust. Thus, the role of analytical chemistry vastly exceeds just satisfying scientific curiosity. Global trade, food safety, clinical chemistry, consumer protection and product quality control are only a few fields where important decisions are based on chemical measurements. The consequences from the results of these measurements can affect the lives of many people - not only today but also in the future. Hence, the science of measurement has become a national sovereignty, institutionalized in the national metrological institutes (NMI's). The science of measurement is subject to international treaties like the Meter Convention (1875) as well as the establishment of the International Organization of Legal Metrology (1955). Hence, metrology is not about esoteric, academic measurements of the highest precision performed in an ivory tower at universities or the NMIs. Metrological rules are driven by the discordant interests of the global trade partners where high quality products should be imported cheaply (if at all) while the own products should be exported with high profits (at best without any limitations). It is evident that promotion of science has no priority in the development of metrology. It is, on the other hand, evident that modern science likewise has developed into a business where competition for funding and publication follows rules not completely different from economic markets. It even may be suspected that political systems could have encouraged experimenters to obtain data to be favorable for a certain politically desirable decision. Measurement values complying with metrological rules in a stable metrological framework will keep their significance over long time periods. Metrology therefore is an important element of sustainability.

Comparability of measurements - the international framework

In order to achieve comparability of measurements on a world-wide level, not only objective criteria and an international agreed system of references must exist. An infrastructure of closely cooperating institutions must exist that ensures and controls the conformity of measurements with criteria and points of reference. The System of Units (système international; SI) represents implements this basis, and by use of traceable measurements provides an international infrastructure for comparable measurements. The institutional infrastructure is realized by the Meter Convention. Even closer ties were formed in 1999 with the Mutual Recognition Agreement (MRA) signed by the directors of 38 member states of the Meter Convention. The objective of the MRA are to: establish the degree of equivalence of measurement standards by NMIs; to provide for the mutual recognition of calibration and measurement certificates issued by NMIs; thereby to provide governments and other parties with a secure technical foundation for wider agreement related to international trade, commerce and regulatory affaires (Wielgosz 2002).

Deficits of existing thermodynamic data

Existing deficits in most contemporary chemical measurements have been discussed elsewhere in some more detail (Chalmers 1993; Thompson 1994; de Bievre 1997; Meinrath and Spitzer 2000). Likewise, fundamental principles of metrology for thermodynamic data as well as concepts for obtaining and applying thermodynamic data by metrological rules have been forwarded (Meinrath et al. 2000a; May and Murray 2001; Ekberg et al. 2002; Meinrath and May 2002).

Serious discrepancies between published thermodynamic parameters of chemical reactions are well-known. Since there are many different causes of these problems (such as experimental error, inadequate theory, and carelessness), they can be very difficult to pinpoint and to eliminate. The situation is made worse because many thermodynamic data persisting in the scientific literature stem from values that are later corrected or become experimentally superseded (May and Murray 2001).

The existing deficits in thermodynamic data of reactions relevant for reactive geochemical transport modeling can be separated in three groups:

- Documentation and communication deficits

- Competence deficits

- Consistency deficits

Documentation and communication deficits mainly limit a critical reassessment of thermodynamic data. Thus, despite the fact that many collections of thermodynamic data are termed 'critical', there is no support for such a claim other than the expectation that the reviewer(s) recognize(s) a 'good' data if he/they see(s) it. The almost complete lack of statistical assessment (including essential issues like the detection and treatment of extraneous data, the discussion of the optimization criterion and the reliability of auxiliary data) of thermodynamic data precludes any post-evaluation. At best, obvious severe errors in the experimental conditions and evaluation of the data can be indicated.

Competence deficits direct to the experimenter performing the experimental work. Not seldom, these are comparatively unexperienced novices to the science of chemical analysis performing the measurements as a part of their academic education.

Next to experimental errors, carelessness and inadequate theory the -sometimes undocumented- use of auxiliary data, the use of different calibration methods and differences in the calibration standards and sometimes even the different equipment generates the well-known differences in chemical thermodynamic data obtained by the nominally same experimental method for the same reaction. Inadequacies in the theory of electrolyte solutions has resulted in a larger number of procedures for extrapolation of experimentally determined formation constants of a reaction to standard conditions, thus adding to the inconsistency of available thermodynamic data.

The role of metrology and statistics

Evaluation and documentation of experimental data serving as a basis of important decisions (= decisions affecting other people) must allow a reassessment by other people. This situation is most obvious if the decision is questioned in a law court. Such a reassessment needs several elements. A primary requirement is a precisely defined terminology. There is an essential difference in using well-defined terms like reproducibility and repeatability (VIM 1994) or undefined terms like validation and standard (De Bievre 1996; 1997a). To provide definitions and to set up criteria for their use is the task of metrology. Metrology is the science of measurement. Thus, it deals with values resulting from measurements.

The requirement of traceable values results both from the request for comparability with related data from sources located at other places but even more from the possible impact of current day decisions on future societies, their socio-economic development and their competitiveness under their living conditions. Once a future society has the impression that the data on which a past decision with long-term effect has been based on (i.e. the construction of a repository for nuclear wastes) have not been interpreted fairly, the affected area may experience drastic losses i.e. in the economic competition, loss of population. The affected society even may have to invest a considerable amount of its GNP to reverse such a past decision.

Statistics is an essential part of metrology. Statistics is searching for patterns in data and tries to find characteristics in groups of data. The major role of statistics is to transfer the message to other people that the data has been treated fairly (Efron 1981). Most people are not natural-born statisticians. Left to their own, human beings are not very good at picking out patterns from a sea of noisy data. To put it the other way, we are all to good at picking out non-existent patterns that happen to suit our purposes and/or prejudices (Efron and Tibshirani 1993). In the past, a major task for statisticians has been to find mathematical approximations in analyzing a given data set in order to make the statistical task tractable. Thus, parametric statistics on basis of some distributions, e.g. normal, Poisson or Student distributions has been common. Nowadays, modern computer-based statistics allow focusing on the question of interest instead of mathematical tractability (Efron and Tibshirani 1991). Complex datasets commonly occurring in chemistry, including chemical thermodynamics, can be treated (Meinrath et al. 2000a; Meinrath 2000; Ekberg et al. 2002a; Meinrath and Lis 2002).

International requirements for comparable data

From the preceding discussion, it is clear that existing thermodynamic data cannot serve as a reliable basis for predictive geochemical modeling because neither the documentation nor the evaluation procedures of individual data are sufficient. A simple value, in some cases even associated with a statement of uncertainty of unknown meaning (standard deviation? marginal confidence interval? Uncertainty limit? Including correlation effects? Considering uncertainty contributions from auxiliary parameters?) is not sufficient for quality assessment.

The enormous and still increasing relevance of chemical analytical data in many areas of daily life has created a demand for a world-wide system of references. The establishment of such a reference system is not without its own difficulties and quite fundamental problems are discussed among metrology scholars. But these aspects are beyond the scope of this discussion. A basis of the international metrological is the document 'Guide to the Expression of Uncertainty in Measurement' (ISO 1993). A major element in this document is the total uncertainty budget as an essential prerequisite for comparability of measurement results inside and outside chemistry. The total uncertainty budget requires the quantitative estimation and inclusion of all contributions of uncertainty to a measurement result. For chemical thermodynamic data, this requirement poses quite a task to the experimenter. It also explains why existing thermodynamic data cannot be comparable. There is no argument what-so-ever that allows to extract reliable values from the heap of doubtful data available in current literature.

It must be stated that at least for the thermodynamic data obtained in pre-1980ies, geochemical modeling was an application out of the scope for an application of chemical thermodynamic data. However, by ignoring important uncertainty contributions in the evaluation of thermodynamic data, the experimenters have invited their own problems. By underestimating the variability of the data forwarded as a result of a metrologically insufficient evaluation process, the differences between data from different sources became significant. Thus they had to be considered as discrepancies (de Bièvre 1996a). The total uncertainty budget, required by metrology as an essential part of the traceability concept, will in most cases lead to larger uncertainties in chemical thermodynamic data. However, some apparent discrepancies will vanish and the real problems will stick out. Thus, the risk to stick with problems caused by seemingly discrepant data will be reduced. Consequently, action to resolve the actual problems will become more efficient. The evaluation of a total uncertainty budget for complex situations is still an open field of discussion. It is clear that such protocols must serve metrological goals: traceability to reference standards and comparability by a reasonable and realistic estimate of a complete uncertainty budget for a given value. Both overinterpretation and underestimation must be avoided. There is no use to forward a value where a small uncertainty is assigned just in order to stick out among a larger number of preceding publications. Since we do not know the motivations having caused the reported reliability estimates in available literature (in the absence of the comparability goal), discarding of these data is possible only in case of obvious errors (Meinrath et al. 1999).

References

Chalmers R.A. (1993) Space age analysis. Talanta 40: 121 - 126.
De Bièvre P. (1996) Where pragmatism and clarity go on collision course. Accred. Qual. Assur. 1: 159.
De Bièvre P. (1996a) Do measurement scientists sometimes invite their own problems? Accred. Qual. Assur. 1: 91.
De Bièvre P. (1997) Measurement results without statements of reliability should not be taken seriously. Accred. Qual. Assur. 2: 269.
De Bièvre P. (1997a) Validation, valid measurement and more about validity. Accred. Qual. Assur. 2: 55.
De Bièvre P. (1998) Uncertainty assessment is an evaluation process. Accred. Qual. Assur. 3: 391.
Efron B. (1981) Nonparametric estimates of standard error: the jackknife, the bootstrap and other methods. 68: 589 - 599.
Efron B, Tibshirani R. (1991) Statistical data analysis in the computer age. Science 253: 390 - 395.
Efron B, Tibshirani R.J. (1993) An introduction to the bootstrap. Monographs on statistics and applied probability 57. Chapman & Hall. London/UK.
Ekberg C., Meinrath G., Strömberg B. (2002) A retraceable method to assess uncertainties in solubility estimations exemplified by a few Americium solids. submitted.
Ekberg C, Börjesson S, Emrén A.T., Samuelsson A (2000a) MINVAR and UNCCON, computer programs for uncertainty analysis of solubility calculations in geological systems. Comput. Geosci. 26: 219 - 226.
Ekberg C, Ödegaard-Jensen A., Meinrath G. (2002) LJUNGSKILE - a computer program for investigation of uncertainties in chemical speciation. REPORT SKI-R--02/xxx. Statens Kärnkraftinspektion. Stockholm/S.
Hefter G (1979) Critical survey of stability constants and related thermodynamic data of fluoride complexes in aqueous solution. IUPAC. Pergamon Press/Oxford.

ISO (1993) Guide to the expression of uncertainty in measurement. International Standard Organistation. Geneva/CH.

May P.M., Murray K. (1991) JESS, a joint expert speciation system I: raison d'être. Talanta 38: 1409 - 1417.

May P.M., Murray K. (2001) Database of chemical reactions designed to achieve thermodynamic consistency automatically. J. Chem. Eng. Data 46: 1035 - 1040.

Meinrath G (2000) Computer-intensive methods for uncertainty estimation in complex situations. Chemom. Intell. Lab. Syst. 51: 175 - 187.

Meinrath G. (2000a) Comparability of thermodynamic data - a metrological point of view. Fresenius J. Anal. Chem. 368: 574 - 584.

Meinrath G (2001) Measurement uncertainty of thermodynamic data. Fresenius J. Anal. Chem. 369: 690 - 697.

Meinrath G., Spitzer P. (2000) Uncertainties in determination of pH. Mikrochim. Acta 135: 155 - 168.

Meinrath G, Lis S (2002) Application of cause-and-effect diagrams to the interpretation of UV-Vis spectroscopic data. Anal. Bioanal. Chem. 372: 333 - 340.

Meinrath G., May P.M. (2002) Thermodynamic Prediction in the mine water environment. Mine Water Environ. 21: 24 - 35.

Meinrath G., Kato Y., Kimura T., Yoshida Z. (1999) Comparative analysis of actinide(VI) carbonate complexation by Monte Carlo resampling methods. Radiochim. Acta 84: 21 - 29.

Meinrath G, Hurst S, Gatzweiler R (2000) Aggravation of licensing procedures by doubtful thermodynamic data. Fresenius J. Anal. Chem. 368: 561 - 566.

Meinrath G, Ekberg C, Landgren A, Liljenzin JO (2000a) Assessment of Uncertainty in Parameter Evaluation and Prediction. Talanta 51: 231 - 246.

Nitzsche O, Meinrath G, Merkel B (2000) Database Uncertainty as a limiting factor in reactive transport prognosis. J. Contam. Hydrol. 44: 223 - 237.

Parkhurst D.L. (1995) User's guide to PHREEQC-a computer program for speciation, reaction, reaction-path, advective transport, and inverse geochemical calculations. Water Resources Investigations Report 95-4277. USGS Lakewood/USA.

Prado P., Draper D., Saltelli S.A., Pereira A., Mendes B., Eguilior S., Cheal R., Tarantola S. (1999) Gesamac: conceptual and computational tools to tackle the long-term risk from nuclear waste disposal in the geosphere. REPORT EUR 19113. European Communities, Luxembourg.

Thompson M. (1994) Statistics - the curse of the analytical classes. Analyst 119: 127N.

VIM (1994) Intenational Vocabulary of Basic and General terms in Metrology. Deutsches Institut für Normung e.V. Beuth Verlag Berlin/FRG

Voss C.I. (1998) Hydrogeol J 6: A4 - A6.

Wielgosz R.I. (2002) International Comparability of chemical measurements. In: Importance of traceable pH measurements in science and technology (Spitzer, Meinrath, eds.). REPORT PTB (xx-2002) Physikalisch-Technische Bundesanstalt Braunschweig, Braunschweig/FRG.

The Influence of Geostatistical Data on the Reliability of the Meshless Method in Transport Modeling

Leopold Vrankar [1], Goran Turk [2] , Franc Runovc [3]

[1]Slovenian Nuclear Safety Administration, Ljubljana, Slovenia
[2]Faculty of Civil and Geodetic Engineering, University of Ljubljana, Ljubljana, Slovenia.
[3]Faculty of Natural Sciences and Engineering, University of Ljubljana, Ljubljana, Slovenia

Abstract. The disposal of radioactive waste in geological formation is of great importance for nuclear safety. A number of key geosphere processes need to be considered when predicting the movement of radionuclides through the geosphere. The main goal of this research is to investigate the influence of geostatistical data on reliability and accuracy of computational modelling. We chose the Kansa meshless method that uses radial basis functions as the mathematical solution technique. The aim of this study is to determine the average and sample variance of radionuclide concentration with regard to spatial variability of hydraulic conductivity modelled by geostatistical approach.

Introduction

The disposal of radioactive waste in geological formation is of great importance for nuclear safety and geosphere. It is considered to be the principal natural barrier, which prevents or inhibits the movement of radionuclides into the biosphere.

A number of important geosphere processes need to be considered when predicting the movement of radionuclide through the geosphere. Computational modelling and simulation based on set of physical processes and geological data can give useful results.

The general reliability and accuracy of transport modelling depend predominantly on the input data such as hydraulic conductivity, water velocity, radioactive inventory, hydrodynamic dispersion, etc. The output data are concentration, pressure, etc. The most important input data are obtained from field measurements, but they are not available for all regions of interest.

For example, the hydraulic conductivity as input parameter varies from one place to another. For such cases geostatistical science offers a variety of spatial estimation procedures (Deutsch and Journel 1998).

One way to study the spatial variability of hydraulic conductivity is stochastic simulation approach (Gomez-Hernandez and Gorelick 1991).

The main goal of this research is to investigate the influence of geostatistical data on reliability and accuracy of computational modelling. The first step of radionuclide transport modelling is to develop geosphere flow model-Darcy's Law (Bear and Verruijt 1987). In the next step, the advection-dispersion equation is used for modelling the transport through the saturated zone with retardation and decay.

The numerical solution of partial differential equations (PDEs) has been usually obtained by either finite difference methods (FDM), finite element methods (FEM), and finite volume methods (FVM). These methods require a mesh to support the localized approximations. Kansa (Kansa 1990) introduced the concept of solving PDEs using radial basic functions (RBFs) for hyperbolic, parabolic and eliptic PDEs. It was also reported (Moridis and Kansa 1994; Moridis and Reddell 1991) that the traditional finite time marching scheme had been replaced with the numerical inversion of the Laplace transforms, which eliminates temporal truncaction errors and the need for numerous time integration steps.

In our case we employ the Kansa meshless method that uses radial basis functions as the mathematical solution technique and the traditional finite difference time marching scheme. In this case no domain discretization and/or domain integration is required. Kansa's method is considered as domain type method and has many features of the finite element method (Hon and Chen 2001).

In this study we determine the average and sample variance of radionuclide concentration with regard to spatial variability of hydraulic conductivity that was modelled by geostatistical approach. By this approach the general applicability, reliability and sensitivity of the Kansa meshless method in transport modelling is verified.

Governing Equations

Advection-dispersion equation with boundary and initial conditions

Dispersion tensor D_{ij} is diagonal in the chosen coordinate system. The advection-dispersion equation for the transport through the saturated porous media zone in macroscopic level with retardation and decay is:

Geochemical and reactive transport modeling

$$R\frac{\partial u}{\partial t} = \left(\frac{D_x}{\omega_e}\frac{\partial^2 u}{\partial x^2} + \frac{D_y}{\omega_e}\frac{\partial^2 u}{\partial y^2}\right) - v_{x_i}\frac{\partial u}{\partial x} - R\lambda u \qquad (x,y)\in\Omega, \quad 0\leq t\leq T \tag{1}$$

$$u\big|_{(x,y)\in\partial\Omega} = g(x,y,t), \qquad 0\leq t\leq T \tag{2}$$

$$u\big|_{t=0} = h(x,y), \qquad (x,y)\in\Omega. \tag{3}$$

where x is the groundwater flow axis, y is the transverse axis, u is the concentration of contaminant in the groundwater [Bqm^{-3}], D_x and D_y are the components of dispersion tensor [m^{2}y^{-1}] in saturated zone, ω_e is the effective porosity of the saturated zone [-], v_{x_i} is Darcy velocity [my^{-1}] at interior points, R is the retardation factor in saturated zone [-] and λ is the radioactive decay constant [y^{-1}].

Meshless method – Kansa method

For the parabolic problem (1–3), we consider the implicit scheme (Hon and Chen 2001):

$$R\frac{u^{n+1}-u^n}{\delta t} = \left(\frac{D_x}{\omega_e}\frac{\partial^2 u^{n+1}}{\partial x^2} + \frac{D_y}{\omega_e}\frac{\partial^2 u^{n+1}}{\partial y^2}\right) - v_{x_i}\frac{\partial u^{n+1}}{\partial x} - R\lambda u^{n+1} \tag{4}$$

where δt is the time step, u^n and u^{n+1} are the contaminant concentration at the time t_n and t_{n+1}, respectively.

The approximate solution can be expressed as:

$$u(x,y,t_{n+1}) = \sum_{j=1}^{N} c_j^{n+1}\varphi_j(x,y) \tag{5}$$

where c_j^{n+1}, $j=1,...,N$ are the unknown coefficients to be determined. Let N be the number of collocation points, N_i is the number of interior points and $N-N_i$ is the number of boundary points. $\varphi_j(x,y)$ is the Hardy's multiquadrics function (Hardy 1990)

$$\varphi_j(x,y) = \sqrt{(x-x_j)^2 + (y-y_j)^2 + c^2} \tag{6}$$

where c is the shape parameter.

By substituting (5) into (1–2), it yields:

$$\sum_{j=1}^{N}\left(R\frac{\varphi_j}{\delta t} - \frac{D_x}{\omega_e}\frac{\partial^2\varphi_j}{\partial x^2} - \frac{D_y}{\omega_e}\frac{\partial^2\varphi_j}{\partial y^2} + v_{x_i}\frac{\partial\varphi_j}{\partial x} + R\lambda\varphi_j\right)(x_i,y_i)\,c_j^{n+1} = R\frac{u^n}{\delta t}(x_i,y_i) \tag{7}$$

where $i = 1,2,...,N_i$

Geochemical and reactive transport modeling

$$\sum_{j=1}^{N} \varphi_j(x_i, y_i) c_j^{n+1} = g(x_i, y_i, t_{n+1}) \qquad i = N_i + 1, N \tag{8}$$

where

$$\frac{\partial \varphi_j}{\partial x} = (x_i - x_j) \varphi_j^{-1}(x_i, y_i); \tag{9}$$

$$\frac{\partial \varphi_j}{\partial y} = (y_i - y_j) \varphi_j^{-1}(x_i, y_i); \tag{10}$$

$$\frac{\partial^2 \varphi_j}{\partial x^2} = \left[1 - (x_i - x_j)^2 \varphi_j^{-2}(x_i, y_i)\right] \varphi_j^{-1}(x_i, y_i); \tag{11}$$

$$\frac{\partial^2 \varphi_j}{\partial y^2} = \left[1 - (y_i - y_j)^2 \varphi_j^{-2}(x_i, y_i)\right] \varphi_j^{-1}(x_i, y_i); \tag{12}$$

from which we can solve the $N \times N$ linear system of (7–8) for the unknown c_j^{n+1}, $j = 1, ..., N$. Then (5) can give us the approximate solution at any point in domain Ω.

The Darcy velocity

Derivation of Laplace equation for the calculation of pressure distribution

The assumption of constant water density is considered. The incompressibility equation is:

$$\nabla \cdot \mathbf{v} = 0 \tag{13}$$

For the description of flow through porous media in gravity field the Navier-Stokes equations (Bear and Verruijt 1987) are used. Neglecting inertial terms, these take the form of:

$$\nabla(p + \gamma z) = \mu \nabla^2 \mathbf{v} \tag{14}$$

where p is the pressure of the fluid, μ is the viscosity of the fluid, and $\mathbf{v}$ is velocity vector, z is elevation and γ is specific weight.

If the motion of the fluid is very slow, Laplace equation is constructed by taking the divergence of Navier-Stokes equation (14). In the case of homogenous and anisotropic porous media we obtain:

$$K_{x_i} \frac{\partial^2 p_i}{\partial x^2} + K_{y_i} \frac{\partial^2 p_i}{\partial y^2} = 0 \tag{15}$$

Solution of the Laplace equation

The Laplace equation is also solved by using the Kansa method. In this case the Neumann pressure boundary conditions will be defined along the whole boundary. Eqs. (15) become:

$$\sum_{j=1}^{N}\left(K_{x_i}\frac{\partial^2\varphi_j}{\partial x^2}+K_{y_i}\frac{\partial^2\varphi_j}{\partial y^2}\right)(x_i,y_i)\,c_j^{n+1}=0 \qquad i=1,2...,N_i, \tag{16}$$

$$\sum_{j=1}^{N}\left(\frac{\partial\varphi_j(x_i,y_i)}{\partial x}s_x+\frac{\partial\varphi_j(x_i,y_i)}{\partial y}s_y\right)c_j^{n+1}=g(x_i,y_i), \qquad i=N_i+1,N, \tag{17}$$

where s_x in s_y are the components of the unit vector normal to the boundary.

The pressure gradient is evaluated by:

$$\frac{\partial p_i}{\partial x}=\sum_{j=1}^{N}c_j^n\frac{\partial\varphi_j(x_i,y_i)}{\partial x} \tag{18}$$

$$\frac{\partial p_i}{\partial y}=\sum_{j=1}^{N}c_j^n\frac{\partial\varphi_j(x_i,y_i)}{\partial y} \tag{19}$$

For the calculation of velocity in principal directions we use:

$$v_{x_i}=-\left(\frac{K_{x_i}}{n\rho g}\right)\frac{\partial p_i}{\partial x} \tag{20}$$

$$v_{y_i}=-\left(\frac{K_{y_i}}{n\rho g}\right)\left(\frac{\partial p_i}{\partial y}+\rho g\right). \tag{21}$$

where ρ is the density of the fluid, g is the acceleration of gravity, n is porosity, K_{x_i} and K_{y_i} are hydraulic conductivity.

A Geostatistical Approach of Determining Spatial Variable Hydraulic Conductivity

The most important input data are obtained from field measurements, but they are not available from all regions of interest. For example, the hydraulic conductivity as input parameter varies from one place to another.

In this study we determine the mean and the sample variance of radionuclide concentration with regard to spatial variability of hydraulic conductivity that was modelled by geostatistical approach.

Spatial variability of hydraulic conductivity has been investigated by several authors (Gomez-Hernandez and Gorelick 1991, Deutsch and Journel 1998). One way to study the spatial variability is the stochastic simulation approach. Each parameter is considered as a random variable at each point in space, and there is a joint probability distribution function at each point in space.

In our case Sequential Gaussian Simulation code was used (based on the program SGSIM, Deutsch and Journel 1998). We chose Gaussian model of random functions. The data that can be changed are semivariogram parameters (nugget effect = 0.2, range, scale, shape and anisotropy). The anisotropy was defined by angles (measured in degrees clockwise from the positive Y). The shape denotes the type of variogram (in our case spherical) with variance 0.8. The range (the horizontal maximum direction) was 600 m.

Solution Procedure for Homogenous Fluid

The solution of a dispersion problem is made up of three independent subproblems. First, the hydraulic conductivity was modelled by SGSIM. Second, the results of simulations and pressure gradient were included in the calculation of the velocity distribution. The resulting velocity v distribution is inserted in the dispersion equation, which is then solved by Kansa method to yield the concentration distribution in the flow domain.

Numerical example

The simulation was implemented for rectangular area which was 600 m long and 300 m deep. The source was Thorium (Th-230) with the activity 1.10^{12} and half life of 77000 years. The source was located on the lower left side of the area. The groundwater flow field is presented for a steady-state conditions. Except for the inflow (left side) and outflow (right side), all boundaries have no-flow condition. The inflow rate was 0.1 m/y. At the outflow side, the time-constant pressures at the boundaries were set. The components of dispersion tensor are approximated by $D_x = a_L v$ and $D_y = a_T v$. Longitudinal dispersivity, a_L, is 10 m and transversal dispersivity, a_T, is 2 m, v is Darcy's velocity. Porosity is $n = 0.25$ and hydraulic conductivity was chosen in different points with geostatistics based on two different sets of input data. In the first one hydraulic conductivity at only four different points is given (values are: 195.52, 230.02, 12.74, 25.41). In the second case the data base of 8 different points is used (values are: 66.00, 71.00, 73.00, 75.00, 76.52, 77.02, 79.74, 83.41). The coordinates of these points are presented in Fig. 1 and 2. For each set 100 simulations were completed. The shape parameter c of the Kansa method was set to 0.01.

In Fig. 1 and 2 the distribution of conductivity is given for both sets of conductivity data. The variability in the first case is much higher, which can be easily seen from these two figures. It is interesting to note how the layer of higher hydraulic conductivity is created.

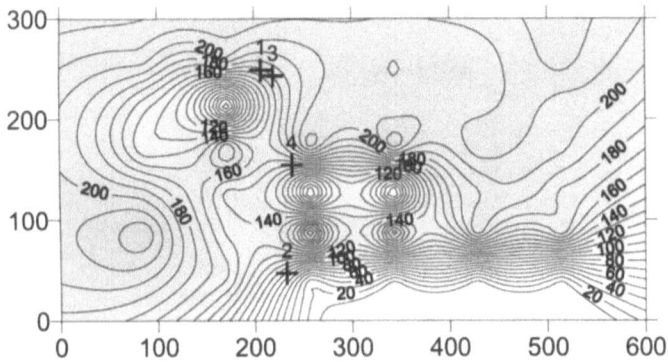

Fig. 1. Distribution of hydraulic conductivity based on 4-point data set

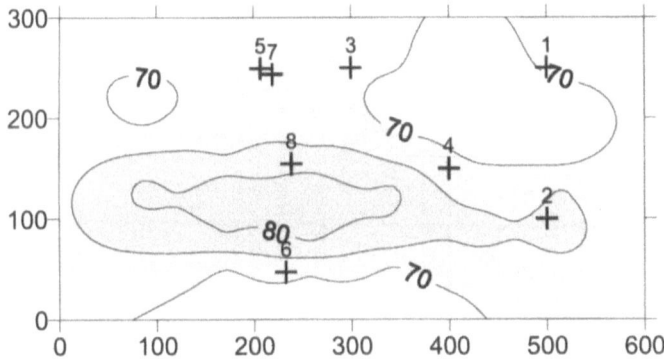

Fig. 2. Distribution of hydraulic conductivity based on 8-point data set

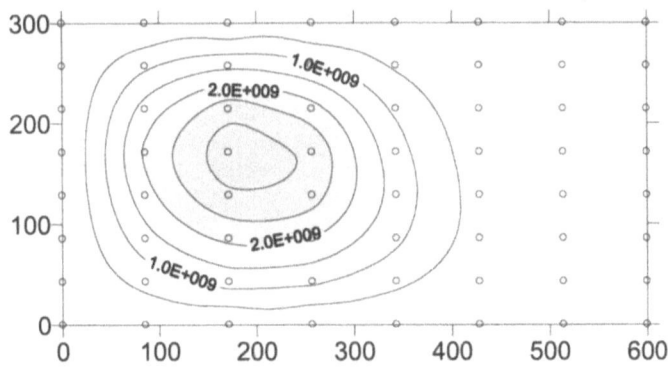

Fig. 3. Distribution of contaminant concentration after 10 years (4 points)

In Fig. 3 and 4 the distribution of contaminant concentration after 10 years is given for one particular simulation. The values are higher in the second case, for which the average hydraulic conductivity is lower.

Geochemical and reactive transport modeling

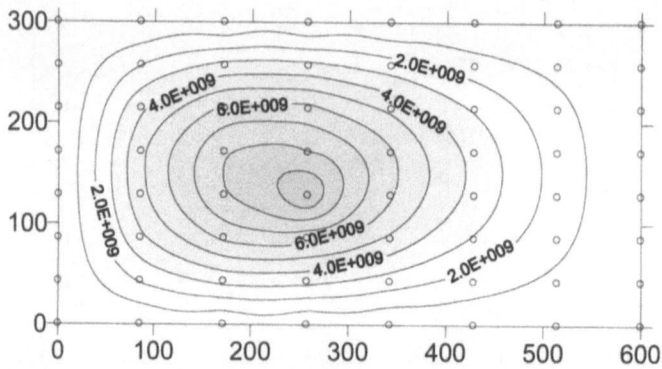

Fig. 4. Distribution of contaminant concentration after 10 years (8 points)

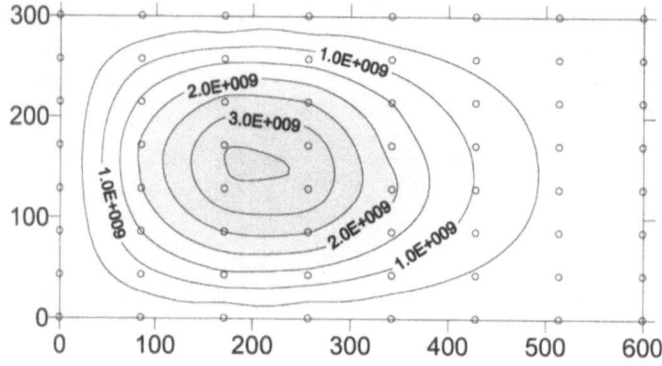

Fig. 5. Distribution of average of contaminant concentration (4 points)

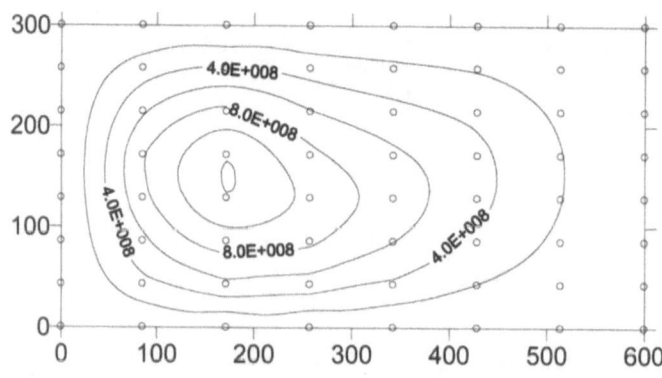

Fig. 6. Distribution of standard deviation of contaminant concentration (4 points)

In Fig. 5 and 7 the distribution of the average value of contaminant concentration after 10 years is given. These values were obtained after repeating 100 simulations. We can note a huge difference between the average value and the distribution for one particular simulation. Obviously, the scatter of the results is very large, which is indicated also in Fig. 6 and 8 showing the standard deviation of contaminant distribution. The standard deviation in the first case is relatively large which yields the coefficient of variation equal to approx.0.3. In the second case we deliberately choose the eight values of hydraulic conductivity with lower variance. As a result the scatter is much lower than in the first case – coefficient of variation is approx. 0.04.

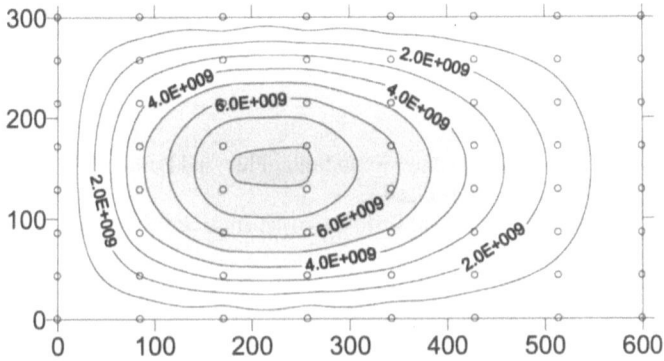

Fig. 7. Distribution of average of contaminant concentration (8 points)

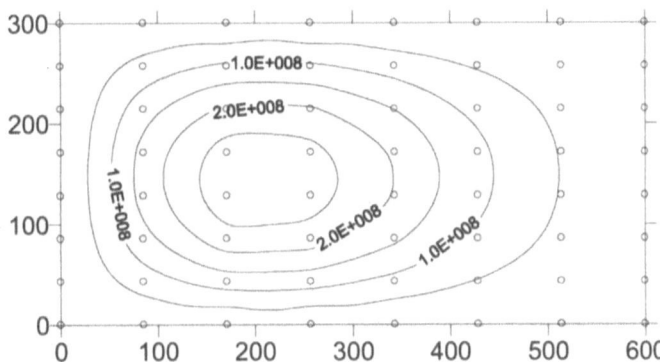

Fig. 8. Distribution of standard deviation of contaminant concentration (8 points)

Conclusions

The influence of geostatistical data on the reliability and accuracy of computational modelling of mass transport problem was investigated. The meshless Kansa method was employed in the determination of velocity distribution which was used as input data in contaminant concentration distribution.

Geochemical and reactive transport modeling

The only physical data modelled as a random field was hydraulic conductivity of the investigated field. In the future research some other parameters such as dispersivity may also be modelled as random field.

In the simulation a very large scatter was observed which is due to very few (only four) very different given values of hydraulic conductivity. We can easily observe that no reliable results can be obtained from such data set. On the other hand, if there are more values with lower scatter, the results are much more reliable.

Reference

Bear J., Verruijt A. (1987) Modeling Groundwater Flow and Pollution. D. Reidel Publishing Company, Dordrecht, Holland.

Deutsch C. V., Journel A. G. (1998) GSLIB Geostatistical Software Library and User's Guide. Oxford University Press.

Hardy R. L. (1990) Theory and Applications of the Multiquadric-Biharmonic Method. Computers Math. Applic. Vol. 19, No. 8/9: 163-208.

Gomez-Hernandez J. Jaime, Gorelick Steven M. (1989) Effective Groundwater Model Parameter Values: Influence of Spatial Variability of Hydraulic Conductivity, Leakance, and Recharge. Water Resources Research, Vol. 25, No. 3: 405-419.

Hon Jichun Li, Y.C. Chen C. S. (2001) Numerical Comparisons of Two Meshless Methods Using Radial Basis Functions, private communication.

Kansa E. J. (1990) Multiquadrics - A Scattered Data Approximation Scheme with Applications to Computational Fluid-Dynamics-II-Solutions to Parabolic, Hyperbolic and Elliptic Partial Differential Equations. Computers Math. Applic. Vol. 19, No. 8/9: 147-161.

Moridis G. J., Kansa E. J. (1994) The Laplace Transform Multiquadrics Method: A Highly Accurate Scheme for the Numerical Solution of Linear Partial Differential Equations. Journal of Applied Science and Computations,Vol. 1., No. 2: 375-407.

Moridis G. J, Reddell Donald L. (1991) The Laplace Transform Finite Difference Method for Simulation of Flow Through Porous Media, Water Resources Research, Vol. 27, No. 8: 1873-1884.

Complex Formation of Uranium(IV) with Phosphate and Arsenate

G. Geipel, G. Bernhard, V. Brendler

Forschungszentrum Rossendorf e.V., Institute of Radiochemistry, P.O.Box 510 119, D-01314 Dresden

Abstract. We studied the complex formation of uranium(IV) with phosphate and arsenate in acid solution using UV-VIS and laser-induced photo-acoustic spectroscopy at various ionic strengths. We found a one to one complex formation between uranium and phosphoric acid and arsenic acid, respectively. During the complex forming reaction one hydrogen ion was released from the acid. Including the protonation constants of the used acids at infinite dilution, the formation constants were derived to be log $\beta^0_{121} = 25.23 \pm 0.13$ for the phosphate system and log $\beta^0_{121} = 23.94 \pm 0.08$ for the arsenate system, respectively.

Introduction

Heavy metals in the aquatic environment are normally transported as a complexed species. Uranium is one of the main potential contaminants in East Germany's former mining areas. Knowledge about the complex formation is therefore an essential constituent in prediction of the migration of these elements. The seepage and mine waters contain several inorganic and organic complex forming agents (G. Bernhard, et. al. 1996),(G. Geipel, et. al. 2000).

Under reducing conditions uranium(IV) is stable and can be complexed by phosphate and arsenate at low pH. Due to their low water solubility no thermodynamic complex formation constants are available for the uranium(IV)-phosphate and arsenate complexes.. For example, the solubility product (I. Grenthe, et. al. 1992) of $U(HPO_4)_2 \cdot 4H_2O$ limits the uranium(IV) concentration in 1 M perchloric acid to less than 1.2×10^{-4} M. Moreover there is a general lack of information on the aqueous chemistry and thermodynamics of tetravalent actinides.

Experimental

We studied the formation of these complexes at uranium concentrations between 2×10^{-6} M and 1×10^{-4} M by Laser-Induced Photoacoustic Spectroscopy and UV-VIS spectroscopy in acid media. The design of LIPAS is described elsewhere (Geipel, et. al. 1998). UV-VIS measurements were carried out using a CARY 5G dual beam spectrophotometer (Varian, Austria).

The U^{4+} was produced by electrochemical reduction of a stock solution of 0.01 M UO_2^{2+} in 0.5 M $HClO_4$. The samples to be measured were prepared by dilution of this solution and adding the calculated amount of phosphoric acid or arsenic acid, respectively. The concentration of hydrogen ions was adjusted with 5 M $HClO_4$ and the final ionic strength was adjusted with 5 M $NaClO_4$. All samples were prepared under inert atmospheric conditions (> 99.9 % N_2, < 10 ppm O_2) in a glove box. Also the cuvettes to be measured were filled and sealed in the glove box.

The measurements were carried out at 1×10^{-4} M and 5×10^{-4} M uranium. The concentration of phosphoric acid or arsenic acid varied between 0 and 1×10^{-2} M. The concentration of hydrogen ions and of the ionic strength was kept constant in a range from 0.1M to 2 M.

To determine the complex formation constant we measured and analyzed the spectra from 600 nm to 690 nm. The spectra were corrected for the absorption of the solvent water. The solvated uranium(IV) shows a very intensive absorption band with a maximum at 650 nm and a smaller maximum at 672 nm. We found a shoulder at 630 nm deconvoluting the absorption spectrum.

In addition, we measured the fluorescence spectra of all prepared solutions to determine the amount of reoxidized uranium(VI). For our calculations we used only spectra where the amount of reoxidized uranium(VI) was less than 5% of the total uranium. The concentration of uranium(IV) was corrected by the data obtained for uranium(VI). The total concentration of uranium in the solutions was determined by ICP-MS.

Results and Discussion

The spectrum changes with increasing phosphate and arsenate concentration. The absorption maxima of the solvated U^{4+} ion are located at 629.5nm, 649.1 nm and 671.7 nm. In the phosphate system the absorption maxima of the formed complex are shifted to be at 645.0nm, 656.6 nm and 667.0 nm. Two isosbestic points were determined at 661.4 nm and 669.4 nm in the phosphate system. In the arsenate system the absorption maxima are found to be 645.1 nm and 662.9 nm. Two isosbestic points were detected at 661.0 nm and 667.5nm. The spectra of free U^{4+}, the phosphate complex and the arsenate complex are shown in (Fig.1).

To obtain the formation constant and the stoichiometry, three parameters had to be varied i) the ionic strength, ii) the concentration of phosphoric acid or arsenic and iii) the concentration of hydrogen ions.

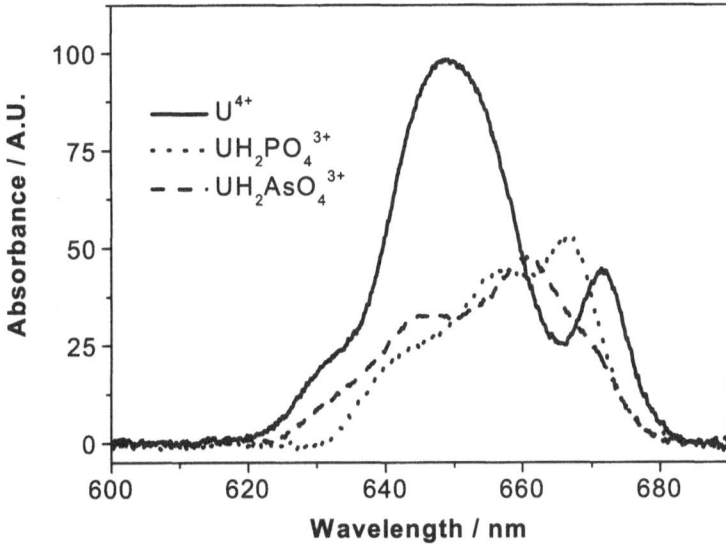

Fig. 1. Absorption spectra of uranium(IV) and dihydrogenphosphate and dihydroge-narsenante complexes

The complex forming reaction can be represented as

$$U^{4+} + x\ H_3ZO_4 \Longleftrightarrow U(H_{3-y}ZO_4)_x^{(4-xy)+} + xy\ H^+ \tag{1}$$

where Z represents phosphorus or arsenic. After rearranging and transforming the mass action law in logarithmic form it becomes

$$\log\frac{[UH_{(3-y)}ZO_4^{(y-x)+}]}{[U^{4+}]} = \log K + x\log[H_3ZO_4] - xy\log[H^+] \tag{2}$$

At a constant hydrogen ion concentration the right side of Equation (2) can be simplified using

$$\log K' = \log K - xy\log[H^+] \tag{3}$$

As example the validation of complex formation for the arsenate system in 0.25 M perchloric acid and at ionic strength of 0.5 M is shown in Fig. 2. By plotting the $\log([complex]/[U^{4+}])$ vs. the $\log[H_3AsO_4]$, we obtain the stoichiometry of the complex reaction from the slope.

Analytical speciation techniques for uranium and related elements

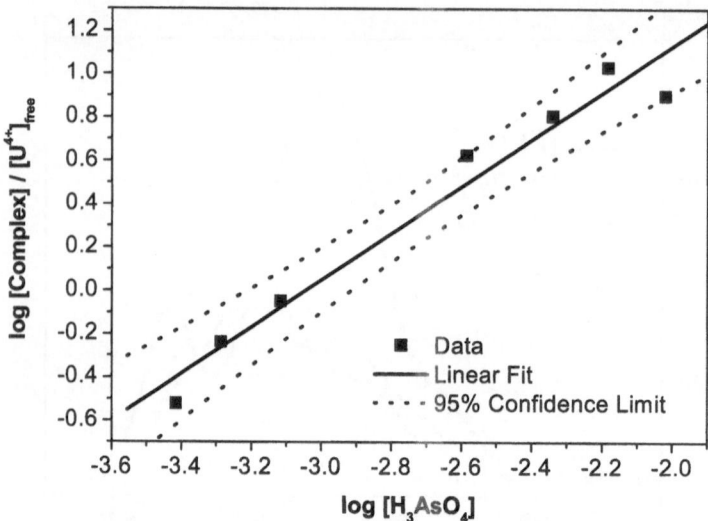

Fig. 2. Slope analysis of the complex formation with arsenic acid

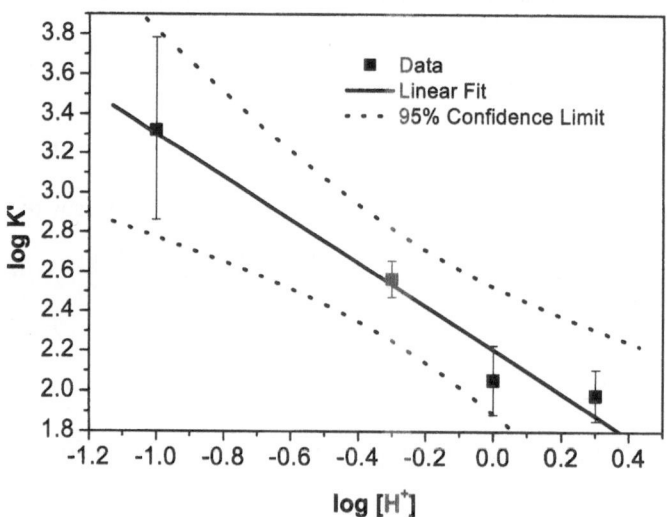

Fig. 3. Slope analysis at an ionic strength of 2.0 M

The value x is found to be 1.08 ± 0.10, which clearly shows a 1:1 complex formation. From the intersection we obtain the formation constant at a fixed hydrogen ion concentration of 0.25 M to be log K' = 3.29 ± 0.27.

In a second step experiments were carried out at various hydrogen ion concentrations. By plotting log K' vs. $\log[H^+]$ we determined the number of hydrogen ions released during the complex forming reaction. From the slope analysis and the intersection we obtained the formation constant at the selected ionic strength. This is shown, for example, in Fig. 3 for an ionic strength of 2 M. We found a slope of -1.03 ± 0.19. This means that one hydrogen ion is released during the first complex formation step. The formation constant was calculated to be log K = 2.24 ± 0.06. The results of all experiments in the phosphate system are summarized in Table 1. The data of the arsenate system are listed in Table 2.

Table 1. Calculated data for the complex formation of uranium(IV) with phosphate

Ionic strength / M	log K	Slope analysis H_3PO_4	Slope analysis H^+
2.00	2.75 ± 0.05	1.25 ± 0.34	-0.82 ± 0.10
1.00	2.52 ± 0.14	1.12 ± 0.26	-0.77 ± 0.14
0.50	2.70 ± 0.08	1.04 ± 0.27	-1.38 ± 0.27
0.25	2.59 ± 0.40	0.98 ± 0.21	-1.08 ± 0.31
0.10	2.65 ± 0.26	1.54 ± 0.16	

Table 2. Calculated data for the complex formation of uranium(IV) with arsenate

Ionic strength M	log K	Slope analysis H_3AsO_4	Slope analysis H^+
2.00	2.24 ± 0.06	1.05 ± 0.17	-1.03 ± 0.19
1.50	2.22 ± 0.10	1.03 ± 0.06	-1.31 ± 0.48
1.00	1.95 ± 0.08	1.01 ± 0.07	-1.31 ± 0.16
0.75	2.31 ± 0.14	1.37 ± 0.21	-
0.50	2.16 ± 0.06	1.14 ± 0.05	-1.40 ± 0.25
0.10	2.18 ± 0.14	0.93 ± 0.17	-

The results suggest a one to one complex formation between uranium(IV) and the phosphoric acid and also for the arsenic acid. During this complex forming reactions one hydrogen ion is released form the acids. The complex forming reaction therefore can be assigned to be

$$U^{4+} + H_3ZO_4 <===> UH_2ZO_4^{3+} + H^+ \qquad (4)$$

To extrapolate the complex formation constant at infinite dilution we used the SIT theory:

$$\log KJ - \Delta z^2 D = \log K^0 - \Delta\varepsilon J \qquad (5)$$

Using the charges of the species in solution, we calculated $\Delta z^2 = -6$. D is a function of the ionic strength:

$$D = \frac{0.5093\sqrt{J}}{1+1.5\sqrt{J}} \qquad (6)$$

Analytical speciation techniques for uranium and related elements

Adding the product of $\Delta z^2 D(J)$ to the formation constants at the various ionic strengths we obtain an expression which should be a linear function of ionic strength. The slope represents the sum of the specific interaction coefficients. The intersection is equal to the formation constant at infinite dilution.

We have to change from the molar to the molal concentration scale. The calculated ionic strengths on the molal scale are listed in Table 3.

Table 3. Data for conversion to molal scale

Ionic strength M	Ionic strength mol/kg	D(J)
2.00	1.74	0.231
1.50	1.35	0.219
1.00	0.95	0.204
0.75	0.72	0.192
0.50	0.48	0.175
0.10	0.098	0.109

Using this extrapolation mechanism, the formation constant for the reaction (6) is found to be log K^0 = 3.53 ± 0.09 for the phosphate system and log K^0 = 3.34 ± 0.08 for the arsenate system. The slopes are calculated to be –0.20 ± 0.08 and -0.16 ± 0.04, respectively. The extrapolation plots are shown in Fig.4 for the phosphate system and in Fig. 5 for the arsenate system.

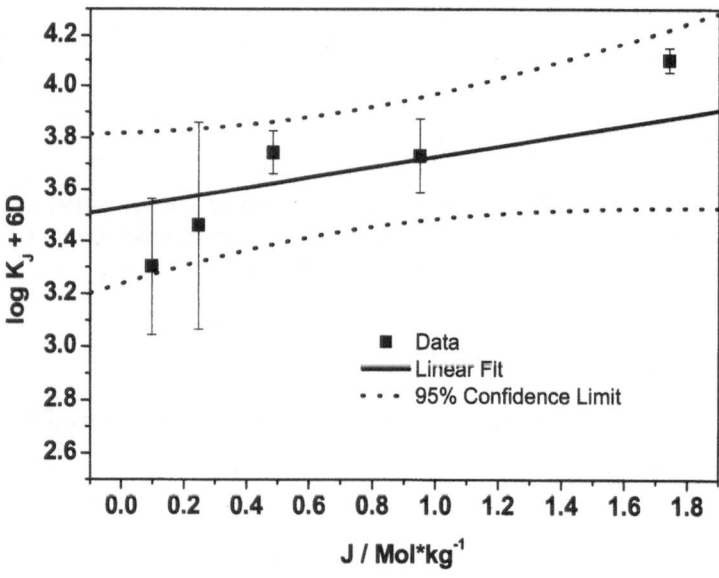

Fig. 4. Extrapolation of the complex formation constant to infinite dilution for the phosphate system

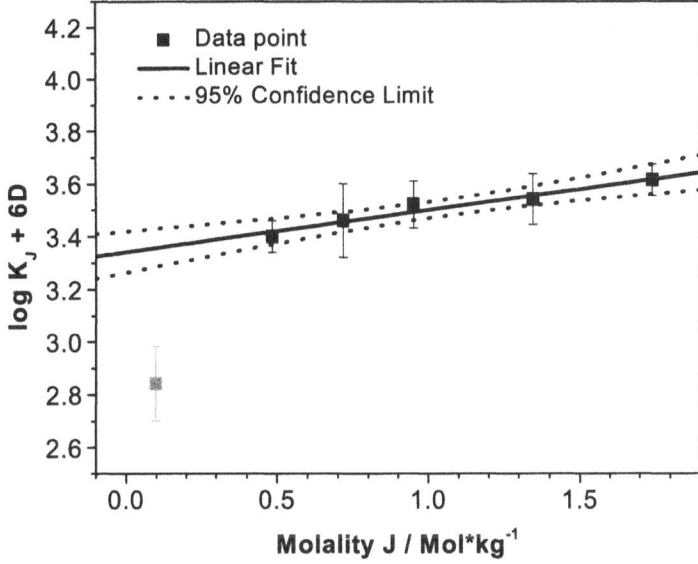

Fig. 5. Extrapolation of the complex formation constant to infinite dilution for the arsenate system

Introducing the protonation of the phosphoric acid and arsenic acid (I. Grenthe et.al., 1992)

$$3 \ H^+ + ZO_4^{3-} \ \text{<===>} \ H_3ZO_4 \qquad (7)$$

with $\log \beta^0_{31} = 21.70 \pm 0.04$ for phosphoric acid and $\log \beta^0_{31} = 20.60$ for arsenic acid, the formation constants of the uranium(IV)dihydrogen complexes can be calculated.

For the reaction

$$U^{4+} + 2 \ H^+ + ZO_4^{3-} \ \text{<===>} \ UH_2ZO_4^{3+} \qquad (8)$$

the formation constant is $\log\beta^0_{121} = 25.23 \pm 0.13$ for the phosphate system and $\log\beta^0_{121} = 23.94 \pm 0.08$ for the arsenate system, respectively. A speciation diagram comparing the complex formation of uranium(IV) with phosphate and arsenate is shown in Fig. 6. As expected, the complex formation of arsenate is somewhat weaker than that of phosphate.

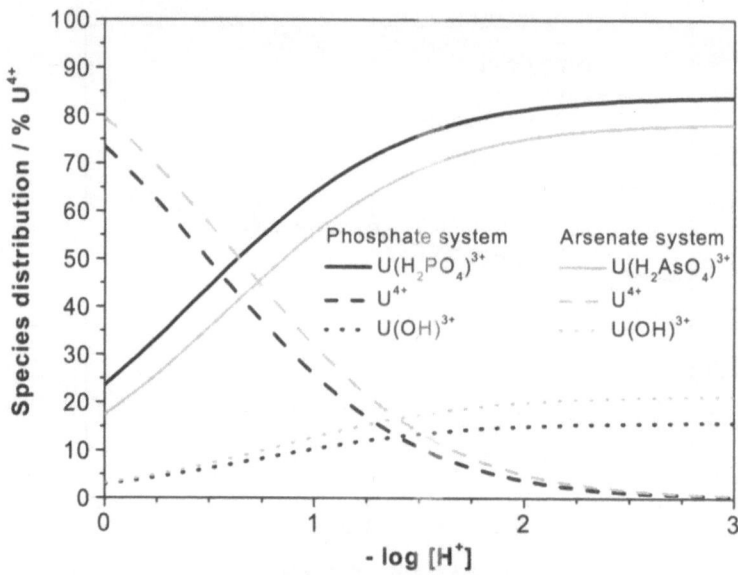

Fig. 6. Speciation of the uranium(IV) in acid phosphoric and arsenic solution

Compared to the data in the uranium(VI) system (I. Grenthe, et. al. 1992), (M. Rutsch, et. al. 1999) the formation constants are about two order of magnitude higher. Also the ion interaction coefficients can be calculated using additional data in (I. Grenthe, et. al. 1992). We assign the ion interaction coefficient ε ($UH_2ZO_4^{3+}$ -ClO_4^-) to be 0.42 in the phosphate system and 0.46 in the arsenate system, respectively. This value is in the same order as for other trivalent charged uranium(IV) species (UOH^{3+} : 0.48 ± 0.08).

References

Bernhard, G. et. al. (1996) Speciation of Uranium in Seepage Waters from a Mine Tailing Pile Studied by Time-Resolved Laser-Induced Fluorescence Spectroscopy (TRLFS); Radiochimica Acta 74, 87.

Geipel, G. et. al. (1998) Complex formation between UO_2^{2+} and CO_3^{2-} studied by laser-induced photoacoustic spectroscopy (LIPAS); Radiochimica Acta 82, 59.

Geipel, G. et.al. (2000) Speciation in Water Released from Mining and Milling Facilities In T.E. Baca and T. Florkowski (eds.), The environmental Challenges of Nuclear Disarmament, Kluwer Academic Publishers, p. 323-332.

Grenthe I., et. al. (1992) Chemical thermodynamics of uranium; North-Holland, Amsterdam-London-New York-Tokyo.

Rutsch, M. et. al. (1999) Interaction of uranium (VI) with arsenate (V) in aqueous solution studied by time-resolved laser-induced fluorescence spectroscopy, Radiochimica Acta 86, 135-141.

Analytical speciation techniques for uranium and related elements

Hyphenated techniques for speciation of Uranium and other actinides

Maria Betti, Laura Aldave de las Heras, Lorenzo Perna, Fabio Bocci, Tania Huber

European Commission, Joint Research Centre, Institute for Transuranium Elements, P.O. Box 2340, 76125 Karlsruhe, Germany

Abstract. Extraction ion chromatography coupled with inductively coupled plasma mass spectrometry has been exploited for studying the kinetic of reaction of neptunium and plutonium under redox conditions. Separations of Th (IV), U(IV) and U(VI) have been studied by capillary electrophoresis employing dynamically modified substrates.

Introduction

Both natural and anthropogenic radionuclides are occurring in the environment through different processes and mechanisms (Warner and Harrison, 1993; Valković, 2000; Betti, 2000). Natural radioactivity has as major sources the very long half-live nuclides that have persisted since the formation of the earth along with their shorted lived daughter nuclides and nuclides produced by cosmic rays in the atmosphere at high altitude. Anthropogenic radionuclides can enter into the environment through a variety of different events. Among the sources of artificial radioactivity, global fallout from atmospheric weapons tests, local fallout accidentally released from nuclear power plants and spent fuel reprocessing plants as well as short-lived nuclides from production of radioisotopes for medical applications, are the most relevant. However, release of radionuclides due to nuclear accidents and storage of nuclear wastes needs also to be considered.

Nowadays, radionuclides constitute an important class of environmental contaminants. In order to assess the impact of uranium and actinides on the environment and human beings, the knowledge of their speciation is essential. In fact, the determination of speciation is of prime importance to understand the mobility, the toxicity and the risk from trace radionuclides in natural systems. It contributes to a better understanding of their fate, in terms of retention, transport and bioavalability, in the framework of environmental and nuclear purpose. For instance, in environmental radiochemistry, isotope speciation is of relevance in order to study transfer/migration mechanisms related to geological (Hoefs, 1987) and cosmic

(Shima, 1986) samples, including age determination (Faure, 1986) as well as bioavalability studies (Mellon and Sandström, 1996; Skipperud et al, 2000). Therefore it is very important to have available accurate, reliable and precise analytical methodologies to check the concentration of their different species.

In our laboratory liquid chromatography has been coupled on-line to an Inductively Coupled Plasma Mass Spectrometer (HPLC-ICP-MS) as well as a capillary electrophoresis (CE) to spectrophotometric detectors to study speciation of actinides. In this presentation the two different analytical techniques proposed for speciation studies are discussed.

Experimental

Instrumentation

An ELAN 5000 ICP-MS instrument (Perkin-Elmer SCIEX, Canada) modified to handle radioactive samples in a glove-box and coupled with a GP50 high-pressure chromatographic pump (Dionex, CA, USA) was used (Betti et al. 1993). The high-pressure pump was located outside the glove-box while the four-ways injection valve (Dionex, Part No 42766, 250 μl loop), guard column and analytical column were placed inside the glove-box (Garcia et al. 1995). The effluent from the chromatographic column was direct to a cross-flow nebulizer. A two ways peristaltic pump (Gilson, France) was used to carry the waste to a vessel inside the glove-box. Standard plasma and ion lens operating conditions are summarised in table 1.

Table 1. Operating conditions of the ICP-MS

ICP ion source	
RF Power	1045 W
Cooling Ar flow rate	$15.0 \, l \, min^{-1}$
Nebuliser Ar flow rate	$1 \, l \, min^{-1}$
Auxiliary Ar flow rate	$0.9 \, l \, min^{-1}$
Lens voltage	
Einzel lens E1	70
Bessel box B	76
Photon stop S2	2
Bessel Box P	45
Data acquisition	
Dwell time	100 ms
Replicate time	100 ms
Scan mode	Peak hop
Reading per replicate	1
Sweep per replicate	1

An Agilent CE system (Agilent Technologies) with diode array detector was used. The wavelength was set at 210 nm. The analysis voltage was 28 kV. Sample injection was carried out with pressure (50 mbar, 0,2 min) at the anodic side. A 56 cm X 50 μm I.D. fused-silica capillary (Agilent technologies) was used. The capillary was flushed with 1M NaOH, water and running buffer each for 10 min at the beginning of each day. A washing step of 2 min with buffer between runs was applied. Automated capillary rinsing, sample injection, and execution of the electrophoretic runs were controlled by a personal computer.

Reagents and materials

Chromatographic actinide separations were performed using UTEVA resin columns. The eluents consisted of 3 M nitric acid, 0.1 M oxalic acid in 2 M HCL and 0.025 M HCl. Nitric and hydrochloric acid were Suprapur grade (Merk, Darmstadt, Germany). Oxalic acid was obtained from Merck (Darmstadt, Germany).

Natural element standards (U and Th) were obtained from SPEX (Grasbrunn, Germany) as 1000 μg ml^{-1} stock standard solutions and diluted as necessary with 1 % nitric acid. Certified enriched isotope ^{237}Np was obtained from "D.I. Mendeleyev" Institute for Metrology (St. Petersburg, Russia Federation); a certified enriched ^{239}Pu from New Brunswick Laboratory (Argonne, IL, USA).

HIBA and 2,6-pyridinedicarboxylic acid were from Fluka (Buchs, Switzerland); Tris from Sigma (St. Louis, MO, USA) and formic acid from Merck (Darmstadt, Germany).

The carrier electrolyte for capillary electrophoresis consisted to 15 mM formic acid, 10 mM HIBA and 0.2 mM 2,6-pyridinedicarboxylic acid or 0.8 mM 2-pyridinecarboxylic acid. pH calibration was made with Tris and the pH set to 4.7. The electrolyte was prepared daily, degasses and filtered through 0.45-μm prior use. All chemicals were of analytical grade, and deionised water was used for the preparation of all the solutions.

Water purified in a Milli-Q system (Millipore, Eschborn, Germany) was used throughout.

All standard solution, spikes and sample were prepared by dilution by mass in polyethylene bottles previously cleaned according to a procedure for trace element analysis (Betti and Papoff, 1988). Radioactive samples and standards were treated in glove-box.

Results and discussion

Kinetics of reaction by extraction chromatography coupled on-line to Inductively Coupled Plasma Mass Spectrometry

The ion exchange separation of high valence metal ions has always posed a notable challenge as for the elution conditions necessary for their separation. In particular, those high valence metal with relatively slow reaction kinetics, which form cationic species in non-complexing acid media. These include Th(IV), U(IV) and U(VI), Pu(III) and Pu (IV), Np(IV) and Np(V). The main problem is the hydrolysis and polymerization can occur unless the pH is quite low.

In this investigation, the UTEVA-Spec. (*Uranium and Tetravalent Actinide Specific*) from Eichrom has been exploited for the chromatographic separation of the Pu and Np in different oxidation states as well as for studying their kinetics of reaction.

UTEVA-spec is a new resin for extraction chromatography. This resin is able to separate and to concentrate uranium and tetravalent actinides from aqueous solutions. The extractant, dyamil amylphosphonate (DAAP), form nitrato complexes with the actinide elements. The formation of these complexes is driven by the concentration of nitrate in the sample solution. Therefore, the uptake of the actinides increases with increasing nitric acid concentration. UTEVA resin has been applied to a variety of analytical challenges: uranium measurements in environmental samples (Adriaens et al. 1992, Corbett et al. 1996, Horwitz et al. 1993) clean-up of uranium content in the samples prior the analysis of other elements (Carney 1995, Good et al. 1995) sequential determination of uranium, plutonium and americium (Hendenson 1996) measurement of actinides in urine (Langstone et al. 1995) and measurement of actinides in high level waste (Maxwell et al 1994). In all these applications the UTEVA resin was always used for on-batch separations. For the purpose of this investigation UTEVA was exploited as stationary chromatographic phase which was coupled on-line with an ICP-MS detector (Perna et al. 2001). Therefore, columns of 5 cm length and 4 mm internal diameter were filled with UTEVA resin consisted of particles with an external diameter of 100-150 μm. In Fig. 1 the kinetic of reaction of Np in a radioactively contaminated natural water sample is reported. Neptunium can occur in solution as Np (IV), Np (V) and Np (VI).

The kinetic was studied in sodium nitrite medium, being a possible environmental contaminant. As can be seen at the beginning, before the addition of nitrite, Np was present in three oxidation states: V, IV and VI, the state V being predominant. After the addition of sodium nitrate, Np can be found predominantly in the oxidation state IV, even though a percent is still in the form V.

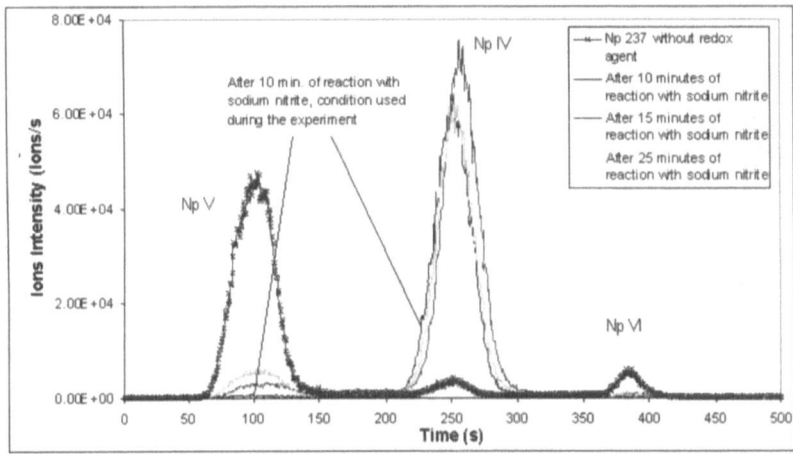

Fig. 1. Reaction kinetics of ^{237}Np with sodium nitrite (redox agent). Column UTEVA, detector ICP-MS.

The same kind of investigation was conducted also for plutonium. This radionuclide can be present in solution as Pu (III), Pu(IV) and Pu (VI). In Fig. 2 its kinetic reaction with sodium nitrite is illustrated.

As can be see, in the original samples chromatographic peaks relevant to the three different oxidation states are present. After 5 minutes of reaction, Pu (III) and Pu (IV) coexist in solution. In 10 minutes all plutonium is converted to the oxidation state (IV).

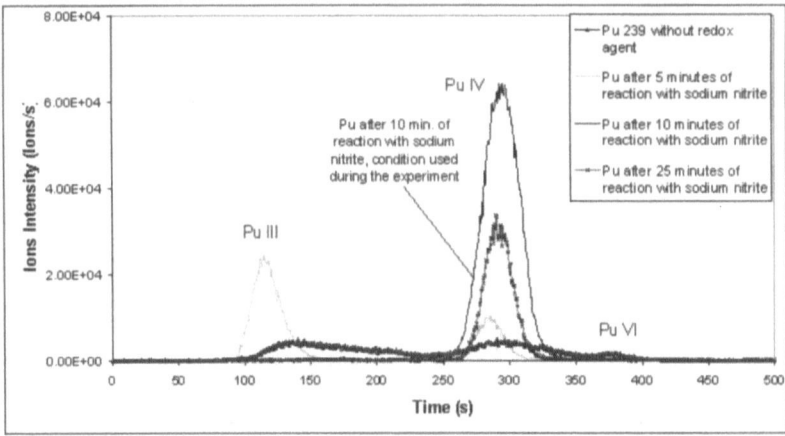

Fig. 2. Reaction kinetics of ^{239}Pu with sodium nitrite (redox agent). Column UTEVA, detector ICP-MS.

Analytical speciation techniques for uranium and related elements

For both Np and Pu, after 25 minutes of reaction in nitrite medium, if a stabilizing agent in not present, a disproportion of the oxidation state IV occurs.

Degrees of complexation by Capillary Electrophoresis

The determination of U and Th is of particular interest with respect to environmental samples and nuclear industry (Barkley et al. 1986, Knight et al. 1984).
In recent years, capillary electrophoresis has been used as method for separation of metal ions. Complex formation is used to differentiate the electrophoretic mobility of hydrolyzed metal ions with equal charge. Thanks to the superior separation efficiency of CE, small differences due to the degrees of complex formation can be revealed and exploited in the separation.

Chelation chromatography utilizing a substrate dynamically modified with 2,6-pyridinedicarboxylic acid (dipicolinic acid) has shown their special separation properties for the determination of uranium and thorium in real samples (Öztekin and Erim 2001, Shaw and Jones 1999, Cowan et al. 2000, Truscott et al. 2001, Sutton et al. 1998). On these basis, separations of Th (IV), U(IV) and U (VI) have being studied by capillary electrophoresis. For this purpose different ligands are used. Among these, hydroxyisobutyric acid (HIBA), picolinic and dipicolinic acid are found to be the most promising for speciation studies. By the use of them, uranium and thorium could be separated and U (IV) could be distinguished from U (IV).

References

Adriaens A. G., Fasset J. D., Kelly W. R., Simons D. S., Adams F. C. (1992), Determination of U and Th concentrations in soil. Comparison of ID-SIMS and ID-TIMS. Anal. Chem. 64: 2945-2950

Barkley D. J., Blanchette M., Cassidy R. M., Elchuk S. (1986), Dynamic chromatographic systems for the determination of rare earths and thorium in samples from uranium ore refining processes, Anal. Chem. 56: 2222-2226

Betti M. (2000), Environmental monitoring of radioisotopes by mass spectrometry and radiochemical methods in urban areas. Microchemical Journal 67: 363-373

Betti M., Garcia Alonso G. I., Arbore Ph., I. Sato, Koch L. (1993), In "Application of Plasma Source Mass Spectrometry II", eds. Holland G. and Eaton A. N., Royal Society of Chemistry, Cambridge. 205-212.

Betti M., Papoff P. (1988), Trace elements: data and information in the characterization of an aqueous ecosystem. CRC Critical review in Analytical Chemistry 19: 271-322

Carney K. P.(1995), The application of micro-column solid phase extraction techniques for the determination of rare earth elements in actinides containing matrices. J. Radioanal. Nucl. Chem., Articles 194: 41-49

Corbett D., Schultz M., Burnett W. (1996), Eichrom Atlanta Users Seminar, Atlanta (USA).

Cowan J., Shaw M. J., Achterberg E. P., Jones P., Nesterenko P. N. (2000), The ion chromatographic separation of high valence metal cations using a neutral polystyrene resin dynamically modified with dipicolinic acid. Analyst 125: 2157-2159

Faure G. (1986), Principles of Isotope Geology, J. Wiley and Sons, New York (USA).

Garcia Alonso G. I., Sena F., Arbore Ph., Betti M., Koch L. (1995), Determination of fission products and actinides in spent nuclear fuels by isotope dilution ion chromatography ICP-MS. J. Anal. Atom. Spectr. 10: 381-393

Good C., Schultz A. (1995), New methods for the determination of transuranics, boron and silicon in uranium hexafluoride, Lockheed Martin, Portsmouth Gaseous Diffusion Plant, POEF-TS-08, prepared for presentation at the Third International Uranium Hexafluoride Conference, Paducah, KY.

Hendenson B.(1996), Determination of selective actinides in water and air filters (Eichrom Resin Column Method with Alpha Spec.), Eichrom Atlanta Users Seminar, Atlanta (USA).

Hoefs J. (1987), Stable Isotopes Geochemistry, Spinger, Berlin (Germany).

Horwitz E. P,. Chiarizia R., Diamond H., Gatrone R. C., Alexandratos S. D., Trochimczuk, Crick D. W. (1993), Uptake of metals ions by a new chelating ion-exchange resin. Part 1: Acids dependencies of actinide ions. Solvent Extraction and Ion Exchange 11(5): 943-958.

Knight C. H., Cassidy R. M., Recoskie B. M., Green L.M. (1984), Dynamic ion exchange chromatography for detrmination of number of fissions in thorium-uranium dioxide fuels, Anal. Chem. 58: 474-478

Langston R., Nguyen S., Simpson T. (1995), A procedure for the determination of Neptunium, Uranium and Thorium in urine, 41st Annual Conference on Bioassay, Analytical and Environmental Radiochemistry, Boston (USA).

Maxwell S. C., Nelson M. R. (1994), Measurement of actinides and strontium 90 in high activity waste, Inst. Of Nuclear Management 35th Annual Meeting (WSRC-MS-94-0163), Naples (USA).

Mellon F. A., Sandström B. (1996), Stable Isotope in Human Nutrition, Academic Press, London (England).

Öztekin N., Erim F. B. (2001), Separation and direct UV detection of lanthanides complexed with pyridine-2-carboxylic acid by capillary electrophoresis. J. Chromatogr. 924: 541-546

Perna L., Betti M., Barrero Moreno J. M., Fuoco R. (2001), Investigation on the use of UTEVA as a stationary phase for chromatographic separation of actinides on-line to ICP-MS. J. Anal. At. Spectr. 16: 26-31.

Shima M. (1986), A summary of extremes of isotopic variations in extra-terrestrial materials. Geochimica et Cosmochimica Acta 50: 557-584

Shaw M. J., Hill S. J., Jones P. (1999), Chelation ion chromatography of metal ions using high performance substrates dynamically modified with heterocyclic carboxylic acids. Anal. Chim. Acta 401: 65-71

Skipperud L., Oughton D., Salbu B. (2000), The impacts of Pu speciation on distribution coefficient in Mayak soil. The Sci. of Tot. Environ. 257: 81-93

Sutton R. C.M., Hill S. J., Jones P., Sanz-Medel A., Garcia-Alonso I.J. (1998), Comparison of the retation behaviour of uranium and thorium on high-efficiency resin substrates impregnated or dynamically coated with metal chelating compounds. J. Chromatogr. A 816: 286-291

Truscott J.B., Jones P., Fairman B.E., Evans E.H. (2001), Determination of actinides in environmental and biological samples using high-performance chelation chromatography coupled to sector-field inductively coupled plasma mass spectrometry. J. Chromatogr. A 928: 91-98.

Warner F., Harrison R.M.(1993), Radioecology after Chernobyl, J. Wiley & Sons, New York.

Valković V. (2000), Radioactivity in the Environment, Elsevier, Amsterdam.

Analytical speciation techniques for uranium and related elements

Raman microspectrometric investigations of UO$_2$ alteration products: Natural analogues to alpha radiolysis effects

Marcus Amme

European Commission, Joint Research Centre, Institute for Transuranium Elements, Postfach 2340, D-76125   Karlsruhe, Germany

Abstract. Disposed spent nuclear fuel can be seen as an artificial counterpart to natural uranium deposits. For investigating safe storage options of this material, its geochemical alterations are examined. Secondary phases which formed during leaching alteration of UO$_2$ surfaces were investigated with Raman microspectrometry and characteristic vibrational spectra of the materials were recorded. U peroxide (UO$_4$) was identified by comparison with a natural studtite sample as the main alteration phase by its characteristic O-O Raman active vibration at 870 cm^{-1}.

Introduction

Directly disposed High-Level Nuclear Waste (like, spent fuel) which is foreseen to be finally disposed in deep geological formations (which is the case in many countries) consists of a variety of fission and activation products, long-lived actinides, and stable isotopes which are present in the matrix of ^{238}UO$_2$ material. The containment stability cannot be guaranteed for geologic timescales and knowledge of the chemistry of the solid-liquid interface, which controls the radionuclide release under these conditions, is important to assess a possible release situation.

Water radiolysis products, which are generated in the liquid phase by strong alpha radiation fields, can alter the redox potential of the solution and lead to the formation of secondary phases. Studies performed with UO$_2$ material under unsaturated test conditions (that is, a small liquid volume contacting a relatively big surface) and in oxic groundwater at 90 °C showed a preferential formation of schoepite during the first two years of the experiment, followed by the formation of uranyl silicates like uranophane and boltwoodite (Wronkiewicz et al. 1992). A more complex paragenetic sequence is observed when spent fuel is used for the experiments. Due to the presence of fission products, alteration phases were found to include rare earth elements (REE) like Ce, Sm, and Nd (Finn et al. 1994).

Uranyl oxihydroxide phases containing Cs and Mo were found in similar investigations (Buck et al. 1997). It was not reported if radiolysis of the contacting water, induced by the strong radiation field present around fresh spent nuclear fuel, changed the alteration phase chemistry. A clear influence of water radiolysis products like H_2O_2 upon the surface development of leached nuclear fuel samples was observed in the case of simulated radiolysis experiments. Samples of UO_2 material, which were exposed to a high dose of alpha particles at the solid/liquid interface using a helium ion beam during the leaching experiment developed a layer of uranium peroxide (UO_4, Meta-Studtite) on the surface (Sattonay et al. 2001). Studies with natural analogue materials let conclude that alpha radiolysis phenomena can govern natural U mineral formation in some cases (Hofmann 1989)

Knowledge of the surface composition of finally disposed spent fuel is the base for solubility and therefore safety considerations. To provide this data, a method has to be found which can determine the complex phase assembly of relatively similar compounds which forms on the surface of the material under final repository conditions, and a spectral database for quick and unambiguous identification would prove useful.

The dissolution chemistry of actinides is strongly dependent on the prevailing redox conditions. In the case of uranium, in solid U phases the oxidation states +IV and +VI are encountered. Most investigations done so far were performed in oxic media and the redox state of resulting secondary products was therefore limited to +6. When U(IV) phases are also present (like it is considered to be possible in the case of spent fuel dissolving under repository vault conditions), a solid analytical method, which can differentiate between U redox states, proves versatile. Raman vibrational spectroscopy can be used to record characteristic spectra of substances which possess unique vibrational bands characteristic for bonds in molecules or crystal lattices. The determination of oxidation states and the differentiation between chemically similar compounds is possible with this method. Furthermore, the technique can be used for microscale determinations , using a microscope measurement accessory. Other advantages are that most samples need no special preparation, no vacuum has to be applied (which might lead to alteration or destruction of the sample). The characterisation of some natural (mineral) and synthetic U compounds (of very similar composition) with Raman spectroscopy and the derivation of their structures using the obtained information was described (Biwer et al. 1990;Hoekstra and Siegel 1973;Sobry 1973). However, not all of the possible alteration phases resulting from the spent fuel corrosion were measured so far.

Modern spectral imaging techniques have broadened the field of applications in materials sciences. Since personal and laboratory computers are able to process great amounts of measured data, methods using spatially resolving spectrometrical measurements (generally known as „mapping") were established for materials investigations. In an combined approach of chemical imaging spectroscopy and optical microscopy (using also IR/Raman and EDX spectroscopy), the determination of the Pu bonding state in heterogeneous environmental samples by functional group analysis was demonstrated (Schoonover et al 1998, 2000).

Experimental

Corrosion experiment

The formation of secondary phases on the surfaces of nuclear fuel samples was induced with leaching experiments using de-ionized water as liquid phase which was spiked with concentrations of the water radiolysis product H$_2$O$_2$ in several concentration steps. The used concentrations of H$_2$O$_2$ (10^{-5}, 10^{-4}, 10^{-3}, 10^{-2} mol/l) were selected in agreement with the amounts of H$_2$O$_2$ which were measured in water previously irradiated with alpha particles (Sunder et al. 1997). The waters were spiked with diluted 30% H$_2$O$_2$ solution (Merck, p. a. grade) and contacted with pellets of UO$_2$ under argon atmosphere. The experiment was ended after 14 days of duration. In the case of the high oxidant concentrations, a yellow alteration layer was visible macroscopically on the originally black sample after 48 hours. After ending the experiment, the solid and the liquid phase were separated by filtration with 450 nm filters.

Synthesis of UO$_2$ alteration products (uranium minerals)

Several uranium minerals were identified as possible alteration products which can form on UO$_2$ surfaces in previous work (Wronkiewicz et al 1996). During the oxidative dissolution of nuclear fuel, it was found in long-term experiments that the original matrix substance, UO$_2$ is changed into different alteration products which appear in certain phases in the timescale of the alteration process, which is a typical indicator for kinetically controlled formation processes. The U minerals compiled in table 1 were found to appear frequently in these alteration sequences (Wronkiewicz et al 1996) and were therefore chosen for setting up a spectral database for the identification of nuclear fuel corrosion products.

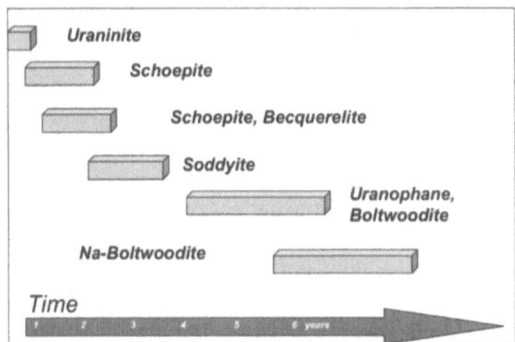

Fig. 1. Sequence of paragenetic mineral formation during (experimentally observed) oxidative alteration of spent nuclear fuel

Analytical speciation techniques for uranium and related elements

Table 1. Uranium minerals which were observed to form as alteration products of UO_2 corrosion during experiments simulating environmental conditions

Mineral	Formula	Main Raman wavenumber measured in this work (cm^{-1})	Reference for synthesis
Becquerelite	$Ca[(UO_2)_6O_4(OH)_6]\cdot 8H_2O$	828	(Protas 1959)
Schoepite	$(UO_2)(OH)_2\cdot x\ H_2O$	843 / 855	(Sandino and Grambow 1994)
Uranophane	$Ca(UO_2)SiO_3(OH)_2\cdot 5H_2O$	793	Natural sample
Na-Boltwoodite	$(H_2O)Na(UO_2)(SiO_4)$	839	(Vochten et al 1997)
Soddyite	$(UO_2)_2SiO_4\cdot 2H_2O$	830	(Moll et al. 1995)
Studtite	$UO_4\ x\ H_2O$	820 / 870	Natural samples
Uraninite	UO_2	445	No special synthesis

Since the groundwaters, which would contact spent nuclear fuel after possible container corrosion and failure in the repository leakage scenario, contain a variety of ions, mixed phases of uranium and ions (like, e.g., Ca, K, Na, Si, CO_3^{2-} and others) must be taken into account to appear in the UO_2 alteration sequence. Numerous minerals can be formed during these processes (Gaines et al 1997;Robault 1960). In this investigation, only phases which contain U, O and H were necessary for spectral identification of alteration products, since the experiments were conducted in de-ionized water and only hydrogen peroxide was added. However, several of the synthesized minerals will comprise the spectral library necessary for future investigations on samples corroded in groundwater.

Synthesis of the solids was performed by following the procedures reported by several authors. Natural samples of studtite (uranium peroxide) and uranophane from the abandoned uranium mine Menzenschwand, Germany (Black Forest) were used to complete the set of uranium standards. About 10 mg of suitable, wellgrown micromounts were removed from the supporting matrix, pulverized and analyzed for the impurity content with SEM-EDX prior to the experiments. The analysis proved a sufficient purity of the minerals.

NIR and VIS laser Raman spectroscopy measurements

Samples of the synthetic U(VI) phases were measured with a Bruker RFS 100/S Fourier Transform (FT) Raman spectrometer with attached Raman microscope (Nikon Optiphot-2 microscope) using an air-cooled near infrared (NIR) Nd:YAG laser with a wavelength of 1064 nm. Several mg of the materials were collected into a capillary glass tube and irradiated with laser light which had a power of 150 mW. The scattered light was collected with a proprietary high-sensitivity Ge diode (cooled with liquid nitrogen). For an average measurement, 128 scans were collected.

Samples of U(IV) phases and, for controlling the method, also U(VI) minerals, were measured with a Dilor XY Raman microspectrometer with attached microscope using an Ar laser with a wavelength of 514 nm. Few mg of the materials were placed on an object slide under the Raman microscope and irradiated with laser light. In the case of corroded pellet samples, a suitable location was selected on the sample surface. For the U(VI) materials, an energy of 0,2 mW was used. The dark U(IV) materials absorb laser light much stronger and were irradiated with an energy of 20 mW in order to enhance the yield of the scattered light. Spectra were recorded with an liquid N$_2$ cooled Silica detector (512 pixels).

Results and Discussion

During the experiment, the solid samples were occasionally examined optically and showed a different behaviour. Samples containing the highest concentration of oxidant showed a pale yellow discolourations on their surfaces which is interpreted with the formation of a layer or coating. The samples which were contacted with solutions containing 10^{-3} M H$_2$O$_2$ showed the same behaviour, but the alterations were much less visible. All other samples showed no visible alterations.

All corrosion samples were examined with EDX analysis for the elemental composition. No other elements besides uranium were found to be present.

Compilation of a spectral library for uranium phases: measurement with VIS and NIR laser Raman spectrometers

Spectra of the synthesized uranium compounds were recorded in order to set up the Raman spectral library before examining the unidentified alteration materials on the corroded samples. Measurements of the synthetic phases were performed with a Near Infrared (NIR) and a Visible Light (VIS) laser. In the case of the NIR-FT Raman measurements, the U(VI) compounds (mostly yellow in colour) yielded spectra with good scatter intensities, due to their high reflectivity. The darkly coloured U(IV) compound (UO$_2$) absorbed the NIR light very strongly which resulted in spectra of poor quality. The bad signal-to-noise ratio of their NIR spectra made an evaluation impossible.

The uranium(VI) compounds delivered spectra which show repeatedly similar features. In each case, a signal is present in the vicinity of 800 cm^{-1}, resulting from the symmetric axial stretch vibration of the uranyl ion. This signal varies in its position in the spectrum, depending from the bonding situation of the uranyl group in the solid (Weidlein et al 1988). A detailed analysis of recorded spectra and a prediction of expected spectra could be performed by the means of a symmetry group analysis. We dispensed with such a theoretical analysis here since the identification of unknown substances by spectra comparison is sufficient for qualitative speciation.

Dark samples were measured with a Ar laser (514 cm^{-1}) to avoid the strong absorption effect of dark material in IR light. Reasonable spectra could be obtained by using this technique. No significant fluorescence disturbed the recording of the spectra. In a first step, the capability of the method to obtain the desired information was tested by measuring a powdered sample of the matrix material UO_2. The resulting spectrum had an intensity sufficient for further evaluation; it showed the typical U-O signals at the position reported by former investigators (Allen et al 1987;Butler et al 1988). The peaks found for the UO_2 oxidation product U_3O_8 between 700 and 800 cm^{-1} could not be detected.

In addition to the dark samples, some of the coloured U(VI) phases were measured with the Ar laser Raman equipment to have direct comparison of the two systems using different measuring wavelengths. Schoepite was measured with the 514 nm equipment to have a comparison of the results delivered by the two methods. The Raman spectrum obtained by this method delivered the major signal at 843 cm^{-1} and hence shows agreement with the NIR result (see Table 1).

Measurement of corroded samples and comparison with synthetic standard spectra

Samples corroded in solutions containing 10^{-3} and 10^{-2} M H_2O_2 were found to be coated by optically visible corrosion layers and were therefore the first specimen selected for Raman microspectrometric measurements.

Spectra which were collected from the unaltered matrix material on the selected samples showed similar characteristic features. In all these spectra, the U-O vibration signal at 450 cm^{-1} was present, identifying the material as UO_2 (Allen et al 1987). Spectra from the yellow areas contained a strong signal at 817 cm^{-1}, typical for the uranyl U-O vibration. The spectra show also a vibration appearing at 870 cm^{-1} which is not present in any of the synthetic standard materials. A comparison with the Raman spectrum of potassium peroxide shows that the O-O vibration of the peroxide group of this substance appears at 890 cm^{-1} (Weidlein et al 1988). It is therefore assumed that the signal found at 870 cm^{-1} in the spectra measured here has its origin in the O-O vibration of uranium peroxide (which has the half-structural formula $UO_2 \cdot O_2$).

The Raman spectrum of studtite can hence be recognized uniquely from that of other uranyl compounds. The spectrum of the corrosion phase shows clear differences from the spectrum of schoepite (which is often observed in nuclear fuel corrosion experiments as an alteration phase (Matzke 1996)), as is shown in Fig. 6. The Raman spectrum which was recorded from a natural sample of studtite (Fig. 5) shows two major signals at 823 and at 869 cm^{-1}, thereby identifying the corrosion layer on the surface of the experimental specimen as studtite phase. A slightly broader signal of the U-O vibration in the spectrum of the natural sample is possibly caused by the presence of further uranyl phases in minor quantity. Some characteristic wave numbers are listed in Table 1; a complete summary of the characteristic wave numbers measured for the substances used in this work and comparison with Raman and IR reference data will be given in an upcoming publication.

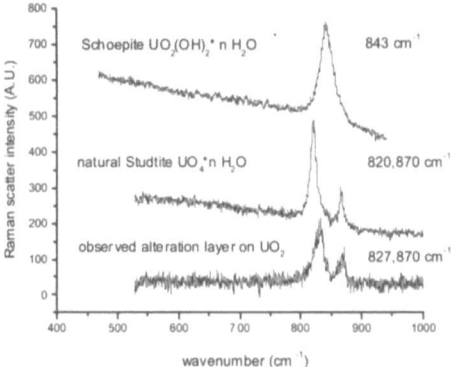

Fig. 2. Comparison of the Raman spectra of the corroded sample surface and the synthetic standards (compounds containing U, O and H). The single peak of the U-O vibration of schoepite cannot be localized in the surface spectrum, the U-O and O-O vibrations of studtite are easily to be identified.

Considerations concerning the geochemical alteration of spent nuclear fuel

The presence of the oxidant hydrogen peroxide directs the corrosion of nuclear fuel material into directions which are different from the normal pathways of oxidative dissolution. The formation of uranium peroxide was observed under laboratory conditions during radiolysis experiments (Sattonay et al 2001), as well as having taken place in nature (Gaines et al 1997; Hofmann 1989). In the latter case, it can be considered to be a relatively rare phenomenon of uraninite alteration, during which Ra, which was formed by U decay and accumulated in Ba minerals, leads to water radiolysis. Raman microspectroscopy constitutes a method which can deliver information about the bonding situation in substances which are composed of the same elements (completing SEM-EDX information); it is applicable for amorphous thin layers as they were observed in the experiments performed here (enhancing the information gathered with XRD); and it does not necessarily change the sample environment during the measurement (therefore avoiding some effects of XPS measurements, e.g., the alteration of samples by a vacuum).

Fission product elements and transuranic actinides are likely to take part in the alteration chemistry of the solid/liquid interface of disposed nuclear fuel after an assumed contact with water. They are present in freshly discharged spent nuclear fuel in a variety of crystalline, amorphous and metallic phases (Kleykamp 1985). The formation of alteration phases on the surface of the UO$_2$ matrix might partly be controlled by the solid phase chemistry of these elements, either by their incorporation into uranium secondary minerals or by separate phase formation. The possibility of a microscopic, highly and spatially resolving scanning application, combined with the software-based identification of the numerous possible fission product secondary materials using a spectral library, seems especially promising for the further investigation of these complex, heterogeneous chemical systems. Future investigations are projected into that direction.

Conclusions

The findings in our work let conclude the following statements:

1. Uranium peroxide (UO_4) is the major phase present on samples of the simulated radiolysis experiment. Its Raman spectrum is almost identical with that of natural studtite (which was most obviously formed by natural water radiolysis). Almost no schoepite was detected on the surface of the specimen .
2. The alteration material is clearly distinguishable from the UO_2 matrix by its features in the Raman spectrum. The Raman microprobe setup allows separate detection of corrosion particles and matrix material which are associated in an area of square-micron dimensions.
3. The examined material can be specified from other very similar U(VI) compounds, since every uranyl compound shows the U-O stretch signal at a unique position in the spectrum. This finding is important for the solid speciation of samples which were generated in an environment with complex chemistry (e.g., corrosion in the presence of groundwater and mineral phases).
4. Even if little facts are known about the long-term stability of U peroxide minerals (which are the only naturally occurring peroxide minerals known) under geologic conditions, observations on natural analogue samples let conclude that these phases will, once formed, be stable for long time.

Acknowledgements

The author acknowledges first of all highly the contribution of Mr. M. Feth and Mr. B. Schmid (University Stuttgart) for their participation in the measurements and evaluation of this work; therefore, they are named as co-authors in regular journal publications. Highly acknowledged is similarly the help of Mr. B. Renker (Research Center Karlsruhe, Institute of solid state physics), without whom many measurements would not have been possible. Dr. Ian Ray, Dr. Thierry Wiss and Mr. Helmut Thiele for help with the SEM imaging technique.

References

Allen G, Butler I, et al (1987) Characterisation of uranium oxides by micro-Raman spectroscopy, J Nucl Mat 144, 17 – 19

Amme M Uranium secondary phase formation during anoxic hydrothermal leaching processes of UO2 nuclear fuel, Geochim Cosmochim Acta (in press)

Biwer B., Ebert W, et al (1990) The Raman spectra of several uranyl-containing minerals using a microprobe, J Nucl Mat 175, 188 – 193

Buck E, Wronkiewicz D et al (1997) A new uranyl oxide hydrate phase derived from spent fuel alteration, J Nucl Mat 249, 70 – 76

Burns P Finch R (Eds) (1999) Uranium: Mineralogy Geochemistry and the environment, Reviews in mineralogy Volume 38 Mineralogical Society of America Washington D C

Butler I, Allen G, et al (1988) Micro-Raman spectrum of triuranium octoxide, Applied Spectroscopy Vol 42, Number 5, 901 – 902

Cejka J Sejkora J Deliens M (1996) Neues JB Miner Mh 3 125-134

Finn P, Bates J et al (1994) Elements present in leach solutions from unsaturated spent fuel tests, Mat Res Soc Symp Proc Vol 333, 399 – 407

Fuchs L, Hoekstra H (1959) The preparation and properties of Uranium(IV)silicate, The American Mineralogist Vol 44, 1057 – 1063)

Gaines R V, Skinner H C W, Foord E E, and Rosenzweig A (1997) Dana's New Mineralogy, 8th Edition John Wiley & Sons, New York

Hoekstra H, Siegel S (1973) The Uranium trioxide – Water system, J Inorg Nucl Chem Vol 35, 761 – 779

Hofmann B (1989), Genese, Alteration und rezentes Fliess-System der Uranlagerstätte Krunkelbach (Menzenschwand, Südschwarzwald), Nagra Technischer Bericht 88-30

Janeczek J Ewing R et al (1996) Uraninite and UO2 in spent nuclear fuel: a comparison, J Nucl Mat 238 121 – 130

Keller C (1964) Über die Festkörperchemie der Actiniden-Oxide, KfK-Rep No 225, Kernforschungszentrum Karlsruhe

Kleykamp H (1985) The chemical state of fission products in oxide fuels, J Nucl Mat 131 221 – 246

Matzke Hj (1996) Analysis of the structure of layers on UO2 leached in H2O, J Nucl Mat 238, 58 – 63

Maya L Begun GM J (1981) Inorg Chem 43 2827-2832

Moll H, Matz W et al (1995) Synthesis and characterisation of Uranyl Orthosilicate, J Nucl Mat 227, 40 – 49

Peters J M (1967), Syntheses et etude radiocristallographique d'uranates synthetiques du type oxyde double d'uranyle, Mem Soc Roy Sci Liege [5] 14, 5 – 57

Protas J (1959), Contributiona létude des oxydes dúranium hydrates, Bull Soc Franc Mineral Crist 82, 239 – 249

Roubault M(Ed) (1960) Les minerais uraniferes francais et leurs gisement, Saclay

Sandino A, Grambow B (1994) Solubility equilibria in the U(VI)-Ca-K-Cl-H2O system: Transformation of Schoepite into Becquerelite and Compreignacite, Radiochimica Acta 66/67, 37 – 43

Sangaletti L, Depero L, et al (1998) Oxidation of Sn thin films to SnO2 – micro-Raman mapping and x-ray diffraction studies, Journal of Materials research Vol 13, No 9, 2457 – 2460

Analytical speciation techniques for uranium and related elements

Sattonnay G, Ardois C et al (2001) Alpha-radiolysis effects on UO2 alteration in water, J Nucl Mat 288, 11 – 19

Schoonover J Saab A, et al (2000) Raman/SEM chemical imaging of a residual gallium phase in a mixed oxide feed surrogate, Appl Spectroscopy Vol 54, No 9, 1362 - 1371

Schoonover J, Weesner F et al (1998) Integration of elemental and molecular imaging to characterize heterogenous inorganic materials, Applied Spectroscopy Vol 52, No 12, 1505 – 1514

Smith D Scheetz B et al (1982) Phase relations in the uranium-oxygen-water system and its significance on the stability of nuclear waste forms, Uranium 1 79 – 110

Sobry R (1973) Etude des „Uranates" hydrates – II Examen des proprietes vibrationelles des Uranates hydrates de cations bivalents, J Inorg Nucl Chem Vol 35, 2753 - 2768

Sunder S, Shoesmith D et al (1997), Oxidation and dissolution of nuclear fuel (UO2) by the products of alpha radiolysis in water, J Nucl Mat 244, 66 – 74

Vochten R, Blaton N, Peters O, van Springel, K, van Haverbeke L (1997) , A new method of synthesis of boltwoodite and of formation of sodium boltwoodite, uranophane, sklodowskite, and kasolite from boltwoodite, Can Mineral 35, 735 – 741

Weidlein, J, Müller U, Dehnicke K (1988) Schwingungsspektroskopie, Georg Thieme, Stuttgart

Wronkiewicz D, Bates J et al (1996) Ten-year results from unsaturated drip tests with UO2 at 90 ° C: implications for the corrosion of spent nuclear fuel, J Nucl Mat 238, 78 - 95

Wronkiewicz D, Bates J, et al (1992) Uranium release and secondary phase formation during unsaturated testing of UO2 at 90 ° C, J Nucl Mat 190, 107 – 127

Sorption of Uranium(VI) onto Schwertmannite - EXAFS investigations

Marcus Walter[1], Thuro Arnold[1], Harald Funke[1], Tobias Reich[1, 2], Gert Bernhard[1]

[1] Forschungszentrum Rossendorf, Institut für Radiochemie, (FZR, IRC), P.O. Box 510119, D–01314 Dresden, Germany
[2] Johannes Gutenberg-Universität Mainz, Institut für Kernchemie Fritz-Straßmann-Weg 2, 55128 Mainz, Germany

Abstract. In acid mine drainage conditions uranium(VI) adsorbs on goethite and schwertmannite, forming surface complexes with the ferric iron polyhedra and sulfate, respectively. The structure of these surface species was identified by EXAFS.

Introduction

The ferric oxide schwertmannite is commonly formed in acid mine drainage waters (pH-range from 2.8 to 4.5) in the presence of high sulfate concentrations (Cornell and Schwertmann 1996). Because of its large specific surface area it may influence the migration of uranium(VI) by sorption processes. By mixing the acid sulfate-rich leachate of the uranium mine Königstein (Saxony) with ground water schwertmannite, as identified by IR–spectroscopy, was precipitated and uranium as a part of the acid sulfate–rich leachate was sorbed on the formed precipitates (Jenk 1996). The aim of this study was to get structural information about the formed uranium(VI) surface complexes by Extended X-ray Absorption Fine Structure (EXAFS) spectroscopy. We include Goethite in our investigations as a reference to compare and interpret the schwertmannite data.

Experimental

Samples for EXAFS analysis (Table 1) were prepared under N_2 atmosphere using 200 mg of schwertmannite or goethite with initial uranium(VI) concentrations of 1×10^{-5} (pH 6.5) to 5×10^{-5} M (pH 4.2). The ionic strength was adjusted to 0.02 N Na_2SO_4 or 0.01 N $NaClO_4$, respectively. Uranium L_{III} edge EXAFS spectra were recorded at room temperature in fluorescence mode at the Rossendorf beamline

(ROBL) at the ESRF in Grenoble. The measured EXAFS oscillations were fitted using EXAFSPAK (George and Pickering 1995) procedure. The theoretical phase shifts and backscattering amplitudes were calculated with FEFF8 (Ankudinov et al. 1998). Following Hudson et al. (1996), only the 4 legged transdioxo multiple scattering path (U - O_{ax} - U - O'_{ax} - U) was included in the fit.

Table 1. Geochemical characterization of the EXAFS samples

U(VI) sample	pH	Ionic strength	U(VI) Concentration in Solution after Sorption	U(VI) Loading on the Sorption Sample
schwertmannite	4.2	0.02 N Na_2SO_4	4.5×10^{-5} M	1,300 mg/kg
schwertmannite	4.2	0.01 N $NaClO_4$	4.0×10^{-5} M	6,900 mg/kg
goethite	4.2	0.02 N Na_2SO_4	4.2×10^{-5} M	6,600 mg/kg
goethite	6.5	0.02 N Na_2SO_4	1.3×10^{-8} M	22,600 mg/kg

Results and Discussion

The structural parameters resulting from the final fit of the EXAFS data (Fig. 1) are summarized in Table 2. All distances (R) of the equatorial oxygens are about 2.37 Å. These are significantly shorter than typical values of 2.41 Å for the aquatic UO_2^{2+} ion (Dent et al. 1992), which indicates strong interactions. The fitted coordination numbers (N) of the equatorial oxygens are between 3.8 and 6.8 (Table 2). The short k–range of the data and the high correlation of coordination number and the Debye-Waller factor (σ^2) may effect this variation. However, the distances of the equatorial oxygens are in good agreement with mean distances of bipentagonal coordinated uranium(VI) in crystal structures (Burns et al. 1997).

An iron backscatterer was clearly found for uranium(VI) sorbed onto goethite at pH 4.2 and 6.5 in a sulfate-rich solution as well as for schwertmannite at pH 4.2 in perchlorate solution. Such a uranium - iron distance is indicative of a bidentate inner-sphere complexation (Reich et al. 1998). In contrast, the EXAFS of uranium(VI) sorbed onto schwertmannite at pH 4.2 in sulfate-rich solutions shows only a sulfur backscatterer at a distance of 3.66 Å, indicating a monodentate uranium(VI) - sulfate coordination (Moll et al. 2000). This could be interpreted either as uranium(VI) surface complexes with the structural sulfate of the schwertmannite lattice ($\equiv$SO–UO_2^+, $\equiv$(SO)$_2$=UO_2) or as a ternary surface complex ($\equiv$FeO–SO$_4$–UO_2). A light backscatterer, like carbon, at a distance of 2.9 Å improves the fit. Although this uranium(VI) - carbon interaction is typical of a bidendate complexation (Bargar et al. 2000), ternary uranium(VI) - carbonate complexes are ruled out by the sorption conditions (pH, high sulfate concentration).

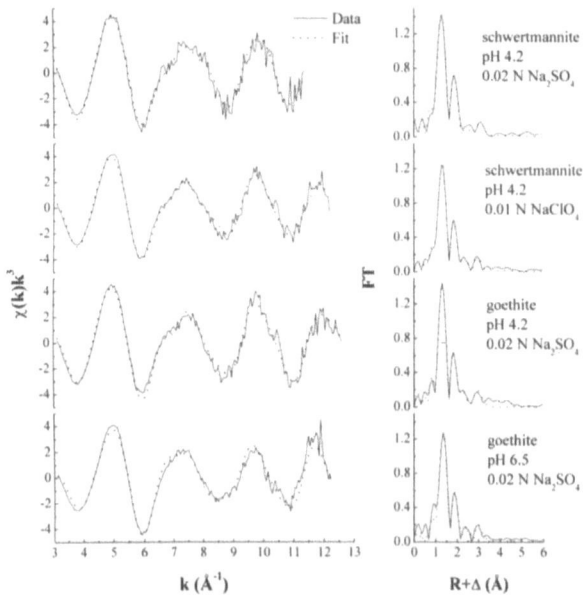

Fig. 1. Uranium L_{III} edge EXAFS and related Fourier Transforms of the uranium(VI) sorption samples.

Table 2. Structural parameters of the uranium L_{III} edge EXAFS samples

U(VI) sample	Shell	N	R (Å)	σ^2 (Å^2)	ΔE_0 (eV)
schwertmannite	O_{ax}	2.3	1.76	0.0029	-14
pH 4.2	O_{eq}	5.6	2.38	0.011	
0.02 N Na$_2$SO$_4$	S	1.3^f	3.66	0.005	
schwertmannite	O_{ax}	2.1	1.77	0.0029	-15
pH 4.2	O_{eq}	4.7	2.37	0.012	
0.01 N NaClO$_4$	Fe	0.7^f	3.46	0.008	
goethite	O_{ax}	1.9	1.76	0.0012	-16
pH 4.2	O_{eq}	6.8	2.36	0.016	
0.02 N Na$_2$SO$_4$	Fe	0.6^f	3.44	0.045	
goethite	O_{ax}	2^f	1.79	0.0022	-12
pH 6.5	O_{eq}	3.8	2.36	0.010	
0.02 N Na$_2$SO$_4$	Fe	0.4	3.43	0.004^f	

^{f} Parameter was fixed during the final fit.
$\Delta N = \pm 25 \%$, $\Delta R = 0.01$ Å

References

Ankudinov A L, Ravel B, Rehr J J, Conradson S D (1998) Real–space multiple–scattering calculations and interpretation of x–ray–absorption near–edge structure. Phys. Rev. B 58: 7565-7576

Bargar J R, Reitmeyer R, Lenhart J J, Davis J A (2000) Characterization of U(VI) – carbonato ternary complexes on hematite: EXAFS an electrophoretic mobility measurements. Geochim. Cosmochim. Acta 64: 2737-2749

Burns P C, Ewing R C, Hawthorne F C (1997) The crystal chemistry of hexavalent uranium: Polyhedron geometries, bond-valence parameters, and polymerisation of polyhedra. Can. Mineral. 35: 1551-1570

Cornell R M, Schwertmann U (1996) The iron oxides. VCH Verlag, Weinheim

Dent A J, Ramsay J D, Swanton S W (1992) An EXAFS study of uranyl ion in solution ans sorbed onto silica and montmorillonite clay colloids. J. Colloid Interface Sci. 150: 45-60

George G N, Pickering I J (1995) EXAFSPAK: A suite of computer programs for analysis of x–ray absorption spectra. Standford Synchrotron Radiation Laboratory. Stanford

Hudson E A, Allen P G, Terminello L J, Denecke M A, Reich T (1996) Polarized x-ray-adsoption spectroscopy of the uranyl ion: Comparsion of experiment and theory. Phys. Rev. B 54: 156-165

Jenk U. (1995) Untersuchungen zu den chemischen und geochemischen Verhältnissen in der Uranlagerstätte Königstein nach Einstellung der Urangewinnung durch schwefelsaure Untertagelaugung. Univ. Diss., Freiberg.

Moll H, Reich T, Hennig C., Roßberg A, Szabó Z, Grenthe I (2000) Solution coordination chemistry of uranium in the binary UO22+ - SO42- and the ternary UO22+ - SO42- - OH- system. Radiochim. Acta 88: 559-566

Reich T, Moll H, Arnold T, Denecke M A, Hennig C, Geipel G, Bernhard G, Nitsche H, Allen P G, Bucher J J, Edelstein N M, Shuh D K (1998) An EXAFS study of uranium(VI) sorption onto silica gel and ferrihydrite J. Elec. Spec. Rel. Phenom. 96: 237-243

Colloid-borne Uranium in Mine Waters

Harald Zänker, Wolfgang Richter, Gudrun Hüttig, Henry Moll

Forschungszentrum Rossendorf, Institute of Radiochemistry, Postfach 510119, D-01314 Dresden, Germany

Abstract. High concentrations (≥ 1 g/l) of ultra fine Fe oxyhydroxy sulphate colloids (< 5 nm) were found in acid rock drainage solutions. The near-neutral "bulk waters" of mines such as tunnel waters, on the other hand, contain lower concentrations (about 1 mg/l) of particles of 100 to 300 nm in size which consist of Fe and Al compounds. The uranium(VI) is normally not colloid-borne in these two water types, either due to the high acidity (acid rock drainage) or due to the formation of dissolved carbonato complexes („bulk waters"). However, most of the U(VI) is colloid-borne in the pH range 4 to 6, a pH range that must be passed through during the flooding of a mine. Scavenging by particles and colloid aggregation and sedimentation can significantly immobilize the U(VI) in this pH range (natural attenuation).

Introduction

There is ample evidence from groundwaters, river waters, lakes or seawater that colloids influence the transport behaviour of contaminants such as As, Pb, Cu, Zn, actinides, fission products, hydrophobic organics etc. Little effort has been made to elucidate the role of colloids in mine waters. We studied mine waters of very different nature for their colloid-chemistry. Emphasis was placed on the question if colloid-borne forms of uranium occur.

Acid Rock Drainage (ARD)

We investigated acid rock drainage solution, so called "Lettenwasser" (pH 2.7, sulphate concentration 411 mmol/l, Fe concentration 93,5 mmol/l), from the Himmelfahrt Fundgrube mine at Freiberg by photon correlation spectroscopy (PCS),

(PCS), centrifugation, filtration, ultrafiltration, scanning electron microscopy (SEM), energy-dispersive X-ray analysis (EDX), ICP-MS, AAS, ion chromatography, TOC analysis and X-ray absorption fine structure (EXAFS) spectroscopy. While filtration through Nuclepore filters of pore sizes of < 50 nm and centrifugation at centrifugal accelerations of up to 40 000 x g were only able to remove a tiny trace (< 20 mg/l) of larger submicron particles, PCS and ultrafiltration revealed the presence of large amounts of ultrafine colloidal particles. A concentration of these ultrafine colloids of ≥ 1 g/l was found. The prevailing particle size was < 5 nm. Iron, arsenic and lead were the metal constituents of the ultrafine particles. Uranium was not colloid-borne. Fig.1 demonstrates the successive unmasking of the ultrafine particles by filtration and their detection by PCS (see also Zänker et al. 2002).

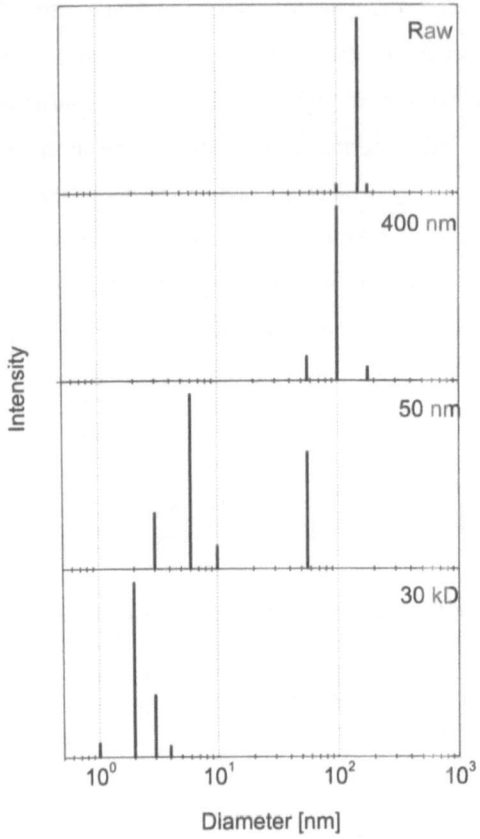

Fig. 1. Light-intensity weighed particle size distribution in the ARD sample according to the CONTIN deconvolution of the autocorrelation functions. Raw sample, 400-nm filtrate, 50-nm filtrate, 30-kD filtrate. Removing the submicron particles results in the appearance of the weakly scattering ultrafine particles (< 5 nm).

Analytical speciation techniques for uranium and related elements

We were not only interested in the size and the chemical composition of the colloid particles but also in their mineralogy and the type of binding of toxic trace elements onto them. X-ray diffractometry (XRD), the "classical" technique of mineral identification, fails with the ultrafine particles: the particles are that small that they appear as "amorphous" to XRD. The method of choice for the investigation of near-order structures of such small objects is X-ray absorption spectroscopy (XAS) with synchrotron radiation, in particular EXAFS spectroscopy. According to EXAFS spectroscopy, the most probable mineralogical composition of the particles is a mixture of hydronium jarosite, $HFe_3(SO_4)_2(OH)_6$, and schwertmannite, ideally $Fe_8O_8(OH)_6SO_4$ (cf. Zänker et al. 2002). A small amount of a relatively coarse precipitate was formed in the ARD solution during the months after sampling. The colloid particles are obviously an intermediate in the precipitate formation process. Our results suggest that the arsenate is bound to the colloids by the formation of a bidentate binuclear inner-sphere surface complex. However, the transformation of the colloidal material to the more aggregated long-term precipitate results in the incorporation of the arsenate into the interior of the iron hydroxy sulphate crystal structures. Pb seems to occur as anglesite.

Bulk Waters

Quite a different type of colloids was found in the oxic "bulk water" of mines as for instance in tunnel waters. The water of the drainage gallery "Rothschönberger Stolln" at Freiberg (pH 7.2, sulphate concentration 1.75 mmol/l, Fe concentration 0.015 mmol/l) was in investigated for its colloid inventories. About 1 mg/l of colloid particles of 100 to 300 nm were found in this adit water. The zeta potential was -10 mV. The particles consist of a matrix of Fe and Al compounds (probably oxyhydroxides). They seem to be formed when anoxic slightly acidic shaft waters mix with oxic near-neutral gallery water. The colloid particles bear toxic trace elements such as As, Pb, and Cu. Almost 100 % of the As and Pb and about 70 % of the Cu are colloid-borne. Dissolved carbonato complexes prevent the uranyl from being adsorbed on the colloids in the unaltered adit water. Decreasing the pH destroys these complexes: up to 50 mass percent of the uranium is attached to the colloids after acidification with HNO_3 to pHs of 4 to 6. Further acidification converts the uranyl again to a 'non-colloidal' form (cf. Zänker et al. 2000). Fig. 2 shows an SEM micrograph as well as an EDX spectrum of typical colloid particles from the "Rothschönberger Stolln" adit water.

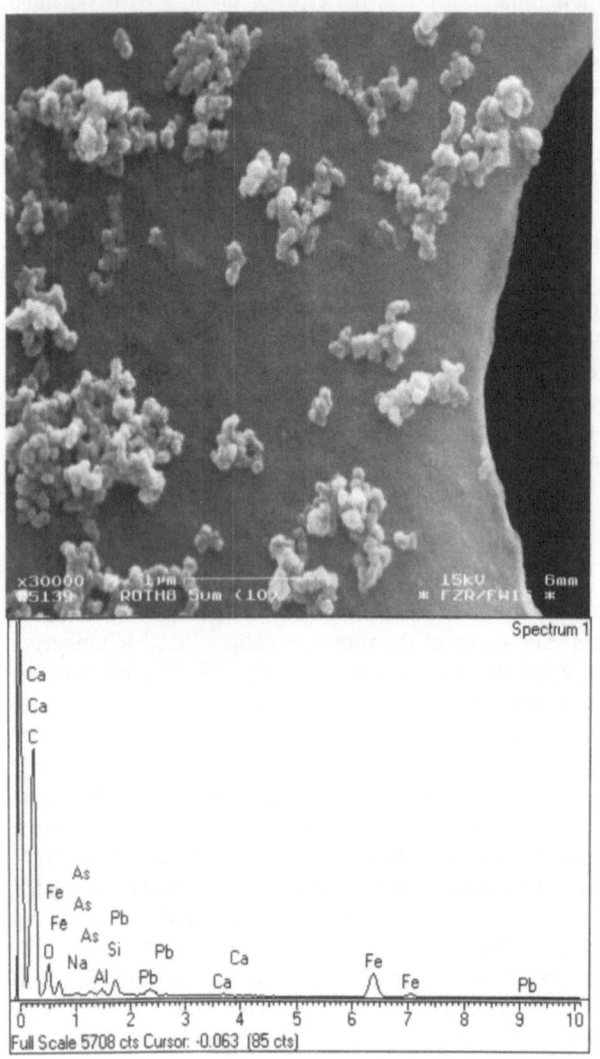

Fig. 2. SEM micrograph (coating with Au) and typical EDX spectrum (coating with C) of colloid particles from the "Rothschönberger Stolln" adit on a 5-µm Nuclepore filter. Filter cake washed three times with Milli-Q water, dried and coated with gold. Magnification: 30 000 : 1. Length of the scale bar: 1 µm. Particles of about 100 nm are visible. Fe, Al, As and Pb are the main constituents of the particles. The high carbon peak in the EDX spectrum is caused by the filter material and thus an artefact.

Analytical speciation techniques for uranium and related elements

Flooding Waters

Acid rock drainage solutions are more or less identical with acidic pore waters. The flooding of a mine can be regarded as the dilution of acidic pore waters to form "bulk water"; flooding waters represent the transition between the two states. We investigated flooding water from the Königstein uranium mine by light scattering, centrifugation, filtration, ultrafiltration, scanning electron microscopy, laser Doppler electrophoresis, ICP-MS, AAS, ion chromatography and diverse radiometric methods. The waters had a pH of 5.6 and an Eh of 450 mV, the concentrations of important constituents were CCa 0.9 mmol/l, CFe 0.3 mmol/l, CU 0.05 mmol/l, Csulfate 1.2 mmol/l, Ccarbonate 1,0 mmol/l and CO_2 0.11 mmol/l. Almost 90 % of the iron was divalent. Because of the Eh value we assume that the uranium was hexavalent. Colloid concentrations in the range of 2 to 3 mg/L were found; they rose rapidly if the access of oxygen was allowed which is attributable to the oxidation of Fe^{2+} and the hydrolysis of Fe^{3+}. The generation of particles is demonstrated in Fig. 3. Fig. 4 gives an SEM micrograph and an EDX spectrum of these particles when lying on a 5-µm Nuclepore filter.

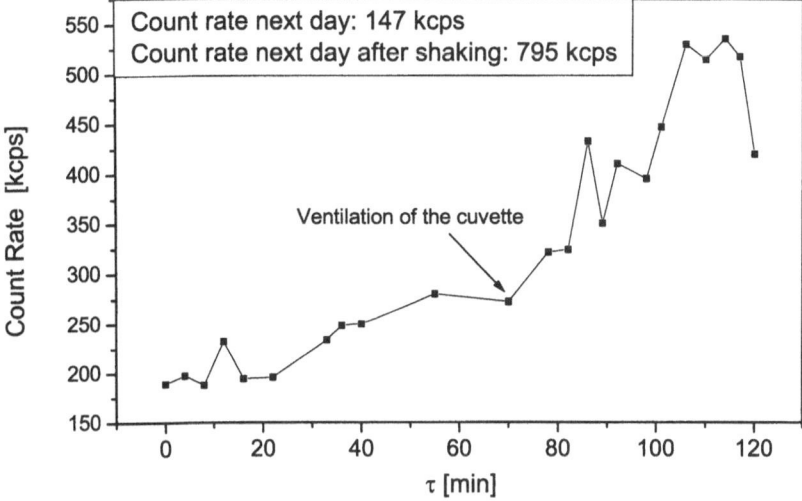

Fig. 3. Time dependence of the intensity of light scattered from a laser beam after centrifugation of the flooding water for 60 min at 1000 rpm (ca. 100 x g). Laser power 400 mW. Measured at 90^0 in a relatively air-tight cuvette. Cuvette opened after 70 min. The increase of the scattered light intensity becomes significantly steeper after ventilation. After about 100 min, a plateau is reached where particle generation rate and particle sedimentation rate are similar. As the ferrous iron was used up, the scattered light intensity dropped down due to sedimentation of the particles.

Analytical speciation techniques for uranium and related elements

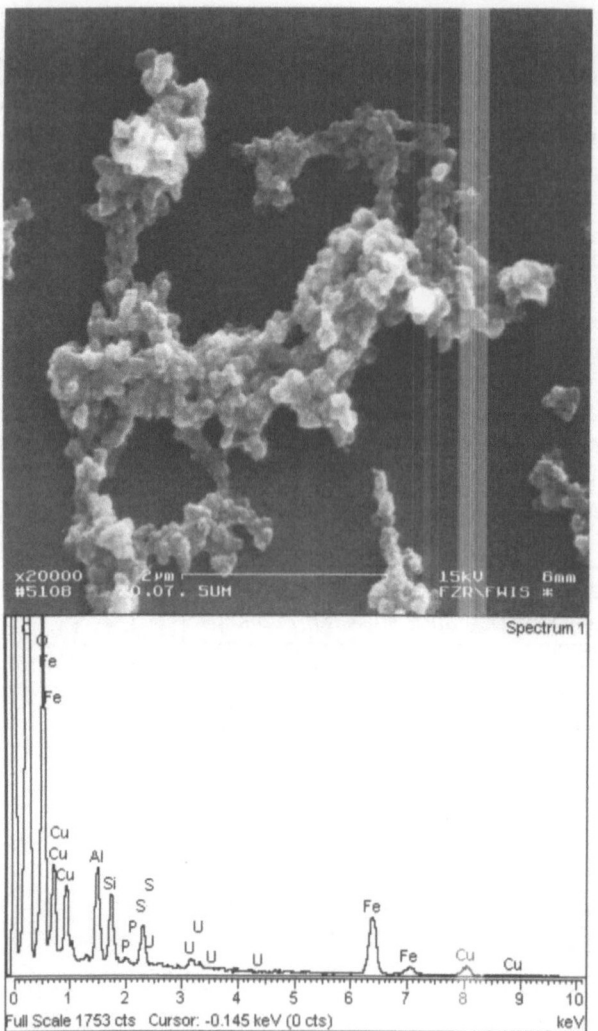

Fig. 4. SEM micrograph (coating with Au) and typical EDX spectrum (coating with C) of colloid particles from flooding water of the Königstein mine on a 5-μm Nuclepore filter. Filter cake washed three times with Milli-Q water, dried and coated with gold. Magnification: 20 000 : 1. Length of the scale bar: 2 μm. Particles of about 100-150 nm are visible. Fe, Al, U, Cu and S are the main constituents of the particles. The high carbon peak in the EDX spectrum is caused by the filter material and thus an artefact.

The micrograph demonstrates the similarity of the flooding water colloids with the "bulk water" colloids depicted in Fig. 2. A particle size of about 100-150 nm can be derived. The particles were primarily composed of iron compounds (Fig. 5).

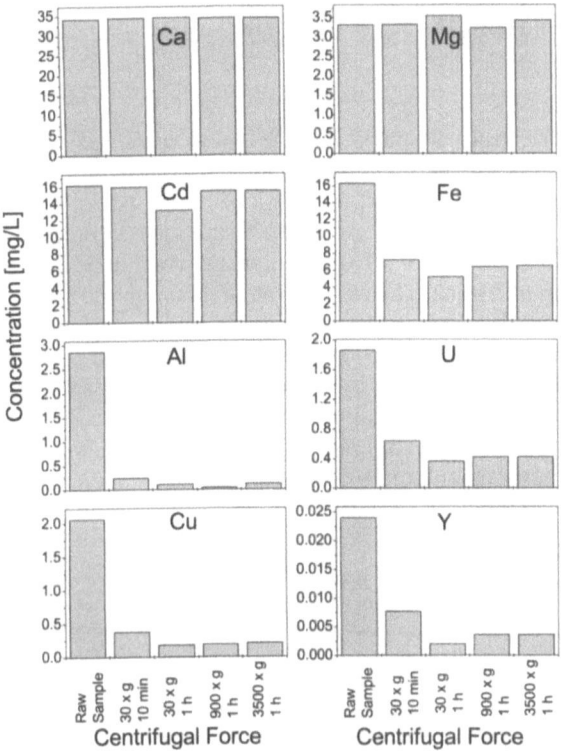

Fig. 5. Concentration of some typical elements in flooding water of the Königstein Uranium mine after centrifugation at varying centrifugal force and centrifugation time. Fe(III) and most of the Al, U, Cu and Y are the colloid-borne components. Part of the Fe is still divalent and thus not removable by centrifugation (centrifugation accomplished in an inert gas glove box).

Further important constituents of the colloids were Al, Cu, and U. Interestingly, the flooding water colloids also contain sulphur which indicates a relationship to the acid rock drainage colloids in addition to the similarities to the "bulk water" colloids. We measured a zeta potential of only –7.5 mV (laser Doppler electrophoresis), i.e. the electrostatic stabilization of the particles was weak. Therefore the particles showed a pronounced tendency to aggregate and sediment. Strongly-sorbing radionuclides such as ^{210}Po and ^{210}Pb, but also significant fractions of the uranium (cf. Fig. 5) were colloid-borne. The amount of the sorbed uranyl was very sensitive to pH changes of only few tenth of pH units. This sensitivity is obviously attributable to the fact that the pH of the water was in the vicinity of the point of inflection of the uranyl - Fe oxyhydroxide "adsorption edge" (adsorption isotherm, cf. Hsi and Langmuir 1985; Waite et al. 1994). The uranium and the other radiotoxic heavy metals studied, with the exception of radium, followed the tendency of the iron(III) particles to aggregate and sediment.

Conclusions

Normally, the uranium(VI) is truly dissolved in mine waters since the adsorption onto colloids is prevented either by high acidity (acidic pore waters) or by carbonate complexation ("bulk waters"). However, there is a pH range where the U(VI) occurs in a colloid-borne form in mine waters, even if carbonate is present. This range is pH 4 to 6 (see also Zänker et al. 2000), a pH range which must be passed through during the flooding of a mine. The assumption of an unretarded migration of uranyl in dissolved forms is conservative for this pH region. Scavenging by Fe(III) particles, which are formed due to O_2 ingress and pH increase, and colloid coagulation plus sedimentation can significantly immobilize the U(VI) in this pH range ("natural attenuation"). Radiotoxic heavy metals like Po, Pb, Th, Ac are also colloid-borne above pH 4. Ra usually exists in truly dissolved forms. Not much is known about colloidal states of uranium in reducing mine waters where it is tetravalent and the iron is divalent.

References

Hsi C-K D, Langmuir D (1985) Adsorption of uranyl onto ferric oxyhydroxides. Application of the surface complexation site-binding model. Geochim. Cosmochim. Acta 49: 1931-1941

Waite T D, Davis J A, Payne T E, Waychunas G A, Xu N (1994) Uranium(VI) adsorption to ferrihydrite: Application of a surface complexation model. Geochim. Cosmochim. Acta 58: 5465-5478

Zänker H, Richter W, Brendler V, Nitsche H (2000) Colloid-borne uranium and other heavy metals in the water of a mine drainage gallery. Radiochim. Acta 88: 619-624

Zänker H, Moll H, Richter W, Brendler V, Hennig C, Reich T, Kluge A, Hüttig G (2002) The colloid chemistry of acid rock drainage solution from an abandoned Zn-Pb-Ag mine. Appl. Geochem. 17: 633-648

Contaminant Release and Carbonate Availability of Sulphide-Bearing Waste Rock derived from Long-term Column Tests

Silvia Jahn, Michael Paul, Delf Baacke

Wismut GmbH, Jagdschänkenstr. 29, 09117 Chemnitz

Abstract. With a view of gaining further knowledge on contaminant release and long-term availability of carbonate waste rock constituents, 11 column tests are ongoing at the Ronneburg site following completion of a comprehensive waste rock pile characterisation program. These tests have been running for about 6 years now. Development of leachate pH is presented and the percentage availability of potential for neutralisation is assessed. Results to date show that for waste rock having a NP:AP ratio of >2 the available potential for neutralisation is sufficient to maintain acid consuming conditions over a protracted test period.

Introduction

Relocation of about 90 % of the sulphide-bearing waste rock accumulated on the surface into the Lichtenberg open pit mine is the most important and cost-intensive single surface restoration project conducted by WISMUT GmbH at the former Ronneburg uranium mining site. The approach used has been extensively described elsewhere (Weise et al. 1996, Jakubick et al. 1997, Hockley et al. 1997, Chapman et al. 1998, Paul 2002).

Extensive investigations were carried out between 1992 and 1996 on waste rock material to determine crucial geochemical and soil physical parameters. In the light of the results achieved, a relocation control program was drawn up which allows a selective relocation of the waste rock material into the Lichtenberg open pit (Hockley et al. 1997, Paul 2002).

The fact that mobility of radionuclides and heavy metals via the aquatic pathway depends essentially on the degree of acid generation of the waste rock makes the prediction of long-term geochemical material behaviour highly significant for remediation planning. The ABA test on drill core samples (Sobek et al. 1978, Coastech 1991) was chosen for the long-term planning of waste dump relocation, since this test allows rapid assessment of the acid mine drainage risk from sulphide-bearing waste. Under this approach, the sulphide sulphur content of a sample representing the acid generation potential AP (c_{S2} in % x 31.25) is compared to the neutralisation potential NP (derived by titration with acid or from the CO_3 content) of the sample, with the NP:AP ratio being a measure of the risk for the generation of acid and highly contaminated seepage. Following is a commonly used interpretation pattern to assess the NP:AP ratio: NP:AP < 1 = potentially acid generating (Class A material), NP:AP 1-3 = behaviour uncertain (Class B material), NP:AP > 3 = potentially acid consuming (Class C material).

With a view of deriving limits for the NP:AP ratio that would be significant and meaningful with respect to the resulting contaminant release, the impact of the NP:AP ratio on the resulting seepage quality was studied in a number of infiltrative column tests which initially varied between 19 and 35 weeks. The results showed clear-cut correlation between the NP:AP ratio and seepage quality. It was also shown that for the Ronneburg waste rock NP:AP ratios of < 1 very probably cause generation of acid drainage while material with NP:AP ratios of > 2 will typically generate neutral seepage which is only slightly contaminated with heavy metals and radionuclides. Therefore, the original interpretation pattern was modified and the following limits used: NP:AP < 1 Class A material, NP:AP 1-2, Class B material; NP:AP < 2 Class C material.

As the spatial variability of material properties cannot be ascertained in sufficient detail from drilling data (obtained on a 100 m grid) for the purposes of waste pile relocation, long-term planning is supported by a program of short-term excavation control. Under this program, test pit samples are taken on a 25 m grid from the excavation face one to three months ahead of relocation. The material is then classified by a combination of paste pH and NAP pH methods (Weise et al. 1996, Hockley et al. 1997). Due to their simplicity, paste and NAP tests, calibrated against ABA data and against kinetic test results, are particularly well suited to provide readily available guidance for mining and rehabilitation planning.

A critical item to be investigated was whether cut-off parameters in use since 1996 were corroborated by the long-term availability of the neutralisation potential. To resolve this problem, 11 column tests were continued following completion of the waste pile investigation program. These tests have been operational for ca. 6 years. Significant findings of these tests are presented below.

Waste rock characterisation

Dumped mine wastes from uranium ore mining at the Ronneburg site of WISMUT amount to ca. 200 million m³. These wastes comprise for the most part shales, limestones and diabases ranging in age from Ordovician to Lower Devonian. Most waste piles contain a mixture of different lithologies (cf. Table 2). Due to variable blending and mixing during mining, mucking and dumping, the mineralisation ratio from the various lithologies is differently reflected in the waste rock material (cf. Table 1). For the most part, waste rock piles consist of rocks which indeed are acid generating but have an excess of neutralising minerals.

According to their mineralogical composition, the Lower (S_1) and the Upper graptolite-bearing shales (S_3) are among acid generating lithologies while leather shale (Og_3), ochre limestone (S_2), nodular tentaculite limestone (D_1^{1-2}) and diabase (ß) count among net acid consuming lithologies.

Silurian graptolite-bearing shales (S_1 and S_2) and ochre limestone (S_2) exhibit exceptionally high percentages of pyrite (up to 7 %). The content of carbonate minerals varies to a great extent and is highest in ochre limestone (S_2) and nodular tentaculite limestone (D_1^{1-2}). These carbonates predominantly occur as dolomites and only secondarily as siderite or calcite.

Table 1. Percentages of rock-forming minerals in lithologies of the Ronneburg deposit (Szurowski 1985)

Minerals	$Og_3^{2+}{}_3$	S_1	S_2	S_3	D_1^{1-2}	D_1^3	ß
Quartz	29	58	13	32	10	24	-
Light mica	43	17	18	34	41	54	3
Kaolinite	10	3	-	3	-	-	-
Chlorite	2.5	1.5	1.5	-	6	15	25
Dolomite	4	1.5	57	7	12	1.5	0.5
Siderite	4	-	-	2	-	-	-
Calcite	-	-	2	-	25	0.5	2.5
Pyrite	3.5	5.5	4	7	1.5	0.5	-
Haematite	2	2	2	3	1	1.5	-
Organic carbon	1.5	9.5	1.5	6	1	0.5	-
Phosphorite	-	2	0.3	3	-	-	-
Feldspars	-	-	-	-	1.5	1.5	45
Pyroxenes	-	-	-	-	-	-	15
Ti-Minerals	-	-	-	-	-	-	6.5
Apatite	-	-	-	-	-	-	0.5

In addition to the differentiated mineralogical and geochemical composition within the various lithologies, sulphides and carbonates are also differently available because of the differing rock structures. As was revealed by summary quantitative analysis and shown in Figs. 1 and 2, unless surrounding carbonates are attacked, availability of sulphides is down by at least 75 % for those contained in leather shale (Og_3), by 90 % for those of the ochre limestone (S_2), by 50 % for those in the tentaculite limestone (D_1), and by 75 % for those of the diabases (ß). This signifies that the overwhelming portion of the acid generating potential which is present in the altogether not acid generating lithologies is embedded in a matrix of neutralising carbonates to a point where it can only become effective once the surrounding neutralisation potential has become available. This analysis is corroborated by comprehensive data material that was collected at the time when the deposit was prospected and is documented in some extensive monographs (Ehrlicher and Paul 1999). Sulphide availability for oxidation is of crucial importance for the acid generating potential.

Analytical speciation techniques for uranium and related elements

Lithology	Ordovi-cian Og₃		Silurian S₁	S₂	S₃	Devonian D₁		Dia b. ß	Q u. Q
Mine dump	1-3	2-3	f.	z.		1-2	3	4	
Innenkippe	3	-	02	2	- 0	1	14	2 0	-
Schm.Balk on	3 8	-	3 4	1	-	3	5	6	1	-
Abset-zerhalde	-	-	9 7	1	5 7	1	8 4	1	-
Nordhalde	-	-	3	1	5 3	1	17 5	2	5 1	2
Halde 4	2 0	-	0 7	1	3	- 5	5 4	-	-	-
Reust	-	66	(	8	-	1	2	-	14	-
Paitzdorf	-	45	(9	1	-	4	13	-	10	-
Halde 377	-	80	-	3	-	-	-	-	17	-
Halde 370	5 0	-	5 0	2	- 5	1	-	-	-	-
Diabashal-de	-	-	(	-	-	6 8	1	3	36 4	1

Legend:

Acid generating lithologies
Neutral lithologies
Low sulphide and low carbonate lithologies
f.: unaltered ochre limestone
z.: decayed ochre limestone: aluminosilicate

"Exfoliation" of shales, which is typical of alum shales from the Lower graptolite-bearing shales (S₁ und S₃) in particular, encourages sulphide weathering. During weathering, the beds separate, inter alia, under the effect of crystallisation pressure from secondary minerals (gypsum) and hence exhibit further pyrite-bearing planes. As a consequence, sulphides from the Lower graptolite-bearing shales are particularly available (Fig. 3).

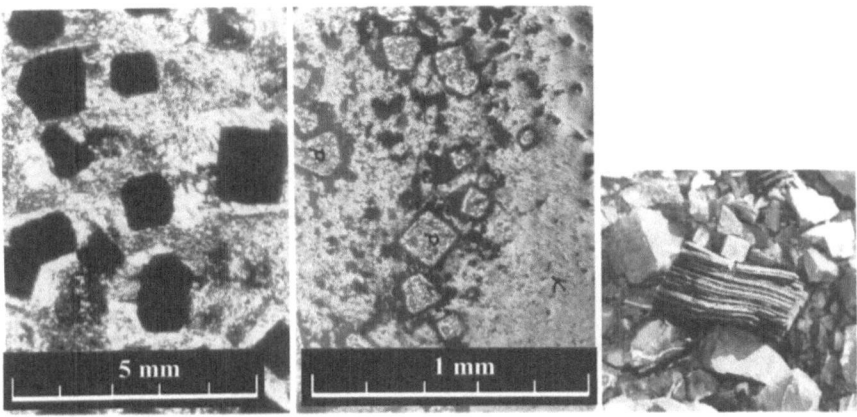

Fig. 1. Micro section of pyrite crystals embedded in a
carbonate matrix

Fig. 2. Pyrite crystals with
quartz fringe in carbonate
rock, polished face
(P=pyrite/K=carbonate)

Fig. 3. Exfoliated weathered alum shale (S₁)

Column testing

Column tests were carried out on test pit samples taken at depths of 4 to 5 m. The
representative sub-samples were ground (to <5 mm) and fractionated. The samples
were characterised by chemical analysis. Grain fraction <63 mm was used in the
column testing

Table 3 lists fundamental material parameters for test pit samples used in
11 long-term column testing. Two samples each were taken from Absetzerhalde
(1.1/5, 1.4/5), Nordhalde (3/6, 3/12), Halde Beerwalde (10/1, 10/5) and Halde
Drosen (11/4, 11/6). Innenkippe (8/13), Halde Reust (2/2E) and Halde Paitzdorf
(6/4) are represented by a single sample each.

With the exception of test pit 1.1/5, classification of test pit samples using the
NP:AP ratio was in relatively good agreement with classification using paste pH,
Nap pH and conductivity tests.

According to the NP:AP ratio, test pit sample 3/6 qualifies as Class B material
(with a NP:AP ratio of 1.0 to 1.4, however, it is in the bottom range of "B") and
when applying the Nap pH value, it qualifies as Class A material. As a consequence, test pit sample 3/6 has to be classified as Class A material. Test pit sample
8/13 is to be classified as Class B material using the NP:AP ratio, while it would
qualify for Class A material when applying the paste pH value. According to the
NP:AP ratio, test pit sample 10/1 is Class C material, but the Nap conductivity
value of >5 mS/cm (5.2) classifies it as Class B material. As this represents another border-line case with regard to classification criteria, this test pit sample has
to be classified as Class C material.

Table 3. Geochemical characterisation of test pit samples by long-term column tests

Sample No.	S^{2-} %	$AP^{a)}$	CO_2 %	$NP^{b)}$	NP:AP	Class $ABA^{c)}$	aste pH	Nap pH	Nap cond.$^{d)}$	Class Field tests$^{e)}$
1.1/5	0.44	13.7	1.6	36.4	2.6	C	4.9	3.1	3.4	A
1.4/5	3.10	96.4	3.1	70.5	0.7	A	4.5	2.5	7.4	A
2/2 E	0.86	26.8	3.3	75.0	2.8	C	7.3	5.6	3.9	C
3/6	0.69	21.5	1.0	22.7	1.0	B	6.3	3.3	-	A
3/12	0.59	18.4	1.8	40.9	2.2	C	5.8	5.8	-	C
6/4	0.41	12.8	1.6	36.4	2.8	C	6.4	6.5	4.8	C
8/13	0.26	8.1	0.55	12.5	1.5	B	4.1	2.8	3.2	A
10/1	1.30	40.6	5.4	122.8	3.0	C	6.4	7.1	5.2	B
10/5	0.93	29.0	4.2	95.5	3.3	C	6.4	7.1	3.2	C
11/4	0.60	18.7	3.1	70.5	3.8	C	6.5	7.2	3.4	C
11/6	0.46	14.4	2.4	54.6	3.8	C	6.8	7.1	3.5	C

a) AP derived from S^{2-} analyses, in kg $CaCO_3$ eq/t
b) NP calculated from carbonate analysis , in kg $CaCO_3$ eq/t
c) NP:AP <1 = A, NP:AP 1...2 = B, NP:AP >2 = Class C material
d) NAP conductivity, in mS/cm
e) according to field test: classification as Class A material when: paste pH ≤4.5 or Nap pH <4, classification as Class B material when: paste pH >4.5 and ≤4 Nap pH <5 or Nap conductivity ≥5 mS/cm, classification as Class C material when: paste pH >4.5 and Nap pH 5 and Nap conductivity <5 mS/cm

Application of the various classification methods to the rest of the test pit samples yielded definite classification with regard to Class A, B or C materials.

Test pit sample 1.1/5 is an exception to the above: it may be classified as Class B material according to its NP:AP ratio. According to the Nap pH value, on the other hand, this sample is definitely Class A material.

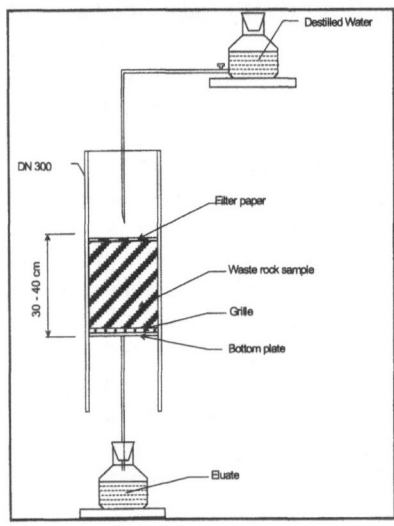

Fig 4. Experimental set-up for infiltrative column testing

Initially, the 11 column tests were carried out under hydrostatic test conditions and in the absence of oxygen in order to identify secondary mineral phases in storage and to determine the chemico-physical leaching control factors. Discussion of these results is not a matter for this paper. Following establishment of constant concentration conditions in the leachate, the columns were turned to an infiltrative test procedure. Fig. 4 shows the experimental set-up of the infiltrative column tests. The columns were

Analytical speciation techniques for uranium and related elements

constructed from PVC piping with an internal diameter of 30 cm and a height of about 45 cm. The cover and bottom plates of the columns were equipped with inlet and outlet valves, respectively. The columns were filled with ca. 43 kg of waste rock (undried material) screened to <63 mm and compacted in several layers.

During the infiltrative test phase columns were operated in a weekly routine. During the first five days, the columns were aerated (rate ca. 6-8 l/h), with the air being humidified by passing through a water bath. During the remaining 2 days of the weekly cycle aeration was stopped and the waste rock was slowly sprayed from the top with 1.5 l of distilled water (over a period of 24 h). The water percolating through the waste rock (leachate) was collected over a period of 24 h at the bottom of the column and analysed.

Results

Development of pH values

The trend of pH development in the leachate from column tests over a period of 307 weeks (5.9 years) confirms the initial classification of test pit samples 3/6 (A, final pH value 3.6), 8/13 (A, final pH value 5.0) and 10/1(C, final pH value 8.2). No definite classification was obtained from the NP:AP ratio, paste pH and Nap pH for test pit sample 1.1/5. Since this sample is rather classified as Class A material according to its Nap pH (3.1) and also in the light of the final pH from the column tests (4.1), the mathematically obtained NP:AP ratio of 2.6 is to be considered as faulty and consequently not representative of the test pit under consideration.

Table 4 below summarises the definite classification of the test pit samples compared with development of pH values from column test leachates.

Test pit samples 1.1/5, 1.4/5, 3/6 and 8/13, classified as acid generating following solid characterisation, exhibit in their respective column eluates a trend of acidic pH in the range from 2.9 to 5.0. After about 5.9 years, pH values show an upward trend.

Table 4. Definite classification of test pit samples into Class A, B or C material compared with development of pH values from column test leachates (infiltrative tests)

Sample No.	NP:AP	Initial pH[a]	Eluate pH 160 weeks	Eluate pH 238 weeks	Eluate pH 307 weeks	Definite classification[b]
1.1/5	2.6	5.7	4.0	4.0	4.1	A
1.4/5	0.7	6.5	2.9	3.6	3.9	A
2/2 E	2.8	6.2	7.3	7.8	8.0	C
3/6	1.0	7.2	3.2	3.4	3.6	A
3/12	2.2	6.5	7.7	7.8	8.0	C
6/4	2.8	6.9	8.1	8.1	8.2	C
8/13	1.5	4.1	4.6	4.6	5.0	A
10/1	3.0	6.3	7.6	7.9	8.2	C
10/5	3.3	7.9	8.0	7.9	8.2	C
11/4	3.8	7.7	8.1	8.2	8.4	C
11/6	3.8	7.7	7.9	8.2	8.4	C

[a]   after 6 to 8 pore volume exchanges during hydrostatic column test
[b] $A \leq pH\ 5;\ C \geq pH\ 7$

Test pit samples 2/2E, 3/12, 6/4, 10/1, 10/5, 11/4 and 11/6, classified as not acid forming following solid characterisation, exhibit in their respective column eluates a trend of neutral pH values in the range from 7.3 to 8.4. Since the beginning of the tests and throughout the period of 5.9 years, the development of pH values has shown a constant upward trend.

Test pit samples with a NP:AP ratio of >2 did not behave acid generating during the long test period (about 5.9 years). The upward trend in pH development is maintained. Test pit sample 1.1/5 is an exception exhibiting a NP:AP ratio of 2.6 and a declining trend in pH (from 5.7 to 4.1) and being definitely acid generating. As stated earlier, determination of the NP:AP ratio is probably faulty, and according to the Nap pH value of 3.1 the sample should be classified as acid generating.

Carbonate availability assessment

The assessment is based on the estimated available neutralisation potential (NP) and on the development of eluate pH. As most of the stored soluble secondary minerals were flushed from the samples during the earlier hydrostatic test procedure, assessment of NP availability was made on the assumption that total concentrations of released calcium and magnesium were produced by ongoing oxidation and neutralisation. This method of consumed neutralisation potential estimation does not take into account gypsum solubility and gypsum precipitation processes, respectively. This means that the calculated cumulatively released amount of calcium is overestimated unless gypsum stored in the test pit samples is completely flushed out during the hydrostatic process; on the other hand, the amount of calcium is underestimated when gypsum precipitation occurs during the testing phase. Table 5 summarises the estimated availability of neutralisation potential as a percentage.

Table 5. Estimated neutralisation potential availability derived from the findings/results of column tests eluates (after 307 weeks), as a percentage

Sample No.	Cumulative Mg release mg/kg	Cumulative Ca release mg/kg	NP consumed to date kg $CaCO_3$ eq/t	NP availability to date %
1.1/5	283	6556	17.6	**48**
1.4/5	3863	5566	29.8	**42**
2/2 E	1012	4919	16.5	**22**
3/6	1511	3660	15.4	**68**
3/12	607	5458	16.1	**39**
6/4	618	5347	15.9	**44**
8/13	167	5686	14.9	119
10/1	1510	5633	20.3	**17**
10/5	645	5180	15.6	**16**
11/4	506	2989	9.6	**14**
11/6	487	2834	9.1	**17**

Analysis and interpretation of results from those tests pit samples that maintain a neutral eluate pH during the test period yield an estimated percentage of neutralisation potential availability in the range from 14 % to 44 %. Comparison of percentage neutralisation potential availability for test pit samples exhibiting neutral eluate pH after 160 weeks (about 3 years) with availability after 307 weeks (about 5.9 years) reveals that NP consumption rose between 2 and 13 % during a period of 147 weeks (ca. 2.9 years).

Conclusions

The results show that the neutralisation potential available in test pit samples exhibiting a NP:AP ratio of >2 is sufficient to maintain acid consuming conditions even during a test period extending over 5.9 years. Solubility of radionuclides and toxic heavy metals which is very high under acidic conditions will remain low for waste rock of these characteristics for long periods of time.

Therefore, more than half of the neutralisation potential is theoretically still available for further consumption. Since pH values from column eluates of test pit samples with a NP:AP ratio of >2 show an upward trend, this may suggest that sulphides as acid formers will be consumed much more rapidly than the considerable excess of available neutralisation potential.

References

Chapman J, Hockley D, Sevick J, Dachel R, Paul M (1998): Pit Backfilling on Two Conti-
nents - Comparison of Recent Experiences in the Wismut and Flambeau Projects.- In:
Tailings and Mine Waste '98. Proceedings of the Fifth Conference Tailings and Mine
Waste '98, Ft. Collins, Co., 26-29 January 1998.- Balkema, Rotterdam - Brookfield,
1998, pp. 55-65

Coastech Research Inc. (1991): Acid Rock Drainage prediction manual.- MEND-Project
1.16.1b, CANMET

Ehrlicher U, Paul M (1999): Analyse der lithologischen Zusammensetzung der Gesteine
des Ronneburger Erzfeldes unter besonderer Berücksichtung der Verteilung von Sulfi-
den und Karbonaten.- Studie, WISMUT GmbH, Chemnitz

Hockley D, Paul M, Chapman J, Jahn S, Weise W (1997): Relocation of waste rock to the
Lichtenberg pit near Ronneburg, Germany. – In: Proceedings of the 4th International
Conference on Acid Rock Drainage, Vancouver, B.C. Canada, May 31-June 6, pp.
1267-1283

Jakubick AT, Gatzweiler R, Mager D, Robertson AMacG (1997): The Wismut Waste rock
pile remediation program of the Ronneburg Mining district, Germany.- In: Proceedings
of the 4th International Conference on Acid Rock Drainage, Vancouver, B.C. Canada,
May 31- June 6, 1997, pp. 1285-1301

Paul M (2002): Geochemische In-situ-Stabilisierung von Bergbaualtlasten.- In: Förstner U,
Grathwohl P, Dahmke, [Hrsg.]: Ingenieurgeochemie – Konzepte und Praxis.- Springer
Verlag, in Vorbereitung

Sobek A, Schuller A, Freemann R, Smith M (1978): Field and laboratory methods applicable to overburden an mine soils, EPA 600/2-78-054, pp. 203 (U.S. Environmental Protection Agency Cincinnati, Ohio)

Szurowski H (1985): Katalog der Gesteinseigenschaften. Teil I+II.- Studie, SDAG Wismut,
Gera.

Weise W, Paul M, Jahn S, Hoepfner U (1996): Geochemische Aspekte der Halden-
sanierung am Standort Ronneburg.- Geowissenschaften, 14 (11), 470-475

Weise W, Zurl R, Hoepfner U, Szymanowsky R, Köhler H (1995): Bewertung Halden-
untersuchungsprogramm, Phase 1.- Studie, Wismut GmbH, Chemnitz.

Separation of uranium from aqueous solution by textile bound calixarenes

Katja Schmeide[1], Gerhard Geipel[1], Dietmar Keil[2], Klaus Jansen[3], Dirk Praschak[4], Karl Heinz Heise[1], Gert Bernhard[1]

[1] Forschungszentrum Rossendorf e.V., Institute of Radiochemistry, P.O. Box 510119, D-01314 Dresden
[2] SynTec Gesellschaft für Chemie und Technologie der Informationsaufzeichnung mbH, Industriepark Wolfen-Thalheim, Werkstattstraße 188, D-06766 Wolfen
[3] Deutsches Textilforschungszentrum Nord-West e.V., Adlerstraße 1, D-47798 Krefeld
[4] Thomas Josef Heimbach GmbH & Co., An Gut Nazareth 73, D-52353 Düren

Abstract. The separation of uranium(VI) from aqueous solution by textile bound calix[6]arenes was studied as a function of pH value and uranium concentration by means of batch experiments. Furthermore, the kinetics of the uranium binding was studied as well as the possibility for remobilization of the bound uranium by solutions of various pH values. Furthermore, the solvent extraction behavior of mono-p-nonyl-penta-p-tert-butyl-calix[6]arene hexacarboxylic acid towards uranium(VI) was compared with that of p-tert-butyl-calix[6]arene hexacarboxylic acid in dependence on pH value and ligand concentration.

Introduction

The remediation of former uranium mining and milling sites of Saxony and Thuringia requires, amongst others, the purification of uranium contaminated seepage and mine waters.

The separation of uranium(VI) from aqueous solution by calix[6]arenes, functionalized with carboxylic or hydroxamic groups at the lower rim, by means of solvent extraction has been described in the literature (e.g., Shinkai et al. 1987, Nagasaki and Shinkai 1991, Schmeide et al. 2001).

However, to facilitate the separation of uranium from uranium containing waters in practice the calixarenes have to be immobilized on certain substrates. For instance, for the separation of uranium from seawater, Aihara et al. (1992) suggested the immobilization of calixarenes on a macroporous polystyrene-divinylbenzene copolymer via a spacer such as polyethyleneimine.

Within our current project 'Integrated environmental conservation in the textile industry – Separation of uranyl ions from seepage and ground waters by uranophile calixarenes' a new procedure for purification of uranium contaminated waters was developed. For this, a uranophile calix[6]arene (p-tert-butyl-calix[6]arene hexacarboxylic acid) is statistically functionalized by spacer groups that allow its permanent fixation onto textile substrates. That means, ideally one tert-butyl group is substituted by one n-nonyl group per calixarene molecule leading to mono-p-nonyl-penta-p-tert-butyl-calix[6]arene hexacarboxylic acid. For fixation, the diffusion of the n-nonyl group into the textile material above the glass temperature is used (Patent No. 10210115.9).

In this paper, we studied the extraction behavior of unfixed mono-p-nonyl-penta-p-tert-butyl-calix[6]arene hexacarboxylic acid (Non^{t}But[6]CH$_2$COOH) towards uranium(VI) by means of solvent extraction and compared it to that of p-tert-butyl-calix[6]arene hexacarboxylic acid (^{t}But[6]CH$_2$COOH) which is used as reference substance. Furthermore, we determined the binding properties of Non-^{t}But[6]CH$_2$COOH fixed onto polyester fabric towards uranium(VI) by means of batch experiments. The results are compared with those obtained for reference material (polyester fabric, not modified with calixarenes).

Experimental

Extractants

In Fig. 1, the formulae of ^{t}But[6]CH$_2$COOH and of Non^{t}But[6]CH$_2$COOH are shown.

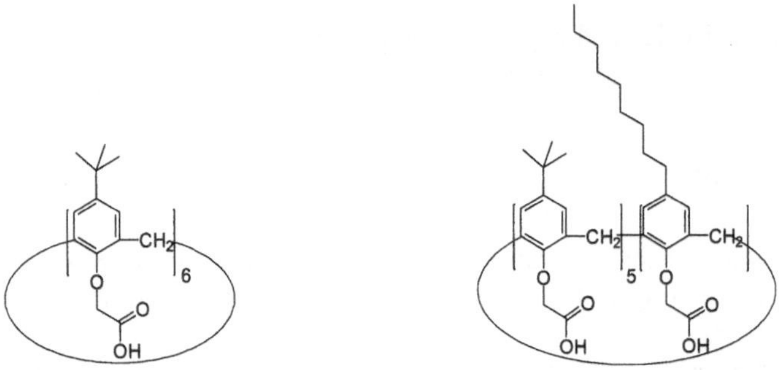

^{t}But[6]CH$_2$COOH   Non^{t}But[6]CH$_2$COOH

Fig. 1. Extractants used in this work.

^{t}But[6]CH$_2$COOH was prepared according to common procedures (Chang and Cho 1986, Shinkai et al. 1989, Ludwig et al. 2000). The synthesis of Non-^{t}But[6]CH$_2$COOH is described in detail in the patent No. 10210115.9. Its synthesis starts from base-catalyzed condensation of p-tert-butyl phenol and p-nonyl phenol with formaldehyde leading to Non^{t}But[6]H which is subsequently treated with sodium hydride and methylbromoacetate in tetrahydrofuran to Non-^{t}But[6]CH$_2$COOCH$_3$. Hydrolysis of the ester to Non^{t}But[6]CH$_2$COOH was achieved applying tetraethylammonium hydroxide.

The extractants were characterized by means of elemental analysis, melting point determination, IR and ^{13}C NMR spectroscopy as well as by mass spectrometry and thin-layer chromatography. The characterization results agree well with literature data (e.g., Chang and Cho 1986, Shinkai et al. 1989) as well as in case of Non^{t}But[6]CH$_2$COOH with theoretical values.

Uranium extraction from aqueous solution by unfixed calix[6]arenes

In solvent extraction experiments, the unfixed calixarenes, dissolved in chloroform, were shaken with the aqueous uranium solution for 6 hours at room temperature. The initial uranium concentration was 1×10^{-5} M and the initial calixarene concentration was 1×10^{-3} M and 3.2×10^{-5} M to 1×10^{-3} M in experiments studying the uranium extraction in dependence on pH value and on ligand concentration, respectively. The phase ratio $V_{(org)}$:$V_{(w)}$ was 1:1. The pH of the aqueous solution was initially adjusted and measured after reaching extraction equilibrium. After phase separation, the initial and the equilibrium uranium concentration in the aqueous phase was determined by ICP-MS and from these values the amount of uranium extracted into the organic phase was calculated.

Uranium separation from aqueous solution by textile bound calix[6]arenes

In separation experiments, 0.5 g calixarene modified polyester fabric and reference material, respectively, were shaken with 20 mL aqueous uranium solution for 6 hours at room temperature. The initial uranium concentration was 1×10^{-6} M and 5.2×10^{-8} M to 2.2×10^{-4} M in experiments studying the uranium separation in dependence on pH value and on uranium concentration, respectively. The pH of the aqueous solution was initially adjusted, readjusted in the course of the experiments and measured after reaching equilibrium. The initial and the equilibrium uranium content in the aqueous solution were determined by ICP-MS. From these values the amount of uranium sorbed onto the material is calculated (results are corrected for uranium sorption onto vial walls).

In remobilization experiments, 0.2 g uranium loaded calixarene modified polyester fabric and reference material, respectively, from the separation experiments were washed 3 times with 7 mL Milli-Q water, 2 times with 7 mL 0.01 M HCl and finally 3 times with 7 mL 0.1 M HCl for 2 hours each time. The uranium content was determined in each washing solution by ICP-MS.

Results and discussion

Uranium extraction from aqueous solution by unfixed calix[6]arenes

Fig. 2 compares the extraction efficiency of $^tBut[6]CH_2COOH$ and Non-$^tBut[6]CH_2COOH$ towards uranium studied by means of solvent extraction.

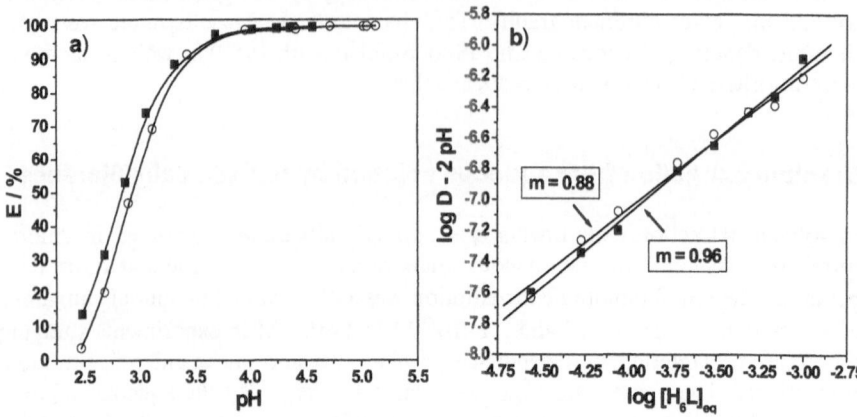

Fig. 2. Uranium extraction from aqueous solution by $^tBut[6]CH_2COOH$ (■) and Non-$^tBut[6]CH_2COOH$ (○) as a function of pH value (a) and ligand concentration (b).

Generally, the equilibrium of the uranium extraction by the calix[6]arene (H_6L) can be written according to the following equation, where the solid line denotes the species in the organic phase:

$$UO_2^{2+} + \overline{H_6L} \leftrightarrow \overline{UO_2(H_4L)} + 2H^+ \tag{1}$$

The extraction constant (K_{ex}) is expressed by the following equation:

$$K_{ex} = \frac{[\overline{UO_2(H_4L)}] \cdot [H^+]^2}{[UO_2^{2+}] \cdot [\overline{H_6L}]} \tag{2}$$

The distribution coefficient (D), defined as the ratio of uranium concentration in the organic and in the aqueous phase at equilibrium, is a measure of uranium extractability.

$$D = \frac{[\overline{UO_2(H_4L)}]}{[UO_2^{2+}]} \tag{3}$$

From the definitions of D and K_{ex} follows:

$$\log D - 2 \cdot pH = \log[\overline{H_6L}] + \log K_{ex} \tag{4}$$

Thus, the extraction constants (K_{ex}) can be calculated according to equation (4) for experiments with variation of ligand concentration.

The plot of the uranium extraction as a function of pH, depicted in Fig. 2a), shows that the extraction efficiency of both calixarenes increases with pH between pH 2.5 and about pH 4.0. Below pH 3.5, the uranium extraction is slightly higher for ^{t}But[6]CH$_2$COOH than for Non^{t}But[6]CH$_2$COOH. At pH values higher than pH 4.0 in the aqueous phase, the maximum of extraction is reached. The plot of log D versus pH, not shown here, shows a curve slope of two which verifies that two carboxylic groups are deprotonated upon extraction of uranium, thereby forming a neutral complex.

The uranium extraction at pH 3.8 in dependence on ligand concentration is shown in Fig. 2b). The slope of both curves is nearly one, which confirms the formation of 1:1 uranium calixarene complexes. The extraction constants (log $K_{ex1,1}$), determined from the intercept of the plots, amount to -3.11±0.08 for ^{t}But[6]CH$_2$COOH and -3.09±0.17 for Non^{t}But[6]CH$_2$COOH. The values correspond to each other very well. That means, that the complexation ability of the calixarene molecule towards uranium ions is not changed when a tert-butyl group is substituted against a n-nonyl group per calixarene molecule. This is important for the application of Non^{t}But[6]CH$_2$COOH for the production of calixarene modified textiles that will be used for purification of uranium contaminated water.

Uranium separation from aqueous solution by textile bound calix[6]arenes

The results of the separation experiments (Fig. 3) show that 97.2% to 98.4% of the uranium present in the initial uranium solution (1×10^{-6} M) is separated by the calixarene modified polyester fabric in the pH range 7.07 to 7.51. Only 27% of the uranium is bound by the reference material (polyester fabric, not modified with calixarenes) under the same experimental conditions. That means, the uranium separation is enhanced by about 70% by the calixarene fixed onto the textile material. Kinetic experiments at pH 7.5 have shown that the uranium binding by calixarene modified polyester fabric reaches equilibrium within 150 min.

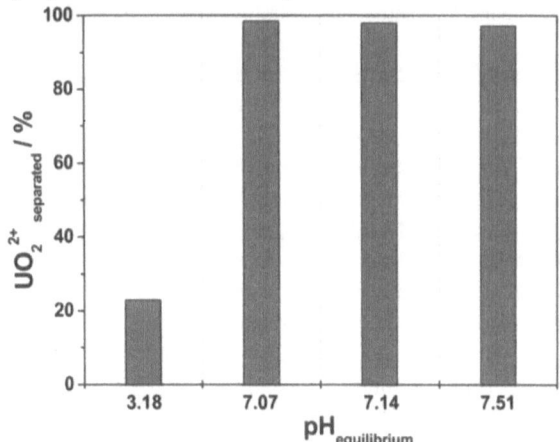

Fig. 3. Separation of uranium from aqueous solution by calixarene modified polyester fabric as a function of pH ([UO$_2$${}^{2+}$]$_{initial}$ = 1×10^{-6} M).

As shown for pH 3.18 in Fig. 3, the uranium separation in the acidic pH range is very low. This indicates a limited applicability of textile bound calixarenes in the acidic pH range.

In Fig. 4, a uranium loading test at pH 5 is shown for the calixarene modified polyester fabric applying increasing uranium concentrations in aqueous solution (5.2×10^{-8} M to 2.2×10^{-4} M). The results show that up to an initial uranium concentration of 9.8×10^{-6} M 92% to 99% of the uranium is separated by the calixarene modified polyester fabric. 7.6×10^{-7} mol uranium is bound at most per 1 g of the calixarene modified polyester fabric under these conditions. At higher initial uranium concentrations the uranium can only partly be separated from the aqueous solution.

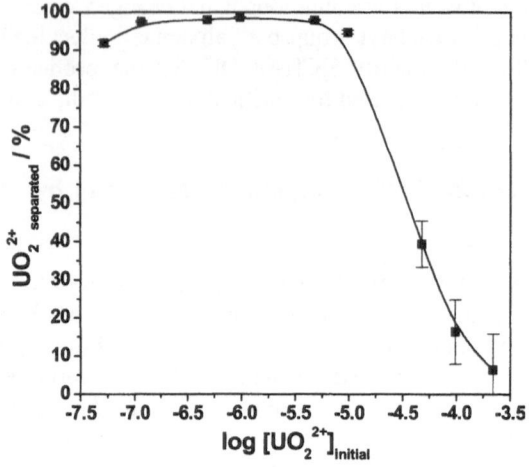

Fig. 4. Separation of uranium from aqueous solution by calixarene modified polyester fabric as a function of the initial uranium concentration (pH 5).

The remobilization of the bound uranium from the uranium loaded calixarene modified polyester fabric as well as from the uranium loaded reference material was studied with the objective to evaluate the binding strength as well as the possibility for remobilizing the uranium bound by the calixarene or by the polyester fabric. The results (Fig. 5) show that from the reference material 47% of the bound uranium is remobilized already by rinsing of the material with water. That means, that a great part of the uranium is weakly bound. In contrast to this, merely 1% of the bound uranium is remobilized from calixarene modified polyester fabric by rinsing with water. However, the uranium can be remobilized by washing this material successively with 0.01 M HCl (82%) and 0.1 M HCl (9%). This indicates that the uranium is strongly bound by the calixarene fixed onto the polyester fabric under environmentally relevant conditions. However, under acidic conditions the textile filter material can be regenerated almost completely.

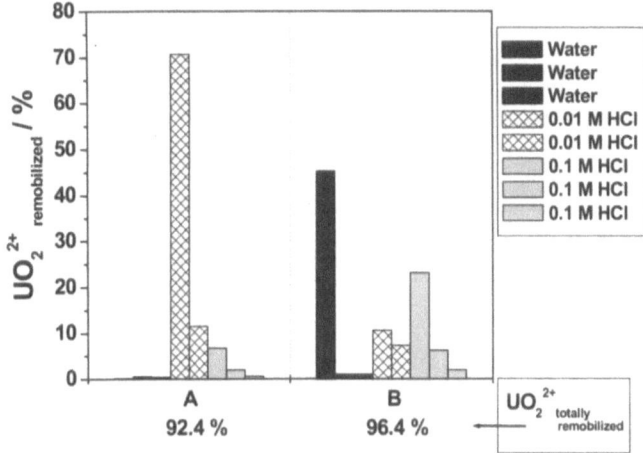

Fig. 5. Remobilization of uranium from calixarene modified polyester fabric (A) and reference material (B) by water, 0.01 M HCl and 0.1 M HCl.

Conclusions

An alternative procedure for purification of uranium contaminated seepage and mine waters was developed. This method, applying calixarene modified polyester fabric, is especially suitable for the separation of uranium from aqueous solution at pH values higher than 5. Its applicability in the acidic pH range is limited. 7.6×10^{-7} mol uranium is bound at most per 1 g of the calixarene modified polyester fabric at pH 5. Under environmentally relevant conditions (neutral pH range) the uranium is strongly bound to the calixarene modified textile and cannot be mobilized. On the other hand, under acidic conditions an almost complete regeneration of the calixarene modified polyester fabric is possible.

The application of this new technique is economically reasonable also for lower contamination levels ($[UO_2^{2+}] \leq 1 \cdot 10^{-6}$ M). The applicability of this separation principle to further actinides is expected.

Acknowledgment

We thank the Bundesministerium für Bildung und Forschung (BMBF) for financial support (contract No. 0339917). Furthermore, we would like to thank B. Barz for her help in performing the experiments and U. Schäfer for ICP-MS analyses.

References

Aihara T, Goto A, Kago T, Kusakabe K, Morooka S (1992) Rate of adsorption of uranium from seawater with a calix[6]arene adsorbent. Separation Science and Technology 27 (12): 1655-1667

Chang SK, Cho I (1986) New metal cation-selective ionophores derived from calixarenes: their syntheses and ion-binding properties. J. Chem. Soc. Perkin Trans. I: 211-214

Ludwig R, Lentz D, Nguyen TKD (2000) Trivalent lanthanide and actinide extraction by calixarenes with different ring sizes and different molecular flexibility. Radiochim. Acta 88: 335-343

Nagasaki T, Shinkai S (1991) Synthesis and solvent extraction studies of novel calixarene-based uranophiles bearing hydroxamic groups. J. Chem. Soc. Perkin Trans. II: 1063-1066

Patent No. 10210115.9 (08.03.2002)

Schmeide K, Barz B, Heise KH, Bernhard G, Gloe K (2001) Solvent extraction of uranium(VI) by calix[6]arene. In: FZR-318, Annual Report 2000. Forschungszentrum Rossendorf, Institute of Radiochemistry: 17

Shinkai S, Koreishi H, Ueda K, Arimura T, Manabe O (1987) Molecular design of calixarene-based uranophiles which exhibit remarkably high stability and selectivity. J. Am. Chem. Soc. 109: 6371-6376

Shinkai S, Shiramama Y, Satoh H, Manabe O, Arimura T, Fujimoto K, Matsuda T (1989) Selective extraction and transport of UO_2^{2+} with calixarene-based uranophiles. J. Chem. Soc. Perkin Trans. II: 1167-1171

The effect of chlorite dissolution on the sorption behavior of U(VI)

Evelyn Krawczyk-Bärsch[1], Thuro Arnold[1], Nils Schmeißer[2], Felix Brandt[3], Dirk Bosbach[3] and Gert Bernhard[1]

[1]Institute of Radiochemistry, Forschungszentrum Rossendorf, P.O. Box 510119, 01314 Dresden, Germany
[2]Experimental Facilities and Information Technology, Forschungszentrum Rossendorf, P.O. Box 510119, 01314 Dresden, Germany
[3]Institute of Nuclear Waste Management, Forschungszentrum Karlsruhe GmbH, P.O. Box 3640, 76021 Karlsruhe, Germany

Abstract. During the dissolution of chlorite the sorption of uranyl(VI) can be described by two different sorption sites. The first site is attributed to a surface site on the chlorite which is occupied after 138 hours of our mixed-flow experiment. Due to the fact that uranyl(VI) is continuously sorbed we conclude that as a second site the sorption sites of the newly forming ferrihydrite colloids are responsible for the additional sorption beyond the available sorption sites on chlorite.

Introduction

The migration of long-lived radionuclides depends on hydrogeological and chemical processes, including mineral dissolution. In uranium tailings associated with the former uranium mining activities in Saxony/Thuringia (Germany) the rock phyllite is commonly encountered. One of the mineralogical component of the phyllite is an iron-rich chlorite. Chlorites are phyllosilicates, commonly found in sedimentary and low-grade metamorphic rocks of the greenschist facies. During acid weathering chlorite tends to dissolve leading to a loss of the octahedral layer cations, i.e. Al^{3+} with Mg^{2+}, Fe^{2+} and Fe^{3+} substitutions. These elements are released in aqueous solution where the dissolved ferrous iron oxidizes to ferric iron and forms iron-oxyhydroxides. In neutral and alkaline media iron-oxyhydroxides are precipitating due to the low solubility of iron in this pH region.

In previous studies (Arnold et al., 1998) the influence of the weathering reactions of the rock phyllite on the migration behavior of contaminants in the vicinity of an abandoned uranium mine environment was discussed. In batch sorption experiments the authors focused that iron-oxyhydroxides, which are forming during the chemical weathering of the phyllite, have a significant influence on the uranium migration behavior. In fact, these iron-oxyhydroxides are dominating the sorption process of uranyl(VI) on phyllite.

The aim of our investigations was to study in mixed-flow experiments the weathering of an iron-rich chlorite and to calculate the uranyl(VI) sorption on the sorption sites on this chlorite.

Experimental

The iron-rich chlorite we used for the experiments was a ripidolite chlorite (CCa-2) from Flagstaff Hill (El Dorado County, California, USA), which was obtained from the Source Clays Repository of the Clay Minerals Society. The chemical formula of the chlorite was determined by Brandt et al. (2002):

$$(Mg_{5.5}Al_{2.48}Fe^{2+}_{3.02}Fe^{3+}_{0.94})[(Si_{5.33}Al_{2.66})O_{20}](OH)_{16}) \tag{1}$$

The individual chlorite grains were broken along their boundaries by deep-freezing with liquid nitrogen and subsequently defrosting at 40 °C. The chlorit surface was therefore more or less left unhurt. After dry sieving, a chlorite fraction of 63 to 200 μm was used for the mixed-flow experiments. A mixed-flow experiment was suited for carrying out mineral dissolution studies since reaction rates can be determined directly. The experiments were performed under oxic conditions and at room temperature using a NaClO₄-solution with an ionic strength of 0.1 M and a pH of 6.5 to approximate natural weathering conditions. In An $UO_2(ClO_4)_2$-solution was added to reach an initial U(VI)-concentration of $1*10^{-6}$ M. During the experiments the solution was pumped through the mixed-flow reactor at 0.3 to 0.5 ml per minute. Half a gram of the chlorite powder was kept between two membrane filters in the reactor. The reacted solution samples were taken two to eight times per day during the experiments. They were analyzed for U, Al, Mg and Si using ICP/MS (*I*nductively *C*oupled *P*lasma-*M*ass *S*pectroscoy) and for Fe using AAS (*A*tomic *A*bsorption *S*pectrometry). The pH of the reactor outflow and the temperature were recorded during the experiments.

Results and discussion

The analyses of the reacted solution samples were used to determined the dissolution rate DR in mol/m²s. The dissolution rate was calculated for Si, Al, Fe and Mg using the following equation after Knauss et al. (1989):

$$DR = (\Delta ppm) \cdot V / M_A \cdot r_A \cdot S \tag{2}$$

where (Δppm) = concentration of element A in reacted solution sample - concentration of element A in blank solution, A = dissolved component (Fe, Si, Al and Mg), V = flow rate (g h^{-1}), M_A = molecular weight of A, r_A = number of moles of A, and S = specific surface area of chlorite (m^2/g).

During the first eight hours of the experiments high initial dissolution rates were determined for Mg, Si, Al and Fe concentrations. They are very likely the result of the dissolving ultrafine particles. After eight hours the calculated rates approache steady state conditions, indicating that the ultrafine particles were completely dissolved and the following dissolution rates are representative for the chlorite fraction of 63-200 µm. Average dissolution rates of $2.73 \cdot 10^{-13}$ and $6.67 \cdot 10^{-12}$ mol/m^{2}s were calculated for Mg and Si, respectively (Fig. 1).

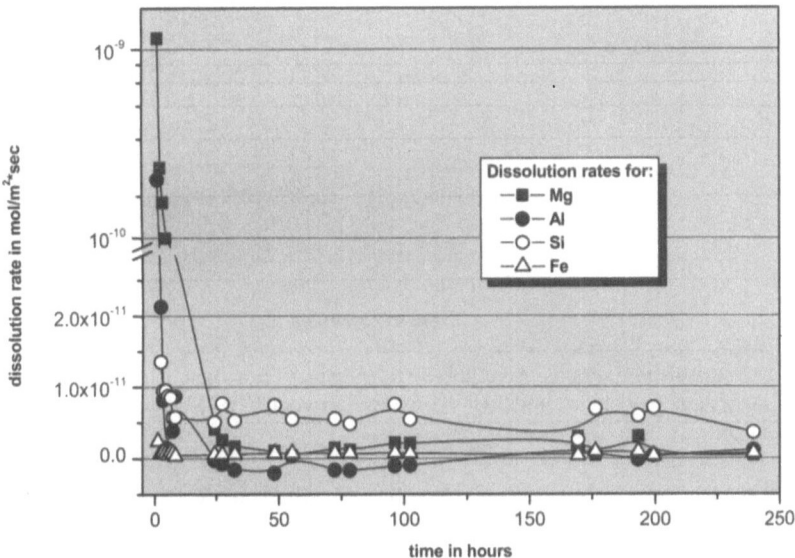

Fig. 1. Dissolution rates of chlorite as a function of time calculated for Mg, Al, Fe and Si in mixed-flow experiments with 0.1 M NaClO$_4$-solution and an uranium concentration of $1 \cdot 10^{-6}$ M.

The average dissolution rates for Fe and Al of $7.00 \cdot 10^{-13}$ and $6.99 \cdot 10^{-13}$ mol/m^{2}s, respectively, seem to be very low. In fact they are primarily higher. However, these rates are based on their aqueous concentration and do not consider precipitated Fe and Al phases. Due to the low solubility of Fe and Al in aqueous solutions of a pH > 5 Fe-oxyhydroxides and Al(OH)-phases have been precipitating. In previous studies (Krawczyk-Bärsch et al., 2002) Fe particles were detected as colloids in the aqueous solution as well as sorbed particles forming iron coatings on the chlorite platelets. The Fe-colloids were spherical in shape, and a few nanometer in diameter indicating that they were ferrihydrite particles.

Analytical speciation techniques for uranium and related elements

In our study the uranium sorption on chlorite was calculated during the mixed-flow experiments. The experimental data of the sorbed uranyl(VI) are shown in Fig. 2 together with the fitted sorption curve. Our calculations revealed that $7,48 \cdot 10^{-7}$ moles uranyl(VI) were sorbed during 239 hours.

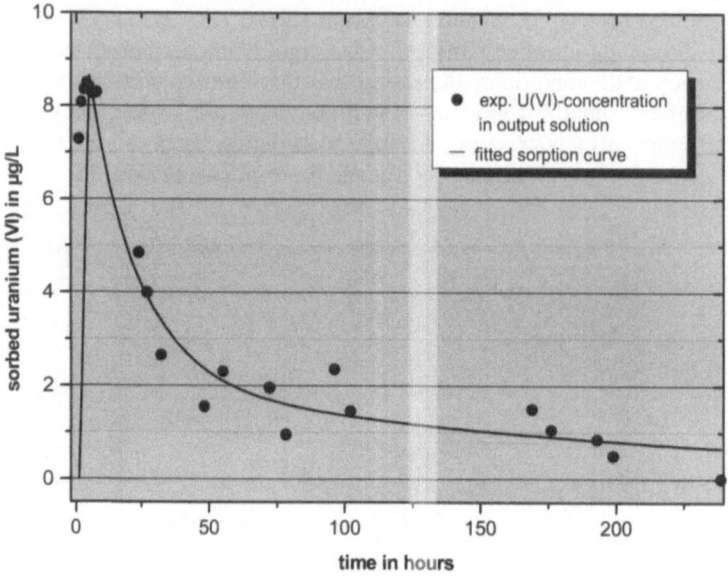

Fig. 2. Experimental results from mixed-flow experiments with chlorite in a 0.1 M NaClO₄-solution with added initial U(VI) concentration of $1 \cdot 10^{-6}$ M together with the fitted sorption curve.

Based on the specific surface area for chlorite, which we determined as 1,1 m^2/g and the surface site density of 1,45 sites/nm^2 (Arnold et al, 2001) the available reactive sites on chlorite were determined to be $7,97 \cdot 10^{17}$. According to EXAFS investigations (Arnold et al., in preparation), uranium is sorbed on the chlorite as a bidentate surface complex, i.e. two mol surface sites bind one mol uranyl(VI). These available reactive sorption sites are occupied after 138 hours of the mixed-flow experiment. However, uranyl(VI) is continously retarded (sorbed) on the chlorite powder indicating that an additional process is responsible for the additional sorption. We conclude that the continously newly forming ferrihydrite colloids, which result of the chlorite dissolution, provide the needed additional sorption sites responsible for uranium sorption beyond the available sorption sites on chlorite. Previous studies (Waite et al., 1994) showed that uranyl(VI) sorbs on ferrihydrite as bidentate surface complexes.

Conclusions

The sorption of uranyl(VI) on chlorite during its dissolution can be described by two different sorption sites. The first site is attributed to a surface site on the chlorite. The second site is attributed to surface sites on continously new forming ferrihydrite colloids which are the result of chemical weathering of chlorite. During the dissolution of chlorite a release of Fe will eventually lead to a precipitation of ferrihydrite. Due to its very high specific surface area and its high affinity to bind heavy metals ferrihydrite is an important sorbent for uranyl(VI). As sorbed particles forming coatings on the chlorite as well as colloids in the aqueous solution ferrihydrite may significantly influence the migration of uranium and other toxic heavy metals in uranium tailings. For risk assessments it is important to distinguish between the Fe-colloids in the aqueous solution and the Fe-particles which may form coatings on the chlorite platelets. Migration of radionuclides through geological formations via the water path distinguishes immobile and mobile contaminants. During the weathering of chlorite the sorbed immobile Fe-particles increase the overall sorption capacity of the chlorite and are thereby retarding the uranium migration. In contrast, the mobile colloids onto which uranium is sorbed are enhancing the rate of contaminant transport.

Acknowledgements

The authors would like to thank U. Schaefer for ICP-MS and AAS analyses and the German Research Council (DFG) for financial support under the contract number BE 2234/4-1.

References

Arnold, T., Zorn, T., Bernhard, G., Nitsche, H., 1998. Sorption of uranium (VI) onto phyllite. Chem. Geol. 151, 129-141.

Arnold, T., Zorn, T., Zänker, H., Bernhard, G., Nitsche, H., 2001. Sorption behavior of U(VI) on phyllite: experiments and modeling. J. Contam. Hydrol. 47, 219-231.

Arnold, T., Walter. M., Krawczyk-Bärsch, E., Bernhard, G. Spectroscopic evidence of a sorbed uranyl(VI) surface species on chlorite by EXAFS (in preparation).

Brandt et al. (2002): Chlorite dissolution in the acid pH-range: A combined microscopic and macroscopic approach. Geochim. Cosmochim. Acta (submitted).

Cornell, R. M., Schwertmann, U., 1996. The iron oxides - structure, properties, reactions, occurrence and uses. VCH Verlagsgesellschaft mbh, Weinheim, 573 pp.

Knauss, K.G., Wolery, T.J., 1989. Muscovite dissolution kinetics as a function of pH and time at 70 °C. Geochim. Cosmochim. Acta 53, 1493-1501.

Krawczyk-Bärsch, E., Arnold, T., Brandt, F., Bosbach, D. and Bernhard, G. (2002): Formation of secondary iron oxyhydroxide phases during the dissolution of chlorite – Effect on uranium sorption. Chem. Geol. (submitted).

Analytical speciation techniques for uranium and related elements

Waite, T.D., Davis, J.A., Payne, T.E., Waychunas; G.A., Xu, N., 1994. Uranium(VI) adsorption to ferrihydrite: application of a surface complexation model. Geochim. Cosmochim. Acta 58, 5465-5478

Major factors of processes forming uranium ore in exogenetic epigenetic deposits of the Chu-Sarysu Province (Southern Kazakhstan)

Vadim Shakhverdov

A. P. Karpinsky All-Russian Geological Research Institute (VSEGEI), 74, Sredny pr., St.Petersburg, 199106, Russia, v-shakh@yandex.ru

Abstract. Uranium ore in Chu-Sarysu province is confined to pinched out stratiform zones of oxidation developed in permeable non-lithified alluvial and coastal-marine Mesozoic and Cenozoic deposits The study of uranium content in Paleogene deposits related to geological history of the development of the region shows that the commercial uranium ore controlled by regional stratiform zones of oxidation is polychron, and the formation of major mineral deposits is caused by favourable integrated complex of interrelated geological processes.

Introduction

In the south of the Commonwealth of Independent States (CIS) there occurs the cis-Tien Shan uranium megaprovince surrounded by the young (N-Q) Tian Shan Orogen. The Chu-Sarysu province lies in the north-east of the megaprovince and is confined to the depression of the same name. The Chu-Sarysu depression is a large negative structure in the NE Turan Plate inherited from the mid-Paleozoic. The depression consists of the basement composed of dislocated metamorphosed Proterozoic-Early Paleozoic formations, the intermediate structural stage represented by lithified distinctly terrigenous and carbonate Middle and Upper Paleozoic formations, terrigenous Jurassic deposits occurring in individual graben-synclinoria, and the upper structural stage composed of weakly lithified or loose Cretaceous, Paleogene and younger sedimentary sequences.

Exogenetic epigenetic uranium deposits are assigned to uranium, selenium-uranium and molybdenum-uranium formation in connection with stratiform zones of oxidation in terrigenous and terrigenous-calcareous sediments. The industrial uranium ore is stratiform and occurs both in Cretaceous and Paleogene deposits. Ore bodies are discontinuous ribbon-shaped from a few hundred meters to 1.5-2.0

km wide. The shape of the ore bodies in the section is intricate and diverse. These are rolls proper, roll-shaped bodies and sheet deposits.

Uranium ore is confined to pinched out stratiform zones of oxidation developed in permeable non-lithified alluvial and coastal-marine Mesozoic and Cenozoic deposits (Fig. 1).

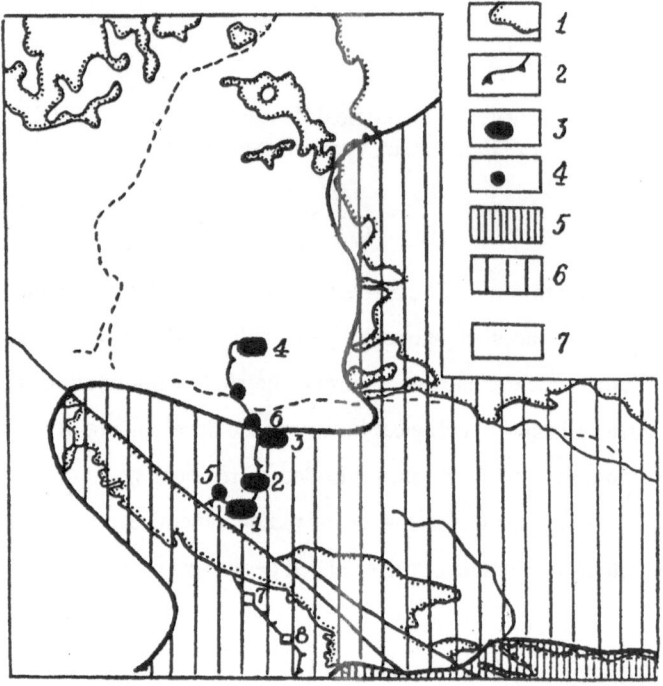

Fig. 1. The Chu-Sarysu province of uranium. 1- wedging out pre-Mesozoic rocks; 2- ore zone in Paleogene deposits; 3 - uranium deposits (1-Kangygan, 2-Moinkym, 3-Tortkydyk, 4-Yvanas); 4 – ore indices of uranium (5-Kaynar, 6-Ylkensor); 5- Tian Shan Orogen; 6 – suborogen area of Tian Shan Orogen; 7 - Turan Plate.

Uranium ore abundance and characteristic features of the mineralization occurring in Paleogene deposits depends on features of a chemical barrier situated at lines of pinched-out stratiform zone of oxidation. The barrier efficiency depends on the following major factors.

1. Facial-paleogeographical factor. Geological studies in the Chu-Sarysu province showed not once that the front of the pinched-out regional stratiform zones of oxidation was situated rather far from the depression frame. Facial-paleogeographical reconstruction and geochemical mapping demonstrated that two types of stratiform zones of oxidation have been developed in the Paleogene deposits of the Chu-Sarysu province: zones of first type, or regional zones, and zones of second type, or local zones. Rear parts of the regional stratiform zones of oxidation are composed of sediments that formed under conditions of eluvial landscapes and therefore deoxidizer-depleted. In the Paleocene these are deposits of proluvial plains and in the Eocene, deposits of dunes and beach ridges. The front of the regional stratiform zones of oxidation is situated at a distance of about 30 km from distribution boundaries of these complexes. Zones of the second type, or local zones, cut the facial-paleogeographical zones and are developed from the Karatau Ridge.

Thus, regional stratiform ore zones of oxidation take appropriate place in landscape-paleogeographical zoning of ore-bearing sediments and in this respect they are "facial". Commercial uranium ore appears only in association with regional stratiform zones of oxidation.

2. Facial-lithological factor. The most extensive ore bodies with stable commercial parameters form only within pinched-out regional zones of oxidation in sediments of particular facial types. In Paleocene horizons these are alluvial channel and marsh-gully deposits, and in the Eocene, deltaic deposits. Amount of uranium ore is controlled by the width of sediment distribution over the area and in the section. Ore-enclosing sediments form in submerged landscapes. The studies showed that these sediments are characterized by high reduction capacitance owing to high organic carbon content (on the average, more than 0.1 %) and ferrous and sulphide iron, as well as higher contents of uranium (on the average, 4.0-10.0 ppm). These geochemical features of the ore-enclosing deposits contribute to the construction of a contrast geochemical barrier on the front of the pinched-out stratiform zone of oxidation. Besides, the removal of uranium and its concentration at the barrier in the process of rock oxidation takes place in most of the deposits.

3. Hydrogeological factor. According to G. M. Shor, deposits in the Paleogene of the Chu-Sarysu province rest both on the high flank and in the centre of the artesian basin of the same name. Their hydrogeological features and directions of the stratiform zones of oxidation are controlled by two flows, which differ in direction, ion-salt composition, mineralization and uranium content.

4. Mineralogical and geochemical factor. Ore-controlling epigenetic redox zonation results from infiltration oxygen-bearing water movement in permeable horizons. The commercial uranium ore occupies strictly determined position in a series of successively alternating zones. There are following zones in the region: zone of rocks, unaltered by oxidizing epigenesis, uranium mineralization dissemination halo, uranium ore zone, and oxidation zone. Studies carried out in co-operation with I. S. Onoshko, N. I. Vetrova, and L. G. Rusinova showed that the distinguished zones differ in mineralogical and geochemical features (Table 1).

The dissemination halo is situated in the frontal part of the mineralization zone and is distinguished by uranium contents in rocks, which exceed background ones. The zone is 1.2-2.0 km long. Besides uranium, iron concentration is also higher in this zone as compared to unaltered rock.

Uranium ore zone. The ores are represented by fine coffinite, pitchblende and polymineral aggregates containing amorphous minerals such as lermontovite. Three subzones are distinguished within the mineralization zone: low-grade ore subzone, ordinary ore subzone, and disintegrating ore subzone.

Distribution of U^{IV} and U^{VI} within the ore bodies and radiological characteristics of uranium ores testify that the primary form of uranium deposition from solutions is sorption form. Further evolution of uranium ore is associated with uranium reduction processes and the development of its mineral form.

Oxidation zone. The oxidation zone is situated in the rear of uranium ores. Almost the whole of reactive ferrous and sulphide iron in rocks is oxidized up to oxide owing to the oxygen-bearing water infiltration. Organic carbon is lacking.

5. Microbiological factor. The ore pinching out (commercial uranium ore) of stratiform zones of oxidation is accompanied by intense activity of element-reducing microorganisms (Method of revealing ore mineralization, author's certificate No 1661208). Barren rocks of the stratiform zone of oxidation and unaltered rocks are characterized by mean value of microbiological activity of element-reducing bacteria of 0.3. This is a background value. The highest microbiological activity has been established in ores and near-ore rocks and has been estimated as 2.2. The microbiological activity increases as an ore body is approached.

Table 1. Mineralogical and geochemical features of ore-controlling epigenetic redox zonation

| Features | Dissemination halo | Uranium ore zone | | | Oxidation zone |
		low-grade ore subzone	ordinary ore subzone	disintegrating ore subzone	
1	2	3	4	5	6
Iron average content, %	0.354	0.933			0.355
Primary form of iron	Ferrous oxide, iron sulphide	Iron sulphide and ferrous oxide			Ferric oxide
Organic carbon content, %	0.022-4.788	0.023-6.2			No
Uranium average content, ppm	2.9-15.2	$< n*1000$	$> n*1000$	$n*100$	1.6-4.5
UO3/UO2	-	2.5	2.1	2.3	-
Uranium solubility (quota U VI), %	-	50-70	14-15	-	-
Uranium minerals	No	amorphous substance with coffinite	coffinite, pitchblende and polymineral aggregates containing amorphous minerals (lermontovite)	pitchblende and coffinite	No
Activity of element-reducing bacteria, number	0.3	2.2			0.3

The study of uranium content in Paleogene deposits related to the geological history of the development of the region shows that the commercial uranium ore con-

Analytical speciation techniques for uranium and related elements

trolled by regional stratiform zones of oxidation is polychron, and the formation of major mineral deposits is caused by favourable integrated complex of interrelated geological processes. It began in the epoch of sedimentation and continued in the stage of exo- and diagenesis. Time of the formation of regional stratiform zones of oxidation (25-30 Ma) is basic in ore genesis, for in this period in addition to internal resources of ore-enclosing rocks (syn- and exodiagenetic uranium concentrations), uranium masses from Oligocene crusts of weathering could be involved into the process. Later on, apparently in the Miocene, crust formation comes to an end and its erosion begins. Therefore, the current infiltration process does not form commercial ore in the Chu-Sarysu province and results in redistribution of earlier generated uranium ores.

Analytical speciation techniques for uranium and related elements

Microbially mediated reduction and immobilization of uranium in groundwater

Werner Lutze[1], Weiliang Gong[2], and H. Eric Nuttall[2]

[1]The Catholic University of America, Washington, D.C. 20063
[2]The University of New Mexico, Albuquerque, NM 87131

Abstract. Aquifers near uranium mines are frequently contaminated with uranium and other metals. Under certain conditions, indigenous microbes can be used to clean up the groundwater. U(VI) can be reduced and precipitated as the mineral uraninite (UO_2). A short summary of published work is given. Experimental results obtained by the authors are presented on immobilization of uranium and other metals in sandstone/groundwater mixtures. The significance of neo-formed iron sulfide (mackinawite) to control and maintain reducing conditions is pointed out.

Introduction

Microbially mediated reduction-oxidation processes have gained increasing interest in environmental remediation during the past decade. This is reflected by the large number of papers presented at international conferences on bioremediation. As an example, approximately 600 papers were presented at 'The Sixth International In Situ and On-Site Bioremediation Symposium' in San Diego, CA, 352 of which were published in 10 proceedings volumes. Volume 6(9) comprises 39 papers on bioremediation of inorganic species and compounds, including uranium.

Let us begin with a short summary of the literature on microbially mediated reduction of uranium. Review articles of the geomicrobiology of uranium and of uranium contamination in the subsurface and related remediation procedures can be found, e.g., in Reviews in Mineralogy, Vol. 38 (1999). Various microorganisms can reduce U(VI) to U(IV) either directly or indirectly under anaerobic conditions. Direct (enzymatic) reduction of uranium is a metabolism dependent accumulation of uranium. Indirect (abiotic) reduction of uranium can take place as a result of enzymatic reduction of other elements or species, e.g. sulfate to sulfide where sulfide can reduce uranium.

Geobacter metallireducens and Shewanella putrefaciens were the first microorganisms found to reduce U(VI) to U(IV) enzymatically (Loveley et al. 1991). These microorganisms used uranium as the sole electron acceptor and hydrogen, acetate, formate, lactate, and pyruvate, respectively, as electron donor. The authors reported that enzymatic reduction of uranium was much faster than abiotic reduction by sulfide. U(IV) precipitated and crystallized as UO_2, the mineral uraninite (Gorby and Lovely 1992). The enzyme cytochrome c3 was shown to be involved in the redox mechanism as electron carrier (Lovely et al. 1993a). Shortly after their work with Geobacter metallireducens and Shewanella putrefaciens Lovely and coworkers found that other microorganisms such as sulfate-reducing bacteria (SRB), Desulfovibrio species (D. sulfuricans, D. vulgaris, and D. baculatum) are also capable of reducing U(VI) (Loveley and Phillips 1992; Lovely et al. 1993b).

Since Lovely et al.'s work in1991, many publications have appeared on microbially mediated reduction of uranium under a wide range of conditions. Suggestions and attempts have been made to use specific microbes or indigenous mixed cultures to clean up sites contaminated with uranium. For example, Francis and coworkers (Francis et al. 1991) studied U(VI) reduction in sediment samples contaminated with uranium processing wastes. These authors showed that U(VI) was reduced and precipitated while Fe(III) and Mn(IV) were reduced and dissolved by the activity of indigenous bacteria. Separation of U(VI) from aqueous solutions by enzymatic reduction was considered more effective than by biosorption. Francis et al. (1994) reported on reduction of uranium by Clostridium sp. Uhrie et al. (1996) studied removal of U(VI) and other metals from aqueous solutions using mixed cultures of sulfate-reducing bacteria from a sulfidogenically active sediment from the Laramie River, Wyoming. The water was amended with sulfate and lactate. Uranium was added at concentrations between 10 and 40 mg/L. Uranium was reduced quantitatively within 3 days. Hard and Babel (1997) isolated strains of sulfate-reducing bacteria from a wastewater pond in Germany and a copper mine in Norway. The isolates were tested under various conditions (pH, concentration of uranium and presence of other heavy metals such as Al, Cd, Cr, Cu, Pb, Ni, Mn, Ag). Ganesh et al. (1997) studied the effect of complexation on uranium reduction and found that reduction rates depended on the types of bacteria and ligands. Reduction by D. desulfuricans of U(VI) complexed with malonate, oxalate, or citrate was slower than with 4,5-dehydroxy-1,3-benzene disulfonic acid. Reduction with S. alga showed the opposite effect.

As long as anaerobic conditions prevail or can be established, uranium and many other heavy metals encounter almost ideal conditions for sequestration in natural aquatic systems. There are many processes possible including biosorption, precipitation of schoepite, a uranyl oxide hydrate (Suzuki and Banfield 1999), adsorption on iron sulfide precipitated by microbes (Watson et al. 2000; Watson and Ellwood 2001) or microbially mediated reduction and precipitation as uraninite (Abdelouas et al. 1998b).

Uranium in groundwater

Redox processes and formation of solid phases

In the present work, we are primarily interested in microbially mediated reduction and precipitation of uranium in contaminated aquatic systems, groundwater in particular. Frequently, such aquatic systems contain sulfate and the surrounding geological material contains plenty of iron, e.g., goethite ($FeOOH$). Sulfate-reducing bacteria are widespread in aquatic systems under anaerobic conditions. Many studies have confirmed ubiquity of these and other microbial species with metal reduction capabilities. Because of limited space only a few references are given here (Abdelouas et al. 1998b, 1999, Arakaki and Morse 1993, Lutze et al. 2001, Revis et al. 1991, Watson et al. 1995, Watson and Ellwood 2001a).

If organic carbon is available, microbial activity will establish reducing conditions. A number of microbially mediated redox reactions will be triggered, depending on water composition. Equations (2) to (6) show how various aqueous species are reduced and what the chemical reactions are. The organic carbon source is lactate in this case. The redox potential decreases in the order of reactions (2) to (6). The equations are written such that the electron for each of the reduction reactions (2) to (6) is provided by equation (1):

Oxidation of organic carbon (lactate)
$$1/12CH_3CH(OH)COO^- + 1/4OH^- = 1/4CO_2 + 3/4H^+ + e^- \qquad (1)$$
Reduction of dissolved oxygen
$$1/2O_2 + 2H^+ + e^- = 2H_2O \quad (2)$$
Reduction of nitrate to nitrogen gas
$$1/5NO_3^- + 6/5H^+ + e^- = 1/10N_2 + 3/5H_2O \quad (3)$$
Reduction of U(VI)* to U(VI) and precipitation of uraninite
$$1/2UO_2^{2+} + H_2O + e^- = 1/2U(OH)_4 \quad (4a)$$
$$1/2U(OH)_4 = 1/2UO_2 + H_2O \qquad (4b)$$
Reduction of bioavailable iron[a] (e.g., goethite)
$$FeO(OH) + 3H^+ + e^- = Fe^{2+} + 2H_2O \qquad (5)$$
Reduction of sulfate to sulfide
$$SO_4^{2-} + 9H^+ + 8e^- = HS^- + 4H_2O \quad (6)$$
Precipitation of iron sulfide, e.g. mackinawite ($FeS_{(1-x)}$ with $x \approx 0.1$)
$$Fe^{2+} + (1-x)HS^- = FeS_{(1-x)} + (1-x)H^+ \qquad (7)$$

In the presence of continuous microbial activity, the sulfur content of mackinawite increases with time and phase transformations produce a series of minerals, e.g. greigite, smythite, and pyrrhotite, $FeS_{(1+x)}$, with $x<1$ (McNeil and Little 1990). The thermodynamically stable phase in a reducing natural environment is pyrite, FeS_2. Pyrrhotite and pyrite have been detected upon formation of mackinawite (Abdelouas et al., 2000).

The most abundant redox sensitive element in aquifer host rock is iron, e.g., in the form of goethite or magnetite. If sulfate is present, its concentration is frequently much higher than that of metal contaminants. If growth of indigenous SRB was stimulated in such systems their chemistry and the redox potential are determined by the reaction products HS^- and FeS_{1-x} (equations (6) and (7)). Fig. 1 shows an Eh-pH diagram for the Fe-S-U-O-H system after Little et al. (1997). According to this diagram, mackinawite is thermodynamically stable only under strongly reducing conditions. If mackinawite forms in an aquifer, the redox potential should be Eh < -0.4 mV at pH ≈ 6. The diagram shows that mackinawite does not form in acid waters (pH<5) but is stable under alkaline conditions.

Fig. 2 shows mackinawite as it is produced by microbially mediated reduction of sulfate and of iron either from minerals contained in sandstone (Abdelouas et al. 1998b) or from freshly precipitated iron hydroxide (Lutze et al. 2001). The SAED pattern in Fig. 2 shows that the material is crystalline. Mackinawite crystals were needle-shaped (≤200 nm long, ≤30 nm thick).

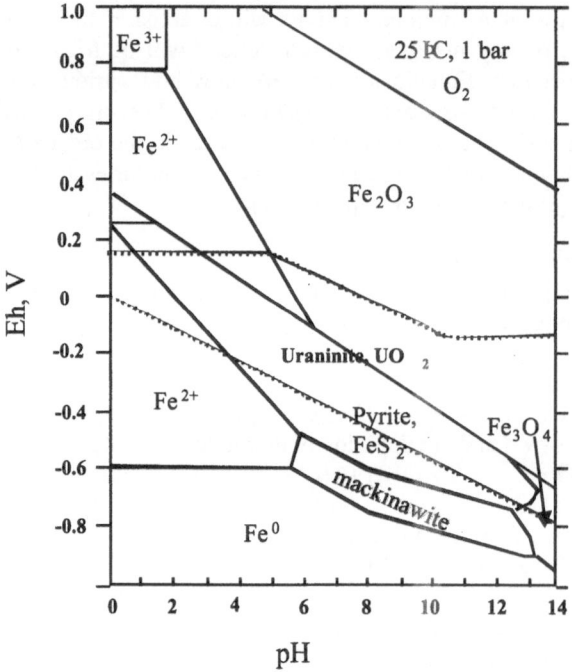

Fig. 1. Eh-pH diagram of Fe-S-U-O-H. (after Little et al. 1997).

As long as mackinawite is present, uranium is only stable in its +4 valence state. Fig. 1 shows the stability field of uraninite. Thermodynamically, uraninite, Fig. 1, would be protected against oxidative dissolution. We have detected uraninite in the presence of mackinawite (Abdelouas et al. 1999). Fig. 3 (a) shows uranium dioxide particles attached to bacteria. Fig. 3(b) is a SAED pattern of the particles in

Fig. 3(a) showing that the material is polycrystalline uraninite. Fig. 3(c) is an EDS spectrum of the material in Figs. 3(a). Uraninite contained some calcium which is not uncommon in natural uraninite. Watson and Ellwood (2001) reported that precipitation of U(IV) as sulfide (US2) can occur down to less than 1 mg/L U. The authors have not given experimental evidence for the presence of US2. Thermodynamically, the formation of US2 cannot be excluded at this low redox potential and at high sulfide concentrations. However, we have not seen any sulfur in precipitated uranium Fig. 3(c).

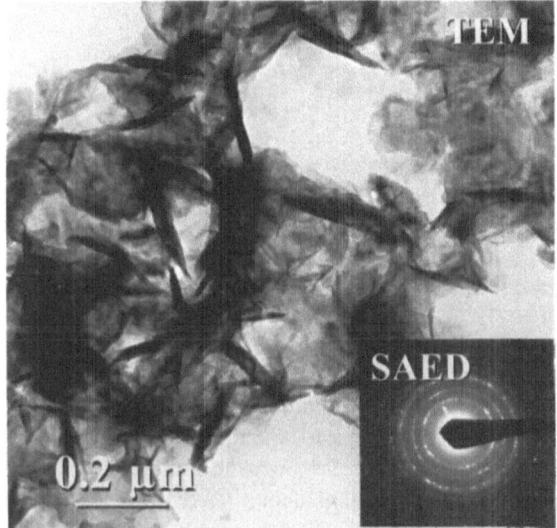

Fig. 2. Mackinawite produced by SRB in a simulated aquifer (groundwater and sandstone from Königstein, Germany). Inlay: selected area electron diffraction (SAED) pattern.

We have studied two aquifers in sandstone as potential candidates for in situ bioremediation, one near Tuba City, Arizona, USA., the other at Königstein, Saxony, Germany. The groundwater at the Tuba City site was contaminated with uranium (up to 1 mg/L) sulfate, and nitrate, whereas at Königstein mine flooding water could become the source of uranium contamination of a nearby aquifer. Using rock and water samples from the two sites, we have studied the processes leading to reduction and immobilization of uranium (Abdelouas et al. 1998b, 1999). Oxidative dissolution of precipitated mackinawite and uraninite was also investigated (Abdelouas et al. 1999; Lutze et al. 2001). Experiments were conducted in closed systems (serum bottles) and in columns filled with sandstone. All microbially mediated reduction reactions were mediated by indigenous microbes residing on core samples of sandstone taken from the aquifers at Tuba City and Königstein. Microbes were detected under the TEM (Fig. 3a) but no effort was made to characterize them. There is ample evidence in the literature that indigenous multicultures are ubiquitous and contain, e.g., dissimilatory Fe3+ reducing or sulfate-reducing species needed to mediate metal reduction.

Microbiology / phytoremediation

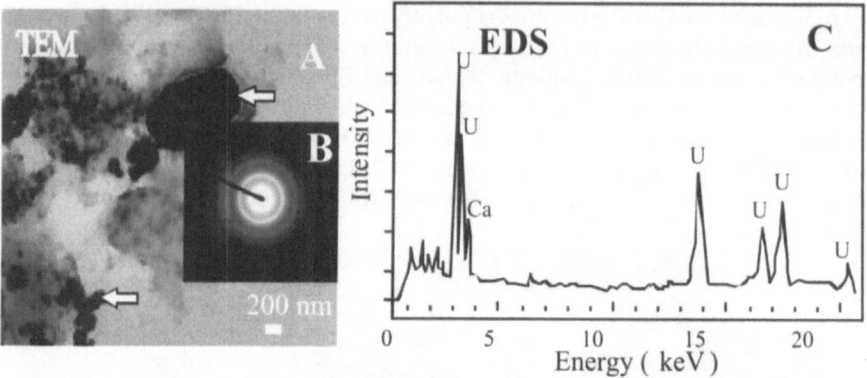

Fig. 3. A: Transmission electron micrograph of uraninite particles (black) attached to bacteria; B: selected area electron diffraction (SAED) pattern of the black material; C: energy dispersive X-ray spectrum (EDS) of black material in A.

Fig. 4 shows new experimental results similar those published previously (Lutze et al. 2001). In the experiment underlying Fig. 4, we have assumed that water from the flooded mine (pH≈3, Fe^{3+}≈0.2g/L, SO_4^{2+}≈1.5g/L, U^{6+}≈17mg/L) would reach the uncontaminated aquifer nearby (pH≈6). Dilution of the mine water by water in the uncontaminated aquifer would increase the pH and decrease the concentrations of dissolved species. Fig. 4 shows what happens to uranium under these conditions. Assuming (somewhat arbitrary) an initial concentration of uranium of 1mg/L after dilution, we see in Fig. 4 an instantaneous decrease of the initial U(VI) concentration by a factor of 25 (circles). This is due to precipitation of $Fe(OH)_3$ and adsorption of uranium as UO_2^{2+}. Without amending the solution, the uranium concentration remained constant at about 40 µg/L for the duration of the experiment. Slight variations of experimental conditions yielded different final uranium concentrations ranging between 15 µg/L and 40 µg/L. In the United States the maximum permissible concentration of uranium in groundwater is currently 44 µg/L (Federal Register 1995) but will be lowered to 20 µg/L in 2002. Analogue behavior could be expected on-site, if the assumed mixing of mine water with water in the uncontaminated aquifer takes place. With amendment (addition of Na-lactate and Na-trimetaphosphate), the initial decrease of U(VI) concentration was the same (Fig. 4, square symbols) due to adsorption on $Fe(OH)_3$. However, after one month, microbial activity became significant. Sulfate was reduced to sulfide (equation (6)) and Fe^{3+} to Fe^{2+} (equation (5)), indicated by visible precipitation of black iron sulfide (equation (7)). Reduction of Fe^{3+} lead to disappearance of $Fe(OH)_3$ and release of adsorbed U(VI) into solution, also seen after about 30 days (Fig. 4). After 50 days the concentration of U(VI) began to decrease reaching a final value of about 1 to 5 µg/L. The evolution of the U(VI) concentration between 30 and 70 days can be explained by the relative rates of uranium desorption and reduction. In the beginning, more U(VI) was desorbed than reduced. Later, desorption ceased because $Fe(OH)_3$ was consumed and reduction of uranium decreased its concentration. With microbial activity the final concentration of uranium is a factor of 20 to 40 lower than without; U(VI) has been reduced and precipitated as an insoluble phase (uraninite). The products mackinawite and uraninite were identified by TEM, SAED and EDX.

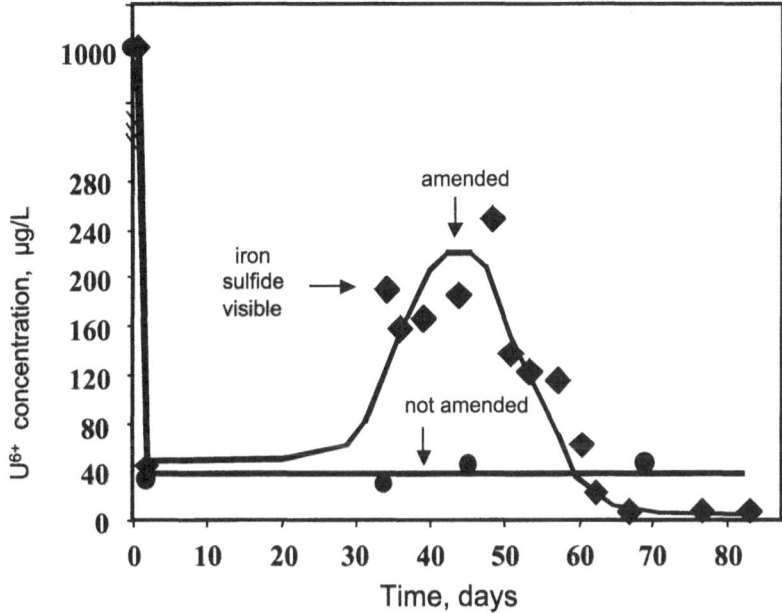

Fig. 4. Adsorption and microbially mediated reduction and immobilization of uranium on sandstone in a serum bottle. Squares: solution amended with sodium lactate and sodium trimetaphosphate; circles: solution not amended; uranium adsorbed on precipitated iron hydroxide (Groundwater and sandstone from Königstein).

We have also studied oxidative dissolution of the two phases mackinawite and uraninite. We have simulated a situation in which dissolved oxygen has access to an aquifer cleaned by in situ bioremediation. Water with known concentration of dissolved oxygen was injected into serum bottles or fed into columns in which uraninite and mackinawite had formed as a result of microbially mediated reduction of U(VI), SO_4^{2-} and Fe^{3+}. Oxygen was consumed by sulfide oxidation and a stoichiometric equivalent of sulfate was measured (Lutze et al. 2001). Uranium concentrations remained below 20 µg/L. Mackinawite was present in much larger quantity than uraninite. Its oxidation consumed all the oxygen. Thus, mackinawite served as a redox buffer protecting uraninite against oxidative dissolution. Formation of goethite (FeOOH) was confirmed experimentally by transmission electron microscopy (Abdelouas et al. 1999).

Comparison of results obtained with material from Tuba City (Abdelouas et al. 1998a, 1998b, 1999) and Königstein (this work and Lutze et al. 2001) yielded very similar results. In both cases uranium was reduced and immobilized as UO_2. Final solution concentrations of uranium were the same (near 1 µg/L). A large excess of mackinawite was formed and reoxidation experiments showed that uraninite was protected by mackinawite from oxidative dissolution. Differences in the results reflect directly the difference in groundwater composition. For example, there was no dissolved iron in the Tuba City water. The pH was close to neutral and no precipitation of $Fe(OH)_3$ and thus no adsorption of uranium was observed.

Other metal contaminants

Many heavy metals form highly insoluble sulfides. Mackinawite forms tiny crystals (Fig. 2) and has a large specific surface area, 400-500 m^2/g (Watson et al. 2001b). Hence, adsorption or chemisorption (Watson et al. 1995, 2001a), formation of separate sulfide phases, e.g. PbS (Fig. 4), and formation of solid solutions, e.g. with Ni, as suggested by the composition of naturally occurring mackinawite $(Fe,Ni)_9S_8$, are all processes sequestering heavy metals.

Fig. 5 shows lead sulfide crystals precipitated from a mixture of contaminated groundwater and sandstone (material from Koenigstein, Germany; for more information on the site and the aquifer see Lutze et al. 2001).

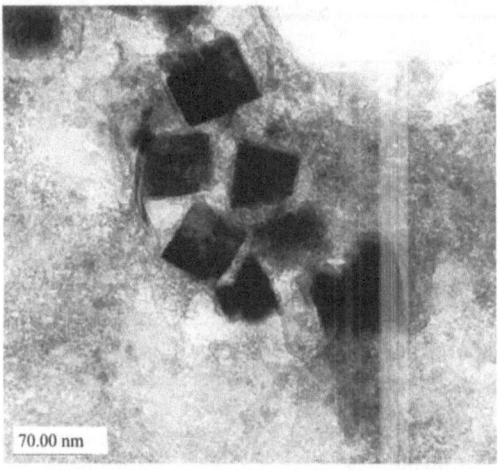

70.00 nm

Fig. 5. Lead sulfide crystals (PbS) embedded in iron sulfide.

The PbS crystals are about 50 nm in size embedded in iron sulfide. The iron sulfide in Fig. 5 is mackinawite but less crystalline than the sample shown Fig. 2. Concentrations of contaminant metals (e.g., U, Cu, Ni, Mo, Zn, Cd, Co) were on the order of 1-10 mg/L before microbially mediated sulfate reduction. After precipitation of iron sulfide we found that the respective metal concentrations had decreased to about or less than 1 µg/L, in good agreement with what was reported by Watson and Ellwood (2001b). Analysis of the mackinawite shown in Fig. 5 by energy-dispersive X-ray spectroscopy allowed us to detect several metals (Co, Zn, Cd) in the iron sulfide phase. Only lead formed a separate phase. The concentration of manganese increased in solution as a result of reductive dissolution of Mn(IV) to Mn(II).

Anaerobic conditions prevail as long as iron sulfide is present. The large quantity of iron sulfide compared with that of precipitated contaminants suggests that the system is well buffered against changes of redox potential, e.g. by small amounts of oxygen in the groundwater after termination of microbial activity. Oxygen is quickly consumed by sulfide oxidation, mostly of mackinawite.

Summary

In summary, it appears that the directive for in situ bioremediation of groundwater contaminated with uranium (and other heavy metals) would be to initiate and support formation of iron sulfide (mackinawite) with the help of indigenous bacteria. The presence of this single phase would support simultaneously several processes (adsorption, chemisorption, precipitation, coprecipitation, reduction and redox buffering) all in favor of sequestering contaminants and stabilizing them over long periods of time. It must be borne in mind that oxygen concentrations are fairly low in groundwater (typically a few milligrams per liter, up to 8 mg/L) and that the amount oxygen with access to a site remediated in situ is also limited by groundwater flow. For the conditions at Tuba City and Königstein total oxidation of the newly formed mackinwite would take on the order 10^3 years, if not longer.

References

Abdelouas A, Lutze W, Nuttall H E (1998a) Chemical reactions of uranium in ground water at a mill tailings site. J. Contam. Hydrology 34: 334-361

Abdelouas A, Yongming Lu, Lutze W, Nuttall HE (1998b) Reduction of U(VI) to U(IV) by indigenous bacteria in contaminated ground water. J. Contam. Hydrology 35, 217-233

Abdelouas A, Lutze W, Nuttall HE (1999) Oxidative dissolution of uraninite precipitated on Navajo sandstone. J. Contam. Hydrology 36: 353-375

Abdelouas A, Lutze W, Gong W, Nuttall HE, Strietelmeier BA, Travis BJ (2000) Biol. reduction of U in groundwater and subsurface soil. Science Total Environ. 250: 21-35

Arakaki T, Morse JW (1993) Coprecipitation and adsorption of Mn(II) with mackinawite FeS under conditions found in anoxic sediments. Geochim. Cosmochim. Acta 57: 9-14

Federal Register (1995) Environmental Protection Agency, CFR 40, Part 192, Groundwater standards for remedial actions at inactive uranium processing sites, Table 1: 2866

Francis AJ, Dodge CJ, Gillow JB, ClineCE (1991) Microbial transformation of uranium in wastes. Radiochimica Acta 52-3: 311-316

Francis AJ, Dodge CJ, Lu F, Halada GP, Clayton CR (1994) XPS and XANES studies of uranium reduction by Clostridium sp. Environ. Science Technol. 228: 636-639

Ganesh R, Robinson KG, Reed GD, Sayler GS (1997) Reduction of hexavalent uranium from organic complexes by sulfate- and iron-reducing bacteria. Applied Environ. Microbiol. 63: 4385-4391

Gorby YA, Loveley DR (1992) Enzymatic U precipitation. Environ.Sci. Techn. 26: 205-207
Hard BC, Babel FW (1997) Bioremediation of facultatively methylotrophic metal-tolerant sulfate-reducing bacteria. Microbiol. Res. 152: 65-73
Little BJ, Wagner PA, Lewandowski Z (1997) Reviews in Mineralogy Vol. 35
Lutze W., Chen Z, Diehl D, Gong W, Nuttall HE, Kieszig G (2001) Microbially mediated reduction and immobilization of uranium in ground water at Königstein. The sixth international conference on in situ and on-site bioremediation, eds. Leeson A, Peyton BM, Means JL, Magar VS, Battelle Press, Columbus; Vol. 9 Bioremediation of inorganic compounds: 155-163
Loveley DR, Phlilips EJP, Gorby YA, Landa ER (1991) Microbialreduction of uranium. Nature 350: 413-416
Loveley DR, Widman PK, Woodward JC, Phillips EJP (1993). Reduction of uranium by cytochrome c3 of Desulfofibrio vulgaris. Applied Environ. Microbiol. 59: 3572-3576
Loveley DR, Rden EE, Phillips EJP, Woodward JC (1993b) Enzymatic iron an uranium reduction by sulfate-reducing bacteria. Marine Geol. 113: 41-53
Loveley DR and Phillips EJP (1992) Reduction of uranium by Desulfofibrio desulfuricans. Environ.Science Technol. 26: 2228-2234.
McNeil MB, Little BJ (1990) Mackinawite formation during microbial corrosion. Corrosion 46: 599-600
Revis NJ, Elmore J, Edenborn H, Osborne T, Holdsworth G, Hadden C, King A (1991) Immobilization of mercury and other heavy metals in soil, sediment, sludge and water by sulfate-reducing bacteria. In Bacterial Processes, eds. Freeman HM, Sferra PM; Vol. 3: 97-105
Reviews in Mineralogy, Vol. 38 (1999) Uranium: Mineralogy, Geochemistry and the Environment; eds. Burns PC, R. Finch R; Mineralogical Society of America, ISSN 0275-0279. Chapter 8: Suzuki Y, Banfield JF, Geomicrobiology of uranium; chapter 9: Abdelouas A, Lutze W, Nuttall HE, Uranium contamination in the subsurface: Characterization and remediation.
Uhrie JL, Drever JI, Colberg PJS, Nesbitt CC (1996) In situ immobilization of heavy metals associated with U leach mines bacterial sulfate reduction. Hydrometallur. 43: 231-239.
Suzuki Y and Banfield JF 1999. Geomicromicrobiology of uranium. In: Reviews in Mineralogy 38: 393-432. Uranium: Mineralogy, Geochemistry and the Environment; eds. P. C. Burns and R. Finch; Mineralogical Society of America, ISSN 0275-0279.
Watson JHP, Ellwood DC, Deng Q, Mikhalovsky S, Hayter CE, Evans J (1995) Heavy metal adsorption on bacterially produced FeS. Minerals Engineering 8(10): 1097-1108
Watson JHP, Cressey AP, Roberts AP, Ellwood DC, Charnock JM, Soper AP (2000) Structural and magnetic studies on heavy-metal-adsorbing iron sulfide nanoparticles produced by sulfate-reducing bacteria. J. Magnetism Magnetic Mater. 214(1-2):13-30
Watson JHP, Croudace IW, Warwick PE, James PAB, Charnock JM, Ellwood DC (2001a) Adsorption of radioactive metals by strongly magnetic iron sulfide nanoparticles produced by sulfate-reducing bacteria. Separation Science Technol. 36 (12): 2571-2607
Watson JHP, Ellwood DC (2001b) Removal of heavy metals and organic compounds from anaerobic sediments. The sixth international conference on in situ and on-site bioremediation, eds. Leeson A, Peyton BM, Means JL, Magar VS, Battelle Press, Columbus; Vol. 9 Bioremediation of inorganic compounds: 61-69

Microbial Transformations of Uranium Complexed with Organic and Inorganic Ligands

Arokiasamy J. Francis

Environmental Sciences Department, Brookhaven National Laboratory, Upton, New York, USA

Abstract. Biotransformation of various chemical forms of uranium present in wastes, contaminated soils and materials by microorganisms under different process conditions such as aerobic and anaerobic (denitrifying, iron-reducing, fermentative, and sulfate-reducing) conditions will affect the solubility, bioavailability, and mobility of uranium in the natural environment. Fundamental understanding of the mechanisms of microbial transformations of uranium under a variety of environmental conditions will be useful in developing appropriate remediation and waste management strategies as well as predicting the microbial impacts on the long-term stewardship of contaminated sites.

Introduction

Uranium exists in several oxidation states U(III), U(IV), U(V), and U(VI), of which U(IV) and U(VI) are the predominant ones in the environment. Uranium may be present in wastes and contaminated soils, and materials as elemental, oxide, coprecipitates, ionic, inorganic-, and organic-complexes, and naturally occurring minerals depending on the process and waste stream. Microbial activity could affect the chemical nature of the uranium by altering the speciation, solubility and sorption properties and thus could increase or decrease the mobility of uranium. Among the radionuclides, biotransformation of uranium has been extensively studied (Francis 1998). In this paper, biotransformation of several chemical forms of uranium under various microbial process conditions such as aerobic, denitrifying, iron-reducing, fermentative, and sulfate-reducing conditions are examined.

Microbial mobilization and immobilization of uranium

Under appropriate conditions, dissolution or immobilization of uranium is brought about by direct enzymatic or indirect non-enzymatic actions of microorganisms. Both aerobic and anaerobic microorganisms are involved in the mobilization and immobilization of various chemical forms of uranium. Dissolution of uranium by autotrophic sulfur- and iron-oxidizing bacteria due to oxidation of inorganic compounds and by heterotorphic bacteria and by fungi due to production of organic acids and chelating agents such as siderophores has been reported.

Immobilization of U is brought about by bioreduction and bioprecipitation reactions. Uranium is reduced by a wide variety of facultative and strict anaerobic bacteria under anoxic conditions in the presence of suitable electron donors.

$$U \, (VI)_{aq} \, \frac{dissimilatory \, metal \, reducers}{fermenters, \, sulfate \, reducers} \rangle \, U(IV)_s$$

Consequently, the potential exists for the use of anaerobic bacteria to concentrate, contain and stabilize U in contaminated groundwaters and in waste with concurrent reduction in waste volume. Reactive barrier technology is based on the activities of these anaerobic bacteria. However, the long-term stability of bacterially immobilized U in the natural environment is poorly understood.

Biosorption and bioaccumulation of uranium has been observed in a wide range of microorganism. It is still one of the intensely investigated areas of research because of the potential use of biomass to remove U from waste streams. Extracellular and intracellular association of U with bacteria was observed but the extent of its accumulation differs greatly with the species of bacteria. Uranium forms complexes with the carboxylate, phosphate, amino, and hydroxyl functional groups present on the cell surface; and intracellularly, by binding to anionic sites or precipitating as dense deposits. Extended X-ray absorption fine structure (EXAFS) analysis of the association of U with halophilic and non-halophilic bacterial cells showed that it was associated predominantly with phosphate as uranyl hydrogen phosphate and as hydroxophosphato or polyphosphate complexes as well as with other ligands such as carboxyl group. These and other studies show that cellular phosphates, including the polyphosphates, bind significant amounts of uranium (Panak et al, 1999, 2000; Sakaguchi, 1996; Francis et al manuscript submitted).

Uranium associated with the bacteria is not very stable, as it was removed completely by Na_2HCO_3 from *Halomonas* sp, from an *Arthrobacter* sp by 0.1 sodium hydrogen carbonate, 0.1M EDTA, and Na_2CO_3, and from Bacillus strains by EDTA (Panak et al, 1999; Sakaguchi, 1996; Francis et al manuscript submitted).

Biotransformations of uranium-organic and -inorganic complexes

Although a wide variety of microorganisms are present in uranium mining wastes and in natural radioactive mineral deposits, the extent to which they regulate the mobility of uranium complexed with organic and inorganic ligands has not been-fully evaluated. Further, there is a paucity of information on the biotransformation of uranium complexed with naturally occurring organic ligands. Biotransformation of radionuclide-complexes may result in precipitation of the released ion as water-insoluble hydroxides, oxides or salts. Such transformations may result in the retardation of uranium migration. Current understanding of the biotransformation of uranium-complexed with organic and -inorganic ligands by aerobic and anaerobic (denitrifying, iron-reducing, fermentative, sulfate-reducing) bacteria is summarized below.

Biotransformation under aerobic conditions

Binary uranyl-citrate complex. Citric acid, a naturally occurring compound, is a multidentate ligand, which forms stable complexes with various metal ions. It forms different types of complexes with transition metals and actinides including formation of a bidentate, tridentate, binuclear, or polynuclear complex species. Biodegradation of metal citrate complexes is dependent upon the type of complex formed between the metal and citric acid; bidentate complexes are readily biodegraded whereas the tridentate complexes are recalcitrant (Francis et al 1992). *Pseudomonas fluorescens* metabolized the bidentate complexes whereas complexes involving the hydroxyl group of citric acid, and the binuclear U-citrate complex are not (Francis et al 1992). The presence of the free hydroxyl group of citric acid is the key determinant in effecting biodegradation of the metal complex. The lack of degradation was not due to their toxicity but was limited by the transport and/or metabolism of the complex by the bacteria. No relationship was observed between biodegradability and stability of the complexes.

Adding excess citric acid to equimolar (0.52mM) uranyl citrate resulted in the metabolism of the excess citric acid; the uranyl citrate complex was not toxic to the bacterium. In the presence of 1-, 2-, and 3-fold excess citric acid, the citric acid remaining in each complex after biodegradation was 0.75, 0.80, and 0.83 mM, respectively. The final stoichiometry of U-citric acid in all three treatments was approximately 2:3, indicating the formation of 2:3 U-citric acid complex.

Ternary Iron- and uranium-citrate complexes. Citric acid forms ternary mixed-metal complexes with various metal ions involving the hydroxyl and carboxyl groups of citric acid. The presence of 1:1:2 Fe:U:citric acid in solution was confirmed by potentiometric titration, UV-vis spectrophotometry, gel-filtration chromatography, and extended X-ray absorption fine structure (EXAFS) analysis (Dodge and Francis, 1997). Comparison of the EXAFS spectra show the 1:1:2 Fe:U:citric acid complex has structural characteristics similar to the 1:1 U:citric acid complex. Biotransformation studies of Fe-U-citrate complex by *P. fluorescens* showed that the ternary 1:1:2 Fe:U:citric acid complex was recalcitrant.

When a one-fold excess citric acid was added to this complex, the excess citric acid was completely degraded with no change in the stoichiometry of the complex. However, with two-fold excess citric acid, a 1:1:1 Fe:U:citric acid complex remained in solution after the excess citric acid was biodegraded. These results suggest that similar to the U-citrate complex, the Fe-U-citrate complex is recalcitrant to biodegradation (Dodge and Francis, 1997). The persistence of mixed-metal-citrate complexes in wastes and contaminated environments may result in the mobilization of uranium.

Biotransformation under Denitrifying Conditions

Metabolism of Uranyl-citrate Complex. Citrate metabolism by *Pseudomonas fluorescens* under denitrifying conditions was facilitated by the enzyme aconitase as in aerobic metabolism. Biodegradation of the various metal-citrate complexes by *P. fluorescens* under denitrifying conditions was similar to that observed aerobically, but the rates were much lower. The bacterium completely degraded bidentate complexes, whereas the tridentate complexes, and the binuclear U-citrate complex, were not metabolized by the bacterium (Joshi-Tope and Francis, unpublished results).

Biotransformation of uranyl-organic and -inorganic complexes. Biotransformations of uranyl nitrate, uranyl-citrate, uranyl-ethylenediaminetetraacetate (EDTA), and uranyl carbonate by a halophilic bacterium, Halomonas sp., isolated from the Waste Isolation Pilot Plant (WIPP) repository site was investigated under denitrifying conditions (Francis et al 2000).

Addition of uranyl nitrate, uranyl-citrate, or uranyl-EDTA to the brine bacterial growth medium containing 4M NaCl resulted in the precipitation of uranium as a uranyl hydroxophosphato species [$K(UO_2)_5(PO_4)_3(OH)_2 \cdot nH_2O$]. Dissolution of the uranium precipitate was concomitant with the growth of the bacterium under denitrifying conditions (Figures 1-3).

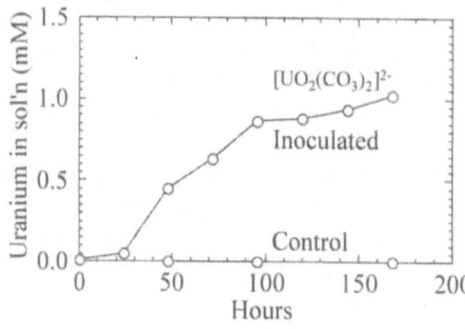

Fig. 1. Addition of uranyl nitrate to bacterial growth medium resulted in the precipitation of U as uranyl phosphate. Growth of *Halomonas* sp. resulted in the solubilization of U as uranyl dicarbonate.

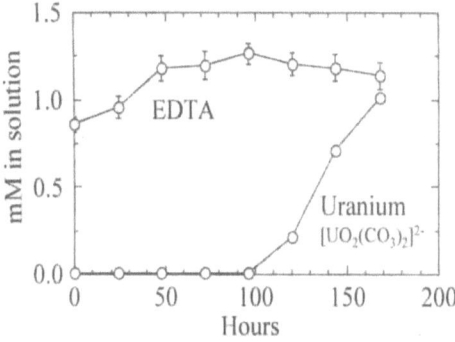

Fig. 2. Addition of uranyl-EDTA to bacterial growth medium resulted in the precipitation of U as uranyl phosphate. *Halomonas* sp. solubilized U as uranyl dicarbonate. EDTA remained in solution and was not metabolized by the bacterium.

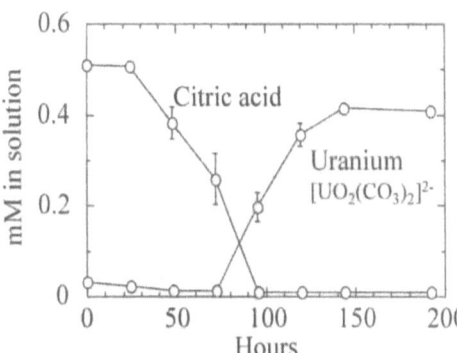

Fig. 3. Addition of uranyl-citrate to bacterial growth medium resulted in the precipitation of U as uranyl phosphate. Growth of *Halomonas* sp. resulted in the dissolution of U as uranyl dicarbonate. Citric acid was metabolized by the bacterium

The bacterium completely metabolized the citrate released from uranyl-citrate complex, but not the EDTA released from the U-EDTA complex. The UV-vis spectra of the culture medium during growth revealed the formation of a uranyl dicarbonate complex $[UO_2(CO_3)_2]^{2-}$ due to CO_2 production from the metabolism of the carbon source. There was no change in the speciation of added uranyl carbonate. These results show that uranium phosphate is readily solubilized by bacterial action with formation of mobile to uranyl carbonate species.

Addition of uranyl nitrate, uranyl-citrate, or uranyl-EDTA to the brine bacterial growth medium containing 4M NaCl resulted in the precipitation of uranium as a uranyl hydroxophosphato species $[K(UO_2)_5(PO_4)_3(OH)_2 \cdot nH_2O]$. Dissolution of the uranium precipitate was concomitant with the growth of the bacterium under denitrifying conditions (Figures 1-3). The bacterium completely metabolized the citrate released from uranyl-citrate complex, but not the EDTA released from the U-EDTA complex. The UV-vis spectra of the culture medium during growth revealed the formation of a uranyl dicarbonate complex $[UO_2(CO_3)_2]^{2-}$ due to CO_2 production from the metabolism of the carbon source. There was no change in the speciation of added uranyl carbonate. These results show that uranium phosphate is readily solubilized by bacterial action with formation of mobile to uranyl carbonate species.

Microbiology / phytoremediation

Biotransformation under Anaerobic Conditions

Biotransformation of uranyl nitrate and uranyl carbonate. A wide variety of facultative and strict anaerobic bacteria reduced U(VI) added as uranyl-nitrate or uranyl carbonate to U(IV) under anaerobic conditions. These include axenic cultures of iron-reducing, fermentative, and sulfate-reducing bacteria. Mixed cultures of bacteria in uranium contaminated ground waters and in wastes also reduced uranium.

Clostridia are strict anaerobic spore-forming fermentative bacteria ubiquitous in soils, sediments, and wastes. They catalyze the reduction of metals from higher to lower oxidation state. Reduction of soluble U(VI) to insoluble U(IV) by *Clostridium sp.* in culture medium was confirmed by X-ray absorption near edge spectroscopy (XANES) and X-ray photoelectron spectroscopy (XPS). Uranium was reduced only in the presence of growing or resting cells. Organic-acid metabolites, the extracellular components of the spent culture medium, and heat-killed cells failed to reduce uranium under anaerobic conditions (Francis et al 1994).

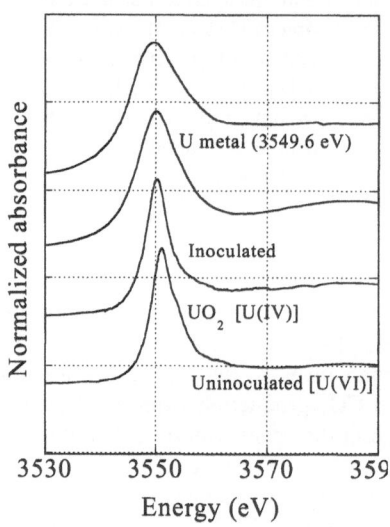

Fig. 4. Uranium reduction by *Clostridium sp.* XANES spectra of U at the M_V absorption edge shows the absorption edge for uninoculated U at 3551.1 eV, which is identical to U(VI) standard. The inoculated sample shows a peak at 3550.1 eV, which is lower than U(IV) but higher than U metal, indicating the presence of a highly reduced form of U, most probably a mixture of U(IV) and U(III).

Metabolism of uranyl–organic complexes by fermentative, iron- and sulfate-reducing bacteria. *Clostridium sphenoides* (ATCC 19403), which utilizes citric acid as the sole carbon source, metabolized equimolar Fe(III)-citrate with the degradation of citric acid and reduction of Fe(III) to Fe(II) but not the U(VI)-citrate complex. However, in the presence of excess citric acid or added glucose U(VI)-citrate was reduced to U(IV)-citrate. *Clostridium sp.* (ATCC 53464) which ferments glucose but not citrate reduced Fe(III)-citrate and U(VI)-citrate only when supplied with glucose. The metals reduced by the Clostridium sp. were present in solution as metal-citrate complexes (Figure 5). These results show that the complexed uranium is readily accessible as an electron acceptor despite the inability of the bacterium to metabolize the organic ligand complexed to the actinide. We have shown that the metabolism of the metal-citrate complex is dependent upon the type of complex formed between the metal and citric acid. Fe(III) forms a bidentate complex with citric acid and was metabolized, whereas U forms a binuclear complex with citric acid and was recalcitrant (Francis et al 1992; Francis and Dodge,1993; Joshi-Tope and Francis 1995).

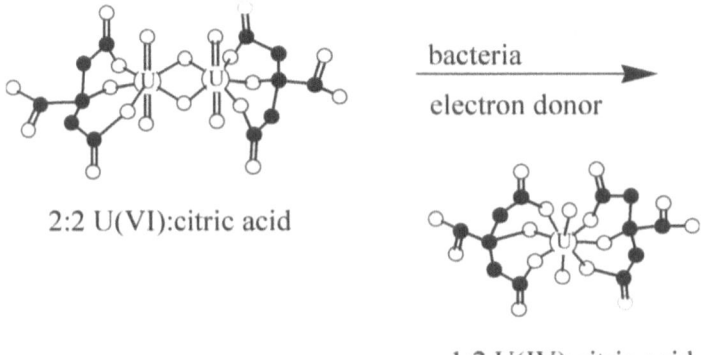

2:2 U(VI):citric acid

bacteria ──────────────▶ electron donor

1:2 U(IV):citric acid

Fig. 5. Proposed mechanism of biotransformation of U(VI)-citrate to U(IV)–citrate by Clostridia.

The resting cells of the sulfate-reducing Desulfovibrio desulfuricans and the facultative iron-reducing Shewanella halotolerans bacteria reduced U(VI) complexed with oxalate or citrate to U(IV) under anaerobic conditions with little precipitation of uranium. The reduced U(IV) remained in solution complexed with oxalate or citrate (Ganesh et al 1997, 1999). These results show that complexed uranium is readily accessible for the microorganisms as an electron acceptor, despite the inability of the bacteria to metabolize the organic ligand complexed to the actinide.

Acknowledgments

This research was funded by the Natural and Accelerated Bioremediation Research (NABIR) Program, Office of Biological and Environmental Research (OBER) Office of Science, U. S. Department of Energy, under contract No. DE-AC02-98CH10886

References

Dodge, C.J. and Francis, A.J. 1997. Environ. Sci. Technol. 31: 3062-3067.
Francis, A.J., C.J. Dodge, and J.B. Gillow. 1992. Nature 356:140-142.

Microbiology / phytoremediation

Francis, A.J. and C.J. Dodge. 1993. Appl. Environ. Microbiol. 59:109-113.

Francis, A.J., C.J. Dodge, F. Lu, G. Halada, and C.R. Clayton. 1994. 28:636-639.

Francis, A.J. 1998. J. Alloys and Compounds 271/273:78-84.

Francis A.J., C.J. Dodge, J.B. Gillow, and H.W. Papenguth. 2000. Environ. Sci. Technol. 34: 2311-2317.

Ganesh R., K.G. Robinson, G.R. Reed, and G. S. Saylor. 1997. Appl. Environ. Microbiol. 63:4385-4391.

Ganesh R., K.G. Robinson, L. Chu, D. Kucsmas, and G.R. Reed. 1999. Wat. Res. 33: 3447-3458.

Joshi-Tope, G. and Francis, A. J. 1995. J. Bacteriol. 177:1989-1993.

Panak, P., S. Selenska-Pobell, S. Kutschke, G. Geipel, G. Bernhard, and H. Nitsche. 1999. Radiochim. Acta 84: 183-190.

Panak, P., J. Raff, S. Selenska-Pobell, G. Geipel, G. Bernhard, and H. Nitsche. 2000. Radiochim.Acta. 88: 71-77.

Sakaguchi, T. 1996. *Bioaccumulation of Uranium*. Kyushu University Press, Hukuoka, Japan. pp 61-95.

Bacterial communities in uranium mining waste piles and their interaction with heavy metals

Sonja Selenska-Pobell, Katrin Flemming, Tzvetelina Tzvetkova, Johannes Raff, Michaela Schnorpfeil, Andrea Geißler

Institute of Radiochemistry, Forschungszentrum Rossendorf, D-01314 Dresden, Germany

Abstract. High diversity and significant differences were found in the structures of bacterial communities present in several U mill tailings and U mining waste piles. Many bacterial strains were successfully cultured from those uranium wastes, most of which are unusually effective in different biotransformations of U. The molecular basis for the selective and reversible binding of U and some other toxic metals by one of the natural bacterial isolates was found to be a novel kind of S-layer protein. Our analysis indicates that uranium wastes are a valuable reservoir for unusual microorganisms prospective for bacteria-based bioremediation.

Introduction

Bacteria play a significant role in migration and detoxification of a large number of heavy metals due to their ability to interact and biotransform them (Lloyd and Lovley 2001; Selenska-Pobell 2002). For this reason, knowledge about the diversity and activity of the indigenous bacteria in the uranium waste piles is of basic importance for understanding the bio-geo-chemical processes occurring in these environments and especially for their remediation by using the high metal-detoxifying potential of the mentioned organisms. Similarly to the situation occurring in other extreme environments, at present most part of the bacterial populations in the uranium wastes are difficult to culture due to our limited knowledge of their life necessities. Possibly, some bacterial groups are even non-culturable at the level of individual strains because of their symbiotic dependence on some other members of the natural microbial communities. The problems of culturability in the analysis of bacterial diversity were overcome during the last decade by the application of direct molecular approaches, such as the 16S rDNA retrieval, which do not require cultivation (Pace 1997, Dojka et al. 2000).

In parallel, many unusual bacterial strains possessing novel and unexpected features were cultured recently from various extreme environments (Kato et al. 1998; Liu et al. 1997; Strous et al. 1999; Takai et al. 2001).

In this work a comparative analysis of bacterial diversity in uranium mining waste piles and mill tailings was performed by using the 16S rDNA retrieval. In addition, particular bacterial isolates were cultured and studied in respect to their interactions with uranium. The molecular basis for the highly effective and reversible accumulation of U by one of the isolates was studied in detail and it was found that the strain possesses a unique surface protein of a very old origin which is responsible for the metal binding.

Materials and Methods

Soil samples

The soil samples studied were collected under sterile conditions from different sites and depths (up to 5 m below the surface) of the following environments: two uranium mill tailings, namely Gittersee/Coschütz located near the city of Dresden, Germany and Shiprock, New Mexico in the USA; one uranium mining waste pile situated near the town of Johanngeorgenstad, Germany; one uranium mining depository site located near Gunnison, Colorado, USA. In the latter case rests of uranium ores after using were transported and stored at a site with appropriate geologic characteristics and with low human density in order to reduce the hazardous consequences due to the migration of radionuclides and other toxic metals. All samples were taken from the corer immediately after the drilling and frozen at -20 °C before analysis.

DNA extraction and manipulations

Total DNA was recovered from the samples according to Selenska-Pobell (1995). 16S rDNA clone libraries and their analysis were performed as described in Selenska-Pobell et al. (2001) and Selenska-Pobell (2002). The analysis of bacterial S-layer genes was performed via PCR and direct sequencing (Raff 2002). The 16S rRNA and the S-layer genes were sequenced by using the fluorescent Dye-Deoxy method of Perkin Elmer and a Model 310 ABI PRISM Genetic Analyser (Applied Biosystems, Perkin Elmer, Forster City, California). They were compared to the sequences available in the GenBank and in the EMBL Nucleotide Sequence Database by BLAST analysis. Phylogenetic trees were generated by using the neighbor-joining algorithm with distance analyses with Jukes-Cantor corrections in the PHYLIP v.3.5 package (Felsenstein 1993).

Results and Discussion

Comparative analysis of the bacteria present in the uranium wastes

The results of the direct molecular analysis of bacterial diversity in the studied uranium mining wastes are summarized in Table 1. As evident from the Table, α-, β-, and γ-Proteobacteria were found in all kinds of wastes. The γ-Proteobacteria were most predominant in the samples collected from the Gitttersee/Coschütz mill tailings (Gitt-MT) and in those from the Gunnison depository site (Gu-DS). In agreement to our previous observations (Selenska-Pobell et al. 2001) the main representatives of this bacterial subdivision were affiliated to *Pseudomonas*. Significant was the part of α-Proteobacteria, in particular of those acting as plant symbionts, in the samples of Johanngeorgenstadt uranium mining waste pile (JG-UM). The α-Proteobacterial populations were also relatively abundant in the samples collected from the Gu-DS. These observations indicate that at these environments a bioremediation in advanced stages occurs.

Table 1. Size of the bacterial populations in % at the studied uranium waste sites.

Origin	Proteobacteria				HA[a]	CFB[b]	Bacilli	Act[c]	N/L[d]	GNS[e]	Plan[f]	Nov[g]
	α	β	γ	δ								
Git-MT[i]	7	15	54	-	-	20	4	-	-	-	-	-
Sh-MT[j]	9	4	17	8	-	-	35	-	10	13	4	-
JG-UM[k]	36	4	16	10	26	1	-	7.5	0.5	1	-	-
Cu-DS[m]	16	7	52	5	9	-	-	-	-	-	-	11

[a] Holophaga/Acidobacterium, [b] Cytophaga/Flavobacterium/Bacteroides, [c] Actinobacteria, [d] *Nitrospira/Leptospirillum*, [e] Green Non-sulfur bacteria, [f] *Planctomycetales*, [g] Novel lineages, [i] Gittersee/Coschütz – mill tailings; [j] Shiprock – mill tailings, [k] Johanngeorgenstadt – uranium mining waste pile, [m] Gunnison –depository site.

δ-Proteobacteria were identified in the soil samples collected from JG-UM, Gu-DS and in the mill tailings of Shiprock, New Mexico (Sh-MT). Interestingly, this bacterial group, consisting mainly of sulfate and metal reducing bacteria, was demonstrated to be predominant in the water samples of these environments (Chang et al. 2001; Selenska-Pobell 2002). The latter is an indirect indication for metal reduction, especially for U-reduction and precipitation from the liquid wastes at these sites. Surprisingly, no δ-Proteobacteria were found in the Gitt-MT, instead, populations of β-Proteobacteria were identified in a higher density than in the other three sites.

Significant differences were found in the distribution of bacteria from the phylums Holophaga/Acidobacterium (HA), Cytophaga/Flavobacterium/Bacteroides (CFB), of Bacilli and of Actinobacteria between the two uranium mill-tailings on one hand and the JG-UM and the the Gu-DS on the other hand (see Table 1 and also the examples given in Figures 1 and 2). Members of the HA phylum seems to be characteristic for the UM and DS. For the two mill tailings Bacilli and CFB are more specific. This observation is not surprising, bearing in mind the high levels of toxic As-compounds in the mill-tailings and the ability of these two groups of bacteria to biotransform them. Members of CFB are also indicative for changes from oligotrophic to organotrophic conditions in the environment as these bacteria are known as most rapidly adaptive from lower to higher nutrient levels (Lebaron at al. 2001). Hence, the significant amount of CFB in the Gitt-MT is an indication for actively running biological processes there. At the Sh-MT site, in contrast, CFB were not found; instead the *Bacillus* population was larger and, in addition, relatively large populations of green non-sulfur bacteria (GNS) and *Nitrospira/Leptospirillum ferrooxidans* (N/L) were identified. The letter are typical for natural and commercial metal-leaching environments (Goebel and Stackebrandt 1994; Selenska-Pobell 2002). On the basis of the above mentioned observations one may consider that Gitt-MT is at present more remediated than the Sh-MT.

Interesting is the presence of *Planktomycetales* (Plan) in the soil samples of the Sh-MT, which are possibly involved in anaerobic oxidation of ammonium to molecular nitrogen. In our previous work representatives of the same bacterial group were found in drain waters of the Gitt-MT (Selenska-Pobell 2002). These bacteria are possibly responsible for the observed strong ammonium fluctuations in the deep waters of the two mentioned mill tailings (Ivanova et al. 2000; Long and Helling, personal communications).

The presence of a relatively large number of representatives from novel not yet characterized bacterial divisions in the samples of the Gu-DS is interesting but not surprizing. Such representatives are possibly also present within the not yet analyzed individual clones of the 16S rDNA libraries of all the sites studied. The letter represent about 50% of the constructed clones and are under investigation in our laboratory.

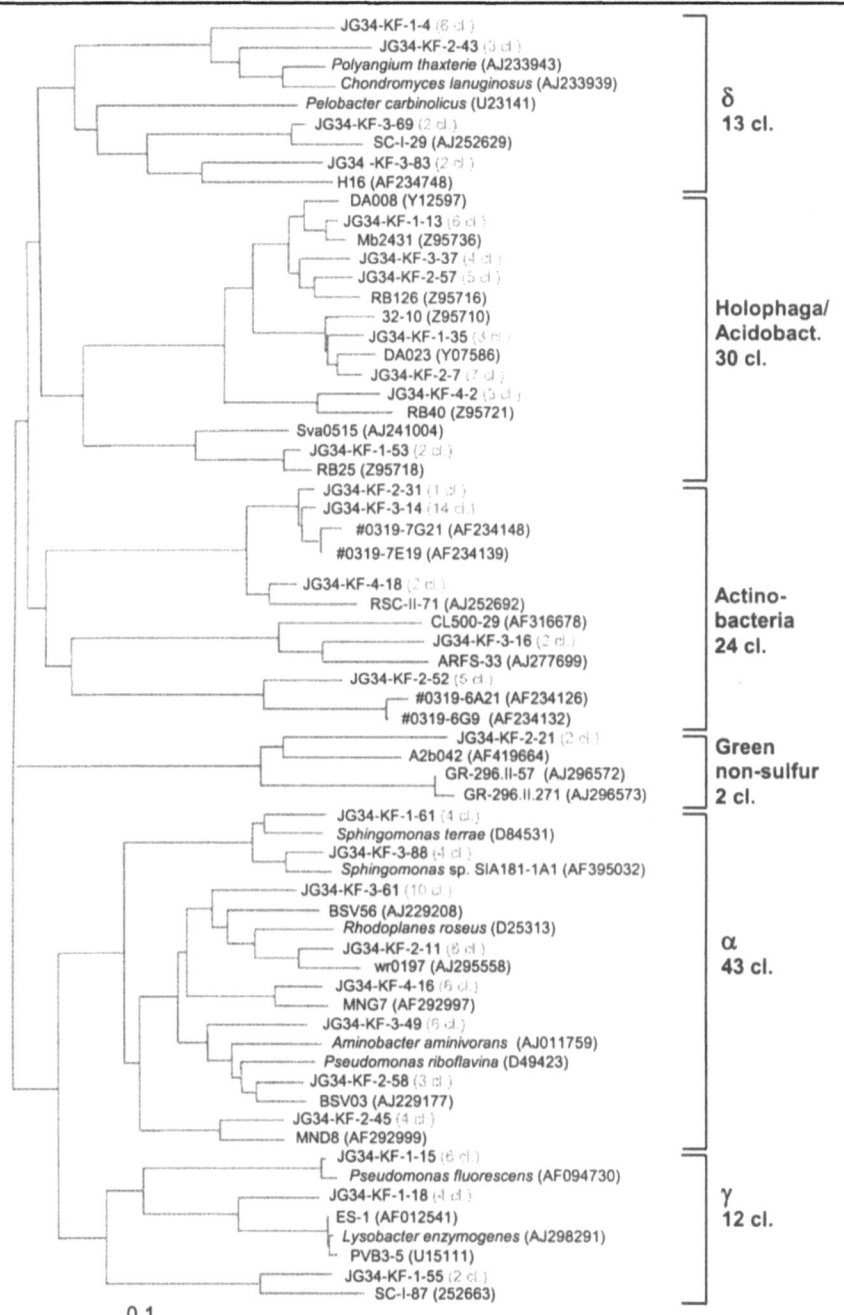

Fig. 1. Phylogenetic tree showing the affiliation of the 16S rDNA clones derived from a soil sample JG34 (2 m below the surface) from the JG-UM.

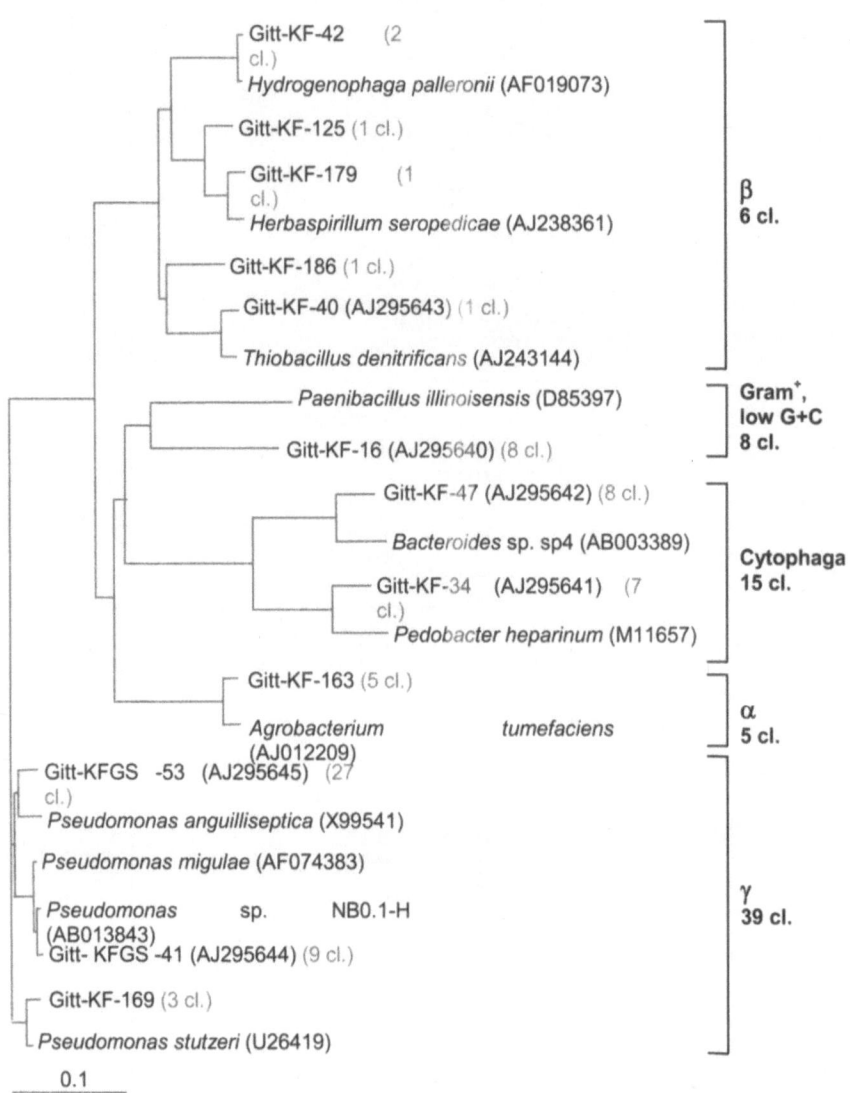

Fig. 2. Phylogenetic tree showing the affiliation of the 16S rDNA clones derived from a soil sample (2.5 m below the surface) from the Gitt-MT.

In parallel to the above described direct molecular analysis we cultured some bacterial groups known to participate in biotransformations of uranium. The main bacterial groups cultivated from the sites most contaminated with U were the chemolithoautotrophic *Acidithiobacillus ferrooxidans* and *Leptospirillum ferrooxidans*. These bacteria play a significant role in the dissolution and mobilization of U and some other metals in the wastes via both direct and indirect oxidation. The following heterotrophic bacteria were cultured from almost all samples studied: strains of the genus *Desulfovibrio*, which reduce sulfate and a large number of metals, strains belonging to γ-*Pseudomonas*, namely *P. stutzerii* and *P. migulae*, which accumulate high amounts of U. Strains related to *Agrobacterium, Rhizobium, Sphingomonas* (Tzvetkova et al. 2001) and a large number of Bacilli, some of which demonstrated an ability to bind selectively and reversibly high amounts of U, Al, Pb, Cu, and Cd (Selenska-Pobell et al. 1999), were cultured from the JG-UM. The failure to identify Bacilli in the samples of the latter environment by using the above described 16S rDNA retrieval indicates that the populations of this bacterial genus are present in JG-UM in a density which is below the detection limit of the method.

Surprisingly, we found that one of the *Bacillus* isolates, called *Bacillus sphaericus* JG-A12, is highly related to the *Bacillus sphaericus* NCTC 9602 reference stain which was isolated from spores of the air almost 50 years ago. The two strains possess almost identical genomic fingerprints and bind the above mentioned metals in a similar way (Selenska-Pobell et al. 1999). In addition, both strains are covered with S-layer protein lattices with similar physical parameters (Raff 2002).

Molecular analysis of the S-layers of *B. sphaericus* JG-A12 and 9602

In order to clear the role of the two above mentioned S-layer proteins in the interactions of the strains *B. sphaericus* JG-A12 and 9602 with U a molecular analysis of their genes was performed. This analysis revealed that the two proteins possess N terminal domains which are almost identical to each other but differ significantly from those of the S-layers of all other *B. sphaericus* strains studied up to date (see Fig. 3). Extremely high is the identity between the two proteins also in the studied part of their central domains. Interestingly, in the latter region of the proteins significant but lower identity occurs to the corresponding parts of the S-layers of two other *B. sphaericus* strains, namely P1 and ATCC4525. No significant identity was found to the central domain of the S-layer of the strain *B. sphaericus* 2362 which, in contrast to the other here mentioned strains, represents a mosquito pathogen.

It is interesting to stress that the minor differences in the amino acid sequences between the S-layer proteins of the strains JG-A12 and 9602 cause dramatic changes in their biochemical properties, especially in their interactions with metals and in their stability. When compared to the S-layer of 9602, the S-layer of JG-A12 binds much faster metals from aqueous solutions and it is much more sensitive to proteolytic digestions by proteases. Additionally, in contrast to the S-layer of 9602, the S-layer of JG-A12 denaturates almost immediately after contact to U and other metals (Raff 2002).

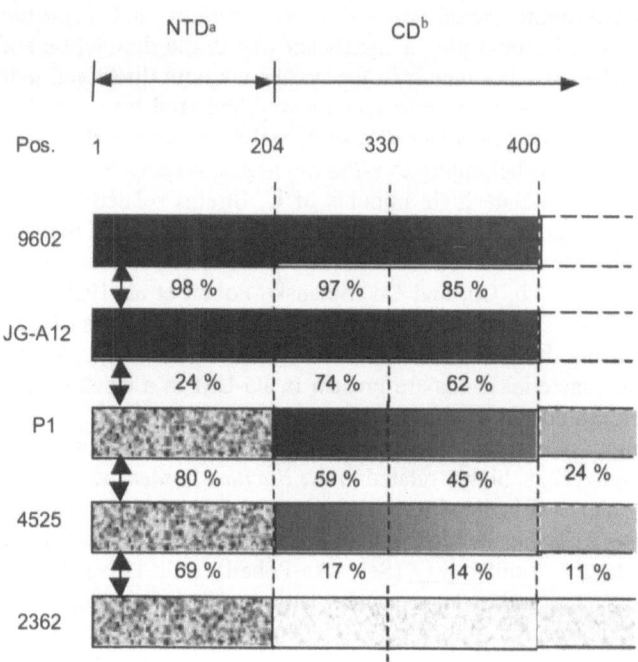

Fig. 3. Schematic comparison of the S-layers of the uranium mining isolate JG-A12 and those of the strains 9602, P1, 2545, and 2362. The identity between the proteins are given in %. [a] N terminal domain, [b] central domain.

The higher affinity of the JG-A12 S-layer to metals can be explained by the fact that this protein seems to be phosphorylated at a larger extend than those of the strain 9602 (contains 6 times more P as measured by ICP-MS, Raff 2002). In agreement to this, the amino acid mismatches found between the two proteins involve mainly amino acids which can be phosphorylated, such as treonin and serin. The abundance of phosphate groups, which are strong metal complexing ligands in the S-layer of JG-A12 is most probably the reason for the fast binding of high amounts of metals. The observed consequent denaturation of the protein is possibly a result of the fast formation of too big metal accumulates or/and it is also based on the described general instability of this protein (Raff 2002). Both the rapid binding of the metals and the fast disruption of the JG-A12 S-layer may be considered as important for the protection of the host bacterial cells from the toxic metals in their natural environment. It seems that the JG-A12 S-layer sheets saturated with heavy metals are continuously replaced by freshly synthesized molecules which ensures a continuous protection function for the bacterial cells. The latter seems to give an advantage of the JG-A12 to survive in the heavily polluted with uranium and other toxic metals environment from which it was recovered.

The abrupt change of the identity in the structures of the S-layers of *B. sphaericus* 4525, P1 and 2362 (see Fig.3) is an indication that these three proteins are products of a recent evolution in which lateral genetic transfer was involved. In contrast, the continuous and very conservative structure of the S-layers of *B. sphaericus* JG-A12 and 9602 indicates that these genes were not subjected to intensive evolution and possibly have a very old origin. Bearing in mind the recent hypotheses about the origin of life in the deep subsurface of the Earth (Pedersen, 1997) one may speculate that the uranium wastes are a reservoir of many of those „old" bacteria which are kept to live in harsh conditions.

Acknowledgements

This work was supported by grants SMWK 7531.50-03-607 and 7531.50-03.0370-01/5 from the Sächsisches Staatsministerium für Wissenschaft und Kunst, and SE 671/7-1 from the Deutsche Forschungsgemeinshaft (DFG),.

References

Chang Y-J, Peacock AD, Long PE, Stephen JR, McKinley JP, Macnaughton SJ, Hussain AKMA, Saxton AM, White DC (2001) Diversity and characterization of sulfate-reducing bacteria in groundwater at a uranium mill tailings site. Appl Environ Microbiol 67: 3149-3160

Dojka MA, Harris JK, Pace NR (2000) Expanding the known diversity and environmental distribution of an uncultured phylogenetic division of bacteria. Appl Env Microbiol 66: 1617-1621

Felsenstein J (1993) PHYLIP (Phylogeny Inference Package) Version 3.5c Department of Genetics, University of Washington, Seatle.

Goebel BM, Stackebrandt E (1994) Cultural and phylogenetic analysis of mixed microbial populations found in natural and commercial bioleaching environments. Appl Environ Microbiol 60: 1614-1621

Ivanova IA, Stephen JR, Chang Y-J, Brüggemann J, Long PEL, McKinley JP, Kowalchuk GA, White DC (2000) A survey of 16S rRNA and amoA genes related to autotrophic ammonia-oxidizing bacteria of the β-subdivision of the class proteobacteria in contaminated ground water. Can J Microbiol 46: 1012-1020

Kato C, Li L, Nogi Y, Nakamura Y, Tamaoka J, Horikoshi K.. 1998. Extremely barophilic bacteria isolated from the Mariana Trench, Challenger Deep at a depth of 11,000 meters. Appl Environ Microbiol 64: 1510-1513

Liu, SV, Zhou J, Zhang C, Cole D, Gajdarziska-Josifovska C, and Phelps T J (1997) Thermophilic Fe(III)-reducing bacteria from the deep subsurface: the evolutionary implications. Science 277:1106-1109.

Lloyd JR, Lovley DR (2001) Microbial detoxification of metals and radionuclides. Curr Opin Biotechn 12: 248-253

Pace N (1997) A molecular view of microbial diversity and the biosphere. Science 276:734-740

Pedersen K (1997) Microbial life in deep granitic rock. FEMS Microbiol Rev 20: 399-414.

Raff J (2002) PhD Thesis, University of Leipzig; March 2002

Selenska-Pobell S (1995) Direct and simultaneous extraction of DNA and RNA from soil. In: Akkermans ADL, van Elsas JD, De Bruijn FJ, eds *Molecular Microbial Ecology Manual 1.5.1*. Dordrecht, Netherlands: Kluwer Academic Publishers, pp.1-17

Selenska-Pobell S, Miteva V, Boudakov I, Panak P, Bernhard G, and Nitsche H (1999) Selective accumulation of heavy metals by three indigenous *Bacillus* isolates, *B. cereus, B. megaterium and B. sphaericus* in drain waters from a uranium waste pile. FEMS Microbiol Ecology 29: 59-67

Selenska-Pobell S, Flemming K, Kampf G, Radeva G, Satchanska G (2001) Bacterial diversity in soil samples from two uranium waste piles as determined by rep-APD, RISA and the 16S rDNA retrieval. Antonie van Leewenhuek 79: 149-161

Selenska-Pobell S (2002) Diversity and activity of bacteria in uranium waste piles, In: Keith-Roach M & Livens F, ed. *Interactions of Microorganisms with Radionuclides*. Oxford, UK; Elsevier Sciences; pp. 225-253

Strous, M, Fuest J A, Kramer E H M, Logemann S, Muyzer G, van de Pas-Schoonen K, Webb R, Kuenen JG, and Jetten K S M (1999). Missing lithotroph identified as new planctomycete. Nature 400: 446-449

Takai, K, Moser D P, Onstott T C, Spoelstra N, Pfiffner S M, Dohnalkova AFredrickson, JK (2001) *Alkaliphilus transvaalensis* gen. nov., sp. nov., an extremely alkaliphilic bacterium isolated from a deep South African gold mine. Int J Syst Evol Microbiol 51: 1245-1256

Tzvetkova I, Tzvetkova Tz, Groudeva V, Selenska-Pobell S (2001) Bacteria cultured from soil samples of uranium mining waste piles. In: Fanghaenel T, ed. *FZR-Report*-318: p. 39

Phytoextraction for clean-up of low-level uranium contaminated soil evaluated

H. Vandenhove, M. Van Hees

Belgian Nuclear Research Centre, SCK•CEN, Radioecology section; Radiation Protection Department, B-2400, Mol, Belgium

Abstract. This study was executed to test if low-level U contaminated sandy soil from a nuclear fuel processing site could be phytoextracted in order to attain the required release limits. Two soils were tested: a control soil (317 Bq ^{238}U kg^{-1}) and the same soil washed with bicarbonate (69 Bq ^{238}U kg^{-1}). Ryegrass (*Lolium perenne cv. Melvina*) and Indian mustard (*Brassica juncea cv. Vitasso*) were used as test plants. The annual removal of the soil activity with the biomass was less than 0.1 %. The addition of citric acid (25 mmol kg^{-1}) one week before the harvest increased U uptake up to 500-fold and up to 3.5 % and 4.6 % of the soil activity could annually be removed with the biomass.

Introduction

The nuclear fuel cycle may be a source of environmental contamination. Uranium exploitation produces large quantities of wastes but also accidental spills at nuclear fuel production, reprocessing or waste treatment plants have led to soil contamination with uranium. U-contaminated soil may be excavated, packaged and removed which is a costly enterprise. Soil washing has also shown promising in removing U from contaminated soil, but results in the generation of liquid wastes and the deterioration of soil properties. In contrast, phytoextraction, the use of plants to remove contaminants from polluted soil, allows for in situ treatment, does not generate liquid wastes and keeps soil properties intact. Furthermore, the contaminated site is covered by plants during phytoextraction and wind and water erosion will be reduced. The draw-back of this option is generally the low extraction efficiency and the generation of large amounts of contaminated biomass, which has to be treated as waste.

The phytoextraction potential depends on the amount of radionuclides extracted and the biomass produced. The annual removal of the contaminant with the biomass is given by

$$\text{Annual removal (\%)} = \frac{\text{TF} \times \text{Yield}}{W_{\text{soil}}} \times 100 \tag{1}$$

with TF, the transfer factor (ratio of Bq g^{-1} plant and Bq g^{-1} soil); Yield, biomass (t ha^{-1}) and W_{soil}, soil weight (t ha^{-1}). Phytoextraction requires several years and evolution of contaminant concentration with time, considering removal by the crop and radioactive decay is given by:

$$C_{\text{soil},t} = C_{\text{soil}_{t=0}} \exp\left\{-\left(\frac{\text{TF} \times \text{yield}}{W_{\text{soil}}} + \frac{0.69}{t_{1/2}}\right) \times t\right\} \tag{2}$$

The U-TF ranges between 0.0002 and 0.1 g g^{-1}. With a TF of 0.1 g g^{-1} and a yield of 20 t ha^{-1}, a soil weight of 1500 t, the annual reduction percentage is 0.13 % and it would take 1650 years to reduce the soil contamination 5-fold.. Amendments to increase the U-availability are hence needed to make the phytoextraction option feasible. Huang *et al.* (1998) observed up to a 1000-fold increase in the U-TF after citric acid addition to the soil.

The objective of present paper is to find out if low-level U contaminated soil from a nuclear fuel processing site can be phytoextracted in order to attain the required release limits.

Materials and methods

The U-contaminated sandy soil was collected from spot contaminations at a nuclear fuel processing site. A portion of the soil was already washed with bicarbonate in order to extract part of the uranium. The total α activity is 1.61 Bq g^{-1} for the control soil and 0.47 Bq g^{-1} for the bicarbonate treated soil. The ^{238}U levels are 0.317 and 0.069 Bq g^{-1}, respectively. The proposed exemption limit for release level is 0.3 Bq g^{-1} alpha activity. Some U-related soil characteristics are given in Table 1.

Two plant species were selected: ryegrass (*Lolium perenne cv. Melvina*) and Indian mustard (*Brassica juncea cv. Vitasso*). Field capacity moist soil was transferred to six 1 L pots, this for each plant species and soil. Seven days before harvest, citric acid (25 mmol kg^{-1} soil, added with 50 ml of 0.5 M solution) was applied to three of the six pots. Plant samples were analysed for ^{238}U content and dry weight production. Yield (t ha^{-1}) was calculated from the yield per pot, culture time and considering a 6 month growing season.

Results and discussion

Following citric acid addition up to 900 times more U was present in the soil solution (Table 1).

Table 1. Some U-related soil characteristics of the control and bicarbonate treated soil

Soil characteristics	Control	Bicarbonate treated
Total ^{238}U (Bq kg^{-1})	317 ± 10	69 ± 4
Soil solution ^{238}U (Bq dm^{-3})	0.258 ± 0.012	9.994 ± 0.096
Ca(NO$_3$)$_2$ extractable ^{238}U (Bq kg^{-1})	5.3 ± 0.3	2.9 ± 0.7
Citrate extractable ^{238}U (Bq kg^{-1})	216 ± 5	48 ± 1

Biomass production was significantly lower on the bicarbonate treated soil. Treatment of the control soil with citric acid resulted in a significant yield decrease of 22 % and 15 % for mustard and ryegrass respectively. Yield decrease was not significant following citric acid treatment of the bicarbonate soil (Fig. 1A and Table 2). Dry weight production in t per ha can be calculated from the dry weight per pot, culture time (1 month for ryegrass, 1.5 months for mustard) and an effective growing season of 6 months.

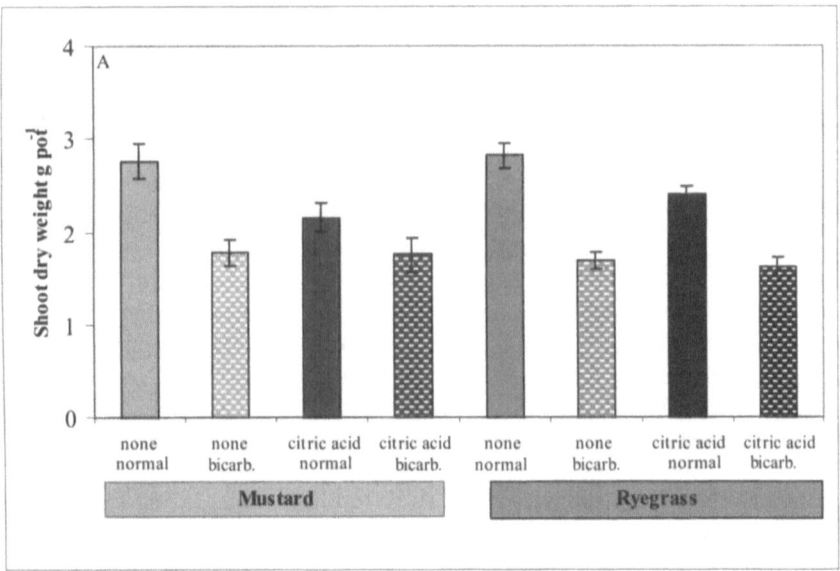

Fig. 1. Dry weight production for ryegrass and mustard grown on control and bicarbonate treated soil, with or without amendment of citric acid

The TF on the control soil was 0.012 g g^{-1} for mustard and 0.0055 for ryegrass. On the bicarbonate treated soil TFs are about a factor 15 higher (Figure 1B and Table 2). Though the citric acid addition decreased biomass production, it increased the U-transfer to the plants up to 500-fold. The increase in TF is higher for the control soil than for the bicarbonate treated soil. The highest TF were observed for mustard with a TF of 7.64 g g^{-1} on the control soil and of 6.27 g g^{-1} on the bicarbonate treated soil. For ryegrass TFs were 2.77 and 5.10 g g^{-1}, respectively.

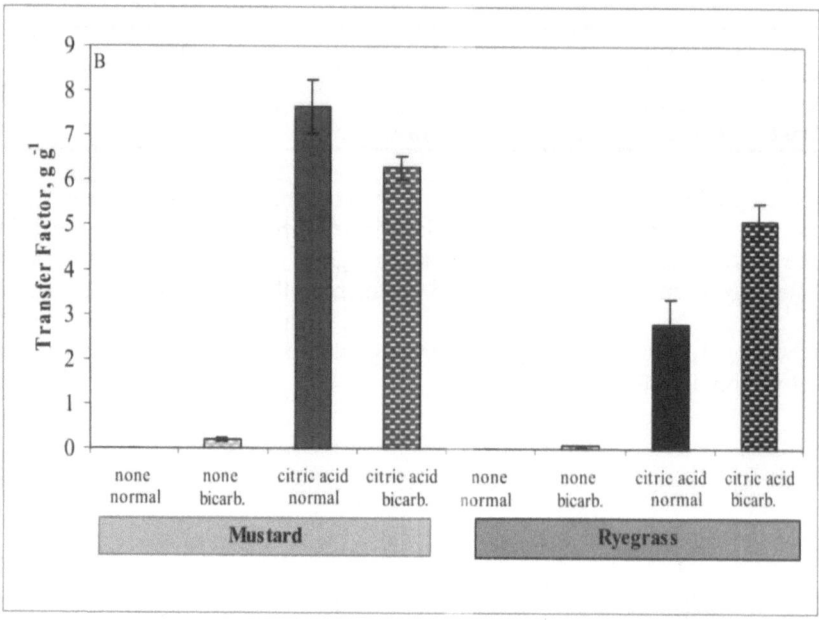

Fig. 2. ^{238}U transfer factors (B) for ryegrass and mustard grown on control and bicarbonate treated soil, with or without amendment of citric acid

Following citric acid addition up to 900 times more U was present in the soil solution (Table 1). A perfect linear agreement between the ^{238}U concentration in the shoots and the soil solution of the control, bicarbonate washed or citric acid treated soil (results not shown), indicates that the U soil solution concentration rules U uptake and that solubilising agents are important to increase plant uptake.

Table 2 shows that (1) without the addition of citric acid phytoextraction levels are <0.1 % and that addition of citric acid results in an up to 500-fold increase in the TF; (2) citric acid treatment resulted in a yield decrease of up to 47 %. Furthermore, dicotylous plants died off completely after citric acid treatment and ryegrass regrowth on the citric acid treated pots was only between 12 and 23 % of the non treated pots (results not shown).

With a desired activity reduction of 1.5 and 5 for the bicarbonate-washed and control soil, respectively, it would take about 10 to 50 years (Eqn. 2) to attain the release limit (0.3 Bq g^{-1}).

Microbiology / phytoremediation

Table 2. Yield (t ha^{-1}), plant activity (mBq g^{-1}) and annual phytoextraction potential (%) of untreated and citric acid treated soils (Soil Depth 10 cm; Soil Density 1.5 kg dm^{-3})

Plant	Soil	Citric acid	Yield	Plant activity	Extraction %
Mustard	Control	No	11.6	3.9±0.5	0.010 ±
	Control	**Yes**	9.1	2422±184	0.002
	Bicarb.	No	7.5	14.3±3.2	**4.618 ±**
	Bicarb.	**Yes**	7.4	438±19	0.384
					0.103 ±
					0.030
					3.284 ±
					0.250
Ryegrass	Control	No	17.8	1.7±1.1	0.007 ±
	Control	**Yes**	15.2	881±184	0.004
	Bicarb.	No	10.7	5.1±1.1	**2.810 ±**
	Bicarb.	**Yes**	10.2	356±29	0.689
					0.052 ±
					0.008
					3.477 ±
					0.474

Conclusions

Following the addition of citric acid one week or a few days before harvest, the soil-to-plant transfer of uranium may increase up to 500-fold and extraction percentages in the range of 2 to 5 % seem achievable even for low level contaminated soils. Depending on the desired contamination reduction factor (e.g. 2-50) it will take roughly between 15 and 200 years before the release limit is attained. Citric acid addition did however result in a decreased dry weight production and crop regrowth (only applicable to ryegrass and potentially to other perennials).

Acknowledgements

This study was partially financed by FBFC, Dessel, Belgium

References

Huang J W, Blaylock, MJ, Kapulnik Y, Ensley BD (1998) Phytoremediation of uranium contaminated soils: Role of organic acids in triggering hyperaccumulation in plants. Environ. Sci. Technol. 32: 2004-2008.

DMT

Wir erobern
die Tiefe

Wir sind dabei:
15.-21. September 2002,
International Conference Uranium
Mining and Hydrogeology, Freiberg

Fragen Sie unsere
Safe Ground Division

- Baugrund
- Bodenmanagement
- Geothermie
- Spezialtiefbau
- Tunnel- und
  Verkehrswegebau
- Hydrogeologie und
  Wasserwirtschaft
- Objektplanung
- Projektsteuerung
- Umweltbiologie
- Wasser- und
  Bodenschutz

Deutsche
Montan Technologie GmbH

Safe Ground Division

Am Technologiepark 1
45307 Essen
Telefon   02 01-1 72 19 00
Telefax   02 01-1 72 17 73
E-Mail   safe-ground@dmt.de
Internet   www.dmt.de

DIN EN ISO
9001
zertifiziert

Long-term differences in transfer and accumulation of potentially toxic trace elements and radionuclides in trees on uranium mining dumps (Erzgebirge, Germany)

Carsten Brackhage, E. Gert Dudel

TU Dresden, Institut für Allgemeine Ökologie und Umweltschutz, Pienner Str.8, 01737 Tharandt

Abstract. Seasonal and long-term variation of radionuclide and heavy metal content in different compartments of trees were investigated on uranium mining dumps in the Erzgebirge, Germany. Seasonal increase in concentration could be substantiated partially for lead and uranium in leaves/needles but for no other element and tree part although variation was high. A major fraction of variation was attributed to the micro-site within each sampling plot. Long-term monitoring comprising a 5-year period showed no differences in radionuclide and heavy metal content.

Introduction

Investigation on radionuclide and heavy metal distribution, transfer and accumulation in the biosphere has found a widespread recognition throughout the scientific society. Since long concentration levels of e.g. uranium in plants have been analyzed (Stoklasa and Penkava 1932). But it was not before dumps and tailings from uranium mining and milling activities remained on a large scale after the second world war that transfer and accumulation became important as potential health risk.

Studies on radionuclide and heavy metal concentration in plants on dumps and tailings are numerous (Sharma et al. 1981; Sheard 1986; Thomas 2000) but often exemplary with small numbers of individuals sampled once in a year. From there transfer factors were calculated (Sheppard and Evenden 1988) complemented by experiments for spotting the factors influencing uptake and accumulation

(Sheppard et al. 1984; Ebbs et al. 1998). However, little is known about seasonal variation in perennial plants and long-term differences.

The present study supplements the data for concentration levels in plants on dumps and tailings and at the same time presents results on time-dependent differences. This is especially important in the course of reclamation including the use of these sites for forestry and other type of land use.

Materials and Methods

Investigation area and sampling

The area of investigation is located within the former uranium mining and milling complex Schlema/Alberoda in the Erzgebirge/Germany. The selected dump site and reference site are found within the catchment area of the Silberbach (480 – 530 m a.s.l.). On dump H310 a sampling plot (10x50m) based on vegetation analysis including minimum area for the present plant community was established and in each case 10 individuals of *Betula pendula* ROTH and *Pinus sylvestris* L. selected. During 1995 and 1996 leaf and twig samples were taken repeatedly added by stem wood and root samples once in 1996. The same trees were sampled again in 2000 together with 4 individuals of *Alnus glutinosa* (L.) GÄRTN.

The soil was characterized as Regosol-Lockersyrosem (German classification) with pH 7,3 – 7,8 and a conductivity of 130 – 180 µS/cm, total uranium concentration being 170 – 420 $\mu mol*kg^{-1}$ (Dudel et al. 1997).

Sample preparation and analysis

Leaf and twig samples as well as root samples were washed very thoroughly with water and with 0,05M EDTA-solution. All plant samples were oven-dried (70°C) to constant weight, milled and homogenized. A small portion of the samples (~200 mg) was then wet digested with HNO_3 (p.a.) in a pressure-controlled microwave system (CEM, MDS 2000). Digested plant samples were analyzed for Co, Ni, Cu, Zn, Cd, Pb and U by ICP-MS (FISONS, PQ2+). The quality of the analytical results was repeatedly checked with standard reference material (NIST No.1575, pine needles; GBW 07604, poplar leaves). The obtained results were in good accordance with certified values (Co 106%; Ni 99%, Cu 89%; Zn 95%; Cd 102%; Pb 90%, U 100%).

Results and Discussion

Species differences and accumulation

As can be seen in Fig.1 differences in uranium concentrations between species are pronounced in above-ground plant parts with *Alnus glutinosa* showing much higher concentrations compared to *Betula pendula* and *Pinus sylvestris*. In contrast no differences can be seen for roots. Although individuals of each tree species are more or less even distributed on the sampling plot in relation to each other a species dependent element distribution would come out only when analyzing the entire rhizosphere for available elements which wasn't done. However, species dependent differences in heavy metal uptake and accumulation are well known, e.g. (Ebbs and Kochian 1998) but differences for uranium are scarcely documented in more detail (Sheppard et al. 1989) requiring experimental investigation.

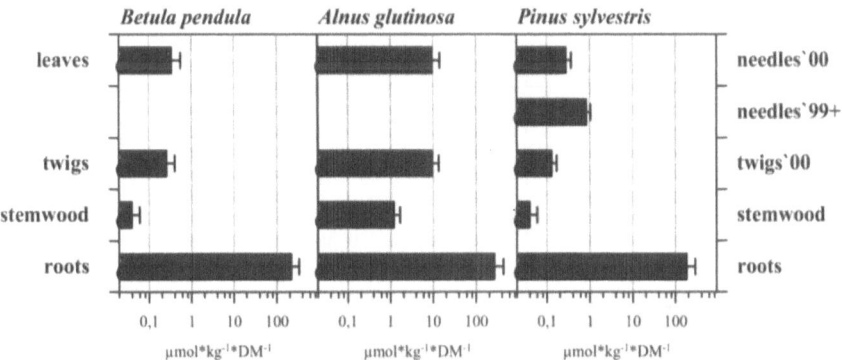

Fig. 1. Uranium concentration in different parts of *Betula pendula* ROTH (n=10), *Alnus glutinosa* (L.) GÄRTN. (n=8), and *Pinus sylvestris* L. (n=10) on the sampling plot dump H310.

Ebbs et al. (1998) could show that under equal conditions in a batch experiment *Beta vulgaris* shows on average much higher uranium concentrations in the shoot compared to *Pisum sativum* and other plant species. No such experiments are yet known for tree species with uranium contamination in the soil or growth medium although the presented results under field conditions indicate species dependence. Clearly element concentrations in plants are influenced by many other factors including availability in soil.

An other important point is the compartment specific distribution of uranium (see Fig.1) as well as other heavy metals. Roots exhibit the highest uranium concentration in all species which is in accordance with other findings (Sheppard et al. 1985). The question remains whether uranium is taken up with subsequent accumulation within the xylem and/or pith of the roots or adsorbed either to the cell wall of periderm or cortex cells or even to mycorrhizal fungi. (Ebbs et al. 1998) come to the conclusion, that uranium uptake is dependent on the uranium species available in the soil which in turn is highly dependent on the pH-value. They found for *Pisum sativum* cv."Sparkle" growing in hydroponic culture with pH = 5 accumulation of uranium in the shoot whereas with pH = 6 and pH = 8 uranium accumulated in the roots. Since the pH-value of the dump material from H310 was found to be between 7 and 8 the high root concentrations in the investigated tree species may be due to precipitation/adsorption of uranium species other than uranyl cations in the root cortex. Further investigation to verify this coherence is required.

Comparing above-ground plant parts it is conspicuous that older needles (Fig.1) and twigs (data not presented) of *Pinus sylvestris* show significant higher uranium and lead concentrations (p<0.01**; Mann-Whitney-U-Test) whereas for other elements no differences could be secured. This is in accordance with findings in other investigations, e.g. for *Pinus banksiana* (Thomas 2000). It shows the accumulation potential of persistent plant parts although the overall concentration level for uranium is 10fold lower compared with other data from contaminated sites (Thomas 2000). No such accumulation pattern is found in stem wood indicating no absorption capacity for uranyl ions or transport in complexed forms.

Seasonal changes

Considering needle samples from *Pinus sylvestris* L. there is a clear tendency for accumulation of uranium and lead (data not shown) during the one year sampling period (Fig.2) which was expected from the analysis of older needles. Not as marked as for needles is the tendency for the uranium concentration in twigs although when comparing first and latest samples a 4fold increase can be seen.

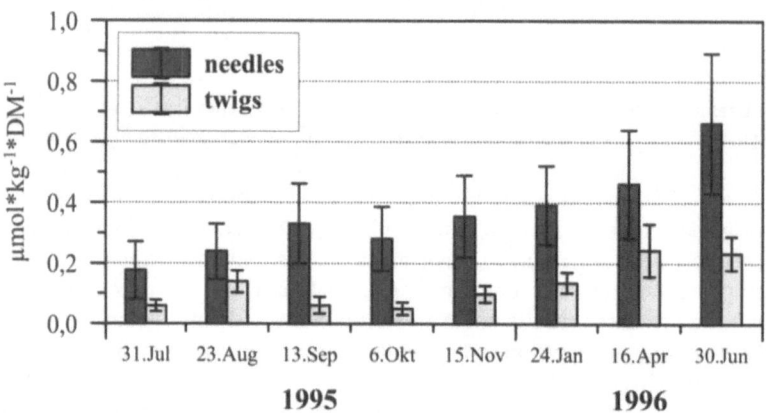

Fig. 2. Uranium concentration in needles and twigs from *Pinus sylvestris* L. on the sampling plot dump H310. Given is the arithmetic mean with standard deviation (n=10).

Seasonal changes could not be verified for all analyzed plant parts and elements in *Betula pendula* ROTH when assuming normal distribution and comparing mean values (see Fig.3, "total"=mean of all 10 individuals). Because three individuals show higher uranium concentrations in leaves and twigs throughout the sampling period these were grouped being "contaminated" and compared with the remaining individuals being "normal" (see also Fig.3). Differences in uranium concentration between these groups were in all cases significant (p<0,05*; Mann-Whitney-U-Test) whereas for all other elements no differences could be substantiated. Focussing on group "contaminated" a slight seasonal increase can be seen where leaf concentrations in samples from August, September and October are higher compared with samples from July.

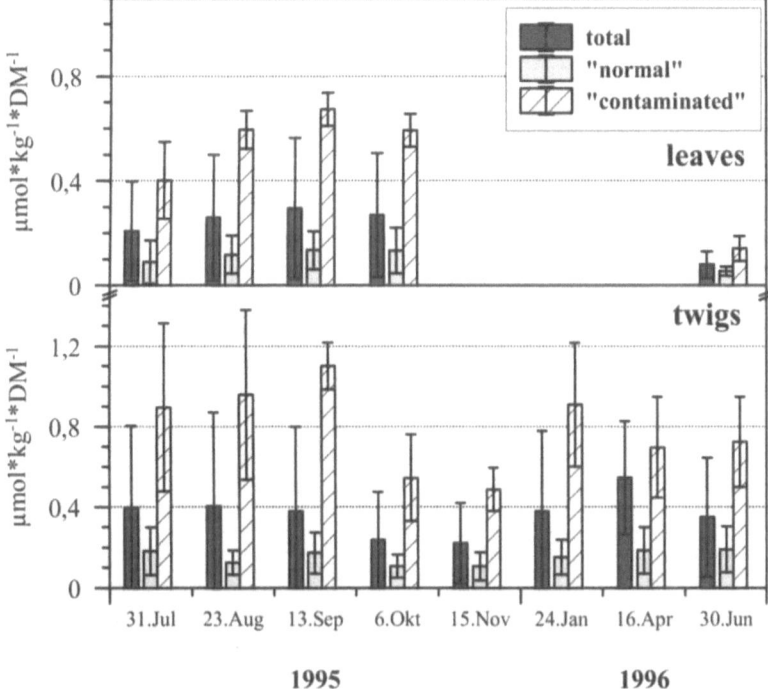

Fig. 3. Uranium concentrations in leaves and twigs of *Betula pendula* Roth (total: n=10) from dump H310. Individuals are grouped as "normal" (n=7) and "contaminated" (n=3) where these 3 individuals exhibit higher uranium concentrations throughout the season.

When thinking about the reason why three of the *Betula pendula* ROTH trees exhibit higher uranium concentrations throughout it is interesting to look at the distribution pattern within the sampling plot (Fig.4) where it can be seen that they grow in close vicinity to each other. It could be presumed that elevated uranium availability on this "micro-site" gives rise to the observed results. This in turn may imply the potential use of birch trees in bio-monitoring programs and at the same time threatens the presumption of homogeneous dump material. Obviously contamination as available for plants can be locally very restricted and therefore impact only very few individuals. Whether tree analysis for bio-monitoring is useful after reclamation of the dump sites has to be evaluated.

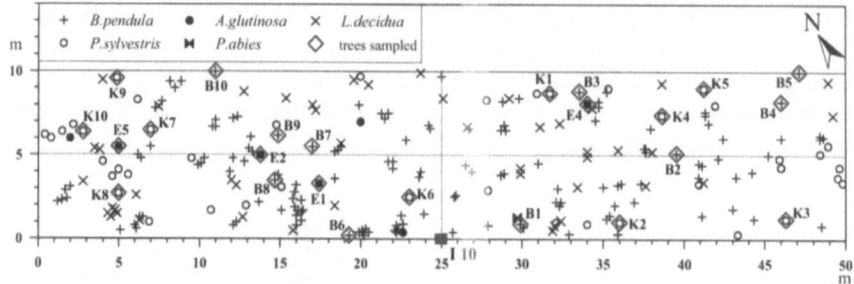

Fig. 4. Location of trees sampled within the sampling plot on dump H310. Individuals 7,8,9 of *Betula pendula* ROTH can be seen to grow in close vicinity.

Long-term differences

Long-term differences could not be found for any of the analyzed elements (Fig.5). Even the difference in uranium concentration between group "contaminated and group "normal" can still be verified. Only lead concentration is significantly lower in 2000 (p<0,05*, Mann-Whitney-U-Test). Although a rigid washing procedure was applied this may be due not to different amounts of uptake and transfer but to different emission of contaminated particles.

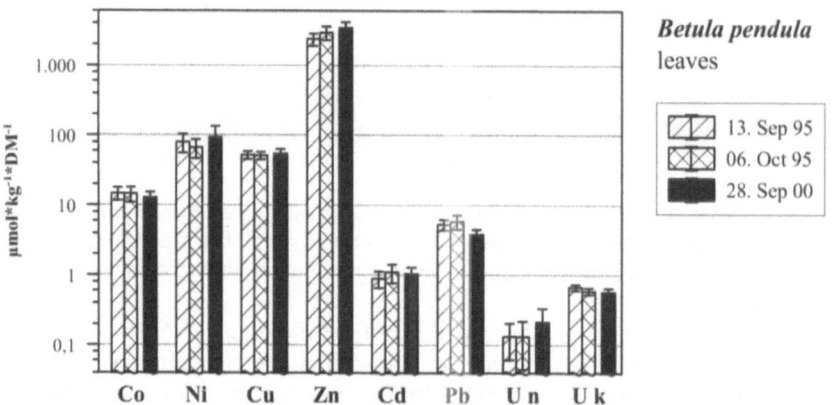

Fig. 5. Mean element concentration in leaves of *Betula pendula* ROTH from dump H310. Samples were taken 13.September 1995, 6.October 1995, and 28.September 2000 (n=10). For uranium it is distinguished between Un = group "normal" (n=7) and Uk = group "contaminated" (n=3) where these 3 individuals exhibit higher uranium concentrations throughout the investigated period.

Conclusions

Tree species on uranium mining dumps in the area investigated show very low concentration levels in above-ground parts. With no long-term accumulation and only slight seasonal changes there seem to be no potential health risk from transfer and accumulation of radionuclides and heavy metals in trees not considering Ra-226 and Pb-210. Moreover very low element concentrations in the stem wood may allow for forestry on reclaimed dump sites not mentioning the poor nutrient supply. In contrast high root concentrations remind to be cautious in case of changes of the availability of these elements in the course of reclamation.

Acknowledgement

Thanks to the Saxonian Ministry for Environment and Agriculture for financial support.

References

Dudel, GE, Brackhage, C, et al. (1997). Freisetzung radioaktiver und nichtradioaktiver E-lemente aus offenen und abgedeckten Bergbauhalden: Einflüsse auf eine land- und forstwirtschftliche Nutzung. Abschlußbericht zum FuE-Vorhaben. Tharandt, TU Dresden, Institut für Allgemeine Ökologie und Umweltschutz.

Ebbs, SD, Brady, DJ, et al. (1998) Role of uranium speciation in the uptake and translocation of uranium by plants. J Exp Bot 49(324): 1183-1190

Ebbs, SD, Kochian, LV (1998) Phytoextraction of zinc by oat (*Avena sativa*), barley (*Hordeum vulgare*), and Indian mustard (*Brassica juncea*). Environ Sci Technol 32(6): 802-806

Sharma, YP, Lal, N, et al. (1981) Trace Content of Uranium in Spices and Condiments. Health Phys 41(4): 680-682

Sheard, JW (1986) Distribution of uranium series radionuclides in upland vegetation of northern Saskatchewan. I. Plant and soil concentrations. Can J Bot 64: 2446-2452

Sheppard, MI, Sheppard, SC, et al. (1984) Uptake by plants and migration of uranium and chromium in field lysimeters. J Environ Qual 13(3): 357-361

Sheppard, MI, Thibault, DH, et al. (1985) Concentrations and concentration ratios of U, As and Co in Scots pine grown in a waste-site soil and an experimentally contaminated soil. Water Air Soil Pollut 26: 85-94

Sheppard, SC, Evenden, WG (1988) Critical compilation and review of plant/soil concentration ratios for uranium, thorium and lead. J Environ Radioact 8: 255-285

Sheppard, SC, Evenden, WG, et al. (1989) Uptake of natural radionuclides by field and garden crops. Can J Soil Sci 69: 751-767

Stoklasa, J,Penkava, J (1932) Biologie des Radiums und Uraniums. Berlin, Paul Parey Verlag.

Thomas, PA (2000) Radionuclides in the terrestrial ecosystem near a Canadian uranium mill - Part I: Distribution and doses. Health Phys 78(6): 614-624

Study courses at the Technical University Freiberg
Akademiestr. 6   09599 Freiberg   Germany
http:/www.tu-freiberg.de   dekan.F3@fggb.tu-freiberg.de

Geo-Informatics

1st – 4th term (undergraduate studies)
* Mathematics
* Applied Computer Science
* Physics and Chemistry
* Geosciences
* oral examination after 4th term

5th – 9th term (graduate studies)
* Applied Mathematics
* Applied Computer Science
* Physics
* Geodata and Geomodeling
* Choice of one out of four modules:
    - Geodynamics
    - Numerical modeling of solid and fluid media
    - Resource Management
    - Geomonitoring
* internship (minimum 4 weeks)
* undergraduate thesis in 7th and 8th term (200 h)
* Masters thesis in the 9th term
* oral Masters examination

Admission
* „Abitur"/Baccalaureate (A-level) or
* specialized college degree or
* equivalent access entitlement

Duration/Degree: 9 semesters/Master of Science in Geoinformatics (Dipl.-Geoinf.)
Start: winter term

Job opportunities
Application of modern information technology in
geo-industries (data management, modelling, presentation,
communication); National and international positions in
resource-related industries, environmental and engineering
consulting, banks, software development, and all levels of
government institutions; Particular option in internet-based
development

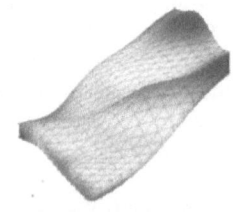

Uranium and thorium uptake on hydrophilic plants

Alexandru Cecal[1], Karin Popa, Ioan Caraus, Iftimie Craciun

[1] "Al.I. Cuza" University, Faculty of Chemistry, 11–Carol I, 6600 – Iasi, Romania

Abstract. The hydrophilic plants were used with the purpose of purifying the radioactive waters containing UO_2^{2+} and Th^{4+} ions. The efficiency of using these plants, rendering by the degree of uptake (bioaccumulation and biosorption) of these plants in the absence of ionic competition was studied in the case of radioactive effluents simulated by $UO_2(NO_3)_2$ and $Th(NO_3)_4$ solutions. Under identical experimental conditions, the degree of decontamination of radioactively wastewater decreases according to the following series: *Riccia fluitans* > *Lemna minor* > *Pistia stratiotes* > *Elodea canadensis*. The influence of certain environmental factors (like temperature, pH, anion nature, etc.) was studied in order to optimize the process. A thin layer radiochromatography study leads to the conclusion that the biochemical fractions in which UO_2^{2+} and Th^{4+} place themselves are of a polysaccharide and lipoid fractions.

Introduction

Biosorption is the property of certain dead organic masses to bind and concentrate various heavy metals, including radionuclides from highly diluted solutions; biomasses having this property, since they act like chemical substances, are ionic exchangers of biological origin. The cellular membrane of certain types of microorganisms, algae, fungus or even plants having separate organs is – mainly – responsible for this phenomenon. Some new studies show that the biosorption process is a very complex one, metallic species being deposited in the solid biosorbent through a multitude of subprocess, among which the most significant ones are sorption, ionic exchange, complexation, etc (Volesky 1994). Biosorption also appears as part of the whole of process of retaining heavy metals and radionuclides on the surface of living cells and on the extracellular polymers (Nakajima et al. 1986; Francis et al. 1991).

The contrary of biosorption is the active, metabolic bioaccumulation on living cells during which the physico-chemical process mentioned above are supplemented by the biological process of transport through membrane. This process depends on the degree of toxicity of the chemical species, on the radioactivity of the isotope accumulated, etc. (Pribil and Marvan 1976).

For the purpose of purifying waters containing α–active radiocations there were used bacteria (Francis et al. 1990; Bouby et al. 1996; Wade Jr. and DiCristina 2000), fungus (Cebotina and Liubimova 1981; Tsezos et al. 1997), algae (Havlie and Robertson 1973; Horikoshi et al. 1979) or even superior plants (Duff et al. 1997). A study on the use of *Lemna minor* in retaining of $^{137}Cs^+$ out of natural waters (Bergamini et al. 1979) lead to the idea of using this plant in purifying waters containing UO_2^{2+} and Th^{4+} ions (Cecal et al. 1999, 2000).

Materials and methods

The starting α–active solutions contain $UO_2(CH_3COO)_2$ 0.1 M (pH = 2.5), $UO_2(NO_3)_2$ 0.1 M (pH = 1.5) and $Th(NO_3)_4$ 0.1 M (pH = 1.5). The reagents used are of "for analyses" quality and they come from MERCK company.

Hydrophilic plants *Elodea canadensis, Lemna minor, Pistia stratiotes* and *Riccia fluitans* were chosen as biocollectors, since they have a high productivity (around 30 %), they develop quickly under natural conditions on any still water surface and they have a high density of active functional groups on the surface (Boyd 1969).

Plants came from the Laboratory of Aquatic Ecology – Piatra Neamt, Romania, and were grown in a "Mytron" thermostatic bath at 20±1 °C, under continuos illumination ($3.6 \cdot 10^{-3}$ J$\cdot$cm$^{-2} \cdot$s^{-1}). Garden soil was used as culture medium (0.5 kg garden soil in 10 L water). Bioaccumulation experiments were conducted under the same conditions. The density attained by the biomass is of approximately 200 g plant per square meter of water surface, being comparable to the one reported under natural conditions (Sutton 1975).

Dry biomass was obtained by separating biological mass from culture medium through filtration, and excessive salts were eliminated by washing with bidistilled water. The plant was then dried at 40 °C for 24 hours.

In order to obtain the "H" form of the biosorbent, dry biomass was treated with an excess of nitric acid 1 N. In this way, all the active groups on the surface of the biosorbent become available for ionic exchange.

In the aim to determine the uptake degree of radioactive cations, equal amounts (0.25 g) of biomass, dry biomass and biosorbent in "H" form are brought into contact with 25 mL radioactive solution. The pH of the solution is maintained constant throughout the experiments by means of an ammonium hydroxide 5 N solution. Temperature was 20±1 °C, and illumination was continuos. At measured time intervals (5, 15, 30, 60, 90, 120, 240, 720 and 1440 minutes), volumes of 0.1 mL of each solution were taken out and poured into porcelain crucibles, then evaporated to dry by means of an IR lamp. Radioactivity of the residue was determined by means of a GMD alpha & beta detector unit connects to the electronic unit of the GMD system. Under similar experimental conditions, several standard samples were prepared, but without biosorbent, for quantitative comparisons (knowing the fact that radioactivity is directly proportional to concentration).

In order to draw the sorption isotherms for different pH values, the uptake degree of radioactive ions in 24 hours on the biosorbent in "H" form was taken into account; acidity determinations were performed by means of a "Radelkis OP-208/1" pH-meter with a glass electrode (the reference being the calomel electrode). Different values of temperature were maintained constant using a waterbath thermostat of the Messgeräte-Werk Lauda" type and a refrigerator.

Besides, in order to optimize the sorption process, the variation of uptake degree of radioactive ions on the biosorbent in "H" form at different temperatures was supervised. For this purpose, the pH was neutral and the conditions were as described above.

A study of thin layer radiochromatography was conducted in order to supervise the placement of UO_2^{2+} and Th^{4+} radiocations in the biochemical components of the plants used. This method was used successfully in tracing the distribution of $\beta+\gamma$–active radiocations like $^{137}Cs^+$, $^{60}Co^{2+}$, $^{65}Zn^{2+}$ and $^{51}Cr^{3+}$ in some algae and hydrophilic plants (Cecal et al. 2002). For this aim, quantities of 0.25 g plant were put in contact with radioactive solutions for 48 hours. Then, the plant was separated through centrifugation and subjected to maceration for five days, in 25 mL acetic acid solution (glacial acetic acid/water = 1/4, v/v). Portions of 0.01 mL extractant were dropped on the starting line of some chromatographic plates 4×8 cm of the Polygram Sil G/UV$_{254}$ type. Elution was performed in covered chromatographic vessels in 10 mL freshly prepared acetic acid solution (glacial acetic acid/water = 1/4, v/v).

In order to determine the areas on the chromatogram where various products of biochemical fraction – radioactive ion interaction spotted, a series of spraying reagents was used, according to indications found in the literature (Krely at al. 1969).

After identifying on the chromatogram the areas where the compounds mentioned above attach themselves, these portions were cut out and subjected to radioactivity measurements.

In the case of retention on the various biochemical fractions, the calculation relation is:

$$G_i = \frac{R_{et} - 2500 \cdot R_{spot}}{R_{et}} \cdot 100 \, (\%)$$

in with R_{spot} is the radioactivity of the spot shown on the chromatogram, characteristic of a certain biochemical fraction, and R_{et} is the radioactivity of the standard sample.

The total uptake degree (calculated) on the extracted components in is given through the following relation:

Microbiology / phytoremediation

$$G_{calculated} = \sum_{\substack{i=1 \\ i \neq 2}}^{9} G_i \ (\%)$$

The total uptake degree (measured) can be determined through the relation:

$$G_{observed} = \frac{R_{et} - R_{sol}}{R_{et}} \cdot 100 \ (\%)$$

Here, R_{sol} is the radioactivity of the sample extracted after 2 days of contact radioactive solution – hydrophilic plant.

Results and discussion

Influence of environmental factors on wastewaters purifying performances using *Lemna minor*

In nitrate solutions (pH = 1.5), UO_2^{2+} is better retained as compared to Th^{4+} (Fig. 1.). This fact can be explained through the fact that Th^{4+} ion hydrates with a bigger number of water molecules, so that the hydrated ion has a bigger ionic radius that UO_2^{2+} (in a linear geometry). UO_2^{2+} ion is bulky, therefore it will take for hydration a small number of water molecules.

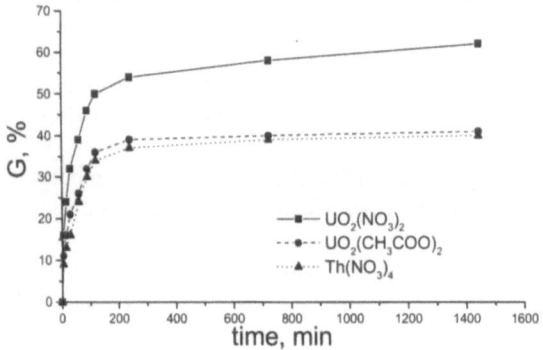

Fig. 1. Variation of UO_2^{2+} and Th^{4+} ions uptake degree, out of nitrate solutions 0.1 M (pH = 1.5) and acetate solutions 0.1 M (pH = 2.5) on the biosorbent obtained from *Lemna minor*, at 22 °C.

UO_2^{2+} ions are better retained in nitrate solutions than in acetate solutions, even though the pH of $UO_2(CH_3COO)_2$ 0.1 M solution (2.5) is superior to the pH of $UO_2(NO_3)$ 0.1 M solution (1.5). The explanation can be given by the fact that UO_2^{2+} in acetate 0.1 M solution is at its limit of solubility. In time,

$UO_2(CH_3COO)_2$ can reprecipitate on the surface of the biosorbent. Therefore a competition appears between the process of biosorption and that of reprecipitation.

As can be observed in Fig. 2., at pH ≥ 5, UO_2^{2+} ions are completely retained, after an interphasic contact time of only a few minutes. This phenomenon is normal, because at pH = 5 there begins the process of precipitation of various uranates and diuranates species (Cotton and Wilkinson 1972) on the surface of the biosorbent (fact visualized by the appearance of the yellow color on the surface of the solid state).

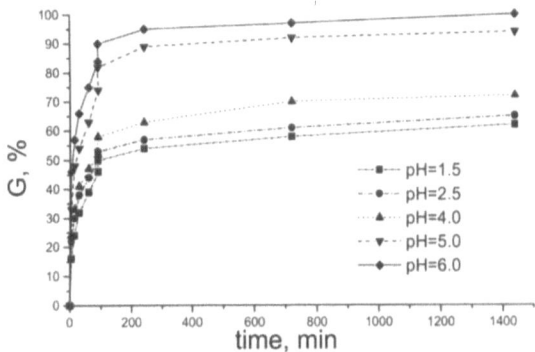

Fig. 2. Variation in time of UO_2^{2+} ion uptake degree from 0.1 M nitrate solutions on the biosorbent obtained from *Lemna minor*, according to pH, at 22 °C.

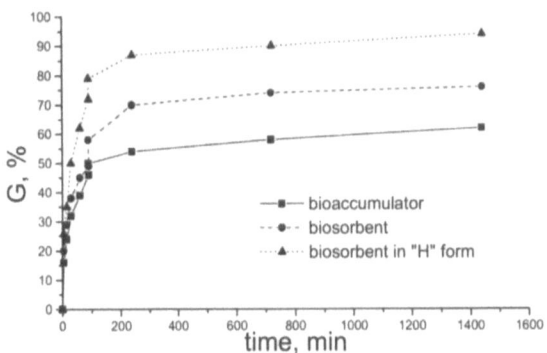

Fig. 3. Variation of UO_2^{2+} uptake degree from 0.1 M nitrate solution (at 22 °C and pH = 1.5) on the various forms of *Lemna minor*.

It can be noticed that the active process (metabolic bioaccumulation on the living plant) is responsible of 50 % of the sorption process approximately (Fig. 3.). UO_2^{2+} ions are better retained on dry biomass. Eliminating the toxic effect of radioactive ions on the plant, the "H" form of the biosorbet is even more active, since all the surface groups became available for chemical bond of radioactive ions, as a result of functionalizing them by means of acids.

The fact that biosorption on cellular membrane and on extracellular polymers plays an important part in the bioaccumulation process is proved by the quick up-take degree during the first minutes of bioaccumulation (when a chelating process of about 80 % of the radioactive ions takes place). Then a plateau is installed, but it is not perfectly parallel to the abscise; the explanation consist of the continuation of active metabolic process among which the most important is the one of transport through membrane.

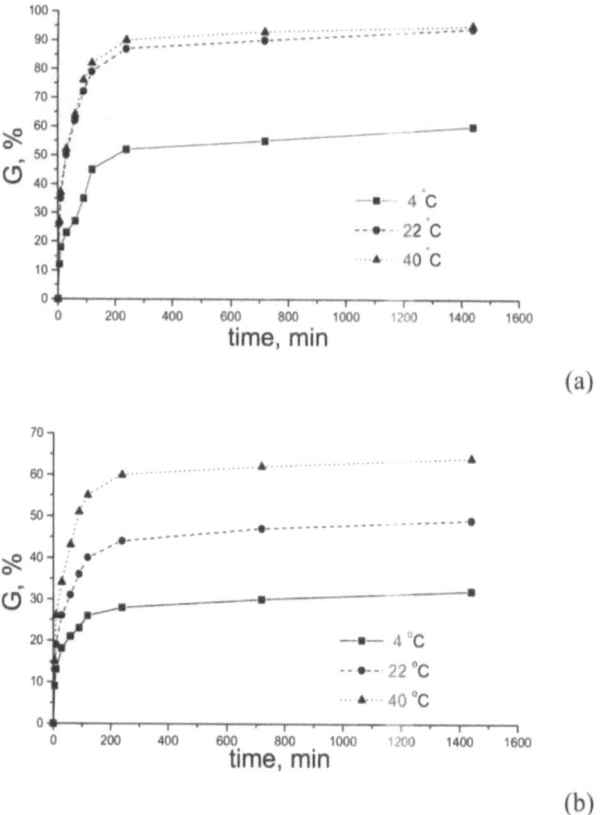

(a)

(b)

Fig. 4. Variation of UO_2^{2+} (a) and Th^{4+} (b) uptake degree from 0.1 M nitrate solution (pH = 1.5) on the biosorbent obtained from *Lemna minor*, according on the temperature.

Since UO_2^{2+} ions are almost completely retained on biosorbent's surface at 22 °C, an increase in temperature over this value becomes useless (Fig. 4.a.). On the contrary, a temperature of 40 °C generates an important increase of the degree of Th^{4+} ions retention (Fig. 4.b.).

Distribution of UO_2^{2+} and Th^{4+} radiocations in the biochemical fractions of hydrophilic plants

Results of the thin layer radiochromatography are summed up in Table 1.

Table 1. Distribution of UO_2^{2+} and Th^{4+} radiocations in the biochemical fractions of hydrophilic plants.

Ion	Uptake degree	Riccia fluitans	Lemna minor	Pistia stratiotes	Elodea canadensis
	Total retention				
	Observed	100.0	74.3	53.2	44.0
	Calculated	97.3	72.2	52.7	43.5
	In lipids, G_1 (%)	30.1	19.0	22.9	18.6
	In sugars and polyalcohols, G_2 (%)	51.7	39.9	22.9	20.9
	In sugars, G_3 (%)	47.8	36.5	22.9	20.3
UO_2^{2+}	In polyalcohols, $G_2 - G_3 = G_4$ (%)	3.9	3.4	0.0	0.6
	In aminoacids, amines and aminosugars, G_5 (%)	7.9	8.1	2.3	2.0
	In aromatic amines, G_6 (%)	3.4	3.6	2.4	2.0
	In phenols, G_7 (%)	1.0	0.0	1.3	0.0
	In aromatic acids, G_8 (%)	2.0	1.1	0.9	0.0
	In vitamins, carotenoids, steroids and terpenes, G_9 (%)	0.8	0.5	0.0	0.0
	Total retention				
	Observed	79.9	56.1	49.9	32.6
	Calculated	79.3	55.5	47.9	32.3
	In lipids, G_1 (%)	31.8	22.1	20.2	13.3
	In sugars and polyalcohols, G_2 (%)	41.3	29.8	19.7	18.1
	In sugars, G_3 (%)	40.0	28.8	17.3	17.3
Th^{4+}	In polyalcohols, $G_2 - G_3 = G_4$ (%)	1.3	1.0	2.4	0.8
	In aminoacids, amines and aminosugars, G_5 (%)	0.8	0.8	2.3	0.6
	In aromatic amines, G_6 (%)	2.3	0.3	3.1	0.0
	In phenols, G_7 (%)	1.1	0.4	1.2	0.0
	In aromatic acids, G_8 (%)	1.2	1.0	1.4	0.3
	In vitamins, carotenoids, steroids and terpenes, G_9 (%)	0.7	0.1	0.0	0.0

One can notice that the values $G_{calculated}$ and $G_{observed}$ are comparable in all the systems radiocation – hydrophilic organism, which leads to the idea that the whole quantity of radioactive ion bioaccumulated placed itself in one of the biochemical fractions extracted according to the mentioned in "Materials and methods" section.

Biomolecules that play a significant role in UO_2^{2+} and Th^{4+} ions uptake are the polysaccharide and the lipoid fractions. The unusually high bioaccumulation activity of lipids is probably due to partial hydrolysis into the fatty acids and triglycerides, with a greater chelating action on the α–active ions.

Comparison of purifying performances of *Elodea canadensis, Lemna minor, Pistia stratiotes* and *Riccia fluitans*

It can be observed that, under identical experimental conditions, the biosorption capacity decreases according to the following series: *Riccia fluitans > Lemna minor > Pistia stratiotes > Elodea canadensis*, for both α–active ions from the simulated radioactive solutions. On can also notice the fact that UO_2^{2+} radiocation is much better retained that Th^{4+}, whatever the plant used as bioaccumulator.

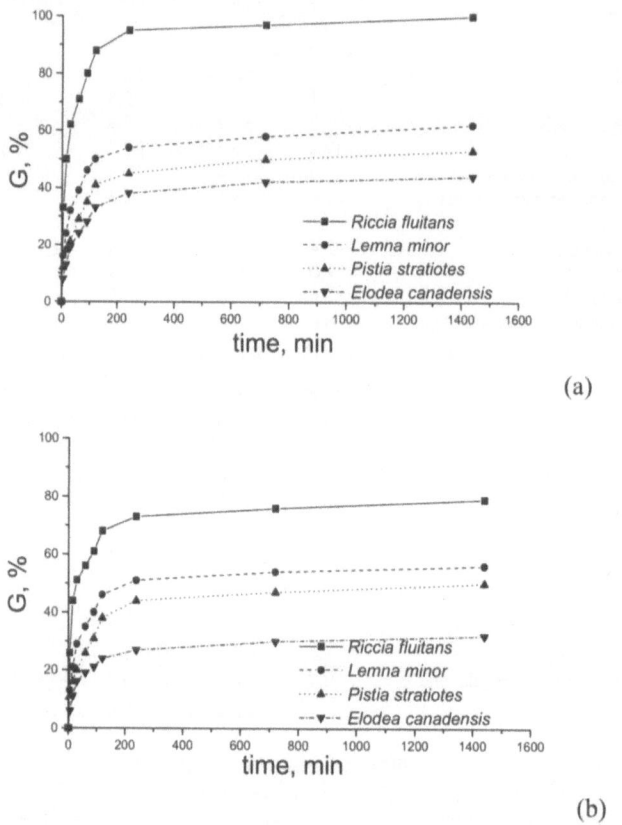

(a)

(b)

Fig. 5. Variation of UO_2^{2+} (a) and Th^{4+} (b) ions uptake degree on hydrophilic plant, at 22°C.

Conclusions

Experiments presented in this study point to the fact that the hydrophilic plants have a high capacity on the UO_2^{2+} and Th^{4+} ions uptake, in the absence of ionic competition. The type and nature of the accumulator (living plant, biosorbent or biosorbent in "H" form), as well as the environmental factors (temperature, pH, anion nature, etc.) have a significant influence on the α–active ions uptake capacity. The studies performed point to the fact that the best results are obtained using *Riccia fluitans* and *Lemna minor*, at 22–30 °C and pH = 5–7.

All the experiments were performed on simulated solutions, in the absence of ionic competition. The next step will be to check whether or not this high uptake degree of UO_2^{2+} and Th^{4+} ions in hydrophilic plants is maintained under natural conditions.

Experiments of thin layer radiochromatography point to the fact that polysaccharide and lipoid fractions are responsible for bioaccumulation. Taking into account this observation and the biochemical composition of the plants chosen for study (Landolt and Kandeler 1987), one can conclude that a high uptake degree is obtained when using plants that have a high percentage of these biochemical components.

References

Bergamini PG, Palmas G, Piantelli F, Sani M, Bantitelli P, Previtera M, Sodi F (1979) Study of ^{137}Cs absorption by Lemna minor. Health Phys 37: 315-321

Boyd CE (1969) The nutritive value of three species of water weeds. Econ Bot 23: 123-127

Bouby M, Billard I, MacCordick J, Rossini I (1998) Complexation of uranium VI with the Siderophore pyoverdine. Radiochim Acta 80: 95-100

Cebotina M, Liubimova C (1981) Sorption dependence of radionuclides in water. Hydrobiol J 17 (5): 101-105

Cecal A, Palamaru I, Popa K, Caraus I, Rudic V, Gulea A (1999) Accumulation of [60]Co[2+] and UO_2^{2+} ions on hydrophilic plants. Isotopes Environ Health Stud 35: 213-219

Cecal A, Paraschivescu A, Palamaru MN, Popa K, Petrescu S, Bontea D (2000) The depollution procedure for rain waters collected after infiltrating the dumps of uranium ores. Ro Patent OSIM A/146

Cecal A, Popa K, Potoroaca V, Puica-Melniciuc N (2002) Decontamination of radioactive liquid wastes by hydrophilic vegetal organisms. J Radioanal Nucl Chem 251 (2): 257-261

Cecal A, Popa K, Caraus I, Potoroaca V (2002) [65]Zn[2+] bioaccumulation on hydrophilic plants. Isotopes Environm Health Stud 38: in print

Cotton FA, Wilkinson G (1972) Advanced inorganic chemistry. Interscience Publ Co, New York, pp. 199

Duff MC, Amrhein C, Bradford G (1997) Nature of uranium contamination in the agricultural drainage water evaporation ponds of the San Joaquin Valley, California, USA. Canad J Soil Sci 77: 459-467

Francis AJ, Dodge CJ, Lu F, Halada GP, Clayton CR (1990) XPS and XANES studies of uranium reduction by Clostridium sp. Environm Sci Technol 24 (3): 373-378

Francis AJ, Dodge CJ, Gillow JB, Cline JE (1991) Microbial transformations of uranium in wastes. Radiochim Acta 52/53: 311-316

Krely KG, Heusser D, Winner H (1969) Spray reagents, in Thin-layer chromatography – A laboratory handbook. Springer Verlag Berlin, pp. 857

Landolt E, Kandeler R (1987) Biosystematic investigations in the family of duckweeds (Lemnaceae). Veroff Geobot Inst ETH, Zurich, vol 2, pp. 42-43

Nakajima A, Sakaguci T (1986) Selective accumulation of heavy metals by microorganisms. Appl Micro Biotech 24: 59-64

Pribil S, Marvan P (1976) Accumulation of uranium by the chlorococcal alga Scenedesmus quadricauda. Arch Hydrobiol S49: 214-225

Sutton DL, Ornes WH (1975) Phosphorus removal from static sewage effluent using duckweed. J Environ Qual 4 (3): 367-371

Tsezos M, Georgouses Z, Remoudaki E (1997) Mechanism of aluminum interference on uranium biosorption by Rhizopus arrhizus. Biotechnol Bioeng 55: 16-27

Volesky B (1994) Advances in biosorption of metals. Biomass types. FEMS Microbiol Rev 14: 291-302

Wade Jr R, DiCristina TJ (2000) Isolation of U(VI) reduction-deficient mutants of Shewanella putrefaciens. FEMS Microbiol Lett 184: 143-148

Soil formation and quality on uranium mining dumps depending on different tree species under special consideration of selected radionuclide contamination

Holger Dienemann, Carsten Brackhage, Anja Dannecker, E. Gert Dudel, Joachim Rotsche

TU Dresden, Institut für Allgemeine Ökologie und Umweltschutz, Pienner Str. 8, 01737 Tharandt

Abstract. Three different sites on an uranium mining dump were investigated, classifying the organic layer as ´raw humus´ on an afforested pine stand and a site with succession and as moder on an afforested alder stand. The radionuclide content (Ra-226, U-238, Pb-210) and radionuclide ratios vary significantly between different humus horizons and between sampling plots within each site. The results show that after 40 years soil along with vegetation development very low concentrations in surface layers can be detected. The nutrient content of the organic layer was analyzed, since it is substantially influenced by the tree species.

Introduction

The discharge of radionuclides from uranium mining dumps and tailings by leaching and gas emission (radon) is often difficult to minimize. Possible solutions are technical measures like multilayered covers. The often used consolidated mineral covers may develop cracks due to dehydration (Melchior 1997).

Other solutions discussed are naturally orientated measures like reforestations where soil formation is one major factor. Within the first decades the development of the organic layer is of major importance. Thus two stands, which were afforested 40 years ago and one area of succession were investigated regarding the organic layer, activity concentrations and distribution of radionuclides.

Material and Methods

Area of investigation

The investigations were carried out on dump H382 in the former uranium mining and milling complex Schlema/Alberoda (Erzgebirge mountains, Germany). The average air temperature is about 8,2 °C (1985-93, climate station Aue). The average yearly precipitation is approx. 883 mm (1951-93, climate station Aue). Dump H382 covers an area of 33,7 ha. A reclamation area with stands of *Pinus sylvestris* L. (Scots pine) and *Alnus glutinosa* (L.) GAERTN. (black alder) was established in the early 60s (1962/63) on the plateau (Ranft 1966). A neighboring part was not afforested and is further referred to as succession area. Dominant species are *Betula pendula* ROTH and *Populus spec.*, the age of the trees varying between 2 and 37 years.

Sampling and Analysis

For sampling of the organic layer from each of three investigated sites 15 samples per horizon were taken and unified to representative composite samples using a rotation contributor. Afterwards the composite samples were homogenized. Due to the high heterogeneity of the succession area two plots of 3 m² each were harvested completely. The air dried material was analyzed by gamma- spectrometry (SILENA, Germany). Plant samples were prepared according to methods by Brackhage and Dudel (2002, this volume).

For nutrient analysis plant and soil samples were homogenized by HF- digestion with HNO_3/H_2O_2 (MDS-2000, CEM). Resulting solutions were analyzed by AAS (VARIAN) and ICP-MS (Pq2+, FISONS).

Results

Organic Layer

The different tree species influenced the formation of the organic layer. Underneath black alder (in symbiosis with *Frankia spec.*) moder developed. Underneath Scots pine a poorly developed (referring to average C-supplies, BMELF 1996) raw humus (mor) emerged. The organic layer beneath the succession area might be labeled as initial raw humus (raw). Differences between the initial raw humus and the poorly developed humus can be seen for example by mass per unit area of each single horizon (see Fig. 1).

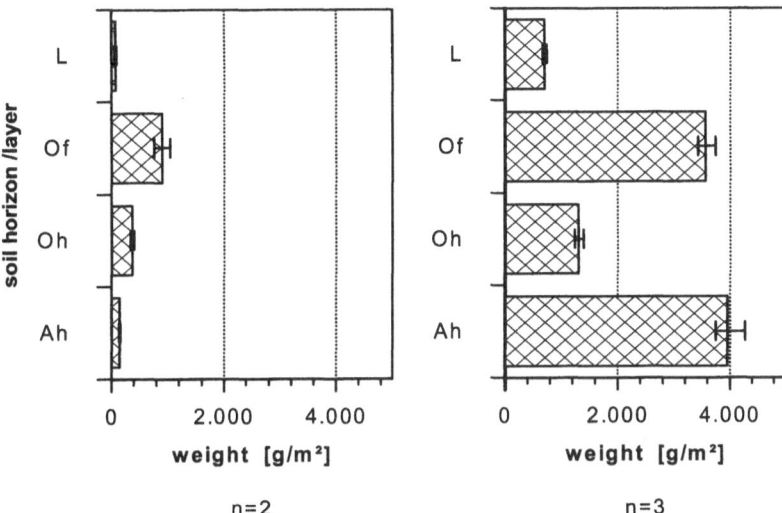

Fig. 1. Weight of each single horizon per m² dry matter (DM) on dump H382 (Schlema). Scots pine afforested approximately 40 years ago, organic layer in both cases raw humus, thickness of the organic layer: succession area 0,5 – 25 cm, beneath Scots pine 4 – 7,5 cm

On the succession area almost no humus developed. The poorly developed raw humus (raw) (underneath Scots pine) shows the highest accumulation compared to the other sites. As expected the Of-horizon is most distinctive in the raw humus. Underneath Scots pine it forms a continuous layer (min. 2 cm thick). On the succession area the Of-horizon is rather thin (max. 1 – 1,5 cm) and disrupted by stones.

The formation of the organic layer on the succession area is more clearly influenced by root litter than by surface litter. Roots on both sites show high mycorrhizal infection rates.

Distribution of radionuclide activity

The activities of U-238, Ra-226 and Pb-210 in individual horizons are shown in Fig. 2. The U-238 and Ra-226 activities (absolute and related to the activity in the yC-horizon) in the Ln-layer and in the L-layer underneath black alder are significantly higher than underneath Scots pine.

The activity concentrations of Ra-226 and Pb-210 in the organic layer are markedly higher than the ones plotted in Fig. 2. The Ra-226/Pb-210-ratio however is much more balanced. The activity concentration of U-238 is very low.

Microbiology / phytoremediation

Site with *Pinus sylvestris L.*   Site with *Alnus glutinosa G.*

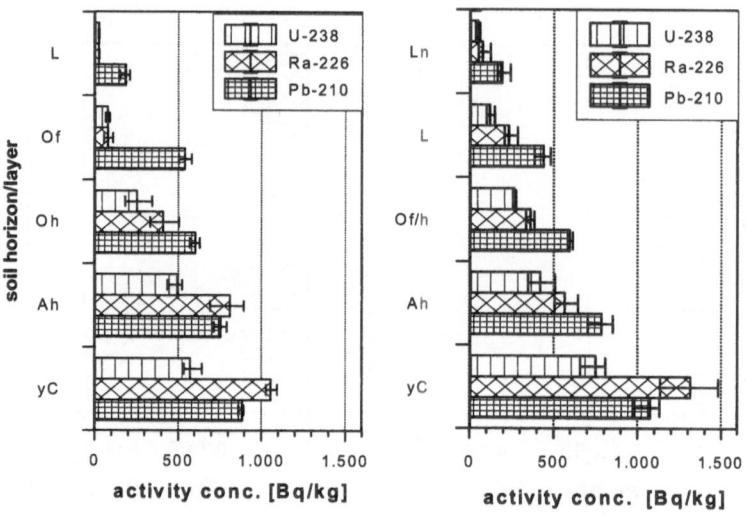

Fig. 2. Activity concentrations (Bq / kg DM) in different horizons under afforested stands (age: approx. 40 years, n=3) of *Alnus glutinosa* G. and *Pinus sylvestris* L.; H382, Schlema

Uranium concentration in plants

Uranium concentrations were measured in different parts of Scots pine and black alder in order to discuss the immediate influence on the soil formation. Therefore plant samples were washed. Uranium concentrations are shown in table 1. Leaves and corresponding twigs show in comparison to needles and corresponding twigs far higher uranium concentrations.

Table 1. Uranium concentrations in different plant parts from trees of the site with Scots pine stand (*Pinus sylvestris* L.) and black alder stand (*Alnus glutinosa* (L.) GAERTN.) on dump H382, Schlema.

Species	Plant part	n	Umin. [µg/kg DM]	∅ U [µg/kg DM]	Umax. [µg/kg DM]
Pinus sylvestris	Needles (1 y old)	5	7	8,6	14
	Needles (2 y old)	4	123	266	393
	Twigs	5	30	45	115
	Roots	5	46000	58500	77000
Alnus glutinosa	Leaves	10	188	315	826
	Twigs	10	611	1077	1673
	Roots	4	3457	28100	67100

Discussion

The investigated sites show different developments of organic layers. The differences (weight, thickness) between the area beneath Scots pine and the succession area might be due to slower colonization with trees on the latter. Furthermore, birch leaves are simply because of their shape more easily blown over than needles. As a result the relatively tight root litter contributes more to the formation of an organic layer on the succession site than on the afforested sites.

Since there are generally higher radionuclide contents in roots than in leaves (Brackhage and Dudel 2002), the shift of the ratio between surface litter to root litter leads to higher radionuclide contents in the organic layer, which is clearly seen on the succession area. However, the different mycorrhizal infection rate might contribute to the higher Pb-210 and Ra-226 activities in the horizons as well.

In comparison to the two afforested areas the development of the organic layer on the succession area over a period of 40 year appears insufficient considering German radiation protection aspects. On the other hand, reforestation appears to be favorable. On the surface layers low activities (< 200 Bq/kg DM) develop. The reason for this is certainly a dilution effect by the surface litter. The differences in uranium concentrations between needles (1 year old), leaves and twigs underneath Scots pine and black alder have an impact on the Of/h horizon. Regarding the activity concentrations the Oh horizon corresponds approximately with the Of/h.

A possible explanation for the differences of the activities in the Of/h resp. Of and Oh horizon regarding the Ln (black alder) resp. L (Scots pine) could be the initial decay of litter, which is faster with alder leaves than with needles (e.g. Cole 1995). Also the influence of soil meso- and macrofauna on rearrangement processes has to be discussed (Scheu 2001).

Biochemically the tree species vary in their Ca- and Mg-requirements which for both elements is higher for black alder (Bergmann 1983, Zimka 1989), which leads to higher (slightly acidic to neutral) pH-values underneath alder. In consequence, heavy metals and especially radionuclides are less soluble, which might be an explanation for the slightly increased radionuclide content in the upper layer of dump material under black alder.

The U-238/Ra-226 ratio is underneath black alder smaller than underneath Scots pine. This might as well result from biochemical-plant physiological reasons. Brackhage et al. (1998) found in leaves of *Alnus glutinosa* higher activities of Ra-226 than U-238.

Much more significant discrepancies than with the U-238/Ra-226 ratio exist for the Pb-210/Ra-226 ratio. In the L and Of horizon underneath Scots pine this ratio varies between 5 and 11,4 (yC horizon underneath Scots pine and black alder ca. 0,85). One reason might be an accumulation of Pb-210 through atmospheric deposition. This supports – like the higher pH-value mentioned above – the reforestation with black alder for reclamation.

Conclusions and Perspectives

The investigations show that reclamation by reforestation with different tree species leads within 40 years to the development of a several cm thick organic layer, which is low contaminated on the surface (Ln, L, Of). The selection of the tree species has significant influence on the activity concentrations of the organic layer as well as on the milieu (pH). In comparison *Alnus glutinosa* appears at the time being to be much more qualified than *Pinus sylvestris L.* (Scots pine) for reclamation purposes when site conditions are comparable. This will be validated by further investigations concerning the emission by the species specific DOC.

Acknowledgement. This study was supported by the the the Saxon State Authority for Environment and Geology (LfUG).

References

Bergmann W (1986) Ernährungsstörrungen bei Kulturpflanzen. 2.erweiterte Auflage.Gustav Fischer Verlag Jena

BMELF (Hrsg.) (1997): Deutscher Waldbodenbericht 1996. Ergebnisse der bundesweiten Bodenzustandserhebung im Wald von 1987-1993 (BZE). Bd. 1

Brackhage C (2002) Long-term differences in transfer and accumulation of potentially toxic trace elements and radionuclides in trees on uranium mining dumps (Erzgebirge, Germany. This volume

Cole DW, Compton JE, Edmonds RL, Homann PS., v.MiegoetT, H. (1995): Comparison of Carbon Accumulation in Douglas Fir and Red Alder Forests. Mc Fee, W.W., Kelly, M. (Hrsg.): Soil Sci. Soc. of Am., Inc. Madison, Wisconsin USA

Dudel GE, Brackhage C, Dabrunz M, Passek U, Rotsche J, Stolz L (1997) Freisetzung radioaktiver und nichtradioaktiver Elemente aus offenen und abgedeckten Bergbauhalden - Einflüsse auf eine land- und forstwirtschaftliche Nutzung - Abschlußbericht 1997. TU Dresden, Institut für Allgemeine Ökologie und Umweltschutz

Melchior S (1997) Die Austrocknungsgefährdung von bindigen mineralischen Dichtungen und Bentonitmatten in der oberflächenabdichtung. In :Workspoh. Sanierunge der Hinterlassenschaften des Uranbergbaus Teil 2. Tagungsband 10.4.1997 Dresden (SMUund LfUG Sachsen)

Ranft H. (1966) Untersuchungen und Versuche zur Rekultivierung von Erzbergbauhalden des sächsischen Erzgebirges. Diss., TU Dresden

Scheu L (2001): Plants and generalist predatore as links between the below-ground and above-ground system. Basic Appl. Ecology, Vol.2, No.1:3-13

Wismut GmbH (1994): Sanierungskonzept für den Haldenkomplex Schlema-Alberoda. Techn. Ressort, Abt. Engineering

Zimka JR (1989) Analysis of processes of element transfer in forest ecosytems- Polish Ecological Studies, Vol. 15, No. 3-4: 135-212

Ecosystem restoration incorporating minerotrophic ecology and Stoneworts that accumulate [226]Ra

Margarete Kalin[1], Martin P. Smith[1] and Mark B. Wittrup[2]

[1]Boojum Research Limited, 468 Queen Street East, Toronto, Ontario, Canada M5A 1T7
[2]Cameco Corporation, Saskatoon, , Saskatchewan, Canada S7N 4M2

Abstract. Uranium removal from shallow lakes ranged from 0.15 kg.d^{-1}.ha^{-1} to 0.65 kg.d^{-1}.ha^{-1}, while [226]Ra removal ranged from 0.05 to 0.2 MBq. d^{-1}.ha^{-1} in the Rabbit Lake drainage basin in northern Saskatchewan, Canada. The removal of U is related to the organic and inorganic components of the TSS in the lakes but [226]Ra is concentrated by stoneworts (Characeae), a family of macrophytic algae. [226]Ra concentrations in algal biomass range from 5 Bq.g dry weight^{-1} (gdw) to as high as 50 Bq.gdw^{-1}, averaging 20 Bq.gdw^{-1}. These alkaline minerotrophic lakes can be integrated into a decommissioning approach.

Introduction

Uranium mining waste management areas require water treatment to remove U and [226]Ra. During operations, this is achieved through conventional wastewater treatment for [226]Ra using barium sulphate and lime. When waste management areas have to be decommissioned, the costs of treatment and sludge storage requirements make conventional chemical treatment both economically and environmentally unattractive. Alternative sustainable decommissioning approaches have to be found to address the long-term water contamination arising from the mining wastes. Such approaches could include integrating minerotrophic conditions for retention of U within the waste management area, leading to the formation of young surficial U ore deposits. The hydrological, topographical and ecological characteristics that contribute to the formation of such deposits are well documented (Culbert et al., 1984). U is adsorbed or integrated into organic and inorganic particulate matter that settles to the sediment, where the U is then mineralized.

Removal of ^{226}Ra in such a sustainable decommissioning approach could be achieved if the waste management area supported the growth of Stoneworts (Characeace), a family of macrophytic algae that accumulate ^{226}Ra. These algae have a high affinity for ^{226}Ra, as their cell walls accumulate Ca, Mg and other carbonates (Price and Whitecross 1983; McConnaughey, 1995) which can comprise as much as 70 % of total biomass. When present, ^{226}Ra is incorporated into these carbonates as they accumulate at locally high pH microzones of the cell walls where Characean cells actively pump out hydroxyl ions. Stoneworts also develop statoliths, organelles at the tips of rhizoid cells which gravitropically direct growth downward (Wang-Cahill and Kiss, 1995). These characteristics of the Stoneworts suggest that, when present in water, ^{226}Ra is likely integrated into Characean biomass. This algal group can be considered as a natural treatment plant for ^{226}Ra, employing both high pH-induced carbonate formation and barium sulphate precipitation.

Combining minerotrophy for U with conditions that support the growth of stoneworts in the waste management area would produce an ecological solution within a decommissioning approach for U mining wastes. To assess this approach to waste water treatment, a research program was initiated in a drainage basin containing U-mining wastes in northern Saskatchewan, Canada, using water quality monitoring for the 1987 to 2000 period.

Location and description of waste management area

Exploration and U mining activities started in late 1960 in the Rabbit Lake drainage basin (995 ha), an area containing a series of lakes located at the southern border of the boreal forest-barren land vegetation zone. The ore body was found under Rabbit Lake, a kettle lake located in the upper reach of the drainage basin. After completion of pit development and the start of mining in 1975, waste rock piles, located to the north of the former Rabbit Lake, started releasing ^{226}Ra and U.

Before 1998, waste rock seepage ran through a drainage ditch into Upper Link Lake (ULL). The first water monitoring station for the drainage basin is referred to as Airport Road (APR), located at the drainage ditch prior to water entering ULL (Fig. 1). ULL (23 ha; 3.0×10^5 m^3) is hourglass shaped. A second station is located at the Narrows, serving as the monitoring point for water quality determination, quantifying the contaminant contribution from the mine slimes to the 16 ha portion of the lake above the Narrows. The upper part of the hourglass includes a small (3.4 ha) delta where mine slimes accumulated during mining of the pit. The water then flows through the third monitoring station at Sedimentation Dam (Sed D), constructed to retain the sediments from the former Rabbit Lake. Below Sed D, water flows through a floating muskeg area (4 ha) into Lower Link Lake (LLL; 22 ha with 2.9×10^5 m^3), finally discharging to Wollaston Lake. Flow monitoring stations are installed at APR and Sed D.

The Link Lakes are shallow (2-4 m deep) with organic-rich sediments. In 1988, ULL was devoid of submerged aquatic vegetation apart from phytoplankton, as the sediments of former Rabbit Lake were discharged during pit development to ULL, greatly attenuating light and eliminating submerged vegetation. Sediments were contained in ULL through the construction of the Sed D (Fig. 1). The muskeg area below the Sed D and the outflow of LLL contains several beaver dams that intermittently control the water level. LLL supports an extensive underwater meadow of stoneworts.

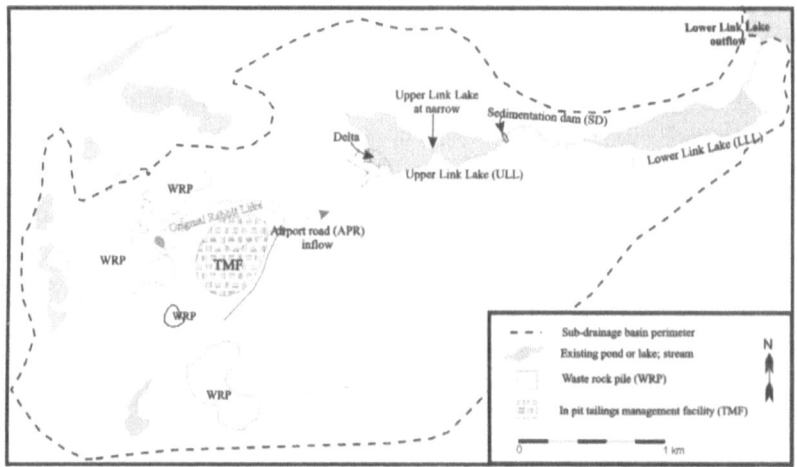

Fig. 1. Rabbit Lake drainage basin existing layout 1999.

Methods and materials

Water sampling and chemical analysis: Water quality monitoring has been carried out since the onset of mining activity. Samples are collected by site staff, then filtered (0.45 μm) and acidified on site. Samples are shipped in coolers to Saskatchewan Research Council (SRC), a certified analytical facility for both elemental and radionuclide determination.

Data refinement: The large volume of water monitoring information was subjected to statistical data refining, transformation, and standardization in order to obtain a reliable data set. The details of the data refinement, yielding the average values used in this presentation, are given in Cao and Kalin (1999).

Stonewort Biomass Sampling: Plant biomass was collected from a known surface area using either an Ekman dredge, or a rake in shallow areas. The biomass was washed free of debris using lake water. The biomass was air dried, ground in a mortar and then subjected to digestion by nitric and hydrochloric acids for determination of elemental (ICP) or radionuclide concentrations.

Transplanting Stoneworts: After small scale lab and field (ULL) growth tests in 1988, twelve tonnes (wet weight) of a species of Stonewort were transported from LLL to ULL in June, 1989 by an all terrain vehicle with a trailer (ARGO). Biomass was then spread by hand from a boat over four zones in ULL (Delta area; northwest corner; Narrows; close to Sed D) in June 1989. Annual direct examination (diving) was used to record the presence / absence and % coverage of stoneworts populations. Annual monitoring of the population continued up to the year 2000.

Results and discussion

The pH of the Link Lakes ranges for all sampling stations from 6.6 to 7.2. The electrical conductivity ranges throughout the system from 94 to 245 $\mu S.cm^{-1}$. For both parameters, the higher values are found in ULL. The bicarbonate concentration ranges from 48 to 57 $mg.l^{-1}$. The Total Suspended Solids (TSS) concentrations are low, ranging from 2.2 $mg.l^{-1}$ to 3.0 $mg.l^{-1}$ along with low sulphur concentrations of 2.5 $mg.l^{-1}$ to 14 $mg.l^{-1}$. There are no metal concentrations in the water, which leaves only ^{226}Ra and U as elements of environmental concern (Table 1a and 1b).

^{226}Ra and U monitoring data are broken up into periods that correspond to the presence or absence of the stonewort populations in the system, or the periods when changes in the drainage basin affected their growth. For ULL, the period between 1987 and 1992 was free of stoneworts, followed by 1993 to 1997 when the population was present. From 1998 to 2000 the effects of the diversion of seepage water from the waste rock seepage to the in pit tailings treatment facility (TMF, Fig. 1) would prevail.

Table 1a. Average total U concentration ($mg.l^{-1}$) in water in the Link Lakes, Avg (N)

		APR inflow	ULL at narrows	Sed D	LLL outflow
1987-92	Before Stoneworts in ULL	3.21 (73)	0.89 (61)	0.85 (71)	0.31 (41)
1993-97	After Stoneworts in ULL	4.04 (62)	1.32 (52)	1.14 (63)	0.37 (38)
1998-2000	After ditch diversion in ULL	1.66 (19)	0.78 (19)	0.57 (29)	0.21 (33)
1987-89	Stoneworts in LLL	2.65 (36)	0.72 (30)	0.65 (32)	0.18 (17)
1990-96	Stoneworts absent in LLL	3.93 (85)	1.18 (72)	1.01 (87)	0.40 (49)
1997-2000	Stoneworts re-establish in LLL	2.63 (33)	1.03 (30)	0.91 (44)	0.25 (46)

This mostly eliminated the flow to ULL from APR (26 $1.s^{-1}$ down to 1.8 $1.s^{-1}$) and as a result reduced the contaminant loading. Neither stoneworts nor any other aquatic macrophytes have colonized the mine slime delta due to the shallow water cover which is subjected to ice scouring each winter and early spring. For LLL, the stonewort population was extensive, covering nearly the entire lake bottom between 1987 and 1989. During the winter of 1990, however, one of the main beaver dams broke at the outflow of LLL, resulting in bottom-bound ice formation and a water level decrease during the following years of nearly 1 m. The large quantity of decomposing biomass inhibited re-establishment of the population until 1997, when the population rebounded from oospores (the "seeds" of stoneworts) to form an extensive population by 2000.

For U, the presence /absence of stoneworts will have less effect on the removal as does its oxidation state. Its removal from the water column is likely related to the suspended solids (organic and inorganic) concentration, rather than incorporation into stonewort biomass. U may be present in the Link Lake system as uranyl carbonate which is anionic and would therefore not be adsorbed easily to organic materials. Several mechanisms have been proposed by which algae interact with U. U^{+6} can be associated with algae by adsorption to cell surfaces (Fisher et al. 1987; Zhang et al. 1997); adsorption to biogenic debris and uptake into cells and on cell walls (Anderson et al. 1989); and finally incorporation into calcium (aragonite) structures associated with some marine and freshwater algae (Edington et al. 1970; Mann and Fyfe 1985). Phytoplankton samples analyzed for U contained from 246 to 1,334 $\mu g.gdw^{-1}$ U, suggesting that phytoplankton does play a role in the transport of U to the sediment in the Link Lakes.

From observations during the study and data not presented here, phytoplankton productivity did not change in the system except after ditch diversion. Hence, the only major change is the presence or absence of stonewort biomass, but not the productivity of the phytoplankton. Average U concentrations should be relatively consistent, slowly decreasing throughout the drainage basin. Between the input concentration and that at the Narrows, a 2.1 to 3.6-fold decrease is noted, followed by no further change in concentration as the water arrives at Sed D. A larger reduction of about 2.5 to 3.7 fold is noted between the dam and the outflow of LLL. Thus the removal of U takes place in ULL above the narrows and between the dam and LLL outflow.

In contrast, the average [226]Ra concentration in the drainage basin exhibits some relation to the presence of absence of the stoneworts in the lake system (Table 1b). The mine slime delta in the ULL contributes [226]Ra to the water, increasing [226]Ra concentrations between the APR station and Narrows for all time periods. However between Narrows and Sed D, [226]Ra concentrations remained the same (0.36 $Bq.l^{-1}$), although concentrations increased in this period at the Narrows from 0.38 $Bq.l^{-1}$ to 0.45 $Bq.l^{-1}$. This apparent reduction was likely due to the establishment of stoneworts. After the ditch diversion, the concentrations of [226]Ra increased further to 0.55 $Bq.l^{-1}$ and the reductions between APR and Sed D (0.28 $Bq.l^{-1}$) are even larger (Table 1b).

Microbiology / phytoremediation

Table 1b. Average total ^{226}Ra concentration (Bq.l^{-1}) in the Link Lakes, Avg (N).

		APR inflow	ULL at Narrows	Sed D	LLL outflow
1987-92	Before Stoneworts in ULL	0.28 (73)	0.38 (61)	0.36 (71)	0.08 (41)
1993-97	After Stoneworts in ULL	0.34 (64)	0.45 (53)	0.36 (63)	0.1 (38)
1998-2000	After ditch diversion in ULL	0.22 (19)	0.55 (18)	0.28 (32)	0.07 (36)
1987-89	Stoneworts in LLL	0.25 (36)	0.32 (30)	0.32 (32)	0.06 (17)
1990-96	Stoneworts absent in LLL	0.29 (85)	0.43 (72)	0.37 (87)	0.10 (49)
1997-2000	Stoneworts re-establish in LLL	0.36 ((35)	0.53 (30)	0.32 (47)	0.07 (49)

Table 2a. Daily average U loads (kg.d-1) in the Link Lakes

		APR inflow	ULL at Narrows	SED D	LLL outflow
	Flow	26.3 l.s^{-1}	42 l.s^{-1}	48 l.s^{-1}	73 l.s^{-1}
1987-92	Before Stoneworts in ULL	7.30	3.22	3.53	1.98
1993-97	After Stoneworts in ULL	9.17	4.78	4.72	2.35
1998-2000	After Stoneworts in ULL[a]	0.26	1.19	1.11	0.86
1987-89	Stoneworts in LLL	6.03	2.61	2.36	0.64
1990-96	Stoneworts absent in LLL	8.93	4.29	3.66	1.44
1997-2000	Stoneworts re-establish in LLL[a]	0.41	1.57	1.77	1.03

a.   After ditch diversion, flow at APR is 1.8 l.s^{-1}, 17.7 l.s^{-1} at Narrows, 22.4 l.s^{-1} at Sed D and 47.4 l.s^{-1} at LLL outflow.

Between Sed D and outflow of LLL, the only factors influencing ^{226}Ra are stoneworts. During the time when stoneworts where absent in LLL, higher concentrations of ^{226}Ra were present at the outflow with 0.10 Bq.l^{-1}, while even higher concentrations were leaving Sed D, suggesting that the floating muskeg area with its permanent stonewort population is very effective. When stoneworts re-established in LLL, the concentrations are further reduced at the outflow to 0.07 Bq.l[-1].

With these trends in [226]Ra concentrations in the drainage basin and the record of the presence and distribution of stoneworts, the removal of the contaminants in relation to load (concentration times flow) for the different time periods will yield relevant information with respect to the effectiveness of these algae in the ecosystem.

The U load to the system from the waste rock seepage (as measured at APR) ranged from 0.26 kg.d[-1] to 9.2 kg.d[-1] (Table 2a).

In Table 2a and 2b, the contaminant load for all time periods are presented throughout the system. It is should be stated that the load at Sed D is that which is affected by populations present in ULL alone and, below Sed D, by its presence in LLL. The U load dropped by half between APR and the Narrows but remained somewhat the same at Sed D. After the diversion ditch was built, LLL's U load was reduced to 0.86 kg.d[-1.] When stoneworts were present in LLL (1987-89), the U load was even lower with 0.64 kg.d[-1]. This pattern of U loads in the drainage basin suggests that the stonewort populations may be exerting some influence. The underwater meadow of stoneworts may well serve as a sieve that removes U-bearing phytoplankton.

Table 2b. Daily average [226]Ra loads (MBq.d[-1]) in the Link Lakes

	Flow	APR inflow 26 l.s[-1]	ULL at Narrows 42 l.s[-1]	SED D flow 48 l.s[-1]	LLL outflow 73 l.s[-1]
1987-92	Before Stoneworts in ULL	0.64	1.37	1.51	0.52
1993-97	After Stoneworts In ULL	0.76	1.64	1.48	0.61
1998-2000	After Stoneworts In ULL[a]	0.03	0.84	0.54	0.27
1987-89	Stoneworts in LLL	0.57	1.17	1.14	0.21
1990-96	Stoneworts absent in LLL	0.66	1.57	1.34	0.37
1997-2000	Stoneworts re-establish in LLL[a]	0.06	0.82	0.62	0.29

After ditch diversion, flow at APR is 1.8 l.s[-1], 17.7 l.s[-1] at Narrows, 22.4 l.s[-1] at Sed D and 47.4 l.s[-1] at LLL outflow.

Mine slimes contributed 0.8 MBq.d[-1] of [226]Ra (Table 2b). Reductions in the [226]Ra load between the Narrows and Sed D are noted throughout all time periods except during the absence of stoneworts (1987-92). In the presence of the stoneworts in LLL in 1987-89, the load leaving LLL was lowest (0.21 MBq.d[-1]).

Microbiology / phytoremediation

Given that the floating muskeg area with its steady population of stoneworts is integrated in the load calculations for the outflow of LLL, the effectiveness of this algal group for [226]Ra removal from the water is evident.

Contaminant removal per day can be partitioned into removals between inflow and the narrows, the narrows and the dam, and the dam and LLL outflow. This partitioning will highlight, for the different periods, the location of sources and sinks in the system, as well as the magnitude of the removal of the contaminant.

In Fig. 2a the U removal and release rates in kg.d^{-1} are plotted for different time periods. ULL becomes a source (i.e. negative removal rates) of U in three intervals, once in the period 1987 to 1992, once after ditch diversion, which overlaps with the period of stonewort establishment in LLL. During all other periods ULL serves as a U sink, removing up to 4.6 kg.d^{-1}. Large algal blooms can explain the change in behaviour from sink to source. U released during the rapid decay of the biomass, after the population crash, could not be retained in the sediments and U was released back into the water.

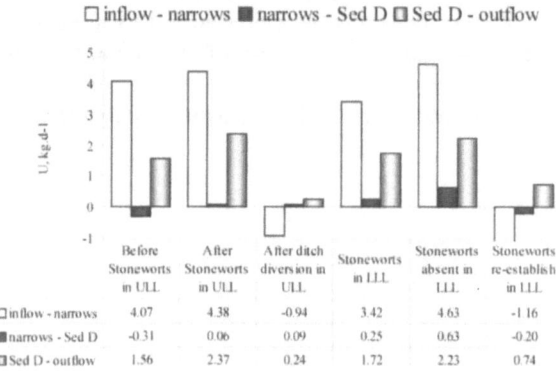

Fig. 2a. Uranium removal from water in the Link Lakes

Removal of [226]Ra was also affected by the ditch diversion (Fig. 2b). [226]Ra removal in the area between Sed D and LLL outflow remained relatively constant at about 1 MBq.d^{-1}, except for the period of the ditch diversion, which overlaps with the period of re-establishment of stoneworts in LLL. However, the system below the mine slime delta remains a sink. In ULL, [226]Ra was not removed before transplant of the stoneworts.

Stoneworts' standing biomass was quantified. In 1992, the standing biomass in the floating muskeg area was about 250 gdw.m^{-2} and in 1998, shortly after ditch diversion, the standing biomass was about 55 gdw.m^{-2}. For ULL and LLL, the standing biomass in 1998 was 61 gdw.m^{-2} and 68 gdw.m^{-2}, respectively. In the biomass from ULL and the muskeg, concentrations of U ranged from 0.13 to 0.27 % (dw), while LLL biomass contained less U (0.08 %). During good growth periods for the ULL population (1988-1997), the concentrations of [226]Ra averaged 22 Bq.gdw^{-1}. In 1998, there was both less stonewort biomass and lower [226]Ra concentrations in the biomass (5.0 to 7.1 Bq.gdw^{-1}). In LLL, the stonewort [226]Ra concentrations ranged from 1.9 to 2.7 Bq.gdw^{-1}.

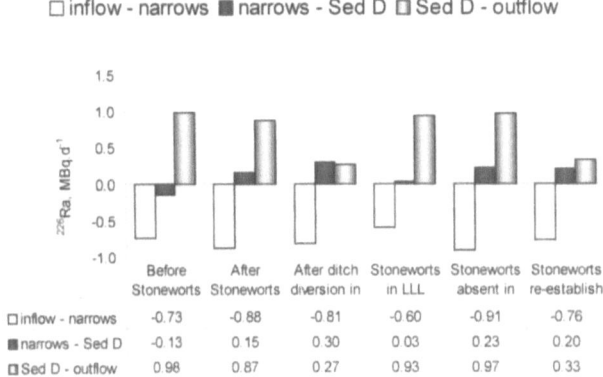

☐ inflow - narrows ■ narrows - Sed D ☐ Sed D - outflow

	Before Stoneworts	After Stoneworts	After ditch diversion in	Stoneworts in LLL	Stoneworts absent in	Stoneworts re-establish
☐ inflow - narrows	-0.73	-0.88	-0.81	-0.60	-0.91	-0.76
■ narrows - Sed D	-0.13	0.15	0.30	0.03	0.23	0.20
☐ Sed D - outflow	0.98	0.87	0.27	0.93	0.97	0.33

Fig. 2b. [226]Ra Removal from water in the Link Lakes.

Conclusions

This study of the Rabbit Lake waste management area was largely driven by observed reductions in [226]Ra concentrations noted as mine wastewater passed through the floating muskeg below the Sed D. It is a descriptive data set, although the transplant of stoneworts from LLL to ULL represents an experimental component of the work. Collection of ecological information about the drainage basin was limited to two to three site visits per year, insufficient to derive a detailed understanding of important ecological parameters pertaining to the ecology of hypereutrophic lakes. While the understanding of the behaviour of the contaminants in the lake system grew, the regulators grew impatient. The experimental intent of transplant was forgotten and the seepage from the waste rock piles was diverted to the treatment plant. The ecological dynamics of the ULL system shifted with some predictable and some unpredictable consequences. In retrospect, some positive conclusions can be drawn. The transplant experiment was successful. The stoneworts survived the ditch diversion. The population rebounded in LLL from oospores. From the ecological perspective, confidence was gained that ecosystems in mine waste management areas respond in a predictable manner.

The data collected over the course of this study support the overall ecological engineering approach to decommissioning. Organic matter and growing algae can sequester U and [226]Ra. U and [226]Ra can be relegated to the sediments. Although the transplant of stonewort was experimental, the spread of the population over ULL along with the collapse of the LLL population provided evidence of the potential capacity of stonewort populations for [226]Ra removal. During the study period, the changes to the system (some engineered, some not) altered the removal processes involved. Ecological engineering solutions to the Rabbit Lake waste management area can be incorporated into decommissioning approaches. Physical parameters for managing the drainage basin to achieve the decommissioning objectives have been derived and are presently under consideration.

Acknowledgements

We would like to thank the environmental staff at the Rabbit Lake Operation. Ongoing support by John Jarrell, Bob Phillips, and originally by Dr. A. Ashbrook, D. Moffat, and G. Vandergaast, was essential throughout the project.

References

Anderson R F, LeHuray A P, Fleisher M Q, Murray J W (1989) Uranium deposition in Saanich Inlet sediments, Vancouver Island. Geoch Cosmoch Acta 53: 2205-2213

Cao Y, Kalin M (1999) Phytoplankton in mine waste water community structure, control factors and biological monitoring. Natural Resources Canada, Biotechnology for Mining. Contract # 23440-8-1016/001/SQ

Culbert R R, Boyle D R, Levison A A (1984) Surficial uranium deposits in Canada. IAEA TecDoc 332: Surficial Uranium Deposits. Report of the Working Group on Uranium Geology. Organized by the International Atomic Energy Agency, 179-192

Edington D N, Gordon S A, Thommes M M, Almodovar L R (1970) The concentration of radium, thorium, and uranium by tropical marine algae. Limnol Ocean 15: 945-955

Fisher N S, Teyssie J L, Krishnaswani S, Baskaran M (1987) Accumulation of Th, Pb, U, and Ra in marine phytoplankton and its geochemical significance. Limnol Ocean 32: 131-142

McConnaughey T (1995) Ion transport and generation of biomineral supersaturation. In: Allemend, D, Cuif J-P (eds) Biomineralisation 93. Bulletin de l'Institut oceanographique, Monaco. Numero specials 14 (2): 1-18

Price G D, Whitecross M I (1983) Cytochemical localization of ATPase activity on the plasmalemma of *Chara corallina*. Protoplasma 116: 65-74

Wang C F, Kiss J Z (1995) The stratolith compartment of *Chara* rhizoids contain carbohydrate and protein. Am J Bot 82 (2): 220-229

Zhang X, Luo S, Yang Q, Zhang H, Li J (1997) Accumulation of uranium at low concentration by the green alga *Scenedesmus obliquus* 34. J Appl Phycol 9: 65-71

Characterization of uranium (VI) complexes formed by different bacteria relevant to uranium mining waste piles

M. Merroun, C. Hennig, A. Rossberg, T. Reich, R. Nicolai, K-H. Heise, S. Selenska-Pobell

Forschungszentrum Rossendorf, Institute of Radiochemistry, D-01314 Dresden

Abstract. A combination of Extended X-ray Absorption Fine Structure (EXAFS) spectroscopy and Infrared (IR) spectroscopy was used to characterize the uranium complexes formed by different bacterial strains of *Acidithiobacillus ferrooxidans* isolated from uranium mining waste piles, and by some reference strains relevant to the predominantly distributed in the wastes *Pseudomonas stutzeri*, and *Pseudomonas migulae*. The results demonstrated that all these bacteria accumulate uranium as phosphate compounds with different structural parameters.

Introduction

The release of radionuclides into the environment is a subject of intense public concern. Although significant quantities of radionuclides were released as a consequence of nuclear weapons testing in the 1959 and 1960s and via accidental release, e.g., from Chernobyl in 1986, the major burden of anthropogenic environmental radioactivity is from the controlled discharge of process effluents produced by industrial activities allied to the generation of nuclear power (Lloyd and Macaskie 2000) and from the mining activities. The intensive uranium mining and milling performed over a period of 40 years (up to 1990) have caused significant pollution with uranium and other toxic metals in the Southeast part of Germany (Saxony and Thuringia). At present a large number of uranium mining waste piles and mill-tailings continue to be hazardous sources for environmental pollution in this region (Bernhard et al. 1996). Removal of uranium from such contaminated sites will require an inexpensive remediation strategy. For effective remediation and stewardship a better understanding is needed of the uranium environmental chemistry and also of the interactions of this metal with bacteria and other biological components present in the above mentioned environments (John et al. 2001). Interactions between uranium and microorganisms have been intensively studied. Microbial cells are known to reduce, oxidize, biosorb, bioaccumulate, and bioprecipitate uranium (Lovely and Phillips 1992; Fredrickson et al. 2000; Merroun and Selenska-Pobell 2001). Living and dead microorganisms possess abundant functional groups, such as carboxyl, hydroxyl and phosphate that bind metal ions. In order to characterize the metal-microbial complexes formed, a variety of physical techniques have been used such as Extended X-ray Absorption Fine Structure (EXAFS) spectroscopy (Hennig et al. 2001), Proton-Induced X-ray Emission (PIXE) analysis (Jeong 1997), nuclear magnetic resonance (NMR) spectroscopy (Andres et al. 1994), Time-Resolved Laser-Induced Fluorescence Spectroscopy (TRLFS) (Panak et al. 2000), Infrared Spectroscopy (IRS) (Yun et al. 2001), etc. In this paper, we present results of U L_{III}-edge Extended X-ray Absorption Fine Structure (EXAFS) spectroscopy and Infrared (IR) spectroscopy of the uranium complexes formed by different bacterial strains isolated from uranium mining wastes such as *Acidithiobacillus ferrooxidans*, and also reference strains of those groups which were demonstrated to be relevant to the uranium wastes, such as *Pseudomonas stutzeri* and *Pseudomonas migulae*. EXAFS measurements provide element-specific short-range structural and chemical information about the U(VI) coordination environments, including identities of, coordination numbers of, and bond distances to neighboring atoms. IR spectroscopy measurements allow identification of bacterial binding sites for uranium.

Material and Methods

Bacterial strains used

The bacterial strains used in this work are: *P. stutzeri* DSMZ 7136, *P. stutzeri* DSMZ 5190, *P. migulae* CIP 105470 and *A. ferrooxidans* D2 recovered from a uranium mining waste (Leduc et al. 1997).

The *Pseudomonas* strains were cultured in nutrient medium containing 5 g of peptone and 3 g of meat extract per liter.

The *A. ferrooxidans* strain was cultured in 9K liquid medium containing 3 g of $(NH_4)_2SO_4$, 0.5 g of K_2HPO_4, 44.2 g of $FeSO_4 \cdot 7H_2O$, 0.5 g of $MgSO_4 \cdot 7H_2O$ and 0.1 g of KCl and 0.014 g of $Ca(NO_3)_2 \cdot 4H_2O$ per liter.

EXAFS and FT-IR spectroscopy measurement

For EXAFS and IRS sample preparation, the cells were resuspended for 48h in 10 ml 0.5 mM solution of uranyl nitrate. After the contact with uranyl, the cells were harvested and washed. The pellets were dried at 70 °C for 24 h and ground (dry samples). Uranium L_{III}–edge X-ray absorption spectra were collected at room temperature in fluorescence mode at the Rossendorf Beamline (ROBL) at the

European Synchrotron Radiation Facility, Grenoble (France) using the Si(III) double-crystal monochromator. The energy was calibrated by measuring the Y K-edge transmission spectrum of a yttrium foil and defining the first infection point set as 17038 eV. Eight spectra for each sample were recorded. The EXAFS oscillations were isolated from the raw, averaged data by removal of the pre-edge background, approximated by a first-order polynomial, followed by μ_0-removal *via* spline fitting techniques and normalization using a Victoreen function. Dead-time correction was applied. The ionization energy for the U L_{III} electron, E_0, was arbitrarily defined as 17185 eV for all averaged spectra. The EXAFS spectra were analyzed according to the standard procedures using the program EXAFSPAK, and the theoretical scattering phase and amplitude functions used in data analysis were calculated using FEFF8. IR spectra of the untreated and uranium treated bacterial biomass were obtained using a Perkin-Elmer FTIR spectrometer model 2000. Spectra from KBr pellets containing equal amounts of sample were recorded at room temperature between 500 and 4000 cm^{-1} at a resolution of 2 cm^{-1}.

Results and discussion

EXAFS

Uranium L_{III}-edge EXAFS spectra of the uranium species formed by *P. stutzeri* DSMZ 7136, *P. stutzeri* DSMZ 5190, *P. migulae* CIP 105470 and *A. ferrooxidans* D2 and their corresponding Fourier Transforms (FT) are shown in Figure 1. The Fourier transforms represent radial distribution functions of the atoms surrounding the uranium atom. The FT' s are not corrected for EXAFS phase-shifts, Δ, causing peaks to appear at shorter distances ($R + \Delta$) relative to the real near neighbour distances (R). The most prominent peak in all samples occurs at 1.3 Å and arises from the backscattering caused by the oxygen atoms which are the nearest to the uranium. They are identified as axial oxygens (O_{ax}) of the linear uranyl group. Backscatters from oxygen atoms laying in the equatorial plane of the uranyl ion appear as a single broad peak at 1.7-1.8 Å. The U-O_{eq1} bond distances of the *A. ferrooxidans* uranium complexes formed at pH 4.5 (2.36 ± 0.02 Å) are similar to of those formed by *Bacillus subtilis* (Kelly et al. 2001) and significantly longer than of those formed by the *Pseudomomas* strains (2.27 ± 0.02 Å) studied in this work. The U-O_{eq1} bond distance of the latter complexes are almost identical to those measured for *Bacillus cereus* and *Bacillus sphaericus* complexes formed also at pH 4.5 (Hennig et al. 2001). As evident from Figure 1 a weak peak appears in all samples at about $R + \Delta R = 3$ Å (not corrected from phase shift) in the FT, which was modeled to the interaction of uranium with phosphorous giving a distance of 3.62 ± 0.02 Å. Therefore, one may conclude that the phosphate groups of the bacterial strains studied are the main uranium binding sites. In addition, the structural parameters of the uranium complexes formed by *A. ferrooxidans* D2 are similar to those of complexes formed by other *A. ferrooxidans* uranium isolates.

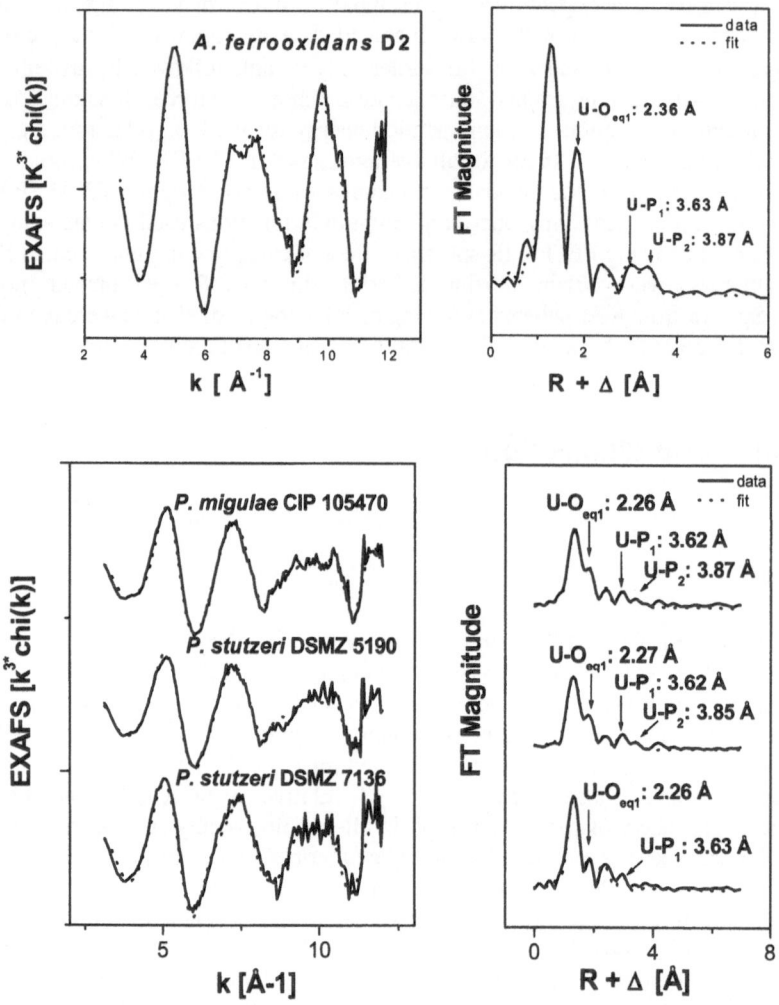

Fig. 1. Uranium L_{III}-edge k^3-weighted EXAFS spectra (left) and corresponding FT (right) of the *A. ferrooxidans* D2, *P. stutzeri* DSMZ 7136, 5190 and *P. migulae* CIP 105470 uranium complexes.

IR Spectroscopy

In order to confirm the type of bacterial functional groups implicated in the inter-action with uranium, the FTIR study was carried out.

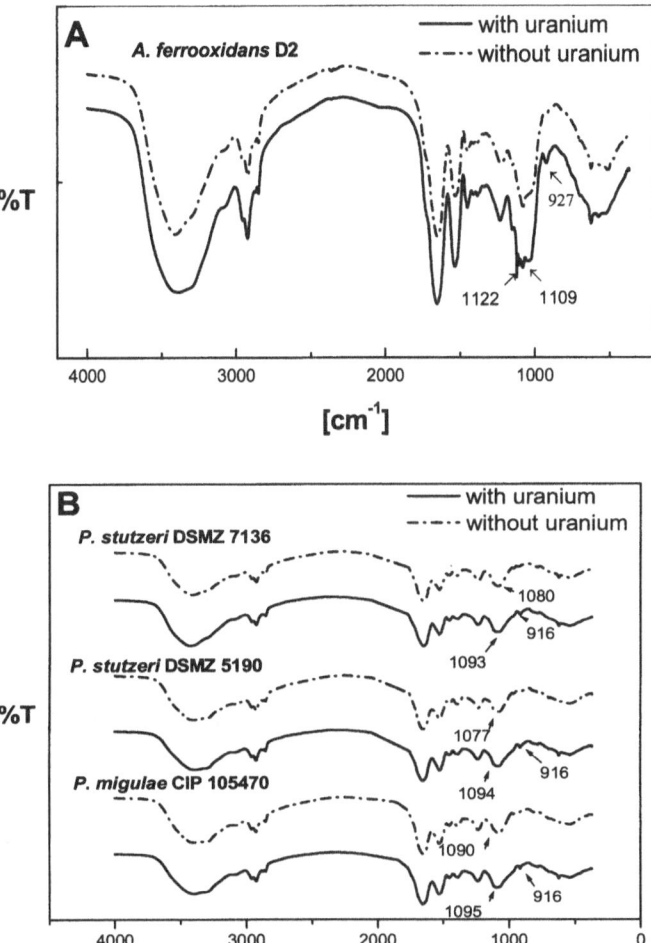

Fig. 2. IR spectra of the native and uranium treated cells of *A. ferrooxidans* D2, *P. stutzeri* DSMZ 7136, 5190 and *P. migulae* CIP 105470.

As shown in Figure 2, the FTIR spectra of the *Pseudomonas* strains and *A. ferrooxidans* (without uranium) are highly complex, reflecting the complex nature of the biomass. Despite this complexity some characteristic peaks can be assigned. The peaks observed at 1647-1652 cm-1 can be assigned to the CO groups. The band at 1530 cm-1 is indicative of the presence of NH bending of the secondary amide groups CONH (amide II band). The spectra also displays absorption peaks in the range from 3000 to 2800 cm-1 which are attributable to CH stretching modes, indicating the presence of the alkyl groups CH_3, CH_2 and CH. In addition the region from 1000 to 1200 cm-1 contain contributions from phosphate groups. These groups belong to various cellular components, such as peptides, phospholipids, peptidoglycan, etc., which are able to complex metals. The absorption spectra of uranium treated biomass were compared with those of untreated biomass (Fig. 2). The treated and untreated biomass were washed, dried and powdered under the same conditions used in the preparation of untreated biomass.

The absorption band at 916 cm^{-1} (*P. migulae* CIP 105470, *P. stutzeri* DSMZ 5190 and *P. stutzeri* DSMZ 7190) and at 927 cm^{-1} (*A. ferrooxidans* D2) originates from the asymmetric stretching vibration of the uranyl unit. The FTIR spectra of the strains treated with uranium revealed a significant shift of absorption peak corresponding to phosphorus residues to higher wave numbers. Similar results were found using other strains of *A. ferrooxidans* isolated from uranium mining wastes.

The IRS results presented here are in agreement with those found by EXAFS measurements.

Summary

EXAFS and IRS measurement of the U(VI) complexes formed by bacteria isolated from uranium mining waste (*A. ferrooxidans* D2) and reference strains (*P. stutzeri* DSZM 7136, 5190 and *P. migulae* CIP 105470) of those which are relevant to this type of wastes demonstrate that phosphorous residues of these bacteria are involved in the interaction with uranium. Further investigations using natural isolates of the *Pseudomonas* reference strains are in progress. The molecular-level structural information obtained by these techniques is essential for understanding the interaction of uranium with the natural bacterial populations in the environment. Such information may be useful for modelling and quantifying the microbial impacts in the fate of uranium and other actinides in nature.

Acknowledgements

This work was supported by grants from the Sächsisches Staatsministerium für Wissenschaft und Kunst (No. 4-7531.50-03-FZR/607) and from the Bundesministerium für Bildung und Forschung, Germany (No.03I4004B).

References

Andres Y, MacCordick HJ, Hubert JC (1994) Binding sites of sorbed uranyl ion in the cell wall of *Mycobacterium smegmatis*. FEMS Microbiol Lett 115: 27-32.

Bernhard G, Geipel G, Brendler V, Nitsche H (1996) Speciation of uranium in seepage waters of a mine tailings pile studied by time-resolved laser-induced fluorescence spectroscopy (TRLFS). Radiochim Acta 74: 87-91.

Fredrickson JK, Kostandarithes HM, Li SW, Plymale AE, Daly MJ (2000) Reduction of Fe(III), Cr(VI), U(VI), and Tc(VII) by *Deinococcus radiodurans* R1. Appl Environ Microbiol 66: 2006-2011.

Hennig C, Panak PJ, Reich T, Roßberg A, Raff J, Selenska-Pobell S, Matz W, Bucher JJ, Bernhard G, Nitsche H (2001) EXAFS investigation of uranium (VI) complexes formed at *Bacillus cereus* and *Bacillus sphaericus* surfaces. Radiochim Acta 89: 625-631.

Jeong BC, Hawes C, Bonthrone KM, Macaskie LE (1997) Localization of enzymically enhanced heavy metal accumulation by *Citrobacter* sp. and metal accumulation in vitro by liposomes containing entrapped enzyme. Microbiology 143: 2497-2507.

John SG, Ruggiero CE, Hersman LE, Tung C-S, Neu MP (2001) Siderophore mediated plutonium accumulation by *Microbacterium flavescens* (JG-9). Environ Sci Technol 35: 2942-2948.

Kelly S D, Boyanov M.I, Bunker B A, Fein J B, Fowle D A, Yee, N, Kemmer, K M. (2001) XAFS determination of the bacterial cell wall functional groups responsible for complexation of Cd and U as a function of pH. J Synchrotron Rad 8: 946.

Leduc LG, Ferroni GD, Trevors JT (1997) Resistance to heavy metals in different strains of *Thiobacillus ferrooxidans*. World J Microbiol Biotechnol 13: 453-455.

Lloyd JR, Macaskie LE (2000) Bioremediation of radionuclide-containing wastewaters, In: Lovley DR, ed. *Environmental Microbe-metal interactions*. Washington, DC: ASM Press; 277-327.

Lovely DR, Phillips EJP (1992) Reduction of uranium by *Desulfovibrio desulfuricans*. Appl Environ Microbiol 58: 850-856.

Merroun ML, Selenska-Pobell S (2001) Interactions of three eco-types of *Acidithiobacillus ferrooxidans* with U(VI). Biometals 14: 171-179.

Panak P, Raff J, Selenska-Pobell S, Geipel G, Bernhard G, Nitsche H (2000) Complex formation of U(VI) with *Bacillus*-isolates from a uranium mining waste pile. Radiochim Acta 88: 71-76.

Yun Y-S, Park D, Park JM, Volesky B (2001) Biosorption of trivalent chromium on the brown seaweed biomass. Environ Sci Technol 35: 4353-4358.

Study courses at the Technical University Freiberg
Akademiestr. 6   09599 Freiberg   Germany
http:/www.tu-freiberg.de   dekan.F3@fggb.tu-freiberg.de

Geology/ Paleontology

1st – 4th term (undergraduate studies)
* Mathematics and Computer Science
* Physics and Chemistry
* Geosciences
* Geological Methods
* Field trips, mapping exercises
* oral examination after 4th term

5th – 6th term (bachelor studies)
* Geoscientific methods (e.g., geological mapping, remote sensing, geophysical prospecting, geo-computing, drilling engineering, environmental geology)
* Bachelor exam (oral exams and bachelor thesis)

5th – 10th term (graduate studies)
* Geoscientific methods (geodynamic/tectonic, petrology/sedimentology, hydrogeology, environmental geology, geomathematics, palaeontology etc.)
* Specialization in selected geological fields
* Field training, mapping exercises, excursions
* 2 months internship
* Masters thesis in the 9th and 10th term, including a mapping task

Admission
* „Abitur"/Baccalaureate (A-level) or
* Specialized college degree or
* equivalent access entitlement

Duration/Degree: 6 semester/Bachelor of Science
10 semester/Master of Science in Geology
(Dipl.-Geol.)
Start: winter term

Job opportunities
Groundwater capture and protection; Landscape planning; Disposal site technology including planning and monitoring; Environmental management; Exploration and use of minerals; Geological information like maps and geo-information systems; Geoscientific research institutes; Regional and federal geological surveys; Hydrological consulting; Geo-industries dealing with exploration and fossil fuels

Uranium speciation in plants

Alix Günther[1], Gert Bernhard[1], Gerhard Geipel[1], Andre Rossberg[1], Tobias Reich[2]

[1]Forschungszentrum Rossendorf/Dresden, Institute of Radiochemistry, P.O. Box 510119, 01314 Dresden, Germany
[2]Johannes Gutenberg University Mainz, Institute of Nuclear Chemistry, Fritz Straßmann-Weg 2, 55128 Mainz, Germany

Abstract. Several spectroscopic methods specially the time-resolved laser-induced fluorescence spectroscopy and the X-ray-absorption spectroscopy were used for the determination of the uranium speciation in plants. Differences between the uranium speciation in the initial solution and inside the plants could be detected. The chemical speciation of uranium is identical in the roots, shoot axis and leaf. It is independent from the uranium speciation in the initial solution and the type of the plant. The results indicate that the uranium is predominantly bound as uranium phosphate (phosphoryl) groups in the plants.

Introduction

Uranium and its natural decay products are being transported by water from the rock piles, flooded mine shafts, and tailing ponds into the environment. To protect the health of the population, knowledge is necessary about the transfer of poison and radioactive heavy metals, like uranium via the food chain soil - water - plant - animal - human. Our investigations are focused on the soil/water-to-plant transfer of uranium.

The overall process of uptake of heavy elements into plant tissues from contaminated soils has been quantified using the soil-to-plant transfer factor (TF). Considering the uranium transfer it is quantified as $TF_{(U)}$ = plant specific activity (fresh mass) (Bq/kg^{-1}) / soil specific activity (dry mass) (Bq/kg^{-1}). Depending on the kind of plants and the experimental conditions, uranium TF's are in the range from 1.0×10^{-3} to 7.2×10^{-5} (Frindik 1986). Up to know, the various species of uranium and other heavy elements have not been considered along the described pathway. The TF is a sum parameter and the value depends of different experimental conditions like kind of sample preparation, soil type, kind and part of the plant and the speciation of the investigated element in the soil/pore water and in the plant itself. The transfer factor has to be unfolded in correct physical-chemical parameters. Knowledge about speciation is a basic prediction for that. In the literature is little information relating uranium speciation to plant uptake.

It was the goal of our study to obtain first results on the change of uranium speciation during the soil to plant transfer. The time-resolved laser-induced fluorescence spectroscopy (TRLFS) and X-ray absorption spectroscopy (XAS) with synchrotron radiation were used as tools for the determination of uranium speciation in pore water of contaminated soils (TRLFS), hydroponic solutions (TRLFS) and the different parts of the plants (TRLFS, XAS). The uranium inside the plant was localized by scanning electron microscopy (SEM), coupled with energy-dispersive X-ray microanalysis (EDX). The reported investigations about the speciation of uranium were focused on lupines.

Materials and Methodes

Cultivation and sample preparation of the plants

Various plants (blue and yellow lupines, dandelion, lamb's lettuce e.g.) were grown in an agricultural test field or in the laboratory in naturally and artificially uranium contaminated soils (1g uranium / kg soil). The plants, which were grown in uncontaminated soil, were transferred to hydroponic (initial) solutions. The uranium concentration of the hydroponic solution was in the range from $1.0 \cdot 10^{-5}$ M to $2.5 \cdot 10^{-2}$ m and the pH-value was adjusted in the range from 3.0 to 8.0. After harvesting, the plants were washed, separated into roots, shoot axis and leaves and cut into small pieces. Furthermore, the fresh samples or dried samples were measured by XAS, TRLFS and SEM. For the XAS measurements, the fresh section samples or pulverized plant samples were placed on an adhesive foil. For scanning electron microscopy at ambient temperature air dried plant samples (fresh cross or longitudinal root or shoot axis sections) were fixed onto conductive adhesive films and coated with carbon or sputtered with gold, respectively.

Laser-induced time-resolved fluorescence spectroscopy (TRLFS)

A Nd-YAG-MOPO pulse laser system (Spectra Physics Mountain View, CA, USA) was used as a light source. The fluorescence of the uranyl complexes is excited with a 266 nm beam of 200 – 500 µJ pulse energy. The emitted fluorescence light of the sample was focused over fiber optics into a spectrograph. The fluorescence signal was measured with a gated intensified diode array (M1492, EG&G). The gate width was set to 2 µs and the fluorescence was recorded with delay times ranging from 10 ns up to 80 µs after application of the laser pulse. Each sample was measured three times on every delay time. 100 laser pulses were collected for each spectrum. All functions of the laser spectrometer are computer controlled. A detailed description of the whole equipment and data analysis is given in Geipel et al. 1996 .

XAS-measurements

The XANES / EXAFS measurements were performed at the Rossendorf Beam Line (ROBL) at the European Synchrotron Radiation Facility (ESRF) in Grenoble, France (Reich et al. 2000). The samples were measured in fluorescence mode using a 4-pixel germanium fluorescence detector (Bucher et al. 1996). The EXAFS spectra were analyzed according to standard procedures using the suite of program EXAFSPAK (George and Pickering 1995). Theoretical scattering phases and amplitudes were calculated with the program FEFF8 (Ankudinov et al.1998).

Scanning electron microscopy (SEM) and energy-dispersive X-ray micronanalysis (EDX)

SEM was carried out with a microscope of type ZEISS DSM 962. For the detection secondary and backscattered electrons were used, whereby acceleration voltages up to 30 kV were applied. The chemical composition of the uranium-containing depositions inside the cells and cell walls were determined by use of the microanalysis system INCA (Oxford Instruments) with integrated Si-detector and S-UTW-window. Principles of methods were described in Schmidt et al. 1994.

Results and discussion

Thermodynamic calculations of the uranium speciation in the pore water and hydroponic solutions were performed with the computer program EQ 3/6 (Wolery 1992) using the NEA data base (Grenthe et al. 1992). The uranium-speciation depends on the uranium concentration and the adjusted pH-values in the initial solutions. The background composition of the hydroponic solution was normal drinking water. In the initial hydroponic solution, the uranium speciation at pH 3.0 is dominated by the free uranyl cation. In the higher pH region hydroxide and carbonate species of uranium exist.

Scanning electron microscopy (SEM) and energy-dispersive X-ray (EDX) microanalysis were used for localization and more detailed characterization of uranyl complexes in uranium-containing cross and longitudinal sections of root and shoot axis samples of plants after the uranium transfer. The uranium-containing depositions are not equally distributed in the samples. There are areas with higher uranium accumulation as well as areas with less or no uranium. The EDX-analyses of the uranium-containing depositions showed the elements uranium, oxygen, phosphorus, calcium, potassium and chlorine, in some cases sulfur and silicon (cell wall).

By determining the fluorescence properties, important information on the coordination sphere of the uranium complexes in the initial solutions and in the plants after the uranium transfer can be obtained. Good fluorescence spectra of the root, shoot axis and leaf samples were recorded, example see Fig. 1.

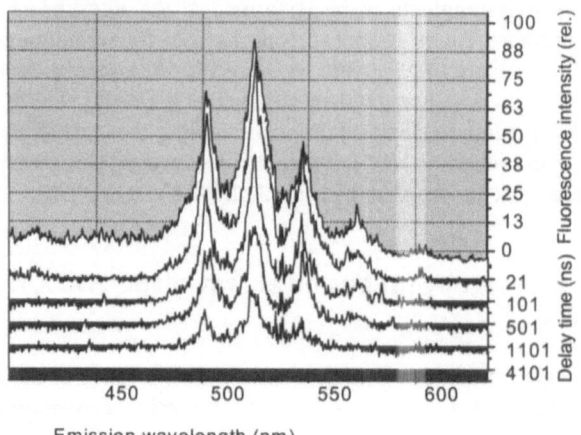

Emission wavelength (nm)

Fig. 1. Time-resolved laser-induced fluorescence spectrum of an uranium containing, fresh shoot axis sample of lupine

Table 1. Uranium fluorescence data of initial solutions and plant samples

Sample	Part of the plant	Main emission bands (nm)					
Pore water, $1 \cdot 10^{-4}$ M U pH 7.9, (1g U/kg soil)		No uranyl fluorescence visible					
Lupine – soil culture	root		489.2	502.6	525.0	547.6	575.6
	shoot axis			500.4	524.5	543.8	571.2
	leaf			503.6	527.7	545.8	
Hydroponic solution $2.5 \cdot 10^{-2}$M U, pH 3.0			482.8	496.5	518.0	542.3	565.4
Lupine - hydroponics	root			501.5	521.9	544.1	572.2
	shoot axis			502.1	524.1	547.7	572.3
	leaf			501.7	530.7	546.9	
Hydroponic solution $2.5 \cdot 10^{-2}$M U, pH 3.2		478.6	496.9		517.5	542.0	567.1
Dandelion - hydroponics	root			500.2	521.2	547.2	568.7
	base of leaf			501.8	521.3	546.5	571.5
	leaf			502.3	524.7	549.8	574.1

We can state that uranium stayed in oxidation state + 6 after uptake into the plants. The main emission bands of uranium in the plant samples are shifted very strong bathochrome in comparison with the bands in the spectra of the initial hydroponic solutions. In all plant samples the main fluorescence wavelengths are quite different to the relevant initial solutions (Table 1). It means the uranium speciation in the initial solutions and inside the plants are different. The fluorescence spectra of dandelion and lamb's lettuce samples agree with the features of the uranium fluorescence spectra obtained from lupines. It seems that the uranium speciation is independent of the kind of the plant under investigation.

Table 2. Fluorescence data of uranyl model compounds or uranyl complex solutions

Species	Main emission bands (nm)					
$(UO_2)_x(PO_4)_y$		488.0	503.0	523.7	546.9	572.5
$Ca(UO_2)_2(PO_4)_2 \cdot 10\ H_2O$		488.6	504.0	524.2	548.0	573.9
$Ca(UO_2)_2(PO_4)_2 \cdot 8H_2O$		491.3	501.8	522.9	546.9	572.2
$Ca_2UO_2(CO_3)_3 \cdot 10\ H_2O$	465.1	483.9	503.5	525.2	548.3	572.6
U(VI)/ Bacillus sphaericus		483.2	502.0	523.5	546.1	570.4

The wavelengths of the emissions bands of the investigated plant samples are almost identical with the wavelengths obtained for liquid or solid uranyl phosphate complexes (Brendler et al. 1996, Geipel et al. 2000), calcium uranyl carbonate (Bernhard et al. 2001) and uranium (VI) bonded to a *Bacillus sphaericus* biomass (Panak et al. 2000) (Table 2). Considering the chemical build-up and the metabolism of the plants, the formation of uranyl phosphato complexes (inorganic and organic) and complexes with other organic compounds is imaginable in plants after the transfer.

Uranium L_{III}-edge X-ray Absorption Near-Edge structure (XANES) confirms the presence of U(VI) in all plant samples investigated. Raw Uranium L_{III}-edge k3-weighted EXAFS spectra and their corresponding Fourier transforms (FT) of uranyl complexes in different parts of lupine are shown in Fig. 2. In the FT four peaks are visible, which were identified with several analysis procedures (Teo 1986).

The first and the second peak corresponds to the scattering contribution of the two axial oxygen atoms $O_{ax(1,2)}$ at a radial distance of 1.77 – 1.80 Å of the uranium(VI) ion and the scattering contribution of the equatorial oxygen atoms $O_{eq(1)}$ at a relatively short radial distance of 2.28 –2.32 Å (Table 3). The small $O_{eq(1)}$ coordination number of 3 and 4 is in agreement with expected value for such a short bond distance. The data are comparable with structural parameters of m-autunite (Makarov and Ivanov 1960, Hennig et al. 2001). In a root sample was found a U-$O_{eq(1)}$ bond distance of 2.37 Å and the larger coordination number. This indicate the formation of another uranyl species, a monodentate coordination between U(VI) and carboxylic group e.g. (Denecke et al. 1997).

We assume for the third peak at a radial distance of 2.81 – 2.88 Å a backscattering oxygen atom $O_{eq(2)}$. Further the identification of uranium-phosphorus-coordination in combination of a multiple scattering path (U-$O_{eq(1)}$-P) was possible. The U-P distance is in the range of 3.59 – 3.63 Å and correspond to those observed for uranyl phosphates (Hennig et al. 2001).

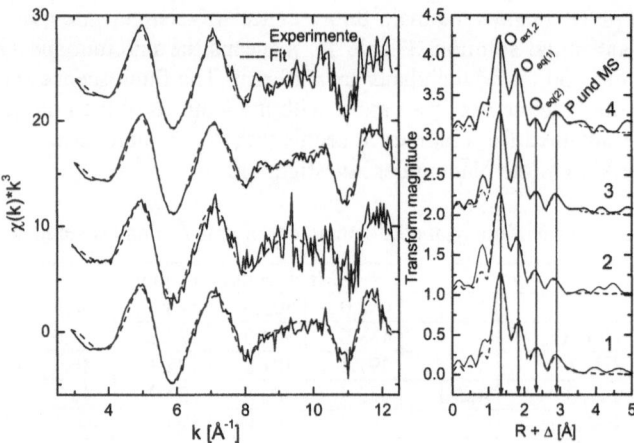

Fig. 2. K³-weighted Uranium L_III-edge EXAFS spectra (left) and the corresponding Fourier transform (right) of different uranium containing plant samples of lupine, 1... root, soil culture, 2... shoot axis, soil culture, 3... root, hydroponics, 10⁻³M U pH3, 4... shoot axis, hydroponics, 10⁻³M U pH3

Table 3. EXAFS structural parameters of the major components (O_{ax}, $O_{eq(1)}$) of the uranium species in lupines

Sample		U - O_{ax}			U – $O_{eq(1)}$	
		R[Å]	$\sigma^2[Å^2]\cdot 10^3$	N	R[Å]	$\sigma^2[Å^2]\cdot 10^3$
Soil culture[a]	r	1.80	2.0	3.4	2.30	3.9
1.0gU(VI)/kg soil, pH 8.1	s	1.78	2.1	3.8	2.27	4.0
Hydroponics[a]	r	1.78	1.3	5.0	2.37	14.0
2.5·10⁻²M U(VI)	s	1.79	1.5	2.7	2.32	6.0
pH 3.0	l	1.79	2.3	3.1	2.31	6.0
Hydroponics[b]	r	1.79	2.1	3.4	2.30	3.1
1.0·10⁻³M U(VI) pH 3.0	s	1.77	2.9	4.7	2.28	5.3
Hydroponics[b] 1.0·10⁻³M U(VI) pH 5.0	r	1.77	2.6	4.2	2.28	4.3

a... dry, b... fresh, r... root, s... shoot axis, l... leaf, N... coordination number, R... radial distance in Å with an uncertainly of± 0.02 Å, σ^2... Debye-Waller factor

We conclude that the uranyl species in plant are different from the species in the solutions after the transfer from solution/soil to the plant. For better identification of the groups of compounds and their functionalities, which are responsible for

Microbiology / phytoremediation

uranyl complexation in plants further model complexation are necessary. Besides the complexation of U(VI) with inorganic phosphates and carbonates we investigate the interaction of U(VI) with organic phosphates (e.g. sugar phosphates, adenosine triphosphate) and other compounds, which play an important role in the chemical build-up and the metabolism of plants. Carboxylic acids, amino acids, peptides, proteines and phospholipides e.g. belong to these compounds.

Acknowledgements

This work was supported by the Deutsche Forschungsgemeinschaft under contact-No. BE2234/1-1, 1-2. The authors would like to thank H. Reuther and E. Christalle for the REM/EDX-measurements. We thank H. Funke and C. Hennig for their support during the EXAFS data collection.

References

Akudinov A L, Ravel B, Rehr J J, Conradson S D (1998) Real-space multiple scattering calculation and interpretation of x-ray absorption near-edge structure. Phys. Rev. B 58: 7565 - 7576

Bernhard G, Geipel G, Reich T, Brendler V, Amayri S, Nitsche H (2001) Uranyl(VI) carbonate complex formation: Validation of the Ca2UO2(CO3)3(aq.) species. Radiochim. Acta 89: 511-518

Brendler V, Geipel G, Bernhard, G, Nitsche H (1996) Complexation in the System UO22+/PO43-/OH-(aq.) Investigations Over Wide Ranges in pH and Concentration. Radiochim. Acta 74 / 75:

Bucher J J, Allen P G, Edelstein N M, Shuh D K, Madden N W, Cork C, Luke P, Pehl D, Malone D (1996) A multichannel monolithic Ge detector system for fluorescence x-ray absorption spectroscopy. Rev. Sci. Instrm. 67: 1

Denecke M A et al. (1997) Differentiating Between Monodentate and Bidentate Carboxylate Ligands Coodinated to Uranyl Ions using EXAFS. J. Phys. IV France 7: C2

Frindik O.(1986) Uranium contents in Soils, Plant and Foods. Landwirtschaftliche Forschung Vol 39, 1-2: 75-86

Geipel G, Brachmann A, Brendler V, Bernhard G, Nitsche H (1996) Uranium (VI) Sulfate Complexation Studies by Time-Resolved Laser-Induced Fluorescence Spectroscopy (TRLFS). Radiochim. Acta 75: 199-204

Geipel G, Bernhard G, Rutsch M, Brendler V, Nitsche H (2000) Spectroscopic properties of Uranium (VI) Minerals studied by time-resolved laser-induced Fluorescence Spectroscopy (TRLFS). Radiochimica Acta 88: 757

George G N, Pickering I J (1995) EXAFSPAK: A Suite of Computer Programs for Analysis of X-Ray Absorption Spectra, Stanford Synchrotron Radiation Laboratory

Grenthe I, Fuger J, Lemire R J, Muller AB, Nguyen-Trung C, Wanner H (1992) Chemical Thermodynamics of Uranium, 1st ed., Elsevier Science Publishers, Amsterdam

Hennig C et al. (2001) EXAFS investigation of uranium, (VI) complexes formed at Bacillus cereus and Bacillus sphaericus surfaces. Radiochim. Acta 89: 625

Microbiology / phytoremediation

Makarov E S, Ivanov V I (1960) The crystal structure of meta-autunite Ca(UO2)2(PO4)2·6H2O. Phys. Rev. B 54: 156

Panak P J, Raff J, Selenska-Pobell S, Geipel G, Bernhard G, Nitsche H (2000) Complex formation of U(VI) with Bacillus-isolates from a uranium mining waste pile. Radiochim. Acta 88: 71-76

Reich T, Bernhard G, Geipel G, Funke H, Rossberg A, Matz W, Schell N, Nitsche H (2000) The Rossendorf Beam Line ROBL- a dedicated experimental station for XAFS measurements of actinides and other radionuclides. Radiochim. Acta 88: 633-637

Schmidt, P F et al. (1994) Praxis der Rasterelektronenmikroskopie und Mikrobereichsanalyse, Kontakt& Studium Band 444, Expert Verlag

Wolery T J (1992) EQ3/6 a software package for the geochemical modeling of aqueous systems, Report UCRL-MA-110662 part1, Lawrence Livermore National Laboratory, California, USA

Mobilization of hazardous metals by plants growing in soils from uranium mining

Gerhard Gramss, Klaus-Dieter Voigt, Hans Bergmann

Friedrich - Schiller - University, Institute of Nutrition and Environment, Dornburger Strasse 24, D - 07743 Jena, Germany

Abstract. Seedlings of rape and Chinese cabbage were grown in potted soil from uranium mining for 36 d. Aqueous extracts of the planted soils contained a total of 183 and 168.7 % the concentrations of As, Cd, Cr, Cu, Ni, Pb, U, and Zn of unplanted soil. This increase in mobility of elements was to 77.7 and 86.6 % the result of increased water-solubility of metal containing humic (HA) and fulvic acids (FA). Their solubility was promoted by carboxylic acids released from plants and microbes.

Introduction

Hazardous elements such as Cd, Cu, Ni. Pb, U, and Zn occur as free cations, complexed with OH^-, Cl^-, oxoanions of C, S, and P, with HA, FA, and as chelates with organic acids, in the soil solution (Griffiths et al. 1994, Schachtschabel et al. 1998). Acidification, i.e., chemical reaction with, and cation exchange by H^+ increases the concentrations of free cations in soil (Sumner et al. 1991). This implies less free cations at alkaline pH when humic substances are more soluble. In soils, production of H^+, phytochelators, HA and FA are predominantly linked with the plant cover and its associated soil microflora feeding on root exudates and dead plant residues (Schinner and Sonnleitner 1996). Significant quantities of H^+ are released during nitrification. Plant roots and microbes also release H^+ in quantities equivalent to the cations taken up (Schachtschabel et al. 1998). Carbon dioxide emitted by microbes and roots is transformed to H_2CO_3 in soils above pH 5.0 and contributes to acidification (Schachtschabel et al. 1998). Soil fungi (Sprecher 1961), bacteria (Schmidt 1986), and roots (Jones 1998, Nigam et al. 2001) release carboxylic acids and amino acids. They form chelates predominantly with 2- to 3-valent metal cations and prevent their immobilization (Nigam et al. 2001). Plants are the major source of organic matter transformed by microbes to humic substances. Some lignin, protein, peptide, cellulose, and polysaccharide residues forming the HA macromolecule are sometimes covalently linked by metal cations or serve as their complexing agents (Stevenson 1994). Low-MW FA form covalent or coordinate bonds with cations via hydroxyl, carboxyl, and amino groups (Klavins and Serzane 2000). Microbial oxidative and hydrolytic enzymes (Gramss et al. 1999a), root-exuded peroxidases (Gramss et al. 1999b), and fumaric and succinic acids (Nardi et al. 1997) decrease the MW of humic substances and enhance their solubility in water.

In the present study, an uranium mine dump (U) soil with high concentrations of hazardous elements (As, 278.2 ± 23.2; Cd, 11.2 ± 1.2; Cr, 22.0 ± 0; Cu, 564.5 ± 44.5; Ni, 102.1 ± 1.9; Pb, 114.3 ± 11.7; U, 105.8 ± 4.2; and Zn, 1112.5 ± 52.5 mg kg^{-1}) was seeded with metal non-hyperaccumulating rape (*Brassica napus* L.) and uranium accumulating Chinese cabbage (*Brassica chinensis* L., Huang et al. 1998). Organic acids in root exudates of rape were quantified and compared with those released by root zone microfungi and bacteria. Mobility increases of toxic elements were determined for water-irrigated, and citric-acid irrigated planted soils, as phyto- and synthetic chelators generally increased the mobility of hazardous elements in soil and their uptake by plants (Huang et al. 1998, Robinson et al. 2000). The changes in the metal content of plant tissue were also determined.

Objective of this paper were estimations of an increased wash-out behaviour of hazardous elements due to the action of plant and rhizosphere microbes, the phytochelators they release, and the humic substances with their associated elements they mobilize. These values were compared with the percentage of metals bound by shoot tissue of plants which reduce the washing-out effect via phytoextraction.

Results and discussion

Carboxylic acids serving as chelators and mobilizers of humic substances were tested under pH-stabilized conditions in phosphate buffer (Table 1). They were likewise efficient in mobilizing cations of 8 elements (ICP-AES data) and humic extract (HE, $r = 0.960$ at $p \leq 0.05$, spectrophotometric data A_{340} / A_{450}) from U soil. Most effective were several 2- to 3-basic oxy acids and 2-basic saturated carboxylic acids, whereas the majority of 1- to 2-basic acids had little or no effect at pH 5.0. Their chelating capacity depends on number and proximity of the carboxyl groups (Jones 1998). The same may apply to their ability to alter the hydrophobic arrangement of the humic structure.

Nearly all of the efficient carboxylic acids were produced in sterile culture of rhizosphere fungi and bacteria on nutrient-rich liquid media (Table 2). In a natural soil, their productivity depends on a continuous supply with root exudates and plant debris (Schinner and Sonnleitner 1996) and may be lower than that of roots.

Table 1. Concentrations of elements (mg kg^{-1} soil DW) and humic extract (HE, g kg^{-1} soil DW) extractable from U soil with 0.2 M potassium phosphate / 0.02 M organic acid buffer within 24 h. Final pH of the soil suspensions 5.0 ± 0.15.

Element	Phosphate	Citric	Oxalic	Isocitric	Malonic	Malic	α-Oxoglutaric	A	B
As	51.24	101.8*	97.95*	68.32*	84.31	81.79*	69.31*	69.11	54.14
Cd	1.342	5.950*	1.277	3.686*	3.253*	3.152*	3.008*	2.352	1.502
Cr	0.022	0.498*	0.353*	0.123*	0.087*	0.108*	0.022	0.012	0.012
Cu	12.86	151.0*	237.8*	42.28*	124.8*	41.34*	29.92*	20.65*	16.12*
Ni	5.243	34.88*	23.28*	23.64*	17.75*	17.59*	11.52*	9.472*	7.781*
Pb	0.094	16.19*	0.332*	1.464*	0.173	0.238	0.137	0.141	0.045
U	1.601	39.51*	13.47*	4.782*	3.303*	3.412*	2.113	1.942	1.810
Zn	49.53	547.8*	337.6*	322.2*	171.9*	228.9*	118.5*	94.24*	94.34*
HE	4.750	11.14*	11.44*	7.435*	6.643*	6.449*	4.816	5.987	4.565

A, *t*-Aconitic, *D*- gluconic, glycolic, oxaloacetic, tartaric, thioglycolic acids.
B, Acetic, adipic, glutaric, pyruvic, succinic acids.
Nonsignificant effects, including increases up to 20 % for a few compounds, in butyric, formic, fumaric, lactic, maleic, and propionic acids. No increases in mobilization of HE.
* Mean values of duplicate reaction mixtures significantly different from the phosphate buffer control at $p \leq 0.05$.

Table 2. Organic acids (mg L^{-1}) formed in culture fluids of sterilely grown rape seedlings (plant DW 10 g L^{-1}), and in axenic liquid cultures of fungi and mixed bacteria, microorganisms which are common to rooted soil (HPLC data).

	Sterile root exudate of rape	Cultures of mitosporic soil fungi		Cultures of mixed soil bacteria	
		Aspergillus niger	*Scytalidium lignicola*	Peptone medium	Sucrose medium[a]
Incubation [b]	45 d	6 d	18 d	14 d	10 d
PH ± SD	5.57 ± 0.07	4.37 ± 0.19	5.90 ± 0.24	8.03 ± 0.31	4.15 ± 0.44
Acids:					
Oxalic	5.61 ± 2.89	33.94	22.44 ± 9.38	0	0
α-Oxoglutar.	0	6.56	11.79 ± 3.33	0	13.44 ± 9.43
Citric	0	61.00	0	0	0
Isocitric	21.47 ± 2.69	0	0	149.25 ± 26	0
D-Gluconic	0	268.83	0	0	0
Pyruvic	0.26 ± 0.03	119.56	12.32 ± 4.93	1.19 ± 0.26	16.81 ± 11.8
Malic	10.33 ± 1.11	0	43.26 ± 1.24	47.13 ± 6.3	313 ± 166.9
Malonic	0	29.15	0	21.3 ± 14.2	0
t-Aconitic	0.08 ± 0.01	0.57	0.61 ± 0.30	0.036 ± 0.01	0.37 ± 0.21
Lactic	4.24 ± 1.34	141.28	666.74 ± 457	28.15 ± 8.23	230.9 ± 36.8
Formic	0	0	0	6.54 ± 2.83	4.97 ± 1.14
Glutaric	0	27.05	6.56 ± 2.13	0	0
Acetic	0	78.34	197.7 ± 17.2	2005 ± 16.05	311.5 ± 55.4
Fumaric	0.62 ± 0.13	3.12	1.81 ± 0	0	27.35 ± 4.03
Propionic	16.16 ± 0.59	82.69	172.91 ± 144	4794 ± 326	10.39 ± 3.06
Isobutyric	0	0	0	446.8 ± 71.8	5.50 ± 2.49
n-Butyric	0	0	0	896 ± 233.3	6.22 ± 1.17

[a] Mixed bacterial populations in part contaminated by mycelia of the mitosporic fungi, *Fusarium* sp., *Penicillium* sp., and *Mucor* sp. [b] Time of maximum acid production.

The present data do not allow to quantify the contribution of plant and microbe to the pool of phytochelators in rooted soil. Unplanted soil with its widely nonvegetative fungal and an altered bacterial flora contained only traces of the nonefficient formic, acetic, and fumaric acids. The close contact of water-irrigated potted rape with U soil nearly doubled the concentration of extractable elements in comparison to unplanted soil (Fig. 1). Irrigation with citric acid as the most efficient chelator, and the most efficient biotic solvent of U in soil (Huang et al. 1998) increased the solubility of most elements further. This fact was stated 6 d after citric acid application and wide degradation when the plants were harvested and soil pH had returned to 7.57 ± 0.08, i.e., 0.1 pH units above that of control soil.

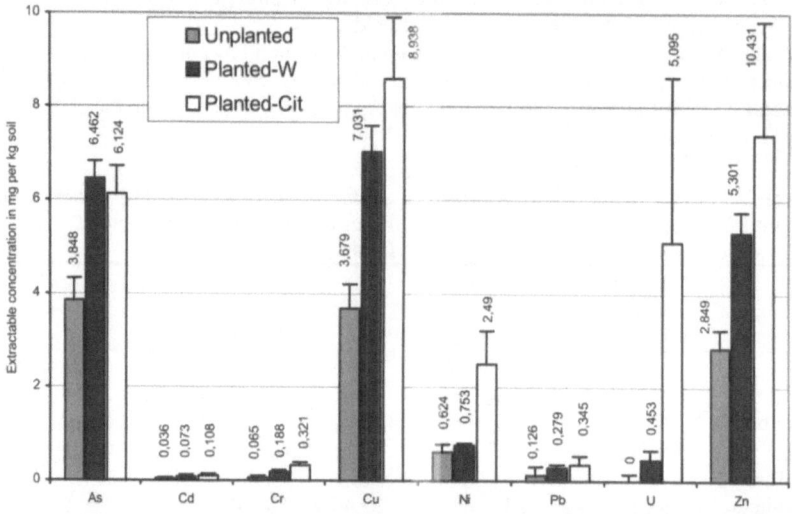

Fig. 1. Elements (mg kg^{-1} soil DW) extracted with water from unplanted and planted U soil of 36-d-old rape pot cultures. Plants irrigated with water (W), or aqueous citric acid monohydrate (Cit, 6 g kg^{-1} soil). Error bars in all figures indicate SD.

Element concentrations in shoots of rape were only in part significantly higher on citric-acid irrigated soil than on water-irrigated soil (Fig. 2). Nevertheless, U concentrations as low as 0.072 mg g^{-1} in rape and 0.201 mg g^{-1} in Chinese cabbage (further data not shown) on the soil with a total U content of 105 mg kg^{-1} are far below the concentrations reported for *Brassica juncea* and Chinese cabbage grown on citric-acid treated soil with 750 mg kg^{-1} of U (Huang et al. 1998). Nevertheless, shoot tissue of rape accumulated phytotoxic concentrations of Cu and Zn, and, in certain treatments, concentrations of As, Cd, and Pb which are hazardous to man but do not exceed the concentrations tolerable to sewage sludge (Schachtschabel et al. 1998).

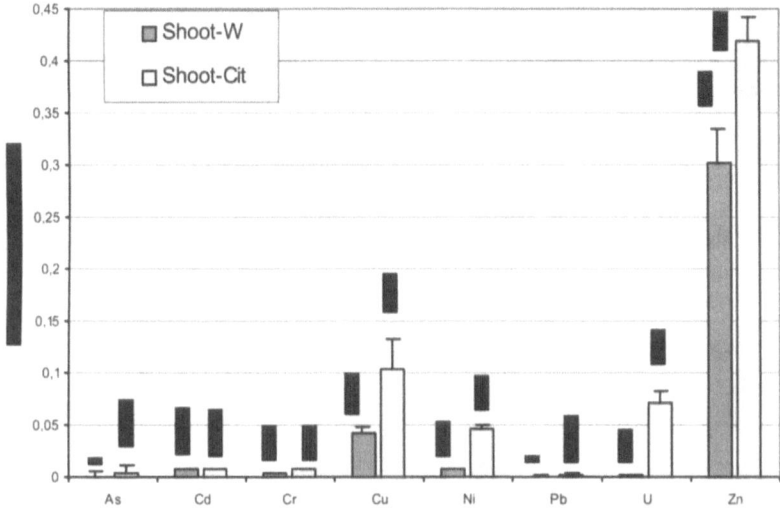

Fig. 2. Elements (mg g^{-1} DW) in shoots of 36-d-old rape seedlings grown on U soil and ir-rigated with water (W), or aqueous citric acid monohydrate (Cit) as indicated in Fig. 1. Concentrations of elements are phytotoxic (*) or exceed the limit tolerable to man (**) (Schachtschabel et al. 1998).

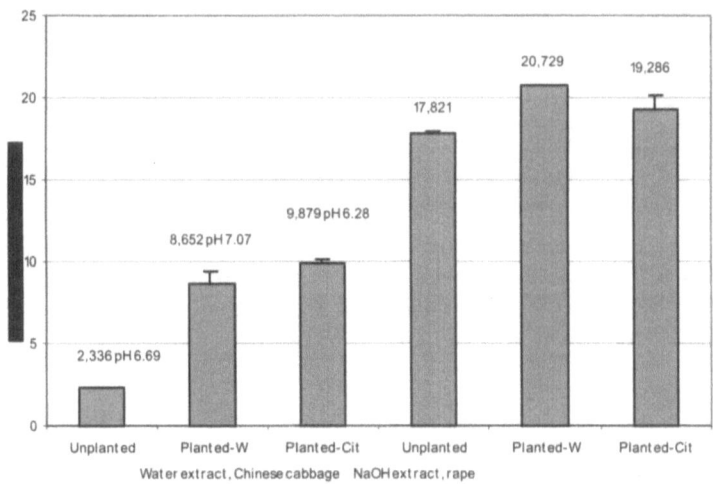

Fig. 3. Sum of humic and fulvic acids (g kg^{-1} soil DW) extracted with water or 0.1 M NaOH solution from unplanted U soil and soils planted to Chinese cabbage or rape for 36 d. Planted samples irrigated with water (W) or citric acid solution (Cit) as indicated in Fig. 1, pH values refer to those of the final extract.

Microbiology / phytoremediation

Planting U soil to rape or Chinese cabbage increased dramatically the quantities of HE soluble in water (widely independent of pH) or 0.1 M NaOH (Fig. 3). It is shown by Figs. 4a and 4b that the majority of elements which dispose of 3-valent oxidation states could be removed from solution to 43.8 to 97.3 % by precipitating HA with HCl at pH 2.0. Contemporarily, HA contained as little as 0 to 15 % the concentrations of the predominantly 2-valent Cd, Cu, and Pb. The remaining portions of free soluble elements may behave as reported by Schachtschabel et al. (1998). In a farmyard soil pH 6.4, 13.3 % (Cu) to 69.5 % (Al) were complexed with FA, whereas in a forest soil pH 3.3, 75.1 % (Cd) to 100 % (Al) were complexed with FA.

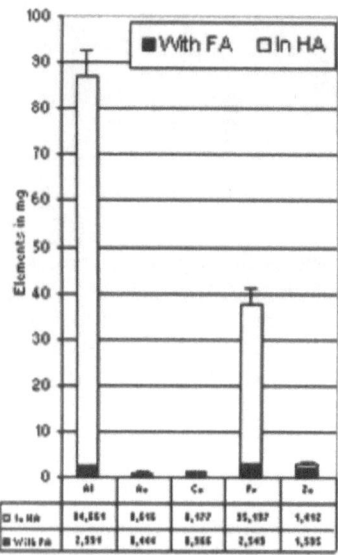 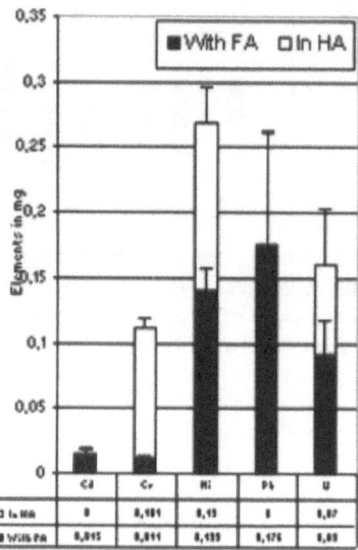

Fig. 4. Elements (mg) extracted with water from 1 kg unplanted U soil diluted to a theoretic content of 1 g kg⁻¹ of extractable humic and fulvic acids. In HA, elements linked, and precipitable with, the portion of HA in 1 g HA + FA. With FA, sum of elements in free solution or linked with the portion of FA in 1 g HA + FA.

Conclusions

Planting U soil increased the mobility of 8 hazardous elements to 183 % in rape and 168.7 % in Chinese cabbage (Fig. 5). This increase was mainly due to an additional mobilization of metal-containing HA and FA in the rhizosphere, possibly promoted by a moderate increase in soil pH and an enzymatic and carboxylic-acid mediated degradation of humic molecules (Gramss et al. 1999a, Nardi et al. 1997). The alternative mobilization of Ni, U, and Zn by direct chelation with citric acid is also documented by Fig. 1, as in citric-acid irrigated soil (with citric acid having been widely degraded) the content of soluble HE was no longer higher than in water-irrigated soil (Fig. 3).

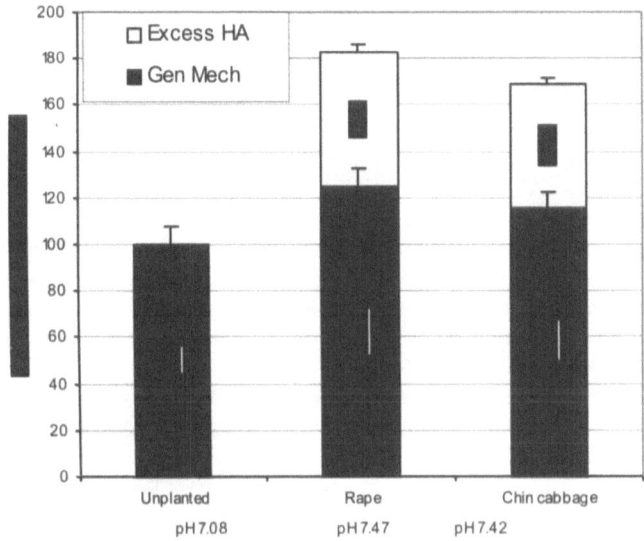

Fig. 5. Sum of increases (%) in water-extractability of As, Cd, Cr, Cu, Ni, Pb, U, and Zn from U soil, unplanted and planted to rape or Chinese cabbage. Plants irrigated with water. Gen Mech, percentage of elements mobilized by the general mechanisms, including complexing with HA and FA. Excess HA, percentage of elements bound to the portion of HA mobilized by plants.

Humic substances extracted from U soil with water (Figs. 4a and 4b) consisted of 89 % precipitable HA and 11 % FA. At pH 7.08, elements such as As, Cr, Ni, U and Zn were strongly associated with HA and were thus subject to increased leaching to surface and ground waters, even in colloidal condition. Cadmium, Pb, and Cu were little or not associated with HA from U soil, but they are known to form complexes with FA (Schachtschabel et al. 1998). If they did so, the planting-induced increase in the mobility of elements was approximately to 77.7 % in rape, and 86.6 % in Chinese cabbage a result of increased mobility of humic extract.

The 36-d-old seedlings of rape / Chinese cabbage had a shoot DW of 3.8 to 5.7 g kg^{-1} soil. Water-irrigated plants accumulated 10.06 / 8.93 % the water-soluble ions of the 8 hazardous elements, and 1.26 / 0.81 % those of U alone. When in citric-acid irrigated soil the mobility of the 8 elements rose to the 8.6-fold, the plants could only accumulate 0.65 / 3.21 % the ions of 8 elements, and 0.61 / 1.91 % those of U alone. An application of chelator to a moderately contaminated soil increased thus far more the leakage than the phytoextraction of the toxicants. Metal ions complexed with dissolved HA may be accessible to the plant. An uptake of Fe dissolved in HA by wheat plants was possible but accompanied by a precipitation of HA from solution (Mackowiak et al. 2001).

Further tests will focus on the promotion of organic matter forming processes in mine dump soils to use organic carbon as a sink of abiotic toxicants, as far as the migration of contaminated humic substances can be controlled.

This work was supported by BMBF, 02S7808 /4.

References

Gramss G, Ziegenhagen D, Sorge S (1999a) Degradation of soil humic extract by wood- and soil-associated fungi, bacteria, and commercial enzymes. Microb Ecol 37: 140-151

Gramss G, Voigt K-D, Kirsche B (1999b) Oxidoreductase enzymes liberated by plant roots and their effects on soil humic material. Chemosphere 38: 1481-1494

Griffiths R P, Baham J E, Caldwell B A (1994) Soil solution chemistry of ectomycorrhizal mats in forest soil. Soil Biol Biochem 26: 331-337

Huang J W, Blaylock M J, Kapulnik Y, Ensley B D (1998) Phytoremediation of uranium-contaminated soils: role of organic acids in triggering uranium hyperaccumulation in plants. Environ Sci Technol 32: 2004-2008

Jones D L (1998) Organic acids in the rhizosphere - a critical review. Plant Soil 205: 25-44

Klavins M, Serzane J (2000) Use of humic substances in remediation of contaminated environments. In Bioremediation of Contaminated Soils. Eds D L Wise, D J Trantolo, E J Cichon, H I Inyang, U Stottmeister. pp 217-235. M. Dekker Inc., New York

Mackowiak C L, Grossl P R, Bugbee B G (2001) Beneficial effects of humic acid on micronutrient availability to wheat. Soil Sci Soc Am J 65: 1744-1750

Nardi S, Reniero F, Concheri G (1997) Soil organic matter mobilization by root exudates of three maize hybrids. Chemosphere 35: 2237-2244

Nigam R, Srivastava S, Prakash S, Srivastava M M (2001) Cadmium mobilisation and plant availability - the impact of organic acids commonly exuded from roots. Plant Soil 230: 107-113

Robinson B H, Mills T M, Petit D, Fung L E, Green S R, Clothier B E (2000) Natural and induced cadmium-accumulation in poplar and willow: implications for phytoremediation. Plant Soil 227: 301-306

Schachtschabel P, Blume H-P, Brümmer G, Hartge K H, Schwertmann U (1998) Lehrbuch der Bodenkunde, 14. Ed. Enke, Stuttgart

Schinner F, Sonnleitner R (1996) Bodenökologie: Mikrobiologie und Bodenenzymatik. Vol. 1. Springer, Berlin

Schmidt O (1986) Investigations on the influence of wood-inhabiting bacteria on the pH value in trees. Eur J For Path 16: 181-189

Sprecher E (1961) Über die Stoffausscheidung bei Pilzen. Arch. Microbiol. 38: 114-155

Stevenson F J (1994) Humus Chemistry. 2nd Ed. J Wiley & Sons, New York

Sumner M E, Fey M V, Noble A D (1991) Nutrient status and toxicity problems in acid soils. In Soil Acidity. Eds B Ulrich, M E Sumner. pp 149-182. Springer, Berlin

Mobility and plant availability of heavy metals in soils of uranium mining dumps

Heiko Schönbuchner[1,3], Matthias Leiterer[2], Bernd Machelett[1], Hans Bergmann[1]

[1]Institute for Nutrition, Friedrich-Schiller-University, Dornburgerstr. 25, D-07743 Jena
[2]Thuringian State Authority for Agriculture, Naumburgerstr. 98, D-07743 Jena
[3]present address: JENA–GEOS-Engineering Ltd, Saalbahnhofstr. 25 c, D-07743 Jena

Abstract. The aim of this investigation was to elucidate the mobility, the availability and the soil to plant transfer of heavy metals in dump soils of the former uranium mining area in Thuringia (Germany). The investigation refers to the elements U, As, Cd, Co, Cu, Cr, Ni, Pb and Zn. For this purpose, soil and plant samples were collected from 34 plots on two different dumps. Pseudo-total contents and mobile portions were investigated in all soil samples. In six selected soil profiles every horizon was analysed by means of a sequential extraction procedure.

Introduction

In the former mining district of Seelingstädt in the eastern part of Thuringia (Germany) uranium ore was mined from 1949 to 1967 in four open-cast workings. The output was 11953 tons of uranium with the ore containing an average uranium concentration of 0.066 percent (Schulze 1993). The deposit developed in two Permian strata of grey sand- and claystone ("Obere Graue Folge" and "Untere Graue Folge") with a maximum of 800 mg/kg uranium in the seam.

In general the mobility of heavy metals in the soil-plant system has been well investigated in Germany in recent years (e.g. Brümmer et al. 1986, Liebe et al. 1997) but only very little information is available concerning mobility and soil to plant transfer of heavy metals in the former East German uranium mining district. The behaviour of uranium in comparison to other elements is of special interest.

Material and Methods

The Seelingstädt uranium mining district includes eight big rock dumps. Two of these dumps are used as pasture land and for mowing while the other six have been afforested. Therefore the "west dump" and the "dump Sorge-Settendorf" were chosen for the investigation of mobility and soil to plant transfer of heavy metals in a meadow ecosystem. Both dumps consist of Permian and Triassic sand- and claystone. Maximum total uranium contamination in the soils is about 100 mg/kg. Other important heavy metal loads exist for cadmium, arsenic, zinc and nickel.

Horizontal soil samples and plant samples were collected from 34 plots on the two dumps. The soil samples were extracted with *aqua regia* for pseudo-total heavy metal concentrations and with ammonium nitrate for the mobile portion. In six selected soil profiles every horizon was analysed by means of a sequential extraction procedure (Zeien and Brümmer 1989) which comprises: (F1) 1 M NH_4NO_3 (readily soluble and exchangeable = mobile), (F2) 1M NH_4O-acetate, pH 6.0 (specifically adsorbed and other weakly bound species), (F3) 0.1 M $NH_2OH·HCl$ + 1 M NH_4O-acetate, pH 6.0 (bound to Mn oxides), (F4) 0.025 M NH_4EDTA, pH 4.6 (bound to organic matter), (F5) 0.2 M NH_4-oxalate, pH 3.25 (bound to poorly crystalline Fe oxides), (F6) 0.1 M ascorbic acid in 0.2 M NH_4-oxalate, pH 3.25, in boiling water (bound to crystalline Fe oxides) and (F7) difference between *aqua regia* extracted content and the sum of all other fractions [F1+F2+F3+F4+F5+F6] (residual, mainly bound in silicates). Plant samples were digested by microwave heated pressure extraction with nitric acid. Measurement of all elements was performed by ICP-MS and ICP-OES.

Results

Soil contamination

The average soil contamination with heavy metals (median values) ranges from 5.25 mg/kg for cadmium to 671 mg/kg for zinc in the uppermost organic horizon (Table 1). Values for all other elements range from 10 to 100 mg/kg.

The mobile portion of all elements is generally low due to the high pH-values (about 6.25), but highest for cadmium and zinc (over 1 percent). Uranium, lead and chromium form a group of heavy metals with low mobility. Besides lead and chromium all elements show mostly significant increases with soil depth. There may be two reasons for this behaviour: First of all there might be a considerable solubilization and leaching of heavy metals due to massive application of liquid animal manure (Japenga et al. 1992). Moreover, heavy metals can be removed from the upper horizons by plant root uptake and can be transferred into plant tissues above ground (Umweltbundesamt 2001).

Heavy metal partitioning on different binding forms

Uranium is scarcely present in the first, mobile fraction (Fig. 1), but there is a considerable amount in the second fraction (specifically adsorbed), which is of high importance for plant uptake. The content of uranium in the organically bound fraction is very low, which contradicts other authors (Edwards et al. 1999).

Table 1: Soil contamination of the uppermost, organic horizon: Median values of all 34 plots (AR = *aqua regia* extraction, AN = ammonium nitrate extraction)

element	AR [mg/kg]	AN [µg/kg]	mobile portion [%]
arsenic	75.8	190	0.25
cadmium	5.25	111	2.11
chromium	22.0	< dl*	< 0.04
cobalt	20.0	83.9	0.42
copper	60.8	440	0.72
lead	58.7	25.6	0.04
nickel	40.4	208	0.51
uranium	20.6	23.8	0.12
zinc	671	7844	1.17

*more than 50 % of values smaller detection limit

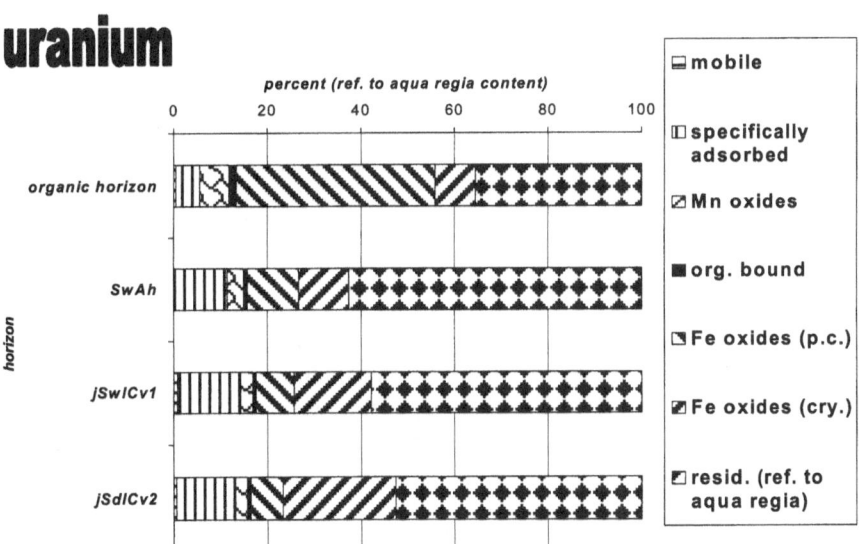

Fig. 1. Distribution of uranium between the fractions of the sequential extraction procedure in a soil profile with changing moisture regime.

EDTA appears to be an unsuitable extractant for uranium; as a consequence organically bound uranium may be found in the next fraction (F5 - bound to poorly crystalline Fe oxides). The portion of uranium in the crystalline Fe oxides increases with soil depth and a lot of uranium is found in the residual fraction generally.

On the other hand cadmium is much more mobile than uranium (Fig. 2). A maximum of eighty percent is bound to fractions F1 plus F2 in the third horizon. In the fourth horizon, the (pseudo)total content of cadmium is only 4.9 mg/kg, but the sum of fractions F1+F2+F3+F4+F5+F6 is higher than this value. So the *aqua regia*-extractable content is exceeded by more than 20 percent. At low total contents the limitations of the procedure become evident.

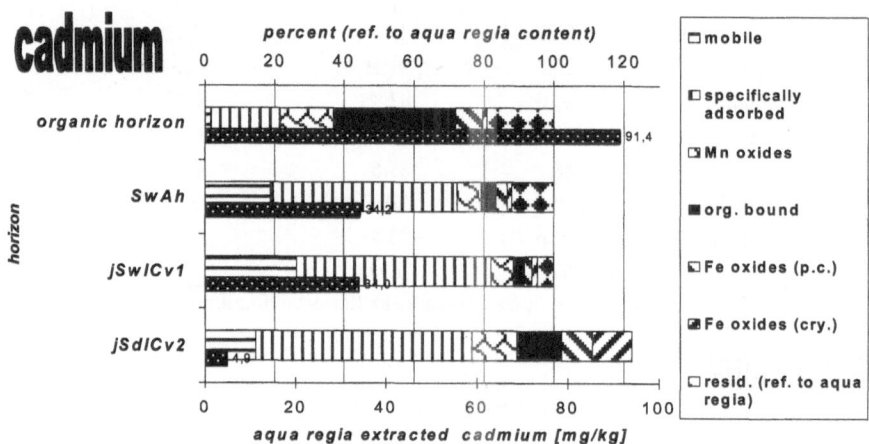

Fig. 2. Distribution of cadmium between the fractions of the sequential extraction procedure in a soil profile with changing moisture regime. Total contents of cadmium are indicated on the lower x-axis.

Arsenic is strongly bound to the Fe oxide fractions and hardly represented in fractions F1 to F4. Cobalt is strongly correlated to the Mn oxide fraction (F3), copper and lead are increased in the EDTA-extractable organic fractions (F4). Nickel behaves in a way similar to that of elements of medium mobility like uranium, zinc is distributed in a way similar to the very mobile element cadmium. Chromium was not investigated by sequential procedure.

Heavy metal contents of plants growing on uranium dumps

The highest heavy metal concentrations in relation to high *aqua regia* extractable soil concentrations and relatively high mobility, were found for zinc (Table 2). Critical cadmium concentrations for pasturage were reached for dandelion and yarrow (> 1 mg/kg). Arsenic and uranium showed very low concentrations in all plant species. This is a consequence of the relatively low mobility in soil and little root to shoot transfer of these elements in soil. Clover showed the highest concentration for five of the nine metals investigated. This is caused by the legumes' special ability to take up mineral nutrients. Concentrations in grass were generally low, because at harvest time in April 1998, the biomass-production of grass was very high compared to the other plants, so heavy metals in grass were subject to a kind of dilution effect. The use of the meadows for pasturage can not be recommended, because cattle may take up a mixture of toxic elements, not to mention the direct uptake of contaminated soil.

Table 2: Median element concentrations in plants from western dump (harvest of April 1998)

element	grass	clover	dandelion	yarrow
	(n = 23)	(n = 23)	(n = 23)	(n=18)
	[mg/kg DS]	[mg/kg DS]	[mg/kg DS]	[mg/kg DS]
arsenic	0.06	0.12	0.11	0.10
cadmium	0.23	0.06	3.41	1.63
chromium	1.53	1.00	0.43	1.28
cobalt	0.04	0.25	0.12	0.20
copper	6.99	8.65	11.46	12.65
lead	0.10	0.58	0.25	0.44
nickel	2.56	4.09	2.76	3.72
uranium	0.01	0.06	0.04	0.03
zinc	87.17	60.83	130.04	146.93

Soil to plant relationships

The best linear correlations between the mobile element-portion in the uppermost, organic horizon and plant parts above ground are found for the mobile elements zinc and cadmium (Fig. 3). For the second harvest in September, heavy metal concentrations in the plants were mostly higher than in the first harvest in April. This is due to a higher biomass production in April and a diminished dry weight of the plant shoots in September. As a consequence the heavy metal concentrations increase in autumn. No correlations exist for more immobile elements like arsenic, chromium and lead. For them, ammonium nitrate-extract is no good indicator for phyto-availability. It is necessary to measure the actual concentration of these elements in plants and no exact predictions can be made by *aqua regia* and ammonium nitrate extraction of soils. Some bad correlations were found for nickel, a few good ones for cobalt. Both heavy metals can be regarded as medium mobile and phyto-available. Copper showed good but logarithmic correlations between mobile soil portion and plant content (Fig. 4). Copper was present in a wide range of concentrations in the soils; with high amounts available saturation seem to take place in the plant tissues.

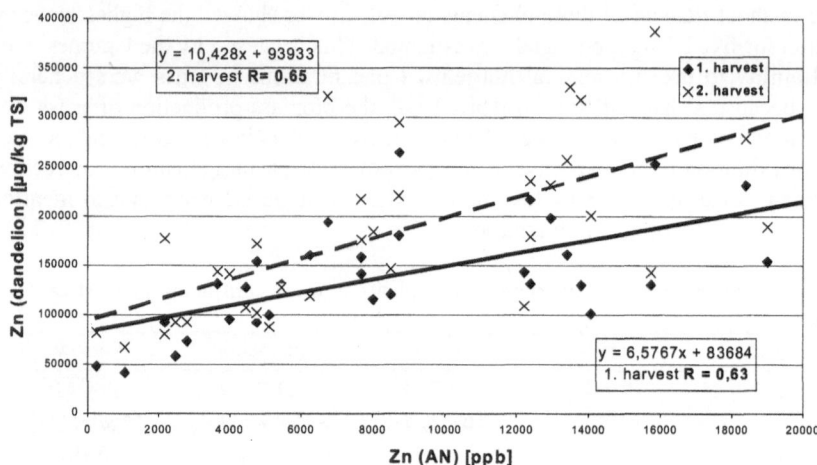

Fig. 3. Correlation between mobile Zn (AN = ammonium nitrate) in the uppermost, organic soil horizon and Zn in aboveground parts of dandelion

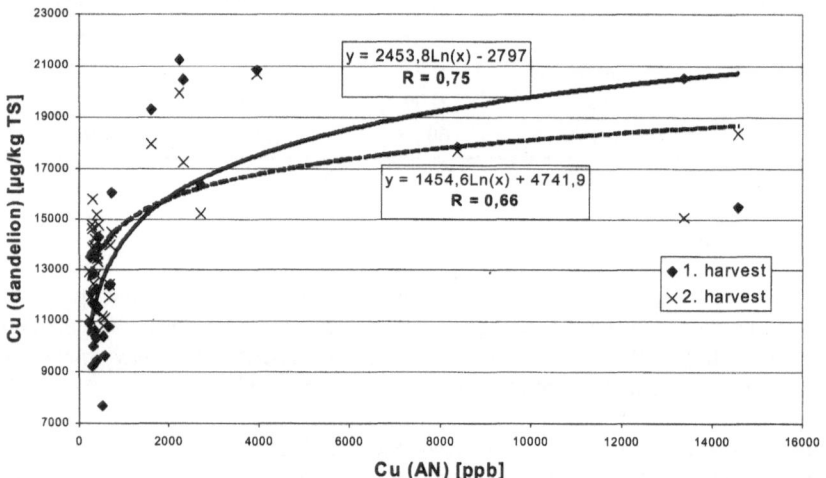

Fig. 4. Correlation between mobile Cu (AN = ammonium nitrate) in the uppermost, organic soil horizon and Cu in aboveground parts of dandelion

Uranium showed a behaviour similar to that of the immobile lead: There were only few correlations between the mobile soil portion and uranium plant tissues. One of these few correlations was found for the first grass harvest (significant on the 0.05 level). But for the second harvest there was again no significant correlation (Fig. 5). It has to be mentioned that the correlations for uranium improved by calculating multiple regressions taking account of pH-values. With lower pH-values the portion of mobile uranyl-cations (Ebbs et al. 1998) increases. These cations can be transferred to the plant shoots more easily than other uranium species.

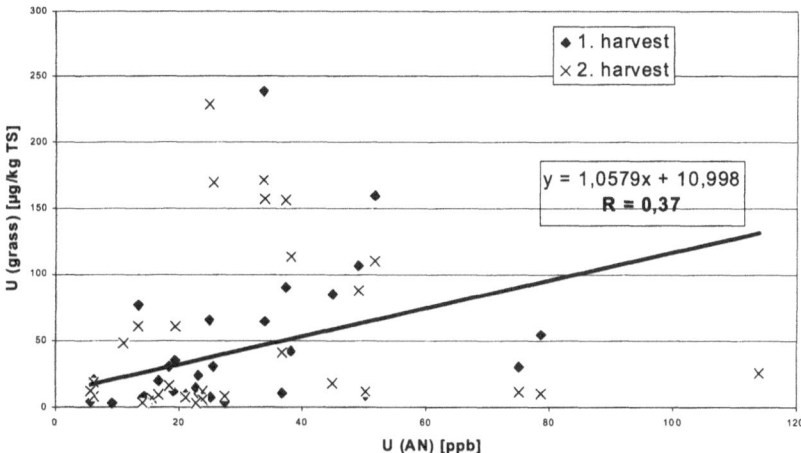

Fig. 5. Correlation between mobile U (AN = ammonium nitrate) in the uppermost, organic soil horizon and U in aboveground parts of grass

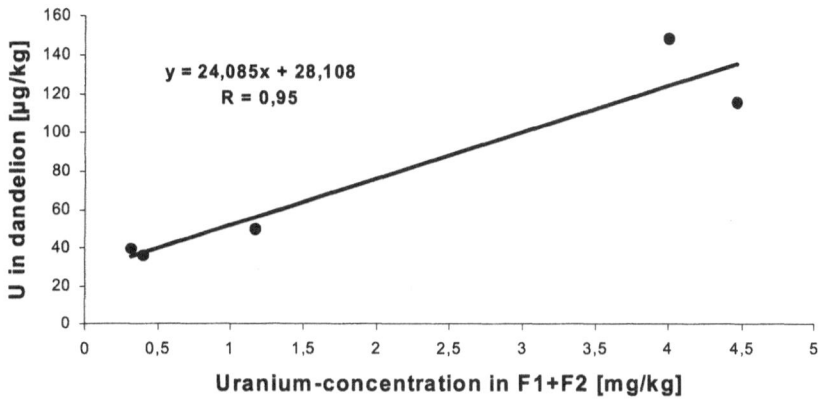

Fig. 6. Correlation between mobile U (F1) + specifically adsorbed U (F2) in the uppermost, organic soil horizon and U in aboveground parts of dandelion

According to Sheppard and Evenden (1992) there should be a correlation between NH_4O-acetate extractable uranium and plant uptake. To test this relationship, fractions F1+F2 were added up and correlated with the uranium content of the plants. A good correlation was obtained for the first harvest of dandelion (Fig. 6), but this needs to be qualified: Due to the lack of further plots investigated by sequential extraction, only five pairs of data could be used; dandelion did not grow on all six plots. Nevertheless, there is a clear trend: Obviously, dandelion from the first harvest used the specifically adsorbed fraction for the uptake of uranium from soil. For the second harvest this correlation vanished; possibly due to a distinct change in the chemical conditions of the soil from April to September. Soil samples were only taken once, in February.

Summary

Zinc and cadmium are fairly mobile, uranium lead and chromium immobile according to NH_4NO_3-extractable contents on the investigated dumps. Reasons for heavy metal increase with soil depth are discussed. Relatively high amounts of uranium are found in the NH_4O-acetate extractable second fraction from the sequential procedure applied. There seems to be some good correlation between shoot content of uranium in plants and the sum of NH_4NO_3- and NH_4O-acetate extractable fractions in the first soil horizon. Cadmium and zinc reach the highest concentrations of all elements in fractions F1 and F2 and are well transferred to plants. In contrast, arsenic for example is largely immobile and hardly represented in fractions F1 to F4 and shows similarly low concentrations in plants as uranium.

References

Brümmer, G W, Gerth, J, Herms, U (1986) Heavy metal species, mobility and availability in soils. Z Pflanzenernähr Bodenk 149: 382-398
Ebbs, S D, Brady, D J, Kochian, L V (1998) Role of uranium speciation in the uptake and translocation of uranium by plants. J Exp Bot 49: 1183-1190
Edwards, R, Lepp, N W, Jones, K C (1999) Weniger häufig vorkommende Elemente mit potentieller Bedeutung für die Umwelt. In: Alloway, B J (Hrsg.): Schwermetalle in Böden – Analytik, Konzentrationen, Wechselwirkungen. Springer-Verlag, Berlin, Heidelberg: 333-385
Japenga, J, Dalenberg, J W, Wiersma, D, Scheltens, S D, Hesterberg, D, Salomons, W (1992) Effect of liquid animal manure application on the solubilization of heavy metals from soil. Intern J Environ Anal Chem 46: 25-39
Liebe, F, Welp, G, Brümmer, G W (1997) Mobilität anorganischer Schadstoffe in Böden Nordrhein-Westfalens. Materialien zur Altlastensanierung und zum Bodenschutz 2, Essen: 383 p.
Schulze, G (1993) Bestandsaufnahme und Charakterisierung der stofflichen Auswirkungen des Uranerzbergbaus und der Uranerzaufbereitung (Standort Seelingstädt) am Beispiel des Wasserpfades. Veröff Museum Gera - Naturwiss Reihe 20: 40-73
Sheppard, S C, Evenden, W G (1992) Bioavailability indices for uranium: Effect of concentration in eleven soils. Arch Environ Contam Toxicol 23: 117-124
Umweltbundesamt (2001) Grundsätze und Maßnahmen für eine vorsorgeorientierte Begrenzung von Schadstoffeinträgen in landbaulich genutzten Böden. Umweltbundesamt Texte, 59/01: 126 p.
Zeien, H & Brümmer, G W (1989) Chemische Extraktion zur Bestimmung von Schwermetallbindungsformen in Böden. Mitteilungen Dt Bodenkundl Gesellsch 59: 505-515

Complex Formation of Uranium(VI) with Glucose 1-Phosphate

Astrid Koban[1], Gerhard Geipel[1], Gert Bernhard[1], Thomas Fanghänel[2]

[1]Forschungszentrum Rossendorf, Institute of Radiochemistry
[2]Forschungszentrum Karlsruhe, Institute for Nuclear Waste Management

Abstract. The complex formation of uranium(VI) with glucose 1-phosphate was determined by time-resolved laser-induced fluorescence spectroscopy (TRLFS) and potentiometric titration. Both measurements show the formation of a 1:1 complex. The complex shows no fluorescence, and the formation constant of $UO_2(C_6H_{11}O_6PO_3)$ was calculated from TRLFS measurements to be log β_{11}=5.60±0.12 at pH=4, and from potentiometric titration log β_{11}=5.37±0.10, respectively.

Introduction

Up to now little is known about the chemical speciation of actinides in biosystems. To compare the obtained spectroscopic data of uranium complexes in several biological systems with model compounds, we investigate the complexation of uranium with relevant bioligands of various functionalities. A very important class of ligands are phosphate esters, which serve as phosphate group and energy transmitters as well as energy storage mediums in biological systems. Heavy metal ions bound to the phosphate esters can be transported into living cells and than deposited.

Therefore, in our study we present the results of uranium complexation with glucose 1-phosphate (G1P) obtained by time-resolved laser-induced fluorescence spectroscopy (TRLFS) and potentiometric titration.

Experimental

The TRLFS-experiments were performed at a fixed uranyl concentration (10^{-5} M) as a function of the ligand concentrations (10^{-5} to 2×10^{-3} M) at pH 4.0 and an ionic strength of 0.1M (NaClO$_4$). A Nd-YAG-MOPO pulse laser system (Spectra Physics, USA) was used as a light source. A laser pulse of 266 nm wavelength and 0.2 – 0.5 mJ pulse energy was used for the excitation of the uranium. The fluorescence signal was measured with a gated intensified diode array (M1492, EG&G). A detailed description of the TRLFS is given elsewhere (Geipel et al. 1996).

The potentiometric titrations were carried out at fixed uranyl (10^{-4} M) and ligand (10^{-3} M) concentrations in a pH range from 3 to 10 and a temperature of 25°C. The experiments were performed with an automatic titrator (736P Titrino, Metrohm, Germany) using the accompanying software (TiNet 2.2). A detailed description of the whole procedure is given elsewhere (Brendler et al. 1996).

Results and discussion

Time-resolved laser-induced fluorescence spectroscopy

Fig.1 shows the TRLFS spectra of uranium(VI) as a function of the ligand concentrations.

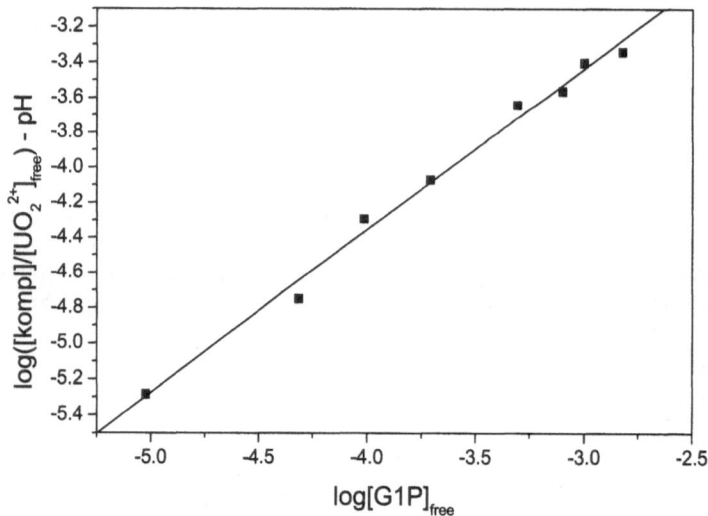

Fig. 1. TRLFS-spectra of uranium(VI) (10^{-5} M) as a function of the glucose 1-phosphate concentration at pH=4.

With increasing ligand concentrations we observed a decrease in fluorescence intensities. At higher ligand concentrations only a small red shift of the emission bands caused by the dominant hydroxide species was detectable.

The TRLFS spectra indicate the presence of the free uranyl ion with a lifetime of 1.3 ± 0.3 μs and a uranyl hydroxide species with a lifetime of 15 ± 3 μs at higher ligand concentrations. We therefore conclude that the complexed uranyl glucose species shows no fluorescence properties. The spectra were corrected for the uranyl hydroxide species and their intensities were eliminated.

The complex formation reaction in general can be written as:

$$UO_2^{2+} + n\,G1P \leftrightarrow \{UO_2(G1P)_n\}^{(2-n)+} + n\,H^+ \tag{1}$$

The associated mass action law can be set up as follows:

$$k = \frac{[\{UO_2(G1P)_n^{(2-n)+}\}][H^+]^n}{[UO_2^{2+}][G1P]^n} \tag{2}$$

This equation can be transferred by forming the logarithm values into the following linear form:

$$\log\frac{\{UO_2(G1P)_n^{(2-n)+}\}}{[UO_2^{2+}]} = n\log[G1P] + \log k + n\,pH \tag{3}$$

To estimate the stoichiometry of the reaction (1) a slope analysis was performed (fig.2).

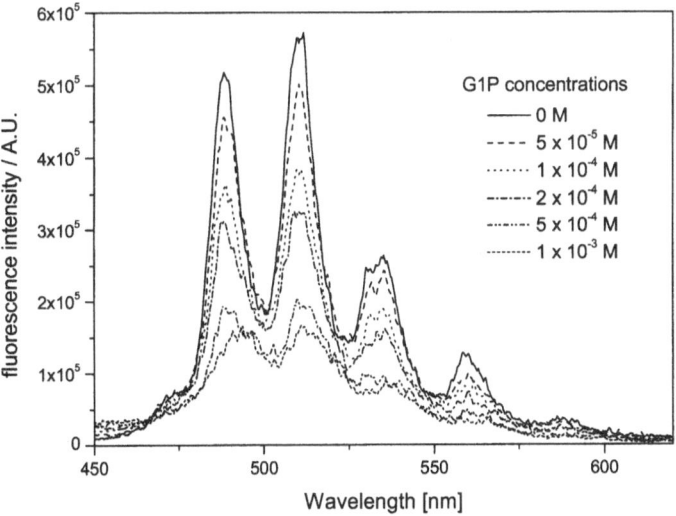

Fig. 2. Slope analysis for reaction (1)

The concentration of the free uranyl ion was determined on the basis of the measured fluorescence spectra. These data were used to calculate the corresponding concentrations of the uranyl glucose phosphate complex and the non complexed ligand.

The slope of 0.90 ± 0.03 indicates a predominant 1:1 complexation.

Under consideration that glucose 1-phosphoric acid ($C_6H_{11}O_6PO_3H_2$) is a two-protonic acid with the dissociation constant values of $pk_{a1}=1.46$ (Murakami and Takagi 1966) and $pk_{a2}=5.98$ (Kramer-Schnabel and Linder 1991), arises for reaction (1) at pH=4 the following equation:

$$UO_2^{2+} + C_6H_{11}O_6PO_3H^- \leftrightarrow UO_2(C_6H_{11}O_6PO_3) + H^+ \tag{4}$$

The constant for reaction (4) was calculated to be log k=-0.38±0.12 at pH=4. Taking into account the second dissociation step of G1P we get the general complex formation reaction:

$$UO_2^{2+} + C_6H_{11}O_6PO_3^{2-} \leftrightarrow UO_2(C_6H_{11}O_6PO_3) \tag{5}$$

with a complex formation constant of log β_{11}=5.60±0.12.

Potentiometric titrations

The examination of the titration data with the software Hyperquad2000NT showed the formation of the 1:1 complex $UO_2(C_6H_{11}O_6PO_3)$ in the pH range between 3 and 7. The complex formation constant was calculated to be log β_{11}=5.37±0.10. This is in good agreement with the results of the TRLFS measurements.

At higher pH-values a 1:2 complex $UO_2(G1P)_2$ appears with a complex formation constant of log β_{12}=8.81±0.05.

Acknowledgement

This work was supported by the Deutsche Forschungsgemeinschaft under contract no. BE 2234/1-1,1-2.

References

Geipel G, Brachmann A, Brendler V, Bernhard G, Nitsche H (1996) Uranium(VI) Sulfate Complexation Studies by Time-Resolved Laser-Induced Fluorescence Spectroscopy (TRLFS). Radiochim Acta 75: 199-204

Brendler V, Geipel G, Bernhard G, Nitsche H (1996) Complexation in the System UO22+/PO43-/OH-(aq): Potentiometric and Spectroscopic Investigations at very Low Ionic Strengths. Radiochim Acta 74: 75-80

Murakami Y, Takagi M (1966) Influence of Metal Ions and of Metal Chelates on the Hydrolysis of α-D-Glucose 1-Phosphate. Bull Chem Soc Japan 39: 122-127

Kramer-Schnabel U, Linder P W (1991) Substituent Effects in the Protonation and Complexation with Copper(II) Ions of Organic Monophosphate Esters. A Potentiometric and Calorimetric Study. Inorg Chem 30: 1248-1254

Performance report on the operations of a pilot plant for the treatment of vanadium in ground water—New Rifle UMTRA Site, Rifle, Colorado, USA

Donald R. Metzler[1], Kenneth E. Karp[2]

[1]U.S. Department of Energy Grand Junction Office, Grand Junction, Colorado
[2]MACTEC Environmental Restoration Services, Grand Junction, Colorado

Abstract. A pilot study conducted at the former uranium- and vanadium-processing site located west of Rifle, Colorado, USA, investigated the feasibility of using an ex situ zero-valent iron (ZVI) through-medium process to reduce vanadium concentrations in the ground water to levels that will naturally flush within the 100-year period allowable by the standards. Prototype testing, consisting of laboratory treatability studies followed by bench-scale field studies, provided performance data and design parameters required to scale up to higher flow rates for ground water remediation. The process was successful in removing vanadium from the ground water to below 0.33 mg/L while also meeting all other chemical cleanup goals.

Introduction

The former New Rifle processing site is located approximately 2.3 miles west of the town of Rifle, Colorado, as shown on Fig. 1. Former vanadium and uranium processing activities at the site have contaminated the alluvial ground water system directly beneath and downgradient of the site. Processing wastes were disposed of in holding ponds on the site, and mill tailings were stored in piles that covered 33 acres. The holding ponds and other surface contamination associated with the processing activities were removed and the site remediated in 1996 in accordance with Title 40 *Code of Federal Regulations* (CFR) Part 192 under the Uranium Mill Tailings Remedial Action (UMTRA) Program administered by the U.S. Department of Energy (DOE).

Ground water modeling has demonstrated that most contaminants of concern (COCs) at the site will flush to applicable standards in the 100 years permitted by natural flushing (DOE 1999). However, vanadium will probably not reach its risk-based concentration of 0.33 milligram per liter (mg/L) (EPA 2001) in 100 years because of its low mobility in the subsurface. In addition, some soils at the site contain highly elevated concentrations of vanadium and may serve as a continuing source of ground water contamination (DOE 2000a). For this reason, a pilot ground water treatment plant was designed, constructed, and operated at the New Rifle site to investigate the feasibility of using an active ex situ method to reduce vanadium concentration.

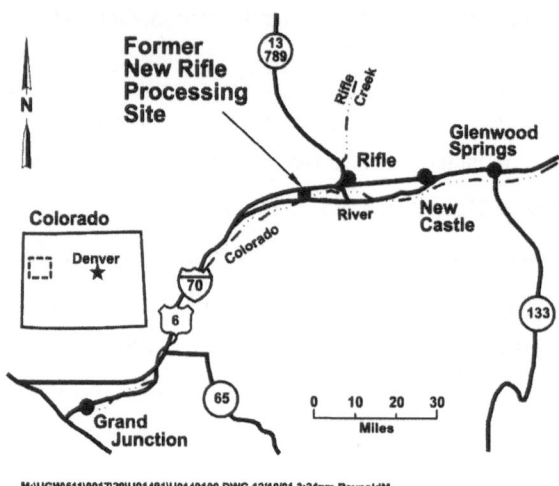

M:\UGW\511\0017\29\U01491\U0149100.DWG 12/10/01 3:24pm ReynoldM

Fig. 1. Location of the New Rifle site.

Background

Laboratory treatability studies were performed at 0.005 gallon per minute (gal/min) to test the feasibility of treating ground water using zero-valent iron (ZVI) in a through-medium process (DOE 2000b). Results of the treatability test were encouraging and demonstrated that ZVI was able to effectively remove uranium, vanadium, and several other constituents from the ground water. Follow-up bench-scale tests were then conducted at 0.5 gal/min under field conditions to (1) evaluate treatment column and settling tank configurations; (2) optimize ZVI mesh size and residence times; (3) address issues expected on a larger scale, such as permitting and "clean" water disposal; and (4) mimic conditions as close as possible to those expected for a full-scale cleanup (DOE 2000b). Prototype test results indicated –60/+100 mesh ZVI was the most effective size to reduce vanadium concentrations; it was also effective in removing arsenic and uranium concentrations. The following design parameters for the 20-gal/min pilot system were determined from prototype tests and were used as the basis for the design of the pilot system:

- Column sizing based on 4-minute residence time:
  15-inch diameter by 3-foot high (4 columns at 5 gal/min each)
  1,008 pounds -60/+100 mesh ZVI (485 pounds per cubic foot, 50% porosity)
- Sludge tank based on 4.3 gallons sludge per 1,000 gallons:
  20 gal/min total inflow
  124 gallons sludge per day
  1,240 gallon tank (10-day capacity)

Pilot System Components

The pilot system consists of three components: (1) extraction system, (2) treatment system, and (3) infiltration systems. Computer flow modeling was used to design the extraction and infiltration systems. Results of the prototype test were used to determine the design parameters for the treatment system. The three components of the pilot system are shown on Fig. 2 and discussed in the following sections.

Extraction and Infiltration System

Results of a detailed characterization of vanadium contamination in the subsurface conducted in March and April 2000 indicate concentrations in the ground water are highest in the middle of the aquifer and are offset approximately 400 feet from the highest soil concentrations in a generally downgradient direction (DOE 2000a). Ground water flow and optimization modeling was performed to determine the optimal number, locations, and pumping rates for the wells that are required to capture the vanadium plume.

Results of the ground water flow and optimization modeling determined the best configuration to meet the extraction design objectives consists of three extraction wells (862, 863, and 864) to remove contaminated ground water and an infiltration gallery to return the treated effluent to the aquifer at a downgradient location. Locations for the extraction wells are shown on Fig. 3. Construction and operating information for the extraction system and the location of the infiltration gallery is presented in the *Operating Manual for the Vanadium Pilot Study* (DOE 2001). The objectives of the extraction and infiltration system are as follows:

- Pumping wells to deliver up to 20 gal/min continuously to the treatment system.
- Extract a total of 2 million gallons of ground water in 8 months from the plume.

Passive insitu treatment techniques, wetlands

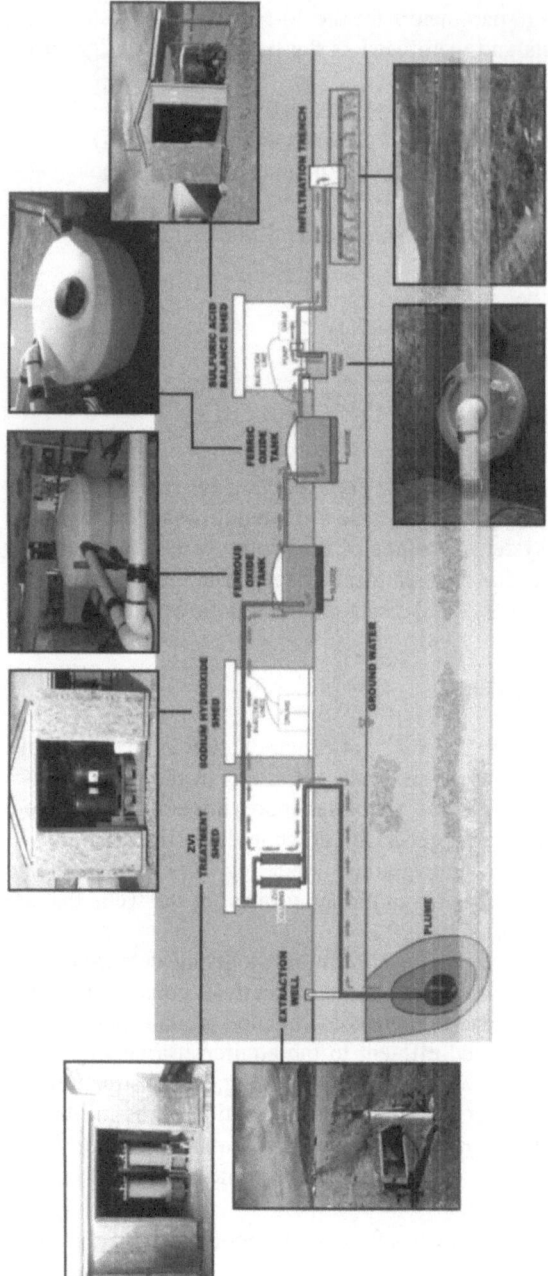

Fig. 2. Pilot system components.

Passive insitu treatment techniques, wetlands

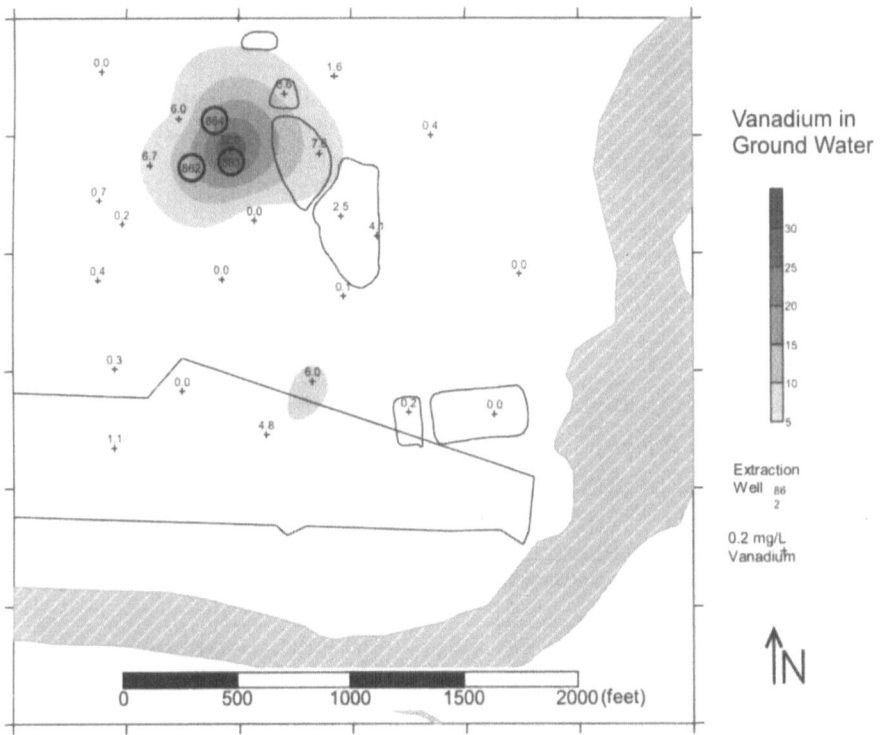

Fig. 3. Extraction well locations at the New Rifle site.

ZVI Treatment System

Contaminated ground water from the 6-inch extraction well 864 shown on Fig. 3 was piped to an aboveground ZVI system for treatment. Effluent from the treatment unit(s) was then piped to a series of 1,550-gallon settling tanks for iron filtration followed by neutralization in a 100-gallon tank before the treated water was returned to the aquifer via the infiltration gallery. The treatment configuration is shown on Fig. 2.

The treatment system was operated to meet the following goals for the effluent water before discharge to the aquifer:
- pH >6<9
- Iron <11 mg/L
- Vanadium <0.33 mg/L

Passive insitu treatment techniques, wetlands

Pilot System Performance Results

Treatment operations were initiated on January 29, 200,1 and continued until November 19, 2001. During the treatment operations, a total of 2,942,350 gallons of contaminated ground water was treated. For details regarding the operation of the plant the reader is referred to the Operating Manual (DOE 2001). Summary results for selected parameters are presented in Table 1.

Table 1. Results of selected operating parameters.

Parameter	Result
Volume treated	>2.9 million gallons
Average treatment rate	9.6 gal/min
Average flow rate per column	2.3 gal/min
Optimized column longevity (ZVI change out)	30 days
Average vanadium influent concentration from plume	8.9 mg/L
Average vanadium concentration from column effluent	1.4 mg/L (6-fold reduction from 8.9 mg/L influent)
Average vanadium concentration from settling tanks effluent	0.08 mg/L (17-fold reduction from 1.4-mg/L influent)
Optimized gallons of water treated per pound of ZVI	>120 gallons per pound
Optimized sludge generated per gallon treated	4 to 5 gallons sludge per 1,000 gallons of water treated

Influent Concentrations

A total of 2,942,350 gallons of contaminated ground water was extracted from pumping well 864 located near the center of the vanadium plume as shown on Fig. 3. Changes in vanadium concentration as a function of volume of water treated is presented on Fig. 4.

It is apparent from data in Fig. 4 that vanadium concentrations in the aquifer were not significantly reduced after removing more than 2.9 million gallons of contaminated ground water. The apparent lack of vanadium reduction is most likely due to the relatively low mobility of vanadium in the subsurface because of its affinity to adsorb onto subsurface materials.

Passive insitu treatment techniques, wetlands

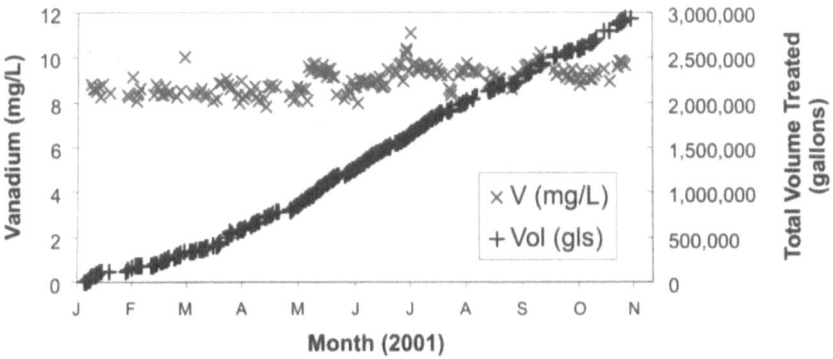

Fig. 4. Influent vanadium concentrations from extraction well 864 (volume in gallons).

Treatment Columns and Settling Tanks

Vanadium and iron concentrations for ground water treated by column 2 (C–2) are presented on Fig. 5 and Fig. 6, respectively. Detailed in each of the referenced figures for C–2 are the concentrations and operational flow rates that correspond with the start of a new batch of ZVI on June 7, 2001, and ending on July 18, 2001, when the medium was spent. The operational duration of this particular run was 38 days. During that time, 109,840 gallons of ground water was treated, 625 pounds of ZVI was used as the treatment medium, and the average flow through the column was 2.1 gal/min.

As shown on Fig. 5 and Fig. 6, the flow rates were decreased from an initial rate of 3 gal/min to 2 gal/min at day 19 and finally to 1 gal/min at day 32. The other columns were operated in a similar manner to extend the longevity of the ZVI.

The amount of water treated per pound of ZVI during the first 2 months of operation ranged from 28 to 91 gallons per pound. As operational experience was gained, the treatment efficiencies improved significantly. Some examples of improved efficiencies include flow rate adjustments as shown on Fig. 5 and Fig. 6, modifications to the columns to minimize preferential flow paths, managing the flow rate through the columns to maximize use of ZVI, and increasing the residence time in the iron-removal settling tanks.

Passive insitu treatment techniques, wetlands

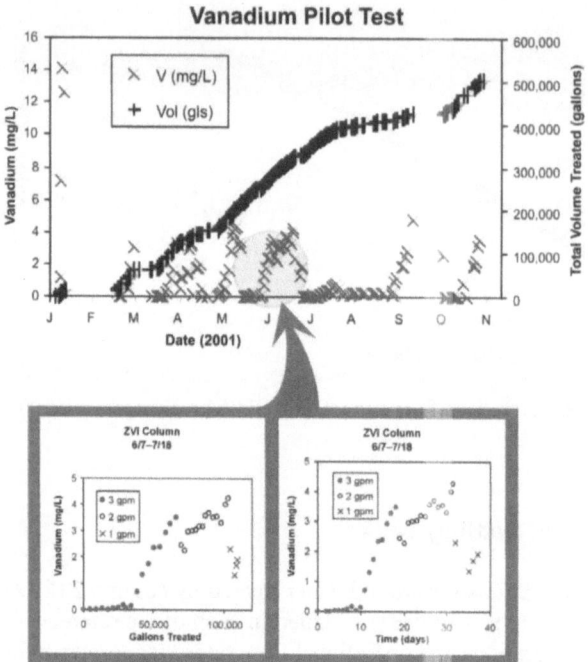

Fig. 5. Vanadium concentrations in treated ground water from C–2.

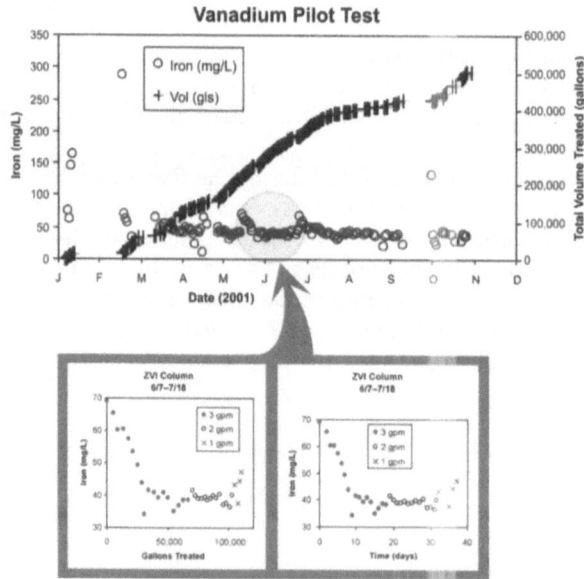

Fig. 6. Iron concentrations in treated ground water from C–2.

Passive insitu treatment techniques, wetlands

Efficiencies were also realized in the amount of sludge generated in the iron-removal settling tanks. This was accomplished by pumping the sludge collected in the settling tanks to an aboveground holding tank so the sludge could be consolidated, thus reducing the volume. The amount of sludge generated per 1,000 gallons of water treated over time is illustrated on Fig. 7.

Sludge Generated in Settling Tanks

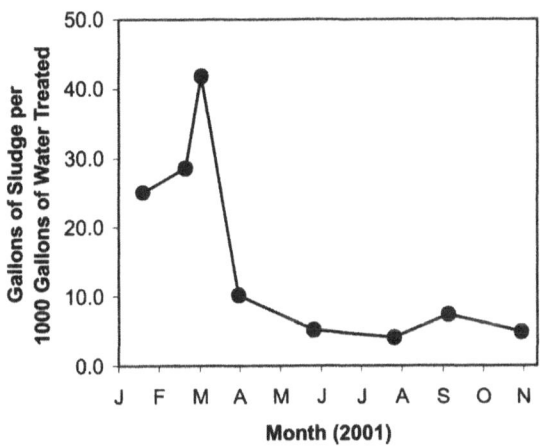

Fig. 7. Sludge generated in the settling tanks over time from effluent concentrations.

The iron removal process also provided additional vanadium reductions. Increasing the residence time for the treated water in the iron-removal settling tanks resulted in greater iron and vanadium reductions. However, as the sludge volume increased in the settling tanks, the residence time for the treated water decreased. This relationship was used to maximize the efficiency of the treatment and to stabilize the effluent concentrations. For example, as the ZVI in the columns decreased in efficiency, the sludge from the settling tanks was pumped to an aboveground holding tank. This transfer allowed maximum consolidation of the sludge while also maximizing the residence time for the treated water. The clean settling tanks and increased residence time decreased both iron and vanadium concentrations. As a result, the effluent concentrations were for the most part relatively stable and stayed within the treatment goals.

Infiltration Gallery

An infiltration gallery was constructed downgradient from the extraction well field to return treated ground water to the aquifer. Before regulatory approval was received, the treated water was reinjected into an on-site monitor well (219) to accelerate startup of the system. Reinjection of treated water into monitor well 219 occurred during the period January 29, 2001, through March 8, 2001. During this time, numerous operational problems resulted because the well was not originally designed for reinjection of water. Most of the problems encountered were associated with screen clogging because of calcium carbonate precipitation and limitations of the relatively small casing diameter and short screen length of the monitor well. Low reinjection rates (1 to 2 gal/min) combined with extensive screen maintenance were required to keep the pilot system operational.

Treated ground water was returned to the aquifer via the infiltration gallery during the period March 8, 2001, through November 19, 2001. The performance of the gallery was monitored by collecting water-elevation measurements with data loggers installed in two monitor wells (PT–680 and PT–682) located 50 feet downgradient of the gallery and in monitor well PT–681 located 250 feet downgradient of the gallery. Water-elevation measurements were collected with a hand-held instrument at the infiltration gallery manhole that provides access to the center of the drainpipe. None of the problems experienced with reinjecting treated water into monitor well 219 was observed with the infiltration gallery during the performance period, even when the pilot plant was operated at relatively high flow rates.

Conclusions

Evaluation of the pilot pump-and-treat system considers both aquifer performance (pump) and treatment performance (treat). Aquifer performance relates to hydrogeologic characteristics that can affect the quantities and rates that ground water can be extracted or injected and geochemical characteristics such as mobility of contaminants in the subsurface that may limit extraction of the contaminant from the aquifer matrix (i.e., aquifer cleanup goals). Treatment performance relates to the adequacy and operation of the technology (e.g., chemical precipitation, ion exchange) to meet the chemical cleanup goals for the extracted ground water (i.e., treatment goals). Summaries of conclusions regarding aquifer and treatment performance of the pilot system are presented in the following sections.

Aquifer Performance

Restoration potential for the aquifer (i.e., aquifer cleanup goals) using conventional extraction wells appears to be limited by the geochemical characteristics of vanadium. Relatively high adsorption of vanadium to the aquifer matrix is evidenced by the apparent lack of significant decreases in vanadium concentrations observed in ground water samples collected from extraction well 864. This limitation is consistent with the relatively high distribution coefficients observed for vanadium at the site (DOE 1999 and 2000a).

Passive insitu treatment techniques, wetlands

Hydraulic parameters of the aquifer, such as permeability and conductivity, did not limit the amount of water or rate that water was extracted from the aquifer for the pilot study; extraction well 864 performed as designed. Similarly, the hydraulic parameters did not limit the amount or rate of water that was infiltrated back into the aquifer via the infiltration gallery. Both extraction goals, (1) to deliver 20 gal/min to the treatment system and (2) to extract 2 million gallons of plume water, were achieved.

Treatment Performance

The treatment process consisted of flowing extracted ground water through a reactor column containing ZVI then removing the iron from the effluent in a series of downstream settling tanks. The ZVI column resulted in an average 6-fold reduction in vanadium concentrations (Table 1). The vanadium removal efficiency was improved by the oxidation and precipitation of iron oxides in the downstream iron settling tanks. The settling tanks provided an additional 17-fold reduction in vanadium concentrations (Table 1) through co-precipitation with the iron oxides. All the chemical cleanup goals for the extracted ground water (i.e. pH, iron, and vanadium) were met, demonstrating the effectiveness of the treatment process.

The ZVI through-medium process and the iron removal system were relatively simple to design, construct, and operate. The system could potentially be simplified further by using only the iron removal portion of the treatment to co-precipitate vanadium with the iron oxides. Iron could be introduced to the treatment stream by injection of ferric chloride.

Total costs to operate the pilot treatment system at 10 gal/min are presented in Table 2 as a function of the volume of water treated. Variable costs include ZVI, chemicals, sludge disposal, labor, and utilities. The fixed cost includes research and development, engineering design, and construction. The estimated cost to treat 5 million gallons of ground water is approximately $0.10 per gallon as shown in Table 2 and graphically illustrated on Fig. 8.

Table 2. Total costs operating at 10 gal/min.

Volume (gal)	Total cost (U.S. dollars) operating at 10 gal/min			
	Variable cost	Fixed cost	Total cost	Total cost (1,000 gal)
1,000,000	18,885	400,000	418,885	419
2,000,000	37,770	400,000	437,770	219
3,000,000	56,656	400,000	456,656	152
4,000,000	75,541	400,000	475,541	119
5,000,000	94,426	400,000	494,426	99

Passive insitu treatment techniques, wetlands

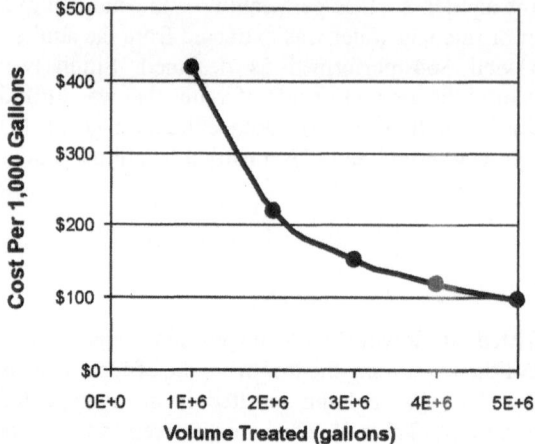

Fig. 8. Total unit cost in U.S. dollars as a function of volume treated.

References

U.S. Department of Energy (November 1999) Site Observational Work Plan for the UMTRA Project New Rifle Site, GJO–99–112–TAR, Rev. 1, prepared for the U.S. Department of Energy, Grand Junction Office, Grand Junction, Colorado

U.S. Department of Energy (July 2000a) Draft Work Plan for Vanadium Pilot Study New Rifle UMTRA Site, Rifle, Colorado, GWRFL10.6.4, prepared for the U.S. Department of Energy, Grand Junction Office, Grand Junction, Colorado

U.S. Department of Energy (April 2000b) Laboratory Treatability Tests To Evaluate Water Treatment Technologies, UMTRA Ground Water Project, Rifle, Colorado, Site, ESL–RPT–2000–03, prepared for the U.S. Department of Energy, Grand Junction Office, Grand Junction, Colorado

U.S. Department of Energy (February 2001) Operating Manual for the Vanadium Pilot Study New Rifle UMTRA Site, Rifle, Colorado, GJO–2001–193–TAR, MAC–GWRFL–10.6.5, prepared for the U.S. Department of Energy, Grand Junction Office, Grand Junction, Colorado, February.

U.S. Environmental Protection Agency (2001) Risk-Based Concentration Table, U.S. EPA Region III. Memorandum from Jennifer Hubbard, Toxicologist, available on the internet at http://www.epa.gov/reg3hwmd/risk/riskmenu.htm.

Passive insitu treatment techniques, wetlands

Inter-disciplinary studies of the impact of gold and uranium mining in the Witwatersrand Goldfield

Henk Coetzee[1], Stéphane Chevrel[2] & Francis Cottard[2]

[1]Council for Geoscience, Private Bag X112, Pretoria, 0001, South Africa, henkc@geoscience.org.za
[2]BRGM – Land use Planning and Natural Hazards, BP 6009 – 45060 Orléans Cedex 2 – France, s.chevrel@brgm.fr, f.cottard@brgm.fr

Abstract. Over the past century, large amounts of gold and uranium have been extracted from sediments of the Witwatersrand Supergroup. Mining activities cover a strike length of more than 300km and the resulting tailings now cover more than 400km^2. In recent years mining activity has declined, with a number of mines closing or scaling down their operations. Furthermore, large areas of the goldfields have been abandoned for many decades, with little or no remediation having been undertaken. We have shown that a combination of airborne geophysics, remote sensing and GIS data integration provides valuable tools for coping with the large scale of the problem.

Introduction

Since the discovery of gold at Johannesburg in the 1880s, an estimated 43 000t of gold and 75 000t of uranium has been produced in the Witwatersrand area. Since gold has been mined along a strike length of more than 300km, a regional approach must be taken to the assessment of environmental impacts.

Assessments of environmental impacts have been undertaken using satellite imagery. Landsat TM, ASTER, SPOT and Ikonos data have been used to create basemaps for environmental assessment, predict geological structures which influence groundwater conditions, identify wetland areas which concentrate pollutants, directly map the areas of infiltration of minewater into the groundwater system and locate old mining activities which influence groundwater flow.

Geophysical data have been used to map geological structures as well as mapping and in some cases quantifying the degree of radionuclide contamination in sediments. Uranium and radium concentrations, which significantly exceed the legally defined regulatory limit, have been measured in sediments downstream of mining activities. Geochemical studies have confirmed these results and demonstrated that uranium and other metals can be remobilised from these sediments, creating a larger source area for potential groundwater contamination with complex regulatory implications.

GIS techniques allow the integration of different data types and sources into layers that can be combined to produce decision support tools. The methods used for the combinations of layers vary from simple arithmetic or Boolean combinations to techniques based on matrix algebra and non-linear simulation techniques, and can be used to project current data into future scenarios.

Application of remote sensing

Traditional space-borne sensors

Traditional satellite based remote sensing techniques (Landsat TM, SPOT XS, SPOT PAN) have been applied to areas in the Witwatersrand goldfield with varying degrees of success. Land cover classification has proved effective in the agricultural and natural areas surrounding mining activities using multispectral Landsat TM data (Chevrel and Coetzee 1997). Classification of imagery has however proved less successful within mining areas and the surrounding dense formal and informal urban areas, particularly because of the wide variety of surface reflectance patterns within small areas. For example, a typical informal settlement plot has an area of approximately 100m², one ninth of the area of a Landsat pixel. Within this area, there will be a mixture of the corrugated iron used as roofing and building material, open soil and vegetation. Traditional sensors also offer sufficient spatial resolution for the visual identification of larger features, such as individual settlements, tailings dams (or zones within the larger tailings dams), fields or wetlands.

In recent years, a number of new sensors have become available to the remote sensing community. The capabilities of some of these sensors are summarised on Table 1. Additional satellites and sensors will be launched in the near future. The major differences between these and traditional remote sensing platforms are greater spatial and/or spectral resolution, better spectral coverage and the ability to perform client-specified missions and data acquisitions.

Passive insitu treatment techniques, wetlands

Table 1. Summary of a few new remote sensing sensors, applicable to environmental remote sensing in mining areas

Sensor	Operator	Spectral coverage	Spatial resolution
Landsat ETM+	NASA	7 Band Multispectral VNIR-TIR + 1 PAN band	15m PAN, 30m multispectral, 90m thermal
ASTER	NASA + Japanese ministry of Industry and Trade	14 Band Multispectral VNIR-SWIR+TIR	15m VNIR, 30m SWIR, 90m TIR
Ikonos	Private Company	3 band multispectral + 1 PAN band	1m PAN, 4m multispectral, resolution enhanced multispectral available
HyMap (Airborne)	Private Company	Hyperspectral	Altitude dependent, typically ~2m

New space-borne sensors - ASTER and Ikonos

Recently, large amounts of data have been made available from the ASTER sensor (the Advanced Space-borne Thermal Emission Radiometer) on board NASA's TERRA satellite. This sensor provides 15m spatial resolution in the visible to near infrared range (VNIR) compared to Landsat TM's 30m resolution and better spectral resolution in the short wave infrared (SWIR), at the same resolution as Landsat TM.

An ASTER image of a portion of the East Rand Goldfield was obtained and used to assess the potential of this type of imagery for environmental analysis of mining areas. Similarly to the previous Landsat TM study, land-cover classification was attempted using the ASTER data. It was found that the greater spectral resolution allowed better separation of land-cover types, but that the limitations imposed by the land-cover types within the area and limited spatial resolution still posed problems. Two classification techniques which show promise are a manual technique, based on the identification of spectral characteristics and the application of simple rules to image and spectral angle mapping, which was found to be extremely effective for assessing the water content of zones within individual tailings dams.

Regarding resolution, the 15m resolution VNIR bands allow the identification of smaller features within a mining district, for example large scale zoning in waste deposits, shallow and deep water and dense vegetation in water seepage zones around tailings dams. Fig. 1 shows a typical example of ASTER VNIR imagery from the East Rand Goldfield. Particular environmental problems encountered here are the proximity of the radioactive and toxic mine wastes to urban areas and local drainage systems.

Passive insitu treatment techniques, wetlands

Fig. 1. Portion of VNIR band 2 (Red) of an ASTER image of a portion of the East Rand. Note the proximity of the tailings dam to urban areas. Also note the drainage to the east of the tailings dam. The tailings dam is approximately 2km from east to west.

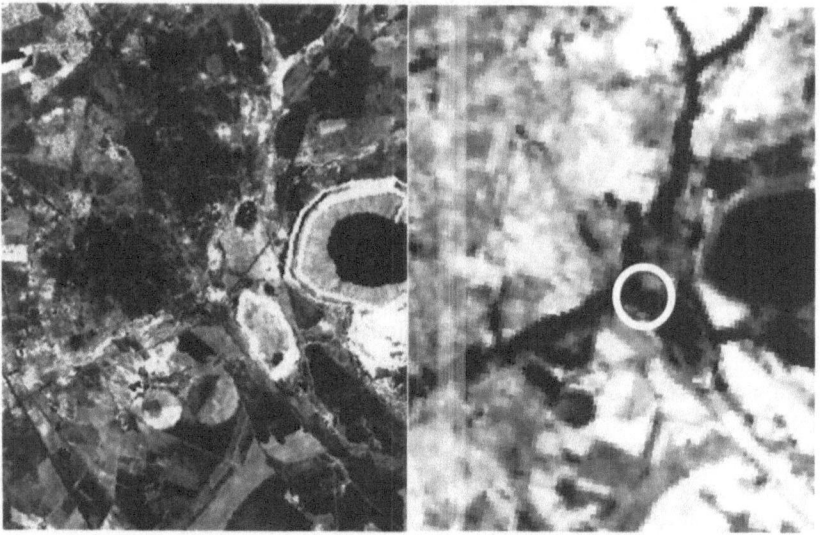

Fig. 2. ASTER VNIR data (Band 3 – NIR) (left) and Thermal (Band 12) (right) images of the area surrounding a tailings dam on the East Rand. Note the seepage area indicated by a white circle on the thermal image.

The thermal data produced by the ASTER sensor is of low spatial resolution (90m pixel size). It does however very clearly detect waterlogged areas. In the study area these include active tailings dams, wetlands and seepage zones where tailings leachate is entering the groundwater system (see Fig. 2).

Data from the Ikonos satellite on the other hand offer less spectral coverage and resolution than ASTER or Landsat, but much higher spatial resolution (see Fig. 3). The standard data products are 1m resolution panchromatic imagery, 4m resolution multispectral (R,G,B,NIR) imagery and resolution enhanced multispectral imagery with a pseudo-resolution of 1m. Such a resolution enhanced multispectral image was used to assess a 100km^2 area within the area studied with ASTER imagery.

Unfortunately, the resolution-enhanced product is unsuitable for quantitative classification. It did however prove invaluable for the planning of a field investigation of environmental conditions within the study area. A large number of sites suspected of being highly contaminated were identified on the image, as were access routes, and contamination sources. These were then visited, investigated and where necessary samples collected and analysed. The use of Ikonos imagery proved highly efficient for the identification of sites needing further investigation, and resulted in considerable cost and time savings over a conventional field investigation. Fig. 4 shows a site where tailings are being eroded and entering the local drainage system.

In addition to contaminated site identification and characterization, subsidence features relating to historical mining activities were also identified on Ikonos imagery. Subsidence due to historical mining has become a particular hazard in the Witwatersrand Goldfield, due to the immediate hazard posed by old mining operations as well as the possibility of the migration of contaminated surface water into the groundwater through these old mining activities.

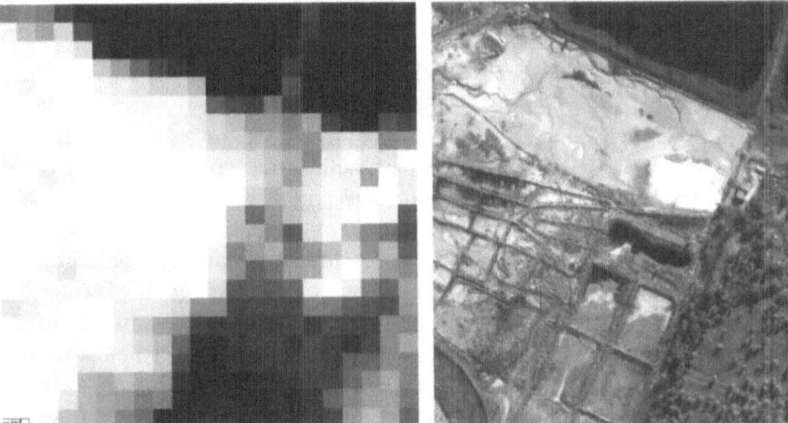

Fig. 3. Comparison of the spatial resolution of Landsat TM (left) and Ikonos (right) imagery. This clearly shows the utility of ultra-high-resolution imagery for the identification of features of environmental interest as well as planning of field investigations.

Passive insitu treatment techniques, wetlands

Fig. 4. Seepage of tailings into a public watercourse as seen on Ikonos imagery.

Airborne radiometric surveying

For several years, the value of airborne radiometric surveys for the characterisation and in some cases quantification of radioactive contamination from Witwatersrand gold/uranium mines has been recognised (Coetzee & Szczesniak, 1993). In addition to clearly showing the increased gamma radiation levels emanating from tailings dams, waste rock dumps, uranium plants and other mining infrastructure, these surveys also show evidence of contamination of watercourses downstream of mining activities, and in some cases wind-blown dust plumes. The downstream contamination is often concentrated in wetlands, some several kilometres downstream of mines and off mine properties. Fig 5 shows an airborne radiometric image of a mining area near the town of Nigel on the East Rand Goldfield. Note the downstream pollution plumes.

Passive insitu treatment techniques, wetlands

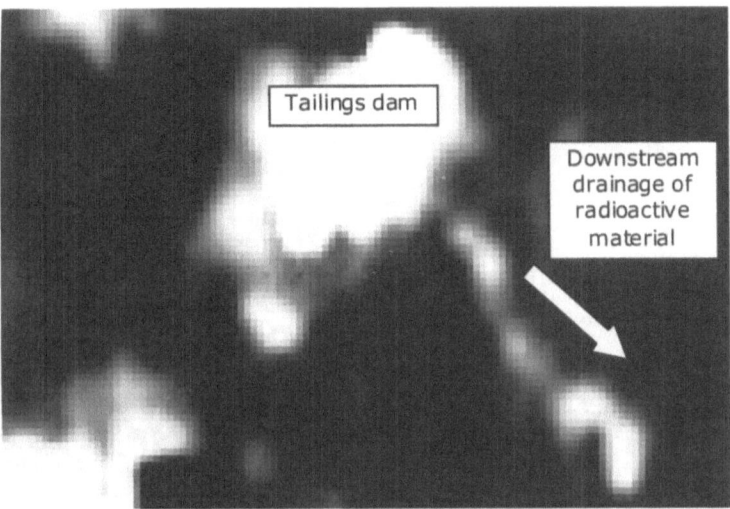

Fig. 5. Airborne radiometric image of a tailings dam and associated downstream plume near the East Rand town of Nigel. Note the significant downstream migration of radioactive material. (Lighter colours indicate higher gamma ray count rates.)

Integration of data using a GIS-based methodology

Such diverse data are often difficult to interrogate together in order to compile a comprehensive environmental analysis. When chemical analytical data, water quality time series and other data with different spatial topologies are included, the process becomes more complex. For this reason, an approach to the assessment of environmental problems has been developed, based on a cycle of conceptual modelling, data collection, data analysis and interpretation, assessment and then either production of final results or iteration of the procedure. This procedure is shown schematically on Fig. 6.

Passive insitu treatment techniques, wetlands

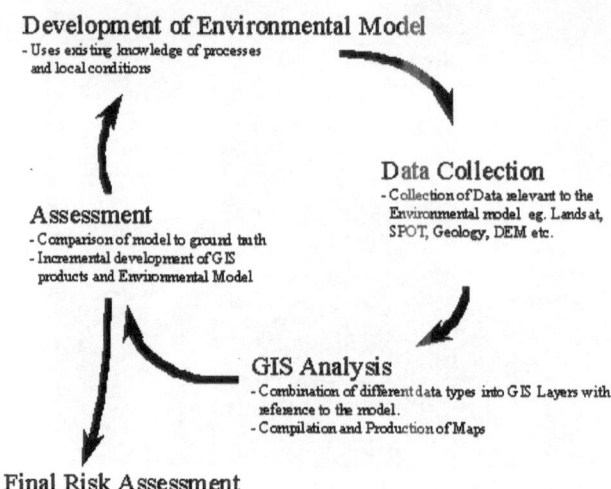

Fig. 6. Schematic representation of the assessment procedure used for multi-parameter data assessment.

Conclusions

The scale of mining and its environmental legacy in the Witwatersrand basin makes detailed site-by-site field studies impracticable and prohibitively expensive. Modern remote sensing and airborne geophysical techniques allow the identification of specific areas requiring remediation and may be also be used to prioritise and monitor remediation.

Geographical Information system methods allow the combination of data from a variety of sources into a common database, that can be intelligently queried to solve specific problems.

References

S Chevrel and H Coetzee (1997) Interdisciplinary study of the impact of gold and uranium mining in South Africa, Council for Geoscience, BRGM, 40pp.

H Coetzee and H Szczesniak (1993) Detection and monitoring of pollution from mine tailings dams along rivers in the Witwatersrand Goldfield using the airborne radiometric method, *16ᵗʰ International Colloquium on African Geology*, 94-96.

Passive insitu treatment techniques, wetlands

Removal of uranium and arsenic from groundwater using six different reactive materials: assessment of removal efficiency

Dirk Mallants[1], Ludo Diels[2], Leen Bastiaens[2], Johan Vos[2], Hugo Moors[1], Lian Wang[1], Norbert Maes[1], and Hildegarde Vandenhove[1]

[1] SCK•CEN, Boeretang 200, B-2400 Mol, Belgium
[2] VITO, Boeretang 200, B-2400 Mol, Belgium

Abstract. Permeable reactive barriers (PRBs) are increasingly being used as a cost-effective technique for *in-situ* treatment of contaminated groundwater. This paper discusses the results from batch tests with six different reactive materials, including zero-valent iron (ZVI), ferric oxyhydroxides, and some composite materials. All materials were tested in their ability to remove uranium and arsenic from groundwater. Results show that fine-grained ZVI was most successful in removing uranium from solution, up to a removal efficiency of 98%. The main mechanism in removing uranium presumably is by reductive precipitation. The results further showed that arsenic removal is most efficient with ferric oxyhydroxides (maximum removal efficiency is 96%). This study illustrates that PRBs using a mixture of fine-grained ZVI and materials containing ferric oxyhydroxides as reactive material may help significantly in removing uranium and arsenic from groundwater.

Introduction

The classical approach to treating contaminated groundwater has been to remove the groundwater by pump-and-treat and to decontaminate the water ex-situ. Alternatively, groundwater may also be treated by using so-called permeable reactive barriers (PRB). These are engineered, subsurface zones of reactive material that can remove contaminants from the pore-water solution, usually by adsorption onto the reactive sites of the materials introduced or by stimulating the formation of sparely soluble minerals (for instance by reductive precipitation in case of redox sensitive elements).

Previous studies have shown the ability of ZVI to remove uranium (Gu et al., 1998; Morisson et al., 2001) and arsenic (Su and Puls, 2001a) from groundwater. However, uranium and arsenic seem to behave differently in the presence of reactive materials. The main mechanism by which uranium (VI) is removed from groundwater is by reduction and precipitation of sparely soluble minerals, including uraninite (UO_2). Adsorption onto reactive surfaces, either generated by corrosion of Fe^0 or initially present as ferric oxyhydroxides, is believed to be another process of uranium removal. In near neutral to alkaline groundwater, adsorption of both As (V) and As (III) is believed to be the most important removal mechanism, rather than reduction. Concerning the effects of inorganic ligands on the mobility of U and As, uranium is clearly more soluble in the presence of carbonate, whereas arsenic desorption is particularly enhanced by phosphate (Su and Puls, 2001b).

The differences observed in the behaviour of U and As in the presence of reactive materials was the motivation to this study. The objective was to test six different materials for their suitability to remove uranium and arsenic from groundwater. The materials tested included iron oxyhydroxides, zero-valent iron (ZVI), composite material, and zeolites.

Materials and Methods

Groundwater characteristics

The groundwater used originated from seepage water collected from a uranium tailings pond in Helmsdorf (Thuringia, Germany). Chemical characteristics of the alkaline groundwater are shown in Table 1 (for a more detailed analysis, see Mallants et al., 2001). The seepage water had an average pH of 9.3, a high concentration of inorganic carbon (670 mg/l on average), and a high sulphate concentration (6500 mg/l). Phosphate concentration was moderately high (between 2 and 10 mg/l).

Table 1. Simplified chemical composition for "Helmsdorf" groundwater

Ion	Helmsdorf		
	GW1, mg/l	GW2, mg/l	GW3, mg/l
Mg^{2+}	46	30	40
Ca^{2+}	4.4	4.6	8.9
U(VI)	11.6	18.8	13.3
total inorganic C	492	844	677
SO_4^{2-}	6180	7420	5970
HPO_4^{2-}	1.8	9.7	8.9
E_h (mV)	~390	~390	~390
pH	9.2	9.4	9.3

GW1 = location 1 at Helmsdorf tailings pond
GW2 = location 2 at Helmsdorf tailings pond
GW3 = location 3 at Helmsdorf tailings pond

Reactive media

Six reactive materials were tested in their ability to remove uranium and arsenic from groundwater: (1) ferric oxyhydroxide (Ferrosorp) with a surface area > 100 m^2/g, (2) coarse-grained zero-valent iron (FeA4) characterized by a surface area of around 1 m^2/g, (3) fine-grained, powdery zero-valent iron (FeX), (4) a composite material made from aluminium and iron oxides (EPV), (5) mordenite, which is a natural zeolite, and (6) a synthetic zeolite, referred-to as Zeolite X.

Batch tests

To test the removal efficiency of the reactive materials batch tests using "Helms-dorf" water were carried out by mixing 0.4 g of reactive material with 40 ml of groundwater. Tests were done for three different uranium and arsenic concentrations, i.e., 11.6 mg/l U and 0.58 mg/l As (named GW1), 18.8 mg/l U and 1.52 mg/l As (named GW2), and 13.3 mg/l U and 1.65 mg/l As (named GW3). Note that these concentrations represent the in situ uranium and arsenic contamination levels observed in the "Helmsdorf" seepage water. Supernatant solution was collected and filtered with a 0.45 μm filter after 12, 24, and 48 h of equilibration. Uranium concentrations were determined by means of ICP-MS at all three times, whereas As was measured only after 48 hours (using ICP-AES).

Results and Discussion

U and As removal efficiency of ZVI

Uranium removal efficiency after 48 h of equilibration was 95% or more for fine-grained zero-valent iron (FeX, see Table 2). Coarse-grained zero-valent iron (FeA4) had an intermediate efficiency between 20 to 71%, depending on groundwater type (i.e., lowest for GW2 and highest for GW1). The composite material (EPV) displays an intermediate efficiency, between 46 and 68 %. Ferric oxyhydroxide (Ferrosorp) had a low efficiency in removing uranium, between 0.5 and 9%. The latter observation may be the result of the limited availability of surface sites and/or the high carbonate content of the groundwater. Zeolites have a very low efficiency. Carbonate is known to be the most important complexing anion in most groundwaters. In this respect it is interesting to note that GW2, which has the highest inorganic carbon content, shows the highest dissolved uranium concentration, irrespective of the reactive material. Similarly, GW1 has the lowest inorganic carbon content and the lowest dissolved uranium concentration, irrespective of reactive material. These results suggest a positive correlation between dissolved uranium and inorganic carbon. This observation is true for all reactive materials.

Table 2. Percentage removal of U and As from contaminated groundwater

Reactive material	Helmsdorf groundwater					
	GW1		GW2		GW3	
	U	As	U	As	U	As
Ferrosorp	9	95	0.5	94	6.7	96
FeA4	71	95	19	80	30	88
FeX	98	88	94	82	97	91
EPV	68	88	46	81	59	85
Mordenite	0	0	0	1	1	0
Zeolite X	1	2	0	1	1	0

The removal efficiency of arsenic is high for Ferrosorp and FeA4, intermediate high for FeX and EPV, and very low for zeolites (Table 2). It is noteworthy that Ferrosorp shows the highest removal efficiency for As (> 94%), whereas it was very inefficient in removing uranium. The low adsorption capacity of ferric oxyhydroxides at high pH (> 9) observed here is in agreement with experimental findings of Davis (2001) when studying adsorption on ferrihydrate surfaces. These observations suggest that different reaction mechanisms and/or reaction rates are operating at the surface of the ZVI. While U(VI) is believed to be predominantly removed by reductive precipitation and less so by adsorption of U on corrosion products of ZVI, As(V) and As(III) surface adsorption seems a more important removal mechanism under alkaline pH and moderate E_h conditions (Su and Puls, 2001). If surface adsorption would indeed be the most important removal mechanism in our study, ferric oxyhydroxide (Ferrosorp) has on average a higher adsorption capacity than the ZVI materials (FeA4) and (FeX) and their corrosion products. Because both UO_2^{2+} and As(V) are present in the solution, competition for the limited number of adsorption sites on the ZVI corrosion products may further explain the lower efficiency of ZVI compared to ferric oxyhydroxide. Although major anions such as phosphate are known to affect adsorption of arsenic (Su and Puls, 2001), results of such analysis were inconclusive for our data.

Batch tests were also used to determine distribution coefficients K_d (l/kg), where $K_d = C_{solid}/C_{liquid}$, C_{solid} is sorbed concentration (mg/kg) and C_{liquid} is porewater concentration (mg/l). Calculated values for K_d based on 48 h of equilibration are given in Table 3. Note that for U reductive precipitation rather than adsorption is the more likely process of removal, hence the term distribution ratio R_d would be more appropriate than K_d. The K_d (or R_d) values confirm the earlier observation that the fine-grained ZVI (FeX) is most efficient for U removal whereas ferric oxyhydroxide is most efficient in terms of As adsorption.

Table 3. Calculated K_d (l/kg) for U and As (after 48 h equilibration)

Reactive material	Helmsdorf groundwater					
	GW1		GW2		GW3	
	U	As	U	As	U	As
Ferrosorp	10	1833	0.5	1589	7	2260
FeA4	247	1834	23	407	42	725
FeX	5799	729	1576	463	3568	1080
EPV	211	729	86	425	142	588
Mordenite	0	0	0	1	1	0
Zeolite X	1	2	0	1	1	0

Figure 1 shows the pore-water concentration of As after 48 h equilibration. Considering the three slightly differing groundwater compositions, groundwater with ferric oxyhydroxide as reactive material has the lowest arsenic concentrations in solution. On average higher concentrations are observed when ZVI is the reactive material, but they are still fairly low. Groundwater containing zeolites as reactive material display the highest concentrations illustrating a removal efficiency which is virtually zero. The WHO recommends a provisional guideline value of 0.01 mg/l arsenic in drinking water (WHO, 1996). The most promising reactive material which may potentially reach the guideline value is ferric oxyhydroxide, although further optimisation would be required (e.g., longer contact time).

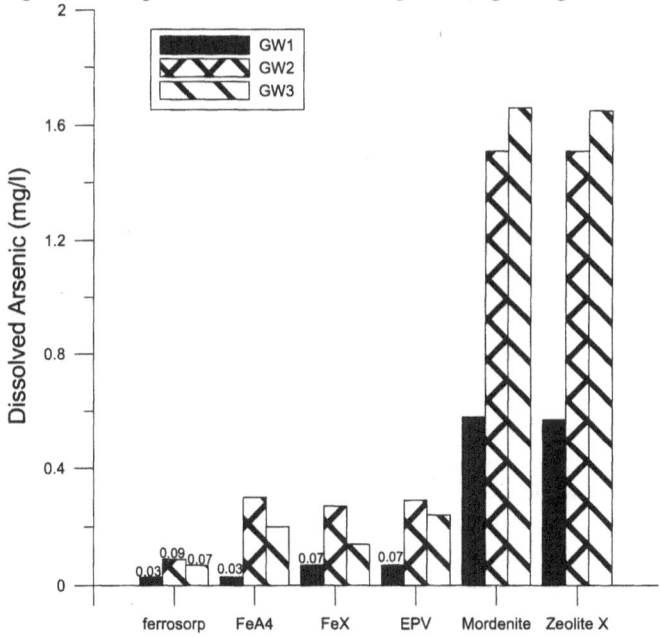

Fig. 1. Results of batch tests using "Helmsdorf" groundwater after 48 h of equilibration. Initial arsenic concentrations for GW1, GW2, and GW3 are, respectively, 0.58 mg/l As, 1.52 mg/l As, and 1.65 mg/l As.

Passive insitu treatment techniques, wetlands

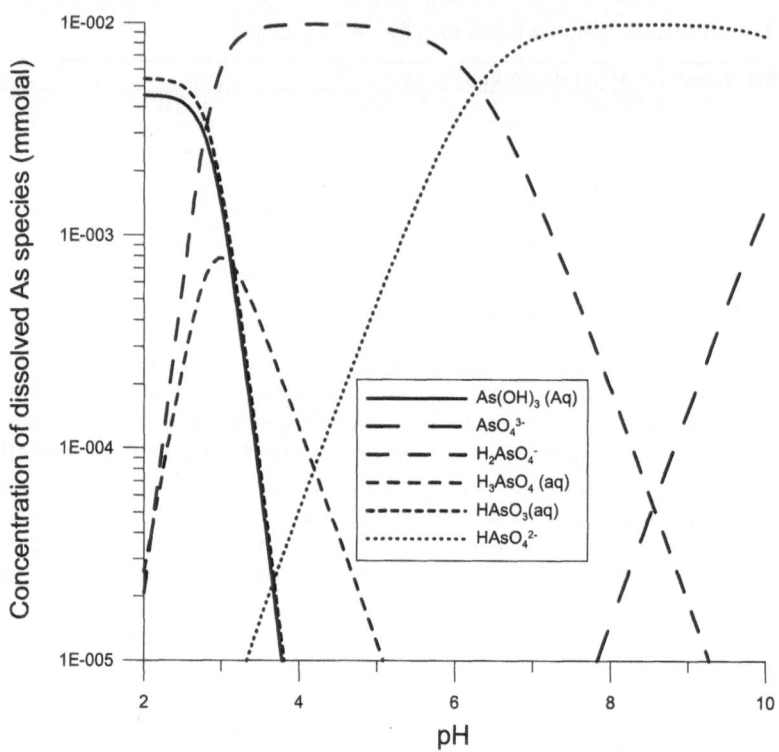

Fig. 2. Speciation of arsenic as function of pH using "Helmsdorf" groundwater (GW1). Total dissolved arsenic concentration is 10^{-5} M. Thermodynamic database is from LLNL, version 8, release 6.

We also carried out geochemical calculations with the aim of identifying the major dissolved arsenic species. The Geochemist's Workbench (Bethke, 2000) was used for this purpose. The thermodynamic database was taken from LLNL, version 8, release 6. The dissolved concentration of various As species as function of pH is given in Figure 2. The calculations are based on the groundwater composition of GW1. The total dissolved arsenic concentration is 10^{-5} M, which is close to the in-situ pore water concentration. The pH of groundwater type GW1 is 9.2. From Figure 2 we note that under alkaline conditions $HAsO_4^{2-}$ is the dominant arsenic (V) species in solution. No As(III) species exist under the observed pH and E_h (approximately +390 mV) conditions. Only at very low pH (or at sufficiently reducing conditions) does the uncharged As(III) species $As(OH)_3$ exist. The successful As removal by means of ferric oxyhydroxide may be explained on the basis of the speciation. The monocharged $H_2AsO_4^-$ is known to have a much higher affinity to the ferric oxyhydroxide surfaces than the uncharged $As(OH)_3$.

Conclusions

Batch tests were carried out to evaluate six reactive materials for their possible use in permeable reactive barriers in case dissolved uranium and arsenic are the primary groundwater contaminants. The materials tested were representative for different removal mechanisms, including surface adsorption onto ferric oxyhydroxides and reductive precipitation in the presence of zero-valent iron.

Removal of uranium (VI) was most successful by using the fine-grained ZVI. The removal efficiency was between 94 and 98% depending on groundwater composition. The ferric oxyhydroxides were very inefficient in removing uranium, with at most 9 % of the initial uranium amount being removed by adsorption.

Unlike uranium, arsenic removal was more successful when ferric oxyhydroxides was used than when ZVI was used. The percentage removed ranged between 94 and 96% for the former compared to 88 to 91% for the latter. This suggests that arsenic removal under alkaline pH is more controlled by surface adsorption than by reduction.

Permeable reactive barriers therefore may have to include different reactive media to address the specificity of chemical reactions when different contaminants are present in groundwater.

Acknowledgements

We gratefully acknowledge May Van Hees and Louis van Ravenstijn for providing technical assistance.

References

Bethke C M (2000). The Geochemist's Workbench, release 3.1, A User's Guide to Rx, Act2, Tact, React, and Gtplot, Hydrogeology Program University of Illinois.

Davis J (2001) Molecular scale observations and models of sorption reactions. In Radionuclide retention in geologic media, NEA Workshop Proceedings Oskarshamn, Sweden, 7-9 May 2001.

Gu B, Liang L, Dickley M J, Yin X, and Dai S (1998) Reductive precipitation of uranium (VI) by zero-valent iron: Environ Sci Technol 32: 3366-3373

Grenthe I, Fuger J, Konings R J M, Lemire R J, Miller A B, Nguyen-Trung C, and Wanner H (1992) Chemical thermodynamics of Uranium, Elsevier Science Publishers B.V.

Mallants D, Diels L, Bastians, Vos J, Moors, H., Wang, L., Maes, N., and Vandenhove Hildegarde (2001) Testing permeable reactive barrier media for remediation of uranium plums in groundwater: Radioactive Waste Management and Environmental Remediation – ASME 2001.

Morisson S J, Metzler D R, Carpenter C E (2001) Uranium precipitation in a permeable reactive barrier by progressive irreversible dissolution of zero-valent iron: Environ Sci Technol 35: 385-390

Passive insitu treatment techniques, wetlands

Su C, Puls R W (2001a) Arsenate and Arsenite Removal by Zerovalent Iron: Kinetics, Redox Transformation, and Implications for in Situ Groundwater Remediation: Environ Sci Technol, 35: 1487-1492

Su C, Puls R W (2001b) Arsenate and arsenite removal by zerovalent iron: Environ Sci Technol 35: 1487-1492

World Health Organisation (WHO) (1996) Guidelines for drinking water quality. Volume 2: Health criteria and other supporting information. Second Edition, Geneva, Switzerland.

Processes in passive mine water remediation with zero-valent iron and lignite as reactive materials

Christoph Klinger[1], Ulf Jenk[2], Jochen Schreyer[2]

[1]Deutsche Montan Technologie GmbH, Safe Ground Division, Am Technologie-park 1, 45307 Essen, Germany. e-mail: Klinger@dmt.de
[2]Wismut GmbH, Jagdschänkenstraße 29, 09117 Chemnitz, Germany.
e-mail: u.jenk@wismut.de, j.schreyer@wismut.de

Abstract. For in-situ decontamination of mine water a reactive material was developed by Wismut GmbH and Deutsche Montan Technologie GmbH. In different scaled tests mixtures of iron chips and lignite proved most effective in terms of all contaminants examined (heavy metals, radionuclides and nitroaromatics). Studies showed different pattern of contaminant accumulation depending on the dominant fixation processes.

Introduction

Within the framework of the flooding concept for the former uranium mine of Königstein (Saxony, Germany), various supportive measures were considered with the aim of attaining the clean-up objective (long-term stable conservation of the mine) more quickly (Schreyer 1996). To minimise the duration and hence the cost of flooding the mine it is generally possible to reduce the mobile pollutant potential by introducing reactive materials. Due to chemical underground leaching using a circulating sulphuric acid solution as main mining method since 1984, mine water contains high concentrations of uranium, radium, acid, sulphate, iron, aluminium, nitroaromatics and various heavy metals.

Reactive materials and test set-up

For acid flood waters as in the Königstein mine (pH 2 - 3) zero-valent iron is effective in hydrochemical terms by leading to an sustainable change in milieu in the water towards rising pH values and falling Eh values. These milieu conditions facilitate or support the fixation of contaminants especially in combination with additives.

In connection with an experimental flooding several test series were conducted by WISMUT GmbH and Deutsche Montan Technologie GmbH with iron and other materials (fine grained lignite ash and limestone) and, with a view of examining combined effects, mixtures of iron with additives (hard coal and lignite, organic residues, barite-containing material) respectively (Klinger et al. 2000).

Performance of these materials was studied in

• underground column tests,
• underground large-scale column test,
• laboratory tests.

The scale of these experimental investigations ranged from laboratory (2 kg) and underground column tests (20 kg and 200 kg) up to a flooded heading (270 t iron only). Percolation direction was varied from vertical to horizontal and water solid ratios up to 120 L/kg were obtained for conditions as close as possible to reality. There were no significant changes in barrier function depending on test set-up or scale.

Contaminant fixation

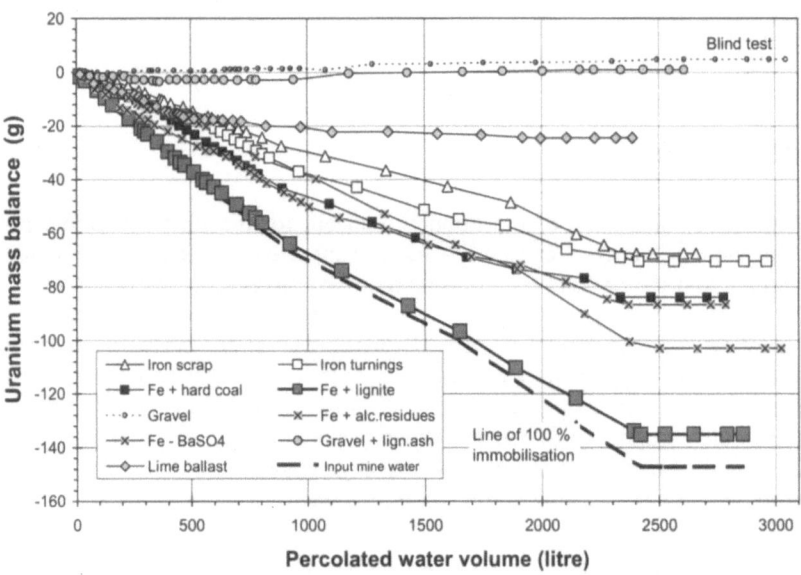

Fig. 1. Uranium mass balance of input and output of vertical columns filled with different reactive materials.

Passive insitu treatment techniques, wetlands

Contaminant input concentrations varied in the different test series representing different flooding stages of a mine: uranium (12-68 ppm), radium (2.5-12 Bq/L), acid (pH 2-3), sulphate (1700-7000 ppm), iron (290-2200 ppm), aluminium (66-270 ppm), nitroaromatics (2.3-60 ppb, predominantly nitrobenzene, nitrotoluenes, dinitrotoluene-2,4) and heavy metals (As 0.2-3.5 ppm, Zn 22-110 ppm).

The mixture consisting of iron and lignite can immobilise practically all potential pollutants. Even the relatively mobile elements like U (Fig.1), Co, Zn, Ni and the nitroaromatics are immobilised substantially better with this mixture than with the other additives examined (e.g. hard coal, lime, fermentation residues). For percolation periods of more than one year a sufficient efficiency without re-release could be verified.

Therefore this reactive material was selected for further tests of hydraulic and geochemical behaviour with the aim of optimisation of material properties. These different test series show the following processes in the reactive material which can potentially cause contaminant fixation:

- acid buffering,
- highly reducing environment,
- degradation of organic substance in combination with sulphate reduction,
- precipitation of hydroxides with potential coprecipitation and sorption,
- sorption on the lignite (can only be assumed).

Analysis of chemical composition of precipitation products and of milieu sensitivity of contaminant mobility should enable identification of the predominant element specific fixation processes.

Iron Corrosion

Reaction of the acid mine water with metallic iron causes generation of hydrogen and buffering of the acid: $Fe + 2 H^+ \Rightarrow Fe^{2+} + H_2 (g)$

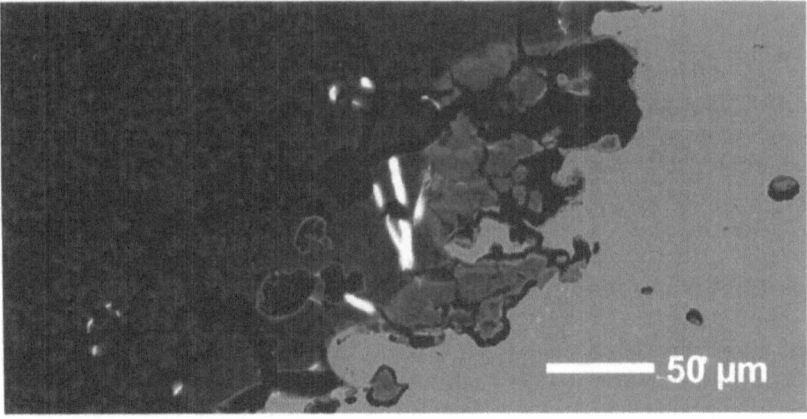

Fig. 2. Scanning electron microscopy of an iron chip thin section with pH 2 mine water corrosion (reactive material consisting of iron chips and hard coal).

Passive insitu treatment techniques, wetlands

Fig.2 shows an iron chip surface in a thin section with solution structures and precipitation products. This iron material was in contact with pH 2 mine water for more than one year.

The generation of the H_2 gas prevents coating if the iron chip surface by precipitation products. This allows high and long term reactivity in spite of a comparatively small surface of this material. The H_2 gas production depends on the acid content of the mine water and could be minimised at pH values of pH 3.2 corresponding with the expected future mine water quality. Then, under atmospheric pressure the percolate is saturated with dissolved hydrogen gas which can completely be discharged with the percolate without generation of a gas phase.

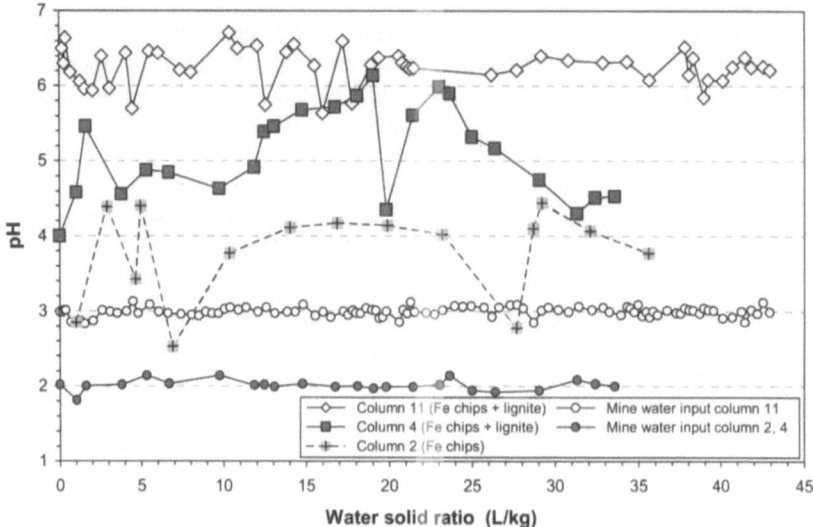

Fig. 3. Buffering of acid mine waters by pure metallic iron and the reactive material mixture with lignite.

The pH buffering level depends on the acid content of the mine water and on the reactive material. Fig.3 shows that pure metallic iron buffers pH 2 mine water up to a pH value of 4. This effect can be enforced by addition of lignite. This reactive mixture produces stable pH values of > 6 for mine water with an original pH 3 level.

Precipitation by buffering

The precipitation products on the surface of the iron chips contain (beside iron residues) predominantly aluminium (maximum 65%), sulphur (maximum 11%) and uranium (maximum 11%). Especially the aluminium phases can be assumed to precipitate because of the buffering process as hydroxides. Fig.4 shows a corrosion structure on an iron chip surface with points of chemical analysis.

Chemical analysis of the SEM thin sections and sequential extractions of the precipitation products show no clear correlation between aluminium and uranium (Table 1). So coprecipitation and sorption on aluminium phases are subordinate processes for uranium fixation.

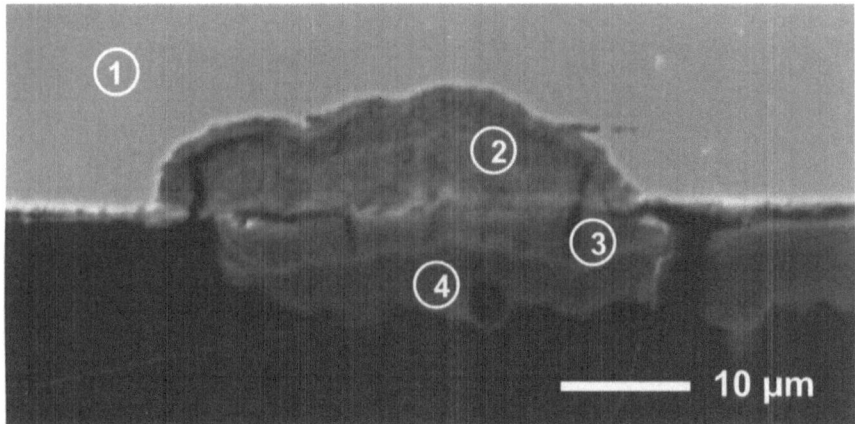

Fig. 4. Scanning electron microscopy of an iron chip thin section with corrosion by contact with pH 2 mine water with points of EDX analysis (reactive mixture consisting of iron chips and hard coal).

Table 1. EDX spot analysis of the SEM thin section in Fig. 4.

Point		① metallic iron	② corroded iron	③ layer 1	④ layer 2
Fe	%	97,5	85,4	30,2	25,1
Mn	%	1,8	3,0	-	0,9
Si	%	0,4	0,4	10,7	7,2
Cr	%	0,3	2,3	0,2	0,4
S	%	-	3,2	6,2	6,7
Cu	%	-	2,2	0,7	1,4
Al	%	-	1,7	48,6	50,6
Ni	%	-	0,7	-	-
U	%	-	-	3,1	3,2

Reduction

Due to the corrosion reaction of the metallic iron the redox environment is shifted towards highly reducing values. Under these conditions precipitation can be attributed to reduction of in this regard sensitive contaminants. For uranium the formation of uraninite (UO_2) by reduction of U^{+6} to U^{+4} can be expected. Similar redox sensitive components in mine water are arsenic and chromium for example. Scanning electron microscope investigations show the precipitation of uranium rich phases on the surface of the iron chip as reactive material without additives

(Fig.5, Table 2). Presumably because of the low arsenic and chromium concentrations in the mine water these elements could not be verified analytically. The low aluminium content results from an insufficient buffering of this material (pH 4, see Fig.3).

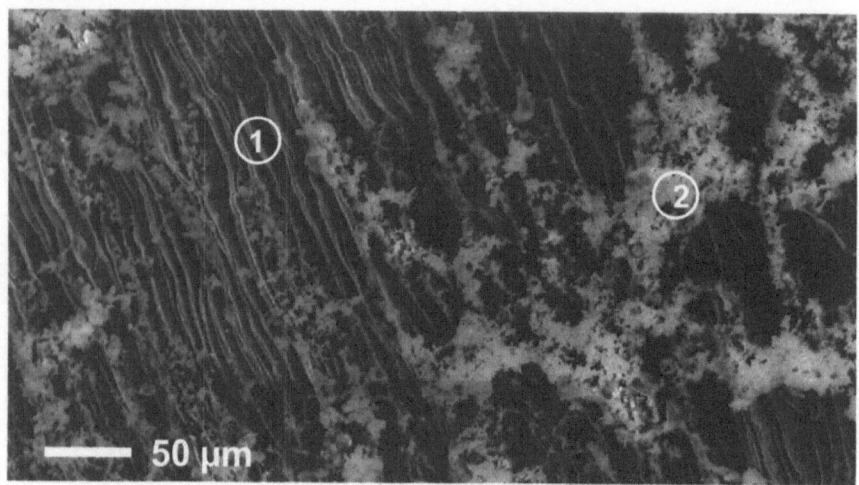

Fig. 5. Scanning electron microscopy of the surface of an iron chip with EDX analysis of the metallic iron and precipitation products (metallic iron as reactive material in contact with pH 2 mine water).

Table 2. EDX spot analysis of the Scanning electron microscopy in Fig.5 (metallic iron) and of a sample of the reactive mixture of metallic iron with lignite as additive.

Sample		Fe Fig.5 ① metallic iron	Fe Fig.5 ② precipitation products	Fe+lignite coating 1	Fe+lignite coating 2
Fe	%	96,2	26,1	64,0	40,5
Mn	%	1,9	0,5	0,6	0,9
Si	%	0,7	2,5	1,0	1,5
Cr	%	0,7	0,5	0,2	0,6
S	%	-	1,0	10,6	8,7
Cu	%	-	0,7	2,4	9,9
Al	%	0,6	6,7	0,8	0,4
Ni	%	-	0,5	0,8	-
U	%	-	56,2	2,9	33,0
Zn	%	-	-	16,7	-

Precipitation products formed in the reducing environment solely caused by iron dissolution (as in Fig.5) do not contain relevant amounts of zinc and copper. The elements as well as cobalt and nickel are not redox sensitive and are mobile as two-valent cations over the entire Eh range and can be fixed under reducing hydrochemical conditions only with a suitable partner for the precipitation reaction.

Sulphate reduction

Precipitation products formed in mixtures of iron chips with lignite are much more heterogeneous and cover nearly the whole iron surface. Associated with increased sulphur concentrations these coatings contain zinc and copper (Table 2). This correlation favours a precipitation as metal sulphides.

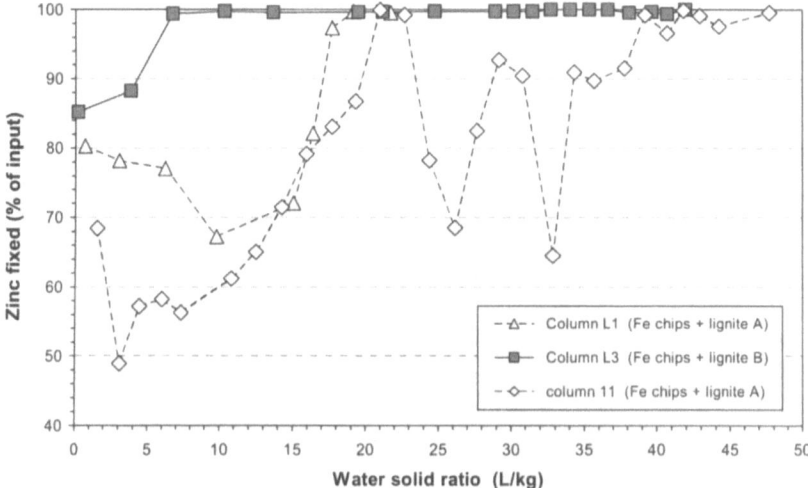

Fig. 6. Fixation rates of zinc in reactive mixtures of metallic iron + lignite (column experiments with pH 3 mine water).

Therefore sulphate reducing bacteria are able to degrade organic substance in the lignite. This bacterial activity causes sufficient sulphate reduction for precipitation of negligible soluble sulphides and nearly complete immobilisation of these otherwise very mobile heavy metals. Eh measurements show that this process is very sensitive to oxygen influx and demands a certain start-up period. Instabilities of the system result in immediate decrease of the fixation rate as shown for zinc in Fig. 6. Experimental results implicate an influence of the lignite type on bacterial availability.

Sorption and decomposition

There are no definite clues for sorption and decomposition of contaminants from the analysis of the reactive materials and the precipitation products. Nevertheless, sorption processes are very likely to take place in the lignite and amorphous aluminium hydroxides.

Very high efficiency for elimination of nitroaromatics has been verified for reactive mixtures containing metallic iron and organic components. There was no decrease in barrier function for percolation periods of more than one year. Because of only slightly minor efficiency of the iron component on its own a combination of inorganic decomposition and sorption / biologic decomposition of nitroaromatics can be assumed.

Conclusion

Mixtures of Fe chips and lignite are capable of efficiently cleaning acid and contaminant-containing mine water in a reactive barrier. Suitable basic conditions ensured all contaminants examined could be fixed almost completely by combination of several processes (Fig.7). This offers different technological applications in situ or in fixed-bed reactors.

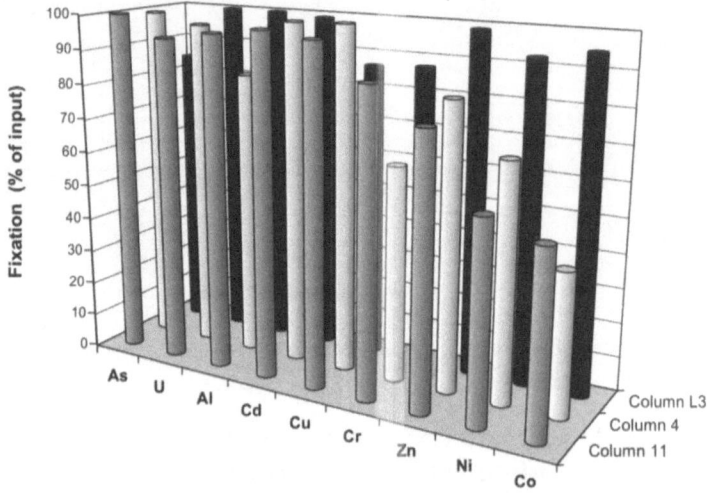

Fig. 7. Efficiency (fixed portions of the input contaminant loads) of the iron chips – lignite mixtures for selected contaminants in different column tests (water solid ratio 22 L/kg).

References

Schreyer J (1996) Sanierung von Bergwerken durch gesteuerte Flutung - Entwicklung und Einsatz eines neuartigen Verfahrens am Beispiel der Grube Königstein. Geowissenschaften, 14 (11): 452-475.

Klinger C, Jenk U, Schreyer J (2000) Investigation of efficacy of reactive materials for reduction of pollutants in acid mine water in the former uranium mine of Königstein (Germany).- In: Różkowski A & Rogoż M [eds.] Proceedings 7th International Mine Water Association Congress, Mine Water and the Environment, Katowice-Ustron, Poland, Sept. 11 - 15: 292-298.

Mechanism of Uranium Fixation by Zero Valent Iron: The Importance of Co-precipitation

C. Noubactep [1,3], G. Meinrath [1,2], P. Volke [1], H.-J. Peter [1], P. Dietrich [1], B. Merkel [1]

[1] Departement for Geology, TU BAF, Gustav Zeuner Straße 12, 09596 Freiberg
[2] RER Consultants, Schießstattweg 3a, D-94032 Passau, Germany
[3] Department for Geology, FSU Jena; Burgweg 11; D-07749 Germany

Abstract. The co-precipitation of U (VI) with iron corrosion products from aqueous solutions by zero valent iron was investigated. The evidence of co-precipitation was demonstrated by conducting experiments with well characterized scrap iron, pyrite and a mixture of both materials with experimental durations of up to four months. Results indicate that under anoxic conditions only less than one tenth of the immobilized U(VI) was associated with the surface of scrap iron, whereas the remaining amount is entrapped in aging corrosion products.

Introduction

Uranium mining activities are sources of contamination for surface and ground waters of worldwide concern (e.g. Meinrath et al. 2002, Morrison et al. 2001). Efficient, applicable and affordable techniques are necessary to mitigate the health risk by eliminating or reducing removal of uranium from the mine waters and contaminated ground waters.

Zero valent iron (ZVI) has been discussed in the literature as a uranium-removing reagent in permeable reactive walls. To be effective in the long term, any remediation technique for uranium must target both mobile aqueous U(VI)-species and U(VI)-precipitates that may be long term sources. Therefore, the remediation with ZVI that possibly reduces mobile U(VI) aqueous species to less soluble U(IV) precipitates is very promising. Furthermore ZVI can maintain reducing conditions in the subsurface, under which beside Fe^0 other electron donors (e.g. organics) may also contribute to the U(VI) reduction.

The results of previous investigations on the U removal mechanism by ZVI are not univocal. Reductive precipitation and adsorption onto iron corrosion products have been shown to govern the U uptake (Cantrell et al. 1995, Farrell et al. 1999,

Fiedor et al. 1998, Qiu et al. 2000). It is believed that under anoxic conditions the U removal will mostly occur through a slow reductive precipitation, whereas the removal will occur through rapid adsorption onto iron corrosion products under oxic conditions. Investigations that came to the conclusion that reductive precipitation is the principal removal mechanism have been conducted under conditions that are very far from the nature (Gu et al. 1998, Abdelouas et al. 1999); i.e. the reaction vessels were shaken for several days or weeks, the initial U solution were over-saturated with respect to the solubility of schoepite, the solid-to-solution ratios of ZVI were very large, up to 200 g/L (Noubactep et al. 2001a and 2002).

Some evidence for co-precipitation in U removal from aqueous solution by ZVI has been shown recently by means of controlling the availability and the reactivity of corrosion products (Noubactep et al. 2001a). Controlling the reactivity of ZVI was achieved by using a pyrite mineral. This mineral is able to lower the pH and to reduce uranium sorption onto corrosion products. Supposedly, pyrite did not exhibit any fixation capacity itself. However, pyrite and other sulfide minerals have been discussed in the literature as potential reductants of U in low-temperature geo-chemical systems ($< 50°C$) (Liger et al. 1999 and references therein). Wersin et al. (1994) have indicated that the reductant for the uranium reduction "is Fe(II) rather than S(-II)". No specific U(IV) minerals could be identified; the conclusions are based on X-ray photoelectron spectroscopy (XPS) observations. This technique however detects only dissolved species (U(IV) or/and U(VI)).

The present study aims on a better characterization of the primary process responsible for the uranium removal from aqueous solution by ZVI. Particular attention was directed at determining the extent to which uranium is associated to the added materials (ZVI and FeS_2) and in-situ generated corrosion products under varying solution chemistry (essentially pH value, iron concentration and speciation).

Theoretical Background

Uranium uptake by ZVI is supposedly based on the electrochemical corrosion of iron resulting in reductive precipitation of U(VI) according to Eq.1 in table 1. This reaction is not the most favorable under natural geochemical conditions (Noubactep et al. 2001b). Competing reactions by local sediment constituents such as MnO_2 may oxidize iron to various hydrous Fe(II) phases, and further to various secondary minerals; e.g. $Fe(OH)_3$, Fe_3O_4, Fe_2O_3, FeOOH (e.g. Ritter et al. 2002).

If, e.g., dissolved Fe(III) is present, UO_2 (resulted from Eq.1) can be re-oxidized according to Eq.2; yielding to increased dissolved Fe(II) and U(VI) concentrations. In the presence of pyrite, there will be a competition for Fe(III) (cf. Eq. 3). Hence, Fe(III) is not available to oxidize UO_2, and U(VI) concentration will remain low. On the other hand, pyrite can reduce U(VI) yielding to UO_2 precipitation according to Eq. 4. Furthermore if any source of Fe(II) exists, the acidification possibility under oxic conditions is increased according to Eq. 5 (Bain et al. 2001).

Table 1. Some relevant reactions for the uranium behavior under experimental conditions. The log K values are from Bain et al. (2001).

Reaction equation			log K	Eq.
$UO_2^{2+} + Fe^\circ$	$\Leftrightarrow$	$UO_{2\,(s)} + Fe^{2+}$	-	(1)
$UO_{2\,(s)} + 2\,Fe^{3+}$	$\Leftrightarrow$	$UO_2^{2+} + 2\,Fe^{2+}$	11.96	(2)
$FeS_2 + 14\,Fe^{3+} + 8\,H_2O$	$\Leftrightarrow$	$15\,Fe^{2+} + SO_4^{2-} + 16\,H^+$	16.78	(3)
$FeS_2 + 7\,UO_2^{2+} + 8\,H_2O$	$\Leftrightarrow$	$7\,UO_2 + 2\,SO_4^{2-} + 16\,H^+$	-20.91	(4)
$2\,Fe^{2+} + \frac{1}{2}\,O_2 + 5\,H_2O$	$\Leftrightarrow$	$2\,Fe(OH)_3 + 4\,H^+$	7.20	(5)

In the presence of ZVI (covered by corrosion products) and pyrite, all the above mentioned reactions (Eq. 1 to 5) are possible. The U(VI) uptake should be governed principally by reductive precipitation. If the reaction vessel is closed, the acidification reaction due to Eq. (5) or the pyrite oxidation through air oxygen will be limited. Then, the removal of U(VI) from the aqueous solution can be due to reductive precipitation by ZVI (Eq. 1) and/or FeS_2 (Eq. 4); sorption on the surface of ZVI, FeS_2 and onto iron corrosion products (iron oxides).

Reductive precipitation trough ZVI will be more favorable when the surface of the material is not covered by corrosion products (especially around pH 4) and the sorption onto corrosion products (iron oxides) will occur favorably at pH > 5 (e.g. Farell et al. 1999). Thus, combining ZVI and FeS_2 into closed vessels is a suitable way to investigate the mechanism of U(VI) uptake by both materials; in particular to understand the mechanism of U(VI) uptake by ZVI. It is expected that various experimental durations will yield various final pH values permitting the characterization of the influence of corrosions products on the removal process.

Experimental Section

Batch experiments without shaking were conducted. The batches consisted in constant amounts of ZVI and a pyrite mineral (FeS_2), respectively. Equilibration times varied from two weeks to four months. A further series of experiment with a mixture of both materials were conducted. Thus, the extend of U fixation by ZVI, FeS_2 and in situ generated iron corrosion products was characterized.

Initial uranium concentration was 20 mg/L (0.084 mM) with a solid:solution ratio of 15 g/L for scrap iron and 0 to 25 g/L for the additives, respectively. The ZVI is a scrap iron from MAZ (Metallaufbereitung Zwickau, Co.) termed internally "Sorte 69". Its elemental composition is given as C: 3.52%; Si: 2.12%; Mn: 0.93%; Cr: 0.66%. The material was fractionated by sieving; the fraction 1.6 - 2.5 mm has been used. The sieved ZVI was used without any further pretreatment. The pyrite mineral was crushed and sieved. The fraction 0.315 to 0.63 mm is used.

Elemental composition is: Fe: 40%; S: 31.4%; Si: 6.7%; Cl: 0.5%; C: 0.15% and Ca <0.01%.

Unless indicated otherwise, 0.3 g of ZVI and 0.5 g of FeS_2 were allowed to react in sealed sample tubes containing 20.0 mL of a uranium solution (20 mg/L or 0.084 mM) at laboratory temperature (about 20° C). The tubes (16 ml graded) were filled to the total volume to reduce the head space in the reaction vessels. All experiments were conducted with the tap water of the city of Freiberg (Saxonia, Germany) of composition (in mg/L) Cl^-: 7.5; NO_3^-: 17.5; SO_4^{2-}: 42; HCO_3^-: 42; Na^+: 7.1; K^+: 1.6; Mg^{2+}: 6.8 and Ca^{2+}: 37.1 (resulting HCO_3^- to U molar ratio: ~8). Initial pH was ~7.2. After equilibration, the supernatant solutions were separated for uranium and iron analysis, pH and E_H measurements.

The contact vessels were turned over-head at the beginning of the experiment and allowed to equilibrate in darkness to avoid photochemical side reactions. The uranium solution were prepared from $UO_2(NO_3)_2 \cdot 6\ H_2O$ in tap water. The samples were filtered through filter paper. Analysis for uranium was performed after reduction to U(IV) with the Asernazo III method (Meinrath et al. 1999 and references therein). Uranium concentrations were determined by a HACH UV-Vis spectrophotometer at a wavelength of 665 nm using cuvettes with 1 cm light path. All chemicals were analytical grade. The pH value and the redox potential were measured by combination glass electrodes (WTW Co., Germany). Electrodes were calibrated with nine standards following a multi-point calibration protocol (Meinrath and Spitzer 2000) in agreement with the new IUPAC recommendation (Buck et al. 2001). Redox potentials are reported relative to the Standard Hydrogen Electrode (SHE). Each experiment was performed in triplicate and averaged results are presented.

Results and Discussion

After the determination of the residual uranium concentration (C) the corresponding total fixation was calculated according to the following equation:

$$P_{tot} = [1 - (C/C_0)] \times 100\%$$

where C_0 is the initial concentration of uranium in solution. To characterize the U(VI) uptake from aqueous solution while taking individual properties of the iron materials into account, three different experiments have been performed over a duration up to 4 months with 15 g/L ZVI and 25 g/L FeS_2: I) ZVI alone, II) FeS_2 alone and III) ZVI + FeS_2 (system I, II and III).

Figure 1 summarizes the results of uranium fixation and table 2 gives the variation of the pH value with the experimental duration in the three systems. Fig. 1 shows the best fixation rate being achieved when ZVI is present alone (> 80%). The efficiency is smallest when FeS_2 is present alone (< 20%). These observations suggest either that the fixation capacity of pyrite for U(VI) is very limited (sorption) or the kinetic of the reductive precipitation by FeS_2 is very slow. The second hypothesis is less probable since the initial fixation rate of 21% (after 14 days, pH 3.5) further decreases to 16% at the end of the experiment (120 days, pH 3.4). Thus the uranium fixation by pyrite for $3.4 < \text{pH} < 3.6$ (table 2) occurs through adsorption. As concerning the system with ZVI alone, it has been shown that the co-precipitation of sorbed U(VI) with corrosions products is the main mechanism of U(VI) removal in the neutral pH-range (Noubactep et al. 2001a).

Table 2. Variations of the pH value with the time in the three systems (initial value: pH ~7.2)

System I: ZVI		System II: FeS$_2$		SystIII: (ZVI + FeS$_2$)	
t (days)	pH	t (days)	pH	t (days)	pH
13	7.6$_2$	15	3.5$_3$	15	4.1$_5$
23	7.5$_5$	25	3.4$_9$	25	4.3$_2$
41	7.6$_3$	43	3.4$_0$	43	3.9$_5$
53	7.5$_8$	55	3.3$_5$	55	3.9$_4$
70	7.5$_8$	72	3.4$_2$	72	4.1$_2$
92	7.5$_9$	94	3.5$_6$	94	4.4$_1$
106	7.6$_2$	108	3.5$_1$	108	4.4$_5$
117	7.5$_0$	119	3.3$_7$	119	4.4$_9$

Investigation of the behaviour of system III (ZVI + FeS$_2$) shows a fixation rate increasing considerably in a very close pH range: from 18% at pH ~3.9 (day 43) to 94% at pH ~4.4 (day 94, table 2 and Fig. 1). It should be pointed out that if reductive precipitation were the dominant removal mechanism, the reduction reaction would be more efficient and rapid around pH 4, where the iron corrosion mostly occurs with or without H$_2$-production depending on the availability of oxygen and the corrosion products mainly remain in the bulk solution, keeping the metal surface free for further reaction. Figure 1 shows the fixation rate for this system first decreasing to a minimum (18%) and subsequent progressive increase to more than 90% after three months. An interpretation of this observation will be given later. The evolution of the pH of this system is depicted on the experimental points. The fact that all final pH values remain below 5 (pH$_{max}$ = ~4.5) suggests a slow reductive precipitation to be responsible for uranium removal in system III. To understand the evolution of this system, it is important to consider also the behaviour of the iron concentration. In system I the final pH was almost constant to an average value of ~7.6, the same observation was made for the system III (final pH: ~3.5).

Beside the pH values, the iron and uranium concentrations, the E$_H$-values and the iron speciation were measured in system III, where variations were expected owing to the evolved possible reactions (Eq.1 to 5). Table 3 summarizes the results. Because of the limited volume of samples the solution parameters (pH and E$_H$) were measured once for each triplicate and after 24 hours.

Passive insitu treatment techniques, wetlands

Table 3 shows a decreasing E_H value with increasing experimental duration. This observation is consistent with the fact both iron corrosion and pyrite oxidation consumes oxygen and care for reducing conditions. This is also confirmed with the predominance of Fe(II) (> 50%) for experimental duration < 100 days, although air oxygen would have oxidized a considerable part since the experiments were conducted under laboratory conditions. Figure 2 depicts the variation of iron concentrations in the three systems.

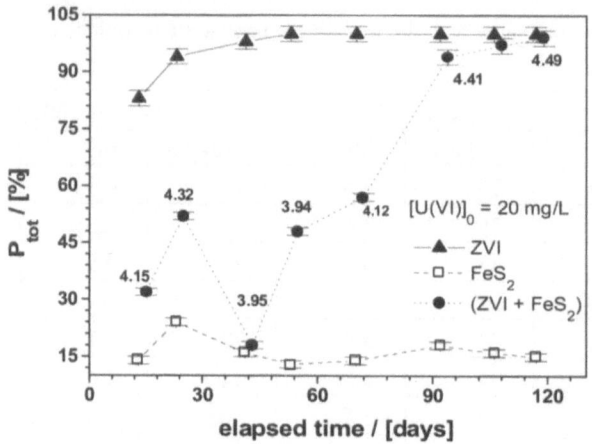

Fig. 1. Uranium fixation by zero valent iron (ZVI: 15 g/L), a pyrite mineral (FeS$_2$: 25 g/L) and the mixture of both materials (ZVI + FeS$_2$) as a function of the time. The reported numbers on the plot (ZVI + FeS$_2$) are the corresponding final pH values. The lines are given to facilitate visualization.

Table 3. Variations of the uranium fixation rate and solution parameters with 15 g/L ZVI and 25 g/L FeS$_2$ for different experimental durations. $P_{Fe(II)}$ is the percentage of Fe(II) in the bulk solution and $P_{uranium}$ represent the total fixed amount of uranium.

Time (days)	pH	$E_{H(SHE)}$ (mV)	[Fe(II)] (mg/L)	[Fe]tot (mg/L)	$P_{Fe(II)}$ (%)	$P_{uranium}$ (%)
15	4.15	85	35.5	58.5	60.7	32
25	4.32	51	38.5	65.0	59.2	52
43	3.95	88	50.5	96.0	52.6	18
55	3.94	77	58.5	104.5	56.0	48
72	4.12	62	53.5	100.5	53.2	57
94	4.41	11	50.0	93.5	53.5	94
108	4.45	8	40.0	92.0	43.5	97
119	4.49	21	39.5	90.0	43.9	99

In system I (pH ~7.5), the iron concentrations remain below 1 mg/L. In system II & III the iron concentrations were essentially higher (up to 100 mg/L or 1.9 mM).

For experimental duration >60 days a concentration decrease was observed in system III. This iron concentration decrease coincides with the decrease of the uranium concentration between pH ~3.9 (18%) and pH ~4.4 (94 %). Co-precipitation is confirmed. Thus the uranium uptake accompanies the precipitation of iron oxides, the sorbed U(VI) is entrapped in the mass of aging corrosion products and is not available for desorption with commonly used carbonated reagents: CO_3^{2-}; HCO_3^- (Gu et al. 1998, Liger et al. 1999). The exact precipitation process is not known, probably several parallel reactions occur, yielding to $Fe(OH)_3$, $FeOOH$, Fe_3O_4 (Ritter et al. 2002).

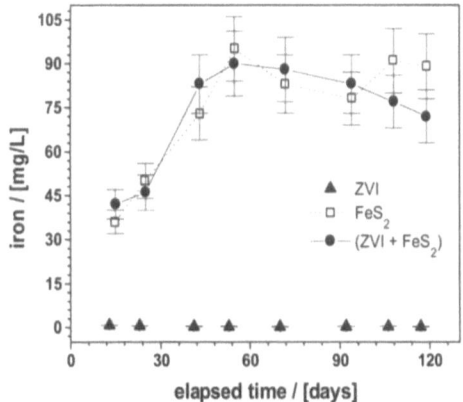

Fig. 2. Variation of the iron concentration in the experimental systems with the time. The represented lines are not fitting functions, they just joint the point to facilitate visualization. The big errors in system II & III resulted from the strong dilution (1/100).

Hence the commonly reported reduction of U(VI) by pyrite and other Fe(II) bearing materials (Charlet et al. 1998, Liger et al. 1999, Wersin et al. 1994) is probably the results of a co-precipitation of U(VI) sorbed onto newly formed Fe-oxides. Noubactep et al. (2001b) have shown that the reversibility of adsorption as measured by desorption with carbonate solutions depends on the age of corrosion products. Freshly formed corrosion products sorb U(VI) by incorporating it in their structure while aging. On the contrary U(VI) sorbed onto surfaces will be readily released in carbonated solutions. The reported U(VI) reduction of by pyrite and other Fe(II) bearing materials will occur by the same mechanism when Fe(II) species are oxidized. Figure 3 compares the total fixation rate for uranium as a function of the final pH value for the three systems. It can be seen that:

- in system I (ZVI) the pH increased from 7.2 to ca. 7.6 and $P_{tot} > 94$ %;
- in system II (FeS_2) the pH decreased and remained < 4 and $P_{tot} < 33\%$;
- in system III (ZVI + FeS_2) the pH evolution (3.95 < pH < 4.49) wasn't monotone and P_{tot} varied considerably; from 18 to 99%.

It is important to note that fixation rate evolution as function the pH value in system III is practically a straight line parallel to the y-axis, indicating that a chemical process at nearly constant pH accompanies the uranium fixation. As discussed above the process is iron oxide precipitation. To gain insight in this phenomenon and better understand the evolution of the fixation curve for system I (Fig. 1) other experiments were conducted with 15 g/L ZVI and varying amount of FeS₂ for 2 and 4 weeks.

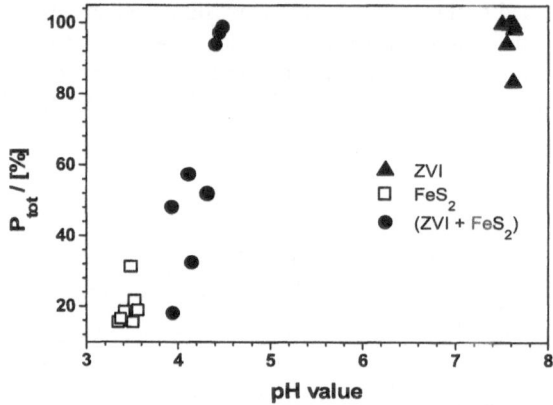

Fig. 3. Comparison of the total fixation rate as function of the final pH value for the three experimental systems.

The results show a decrease of the fixation rate (Noubactep 2002) with the amount of pyrite (decreasing pH). For low pyrite doses (< 10 g/L), the fixation rate increases for a longer reaction duration (4 weeks). This is attributed to the predominance of iron corrosion on the pyrite dissolution. The final pH values were > 4.6. For larger pyrite doses (> 10 g/L) on the contrary, the iron corrosion was not able to consume the acidity produced by the pyrite dissolution and the final pH value was < 4.20. Adsorbed uranium (after two weeks) was partly released into the solution. Thus an explanation of the behaviour of the curve for system III in figure 1 can be given. It is obvious that U(VI) first adsorbs onto ZVI, FeS₂ and iron corrosion products. During the first stage of the experiment, pyrite dissolution predominates over iron corrosion and determines the pH of the system. At this stage pH decreases, the iron concentration increases and the fixation rate decreases. Between day 25 and day 43 the minimum is attained due to a lack of oxygen for further pyrite oxidation. The evolution of the system is determined by anaerobic iron corrosion with production of H_2 (gas bubbles were observed in the reaction tubes), the pH increased progressively and eventually reached a value of ~4.5 (120 days). Thus the primary mechanism of uranium fixation by zero valent iron even under anoxic conditions is the co-precipitation with corrosion products.

As discussed above the fixation of uranium by ZVI is strongly dependent on the pH value. For example, the total amount of uranium fixed in the experiment with (ZVI + FeS₂) and FeS₂ alone are identical after 43 days. This observation suggests that the uranium fixation by ZVI itself in this pH range (3.40 - 3.95) is negligible. To better understand this observation, another experience was conducted with the same amount of FeS₂ (25 g/L) and 0, 3, 8 and 15 g/L ZVI for 1 month. The corresponding fixation rates were: 18, 20, 20 and 28 % respectively, and the pH value varies from 3.6 to 4.2. The maximal efficiency difference was 8% when the ZVI dose was quintupled (3 to 15 g/L). These Results confirmed the hypothesis that uranium reduction didn't play any important role in the mechanism of the U(VI) fixation by ZVI. On the other hand the efficiency difference between 0 and 15 g/L ZVI was 10 %, indicating that maximal 10 % of the fixed uranium is associated with the ZVI surface (not necessarily reduced).

Conclusions

U removal from the aqueous solution by ZVI in the pH range 3.8 to 7.6 is mostly due to the co-precipitation of adsorbed U(VI) with aging iron corrosion products. This mechanism is predominant both under oxic (system I: ZVI alone) and anoxic conditions (system III: ZVI and FeS_2).

Pyrite and other iron (II) bearing minerals also fix U(VI) by co-precipitation with newly formed iron (II, III) oxides, even under anoxic conditions. Thus Fe(III) and Fe(II, III) oxides are formed. Co-precipitates with U(VI) enclosing uranium into their matrix, making them unavailable for any resolubilization so far these iron oxides remain stable. In-situ iron oxide barriers will evidently have a limited remediation capacity. In contrast, the use of ZVI has the advantage of continuous production of fresh and very active corrosion products that may incorporate U(VI) into their structure while aging. However both the limited volume of pore spaces in the reactive barrier (for further corrosion products, whose volumes are at least 2.3 times larger than that of Fe in the ZVI-material) and the potential inhibition of the electrochemical dissolution of ZVI through corrosion products have been recognized but not yet solved (Noubactep 2002).

References

Abdelouas A.; Lutze W.; Nutall H.E., and Gong W. (1999): Réduction de l'U(VI) par le fer métallique: application à la dépollution des eaux. C. R. Acad. Sci. Paris, Sciences de la terre et des planètes / Earth & Planetary Sciences. 328, 315-319.

Bain J.G., Mayer K.U., Blowes D.W., Frind E. O., Molson J.W.H., Kahnt R. and Jenk U. (2001): Modelling the closure-related geochemical evolution of groundwater at a former uranium mine. J. Cont. Hydrol., 52, 109-135.

Buck R.P, Rondinini S., Covington A.K., Baucke F.G.K., Brett C.M.A, Camoes M.F., Milton M.J.T., Mussini T., Naumann R., Pratt K.W., Spitzer P., and Wilson G.S. (Draft, 6 July 2001): The measurement of pH - definition, standards and procedures. Report of the working party on pH (34 pages).

Cantrell K.J., Kaplan D.I., and Wietsma T.W. (1995): Zero-valent iron for the in situ remediation of selected metals in groundwater. Jour. of Hazard. Mat. 42, 201-212.

Charlet L., Liger E., and Gerasimo P. (1998): Decontamination of TCE- and U-rich waters by granular iron: role of sorbed Fe (II). Journal Environ. Eng.,124 (1) , 25-30.

Farrell J., Bostick W.D, Jarabeck R.J., and Fiedor J.N. (1999): Uranium removal from ground water using zero valent iron media. Ground water 34, 618-624.

Passive insitu treatment techniques, wetlands

Fiedor J.N., Bostick W. D. Jarabek R.J and Farrel J. (1998): Understanding the mechanism, of uranium removal from groundwater by zero-valent iron using X-ray photoelectron spectroscopy. Environ. Sci. Technol., 32, 1466-1473.

Gu B., Liang Liyuan, Dickey M.J., Yin X. and Dai S. (1998): Reductive precipitation of uranium (VI) by zero-valent iron. Environ. Sci. Technol., 32, 3366-3373.

Liger E., Charlet L. & Van Cappellen (1999): Surface catalysis of uranium(VI) reduction by iron (II). Geochim. Cosmochim. Acta 63, 2939-2955.

Meinrath A., Schneider P. and Meinrath G. (2002): Uranium ores and depleted uranium in the environment - With a reference to the biosphere from the Erzgebirge/Sax´chsen, Germany. J. Environ. Radioactivity (in press)

Meinrath G. and Spitzer P. (2000): Uncertainties in determination of pH. Mikrochim. Acta 135: 155 - 168

Meinrath G., Volke P., Helling C., Dudel, E.G. and Merkel P. (1999): Determination and interpretation of environmental water samples contaminated by uranium mining activities. Fresenius J Anal Chem. 364: 191 - 202

Morrison J.S, Metzler, R. D. and Carpenter, E. C. (2001): Uranium precipitation in a permeable reactive barrier by progressive irreversible dissolution of zerovalent iron. Environ. Sci. & Technol., 35, 385-390.

Noubactep C. (2002): Untersuchungen zur passiven in-situ-Immobilisierung von U(VI) aus Wasser. Ph.D. Thesis, TU Bergakademie Freiberg.

Noubactep C. Meinrath G., Volke P.; Peter H.-J., Dietrich P. and B. Merkel (2001a): Understanding the mechanism of the uranium mitigation by zero valent iron in effluents. . Wiss. Mitt. Institut für Geologie der TU Bergakademie Freiberg, Band 18, pp.37-44. ISSN 1433-1284.

Noubactep C., Meinrath G., Dietrich P. and B. Merkel B (2002) Mitigating of uranium in ground water: prospects and limitations; Environ. Sci. & Technol. (submitted)

Noubactep C., P. Volke; G. Meinrath and B. Merkel (2001b): Mitigation of uranium in effluents by zero valent iron: the role of iron corrosion products; paper presented at the ICEM'01 Conference, September 30 - October 4, 2001, in Brugge Belgium. published on CD-ROM, December, 2001

Qiu S.R., Lai H.-F., Roberson M.J., Hunt M.L., Amrhein C., Giancarlo L.C., Flynn G.W., and Yarmoff (2000): Removal of contaminants from aqueous solution by reaction with iron surfaces. Langmuir 16, 2230-2236.

Ritter K, Odziemkowski and Gillham R.W. (2002): An in situ study of the role of surface films on granular iron in the permeable iron wall technology. J. Cont. Hydrol., 55, 87-111.

Wersin P., Hochella Jr. M.F., Per Person, Redden G., Leckie J.O. and Harris W.D. (1994): Interaction between aqueous uranium (VI) and sulfide minerals: spectroscopic evidence for sorption and reduction. Geochim. Cosmochim. Acta 58, 2829-2843.

Passive insitu treatment techniques, wetlands

Ecological water treatment processes for underground uranium mine water: Progress after three years of operating a constructed wetland

Margarete Kalin[1], Gunter Kießig[2] Annette Küchler[2]

[1] Boojum Research Ltd. Toronto, Ontario, M5A 1T7 Canada
[2] Wismut GmbH, Jagdschänkenstrasse 29, Chemnitz, 09117, Saxony, Germany

Abstract. Sustainable treatment approaches are sought for U mine waste water utilizing constructed wetlands. In 1998 a pilot system was constructed to remove U, ^{226}Ra, As, Fe and Mn from the effluents of the flooded workings of the Pöhla-Tellerhäuser mine of Wismut in Germany. Gravel beds and their biofilms in the system concentrate ^{226}Ra. As is co-precipitated/adsorbed onto iron-oxyhydroxide particles collecting in treatment cells. U occurs as uranyl carbonate at pH 7.3-8.0 and is not removed. An ecological system oxidizes iron, provides particulates for adsorption and organic matter to support bio-mineralization in anaerobic sediments.

Introduction

Metal and radionuclide-contaminated waters from underground mine workings, waste rock piles and tailings deposits represent long-term environmental and financial liabilities. Low cost, environmentally sustainable approaches to waste water treatment are sought by utilizing wetlands, sometimes constructed specifically for this purpose. Wetlands as treatment systems are very effective in addressing organic water pollution (Hammer 1989; Moshiri 1993; Kadlec and Knight 1996). Similar passive treatment techniques have begun to show value to the mining industry with a number of successful applications over the past two decades. However, in many systems, metal adsorption onto organic/ inorganic material used in the construction of the wetland will ultimately cease, and so will the capacity of the wetland to generate sufficient new organic material (Fyson et al. 1995). These approaches have certainly assisted in advancing knowledge and awareness of the potential of natural systems, and contribute to the resolution of these industrial challenges. However, they fall short of providing a sustainable ecological solution to water treatment, as such systems are likely to have a limited functional life span.

A long-term, sustainable wetland is composed of three integrated but critical parts; an oxidation pond, a biological section, and a settling pond. Metals must be reduced in an oxidation pond, before the waste stream is allowed to enter the biological section, to prevent them from coating the organics that would then cease to function as adsorption sites and serve as microbial carbon source. In the biological section, productivity must be high to produce organic matter at a rate equal to or higher than adsorption sites are needed to adsorb metals. Finally, settling ponds must provide anaerobic conditions in the sediments in which the metals can be bio-mineralized by reducing microbial consortia.

When a pilot wetland test system was constructed in 1998 to treat passively underground water from the Pöhla-Tellerhäuser mine it was believed that U, ^{226}Ra, Fe, As and Mn would be removed primarily by the actions of plants (Gerth and Kießig, 2001). Three years of monitoring data from each section of the system are assessed. It is suggested that in fact plants play a minor role in contaminant removal.

Pilot test system description

In Fig.1 a schematic overview is presented of the pilot system components that were constructed from a former storm water retention pond. The overall system has a volume of $415m^3$ covering a surface area of 474 m^2 and is divided into 5 cells, constituting the passive pilot system. Two small additional cells are used to test reactive materials after the passive treatment, which are not discussed here. After it has passed over a 21 m long cascade, the mine effluent arrives through a pipe at the bottom of the first cell, a sludge collection cell (2.5x5.8x2.1 m). From the sludge cell the water travels along a trough to the entrance of a larger settling cell (21.4x5.8x2.1 m). The water overflows into a gravel bed installed in cell 3 (16.7x5.8.2.18 m) where it leaves through 5 drainpipes at the bottom and flows upward into a second gravel bed into cell 4 (17.3x5.8x2.8 m). This third cell consists of 0.8 m of gravel, a thin layer of sewage sludge and straw to supply nutrients for plant and microbial growth, covered by an additional layer of 0.4 m of gravel. This 3rd cell serves as a source of nutrients to the fourth cell, where a 1.9 m deep gravel and sand bed is planted with reeds, rushes and cattails., representing the first "wetland" component of the pilot system. The water overflows into the final treatment cell, #5 (23.6x5.8x2.8 m) where swamp iris, rushes, reeds and cattails grow in a substrate consisting from top to bottom of a 0.4 m layer of gravel, a 1.16 m layer of soil, followed by a 0.25 m layer of compost.

Flow through the system was 3.5 $m^3.h^{-1}$ from January 1999 to August 2000. A reduced flow of 1 $m^3.h^{-1}$ passed through the system between September and December 2000. From January 2001 to the beginning of March 2001 the system was operated at 2.0 $m^3.h^{-1}$ when the flow was again reduced to 1.0 $m^3.h^{-1}$ until November 2001. From November until the end of monitoring period reported here, December 19th 2001, the system was operating at 2.0 m^3.h^{-1}.

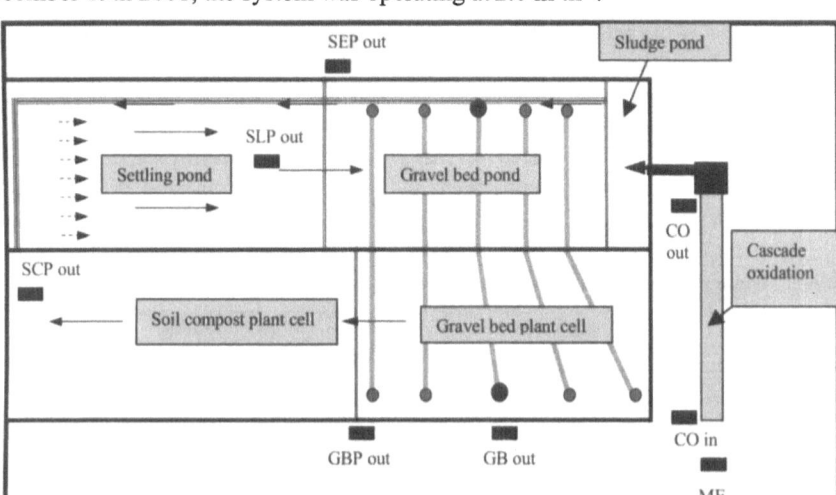

Fig. 1. Overview of the pilot system components and sampling locations.

Water sampling and analysis

Water quality was measured weekly at 9 sampling stations located throughout the pilot system, but only 8 are relevant to the discussion of the system performance. The sampling stations are: 1) mine effluent (ME) sampled before it enters the oxidation cascade 2) (CO in) and as it is leaving the cascade 3) (CO out) to enter the sludge pond. The overflow from the sludge pond after the water has traveled through a trough is sampling point 4) (SLP out), where the flow enters the settling cell. The water is sampled at the overflow to the first gravel cell 5) (SEP out) entering the first gravel bed in cell 3 from the top. After flowing through the gravel bed and before it enters the next gravel bed 6) (GB out) is located, followed by station 7) (GBP out), as the water is leaving the first cell with wetland plants. The final outflow 8) (SCP out) is collected after the water flows over surface of the soil, compost cell through the plants.

Water samples were filtered on site through a 0.45 μm filter, acidified for the determination of As, Fe, Mn and U and are shipped to the laboratory of Wismut for chemical analysis by ICP-OES for As, Fe and Mn. U is determined by the KPA method (Kinetic-Phosphorescence). The values of ^{226}Ra are determined on unfiltered samples by Alpha spectrometry. pH, Eh, electrical conductivity and dissolved oxygen concentration are measured in the field with a combination probe on a WTW Multi Line meter.

Passive insitu treatment techniques, wetlands

Results and discussion

The pH values of the mine effluent ranged from 6.8 to 7.6, with an average value of 7.3. The pH leaving the sludge cell (SLP) ranged from 7.6 to 8.4 with an average pH of 8.0. Overall, the average pH of the system was 7.6. The bicarbonate concentrations in the effluent averaged 416 mg.L^{-1}.

Uranium: Concentrations of U in the pilot system, covering three years of weekly sampling have been summarized in Table 1. Both total and filtered concentrations are shown.

Table 1. Average uranium concentration in water at Pöhla system, 1999-2001

	ME	CO in	CO out	SLP out	SEP out	GB out	GBP out	SCP out
			Total	U	mg.L^{-1}			
N	141	142	92	92	150	149	140	144
Avg	0.09	0.08	0.07	0.07	0.09	0.09	0.09	0.08
Stdev	0.02	0.03	0.01	0.05	0.03	0.02	0.02	0.02
			Filtered	U	mg.L^{-1}			
N	149	19	19	134	148	138	143	141
Avg	0.09	0.06	0.06	0.09	0.08	0.09	0.08	0.08
Stdev	0.03	0.01	0.01	0.02	0.02	0.02	0.02	0.03

In freshwater, U is present in its oxidized form as UO_2, uranyl, or uranyl hydroxide as a cation. In the presence of carbonate, it forms uranyl carbonate $UO_2(CO_3)_3^{4-}$. With the high bicarbonate concentration, U is most likely in the uranyl carbonate form, which is a stable anion (Langmuir 1978). As an anion, U is less attracted to the negatively charged cell walls of plants, algae and microbes (Myers et al. 1973; Neihof and Loeb 1972). This may explain why uranium concentrations are unaffected throughout the system.

If pH were lowered, more U would be removed by the biomass in the system. In several studies of U uptake by algae, the optimum pH for U removal was between 3 and 7, with most having pH optima around 5 (Liu and Wu 1993; Pribil and Marvan 1976; Yang and Volesky 1999; Franklin et al. 2000). In another example, pond algae accumulated U favourably in waters with higher calcium carbonate ratios, which had lower pH values, than waters with higher calcium carbonate ratios (Duff et al. 1997).

Iron and Arsenic: The concentrations in total and filtered Fe and As are given in Fig. 2a and 2b. Both iron and arsenic concentrations decrease significantly in the system. Most of the iron and arsenic concentrations are reduced as the water oxidizes passing over the cascade (CO out). A further reduction in total As concentration is evident after the water leaves the sludge pond (SLP out) but the filtered As remains at the same concentration of 444 µg.L^{-1}. After the settling pond (SEP out) both total and filtered As concentrations are essentially the same, suggesting little removal through the planted cells and on the biofilms on the gravel. The average concentration of As leaving the system is 390 µg.L^{-1}.

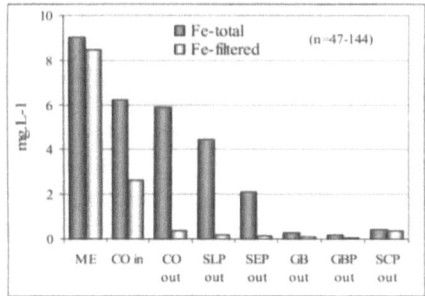 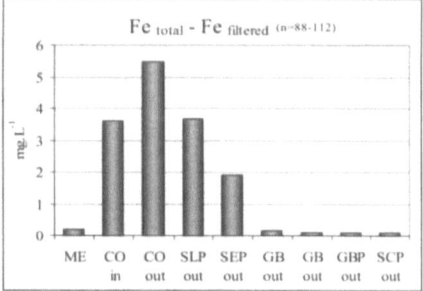

Fig. 2a. Average [Fe] and Fe particulates in water at Pöhla system, 1999-2001

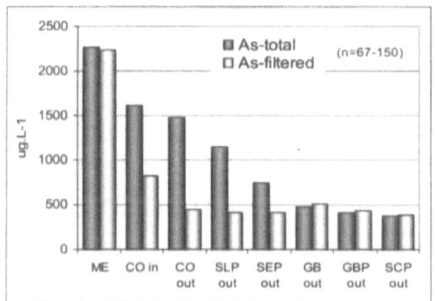 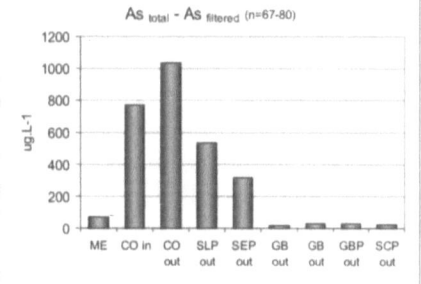

Fig. 2b. Average [As] and As particulates in water at Pöhla system, 1999-2001

The total Fe concentration leaving the cascade (CO out) has the same pattern as As, being reduced as the water passes through the sludge pond and the settling pond. The smaller particles of iron in the filtered water further decrease in concentration coming out of the cascade at 0.37 mg.L^{-1} (CO out) reducing to 0.04 mg.L^{-1} in the first planted cell (GBP out) only to increase again to an average concentration of 0.37 mg.L^{-1} at the outflow (SCP out).

Iron is found in the mine effluent as reduced Fe^{+2}. In the cascade and settling sections, most of the Fe^{+2} is oxidized to Fe^{+3}, with a concomitant hydrolysis forming Fe$_3$(OH)$_2$, and settling out of the water column. Subtracting the filtered concentrations from total concentrations, gives the amount of particulate Fe and As (Fig. 2a and Fig. 2b). The distribution of the particulates is identical for both Fe and As suggesting clearly that they are associated in the removal process. Particulates of Fe are present in sizes as low as 1 nm (Buffle et al. 1992). In fact, iron oxyhydroxide is considered a colloid (Stumm and Morgan, 1995 Pizarro et al. 1995) Arsenic readily adsorbs to iron hydroxide colloids and particulates, removing it as well. As adsorption capacity of iron hydroxides is an extensive field of study and is not discussed in detail here (Paige et al 1996; Pierce and Moore 1982). Equally active is the field of microbial interactions with the iron-hydroxide particulates and the liberation of adsorbed As (Niggemeyer et al. 2001; Glasauer et al 2001; Oremland et al. 2002).

Passive insitu treatment techniques, wetlands

The pilot system was operated under different flow regimes ranging from 1 $m^3.h^{-1}$, to 2 $m^3.h^{-1}$ and 3.5 $m^3.h^{-1}$. Multiplying the flow by the concentration gives contaminant load, the unit that is needed to assess removal in relation to the retention time of the water in the system, or the rate of particulate formation. In Fig. 2c the % of the daily Fe load that did get removed is plotted for the different flow rates in relation to the load entering the cascade from the mine. The flow regime does not largely influence removal as the water passes over the cascade, but the retention time reflects removal in the sludge pond and the settling pond (SLP out and SEP out) when total iron load is considered. The filtered iron load removal suggests that the cascade affects particulate formation. Burlap mats have been installed in the settling pond to test if Fe and As removal can be improved by increasing surface area. From the first gravel bed to the outflow, essentially 100 % iron is removed (Fig.2c).

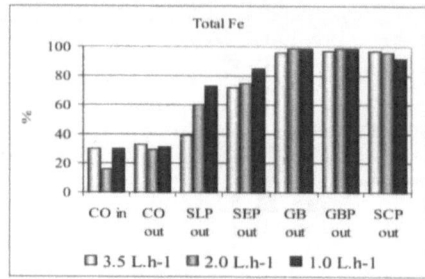 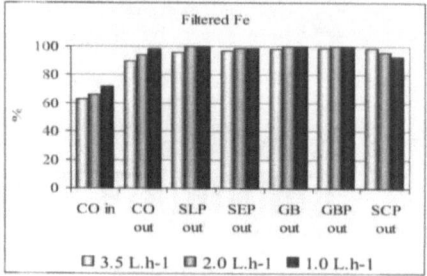

Fig. 2c. Fe removal from mine effluent at Pöhla system, 1999-2001

The same evaluation for As total load and filtered load indicates the stepwise increase of total As removal increasing only in the sludge pond (SLP out). The high flow rate with 40 % removal and 70 % at low flow rate. For the filtered load, the smaller particles, flow rate has a minimal effect on the removal (Fig.2d). As may be present as As III and remains in the water (Driehaus et al, 1999).

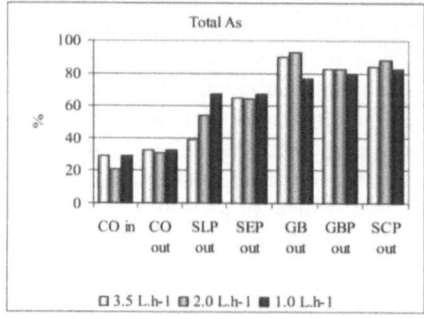 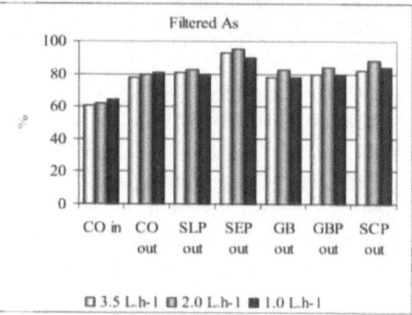

Fig. 2d. As removal from mine effluent at Pöhla system, 1999-2001

^{226}Radium: Concentrations of ^{226}Ra in $mBq.L^{-1}$ are shown in Fig.3a. ^{226}Ra concentrations decreased from 4500 $mBq.L^{-1}$ to about 1325 $mBq.L^{-1}$, a 70.6 % decrease, over the entire system. Each successive sampling contained lower concentrations of radium. Unlike iron and arsenic, a fraction of the radium was removed in each section of the system. Between the mine effluent (ME) and the settling pond (SEP out) the reduction in ^{226}Ra concentration is 31 % followed by a 58 % reduction in the gravel and plant cells (SEP out). This suggests that the removal process is related to a factor or media, which are present throughout the system, but more prevalent in the gravel bed. The largest reduction between two components of the system occurs between the settling pond and the gravel bed.

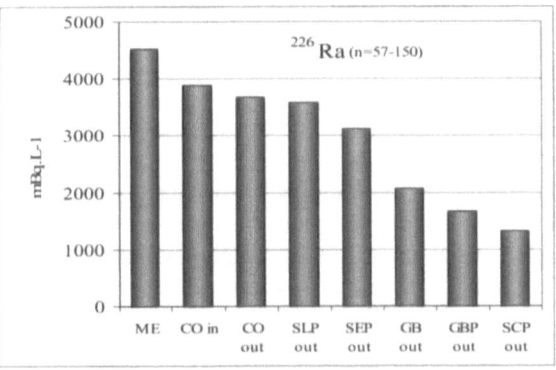

Fig. 3. Average [^{226}Ra] in water at Pöhla system, 1999-2001

Geochemical processes that influence radium removal include co-precipitation and adsorption to particulates. Iron hydroxide sludge analyzed from the sludge cell contained 73 Bq.g^{-1} of ^{226}Ra. When gravel from the first gravel bed was removed for analysis of ^{226}Ra and washed free of sludge and biofilms, it contained 140 Bq.g^{-1} of gravel, compared to its original concentration of 5 Bq.g^{-1}. The biofilm and sludge washed off the gravel contained 5.9 Bq.g^{-1}. Beneš (1982) compared the adsorption of ^{226}Ra to ferric hydroxide, koalinite, quartz sand and silica over a broad range of conditions. He concluded that adsorption increases in the order of Kaolinite> ferric hydroxide> quartz sand> silica at pH 6 and ferric hydroxide is the highest adsorbent at pH 8. Pradel (1976) concluded that co-precipitation with ferric hydroxide can account for a rapid decrease in the concentration of dissolved radium in river water receiving untreated iron rich mine drainage water. This explains the large reductions of ^{226}Ra concentration between the settling pond and the gravel beds and the high concentrations in the hydroxide sludge.

In water, radium exists primarily as a divalent ion (Ra^{2+}) and has chemical properties that are similar to barium, calcium and strontium. Radium favours coordination with oxygen donors and does not usually form complex ionic species (Kabata-Pendias and Pendias 1984). Radium itself cannot form precipitates, as its mass concentrations are low, measured in Bq where 1 Bq is equivalent to $2.7.10^{-9}$ mg. However it can co-precipitate with suitable solids. Sebesta et al. (1980) found that radium can co-precipitate with barium sulphate. Barium concentrations in the mine effluent are around 1.4 mg.l^{-1} but sulphate concentrations were reported as less than 5 mg.l^{-1}. As both sulphate and barium are low in this water, barium sulphate precipitation is an unlikely removal process for ^{226}Ra.

Passive insitu treatment techniques, wetlands

The chemical properties of ^{226}Ra suggest that it might be taken up by living cells via the same mechanism as other alkaline earth elements (Mg, Sr, Ba), but the evidence on such relations is contradictory. Williams (1982) analysed 19 species of aquatic and semi-aquatic macrophytes and found that the radium distribution was not significantly correlated with that of Ca, Mg, Na, Cu, Zn, Fe or S but was positively correlated with that of Mn. In 18 samples of five organs (lamina, petiole, fruit, peduncle, rhizome), from a single species, *Nymphaea gigantea*, radium distribution was not correlated with that of Ca, Mg, Na, Zn, U or S but was positively correlated with that of Cu, Fe and Mn and negatively correlated with that of K and P.

Radium can also be incorporated into calcium carbonate as it is being laid down by some marine and freshwater algae, such as the stonewort (Smith and Kalin, 1989). Given the documented adsorption of ^{226}Ra to gravel, it is postulated that diatoms incorporate ^{226}Ra into silicaceous structures. As diatoms are present in the system as biofilms throughout the gravel beds it is reasonable to suggest that the latter may play a role in the removal throughout the system.

Manganese: Average concentrations of Mn in different sections of the pilot system are shown in Fig. 3. Initial concentrations entering the system averaged a little over 0.6 mg.L^{-1}. This dropped slightly to about 0.5 mg.L^{-1} as the water passed through the cascade, and settling cells. The initial loss of Mn is probably due to co-precipitation with iron hydroxide in the sludge cell. The major reduction of the concentration of Mn occurs for both the total and the filtered as the water pass through the gravel bed. Mn oxidizes slower than Fe and this facilitated by microbes, which is well documented (Ehrlich 1990). Mn nodules, streaks or varnish as the black precipitates are referred to, where abundant on the gravel bed surfaces.

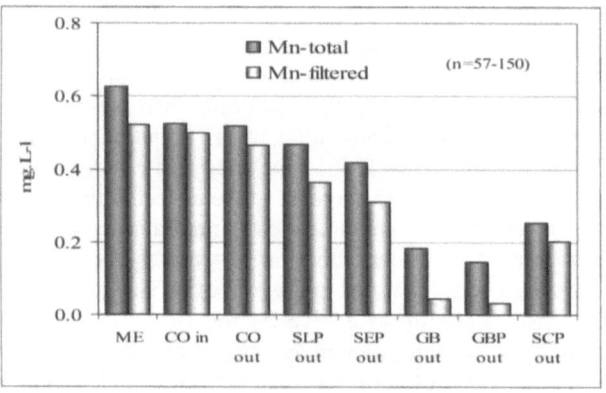

Fig. 4. Average [Mn] in water at Pöhla system, 1999-2001

Conclusions

After three years of operations the pilot system constructed to treat the mine water from the water-flooded Pöhla-Tellerhäuser mine of Wismut in Germany has produced some very interesting results. The data show that the system does remove iron, arsenic, manganese and ^{226}Ra and that the removal processes are based on geochemical characteristics of the contaminants. For Mn and ^{226}Ra some biologically facilitated removal is suggested through biofilm formation. Uranium, at the pH value and in the presence of high bicarbonate concentrations in this mine water, is not removed. The monitoring data generated from the Pohla pilot system facilitate greatly the understanding of the natural removal processes at work and have laid the foundation for scale-up following the principles of ecological engineering.

References

Beneš P (1982) Physico-chemical forms and migration in continental waters of radium from uranium mining and milling (IAEA-SM-257/84). In: Environmental Migration of Long-Lived Radionuclides. IAEA, Vienna, 3-23

Buffle J, Perret D, Newman M (1992) The use of filtration and ultra filtration for size fractionation of aquatic particles, Colloids and macromolecules. In Environmental Particles, Vol. 1, Buffle and R.R. de Vitre Eds. Lewis, Chelsea, MI

Driehaus W, Seith R, Jekel M (1994) Oxidation of Arsenate (III) with Manganese Oxides in water treatment. Wat Res 29:1 297-305

Duff M, Amrhein C, Bradford G (1997) Nature of uranium contamination in the agricultural drainage water evaporation ponds of the San Joaquin Valley, California, USA. Can J Soil Sci 77:459-467

Ehlrich H L (1990) Geomicrobiology. 2nd edition, New York, Marcel Dekker Inc.

Franklin N, Stauber J, Markich S, Lim R (2000) pH-dependent toxicity of copper and uranium to a tropical freshwater alga (*Chlorella* sp.). Aquat Tox (Amsterdam) 48:275-289

Gerth A, Kießig G (2001) Passive / biological treatment of waters contaminated by uranium mining In: Leeson A, Foote E A, Banks M K, Magar V S: Phytoremediation, Wetlands and Sediments. 6th International In-situ and on-site bioremediation symposium Battelle Press (6) 5 173 -180

Glasauer S, Langley S, Beveridge T J (2001) Sorption of Fe-(hydr) oxides to the *Shewanell putrefaciens*: Cell-bound Fine grained minerals are not always formed de novo. Appl Env Microb 67 (12): 5544-5550

Jensen B (1980) The geochemistry of radionuclides with long half-lives-Their expected migration behavior. Risø National Laboratory

Kabata-Pendias A, Pendias H (1984) Trace elements in soils and plants. CRC Press Inc. Boca Raton Florida.

Kadlec R.H, Knight R L (1996) Treatment Wetlands. CRC Lewis Publishers

Kalin M, Fyson A, Smith M P (1995) Passive Treatment Processes for the Mineral Sector. Proceedings of the 34th Annual Conference of Metallurgists: 'International Symposium on Waste Processing and Recycling in Mineral and Metallurgical Industries', Vancouver, B.C., August 19-24. 363-376

Langmuir, D (1978) Uranium solution-mineral equilibria at low temperatures with applications to sedimentary ore deposits. Geochim Cosmochim Acta 42:547-569

Lui H, Wu T (1993) Uptake and recovery of americium and uranium by *Anacytis* biomass. J Envir Sci Health Part A: Env Sci Eng 28:491-504

Moshiri G A (1993) Constructed Wetlands For Water Quality Improvement. Lewis Publishers, Ann Arbor , MI

Myers V B, Iverson R L, Harris R C (1973) The effect of salinity and dissolved organic matter on the surface charge characteristics of some euryhaline phytoplankton. J Exp Mar Biol Ecol 17: 59-68

Neihof R A, Loeb G L (1972) The surface charge of particulate matter in seawater. Limnol Ocean 17: 7-16

Niggemeyer A, Spring S, Stackebrandt E, Rosenzweig R F (2001) Isolation and Characterisation of a Novel As(V) reducing bacterium: Implication for Arsenic Mobilisation and the Genus *Desolfitobacterium*. Appl Env Microb 67(12): 5568-5580

Paige C R, Snodgrass W J, Nicholson R V, Scharer J M (1996) The crystallization of arsenate-contaminated iron hydroxide solids at high pH. Water Env Res 68: 981-987

Pierce M L, Moore C B (1982) Adsorption of arsenite and arsenate on amorphous iron hydroxide. Water Res 16: 1247 1253

Pizarro J, Belzize N, Filella M, Leppard G, Negre J-C, Perret D, Buffle J (1994) Coaguation/sedimentation of submicron iron particles in a eutrophic lake. Wat Res 29 (2): 617-632

Oremland R S, Newman D K, Kail B W, Stolz J F (2002) Bacterial respiration of arsenate and its significance in nature. In Frankenberger W T Jr. ed. Environmental chemistry of arsenic. New York. Marcel Dekker in press.

Pradel J (1976) French Contribution to the IAEA Panel on the Distribution, Movement and Deposition of Radium in Inland Waterways and Aquifers

Pribil S, Marvan P (1976). Accumulation of uranium by the chlorococcal alga *Scenedesmus quadricauda*. Arch Hydrobiol, Supplement 49: 214-225

Sebesta F, Sedlack, J, Obdrazlek M, Sandrick R, Benes P (1978) Studies on the source, distribution, movement and deposition of radium in inland waterways and aquifers. Progress Report of the IAEA Research Contract No1729/RB, Technical University of Prague

Stumm W, Morgan J J (1995) Aquatic Chemistry. Chemical Equilibria and Rates in Natural Waters. 3rd Edition, John Wiley & Sons

Smith M P, Kalin M (1989) Biological polishing of mining waste water. Bioaccumulation by the Characeae. In Salley J, McCready R G L, Wichlacz P L, Biohydrometallurgy, Proc. of the Int. Sym. Jackson Hole CANMET SP89-10 659-668

Williams A R (1982) Biological uptake and transport. Ch. 6. In: The Behavior of Radium in Waterways and Aquifers. IAEA Report 1982-04-15

Yang J, Volesky B (1999) Modeling uranium-proton ion exchange in biosorption. Env Sci Tech 33:4079-4085

Long-term Stability and Resilience of Passive Mine Water Treatment Facilities: A Joint Experimental and Simulation Approach

Christian Kunze[1], Franz Glombitza[2], André Gerth[3], Gunter Kiessig[4], Annett Küchler[4]

[1]B.P.S. Engineering GmbH, Reinsdorfer Str. 29, 08066 Zwickau, Germany
[2]G.E.O.S. Freiberg Ingeniergesellschaft mbH, PF 1162, 09581 Freiberg, Germany
[3] BioPlanta GmbH, Benndorfer Landstrasse 2, 04509 Delitzsch, Germany
[4] WISMUT GmbH, Jagdschänkenstr. 29, 09117 Chemnitz

Abstract. Passive water treatment technologies such as wetlands at abandoned mining sites are an attractive and economically sensible alternative to conventional technologies for long time-scales and relatively small contaminant loads. However, long-term stability and resilience with respect to external perturbations are a major concern for both wetland operators and regulators. In this paper, we outline a research project which addresses questions such as resilience and recovery behaviour of passive biological water treatment systems.

Introduction

Passive water treatment technologies such as wetlands at abandoned mining sites are an attractive and economically sensible alternative to conventional technologies for long time-scales and relatively small contaminant loads. However, long-term stability and resilience with respect to external perturbations are a major concern for both wetland operators and regulators. Apart from the investigation of potential failure scenarios that lead to a full or partial breakdown of a wetland's function for a certain time, it is the time a water treatment system needs to restore its function after a breakdown and the question whether it will return to its designed state of operation at all which must be considered before approval will be given by regulators.

In WISMUT's remediation activities at uranium mining and milling sites in Eastern Germany, water treatment plays an eminent role. Apart from mining water and free water from tailings ponds which are treated in conventional facilities, there are a number of seepage waters from waste rock piles, tailings dams, and smaller mines which require treatment before they can be released into streams but do not warrant conventional treatment due to their low flow rate, and due to the long-term nature of their contamination. In most cases, it is Uranium, Radium, and Arsenic which are the main contaminants, while at some sites, Nickel and other non-radioactive metals are present, too.

For these waters, passive systems are an attractive alternative to conventional treatment facilities. Numerous approaches have meanwhile been developed and discussed in the literature which use natural processes to remove metallic contaminants from mine and seepage water. Plants and microbes create hydrochemical conditions which lead to a shift of pH or redox potential; in other cases adsorption or the incorporation of metals in the microbial and/or plant metabolism are possible mechanisms to reduce the contaminant concentration in the outflow. Theoretical explanations for all these chemical and physical processes have been developed, and the basics seem to be well-understood, in principle. However, implementing biological water treatment systems are still regarded as "tricky", and scepticism as to their effectiveness and stability is still prevalent among practitioners, the public and, perhaps most importantly, regulators (Suthersan 2002).

A necessary precondition for the approval of wetlands by regulators, particularly if radioactive components in the water attract enhanced attention from the public, is that safe operation can be guaranteed over a long time span. Here we are faced with a dilemma: on one side passive biological water treatment systems derive their attractiveness from their low level of maintenance required which leads to low costs, on the other side a certain degree of reliability must be proven before they can be left unattended. This dilemma, combined with the fact that passive biological treatment systems are no "plug-and-play" technology but need careful adjustment to site conditions and sometimes show inexplicable fluctuations of performance, is still a barrier to the widespread use of passive systems.

The agenda on the way to better acceptance of passive biological treatment systems for mine and seepage waters is therefore twofold: (1) the chemical, biological and physical processes, as well as their complex interplay, need to be better understood, and (2) reactions of the system to external perturbations, and its ability to return to the normal operating state (i.e., its resilience), must be investigated and optimised. It is primarily the second question which will be addressed in this paper.

Our paper reports on the R&D project "BioRobust" carried out by a consortium of firms. Its objective is to develop a framework to assess the robustness of constructed wetlands and to identify and compare measures to increase the robustness by suitable design and combination of several stages with distinct functionality. Two different types of wetlands are currently being installed at WISMUT's Schlema mining site: a microbiological system based on biofilms on a gravel-bed (G.E.O.S. GmbH), and a plant root system (BioPlanta GmbH). The design of both systems is based on parameters from laboratory experiments in both firms. The data from lab (and, later on, field) experiments are used to develop a system model (B.P.S. Engineering GmbH) which allows to simulate the dynamic behaviour of wetlands.

Passive insitu treatment techniques, wetlands

As of the time of writing this paper, the project is still in progress. This means that there are no final conclusions but a number of promising approaches to be pursued in the further course of our work. This paper does not attempt to give any final answers but to outline our approach.

Concepts of Robustness, Resistance and Resilience

There are many approaches to the question of what constitutes a robust system. Definitions and concepts have mainly originated in engineering, biology or sociology, but they are too numerous to be discussed here in detail. With respect to ecosystems, the interested reader may be referred to Jorgensen (2000) which contains a number of interesting concepts. We will confine ourselves to the narrow but practicable terminology of Gunderson (2000) who uses the terms resistance and resilience as constituents of the broader concept of robustness. Using a physical analogue (potential well model), resilience and resistance can be represented as shown in Fig. 1.

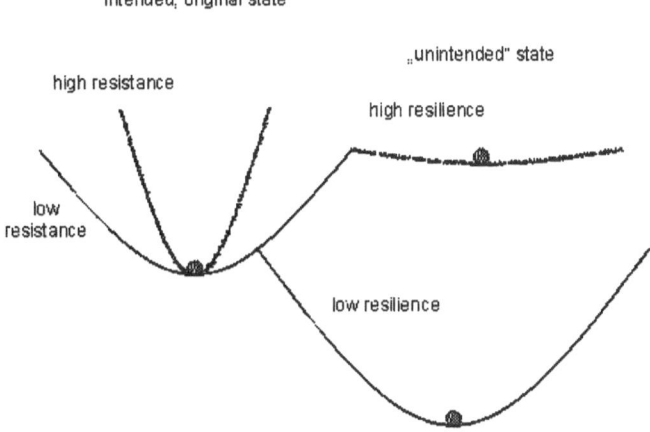

Fig. 1. Schematic representation of the concepts of resilience and resistance (adapted from Gunderson 2000)

It is important to distinguish the concepts of robustness and resilience from that of failure risk. Classical concepts of risk analysis (Renn 1985) ask two questions: (1) how likely is an event, and (2) how severe are its consequences. By contrast, robustness is a term used to characterise the ability of a system to withstand external perturbations and/or to return to the desired operational state after such a perturbation. According to the terminology introduced by Gunderson (2000), robustness comprises the concepts of resistance and resilience. Resistance is a term used to describe the reaction of a system to changes of external conditions, or more precisely the extent to which a system deviates from its normal operating state under the influence of an external force. Resilience, on the other hand, describes the dynamic behaviour of a system after an external shock, and how the system returns to its normal operational state. In the resilience concept, the questions asked are (1) can the system recover after the perturbation is over (or into which new stable state will it move), and (2) how long does it take for the system to recover. Resilience is thus an inherently dynamic concept, and attempts to deal with a system's resilience must start from a dynamic model.

Returning to the issue mentioned in the Introduction that public and regulatory concerns over the long-term behaviour of passive biological water treatment systems are an obstacle to their broader application, we may invoke the resilience concept. Nobody in their right mind would require a passive, unattended water treatment system to function without any fluctuations over many decades. What is required, however, is that the periods during which the system will under perform are short compared to the total operating time, and that the system is able to recover from a wide range of perturbations without human intervention in sufficiently short time.

A practical way of characterising the resistance and resilience of passive biological water treatment systems is thus to provide the following sets of quantities and their interrelationship: the extent of a deviation of performance from the normal state during an external perturbation, the time to recover and the end-point(s) of the recovery process (original state or "lock-in" state other than original). These parameters will often depend on the duration and strength of an external perturbation.

The wetland system can be perturbed by external factors such as
- flow rate fluctuations (dry-out during extended draught periods, or extremely high flow rate due to extreme precipitation events)
- temperature fluctuation (seasonal temperature curve ranges from extremely hot seasons to harsh winters)

These two parameters are likely to have the largest impact on the systems performance. Both flow rate and temperature are subject to seasonal changes and overlaid stochastic fluctuations within a range that can be regarded as normal. In rare events, however, these ranges can be exceeded so that an extreme event occurs. The dynamic system model must be able to describe the behaviour under both situations (i.e., normal and extreme fluctuations), each with its probability distribution or probability of occurrence, respectively.

Other perturbations such as deliberate malevolent destruction may be of theoretical interest but are not part of our investigations because of the breadth of possible scenarios and conceptual difficulty of including them in a coherent framework.

Model Concept

In order to describe the dynamics of a biological system, we have developed a simple model framework which consists of 3 parts or submodels which are interlaced: (1) a model for the physical quantities such as temperature, flow rate etc., (2) a model describing the removal of contaminants by biological activity such as microbial growth and microbiologically induced redox reactions, and physical/chemical conditions such as precipitation, sorption etc., and (3) a model for the biological processes including a very simple concept of hydrochemical processes, metabolism and nutrient cycle. The concept is graphically represented in Figure 2.

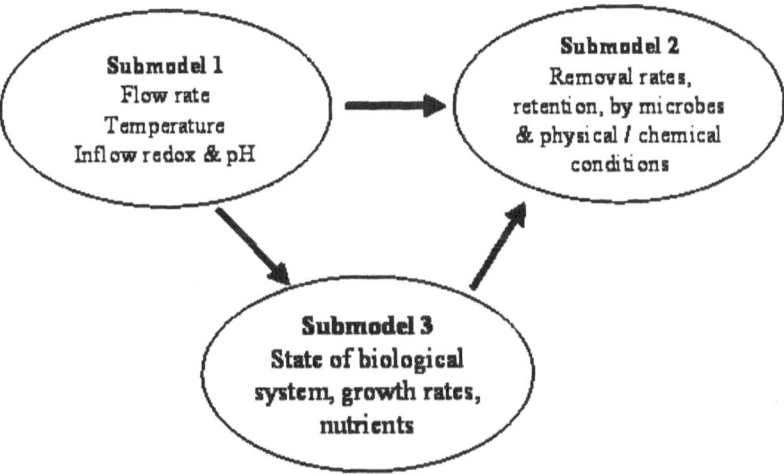

Fig. 2. Schematic representation of the model framework used in the project

It should be noted that the model does not intend to map the whole complexity of the real world. Rather, it aims at giving an understanding of the most important processes taking place in a passive biological water treatment system. With respect to the biological processes, we must restrict ourselves to the most basic relationships.

Submodel 1 primarily aims at replicating the environmental conditions and allows for the simulation of different scenarios. Monitoring data of the site have been statistically analysed in order to generate realistic time series for temperature and flow rate, including extreme events. Alternatively, re-sampling of measured time series and constant values for all relevant parameters can be chosen by the user in order to investigate a wide variety of scenarios and improve the understanding of the complex processes. Some chemical parameters of the inflow water are strongly correlated with the flow rate (e.g., sulphate and nitrate concentrations at the inflow into the passive wetland system). This is due to the fact that different types of water can mix (e.g., different flow paths in a waste rock pile). The model must take this dependency into account. Furthermore, oxygen supply is an important factor which depends on both the oxygen saturation of the inflow water (which in turn is pH dependent) and additional oxygen supply via plant roots.

Submodel 2 links microbial activity and macro parameters such as redox and pH to removal rates. For example, it is now known that Uranium reduction is a process concurrent to sulphate reduction, both processes being mediated by anaerobic microbial activity. Remobilisation is also covered by Submodel 2, for example if anaerobic microbial activity slows down or stops altogether, or redox conditions change due to a breakdown of oxygen consumption in the aerobic zone, so that the solubility of certain contaminant species increases.

Submodel 3 is the central component of the model. It describes the dynamics of the microbiological system as a whole. It is outlined in more detail in the next section.

The dynamic system model is being implemented under iThink, a visual simulation environment similar to the much better known system Stella, both produced by HPS Inc., Hanover NH. It provides the opportunity to create a user-friendly graphical interface which will be a great advantage when communicating the simulation results to the public, regulators or other stakeholders.

Microbiological Submodel

This submodel describes microbial growth and decay. Apart from the biological processes, it must take account of the nutrient balance. Organic carbon, in particular, plays an important role.

We take account of the following processes, in order of their preferred redox range: oxygen consumption by aerobic bacteria, nitrate reduction, sulphate reduction. As an option, complementary processes such as the fermentation of organic matter into alcohol and methanogenesis can be included. Each of these processes is governed by a growth equation of the Michaelis-Menten type, i.e.,

$$dX/dt = X \; \mu_{max} \; c/(k+c) - X/T$$

and a nutrient balance equation or the type

$$dc/dt = -[X \; y + dX/dt \; y']$$

where X is the biomass of a microbial species, μ_{max} is the maximum growth rate, c is the nutrient concentration available to the microbes, k is the Michaelis-Menten constant, and T is the average lifetime of the microbes. y and y' are the consumption coefficients for living biomass and biomass growth, respectively.

Microbial growth of a species occurs until the component reduced by this process is used up. So, if the dissolved oxygen is used up by aerobic oxygen consumption, nitrate reduction sets in until nitrate is completely used up. Only then the sulphate reduction can start, and finally Uranium is reduced to an insoluble sulfidic form.

Passive insitu treatment techniques, wetlands

Each of these processes consumes carbon to build up biomass. It is also partly transformed into carbon dioxide, as is the case in the aerobic zone. Microbial autolysis, on the other hand, sets carbon free which is available for the next cycle.

It must be noted that some of the model parameters (k, T, μ_{max}) are difficult to obtain. Laboratory and literature data can help here, but it is critical that they are valid under site conditions, too. Much work is still needed to establish a reliable dataset. Another problem is the oversimplification of most processes. It would be possible, of course to develop a more detailed multi-stage model of the nitrate reduction, for example. On the other hand it seems questionable that the data required to run such a model would be available. After all, what the model attempts to achieve is a rather general understanding of the basic processes and their dynamics and interconnectedness.

An extension of the model concept described above are macrophytes on the wetland surface. They contribute to the oxygen supply, release carbon compounds via the rhizosphere, and form a basis for microbial growth and adsorption of contaminants.

Summary and Outlook

In a current R&D project ("BioRobust"), the robustness of passive biological water treatment systems for mine and seepage water in terms of resilience and resistance is investigated. Starting from definitions of what we mean by these terms, we have developed a systems simulation model to describe the dynamic behaviour of a passive water treatment system (wetland).

Although we believe that we have reached a good compromise between the complexity of the real world processes and the limits set by a modelling approach, be it the model itself or the available data needed to run the model, much remains to be done both theoretically and experimentally. First of all, two pilot-scale wetlands will be finished at WISMUT's Schlema site. On the theoretical side, macrophytes will have to be included in the model and other refinements made.

We will report on the progress of the project and preliminary results in forthcoming papers.

Acknowledgements

This work is supported by a grant from the Federal Ministry of Education and Research (BMBF).

Passive insitu treatment techniques, wetlands

References

Gunderson L.H. et al. (2000) Resilience in Ecological Systems. In Jorgensen and Müller (2000, pp. 385)

Jorgensen, S.E., Müller, F (2000) Handbook of Ecosystems Theories and Management, ed. by S.E.Jorgensen and F.Müller, Lewis Publishers, Boca Raton, London, New York, Washington DC

Suthersan S.S. (2002) Natural and Enhanced Remediation Systems, Lewis Publishers, Boca Raton

Renn, O. (1985) Risk Analysis: Scope and Limitations, In: Regulating Industrial Risk (Ed. by. Otway, H. and Peltu, M.), London 1985

Passive insitu treatment techniques, wetlands

Retention of Radionuclides and Arsenic by Algae Downstream of U mining Tailings

Claudia Dienemann[1], Gert E. Dudel, Holger Dienemann and Lieselotte Stolz

[1]Technische Universität Dresden, Institut für Allgemeine Ökologie und Umweltschutz, Pienner Straße 8, 01737 Tharandt, Germany

Abstract. Algae including cyanobacteria function as efficient temporary sinks in wetlands below uranium mining tailings and dumps. As far as investigated their accumulation potential exceeds in parts higher plants and soils, microphytic and epiphytic algae each being superior to macrophytic algae and vascular plants. The enrichment compared to the surrounding water is high. In the living biomass uranium can be accumulated up to 300 mg/kg dry matter (DM), lead up to 250 mg/kg DM and arsenic even up to 4000 mg/kg DM. On average approx. half of the contamination is loosely adsorbed on the cell surfaces. The dimension of contaminant remobilization is yet unknown.

Introduction

Heavy metals and arsenic are – with few exceptions – very efficient accumulated in algae and aquatic plants (literature overview see Vymazal 1994). In photoautotrophic algae and cyanobacteria enrichment up to several percent per gram dry matter can be expected due to relatively high surface/volume ratio, mucilage (metal complexing) sheaths (e.g. Yen et al. 2001), fast growth rates and the ability to store phosphate actively as polyphosphate at growth limitation caused by the lack of other resources (Vymazal 1994; Franklin et al. 2000). A P like behavior of As and interactions with Uranium are assumed (Planas and Healy 1978; change in chemical speciation: Knauer et al. 1999; Meinrath et al.1999; Mkandawire and Dudel, this volume).

The specific performance of photoautotrophic algae or chemoautotrophic bacteria has so far been sparsely considered in field investigations. However, concentration factors resp. transfer factors for uranium and especially for radium have been calculated already decades ago (for summary see Williams 1990).

This study focuses on the content of various radionuclides in algae and aquatic plants in relation to their milieu below uranium mining dumps and tailings in Saxony. They were particularly investigated and discussed along two flow paths.

Material and methods

Sampling and sample preparation

Sampling sites in Saxony (Tailing Lengenfeld/Vogtland; Tailing Mechelgrün/ Vogtland) were chosen based on the German Altlastenkataster (ALASKA; Altlastenkataster = Register of old neglected deposits of toxic waste) as well as on hydrological and geological maps showing local activity concentrations higher than 1 kBq/kg DM in sediments and hints in the water path higher than 20 µg /l Uranium.

Sampling took place in the period between August and October 2001 along two contaminated water courses (2500 m in Lengenfeld, 250 m in Mechelgrün) and by some stagnant surface waters at seepage spats (directly downstream of the tailings in Lengenfeld and Mechelgrün). Plant samples were taken as mixed samples from the open water zone as well as from the banks. Sampling was repeated several times within a few weeks until the end of the vegetation period.

The plant samples were aliquoted and one part was washed for 15 minutes in a defined amount (2 resp. 6 l) of Aqua dest.. This procedure was repeated up to four times to remove attached and roughly adsorbed suspended fine particles. All samples were oven dried to constant weight at 65 °C, ground and homogenized (ZM 1000, RETSCH).

Together with plant sampling water samples were taken and immediately prefiltered (12 µm pre-washed nylon filter). Within 24 hours all water samples were filtered in the lab (0.45 µm-cellulose-nitrate-filters, SARTORIUS, Göttingen). For comparison samples were analysed unfiltered (only 12 µm filtration in the field).

Chemical analysis

Uranium was mainly analyzed by ICP-MS (PQ2+, Fisons, U.K). For this purpose solid parts of the samples (dried and ground, filter residues) were mineralized with with HNO_3/H_2O_2 in a microwave digestion process (MARS 5, CEM, U.S.A.).

Radionuclide were analyzed by gamma-spectrometry (IAF Radiooekologie, Dresden). Dried samples were enclosed into radon-tight 20 ml flat trays. Samples were measured twice with a temporal off-set. Time of measurement varied from 5 to 24 hours. For quality assurance replicate measurements were made and results were compared with the results from ICP-MS by regression analysis (r^2=0.99).

Passive insitu treatment techniques, wetlands

Results

Element and radionuclide distribution in water pathway

Uranium, Arsenic, Lead

Uranium, arsenic and lead load can be clearly distinguished between the different sampling sites (Table 1). As expected, concentrations are high near the contamination sources and decrease with distance. Depending on the element, the overall contamination is always less than 1 mg/l.

The tailings near Lengenfeld and Mechelgrün are different not only with respect to uranium and arsenic concentrations. Their element ratios are quite different as well. In site "Lengenfeld" the arsenic concentration is rather high compared to uranium. In "Mechelgrün" uranium is dominant.

Table 1. Concentrations of arsenic, lead and uranium in water samples from different sites in varying distance from the contamination source; n= number of sample replicates; standard deviation in %.

Location		As [µg/l]	Pb $_{tot.}$[µg/l]	U $_{tot.}$[µg/l]
Lengenfeld				
near source – Covered tailing (ore processing)				
0...1 m from seepage influx site	(n = 6)	31,5	0,2	131,9
		27 %	25 %	24 %
flow off pathway about 25 m	(n = 20)	12,5	0,03	54,6
		157 %	164 %	121 %
source-distant – Stauteich				
"free" running water		2,1	1,0	5,1
flow off pathway about 500 m	(n = 9)	7 %	46 %	18 %
interstitial water	(n = 16)	12,2	2,2	9,3
		107 %	163 %	92 %
Mechelgrün				
near source –tailings (ore milling)&mine water		25,7	2,9	330,1
0...1 m from influx site	(n = 3)	50 %	85 %	98 %
source-distant – Pond ("Forellenteich")		8,3	1,8	198,4
flow off pathway about 200 m	(n = 4)	6 %	100 %	3 %
reference brook (in catchment of Plohnbach)		1,4	2,2	2,9
	(n = 9)	53 %	116 %	74 %

A significant reduction of the uranium concentration already exists during the relatively short flow path (about 25 m) between seepage influx site and downstream of the tailing "Lengenfeld". In contrast, below the tailings in Mechelgrün a reduction of approx. 330 to 200 µg/l U was found, although the total uranium load is comparatively high.

Passive insitu treatment techniques, wetlands

Radionuclides (^{238}U, ^{226}Ra, ^{210}Pb resp. ^{228}Ra, ^{224}Ra)

The decay products of ^{238}U as well as of ^{235}U- and ^{232}Th do not reach the activity concentrations of uranium (Fig. 1). Although the contamination at both sampling sites is based on very different sources, the activity concentrations vary only slightly. The U/Ra-ratio is shifted more clearly towards uranium (max. 175) in "Mechelgrün" (in "Lengenfeld" water: about 20). The outcome of this is a minor shift towards radium in the U/Ra-ratio in the Lengenfeld area. Reasons can be seen in the uranium milling as well as the uranium-rich ores in Mechelgrün and the specific geochemical behavior of U in mineral rich surface water, respectively.

Along the flow off pathway the activity concentration of uranium decreases more significantly than that of the decay products resulting in a shift of the U/Ra-ratio. The reduction of the activity concentration during the flow off pathway is significant (P<0,01).

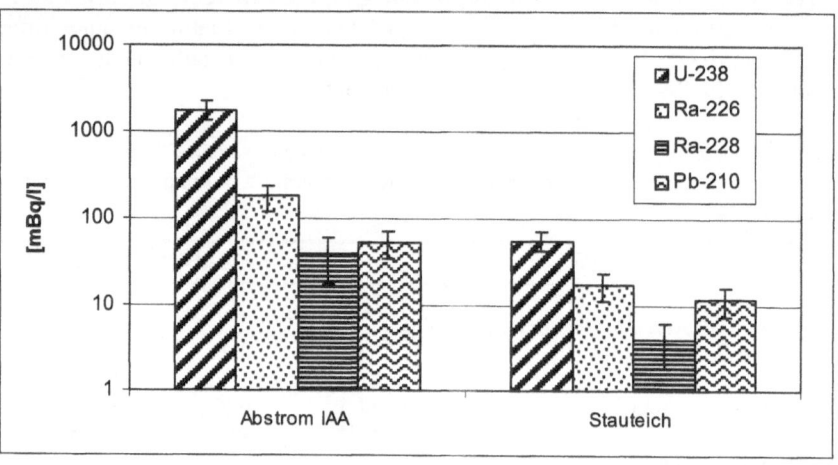

Fig. 1. Activity concentration of ^{238}U, ^{226}Ra, ^{228}Ra, ^{210}Pb [mBq/l] in water sampled near seepage source (left columns: below tailing dam "Abstrom IAA") and about 400 m downstream (right columns: run off below pond "Stauteich"); n = 5; all differences between the two location are significant (P<0,01, pair-wise comparison of each isotope).

Enrichment in macrophytic algae and microphytic periphyton (Transfer coefficients)

The enrichment (biosorption/accumulation) in comparison to the surrounding water exceeds in most cases one order of magnitude (Fig. 2). It is only slightly reduced with distance from the source, although the contaminants are diluted (see Fig. 1). The major part of contamination appears to be only loosely attached to the cell surfaces (see below).

Transfer coefficients (water/algae) were calculated as 250 ($\pm$ 35 %) for ^{238}U, 111'000 ($\pm$ 65 %) for As and 73'000 ($\pm$ 60 %) for ^{208}Pb.

In the relatively fast sedimenting detritus following factors were calculated: 2000 ($\pm$ 10%) for U-238, 210'000 ($\pm$ 25%) for As and 232000 ($\pm$ 80%) for Pb-208.

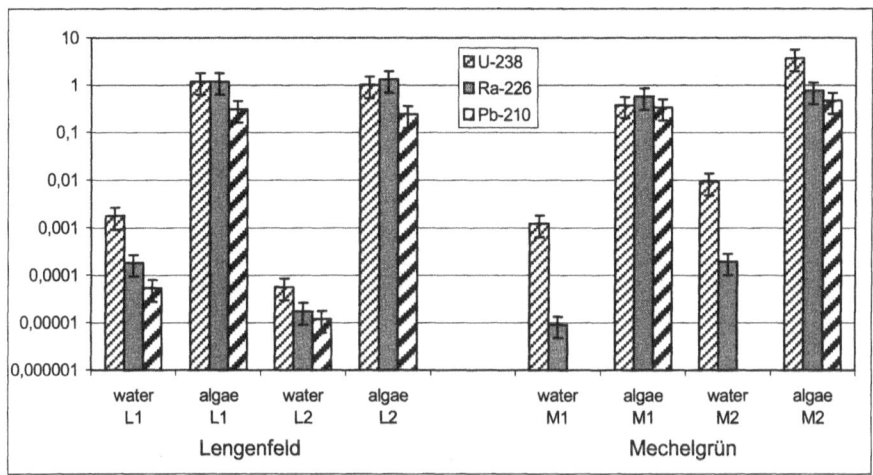

Fig. 2. ^{238}U-, ^{226}Ra-, ^{210}Pb- activity concentrations [y-axis unit: Bq/g dry matter for algae and Bq/l for water] in algae and surrounding water from different sampling sites below tailing Lengenfeld and Mechelgrün (downstream of the tailings) ($4 \leq n \leq 9$; error bars: standard deviation)

Local differences related to the algae community structure

The concentrations vary only slightly within the sampling period despite changing weather conditions from August to October. However, local distributions on both tailing sites are much more differentiated (Table 2). Depending on algae structure (functional groups) there are locally high enrichments in the surrounding water as well as in the biota.

This was pronounced at the sampling site "Abstrom IAA" (directly below tailing dam Lengenfeld). Here, due to aeration and biological activity, the chemical characteristics of the water change drastically. In the shadowed areas with varying mixing conditions of the more or less stagnant water a variety of algae or (cyano-)bacteria communities have developed. Also the contaminant concentration and accumulation pattern within algae and bacteria communities differ significantly from sampling point to sampling point.

The samples were washed in order to differentiate between biosorption (incl. uptake and accumulation into the cells) and loose sorbed or simple aggregated material (fine particular matter and detrical particles) on the large algae surfaces the samples were washed. The total enrichment of U is not so high when compared

Passive insitu treatment techniques, wetlands

with As although depending on the site - due to washing - only a small part of U can be removed (see Table 2). It is suggested, that in general U is fixed tight on cell surfaces or is taken up (accumulation within cells). Extraordinary high enrichment (gram range!), especially of As, was effected additionally by aggregation of fine particulate (humic?) matter with (filter effect). The contaminant content of detritus was in all cases higher than in algae. The "filter" effect depends on local conditions and characteristics of algae. Macrophytic species (e.g. *Cladophora* sp.) are more effective than small cyanobacteria mats.

Table 2. Arsenic, lead and uranium concentrations in algae of dominant life forms and functional groups, resp. from different locations (Cyanobacteria: genus *Oscillatoria sp.. Phormidium* sp., Chlorophyceae: *Cladophora crispa, Rhizoclonium hieroglyphicum;* Periphyton: epiphytic algae on vascular plants and macrophytic algae, mainly Naviculaceae)

Location	Functional group	Treatment	As [mg/kg]	$Pb_{tot.}$ [mg/kg]	$U_{tot.}$ [mg/kg]
Lengenfeld					
Below tailing dam	*Cyanobacteria*	unwashed n = 2	1382,0	4,2	119,0
"Abstrom IAA"	*Chlorophyceae*	washed n = 1	910,3	0,9	39,3
		unwashed n = 1	4150,3	3,5	78,6
Upstreams 500m below tailing dam	*Chlorophyceae*	unwashed n = 3	151,4 3 %	37,9 4 %	91,4 5 %
Stauteich"	Periphyton	washed n = 2	72,6	33,2	152,0
		unwashed n = 2	166,9	181,0	131,7
Mechelgrün					
Tailing	*Chlorophyceae*	washed n = 9	44,2 20 %	6,4 37 %	20,1 32 %
		unwashed n = 19	61,3 4 %	3,7 7 %	31,0 7 %
	Periphyton	n = 2	243,8	28,0	42,4
Reference	*Chlorophyceae*	washed n = 8	33,4 29 %	3,4 81 %	4,0 32 %
		unwashed n = 4	44,9 30 %	5,6 44 %	8,9 76 %

Discussion

The transfer coefficients calculated in this study exceed one order of magnitude in particular regarding arsenic and radium. Such extraordinary enrichment was already published, mainly from laboratory experiments and common heavy metals, respectively (summary by Vymazal 1994). However, information from field studies including radionuclides are limited (Kalin and Smith 1986, 1996; Kalin 1989; Kalin and Wheeler1992). Compared with higher plants algae are very efficient (Dudel et al. 2001; Brackhage et al. 1999, 2001).

It must be taken into account that all macrophytic algae were covered more or less densely by epiphytic algae, particularly Bacilliarophyceae (Pennales). According to Emerson and Hesslein (1973) epiphytic organisms (algae and bacteria) can increase radium accumulation (biosorption) significantly. The radium content depends directly on the biomass of algae as summarized by Williams (1990) and again shown in this investigation (0.5-2.3 Bq/g dry matter). This concentration range was relatively independent on the location of contaminated site and sampling time.

The biosorption of uranium has so far been investigated mainly with bacteria and various fungi (e.g. Golab et al. 1991; Bustard and McHale 1997; Hennig et al. 2000; Yen et al. 2001). The average uranium concentration in algae within the direct surroundings of both tailings is higher than 20 mg/kg dry matter, relatively independent from the variable concentration in the water in the range of some hundred µg/l. In many field studies a comparable transfer was shown already for other heavy metals (Cd, Cr, Cu, Pb, Ni und Zn: summarized by de Fillipis and Pallaghy 1994) and in some cases for U (Kalin and Smith 1986; Goldberg et al. 1998; Kalin this volume).

Regarding the stability of the immobilization it is particularly interesting, which part of the surface sorbed uranium (e.g. complexed with polygalacturonic acids) can be removed by water in the course of changing chemical conditions (pH, Eh, ion strength) and which part has been actively accumulated in the cells. As shown by Macaskie et al. (1988, 1992) U is taken up fast from water - independent of concentration in milieu but, however, dependent on biomass of bacteria (*Citrobacter* sp.). Above pH 4 U was fixed (85-99%). For the green algae *Scenedesmus subspicatus L.* Clemens (2000) proofed that under controlled water chemical and growth conditions (laboratory cultures) even with organic acids and/or other complexing agents (EDTA) only approximately 50 % of immobilized uranium can be removed.

Conclusions

Despite considerable differences between the two sampling locations the investigated algae communities were found to be generally suited for contaminant retention. Thus it may be possible to immobilize radionuclides in situ in seepage resp. surface waters or to extend and design (artificial) retention systems. However, the question of the quantity of the long-term biosorption resp. re-mobilized proportion (e.g. by fungal or bacterial mineralization) yet remains. Further investigations focusing on this subject are in progress.

Passive insitu treatment techniques, wetlands

Acknowledgement

This project was supported by the Saxon State Authority for Environment and Geology (LfUG).

References

Bustard M, McHale PM (1997) Biosorption of uranium by cross-linked and alginate immobilized residual biomass from distillery spent wash. Bioprocess Eng 17: 127-130

Clemens C (2000) Immobilisierung von Uran durch *Scenedesmus subspicatus* L. unter Berücksichtigung der spezifischen Wachstumsrate, der Phosphatkonzentration und der Biomassedichte. Diplomarbeit, Institut für Allgemeine Ökologie und Umweltschutz, Technische Universität Dresden, Fakultät Forst-, Geo-,Hydrowissenschaften

Brackhage C, Dienemann H, Dannecker A, Dudel EG (1999): Radionuclide and heavy metal distribution in soil and plants from a 35-year old reclaimed uranium mining dump site. Proceedings of 5th Int. Conf. On the Biogeochemistry of Trace Elements, Vienna, Vol II., p. 1174-1176.

Dudel GE, Brackhage C, Clemens C, Dienemann H, Mkandawire M, Rotsche J, Weiske A (2001) Principles and limitations for natural attenuation of radionuclides in former uranium mining and milling sites. Proceedings of the 8th International Conference of Environmental and Radioactive Waste Management (ICEM 01) Brugge (Belgium), American Society of Mechanical Engineers (ASME) 8 p.

Dudel GE, Brackhage C, Clemens C, Stolz L, Dienemann H, Weiske A (2002) Rückhaltevermögen von Radionukliden durch Algen im Abstrom Industrieller Absetzanalgen. FuE-Abschlussbericht, Sächs. Staatsministerium für Umwelt und Landesentwicklung, Landesamt für Umwelt und Geologie.

Emerson S, Hesslein R (1973) Distribution and uptake of artificilly introduced radium-226 in a small lake. J Fish Res Board Can 30: 1485-1490

De Filippis LF, Pallaghy CK (1999) Heavy Metals: Sources and Biological Effects, in: Rai, LC, Gaur JP, Soeder CJ (eds.) Algae and Water Polution. Stuttgart: 32-77

Franklin N, Stauber J, Markich S, Lim R (2000) pH-dependent toxicity of copper and uranium to a tropical freshwater alga (*Chlorella sp.*). Aquat Toxicol 48, 275-289

Golab DR, Orlowska B, Smith RW (1991) Biosorption of Lead and Uranium by *Streptomyces Sp.*. Water Air Soil Pollut 60: 99-106

Goldberg EL; Gracher, MA, Bobrov UA (1998) Do diatom algae structures accumulate uranium ?, Nucl Instr Meth A, 405 (2-3), 584 - 589

Hennig C, Panak PJ, Reich T, Raff J, Selenska-Pobell S, Bernhard G, Nitsche H (2000) Uranium(VI) complexation by *Bacillus species* isolated from uranium mining waste pile – a Comparative EXAFS Study

Kalin M, Smith MP (1986) Biological Polishing Agents for Mill Waste Water: An Example Chara, In: Lawrence, RW, Branion RMR, Ebner HG (eds.) Fundamental and Applied Biohydrometallurgy, p. 491

Kalin, M and Smith MP (1996) Mining Waste Management Areas : Converting Waste Sites to Ecosystems with Biologically Mediated Processes, Water Environment of Ontario Conference 1996, Design Criteria for Passive Treatment Processes for Mining Wastes, Toronto, Ontario 8 p.

Passive insitu treatment techniques, wetlands

Kalin M and Wheeler WN (1992) Periphyton Growth and Zinc Sequestration. Proceedings of the Ninth Annual General Meeting of BIOMINET, Edmonton, Alberta, pp. 49-64

Knauer K, Behra R, Hemond H (1999) Toxicity of inorganic and methylated arsenic to algal communities from lakes along arsenic contamination gradient. Aquat Toxicol 46: 221-230

Macaskie LE, Blackmore JD, Empson RM (1988) Phosphatase overproduction and enhanced uranium accumulation by a stable mutant of a *Citrobacter sp.* isolated by a novel method. FEMS Microbiology Letters 55: 157-162

Macaskie LE, Empson RM, Cheetham AK, Grey CP, Skarnulis AJ (1992) Uranium Bioaccumulation by a *Citrobacter sp.* as a Result of Enzymically Mediated Growth of Polycrystalline HUO_2PO_4. Science 257: 782 - 784

Meinrath G, Volke P, Merkel B, Dudel EG (1999) Determination and interpretation of uranium contents in environmental samples. Fres Anal Chem 364: 191-202

Planas D; Healy F (1978) Effects of Arsenate on growth and phosphorus metabolism of phytoplankton. J Phycol 15: 424 - 428

Vymazal J (1994) Algae and Element Cycling in Wetlands. Lewis Publishers, USA

Williams AR (1990) Radium uptake by freshwater plants. in: The Environmental Behaviour Of Radium. Vol 1 - Technical Reports Series 310; International Atomic Energy Agency, Vienna: 487-508

Yen P, Yu Q, Lin Z, Kaewsarn P (2001) Biosorption and desorption of Cadmium(II) by biomass of *Laminaria japonica*. Environ Technol 22(5): 509-514

Passive insitu treatment techniques, wetlands

Geophysics

1st – 4th term
(undergraduate studies)
* Mathematics, Computer Science
* Experimental and Theoretical Physics
* Earth Sciences
* Economic Geology
* Geophysics
* Technical basics
* Geological/geophysical fieldwork after the 2nd term
* Additional topic of choice
* Practical training in Computer Science after the 3rd term
* oral examination after 4th term

5th – 6th term (bachelor studies)
* Mathematics, Physics, Chemistry
* Applied Geophysics
* General Geophysics
* Additional topic of choice
* Bachelor examination

5th – 10th term (graduate studies)
* Mathematics, Physics, Chemistry
* Geophysics
* Geophysical methods
* Geophysical field trip and internship
* Masters thesis in the 10th term

Admission
* „Abitur"/Baccalaureate (A-level) or
* Specialized college degree or
* equivalent access entitlement

Duration/Degree: 6 semesters/Bachelor of Science
10 semesters/Master of Science in Geophysics (Dipl.-Geophys.)
Start: winter term

Job opportunities
Research and development; National and international field work; Scientific organizations; Consultancy and planning; Companies; Regional and federal institutions; Geophysical observatories and research centers; Engineering companies for environmental analysis or geotechnics; Geo-scientific research institutions and public services

Bacterial-based bioremediation of uranium mining waste waters by using sol-gel ceramics

J. Raff[1], U. Soltmann[2], S. Matys[3], M. Schnorpfeil[1], H. Böttcher[2], W. Pompe[3], S. Selenska-Pobell[1]

[1]Institut für Radiochemie, Forschungszentrum Rossendorf, 01314 Dresden
[2]Arbeitsgruppe Funktionelle Schichten, GMBU, 01317 Dresden
[3]Institut für Materialwissenschaft, Technische Universität Dresden, 01062 Dresden

Abstract. Cells, spores and the surface layer protein (S-layer) of the *Bacillus sphaericus* strain JG-A12, recovered from a uranium mining waste pile, were embedded in SiO_2 matrices by using sol-gel techniques. Sorption and desorption of uranium and copper by the free biocomponents and the obtained biological ceramics (biocers) were investigated. The biocer with cells possesses the highest binding capacity, followed by the S-layer biocer and the spore biocer. For renewed use of the biocers, the bound uranium and copper can be completely removed by washing with aqueous citric acid.

Introduction

During the long period of uranium mining and production in Saxony large amounts of ores were treated and leached. Waste piles and waste waters stemming from that procedures are still hazardously polluted with heavy metals and radionuclides. Some bacteria living in these environments are good candidates for construction of bacteria-based biological ceramics (biocers) for *in situ* bioremediation of the liquid wastes in that environments. The immobilization of the biological component in such bioceramics is of fundamental importance. It was demonstrated that by using the so called sol-gel technique various biomolecules or microbial cells can be embedded in SiO_2 ceramics without loosing their activity (Carturan et al. 1989, Braun et al. 1992, Böttcher 2000, Livage et al. 2001). The sol-gel ceramics possess further advantages such as (a) high mechanical, thermal and photochemical stability, (b) biological and toxical inertness and (c) controlled matrix porosity.

In this work *B. sphaericus* JG-A12, recovered from the uranium mining waste pile "Haberland" situated near the town of Johanngeorgenstadt (Selenska-Pobell et al. 1999), was used for the preparation of biological ceramics. Different biocers were prepared by embedding vegetative cells, spores and S-layer of *B. sphaericus* JG-A12 in porous matrices. The binding of uranium and copper by these biocers were studied by using ICP-MS, scanning electron microscopy and EDX analyses.

Material and Methods

Organism, growth, and preparation of spores and S-layer

Bacillus sphaericus JG-A12 cells were grown on nutrient broth (8 g l^{-1}), pH 7.0. All fermentations were performed as batch cultures at 30 °C, a stirring speed of 500 rpm using a magnetic stirrer and flow rates of 3 l air per min. Cells were harvested at the late exponential growth phase by centrifugation at 10000 g for 20 min.

For sporulation the nutrient broth medium was supplemented with 10 mg l^{-1} $MnSO_4$ x H_2O and the fermentation was carried out until the vegetative cells were completely sporulated. Spores were harvested by centrifugation at 10000 g for 20 min, resuspended in cold ultra pure water and treated alternately with 0.2 mg l^{-1} lysozyme and 0.1 mg l^{-1} trypsin at 4 °C for the complete lysis of the intact cells and cell wall fragments.

S-layer was prepared after the cells were suspended in a buffer solution (50 mM TRIS-HCl, 1 mM $MgCl_2$, 3 mM NaN_3, pH 7.5) and homogenized in a rotating-blade bender at maximum speed for 10 min on ice. By this procedure bacterial flagella were removed. After centrifugation at 6000 g for 10 min at 4 °C, cells were suspended 1:1 with the above buffer and mixed with a few crystals of DNAse II and RNAse. The cells were disintegrated in a mixer mill at 4 °C with glass beads with a diameter of 0.1 mm. Glass beads and unbroken cells were removed by differential centrifugation. Plasma membranes were solubilized in 1 % Triton X-100 in the buffer solution for 10 min at RT. The remaining peptidoglycan was lysed by incubating the samples in the buffer solution containing 0,2 mg ml^{-1} lysozyme for 6 h at 30 °C. The S-layer fraction was washed several times and suspended in the buffer solution. Isolated native S-layer sheets were stabilized by incubating with 1-Ethyl-3-(N,N'-dimethyl-aminopropyl)-carbodiimid (EDC) at a concentration of 30 mg ml^{-1} in buffer solution for 48 h at 20 °C. Crosslinked S-layer were harvested by centrifugation at 12400 g for 30 min at 4 °C, washed three times in distilled water and stored at 4 °C.

Transmission electron microscopy (TEM) investigations were carried out using a Philips CM200 with tungsten cathode at 160 kV or 200 kV. A droplet of sample solutions was placed onto a carbon coated TEM grid for about 1 min and then removed using a filter paper. Immediately after that, a droplet of 2% uranyl acetate was placed on the grid for 30 to 60 s for negative staining and again removed by using a filter paper. Then the samples were air-dried.

Preparation of aqueous silica nanosols and biocers

The sol was prepared by stirring 10 ml tetraethyl orthosilicate (TEOS), 40 ml ethanol and 20 ml 0.01N HCl catalyst for 20 h, 20 °C. For aqueous silica-sols the ethanol was evaporated by leading air through the solution. Successive volumes of evaporated ethanol were substituted by water. The resulting mean silica particle size in the aqueous nanosol was 6 nm.

The pH of the aqueous silica sol was increased up to about pH 7 by adding NaOH before mixing with *B. sphaericus* cells, spores or S-layer. 20 % (w/w SiO_2) sorbitol was added to the silica sols to get a higher porosity. Gelling occurs a short time after neutralization and addition of the biocomponent. The gels were aged for 3 days at 4°C, cut into small pieces and dried at room temperature or by freeze drying. The dry gels were sieved to particles with a size of 355-500 µm.

Sorption and desorption of uranium and copper

Binding of uranium and copper was investigated using 200 mg dry weight of biocers or silica gel particles with a size of 355-500 µm containing 36.4 mg cell biomass (2.6×10^{10} particles), 17.23 mg spore biomass (2.8×10^{10} particles) or 36.4 mg S-layer protein. The metal binding capacity of the same amounts of the free biocomponents were measured as well. All samples were shaken in 35 ml 0.9% $NaClO_4$, pH 4.5 with 9×10^{-4} M $UO_2(NO_3)_2 \times 6\ H_2O$ or $CuCl_2$ at 30 °C for 48 h. The amount of metal in the water solutions was determined by inductive coupled plasma mass spectroscopy (ICP-MS). Desorption experiments were carried out after washing with 0.9 % $NaClO_4$, pH 4.5 and 35 ml 0.5 M citric acid, trisodium salt, pH 4.5 at 30 °C for 24 h.

After the binding experiments biocers with bound uranium or copper were embedded in liquid colloidal silver on conductive carbon sheets. Samples were shadow casted with carbon and investigated with a scanning electron microscope type Gemini 982 (LEO) with a energy dispersive X-ray analyzer.

Results and discussion

The S-layer of *Bacillus sphaericus* JG-A12

Bacillus sphaericus JG-A12 cells possess as outermost component of its cell wall a paracrystalline protein layer (S-layer). This S-layer shows a square symmetry with a lattice constant of 12.5 nm (Fig. 1). The molecular weight of the *B. sphaericus* JG-A12 S-layer protein was estimated to be (135 ± 5) kDa. Its N-terminal domain differs significantly from these of the other *B. sphaericus* S-layers studied up to date. The natural function of the protein envelopes, which may form also other kinds of symmetry (oblique or hexagonally) is not uniform. The S-layers may function as molecular sieve or ion trap (Beveridge 1979, Sàra and Sleytr 1987, Sàra et al. 1992) and interact with several metals by forming nanoclusters (Shenton et al. 1997, Pompe et al. 1999, Wahl et al. 2001).

Fig. 1. TEM micrograph of uranyl-acetate negative stained native S-layer sheet of *B. sphaericus* JG-A12. The insert shows a correlation averaged reconstruction of a part of the TEM micrograph.

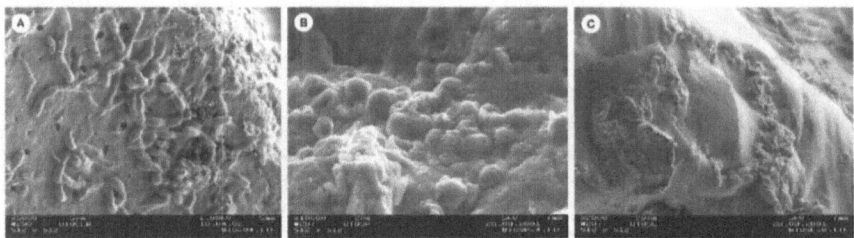

Fig. 2. SEM micrographs of biocers with embedded *B. sphaericus* cells (A), spores (B) and S-layer (C).

The metal binding makes the S-layers, beside the intact cells and spores of *B. sphaericus* JG-A12, a good candidate for embedding in sol-gel ceramics and use as selective filter material for remediation of heavy metal and radionuclide polluted waters.

Preparation of biological ceramics (biocer)

A complete and stable entrapment of the biological components is essential for their effective use in bioremediation technologies. Furthermore, the structure of the biocers especially their porosity strongly influences both the sorption and desorption kinetics and the binding capacity of the constructed particles.

To retain the high biosorption capacity known for vegetative cells and spores of *B. sphaericus* JG-A12 (Selenska-Pobell et al. 1999) the composite material was prepared using a aqueous sol-gel protocol. Thus cell-, spore- and S-layer-biocers were prepared by mixing with nanosols, gelling and air drying (Fig.2).

As shown in Fig.3 the biological components of air dried biocers were completely immobilized also on the surface after several washing steps. The biocers possess high stability but only a few pores and channels. An improvement of the fine structure increases the inner surface and, in this way, the binding rate and capacity. Adding water soluble compounds as sorbitol leads to a higher porosity (Fig. 3B). More efficient for the obtaining of better fine structure is the freeze drying of the silica gel (Fig. 3C).

Fig. 3. SEM micrographs of biocers with embedded *B. sphaericus* cells after air drying (A), with sorbitol (B) and after freeze drying (C).

In contrast to the air dried biocers or biocers with sorbitol in the biocers obtained by freeze drying no shrinkage of the material occurs, but they possess lower stability of the silica network.

Sorption and desorption of uranium and copper

To estimate the influence of the biocomponents immobilization on the metal binding the sorption of uranium and copper was investigated for each of the used biocomponents individually and for the corresponding biocer (Table 1).

Table 1. Metal binding capacity of constituents used for the biocers (dw = dry weight) and prepared biocers. One mg dry weight of biocer contains either 182 µg cells, 86.15 µg spores or 182 µg S-layer.

		uranium	copper
cells	µg/mg dw	64	6
	µg/10^9 cells	145	10
spores	µg/mg dw	110	8
	µg/10^9 spores	90	3
S-layer	µg/mg dw	19	2
silica gel	µg/mg dw	14	0.1
cell-biocer	µg/mg dw	18.5	1.3
spore-biocer	µg/ mg dw	13	0.35
S-layer biocer	µg/mg dw	14	0.35

Based on dry weight, spores show the highest binding capacity for uranium and copper followed by cells, S-layer and the silica gel. The binding of copper is more specific to the biological component, i.e. silica gel bounds uranium but not copper (Fig. 4).

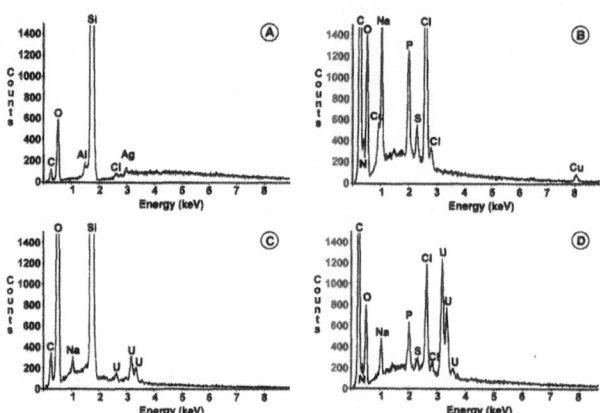

Fig. 4. EDX spectra of xerogel samples with copper (A) or uranium (C) and of *B. sphaericus* JG-A12 cells incubated with copper (B) or uranium (D).

In the case of the biological ceramics, cell-biocer binds more uranium and copper than the S-layer-biocer or the spore-biocer (Table 1). Determined binding capacities for the latter were noticeable lower than those of the xerogel and the spores. In the sorption experiments binding capacities of the spore-biocer reaches only 58 % for uranium and 47 % for copper of the theoretical values. Probably this is connected to the drying effects. During the drying steps the cells, the S-layer, and the xerogel matrix are shrinking due to their high water content. In contrast, spores are nearly water free and keep their original structure. For this reason only the surrounding xerogel is shrinking and leads to a loss in pores and in a decrease of the inner surface and binding sites.

The binding kinetics and capacities were positively influenced by freeze drying instead of air drying of the material or by adding sorbitol during the preparation of the biological ceramic (Fig. 5). In both cases the assignable cause is an improvement of the matrix porosity.

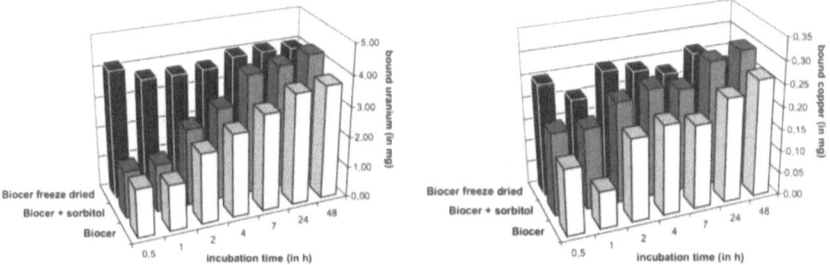

Fig. 5 Uranium (left) and copper (right) binding of biocers with different porosity. In all cases *B. sphaericus* cells were embedded.

The maximal amounts of both metals were bound by freeze dried biocers after half an hour, whereas the air dried and the sorbitol biocers reach their maximal binding capacity after 24 h for uranium and 48 h for copper. Furthermore, copper binding by freeze dried biocer is faster than by free cells, i.e. the structure of the modified biocer supports the sorption process.

Complete desorption of uranium and copper was possible by washing in 0.5 M citric acid. The binding of uranium is stronger than the binding of copper. Washing two times with 0.9 % $NaClO_4$ resulted in desorption of 19 % of the bound uranium and nearly 50 % of the bound copper.

Acknowledgements

This study was supported by grant DFG PO 392/15-1, SE 671/7-1, BO 1070/4-1 from Deutsche Forschungsgemeinschaft, Bonn, Germany. We thank R. Wahl (Institute of Materials Science, Technical University Dresden) for providing us the TEM micrograph.

References

Beveridge T J (1979) Surface Arrays on the wall of *Sporosarcina ureae*. J. Bacteriol. 139 (3): 1039-1048
Böttcher H (2000) Bioactive Sol-Gel Coatings. J. Prakt. Chem. 342: 427-436
Braun S, Shtelzer S, Rappoport S, Avnir D and Ottolenghi M (1992) Biocatalysis by sol-gel entrapped enzymes. Journal of Non-Crystalline Solids, 147&148: 739-743
Carturan G, Campostrini R, Dirè S, Scardi V and De Alteriis E (1989) Inorganic gels for immobilization of biocatalysts: Inclusion of invertase-active whole cells of yeast (Sac-

charomyces cerevisiae) into thin layers of SiO₂ gel deposited on glass sheets. Journal of Molecular Catalysis, 57: L13-L16

Livage J, Coradin T and Roux C (2001) Encapsulation of biomolecules in silica gels. J. Phys.: Condens. Matter, 13: R673-R691

Pompe W, Mertig M, Kirsch R, Wahl R, Ciacchi L C, Richter J, Seidel R and Vinzelberg H (1999) Formation of Metallic Nanostructures on Biomolecular Templates. Z. Metallkd. 90 (12): 1085-1091

Sára M, Pum D and Sleytr U B (1992) Permeability and charge-dependent adsorption properties of the S-layer lattice from *Bacillus coagulans* E38-66. J. Bacteriol. 174 (11): 3487-3493

Sára M and Sleytr U B (1987) Charge distribution on the S-layer of *Bacillus stearothermophilus* NRS 1536/3c and importance of charged groups for morphogenesis and function. J. Bacteriol. 169 (6): 2804-2809

Selenska-Pobell S, Panak P, Miteva V, Boudakov I, Bernhard G, Nitsche H (1999) Selective accumulation of heavy metals by three indigenous *Bacillus* strains *B. cereus, B. megaterium* and *B. sphaericus* from drain waters of a uranium waste pile. FEMS Microbiol. Ecol. 29:59-67

Shenton W, Pum D, Sleytr U B and Mann S (1997) Synthesis of cadmium sulphide superlattices using self-assembled bacterial S-layers. Nature 389: 585-587

Wahl R, Mertig M, Raff J, Selenska-Pobell S and Pompe W (2001): Electron-Beam Induced Formation of Highly Ordered Palladium and Platinun Nanoparticle Arrays on the S layer of *Bacillus sphaericus* NCTC 9602. Adv. Materials 13 (10): 736-740

Passive insitu treatment techniques, wetlands

Uranium attenuation from tailing waters by floating macrophyte Lemna gibba L.

Martin Mkandawire, Gert E. Dudel

Dresden University of Technology, Institute of General Ecology and Environmental Protection, Pienner Straße 19, 01737 Tharandt. Germany. Tel: 49 351 463 31393. Fax: 49 351 463 31399. E-mail: mkanda@forst.tu-dresden.de

Abstract. Biosorption of uranium from mine tailing waters by floating macrophyte *Lemna gibba* L. was investigated in nutrition solutions. Changes in pH (from average 7.0 to 5.5), speciation and removal of U from the solutions and accumulation in the plant directly corresponded to specific growth rate, Fe and PO_4 speciation. Macrophytes (e.g. *Lemna* sp.) are reported to produce a range of 5-1000µg kg^{-1} day^{-1} dry biomass of organic compounds particularly oxalic acids and proteins which are triggered by Fe, or PO_4 deficiency. Modelling predicts specation change of Fe and PO_4, and U-oxalic comlexation which influence U speciation and its bioavailability.

Introduction

Capacity and applicability of plant and microbe mediated remediation processes of heavy metal (HM's) and radionuclide (RN's) contamination are well documented e.g. Boniardi, et al., 1994; Lewis and Wang, 1997; Macaskie, et al., 1997; Salt, et al., 1995. This has drawn increasing attention from ecological and environmental engineers to salvage the knowledge for use in cost-effective remediation and restoration of former uranium mining, dumps and tailings. The interest is gradually shifting from adoption to identifying mechanism employed in plant remediation processes that can be engineered for more effectiveness and efficiency.

Knowledge of the physiological and molecular mechanisms of phytoremediation have emerged together with biological and engineering strategies for optimisation and improvement (Miller, 1996). These benefit from: (i) the ability of some plant species to avoid toxic substances from transferring and accumulating in the shoots through rhizofiltration and sorption, and/or complexation on organic surfaces on root or bacteria cell surfaces and in rhizodermis, or production of counteractive biochemical which may transform toxicant to non bio-available state (e.g. change of redox state); and, (ii) species with exceptional high turn over and capable to accumulate and tolerate high levels of toxic substances. In a longer run of carbon fixation (primary production), dilution in accumulated recalcitrant organic matter occurs in relation to the concentration in mining debris and tailings (Dudel, et al., 2001). The mechanisms involved are mostly under investigation, but likely depend on the RM and HM bioavailability and mobility dictated by their speciation. The most important speciation parameters are the concentrations of ions and the extent of complexation by organic ligands, because they provide information on free metal ion concentrations, conditional stability constants of metal complexes, and ligand concentrations.

From this, it is hypothesised that plants produce organic compounds (ligands or chelatins) to counter respond toxic substance in their environment, which change speciation of essential elements, is principle mechanism by which plants acquire metal ions (Treeby, et al., 1989). Hence the chemistry of biogenic organic matter is one key to understand mechanisms of bioavailability for remediation purposes. This was investigated with a floating macrophyte, *Lemna gibba* L. in nutrition solutions whose properties were adapted to natural conditions based on field data from uranium tailing waters of Lengefeld and Mechelgrun in eastern Germany. Speciation of components of the solution were calculated and processes of speciation modelled with computer programme PhreeqC.

Methods

Chemicals and nutrient solutions

The experiments were carried out with modified standard Hutner solutions (table 1) in batch culture, factorial arranged and replicated four times. Modifications were in respect to PO_4 content to produce nutrient solution I and II and control. 100 mg L^{-1} of uranium stock solution was prepared from analytical grade $UO_2(NO_3)_2 \cdot 6H_2O$, and diluted to the experimental required concentrations of 20, 50, 100, 250, 500, and 1000 μgL^{-1} in nutrient media.

Test plant

Lemna gibba L was obtained from the Humboldt University Berlin Arboretum Baumschulenweg and stock cultured in standard Hutner solution (Table 1) adjusted to pH 7.0 ± 0.5. The culture was maintained by selecting 30 fronds for re-culturing every two weeks repeatedly for two months.

Table 1. Nutrition solution based standard Hutner for hydroponic culture

Macro- elements (mgl^{-1})	$CaCl_2 \cdot 2H_2O$	12.2
	K_2HPO_4*	40.0
	$MgSO_4 \cdot 7H_2O$	50.0
	NH_4NO_3	20.0
Micro-elements (µgl^{-1})	H_3BO_3	1500.0
	$ZnSO_4 \cdot 7H_2O$	6500.0
	$CuSO_4 \cdot 5H_2O$	400.0
	$CoSO_4 \cdot 7H_2O$	20.0
	$MnCl_2 \cdot 4H_2O$	3514.8
	$FeSO_4 \cdot 7H_2O$	2500.0
	$Na_2MoO_4 \cdot 2H_2O$	2500.0
Chelate	Na_2 EDTA$\cdot 2H_2O$	50.0

* In the experiments when PO_4^{3-} was a variable, other PO_4^{3-} concentrations were as follows: 0 µgl^{-1} as control; 136.1µgL^{-1} = 1 µMol; and 13.61mgL^{-1} = 10 µMol

Speciation simulations and calculations

The speciation of components in the nutrient solution in the course of experiment in milieu of Lemna was calculated with the PhreeqC computer program (USGS, 2001). A modified database which include species elements such as Mo, Co and a number of organic ligands e.g. oxalic acid of interest in the study was used in the procedure (Mkandawire, 2001).

Bioaccumulation and removal experiments

30 fronds (2 or 3 fronds per plant) of similar size satisfying a seven-day pre-culture conditions were systematically inoculated into two parallel set of experiments, placed one at a time into each of 250 ml Erlenmeyer vessel containing 200 ml of test solutions until each contained 30 fronds. The vessels placed in a phyto-chamber (NEMA, Netzschkan, Germany) programmed at temperatures 24 °C (±1 °C) during the 14 hour photo-period (day) of light intensity (PAR) of 85-125 µMol m^{-2} s^{-1} and 16 °C (±1 °C) in the night. The experiment period was 21 days and growth was monitored through increase in number of fronds and frond area at every 24 hours. Lemna biomass and back ground U content in plant and media were quantified at the start of the experiment.

In the first set of experiment, growth in relation to U concentration was observed. Media and plant samples were collected at the end of the experiment. In second set, samples were collected every 5 days. In both sets biosorption was monitored by collecting 10 ml media samples every 30 minutes for the first 3 five hours. All samples were processed for chemical analysis as described below.

Element analysis

Background content of Uranium in *L. gibba* and nutrient media were quantified at the start. Samples of Lemna were collected wholly, washed, freeze dried (Christ freeze dryer ALPHA 1-2/LD, Osterode, Germany), weighted, and digested by HNO_3 in microwave digester (CEM Model MDS 200, USA). All samples were diluted with 2 % HNO_3 to required concentrations for determination by inductive couple plasma – mass spectroscopy (ICP-MS PQ2+ VG Elemental, UK) and atomic absorption spectroscopy (AAS GF95, Unicam, UK). Total phosphorus was determined with inductive couple plasma - optical emission spectroscopy (ICP-OES Perkin Elmer Plasma 2, USA).

Data analysis

Specific growth rate (μ) was defined as:

$\mu = (\ln N_{tn} - \ln N_{t0})/ (t_n - t_0)$

where N_{tn} is number of fronds at time n (t_n) and N_{t0} is number of fronds at time 0 (t_0). Dry biomass has been calculated from proportionality calibration of frond number-biomass relationship.

Bioaccumulation values (φ) (in g kg^{-1}) of Uranium in *L. gibba* were defined as:

$\varphi = g_u / g_L$

where (g_u) is mass U in frond material and (g_L) is dry Lemna fronds biomass.

Uptake was defined as:

$q = (Cf - Ci)V/g_L$

where Ci is initial U concentration and correspond to the control samples, Cf is final U (residual) concentration, V is media volume.

The results were subjected to Analysis of Variation (ANOVA) or student t-test for pair wise analysis, and regarded significant at $p \leq 0.05$. Statistical package SPSS version 10.0 for MS Windows was used.

Results and discussion

U effect on growth

Lemnaceae species are widely used macrophytes in ecotoxicological and phytoremediation studies due to their wide distribution, fast growth, short life span, simple physiological formation and their sensitivity to environmental change (Landolt 1980; Lemon et al. 2001). They have been found growing in relative high heavy metal content environment (Fe 300 mg l^{-1}; U 30µg - 300 mg l^{-1}) in uranium tailing waters of mines in eastern Germany (unpublished results).

The effects of U Concentration on μ in *L. gibba* are illustrated in Figure 1. Maximum μ was observed at initial 8 to 10 days of growth after inoculation in the nutrition medium. High U concentrations (500 and 1000 µg/l) in the medium inhibited significantly the specific growth rate (student t-test $p = 0.05$).

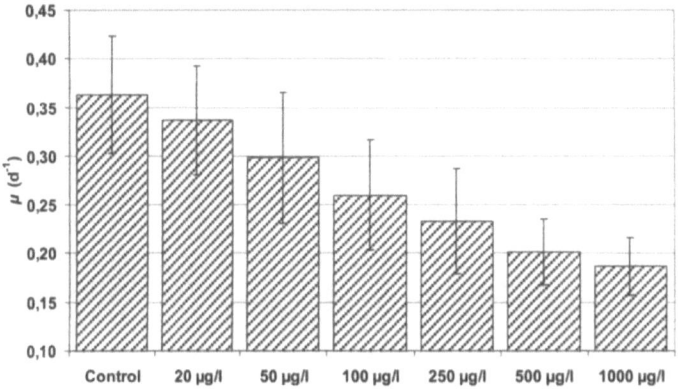

Fig. 1. μ (d^{-1}) of *L. gibba* grown on modified Hutner medium with different concentrations of U. The presented μ is mean of 4 replicate, and calculated from frond number.

Biomass dependent U biosorption

The rate of surface absorption of U was found to be maximum at between 70 and 120 hr of growth. However, significant U uptake was noticed during the initial 24 hr of growth. From Figure 2 it is evident that the rate of U absorption by *L. gibba* is dependent on the concentration, and is related to growth increment. Reduced frond size and chlorosis were observed at the concentration of 500 and 1000 µg/l of U.

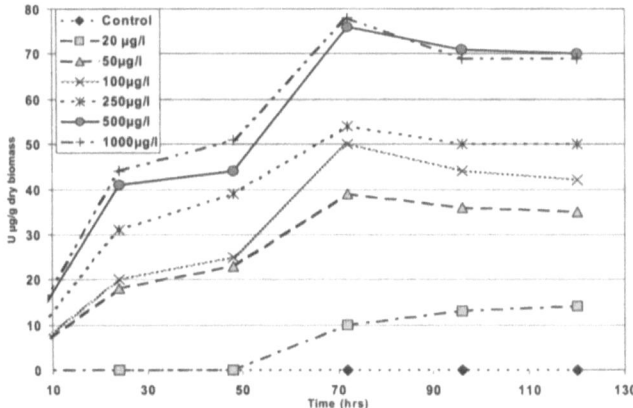

Fig. 2. Rate of φ in *L. gibba* L. after batch culture in modified standard Hutner.

Passive insitu treatment techniques, wetlands

Table 2. Species of U, PO_4, Fe, with calculated oxalic acids at the beginning and the end of the experiment. A modified database is used in the PhreeqC procedure

PO4 conc.	Elements	Uranium concentration (μgL^{-1})			
		100		1000	
		No. Species	Molality	No. Species	Molality
1.	At the beginning of the experiment, oxalic acid = 0.00 μgL^{-1}				
13.6	Fe(2)	10	1.00×10^{-10}	11	1.00×10^{-10}
	Fe(3)	16	8.99×10^{-6}	18	8.99×10^{-6}
	P*	26	1.00×10^{-6}	24	$1.00 \times 10^{-6*}$
	U(4)	14	0.00	13	0.00
	U(5)	1	0.00	1	0.00
	U(6)	12	4.20×10^{-7}	12	4.20×10^{-6}
40.0	Fe(3)	19	8.99×10^{-6}	19	8.99×10^{-6}
	P	24	2.32×10^{-4}	25	$2.32 \times 10^{-4**}$
	U(4)	15	0.00	15	0.00
	U(5)	1	0.00	1	0.00
	U(6)	13	4.20×10^{-7}	13	4.20×10^{-6}
2.	At the end of experiment. oxalic acid = 100.0 μgL^{-1}				
13.6	Fe(2)	10	1.06×10^{-10}	10	1.15×10^{-10}
	Fe(3)	16	3.57×10^{-8}	16	3.57×10^{-8}
	P		n.d.		n.d.
	U(4)	14	0.00	5	0.00
	U(5)	1	1.62×10^{-9}	1	1.55×10^{-8}
	U(6)	10	1.29×10^{-7}	9	1.49×10^{-6}
40.0	Fe(2)	11	1.08×10^{-10}	10	1.15×10^{-9}
	Fe(3)	18	3.57×10^{-8}	16	3.57×10^{-7}
	P	19	$2.15 \ 10^{-5}$	17	4.53×10^{-7}
	U(4)	13	0.00	11	0.00
	U(5)	1	0.00	1	1.36×10^{-8}
	U(6)	12	$1.31 \times 10^{-7***}$	9	$1.50 \times 10^{-6***}$

* predicted precipitation UO2(OH)2:H2O saturation index (SI) = 0.14;**MnHPO4 SI = 0.96;*** U3O8(C) SI = 1.11; U3O9 SI = 0.84; Fe2O3 SI = 2.93

Estimated effects on U speciation

Change in speciation of solution components corresponded to *L. gibba* L. specific growth rate (table 2). The change strongly influenced removal of uranium from the solution, and accumulation in *L. gibba*, and were significantly correlated to calculated biomass yield and specific growth rate ($r^2 = 0.87$; p= 0.05). Specifically, the correspondence was in respect to Fe, PO_4 species in the nutrient solution. *Lemnaceae sp.* can produce a range of 5-1000μg kg^{-1} dry biomass day^{-1} of organic acids, saccharides, or/and proteins when toxic metals are introduced into the environment (Golovchenko, et al., 2002; Miranda and llangovan, 1996; Ovodova, et al., 2000). *L. gibba* increases production of organics with increasing Pb concentration in nutrient solution until when specific growth rate is significantly inhibited (Miranda and llangovan, 1996). Production of oxalic acids in *Lemna minor* L increased with biomass production (Baker and Farr, 1987; Bornkamm, 1965). Physiological and biochemical formation in *L. minor* for Ca-oxalate formation have been studied (Li and Franceschi, 1990). Inverse modelling of the residual solution at the end of experiment to the original with PhreeqC predicted U-oxalate species models at an allowable uncertainty of 0.25.

U-Oxalic acid complexation

U in plant material and residual nutrient were far from equilibrium. A range of 6 to 10 % in low, and 15 to 18 % in high concentration U loss were observed. The trend was typical of sorption isotherms. Such losses may result from dilution and filtered-out precipitation (on 0.2 μm membrane filter, Satorius Göttngen) from residual solution. Comparison with control (loss range of 2.3 ± 1.7 to 4.2 ± 1.3 %), reveals that U loss was mostly through filtration of U complexed on oxalic acids or derivatives. This agreed with observed pH decreased which corresponded to specific growth rate and the U concentration i.e. least pH changes observed in U 20 $\mu g\ L^{-1}$ (pH 6.95 ± 0.04 to 6.13 ± 0.24) and 1000 $\mu g\ L^{-1}$ (pH 6.95 ± 0.19 to 5.9 ± 0.18), and highest in 250 $\mu g\ L^{-1}$ (pH 6.91 ± 0.10 to 5.49 ± 0.23). It can be concluded with assumption from literature (Bornkamm, 1965; Miranda and llangovan, 1996; Treeby, et al., 1989) that the decrease and change of speciation was a result of oxalic acid exudates from *L. gibba*. P and Fe had an important role, even though S is not ruled out for its properties close to P.

Conclusion

These suggest that *L. gibba* L. and lemnaceae in general potentially immobilise or remove U efficiently from tailing waters through biosorption and exudation of oxalic acids and that modification of P, Fe and other essential elements can optimise the system's efficiency. Exudation of oxalic acids by *L. gibba* is deductive from inverse modelling, secondary analytic and literature e.g. (Bornkamm, 1965) and (Treeby, et al., 1989). Modelling may be restricted by thermodynamic database used in PhreeqC procedure. Hence, further investigation and analytical determination of oxalic acid and other organic substance from *L. gibba* in tailing waters is recommended.

Acknowledgement

The Project was supported by Bundesministrium für Bildung und Forschung (BMBF). Martin Mkandawire is supported in Germany by DAAD. Frau Agnes

Franke, Karin Krinzmann and Annet Jost assisted in laboratory procedures. Dip. Chem. Arndt Weiske did all analyses and Dip. Min. Holger Dienemann worked in hand in modifying thermodynamics databases for PhreeqC.

References

Baker, J. H. and Farr, I. S. (1987). Importance of organic matter produced by duckweed (Lemna minor) in southern English river. *Freshwater Biology* 17, 325-330.

Boniardi, N., Vatta, G., Rotta, R., Nano, G. and Carra, S. (1994). Removal of water pollutants by Lemna gibba. *Biochemical Engineering Journal* 54, 41-48.

Bornkamm (1965). Die Rolle des Oxalats im Stoffwechsel höherer grüner Pflanzen: Untersuchungen an *Lemna minor* L. *Flora, Abt. A* 156, 139-171.

Dudel, E. G., Brackhage, C., Dienemann, H., Mkandawire, M., Rotsche, J. and Weiske, A. (2001). Principles and limitations for natural attenuation of radionuclides in former uranium mining and milling sites. *The 8th Int. Conference of Radioactive Waste and Environmental Management*, pp. 8.

Golovchenko, V. V., Ovodova, R. G., Shashkov, A. S. and Ovodov, Y. S. (2002). Structural studies of the pectic polysaccharide from duckweed Lemna minor L. *Phytochemistry* 60, 89-97.

Lewis, M. A. and Wang, W. (1997). Water quality and aquatic plants in Raton, B. (Ed), *Plants for environmental studies*, CRC Press.

Macaskie, L. E., Yong, P., Doyle, T. C., Diaz, M. and Monzas, T. (1997). Bioremediation of uranium-bearing wastewaters: Biochemical and chemical factors influencing bioprocess application. *Biotechnol.-Bioeng.* 53, 100-109.

Miller, R. R. (1996). Phytoremediation, Ground-water Remediation Technology Analysis Centre in cooperation with United States Environmental Protection Agency's (EPA) and Technology Innovation Office (TIO).

Miranda, M. G. and llangovan, K. (1996). Uptake of Lead by *Lemna gibba* L.: Influence on Specific Growth Rate and Basic Biochemical Changes. *Bull. Environ. Contam. Toxicol.* 56, 1000-1007.

Mkandawire, M. (2001). PHREEQC (or PHAST) data base, organic ligands and trace elements, USGS.

Ovodova, R. G., Golovchenko, V. V., Shashkov, A. S., Popov, S. V. and Ovodov, Y. S. (2000). Structural studies and physiological activity of lemnan, a pectin from Lemna minor L. *Bioorganicheskaya Khimiya* 26, 743-751.

Salt, D. E., Blaylock, Kumar, P. B. A. N., Dushenkov, V., Ensley, B. D. and Chet, I. (1995). Phytoremediation: A novel strategy for removal of toxic metals from the environment using plants. *Biotechnology* 13, 468-475.

Treeby, M., Marschner, H. and Römheld, V. (1989). Mobilization of iron and other micronutrient cations from a calcareous soil by plant-borne, microbial, and synthetic metal chelators. *Plant and Soil* 114, 217-226.

USGS (2001). PHREEQC - A Computer Program for Speciation, Batch-Reaction, One-Dimensional Transport, and Inverse Geochemical Calculations, United States Geological Survay.

Passive insitu treatment techniques, wetlands

Exploring the potential for natural attenuation to ground water cleanup goals at select former uranium mill tailings sites in the western United States

Donald R. Metzler

U.S. Department of Energy Grand Junction Office, Grand Junction, Colorado

Abstract. One of the missions of the U.S. Department Energy (DOE) is to plan, implement, and complete the restoration of the 22 former uranium-ore processing sites identified as Title I sites in the Uranium Mill Tailings Radiation Control Act. DOE controlled and stabilized the tailings and associated contaminated material at these sites in engineered disposal cells that have design longevities of 1,000 years. All the disposal cells and associated surface remediation activities were completed by 1998 under the Uranium Mill Tailings Remedial Action (UMTRA) Project. Passive ground water cleanup with the application of natural attenuation (flushing), coupled with monitoring and institutional controls, has been proposed for 8 of the 22 sites. Natural attenuation allows the natural ground water movement and geochemical processes to decrease contaminant concentrations. Establishment of compliance targets by identifying point-of-compliance wells and proposed concentration limits gives confidence that performance measures are in place to determine the adequacy and effectiveness of the natural attenuation process. The U.S. Environmental Protection Agency requires that natural attenuation must reduce contamination to levels within regulatory limits within 100 years. DOE has been working toward developing ground water compliance at these sites since 1991. The U.S. Nuclear Regulatory Commission (NRC) has approved natural attenuation for one of the eight sites.

Introduction

Natural attenuation with monitoring, coupled with viable institutional controls, can be an effective solution for addressing residual contaminated ground water resulting from past milling practices (International Atomic Energy Agency 1999). Most of the former uranium mills in the United States operated in the 1950s and 1960s. Many of these mills were sited over shallow alluvial systems adjacent to surface water, such as a river. Practices in those early years lead to contamination of the underlying aquifers by seepage of contaminants from milling activities and tailings.

One of the missions of the U.S. Department of Energy (DOE) is to plan, implement, and complete the restoration of the 22 former uranium-ore processing sites identified as Title I sites in the Uranium Mill Tailings Radiation Control Act (UMTRCA). DOE controlled and stabilized the tailings and associated contaminated material at these sites in engineered disposal cells that have design longevities of 1,000 years. All the disposal cells and associated surface remediation activities were completed by 1998 under the Uranium Mill Tailings Remedial Action (UMTRA) Project. The UMTRA Ground Water Project was assigned to the DOE Grand Junction Office in 1996 with responsibility to achieve compliance with ground-water standards at the 22 sites.

Passive ground water cleanup with the application of natural attenuation, coupled with monitoring and institutional controls, has been proposed for 8 of the 22 sites. Natural attenuation allows the natural ground water movement and geochemical processes to decrease contaminant concentrations. The U.S. Environmental Protection Agency (EPA) requires that natural attenuation (flushing) must reduce contamination to levels within regulatory limits within 100 years. DOE has been working toward developing ground water compliance at these sites since 1991. The U.S. Nuclear Regulatory Commission (NRC) has approved natural attenuation for one of the eight sites.

A natural attenuation strategy will be effective if (1) the source can be completely decoupled from the ground water; (2) advective, dispersive, and geochemical processes will be active enough to reduce or dilute contaminant particles within the allowed time frame; (3) the ground water resource is neither currently nor projected to be a public drinking water supply; (4) monitoring and enforcement of an effective institutional control will be in place for the necessary time period; and (5) adequate site characterization, uncertainty analysis, and numerical flow-and-transport modeling have been completed and approved by NRC (U.S. Department of Energy 1999). Establishment of compliance targets by identifying point-of-compliance wells and proposed concentration limits gives confidence that performance measures are in place to determine the adequacy and effectiveness of the natural attenuation process.

Problem statement

In the early days of siting uranium-ore processing mills, environmental factors were generally low-priority considerations. Only in the later years of uranium-ore milling did air and water pollution problems, accentuated by proximity of population centers, streams, and rivers, affect decisions on mill locations and processing methods (see Fig. 1).

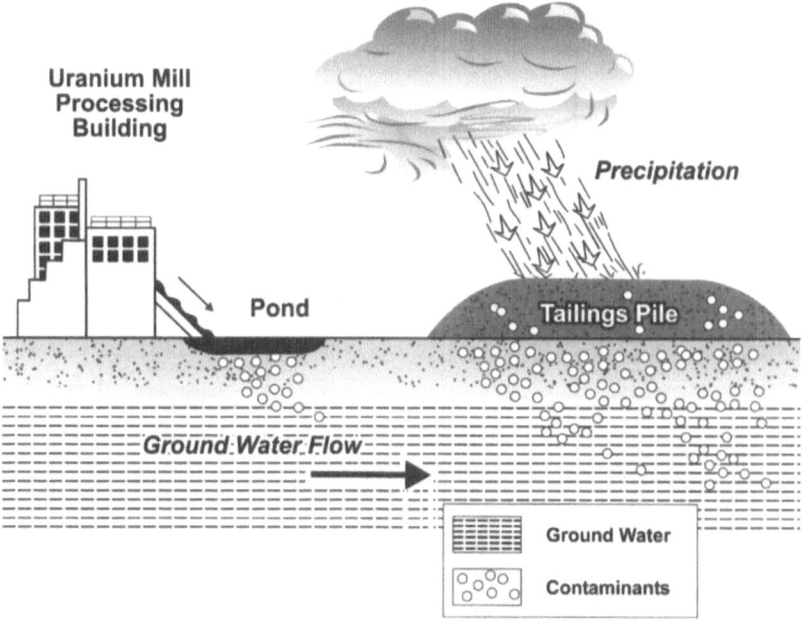

Fig. 1. Uranium mill processing site (U.S. Department of Energy 1996).

The extraction phase of the milling operation involved extractive metallurgy to obtain solutions containing the dissolved uranium. In some cases, a roasting step was necessary. The leaching step was a key operation in the mill process with the choice of acid or alkaline leach reagents. Sulfuric acid and sodium carbonate were the common reagents used. Acid was used at a majority of the mills (see Fig. 2). Solid or liquid oxidants were added during leaching if the uranium minerals were in a reduced state. A further step in the extraction phase of the milling operation was the separation of the pregnant solution from waste solids. This process usually involved resin-ion exchange and solvent extraction. The final step in the process was chemical precipitation of uranium from the solution. In alkaline process operations, this step involved the addition of caustic carbonate-bicarbonate liquors.

Case studies: active and abandoned Uranium mines

Processing Operations

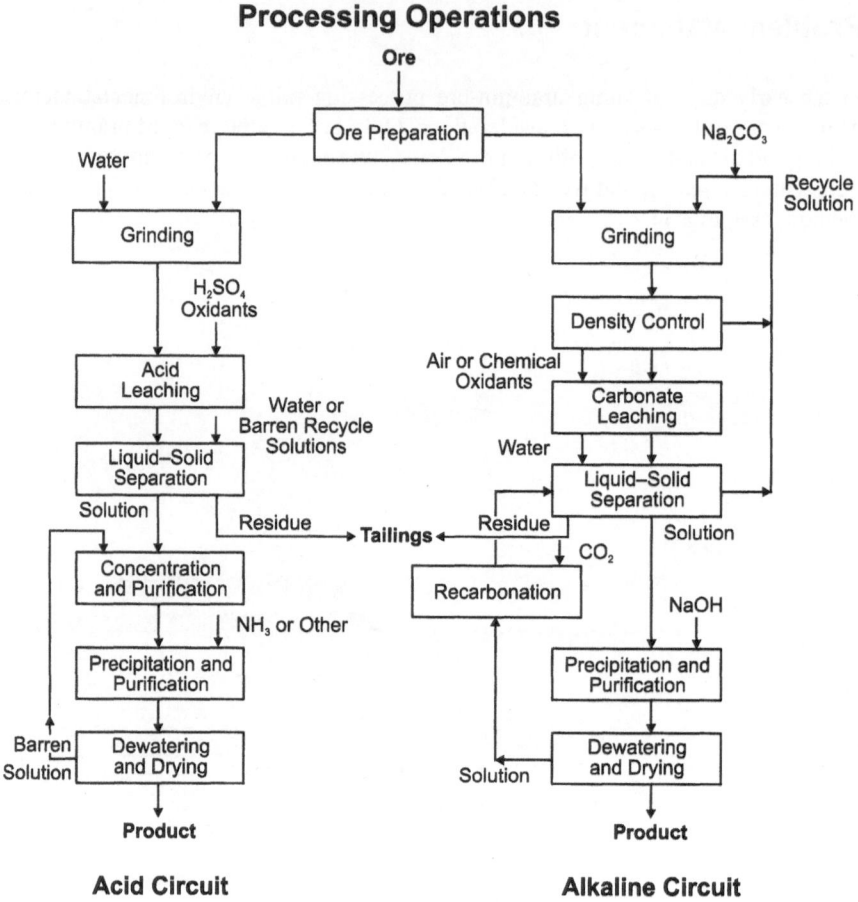

Fig. 2 Acid and alkaline circuits used to extract dissolved uranium (Merritt 1971).

In acid process plants, the uranium was precipitated by neutralization with a base such as lime, magnesia, or ammonia. Final stages of the milling involved dewatering (Merritt 1971).

Many of the mills that used these extraction processes were located over shallow alluvial systems adjacent to surface water, such as streams and rivers (see Fig. 3). These milling practices resulted in large amounts of contaminated ground water.

Fig. 3. UMTRA Project site locations.

The UMTRA Ground Water Project sites have an estimated total of more than 40,000,000 cubic meters of contaminated ground water. Contaminants in the ground water include arsenic, cadmium, chromium, lead, molybdenum gross alpha, nitrate, radium, selenium, uranium, manganese, and ammonium.

Meeting the ground water standards

DOE is required to meet ground-water cleanup standards set by EPA in 60 *Federal Register* (FR) 2854 (Federal Register 1995). The standards in Title 40 *Code of Federal Regulations* (CFR) Part 192, Subpart B, are a combination of prescriptive and risk-based provisions (Code of Federal Regulations 1993). Ground-water remedial actions are required to meet, at a minimum, one of the following provisions:

- **Background levels** for contaminants of concern.
- **Maximum concentration limits** (MCLs) that are EPA's maximum concentrations of certain hazardous constituents for ground water protection. Hazardous constituents with MCLs that may be present in contaminated ground water at UMTRA Project sites include arsenic, barium, cadmium, chromium, lead, mercury, molybdenum, nitrate, radium, selenium, silver, and uranium.
- **Alternate concentration limits,** which are concentrations of contaminants that may exceed the MCLs or limits for those constituents without MCLs. If

Case studies: active and abandoned Uranium mines

DOE determines and NRC concurs that human health and the environment would not be adversely affected, DOE may meet an alternate concentration limit to meet the EPA ground water standards.

The standards also define limited-use ground water. Ground water may be classified as limited use if the total dissolved solids exceed 10,000 milligrams per liter, if there is widespread surrounding contamination that cannot be cleaned up using treatment methods reasonably employed in public water supply systems, or if the quantity of available ground water is less than 570 liters per day.

The standards also have provisions that allow natural flushing to meet the EPA ground water standards. Natural flushing means natural ground water processes reduce the contamination in ground water to background levels, to below MCLs, or to alternate concentration limits. The following conditions must be met before natural flushing can be implemented:

- Natural flushing must allow standards (i.e., background levels, MCLs, or alternate concentration limits) to be met within 100 years.
- Institutional controls with a high degree of permanence that will effectively protect public health and the environment and satisfy beneficial uses of ground water must be viable and enforceable.
- Ground water must not be a current or projected source for a public water system during the period of natural flushing. A public water system is defined in Title 40 CFR Part 125.58 as a "...system for the provision to the public of piped water for human consumption, if such a system has at least fifteen (15) service connections or regularly serves at least twenty-five (25) individuals. This term (public water system) includes (1) any collection, treatment, storage, and distribution facilities under the control of the operator of the system and used primarily in connection with the system; and (2) any collection of pretreatment storage facilities not under the control of the operator of the system which are used primarily in connection with the system."

The standard also requires that DOE monitor the concentrations in ground water contaminants for compliance with Subpart B standards and define the extent of ground water contamination so that measures can be taken, if necessary, to protect human health and the environment.

The EPA standards specify a point of compliance for disposal of the surface contamination but indicate that this does not suffice for the cleanup of contaminated ground water. For the UMTRA Ground Water Project, "...compliance must be achieved anywhere contamination above the levels established by these standards is found or projected to be found in ground water outside the disposal area and its cover" (Federal Register 1995).

Institutional controls

Institutional controls are controls that effectively protect public health and the environment. They typically depend on some social order to ensure that protection is effective. For the UMTRA Ground Water Project, institutional controls would reduce exposure to or mitigate health risks by (1) preventing intrusion into contaminated ground water or (2) restricting access to or use of contaminated ground water for unacceptable purposes. As a last resort, institutional controls could limit human access to the land above the contaminated ground water. The EPA standards allow the use of institutional controls in place of remediation only if their effectiveness can be verified and maintained. The EPA standards permit the use of institutional controls at sites where remediation can occur through natural flushing of the aquifer within 100 years. However, the standards do not limit the use of institutional controls to the sites that can meet the standards through natural flushing. Institutional controls can also be used to protect public health or the environment when DOE finds them necessary and appropriate before the start of active remedial action, during active remedial action, or during implementation of other compliance strategies.

The EPA standards require that institutional controls

- Have a high degree of permanence.
- Protect public health and the environment.
- Satisfy beneficial uses of ground water.
- Are enforceable by administrative or judicial branches of government entities.
- Can be effectively maintained and verified.

An example of acceptable institutional controls cited in the EPA standards is a deed restriction that can be enforced by a unit of government (either administratively or through judicial processes). Another example is federal or state ownership of land containing contaminated ground water. EPA recognizes that a combination of controls may be needed to protect public health and safety adequately. Measures such as signs, health advisories, or other measures that require voluntary cooperation of private parties can be used to complement other enforceable institutional controls but cannot be considered as primary protective measures. In addition, the use of an alternate water supply in conjunction with institutional controls that would prevent human contact with contaminated ground water would be a viable institutional control.

Key to identifying, implementing, and enforcing institutional controls is participation by tribal, state, and local governments. While DOE is responsible for compliance with the EPA standards at UMTRA Ground Water Project sites, its authority to implement and enforce institutional controls may be limited, particularly where tailings are disposed of in repositories that are not located on the processing site and the land is privately owned or is owned or controlled by tribal, state, or other public agencies. Similarly, ground water contamination resulting from uranium-ore processing may have moved beyond the processing site to areas that are not within the DOE jurisdiction.

The need for and duration of institutional controls depend on the compliance strategy selected for a site, the type and level of risk, and existing site conditions. As risks decrease over time, so should the restrictiveness of institutional controls. Contaminated plume movement might require applying the restrictions to an extended area over time. Therefore, the effectiveness of institutional controls

Case studies: active and abandoned Uranium mines

needs to be verified and modified as necessary to ensure extended protection of public health, the environment, and beneficial uses the water could have satisfied.

Institutional controls, if any, will be selected in cooperation with Indian tribes, states, and local governments. DOE will verify that the institutional controls are effective. Site-specific institutional controls will not be selected and implemented without DOE and NRC concurrence.

Natural attenuation processes

Natural attenuation processes occur at all sites but to varying degrees, depending on the types and concentrations of contaminants and the hydrologic and geologic characteristics of the site. Natural attenuation may reduce the potential risks posed by site contaminants in a number of ways, depending upon the type of contaminants:

- Contaminants may be transformed to a less toxic form through destructive processes such as radioactive decay or biodegradation.
- Potential exposure levels may be reduced by lowering concentration levels through processes such as dilution and dispersion.
- Contaminant mobility may be reduced by geochemical processes that involve contaminant interactions with the ground water and aquifer matrix.

Table 1 identifies the most common geochemical processes that control contaminant migration through an aquifer.

DOE's monitored natural attenuation definition

Monitored natural attenuation (MNA) may be defined as the reliance on natural attenuation processes, within the context of a carefully controlled and monitored site cleanup, to achieve site-specific remedial objectives within a time frame that is reasonable compared to that offered by more active methods. Monitoring, therefore, is the critical component of any remediation by natural attenuation. Monitoring is imperative (1) to ensure that performance objectives are being achieved as expected and (2) to detect unacceptable migration of contamination so that contingency measures can be implemented to prevent any unacceptable risks to human health and the environment. Figure 4 presents the conditions that would be considered when evaluating MNA as a remedial alternative.

Table 1. Geochemical processes that control contaminant migration through an aquifer.

Process	Definition
Dissolution	The process of dissolving minerals from the aquifer matrix.
Precipitation	The separation of chemical constituents from ground water to form new minerals on the aquifer matrix.
Adsorption	The adhesion of chemical constituents on minerals within the aquifer matrix.
Desorption	The removal of a chemical constituent from the aquifer matrix by the reverse of adsorption.
Ion Exchange	The replacement of adsorbed chemical constituents by constituents in the ground water.
Biological	The process of transforming chemical compounds into different chemical compounds.

Case history for a successful deployment of a natural attenuation site

Conceptual model for a Wyoming Title I site

Uranium-ore milling activities at the Riverton, Wyoming, tailings site operated from 1958 to 1963. During these 4 years of operation, the mill processed approximately 800,000 metric tons of ore. Waste solids from the uranium ores were transferred to a tailings pile. The rectangular pile covered 28.5 hectares. Between 1988 and 1990, the uranium mill tailings pile and contaminated soils were removed from the site and were relocated to another site for disposal. The excavated areas were backfilled with clean fill, graded to form a crown, and seeded.

The area is arid with 20 centimeters of average annual precipitation. The site lies on a nearly level surficial terrace in a major alluvial basin, a few miles upstream of the confluence of two large rivers.

A system of unlined irrigation canals that exists along the northern and eastern sides of the site is operational for 5 months each year. The uppermost aquifer consists of two hydrogeologic systems: (1) an unconfined surficial aquifer composed of alluvial sands and gravels 5 to 6 meters thick and (2) a hydraulically connected, semiconfined shale system 12 meters thick that acts as a leaky confined aquitard. Deeper systems are unaffected by tailings seepage.

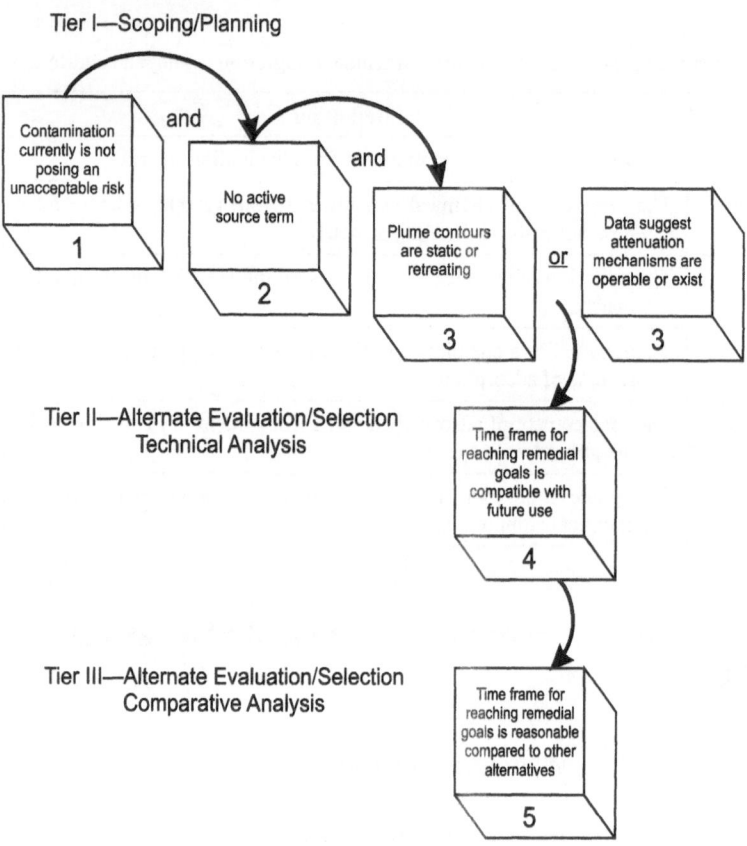

Fig. 4. Favorable conditions for evaluating monitored natural attenuation as a remedial alternative.

The water table in the surficial system is approximately 2.3 meters below the ground surface. Ground water flow direction is to the east-southeast with a horizontal gradient of 0.0024. Seasonal fluctuations to 1 meter occur in ground water elevations during the year. Ground water in the surficial system discharges predominately to a gaining river along approximately 1.6 kilometers of its course. Hydraulic conductivity is estimated to be approximately 17 meters per day. The calculated average linear ground water velocity is approximately 50 meters per year. The leaky aquitard does not contain a significant amount of contamination for modeling interpretations.

Case studies: active and abandoned Uranium mines

Model overview

GANDT software is used to simulate the flow and transport of a contaminant in a ground water system. GANDT uses external models to implement analytical and numerical solutions to aqueous and vapor-phase problems and supports single and unconditional or conditional Monte Carlo simulations. GANDT uses two variants of the Multimed code developed by Salhotra et al. (1993) to simulate the migration of a contaminant from its source to the water table. Contaminant transport in the saturated zone is modeled using subroutines from FTWORKS software by Faust et al. (1994). The interface between the two groups of code is accomplished by applying the contaminant flux from the unsaturated zone to the appropriate finite difference blocks of the saturated surface layer. GANDT software supports six model types: aqueous chemical, aqueous radioactive, and vapor in both analytical and numerical solution phases.

In modeling the candidate site, the conceptual model of the site was developed and model input parameters and justifications were created. Next, Monte Carlo simulations were run to translate site uncertainties quantitatively into model output uncertainties. A Monte Carlo simulation is a series of runs in which various parameter values are randomly generated from user-defined distributions. Parameters were defined as uncertain and represent areas where site data may not be consistent for the whole site. Monte Carlo simulations were post-conditioned to observed concentration data. After running Monte Carlo simulations with multiple layers generated, the realization can be presented in a visualization. The ultimate purpose of the visualization is to predict the scenario in which the site will naturally flush.

Model Simulation Results

To demonstrate the potential for meeting MCLs using the natural flushing strategy, the GANDT software was used to model the Riverton, Wyoming, Title I site. The contaminant of concern, uranium, represents the aqueous chemical contaminant phase. Time zero is when the milling operation started in 1958. Source removal was completed in 1990. Both the analytical and numerical solution phases were simulated. The analytical solution does not account for boundary effects and, therefore, will not account for surface water–ground water interactions. Figure 5 presents the predicted probability distribution of average uranium concentrations in 2072 (operation start-up was in 1958). As depicted in Fig. 5, the analytical solution does not accurately account for recharge and discharge phenomena.

The numerical solution that accounts for recharge and stream interactions was used to produce the probability distribution for average uranium concentrations in 1997, 2022, and 2072 (operation start-up was in 1958) as shown in Fig. 6.

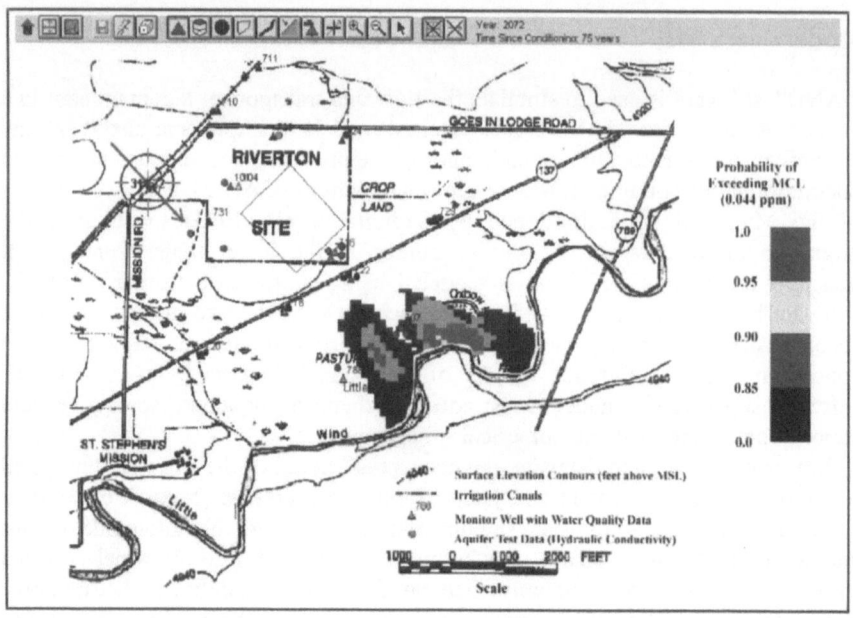

Fig. 5. Analytical solution of probability distributions of average uranium concentrations in 2072.

Summary and conclusions

The numerical solution more accurately accounts for hydraulic and chemical processes that affect natural flushing. Simulations at time step 35 years (1993) show a relatively good correlation with observed data. This calibration gives confidence that the conceptual model represents the system and source term fairly accurately. The 70-year time step results in average uranium concentrations still exceeding the EPA ground water MCL (0.044 milligram per liter). At the 90-year time step, the probability distribution for the average uranium concentration is below the MCL. The results of the GANDT numerical simulations support the prediction that this Title I site is a strong candidate for natural flushing. NRC approved this site for natural attenuation with monitoring in May 1999. The DOE UMTRA Ground Water Project has completed 11 of the 22 sites by meeting the EPA ground water standards.

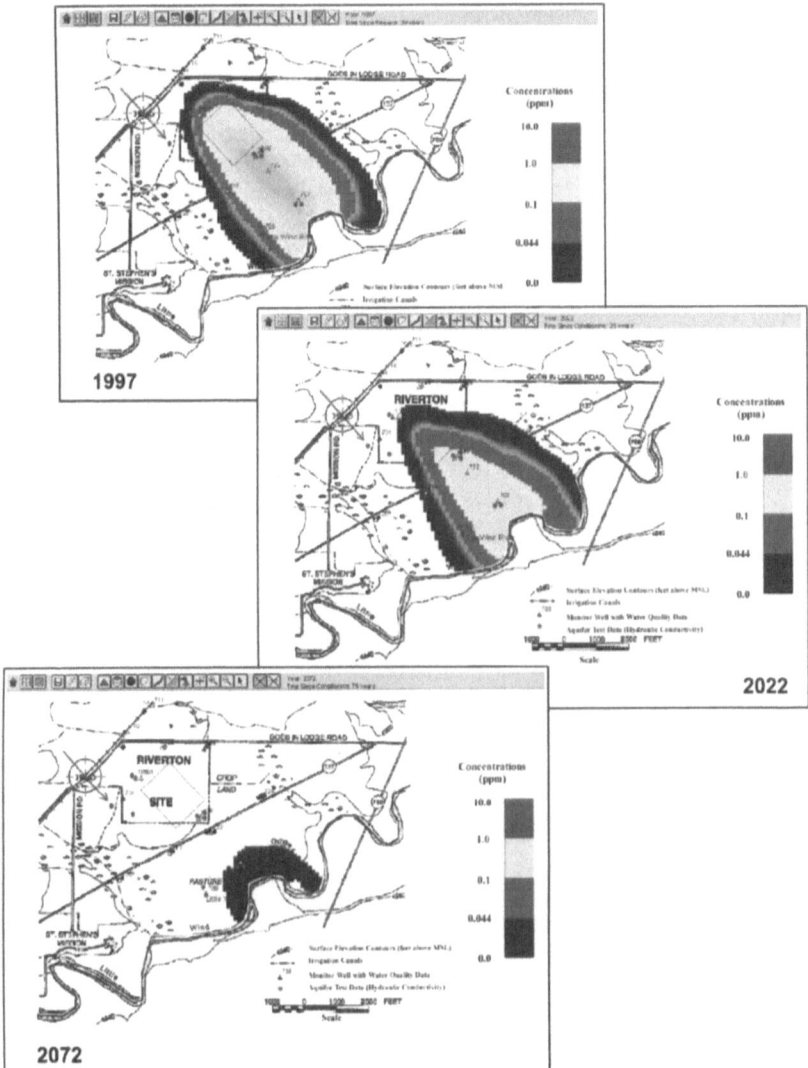

Fig. 6. Numerical solution of probability distributions of average uranium concentrations in 1997, 2022, and 2072.

References

Merritt, Robert C. (1971) "The Extractive Metallurgy of Uranium," Colorado School of Mines Research Institute, Golden, Colorado.

Code of Federal Regulations. Title 40 CFR Part 192, Subpart B, "Protection of the Environment."

Case studies: active and abandoned Uranium mines

Faust, C.R., et al. (1994) "FTWORKS: Ground Water Flow And Transport in Three Dimensions," Geotrans.

Federal Register (1995) 60 FR 2854, "Ground Water Standards for Remedial Actions at Inactive Uranium Processing Sites."

International Atomic Energy Agency (1999) "Technical Options for the Remediation of Contaminated Ground Water," IAEA–TECHDOC–1088, Vienna, Austria.

Salhotra, A.M., et al. (1993) "Multimedia Exposure Assessment model (MULTIMED) for Evaluating the Land Disposal of Wastes—Model Theory."

U.S. Department of Energy (1996) "Final Programmatic Environmental Impact Statement for the UMTRA Ground Water Project," Volumes I and II, DOE/EIS–0198, Grand Junction Projects Office, Grand Junction, Colorado.

U.S. Department of Energy (1999) "Decision-Making Framework Guide for the Evaluation and Selection of Monitored Natural Attenuation Remedies at Department of Energy Sites," Office of Environmental Restoration, Washington, DC.

Case studies: active and abandoned Uranium mines

Changing standards and continuous improvement: A history of uranium mine rehabilitation in northern Australia

Peter Waggitt

Office of the Supervising Scientist, GPO Box 461, Darwin NT 0801, Australia

Abstract. Uranium has been mined in northern Australia since 1949. In the early days exhausted mines were simply abandoned. Over time community concern with environmental issues has increased to the point where minesite rehabilitation is now mandatory. The paper reviews the changing standards that have been applied to the uranium mining industry in the north Australian region, in particular the ways in which the quality of rehabilitation and standards of long term environmental protection have been continuously improving.

Introduction

Minesite rehabilitation is not a new issue. Since mining first began there has been community concern about adverse environmental impacts and the state of the site once the project has been completed. The father of modern metallurgy, Agricola (1556), drew attention to these matters in his book *De Re Metallica*. Even then there was concern that mineral resource development was at the expense of agriculture and that land would be lost to production after mining. This is perhaps the earliest reference to sustainable development. The level of concern is also affected by the nature of the mineral being mined. Whilst mining is regarded with some acceptance by society, albeit grudgingly, radioactive minerals, especially uranium, give rise to a whole new range of community concerns. As a result, the need for effective, lasting and safe rehabilitation is now regarded as essential at locations where uranium has been, is being, or will be mined.

For many years minesites were simply abandoned as operations ended. There was rarely any attempt at rehabilitation, and usually no regulatory requirement for any to be done. The modern community's attitude to environmental issues has changed. Today a new mine cannot start until all aspects of the environmental management of the operation, including rehabilitation, have been subject to scrutiny and approval by a process that includes substantial community consultation.

In effect, the community is "licensing" new operations just as much as the regulators (Ellis, 1995).

Uranium mining and rehabilitation in northern Australia

Uranium mining in the Northern Territory began with the operations at Rum Jungle from 1949 until 1958 (Annabel, 1977) followed by the mines of the South Alligator Valley (1953 to 1964) which marked the end of the "first phase' (Waggitt, 2000a). Renewed exploration activity in the late 1960s saw two new operations commence after the introduction of the Commonwealth Government's *Environment Protection (Impact of Proposals) Act* in 1974. These were Nabarlek in 1979 and Ranger in 1980. Ranger was the subject of a major environmental inquiry (Fox et al, 1977). These mines may be considered the "second phase". Between these phases was another significant uranium mine at Mary Kathleen in Queensland. This mine operated in both "phases", from 1958 until 1963, and then again from 1976 until 1982, after EIA legislation was in force (MKU Ltd, 1986).

The third phase of mining is represented by the Jabiluka project, which was begun in 1998 but has not progressed beyond the initial development stage. Jabiluka has been subjected to the most stringent processes of environmental impact assessment in Australia. All mining projects begun since 1974, the date of introduction of EIA legislation, have had to provide details of their rehabilitation plans before being given approval to proceed. A location map for all these sites may be found at: http://www.uic.com.au/fmine.htm.

Phase One operations

Rum Jungle

A copper mine since 1905 uranium was first identified at Rum Jungle in 1912 (Kraatz, 1992). Following a 'new' discovery in 1949 uranium mining operated from 1954 until 1958 and copper mining continued until 1965, the mine finally closing in 1971. The uranium was mined specifically on behalf of the Commonwealth (Federal) Government who had no obligation to rehabilitate the site, which was simply abandoned. Buildings and machinery remained on site decaying, tailings frequently washed away into the Finniss River and sulphides in the stockpiles weathered to sulphuric acid leading to releases of substantial amounts of metals and other pollutants into the Finniss River. A clean-up operation in 1977-78 was not wholly successful, achieving some revegetation, but completely ignoring the issue of acid rock drainage and the associated pollution from the waste stockpiles (Kraatz,1992).

In 1980 due to growing public concern about the impacts on the Finniss River, by now virtually devoid of aquatic life for about 14 km downstream of the site, the Commonwealth Government agreed to fund a rehabilitation program which was undertaken on it's behalf by the Northern Territory Government. The program ran from 1982 until 1986 at a cost of AUS\$18.6M (Allen & Verhoeven, 1986).

The major objectives of the program (Kraatz,1992) were to:

- reduce the pollutants leaving the site by specific amounts (Cu by 70%, Zn by 70% and Mn by 56%),
- restrict water infiltration in waste rock piles to 5% of incident rainfall,
- to contain and reduce pollution in the water-filled open cuts,
- to reduce radiation levels to suitable levels, and
- to make the area safe and to improve the site's visual appearance.

Finally, it was required that the works had a structural life of 100 years. There was no intention that the site should be rehabilitated such that the public could have unrestricted access. In the works program tailings were contained, waste rock was encapsulated in new landforms to restrict the ingress of air and water, waste-waters were treated to raise pH and remove heavy metals, and the site was revege-tated. The grass species used were non-indigenous and needed fertilising and mowing to maintain the effectiveness of the cover. The result was a site quaran-tined from any future use and which required ongoing management for the reha-bilitation to remain effective. Such outcomes would be unacceptable today.

Whilst such an outcome was apparently acceptable in the 1980s, modern com-munities expect former minesites to be restored to a productive use. This became apparent when the outlying mine at Rum Jungle Creek South was rehabilitated in 1990-91. The program there required the former open cut, waste rock dump and surrounding area be rehabilitated to become a recreational lake and picnic area with unrestricted public access (Bastias, 1992). The significant change in commu-nity expectations over a period of less than 15 years is clearly demonstrated in the widely differing ways two sites only 5 km apart have been rehabilitated.

South Alligator Valley mining field

Thirteen small uranium mines and a mill operated in the South Alligator from 1955 to 1964 and all were abandoned when mining activities ceased. The area re-mained untouched until concern about public safety issues arose in the mid 1980s when the valley was included in a National Park. In 1986 tailings were removed from the millsite to a location outside the Park and processed to extract gold. Be-tween 1990 and 1992 the Commonwealth Government undertook a program of hazard reduction works at all the minesites, including the mill. The objectives of the program were to reduce physical hazards at old workings and to reduce radia-tion exposure for park visitors and traditional landowners to levels compatible with the new land use (Waggitt, 2000a). The work included revegetation of dis-turbed areas with native species to match the surrounding vegetation. No attempt was made to fully rehabilitate the sites. In 1996 the land was the subject of a suc-cessful native title claim. As the land was handed over to the traditional owners it was immediately leased back to remain as a National Park. However, the lease re-quired that all mine sites be fully rehabilitated by 2015. The negotiations for the design of a suitable program have taken several years but the task is nearly complete (Waggitt, 2002). A major objective of the rehabilitation is to ensure that the sites blend in with the surrounding countryside and do not require any special management.

Mary Kathleen

From 1958 to 1963 and again from 1976 to 1982 the Mary Kathleen uranium mine operated in Queensland. After the first phase, no rehabilitation work was undertaken, none being required by legislation at that time. However, after the second operational life ended in 1982, there were expectations from the community and requirements of the regulators that the site would be made safe. New environmental legislation had been promulgated between 1963 and 1976, and consequently the mining company had prepared a complete Environmental Impact Statement, including a rehabilitation plan, prior to the operation's restart in 1976. The rehabilitation plan was updated prior to implementation in 1982 to bring it into line with the Code of Practice published by the Commonwealth Government (DASETT, 1982). The main objectives of the plan were: to make all areas safe for public access in terms of radiation hazards and physical risks; to remove all structures that could degrade or become hazards; to make remaining ground surfaces as erosion resistant as possible; and revegetate them with native species.

The work program, which took nearly three years to complete, included flooding the mined-out pit, covering, contouring and revegetating of the rock stockpiles, revegetating borrow pits, decommissioning and removing of the processing plant and township and revegetating those areas, and ensuring secure containment of tailings in the existing above-ground facilities. The standards of rehabilitation were basic by today's criteria, with limited reshaping of stockpiles and little effort to match the surrounding landforms. Concrete slabs and exotic flora are still visible in the former township areas today. Whilst the rehabilitation outcomes were apparently satisfactory at the time, it is unlikely that the site would meet the expectations of either the community or a regulator today.

Phase two operations

Nabarlek

Nabarlek is the most recent example of uranium mine rehabilitation in Australia. (Klessa, 2001) The ore body was discovered in 1973 and its development was subjected to what was then considered intense scrutiny through a Public Inquiry (Fox et al, 1977) and an EIA process (Waggitt, 2000b) The disturbed area was relatively small as the nature of the ore-body allowed mining to be completed in one 143-day campaign in 1979. The ore was stockpiled whilst the mill was built. Milling continued until 1989 when the site was "mothballed" for 5 years whilst the mining company explored for new reserves. In 1994 the Supervising Authorities directed the mining company to rehabilitate the site, requiring that all site work should be completed by 31 December 1995. A full description of the rehabilitation program has been given elsewhere (Waggitt, 2000b; Klessa, 2001).

During the pre-mining negotiations with the Aboriginal traditional land owners it was agreed that the site was generally to be landscaped to match the original contours to the greatest extent practicable and revegetated to "blend in with the surrounding vegetation". Also, in accordance with the Environmental Requirements of the Commonwealth Government, all tailings were required to be returned to the mined-out pit. At Nabarlek, this was not a problem as the tailings had been placed in the pit as the stockpiled ore was milled (Waggitt, 1994).

Other criteria to be met for successful rehabilitation were that the site should be safe to enable Traditional Owners to follow a traditional lifestyle without limitation on access. This would include hunting and food gathering across the site with occasional overnight camping. The issue of improving physical safety to meet the criteria was addressed by the dismantling, decontamination and removal of the mill and the associated infrastructure. All residual plant and machinery items that could not be sold or satisfactorily decontaminated were placed in the pit together with all other contaminated materials. Landforming was completed in early December 1995, immediately before the anticipated onset of the wet season, thus allowing seeding to proceed at the optimum time By mid-1996 there was an initial vegetation cover across the site. Since then tree growth has become uneven and introduced grass species are dominating some areas (Klessa, 2001).

Modern standards are exacting and the stakeholder community, including both regulators and landowners, is requiring that the satisfactory development of an "appropriate" climax vegetation association be adequately demonstrated before the mining company can be released from it's obligations for the site (Klessa. 2001).

Ranger

Ranger is the only uranium mine currently operating in the Region. It is a conventional open cut operation with one active pit and one mined-out pit, which is the active tailings repository. Rehabilitation planning at Ranger is seen by the stakeholders to be an example of best practice that stands as a benchmark for similar operations elsewhere in the region and the world. The five major elements of the program are:

- a clear understanding of the goal and objectives of rehabilitation agreed by the stakeholders,
- stakeholder participation in the planning and updating processes,
- an approved plan for rehabilitation that is revised annually,
- a process which ensures that the finance for rehabilitation is completely secure, and
- implementation of progressive rehabilitation wherever possible.

The rehabilitation plan of the Ranger mine was set out initially in the EIS and was specifically written into the various agreements between the Traditional Owners, the Government and the mining company at the time development was approved. It was agreed that the final goal and objectives would be set down and agreed by the main stakeholders within a set period of time. This was finally achieved in 1990 (OSS, 1990) The goal is for the project area be rehabilitated to establish an environment which reflects that existing in the surrounding country and could permit the incorporation of the former site into the surrounding Kakadu National Park without detracting from the Park values or needing special management.

The mining company is required to revise the rehabilitation plan annually for approval by the Supervising Authorities. The plan has to be sufficiently detailed for stakeholders to be confident that the goal and objectives would be met should the plan be implemented during its lifetime. The detail provided must be sufficient to enable a complete costing to be made. The mining company is then required to place a cash deposit in a rehabilitation trust fund sufficient to pay for the approved plan. The annual plan is currently in version #27 but a complete blueprint for the total decommissioning of the site is currently being prepared by the company.

Phase Three operations

Jabiluka

The system in place at the Jabiluka project is less complex due to the relatively small amount of development undertaken to date. However, the project has been the subject of a thorough EIA process and subsequent assessment by an independent scientific panel of the World Heritage Committee. The company has provided a detailed rehabilitation plan, which has been assessed by stakeholders and approved by the Supervising Authorities in the same way as for the plan at Ranger. This approved plan has been costed and the company has posted a guarantee for the entire rehabilitation costs. Whenever the project moves into the next stage of development the plan will be revised. In the event that full development proceeds it is likely that the process of rehabilitation bonding to be imposed will be similar to that employed at Ranger. Again a complete blueprint for decommissioning has yet to be prepared.

Long Term Surveillance and Monitoring

A significant issue that remains unresolved is the stewardship or long term care of rehabilitated uranium mine sites (Needham & Waggitt, 1998). There is a real concern that the systems in place at present may rely too much on institutional controls to remain effective for the long term. Many members of the community consider that former uranium mine sites can never be regarded as completely safe, no matter how well controlled the rehabilitation process has been. Thus all rehabilitation situations now require a plan for the site's long-term stewardship. The major elements of stewardship are: appropriate monitoring and surveillance for as long as required (in perpetuity if necessary); provision of maintenance as required; ability to undertake further remedial actions if needed; and, communication and consultation with stakeholders. Few organisations other than central governments are likely to have the capability to provide adequate resources to manage the situation effectively and to the degree expected by the community.

Finally, the program of stewardship must include an element for consultation and information exchange with the stakeholders. Local communities must feel that they are being kept informed of the hazards and risks associated with a rehabilitated mine in their district and that they have real opportunities to contribute to decision making, especially in the stewardship period. The ultimate goal of stewardship must be to ensure that environmental protection is paramount and maintained at the required level.

Conclusions

Comparison of the rehabilitation outcomes at Mary Kathleen and Nabarlek show how much standards have changed over time. A significant difference was that the Nabarlek EIS was drawn up after the Fox Report (Fox et al, 1977; Waggitt, 2000b). Over the inquiry period (1975–77) it became apparent that society was no longer prepared to accept mining as a one-time user of land and former sites would in future need to be returned to some form of alternative, and acceptable, land use. Furthermore, the Nabarlek plan was implemented long after it was drawn up and community expectations and technical standards had been raised considerably in the intervening years. The plan was continually updated in light of these developments (Klessa, 2001). The outcomes being planned for, and expected, at Ranger and Jabiluka represent a further shift and improvement in standards.

The modern mining industry has developed a better understanding of the importance to consult frequently, extensively and effectively with stakeholders and systems are in place that ensure funding for rehabilitation is guaranteed. Rehabilitation requirements and standards take account of community concerns and take account of concepts such as continuous improvement, sustainable development and inter- and intra-generational equity. Proponents who fail to consult with their stakeholders risk having their programs held up or even stopped in an atmosphere of conflict. As society has expressed its desire for ever better levels of environmental protection and rehabilitation, so regulators have responded and the mining industry, in particular the northern Australian uranium mining industry, has acted to address those concerns through a process of continuous improvement.

Acknowledgements

The author wishes to thank his colleagues in OSS and other professional associates for their counsel and discussions during the preparation and internal refereeing of this paper.

Case studies: active and abandoned Uranium mines

References

Agricola, G 1556. *De Re Metallica*. Translated Hoover, H.C. & Hoover, L.H., pub 1950: Dover Publications Inc., New York.

Allen CG & Verhoeven TJ 1986. *Final Project Report-The Rum Jungle Rehabilitation Project*. Northern Territory Department of Mines and Energy, Darwin.

Annabel R 1977. *The Uranium Hunters*. Rigby, Adelaide

Bastias J 1992. The Rum Jungle Creek South rehabilitation project. Aus.J. Mining 7(66) March p40.

DASETT 1982. *Australian Code of Practice on the Management of Radioactive Wastes from the Mining and Milling of Radioactive Ores*. Department of Arts, Sport, the Environment, Tourism and Territories. AGPS, Canberra.

Ellis J 1995. Meeting the Challenges. *Groundwork* No. 1 Vol. 1 Australian Minerals & Energy Environment Foundation, Melbourne.

Fox R.W, Kelleher GG & Kerr CB. 1977. *Ranger Uranium Environmental Inquiry-Second Report*. AGPS, Canberra

Klessa DA(ed). 2001. *The Rehabilitation of Nabarlek Uranium Mine: Proceedings of a Workshop*. Darwin NT Australia, 18-19 April 2000, Supervising Scientist report 160, Supervising Scientist, Darwin NT.

Kraatz M 1992. *The Rum Jungle minesite. A case study in mine rehabilitation*. Landcare Fact Sheet #2. Conservation Commission of the Northern Territory, Darwin,Australia

Mary Kathleen Uranium Limited, 1986. Rehabilitation of the Mary Kathleen Uranium Mine-Review Report, June 1986 (unpublished).

Needham RS & Waggitt PW, 1998. Planning Mine Closure and Stewardship in a World Heritage Area-Alligator Rivers Region, Northern Territory, Australia. in Proceedings of the *Long Term Stewardship Workshop*, Denver, CO, 2-3 June 1998. US Department of Energy, Grand Junction Office (CONF-980652).

OSS 1990. Annual Report of the Supervising Scientist for the Alligator Rivers Region. 1989-1990. Australian Government Printing Service, Canberra.

Waggitt, 1994. *A worldwide review of uranium mill tailings disposal systems*. Technical Memorandum 48, Supervising Scientist for the Alligator Rivers Region, AGPS, Canberra.

Waggitt PW 2000a. The South Alligator Valley, Northern Australia, Then and Now: Rehabilitating 60's uranium mines to 2000 standards. in *proceedings of the SWEMP 2000 Conference, Calgary, Canada. May30 – June 2, 2000*. pub: Balkema, Netherlands

Waggitt PW 2000b. Nabarlek uranium mine: From EIS to decommissioning. in *Proceedings, URANIUM 2000 - International Symposium on the Process Metallurgy of Uranium. Saskatoon, Canada 9-15 September 2000*. Pub. Canadian Institute of Mining, Metallurgy and Petroleum, Montreal, Canada.

Waggitt PW 2002. Community Consultation in the rehabilitation of the South Alligator Valley Uranium Mines. *In proceedings of the conference* Tailings and Mine Waste '02. 27-30 January 2002. Colorado State University, Fort Collins, CO. Balkema, Netherlands.

A New Approach to Water Management and Pollution Control at Rössing Uranium Mine

Sandra Müller

Rössing Uranium Limited, Private Bag 5005, Swakopmund, Namibia

Abstract. At Rössing Mine pegmatitic granite is mined, from which uranium is extracted by sulfuric acid leaching. The tailings contain residual carbonates and can be discarded without neutralisation. A new approach to seepage control was introduced recently by installing boreholes on the tailings dam to recover water for re-use and control seepage at source. This method is feasible because the tailings are coarse-grained and the boreholes have reasonable yields. However, the chemical composition of the water is causing operational problems due to the precipitation of jarosite and iron hydroxides.

Introduction and Geology

Rössing Mine is an open cast uranium mine situated 65 km inland from the coast of Namibia in southern Africa (Fig. 1). The mine is located in the Namib Desert, which experiences low and erratic rainfall, high temperatures, and strong seasonal winds resulting in high evaporation rates. The landscape varies from gravel plains with saline soils to bare rocky outcrops with some spectacular geology.

Geological formations in the mine area belong to the Precambrian Damara orogen. They consist mainly of folded, steeply-dipping metasediments (gneiss, schist, quartzite and marble) arranged in a NE-SW striking belt. The uranium deposit is associated with a migmatitic dome structure and occurs in a pegmatitic granite know as alaskite, which intruded into the metasediments. Several tectonic phases caused intensive folding and jointing, but due to their predominantly compressive nature, open fractures are rare.

Hydrogeology

The regional groundwater flow pattern shows a gradient from NE to SW in accordance with the topography (Fig. 1). Groundwater resources in this part of the Namib are concentrated in the alluvium of ephemeral rivers. The Khan River supplies brackish groundwater (TDS 5000 mg/L) to the mine, which is mainly used for dust suppression in the open pit. The alluvium in the Khan River reaches a thickness of 20 m, while its tributaries crossing the mine site contain 5-10 m thick sediments of ill-sorted, silty coarse sand and gravel. Boreholes in alluvium have yields of 20-50 m^3/h.

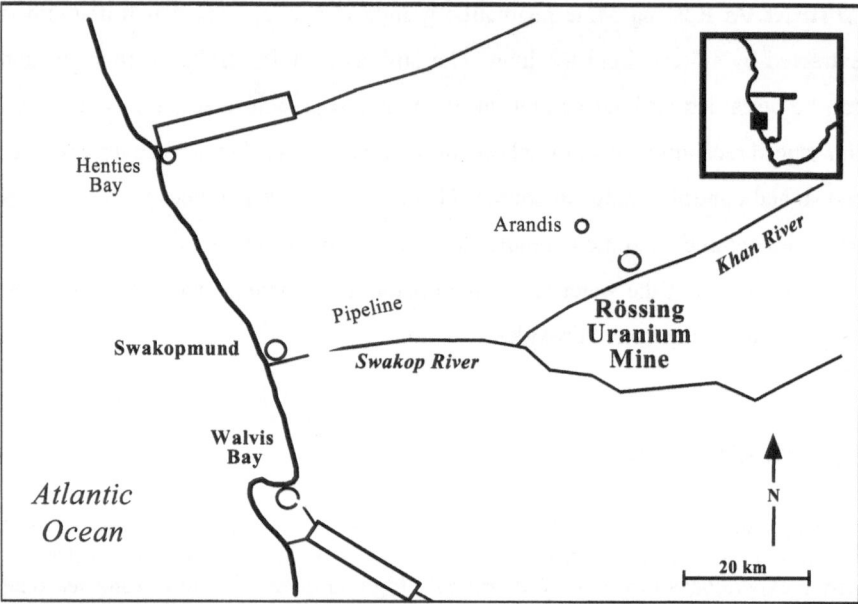

Fig. 1. Location of Rössing Mine

The hydraulic conductivity of the metasediments is generally low and typical yields are under 1 m^3/h, but boreholes on fracture zones can yield up to 5 m^3/h. Groundwater encountered in secondary aquifers is saline with TDS around 15 000 mg/L. Table 1 shows the hydrogeological parameters of the geological formations at Rössing Mine.

Superimposed on the natural groundwater system are sources and sinks created by mining. The open pit is nearly 300 m deep and cuts off flow from the uranium mill towards the Khan River. The main source of potential groundwater pollution is the tailings dam. It covers an area of 650 ha and contains over 200 million tons of tailings. Due to the acid-leaching process employed at Rössing the tailings solution is acidic and contains traces of heavy metals and radionuclides.

Case studies: active and abandoned Uranium mines

Table 1. Summary of hydrogeological information

Period	Formation	Lithology	Property	K (m/d)	S
Recent		Alluvium	Aquifer	2-80	0.1-0.2
Damara System	Intrusives	Alaskite	Aquitard	0.01-0.1	0.001
		Granite-gneiss	Aquitard	0.005	0.001
	Karibib Fm	Metasediments	Aquifer	0.15	0.01
	Chuos Fm	Meta-tillite	Aquitard	0.005	0.001
	Rössing Fm	Metasediments	Aquifer	0.1-1.0	0.01-0.15
	Khan Fm	Schist	Aquifer	0.1-0.5	0.01
		Gneiss	Aquitard	0.01-0.1	0.001

Pollution Control

The tailings dam overlies the alluvium of a tributary that drains towards the Khan River. Seepage from the tailings dam flows into a seepage collection dam 1 km downstream of the tailings dam, from where it is recycled. To reduce evaporation from the surface water streams and the seepage dam, cut-off trenches were constructed at the toe of the tailings dam in 2000 and seepage is now pumped directly from these trenches into the recycle system. Seepage control installations such as boreholes and cut-off trenches are in place around the tailings facility to prevent contamination of groundwater. Water recovered by these systems is returned to the processing plant. To detect the impact of seepage on groundwater 80 boreholes on the mine and in the Khan River are monitored on a regular basis.

Table 2. Borehole data

	Depth (m b.g.)	Diameter (mm)	Water level (m b.g.)	Transmissivity (m²/d)
TX1	60.0	63	32.7	not tested
TX2	71.5	200	32.9	3
TX3	61.0	200	38.2	35
TX4	71.0	200	46.3	350
TX5	61.0	200	22.0	40
TX6	57.0	63	34.6	not tested
TX7	52.0	200	39.6	not tested
TX8	44.0	200	26.5	not tested
TX9	50.0	200	31.9	not tested
TX10	44.0	200	29.0	not tested
TX11	53.0	200	38.0	not tested

Tailings Dam Extraction System

A new approach to seepage control was introduced in 1999 with the commissioning of a number of production boreholes extracting water directly from the tailings dam. The volume of water stored in the tailings dam was determined as 36 000 ML, of which 10 000 ML are extractable from coarse to medium-grained tailings. The purpose of the wellfield was not only to control seepage, but also reduce the mine's freshwater consumption and shorten the drainage time of the tailings dam after mine closure.

Table 3. Water quality data

	Tailings solution	Surface seepage	TX2	TX3	TX4	TX11
pH	2.0	4.3	6.4	5.9	6.0	5.3
Aluminium (ppm Al)	600	71		25	12	56
Conductivity (μS/cm)	20700	15900	18200	14400	13500	13200
Total dissolved solids (ppm)	37994	22713	30940	24392	20771	21174
Calcium (ppm Ca)	574	530	530	484	489	495
Magnesium (ppm Mg)	1560	1345	1889	1411	1173	1340
Total Alkalinity (ppm $CaCO_3$)	<0.1	15	1057	386	394	104
Chloride (ppm Cl)	1920	2100	2105	1285	1560	1700
Sulphate (ppm SO_4)	19223	10655	10192	14591	8845	9899
Fluoride (ppm F)	61	47	21	60	72	
Nitrate (ppm NO_3)	77	46	<2	25	<2	5
Ammonia (ppm NH_3-N)	280	132	136	191	142	163
Potassium (ppm K)	115	75	112	54	68	24
Sodium (ppm Na)	2295	1725	2233	1271	1491	1360
Manganese (ppm Mn)	2190	798	120	1016	617	805
Total Iron (ppm Fe)		333	115	421	347	405

Boreholes were drilled on the tailings dam with rotary mud flushing and installed with 200 mm uPVC casing, which is slotted below the water table (see Table 2 on the next page for borehole data). The thickness of the tailings in the target area varied from 20-70 m, while the saturated thickness was 10-20 m. The tailings are medium to coarse-grained with a d10 of 0.015 mm and the boreholes have yields of 2-15 m^3/h. The first five test boreholes have been in production for over two years. They have confirmed that the system can run effectively even if some clogging was experienced, which was removed by hydrochloric acid treatment. Eight more boreholes were established in 2001 and the first production figures indicate a combined yield of up to 1000 ML/year. Table 3 shows the chemical composition of water from boreholes (TX2-11) compared to the original tailings solution and surface seepage emanating at the toe of the tailings dam. There is an increase in pH due to neutralization accompanied by an increase in total alkalinity and a reduction in sulfate and manganese concentrations in seepage and borehole water.

Water Management

Water is a scarce commodity in Namibia and water management is of prime importance at Rössing Mine. This includes the reduction of freshwater consumption by reuse of process solutions and recycling of water wherever possible. Currently the mining and milling process requires 2400 ML of fresh water per year. This water is bought from the Namibian bulk water supplier who abstracts it from wellfields in the ephemeral Omaruru and Kuiseb rivers (Fig. 1). The mine and the growing coastal towns of Walvis Bay, Swakopmund, Henties Bay and Arandis share these water resources of which the mine uses about 25%.

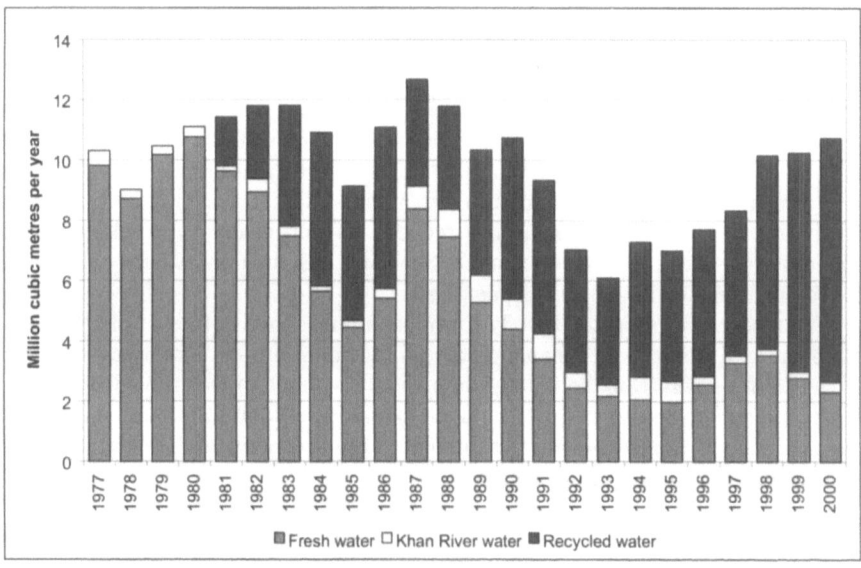

Fig. 2. Trend of Fresh and Recycled Water Use at Rössing

The limited and partly non-renewable potable water resources have been put under increasing pressure during the last decade due to low recharge and over-extraction (Fig. 2). Construction of a seawater desalination plant has been proposed, but postponed in favour of initiatives to manage demand. Demand management has been introduced by the mine, the municipalities and the water supplier. Rössing Uranium has contributed to the water conservation initiatives by creating awareness among employees, using alternative lower quality water sources such as recycled wastewater and minimizing evaporative water losses.

Recycling

The basic tool for water management at Rössing is the process water balance, which shows that input of fresh water balances losses due to evaporation and storage in the tailings material. The bulk of water currently used in the process (~60%) is recycled. Between 1976 and 1980, only fresh water was used and no recycling took place. Used water accumulated on the tailings facility in an evaporation pond of 150 ha. Recycling of tailings solution for use in the mills began in 1980 and in 1988 the paddock deposition method was introduced. This entailed the subdivision of the tailings facility into smaller areas of typically 40 ha, which were successively used for deposition. Decanting pumps were installed in each operational paddock to return the solution to the mills before much evaporation could take place.

Benefits

Modification of the tailings disposal facility and recycling of wastewater reduced the mine's freshwater demand from 0.66 m^3 per tonne of ore milled in 1980 to 0.21 m^3 per tonne of ore milled in 2002. This ensures that the mine's operations will remain sustainable with regard to water availability. Through this reduction in freshwater consumption Rössing has reduced its impact on the coastal aquifers and other users of this scarce resource. Seawater desalination, which would have had a great financial impact, was postponed.

The recovery of seepage directly from the tailings dam is of benefit to the environment as it reduces the impact of the tailings dam on the underlying aquifers. The piezometric head in the tailings dam and the volume of water available for infiltration is kept as low as possible. This method was found to be more effective than seepage control by means of boreholes in the low-permeability secondary aquifers surrounding the tailings facility.

Application of a Dry Cover to Remediate the Acid Rock Drainage Generation in the Uranium Mining and Milling Site of Poços de Caldas - Brazil

Horst Monken Fernandes, Mariza Ramalho Franklin

Instituto de Radioproteção e Dosimetria – Av. Salvador Allende s/n – Rio de Janeiro – RJ – CEP 22780-160 – RJ - Brazil

Abstract. This paper addresses the remediation of acid rock drainage originated in one of the waste rock piles of the mining and milling site of Poços de Caldas. The dump consists of a total amount of 29.8×10^6 tons corresponding to a volume of 12.4×10^6 m^3. It was estimated that acid drainage generation will last for 600 years and because of that permanent remedial solutions will have to be considered. In this paper the efficiency and costs of a dry cover system was assessed. It was found that the application of a material with a diffusion coefficient to oxygen of 10^{-9} m^{-2} and with a thickness of 1.0 m would reduce the pollutant loads to acceptable values. The overall costs of this remediation scheme would amount US\$ 10 million

Introduction

One of the main issues related to mining activities is the occurrence of Acid Mine/Rock Drainage (AM-RD). The term is used to refer to drainage of polluted water from sulfidic mine wastes and waste-rock dumps. This problem is common at mine sites because pyrite is frequently associated with minerals of economic significance, including uranium, coal and gold. The driving mechanism that leads to pollution generation is the oxidation of sulfides, that occurs when the reactive pyrite present in the wastes comes into contact with oxygen and water producing acid waters. In the beginning, pyrite is oxidized by oxygen in the presence of water and ferrous iron, Fe^{+2}. Sulfate and H^+ ions are then produced according to equation 1:

$$FeS_2 + 7/2\ O_2 + H_2O \rightarrow Fe^{2+} + 2SO_4^{2-} + H^+ \text{ (equation 1)}$$

The chemical equation of pyrite oxidation shows that the reaction is sustained by the supply of water and oxygen. In the vast majority of studied situations there will be plenty of water contained within the pore spaces of a dump, resulting from rainfall infiltration, while the supply of oxygen becomes a limiting factor for maintenance of the pyrite oxidation process.

Oxygen transport into the dump occurs by a combination of processes including diffusion, through the pore spaces of the rock, advection, driven by external pressure variations, and thermal convection, produced by the elevated temperatures resulting from the exothermic oxidation reaction, and finally, as dissolved oxygen in the infiltrating water. Ritchie (1995) shows that the only effective mechanism for oxygen transport into the dump is by diffusion through the rock pore spaces, the process being driven by the oxygen gradient generated by the consumption of oxygen within the dump. Thermal convection becomes a significant process in dumps, whenever the material shows gas permeability coefficients greater than 10^{-10} m^{-2}. Recently, the relative importance of the dissolved oxygen in the infiltrating water in the dump has been argued. Even at high infiltration rates, the amount of dissolved oxygen in the infiltrating water is far too low to support the oxidation rates which have been measured in the field, which range from 0.3 to 8.8 x 10^{-8}kg.m^{-3}s^{-1} (Ritchie, 1995).

The rate of oxidation of a pyritic material in a dump is governed by the transport of reactants to oxidation sites, as well by the Intrinsic Oxidation Rate (IOR). The Intrinsic Oxidation Rate is a key parameter to understand how oxidation is taking place in a dump. It represents the rate of oxygen consumption by the material deposited in the dump under the conditions applicable to the particular case. It is a function of a large number of variables including pore-gas oxygen concentration, particle-size distribution, mineral surface area, bacterial population, temperature, pH, and ferric ion concentration.

Values of IOR in the range of 10^{-9} to 10^{-8} kg (O$_2$)m^{-3}.s^{-1} are typical for a large number of waste rock piles Ritchie (1995). Values greater than 10^{-7} can be considered extremely high, whereas values around 10^{-10} are related to marginal acid drainage environmental problems.

The objective of remediation strategies is to reduce the concentration of pollutants in the drainage to acceptable levels over the lifetime of oxidation in the mine wastes. The pollution levels are correlated with the oxidation rates (IOR) in the pyritic wastes. There are two methods for reducing ARD. First by reducing IOR and secondly by reducing oxygen flux. The reduction of IOR has been tried by means of chemical modifications of the surface of the pyritic material, and by reducing the bacterial population. The first option is in the experimental phase and does not provide solid basis for consistent remediation strategies at this stage of our knowledge. The second will play an important role if bacterial catalysis is a significant processes in determining the oxidation rate. Reduction in oxygen flows is clearly a way to reduce the pollutant generation rate in waste rock dumps. The practicality of doing so depends very much on the mine site location and the availability of suitable materials. In the case of AMD remediation at Woodlawn Mines, Australia, it was observed that the conventional rehabilitation technique of a layered cover including an impervious clay lining over the waste was considered im-

Case studies: active and abandoned Uranium mines

practical. A conventional 30 cm layer of clay would have required the provision of at least 300,000 m3 of clay. It was determined that such source of clay would probably not be available at the site.

Methods to reduce oxygen flow can be divided into three categories: i) wet covers, ii) dry cover and iii) vegetation cover. In general, the success of a cover design is a function of the hydraulic characteristics of the cover material, the thickness of the cover system, the ability of cover system to sustain vegetation and the response of the cover system to climatic forcing. But it is the determination of the characteristics of the cover layer that must be assessed properly to orientate the engineering works to put the cover in place

The objective of this paper is to propose a remediation solution to be implemented based on cost x effectiveness analysis to one of the waste-rock piles of the Pocos de Caldas mining site. The WRP-4 as the object of this study because most of the infiltrating water is collected in a single holding pond. This situation represents a unique opportunity that is not generally available in worldwide mining sites and has not been previously exploited. This allows for a better assessment of the cover

Study Area

The uranium mine and mill site is located at the Poços de Caldas plateau, in the southeast region of Brazil (figure 1). The alkaline complex corresponds to a circular volcanic structure which formation began in the upper Cretaceous (87 ma) and evolved in successive steps until 60 ma. This intrusion is rounded by the leveling of bed rocks, consisting of granites and gneisses. These rocks are frequently cut by diabase dykes , amphibolites and gneisses.

The igneous-policyclic activities, of alkaline nature, associated to intense metassomatic processes and a strong weathering, gave rise to a variety of rock types belonging to the Nepheline-Syenite family and to uranium mineralization.The uranium enrichment in Poços de Caldas mine is related to hydrothermal events (primary mineralization) and to latter weathering processes (secondary mineralization). The mine covers an area of about 2.5 km^2 and is divided into three mineralized units designated as ore bodies A, B and E for mining purposes. The mining and milling facilities began commercial operation in 1982. However, the original intended production of 500 ton of U_3O_8 per year was never reached. As of 1995, 1,172 tons of U_3O_8 were produced. In the development of the mine, $44.8 \times 10^6 \text{ m}^3$ of rock were removed. From this amount, 10 million tons were used as building material (roads, ponds, etc). The rest was disposed of in two major rock piles, i.e., waste rock pile 8 (WRP-8) and 4 (WRP-4). In contrast to WRP-8, all the drainage from WRP-4 is collected into a single holding pond.

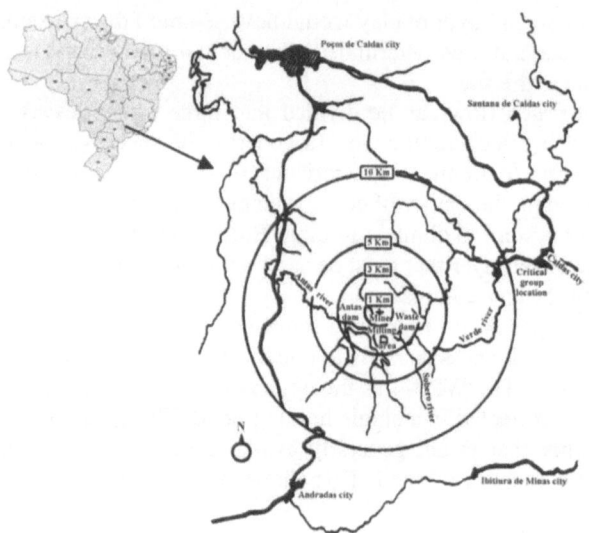

Fig. 1. Site location

Assessment of Generation of Acid Rock Drainage

For the calculation of IOR, it was assumed that all sulfate in the drainage results from pyrite oxidation and that no major precipitation reactions involving sulfate take place. Local precipitation averages 1.7 m.a^{-1}. It can be estimated that the average volume of precipitation collected by the dump area in one year is about 9.67×10^5 m^3. The infiltration rate for the undisturbed rock at the mining area is 0.1 mm^3 m^{-1} year^{-1} according to Cross et al. (1991). This figure is 5% of the local precipitation rate. However, the dump surface is completely altered in relation to the original surface. It is suggested that 40 to 70% of the precipitation will infiltrate into the dump (Morin and Hutt, 1994). It is reported by Nordstrom (1990) that infiltration in different dumps averages 50% of the total precipitation. Since the infiltration rate for the current site is not known, we decided to use the upper bound value of 70% as suggested by Morin and Hutt (1994). The mine operator reports that 1.5×10^5 m^3 and 1.98×10^5 m^3 have been pumped from the holding pond in the years 1996 and 1997 respectively. These figures correspond to 16% and 20% of the incident rain over the dump. The average of these two values, i.e., 1.74×10^5 m^3 will be used in further calculations.

If the average sulfate concentration in the drainage water (1,010 mg.L^{-1}) is multiplied by the average amount of water pumped from the holding pond in a year (1.74×10^5 m^3) an amount of, 1.76×10^8 g sulfate will be recovered.

Now, if the total area of the dump is taken into account a Global Sulphate Generation Rate (GOR) of 7.9×10^{-7} kg(SO$_4$).m^{-2}.s^{-1} is calculated. IOR is related to GOR according to equation 2:

$$GOR = IOR * L \text{ (equation 2)}$$

Where L is the height of the dump.

Fernandes (1997) reports that the maximum height of WRP-4 is 90m. As a result, a maximum IOR value of 2.72×10^{-9} kg (O_2) $m^{-3}.s^{-1}$ can be applied to the dump. This result is near the low end of values that are typical of waste rock piles (Ritchie 1995). If it is also assumed that oxygen diffusion is the major mode of oxygen transport into the dump the average depth of the surface region under oxidation may be calculated according to equation 3 (Ritchie, 1994):

$$X = \sqrt{\frac{2DC_0}{IOR}} \, (equation 3)$$

Where C_0 ($mg.kg^{-1}$) is the concentration of oxygen in the air (0.265 $kg.m^{-3}$), D ($m^2.s^{-1}$) is the diffusion coefficient of oxygen in the pores of the rock. Ritchie (1995) reports that a typical value of D is 5.0×10^{-6} $m^2.s^{-1}$. Applying these parameter values to the above equation, the thickness of the oxidizing region will be about 31.2 m.

The time needed for a region of a given thickness to be oxidized is given by equation 4 (Ritchie, 1995):

$$T = \frac{\varepsilon \delta_{rs}}{IOR} \, (equation 4)$$

Where ε is the mass of oxygen used per unit mass of sulfur oxidized and assumes a value equal to 1.75; δ_{rs} is the sulfur bulk density as pyrite (30 $kg.m^{-3}$.). As a result of these calculations the time needed to the region under oxidation to be consumed will be approximately 600 years.

The previous results rely on the assumptions that the dump is homogeneous in respect to the concentration of pyritic material; diffusion is the dominant mechanism of oxygen transport into the dump and no major precipitation reaction (that would cause significant losses of sulfate from the solution) is taking place.

These calculations have also to consider the complexity of a mine waste-rock pile in respect to the preferential channels for water flow. For large piles, only 5 to 20% of the rock surfaces may be regularly flushed by infiltration events (Morin & Hutt, 1994). As a result of the above discussion, the calculations done must be regarded as a general trend and used as an initial guidance for decision making.

Case studies: active and abandoned Uranium mines

Remediation Strategy Scheme

The primary objective of a remediation scheme to be used in the waste rock piles is to reduce the pollutant loads that are introduced into the environment along with the acid drainage. One possible way to achieve this goal, as mentioned before, is by reducing the oxygen flux into the dump. The use of dry covers suits this objective pretty well. However, its effectiveness need to be assessed. The cover consists of a layer of an appropriate material that has an oxygen diffusion coefficient lower than the rocks forming the dump. Once this condition is satisfied a reduction in the global oxidation rate by a significant factor will be achieved. It has to be noticed that the reduction factor will depend much more on the cover properties than on the waste rock pile properties themselves. The practical problem is to design a cover that remains relatively intact over a long period of time. The Global Oxidation Rate of the entity may be related to the cover thickness X_c and oxygen diffusion coefficient of the cover Dc by.

(equation 5), where

$$GOR = \sqrt{2C_0 DS^*} \left(\sqrt{\alpha+n} - \sqrt{\alpha+n-1} \right)$$

$$\alpha = \sqrt{\frac{X_c}{D_c} \left(\frac{S^*D}{2C_0} \right)}$$

Table 1. Efficiency of the different remediation schemes applied to the dump

Dif. Coef. $(m^2.s^{-1})$	10^{-8}	10^{-8}	10^{-8}	10^{-9}	10^{-9}	10^{-9}	10^{-10}	10^{-10}	10^{-10}
Thick (m)	0.5	1.0	2.0	0.5	1.0	2.0	0.5	1.0	2.0
GOR $(kg.m^{-2})$	1.8×10^{-8}	2.2×10^{-10}	1.1×10^{-10}	2.0×10^{-9}	1.0×10^{-9}	2.0×10^{-11}	2.0×10^{-10}	1.0×10^{-10}	5.0×10^{-11}
Load $(ton.y^{-1})$	139	75	38	15	7.7	3.9	1.8	0.76	0.34
SO_4^{2-} $(mg.L^{-1})$	287	154	78	32	16	8.0	3.7	1.6	0.79
^{238}U $(Bq.L^{-1})$	49	26	13	5.4	2.7	1.3	0.63	0.27	0.13

Case studies: active and abandoned Uranium mines

Table 2. Covering system associated costs

Layer Type	US\$.m^{-2}	m^2	Total (US\$)
Lower Layer	3.5	56.5 x 10^4	1.98 million
Intermediate Layer- Low Permeability	11	56.5 x 10^4	6.22 million
External Layer (gravel)	3.0	56.5 x 10^4	1.70 million
Total			9.9 million

If it is expected that the cover shows some efficiency, α will have to be greater than unity. With the aid of equation 5 the sulfate concentration and load arising from the dump drainage was simulated for different cover thickness and oxygen diffusion coefficients. The results are shown in table 1. It can be observed that from the option 4 on, the reductions in the sulfate and uranium concentrations begins to be negligible.

It has been mentioned that the integrity of the covering system has to be assured in the long term. As a result a three layered cover system is usually employed. The system comprises a lower granular layer, an intermediate one, of low gas permeability and an external one of sand and gravel to avoid the erosion of the system by rain and wind. The costs of a system like that, adapted to the dimensions of the WRP-4 is represented in table 2. It can be seen that US\$ 10 million approximately would be necessary for the application of a covering system like that to the entity. Other possibilities were considered like covering the entity with plastic material or backfilling the open pit. The first option would amount US\$ 12.7 million while the second one would amount about US\$ 70 million. It must be emphasized that these costs refer only to one of the existing waste rock piles at the site.

Conclusions

Acid drainage generation is a long-term problem at mining sites where sulfidic material is present in the rock. Remediation schemes to be adopted in such situations should concentrate on the reduction of oxygen diffusion into the dump. This may be achieved by covering the dump with some sort of material (e.g. clay or compacted clay) that has a lower oxygen diffusion and lower permeability to water than those of the dump. It was demonstrated that in the studied situation a cover with a thickness of 1.0m and a diffusion coefficient of 1 x 10^{-9} m^2 would be an efficient solution to the problem. However, monitoring programs, to be initiated after the cover installation, have to be undertaken. They should address the assessment of the cover efficiency. They must be focused at evaluating the rainfall infiltration rates – these data may be obtained by the installation of lysimeters (work in progress) - and oxygen profiles in the dump –achieved by the installation of probes to measure oxygen concentration at different depths inside the dump (work to be developed).

References

Cross, J.E., Haworth, A., Neretnieks, I., Sharland, S.M. & Tweed, C.J. (1991). Modelling of Redox Front and Uranium Movement in a Uranium Mine at Poços de Caldas. Radiochimica Acta. 53/53, p. 445 - 451. (1991).

Fernandes, H.M. (1997) Subsídios ao Descomissionamento ad Primeira Indústria e Mineração e Beneficiamento de Urânio no Brasil – O Caso do Complexo Mínero Industrial de Poços de Caldas – Niterói.   (D.Sc. Thesis) Departamento de Geoquímica da Universidade Federal Fluminense. pp. 250. (in portuguese)

Morin, K. & Hutt (1994) N., An Empirical Technique for Predicting the Chemistry of Water Seeping from Mine-Rock Piles. In: International Land Reclamation and Mine Drainage Conference on the Abatement of Acid Drainage. Pittsburgh. 12 – 19.

Nordstrom, D.K., Puigdomènech, I. & McNutt, R.H. (1990) Geochemical Modelling of Water-Rock Interactions at the Osamu Utsumi Mine and Morro do Ferro Analogue Study Sites, Poços de Caldas, Brazil Swedish Nuclear Fuel and Waste Management Co. SKB Technical Report 90-23. 733p.

Ritchie, A.I.M. (1995) Application of Oxidation Rates in Rehabilitation Design. In: Second Australian Acid Mine Drainage Workshop. Eds. N.J. Grundon & L.C. Bell, Charles Town, p. 101 - 116.

Groundwater rebound compatible with the aquatic environment – technical solutions at WISMUT's Ronneburg mine

Werner Unland[1], Michael Eckart[2], Michael Paul[3], Werner Kuhn[4], Ralf Ostermann[5]

[1] Dr. Werner Unland, Deutsche Montan Technologie GmbH,
Am Technologiepark 1, 45307 Essen, Unland@dmt.de
[2] Dr. Michael Eckart, Harress Pickel Consult GmbH Niederlassung Gera, Otto-Dix-Str. 9, 07548 Gera, HPC-Gera@t-online.de
[3] Dr. Michael Paul, WISMUT GmbH, Jagdschänkenstr. 29, 09117 Chemnitz, m.paul@wismut.de
[4] Werner Kuhn, WISMUT GmbH Sanierungsbetrieb Ronneburg, Postfach 41, 07576 Ronneburg, w.kuhn@wismut.de
[5] Ralf Ostermann, Franz Fischer Ingenieurbüro GmbH, Wilhelmstr. 26, 42697 Solingen, ralf.ostermann@fischer-teamplan.de

Abstract. Flooding of the Ronneburg mine would result in 140 to 330 m³/h of mine influenced water to seep into the Gessen valley from below and to finally drain into local small creeks if no action were taken. A sophisticated engineering design provides for a safe and reliable system utilising the inherent hydraulic head of the groundwater and gravity as the only driving forces to collect the water within the uppermost aquifer. The Fe(II) content of the seepage water demands special features to prevent oxidation and subsequent clogging of intake structures and the piping system.

Introduction

The WISMUT GmbH is in charge of remediating, among others, the former uranium mining district in the Ronneburg area. One important objective of the remediation strategy is to flood the underground galleries and the Lichtenberg open pit presently being backfilled. The WISMUT approach is presented in general terms by Gatzweiler et al. (2002) and a comprehensive status report on the Ronneburg mine flooding is given by Paul et al. (2002).

Our contribution is dedicated to the technical implications involved in pursuing the strategy to catch the rising groundwater as close to ground level as possible without detriment to the environment. Preliminary design work started in March 2001. The following summarises the engineering design submitted by WISMUT to the permitting agency for approval in April 2002.

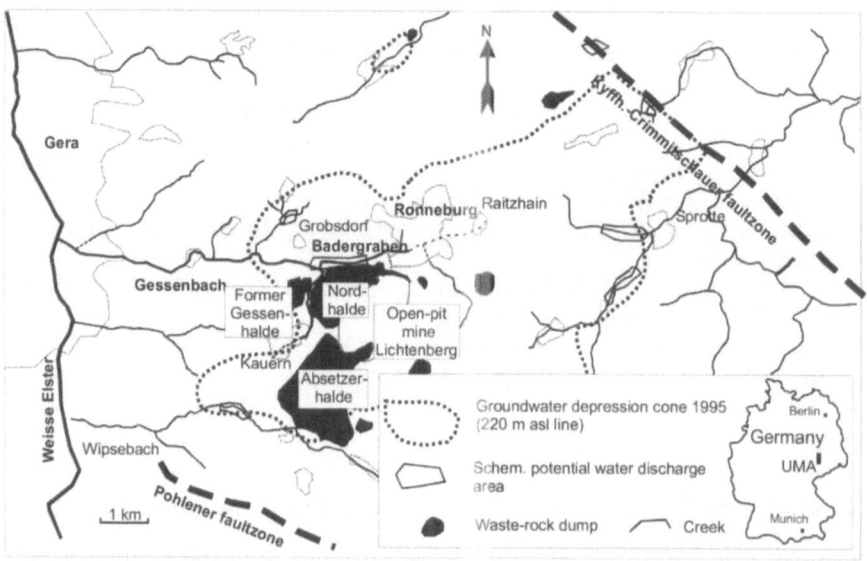

Fig. 1. Location of the Ronneburg mining area and the Gessen valley (from Geletneky et al. 2002). UMA – Uranium mining area of Ronneburg

Hydrogeological background

Preceding studies confirmed that the Gessen valley will be the location (Fig. 1) where rising mine water levels could firstly exceed ground elevations. This valley is located west of the city of Ronneburg, close to the former Lichtenberg open pit. It stretches about 2000 m in east – west direction at elevations from 250 down to 235 m a.s.l. and is about 50 m wide in its upper part. The most permeable country rocks, namely Silurian and Ordovician formations, which were also subjected to mining, crop out in the Gessen valley at their lowest elevation. Geological conditions are well defined by mining and exploration drilling and special drilling campaigns involving about 200 perforations of the Quaternary cover. The thickness of the unconsolidated sediment cover is generally 3 to 5 m at the fairly flat valley bottom and approaches zero at the valley slopes. The bottom part of the Quaternary cover varies laterally in granulometric composition from fine to coarse, often with stones in a loamy matrix, at the lowermost part. Hydraulic conductivities were determined to range in orders from 10^{-4} to 10^{-8} m/s. The upper part is built by a more or less uniform loamy layer, generally more than 2 m thick with a hydraulic conductivity of about 1 x 10^{-8} m/s (see also Fig. 3).

The valley is drained by two small creeks, the Gessenbach and the Badergraben which naturally cut to some extent into the loamy top layer and were both modified by WISMUT surface operations.

The present conditions are such that the area of the future groundwater discharge can be well delineated but due to the presently unsaturated conditions in the bedrock the avenues of fracture flow can not be explored in detail beforehand. At an advanced rebound stage and once saturated, the bedrock flow will be confined by the loamy top of the quartenary cover, which will act like a lid. At a zero action scenario, the groundwater would migrate through the more permeable Quaternary bottom sediments until it finds a point of relief to ground level. The impact were unforeseeable soil instability, unexpected generation of wetlands and uncontrolled surface flow of possibly contaminated water into the surface water environment.

Expected groundwater quantity and quality

To describe the flooding process, a numerical 3-D box model was developed for the entire Ronneburg mining district which is able to describe also the vertical velocity component of groundwater flow through the extended network of mining voids. The box model is integrated with a 3-D finite difference groundwater flow model which represents the groundwater flow in the regional aquifer system. A further detailed groundwater flow model was designed for the area of the Gessen valley. This model was refined by our studies and also integrated with the 3-D box model. Our results confirmed previous assessments and concluded that depending mainly on the hydraulic conductivity of the Silurian/Ordovician bedrock the long term upward groundwater flow into the Gessen valley will average approximately 140 m³/h. This value is based on what we consider a realistic hydraulic bedrock conductivity of 2 x 10^{-7} m/s equivalent to the average value of previous permeability field tests. Since the Gessen valley is embedded in the regional groundwater flow system the regional model allowed a sensitivity analysis for significantly different values of the bedrock hydraulic conductivity. Following implications were derived:

Lower hydraulic conductivity: lower flow rate into the Gessen valley but higher hydraulic heads at final rebound so that further groundwater discharge points at other locations would be triggered.

Higher hydraulic conductivity: lower hydraulic heads but higher flow rates into the Gessen valley which eventually would become the only relevant discharge point with outflow being limited by water balance considerations.

This situation demanded to take a conservative approach. Taking also earlier studies on variations of annual water balances into account we concluded that a design capacity of 330 m³/h would represent an upper limit at reasonable assumptions. However, the overall system would also allow for larger rates in cases of theoretical emergencies.

The expected quality of the groundwater was based on experience to date, laboratory simulations, and theoretical calculations. The range for the most important parameters is given in Table 1. The expected quality requires treatment prior to discharge into the environment. From the technical point of view, especial concern is directed to the content in Fe, Mn and Al and to some extent to the acidity of the groundwater while sulphate is not expected to precipitate.

Table 1. Expected groundwater quality at the Gessen valley. The values do not account for additional contaminated seepage from the former Nordhalde waste dump area.

Parameter	Unit	Minimum	Maximum
SO_4	mg/l	1390	3000
Cl	mg/l	35	100
U	mg/l	0,52	0,75
Ca	mg/l	239	251
Mg	mg/l	199	590
hardness	°dH	79	171
Fe_{total}	mg/l	10	125
Fe^{2+}	mg/l	8	80
pH		4,1	5,5
Eh	MV	300	439
Mn	mg/l	3,9	15
Al	mg/l	7,2	29
Ni	mg/l	0,8	3
HCO_3	mg/l	180	310
K	mg/l	9	55
Na	mg/l	42	80
Ba	mg/l	0,12	0,12
Cu	mg/l	0,08	1,8
Zn	mg/l	0,65	0,9
Ra	mBq/l	100	220

Objectives

The time groundwater of the described quantity and quality could surface at the Gessen valley depends on natural factors, especially groundwater recharge by precipitation, as well as on progress in the overall permitting process. A realistic schedule suggests 2004 or delay up to 2005. This means the groundwater catchment system in the Gessen valley has to be in place by the end of 2003.

It is an accepted fact that the approvals to proceed with the flooding process will require practically all groundwater to be contained and treated before being discharged into the environment until the groundwater quality is significantly improved above what is displayed in Table 1. On the basis of present knowledge, this process is expected to last some 25 to 30 years. This dictates the minimum design life time of any technical system catching the contaminated groundwater flow. Especially the low surface water flow rates in the Gessenbach and Badergraben will not tolerate exceptions. In addition, the lower Gessen valley is a protected area already now. For this reason and, additionally, for compliance with the EU Water Framework Directive no deterioration of the surface water quality will be permissible. Any construction work will have to respect the vegetation under protection and any unavoidable damage will require full compensation.

The very same area has also been designated to become a show case for the 2007 National Horticultural Exhibition. There is an agreement between approval authorities, the event organisers and WISMUT that surface installations for groundwater shall be reduced to a minimum and necessary structures will have to blend into the exhibition ground concept.

Technical concept and engineering design

The scope of the engineering work is easily defined: catch the uprising contaminated groundwater prior to entering into the aquatic environment and convey it to a water treatment facility located about 90 m higher and about 3,800 m away.

While the delivery end is a straight forward design job, the uptake and collection of the groundwater requires joint creative thinking of hydrogeological sciences and engineering skills. On a pre-feasibility level theoretical alternatives like well point systems, sheet piling, slurry walls or extended filter blankets were eliminated for various reasons. To minimise needs for landscape destructions and to find an acceptable balance between environmental safety and monetary expenses it was agreed to limit construction work to where needs will arise. Knowing the area but not the definite spots where groundwater will surface led to a general concept made up by a basic collection system to be amended by flexible and quickly to be installed satellite systems on an as required basis. The backbone of the groundwater collection system in the Gessen valley are strings of combined drainage and collection pipes finally entering a pumping station from where the collected groundwater will be pumped up to the water treatment plant. The location and depth of the drainage and collector pipes were the result of a detailed hydrogeological analysis considering:
- bedrock contours and lithology
- high transmissivity areas within the Quaternary cover
- thickness of the upper loamy layer
- maintenance of saturated semi-confined conditions in the Quaternary aquifer

in synthesis with technical requirements like:
- all way through gravity drainage
- possible takeover of additional water from the NW part of the backfilled Lichtenberg pit
- distance of > 5 m from a public storm sewer line

- distance of > 10 m from the present Gessenbach and Badergraben
- a decent grid of service roads.

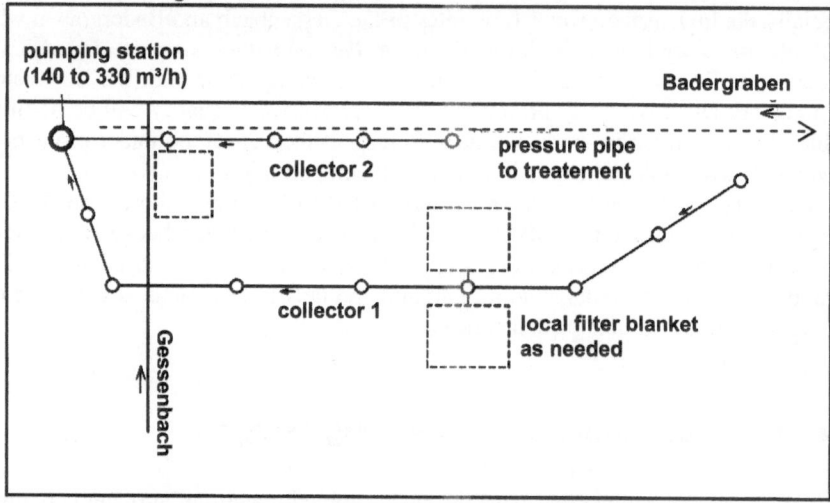

Fig. 2. Principal layout of the groundwater catchment and transportation system

Collector pipes and drainage lines are designed in parallel, buried in one trench and total in length about 1,640 m each. The lines are split in segments by a total of 23 manholes providing access and control as well as connecting points to satellite drainage systems. By comparative analysis and evaluation a DN 350 HDPE collector pipe system with a parallel DN 200 HDPE drainage pipe system was favoured. The parallel lines provide for an additional safety feature in case of repairs or pipe maintenance. Fittings in the adjacent manholes allow exclusion of pipe sections from the normal straight through flow. A typical cross section of the collector and drainage line is shown in Fig. 3.

In addition to the drainage lines representing linear elements along the Gessen valley there are two satellite drainage systems foreseen to date:

- buried vertical drains where hydrogeological conditions facilitate these well type installations.
- local filter blankets augmented with drainage pipes and with a top seal of local loamy sediments.

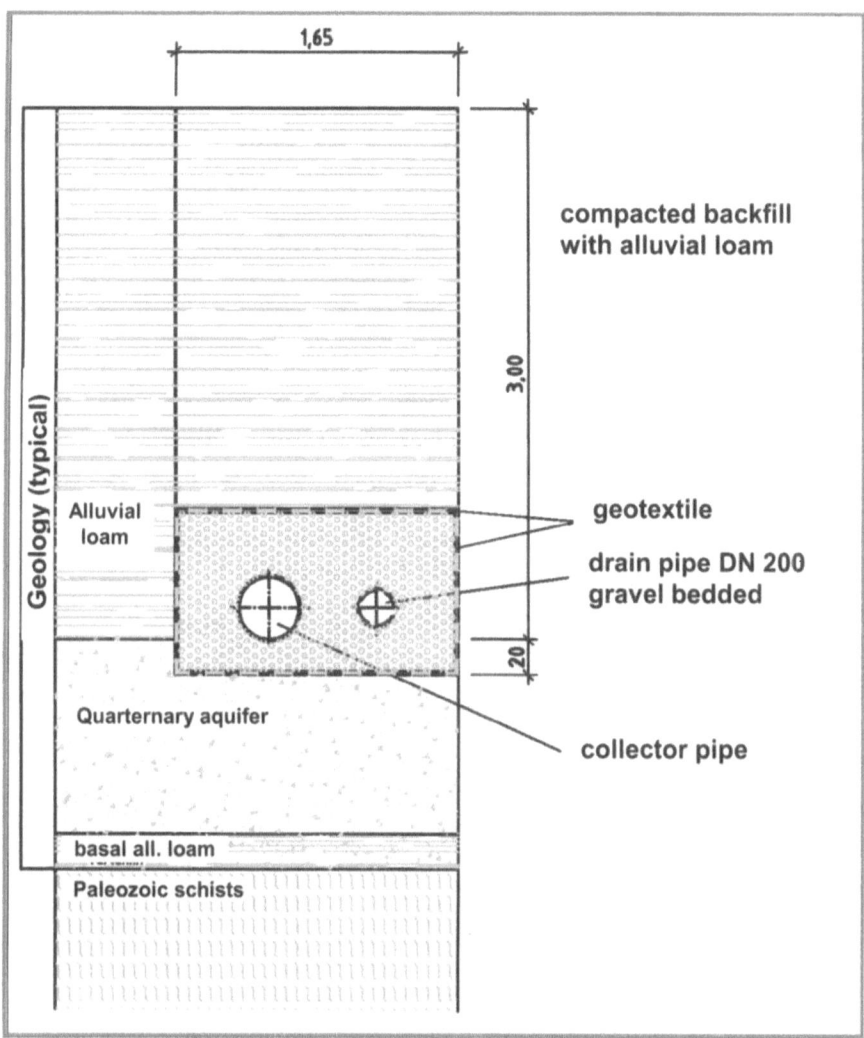

Fig. 3. Cross section showing parallel placement of drainage and collector pipes in a typical geological profile of the Quaternary cover.

A total of 8 vertical drains could be identified on the basis of the extensive drilling program executed and are integrated in the detailed plans already. Local filter blankets will be installed on an as required basis when a need can be spotted during the comprehensive monitoring program. The operational procedures call for a temporary catchment of the upward seepage and pumping into the nearest manhole of the collection system while construction is underway. The satellite drainage systems will not require any moving parts and will deliver the water by gravity only to the main collector. The only driving force will be the hydraulic head building up during the groundwater rebound. This hydraulic head can be controlled to some extent by a riser pipe installation at each manhole (Fig. 5). The overflow system serves two purposes:

-    to maintain a minimum hydraulic head in the Quaternary cover so that semi-confined conditions prevail reducing groundwater recharge by precipitation and preventing air to invade the unsaturated aquifer system.
-    to minimise contact of air with groundwater also in the gravity lines, with the objective to prevent Fe^{2+} from becoming oxidised and forming incrustations in the downstream piping and pump system.

The overflow system is shown schematically in Fig. 4. The system is simple and will need some attention only in the initial stages of the groundwater rebound for proper balancing.

The groundwater catchment system is dimensioned in excess of the design flow rate of max. 330 m³/h. The pumping system at the end of the gravity line consists of 3 identical pumps designed to switch regularly between stand-by and operation mode. The maximum total pumping capacity installed right from the beginning will be 400 m³/h. The pump house is designed for quick machinery changes in case of repair and maintenance.

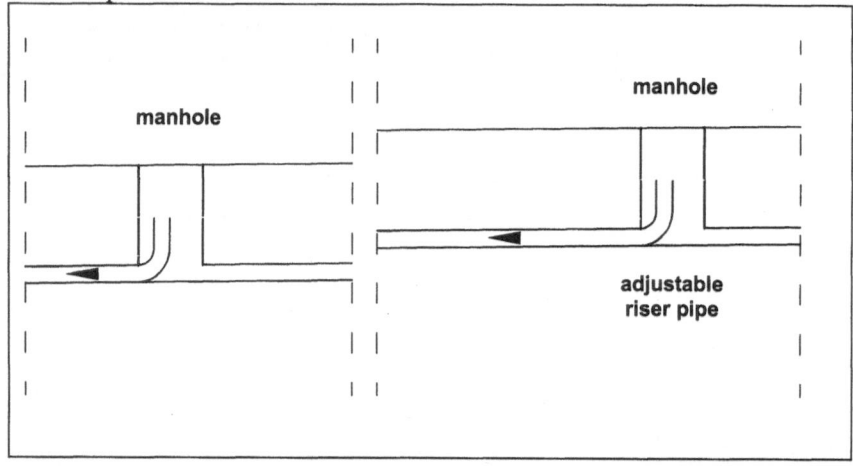

Fig. 4. Scheme of overflow system preventing air from entering the collector pipes.

The system is designed for continuous flow through and does not provide for any significant storage. For cases of emergency and failures of the pumping system an automatic overflow into a 1,600 m³ capacity storage basin is provided. For an ultimate event the storage basin features a controlled spillway into the Gessenbach.

While the design endeavours excellent reliability and redundancy of the mechanical and electrical components significant concern remained as to the impact of iron precipitation and subsequent clogging within the system. Therefore, some additional efforts were engineered to prevent oxygen to get into contact with the groundwater. All manholes have been provided with gas tight covers and will be positively pressurised by a nitrogen atmosphere (Fig. 5). A central nitrogen supply station is installed within the pumping station. Nitrogen is generated from the ambient atmosphere by molecular sieve and adsorber techniques at reasonably low investment and operating costs. Nitrogen is supplied through a 2" distribution line to each manhole. The inert atmosphere and pressure will be controlled in the course of the regular inspection program.

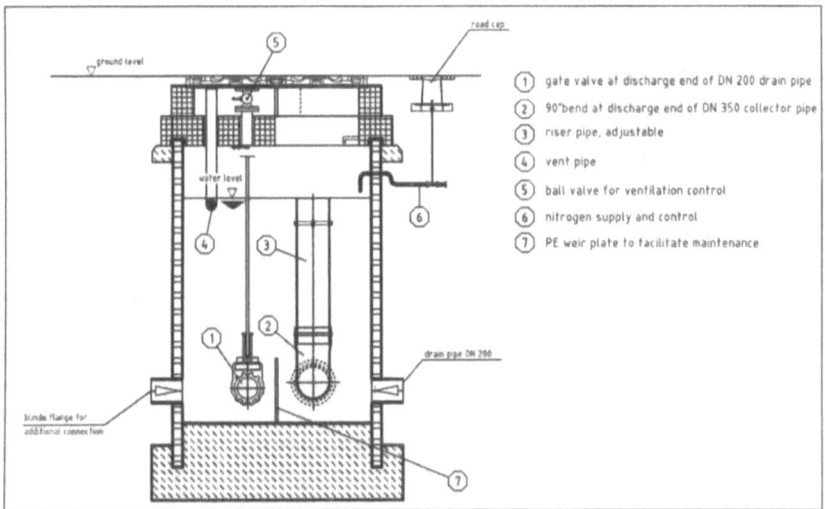

Fig. 5. Cross section of a typical manhole featuring water level and inert atmosphere controls.

Monitoring system

Today a comprehensive monitoring system exists already which allows adequate control over surface water flow and especially groundwater levels and quality. A total of 8 additional bedrock piezometers was placed in 2000/2001 at strategic positions in the Gessen valley to observe the general groundwater level increase. Herewith, the time the catchment system will become operational can be well predicted. The vertical drains will act as excellent indicators for the pressure build up since they are installed at the most permeable low points within the Quaternary cover and can be expected to become the preferred collectors prior to discharge into the drainage piping system. Quantity and quality of the groundwater captured can be identified by measurements and sampling at the manholes defining begin and end of each drainage section. Overall quantity and quality will be monitored at the pumping station.

Surface water quality at the Gessenbach and Badergraben will be monitored upstream and downstream at regular intervals. Detailed monitoring schedules are subject to the approval process.

Case studies: active and abandoned Uranium mines

The performance of the technical system will be remotely controlled, with data on pumping rates, pressure rates and current uptake being transmitted on a continuous basis to a central WISMUT control room.

Regular visual inspections of the system are mandatory. Especially during the first phase of groundwater entering the Gessen valley ground observation is important to quickly spot any unexpected seepage to the surface or groundwater induced change to the vegetation so that counteraction can be initiated. Manholes will be inspected on a regular basis to identify possible build up of incrustations and needs for pipe cleaning.

Conclusions

An important step is finished towards preparation for the groundwater rebound in the Ronneburg mining district. A safe and flexible system is designed by joint efforts of engineers and scientists to cope with the groundwater quantity and quality which can reasonably be expected. The basic system entailing drainage lines at strategically important longitudinal axis is flexibly supplemented by additional drainage elements on an as needed basis. The entire system works by gravity forces towards a central pumping station. Redundancy and emergency features are included. Concerns about iron precipitation and clogging of the system could be alleviated to the extent possible. Total investment costs will amount to about 2.8 million € and hence, remain tolerable expenditures for the environmental protection of the aquatic system in the Gessen valley. Provided the approval process will not exhibit any unduly delay construction work will be completed reasonably ahead of the predicted groundwater rebound.

References

Gatzweiler R, Jakubick AT, Paul M, Schreyer J, Meyer J (2002): Flooding the WISMUT mines - Learning by doing or applying a comprehensive systematic approach?- UMH III, Freiberg, September 2002

Geletneky J, Büchel G, Paul M (2002): Impact of acid rock drainage in a discrete catchment area of the former uranium mining site of Ronneburg (Germany).- Tailings and Mine Waste '02. Proceedings of the Ninth International Conference on Tailings and Mine Waste, Ft. Collins, Co., 27-30 January 2002.- Balkema, Lisse, 67-73

Paul M, Gengnagel M, Vogel D, Kuhn W (2002): Four years of flooding WISMUT's Ronneburg uranium mine - a status report- UMH III, Freiberg, September 2002

Water protection by long-term efficient surface contouring and covering of uranium mining dumps and tailings ponds –results of ten years of research and develop

Roland Hähne, Klaus-Dieter Oswald, Michael Schöpe

C&E Consulting und Engineering GmbH, Jagdschänkenstr. 52, D-09117 Chemnitz

Abstract. The way of closure, remediation and re-cultivation of dumps, tailings ponds and sites contaminated by radioactivity depends primarily on the need for reducing the discharge of contaminants via the water pathway. In combination with optimization requirements within radiation protection laws for uranium mining liabilities special contours and cover systems were developed, tested and optimized during the last ten years, especially on the basis of complex water balance assessments.

Principles

A long-term efficient remediation of waste rock dumps and tailings ponds from the former uranium mining is necessary as for reasons of reducing the contaminant release via air and water pathway as well as the human contact with radioactive wastes, the re-establishing of the natural water balance, the ensuring of the geotechnical long-term stability and for reasons of landscaping the effected areas.

There, where especially because of radiation protection legislation and of water laws a surface covering is necessary, technical feasible, site specific, self regulating natural solutions will be preferred instead of complicated multi barrier systems with a high expenditure for repair and attention.

Compared to current conditions of high infiltration and contaminant release rates, these approaches must be characterised by balanced water regimes including elevated storage capacities and permanent availability of soil moisture to plants, by diminished percolation into the waste rock dump and tailings bodies as well as by their resistance to erosion forces and by their stability. And in the end they are measures rising the value of formerly devastated properties.

By a detailed comparison of technical options of contouring and surface safeguarding, of their ecological and landscaping effects and economical consequences the selection of the option with the biggest netto-effect in relationship with the current situation is possible. For the land owner or the authorities the decision about the technical measures needed and the costs to ensure this netto-effect will be easier.

The technical dimensioning of the final covering has a special importance because it represents an important element of function of the whole remediation. By the final cover a long-term save deposition of the residues of the uranium mining and milling has to be ensured.

For this purpose a period of time of 200 to 1000 years has to take into consideration in connection with the selection of cover materials, of vegetation and of the evaluation of the options concerning the environmental impact.

Demands on surface contour and cover system:
- Securing erosion protection (wind and water)
- Securing long-term stability of slopes
- Drainage of surface runoff and interflow by suitable surface relief structures in the plateau area
- Avoiding deformation of the surface by consolidation
- Use of soil materials, for which the long-term functionality is assessable
- Following the criterions of the soil protection law (BBodSchV)
- Minimal expenditure for attention and minimal long-term tasks

Demands on the storage layer and the re-cultivation layer to ensure a long-term stable vegetation
- Sufficient air capacity of soils
- Good roots

Demands on the cover system for safeguarding the sealing layer against rooting, frost penetration and drying
- Securing a minimal thickness of the storage layer
- Securing a compacted installation of the sealing layer

Demands on the vegetation
- Drought resistance and insensitivity against wind and biological pests
- High evapotranspiration
- Low expenditure for attention
- Integration in the landscape
- Efficient protection against erosion by wind and water
- Site suitable selection of plants

Additionally to the above mentioned assessment criterions to reach the goal of the remediation and to ensure the long-term performance the costs are an important criterion, too. Construction costs for the covering and the long-term costs to ensure the long-term functionality of the cover system have to be distinguished.

The following explanations are focused at the selection of suitable cover options as on the basis of water balance investigations as well as on the needed proof of the erosion stability and slope stability. Without this long-term water protection is impossible.

Case studies: active and abandoned Uranium mines

Selection of suitable cover options by water balance investigations

Cover systems differ concerning their water balance relevance by

- Different systems of cover layers
- Use of different soil materials
- Variation of layer thickness
- Various construction geometries and
- Various kinds of vegetation

Typical cover systems, which were analysed by sensitivity analysis, are:

1. One layer System   Recultivation/ Storage layer
2. Two layer System   Storage layer+ Sealing layer
3. Three layer System   Storage layer + Drainage layer + Sealing layer

The following characterisation is the result of sensitivity investigations of various objects in the region of Saxony, Thuringia and Saxony-Anhalt.

Storage layer: The variation of the hydraulic conductivity from about $1 \cdot 10^{-6}$ m/s to $1 \cdot 10^{-8}$ m/s doesn't have any important influence on the water balance. However, the conductivity should not fall below the above mentioned ones. The thickness should be higher than 1 m and lower than 1,5 m especially for plateau areas because of percolation and evapotranspiration. In the case of reforestation the thickness should be about 1,5 m to ensure the suitable space for rooting.

Sealing layer: The hydraulic conductivity of the sealing material should not be better than $1 \cdot 10^{-9}$ m/s. The thickness of the sealing layer has no important influence on the permeability of the cover system. Therefore the thickness should be estimated in accordance to technological and geotechnical points of view.

Drainage layer: At surfaces with very low inclination the hydraulic conductivity should not be lower than $1 \cdot 10^{-4}$ m/s in order to prevent a high swell of water and related to this an increasing percolation of the sealing layer.

The percolation rate is an important factor to estimate the seepage water discharge and the contaminant release. The dimension of the water swell above the sealing layer is important to calculate the geotechnical stability of the slopes.

Results and examples of the water balance based dimensioning of cover systems

Since the end of the eighties investigations to the effect of surface cover systems were realised in preparation of the final remediation of residues from the uranium mining in Saxony and Thuringia. In 1988 the first test fields with big field lysimeters (infiltrometers) for various cover systems were installed in the region of

Case studies: active and abandoned Uranium mines

Ronneburg. Later since 1993 the installation of more than hundred small lysimeters followed for waste dumps and their cover material of the region of Schlema-Alberoda to find out the optimal cover configuration.

Since the end of the nineties extensive measurement equipment for the investigation of water balance and of percolation process was installed at the waste rock dumps Beerwalde/ Thuringia, Borbach und 371/ Saxony as well as the backfill cover of the Ronneburg open pit. Last year a big lysimeter facility with 10 large test fields various cover systems for tailings ponds was established in Seelingstädt. A lot of column tests to investigate the seepage process and the discharge of contaminants complete the optimisation of cover configurations. It is not always possible to extend the site specific experiments because of time and cost limitations. So in 1994 the development of special software began, in order to force the prognosis possibilities of water balance effects.

The one dimensional balance model BOWAHALD was developed in co-operation by the Mining Academy Freiberg and C&E. This model was modified over the last years and partially validated. The software is suitable for estimation of the infiltration rate taking into account different vegetation. In the meantime the model HELP of the US-EPA was modified for European conditions and was used for various cases in the region. Comparisons with field tests and with other models showed that suitable results were achieved concerning vertical and lateral runoff in the aeration zone as well as concerning the swell of water on sealing layers.

HYDRUS-2D is suitable for the modeling of the two dimensional runoff into the cover, of the estimation of the capillary rise and of the drying of soil in connection with special meteorological situations.

By means of the model SOIL and the own development EVAB the evaluation of the influence of specific vegetation at the water balance of cover systems is possible.

During the last years the development of a combination of model compartments was forced which includes water movement in the unsaturated zone, runoff in the saturation zone, discharge and transport of contaminants. Examples are the models AERA and TENSIC.

Realistic prognosis supporting the decision of contouring, specific covering, cover material and kind of vegetation are possible in combination of model output and   results of field tests.

Practical examples:

1.) Re-cultivation of slopes of dumps of ore mountains / Saxony (inclination 34 %, E-exposition)

	0,3 m re-cultivation layer			k_f 1E-06 m/s	
Vegeta-tion	Pre-cipitation [mm/a]	Evapotran-spiration [% P]	Surface runoff [% P]	Interflow (drainage) [% P]	Percola-tion [% P]
grass	869	59	0	0	41
forest	869	67	0	0	33

Case studies: active and abandoned Uranium mines

2.) Re-cultivation plateau TPF Lengenfeld (Surface slope inclination ≤ 5 %)

1,0 m recultivation layer	k_f 5E-06 m/s

Vegeta-tion	Pre-cipitation [mm/a]	Evapotran-spiration [% P]	Surface runoff [%. P]	Interflow (drainage) [% P]	Percola-tion [% P]
grass	913	60	5	2	33

3.) Monolayer cover plateau TPF Trünzig (Surface slope inclination ≤ 5 %)

0,3 m re-cultivation layer	k_f 1E-06 m/s
1,2 m storage layer	k_f 5E-07 m/s
	k_f 1E-08 m/s

Vegeta-tion	Pre-cipitation [mm/a]	Evapotran-spiration [% P]	Surface runoff [% P]	Interflow (drainage) [% P]	Percola-tion [% P]
grass	680	68	5	0	27
forest	680	68	4	0	14

4.) Twolayer cover dumps Aue/ Saxony (surface slope inclination 34 %, E-exposition)

0,2 m Upper layer	k_f 1E-05 m/s
0,8 m Lower layer	k_f 1E-06 m/s

Vegeta-tion	Pre-cipitation [mm/a]	Evapo-transpiration [% P]	Surface runoff [% P]	Interflow (drainage) [% P]	Percola-tion [% P]
grass	865	62	1	6	31
forest	865	70	0	0	30

5.) Multilayer cover dump Beerwalde / Thuringia (surface slope inclination 34 %, E-exposition)

1,0 m recultivation/ storage layer	k_f 1E-05 m/s
	k_f 5E-06 m/s
0,4 m sealing layer	k_f 1E-09 m/s

Case studies: active and abandoned Uranium mines

Vegeta-tion	Pre-cipitation [mm/a]	Evapo-transpiration [% P]	Surface runoff [% P]	Interflow (drainage) [% P]	Percola-tion [% P]
grass	666	69	11	16	4
forest	666	84	1	11	4

6.) Multilayer cover plateau TPF /option (surface slope inclination ≤ 5 %)

	0,3 m re-cultivation layer	k_f 1E-05 m/s
	1,0 m storage layer	k_f 1E-06 m/s
	0,2 m drainage layer	k_f 1E-03 m/s
	0,4 m sealing layer	k_f 1E-09 m/s

Vegeta-tion	Pre-cipitation [mm/a]	Evapo-transpiration [% P]	Surface runoff [% v. P]	Interflow (drainage) [% P]	Percola-tion [% P]
grass	666	70	1	25	4
forest	666	73	1	23	3

7.) Multilayer cover Theissenschlamm-deposit Helbra / Saxony-Anhalt (surface slope inclination 5 %, E-exposition)

	1,0 m re-cultivation layer	k_f 1E-04 m/s
	0,3 m drainage layer	k_f 1E-03 m/s
	KDB (HDPE foil 2,5 mm)	-
	0,5 sealing layer	k_f 5E-10 m/s

Vegeta-tion	Pre-cipitation [mm/a]	Evapo-transpiration [% P]	Surface runoff [% P]	Interflow (drainage) [% P]	Percola-tion [% P]
grass	588	55	0	45	ca. 0
forest	588	69	0	31	ca. 0

Proof of stability

An essential aspect of the assessment of suitability of covering systems reducing the distribution of harmful substances is the proof of long term stability.

Because of the low thickness of the cover in relation to the length of dam slopes in the framework of local stability assessment of final covering the investigation of polygonal slide lines in the strata of final covering without consideration of supporting effects of bermes is the matter of interest.

The investigations of stability were performed in view of quasi static and dynamic cases with analytic methods (e.g. JANBU) and numeric methods (PLAXIS 8.0 / FLAC). Using numeric methods only the stability of the covering system themselves was examined.

In addition to the geometric data input of the stability calculations for all materials should be geotechnical parameters (friction angle, cohesion, density) estimated on the basis of soil mechanical investigations of the appointed materials.

Data of the possible water saturation in the storage layer are an additional prerequisite of the stability proof. The calculation of storage level can be performed with the calculation programs HELP or HYDRUS, e.g.

The framework of static safety assessment in a sufficient conservative way should assume the complete saturation of the covering system beforehand. On the other hand for the calculations of dynamic stability an earthquake with an interval of reoccurrence of 10 000 years including the consideration of a „probable" saturation was assumed.

The stability proof for slope inclination degrees of 1 : 4 shows that in general a wide range of covering systems confirm the requirements of long term stability. For slope rates of 1: 3 and steeper restrictions for the covering design are resulting. Taking into consideration additional boundary conditions the demand of the flattening of dam slopes can be required.

Proof of safety against erosion

Beside the stability in long terms a prerequisite to get a license for a covering system is the proof of safety against erosion. This proof is based on the comparison of occurring shear stresses τ_0 with each (critical) drag τ_{Cr}.

The extreme discharge coefficient ψ_S describes the degree of earth surface sealing and thereby that part of surface runoff that is valid for rainfalls events having a duration of a 15 minutes.

Furthermore the consideration of rainfall duration in relation to the case of assessment is necessary for the calculations. In the framework of proofing long term safety against erosion have to considered a 1000-year-precipitation-occurance with a rainfall duration of 15 minutes. This is an appropriate value.

The calculations of safety against erosion always should be performed for the steepest part of the slope or the part with the longest slope surface. These parts of covered dam slopes are endangered to erosion at most. It must be considered that

Case studies: active and abandoned Uranium mines

in practice surface runoff isn't homogeneously and leads to the concentration in grooves on the slope surface. To simulate this phenomenon it is common to multiply surface runoff with a concentration factor of 3 (NUREG 1623).

On the basis of discharge calculation in the case assessment of investigation with help of the MANNING-STRICKLER formula is resulting the depth of flow. Finally shear stress has to be calculated.

In the technical literature for grass covered areas a (critical) drag of $\tau_{Cr} = 30$ N/m^2 is listed. The safety proofs against erosion of a slope are performed by means of a comparison of occurring shear stresses and the drag τ_{Cr}. The proofs are given if results of calculations show that in situ shear stresses τ_0 are smaller than (critical) drag τ_{Cr} For sites in Saxony and Thuringia with slope rates of 1 : 4 and moderate lengths of the slope surfaces it was shown, that a grass covered surface confirms the requirements of safety against erosion.

Resume

The design of the final cover of dumps and tailings is the result of an expending multidisciplinary optimisation. The selection of site specific cover configurations is primarily determined by the long-term effect of the water balance under consideration of stability and safety against erosion.

During the last 10 years many field tests were realised, specified water balance models and geotechnical models were used and developed, which used in combination contributed to the definition and the reason of cover measures.

So costs and time was saved and the certainty of the prognosis of the development of the discharge of contaminants as well as the certainty of the safeguarding of the functionality of the cover could be increased.

Delineation of contamination flows produced by acid mine drainage in a former uranium mining site (Ronneburg, Eastern Thuringia, Germany) by the use of rare earth elements as tracers

Dirk Merten, Jörn Geletneky, Kathrin Lahl, Georg Büchel

Institute of Earth Science, Friedrich-Schiller-University, Burgweg 11, D-07749 Jena, Germany

Abstract. Between 1950 and 1990 the area of Ronneburg (Eastern Thuringia, Germany) was the largest uranium mining site in black shales in the world. Sulfite oxidation still leads to pH-values in seepage waters as low as 2.5 and subsequent enrichment of heavy metals, including uranium and rare earth elements (REE). For seepage water sampled over a period of up to two years the shale normalized REE patterns show only minor variations, although the absolute concentrations differ. It is shown that not only the identification of contaminant sources can be performed using REE patterns, but it is also possible to follow the flow of contamination for both surface and groundwater. Percolation in batch experiments shows that the unique heavy REE enriched patterns are due to preferential leaching. The REE patterns in the water samples do not reflect the source rock pattern as can be demonstrated by laser ablation ICP-MS experiments.

Introduction

Background

The Ronneburg mining district in the eastern part of Thuringia (Germany) was one of the largest uranium mining sites in the world. Between 1950 and 1990 about 216 kt of uranium were produced in the eastern part of Germany. After the USA and Canada the former GDR was the third largest uranium producer in the world

(Barthel 1993). About half of the production (113 kt) originated from the underground and the open pit operations near the city of Ronneburg.

One of the remediation topics is the passive flooding of the mine initiated in
1998. The flooding is expected to be completed between 2003 and 2005. The investigation area "Gessental" valley is one of the main discharge areas for the
flooding water in the future.

Rare earth elements

REE (La-Lu) show smooth but continuous variations of their chemical behaviour
as a function of their atomic number. After normalization to PAAS (Post Archean
Australian Shale) (Taylor and McLennan 1985) they can be used as tracers in
ground water and surface water (Johannesson and Lyons 2000). Furthermore, they
are suited to study processes such as dissolution (Sholkovitz and Hannigan 2001),
sorption (Ohta and Kawabe 2001; Coppin et al. 2002), complexation (Johannesson
et al. 1996; Schijf et al. 2001), (co)precipitation (Byrne and Kim 1993) and especially water-rock interaction (Worrall et al. 2001).

Geology

The uranium deposit of Ronneburg, Thuringia, Germany, is a strata-controlled,
structure bound deposit. It consists of uranium concentrations in small scale brittle
structures which form stock works within or immediately adjacent to carbonaceous, pyritic black shales. The Paleozoic host rocks mainly consist of argillaceous and siliceous black shales with intercalations of dolomitic and phosporite
nodule beds (Silurian "Graptolithenschiefer"). The main black-shale horizon lies
below Ordovician carbonaceous sandy shales and overlies Silurian carbonate
rocks ("Ockerkalk"). Some Devonian metabasaltic dikes and sills cut the meta-
sedimentary rocks (Dahlkamp 1993). The rocks contain up to 7 wt% sulfides, 5-9
wt% organic carbon, 40-60 ppm uranium and a series of trace elements.

Investigation Area

One of the most important drainage systems of the region is the catchment area of
the creek "Gessenbach". The Gessental valley is located in the western part of the
district between the cities of Ronneburg and Gera. In its Eastern part near the city
of Ronneburg it is influenced by two (former) waste rock dumps called Gessenhalde and Nordhalde. The Gessenbach creek and its tributary Badergraben follow
the valley in western direction (Fig. 1). The Gessental valley will be one of the
main discharge areas of mine water in the post-flooding situation.

Case studies: active and abandoned Uranium mines

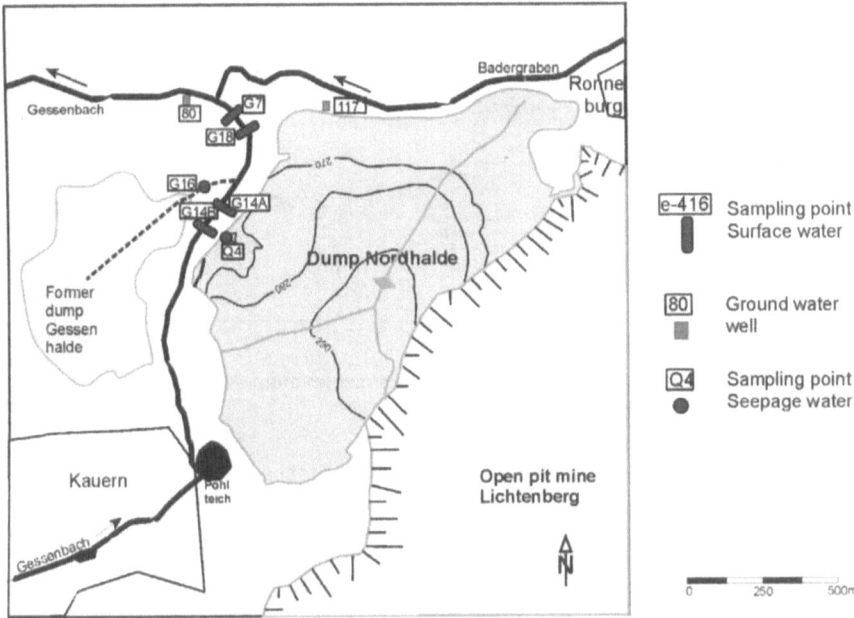

Fig. 1. Sampling points of seepage water, surface water and groundwater around the waste rock dump "Nordhalde".

Results

The hydrological settings of the Gessental valley are dominated by the existence of the (former) dumps and by the groundwater depression cone of the underground mine.

The investigated seepage waters (Q4, G16, Fig. 1) are highly mineralized, have a low pH-value (2.5-5.4) and high redox potentials. This water has high concentrations of Fe, Ca, Mg, Al, Mn, SO_4, Si, Cu, Zn, Ni, Cd, Cr, U and REE (Acid rock drainage: ARD).

Using REE-patterns as tracers for ARD

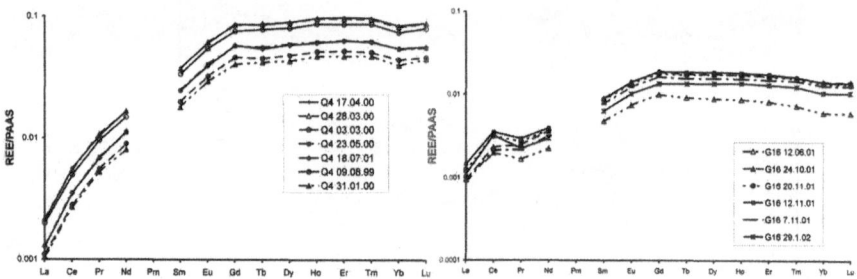

Fig. 2. REE patterns for seepage waters in the former uranium mining region of Ronneburg, Germany (for sampling points see Fig. 1)

When normalizing REE concentrations to Post Archean Australian Shale (PAAS) enrichment of middle REE (Sm to Dy) and especially of heavy (Ho to Lu) REE as compared to the light ones (La-Nd) is observed in the investigated seepage waters. For the investigated seepage waters (Q16, Q4) significantly different REE patterns are obtained (Fig. 2). Furthermore, the presence and absence of positive Ce anomalies could be observed. Nevertheless, for both seepage waters - sampled over a period of two years and half a year, respectively- the shale normalized REE patterns show only minor variations, although the concentration differs. Thus, REE patterns are independent of REE concentrations and can be used to distinguish between different dumps as sources of contamination.

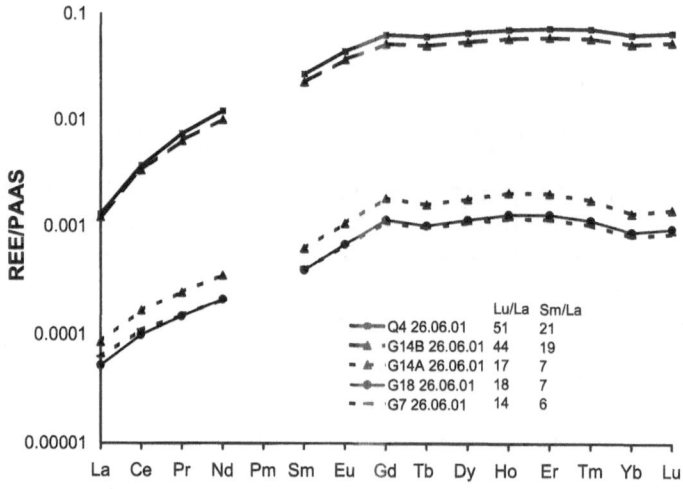

Fig. 3. REE concentrations normalized to PAAS with enrichments in middle (expressed as Sm/La) and heavy REE (expressed as Sm/La) along a flowpath in the Gessenbach creek

Case studies: active and abandoned Uranium mines

At the sampling points in the surface water (G14A, G14B, G18, G7, Fig. 1) nearly the same REE patterns were observed as in the seepage water Q4 (Fig. 3). This represents a diffuse inflow of REE-rich ARD of the dumps into the creek. The absolute concentrations of REE in the creek are up to 100 times less than in seepage water due to mixing and (co)precipitation of REE. Lu/La and Sm/La relations show a significant decrease with increasing distance from the dump. This is caused by preferential (co)precipitation of heavy REE with amorphous Fe-hydroxides along the Gessenbach creek (Astrom 2001).

The groundwater situation in the valley is strongly influenced by the groundwater depression cone. The alluvial sediments north of the waste rock dumps are only partly saturated with groundwater. However, groundwater in wells 117/00, 80/00 (see Fig. 1) could be sampled in summer 2000. The water can be classified as Mg-Ca-SO$_4$-type with a pH varying between 5.3 to 6.4. The general shape of the REE pattern of the dump Nordhalde can also be found in ground water wells (see Fig. 4). However, the degree of enrichment of heavy REE is less pronounced. Again, the REE patterns of the groundwater samples as a function of time are very similar. The absolute concentrations of REE in groundwater are 1000 times lower than in seepage. A negative Ce anomaly is caused by precipitation of Cerianite due to the Eh-pH conditions in the groundwater (Leybourne 2000).

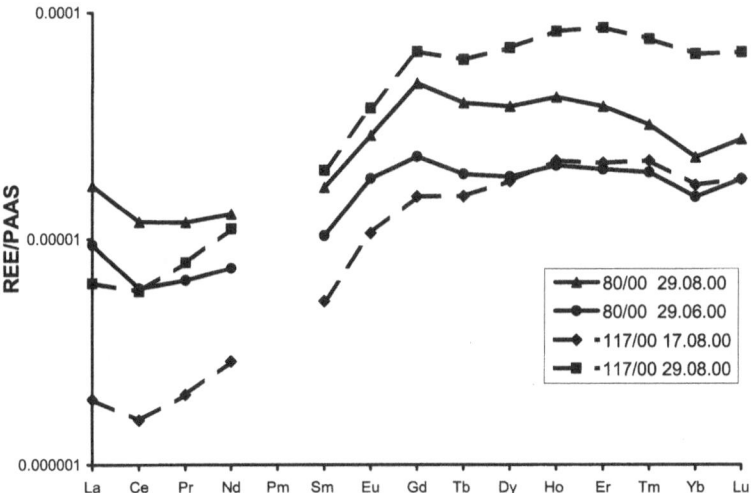

Fig. 4. REE patterns in groundwater samples

In order to investigate whether the REE patterns in the seepage water resemble the REE concentrations in the source rocks LA-ICP-MS measurements were performed using glass beads of the most important lithologies Silurian carbonate rock "Ockerkalk" and fine grained alluvial sediments. The results are compared to those obtained when eluting the lithologies with demonized waters in batch experiments (see Fig. 5).

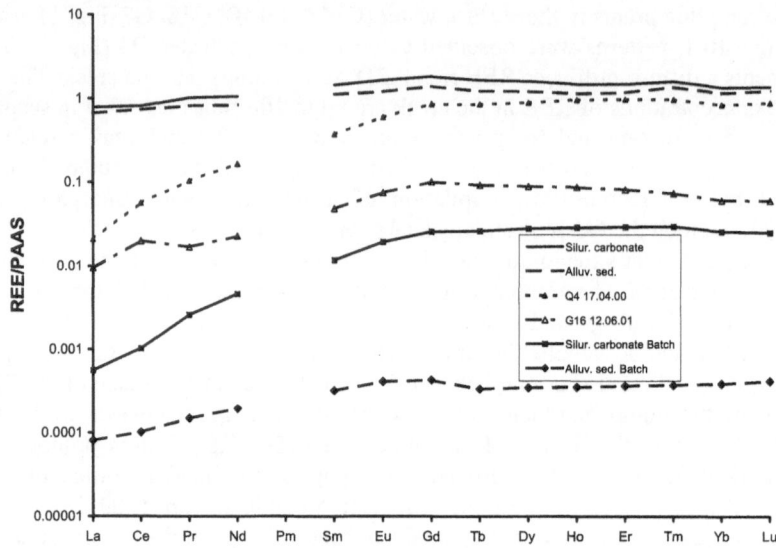

Fig. 5. REE patterns of source rocks (Alluvial fine grained sediments and Silurian Carbonate rocks) as compared to to eluates (Batch) of the lithologies and REE-patterns of the investigated seepage waters

In contrast to the patterns of the seepage, the REE patterns of the Silurian carbonate rock and the alluvial sediments are featured by rather flat patterns with enrichment of middle REE (Sm – Dy). The concentrations of REE in the lithologies investigated differ only by a factor of 2.

The REE concentrations eluted from the lithologies differ by a factor of 10 and are at least one order of magnitude lower than in the source lithologies. The results indicate a preferentially leaching of heavy REE from the investigated lithologies. The REE pattern of the eluate of the Silurian "Ockerkalk"closely reflects the pattern found in the seepage water Q4. Therefore, it is assumed to be the most important source for the occurrence of this special REE pattern.

Conclusions

It is shown that the identification of contaminant sources can be performed using shale normalized REE patterns. It is also possible to follow footpaths of contamination in surface water and groundwater by using REE patterns.

Both the chemical composition and the REE pattern show an influence of acidic seepage water on surface and groundwater, which is originated from the dumps. Both surface and groundwater show enrichment in the middle and in the heavy REE.

The REE patterns in the water samples do not reflect the source rock pattern as was demonstrated by laser ablation ICP-MS experiments. Percolation in batch experiments shows that the unique heavy REE enriched patterns are due to preferential leaching especially from Silurian "Ockerkalk".

References

Astrom, M (2001) Abundance and fractionation patterns of rare earth elements in streams affected by acid sulphate soils. Chem Geol 175: 249-258

Barthel FH (1993) Die Urangewinnung auf dem Gebiet der ehemaligen DDR von 1945 bis 1990. Geol Jb A142: 335-346

Byrne RH, Kim K-H (1993) Rare earth precipitation and coprecipitation behavior: The limiting role of PO43- on dissolved rare earth concentrations in seawater. Geochim Cosmochim Acta 57: 519-526.

Coppin F, Berger G, Bauer A, Castet S, Loubet M (2002) Sorption of lanthanides on smectite and kaolinite. Chem Geol 182 : 57-68

Dahlkamp FJ, (1993) Uranium ore deposits. Berlin: Springer.

Gatzweiler R, Hähne R, Eckart M, Meyer J & Snagovsky S (1997) Prognosis of the flooding of uranium mining sites in east Germany with the help of the numerical box-modeling. In: Veselic M & Norton JPJ (eds). Mine water and the environment. Proc 6th IMWA congr, Bled, 8-12 September 1997

Johannesson KH; Lyons WB; Yelken MA; Gaudette HE; Stetzenbach KJ (1996) Geochemistry of the rare-earth elements in hypersaline and dilute acidic natural terrestrial waters: Complexation behavior and middle rare-earth element enrichment. Chem Geol 133 : 125-144

Johannesson KH, Lyons WB (2000) Rare earth elements in groundwater in: P. Cook and A. Herczeg (eds.) Environmental Tracers in Subsurface Hydrology: 85-492. Dordrecht: Kluwer

Leybourne MI, Goodfellow WD, Boyle DR, Hall GW (2000) Rapid development of negative Ce anomalies in surface waters and contrasting patterns in groundwaters associated with Zn-Pb massive sulphide deposits. Appl Geochem 15: 695-723

Ohta A, Kawabe I (2001) REE(III) adsorption onto Mn dioxide (d-MnO2) and Fe oxyhydroxide: Ce(III) oxidation by d-MnO2. Geochim Cosmochim Acta 65 : 695-703

Schijf J, Byrne RH (2001) Stability constants for mono- and dioxalato-complexes of Y and the REE, potentially important species in groundwaters and surface freshwaters. Geochim Cosmochim Acta 65 : 1037-1046

Sholkovitz RE, Hannigan ER (2001) The development of middle rare earth element enrichments in freshwaters: weathering of phosphate minerals. Chem Geol 175 : 495-508

Taylor SR, McLennan SM (1985) The continental Crust: Its Composition and Evolution. Blackwell: Oxford

Worrall F, Pearson DG (2001) Water-rock interaction in an acidic mine discharge as indicated by rare earth element patterns. Geochim Cosmochim Acta 65: 3027-3040

Case studies: active and abandoned Uranium mines

◄ fieldSPEC

Digital hand held spectrometer and dose rate meter with nuclide identification.

nanoPROBE ►

Integrated Multi-Channel-Analyzer with NaI(Tl) for on-line spectroscopy and for data aquisition networks. Shock resistant and water protected (IP67).

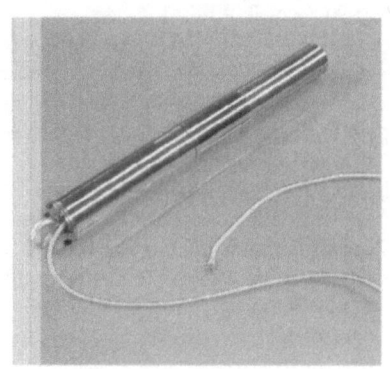

◄ nanoSPEC pro

Portable MCA for NaI(Tl)-Spectroscopy with Pocket PC.

42651 solingen · germany
phone +49 (0) 212 / 22 20 90
fax +49 (0) 212 / 20 10 45
target@target-systems-gmbh.de
www.target-systems-gmbh.de

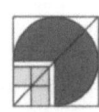

target

systemelectronic gmbh

Using acid rock drainage (ARD) of an uranium mining dump as tracer in surface and groundwater: Case study at the Ronneburg uranium mining site

Jörn Geletneky

Institute of Earth Science, Friedrich-Schiller-University, Burgweg 11, D-07749 Jena, Germany

Abstract. Natural tracers (e.g. sulphate, heavy metals, rare earth elements) in acid rock drainage can be used to identify sources and sinks for heavy metals as well as processes in hydrological systems. High contents of these tracers are available at the former uranium mining site of Ronneburg. The Eastern part of the Gessental valley near Ronneburg is dominated by (former) uranium mining dumps and by a groundwater depression cone of the underground mine of Ronneburg. The influence of highly mineralised, acidic seepage water (ARD) on the unsaturated valley sediments as well as on surface and groundwater has been investigated in situ and with geochemical modelling.

Introduction

Investigation Area

The Ronneburg mining district in Eastern-Thuringia (Germany) was one of the largest uranium mining sites in the world. Between 1950 and 1990 about 231 kt of uranium were produced in the Eastern part of Germany. After the USA and Canada the former GDR was the third largest uranium producer in the world (Barthel 1993, Gatzweiler et al. 1997). About half of the production (113 kt) originated from the underground and the open pit operations near the city of Ronneburg.

One of the remediation topics is the passive flooding of the mine initiated in 1998. The flooding is expected to be completed between 2003 and 2005. The in-

vestigation area "Gessental-valley" with the creeks "Gessenbach" and „Badergraben" is the major prognostic discharge area for flooding water in the near future.
The Gessental valley is located in the western part of the district between the cities
of Ronneburg and Gera. In its Eastern part near the city of Ronneburg it is influenced by two (former) waste rock dumps „Gessenhalde" and „Nordhalde". The
Gessenbach creek and its tributary Badergraben follow the valley in western direction (Fig. 1).

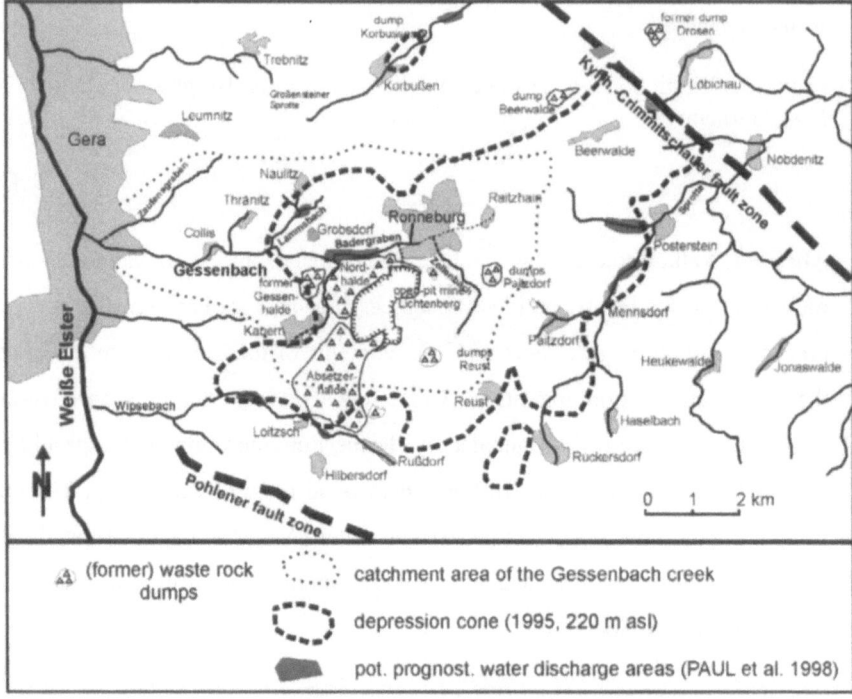

Fig. 1. The catchment area of the Gessenbach creek is located in the Western part of the
Northern former uranium mining site Ronneburg, Germany (1950-1990). A noticeable area
was inside the depression cone of 1995, which is now in flooding progress. The major
prognostic water discharge area in the post-flooding-stage is situated inside the Gessental
valley.

Geology

The uranium deposit of Ronneburg, Thuringia, Germany, is a strata-controlled,
structure bound deposit. It consists of uranium concentrations in small scale brittle
structures which form stockworks within or immediately adjacent to carbonaceous, pyritic black shales. The Paleozoic host rocks mainly consist of argillaceous and siliceous black shales with intercalations of dolomitic and phosporite

nodule beds (Silurian "Graptolithenschiefer"). The main black-shale horizon lies below Ordovician carbonaceous sandy shales and overlies Silurian carbonate rocks ("Ockerkalk"). Some Devonian metabasaltic dikes and sills cut the meta-sedimentary rocks (Dahlkamp 1993). The rocks contain up to 7 wt% sulfides, 5-9 wt% organic carbon, 40-60 ppm uranium and a series of trace elements (Szurowski 1985).

The intensively folded and faulted, incompetent and competent rocks have high density small scale brittle structures (fissures, joints, faults). Permian and Tertiary supergene oxidation processes associated with mobilization and precipitation of trace elements created an oxidation and cementation zone. The irregular distribution and size of the ore bodies is controlled by major and minor faults. The uranium ore appeared near the surface in the southern part and down to 1000 m deep in the northern part of the Ronneburg mining district.

Methods

A monitoring system with 8 sampling points was installed in the creeks along the Gessental valley and near the uranium mining dump Nordhalde (Fig. 2).

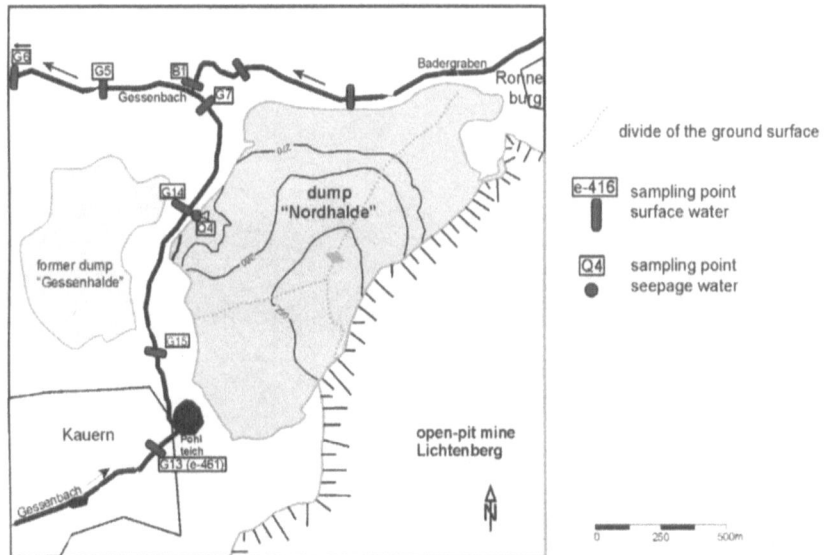

Fig. 2. Sampling points of seepage water, surface water and groundwater around the waste rock dump "Nordhalde".

The average discharge, hydrochemical parameters (pH, Eh, electr. cond.) and the chemical composition of the water were analysed with AAS, ICP-MS, IC to describe the recent processes and the flux. Wells were drilled in the valley sediments and sampled for analysing chemical composition of the groundwater together with the Wismut GmbH (Geletneky et al. 2001).

Results

Seepage, surface water and groundwater

The seepage water in the Western part of the „Nordhalde" (Q4 in Fig. 2) is highly mineralised, has a low pH (2-3.5) and high redox potentials. This water has high concentrations of Fe, Ca, Mg, Al, Mn, SO_4, Si, Cu, Zn, Ni, Cd, Cr, U, REE (Fig 3).

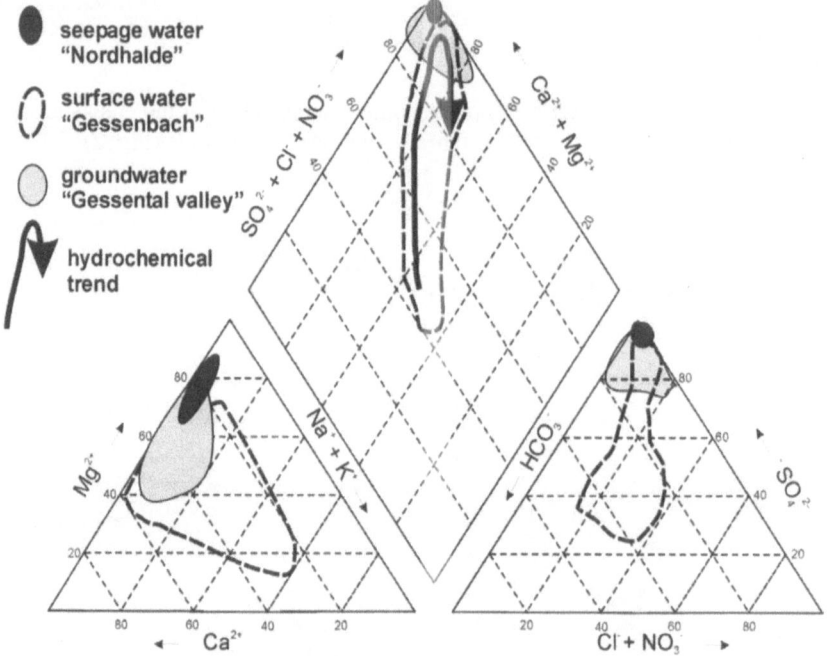

Fig. 3. Piper-plot of the hydrochemistry of the seepage and surface water and the groundwater of the Gessental valley.

The amount of diffuse discharge at Q4 could only be estimated. Discharge measurements in the Gessenbach creek showed that the amount of seepage water flowing into the creek is less than 1 l/s.

In the Gessenbach creek the chemical composition of the water changes from an alkaline $Ca-Mg-SO_4-HCO_3$-type to an acidic $Mg-Ca-SO_4$-type. The heavy metal (e.g. Ni, Zn, U) contents of the water increases by diffuse influx of ARD (Fig. 3).

After the dump passage the surface water chemistry changes quickly in flow direction due to mixing, dilution and (co)precipitation back to $Ca-Mg-SO_4-HCO_3$ type with low metal loads. Iron-rich ochreous precipitates with high contents of natural tracers like Ni, Al, Zn and U occur temporally in the creek. Hydrogeochemical modelling showed, that the phases mainly consist of poorly cristalline $Fe-SO_4$-minerals (e.g. ferrihydrite, schwertmannite), Al-(hydr)oxides, and jarosite. Investigations collected from the creek showed enrichments in most of the heavy metals, which are also enriched in ARD (Geletneky 2002).

The inflow of seepage water can also be detected by using natural tracers in discrete footpaths of the partly saturated alluvial sediments of the Gessental valley. The edge of the depression cone can be seen by comparing the chemical composition and the hydraulic heads of the shallow groundwater in the valley sediments. The groundwater chemistry is characterised as $Mg-Ca-SO_4$ type (Fig. 2) by the influence of the dumps inside the depression con and low mineralised $Ca-Mg-HCO_3$ type outside the depression cone.

Conclusions

The amount of TDS and the water type in the surface water changes when the Gessenbach creek passes the dump areas. The amount of TDS in the creek decreases rapidly after the „Nordhalde". This is caused by a combination of iron-hydroxide/sulphate precipitation, co-precipitation of heavy metals and dilution.

More than 50% of the seepage water of the Nordhalde discharged directly or indirectly into the former underground mine (C&E 1995). Only a minor part seeps diffusively into the creeks and into the valley sediments. Using the seepage water as tracer, the existence of discrete flow paths within the valley sediments is suggested. This information can be used for the implementation of effective remediation strategies.

References

Barthel FH (1993) Die Urangewinnung auf dem Gebiet der ehemaligen DDR von 1945 bis 1990. Geol Jb A142: 335-346

C&E (1995) Wasserhausrechtliche Untersuchungen im Bereich der Nordhalde, Teil I: Untersuchung zum Istzustand. C&E Consulting und Engineering GmbH internal report prepared for Wismut GmbH

Dahlkamp FJ, (1993) Uranium ore deposits. Berlin: Springer

Gatzweiler R, Hähne R, Eckart M, Meyer J & Snagovsky S (1997) Prognosis of the flooding of uranium mining sites in east Germany with the help of the numerical box-modeling. In: Veselic M & Norton JPJ (eds). Mine water and the environment. Proc 6th IMWA congr, Bled, 8-12 September 1997

Geletneky J (2002) Hydrogeologische/Hydrologische Untersuchung einer Prä-Flutungssituation am Beispiel des Gessentals im ehemaligen ostthüringischen Uran-bergbaugebiet. PhD-work Friedrich-Schiller-University Jena: 261 S

Geletneky J W, Merten D, Büchel G (2001) Seepage water from uranium mining dumps in Eastern Thuringia, Germany: A hydrogeochemical study. EUG 11, Strasbourg, J. Conf. Abs, 6: 44

Szurowski H (1985) Katalog der Gesteinseigenschaften (eine kurzgefaßte petrographische, mineralogische, geochemische und physiko-mechanische Charakteristik der Gesteine im Ronneburger Raum). Internal report of Wismut GmbH

Environmental Impact of Uranium ISL in Northern Bohemia

Radomír Smetana[1], Jiří Mužák, Jan Novák[2]

[1]DIAMO s.p., o.z. TÚU, Máchova 201, 471 27 Stráž pod Ralskem, Czech republic
[2]Technical university, department of modelling of processes, Hálkova 6, 461 17
Liberec, Czech Republic

Abstract. The acid mining of uranium in the Stráž deposit in Northbohemian Cretaceous basin was in 70's and 80's one of the biggest application of „in situ leaching" in the world. This paper describes origin and character of the contamination, its areal and vertical distribution. Evaluation of the risk of penetration of separate contaminants into the environment has been done by the transport-reactive model of inter-collector transport. The paper discusses advantages and disadvantages of the main methods of cleaning. In conclusion the concept accepted by the Czech Republic government is presented.

Chemical mining in the Stráž deposit

In the Stráž deposit the acid leaching of uranium ore was carried out in 1967-1996. The Stráž deposit and several other deposits are located in the Stráž block of North-bohemian Cretaceous basin. There are two sandstone aquifers in the block – lower Cenomanian and upper Turonian aquifers – isolated by 60 m thick aquitard created of muddy limestones, marlstones and sandy siltstones (see Fig. 1). The ore mineralization is embedded on the base of the Cenomanian aquifer, upper Turonian aquifer represents the source of drinkable water for wider region.

The deposit has been opened from the surface with the wells. During the first phase of mining (until 1975) the wells were fitted only with polyethylene casing, later a protection of Turonian water was ensured by the outer steel casing (till the aquitard) and by the cementing of inner space. There were completely drilled more than 8000 technological wells and the area of leaching fields reached 628 ha. In seventies it was the biggest application of acid leaching in the world. Annual production of uranium was about 700 tons.

Separation of uranium from the leachate has been made (and still is made) in the chemical plant. Extracted solution, which contains together with uranium also another products of leaching and non-reacted acid, is led through the sorption columns filled with ion exchange resin (ionex). Uranium is washed out from the saturated ionex by nitric acid (elution), regenerated ionex is returned into the sorption. From the eluate the ammonia diuranite (yellow cake) is precipitated. Remaining solution after the precipitation containing ions NO_3^- and NH_4^+ has been added into the solution, passing through the sorption columns. After a completion of the main reactive agent – sulphuric acid – this solution was again injected. Scheme of separation is in Fig. 2.

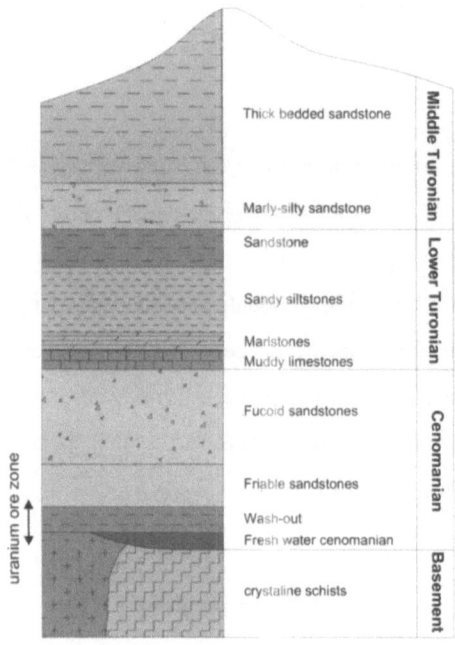

Fig. 1. Stratigraphic structure of the Stráž deposit

The solution circulation was running in the closed circle. Because the separation of uranium was selective, the others substances leached from the rock remained in the solutions and their concentrations gradually increased. It was mainly Al and Fe which are now considered as majority components of solutions. The most important minor component is As from oxidated arsenopyrite. As and NH_4^+ belong between the most risk contaminants.

Conditions for chemical mining were not ideal in the Stráž deposit. Part of mineralization was embedded in less permeable basalt sediments of wash-out horizon where mainly diffusion leaching has run. Considerable part of ore was in badly leachable uranium-zirconium minerals of crandallite group. Therefore it was necessary to use higher dosage of the acid and longer time for leaching of individual block in comparison with ISL applications in other world deposits.

Case studies: active and abandoned Uranium mines

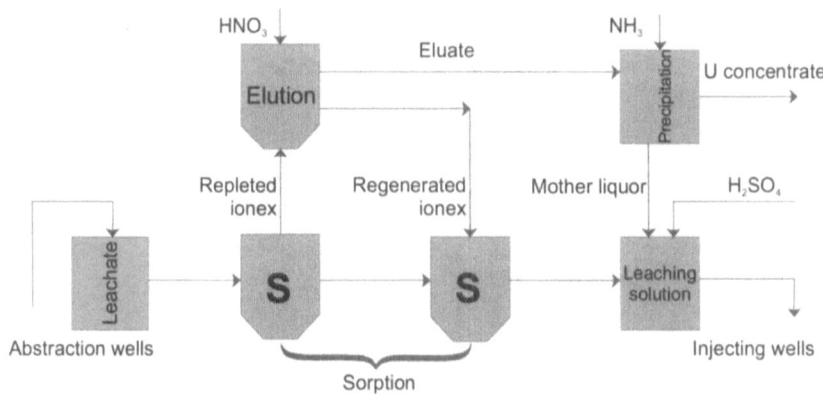

Fig. 2. Scheme of the surface reprocessing of uranium leachate

Extent of contamination

Actual extent of contamination is contingent on the process of chemical mining and on complete mining activity in the Stráž block. Extent of influenced area in the Cenomanian aquifer is 24 km². The total volume of contaminated water is 180 million m³, total amount of dissolved solids is 4,8 mil. tons, from that 3,7 mil. tons of SO_4^{2-}, 420 thousand tons of Al, 110 thousand tons of Fe, 90 thousand tons of NH_4^+ and 50 thousand tons of NO_3^-.

In the area of leaching fields almost whole thickness of the Cenomanian aquifer is affected. Here is located about 50 % of contaminated solutions and almost 90 % of dissolved solids. Acidity of the solutions is from 30 to 100 g/l. The contamination of less permeable upper fucoid sandstones is caused by the high gradients in vertical direction in neighbourhood of active wells.

Actual extent of solutions is shown in Fig. 3. The highest concentrations are in the area of leaching fields, where is 100-600 kg of contaminants per m². Large outer area is affected considerably less. Scheme of escape of solutions outside of leaching fields area shows Fig. 4. Content of dissolved solids in these solutions is 1-10 g/l, density of contamination is mostly 10 – 50 kg/m².

Rise of large areal contamination and its expected development is shown in Fig. 5. Before building of hydraulic barrier (HB) the solutions moved from the area of leaching fields into the depression cone of drainage mine. After building of HB in 1977 – 1980 were solutions caught between the HB and the mine partly cleaned and discharged into river, partly again injected into HB (see Fig. 6). Injection up to 18 m³/min was completed with clean water. As a result of increased water level on east edge of leaching fields area the solutions escaped toward south. The escape was stopped by the Svébořice barrier where 8 m³/min of Turonian water was injected. After 1985 the extent of the solutions was stabilized.

Case studies: active and abandoned Uranium mines

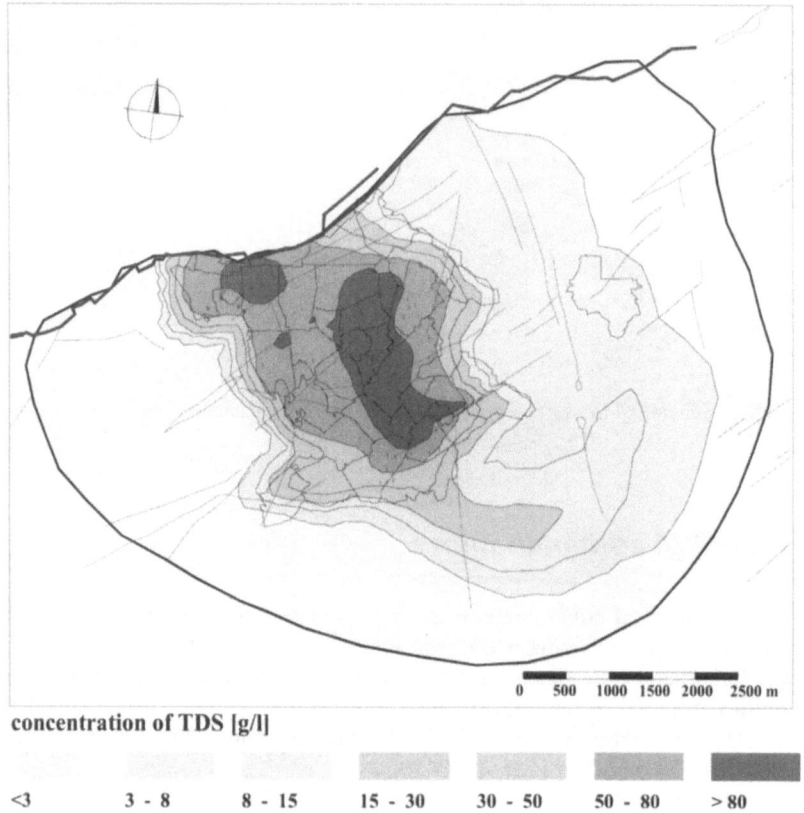

concentration of TDS [g/l]

| <3 | 3 - 8 | 8 - 15 | 15 - 30 | 30 - 50 | 50 - 80 | > 80 |

Fig. 3. Actual extent of solutions in the area of the Stráž deposit – concentration of total disolved solids (TDS)

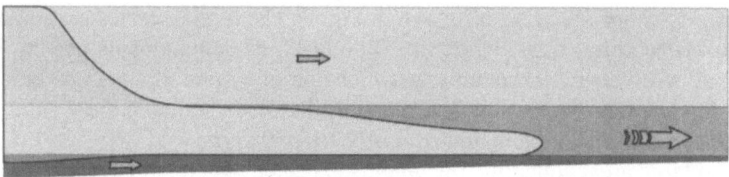

Fig. 4. Scheme of escape of solutions from the area of leaching fields

In 2001 a flooding of the deep mine Hamr has begun. After a water level stabilization the center of hydraulic depression will move into area of leaching fields. The depression will be caused by the activity of evaporation plant, which represents the main element of the remediation technologies chain. A contour of solutions extant will decrease. After the finish of remediation the remaining solutions will move in direction of natural hydraulic flow toward south-west.

Case studies: active and abandoned Uranium mines

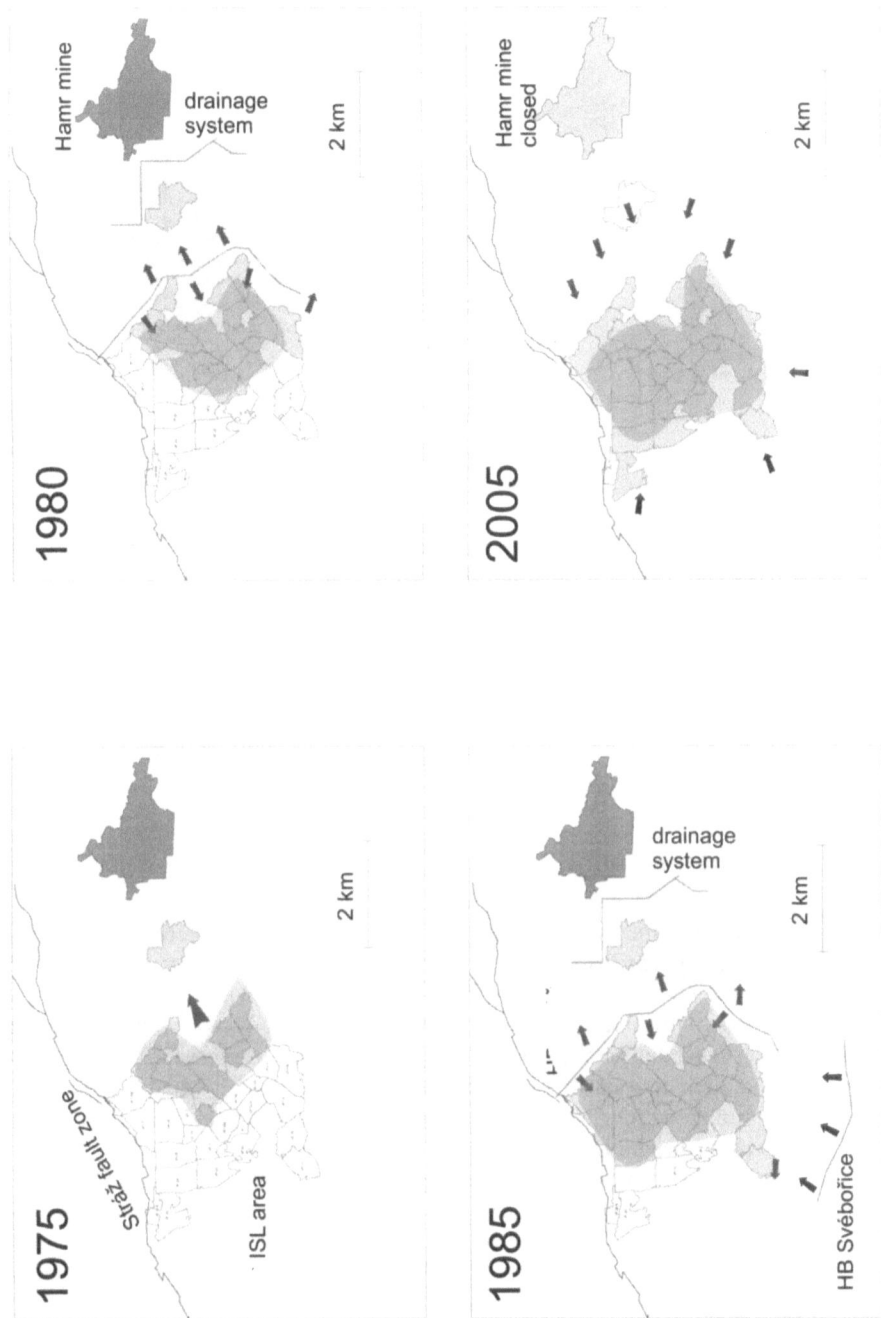

Fig. 5. Development of contamination in neighbourhood of the area of leaching fields in period of uranium mining and in the first phase of remediation

Case studies: active and abandoned Uranium mines

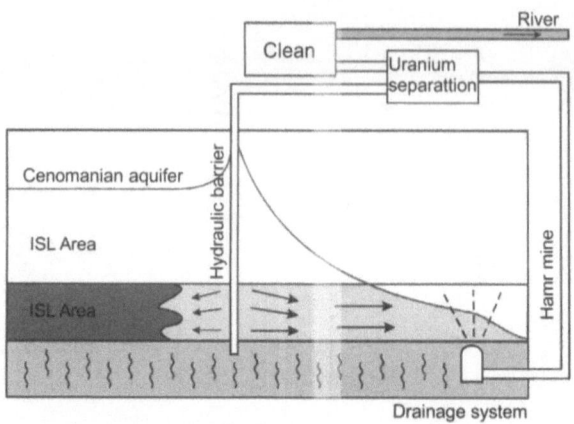

Fig. 6. Scheme of the hydraulic barrier activity

Contamination of Turonian aquifer is relatively separated problem. Origin of the contamination is in escape of solutions from the surface technologies and mainly in the untightness of cementing of the wells without double-casing through the Turonian aquifer. There is about 30 thousand tons of dissolved solids in Turonian aquifer, partly in smaller lens with concentrations 5 – 30 g/l, partly in larger volumes with concentrations about 1 g/l. Here the solutions are extracted, cleaned by the membrane technologies or by neutralization and discharged into river.

Risks of the actual state

At present the piezometric head of the Cenomanian aquifer id decreased as consequence of unfilled depression in the area of Hamr mine and activity of evaporation plant. After the end of remediation the piezometric head will rise into the initial level, which is in the area of leaching fields and in area towards south-west higher than the Turonian aquifer water level. Then the solutions can escape from the Cenomanian aquifer through low Turonian aquitard, damaged by quantity of wells, into the Turonian aquifer and endanger the sources of drinkable water for towns Mimoň, Česká Lípa and Doksy.

For the risks evaluation the 3D reactive transport model has been developed. The model covers an area of the length of 20 km from the Stráž deposit towards south-west. Flow has been solved by the finite element method. The mix-hybrid element has been used in the model, this element gives as a solution the transfer of water through sides of the element. The elements have shape of the vertical trilateral prism with arbitrary inclined bases. This shape is convenient for description of the geological environment. It enables describe both downy boundary of layers and tectonics and tokens of tectonics and volcanic activities.

Case studies: active and abandoned Uranium mines

The model mesh consists of 35000 space elements in 10 model layers. Lower 2 layers represent friable sandstones, next 2 ones a fucoid sandstones, next 2 ones represent lower-Turonian aquitard and upper 4 layers represent middle Turonian aquifer.

Fig. 7 shows extent of solutions in friable sandstones in time 1000 years after the end of remediation, middle part of figure shows situation in the thick-bedded sandstones, lower part of figure shows the cross-section of modeled area. As a characteristic risk component has been chosen ammonium ion NH_4^+. In the Cenomanian aquifer the head of solution with concentration $5 - 10$ mg/l moved cca 15 km toward the deposit to south-west. The transfer into the Turonian aquifer is intensive in the places where the aquitard is weakened by drilling works or tokens of volcanic activities near Ralsko and where the Cenomanian piezometric head is higher than Turonian water level. The biggest elevation is caused by decreasing of Turonian water level along the river Ploučnice. One of the most important water supply areas, water source Mimoň, is also struck . Concentration of NH_4^+ is overreaching for several hundred years the limits for drinkable water $5 - 10$ times. Therefor it is not possible to let the contamination in the underground at present extent and it is necessary to carry out the remediation.

Also preliminary target parameters of remediation, i.e. quantity and concentrations of the substances which could stay in the aquifer after remediation, have been defined by means of the mentioned model. Acceptable quantity represents about 30 % of present state. The most risk compounds of solutions are NH_4^+ and As. Radioactivity is relatively low, because Ra is not mobile in the acid medium. The main radioactive components are ^{230}Th and ^{234}Th.

Possibilities of remediation

The remediation of rock environment of the Cenomanian aquifer is a long-term and technologically sophisticated activity. Three main processes come into account:

a)   extraction of solutions and their neutralization on surface,
b)   evaporation of the solution and separation of ammonia alum,
c)   immobilization of the solutions in the underground.

Each of these processes has its advantages and its risks.

Neutralization of solutions on surface

Neutralization technology is relatively cheap and simple, its disadvantage is however following ecological impact of the area. Neutralization with the lime brings a large volume of neutralization sludge, which must be stored in setting pit. Volume of the sludge would be about 16 mil. m^3, i.e. almost the same volume as volume of sludge from former uranium ore reprocessing plant from the Hamr mine.

Case studies: active and abandoned Uranium mines

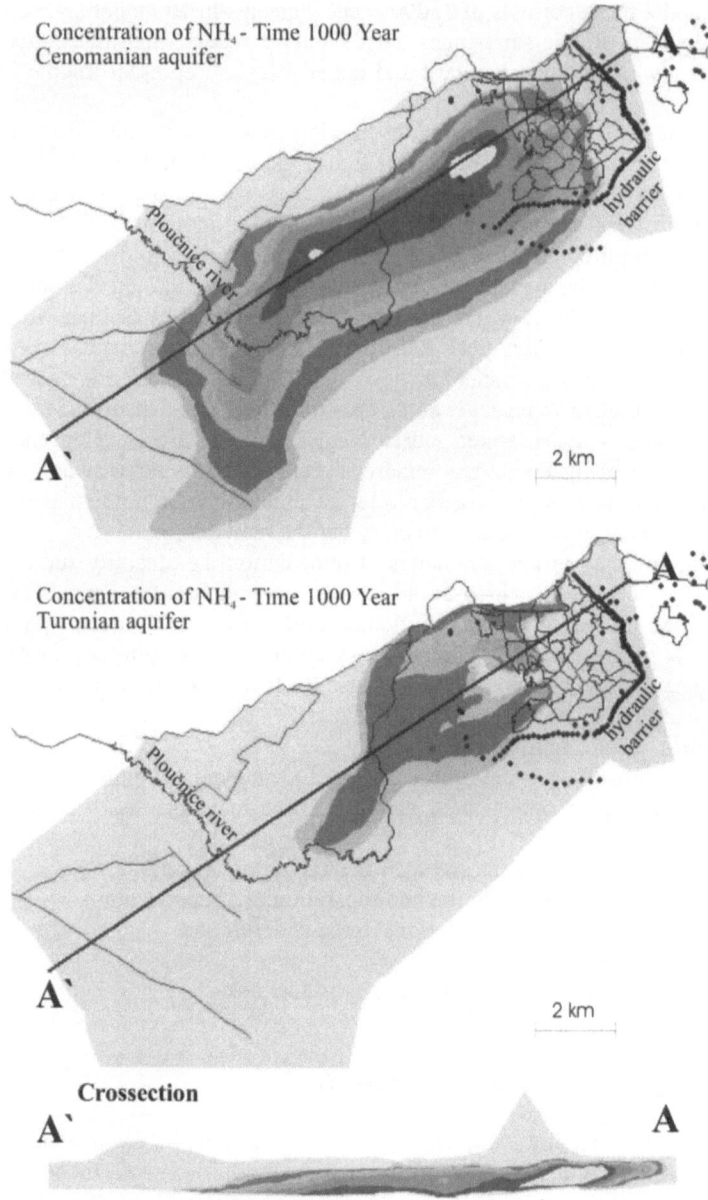

Fig. 7. Extent of contamination after 1000 years after the end of remediation – modelling results

The main component of the sludge would be gypsum $CaSO_4$. Production of gypsum is also possible but the marketing studies say that there is not a consumption for several millions tons.

Case studies: active and abandoned Uranium mines

Solutions evaporation and production of alum

This process is ecologically cleaner but capital and operative intensive. Solutions are extracted into the evaporation plant and thickened. The condensate from the process is discharged into river and it provides a necessary under-balance, which prevent escape of the solutions in the underground.

Concentrate is completed with ammonia to keep stoichiometric ratio NH_4^+ : Al (2:3) in alum. Ammonia alum $NH_4AL(SO_2)_2$ crystallizes from cooled concentrate. After a re-cleaning by re-crystallization it is possible to use it for production of economic usable products, e.g. aluminum sulphate, ammonia sulphate, various Al oxides, heat resistant and constructive materials etc.

Immobilization of contaminants in the underground

This inviting alternative requires both a good theoretical preparation and a realization of pilot plant experiments. It is necessary to verify experimentally an efficiency of particular methods, to define their technical and technological parameters and after it to carry out an economic evaluation.

For the present the most developed method is injection of lime milk into the underground. Reaction of milk with the acid solutions implies a precipitation of insoluble $CaSO_4$ and as a result of decreasing of the acidity also precipitation of Fe^{III} and Al hydroxides. Colmation of porous space of rock with the precipitates has only secondary importance, because the lab test results show that it is possible to inject a lime suspension only in low concentrations until 3 % CaO.

But in means that it is necessary to infuse into the deposit a large volume of water and to keep an under-balance by extraction of appropriate volume through the evaporation plant or by another way. An efficiency of injection and spreading of suspension in the underground is not still enough known. An experimental verification implies the requirements on number of injection wells and their construction.

Accepted remediation concept

Valid concept of remediation according to the decree of the government of Czech Republic is based on extraction of contaminants from the underground, evaporation of solutions and reprocessing of alum.

Parallelly with the evaporation the part of weaker solutions from boundary parts of the deposit will be neutralized.

Problems of the remediation course and its optimization is discussed in paper of H. Čermáková "Optimization of the rock environment remediation process in the Stráž deposit".

Case studies: active and abandoned Uranium mines

Conclusions

The result of the chemical mining of uranium in the Stráž deposit is large contamination of the rock environment.

In the Cenomanian aquifer there is about 180 million m^3 of solutions containing 7,8 million tons of dissolved solids.

The main component of the solutions is SO_4^{2-}, the most dangerous substances are NH_4^+, As and Be.

During the whole period of remediation the hydraulic depression will be keeping in the deposit. The depression will inhibit spreading of solutions in horizontal and vertical direction. After the end of remediation the Cenomanian water level will rise on its initial level and the overflow into the Turonoan aquifer, the source of drinkable water for a wide region, will take place.

For calculation of spreading of solutions in long-term time horizon, risk evaluation and definition of requirements on target parameters of remediation the mathematical models are used.

References

Novák J, Mareček P, Smetana R (1994) Modelling of the ISL in the Stráž deposit. In: Technical Committee Meeting IAEA on Computer application in uranium exploration, Vienna, Austria

Novák J, Smetana R (1994) Modelling of restoration after ISL in the Stráž deposit. In: Technical Committee Meeting IAEA on Computer application in uranium exploration, Vienna, Austria

Novák J, Mareček P, Smetana R (1995) System of the model solutions of the physical-chemical processes in the Stráž deposit used for rock environment restoration. In: Proceedings of the International Conference Uranium-Mining and Hydrogeology, Freiberg, Germany

Kazda I, Novák J, Vaníček I, Vaníček M (1997 Environmental impact of uranium mining in Northern Bohemia. In: Yong, R. N. –Thomas, H. R. (eds.): Geoenvironmental engineering-Contaminated ground: fate of pollutants and remediation, pp. 437-442, Thomas Telford, London

Novák J, Smetana R, Šlosar J (1998) Remediation of a sandstone aquifer following chemical mining of uranium in the Stráž deposit, Czech Republic. Proceedings of the 9th International Symposium on Water-rock interaction, Taupo, New Zealand

Novák J (2000) Groundwater Remediation in the Straz Leaching Operation. In: Tagungsband Internationale Konferenz Wismut 2000 - Bergbausanierung in Schlema, Germany

Environmental Issues and Remedial Actions of the Abandoned Linchang Uranium Mine in China

Lechang Xu [1, 2], Yanxin Wang [1]

[1]Department of Environmental Engineering, China University of Geosciences, Wuhan 430074, China
[2]Research Institute of Uranium Mining, China National Nuclear Corporation, P.O. Box 48, Hengyang, 421001, China

Abstract. Abandonment of Linchang Uranium Mine left one tailings pile, twenty waste piles and eighteen adits and inclined shafts, with severe erosion of tailings and mining debris. Mine waters contain contaminants which include U, Ra, Cd, Cr^{+6}, As, Pb, Cu, Zn, Mn, SO_4^{2-} and Tp. Remediation of solid wastes entails on-site and off-site disposal. Tailings piles will be stabilized with multi-covers consisting of white lime, other solid wastes, loess, clay containing gravel and revegetation taking into account cost-effectiveness and the higher radon flux of the tailings. Abandonment of adits and inclined shafts entails flooding and backfill. After remediation, radon release will reduce to 1.08 TBq/a from 11.01 TBq/a.

Introduction

Uranium mining/milling in China began in the late 1950s. Thirty four mining/milling facilities were successively built to the date. During 1980s and early 1990s, twenty three of these were closed and began to be decommissioned in succession due to a decrease in the demand for uranium and an increase in the overall supply or depleted and barren reserves. Remediation of these closed facilities commenced in 1990. Remedial works have been completed at ten facilities with work at seven more sites to commence by the end of 2002. In the past 10 years, China issued policies and regulations and gained some valuable experience on the decommissioning and remediation of uranium production facilities. This paper reports relevant environmental issues and remediation acts of the abandoned Linchang uranium mine/mill complex, Yunnan Province, China, which is being remedied.

Site characterization

The Linchang uranium mine is a small mine/mill complex situated about 12 km southwest of Linchang County, Yunnan Province, China. It began production in 1970 and was closed in 1994. The final product was $(NH_4)_4UO_2(CO_3)_3$. The ores were mined by conventional open pit and underground methods. The mill process adopted filtration leaching of non-crushed ore by H_2SO_4 and MnO_2 and extraction of settled leachate by organic phase. The mine extracted ores of No. II, III, V/VI and VIII mining areas and left one pile of dry uranium mill tailings and twenty piles of mining debris with little or no on-site environmental control. The streams and rice fields nearby accumulated some uranium mill tailings and mining debris resulted from erosion.

The mining areas and the tailings pile are located in the NE-SW oriented Meng-tuo basin. The basin has two river systems, one in the north and another in the south. The headwater of the southern watercourse is one of the headwaters of drinking water sources in Linchang County. The tailings pile is at the watershed. Overflowing waters from tailings pile and No. II , III mining areas flow into the northern and southern systems.

In this region, annual average temperature is 16.8 °C—17.7 °C; average precipitation and evaporation is 1163.9 mm and 1580.9 mm, respectively. Precipitation concentrates May to October (rainy season), occupying 88.3% of annual total. Bared soil is subjective to dry cracking in the dry season and erosion in the rainy season. Drought and flood happen frequently. The earthquake intensity of the study area is VIII rank.

Objectives and regulatory requirements for remediation

Objectives

The objectives of the remediation for Linchang uranium mine/mill complex are:
- Protection of water catchment of Linchang county;
- Reduction of the area of restricted use as far as reasonably possible;
- Long-term stability of the remediation area in terms of safety and public health, especially, limitation of soil erosion;
- Limitation as far as reasonably possible of the residual impacts;
- Prevention of risk as consequence of intrusion.

Regulatory limits

Regulatory limits for remediation involve safety stability, individual dose, surface contamination, water, air and soil.

- Individual dose
  Annual average individual effective dose over the lifetime from remedied uranium mine/mill does not exceed 0.25 mSv/a (Li 2001).
- Safety stability
  Stability period of debris barrier, slope protection and aslet of mining debris is proposed to be at least 100 years; stability period of the tailings dam is recommended to be at least 200 years (Li 2001).
- Radon flux
  Annual average radon flux over the surface of remedied tailings, mining debris, open mining pit and heap leaching pile is not over 0.74 Bq/(m^2.s) (GB 14586-93).
- ^{226}Ra in soil
  The average concentration of ^{226}Ra in land averaged over any area of 100m^2 shall not exceed the background level by more than 0.18 Bq/g over the first 15 cm and 0.56 Bq/g over the second 15cm of the soil below the surface (GB 14586-93).
- Uranium in surface water
  Uranium concentration of remedied surface water shall not exceed 0.05 mg/L (Li 2001).
- α-Surface contamination
  Equipment, material and structure may be permitted reutilizations in other general industry (except in food processing) as their non-fixed α-surface contamination levels does not exceed 0.08 Bq/cm^2; steel/iron may be permitted unrestricted reuse or recycle as their surface contamination levels does not exceed 0.04 Bq/cm^2 after decontaminated (GB 8703-88).
- Other non-radioactive constituents
  II rank standard of Environmental Quality Standard For Surface Water (GHZB 1—1999) is applied to overflowing mine water as maximum allowable release limits (Table 1).

Table 1. Some effluent concentration limits of overflowing mine water, mg/L

Cd	Cr^{6+}	As	Pb	Cu	Zn	Mn	SO$_4^{2-}$	F$^-$	Tp	Ph
0.005	0.05	0.05	0.05	1.0	1.0	0.1	250	1.0	0.1	6.5—8.5

Environmental issues

Tailings, mining debris, open mining pit, surface subsidence pit and patio

One pile of tailings and twenty piles of mining debris (including open mining pits) of Linchang uranium mine were left on-site with little or no control and steep slopes, resulting in extensive erosion of tailings and mining debris. It was estimated that 20% volume of tailings were washed into the nearby streams (Xu 1999).

Average radon flux and gamma air dose rate from the tailings is up to 10.64 $Bq/(m^2.s)$ and 5.43 μGy/h, respectively, much more than regulatory limit and more than from mining debris, most of which exceed radon flux limit (Table 2). The total radon release was 11.01TBq/a.

Seepage from the tailings contains radionuclides and non-radioactive constituents, such as, Cd, Cr^{6+}, As, Pb, Cu, Zn, Mn, SO_4^{2-}, and Tp, exceeding relevant regulatory limits (Table 3). Average annual flow of seepage from the tailings was $30.76m^3/d$. Some of the mining debris has been reused as cultivating crop by local farmers.

There were thirty nine surface subsidence pits and four patios, all in No. V mining area. The subsidence pits have 20—765 m^2 areas and 2—6.5 m depths. The subsidence pits and patios are steeper and tend to result in soil erosion, safety risk (e.g. man and livestock falling into subsidence pits), radon release and further contamination of underground water as a result of surface water infiltration.

Table 2. Some parameters of mining and milling wastes

Name of waste pile	Mass (kt)	Mean depth (m)	Bare area (ha)	Radon flux (Bq/(m².s))		γ dose rate (μGy/h)		Radon release (TBq/a)
				Range	Mean	Range	Mean	
Tailings	167	3.50	2.69	0.89-38.5	10.64	130-1000	5.43	9.0261
Dihua waste	2.03	0.46	0.246	0.23-8.18	2.94	70-280	1.89	0.2281
II open pit	1.39	0.40	0.254	0.12-1.77	0.35	36-160	0.76	0.0288
II 1972	2.10	2.47	0.234	0.54-1.00	0.84	74-200	1.28	0.0620
II 1950	0.86	0.38						
II 1948	2.37	0.94	0.181	0.26-0.84	0.44	48-125	0.77	0.0251
II 1925	5.01	1.86	0.195	0.37-1.35	0.83	40-200	1.05	0.0510
III 1987	3.64	1.77	0.207	1.28-5.36	3.04	45-265	1.12	0.1984
III 1964	2.39	0.66	0.301	0.32-2.93	1.11	47-300	1.14	0.1054
III 1942	6.01	0.90	0.458	0.12-3.40	1.09	25-420	1.10	0.1574
III 1933	1.88	1.16	0.105	0.02-0.20	0.13	17-21	0.19	0.0043
III 1934	0.77	0.28	0.225	0.04-0.17	0.10	19-22	0.21	0.0071
III 1925	7.08	1.13	0.431	0.64-3.64	1.18	34-180	0.68	0.1604
V 1947	11.3	2.29	0.354	0.24-1.63	0.72	30-100	0.58	0.0804
V 1934	34.0	3.89	0.607	0.70-3.25	1.85	35-80	0.91	0.3541
V 1900-1	5.87	1.44	0.278	0.25-2.22	0.94	18-480	1.56	0.0824
V 1900-2	17.6	3.42	0.361	0.85-2.46	1.48	56-192	0.93	0.1685
V 1900-3	21.9	3.75	0.455	0.47-2.28	1.31	50-310	1.09	0.1880
VI 1880	17.6	6.35	0.189	0.06-1.38	0.41	40-300	1.42	0.0244
VI open pit	1.99	2.23	0.530	0.07-1.03	0.34	35-170	0.84	0.0568
1820	2.42	0.67	0.234			23-86	0.34	

Case studies: active and abandoned Uranium mines

Table 3. Effluent concentrations of overflowing water, mg/L (except 226Ra Bq/l, and pH)a

Location	U	^{226}Ra	Cd	Cr^{+6}	As	Pb	Cu	Zn	Mn	SO$_4^{-2}$	F$^-$	Tp	pH
II 1925	31.8	0.782	1.00	5.700	8.478	3.125	0.88	1.78	2.67	2455	2.90	0.21	2.47
Tailings	12.2	0.297	0.016	0.932	0.008	1.250	0.25	0.21	3.07	3404	2.28	0.13	3.45
V 1850	0.39	6.16	0.015	1.667	0.095	0.277	0.50	0.14	2.76	1381	2.02	0.11	4.95
III 1925	0.01	1.40	0.001	0.733	0.015	0.025	0.55	0.17	1.11	104	0.57	0.30	6.10
III 1942	0.005	0.035	0.016	0.333	0.094	0.025	0.85	0.31	0.42	20.2	0.24	0.77	5.95

[a] U and ^{226}Ra: annual mean; non-radioactive constituents: single monitoring in dry season.
[b] U, ^{226}Ra and flow: annual mean before flooding.

Adits and inclined shaft

Linchang left seventeen adits and one inclined shaft (V 1850) after closure. Among them, five adits and the inclined shaft overflowed water. Average annual flow of overflowing water from II 1925 adit, the inclined shaft (V 1850), III 1925 adit, III 1942 adit was 15.12, 104.72, 52.96 and 9.68 m^3/d, respectively. Overflowing water contains radioactive and non-radioactive constituents mentioned above, some of which exceed regulatory limits (Table 3). The rivers receiving overflowing water from adits and seepage from the tailings pile indicated radioactive abnormity. Additionally, the adits and the inclined shaft were releasing radon. Radon concentrations at the exits were 126—14900 Bq/m^3.

Soil/sediment contamination

The erosions of tailings and mining debris are intensive because of flooding and lack of containment. Rice fields of about 2.86 ha and stream sediment downstream for 1.59 km are contaminated. The mean depth of tailings depositing on the stream is up to 0.60m. The rice fields close to the tailings pile, No. III and No. VIII mining area is 2.51, 0.12, 0.23 ha, respectively; and the stream sediment close to the tailings pile and III mining area is 0.79 and 0.80 km, respectively.

^{238}U and ^{226}Ra are anomalous (Table 4 and Table 5) in soils/sediments as compared to the background (^{238}U 6.84—306.5 and ^{226}Ra 7.84—421.8 Bq/kg in soils of Yunnan province) (Li 1991). The anomalies are as deep as 0.90m in the stream sediment and 0.40 m deep in the soil under the base of the mining debris (Table 6).

Table 4. Average contents of radionuclides of contaminated rice fields and stream sediments/alluvial soils (<30 cm depth) of Linchang uranium mine

Parameters	Rice field			Sediment/ soil	
Location/near to	Tailings	III mining area	VIII mining area	Tailings	III mining area
^{238}U (Bq/g)	0.38	1.08	0.56	0.58	1.19
^{226}Ra (Bq/g)	2.58	0.94	0.74	5.01	0.89

Table 5. Radioactive contents at different depth and distance from the source in the stream close to the tailings pile in Linchang uranium mine

Distance from the source (m)	^{238}U (Bq/g)					^{226}Ra (Bq/g)				
	20^a	40	65	90	115	20	40	65	90	115
200	1.04	0.294	0.073	1.10	0.133	5.81	3.80	9.07	9.20	0.422
540	0.361	0.279	0.38	0.846	0.398	2.13	2.29	4.04	6.21	1.21
790	0.435	0.174	0.153	0.161	0.198	6.66	0.822	1.20	0.255	0.44

^{a} 20, 40, 65, 90 and 115 are sample depths (m).

Table 6. Average radioactive contents of the soil under the base of the No. II mining debris

Sample depth (m)	30	40	50	60	80	100
^{238}U (Bq/g)	0.548	0473	0.418	0.310	0.407	0.363
^{226}Ra (Bq/g)	0.955	0.512	0.172	0.118	0.124	0.100

Equipment, material and structure

The areas of contaminated structures were 6198 m^2, where the gamma air dose rates were 0.51—1.67 µGy/h. There were 37,849 m of contaminated rails, pipes, cables and electrical wire, 107 contaminated valves and 541 pieces of contaminated equipment. These equipment and materials contained 0.044—4.213 Bq/cm^2 α surface activity mean, exceeding regulatory limit.

Radiation safety

It was estimated that individual effective dose of the critical group resulting from Linchang uranium mine was 0.24 mSv/a, which was close to the regulatory limit. However, the potential individual effective dose was as high as 0.64 mSv/a, which exceeded the regulatory limit (Xu 1999).

Remedial actions

Remediation of the tailings, mining debris, open mining pit, surface subsidence pit, patio and contaminated soil/sediment

Considering factors such as disperse distribution of mining debris piles, their lower radon flux and gamma air dose rate than tailings piles, and cost-effectiveness of remediation measures to be taken, a combination of on-site and off-site disposal will be applied.

Contaminated rice field and stream sediment, Dihua waste, II 1948, II 1925, III 1987, III 1964, III 1942, III 1925 mining debris, contaminated soils under these waste rocks and other contaminated soils and dismantled building rubbles will all be relocated to the tailings pile. V 1900-1 waste rock pile will be cleared up and removed to V 1900-2 waste rock pile and V 1947 removed to surface subsidence pits and patios of V mining area. Cleaned rice fields will be restored after covered with loess. After waste rock relocation, the sites will be revegetated and may have unrestricted used.

III 1933, III 1934 mining debris will be revegetated only since their radon flux and gamma air dose rate were not over regulatory limits.

The other waste rock piles and open mining pit will be left in place to be rehabilitated in-situ. The piles will be covered with 0.3—0.4 m depth loess depending on their radon flux and 0.3 m clay containing gravel from bottom to above and then revegetated to prevent erosion and limit radon flux and gamma radiation exposure after the slope regraded to a stable ratio of 1:2. Additionally, ditches and astels will be constructed to catch and control surface water run-off and to further prevent erosion.

Surface subsidence pits and patios will be backfilled with V 1947 mining debris and then covered with loess and revegetated.

The tailings pile will be reshaped into multilevel steps and slopes regraded to a ratio of 1:4. The tailings dam will be reconstructed to control the tailings considering VIII rank earthquake intensity. 0.4 m thick white lime over the reshaped tailings pile and 3m thick white lime between the dam and tailings will be applied to prevent acid seepage from overflowing before receiving the waste rocks mentioned above. After that, the tailings pile will be covered with 2.34 m thick removed waste rocks, 0.9 m thick loess and 0.3m thick clay containing gravel from bottom to above and then revegetated. Clay cover containing gravel and revegetation are aimed to prevent soil erosion in rainy season and dry cracking in dry season. The ditches outside and inside the tailings pile will be constructed to catch and control surface water runoff. These measures are enough to keep long-term stability of the tailings pile and limit radon flux and gamma radiation exposure.

All above facilities after remediation will meet regulatory limits. It is estimated that radon release will reduce to 1.08 TBq/a, reduced to 1% of original; individual effective dose of the critical group and potential individual effective dose will decrease to 0.017mSv/a and 0.25mSv/a, respectively.

Adits and inclined shaft

The adits and inclined shaft (V 1850) will be flooded and closed down considering radiation safety, general safety, and overflowing or potential overflowing water from adits.

Flooding is the most environmentally friendly, technically safest option to limit contaminated water from the underground mine to overflow. VI 1880, V 1880, V 1990, III 1925, III 1942, II 1925 adits and the inclined shaft will be flooded by adopting heavy curtain flooding with thick liquid, which mainly consists of cement and PCC (a water-tight patent material). The 0.2 m gap between the two water walls constructed in a flooding adit and some heavy curtain boreholes around the adit will be used to flood with thick liquid. The rock walls in the other adits will be treated with grout. All adits and inclined shaft will be closed down after backfilled with loess outside the water walls and rock walls.

The flooding and shutdown of the inclined shaft was tested from October to November in 1995. Wastewater overflowed from VI 1880 and V 1880 adits rather than the inclined shaft. Overflowing water volume reduced to 0.13—1.40 m³/d from 96.77—115.78 m³/d; Uranium and ^{226}Ra of the overflowing water attenuated to 0.016—0.02 mg/L from 0.25—0.80 mg/L and 0.18—0.23 Bq/L from 4.16—7.95 Bq/L, respectively. All adits and the inclined shaft will not release radon and waste water after flooding and close down.

Equipments, materials, and structures

All the equipments and materials, which will no longer be used, and the contaminated structures will be dismantled, demolished, crushed and then used as interim cover on the tailings pile. Uncontaminated or decontaminated equipments and materials, which have value of service, will be reused and recycled, for example, in other active uranium mine. Decontamination reagents are water and 10% $H_2SO_4^{2-}$.

Water treatment

Wastewater from flooding and equipments/materials decontamination will be treated via neutralizing with lime. The neutralized sediment will be removed to the tailings pile.

Conclusions

Linchang uranium mine left one tailings pile, twenty mining debris piles, thirty nine surface subsidence pits, four patios, seventeen adits and one inclined shaft. The mine brought about environmental issues such as radiation exposure to public, soil erosion, soil/sediment contamination, water contamination, radon release into air and safety risk.

The tailings and mining debris erosion were serious. About 20% of the tailings were washed into and deposited on the nearby streams and rice fields. Contaminated rice fields and stream sediments resulting from erosion were up to 2.86 ha and 1.59 km, respectively. Seepage from the tailings and overflowing water form adits and the inclined shaft contained radionuclides and non-radioactive constituents including Cd, Cr^{+6}, As, Pb, Cu, Zn, Mn, SO_4^{2-}, Tp. Radon concentrations at the exits of the adits/inclined shaft were 126—14900 Bq/m^3. Average radon flux and gamma air dose rate from the tailings is up to 10.64 $Bq/(m^2.s)$ and 5.43 $\mu Gy/h$, respectively, much more than regulatory limit or from mining debris.

On the basis of the specific environmental issues and site characterization, remediation of the tailings, mining debris and contaminated soils/sediments combine on-site and off-site disposal, i.e. contaminated soils/sediments and some small and disperse mining debris piles are removed to the tailings pile; adits and inclined shaft combines flooding and backfill. The tailings pile will be stabilized with multi-covers consisting of white lime, waste rock, loess, clay containing gravel and revegetation from bottom to above.

The remediation execution of Linchang uranium mine began in 2001 and is anticipated to terminate in 2005. Remediation effects are obvious. Remediation objective and regulatory limits after remediation will all be met. Soil erosion will be reduced to a minimum. Radon release will reduce to 1.08 TBq/a from 11.01 TBq/a; individual effective dose of the critical group and potential individual effective dose will decrease to 0.017 mSv/a from 0.24 mSv/a and 0.25 mSv/a from 0.64 mSv/a, respectively. Any adit and the inclined shaft will not release radon and waste water after flooding and close down.

In addition, it will have the following advantages that contaminated soils/sediments and some small and disperse mining debris piles are removed to the tailings pile:

- Reduce radon flux and gamma radiation dose rate from the tailings pile; consequently decrease loess volume covering the tailings.
- Reduce the contamination sources; consequently enhance utilizable value of soil.
- Facilitate long term post-decommission maintenance and surveillance.

Acknowledgements

The research work was funded by National Natural Science Foundation of China (NSFC-49832005) and the Teaching and Research Award Program for Outstanding Young Teachers in Higher Education Institutions of Ministry of Education, P. R. China.

References

Li R (2001) Decommissioning and Environmental Disposal of Uranium Mining and Milling Facilities (in Chinese). Beijing: Atomic Energy Publishing House, p52.

GB14586-93 Technical Regulatory for the Environmental Management of Decommissioning of Uranium Mining and Milling Facilities (in Chinese).

GB 8703-88 Regulations for Radiation Protection (in Chinese).

GHZB 1-1999 Environmental Quality Standard for Surface Water (in Chinese).

Xu L (1999) Environment Impact Report on Environmental Treatment Engineering of Decommissioning Linchang Uranium Mine (Feasibility studies stages) (in Chinese).Linchang: No. 765 mine of China National Nuclear Corporation.

Li Y (1991) Investigation of Environmental Natural Radiological Level in Yunnan Province (in Chinese). Kunming: Yunnan Science and Technology Press.

Water balance and hydrogeological modeling in the case of the TPF Rozna/ Czech Republic

Roland Hähne[1], Michael Schöpe[1], Michael Paul[2], Jan Slezak[3]

[1]C&E Consulting und Engineering GmbH, Jagdschänkenstr. 52, D-09117 Chemnitz
[2]WISMUT GmbH, Jagdschänkenstr.29, D-09117 Chemnitz Chemnitz
[3]DIAMO s.p., Straz p.R., Czech Republic

Abstract. The still operative Tailings Pond Facility (TPF) Rozna is situated near to Brno. It includes an area of 40 ha containing 15 Mio m³ of tailings. To reduce the exposure of contaminants and to modify the water treatment technology before the closure of the uranium mining and milling in Rozna, a PHARE project was realized in 1999 to resolve the water balance by numerical modeling, both for the present and the future. For this purposes, HELP was used to clarify the percolation, MODFLOW to model the groundwater movement and PHREEQE to simulate the hydro chemical discharge of contaminants for different remediation options and phases.

Introduction

The TPF Rozna is situated at the uranium mining and milling site Dolni Rozinka / Rozna near Brno and is still in operation. Actually a non acceptable environmental impact via the water pathway cannot be observed because contaminated seepage waters are collected and pumped back in the tailings pond. But preparing the closure of the uranium production, prognosis of the release of contaminants is important and will give recommendations to the closing and remediation strategy including the water treatment needed.

Therefore C&E in co-operation with WISMUT got a contract from the EC to clarify the „Development of the tailings ponds water balance". This project was part of the Multi-Country Project „Tailings Ponds Remediation Planning in the CEEC". The results of the a.m. pilot project should be used for other cases in Central- and Eastern Europe, too. Therefore experts from Bulgaria, Czech Republic, Estonia, Hungary, Poland, Romania, Slovenia were involved.

General Methodology

The methodological procedure includes the following working steps:
1. definition of site specific goals for a water balance (recent status, remediation status, long-term status; prediction of seepage amount, chemical load and/or groundwater contamination etc.)
2. development of a conceptual model describing the relevant subjects (tailings pond body, dams, waste rock disposals, catchment areas) and processes (infiltration, percolation, runoff, influences from outside and to outside, internal interaction processes, transport processes etc.)
3. identification of key parameters which are important for the prediction of the development of the water balance and selection of special criteria which have major influence on the water balance
4. collection and evaluation of existing data as well as special investigations and measurements to obtain additional data
5. design of special data bases for various model compartments of the water balance and the modeling (data base for water budget, pore water release, chemical interaction processes and transport)
6. design of a site specific model concept under consideration of the objectives, the available data and the possibility of comparison with results of an analytical balance
7. modeling (relevant compartments) including calibration and sensitivity analysis
8. establishment of a water balance for recent conditions and the state during and after the remediation using modeling results and results of the analytical balance based on monitoring
9. recommendations for the water management and necessary water treatment depending on the development of the water balance (water quantity and quality)

The prognosis of the development of the tailings pond water balance followed this procedure. The recent water balance was been investigated by analytical methods, too. The prediction for the remediation state and the long-term state required numerical modeling methods.

The application of numerical modeling is taken into consideration:
- water budget (calculation of infiltration and percolation rates for the pond, the dams and the environs and for various scenarios of development, calculation of the pore water release by tailings settling/ consolidation)
- water flow (investigation of water movement through saturated and unsaturated zones e.g. tailings body, dams, aquifers)
- contaminant release (interaction between tailings, waste rock, basement, infiltration water, pore water, groundwater; transport processes to the environs).

Case studies: active and abandoned Uranium mines

Site investigations

Additional field and laboratory investigations were realized in 1998. The main objective was the capture of information about the properties and structure of the tailings, the dam and the underlying rocks.

6 boreholes were drilled of the total length of 202.8 m into the K1 pond Rozná (Fig.1, 2).

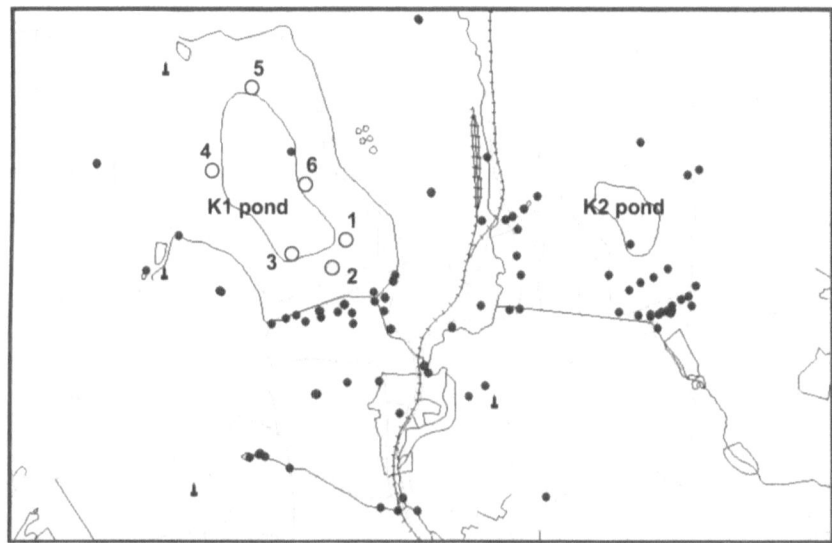

Fig. 1. Position of the boreholes for the site investigation of the Rozna TPF

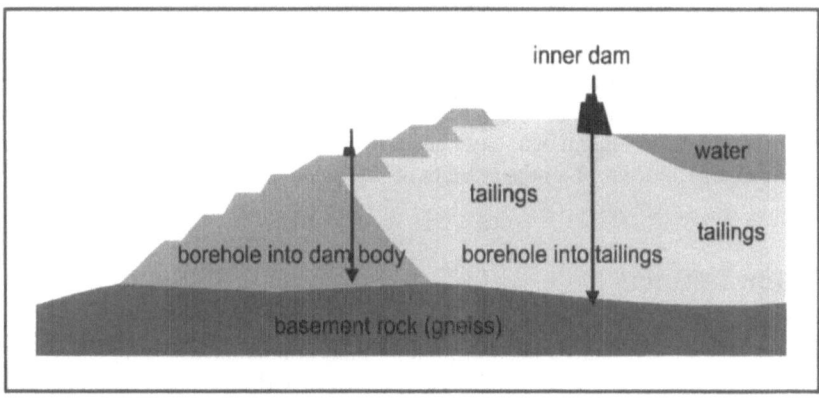

Fig. 2. Scheme of the interior of the Rozna TPF and of the position of the boreholes

Case studies: active and abandoned Uranium mines

Following values of filtration coefficient have been determined:
⇒ $5 \cdot 10^{-7}$ to $1 \cdot 10^{-6}$ m/s for coarse-sandy tailings,
⇒ $5 \cdot 10^{-9}$ to $2 \cdot 10^{-7}$ m/s for fine-sandy tailings,
⇒ $1 \cdot 10^{-9}$ to $5 \cdot 10^{-8}$ m/s for very fine (plastic) tailings.

As the results of pore water chemistry investigations the following conclusions have been drawn:

- The high salinity (~30-40 g/l TDS with corresponding SO_4^{2-} concentration ~20-27 g/l) in the whole tailings depth profile obviously affects basement soils too.
- Na_2SO_4 is the major pore water component; it takes up about 80-95 % of the total salinity. Na-concentrations reach values ~7-9 g/l in the upper part of the boreholes, while they are much higher in the deeper part (~11-13 g/l).
- The pH value varies in a relatively short interval 7.8-8.6.
- The total alkalinity (as HCO_3^-) increases quickly with the depth, the concentration maximum reaches a value of 5.5 g/l.
- The U-concentration profiles show the similar trend as the total alkalinity. Maximum values (up to ~135 mg/l) have been found in the deepest boreholes KI-1 and KI-3 (older tailings). On the other hand, the determined maximum in the "shallow" zone, filled with „younger" tailings is lower (55-70 mg/l).
- The Ca- and Mg-concentrations show the opposite trend in comparison with the concentration of HCO_3^- in consequence of precipitation of carbonate minerals. The Ca-concentration reaches the maximum values (~400-600 mg/l) in the dam material (gypsum equilibrium).
- Significantly higher concentration levels of further components were found in the deepest and oldest part of tailings pond, such as NO_3^- (~3.5 g/l in maximum) and NH_4^+ (1.1 g/l in maximum).
- The initial Cl-concentration of ~350-400 mg/l at the top regularly decreases to ~200-250 mg/l on the bottom of boreholes.
- The K-concentration varies between 100-250 mg/l predominantly all over the borehole profiles, and is rarely reaching values of ~300 mg/l.
- The Mo-concentration varies between 7-10 mg/l in all depth profiles. Other heavy metals, such as V, Cr, Ni, Pb, Cd, As were not present, only trace concentrations of Cu and Mn have been determined by semi-quantitative spectral analysis (in order of tenths to units of mg/l).

Water balance

The results of the investigation from 6 boreholes and about 200 samples as well as data on the hydrogeological situation, geomechanical properties and structure of tailings, chemical composition of tailings, rocks of basement and sediments of the catchment area, pore water, drainage water, seepage water and groundwater, data on the meteorological conditions, on the foreseen contour and cover system and water management data was the basis for the recent water balance and the prognosis of it's development after the remediation.

The recent status of the Rozna site shows an area of the tailings pond surface of 62.8 ha, including the area inside the dam crown of 34.5 ha with 20.4 ha freewater. The catchment's area is about 176 ha. Two ditch systems are installed around the whole pond collecting drainage water from the pond together with surface water and partly also groundwater from the catchment's area.

The whole amount of drainage water in the inner ditch is in average about 500,000 m³/a (Fig. 3).

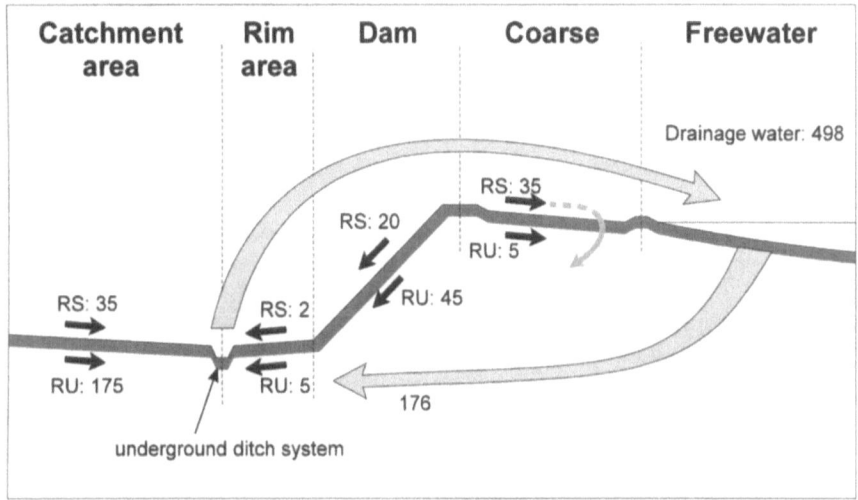

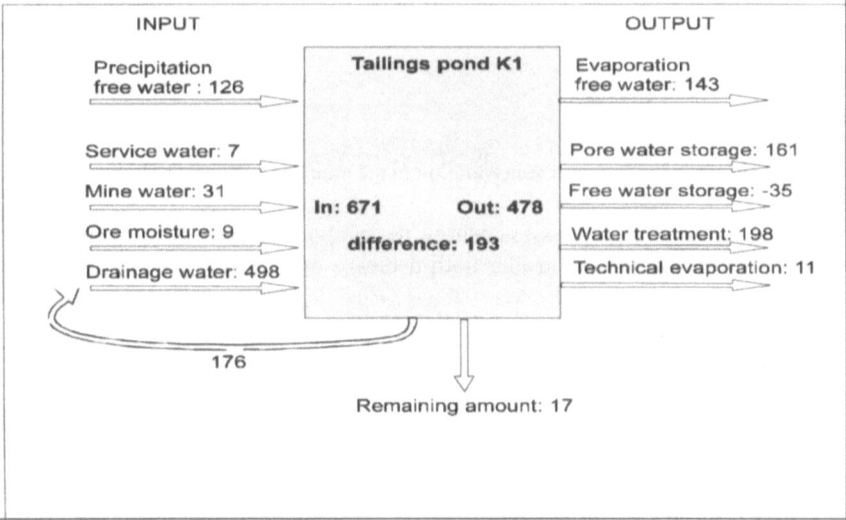

Fig. 3. Water balance (Tm³/a) of the Rozna K1 TPF

Case studies: active and abandoned Uranium mines

The concentration of contaminants in 1997 was 16 g/l SO_4, 27 mg/l U, 0.1 Bq/l Ra, 0.6 g/l NO_3, and 80 mg/l NH_4.

As result of remediation of the tailings ponds the amount of water discharge decreases for an average case to 370,000 m³/a after the interim covering, to 170,000 m³/a after contouring and final covering, to 140,000 m³/a 5 years after finalizing the remediation, 130,000 m³/a 10 years after the covering and 110,000 m³/a 50 years after covering.

The concentration of contaminants will be changed for SO_4 from actually 16 g/l to 8 g/l after the final covering and 3.5 g/l after 50 years and for uranium from actually 27 mg/l to 22 mg/l after the final covering and 10 mg/l after 50 years.

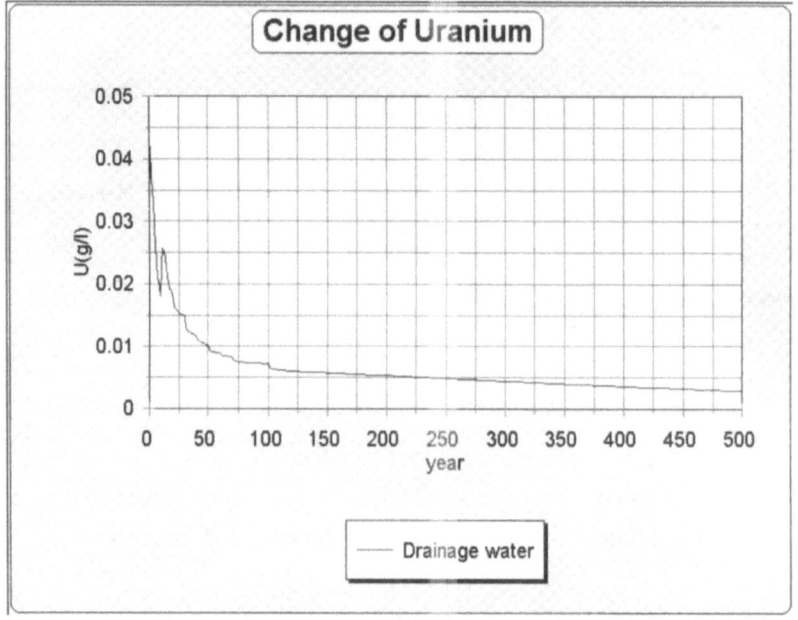

Fig. 4. Prognosis of uranium-concentration in the drainage water of the Rozna K1 TPF

In combination with the assessment of the temporal development of the water balance one may have to consider both decrease of drainage water amount and decrease of concentration.

The results of modeling show that the amount of water discharge can be reduced by modified cover systems and changed contour, too. A higher thickness of the storage layer without a separate drainage layer at the flat tailings surface ensures the lifetime of the underlying sealing layer and decreases the percolation rate.

A flat final contour of the valley type reduces the contribution of pore water release by settling and consolidation and in this case the contaminant's load into the seepage water.

Case studies: active and abandoned Uranium mines

We proposed to realize a cost-benefit-analyses for several remediation options including the changed influence on the water balance and the influences on the technology and working duration of a water treatment plant and including the remediation of the tailings pond K2 and the remediation and flooding of the mine.

Resume

The results of the prognosis of the K1 tailings pond water balance for the remedial state show the need of a changed water management and of a water treatment. The procedure of the water management is depending on the remediation concept and the time schedule of interim covering, contouring and final covering. The pumping of free water is possible together with the progress of the interim covering and has to consider the time which is necessary for the first consolidation of the upper part of the tailings. This has influence on the water treatment too, as well as the amount of pore water release is depending on the kind and time schedule of the contouring work.

The duration of the water treatment depends on the boundary conditions and the limits for the seepage water discharge to the creeks. It depends on the cover system and the long-term functionality, on the functionality of the drainage system. But this is linked with the need of mine water treatment after the flooding of the Rozna mine, too.

On the basis of the recently given limits it should be necessary to treat the discharged water from the tailings pond K1 over a period of about 50 years or longer.

During this time the water amount will decrease from 400,000 m^3/a to about 120,000 m^3/a.

The development of the chemical load of discharged water from the tailings pond will be influenced especially by

- the concept and the time schedule of interim covering, contouring, final covering and therefore the contribution of pore water release
- the concept of a long-term stable cover system
- interactions between mine flooding and remediation of K1 and K2.

It can be concluded that the concept of DIAMO for the contour work, for the dewatering and the cover system was not optimized yet and has to be modified after a site specific cost-benefit-analysis of alternative options. For this optimization additional investigations are necessary especially concerning the internal structure and the geomechanical and hydrochemical interior of the tailings pond.

Study courses at the Technical University Freiberg
Akademiestr. 6   09599 Freiberg   Germany
http:/www.tu-freiberg.de   dekan.F3@fggb.tu-freiberg.de

Geotechnical Engineering and Mining

1st – 4th term (undergraduate studies)
* Mathematics, Physics, Chemistry
* Earth Sciences
* Geo-Engineering
* Economics and Public Law
* Practical Training (60 days)
* oral examination after 4th term

5th – 9th term (graduate studies)
* Choice of one out of five modules:
    - Mining
    - Soil Protection
    - Drilling and Fluid mining
    - Geotechnical Engineering
    - Special civil Engineering
* Excursions (minimum 10 days)
* Subject specific internships (60 days)
* Literature review in the 7th and 8th term
* Masters thesis in the 9th term

Admission
* „Abitur"/Baccalaureate (A-level) or
* Specialized college degree or
* equivalent access entitlement

Duration/Degree: 9 semesters/Master of Science in Engineering
                  (Dipl.-Ing.)
Start: winter or summer term

Job opportunities
Planning, Organization, Management, National and international
business supervision in the areas of Civil engineering; Underground
and surface mining; Construction of tunnels, galleries, cavities and
caverns; Extraction and conditioning of raw materials, Drilling
techniques; Oil and gas industries; Disposal sites; Engineering and
environmental consultancy; State control organizations; Research
institutes; Insurance companies; Banks

Rehabilitation of the Nabarlek uranium mine – Will close out ever be achieved?

RA McGill[1], RE Fox[2] and AR Hughes[1]

[1]Mines Division, Northern Territory Department of Business Industry and Resource Development (previously NT Department of Mines and Energy), GPO Box 3000, Darwin NT, Australia 0801
[2]Ecofox Enterprises Pty Ltd, PO Box 2580, Palmerston NT, 0831, Australia

Abstract. The Nabarlek uranium mine commenced mining operations in May 1979 and ceased milling in 1988. It was retained on care and maintenance status from 1988 until decommissioning earthworks were completed during 1994 and 1995. The company gave only generalised commitments to rehabilitation in its EIS, and although this was noted, it was not identified as a serious problem by those who provided comments. A decommissioning working group was unable to agree on a revegetation standard for the mine and an independent expert was appointed be the arbiter of revegetation success. The expert's 1999 report was rejected by some stakeholders and in 2002 the company remains unaware of what will be required to achieve close out and relinquishment of its lease.

Background

In Australia the Commonwealth Government controls the mining of uranium and day to day regulation of approved uranium projects is the responsibility of State or Territory Governments. The Northern Territory achieved self-government in 1978. The Nabarlek uranium mine in northwest Arnhem Land (Fig.1) was one of the first major mining projects to be regulated by the Northern Territory. Commonwealth Government approval of the Nabarlek project was contingent on a set of Environmental Requirements (ERs) drawn up in consultation with stakeholders. The ERs formed the basis of the Northern Territory regulation of the mining operations. The Commonwealth maintains an overseeing body, the Supervising Scientist, which reports to the Commonwealth on the administration of uranium mining in the Northern Territory.

Planning for decommissioning was an integral part of the mine design outlined in the Queensland Mines Limited (QML) Environmental Impact Statement (EIS) in 1979 (QML 1979). The section in the EIS on rehabilitation did not include much detail on revegetation but the public comment process did not identify this as a major concern. From the commencement of the project the company employed environmental rehabilitation officers and set up a nursery to germinate and produce local seedlings for revegetation. Consultants were engaged to undertake flora and fauna studies and revegetation trials were commenced on waste rock, topsoil and surplus materials stockpiles.

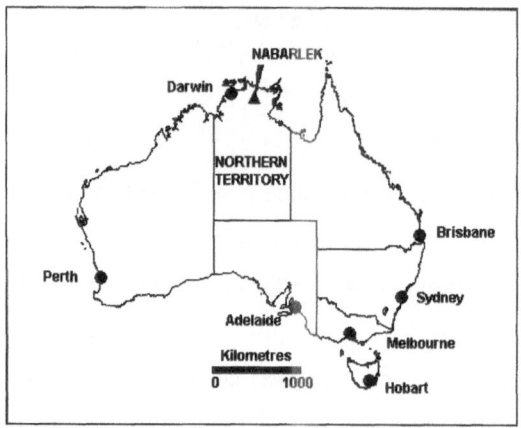

Fig. 1. Location plan Nabarlek uranium mine.

After about five years of operations, the Nabarlek Decommissioning Group was established. The Group included representatives from QML (by then Queensland Mines Pty Ltd – QMPL), the Northern Land Council (NLC), the Office of the Supervising Scientist (OSS), the then Northern Territory Department of Mines and Energy (DME) which is now known as the Department of Business Industry and Resource Development (DBIRD), the Conservation Commission of the NT, and the Water Division of the NT Department of Transport and Works. The functions of the Group as agreed at its first meeting on 8 February 1984 were:
- To ensure best practicable technology is used in the decommissioning plan;
- That QMPL would formulate the decommissioning plan;
- That the other members of the Group would provide technical advice to the statutory approvers relating to the plan.

Also agreed at that meeting were the operations of the Group, including the statement that paper work was to be kept minimal (DME/DBIRD files). With hindsight, this proved to be a mistake as the poor record keeping allowed stakeholders to disavow commitments made.

Progress was slow and the minutes of a meeting on 26 February 1987 recorded: 'The meeting could not agree on revegetation standards for Nabarlek'.

In 1988 the mine was mothballed and the first stage of decommissioning of running down the mill, cleaning out the pipes etc was completed by the end of

1989 (Waggitt 2000). QMPL attempted to sell the mine on an 'as is where is basis' but this was unsuccessful and the final decommissioning and rehabilitation works were undertaken during 1994 /95.

Nabarlek is situated on Aboriginal Freehold land in the Northern Territory of tropical northern Australia. During the late 1980s there was significant Aboriginal opposition to further mining or exploration in the region. After it became clear that, as a result, the company was unlikely to be able undertake any further mining in the area, development of a decommissioning proposal recommenced with a view to achieving close out in the shortest possible time. Disposal of the excess water on the site was a major problem and QMPL began exploring alternatives for supplementing natural evaporation by release within surrounding catchment including:

- Flood irrigation adjacent to the airstrip;
- Spray irrigation adjacent to the airstrip;
- Irrigation of bushland areas adjacent to Kadjirrikarmarnda Creek.

The locations of these areas are shown in Fig. 2. The results of the irrigation were less than satisfactory because the natural vegetation had difficulty coping with the overzealous application of mildly saline water, leading to tree deaths.

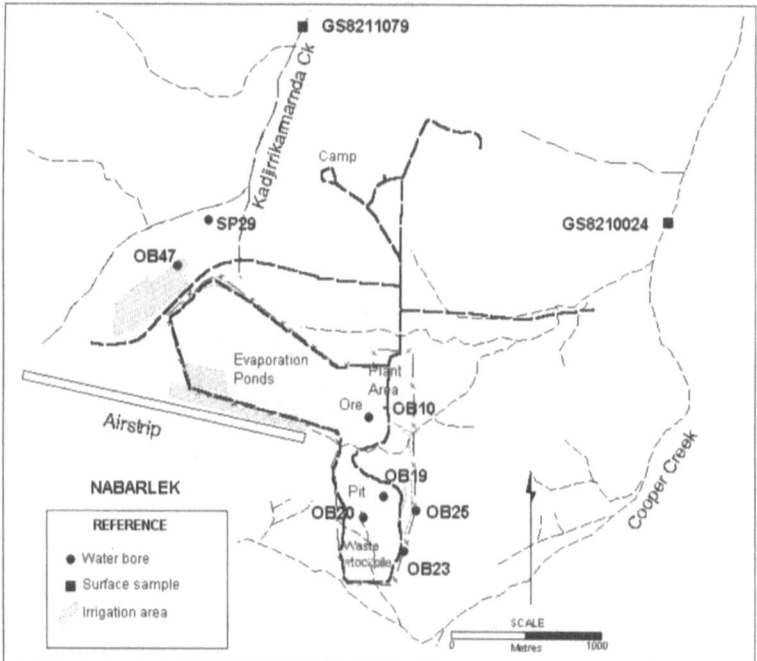

Fig. 2. Nabarlek site plan

The issue of close out criteria for Nabarlek

Early in the Nabarlek project, QMPL set up an extensive nursery and employed a revegetation officer and supporting staff to undertake revegetation studies within the project area (Buckley 1987). Disused plant and temporary campsites were revegetated fairly successfully in the early 1980's.

In 1984 it was expected that the newly formed Decommissioning Working Group would establish close out criteria for the whole site. The failure to achieve this goal was largely due to the fact that a simple index of rehabilitation success could not be agreed as each scientific expert consulted had a different opinion and there were as many opinions as there were experts. Nevertheless, engineering plans had to be drafted and there was a significant cost involved in maintaining a presence on the site.

The company endeavoured to break the impasse with a proposal to complete a decommissioning program to be approved by the supervising authority (DME/DBIRD). The outcome of that proposal was that the program should also be the responsibility of the approver and that the company should be given close out upon implementation of the approved work program. Since there were a number of stakeholders who believed that they may be potentially disenfranchised, this proposal was rejected.

In another effort to break the impasse, a general understanding was reached whereby a consultant, or panel of consultants, would be chosen by the four principal stakeholders (QMPL, DME/DBIRD as representative of the NT, OSS as representative of the Commonwealth and NLC as representative of the landowners). If this independent arbiter assessed that the site was stable and would ultimately reach the design criteria, all parties would respect that assessment and a close out certificate could be issued. It was accepted that this might allow some subjectivity to enter the assessment but had the advantage of limiting the number of conflicting opinions. It is not clear exactly when this understanding was reached, largely because of the Group's maxim to minimise paperwork.

Although recording of minutes of the Group's meetings was sparse, the following is known and verifiable from DME's file records:

- Agreements were made between QMPL and the NLC that defined rehabilitation criteria in at least generic terms.
- Site-specific revegetation goals were agreed between QMPL and NLC in October 1987 and this agreement was progressively modified through an exchange of letters between the two parties until at least February 1988.
- A settlement deed of 23 November 1993 between the same two parties required the company to rehabilitate the site in accordance with the standards, principles and objectives set out in the Nabarlek Revegetation Standards. This would continue until the supervising authority (DME/DBIRD) issues a certificate of revegetation pursuant to paragraph 27(c) of the company's statutory environmental requirements.
- The summary record of the 12th meeting of the Group on 12 June 1992 included the statement that 'Agreement in principle was reached that a group of

experts could be found who would pronounce on the efficiency of the final rehabilitation.'

- A letter dated 26 July 1994 from DME's Director of Mines stated *inter alia* that it had already been agreed that a panel of experts will certify successful rehabilitation, it had been accepted that numerical standards would not be imposed and there seemed to be no reason to revise that position.

- In July 1996 QMPL agreed with the NLC to appoint Dr Mark Adams of Adams Ecological Services to '…act as an expert with regard to the determination of satisfactory revegetation of the Nabarlek mine site'. Also within the agreement was the statement that '…both QMPL and NLC will accept as binding the determination of satisfactory revegetation at Nabarlek by Dr Adams acting as an expert'.

- The summary report of the Environmental Performance Review of Nabarlek in July 1996 (EPR5) recorded *inter alia* that Dr Mark Adams was to visit the site in July of that year with a team of four to develop a program of studies in revegetation, biological monitoring and soil development. It was also confirmed that both NLC and QMPL had 'agreed to abide by Dr Adams' assessment of revegetation success'. Dr Adams produced a report in which the revegetation was assessed as satisfactory (Adams 1999).

- On 11 April 2000 QMPL made application to DME for a certificate of revegetation, which is a prerequisite for the company seeking relinquishment of its mineral lease.

- OSS convened a workshop on 18 and 19 April 2000 to discuss the issue of rehabilitation, particularly revegetation and the consultant's report. Based on the workshop OSS concluded that some 24 requirements remained to be addressed in fields including revegetation, monitoring, fire management, erosion, weed and feral animal control. OSS advised DME that it would not support any proposal to issue a certificate of revegetation until the above issues were satisfied.

- NLC advised DME that the Deed of Settlement between NLC and QMPL prevents the company from applying for revegetation closure without the consent of the NLC.

The current status of the site

There has been no caretaker on the site for the past seven years and DME/DBIRD has been undertaking a limited monitoring program of surface and ground waters. QMPL has continued ongoing maintenance of the site through contractors for fire management, weed control, fencing, erosion control and other associated matters.

Surface water

Indications are that there has been little effect, as a result of mining activities, on water quality of the ephemeral streams downstream of the mine site (Kadjirrikarmarnda Creek at sampling location GS8211079 and Cooper Creek at GS8210024).

Fig. 3 depicts the concentration of sulfate in surface water at sites within these catchment and their locations are shown in Fig. 2. Sulfate was chosen as an indicator of mine-related contamination because of very low background levels and its conservative behaviour in the environment. Contamination evident in Kadjirrikarmarnda Creek is the result of irrigation in the mid 80's. From monitoring results there is no evidence of any significant contamination of surface water in Cooper Creek having occurred.

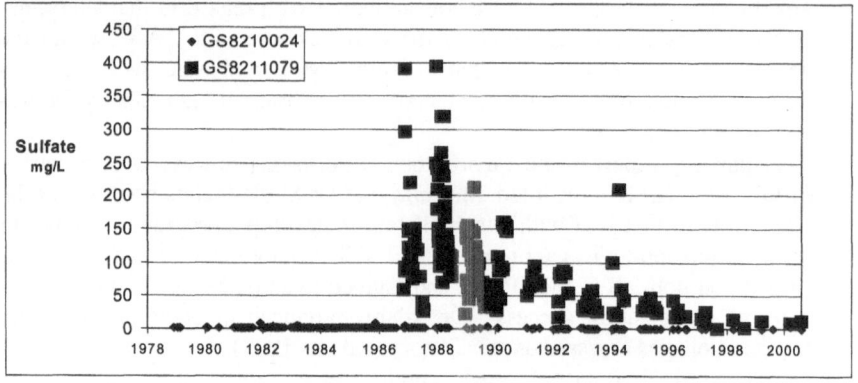

Fig. 3. Sulfate concentrations in surface water, Cooper and Kadjirrikarmarnda Creeks.

Groundwater

Overall, ground water contamination plumes as identified in bore samples appear to be improving in quality and dispersing slowly. There is no indication that these plumes will significantly contaminate the Cooper Creek surface water system. However the ground water situation is more complex than that of surface waters due to the presence of more than one principal aquifer, three main rock types with differing hydrological characteristics and various contaminant sources.

Generally there is a shallow aquifer, typically within the regolith, which has relatively isotropic lateral transmissivity and gradational vertical transmissivity. In contrast, a second, deep aquifer in unweathered rock relies largely on preferential pathways such as faults and fractures for groundwater movement (Salama and Verma 1985).

Geologically, a variably lateritised regolith is underlain by sandstone to the north of the mine pit, plant area and evaporation ponds. Dolerite underlies the plant and evaporation ponds areas and the mine pit is situated within metamorphic rocks, mainly schists. Where the laterite profile is more completely developed, and within the weathered zone of the dolerite, a clayey layer acts as an aquitard, separating the shallow and deep aquifers.

Contaminant sources include areas where irrigation was conducted, locations of previous evaporation and runoff ponds, soils and hard stand in the plant and mill area, tailings which are buried in the backfilled pit and, possibly, residual pockets of natural, unmined uranium mineralisation that may occur in the area.

Case studies: active and abandoned Uranium mines

Fig. 4 depicts the trend of groundwater quality returning to background levels in deep and shallow aquifers following irrigation in the forest irrigation area. Bore locations and their relationship to the irrigation area are shown in Fig. 2. In this area the substrate is sandstone and transition between the two aquifers is gradational. A first peak in OB47 data corresponds to initial irrigation. The corresponding peak in SP29 is delayed as SP29 is down gradient from the irrigation area. During the period from 1988 to 1994 the shallow aquifer can be seen to recover more quickly than the deep aquifer due to flushing and dilution from seasonal heavy rains. Quite clearly, no further intervention or rehabilitation is required in this area.

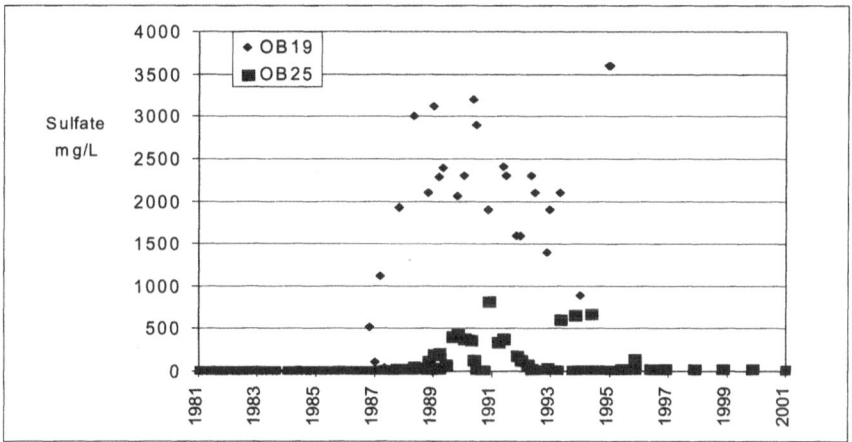

Fig. 4. Status of sulfate contamination in shallow (SP29) and deep (OB47) aquifers due to irrigation.

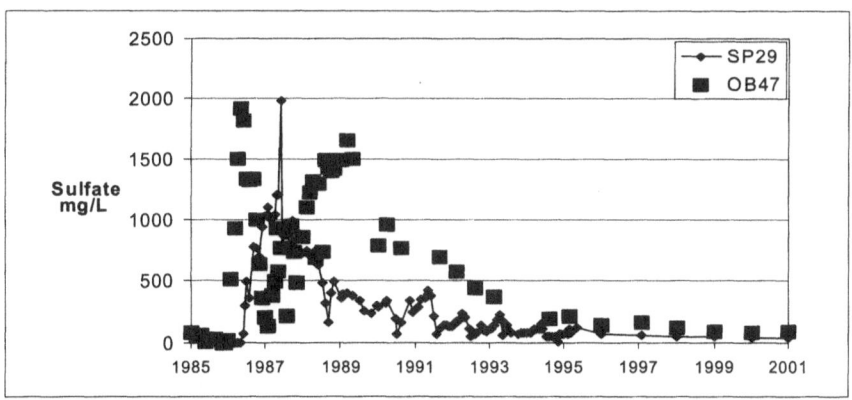

Fig. 5. Sulfate levels in groundwater – Nabarlek pit area

Fig. 5 shows sulfate levels in deep aquifer groundwater from bores OB19 and OB25 adjacent to, and down gradient from the backfilled pit which is used as the final repository for mill tailings. OB19 intercepted contamination ahead of that in

Case studies: active and abandoned Uranium mines

the more distant OB25 by approximately two years. Contaminant levels were significantly higher in OB19, presumably due to dilution and attenuation which occurred during the time and distance required to traverse the intervening 230 metres. The delay and attenuation are even more apparent between OB 20 and OB23 which are separated by 350 metres, however the low order peak in OB23 may have not yet passed (Fig. 6).

The buried tailings, which comprise the principal ongoing source of contamination due to mining, do not appear to be continuing to contaminate the ground water as evidenced in OB19, OB20 and OB25. It is interpreted that after the pit was finally filled and capped, rainwater has been prevented from entering in large volumes during the tropical wet season. Since that time the low permeability and transmissivity of both the tailings mass itself and the host schist has resulted in minimal through flow and stabilisation of ground water in and around the pit appears to have occurred. What little outward movement that occurs undergoes attenuation and dilution and it is therefore unlikely that any significant concentrations of contaminant from the process residues will migrate beyond the immediate mine area or reach the surface drainage system.

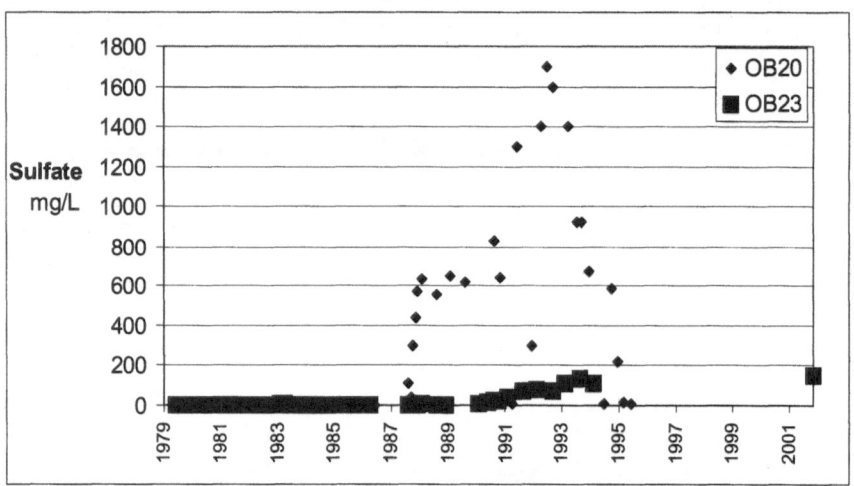

Fig. 6. Attenuation of sulfate levels – Nabarlek pit area

A potential risk identified during the planning stages was that a larger mass of contaminated groundwater would develop as a mound beneath the extensive evaporation ponds during operations. Consequently the ponds were constructed over weathered dolerite where a substantial clay layer was present. The ponds and the water in them have now been completely removed as part of the decommissioning program in 1994 and 1995. Since that time most of the potential contaminants in the shallow aquifer have been flushed and diluted in a similar fashion to that seen in shallow bore SP29. What has seeped through to the deep aquifer appears to be slow moving, along preferred pathways. As the source has been removed, this water will also continue to improve in quality through dilution and attenuation processes. Again, it is difficult to see how any significant contamination from this source could reach the surface environment.

Terrestrial Radiation

A detailed airborne gamma radiation survey was flown over the site at 100m line spacing and 50m mean terrain clearance in 1997 following completion of restorative earthmoving. Detailed ground traversing along a number of transects was completed in 1998 to enable estimation of surface gamma radiation levels in the area (Martin et al 2001). Preliminary conclusions from this work estimate that the average terrestrial-origin absorbed dose rate from gamma radiation of 0.25 microGrays per hour with a pre-mining component of 0.10 microGrays per hour.

Normal expected land use in this area, comprises hunting and camping with residence time of less than 50 days per year. It is calculated that the resultant above pre-mining effective dose rate will be up to 0.14 milliSieverts (mSv) per year compared with a public dose limit of 1 mSv per year. There are, however, assumed to be small areas on the site, for example over the backfilled pit, where dose limits might be exceeded by a member of the public taking up permanent residence. The risk of this occurring is low as the site is less attractive for occupation than other nearby areas with lower background contaminant levels.

The DBIRD perspective

The DBIRD perspective in relation to close out of the Nabarlek site is that it accepts the position of the landowners of the site as has been presented by their representative, the NLC. The intended sequential land use of the site was to not restrict the enjoyment of the traditional owners in their use of the rehabilitated site and the close out criteria were to be the report of the independent assessor.

It is acknowledged that members of the Minesite Technical Committee (MTC) comprising representatives of the four principal stakeholders, expressed reservations about the final report produced by Dr Adams (Adams 1999). However, with the advantage of hindsight, it would appear that the formal role of the MTC in considering the report was possibly questionable, as the report was a consequence of an agreement between QMPL, NLC and Dr Adams. It is presumed that the company tabled the report in good faith and in the spirit of transparency of process. In any event, the only consistent criticism from the MTC was not of the conclusions reached but of the relative lack of documented evidence presented to support those conclusions.

A group of scientists could not reach agreement in over ten years on the basic issue of close out criteria for what is a relatively simple mine site. DME officers working independently produced a set of simple, generic yet powerful close out criteria in a matter of weeks (Norris et al 1997). The then NT Minister for Mines and Energy approved a set of generic Mine Closure Criteria for use on NT mines in 1997 and, with or without the Adams report, it would be difficult to argue that the rehabilitation at Nabarlek does not fulfil those criteria. DBIRD has sympathy with the company's position in that it has fulfilled its obligations of an agreement with the traditional owners. An independent arbiter was sought, the parties agreed to abide by his assessment and now it appears that there may be doubts as to whether the agreement will be honoured.

Conclusion

It would be unfair and unreasonable to expect the company to reopen all the uncertainties with respect to close out when the rehabilitation has been completed. The question remains as to when this has been achieved. DBIRD's closure guidelines require that the landowners be satisfied with the state of rehabilitation. For Nabarlek, this is formalised by the Deed of Settlement between QMPL and NLC. Arguably, the 24 conditions proposed at the April 2000 workshop convened by OSS also remain to be satisfied.

The respective roles of the various stakeholders are, of course, acknowledged and DBIRD would prefer that the final close out of Nabarlek be reached with the consensus of all parties. Although there are a number of interested stakeholders, relinquishment of the Nabarlek Mineral Lease is primarily a matter between the leaseholder, the landholder and the regulator's representatives. Formal sign off to give effect to relinquishment is the responsibility of the Northern Territory Government, but is also contingent on final agreement between the NLC and QMPL.

References

Adams Ecological Consultants (1999) Revegetation of the Nabarlek uranium mine, Final Report. Adams Ecological Consultants, Subiaco, Perth, WA. Unpublished confidential report.

Buckley R (1987) Revegetation strategy, Nabarlek uranium mine, NT. Amdel report No 1656 for Northern Territory Department of Mines and Energy.

Martin P, Tims S, McGill R, Ryan B and Prendergast B (2001) Use of airborne gamma ray spectrometry for environmental assessment of the Nabarlek uranium mine. Unpublished report in progress.

Norris V, Jan G, Fox R, Sedman J (1997) Mine closure criteria. In Proceedings of the 22nd Annual Minerals Council of Australia Environmental Workshop: Demonstrating Environmental Excellence 97, 12-17 October 1997, Adelaide. Minerals Council of Australia, Canberra, 121-132.

Salama R and Verma M (1985) Nabarlek Deep Aquifer Models. Water Division of Department of Transport and Works, Northern Territory. Unpublished report.

Queensland Mines Limited (1979) Final Environmental Impact Statement; Nabarlek uranium project, Arnhem Land, Northern Territory.

Waggitt PW (2000) The decommissioning and rehabilitation of Nabarlek uranium mine, Northern Australia, *The Rehabilitation of Nabarlek Uranium Mine*, Proceedings of a workshop, Darwin NT Australia 18-19 April 2000, Supervising Scientist Report 160. Environment Australia, Commonwealth of Australia.

Restoration status of the first abandoned uranium mine in Brazil

Elisabete A. De Nadai Fernandes[1], Hélio Antonio Scalvi[2]

[1]Centro de Energia Nuclear na Agricultura, Universidade de São Paulo – CENA/USP, CP 96, 13400-970 Piracicaba SP Brazil
[2]Indústrias Nucleares do Brasil – INB, CP 913, 37701-970 Rodovia Andradas-Caldas MG Brazil

Abstract. This work describes the planned actions for the mine site rehabilitation, comprehending impermeability of berms and embankments, re-vegetation, surface and groundwater hydrogeological studies, containment and chemical treatment of drainage waters to comply with the standards prescribed by the federal agency responsible for implementing regulations applicable to uranium mine site.

Introduction

Brazil has one of the largest uranium reserves worldwide, which backs the supply of the long-term domestic needs and makes the surplus available to the international market. Prospecting studies carried out in one-fourth of the Brazilian territory indicated approximately 300,000 tons of uranium with important occurrences in the states of Bahia (Lagoa Real and Caetité) and Ceará (Itataia). The country also holds uranium occurrences associated to other minerals, which contribute to an additional amount of some 150,000 tons of U_3O_8.

The Indústrias Nucleares do Brasil (INB) is the company responsible in Brazil for the nuclear fuel cycle related products and services. Supplying 60% of the enrichment of uranium services demand of the nuclear power plants Angra 1 and Angra 2, INB will be increasing the fuel cycle nationalization index and also opening new perspectives for a technological improvement and for the competitiveness of its products. Due to the technological achievement reached throughout the various phases of this cycle, INB has become a member of the select group of highly qualified companies acknowledged for their technological expertise in the nuclear field. Its activities are backed by last-generation facilities and equipment and specialized technical staff. Having obtained an ISO 9001 certification granted by the German TÜV organization, INB maintains a rigorous control of all its processes and has adopted quality assurance procedures in every production phase. Thereby the company is qualified to supply products that demand strict control during their fabrication process and full reliability in the final results obtained.

INB operates cutting-edge liquid and gaseous effluent treatment systems, as well as those used for compacting and segregating solid wastes, thus minimizing the risk of causing an environmental impact. It exercises full control of its activities and staff health through the implementation of permanent monitoring programs to protect people and the environment. It also performs modeled actions to ensure the preservation and restoration of the ecosystem wherever its units are established. Fully aware of its social responsibility, the company appropriately adjusts its activities in the nuclear field by fostering education, culture and well-being of all communities settled nearby its operating units.

With this philosophy in mind, INB is running the decommissioning of the Caldas unit which first operated in the country for the production of uranium concentrate. The unit is located in an area of about 15 km^2 in the Poços de Caldas Plateau, Minas Gerais state, formed by a 35 km diameter caldera, the largest alkaline volcanic complex in South America. It lies within the watershed boundaries of the Antas river and the Verde river, both intensively used for crop irrigation and cattle watering.

Present status of the first uranium mine

The Caldas unit has produced from the onset of its activities in 1982 uranium concentrate that basically supplied the demands of Angra I reactor reloads and technical development programs. Besides its importance as an ore deposit, it was in this unit where the fuel cycle technology development and the uranium concentrate (yellowcake) production as ammonium diuranate (ADU) have begun. The open pit mine was exploited down to a depth of 120 m. About $47 \times 10^6 m^3$ of cover material, ore and waste rock were taken out of the pit. The mining activities from 1977 to 1996 generated $45 \times 10^6 m^3$ of waste rocks characterized by the presence of low-grade uranium below 200 ppm of U_3O_8 (Wiikman et al. 1995, Wiikman 1998). The plant running at full was able to process 2,500 tons of ore per day and produce 400 tons per year of concentrate. Solid and liquid wastes were sent to a tailings pond. Since the beginning of operation, control actions were taken to minimize environment impact, such as:

- Treatment of acidic water and adequate disposal of waste rocks based on the distance from the pit and local topography;
- Ditches for waste rock disposal have been formerly prepared by construction of drains for water contention using waste rock, transition and clay materials;
- Lining of the waste rock piles surface with 30 cm layer of compacted clay in order to avoid percolation of rain water;
- Alternatives of re-vegetation aiming at stabilization of the deposit against erosion due to wind and water penetration into the waste rock piles.

Nowadays the uranium concentrate is being produced in Caetité, Bahia state, due to the rise of production costs in the Caldas unit. The mine is under decommissioning and the industrial facilities will be used for other projects like the monazite chemical processing and rare-earth production.

Decommissioning action plan

In 1997 a systematic action approach for the decommissioning of the area had started in the potential pollutant sources - mine pit, waste rock piles and tailings pond (Scalvi 2000). The techniques applied for the environmental restoration of the uranium ore treatment unit (UTM-Caldas) are largely used worldwide and have been authorized by the Brazilian National Nuclear Energy Commission - CNEN (094/SLC, 14 October 1997). Mostly the actions are focused on enhancing the impermeability of berms and embankments, re-vegetation, development of a drainage system for pluvial waters and hydrogeological assessments aiming at reducing the generation of acid drainage water (Table 1).

The mining derived waste rocks disposed surrounding the mine pit gave rise to several piles, being the BF04 and BF08 the most significant ones in terms of interaction with the environment due to location and quantity/quality of material deposited. BF04 was constructed over the Consulta creek valley of the Rio Verde basin with 12.4×10^6 m^3 of waste rocks from body B and mine cover material distributed on a surface area of 56.9×10^4 m^2 with maximum embankment height of 90 m. Much of the waste is sulfidic containing approximately 2% pyrite (FeS$_2$) and water infiltrating into the pyritic material is acidified leading to high concentrations of metals and sulfate, a process catalyzed by the bacteria *Thiobacillus ferrooxidans* (Schorscher et al. 1997).

Monitoring of Consulta waters alerted for the need to control the contents of uranium and stable elements to acceptable levels, thereby piping such waters to a lime treatment station. In 1997, the INB implemented measures to decrease the generation of such huge volume of drainage water sent for treatment and to reduce associated costs. The acid drainage water flow was 90 m^3 h^{-1} in the rain period and 65 m^3 h^{-1} in the dry period. After the interventions were performed such flows decreased to 65 m^3 h^{-1} and 55 m^3 h^{-1}, respectively. Thus a decrease of about 220,000 m^3 y^{-1} in acid drainage water treatment was accomplished in the last 4 years (Table 2).

The investments for the next 5 years are estimated to be about US$ 250,000 and the actions to be taken for the full decommissioning will involve:

- improve embankment's permeability
- Consulta creek dam filling
- Consulta creek deviating canal (sealing work)
- Complementation of superficial drainage system
- re-vegetation of berms and embankments
- other activities

Case studies: active and abandoned Uranium mines

BF08 was formed with 15.0×10^6 m^3 of waste rocks originated from bodies B and E and mine cover material on a surface area of 64.4×10^4 m^2. Since the material was deposited on the Cercado creek watercourse, a lateral deviation had to be constructed, which sometimes contacts the pile loosing part of the water that should be delivered directly to the environment without treatment. This water after infiltration in the pile has to be collected with the other waters and sent for lime neutralization. The same problems of acid drainage water generation are faced because the pyrite is a common occurrence for both areas. Actions to recover the superficial draining and concrete canals were implemented in 1998, while clay covering of berms and embankments and posterior re-vegetation begun later in 2001. The results are still incipient since a small area of the deposit has been treated. The operational costs for this pile are higher than for the BF04 because the bed of the deviating canal of the Cercado creek has to be recovered and lowered, and also it has to be extended for about 1200 m to deliver the waters directly into the Antas river, in the Águas Claras basin. It is expected that much less water will infiltrate through the pile and consequently less acid drainage will be generated.

Costs will reach about US$750,000 for a period of six years, encompassing the following actions:

- recovering and lowering the existent deviating canal
- construction of the second stretch of the deviating canal
- improve impermeability of berms and embankments
- re-vegetation of berms and embankments
- superficial drainage recovering
- construction of underdrain system
- consultancy and technical support

The remaining waste rock piles (BF01, BF03 and BF07) are not presenting chemical problems since the pyrite is already oxidized. Nevertheless, physical stabilization is mostly required because the pile structures have been affected by erosion phenomena in the last 20 years. Main work has to be done for the stabilization of collapses that are carrying the fine sediments to the watercourses and basins within the UTM. Estimated budget needed for such actions are US$ 300,000 mainly related to the heavy equipments to be procured.

It shall be emphasized that the studies linked to reducing generation of acid drainage water have been performed in a fully coordinated action among several research institutes and universities and (inter)national consultancy (Rodrigues 2001; MDGEO 2000, 2001; ENGRAD 2001). In brief, these studies are related to an inventory of the groundwater springs in the Osamu Utsumi mine surrounding area, hydrogeological and hydrochemical aspects of the waste rock piles, and geophysical prospecting to evaluate water percolation in soil and rock fractures with the purpose of filling-up the mine pit with rainwater to the elevation of 1332 m.

Table 1. Environmental restoration data for the activities performed in the Caldas unit during the period 1997-2001.

Activities	1997	1998	1999	2000	2001	Total
Impermeability (m²)	4000	246000	28000	38000	114000	430000
Re-vegetation (m²)	3800	136200	22500	185000	85000	432500
Native and fruit trees (n°)	-	10200	12100	54650	68340	145290
Eucalyptus trees (n°)	850	3100	3800	1200	1114	10064
Underdrain (m)	650	4350	3900	1860	8600	19360
Concrete canals (m)	200	3100	2300	1120	2.100	8820
Cleaning of canals (m)	1200	2800	2600	3420	3920	13940
Seedlings (n°)	-	16500	52000	63240	126400	258140
Total revegetated area (m²)	6350	202700	102000	464250	432270	1207570

Table 2. Acid drainage water treatment in the Caldas unit during the period 1998-2001.

Marginal Water Treatment AA-440					
Year	Treated water volume (m³)	Annual consumption AA-440			
		Lime		Polielectrolyte	
		Ton	kg m⁻³	kg	g m⁻³
1998	1163670	2573	2.21	1.77	1.50
1999	1232616	2249	1.82	2.94	2.40
2000	1604546	2907	1.81	1.22	0.76
2001	1414851	1966	1.39	1.44	1.02
Tailings Pond					
Year	Treated water volume (m³)	Annual consumption AA's- 540/570/580			
		Lime		Barium chloride	
		Ton	kg m⁻³	kg	kg m⁻³
1998	746370	427	0.57	7.55	0.0101
1999	541039	247	0.46	3.18	0.0059
2000	743400	*434	0.58	4.13	0.0056
2001	456023	96	0.21	1.62	0.0036

* January 2000 consumption of lime was 239 tons due to heavy rain

According to the "Plan for Re-vegetation and Environmental Restoration of the Waste Rock Piles and Waters Associated with the Caldas Mine", a tree farm was implemented in 1998 occupying an area of 10 ha within the UTM. Presently the INB tree farm is producing some 300,000 plants of native, fruit, gramineous, eucalyptus and ornamental species per year, a sufficient quantity to supply the demand for re-vegetation of all areas degraded during the operational phase. Besides, the INB tree farm has been serving as a regional reference center to promote education of students and public in general interested in the reforestation program of the mined area. It is also helping the municipalities of Caldas, Andradas and Poços de Caldas in recovering extensive devastated areas along the rivers and creeks due to the uncontrolled predatory agriculture land uses.

Case studies: active and abandoned Uranium mines

Future compromises for environmental restoration

Mine and waste rock piles

In 1999, the INB submitted the Program for Decommissioning of Mine and Waste Rock Piles to the Brazilian National Nuclear Energy Commission (CNEN) and to the Brazilian Institute for Environment and Renewable Natural Resources (IBAMA). This program consisted in the first approach for the decommissioning work to be realized in the UTM in the next 15 years, a pioneer activity in the Brazilian territory. After revisions and inclusions by national and international bodies, the outlines of the program with the studies and respective objectives to be developed in the mine pit are:

- Consolidation of the preliminary studies performed in the area: subsidize the actions for the mine pit decommissioning
- Chemical characterization of the mine pit water: mapping the quality of the water which drains into the mine pit
- Geological, geotechnical and geophysical characterization of the rock bed in the mine pit region: identify the structural features
- Hydrogeological and hydraulic characterization of the mine pit: determine the quality and quantity of underground water; flow direction and rate, depth of the hydrostatic level; charge and recharge areas
- Definition of the mine pit closure plan: physical stability of the mine pit benches; prevention of contaminants dispersion into the environment; prevention of acid drainage water generation; definition of the future use of the area; program for rehabilitation of the area; definition of future scenarios (pos-remediation and rehabilitation)

For the waste rock piles, the proposed studies and related objectives are:

- Consolidation of existent data: subside the future actions
- Petrographic, geochemical and geotechnical characterization of the waste rocks: knowledge of composition and disposal status
- Hydrological balance of the waste rock piles: quantify the water sources and drains
- Plan for stabilization of the waste rock piles: containment of waste rocks to prevent their dispersion; physical stabilization of the piles
- Waste rock piles: prevent generation of acid drainage water; implement remediation strategies; elaborate rehabilitation program; define future uses of the area; identify and analyze consequences (risks)
- Definition of legal and environmental requirements: define type and quantity of radioactive material potentially available for liberation

Industrial area

With the uranium production stop since 1995, the INB has been searching for alternatives for its industrial facilities in order to make good use of material resources, infrastructure and personnel specialized in the nuclear fuel cycle. Besides some favorable perspectives were identified, there is still the need to dismantle the sulfuric acid plant, the crushing unit and the filtration system.

Tailings pond

During the year 2000, a contention ditch was constructed on the NE part of the tailings pond, in the place named Asa da Andorinha, to deviate the pluvial waters direct to the environment, thus avoiding the silting up of the tailings pond as well as the abrupt increasing of water level and acid drainage water, therefore reducing significantly the lime consumption. Concomitantly, a 20000 m^2 area downstream to the new ditch was re-vegetated to settle the existing fine sediments. Interesting results can already be attributed to such countermeasures:

- Reduction of re-suspension of deposited sediments during heavy rain periods
- Significant reduction of the treated water volume at the site 23-E (456023 m^3) in relation to the year 2000 (743400 m^3)
- Installation of about 1000 m of pipes (PEAD, 6") for the transfer of water from Barraginha (drain well P.27) to the Area 540/570 which receives lime treatment up to the pH 11 before being discarded to the tailings pond. In former times, this discharge was done without any treatment
- Installation of a platform to discharge effluents and limewater into the deeper part of the tailings pond thus allowing better reagent distribution. The main response of such changing was better water quality at the site 23-E, improved operational control of the water pH in tailings pond and significant reduction of lime consumption, from 434 tons in 2000 to 96 tons in 2001

The studies to be performed are dictated by the only in force Brazilian norm for radioactive tailings pond and by the CNEN, and can be summarized as follows:

- Geological and hydrogeological evaluation of the area
- Compactness of the material deposited in the tailings pond
- Occurrence of acid leaching in the tailings pond
- Tailings physico-chemical and radioactive characterization
- Inventory of the material deposited in the tailings pond
- Composition and mobility in relation to compactness
- Material qualification for tailings covering
- Engineering definition for tailings pond landfill
- Draining features for the contention system and pluvial water deviation
- Selection of appropriate types of vegetation for area covering and reforestation

Case studies: active and abandoned Uranium mines

Final remarks

Parallel to the implementation of the above discussed uranium mine decommissioning plan, part of the INB installations has been adapted for the monazite chemical treatment plant (TQM), which is under commissioning phase to start the operation with monazite concentrate originated from the Buena unit, located in the Rio de Janeiro state. INB's main objective for its UTM-Caldas is to creating the technological basis for the chemical processing of monazite mineral to obtain rare earth compounds, becoming the only national producer to attend the Brazilian market demand still dependent on imported products.

References

Inventário de pontos d'água no entorno da Mina de Caldas e demais instalações das Indústrias Nucleares do Brasil, MDGEO – Serviços de Hidrogeologia Ltda., Relatório INB (2001).

Projeto conceitual da drenagem subterrânea do Bota-fora 4 na Mina de Caldas-MG, MDGEO – Serviços de Hidrogeologia Ltda., Relatório INB (2000).

Relatório de Prospecção Geofísica, ENGRAD – Consultoria, Geociências e Radioproteção Ltda. (2001).

Rodrigues JA (2001) Drenagem ácida do Bota-fora 4 (Mina de Urânio de Caldas-MG): Aspectos hidrogeoquímicos e hidrogeológicos. Dissertação de Mestrado, Universidade Federal de Ouro Preto MG 145 p.

Scalvi HA (2000) Plano diretor de descomissionamento da Mina de Caldas e depósitos de estéreis. Relatório Interno INB.

Schorscher HD, Capovilla MMGM, Fallick AE (1997) Studies of pyrit generations related to uranium mineralization at Osamu Utsumi Mine, Poços de Caldas, Alkaline Complex, Brasil. In : SSAGI, Brasil. Proceedings, 289-290.

Wiikmann LO (1998) Caracterização química e radiológica dos estéreis provenientes da mineração de urânio do planalto de Poços de Caldas. MSc Thesis CENA/USP, Piracicaba SP 98 p.

Wiikmann LO, Figueiredo N, Taddei JF, Valente SMC, Chilelli V, Souza VP (1995) Gerenciamento de rejeitos da lavra de minério de urânio do CIPC. In Symposium on Regional Integration in Nuclear Energy. Rio de Janeiro. Proc Latin American Section of the American Nuclear Society p 117-131.

Flooding of the WISMUT mines - learning by doing or applying a comprehensive systematic approach ?

Rimbert Gatzweiler, Alexander T. Jakubick, Jürgen Meyer, Michael Paul and Jochen Schreyer

WISMUT GmbH, Chemnitz, Germany

Abstract. In mine closure and particularly in mine remediation flooding is the most important step because of its long-term consequences and its strong relevance to costs. Also, mine flooding can be strongly interrelated to other measures in mine remediation. Furthermore, the environmental impacts of mine flooding on the water, soil and even on the air path can be substantial. Forecasts based on models initially imply major uncertainties while representative monitoring data is not yet at hand. The WISMUT project since 1991 comprises the flooding of five underground mines and one open pit. The experience gained so far underlines the need for comprehensive planning for flooding and for a flexible approach as part of a mine closure management plan which allows continuous updating of initial planning and corrective actions on the basis of monitoring data recovered during the flooding process and additional findings.

Introduction

In the past, the usual way in closing a mine was to pull out all valuable materials, to seal the mine openings and then to stop pumping and let the groundwater rebound within the open mine space. With increasing environmental awareness in the mining industry and the introduction of regulatory measures for improved environmental and, especially, protection of surface and groundwater, comprehensive conceptual planning for mine closure is nowadays required as part of Environmental Impact Studies for licensing new mines. In mine remediation the closure of a mine and especially the flooding of a mine requires optimised conceptual planning and continuous comprehensive monitoring to control the flooding process and to facilitate the calibration of models to predict quantities and qualities of waters which might need pump and treat measures to avoid contamination of surface and groundwaters. Besides the requirement of controlling the flooding process often measures need to be developed and taken to steer the flooding process including the provision of fall-back options. Since the flooding process is interrelated to several other remediation measures and initially characterized by uncertainties, it needs optimisation and is best developed in a systems approach including risk and sensitivity analyses. An optimised flooding concept includes the development of a water management plan and the evaluation of potentially applicable water treatment processes and possible alternatives such as in situ or ex situ passive water treatment systems (Gatzweiler et al. 2002).

Flooding concepts

Flooding is an integral part of mine remediation. The general aim of mine remediation is to exclude hazards and unacceptable risks for life and health of the population and the environment concerned and thus allow further utilisation of the mine lands without inappropriate restrictions. In practice, this means to restore the mine field as far as economically and ecologically justified (surface and underground) and to stabilize it geomechanically, hydrologically and geochemically (Gatzweiler and Meyer 2000). While in modern in situ leach (ISL) mining the restoration of the groundwater aquifers in which leaching took place is a legal requirement, in remediation of old mines which were started under less stringent environmental legislation the prevention of hazards for man and environment must be the primary aim since total restoration would be unaffordable. The geomechanical stabilization of a mine and the geomechanical effects of flooding a mine are well understood and the respective remediation measures are largely prescribed by the mining ordinances and are part of good mining practise. The necessary expenses are calculable with sufficient precision. Forecasts of the impacts of flooding a mine on the local groundwater and surface water regime are more difficult and less precise. This is due to a variety of uncertainties relating to e.g. physical properties of the mine voids, hydraulic conditions and, in particular, external natural influences like meteorological conditions which all influence the flooding process. Because of these uncertainties, which can be reduced but not totally excluded through careful planning before flooding starts, remediation measures to minimize the risks related to the flooding of a mine are largely of a precautionary nature and at least on the side of the licensing authorities conservative judgement predominates which mostly corresponds directly with costs.

The essential components of a flooding concept are the timely installation of water treatment, pumping, discharge and residue management facilities and the establishment of a monitoring system. Furthermore, the concept must include a suitable organisational structure which on one hand requires the availability of an interdisciplinary team of experts equipped with the appropriate tools for evaluation, control and preparation of decision making and on the other hand requires the clear fixing of managerial decision lines. At WISMUT, the installation of management units being especially responsible for mine flooding proved to be successful.

As a general rule, the two most important preconditions for an optimised flooding concept are, firstly, a quick start of flooding and, secondly, an uninterrupted flooding process to arrive as early as possible at the final flooding level where quasi-stationary conditions can be expected. A quick start implies that the mine workings can be abandoned and mine ventilation and pumps can be switched off. This, on one hand, means substantial cost savings and, on the other hand, decreases oxidation of residual ores and wall rock and, as a consequence, diminishes the mobilisation of contaminants.

However, absolute priority must be given to preparatory measures as e.g. comprehensive clarification and documentation of the physical, hydraulic, hydrogeological, geochemical and geomechanical conditions of the mine respectively the ore deposit. Those measures, where applicable, must include already abandoned parts of the mine. The endeavour to abandon the mine workings as fast as possible and the careful preparation of the flooding process including e.g. the construction of hydraulic barriers to avoid mixing of differently contaminated mine waters in different mine fields or the installation of adequate monitoring facilities or the installation of facilities for alternative pumping options demand both compromises and the elaboration of a sound technological concept for abandoning individual mine fields and levels which allows to start with flooding at the lower levels while underground work is still carried out at the upper levels. The general aim to flood a mine continuously and as fast as possible can be inconsistent with avoiding negative geomechanical effects at the surface. Furthermore, it is often questionable whether all available waters in a mine should be used for flooding or whether it would be more advantageous to discharge little or not contaminated infiltration waters from higher levels within the mine without treatment directly at surface. A similar problem exists regarding the start of a pump and treat operation: To select an optimal mine water treatment process it is important to have reliable data on quantity and quality of waters to be treated available. In the course of flooding monitoring data allows the calibration of models and the respective predictions increase in accuracy. An early installation of a treatment facility bears the risk of not choosing the optimal treatment process and can involve costly changes to the treatment facilities later on (Gatzweiler et al. 2000). More reliable predictions on the development of the mine water quality and quantity are possible in the final stages of flooding when near stationary conditions prevail. With the raising flooding level generally oxygen is depleted in the flooding waters which reduces mobilisation and enhances the potential for precipitation, specifically of uranium. Also natural attenuation processes are enhanced. Conceptionally, an early start of pump and treat operation could potentially reduce the overall duration of necessary water treatment if strong inhomogenities exist within the mine waters and water with a high contaminant load could be tapped first by an appropriate positioning of pumps. However, careful optimisation and consideration of all cost relevant parts of the treatment chain is necessary to justify such a decision.

Case studies: Uranium mine flooding

The decision on if and when to prepare for the installation of pump and treat facilities and whether to allow the flooding level continuously to raise to its final level is strongly influenced by the perception of risks connected with such action. Due to mining over longer periods the original hydraulic conditions of the mine field are disturbed and normally cannot be totally restored. A specific problem often are boreholes which in former years were not or insufficiently sealed. Contaminated flooding waters therefore potentially intrude into adjacent aquifers, or outflow to surface occurs via mine workings or hydraulically conductive zones. Licensing authorities tend to demand extensive monitoring and the early installation of pump and treat facilities capable to lower the flooding level. The cost-conscious remediation operator on the other side tends to prepare for collection and discharge systems and a „just in time – if needed" option for treatment of seepage waters. The seepage collection respectively overflow collection option promises economic advantages and also a better potential for natural attenuation processes. It also eases the change from conventional water treatment to passive treatment systems as e.g. wetlands.

Flooding concepts and strategies have to be site specific though general routines can be applied (Gatzweiler and Meyer 2000). Continuous updating is required .Uncertainties can be reduced by flooding tests, high quality monitoring and evaluation of available data by a competent expert team guiding the flooding process. Well founded and documented management decisions will promote the confidence of the licensing authorities to arrive at well balanced decisions.

Key aspects of flooding the WISMUT mines

The WISMUT remediation project comprises five underground mines and one large open pit, which are being flooded. Approx. 80 per cent of the total historic production of about 230,000 t of uranium originated out of these mines. The most important mines, from a production point of view, are those at Ronneburg and at Schlema. With regard to their environmental impact especially on the local hydrosphere the Ronneburg and the Königstein mines take the first rank. The geographical, geological and hydrological environments, ore deposits and wall rocks of all five mines differ substantially. Since 1991 Wismut GmbH is undertaking remediation actions. Mine remediation is in an advanced stage. Flooding is completed at the Pöhla mine and in progress at the other mines. At all five sites water treatment facilities have been installed (Hagen et al. 2000; Gatzweiler 2000).

Pöhla

At the Pöhla mine flooding started in 1991 and the final flooding level was reached at the end of 1995. Since then, an annual average of 0.15 million m³ of flooding water flows out of the mine through the former main haulage adit and is treated for arsenic, radium, iron and manganese. An additional annual 0.15 million m³ of little contaminated infiltration water is discharged directly without treatment. The average annual mine water discharge during production reached 0.5

million m³. The mine water flow rate declined substantially as a result of flooding. The flooding water is consistently neutral in pH and shows slightly reducing redox conditions. Contaminant concentrations initially increased and later on sharply dropped (uranium), stagnated (arsenic) or increased (radium). The physico-chemical conditions responsible for this development are still not completely understood. Treatment for uranium could be abandoned in 1999 when concentrations fell below the discharge limit of 0.1mg/l. In 2002, tests are undertaken to precipitate part of the contaminant load within the mine. A lower cost alternative water treatment system has been tested successfully and will be put into operation in 2004.

Schlema

The very large Schlema mine has a total open mine space of about 35 million m³. The mine follows the eastward plunging ore body in four "cascades" down to a depth of about 2000 m below surface. The older shallow part of the mine is densely developed at surface by the town of Schlema. Extensive remediation work from surface and from underground aims to stabilize the surface geomechanically. An unique seismic monitoring system registers flooding induced rock bursts. Flooding of the upper levels of the mine is expected to result in major impacts at surface. Flooding started in 1991 (Meyer et al. 2002, this volume). By the middle of 2001 about 93 % of the mine workings had been flooded. Since then, the flooding level is kept between 120 m and 90 m below the level of the Markus-Semmler-Adit through which discharge into the river Mulde will occur at the final flooding stage. The delay in the flooding process was necessary to complete underground remediation work for stabilizing the surface. The amount of infiltration water into the mine is highly dependant on the meteorological conditions. A historical analysis shows annual variations between 5 million and more than 10 million m³. Indications for a decrease of infiltration quantity into the mine once the flooding level had risen into the upper levels of the mine, as originally expected, cannot be pinpointed so far. The water treatment capacity has recently been increased to 1000 m³ per hour. Extended periods of higher infiltration, which cannot be excluded, are buffered by 1 to 2 million m³ of open mine space below the overflow level. The flooding waters are pH neutral and show a slightly reducing redox potential. The main contaminants are uranium, arsenic, and radium as well as iron and manganese. Uranium and Radium concentrations show a declining trend while arsenic concentrations are stable. Due to the great depth of the mine, with rock temperatures of more than 50 degrees Celsius at the lower levels, strong convection flows prevail within the mine water body, thus enhancing mixing. This is probably further supported by pumping the mine water for treatment from below the 540 m level. Dilution by infiltration and treatment and discharge of flooding water appears to be the main reason for the declining trend in the concentrations of uranium and radium. Water treatment probably needs to be extended for about 25 years. Studies on applying in situ treatment did not provide sufficient encouragement for justifying pilot testing.

Case studies: Uranium mine flooding

Dresden-Gittersee

The former uranium coal mine Dresden-Gittersee is part of a small historic coal mining district. Flooding started in October 1995. The initial conceptual model was based on a hydraulic connection between the uranium mine and old dewatering adits which would drain the overflowing flooding water into the river Elbe. However, the overflow did not yet occur as forecasted. The alternative concept aims to let the flooding level raise up to its maximum level (after a toxic waste deposit has been excavated). Discharge of the flooding waters is then expected to occur via old dewatering adits connecting the mine field with the rivers Weißeritz and Elbe or via diffuse seepage into the river Weißeritz. An extensive monitoring system has been put in place to register the impacts of this procedure at the densely developed surface. Whenever required, the flooding level can be lowered again through pumping from two wells. Should the impacts of this procedure require pump and treat action for an incalculable time it is anticipated to establish an artificial underground hydraulic connection between the flooded mine field and an existing pervious dewatering adit with discharge to the river Elbe.

The flooding waters at Dresden-Gittersee carry a high iron and sulphate load. They are treated for iron. The infiltration water is clean. The pumping action enhances mixing and likely prevents the formation of a stratified water body with iron and sulphate rich water at the deeper parts of the mine overlain by clean infiltration water. At the final stage of the presently pursued concept no treatment is expected to be necessary.

Ronneburg

The very large mining field at Ronneburg consists of six formerly separate mines which were interconnected in the course of mining that started at the Schmirchau mine in 1951 and ceased in 1990 in the whole district. The open mine space has a total volume of about 27 million m^3. In addition, the Lichtenberg open pit with a total volume of 160 million m^3 and a maximum depth of 240 m being presently refilled with mine dump rock is located in the centre of the mine field. Flooding was initiated at the beginning of 1998. In preparation of flooding about 120 hydraulic barriers were installed within the main drifts at various levels in order to separate the southern part of the mine field where highly contaminated flood waters had to be expected from the northern part where mine waters contained low contaminant loads. Furthermore, barriers were erected to isolate individual mine fields with strong differences in the mine water quality. Despite major efforts in establishing a comprehensive flooding model and extensive monitoring forecasts on the duration of flooding and quantity and quality of flood waters to be treated are complicated by the complex mine structure, size and nature of the influenced groundwater regime and changing infiltration water quality due to ongoing remediation work at the surface. Currently, the final flooding level is forecasted to be reached within the period 2003 to 2005 depending on interim pump and treat action (Paul et al. 1998 and 2002).

Case studies: Uranium mine flooding

Due to the topography of the mine field and the sealing of all mine openings it must be anticipated that the flooding level will raise until natural discharge will occur at various topographic low points. The most important of these seepages will be in the Gessenbach valley. The preferred strategy is to bring the flooding level up close to surface and then to collect the flooding waters for treatment and discharge. As it is uncertain where exactly future seepage will occur a phased construction of a two level flooding water catchment system is foreseen starting with monitoring wells with a depth of about 30 m which can be turned into pumping wells. The final flooding level will be determined by a stepwise approach. While generally a maximum possible inundation is aimed at in order to reduce the thickness of the unsaturated zone specifically of the refilled open pit thus minimizing the generation of acid drainage a high inundation level will increase potentially diffuse seepage. A critical level of 240 m above sea level. has been defined which corresponds to the lowest point in the Gessenbach valley with potential contact of contaminated flooding water with surface water. To control the raise of the flooding water and as a fall-back option a 600 mm pumping well with a capacity equivalent to the inflow into the mine has been installed in the centre of the southern minefield. It is presently used to pump water to the recently completed water treatment plant during commissioning of the plant. Collect and treat action after arriving at the final flooding level is expected to last at least 20 years. The final flow field without intervention by pumping will enable the formation of a density stratification within the flooded mine. This will potentially reduce the long-term treatment requirements. Meanwhile treatment is needed for uranium and radium, iron, arsenic and base metals. Sulphate reduction is demanded by the permitting agency in case of discharge into minor streams.

Flooding of the northern part of the Ronneburg mine field has a much smaller impact on ground and surface waters. As in the southern part the concept aims at raising the flooding level to its natural maximum level and to install a combined monitoring and collection system. The flooding water quality probably does not demand treatment. However, as part of a fall-back option the main shaft of the Beerwalde mine has been equipped with a well and plans are at hand to install a water treatment facility should this be required.

Königstein

The sandstone-type uranium deposit at Königstein was initially mined by conventional underground methods, and at a later stage by underground block leaching using dilute sulphuric acid. The ore deposit is located in the lower most of four aquifers at the margin of a Cretaceous basin with varying fluviatile to marine depositional conditions. The primary aim of the flooding concept is to protect the 3rd aquifer which represents an important drinking water resource for the densely populated region. It is separated from the mined deposit by a 10 to 30 m thick aquitard, which itself is transected by one major and several minor fault zones, drill holes and shafts (Schreyer and Zimmerman 1998). Due to the acid leaching, the geochemical status of the deposit has been substantially changed. The pollutant potential of the mine comprises approx. 2 million m^3 of acid and strongly mineralised pore solutions with high concentrations of uranium, radium, iron, manganese and base metals. The flooding concept comprises control as well as steering components. An important control element is a drift surrounding the mined deposit at its hydraulically open sides at a distance of 30 to 50 m. This allows the collection of contaminated flooding waters draining through the pillar between the mine and the control drift while the mine is being flooded. Two extensive in situ flooding experiments, several laboratory studies especially with regards to pollutant output and comprehensive modelling efforts support the flooding strategy presently applied. The step by step approach includes an initial phase during which the flooding level is raised to the 140 m mine level. During this phase which will last approx. 4 years, the flooding water draining into the control drift is collected, pumped to surface and treated by a conventional high density sludge lime precipitation process before discharge. Due to the initial flushing or sweeping of the easily accessible open mine voids the contaminant load of the initial drainage will be extremely high. At a later stage, the contaminant release is expected to be lower since it will largely relate to the pore solutions. According to the experiments and studies this will then lead to a sharp fall off of the flooding waters' contaminant load (Schreyer and Zimmerman 2002). During this phase the control drift will be flooded to establish almost natural hydraulic conditions within the mine. The flooded control drift can then be used as a horizontal well from which contaminated flooding waters can be pumped to surface and treated.

A variety of options to support respectively optimise the flooding process have been developed and partly pilot tested. This includes the separate flushing of mineralised ore blocks which were already prepared for leaching at the end of production, the immobilization of contaminants in mineralised sandstone by infiltration with barium sulphate solution and controlled crystallization and the construction of a reactive barrier within the northern control drift. Decisions on whether these measures are applied or not will be taken after careful evaluation with respect to their ecological benefits and their technical and economic feasibility.

A decision on the final flooding of the Königstein mine has not been taken yet. Options include

- the continuation of controlled flooding up to the highest level naturally achievable and further collect and treat action by using the open control drift,
- decommissioning of the underground mine and use of the control drift as a horizontal well with continuation of pump and treat action. Treatment could be continued either conventionally or later on change to a passive treatment system using reactive materials in a solid bed reactor;
- construction of a reactive barrier within the control drift and flooding of the control drift. Besides the monitoring system to control the reactive barrier wells will have to be installed and the treatment plant would need to be kept on stand-by as a fall-back option;
- construction of a dewatering adit connecting the mine at 140 m above sea level with the river Elbe and discharge without treatment.

Flooding of the Königstein mine started in January 2001. By March 2002 the flooding level has been brought up to 80 m above sea level filling approx. 2.7 million m³ of mine voids. Predictions and reality are dealt with by Schreyer and Zimmerman in this volume (Schreyer and Zimmerman 2002).

Summary

Up to now, the flooding of four underground mines and a large open pit mine has been carried out successfully at WISMUT and yet no uncontrolled and negative impacts on the environments concerned have been observed. The flooding procedures also provided extensive experience. Key factors of the approaches taken and lessons learned are as follows: careful evaluation of all relevant available data and the definition of critical data gaps; evaluation of available options and decision points as part of the initial flooding concept before the start of flooding; the timely design and installation of a monitoring system; continuous updating of the initial concept with the recovery and evaluation of monitoring data and experience gained; conductance of the flooding process by an interdisciplinary team of experts with managerial competency; clear definition of uncertainties in modelling and forecasting; timely preparation of sound documentation for permit application and sufficient discussion for clarification of relevant scientific and technical problems between permit applicant and authority; and an effective reporting to build up confidence and thus ease decision making by the permit authorities.

References

Gatzweiler R., Jakubick A.T. et all (2002) Flooding strategies for decommissioning of uranium mines - a systems approach. Simon F.G., Meggyes T. and McDonald C. (eds): Advanced Groundwater Remediation - Active and Passive Techniques, Thomas Telford, London (in press).

Gatzweiler R. and Meyer J. (2000) Umweltverträgliche Stillegung und Verwahrung von Uranerzbergwerken - Fallbeispiel WISMUT. Tagungsband Internationale Konferenz - WISMUT 2000, 11. - 14.7.2000 in Schlema, WISMUT GmbH, Chemnitz.

Gatzweiler,R.; Jakubick,A.T. and Kiessig,G. (2000) Remediation options and the significance of water treatment at former uranium production sites in Eastern Germany. In: URANIUM 2000 - Proceedings of an International Symposium on the Metallurgy of Uranium, Saskatoon, Canada.

Gatzweiler,R. (2000) Remediation of former uranium mining and milling facilities in Germany. In: Restoration of Environments with Radioactive Residue, Proceedings of an International Symposium at Arlington,Virginia,USA (1999), pp 477 - 501, IAEA Vienna.

Hagen, M., Gatzweiler, R. et all (2000) Status and Outlook for the WISMUT Remediation Project in the States of Thuringia and Saxony, Germany. Proceedings of the NEA/IAEA Workshop RADLEG - 142-0, Moskov, November 2000, IAEA Vienna (in preparation).

Meyer, J, Jenk, U. et all (2002) Characterisation of final mine flooding at the Schlema-Alberoda site of WISMUT GmbH with particular emphasis on flooding water quality evolution. UHH III, Freiberg ,September 2002, this volume.

Paul, M., Sänger,H.-J. et all (1998) Prediction of the flooding process at the Ronneburg site - Results of an integrated approach. Merkel,B. and Helling,C. (eds.) Uranium Mining and Hydrology II, Proceedings of the Intern. Conference and Workshop, Freiberg, Germany, September 1998, pp 130 - 139, Sven von Loga, Köln.

Paul, M., Gengnagel,M. et all (2002) Four years of flooding the Ronneburg uranium mine - a status report. In: UMH III, Freiberg, September 2002, this volume.

Schreyer,J. and Zimmerman,U. (1998) The Königstein flooding concept - Status and Outlook. Merkel,B. and Helling,C. (eds.) Uranium Mining and Hydrology II, Proceedings of the International Conference and Workshop, Freiberg,Germany, September 1998, pp 140 - 151, Sven von Loga, Köln.

Schreyer,J. and Zimmerman,U. (2002) Flooding of the Königstein mine up to 80 m a.s.l. - Prediction and Reality. In: UMH III, Freiberg, September 2002, this volume.

Case studies: Uranium mine flooding

Flooding of the Königstein mine up to 80 m above sea level – Prediction and Reality

Jochen Schreyer[1], Udo Zimmermann[2], Ulf Jenk[1]

[1]WISMUT GmbH, Chemnitz, Germany
[2]WISMUT GmbH, Niederlassung Königstein, Germany

Abstract. Decommissioning of the Königstein mine is a special case in uranium mine flooding. This is related to the mine's location in the depths of four aquifers and to the fact that uranium mining was both by conventional methods and underground in-situ leaching using sulphuric acid. The area potentially affected by flooding of the mine workings is densely populated and the surrounding groundwater is an important drinking water resource. Following more than seven years of flooding experiment, the completion of numerous scientific, experimental and technical studies as well as of mining and engineering requirements, the competent authorities approved of the controlled flooding to start from January 2001. In a first step the mine was flooded up to a level of 50 m a.s.l. (above sea level) involving roughly 1.2 million m^3 of water. Between October 2001 and April 2002 the flooding level was raised to 80 m a.s.l., flooding ca. 2.7 million m^3 of mine voids. Flooding went trouble-free and provided satisfactory evidence of the technical controllability of the flood rise. Hydrochemical parameters of the flood water developed within predictions. The key issue of how drainage volume would develop could be answered as a result of flooding implementation. Prediction tools were validated and provide a valuable basis for further flood level rise and decision-making on final flooding stages.

Introduction

Remediation of the Königstein uranium mine south of Dresden/Saxony has some highly specific features. The uranium was extracted from the 4th sandstone aquifer initially using conventional mining methods and later an underground in situ leaching method using sulphuric acid. The mine is located in a ecologically sensitive and highly populated area (Fig. 1).

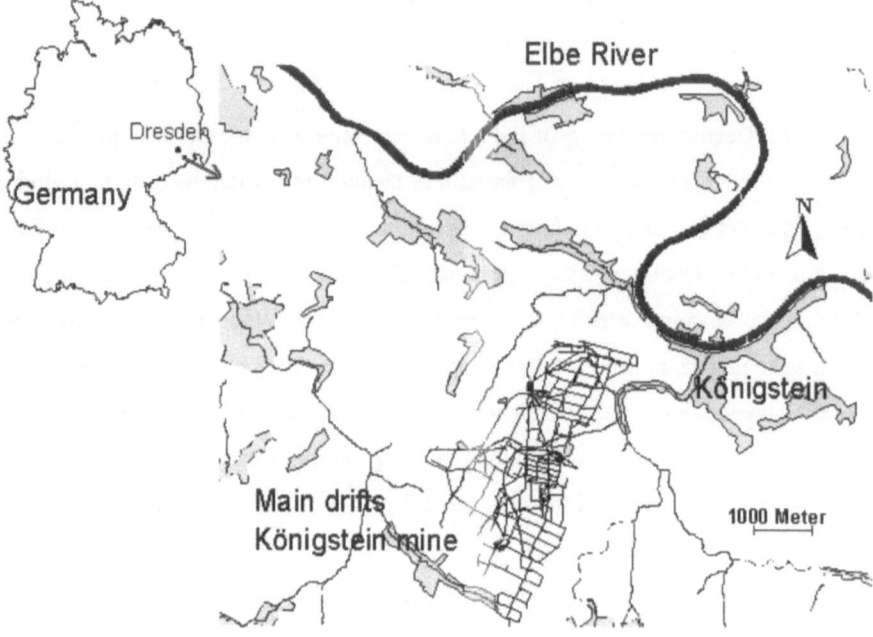

Fig. 1. Location of the Königstein mine

Experimental flooding of a small part of the mine was initiated in 1993 to gain hydrological, hydrogeological, geochemical, and rock mechanical data which will be used to predict conditions during the flooding process. The experimental area, being representative for the mine, is situated in its deepest parts and includes several former leaching blocks. Until October 1997 the experimental flooding level was at 33.4 m a.s.l., thereafter the level was raised to 40.0 m a.s.l. Quality of the flooding water is determined by the former leaching process and corresponding geochemical processes:

– pH 2 – 3, 10,000 mg/L solved substances
– 3,000 mg/L sulphate
– 500 mg/L iron
– 10 – 100 mg/L uranium.

WISMUT used the concept of controlled flooding (Schreyer and Zimmermann 1998), a major element of this approach being the control drift system which allows the collection of the draining flooding water. The controlled flooding will allow the reduction of pollutant concentrations to acceptable levels, to restore hydraulic conditions to nearly pre-mining conditions, and to prevent pollutant migration into the aquifer lying above and downstream of the mine. The flooding water, collected in the control drift system will be treated and then discharged to the Elbe river (Fig. 2). No relevant environmental impact is expected to occur with the control drift system and water treatment functioning.

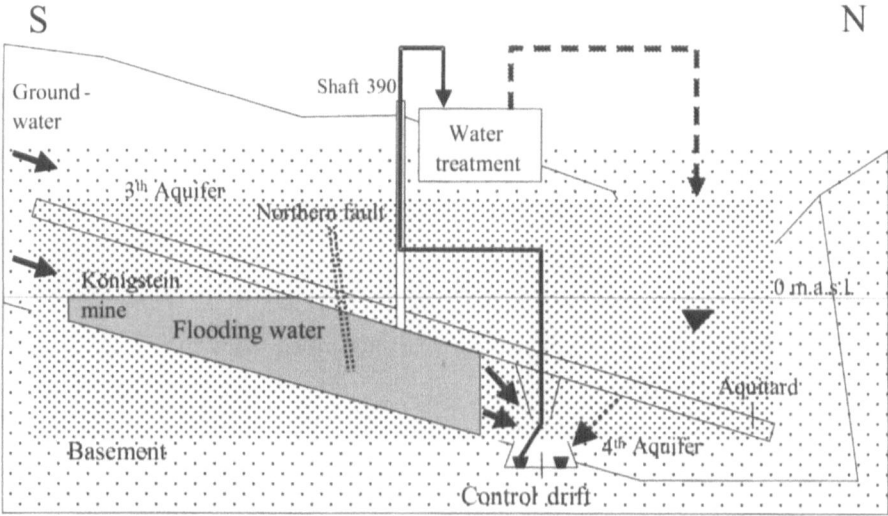

Fig. 2. Schematic cross section of the Königstein mine and the downstream region

Based on the experience gained from the flooding experiment and several scientific and engineering investigations the authorities permitted flooding of the Königstein mine to start in January 2001. Rise of the flooding water is limited to the level of 140 m a.s.l. to protect the groundwater aquifer. The results of the successful flooding are a good basis to make the decision on final flooding.

Results of the Flooding

Since flooding was initiated on January 29th 2001, ca. 3.7 million m³ of water were conducted into the flooding space by April 2002. Off that total, approximately 670,000 m³ drained from the flooding space into the control drift. 950,000 m³ were released via dams. Given that about 320,000 m³ of water (from the flooding experiment) were already stored in the flooding space when flooding of the mine was initiated, the water volume stored in the flooding space amounted to. 2.4 million m³ by the time the water level had risen to 80 m a.s.l. (Fig. 3).

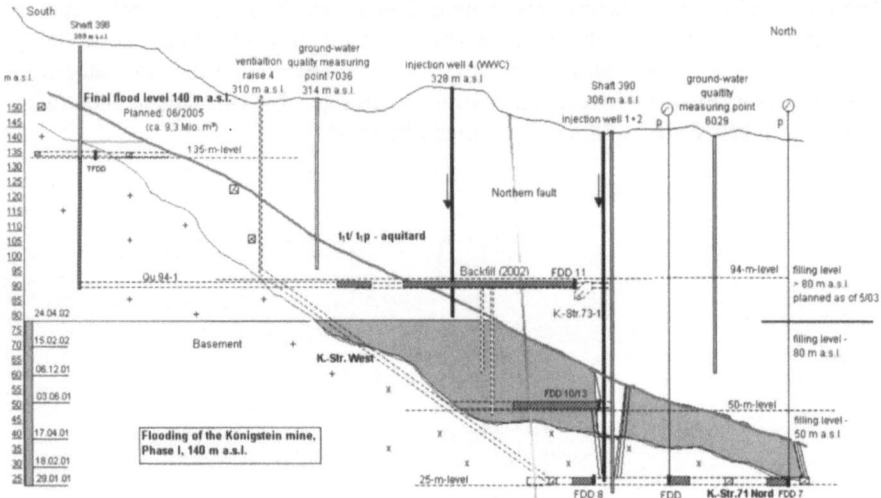

Fig. 3. Schematic diagram of the Königstein mine and flooding stages

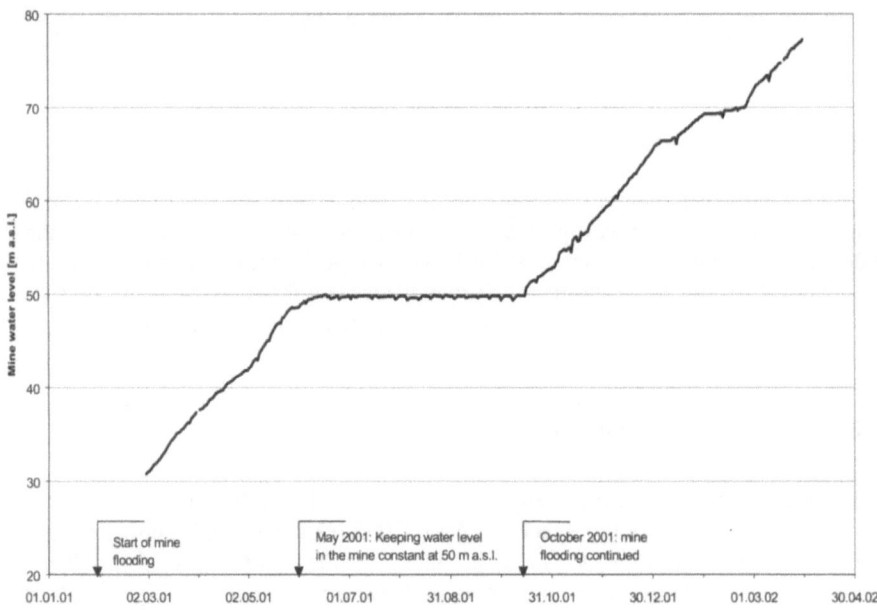

Fig. 4. Development of mine water level

Case studies: Uranium mine flooding

As the flooding was proceeding, rise of the flood level was subject to active control by the controlled addition of groundwater and the release of flood water. Between May and October 2001, for example, the flood level was kept constant at 50 m a.s.l. This halt was conditioned by the necessity to finalize dam construction work at the 50 m level. Filling was resumed in October 2001 and the flood level allowed to rise to 80 m a.s.l. by April 2002 (Fig. 4).

The flood level will be kept at this elevation for about one year which will allow to implement construction work prior to flooding of the 94 m level (Fig. 5).

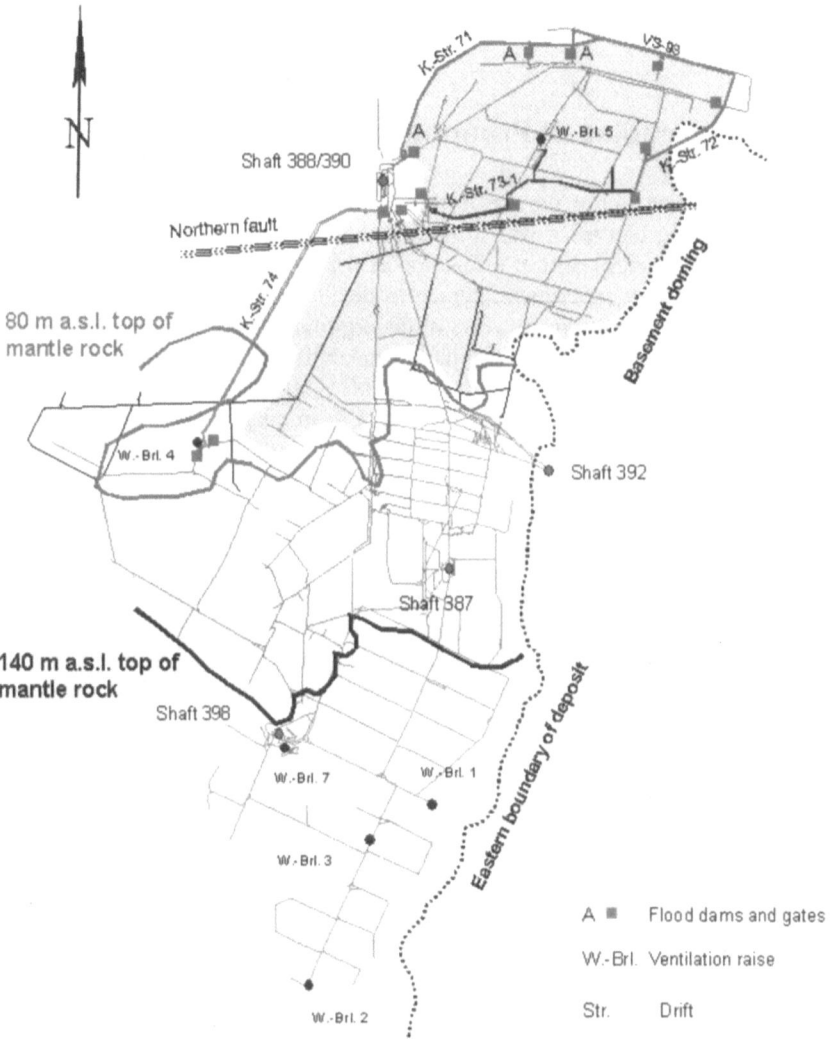

Fig. 5. Contour of flood water raise by April 2002, ca. 79 m a.s.l.

Case studies: Uranium mine flooding

Flooding progress has established proof that the existing tools allow very flexible control of the flooding process. Rise of the flood level was regulated in a range of 0 m/d to ca. 40 cm/d. If need be, the flood level might also be lowered.

Comprehensive and sophisticated systems to monitor the flooding process function flawlessly, ensuring safety of the flooding process and preventing any environmental hazards.

The course which the flooding has taken until now indicates that the prognoses underlying the design and sizing of the technical systems were sufficiently accurate. But it demonstrated at the same time that prognoses made for the evolution of a natural system are afflicted with uncertainties and that local phenomena cannot in any case be readily predicted.

Selected Examples of Prognoses and Reality

Drainage volumes: one of the greatest difficulties in establishing prognoses for flooding the Königstein mine was to predict the drainage volumes which would accumulate in the control drift as a function of the flood level. Based on the findings from the flooding experiment, numerous detailed studies and in particular on the experience gained from many years of underground leaching it was possible to narrow down the probable range of pillar permeability. For the flood level rise to 140 m a.s.l. it was assumed that the available groundwater volume of about 450 m^3/h would suffice to reach the level of 140 m a.s.l., i.e. that the drainage would not exceed the flood water volume of 450 m^3/h. The capacity of water treatment installations was designed accordingly to handle 500 m^3/h.

When flooding was underway, there was noticeable drainage in the control drift as early as in March 2001. At a flood level of 50 m a.s.l. drainage rose to ca. 40 m^3/h. With the rise of the flood level from 50 m a.s.l. to 80 m a.s.l. the entire northern pillar in the control drift zone was submerged and drainage volumes rose. In April 2002, ca. 180 m^3/h drainage water were collected in the control drift and pumped for treatment. As this drained volume does not yet correspond to steady-state conditions, one has to anticipated a further increase with stationary flood water level. In the light of developments until now it is assumed that the design flood water level of 140 m a.s.l. may be reached with the available groundwater quantities. A more accurate assessment can only be made once steady-state conditions have established at a flood level of 80 m a.s.l. Uncertainties will nevertheless remain since the rise of the flood water level to 140 m a.s.l. will affect new pillar areas of unknown drainage behaviour.

Hydrochemistry of the flood water: Fundamental processes which determine the flood water composition and its development over time as the flooding proceeds could be established by the flooding experiment and accompanying studies (Jenk and Schreyer 2001).

During the flooding experiment it was already found that the obtained data would only allow summary assessments of the total system behaviour. A major shortcoming of the flooding experiment was the failure to establish by experiment or to narrow down the impact of the drainage pillar on flooding water composition. Therefore, there was no appropriate data for the calibration of prognosis tools.

With the flooding level reaching 140 m a.s.l. the prognosis predicted e.g. discharge with the flooding water of about 800 t uranium, up to 40,000 t sulphate and up to 10,000 t iron. With the flooding running for 15 months now and a flood water level at 80 m a.s.l., discharge up to now amounts to ca. 120 t uranium, 4,000 t sulphate and 1,000 t iron. This means that contaminant discharge on the whole is within the anticipated range. But it also turned out that some locations of the drainage pillar showed concentrations which strongly deviate from predictions. For instance, flood water draining from the pile at the location of the former main shafts showed uranium concentrations of up to 7 g/l und pH-values of below 1. This resulted in temporary uranium concentrations in the pumped off flooding water of about 250 mg/l. Anticipated average uranium concentrations were in the order of 60 mg/l.

Altogether, the flood water rise to 80 m a.s.l. showed sufficiently good matching with prognoses. On the other hand, it was obvious that a complex natural system like flooding of the Königstein mine is no obvious subject for detailed predictions. Presently available data does not allow to check prognoses for the temporal behaviour of the overall system.

Summary

The concept of controlled flooding was definitely confirmed by the course flooding has taken. On the other hand, there was no trend in the flood process outside the subjects/items covered by the predictions. Local deviations are readily detected by the underground monitoring system. In such case, flood control by means of the unflooded control drift allows appropriate intervention into the overall process and prevents that such local deviations might cause any environmental impact.

Establishment of stationary conditions with a flood water level at 80 m a.s.l. in late 2002 will provide a database that will allow to further hone the existing prediction tools. It will then become possible to provide verifiable prognoses for a number of flooding options beyond the level of 140 m a.s.l. Such prognoses will then serve as a decisive basis for decisions on the final flooding phase.

References

Schreyer J. and Zimmerman U. (1998) The Königstein Flooding Concept - Status and Outlook. Editors Merkel B. and Helling C.: Uranium Mining and Hydrology II, Proceedings of the International Conference and Workshop, Freiberg, Germany, September 1998. Sven von Loga, Köln.

Jenk U. and Schreyer J. (2001) Pollutant Release Level Prognosis – A Major Input into the Flooding Concept for the Former ISL Uranium Mine at Königstein (WISMUT Germany). Proceedings 8th International Conference on Environmental Management, September 30 – October 4, 2001 Bruges, Belgium

Case studies: Uranium mine flooding

Uranium Removal from Flood Water at Königstein Mine

Horst Märten[1], Jochen Schreyer[2], Dieter Seeliger[1], Klaus Sommer[1]

[1]UIT GmbH Dresden, Zum Windkanal 21, D-01109 Dresden, Germany
[2]Wismut GmbH, Jagdschänkenstraße 29, D-09117 Chemnitz, Germany

Abstract. The acidic flood water from the former in-situ leaching (ISL) mine Königstein is characterized by considerable U concentrations (about 10 to 60 mg/l). Flood water treatment includes (i) U removal by ion exchange (IX) and further U processing by multi-stage elution, H_2O_2 precipitation of U, and U oxide washing and de-watering and (ii) water cleaning in a HDS plant. The former U plant at the site was converted and technologically adapted to meet the specific requirements determined by controlled flooding, by disposal/usability criteria for U as well as by legislation and authorities.

Uranium plant conversion – from former U production to U removal technology within remediation

ISL technology had been applied in Königstein (Saxony) within open mine works to recover uranium from ore blocks prepared with leachant inlet and leachate drainage systems. The ore bodies are located within a Cretacious sandstone aquifer. The sulfuric acid leaching led to a considerable contamination of the pore water, in particular with uranium in a strongly acidic milieu. In order to avoid the wide-spread migration of hazardous material into neighbored aquifer regions, flooding of the hydraulically isolated mine has been arranged in a controlled manner. Controlled flooding (Schreyer and Zimmermann 1998) includes

(i)   Collection of the contaminated flood water by a sophisticated drainage system via a control tunnel at the lower border line of the sloping mine aquifer

(ii) Pumping of the flood water to the surface and treating it in two process steps: uranium removal and water cleaning (Heinze et al. 2002)

(iii) Automatic monitoring and process control (Märten et al. 1998), in particular control of flooding level

The conversion of the former Königstein mine operation (including mine water treatment and bleeding from the leachant/leachate circle) to the present controlled flooding and flood water treatment operation is illustrated in Fig. 1. Whereas the old mine water and bleed treatment facilities had to be replaced by a new HDS plant for efficient flood water treatment (Heinze et al. 2002), the former uranium recovery plant could be adapted by technological and technical changes to meet the requirements of flood water pre-treatment by removal of U. The necessity for U removal from flood water became obvious by geo-chemical forecasts of the flooding process and has been clearly confirmed within the first stages of flooding (U concentrations of about 10 - 60 mg/l and locally beyond).

The former U recovery by IX and further processing steps (elution, precipitation both with considerable consumption of chemicals) resulted in an intermediate product (sodiumdiuranate). In order to meet usability criteria, the former U processing technology had to be changed to prepare a clean U oxide product (in fact, as a by-product of remediation).

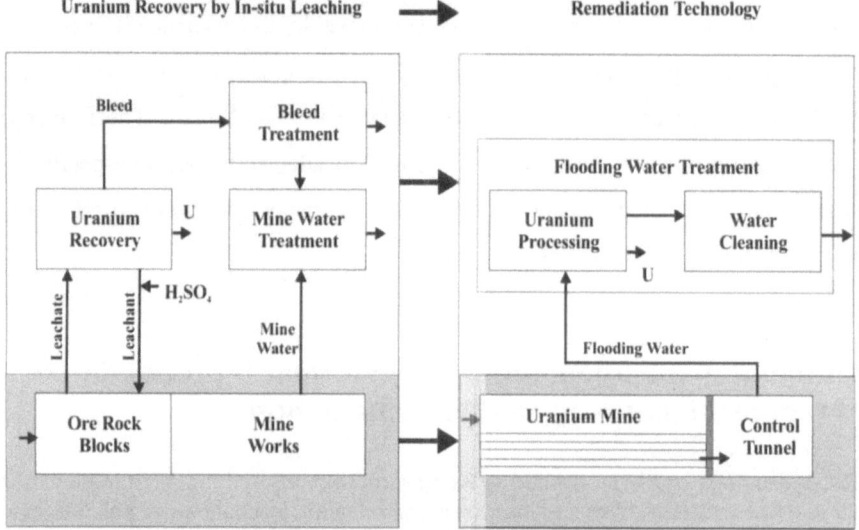

Fig. 1. Schematic representation of the former Königstein mine operation in comparison with the present controlled flooding and flood water treatment processing

The new U processing technology at Königstein is also based on sorption on an anionic ion exchange resin, but elution and precipitation had to be modified considerably. Because of the remarkable iron concentration in the flood water (typical molar Fe : U ratio in the order of 50 : 1) and in view of the incomplete selectivity of the IX resin (molar load ratio Fe : U of about 5 : 1), in particular against the sorption of $Fe(SO_4)_2^-$ ions, special emphasis had to be laid to the suppression of iron in the final U product (The product standard of iron impurity is < 0.15 %, this corresponds to a molar ratio Fe : U in the product of about 1 : 200).

The scheme of the new U processing is shown in Fig. 2. It is described in more detail in the following paragraphs.

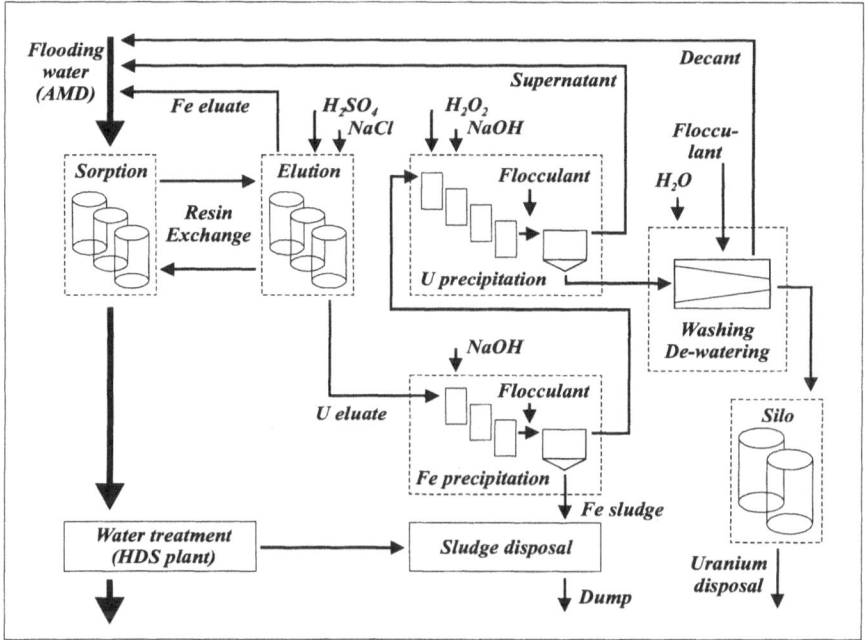

Fig. 2. Schematic representation of the former Königstein mine operation in comparison with the present controlled flooding and flood water treatment processing

Basic IX chemistry

For a better understanding, the principals of IX chemistry are briefly summarized. In sulfuric acid solutions with U and Fe, the following species have to be mainly considered in describing ion exchange:

- Cl^-, $Fe(SO_4)_2^-$, HSO_4^-, SO_4^{2-}, $UO_2(SO_4)_2^{2-}$, and $UO_2(SO_4)_3^{4-}$

The dominant ion exchange reactions at the resin (abbreviated by R^+) include:

$R^+ + Cl^- = RCl$	$\log\_K = 0.854$
$R^+ + Fe(SO_4)_2^- = RFe(SO_4)_2$	$\log\_K = 1.4$
$R^+ + HSO_4^- = RHSO_4$	$\log\_K = 0.354$
$2\,R^+ + SO_4^{2-} = R_2SO_4$	$\log\_K = 0$
$2\,R^+ + UO_2(SO_4)_2^{2-} = R_2UO_2(SO_4)_2$	$\log\_K = 1.6 - 1.9$
$4\,R^+ + UO_2(SO_4)_3^{4-} = R_4UO_2(SO_4)_3$	$\log\_K = 3.9 - 4.6$

(The thermodynamic constants $\log\_K$ depend on resin sort. Approximated values are given for the WOFATIT SBT resin used in Königstein.)

Sorption and elution conditions are determined by concentration ratios in the inflow to the IX columns. Elution means replacing of metal-sulfato complexes by high-ionic solutions either with sulfate or with chloride (or alternatives not considered here).

Multi-stage elution

Using eluants with suitable cations (Cl^-, SO_4^- or others) of about 1 M concentration, an efficient elution of cationic uranyl-sulfato complexes U from the loaded resin can be achieved. However, the $Fe(SO_4)_2^-$ complexes are also eluted. Based on pilot tests together with 1D transport model simulations using the unpublished code KIN (UIT development) including PhreeqC (Parkhurst and Appelo 1999), a multi-stage elution has been developed. The elution operation consists of the following stages:

(i)   Fe elution by a 0.1 M H_2SO_4 solution followed by a water flushing period

(ii)   U elution by a 1 M NaCl solution prepared from the recycled conversion outflow (step iii)

(iii)   Resin conversion from chloride to sulfate load by a 0.1 M H_2SO_4 solution, which is recycled to the eluant preparation for step (ii)

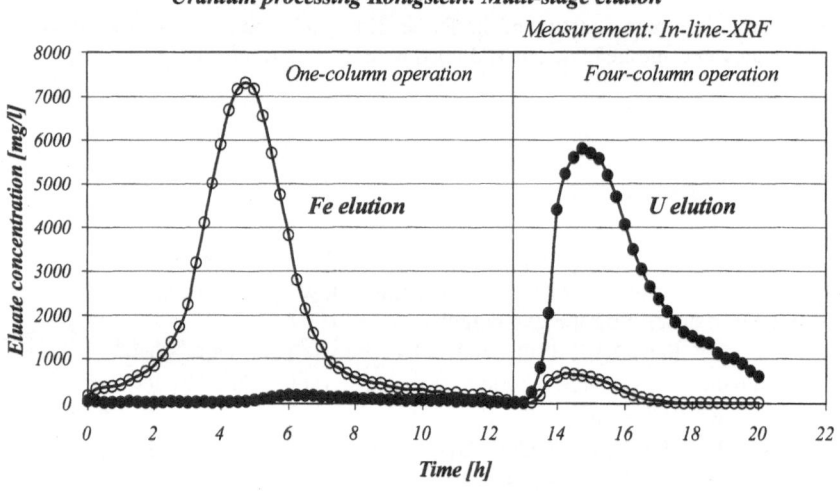

Uranium processing Königstein: Multi-stage elution

Measurement: In-line-XRF

Fig. 3. Elution curves for uranium and iron, separated for the two main elution stages applied (measurement by in-line X-ray fluorescence analysis)

As shown in Fig. 3, the first elution step results in an efficient de-loading of the ferric sulfato complex, whereas uranyl sulfato complexes are kept on the resin. The Fe eluate with about 2 g/l Fe and 30 mg/l U is recycled to sorption (Fig. 2). The typical metal concentrations in the U eluate from the second elution step are 2

Case studies: Uranium mine flooding

- 10 g/l U and 200 – 400 mg/l Fe, i. e. the molar ratio Fe : U is already reduced to about 1 : 5.

Note that chloride elution had to be favored against sulfate elution, which works with comparible efficiency, for two reasons:

- U precipitation with hydrogen peroxide (cf. next paragraph) is less efficient at high sulfate concentrations.
- Less total costs of chemicals are achieved in the case of chloride elution.

Continuous U precipitation

In order to eliminate the iron from the U eluate to an acceptable level, a first continuously operating precipitation cascade with 3 reaction tanks is used. The pH value of the acidic U eluate (pH $\approx$ 1) is enhanced by controlled dosage of NaOH to about pH $\approx$ 3.6, thus, eliminating iron from the solution by precipitation of ferric hydroxide. The solids are separated in a thickener and disposed within the sludge disposal unit of the HDS plant (Fig. 2).

The thickener overflow is pumped to a reaction tank cascade with 4 reaction tanks to precipitate the uranium in continuous operation. The precipitation reaction with hydrogen peroxide reads:

- $UO_2^{2+} + H_2O_2 + 2\ H_2O = UO_4 \cdot 2H_2O + 2\ H^+$

Hydrogen peroxide is added to the first 3 reaction tanks with a total stoichiometric access of about 100 %, i. e. dosage has to be accordingly controlled by the U mass flow to the precipitation cascade. In order to compensate the acid production from U precipitation and to achieve the optimum U precipitation pH value of about 3.5 – 4.0 (Merritt 1971), NaOH is added by pH control.

The U concentrate is separated from the suspension in a second thickener. The U rest concentration in the supernatant amounts to about 10 mg/l typically. Due to the suppression of iron by staged elution and rest iron precipitation in a first reaction cascade, the U concentrate meets the product requirements concerning Fe impurity. However, the liquid phase of the wet concentrate contains a considerable amount of chloride and sulfate to be eliminated in a next processing step.

Washing and de-watering

A multi-step washing (by dilution with pure water) and de-watering (by centrifuge) facility has been built up to remove soluble impurities from the U concentrate. The number of necessary steps to meet impurity criteria is controlled by the on-line measurement of conductivity (as a measure of ionic strength) in the decant. The decant from the centrifuge is recycled to sorption inflow (cf. Fig. 2). Two silos with 200 t capacity store are used for U concentrate storage.

The effect of washing / de-watering is illustrated in Fig. 4 representing two microscope views of the U product before / after treatment. Whereas the washed product shows pure U oxide particles, the initial U concentrate includes salt crystals.

Fig. 4. Electron microscope views of the uranium product before (left) and after (right) washing and de-watering

Summary

The Königstein U plant, which has been converted from the former U recovery within ISL operation to new elution and precipitation technologies, meets all requirements for its application in conjunction with mine flooding to remove U from the flooding water and to process it to a pure U concentrate. The total capacity of U processing is in the order of 1 t/d U or 500 m³/h flood water inflow to sorption (reserve capacity up to 1000 m³/h). Based on multi-stage elution and two-step (continuous) precipitation followed by washing / de-watering of the concentrate the U by-product of remediation complies with impurity criteria for utilization.

References

Heinze G, Märten H, Schreyer J, Seeliger D, Sommer K, Vogel D (2002) these proceedings
Märten H, Neubert F, Schneider P, Schramm H (1998) Proc. Int. Conf. On Uranium Mining and Hydrogeology II, Freiberg, Germany, September 1998, Ed. Merkel B and Helling C, Verlag Sven von Loga, Köln: 159-167
Merritt R C (1971) The Extractive Metallurgy of Uranium, Colorado School of Mines Research Institute
Parkhurst D L, Appelo C A J (1999) User's Guide to PhreeqC (Version 2). Water-Resources Investigations Report 99-4259, Denver, Colorado
Schreyer J and Zimmermann U (1998) Proc. Int. Conf. On Uranium Mining and Hydrogeology II, Freiberg, Germany, September 1998, Ed. Merkel B and Helling C, Verlag Sven von Loga, Köln: 140-151

Characterisation of final mine flooding at the Schlema-Alberoda site of WISMUT GmbH with particular emphasis on flooding water quality evolution

Jürgen Meyer[1], Ulf Jenk[1], Andrea Göhrs[1], Werner Schuppan[2]

[1]WISMUT GmbH, Chemnitz, Germany
[2]WISMUT GmbH, Niederlassung Aue, Germany

Abstract. Controlled flooding of the Schlema-Alberoda uranium mining site comprising ca. 38 million m^3 of voids and reaching down to a depth of 1,800 m began in 1991 under the mine closure program. The flooding progress is monitored for its environmental effects in terms of geomechanics, radiology and hydrochemistry. By mid-2001 ca. 93 % of the floodable voids at the site were filled up and the flood water level rose to some 100 m below the natural spillover to the Zwickauer Mulde river. This was accompanied by a slowdown in the continuous rapid rise and the transition to a discontinued, step-by-step flooding progress. Flooding of the remaining near-surface-voids will have to be closely monitored with a view to ensuring targeted control of flooding impacts on this densely populated area. Slowed down final flooding is a good prospect for getting first-hand information from the analysis of contaminated flood water and for establishing detailed prognoses of long-term water characteristics. In the medium term, such prognosis will help optimise cost-intensive operation of the water treatment plant which came on line in 1999. Current trends in water quality are characterised by predominantly stagnating and/or downward concentrations of major components in a neutral and reducing hydrochemical environment. Convection-driven transfer of constituents still continues in the flooded mine.

State of mine flooding

Details regarding mine characteristics and implementation of the flooding process have already been published (Meyer et al. 1998, 2000). This paper will focus on current results.

Initiated in 1991, flooding of the mine relies essentially on mine water inflow. In hydraulic terms, flooding is well advanced. By June 2001, the flood level had reached the maximum elevation of 135 m a.s.l. (above sea level) corresponding to the -120-m mine level. The natural spillover from the flooded mine workings to the Zwickauer Mulde river is at 223 m a.s.l., the greatest mine depth at 1475 m below sea level. The flood water level is to a large extent uniform across the entire mine field.

Flood progress is controlled for a number of reasons, e.g. the precipitation-dependent volume of water inflow into the mine, completion of work to ensure safe mine closure and control of geomechanical effects of flooding. Partial pumping of flood water is the main control feature since late 1999. Prior to discharge into the receiving stream, the pumped flood water passes through a conventional water treatment plant (WTP) of 1000 m³/h design capacity. The multi-stage process removes U, Ra, As, Fe, and Mn down to acceptable residual concentrations.

Annual flooding progress was in the order of 1.2 to 5.2 million m³ of flooded mine voids. By mid-2001, the cumulative flood water volume amounted to ca. 33 million m³. Yet, there is no indication of major additional cavities.

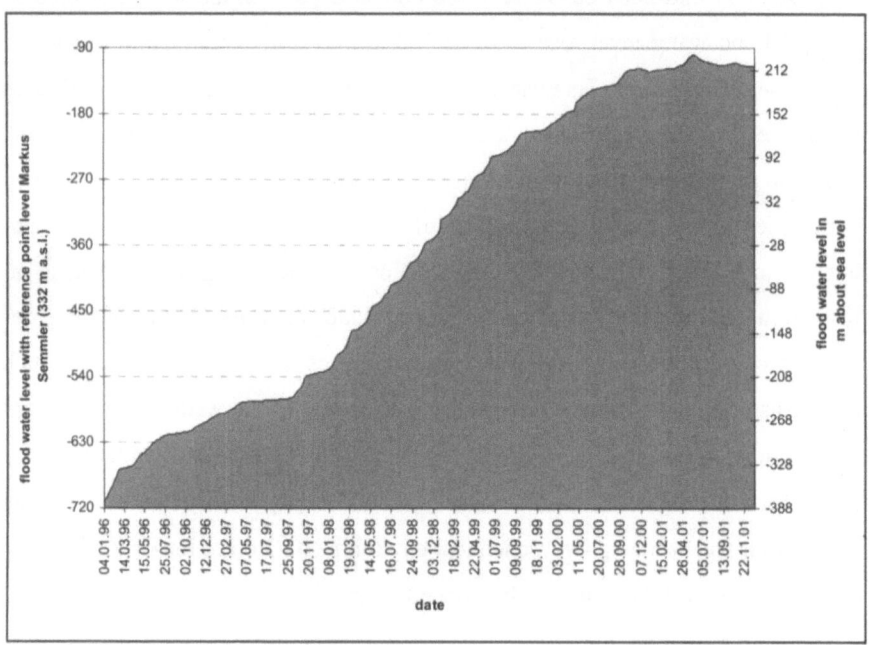

Fig. 1. Development of flood water level in Schlema-Alberoda-Mine since 1996

Case studies: Uranium mine flooding

Total annual inflow into the mine varies from 5.9 to 9.5 million m^3. Short-term inflow may peak at 1200 m^3/h. Until now, there is no flood-induced decline in inflow volume as was found at other mine floodings. This is mainly due to the dominance of near-surface inflow over inflow from deep-seated aquifers.

To the extent where it can be monitored by measurements, the flooded mine space shows a largely homogeneous composition of water. Relatively intensive convective water exchange within the hydraulically well connected mine workings accounts for that phenomenon.

Final flooding approach

The final flooding volume of ca. 3,5 million m^3 will remain suspended until completion of all closure operations in the floodable near-surface area. These operations are to ensure:
–   geomechanical stability of the mine site environs
–   ventilation of the non-floodable near-surface mine space (for radiation protection purposes)
–   sustained flood water feed to water treatment plant/spillover point
Furthermore, cost-effective operation of the cost-intensive water treatment process together with residue disposal (operational optimisation) have to be implemented. In accordance with current estimates, water treatment may be required for up to 30 years.

The final flooding will require ongoing control and monitoring for a number of environmental aspects. With this in mind, the final flooding is designed to proceed in a step-by-step manner with relatively small annual progress. Current considerations are based on the need for a hydraulic storage facility of the WTP in the near-surface mine area. Such a buffer storage would compensate peak inflow exceeding 1000 m^3/h and a possible WTP failure.

For the overall flooding project, the year 2001 marks a fundamental transition from rapid to slowed down or even stagnating flooding.

Retrospective development of flood water condition

Development

The rising flood water was routinely monitored by sampling/analyses and a number of in situ measurements. In view of the above-mentioned flood water homogenisation and despite objectively limited measurement possibilities, the observed data are of sufficiently good quality.

Quality development was characterised by continuous trends. Major discontinuities occurred essentially during the period before 1994.

The chemical environment of the flood water (and of the inflow as well) is in general characterised by neutral pH conditions. Redox conditions are intermediary to reducing. The inflow, in contrast, shows oxidising to intermediate redox conditions. Temperature of the flood water shows a downward trend and had fallen to approx. 30 °C in late 2001.

Total mineralisation had increased to ca. 4.5 g/l by 1999 and then started to recede. Principal components are cationic Ca, Mg, Na, also K and anionic SO_4, HCO_3, also Cl. Development of those parameters is similar to that of overall mineralisation. For some parameters, however, stagnation was observed following the end of concentration rise or before the onset of recession.

The redox sensitive parameters Fe and Mn show on the whole ascending to stagnating concentrations at a low level (together < 20 mg/l).

Parameters U, As and Ra-226 show environmentally relevant concentrations. Other nuclides and heavy metals are scarcely mobilised. Steady increase was noted for U until 1998, followed by stagnation (8 mg/l), and since 2000 there is a clear-cut recession. As for Ra, the increase came to an end as early as 1997 to be followed by several years of stagnation (4 Bq/l) and a decrease from 2000 onwards. Arsenic concentration peaked already in 1993 (at 8 mg/l) and then decreased until 2000. Current signs point to stagnation.

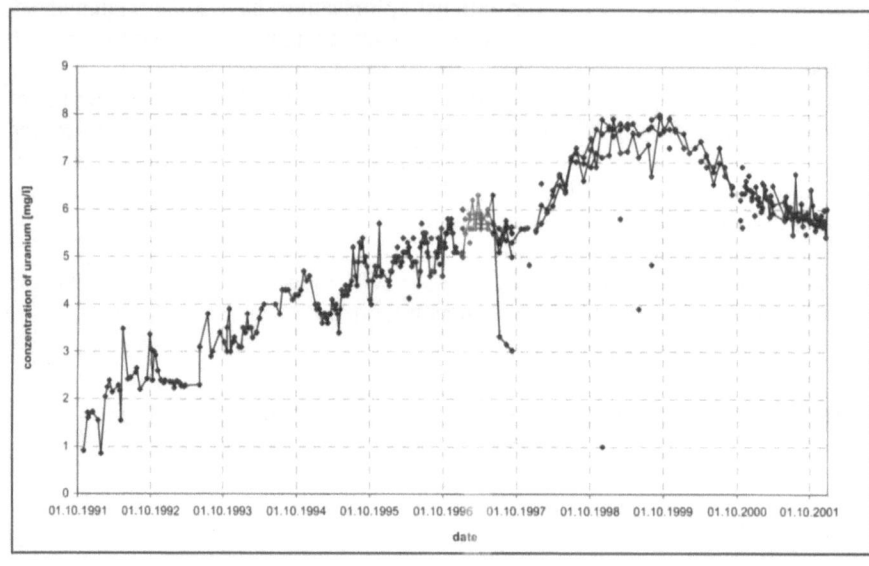

Fig. 2. Concentration of uranium in Schlema-Alberoda-Mine

For the most part, flood water is free of solids. Parameters which are environmentally relevant and redox sensitive occur predominantly in dissolved and partly also in colloidal form.

The additional consideration of mobilised mass quantities, basically, shows mainly a pattern that is analogous to that of mass concentrations.

Tentative Interpretation

The flooding process triggered a significant mobilisation of macro chemical and micro chemical substances into inflowing and rising flood waters, respectively. Sources of mobilisation include rocks and minerals in the area of the deposit, in particular backfilled stopes and fine-grained mine sludges as well as propping, supporting and lining materials. These sources were already subject to natural weathering during the decades of exploitation.

Lack of sufficient geochemical data does not allow quantification of relevant contaminants and weathered products, respectively, for the various mine sections. Information on vertical distribution is available only for uranium, the resource exploited. Independently of the data situation, it may be assumed as plausible that weathering was more intense in near-surface, humid areas of the mine that in deep-seated dry areas.

The rising flood water level mobilises weathered products in dissolved, partly colloidal and only marginally in solid form. In modelling terms, one may assume the formation of a contaminant wave in the top section of the rising flood water column. As a result of the convective exchange, the mobilised substances are distributed across the entire flooded space (homogenisation).

Concentration of substances in the flood water was governed by the available quantities of soluble substances and the specific flood water solubility. With regard to solid phases of gypsum and a number of carbonates, there are temporarily specific solubility limits for SO_4 and HCO_3 as the quantitatively dominating parameters. With regard to partial atmospheric pressure, the flood water is definitely oversaturated with CO_2.

On the analogy of the above, rapid release of uranium from mine-related weathered rock is evident (carbonato complexes). Given the prevailing chemical conditions, further mobilisation from primary mineralisations might be of minor importance only. As for As it is assumed that mobilisation is controlled by kinetics and hence only a small portion of the theoretically available mass potential will be released.

Due to pH-neutral conditions of the flood water, mobilisation of environmentally relevant metals is not stimulated by free acids. Their chemical neutralisation occurs at a number of carbonate, oxidic, and silicate substances contained in the mine rock and the flood water, respectively.

Intermediate to reducing conditions trigger the mobilisation of Fe and Mn. For the time being, there is no definite indication of uranium being chemically reduced, as was learnt from other mine flooding projects. Occurrence of such a demobilisation is thought possible.

The decline in some parameters which was first noticed in 1999/2000 is at first sight caused by the pumping of flood water to the WTP which began in 1999 and was stepped up in 2001 as well as by natural dilution of water in the flooded space by the inflow of low mineralised and low polluted water.

Water quality prognosis/conclusions

It is anticipated that stable hydraulic and hydrochemical conditions will develop as early as the final flooding stage is under way. In the mid-term however, there will be continued flood control in accordance with the flooding concept (e.g. deep-seated flood water pick-up, buffer storage operation).

As for the future chemical environment of the flood water, one has to assume the continuance of pH-neutral conditions and the stabilisation of reducing redox conditions. It is therefore anticipated for the final flooding stage that there will be only limited mobilisation from the newly flooded mine areas.

According to current knowledge, convective exchange will diminish and continue in mid- and long-term. Further flood monitoring and specific in-situ measurements shall document to what extent the flood water will trend towards stratification.

In terms of exponential time functions, continued pumping of flood water to feed the water treatment plant will therefore ensure continued fall in most contaminant concentrations. Mid-term monitoring will provide data for more detailed quantification to be completed in late 2003. Tentative deduction will show the extent to which mere dilution or more complex chemical processes will be involved for the various parameters.

Operation of the water treatment plant will have to be adjusted/optimised in the light of changing water quality.

These monitoring data should also allow to further improve understanding of the chemical interactions in the flooded mine space. In that context, search shall also be continued with a view to identify ways for in-situ flood water treatment. In case of positive results, this may lead to measures that would supplement conventional water treatment by a more rapid demobilisation of environmentally relevant substances.

References

Meyer, J. et al. (1998) Hydrochemische Aspekte und Interpretationen der Grubenflutungen des Sanierungsbetriebes Aue der WISMUT. Proceedings of the International Conference and Workshop "Uranium Mining and Hydrogeology", Freiberg, September 1998

Meyer, J. et al. (2000) Prognose und Realität der Flutung von Urangruben der WISMUT GmbH im Westerzgebirge. Tagungsband 5. Workshop zur Sanierung der Hinterlassenschaften des Uranerzbergbaus "Chancen und Grenzen der geochemischen Transportmodellierung bei der Verwahrung von Uranbergwerken und bei der Endlagerung radioaktiver Abfälle", Dresden, Mai 2000

Four years of flooding WISMUT's Ronneburg uranium mine – a status report

Michael Paul[1], Manfred Gengnagel[1], Dietmar Vogel[1], Werner Kuhn[2]

[1] WISMUT GmbH, Unternehmensleitung, Jagdschänkenstr. 29, 09117 Chemnitz
[2] WISMUT GmbH, Niederlassung Ronneburg, 07580 Ronneburg

Abstract. The Ronneburg uranium mine is flooding since the turn of 1997/1998. Until May 2002 the mine water rise has reached levels of about 148 m a.s.l. in the southern mine fields and 90 m a.s.l. in the northern mine fields. For the southern mine fields a water treatment plant proved to be necessary. Test runs began in January 2002. As flooding of the northern mine fields will have much less impact on ground and surface water bodies, operation of a water treatment plant is not deemed necessary for that area. The flooding operation is expected to complete between 2004 and 2005.

Introduction

WISMUT's Ronneburg uranium mine is flooding since the turn of 1997/1998. By May 2002 the mine water rise has reached levels of about 148 m a.s.l. (above sea level) in the southern mine fields and about 90 m a.s.l. in the northern mine fields of Drosen and Beerwalde.

This paper outlines important features of the flooding process observed so far, describes the technical measures undertaken for controlling the final steps of flooding, and gives an outlook on the further flooding process.

Site characteristics

The Ronneburg mine which is located about 10 km east of Gera/Thuringia was the largest WISMUT mine site. It was active between 1951 and 1990. Its 3,000 km of mine workings covered an area of about 60 km^2 and caused a corresponding groundwater depression cone. Mining extended over 22 levels in depth between 30 m and 940 m.

The deposit is hosted in a blackshale-limestone-metasedimetary series of Ordovician to Devonian age. A more detailed description of the geological site features has been given elsewhere (Lange and Freyhoff 1991, Paul et. al. 1998).

The Ronneburg mine consists of six formerly separate mines comprising 14 mine fields with 40 shafts. These mines which accounted for about 50 percent of WISMUT's total uranium production of about 230,000 tonnes are (from southwest to northeast): Lichtenberg, Reust, Schmirchau, Paitzdorf, Beerwalde, and Drosen. The mine levels were numbered according to their vertical distance to the surface. The main activities were concentrated between the 120-m- and the 390-m-level.

The overall volume of the mine workings at the Ronneburg site amounted to about 68 million m^3 of which about 44 million m^3 were backfilled mainly with a special fly-ash cement as a characteristic feature of the cut and fill mining technology performed at Ronneburg since the late 1960s. About 4 million m^3 arose from caving which was practiced in the early mining period of the 1950s and 1960s.

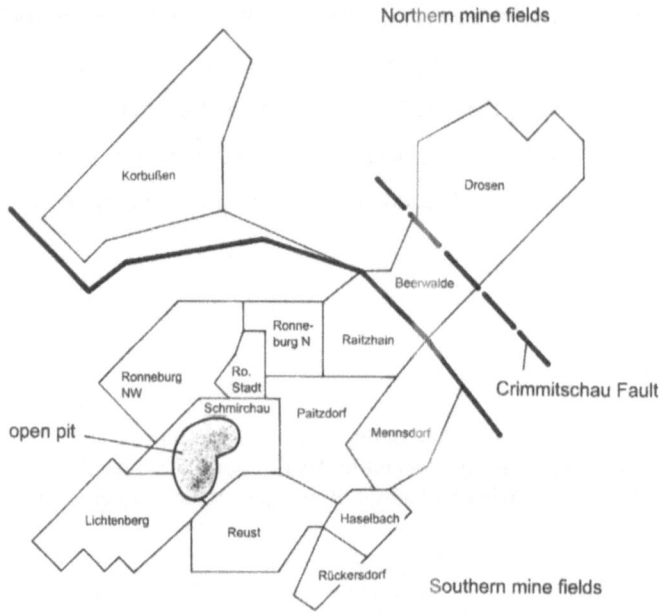

Fig. 1. Schematics of the Ronneburg mine site showing the different mine fields

In addition to the underground mine the Lichtenberg open pit with a total volume of 160 million m³ and a maximum depth of 240 m is located in the centre of the mine field. The open pit is in the process of backfilling with waste rock from Ronneburgs waste rock dumps which is planned to be finished in 2007. Inundation of the open pits so-called A-Zone is part of the rehabilitation concept (Paul 2002).

The flooding process of the Ronneburg mine is taking place within two hydraulically isolated areas (see Fig.1):

(1) Mine fields south of the federal motorway A 4 which comprises the former mines of Schmirchau, Reust, Lichtenberg and Paitzdorf,

(2) Mine fields north of the motorway A 4 including the Beerwalde and Drosen mines.

Both parts of the mine are separated hydraulically by six concrete plugs built at five mine levels which formerly connected the southern and northern mine fields.

Flooding of the southern mine fields

Before the complete mine flooding started, deep and isolated mine sections had already been partially flooded since 1989 (Table 1).

Table 1. Mine sections of the southern mine fields flooded before 1997.

Mine section	Mine levels	Flooding time	Open space flooded [Tm^3]
Reust, blind shaft 4	485-m to 390-m-level	12/89 – 02/91	347
Lichtenberg, blind shaft 6	525-m to 345-m-level	09/91 – 09/92	275
Ronneburg NW, shaft 381	570-m to 480-m-level	07/93 – 08/95	63
Ronneburg N, shaft 396	570-m to 435-m-level	04/95 – 03/96	100

Fully fledged flooding of the southern mine fields was initiated at the turn of 1997/1998 after a four-year-permitting and preparation process. The flooding operation is accompanied by a comprehensive monitoring program which provides detailed information on the time history of flooding, on the impact of hydrometeorological events on flooding dynamics and on water quality trends. Currently, about 20 wells are used for monitoring purposes.

The flooding process started at a water level of -168 m a.s.l. in the Ronneburg NW mine field whereas in Paitzdorf (-126 m a.s.l.) and Lichtenberg (-45 m a.s.l.) higher inundation levels had already been reached (see Table 1). Since 1999 all southern mine fields show the same water level. Meanwhile, the flooding water has reached a level of about 148 m a.s.l. Hydraulic consistent conditions in this part of the Ronneburg mine can be demonstrated by the water level differences between the various mine fields, which reach about 1 or 2 meters in maximum. A significant deviation from that general behaviour could only be detected for the mine field of Paitzdorf during 1999 as shown in Fig. 2.

During a particular period, the flooding process was faster than originally predicted. A revision of the floodable volumes and the inflow rates showed that a major reason for this wrong prediction was the overestimation of the various pore volumes and their dewatering/rewetting rates (Hähne et al. 2000, Table 2).

Table 2. Categorization of floodable mine space and pore volumes, southern mine fields, revised estimates, up to 270 m a.s.l.

Category	volume [million m³]
Open volume (mine workings)	15.7 [a]
Floodable pore volume in the backfill	4.1 ... 4.3
Floodable pore volume in the caving area	1.6 ... 2.3
Floodable pore volume in the backfilled open pit	1.8 ... 2.7
Dewatered pore volume in the host rock (depression cone)	3.8 ... 5.1

[a] Volumes of mine sections completely or partially flooded before 1997 (see also Table 1) are not included

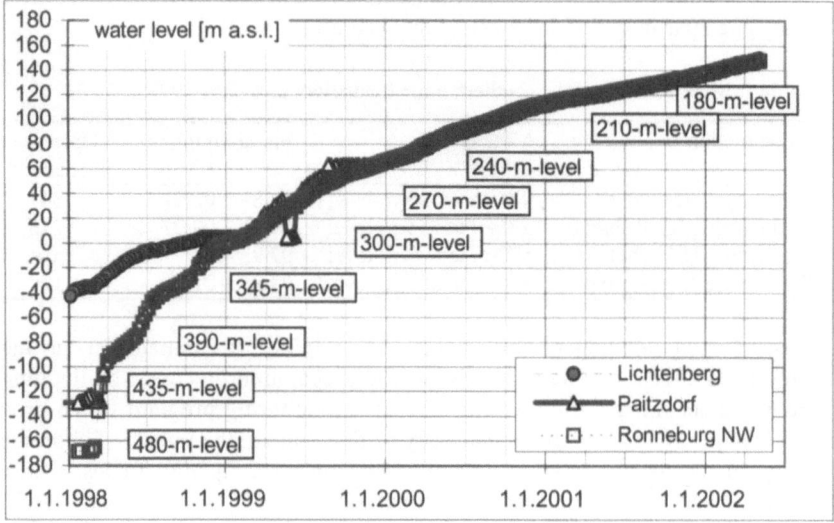

Fig. 2. Flooding curves for the mine fields of Lichtenberg, Ronneburg NW (belonging to the Schmirchau mine) and Paitzdorf

The velocity of the water table development as represented in Fig. 2 exhibits two significant phenomena:

(1) The annual rate of water rise shows a downward trend from 1998 to 2001 (1998: about 170 m p.a., 1999: 67 m p.a., 2000: 46 m p.a., 2001: 24 m p.a.). This is in general compatible with the vertical distribution of open mine space: in 1998 deep mine levels (480 up to 345) with only relative small amounts of open volume have been flooded whereas since 1999 the main levels (300, 240, 180) with more than 2 million m³ each were inundated (Table 3).

(2) The steplike shape of the curve as typical between -160 and 100 m a.s.l. flattens more and more above 100 m a.s.l. since the various mine levels are vertically overlapping in the upper part of the mine. Additionally, the effect of several types of pore volumes is increasing.

Case studies: Uranium mine flooding

Table 3. Ronneburg mine, southern mine fields, open volume by mine level [a]

Mine level	Open volume [Tm^3]	Mine level	Open volume [Tm^3]
30 m	-	270 m	931
60 m	32	300 m	3065
90 m	87	345 m	1649
120 m	1118	390 m	1368
150 m	1443	435 m	622
180 m	2447	480 m	388
210 m	1093	525 m	36
240 m	2503	570 m	74

[a] Volumes of mine sections flooded before 1997 (Table 1) are included

Mean annual precipitation in the Ronneburg region can be quantified at about 680°mm/a based upon a time serie from 1960 to present (DWD-German Weather Service). Annual variations can be characterized using years with extreme conditions: Low precipitation rates at about 400 mm/a are contrasting years with high precipitation rates at nearly 1,000 mm. Compared to those variations in the general hydro-meteorological regime the conditions in the four-year-period from 1998 to 2001 can be characterized with nearly medium precipitation rates; the measured values in this period exceed the long term average by about 3%.

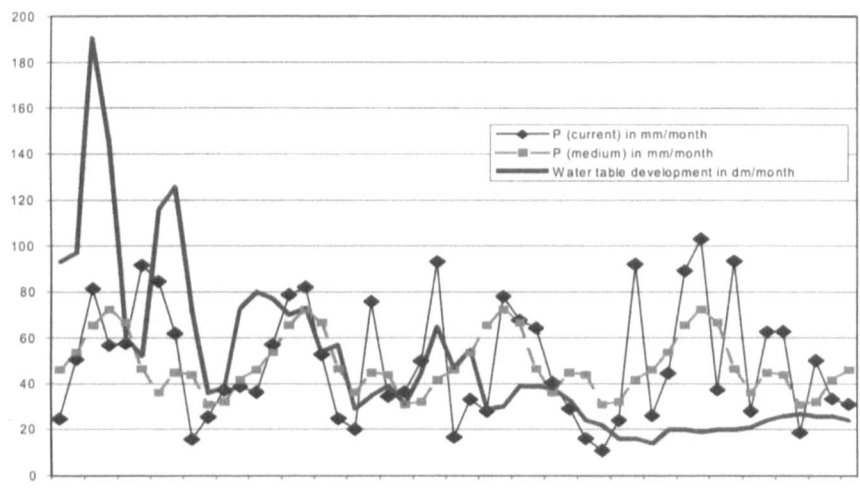

Fig. 3.   Time series of non-corrected precipitation (P) and water level development in the southern mine fields

A more detailed description of the hydro-meteorological regime is possible using the monthly precipitation rates as shown as a time series diagram in Fig. 3 which also contains a time series of the monthly water level development data for the southern mine fields. The general retrograde trend represents the increasing values of floodable mining volume with the rising flooding water level as explained above. The variation of monthly precipitation values seems to entail an analogous variation in the monthly mine water flooding level developments with a typical time delay of two or three months especially during 1998 and 1999. Nevertheless, a simple correlation between the monthly rainfall and the rates of water rise cannot be derived, as the inundation rates are strongly related to the vertical distribution of mining voids.

In contrast to the long term trend the period from September to November 1998 was very wet and caused a high groundwater recharge rate. This is believed to be the reason for the only weak reduction of the flooding rate during the inundation of the 300-m-level which is the mine level with the highest open volume of all mine levels (Table 3).

An important question to be answered is to what degree the inflow rate into the depression cone will decrease with rising water table. Before flooding the average inflow into the southern mine fields amounted to about 650 m^3/h with variations from < 500 m^3/h up to 800 m^3/h (based on monthly values, time series between 1991 and 1999; mine water flow rates from the period of active mining (before 1991) have been influenced by technical waters (dust control, drilling, mine back fill by concrete) and are suitable for plausibility considerations only). There are strong indications derived from monitoring and modelling of the regional water balance for a significant decrease of the inflow rates with higher inundation levels. For a flooding level of 232 m a.s.l. ("security level", see the last chapter) a decrease to some 450 m^3/h is expected under average conditions which was important for the design of the water treatment plant.

In terms of water quality conditions it has to be mentioned that at Ronneburg a very serious acid mine drainage problem exists. During active mining the mine waters were characterized by acid to neutral pH and high concentrations of radionuclides, heavy metals, iron, sulphate, and magnesium. Because of the huge area of the mine and its very complicated structure with different geochemical and hydraulic conditions water quality differed substantially in the various mine fields and mine levels.

Monitoring data of the flooding process confirm the significant water quality differences between the various mine fields. Most contaminated waters are concentrated in the mine fields of Schmirchau and Lichtenberg. Mixing and horizontal flow in the course of flooding led to variations in the water quality of some of the monitoring wells over time.

Furthermore, internal hydraulic processes and conditions have to be considered. As an example for these Fig. 4 shows the results of conductivity- and temperature-logs of the flooding water body in the mine workings. The increasing temperature trend with depth refers to the regional geothermal conditions in the Ronneburg area. The conductivity profiles exhibit a density stratification in the flooding water body. These conditions could hitherto be observed as a stable situation in some of the monitoring wells, which was neither disturbed by the relatively weak geothermal gradient nor by other influences. Variabilities in the log measurement results over time as shown in Fig. 4 are considered to be relatively small and do not influence the general trend of density stratification. Nevertheless continued monitoring activities and further investigations of the observed stratification effects in the flooding water and of their implications for water quality seem to be important.

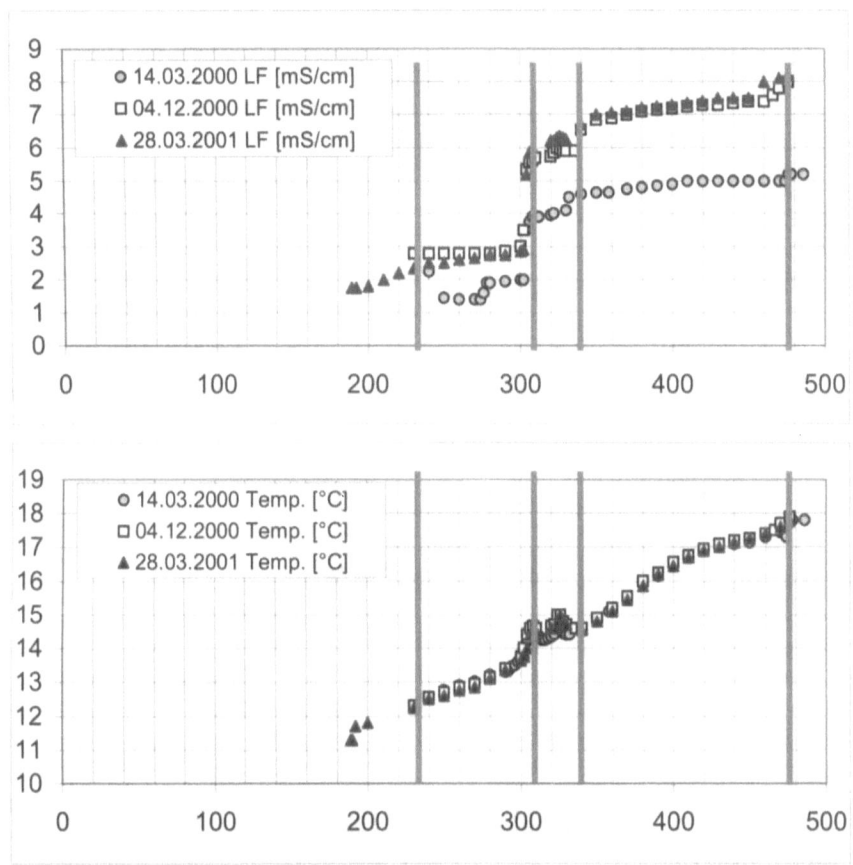

Fig. 4. Temperature and conductivity profiles (LF in mS/cm) at monitoring station e-1260 in the mine field of Ronneburg NW. Depth in meters below surface. Grey vertical bars indicating mine levels which are interconnected with the well.

The amount and degree of mine water contamination in the southern mine fields required the construction of a water treatment plant for the separation of iron, heavy metals, radionuclides, and arsenic. The water treatment plant is under test operation since January 2002. It is operating according to the HDS-lime precipitation scheme. Mine water supply is realized by pumping from a deep well situated in the Schmirchau mine field which is connected to the mine workings of the 240-m- and 180-m-level. The plant is constructed for a throughput of 600 m³/h, an optional capacity extension is possible.

The mine water which is pumped from the Schmirchau mine field is weakly acid but very high in sulphate, iron, and magnesium (Table 4). Nevertheless the trial runs showed a sufficient removal rate for iron, heavy metals, radionuclides, and arsenic. Continued operation should be used to quantify the actual inflow rate into the depression cone of the mine and to examine the water quality trend.

Table 4. Results of the water treatment plant trial runs, Jan. to April 2002, average data

		Inflow water treatment plant	Outflow water treatment plant
Magnesium	[mg/l]	725	700
Calcium	[mg/l]	550	900
Total iron	[mg/l]	775	< 2
Manganese	[mg/l]	20	< 1
Aluminum	[mg/l]	45	< 2
Sulphate	[mg/l]	5750	5200
Uranium	[mg/l]	2.0	< 0.5
Radium-226	[mBq/l]	≈ 500	< 400
Cobalt	[µg/l]	2000	< 100
Nickel	[µg/l]	6500	< 50
Zinc	[µg/l]	2300	< 200
Arsenic	[µg/l]	130	< 20
pH	-	4.7	6.5 - 8.5

Flooding of the northern mine fields

The Beerwalde and Drosen mines north of the federal motorway A 4 are flooding since mid 2000 comprising the mine field of Korbußen. As already explained, the northern mine fields are hydraulically separated from the southern ones by six concrete plugs. The maximum water level difference between the northern and the southern fields of about 160 m was reached during the year 2000.

The mine fields of Beerwalde and Drosen are separated by the Crimmitschau fault, a supra-regional fault zone. Mine workings which connected both mine fields have been backfilled prior to flooding. In contrast to the rest of the mine site at the Drosen mine field the Palaeozoic host rock is covered by a series of platform sediments of Permian to Triassic age.

Flooding at Beerwalde and Drosen develops as expected. At Drosen the mine workings are already completely flooded and the depression cone is filling up. The water level differences between the two mine fields do not exceed 10 to 20m.

The Korbußen mine field is totally isolated from the rest of the mine. It is connected to the Beerwalde mine by only one working which has been sealed by a concrete plug. As a consequence of that the water level rose quickly from July 2000 to the spring of 2001. Since march 2001 the water level dropped from about 170 m a.s.l. to about 35 m a.s.l. before it began to rise again since August 2001. The reason for that phenomenon is under discussion but has no consequence for the general flooding strategy of the northern mine fields. At present the water levels in Beerwalde and Drosen are 50 to 70 m deeper as in Korbußen.

Since the platform strata northwest of the Crimmitschau fault are comprising three productive aquifers partially used for drinking water supply downstream the

Drosen mine field (distance 1 to 10 km) the effects of flooding on groundwater quality had to be considered. The investigations led to the conclusion that a significant impact of rising mine waters on groundwater quality can be excluded since the mine waters in the northern mine fields are much less contaminated than the mine waters of the southern fields. In addition preferred flow paths from the mine to the overlying aquifers could be sealed to a significant degree by concrete injections prior to flooding which minimize the post-flooding leakage rates. In the case of the Beerwalde and Korbußen mine fields it is also predicted that exfiltrating mine waters will have no or a just small impact on surface waters and local groundwater bodies so that the construction of a water treatment plant for the northern mine fields is not foreseen. Nevertheless, fall-back options are being drawn up should unacceptable situations occur, which would deviate substantially from the predicted scenarios. So a well has been established within the main shaft of the Beerwalde mine, and conceptual plans are at hand for water treatment should it become necessary.

Conclusions for the further strategy of controlled flooding

Since the mine is completely backfilled in its uppermost 100 meters and has no dewatering adits rising flood waters are expected to eventually discharge "naturally" into local receiving streams at topographic low points, unless pumping or collection measures are taken during the final stages of the flooding process. As these low points have altitudes of 240 m a.s.l. or higher the flooding process is permitted so far up to "security levels" of 232 m a.s.l. in the south and of 230 m a.s.l. in the north. Strategies for the final steps are in the permitting process.

The flooding operation is expected to complete between 2004 and 2005 by reaching a quasi-steady state. There is much uncertainty on how fast the final water rise will happen especially in the south: After reaching an inundation level of about 200 m a.s.l. flooding should accelerate since there are no open mine workings above this level. On the other hand inundation of the dewatered weathering zone of the bed rock should provide considerable pore volume, and water inflow is expected to decrease with rising water table; both phenomena would delay the water rise. Moreover the operation of the water treatment plant will affect the final flooding phase for the southern fields substantially.

As discussed by Gatzweiler et al. (2002) the flooding strategy at the Ronneburg mine site aims at a high inundation level in order to minimize the catchment area of the mine, to limit the thickness of the unsaturated zone which is subject to further acid mine drainage (AMD) generation and to lower operational costs for water management, including water treatment and sludge disposal. For this purpose the pump-and-treat-strategy has to be replaced by a collect-and-treat-approach. Water catchment systems have to be installed at locations, where contaminated groundwaters are expected to exfiltrate to the local receiving streams. Since the Gessental valley west of Ronneburg is expected to be the area of highest priority a catchment system comprising linear strings of combined drainage and collection pipes has been designed and submitted for approval (see Unland et al. 2002). It is planned to build this system in 2003. Other potential exfiltration areas are subject of further investigation and monitoring.

The optimum final flooding/groundwater level has to be determined by a stepwise approach. Water management by shallow catchment systems, but without intervention by pumping from the flooded mine, will allow density stratification to stabilize within the mine water column which will minimize the contaminant loads leaving the mine over the long term.

After four years of flooding it is evident that this operation will be a major contribution to restoring the Ronneburg region to good environmental conditions.

References

Gatzweiler R, Jakubick AT, Paul M, Schreyer J, Meyer J (2002): Flooding the WISMUT mines - Learning by doing or applying a comprehensive systematic approach?- UMH III, Freiberg, September 2002

Hähne R, Paul M, Schöpe M, Gengnagel M (2000): Abhängigkeit des Verlaufs der Grubenflutung des Ronneburger Uranerzbergbaureviers von der Beschaffenheit des Flutungsraumes und von den Neubildungsbedingungen.- Internationale Konferenz Bergbausanierung WISMUT 2000, Schlema, 11.-14. Juli 2000

Lange G, Freyhoff G (1991): Geologie und Bergbau in der Uranlagerstätte Ronneburg/ Thüringen.- Erzmetall 44: 264-269

Paul M (2002): Geochemische In-situ-Stabilisierung von Bergbaualtlasten.- In: Förstner U, Grathwohl P, Dahmke [Hrsg.]: Ingenieurgeochemie.- Springer, in Vorbereitung

Paul M, Sänger HJ, Snagowski S, Märten H, Eckart M (1998): Flutungsprognose am Standort Ronneburg - Ergebnisse eines integrierten Modellansatzes.- Uranium Mining and Hydrogeology II, Proceedings of the International Conference and Workshop, Freiberg, Germany, September 1998.- GeoCongress 5, Verlag S. v. Loga, 130-139

Unland W, Eckart M, Paul M, Kuhn W, Ostermann R (2002): Groundwater rebound compatible with the aquatic environment – technical solutions at WISMUT's Ronneburg mine.- UMH III, Freiberg, September 2002

Flood Water Treatment by Improved HDS Technology

Grit Heinze[1], Horst Märten[1], Jochen Schreyer[2], Dieter Seeliger[1], Klaus Sommer[1], Dietmar Vogel[2]

[1]UIT GmbH Dresden, Zum Windkanal 21, D-01109 Dresden, Germany
[2]Wismut GmbH, Jagdschänkenstraße 29, D-09117 Chemnitz, Germany

Abstract. HDS (high density sludge) technology is widely used for the treatment of acid mine drainage (AMD). Treatment efficiency and sludge de-watering properties can be considerably improved by applying staged neutralization to control super-saturation with reference to relevant solid phases, thus, increasing the precipitation rate of crystalline solids in significantly larger particles. Adsorption on Fe phases is essential to remove various metals and radionuclides efficiently. Two flood water treatment plants, both built on the base of improved HDS technology, are used at WISMUT in special modifications according to inflow characteristics.

Improved HDS technology

The HDS technology is a lime neutralization method with partial sludge recycling used for increasing alkalinity and for efficiently removing heavy metals from AMD. Since the 1950's it has been widely used and is generally accepted as the best available neutralization technology for the treatment of contaminated mine water which is typically characterized by its low pH value as well as its considerable sulfate and heavy metals (mainly Fe^{2+}) concentrations.

The HDS process is schematically represented in Fig. 1. The treatment includes neutralization, flocculation and sedimentation. The sludge from the thickener is partly recycled with typical rates between 80 and 95 % and afterwards conditioned by lime addition controlled by the final pH value set in the neutralization tank(s). Commonly, aeration in the neutralization tank(s) is applied to improve the oxidation of metals (in particular the oxidation of ferrous iron Fe^{2+} to ferric iron Fe^{3+}).

A novel lime treatment process based on staged neutralization has been proposed and tested in laboratory scale by Demopoulos et al. (1995) by using NaOH instead of lime. In the present paper, two new industrial-scale plants are presented.

The main idea of the improved HDS technology which is based crystallization principles is to limit the supersaturation of the neutralized suspension and, hence, to suppress the amorphous / colloidal precipitation, in particular of ferric iron in form of (oxy)hydroxide, and to favor the precipitation of crystalline solid phases like basic ferric sulfate $FeOHSO_4$ or more complex mixed phases like Schwertmannite $Fe_8O_8(OH)_6SO_4$ (or similar formulae).

The improved HDS technology with staged neutralization is represented in Fig. 1 (bottom). The neutralization circuit is designed to consist of a 3- to 5-step cascade with increasing pH value. This means that pH-controlled lime dosage has to be applied to all tanks (or reaction tank chambers) separately. The application of a high-rate thickener is preferred because of improved separation properties by feed injection into the pulp bed.

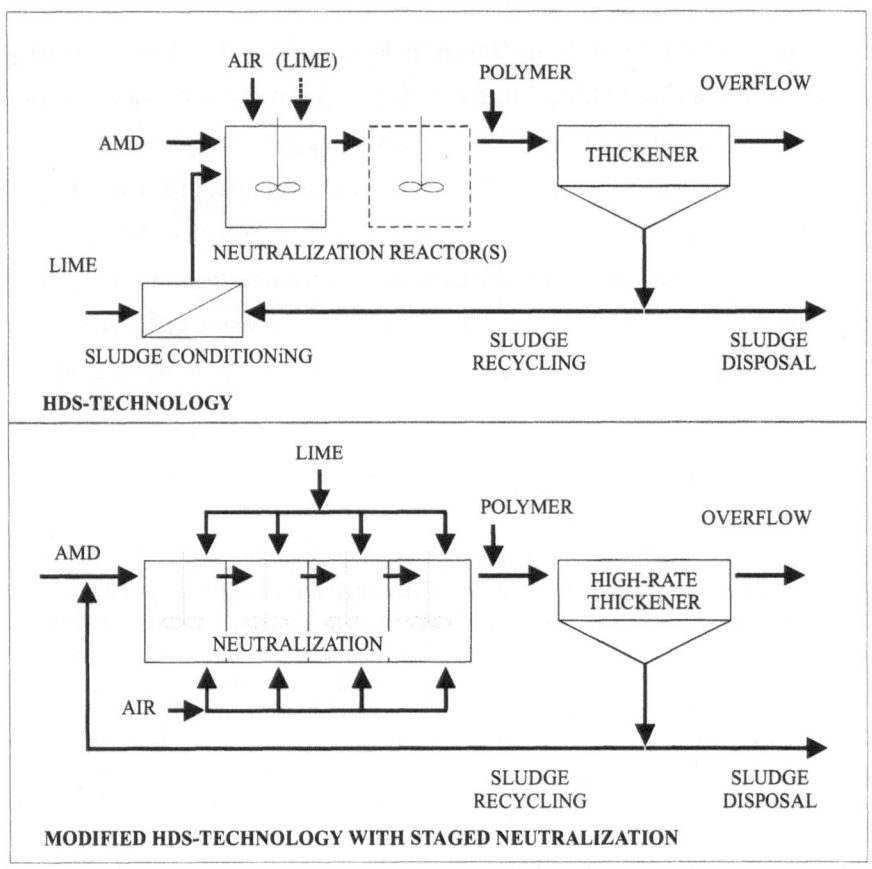

Fig. 1. Schemes of HDS and improved HDS technology

Case studies: Uranium mine flooding

The main advantages of the improved HDS technology include:

- Preferred precipitation of ferric iron in crystalline (opposed to completely amorphous) phases, thus, improving particle growth and increasing the sludge density considerably
- Decrease of lime consumption due to the fact that OH^--groups in the ferric iron phases are replaced by sulfate ions (limited hydrolysis)
- Increased settling rate of the sludge leading to a smaller necessary clarification area for solid/liquid-separation in the (high-rate) thickener (reduced construction costs due to smaller thickener dimensions)
- Increased sludge density in the underflow of the thickener
- Much better sludge de-watering properties
- Efficient removal of heavy metals, in particular radionuclides, from the AMD by improved precipitation in a complex hydroxide-gypsum-matrix (in particular adsorption on ferric iron phases)

The conceptual design and basic engineering of AMD (flood water) treatment plants for the Wismut sites Königstein and Ronneburg (to be represented in more detail in the following paragraphs) have been based on both pilot tests and chemical model calculations by the use of the code aquaC (Kalka 2001), which includes PhreeqC (Parkhurst and Appelo, 1999) as subroutine.

Application case Königstein

The Königstein uranium mine (Schreyer and Zimmermann 1998) is exceptional because of the combination of conventional mining (from 1971 to 1984) with in-situ leaching (up to 1990), in total 18,000 t. The geology consists of sedimentary strata (sandstone, chalk, and clay) with 3 ore horizons within the 4th aquifer consisting of Cretaceous sandstone overlaying granitic basement rocks. In-situ leaching by the use of sulfuric acid leachant led to a high contamination of the pore water. Remediation by controlled flooding of the hydraulically isolated mine includes the collection of the contaminated flood water via drainage, pumping it to the surface and treating it by uranium removal and improved HDS process.

The typical composition of the Königstein flood water is characterized by the following parameters: pH = 2 to 3, SO_4 = 1 to 6 g/l, Fe = 0.2 to 1 g/l, U = 10 to 60 mg/l (temporarily higher concentration values possible). The flood water treatment plant at Königstein has been designed for a capacity of 650 m^3/h (temporarily up to 1000 m^3/h). Uranium removal by ion exchange leads to a decrease of the uranium grade in the flood water down to about 1 - 3 % of the initial value (depending on initial value itself). The uranium removal and processing is described elsewhere (Märten et al. 2002).

A scheme of the flood water treatment plant at Königstein is represented in Fig. 2. The main treatment steps include:

- Uranium removal by ion exchange
- Staged neutralization by lime dosage in a concrete pond separated into 2 trains each with 4 chambers (neutralization steps), cf. Fig. 3

Case studies: Uranium mine flooding

- Intense aeration for oxidation and agitation of the suspension
- Dosage of $BaCl_2$ for Ra precipitation
- Flocculation and high-rate thickening for efficient sedimentation
- Sludge recycling and conditioning with lime (controlled by first pH)
- Continuous sludge disposal via de-watering in centrifuges
- Water polishing pond and final multi-layer filtration

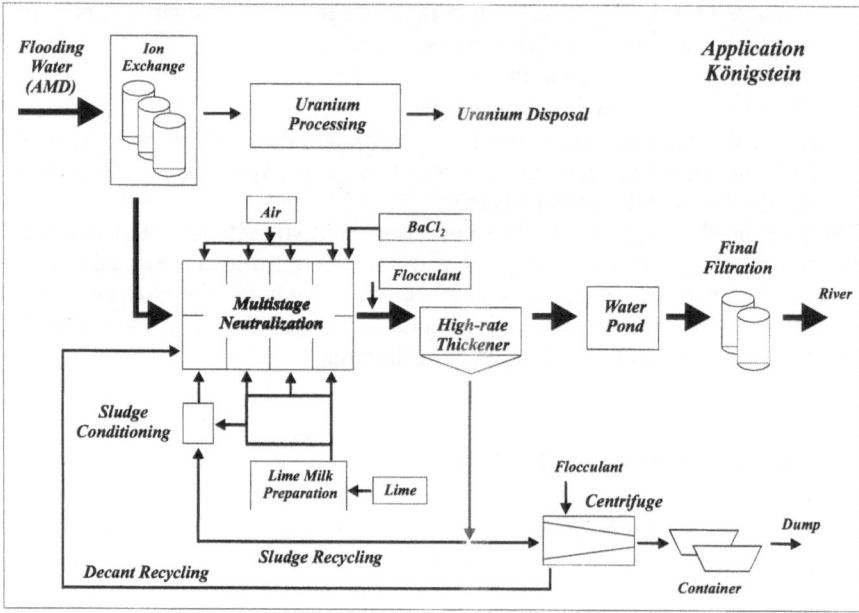

Fig. 2. Scheme of the HDS plant Königstein

As listed in Table 1, the main contaminants are removed from the AMD raw water efficiently.

Table 1. Water treatment results for selected elements (results of test period 10/00 to 02/01)

Element	Unit	AMD inflow	Outflow	Limit
Uranium	mg/l	35 → 1.5 (ion exchange)	< 0.05	0.5
Radium	mBq/l	1253	31	400
Iron	mg/l	363	< 0.24	3
Aluminum	mg/l	75	0.24	3
Nickel	µg/l	1770	< 41	500
Chromium	µg/l	239	< 8	500
Zinc	µg/l	31826	< 142	2000
Lead	µg/l	338	< 20	200
Arsenic	µg/l	462	< 50	100
Cadmium	µg/l	239	< 10	200
Copper	µg/l	503	< 12	500

Fig. 3. View of the 2-train / 4-stage neutralization circuit of the Königstein HDS plant

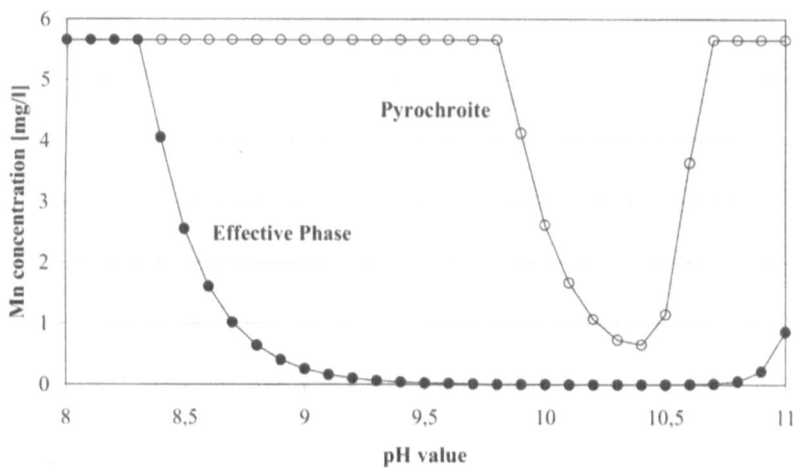

Fig. 4. The effect of adsorptive phase precipitation on Mn concentration in the treated water (aquaC / PhreeqC simulation)

Case studies: Uranium mine flooding

The radionuclide concentrations are reduced to 1.5 % (U isotopes), 3 - 6 % (Ra isotopes), 0.1 - 0.4 % (Th isotopes), and 0.2 - 0.3 % (Ac and Pb isotopes).

The effect of adsorptive precipitation with ferric iron phases is demonstrated for Mn on the basis of aquaC calculations. Fig. 4 shows the model result for precipitation in a pure Mn phase (pyrochroite) compared with an effectice phase approach (Kalka 2001). This effect has been clearly confirmed by plant operation.

The HDS plant in Königstein has already been used for a wide range of raw water compositions (Fe from 50 mg/l to 1000 mg/l) without limitation of treatment efficiency. In addition, contaminated (neutral) surface water is treated in the HDS plant by inlet into the 3rd chamber of the neutralization circuit.

The improved HDS technology gives a sludge with excellent de-watering properties. This has been demonstrated in an operation period with low sulfate (hence, vanishing gypsum in the sludge) and considerable Fe (300 mg/l) and Al (80 mg/l) grade in the flood water. In conventional lime treatment plants, this would lead to an amorphous / colloidal ferrous hydroxide sludge with very limited de-watering properties. Under the above operational conditions, the sludge underflow of the high-rate thickener had a solids content of about 20 % (mass). Even with centrifuges, a de-watering up to 45 - 48 % solids has been achieved in routine operation. The geophysical and chemical properties of de-watered sludge allow a direct disposal on the heap at the site without further conditioning.

Application case Ronneburg

The Wismut Ronneburg uranium mine area consists of 14 hydraulically coupled mines together with an open pit. The flood water to be treated either by pumping from underground levels or by transmitting the flood water outlet at the Gessental valley to the HDS plant has the following characteristic composition: pH = 4 to 6, SO_4 = 1 to 7 g/l, Fe = 10 to 200 mg/l, HCO_3 up to 700 mg/l, U = 0.5 to 2 mg/l. It is quite different from the Königstein flood water due to the higher neutralization potential in the underground. Consequently, the HDS technology for the Ronneburg plant had to be modified in the following items:

- *Optional* acidification by H_2SO_4 and de-carbonization to reduce the bicarbonate concentration in the raw water (to reduce uranyl-carbonate complexation for efficient U precipitation and for reducing sludge volume)
- *Optional* $FeCl_3$ dosage for efficient coagulation in the case of low iron grades in the AMD raw water
- *Optional* HCl dosage to the treated water before filtering
- Immobilization of the de-watered sludge by adding cement / lime as well as inhibitor to retard setting (required by disposal conditions)

A scheme of the modified Ronneburg HDS plant is shown in Fig. 5. The plant capacity is 450 m^3/h. The plant has been completed at the end of 2001. Test operation started in January 2002.

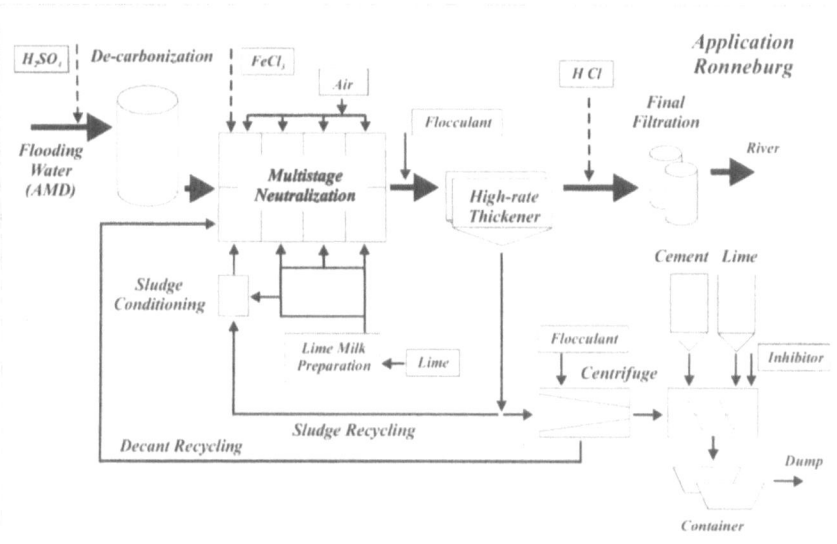

Fig. 5. Scheme of the HDS plant at Ronneburg

Fig. 6. View of the HDS plant at Ronneburg (2-train / 4-step neutralization and de-carbonization tower on the left-hand side)

Case studies: Uranium mine flooding

Summary

Two plants for the treatment of AMD flooding water based on the improved HDS technology have been implemented at Wismut Königstein and Ronneburg in two modifications with capacities of 650 (1000) and 450 m^3/h, respectively.

The essential process is the staged neutralization combined with the partial sludge recycling according to the classical HDS principle. In this way, the precipitation of ferric iron in crystalline phases is dominant over amorphous / colloidal ferric hydroxide phases. Accordingly, the main advantages of the improved HDS technology include:

(i)   Improved treatment efficiency
(ii)   Reduced lime consumption
(iii)   Increased settling rate and, hence, smaller separation area / lower construction costs
(iv)   Increased average particle size and, consequently, improved sludge de-watering properties

The operational parameters of the plant, in particular sludge recycling rate and pH cascade within the staged neutralization, must be adjusted to the chemical AMD inflow conditions. Besides empirical know-how, this adjustment is supported by chemical model simulation (Kalka 2001) performed for all operational steps and for the relevant suspensions. The treatment efficiency for heavy metals can only be reproduced by considering effective model approaches for adsorptive processes on ferric iron solids.

References

Demopoulos G P, Zinck J M, Kondos P D (1995) Proc. Int. Symp. On Waste Processing and Recycling in Mineral and Metallurgical Industries II, Vancouver, British Columbia, August 20-24, 1995, Ed. Rao S R et al.: 401-411

Kalka H (2001) aquaC Program Manual, UIT Report (cf. www.aquac.de)

Märten H, Schreyer J, Seeliger D, Sommer K (2002) these proceedings

Parkhurst D L, Appelo C A J (1999) User's Guide to PhreeqC (Version 2). Water-Resources Investigations Report 99-4259, Denver, Colorado

Schreyer J and Zimmermann U (1998) Proc. Int. Conf. On Uranium Mining and Hydrogeology II, Freiberg, Germany, September 1998, Ed. Merkel B and Helling C, Verlag Sven von Loga, Köln: 140-151

Optimizing the remediation of the subsurface environment in the Stráž deposit

Hana Čermáková

Technical University of Liberec, Czech Republic

Abstract. The remediation of subsurface environment in the Stráž deposit after the uranium mining is realized by consecutive pumping of contaminated groundwater. The pumped water is treated in a conventional plant. The simulation model presents information about the changes of contamination in the underground during remediation. One subject of the optimization model is to identify sites for pumping and injection wells. Pumped groundwater must meet the capacity of technological devices and technological demands. The injection has to prevent large hydraulic depression which would make further pumping impossible.

Introduction

There is a great residue of contaminated groundwater in the Cenomanian aquifer after chemical mining at the Stráž deposit located in the northern-Bohemia in the Northbohemian cretaceous table The volume of contaminated groundwater is about 180 mil. m^3 containing 4,8 tons of solvents. The afflicted area has a size of about 24 km^2. The main contamination components are SO_4^{2-}, Al^{3+}, NH_4^+, Fe^{3+}, Fe^{2+}, NO_3^-.

Remediation concept

The remediation is realized by the consecutive pumping out of the contaminated water. About 5,5 mil. tons TDS in total have to be pumped form the subsurface. Since 1996, an evaporation station with a capacity of 4 m^3/min of evaporated water is in operation. The concentrate is then supplied with the ammonia. The ammonium-aluminium sulphate (alum) crystallizes from the depleted concentrate. The evaporated water is drained to the watercourse thus implicating the necessary low balance, which prevents any further solution diffusion. The alum can be used for manufacturing of commercial products. The evaporation station capacity enables production of 200 thousands tons of alum per year. In 2001, a new technology of aluminum sulphate production was introduced processing 30 thousands tons of the alum a year. New technologies are planned and they will be established successively.

Meanwhile, the remaining solution is injected back to the deposit after the crystallization. New processes of thickening in a separate evaporator, crystallization of additional amount of aluminum and calcinations of the reprocessing of the remaining solution are planned after constructing of appropriate devices.

Groundwater concentrations will decrease with time. Since the evaporation of weak solutions is uneconomic, they will be thickened using the membrane technology before being evaporated. Thus the volume of withdrawn groundwater increases and it will be necessary to inject a part of the cleaned water back into the underground, to avoid a severe drawdown of teh water level in the deposit. Also, the aim of the injection is to remove the contamination from most worse permeable layers.

Concurrently with the evaporating, a part of the less contaminated solutions from the boundary area will be neutralized. Also in this case, the injecting of a part of the neutralized water is planned. A part of the neutralized water, the dilute from the membranes and the distillate can be drained to the watercourse in appropriate amount.

Remediation model

To evaluate remediation procedures, a 3D transport-reaction model was developed. This model first solves groundwater flow by the finite element method. The results of this first step are the fluxes of the solutions through the boundaries elements. This method enables to balance mass transfer in the time steps of the transport model and to simulate the chemical reactions. The finite elements are trilateral prismatic with vertical sides and a common slope of bases. This mesh consists of 1400 plane elements and 16000 spatial elements. An example mesh covering about 40 km^2 is shown in Fig. 1.

The chemical reactions in the solutions are solved by using the principle of thermodynamic equilibrium. The interactions between the solutions and the rocks (dissolving and coagulation of solid phase) are governed by the relations of chemical kinetics. The calculation of the chemical transformations is performed in each time step, individually in each element of the model mesh.

The simulation model of the underground processes presents information about the changes of contamination distribution in the underground during the remediation. The changes depend on the performance of the surface technologies and on a choice of the wells, included to the process in the particular periods.

Case studies: Uranium mine flooding

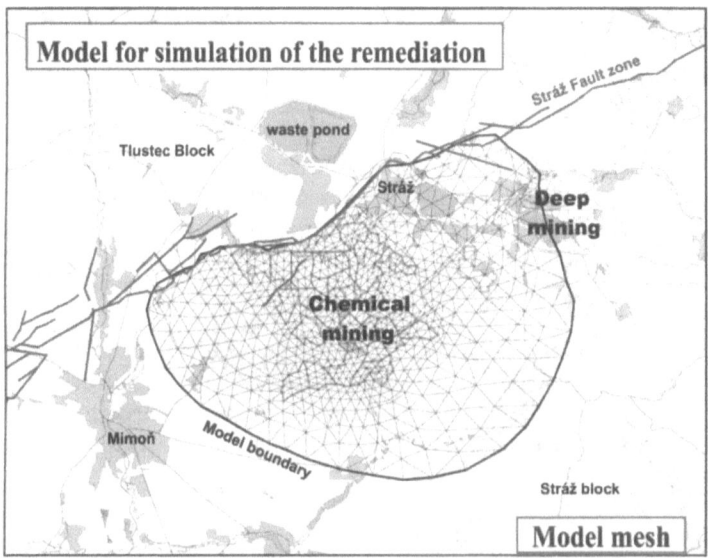

Fig. 1. The discretization element mesh

The optimization model

The complicated process of the remediation technologies, requiring parallel pumping and injecting of the several solutions streams of various properties, poses a great demand on the remediation operation management. For this purpose, a special optimization model was built, based on the existing simulation model.

The scheme of the optimization model can be described by the following: The remediated area is divided into sub-domains – element, whose centers represent places of potential wells. For each element, at the beginning of each modeled period (one year of remediation), an estimate of drawn concentrations of the monitored components for the next period is at disposal. The solutions drawn from the underground are processed by the surface technologies. A continuous and efficient process requires continuous supply of solutions. The cleaned solutions are partially injected back to the underground providing the necessary underbalance (volume of abstracted water is larger then volume of injected water) in the area and a partially drained to the watercourse (complying the rules for a pollutant content in the drained water).

The problem is solved by linear programming. As the variables, the linear model applies the elements and their properties (concentrations of the individual contaminants, properties of the element's vicinity, existence of a well in the element, etc.). The constrain conditions of the model are implied by the given technological concept and they describe technological and technical dispositions of the cleaned area (performance of the individual devices, parameters of suitable and efficient regimes of the surface technologies). Additional constrains of the model can be constructed by a transformation of suitable strategies of remediation control to a mathematical form.

The linear model for the optimization of the selection of element for remediation is a mathematical formulation of the described problem. The variables of the model consist of available pumped volumes from potential pumping and injecting sites. The set of technical and technological dispositions of the surface remediation technologies generates the constrain condition of the model. The constraining demands are:

- providing an efficient use of the surface technology capacities complying with the technological dispositions and the parameters for trouble-free and efficient operation of all process devices included in the remediation
- providing a necessary amount of elements in the pumped solutions for alum production
- providing a uniform and for practical needs suitable distribution of the remediation process intensity in the area
- complying the technical requirements of the pumping and injecting operations the area
- providing the area distribution of the needed lo
- complying the adjusted total volume underbalance in the area
- complying the limits of the contaminant draining to the watercourse

The optimization calculation results a list of sites for pumping and injecting of the individual solution streams. The list is an input to the underground processes simulation model which calculates the conditions for the beginning of the next period. Simultaneously, the program module of surface technologies monitors the amount of matters drawn from the underground and the amount of manufactured products. After the evaluation and comparison of variants, a remediation strategy is chosen, respecting the given technological concept. The strategy consists in spatial, time and dispositional organization of pumping and injecting in the particular remediation periods.

The procedure of optimization calculation

The optimization system is implemented as a user-friendly interactive software tool. User inserts input data into dialog boxes and simultaneously can see both the feedback and partial results on the screen. Considering the variability of the underground situation and progressive construction of the surface technologies, the input conditions and remediation requirements change. Thus the calculation is performed in temporal sub-steps until reaching the objective remediation parameter. The system allows to go back from any time step of the problem processing to previous positions, with full reproduction of the adjusted values of the input parameters.

Scenario of the technological concept of the remediation

The scenario defines the types of technological devices for processing the remediation solutions (evaporator, crystallization, thickening of stock liquors, disposal of the residual stock liquors, neutralization of weak solutions, thickening of weak solutions on membranes), their capacities, operation schedules and basic conditions and requirements of their operation. For the next periods, the scenario is largely determined by the schedule of construction of the technological nodes. In the later periods, we can alternatively choose a number of units for processing of diluted solutions, terms and conditions of their start-up. The scenario of the technological concept is the basic tool for remediation control (Fig. 2).

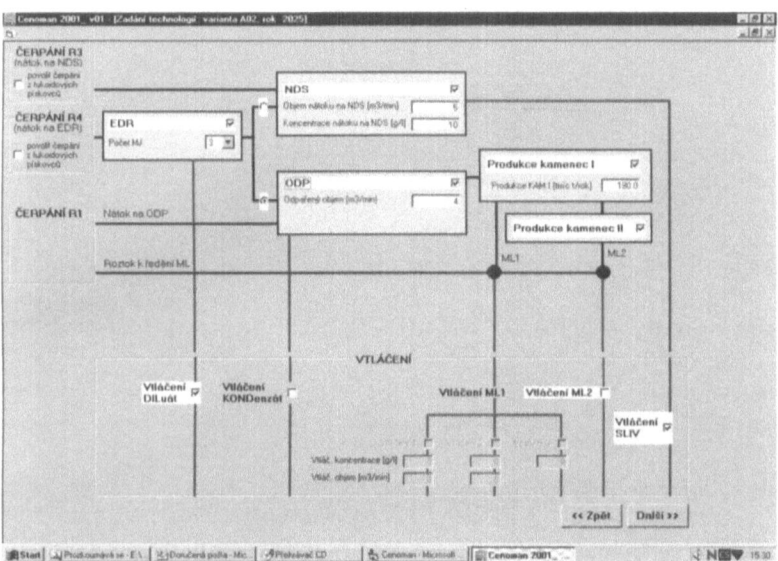

Fig. 2. The setup of a technological scenario of the remediation

Availability of the particular devices is the limiting factor for the amount of matters, which can be recovered in the considered period.

Choice of a model of the leaching fields

The control of pumping and injecting of the solutions during the remediation progress is very complex. It is necessary to prevent mixing of solutions with different contents. The system conditions vary relatively quickly in time. For each period, we can specify the following requirements:

- localization of drawing and injecting of individual solution streams on the surface of the deposit (according to the proper leaching field)
- vertical localization of pumping and injecting in the strata of the friable or the fucoid sandstone (the position of well filter)
- concentration range of a chosen contamination component in each stream of the pumped solutions
- required range of the concentration of a chosen contamination component in sites of injection
- distribution of the solutions' low balance (the difference between pumping and injecting) in the deposit with respect to provide the necessary piezometric level

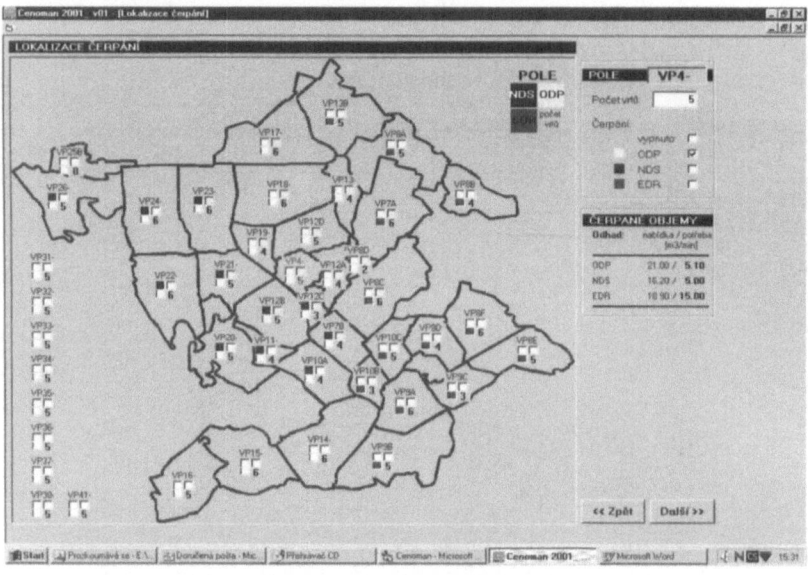

Fig. 3. The setting of the pumping localization demands

The search for a suitable regime of the leaching field belongs to the most difficult parts of the system. A choice of a large number of parameters is concerned and their influence is dependent on each other and can appear as late as after several years. An example of a screen for setting the requirements on pumping localization is given in Fig. 3.

Setting demands for the water quality

The device capacity is a limiting factor but it is not sufficient for its description. For example the aluminum production can be provided by solutions of various quality at different degrees of thickening in the evaporator. The regime variability (upon the maximal capacity exploitation) is necessary during the long-term fluctu-

Case studies: Uranium mine flooding

ating of the solutions composition in the underground. The following parameters can be changed in the individual periods:

- acceptable contamination range in the pumped solutions,
- NH_4:Al ratio in the pumped water,
- degree of solutions thickening in the evaporator,
- demand on the bore holes existing,
- demands on the contamination distribution in the surroundings.

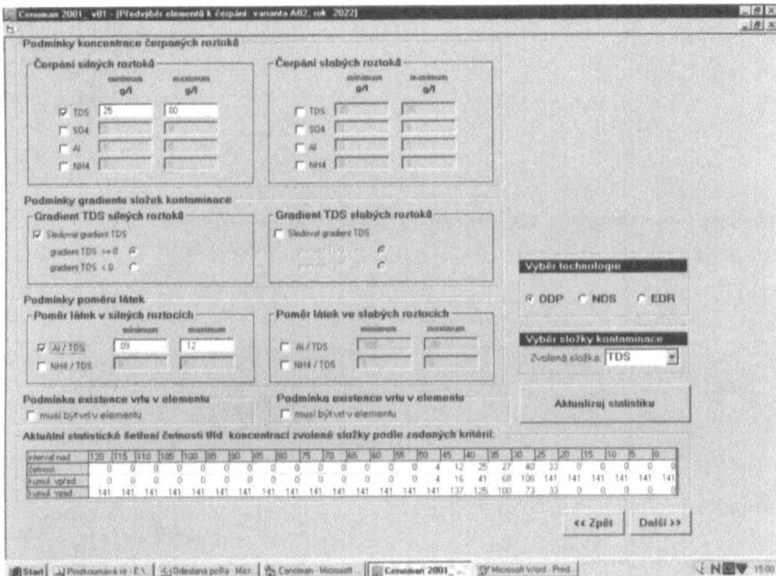

Fig. 4. The setting of the demands on the contamination's concentration in the individual process streams

These parameters can be set separately for each pumping stream. Data are not independent and it is necessary to set their acceptable combination during the processing. Demands on the individual component concentration range in the pumping streams are selected on the screen shown on Fig. 4. Reduction of the elements, which are suitable for pumping, are carried out by the demands specification on the pumped solutions. Therefore the screen with demand setting is completed by information about the selected elements as a statistic overview.

Optimal solution

The optimal solution of the model contains the overview of the model's limiting conditions and the list of the selected pumping point. Besides this, there is an overview of the particular streams of the remediation solutions in the lower part of the form, which includes both the summary on the processing technologies and partial and also final outputs of the surface processes. The data are completed by the overview of the impact of pumping. The valuation of the aluminum production, the amount of extracted contaminants and the necessary addition of ammonium and sulfate for providing the distinguished aluminum production are presented here. The optimal solution results are documented in Fig. 5.

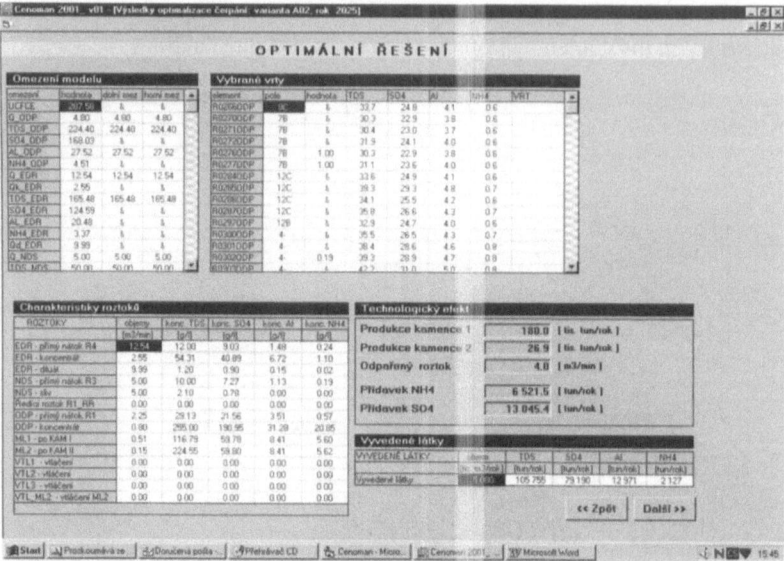

Fig. 5. The optimal solution of the pumping model

Optimization of injections

The injecting model is solved in consequence from the pumping results. The amounts of solution available for injecting and their parameters (concentration of contaminants) are known in this moment. The form and the content of the limiting injecting parameters are similar to the pumping model and they comprehend demands on the injecting sites in vertical and horizontal direction and requirement of injection site properties, which are formed as contamination data in potential injecting points. These data are added to the model with individual injecting streams resolution. The input data is completed with information about necessary underbalance in the remediation area and its distribution in the area, and further information required for monitoring of limits of contaminating substances drained into the watercourse.

Visualization of the solution

The general result of solved optimization task for modeled periods - selected pumping and injecting sites - can be displayed on a remediation map. The visualization system distinguish between the particular pumping and injecting streams with respect to technologies, on which they are reprocessed. Optionally the final solutions of the reprocessing technologies shows, which solutions will be injected into underground. The drawing and injecting streams are distinguished by color and numeric signs. The background of a map includes contour lines of contamination distribution in modeled periods. Thereby the progress of cleaning the underground can be monitored in consequence of a selected cleaning strategy. Reactions of the model on the selection of limited pumping and injecting conditions can be controlled and suitable methods and strategies for the next period can be simulated with respect to special conditions in the underground. The result of the optimizing model is shown on Fig. 6.

Fig. 6. The pumping and injecting well localization according to selected optimal solution

The comprehensive results of the remediation optimization

The results of remediation partial periods are summarized in overviews documenting the global course of remediation according to proceeding variants. Part of the results is a survey of economic parameters calculated by the model. The global economic effect of proceeded remediation variant is expressed by value of global remediation costs up to the moment of its termination. The time of the remediation termination is given by the period of reaching the final remediation parameter, which means, to clean the area to the acceptable residue level of contamination. The most favorable, effective and practical methods are chosen by comparison of global remedial parameters (total costs, time of termination, remediation course)

Case studies: Uranium mine flooding

for proceeded variants. The remediation course is documented by many graphic outputs illustrating evolution of selected remediation characteristics. An example of amounts of extracted substances by individual technologies is shown in Fig. 7.

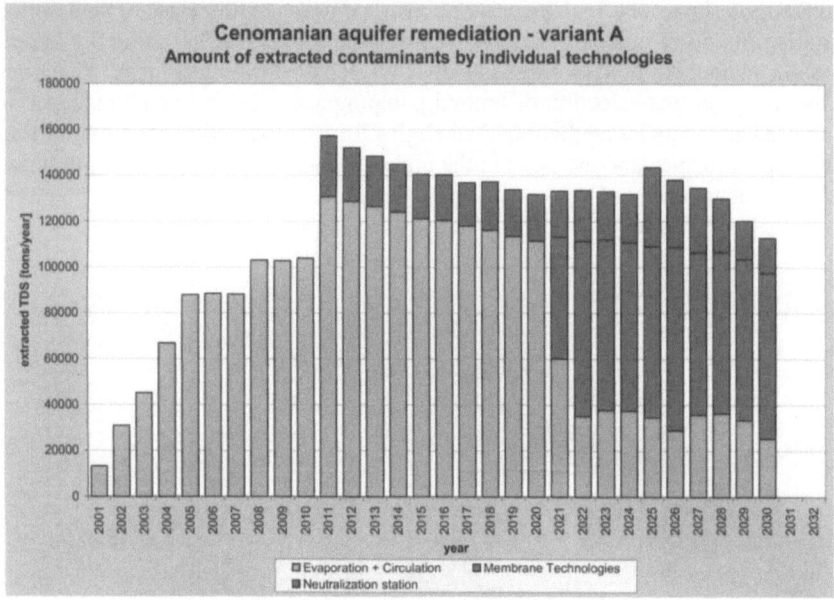

Fig. 7. The amounts of extracted substances according to remediation technologies

Conclusions

The optimizing system of remediation modeling makes it possible to simulate the remediation control by selecting pumping and injection wells for various scenarios of processing and remediation actions. Using the model, the estimated remediation time was optimized from 34 years to 30 years, thus saving 9 % of the global remediation costs.

References

Novák J (2000) Groundwater Remediation in the Straz Leaching , Im Tagungsband Internationale Konferenz Wismut 2000, Bergbausanierung. in Schlema, Germany 2000

Case studies: Uranium mine flooding

Uranium contamination of streams by tailings deposits – case studies in the Witwatersrand gold mining area (South Africa)

Frank Winde, Abraham Barend de Villiers

Potchefstroom University (South Africa)

Abstract. Many mines in the Witwatersrand basin are located at the outcrop of the auriferous ore, right on the watersheds of drainage basins. This resulted in the concentration of waste deposits in sensitive headwater regions of streams. With falling prices for uranium (U) the recovery of this commodity virtually stopped increasing the U-levels in active slimes dams. The following questions are addressed: Is U transported from slimes dams into the environment? To what extent are aquifers and adjacent streams affected? What are the main transport mechanisms and are there common patterns of contamination discernable?

Introduction

While U is normally only found as traces in the natural environment, mining of U bearing ore often transfers high concentrations of the radio-nuclide from the "safe" geological underground into the biosphere. In SA U was mainly mined as a by-product of gold, since both metals are commonly associated in the mined ore bodies. In terms of U the ore mined in SA is regarded as low-grade ore, ranging from 20ppm to 1000ppm, with an average of some 100ppm. After milling and leaching of the ore an average of approximately 10% of the original U concentration remains in the tailings (Which are termed "slimes" in SA) (Ford 1993).

During the first 66 years of gold mining on the Witwatersrand, which started in 1886, only gold were extracted from the ore. U and other heavy metals formed part of the waste products dumped initially as sand dumps and later on in slimes dams. In 1952 the first U-recovery plant started producing U. At the peak of production in 1980, 20 mines produced 6002 tons of U_3O_8 from 18 U plants. The production dropped to only 3 mines, producing 1640t of U_3O_8 at four plants, during 1994 (CfG 1998). Currently less than 1000t/a of yellow cake are produced in SA (Venter 2001).

Since U was extracted to about 90% from milled ore, the stop of U-extraction results in a tenfold increase in the U content in tailings produced since than. If the current production of U of less than 1000t/a is compared to the peak in 1980, when more than 6000t was produced, an estimated additional 5000t of U is dumped on slimes dams every year (constant concentration of U in the ore and stabile rate of gold production provided). In addition to this, all the slimes produced before the WW II contain fairly large amounts of non-recovered U. Taking the vast area, which is at present covered by slimes dams into account as well as the ecological sensitivity of affected watersheds in the semi-arid country, the problem of tailing related contamination is significant. It is estimated that slimes dams in the Witwatersrand goldfields cover an area of 400km^2 (CfG 1998). With about 6 billion tons of gold tailings produced since 1886 and an average U concentration of 100g/t (ppm) an amount of 600,000t of unrecovered U_{nat} results (Winde 2001).

Mining in the Witwatersrand basin originally exploited the easily accessible outcrops near the top of the hills containing the auriferous sediments, shafts and slimes dams were located right on the rims of the drainage basins. This had immediate effects on the headwater regions of many streams. These regions, which are usually the most pristine ones along a watercourse, are particular vulnerable to pollution since the water yield from such small catchments is often insufficient to dilute contaminants to harmless levels. Contamination of those regions therefore often has far reaching effects on downstream users. With respect to the latter the waterborne transport of solute contaminants is of particular concern, since dissolved metals display a higher mobility and bioavailability.

The emphasis of current environmental management of slimes dams is placed on the prevention of the erosion of slimes dams by wind- and water , and reduction of point sources of stream pollution such as down falls of contaminated water from underground mine-workings and processes water into nearby streams. The significance of tailings deposits (which in SA are termed "slimes dams") as sources of non-point pollution for streams and, however, is widely unknown.

Study sites and methods

The three study sites are all located in the Witwatersrand basin. While at the first site (Boetrand) only an unnamed non-perennial streamlet is affected, at the two remaining sites perennial streams, the Koekemoerspruit and the Natalspruit are impacted on.

To quantify impacts of slimes dams with different U-concentrations on the extent of contamination of surrounding environments, goldfields with distinct differences in the U-content of the mined ore have been chosen. For each of the sites field measurements of electrical conductivity (EC) and pH in relevant waters such as seepage, streams and groundwater were taken as an indicator for mining-related pollution. EC were measured were-ever possible for watercourses both above and below the slimes dams as well as for the seepage originating within the dam itself. In addition to this, the concentration of U in slimes, stream sediments, soils and salt-crust as well in said waters was determined by optical emission spectrometry (ICP-OES) (after leaching solid samples in *aqua regia*) and laser phosphorescence (liquids). All references to background values for heavy metals refer to global mean concentration in uncontaminated shales (Turekian and Wedepohl 1961).

Results

Alberton (Central Rand, near Johannesburg)

Background

This area has been largely mined out by 1994 and many decommissioned slimes dams that are several decades old occur. While some of them are partly vegetated the majority is severely eroded. Wymer (1999) estimates the average U-concentration of the slimes dams at 16ppm. The rather low concentration is due to a relatively low U-content in the mined ore. Some of the slimes dams are currently reworked to extract remaining gold. The catchment of the upper part of the Natal-spruit occupies approximately 11km^2. With 9 slimes dams covering an area of 0.6km^2 and 12 reclaimed sites (1.6km^2) more than 20% of the total area are covered by mining residues.

U-contamination and stream pollution

The location of slimes dams, stream and sampling sites including the U-concentration found are shown in Fig. 1.

The measured U-concentration of 3ppm in slimes material is lower than estimated by Wymer (1999) and close to the natural background. Significant higher U-levels, however, were found in sediments of the adjacent floodplain and in the stream channel, comprising large amounts of eroded slimes (Fig.1). Since slimes are rather low in U the elevated levels are likely to be caused by crusts of sulphates and iron-hydroxides, which cover the floodplain sediments extensively. Sulphate crusts occur as thin white layers on top of floodplain sediments. With the majority of the sample constituting of underlying material, crust-covered sediments contain the highest U-concentration at the study site (22ppm). In contrast to this, iron hydroxide forms thick brownish crust floating on groundwater puddles at the surface.

A pH of only 3.31 and elevated EC (6.3mS/cm; 2400ppm SO$_4$) in the alluvial groundwater point to contamination by acid mine drainage in adjacent slimes dams. Despite the excessive precipitation of iron as hydroxide-crusts the groundwater still contains large amounts of the metal (125ppm) together with elevated concentration of other heavy metals, such as Mn (214ppm), Ni (13.6ppm), Zn (9.5ppm) and U (0.48ppm). Elevated groundwater-levels in the floodplain allow for capillary ascent and subsequent evaporation of groundwater from the sediment surface during dry periods. That leaves sulphate-crusts behind in which U is re-concentrated by more than 40 times of its solute concentration and 7 times its concentration in the slimes. Those crusts are easily re-soluted as elevated sulphate levels in rainwater run-off suggest (240ppm). While sulphate crusts only form during dry conditions, iron-hydroxide precipitation takes place independently from weather conditions as soon groundwater is exposed to free oxygen. Since such crusts were found in sub-aquatic sediments of the stream channel, which therefore display a significantly lower U-concentration (9ppm) (Fig.1).

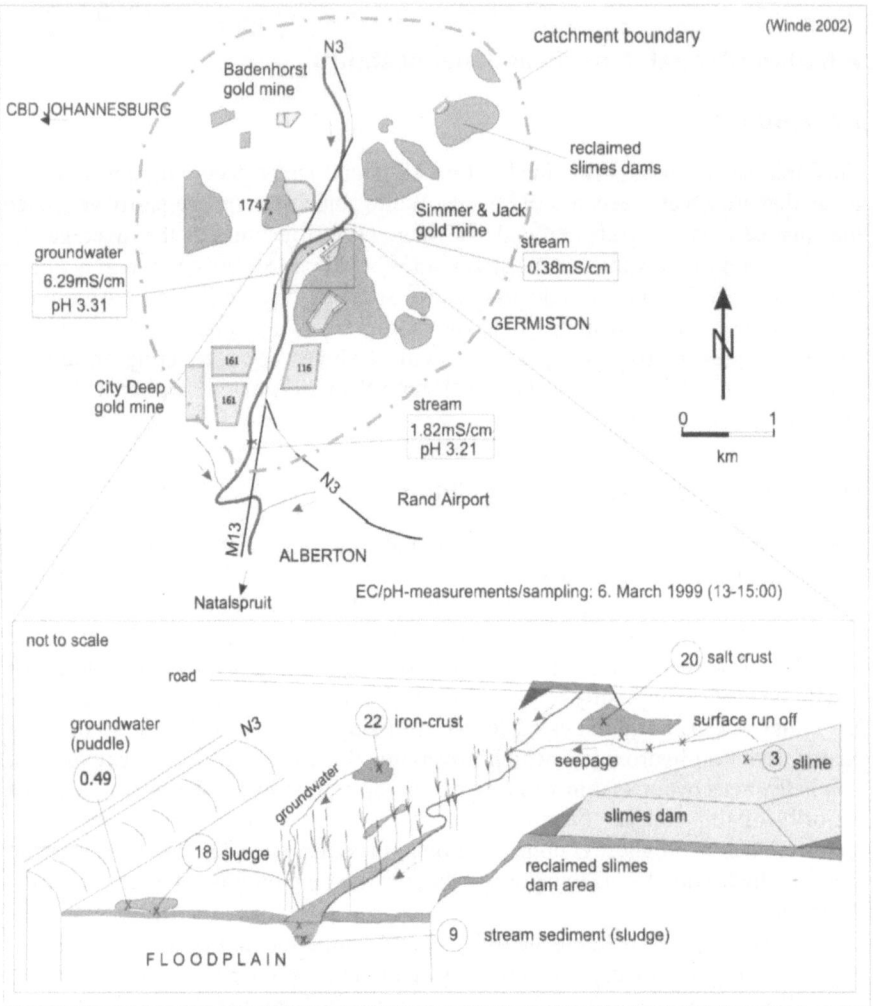

Fig. 1. Location and U-concentration [ppm] in water and sediments at the Alberton site

The drastic increase of the EC ($0.38 \rightarrow 1.82$mS/cm) and a simultaneous drop of the pH to 3.21 in the stream water after passing the slimes dams, indicate that the contaminated groundwater accounts for a significant portion of the stream flow. Using an EC-based volume-concentration balance it was calculated that some 24l/s of seepage (with known EC) enters the stream at an approximately 2km long reach that runs between several slimes dams on both sides. At an estimated flow rate of about 100l/s this accounts for about 24% of the total stream flow and reflects (incidentally?) the proportion of the catchment area that covered by mining residues. With an U-concentration in seepage of about 0.5ppm the stream receives nearly 400kg U per year. After diluting to a fourth of the seepage-concentration the U-content in the stream is estimated to about 0.12ppm.

Site-summary

The site is a typical example for effects of old, decommissioned slimes dams on diffuse pollution of adjacent streams. Although slimes material is washed into the stream the associated U-contamination is rather small because of low U-concentration in the tailings. Much more important is the solute transport of U from tailing deposits via the aqueous pathway. This leads to off-site re-concentrations of U and other heavy metals in sulphate and iron hydroxide-crusts, which eventually display higher concentrations than the slimes dams as actual source. The leaching of heavy metals from tailing material is enhanced by acid mine drainage which typically occurs in old slimes dams that were never neutralised. It results in acidic groundwater with high concentrations of dissolved U that seeps diffusely into adjacent streams. Due to the large proportion of the catchment covered by mining residues and accelerated subsurface-flow from topographically elevated slimes dams, seepage generates a significant part of stream flow. The fact, that the pH of the stream eventually is lower than in the contaminated groundwater near the slimes dams points to a low buffer capacity of the stream. This, in turn, allows for dissolved heavy metals to stay in solution and being transported downstream into where the polluted streamwater may be used for human consumption.

Boetrand (Klerksdorp goldfield, near Klerksdorp)

Background

Large-scale gold production started in this area at the beginning of the 1950's. Due to comparably high U-concentrations (average 500ppm) in the Vaal Reef, as most important ore body mined in the region, U production formed an important part of the mining activity. From 1953 to 1995 some 72,879t U was produced (1695t U/a). With Vaal Reefs and Hartebeestfontein Gold Mine two of the three mines still producing U in South Africa are located here. Current average U-levels in the ore range from 224ppm (Vaal Reefs) to 106ppm (Hartebeestfontein) (CfG, 1998).

Case studies: Uranium tailings

U-contamination and stream pollution

The two slimes dams investigated are located SW of Boetrand, which is an extention of the mining town Klerksdorp. They are on either side of a non-perennial
streamlet, which occasionally (after heavy rain fall in summer) may discharge into
the (perennial) Schoonspruit, some 6km away. Both slimes dams are decommissioned and unvegetated. The western slimes dam is fenced and equipped with so
called "toe dams" consisting of small walls made of slimes material. They contain
red to dark-red seepage and eroded slimes material. With a pH<3 and an elevated
EC (up to 4.5mS/cm) the reddish seepage indicates acid mine drainage. Accordingly, high concentrations of dissolved metals in the seepage were found, in particular for iron (1162ppm) and U, which with 30,1ppm displayed the highest solute concentration (except of process water) encountered during the study. It equals
almost 30% of the U-concentration in the tailings and illustrates how mobile U
under acid mine drainage conditions can be. This is of particular concern since the
U is concentrated far above background levels and therefore able to contaminate
the environment. Although all other heavy metals in the slimes show no or only
moderate enrichment against background concentrations, some of them are highly
concentrated in the acidic seepage, e.g. As (36.3ppm), Zn (31,0ppm), Ni
(21,6ppm) and Cu (7,3ppm).

No toe dams were found at the eastern slimes dam, which is also severely
eroded. The yellowish slimes material has been distributed over several hundred
meters from the site, preferably along a depression line. The concentration of U on
the surface diminishes rapidly from the foot of the slimes dam in the direction of
transport decreasing from 47ppm to 22ppm within the first 10 meters. This is
mainly due to dilution of the slimes by uncontaminated soil that mixes in during
the transport (Fig.2).

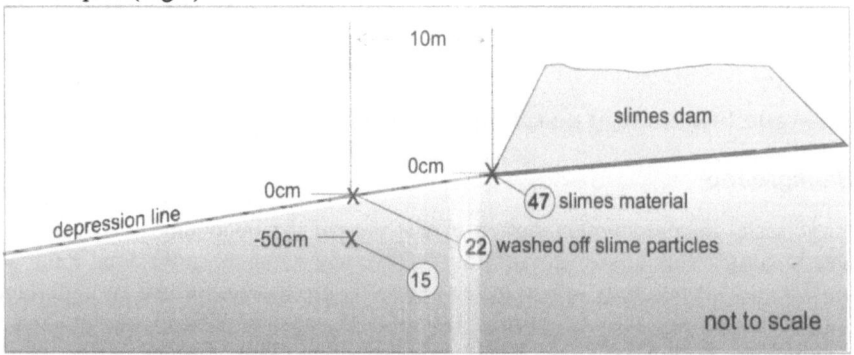

Fig. 2. U-conc. [ppm] in sediments next to the Eastern slims dam at the Boetrand site

The elevated U-concentration in soil well below the bottom of the depression line
where no slimes particles indicate that U is also transported in dissolved form.
Most likely U is dissolved in run-off that either flows along the depression line
and infiltrates into the underlying soil or migrates as subsurface-flow. Contaminated groundwater as medium of transport, however, is unlikely, since no signs for
a reduction zone within the soil profile were observed.

Site-summary

Although no direct stream contamination takes place, slimes dams in the area contain relatively high U-concentrations and act as sources of contamination. This is facilitated by the transport of U bearing slimes particles along with run-off after rain events. Due to mixing with uncontaminated material U-concentration rapidly decreases along the way of transport. However, rain events also cause waterborne contamination by dissolved U, which under acid conditions can be leached to almost 30% of its concentration in slimes particles. Due to high U-levels in slimes dams of this area the high leaching-rate is of particular concern.

Grootspruit (Evander goldfield, near Evander)

Background

The Evander goldfield is the smallest of the Witwatersrand basin and mined since the late 1950's. Due to rather low U-concentration in the mined Kimberley reef U never was produced in this area. The investigated slimes dam is still active and located some 500m distant to the Grootspruit. After flowing through a wetland the small perennial stream partly feeds into a dam that also receives mine-water from a settling pond in which excess water from the dam is collected (Fig. 3).

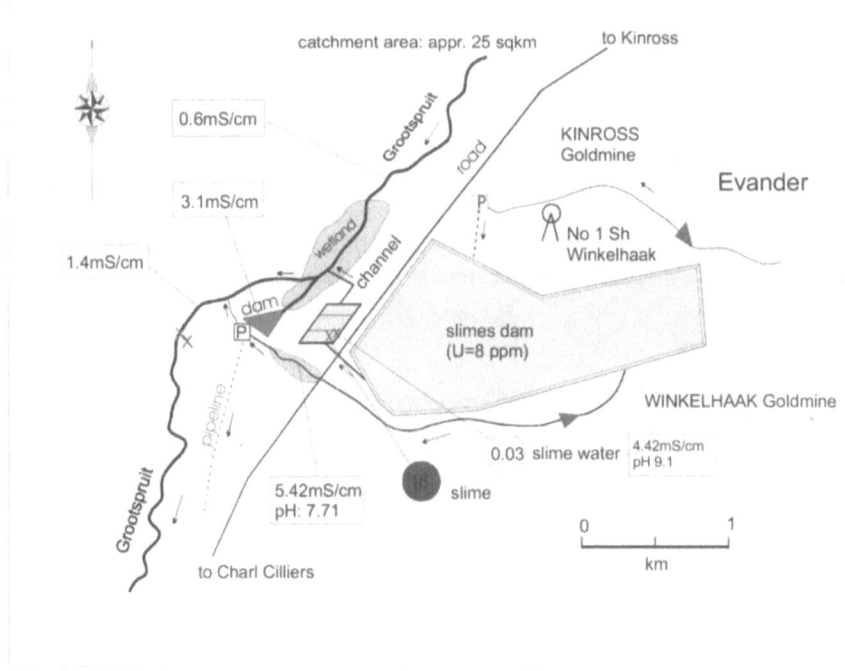

Fig. 3. Location and U-concentration in water and sediments at the Evander site

U-contamination and stream pollution

With 16ppm the U-concentration in the slimes is comparably low, although double as high as estimated by Wymer (1999) and still significantly above the natural background, in fact, more than all other heavy metals. The water in which the slimes are transported to the settling ponds is alkaline (pH>9) with a high EC (4.42mS/cm). The latter is mainly due to high concentrations of NO_3 (1242ppm) and Cl (910ppm) while SO_4, which usually dominates, is of much lower importance (410ppm). High Ca-concentrations (258ppm) point to neutralisation as reason for the alkalinity. The concentration of dissolved U in the water is comparably low (0.03ppm).

Increasing EC in the stream after it passes the wetland (0.6→1.4mS/cm) points to infiltrating mine-water. Since wetlands often occur in vicinity of slimes dams, it is likely that they are caused by seepage-related groundwater-elevation allowing for reeds to grow throughout the year. The rising EC in the stream indicating seepage influence while passing the wetland suggest that this is also the case here.

Compared to the dam, which receives water directly from the settling pond that leads to an increase of the EC to 3,1mS/cm, the non-point stream pollution by contaminated groundwater in the wetland is comparably moderate, resulting in a U-concentration of 0.015ppm. The high salt-load of the stream leads to the precipitation of white crusts in capillary fringe of the bank sediments. Due to the high Cl and Na content these crust are likely to consist of NaCl rather than sulphate, which is known for re-concentrations of U. Although the U-concentration of 4ppm in the bank sediment is rather low, it still represents a >100 times enrichment from the water column (assumed regional background of U of about 2ppm). While all other analysed heavy metals in slimes and sediments display concentrations below background values, the maximum concentrations for manganese (922ppm) and U (16ppm) are well above (2 resp. 4times). High Mn-concentrations in the bank sediments point to precipitation of Mn-oxide that preferably occurs in alkaline water (pH>8.2), as the Grootspruit is, after passing the wetland. By co-precipitation it may contribute to the immobilisation of U in the bank sediments.

Site-summary

In terms U-pollution the site is an example for a low risk area since concentrations in tailings as potential source of contamination are comparably low. In addition to this, the slimes dam is in active use and proper maintenance reduces stream pollution by erosion or acid mine drainage. However, that does not prevent seepage from diffusely contaminating the stream with dissolved metals and salts. Due to low source concentrations and an alkaline environment that reduces metal mobility, no significant off-site accumulation of U or other heavy metals were found.

Conclusions

In two of the three cases it was evident that nearby streams were contaminated by slimes dams. At the third site (Boetrand) stream contamination could not be established due to the absence of a perennial stream, rather than the absence of contamination. In fact, the site is to be termed as worst-case scenario in terms of metal mobility from tailings deposits. As an extreme example for acid mine drainage it illustrates to what large extend U can be re-soluted from tailings and transported into the environment. Together with the Alberton-site it suggests that old, decommissioned slimes dams are particular prone to acid mine drainage, more than actively used and maintained ones. This is mainly due to weathering fronts and pyrite-oxidation that move deeply into old slimes deposits increasingly acidifying the porewater.

Despite the negative annual water balance it was found that also in old slimes dams, on which water-saturated slime (slurry) is no longer applied, a phreatic surface develops. Due to the elevated heights of slimes dams above their surrounding ground steep hydraulic gradients between the phreatic surface in the slimes dam and adjacent streams or underlying aquifers result, which drive seepage-flow into the environment. The spontaneous formation of wetlands between slimes dams and adjacent streams indicates the permanent outflow of seepage, thereby increasing groundwater levels and allowing for the formation of wetlands. Since the latter are known to be sinks for dissolved metals it simultaneously reduces associated stream pollution (Payne 1998).

Although the concentrations of dissolved heavy metals in migrating waters are rather low significant re-concentrations were found in sediments well away from the from the sources of contamination, frequently even exceeding the concentration in the source. This particular was observed for sulphate crusts, which contained up to 1200ppm U compared to 130ppm in the contaminating slimes dams (Winde 2001). As shown at the Alberton site also precipitated iron-hydroxide tends to re-concentrate U and other heavy metals. While the retention of U in sulphate crust is only temporal, since they are easily re-soluted by rainwater, iron-hydroxide crusts are long-term sinks. However, they need free oxygen and certain pH-values to form, which restricts their efficiency for U immobilisation.

Apart from solute transport also particle-bound transport of U into stream was observed. This is mainly due to eroded particles from tailings deposits that are washed into adjacent streams or riparian zones. Due to mixing with uncontaminated soil and sediments along the way the U concentration in the environment steadily declines with increasing transport distance from the source. Although the pollution of streams with slimes particles adversely affects aquatic ecosystems, it is of lesser concern than the solute transport of U, which leads to longer transport-distances and higher degrees of off-site contamination.

Acknowledgement

This study is part of research project funded by the German Academy of Natural Scientists Leopoldina (BMBF-LPD 9801-17).

References

CfG, Council for Geoscience, (1998): The Mineral Resources of South Africa. Handbook 16. CTP Bookprinters, Cape Town, 740pp.

Ford MA (1993): Uranium in South Africa. Journal SA Inst. Mining and Metallurgy, 93.2, 37-58

Payne TE (1998): Uranium sorption on tropical wetland sediments. Abstract in http://www.geo.tu-freiberg.de/umh/abstracts.doc.html, accessed: 24.08.1999

Turekian T, Wedepohl H (1961): Distribution of the elements in some major units of the earth's crust. Geological Society of America Bulletin 72, 175-192.

Venter I (2001): South African mining sector facing radiation challenge. Mining Weekly, vol. 7 no 43 (November 9-15 2001), 2-3.

Winde F (2001): Fluvial processes and uranium transport – pitfalls from the Wismut region (Germany) and the Klerksdorp gold field (South Africa). 4th International Conference of the Society of South African Geographers., 2-5 July 2001 in Goudino Spa (South Africa), Conference proceedings (CD), 5pp

Wymer D (1999): Compilation of volumes and uranium concentration of milled ore and tailings of South African goldmines. Chamber of Mines, Johannesburg, South Africa (unpublished).

An IAEA Co-Ordinated Research Project (CRP) on Technologies and Methods for Long-Term Stabilization and Isolation of Uranium Mill Tailings

W. Eberhard Falck

Waste Technology Section, International Atomic Energy Agency, Wagramer Str. 5, P.O. Box 100, A-1400 Vienna

Abstract. This paper provides an overview over the work undertaken within an IAEA Co-ordinated Research Project (CRP) that is intend to address some issues related to the long-term stability of uranium mill tailing disposal facilities. The long-term stability is of concern since all engineered infrastructure will only have a limited lifespan after which the confinement of the long-lived radionuclides in the tailings cannot be guaranteed anymore. The project began in 2000 and will run to 2004 and addresses both technical and managerial measures.

Introduction

IAEA Co-ordinated Research Projects (CRPs) are intended to bring together re-searchers from different Member States with the view to share and disseminate the experience in solving problems of common interest.

Mining and milling of uranium ores has been undertaken at many places around the world, resulting in large volumes of mining/milling residues with low activity concentrations of long-lived nuclides. The common mode of disposal of mill tail-ings was in near-surface impoundments in the vicinity of the respective mill, often in a haphazard fashion, utilizing geomorphological depressions or by filling-in valleys. As a result, there often was (is) little or no care taken to isolate the tailing materials from their environment. Cost and convenience were the main drivers in selecting disposal options.

Over the last couple of decades or so regulations and standards were put in place requiring practices to minimise environmental impact, although radiation dose to the public has continued to be used as the yardstick by which the level of protection of the environment has been measured. Only in recent years have the deleterious impacts from non-radiological components of uranium mill tailings become a significant issue, and it is now well documented that impacts from such sources are more limiting on operational practice than are "conventional" concerns over ionising radiation.

While geomechanical aspects, such as the stability of pile slopes, dikes and retaining dams, are standard engineering problems, for which in most countries provisions are made in the relevant building or mining regulations, environmental and radiological impacts are often neglected. Typical environmental problems arising from mill tailings are radon emanation, the leaching of contaminants into surface and groundwaters and dust from drying out beaches in tailing impoundments.

A range of geotechnical engineering measures can be employed to prevent or reduce the extent of these processes. However, they require maintenance activities, and hence active institutional control, over a prolonged period of time, extending over decades if not longer. Whilst the engineering standards may appear adequate under normal operations, they may be inadequate under extreme events such as flooding or earthquake. In addition, risk estimation for uranium tailings impoundments should consider the very long time frame over which safe containment is required due to the long half-lifes of the relevant radionuclides.

The aim of the CRP, therefore, is to develop conceptual and technical solutions that

- render tailings more inert over prolonged time-spans,
- render impounded materials and engineered structures stable over prolonged time spans,
- minimize the need for active maintenance,
- and that are technically and economically feasible.

The emphasis is on solutions that can be applied in retrospect, i.e. in a restoration/remediation context. It is recognized, however, that these objectives cannot exclusively be met by engineering design, but must involve an integrated approach, including appropriate management and planning procedures. Solutions that aim at *incorporation* of the tailings within the natural environment in such a way that release of contaminants is prevented or below levels of concern, rather than *isolation* from it seem to be more promising.

This IAEA Co-Ordinated Research was initiated in 1999, bringing together scientists and industry people from 13 countries. The countries participating are Brazil, China, Czech Republic, France, Germany, Kazakhstan, Korea, Poland, Russian Federation, Slovenia, Ukraine and USA.

The work has been grouped into four topical areas:

- Tailings remediation case studies.
- Capping of tailings
- In situ conditioning of materials
- Management of tailings in remediation situations

Case studies: Uranium tailings

Tailings remediation case studies

There are, of course, numerous examples around the world of successful, and not so successful, attempts to remediate tailing impoundments. Three examples were chosen, representing different operational states and different socio-economical circumstances:

- A case study on the uranium tailings dam of the Poços de Caldas uranium mining and milling site (H.M. Fernandes, Instituto de Radioprotecaô e Dosimetrica, Rio de Janeiro, Brazil)
- Cameco Research and Development Projects for Tailings Disposal Technology (P. Landine, CAMECO Co., Saskatoon, Canada)
- Remediation and Long-Term Stabilization of Waste Disposal Facilities at the Closed Uranium Mine and Mill at Zirovski Vrh, Slovenia (B. Kontic, University of Ljubljana, Slovenia).

In the Poços de Caldas case study a conceptual model for the mass flow of both water and contaminants was developed. This model will be used to assess the long-term environmental impacts and to design appropriate countermeasures. Tailings disposal in Canada faces different problems in the face of extreme climatic conditions. The freezing of tailings has prevented their consolidation in some places, e.g. in the in-pit disposal facility at Rabbit Lake. The injection of warm tailings from the production cycle is intended to remedy the situation. The expelled porewaters are of environmental concern and are investigated in detail. The tailings impoundment from the Zirovski Vrh mill in Slovenia suffers from serious geotechnical instabilities, being situated above ground on a gentle slope. Below-grade disposal offers, in principle, the greatest protection agains catastrophic failure and, hence, backfilling into the mine is being considered. Using multi-attribute analyses the project evalutates the various environmental implication arising from such action, such as groundwater contamination.

Capping of tailings

Cappings aim to reduce erosion rates, both from water and wind, and, hence, airborne dispersion of radioactive materials. The other main objective is to delay air exchange with the ambient atmosphere in order to reduce radon exhalation. Low-permeability cappings also reduce infiltration of atmospheric precipitation and thus the leaching of radionuclides into underlying groundwaters. This set of projects is concerned with the improvement of geomechanical and geochemical properties of capping materials. Four projects were grouped into this topical area:

- Studies of bentonite and red soils as capping of the uranium mill tailing impoundments (Z. Wen, Beijing Res. Inst. of Uranium Geology, China)

Case studies: Uranium tailings

- Development of Method of Covering Raising Dust Beaches of Radioactive Wastes Storage Out of Operation (A. Gagarin, Ulba Metallurgical Plant OSC, Ust-Kamenogorsk, Kazachstan)
- Polymeric Coats for Contaminated Surfaces Localization (S. Mikheykin, A.A. Bochvar Institute, Moscow, Russian Federation)
- Improvement of Soil Properties Applied to Capping and Multi-Layer Barriers (J. Koszela, Wroclaw University of Technology, Poland)

Various conditioning and capping materials are being studied, such as natural clays and man-made organic polymers. The latter in particular will increase the cohesion between particles and thus increase the erosion resistance. Reducing the erosion rates by either increasing the cohesive strengths of materials already in place, or by adding layers of higher cohesive strength also enhances the revegetation probability. A stable vegetation cover in turn will reduce infiltration rates and thus contaminants leaching rates. The materials are being tested under the countries' specific climatic conditions, e.g. under semi-arid or subtropical climates, for their applicability and performance. First results are very promising. The long-term stability of synthetic organic polymers, however, is completely unknown and for this reason, their application can only be viewed in support of other technical measures.

The use of cohesive materials, such as clays has been well studied in a number of countries, starting with the UMTRA projects in the USA. However, the design parameters have to be modified for each site in order to take account of the local availability of materials. The Chinese project investigates the performance of local red-clays and additives required.

Conditioning of impounded materials

The main problem arising from liquid tailings is the expelled water and the possible radionuclide contamination associated with it. Leaching of contaminants from the solid phase is also a concern. For this reason various technologies to thicken tailings ('paste technologies') before emplacement are being explored world-wide. Also, research is being undertaken to make tailings materials more leach-resistant by reducing their permeability or by encapsulation. Four project in the CRP are directly concerned with the conditioning of tailings materials:

- Remediation of Uranium Mill Tailings Using Natural and Organo-Clays (S. Choi, Kyung Pook National Unversity, Taegu, Korea)
- Research and Development of Measures to be Taken for Long Term Stabilization of Uranium Liquid Wastes (G. Maslyakov, Ukrainian Research and Design Institute for Industrial Technology, Zhovty Vody, Ukraine)
- Development of Technologies for In-Situ Remediation of Contaminated Sites by Directed Formation of Naturally Occurring Slightly Soluble Minerals (G. Ziegenbalg, TU Bergakademie Freiberg, Germany)

Case studies: Uranium tailings

- Low Temperature Ceramics, the Breakthrough Material for Long Term Stabilisation and Isolation of Low-Waste Uranium (A. Piestrynzki, University of Mining and Metallurgy, Krakow, Poland)

In many instances financial and operative resources are not available to relocate tailings into new and better-engineered impoundments. Under such circumstances an *in situ* improvement of the materials' performance may be considered. The methods and materials under study are intended to reduce the source term for radionuclides contained in the tailings materials. This is being achieved by increasing the sorption capacity, by changing the chemical binding forms and by reducing the permeability. For instance, additives are being used to precipitate sulfates of low-solubility that will bind radionuclides such as radium. This technique has already been tested successfully in underground mine workings and is now being extended to the conditioning of tailings. A major problem currently addressed is the controllability of the chemical reaction under field conditions. The addition of binders or precipitating agents will entail significant costs, and consequently research also focuses on minimising the amounts needed, while still achieving the desired geomechanical and geochemical performance.

Management of tailings in remediation situations

It is clear that in cost terms, prevention of contamination is much more effective than cleaning it up afterwards. In relation to the general performance of uranium mill tailings impoundments, the main current concerns relate to longevity of containment, and seepage to groundwater. It is recognized that the objective of long-term stability cannot exclusively be achieved by engineering design, but must involve also adequate planning and management procedures, and stewardship measures, involving a thorough understanding of the surrounding environment:

- Predicting the Long-Term Stabilisation of Uranium Mill Tailings (J. Trojacek, DIAMO s.p., Stráž pod Ralskem, Czech Republic).
- Harmonization of Radiological Impact Assessment Methodologies of Uranium Mill Tailings Repositories (A.-C. Servant, Institute de Radioprotection et de Sûreté Nucléaire, Fontenay-aux-Roses, France).
- Holistic Approach to Remediating Uranium Mill Tailings and Contaminated Groundwater (D. Metzler, U.S. DOE, Grand Junction Office, USA).

The three projects in this group address the performance and impact assessment criteria, by which a tailings disposal site would be evaluated with respect to its long-term stability. Predicting the consolidation behaviour of tailings impoundments in response to engineering measures is an essential element of the long-term management strategy. Even though measures to contain the radioactivity within the impoundment are taken, over long periods of time releases cannot be completely ruled out and the likely effect has to be assessed. Indeed, as has been pointed out above, a design that aims at incorporating the tailings into the natural environment without releases of concern, rather than to isolate them from it, is more likely to exhibit the desired performance over very long time-scales, based on the current knowledge.

The French project focuses on intercomparing the various methods by which the radiological impact and impoundment performance can be assessed. It is hereby recognised that tailings not only impact the environment from a radiological perspective, but they may also adversely influence the quality of adjacent environmental compartments, such as the groundwater.

While a given solution may be technically acceptable or even desirable, its implementability will depend on a variety of non-technical factors also. Public acceptance can be one of those factors. Experience from the USA and world-wide tells that a holistic approach integrating public involvement, planning the conceptual and final disposal design, alternatives analysis, providing for institutional controls, long-term monitoring and surveillance, and for stewardship is essential to achieve the long-term goals and objectives of preventing adverse environmental impacts from tailings impoundments.

Summary and Outlook

The outstanding issues relating to stabilisation and isolation of uranium mill tailings concern the confidence for long-term secure containment, prevention of seepage from the containments, chemical mobility of contaminants in the short and long term, and air-born releases in the form of dusts and radon gas. New concepts are emerging that seeks to mimic natural processes as much as possible in order to improve physical and chemical stability by understanding and taking advantage of natural assimilation and attenuation. This could lead to a fundamental change in the way that impoundments are designed, away from structures engineered to keep their contents out of contact with the environment (i.e. isolation), to structures that somehow utilise natural processes to assimilate the entire structure and its contents with the surrounding environment over the long term. Improvement of the geochemical and/or geomechanical properties may be required, at least for a limited period of time.

Hence, the main foci of current research are methods to stabilise tailings. Whilst the above discussion has divided the work into physical and chemical stabilisation of tailings piles, and the stabilisation of covers, much of the work being done cuts across all those areas. This is because the majority of the work is on methods to "glue" the tailings together with materials that are able to bind contaminants (with emphasis on the metals) as a result of redox reactions or modifications to pH. Therefore both chemical and physical stability result from the one process. The agents vary from tried and tested Portland cements, to inorganic and organic reagents, and high-tech polymers.

By the time the UMH III conference takes place, the second progress meeting will have been held in Rio de Janeiro (6 to 10 May 2002), where the efforts of the first two project years will have been summarised. The experience gathered will be published in an IAEA technical document. This will also encompass a review of current and past tailings management technologies, which was contracted out to a consultant.

Acknowledgments

This Co-ordinated Research Project represents the concerted effort of a large number of people. The author wishes to thank the chief scientific investigators in this project: Sang-June Choi (Korea); Horst Fernandes (Brazil); Alexander Gagarin (Kazachstan); Branco Kontič (Slovenia); Josef Koszela (Poland) ; Patrick Landine (Canada); Gennadi Maslyakov (Ukraine); Donald Metzler (USA); Sergey Mikheykin (Russian Federation); Adam Piestryñzki (Poland); Anne-Christine Servant (France); Jan Trojaæek (Czech Republic); Z. Wen (China); Gerald Ziegenbalg (Germany). Special thanks are also due to Stewart Needham (Australia) for carrying out the review of past and current tailings disposal technologies.

Mailuu-Suu Tailings problems and options for remediation

Hildegarde Vandenhove[1], Jean Jacques Clerc[1], Holger Quarch[3], Johan Paridaens[1], Abdilamit Aitkulov[4], Salamat Imanakunov[4], Alexander Naravagov[5], Maxim Savosin[5], Isakbek Torgoev[6], Muradil Mirzachev[7], Omor Mombekov[7], Theo Zeevaert[1], Hans Vanmarcke[1].

1 Belgian Nuclear Research Centre, Boeretang 200, 2400 Mol, Belgium
2 GESTER, Techniparc, 24 rue Lavoisier, 92000 Nanterre, France
3 Allmendweg 6, 66453 Gersheim
4 Alex Stewart Assay and Environmental Laboratories, Kara-Balta, Kyrgyzstan
5 CHU-Ecological Laboratory, Kara Balta, Kyrgyzstan
6 Geopribor, Bishkek, Kyrgyzstan
7 KyrgyzGIIS, Osh, Kyrgyzstan

Abstract. The area of the town of Mailuu Suu, Kyrgyzstan, is polluted by radionuclides and heavy metals in tailing dumps and heaps resulting from the historic exploitation of U-mines in the Mailuu Suu area in Kyrgyzstan. Radioactive substances are stored in twenty three tailings and ten heaps situated along the Mailuu Suu River. The actual exposure at the tailings is of no concern even when considering rather conservative utilization scenarios. The actual stability of tailing 3, considered the most risky one, is low. Effects of external influences and remedial measures on tailing stability are evaluated.

Introduction

The area of the town of Mailuu Suu, Kyrgyzstan, is polluted by radionuclides and heavy metals in tailing dumps and heaps resulting from the historic exploitation of U-mines in the Mailuu Suu area in Kyrgyzstan. Radioactive substances are stored in twenty three tailings and ten heaps situated along the Mailuu Suu River. The stability of many tailings is at risk. Some of those tailings are already damaged by landslides, mudslides and floods, and some others are in high-risk areas where major landslides are expected. High risk for potential future exposure is present at deposits located along the riverbank in the narrow gap of the Mailuu-Suu river immediately south of the village of Kara Agach. Tailing 3 is considered as being the more risky one, because of its important radionuclide inventory and since threatened by a major landslide.

Major landslides potentially affecting directly or indirectly uranium ore processing tailings deposited along the Mailuu Suu River valley (i.e. tailings deposits No. 3, 9, 10, 8, 5 and 7 are the Tectonic,Technicum, Isolit and Koetash landslides which partly already caused river blockages in previous times (such as Tectonic Landslide in 1992). A recent reactivation of the southern limb of the Tectonic landslide again caused partial blockage of the Mailuu Suu River during spring 2002.

Earthquakes may have a direct or indirect effect on mudflows or landslides. Kyrgyzstan has been affected by a series of seismic disasters, the strongest being the Ms=8.2 Kemin earthquake in 1911. Recently, the Ms=7.3 Suusamyr earthquake affected the Northern and central Tien Shan mountain regions.

The objective of the EC-TACIS funded project is to evaluate and prepare measures to be taken by the authorities to reduce the radiological exposure of the population and to prevent environmental pollution by radionuclides and heavy metals in case of loss of tightness of dams and damage to dumps and heaps from mining and milling by land and mudslides and to propose sustainable remedial options, accepted by the public. This is achieved through setting up monitoring networks, conducting impact and risk assessments, feasibility study of a number of remedial options and the study and evaluation of rehabilitation plans to decrease the impact of a disaster scenario. In the present paper the gamma monitoring results from all tailings deposits in the Mailuu Suu area is discussed. In a second part, relying on field observation, the actual stability of tailing 3 is assessed and in situ remedial actions are evaluated.

Gamma monitoring campaign

Monitoring results

In August 2001, tailings were screened for external exposure in order to indicate areas of concern and set up a priority list of tailings to be fenced. Tailings were overlaid with a 10m by10 m mesh and the gamma dose was measured using highly sensitive gamma AUTOMESS AD6 with plastic detector Automess Adb. The readings were recorded (nSv/h), entered into a GIS, and plotted. The background gamma exposure is in range of 60-110 nSv/h. For most tailings and for most of the surfaces the exposure was around background or increased by a factor of two. At tailing 3 (Fig. 1), the borders of tailing 4, tailing 5, 11, 13, 18 and 21 higher dose rates were noted. At the waste dump bordering T13 dose rates up to 19300 nSv/h were recorded. Summary statistics on external dose is given in Table 1.

Most relevant dose contributions to the local population from the tailings are direct irradiation from radium and its daughter products, and inhalation of re-suspended dust and of emanated radon, during occasional residence on the tailings. To assess the contribution of the different exposure pathways, we considered the example case at the waste dump of Tailing 13, a location where a maximum dose rate of 19300 nSv/h was measured and the ^{226}Ra concentration amounted to 42 Bq/g (^{226}Ra is expected to account for > 95% of the global activity). When assuming a suspended dust concentration of 5 10^{-8} kg/m³, inhalation of ^{226}Ra at a rate of 1.2 m³/h yields an inhalation dose rate of 24 nSv/h, which is negligible as compared with the external dose rate. This is also the case for inhalation of ^{230}Th in re-suspended dust, being in equilibrium with its daughter radionuclide, ^{226}Ra. Also the inhalation of emanated radon originating from that ^{226}Ra concentration in soil leads to relatively low doses: adopting a conversion factor of 9 10^{-3} Bq/m³ ^{222}Rn in air per Bq/kg ^{226}Ra in soil gives a concentration of 360 Bq/m³ ^{222}Rn in inhalation air or an inhalation dose rate of 1170 nSv/h, which is also an order of magnitude lower than the external dose rate. As a consequence we may conclude that only the measured external dose rate needs to be considered when evaluating present radiation exposure from the tailings.

Table 1. Measured dose rates at the tailings at Mailuu Suu (nSv/h)

	Averages	St. Dev.	Median	95th Perc.	Max.	Min.
TOTAL*	**186**	**410**	**130**	**410**	**19300**	**41**
T1	180	53	175	260	490	90
T2	150	37	145	206	285	41
T3	245	291	180	516	3000	99
T4	178	93	150	324	1150	80
T4_Terraces	297	279	195	759	1920	75
T5	305	625	120	820	6000	80
T6	100	22	95	135	190	55
T7	111	58	100	180	850	60
T8	106	19	100	140	170	70
T9	118	22	120	153	180	70
T10	129	42	110	210	260	70
T11	220	102	200	404	700	95
T13	542	1282	265	1700	19300	100
T14	150	51	140	200	580	90
T15	105	25	100	145	420	45
T16	137	40	130	185	430	45
T18	377	878	200	490	5000	130
T19	145	35	140	188	230	90
T20-21	248	169	175	641	800	115
Mean**	202	217	149	403	2208	76

TOTAL: Values calculated on all measurements on all tailings; ** Mean: Values obtained from values presented in table

Case studies: Uranium tailings

When assuming that an individual is randomly staying at the mean dose rate on the tailings considered, even exposures of 4h a day will not lead to a dose of 1 mSv/y, for none of the tailings. When considering exposures at the 95th percentile of the dose rates for 2h a day, doses higher than 1 mSv/y will be reached only at tailing 13 (and especially at small dump next to T13) (1.2 mSv/y). If we would consider an extremely conservative scenario, someone staying at the place of maximum exposure during 2h/day, the exposure rates would be higher than 1 mSv/y for 5 tailings: Tailing 3: 2.12; the dam of Tailing 4: 1.36; Tailing 5: 4.2; Tailing 13: 13.62 and Tailing 18: 3.53.

To conclude, we may state that one tailing certainly needs fencing off: the dump next to tailing 13. It is however advisable to fence off the tailings since they contain radioactive material. In case of funding limitations, the fencing off can be dealt with in the order of descending dose rates, indicated in Table 1.

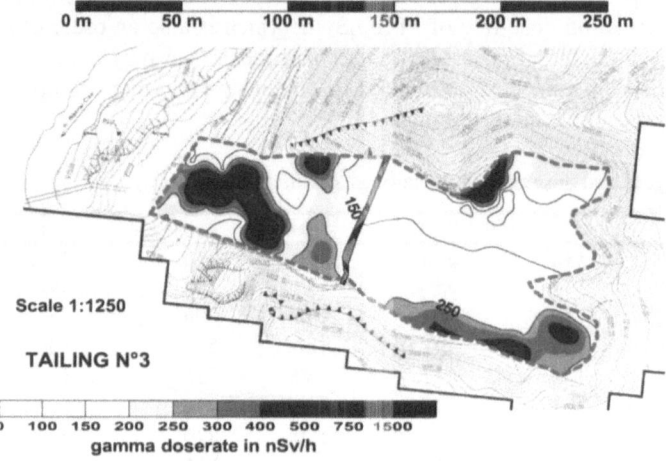

Fig. 1. External exposure at tailing 3 in Mailuu Suu

Suggestions for stabilization of Tailing 3

Under the TACIS project different remedial options will be evaluated for tailing 3 including in situ stabilization and tailing translocation. Only the first option will be discussed here.

Location, geometry

Tailing 3 is located at the foot of a steep slope (close to 45°) which overhangs the tailing 400 m to the East. It is restricted to the North and to the South by two small

20 m high spurs which are perpendicular to the slope foothills of which they are made of. The tailing's surface presents a top platform of variable width (between 70 and 100 m) and a 100 m length; to the West, the tailing presents a sloping downhill embankment towards a road which travels at the slopes foot and which was set up on the high bank of the Mailuu Suu river.

Under the downhill embankment (slope tgβ = 31 % slope) there is a 1.5 to 2.1 m thick infill which is rocky/stony then silty; the treatment residues (pulpa), a hardly thick grey mud with thickness expanding from 0 m downhill to about 13 m uphill; the altered and reshaped substratum made up of silty sands on its upper part or with sandy silt with rocky debris. Under the top shelf (slope tgβ = 8 à 9 %) we encounter from top to bottom: sandy and silty deposits originating from the erosion of the cliff with thickness increasing uphill from 2,3 to around 7 meters. In two boreholes a 0,2-0,3 m concrete layer showed (impossible to say if this is a continuous concrete slab or to debris from demolition practices; sand and gravel materials were hence identified at 0.5 or 1.5; the tailing material, around 13 meters thick in its central section, with a thickness decreasing to zero towards the cliff; the altered and reshaped substratum made up of silty sands on its upper part or with sandy silt with rocky debris. In the drillings equipped with piezometers, the water levels were measured between 4,3 and 8,3 meters in the beginning of September 2001.

Laboratory results of mechanical testing of soil samples

In order to be able to make stability calculations of the tailing disposal site, resistance tests were performed. To determine the short-term stability, 5 quick shear tests were done on tailing material showing that the undrained cohesion C_u increases with the preconsolidation strain σ' according to the formula: $C_u = 55 + 0,26\ \sigma'$ (C_u and σ' are in kPa). To this cohesion C_u, we associate a friction angle $\varphi u = 0$. In the tailing, the preconsolidation strain σ' increases with depth z according to density $\gamma = 20$ kN/m^3 above ground water and the density degauged $\gamma' = 10$ kN/m^3 below groundwater:

$$\left. \begin{array}{l} \sigma' = 5 \times 20 + (z-5)\ 10 = 50 + 10z \quad \text{in kPa with z in meters} \\ C_u = 25 + 2,6z \qquad\qquad\qquad\qquad C_u \text{ in kPa with z in meters} \\ \varphi u = 0 \end{array} \right\}$$

Hence,

For the long-term stability calculations, slow consolidated shear tests were done; test results in $\varphi' = 27°$ and $c' = 22$ kPa.

Screening of some available stabilization techniques

Different techniques could be available to stabilize the pulpa. Some available techniques are based on the structure of the tailing itself, especially the surface of the tailing; other techniques aim at improving the tailing material geotechnical characteristics.

Case studies: Uranium tailings

Top platform discharge and creation of a foothill barrier of the downhill embankment: This solution, well adapted to the problem, can be easily applied using available regional material and local people. It is described in the next paragraph.

Diversion of the slope runoff water and impervious covering of the tailing: The cover should consist of an impervious clay layer and a PEHD or tar geomembrane; a geotechnical protection using geotextiles and a layer of infill should be used on the top platform but would not be stable enough on the downhill embankment. Stabilization of the tailing using resistive columns treated with lime and cement and strengthened with metallic girders is another alternative as well as improvement of the global characteristics of the residues by ballast columns used by vibration. This technique should help annihilate the pulpa probable risk of liquefaction during earthquakes. Calculation of the improvement of residue characteristics between the columns requires additional studies.

Horizontal drainage using wells: waters levels in the tailing are located between 4,3 and 8,3 meters; each well should be equipped with a pump in order to reduce the water level. Given the limited pervious ness of the tailings, wells should be set up close to each other. Extracted waters must be collected and treated before discharging to the environment. This option is expected difficult and costly to set up.

Horizontal drainage using wick wells: the wick wells can not be equipped with pumps. In order for them to extract water, the platform should be loaded with infill, hence creating a pressure point, or it should be covered by a membrane underneath with a depression created by air extraction. That overload would reduce the tailing's stability; thus solution should not be advised.

Sub horizontal drainage: The density of the sub horizontal drains starting downhill of the slope should be high because of the residues' weak pervious ness. The collected waters must be treated before being rejected back into the environment.

Stability study of different case studies – practical consequences

Actual stability of the tailing

The program used for the stability calculations was TALREN. The calculation method is that of BISHOP where the examined rupture surfaces are cylinders (or circles when represented in the vertical axis); it examines all the potential landslide circles. Only the most unfavorable circle and the safety coefficient (F) of the rupture are given by the summary figure of the case study. Regarding the other circles, the safety coefficient is given with the circle centre and circles are not drawn. With the value $C_u = 25 + 2{,}6z$ deducted from the laboratory testing, the safety coefficient of the current situation is $F = 0.96$ (Fig. 2).

The current situation is apparently at the state of limited stability. We admit that, as a reference only, the actual safety coefficient is $F = 1{,}00$ which corresponds to the limit of equilibrium. The corresponding undrained cohesiveness is:

$C_u = 27 + 2{,}6z$ C_u in kPa with z in meters

Case studies: Uranium tailings

This value is used in each of the following examples evaluated.

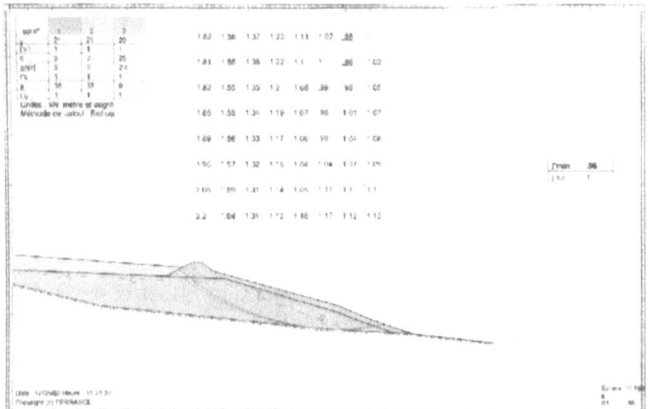

Fig. 2. Actual stability of Tailing 3 determined with Talreg

Seismic influence

Only the influence of vertical and horizontal accelerations induced by the potential seismic activity inside the tailing is examined. The acceleration values used for these calculations are purely hypothetical; the probable values, function of the site's distance to the earthquake's epicenter whose magnitude needs to be clarified according to its probability of occurrence, should be obtainable from a recent regional tectonic study. In presence of an earthquake, assuming a horizontal acceleration of $\gamma h = 0,1$ g and a vertical acceleration of $\gamma v = +$ or $- 0,05$g, the safety coefficient F is 0.71. Under the influence of a strong earthquake applying a horizontal acceleration of $\gamma h = 0,2$ g and a vertical acceleration of $\gamma v = +$ or $- 0,1$ g, the safety coefficient F falls down to 0,54.

Influence of changes on the top platform

If we add a 2 m thick additional layer (accumulation of erosion depot originating from the slope) on the entire surface of the top shelf, the safety coefficient decreases slightly: F = 0.97. If the thickness of that additional layer increases uphill at a value of 3 m against the small embankment (little dam) located on the crest at the junction of the flat area and the downhill slope, at a 10 m value towards the foot of the slope (cliff) dominating the tailing, the safety coefficient also decreases slightly (F = 0.96) If we reduce the thickness of the existing cover layer, at an increasing 2 m value uphill near the embankment (little dam) and 5 m to the foot of the slope in such a way that the left-over thickness of the depot located on the platform is 1 m thick everywhere, the safety coefficient stays unchanged. F = 1.00.

What precedes shows that the exerted overload close to the junction between the shelf and the downhill slope induces a predominant instability effect. The infill located on the main part of the top platform does not have any incidence on the local stability. Therefore, it is suggested to retrieve the small embankment (little dam) uphill (towards the East) over a 20 m distance, the safety coefficient increases to F = 1.08, even though the thickness of the infill appears to be high. The local stability of a 5 m high embankment infill, which is set upon a platform, made of erosion depots, whose thickness was reduced to 1 m (presently 3 to 7 m), presents a strong safety coefficient: F = 1,59

Influence of an infill on the downhill embankment

If we add a 2 m thick layer over the entire surface of the downhill embankment, the safety coefficient decreases to 0.90. The destabilizing effect of the infill on a higher part of the embankment exceeds the stabilizing effect of the infill at the bottom of the embankment. If we just add the infill at the foot of the embankment (little dam), e.g. 20 m width and up to mid-height, the safety coefficient increases to F = 1.50.

Influence of a reinforcement by resistive columns (COLMIX)

The reinforcement studied consists of introducing resistive columns inside the treatment residues by an incorporation of cement and lime using a system of continuous bores which mix the material. These columns have many advantages: they consolidate the residues and make them chemically inert, they limit their displacement and strengthen the resistance of the materials. Recommended columns are triple columns, built using three bores (diam. 0,50 m) slightly linked. In each column a steel girder HEB 200 is thrust by vibration. With one triple column per 20 m², 10 m² and 7 m², the safety coefficient F is, 1.12, 1.29 and 1.48, respectively in absence of an earthquake. With an earthquake of an horizontal acceleration of 0.1 or 0.2 g, the stability coefficient reduces to 0.96 and 0.67, respectively (for 1 column per 10 m²).

Conclusions regarding improvement of stability of tailing 3

The overloads (additional material induced by hill erosion or landslides) accumulating on the top platform have an unfavorable effect when located close to the existing embankment (small dam) which is located on the edge of the downhill embankment. Extracting the overloads (scrape the existing embankment on the downhill shelf of the platform) has a positive effect. The overloads settled on the most part of the platform only have a slight impact.

The additional embankment (infill) located on the lower part of the downhill embankment of the tailing have an extremely positive effect. On the other hand, the additional infill located on the upper part of the downhill embankment have an extremely negative effect. A total cover should not be applied on the entire downhill embankment but only on the bottom.

The creation of reinforcements using the Colmix columns enforced by metallic girders will considerably increase the stability. They will also turn the tailing material chemically and physically inert and limit the possibility of dispersion of tailing contaminants or residues to the river.

An earthquake has a direct averse effect on tailing stability and only with a moderately high seismicity the COLMIX improved tailings will fail. The second possible strain is that of a landslide, which would create an overload on the tailing and could provoke dam rupture. A concomitant effect, is the risk of a liquefaction of the saturated residues, which is linked with the limited density and tailing granulation. The effect of these actors on tailing stability will be assessed at a later stage.

For the stabilization option considered most appropriate, overall suitability of the option will be evaluated in terms of costs, long-term effectiveness, dose to workers and population, remaining risks (also linked with the landslide and seismic activity also studied under present project) and compared with an alternative option which is the one of translocation of the tailing.

Case studies: Uranium tailings

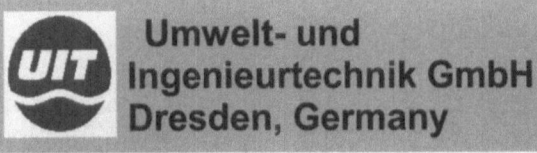

Umwelt- und Ingenieurtechnik GmbH Dresden, Germany

Zum Windkanal 21, D-01109 Dresden
Postfach 80 01 40, D-01101 Dresden
Phone: (+49) 3 51 8 86 46 00
Fax: (+49) 3 51 8 86 57 73
e-mail: uit.dd@t-online.de
www.uit-gmbh.de

Umwelt- und Ingenieurtechnik GmbH Dresden is a medium-sized German company which is specialized in

- development and fabrication of environmental equipment and facilities,
- development and application of new remediation technologies,
- construction of disposal and water treatment plants as well as
- relevant engineering and consulting

Environmental Technology
- Consulting, Planning, Management -

Engineering Services
- Process engineering for remediation and waste disposal (incl. radioactively contaminated sites)
- Water treatment (incl. mine water)
- Waste management and disposal
- Process automation and monitoring systems

Project Management
Modeling/Software Development
- Geo-chemical and geo-hydrological modeling of mine flooding
- Modeling of chemical processes
- Modeling and optimization of water treatment plants

Environmental Technique
- Development, Manufacturing, Installation -

Laboratory Systems

Measuring and Sampling
Technique (Water, Soil, Air)
- Multi-Sensor Probes (Water)
- Data Logger
- Soil Moisture/Temperature Measurement
- Tensiometer, Soil Water Samplers

Field Monitoring and Sampling
Systems
- (Mobile) Water Treatment Stations
- Field Monitoring Systems
- Water Quality Monitoring Stations

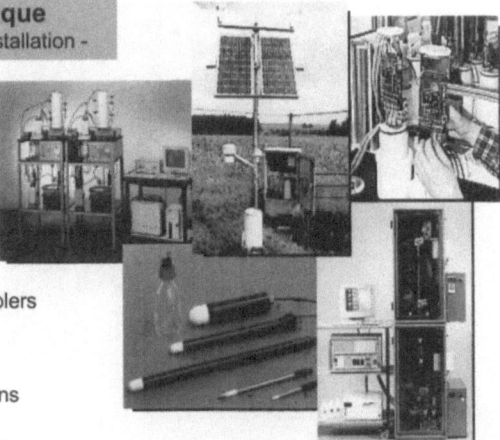

Uranium mill tailings in the Pine Creek Geosyncline, northern Australia : past, present and future hydrogeological impacts

Gavin M. Mudd

[1] Formerly Research Fellow, Dept. Civil Eng., Uni. of Queensland, St Lucia, QLD Australia (now P.O. Box 6405, St Lucia, QLD 4067 angelb@netspace.net.au)

Abstract. Uranium mining has been a principal part of Northern Australia since mineralisation was discovered at Rum Jungle in 1949. To date, 5 mills have operated at Rum Jungle, Rockhole, Moline, Ranger and Nabarlek supported by 27 adjacent or other NT mines. The tailings management at 1950s mills was through discharge to adjacent lowlands with in-pit tailings at Rum Jungle in the 1960s. Due to higher environmental expectations in modern times, Ranger and Nabarlek have used in-pit plus above ground interim storage of tailings. The environmental and hydrogeological impacts of these sites is reviewed.

Uranium mining history in the Pine Creek Geosyncline

The identification of uranium minerals at Rum Jungle in August 1949 heralded the start of a long history of uranium mining and milling in the Pine Creek Geosyncline, at the top of the Northern Territory (NT), Australia (Fig. 1). The geology is described in Ferguson and Goleby (1979) and Needham and De Ross (1990).

The 1950s saw a major mining and milling project at Rum Jungle, with additional ore sourced small mines across the NT. In the South Alligator Valley, 12 U mines were operated to support the Rockhole and Moline mills. Exploration in the early 1970s led to the discovery of new U deposits at Ranger, Jabiluka, Nabarlek and Koongarra. Due to community concerns, only the Ranger and Nabarlek projects were developed and operating by the early 1980s.

The main issues relating to uranium tailings management are climatic extremes (wet/dry), radon, permeable carbonate units and/or fractures in groundwater flowpaths and links from shallow aquifers to surface water systems (Mudd 2002).

The approach to tailings management has developed together with improved scientific knowledge of their associated hazards, such as gamma radiation, radon flux, erosion, heavy metal and water quality impacts (Mudd, 2002). The impacts from these sites will be reviewed (data in Table 1), thereby developing thematic issues for uranium mill tailings management in northern Australia.

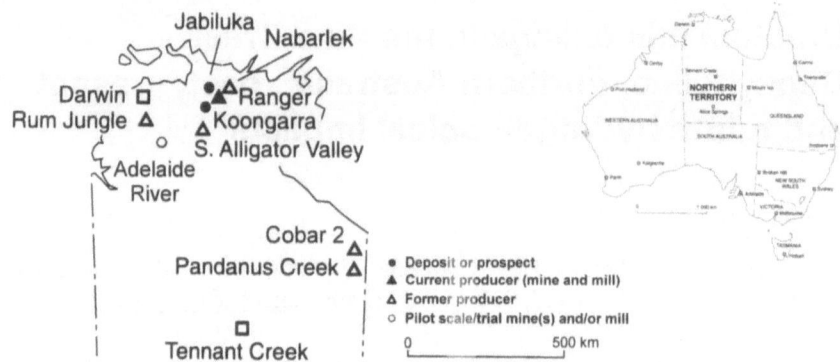

Fig. 1. Location of uranium projects in the Northern Territory, Australia

Table 1. Uranium mill tailings and waste data, Pine Creek Geosyncline (Mudd 2000, 2002)

	Rum Jungle	Moline	Rockhole	Nabarlek	Ranger [a]
Mill operating	1954-71	1956-64	1959-62	1980-88	1981-??
Ore milled (t)	1.50 Mt [b]	135,444 [b]	13,418	0.76 Mt [c]	22.91 Mt
Waste rock (Mt)	14.28	??	??	2.3	88.68
U_3O_8 t Produced	3,530	716	139.7	10.955	65,468
U_3O_8 Ore (%)	0.32%	0.46%	1.11%	1.84%	0.319%
U_3O_8 Tail's (%)	0.086%	0.065%	0.048%	0.034%	0.033%
Ore (Bq/kg)	488,868	689,272	1,672,769	2,772,878	480,696
Tail's (Bq/kg)	438,127	605,969	1,447,014	2,389,005	419,957
Initial (Bq)	6.6×10^{14}	8.2×10^{13}	1.9×10^{13}	1.4×10^{15}	9.6×10^{15}
Tail's (Bq/kg) ^{226}Ra	33,686	47,495	115,264	191,068	33,123
Tail's (Bq/kg) 10^3 yrs	383,519	517,447	1,209,639	1,985,616	355,879
Tail's (Bq/kg) 10^6 yrs	128,879	98,466	71,916	49,903	49,983

[a] data to March 31, 2002. [b] not including 1.11 Mt / 152,600 t of base metal ores milled.
[c] includes 157,000 t of low grade ore heap leached (not included in radioactivity calc's).

Rum Jungle uranium/copper project

The Rum Jungle U/Cu project was developed between its discovery in 1949 and 1954 when commercial operations began. All uranium was sold to the Combined Development Agency for the USA-UK nuclear weapons programs until 1963. The project was owned by the Commonwealth Government through the Australian Atomic Energy Commission (AAEC), with the project being operated on a special contract by a dedicated subsidiary of CRA Ltd (now Rio Tinto Ltd). The uranium produced from 1963-71 was stockpiled by the AAEC in Sydney, NSW.

The area was discovered to contain several small to moderate U deposits, with base metal ores (Cu, Pb) often present (about 20,000 t Cu was produced). Mining data for all deposits is given in Mudd (2002). Rum Jungle also processed custom ores from numerous small U mines across the NT.

The Rum Jungle site has become a major source of acid mine drainage pollution of the Finniss River system to the west, with significant quantities of radium and heavy metals (eg. Cu, Mn, U, Zn) impacting an area of some 100 km^2 (Kraatz 1998). The ore and waste rock contained about 3% sulfides (Richards et al. 1996).

Between 1954-61 tailings were discharged onto lowlands adjacent to the mill, later known as Old Tailings Creek, and proved to be highly erodible (Richards et al. 1996). About 1 million L of liquid wastes were discharged daily, at a pH of 1.5 containing heavy metals and radionuclides. The 640,000 t of tailings that settled out covered 35 hectares. By 1984, about 10% or 10 mm per year (~3,000 t/yr) had been eroded into the Finniss River (Mudd 2002). In 1961-62, a hole appeared in the ground near the copper cementation launders (presumably due to dissolution of dolomite by acidic liquid wastes) (see Davy 1975).

In 1961 tailings were dumped in the former Dyson's open cut, and to White's open cut from 1965 until closure in 1971. The use of pits and adjacent flood control dams were part of an attempt to improve water quality in the Finniss, though later estimates in the mid-1970s showed that the dilution ratios were not sufficient.

The radon flux from the Old Tailings area averaged about 2 Bq/m^2/s (Ritchie, 1985). Other sources of radon include waste rock dumps, which were studied prior to rehabilitation in the 1980s. For the White's and Rum Jungle Creek South (RJCS) dumps, the average radon fluxes were 1.1 and 2.7 Bq/m^2/s with uranium grades of 0.01% and 0.054% U$_3$O$_8$, respectively (Mason et al. 1982). Given that the RJCS deposit was 'blind', the additional radon load from the dump would be well in excess of pre-mining conditions (Mudd 2002). The target set for radon flux during rehabilitation was 0.14 Bq/m^2/s (Allen and Verhoeven 1986).

Due to public pressure, the Rum Jungle site was rehabilitated at a cost of ~$20 million between 1983-86, with the open cut at RJCS rehabilitated over 1991-92. The main works involved relocation of the 'Old Tailings' and unsuccessful copper heap leach pile to Dyson's open cut, re-diversion of some surface water features, water treatment, covering dumps and re-vegetation.

The main approach used for reducing acid mine drainage was the construction of multi-layered soil covers, designed to limit infiltration of rainfall and thereby oxygen ingress. As the site was among one of the earliest to employ this technique, the ongoing monitoring is important in assessing its efficacy.

The annual loads of heavy metals and radium prior, during and after rehabilitation are summarized in Table 2. The water quality downstream in the Finniss River, as measured over the 1992/93 wet season, is summarized in Table 3.

Prior to rehabilitation, up to 50% of rainfall would infiltrate into the various dumps. The post-rehabilitation target was set at 5%. After rehabilitation the rates of infiltration into White's dump was reduced to 1.4-3.7% (1984-93). For the Intermediate dump, infiltration rates were reduced to 3.5-5.9% (1985-89). Recent monitoring has shown infiltration at both dumps increasing to about 10%, which is likely to lead to increased metal loads (Mudd 2002).

Although water quality and metal load reduction targets were set for the rehabilitation works, they were not based on eco-toxicological criteria and still allow metal concentrations downstream many times higher than upstream (Mudd 2002).

Table 2. Range of annual heavy metal and radium loads from Rum Jungle (Mudd 2002)

Period	Rain (m)	Flow (GL)	Cu (t)	Zn (t)	Mn (t)	SO$_4$ (kt)	^{226}Ra [a] (GBq)
54-61	nd	nd	~43	nd	~114	nd	~914
61-67	nd	nd	~100	~6.7	~167	nd	~1,011
67-74	0.90-2.00	7.0-69.0	40-106	20-30	46-110	3.3-13.0	~703
82-84	1.12-1.70	9.5-48.0	23-28	5-9	6-21	1.52-3.6	~23
84-90	0.90-1.60	3.1-35.0	1.8-9.1	1.6-4.4	3.9-19.2	0.76-4.4	0.35-1.6
90-93	1.00-1.56	7.1-40.5	3.8-14.9	2.7-7.4	9.1-30.5	1.26-4.0	nd

[a] Estimates based on ^{226}Ra in liquid wastes of 1,480 GBq (1954-61); 1,850 GBq (1961-67); and ^{226}Ra in tailings of 14,060 GBq (an annual average of 703 GBq). The data from 1954-74 does not distinguish between dissolved and sediment-bound radium. (nd - no data).

Table 3. Finniss River water quality, downstream of Rum Jungle, 1992/93 (Kraatz 1998)

	mg/L			μg/L								
	Al	Ca	Fe	As	Ba	Co	Cr	Cu	Ni	Pb	Th	U
Ave.	3.6	9.9	1.71	4.1	37	176	5	485	169	76	3.3	33
Min.	0.21	4.2	0.096	0.6	21	53	0.7	180	53	2	0.02	6
Max.	9	29	14	41	120	480	33	1100	430	880	26	63

The rehabilitation design life was 100 years, although as the water quality and in-filtration data show, deficiencies are evident after 15 years (likely to be related to either the construction or design thickness of the covers). A major failure of the rehabilitation works was to not remediate groundwater, which receives saline seepage from the dumps and contributes significant metal loads to the Finniss River, especially during the dry season and early wet season (Parker, 1999).

The tailings, now all below-grade in White's and Dyson's open cuts, appear to have been stabilized, though the extent of groundwater contamination remains uncertain. If the covers continue to deteriorate, then it is likely that pollution loads will increase again in the Finniss River. It is clear that the Rum Jungle site is far from stable environmentally, with continuing metal and radionuclide loads (eg. U) discharging into the Finniss River every wet season derived mainly through seepage from the various dumps and groundwater.

Upper South Alligator Valley - Rockhole, Moline projects

The 1950s saw considerable exploration in the Upper South Alligator Valley (SAV) leading to the development of numerous U mines (0.13-2.5% U$_3$O$_8$) which supplied two mills at Rockhole within the SAV and Moline 60 km to the south-west (some ores also had gold). The practices of both were similar and are presented jointly (see Mudd (2002) for detailed data and analysis).

The discharge of non neutralized acidic tailings was onto lowlands adjacent to and downhill from each mill, with no serious attempt to retain tailings (Waggitt 1994). Observers at the time noticed that tailings were easily eroded into the river during the intense rain and flooding of the wet season (Waggitt 1994). The metal concentrations and radiation rates from the tailings sites are given in Table 4. By the mid-1980s, up to 10% (~1,500 t) and 25% (~63,000 t, including later base metal tailings) of the Rockhole and Moline tailings had eroded into creeks and the South Alligator and Mary rivers, respectively (Fig. 2) (Mudd 2002).

Table 4. Radiation rates and heavy metals in Rockhole/Moline tailings (Mudd 2000, 2002)

		Rockhole (R)		Moline (M)	
		Ave.	Range	Ave.	Range
Gamma [a]	μGy/hr	18	<2.5 to 65	~5	<2.5 to 24
Radon	$Bq/m^2/s$	~6	<5 to 21.1	~2	<1 to 17.9
	Bq/m^3	??	52 to 9,586	??	50 to 1,654
	Bq/m^3	190	wind >2 m/s [b]	135	wind >2 m/s [b]
	Bq/m^3	1,895	wind <2 m/s [b]	395	wind <2 m/s [b]

	(mg/kg)	As	Au	Ag	Cu	Pb	Zn	^{226}Ra [c]	U
M	Min.	88.8	0.026	<1	40.5	205	32	270	3.7
M	Max.	9,800	9.49	40	7,770	5,800 [d]	6,300	87,900	740
M	Ave.	1,438	1.5	12.5	661	1,227	1,107	40,290	88
M	Area [e] (18 ha)	30% >500	20% >2	20% >10	20% >500	40% >500	40% >500	-	similar
R	Min.	12	0.107	-	13	28	10	28,500	3
R	Max.	570	23.7	-	1,030	1,090	87	222,000	830
R	Area (~2 ha)	30% >100	60% >2	-	1% >500	30% >500	uni-form	(2 sam-ples)	60% >100

[a] Typical background gamma radiation is about 0.1 μGy/hr. [b] Wind speed.
[c] Bq/kg. [d] El Sherana ore had up to 5% Pb in places. [e] Area greater than conc'n.

By the 1980s, tourist numbers had increased in the former mining province, and the SAV was being considered for inclusion in Stage 3 of Kakadu National Park. As the gold industry was going through a major expansion, the tailings were tested for reprocessing. After assessment, approval was given for excavation of all the Rockhole tailings to be transported to Moline for gold extraction. Waggitt (1994), however, only lists 6,000 t or about half being transported.

The reprocessed Moline/Rockhole tailings (producing ~10,000 oz Au) were discharged into a new engineered dam (and later became the site of a major dam for 3 Mt of gold tailings from the Moline gold mine, 1988-92; see Mudd 2002).

In 1991-92, a series of 'Hazard Reduction Works' were undertaken throughout the SAV, including the erection of fences, warning signs and the redirection of roads (Waggitt 1996). Although localized areas of gamma radiation could be found (eg. El Sherana camp), radon progeny were considered low at 4.6 mWL, although a gully area of elevated gamma and radon was detected and not included in works at this time (the site was monitored).

Fig. 2. Erosion of Moline U/Au tailings (Cull et al. 1986).

In 1999, the gully containing elevated radiation was found to be the source of uranium tailings dispersed onto the SAV tourist road. It is likely that this is the remaining 6,000 t of tailings not transported to Moline during the mid-1980s.

There are many continuing management issues with the former uranium mines of the SAV and their associated tailings. This includes acid drainage from adits at Rockhole, erosion problems at shallow burial and most mine sites, bat colonies in underground shafts, re-vegetation difficulties and further areas of elevated gamma rates. The continuing acid drainage in particular highlights the difficulties of interpreting and predicting groundwater flowpaths and the efficacy of rehabiliaition.

Nabarlek uranium project

The Nabarlek uranium deposit was discovered in 1970 and was found to be ~1.9% U_3O_8 in 606,000 t of ore. After the environmental, land rights and nuclear debates of the mid-1970s (the Ranger Uranium Environmental or 'Fox' Inquiry; Fox et al. 1977), the project was approved and began in 1979.

The ore was mined in 4 months of the mid-1979 dry season with the high and low grade ores and waste rock stockpiled separately. The mill began operation in June 1980, allowing the unique situation where the tailings, neutralized with lime, could be discharged to the former open cut. Between 1984-88, ~157,000 t of low grade ore (~0.05% U_3O_8) was heap leached. After the exhaustion of primary ore, the mill formally closed in late 1988.

The pre-mining radon fluxes were measured in September 1978 and June 1979 to range from 3.7 to 44.0 and 11.5 to 164.0 $Bq/m^2/s$, respectively, with the latter including the effects of cleared vegetation due to mining (Clark et al. 1981).

The principal issue affecting tailings during operation was water management, due to the need to store large quantities of contaminated waters in the former pit. This led to low settled tailings densities, which would create difficulty at the time of rehabilitation. In mid 1985, the use of sub-aerial deposition was authorised, which led to higher settled densities. After mill closure in 1988, a series of vertical drainage wicks were installed to aid tailings consolidation, the first use of this approach in the mining industry and thought successful (Waggitt and Woods 1998).

During operation, it was decided to trial the irrigation of contaminated pond waters in 1984 on a small region adjacent to the airstrip (see Akber 1991). The main solutes of concern were perceived to be NH_4 and SO_4. After further trials in 1985, the irrigation was expanded to a nearby forest area. The total water irrigated by the end of 1987 was 701 ML. The high NH_4/SO_4 concentrations in the pond water (~1,040 / 5,340 mg/L) led to significant impacts on groundwater quality, exacerbated by oxidation of the NH_4 to NO_3 and the release of acidity, as well as tree deaths in the forest irrigation area (Mudd 1999). The impacts on shallow ground waters were detected within months of irrigation, which discharged to the adjacent creek (see Akber 1991; Waggitt and Woods 1998).

Further complexity in groundwater behavior was demonstrated by monitoring down gradient of evaporation pond 2 (EP2). Three bores each 50 m from EP2 showed SO_4 concentrations of 488, 12 and 1,260 mg/L, respectively, demonstrating the importance of fracture zones or high permeability preferential flow paths (Waggit and Woods 1998; Mudd 1999).

After prompting by government authorities, decommissioning began in late 1994 and the site was rehabilitated by late 1995. The main works involved disposal of the remaining contaminated water and removal of the evaporation ponds, backfilling of the former pit/tailings with wastes from the mill and other contaminated materials and re-contouring and re-vegetation of the area. A review of Nabarlek's rehabilitation is given in Klesaa (2001) and Mudd (2002). Environmental and some radiological monitoring has been ongoing since 1995.

A major radiological study of the rehabilitated Nabarlek site is given by Martin (2000). The gamma surveys showed that current dose rates average 0.27 µGy/hr over the 98 hectares of the former site, and clearly outlines features such as the evaporation ponds, waste rock dumps and the pit/tailings area. The pre-mining gamma dose rate was estimated to average 0.18 µGy/hr over the same area.

The radon fluxes from the various areas at Nabarlek have also been studied since rehabilitation, reviewed in Martin et al. (2002) and (Mudd 2002). Although pre- rehabilitation estimates suggested a reduction of some 10^{22} in the radon flux from the former ore zone, recent field measurements have shown that the cover over the tailings and former pit has an average radon flux of 1.03 $Bq/m^2/s$, which is only 10-100 times reduced. This was thought to be related to the use of waste rock in the final cover instead of soil, since the topsoil had become sterile during the 16 years since mining (Mudd, 2002).

Ranger uranium project

The Ranger uranium project has been a controversial focal point for the complex issues of Aboriginal land rights, uranium mining (reflecting community concerns on nuclear issues) and national parks / environmental protection. The deposits were discovered by Noranda in 1969 on Peko-EZ leases and proved by mid-1970.

The Ranger project was located within land originally intended for a national park proposed in 1965, and spirited national debate led to the Ranger Uranium Environmental Inquiry being established in July 1975. It handed down 2 reports, the final in May 1977 concentrating on the Alligator Rivers Region and the Ranger Project (in which the Commonwealth held a 50% stake) (Fox et al. 1977). It remains one of the most quoted inquiries on nuclear issues and uranium mining and made several wide-ranging recommendations regarding Ranger (as well as other proposed mines at Nabarlek, Jabiluka and Koongarra).

Primarily it supported land rights for Aboriginal people, the creation of Kakadu National Park and urged caution and best practice for all proposed uranium mines, including final below-grade tailings and 'no-release' water management systems as well as different waste rock and stockpile locations. The distinction was also made that low grade ore be considered as the same environmental risk as tailings and also be placed below-grade. When the Ranger project was authorized by the federal government in January 1979, 'Environmental Requirements' (ER's) were included covering water management, radiation protection, tailings and the like.

In the short-term, until pit #1 was mined out, an above ground storage dam has been used for tailings management, built to the west of pit #1 and the mill. The original ER's modified the Fox Inquiry recommendation to state that, after 10 years from the date of approval, Ranger may apply for an alternative to below-grade tailings management if it can be demonstrated that the environment is no less well protected. Despite considerable research, however, Ranger finally stated publicly their commitment to final below-grade tailings management in late 1997.

The hydrogeology of the Ranger site is described by Haylen (1981) and Ahmad and Green (1986). There are three main aquifer types - sands and gravels ('Type A'), weathered soils ('Type B') and fractured rocks ('Type C'), shown in Fig. 3. The relationship between the aquifers is often complex. The water quality is generally of low salinity, up to 300 mg/L, with low heavy metal and radionuclide content (except ore zones).

In 1996 Ranger began the deposition of tailings into pit #1, although no lining was constructed. The current height limit is 'RL 0 m', or the base of the weathered zone, where the tailings would come into direct contact with the permeable surficial aquifers which can connect directly to surface water systems.

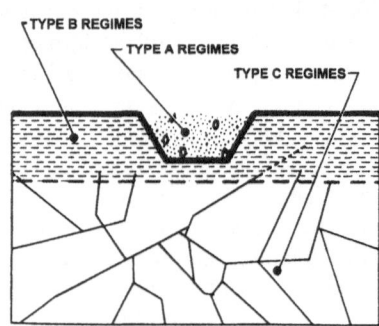

Fig. 3. Groundwater regimes at Ranger (Ahmad and Green 1986)

A detailed analysis of the potential seepage pathways from tailings stored in both the above ground dam and pit #1 was presented by Haylen (1981), shown in Fig. 4. Essentially the critical role of permeable carbonate units, such as cherts, as well as fracture zones, are clearly highlighted as being major and potentially rapid pathways for contaminants from tailings. Haylen (1981) noted that the chert units were present downstream in the Magela Creek, which flows into world heritage-listed Kakadu National Park. Given the environmental senstivity of the region and the concerns of Aboriginal people, it is disappointing that there appears to be little follow-up research on these issues for long-term tailings management at Ranger.

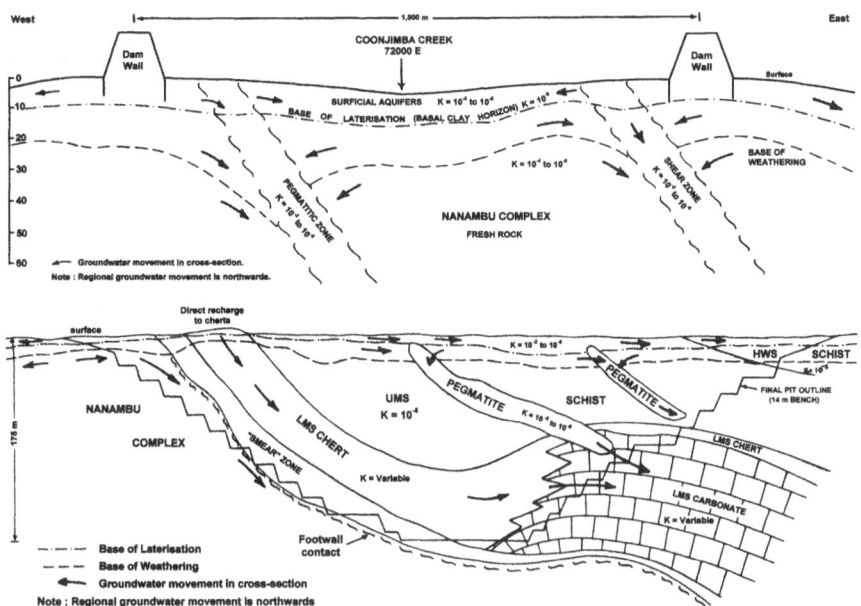

Fig. 4. Permeability (K, cm/s) zones and inferred groundwater flowpaths for the above ground tailings dam and pit #1 at Ranger (adapted from Haylen 1981)

Discussion and conclusions

The management of uranium tailings and the minimization of associated environmental and radiological impacts has improved over the last 50 years in northern Australia. The acid mine drainage problems at Rum Jungle continue to lead to groundwater pollution and metal loads to surface water systems. For Moline and Rockhole, most of the tailings appear to have been stabilized except for the residual remaining within Kakadu National Park. The use of in-pit tailings at Nabarlek was innovative, though concerns remain about the potential for fractures to transmit solutes. Ranger has managed its tailings through use of an interim above ground dam and is now using pit #1. All sites show a clear case for final below-grade tailings, links with water management and better emphasis on groundwater.

References

Ahmad M and Green DC (1986) Groundwater regimes and isotopic studies, Ranger uranium mine area, Northern Territory. Australian J Earth Sciences 33: 391-399

Akber RA (Co-ord.) (1991) Land application of effluent water from uranium mines in the Alligator Rivers Region. OSS Workshop, Jabiru, NT, September 1991 (Pub. 1992)

Allen CG and Verhoeven TJ (Ed's) (1986) The Rum Jungle rehabilitation project - final project report. NT Dept. of Mines & Energy, Darwin, NT, June 1986

Clark GH, Davy DR, Bendun EOK and O'Brien B (1981) Meteorological and radiation measurements at Nabarlek, Northern Territory, June to July 1979. AAEC/E505

Cull RF and others (1986) Tailings transport and deposition downstream of the Northern Hercules (Moline) mine in the catchment of the Mary River, NT. In: 10TH North Australian Mine Rehabilitation Workshop, Darwin, NT, June 1986, 199-216

Davy DR (Ed.) (1975) Rum Jungle environmental studies. Australian Atomic Energy Commission, AAEC/E365, Lucas Heights, NSW, September 1975, 322 p

Ferguson J and Goleby AB (Ed's) (1979) International Symposium on Uranium in the Pine Creek Geosyncline, BMR-CSIRO-IAEA, June 1979, Sydney, NSW (Pub. 1980)

Fox RW, Kelleher GG and Kerr CB (1977) Ranger uranium environmental inquiry - second report. Tabled May 17, 1977, Canberra, ACT, 425 p

Haylen ME (1981) *Uranium Tailing Disposal : Ranger Project - A Rationale.* M Env Stud Thesis, Macquarie University, 181 p

Klessa DA (Ed.) (2001) The rehabilitation of Nabarlek uranium mine. Workshop Proc., Darwin, NT, April 2000, Supervising Scientist Report 160 (Pub. August 2001)

Kraatz M (Ed.) (1998) Rum Jungle rehabilitation project - monitoring report 1988-1993. Dept. Lands, Planning & Environment, Darwin, NT, March 1998, 180 p

Martin PC (2000) Radiological impact assessment of uranium mining and milling. PhD Thesis, Queensland University of Technology, Brisbane, QLD, February 2000, 226 p

Martin P, Tims S and Ryan B (2002) Radon exhalation rate from the rehabilitated Nabarlek surface. In: Supervising Scientist Research Summary 1995-2000, Ed's J Rovis-Herman et al., Supervising Scientist Report (In press), Darwin, NT, 15-19

Mason GC, Elliott G and Gan TH (1982) A study of radon emanation from waste rock at Northern Territory uranium mines. Australian Radiation Laboratory, ARL/TR044

Mudd GM (1999) A review of Australia's uranium mining and the proposed Jabiluka uranium mine : a scientific case for placing Kakadu as world heritage in danger. Technical Submission to the World Heritage Committee, May 1999, Melbourne, VIC, 88 p

Mudd GM (2000) Remediation of uranium mill tailings in Australia : a critical review. In: Contaminated site remediation : from source zones to ecosystems. Dec., 2: 777-784

Mudd GM (2002) Uranium mining and milling wastes in Australia : past, present and future management. Research report, in preparation, ~140 p (angelb@netspace.net.au)

Needham RS and De Ross GJ (1990) Pine Creek Inlier - regional geology and mineralisation. In: Geology of the mineral deposits of Australia and Papua New Guinea", Ed. FE Hughes, AusIMM, Mono. 14, Melbourne, VIC, Vol. 1: 727-737

Parker G (1999) A critical review of acid generation resulting from sulfide oxidation : processes, treatment and control. Occassional Paper No. 11, AMEEF, Melbourne, VIC

Richards RJ, Applegate RJ and Ritchie AIM (1996) The Rum Jungle rehabilitation project. In: Environmental Management in the Australian Minerals and Energy Industries - Principles and Practice, Ed. DR Mulligan, UNSW Press & AMEEF, 530-553

Ritchie I (1985) The Rum Jungle experience - a retrospective view. Aust. Science, 4: 22-30

Waggitt PW (1994) A review of worldwide practices for disposal of uranium mill tailings. Technical Memorandum 48, Office of the Supervising Scientist, AGPS

Waggitt PW (1996) Hazard reduction works at abandoned uranium mines in the Upper South Alligator Valley, Northern Territory. In: Radiological Aspects of the Rehabilitation of Contaminated Sites Workshop, Darwin, NT, June 1996, 70-78 (Pub. 1998)

Waggitt PW and Woods PH (1998) Nabarlek uranium mine, northern Australia : history, rehabilitation and groundwater studies. In: Uranium mining & hydrogeology II, Freiberg, Germany, GeoCongress 5, September 1998, Vol. 1: 603-612

Radioactive Contaminant Transport of Aue Mine Dump 371: A Hydrogeochemical and Isotopic Case Study

Schneider, P.[1]; Voerkelius, S.[1]; Osenbrück, K.[1]; Meyer, J.[2]

[1]Hydroisotop-Piewak GmbH, Oberfrohnaer Strasse 84, D-09117 Chemnitz,
[2]WISMUT GmbH, Jagdschänkenstrasse 29, D-09117 Chemnitz

Abstract. In order to evaluate the current and future environmental impact of the disposal site, investigations of environmental and radioactive isotopes have been performed as part of a hydro geological and hydro chemical case study at Aue Mine Dump 371. The results were summarised in a hydro geological model of the dump. The main contaminants reaching the nearby rivers are of uranium and arsenic. The mean residence time of those waters percolating the dump is about 2 to 5 months. The ^{3}H contents in the ground water of the first aquifer yield a mean residence time of about 10 to 20 years.

Introduction

The Aue Uranium Mine in Saxony, East Germany, was shut down in 1990 due to the end of uranium production in the former GDR. As a result of the exploitation of the uranium mine 371 about 40 dumps were left in the Aue area. The major dump is the Mine Dump system 371, which consists of two dumps of about 45 ha (Aue Mine Dump 371/I) and 22 ha (Aue Mine Dump 371/II). After in-situ remediation of the dumps, the ground and pore water, which are released in the rivers nearby, will become the most important aqueous pathway for the migration of radioactive and toxic contaminants. In order to evaluate the radionuclide potential and to specify the current and future environmental impact of the disposal site, investigations of environmental and radioactive isotopes have been performed as part of a hydro geological and hydro chemical case study.

Description of Aue Mine Dump

Site characterisation

Aue Mine Dump 371 is situated in the south west of Saxonia, Germany, next to the city of Aue. The study site (see Figure 1) receives an average annual precipitation of 800 mm with maxima in summer.

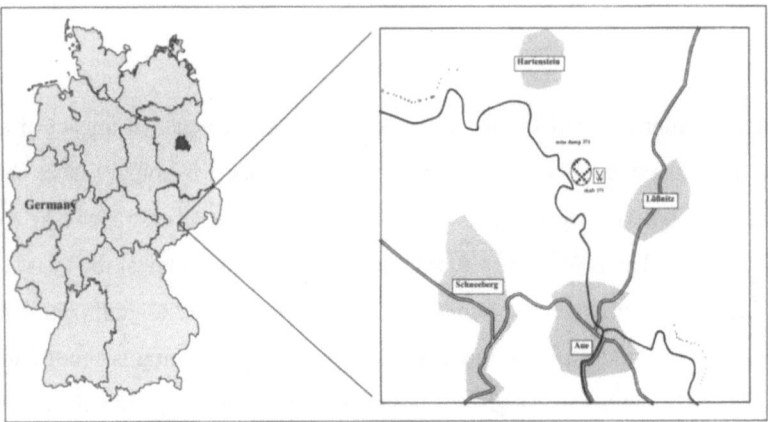

Fig. 1. Map showing the location of Aue Mine Dump.

Aue Mine Dump 371 is the most voluminous dump of the Aue dump complex and reaches a thickness up to 30 m, in maximum 60 m. It was filled from 1946 to 1990 with rocky mining waste of the shaft 371. Aue Mine Dump 371/I contains 9,9 mil m³, Aue Mine Dump 371/II 3,7 mil m³ mining waste consisting of waste rock from the mining process and mining residues. The deposited mining waste includes 642 t uranium, 10.692 GBq radium and 7036 t arsenic. The 371 Mine was closed in 1990 (Meyer et al 1998). The flooding of the pit started in 1991. All dumps will be covered with mineral soil and replanted.

Geological and hydro geological setting

The Aue area is characterised by paleozoic micas and phyllites, which were formed during the variscic orogenesis. The rock is fractured and clayey weathered from the surface down to a depth of 2 to 30 m. Monitoring wells were installed upstream and downstream of the dumps. These wells monitor the water flow in the unweathered rock aquifer and the weathering zone of the phyllites. Referring to pumping tests, the mean hydraulic conductivities of the unweathered rock zone

Case studies: Uranium tailings

range from 10^{-7} to 10^{-8} m/s, with elevated values of 10^{-5} m/s in tectonically altered areas. The weathered loamy zone has a hydraulic conductivity of 10^{-6} m/s. The heap material shows high hydraulic conductivities up to 10^{-2} m/s, locally. The phyllite aquifer can be seen as fracture aquifer. Recharge takes place throughout the study area by infiltration of precipitation and vertical leakage. An artesian ground water table has formed in the loamy weathering zone, locally.

Aue Dump 371/I and II are located in former river valleys (Kohlungbach, Wiesenbach) and the water courses are flowing underneath of the heaps. About 500 m downstream of the dumps the rivers are joining the main river Zwickauer Mulde. Water saturation within the dump is maintained only at the bottom with an extension of about 2 m. In most parts of the dump, an impermeable basement is missing. Consequently, percolating dump waters can directly enter the phyllite.

Investigations Methods

Following topics were investigated in detail at Aue Mine Dump 371:
*   hydrological and hydro chemical characterisation of surface and ground water including water balance of the dump and dump catchments,
*   isotopic and hydro chemical composition of the pore, surface and ground water to characterise the long term contaminant potential,
*   hydraulic and hydro geological characterisation of the dump, the underlying weathering zone, the fracture aquifer und the nearby rivers,
*   residence times and recharge rates for water.

The water balance was calculated based on hydrological data of field measurements. Following equipment was installed:
*   monitoring wells,
*   various runoff weirs using continuous data logging units,
*   climatologically station,
*   lysimeters for the evaluation of pore water recharge.

According to the specific hydraulic situation of the site the following isotope analysis were performed:
*   oxygen-18 analysis of the surface and ground water to analyse the annual variation of mean residence time and evaporation effects,
*   tritium (^{3}H,) analysis of ground water to evaluate the residence time of the older component,
*   investigation of heavy runoff events (including measurements of runoff, conductivity, oxygen-18).

The neutralisation potential towards acid mine drainage was investigated by the $^{32/34}$S method.

Results

Hydrochemical analysis of ground and surface waters

The hydro chemical data base of about 40 surface and ground water monitoring locations contains about 800 analytical data sets. All data were analysed using statistical methods.

The natural ground and surface waters refer to the HCO_3-SO_4-type. Due to the natural HCO_3-content, the dump waters are characterised by neutral pH-values and no acidification potential. The natural mineralization reaches 0.2...0.4 mS/cm. The main contaminants of the pore waters are uranium (0,8-2,5 mg/L) and arsenic (28-270 µg/L). In addition, contaminants such as zinc and sulphate are present in the pore water. In contaminated surface waters uranium concentrations range from 0,5-1,5 mg/L, arsenic concentrations from 80-560 µg/L. Radium-226 activity reaches 90 mBq/L, locally.

The radionuclide concentrations in the dump pore waters are 10 times higher than at the natural inflow of Kohlungbach (MP 067) and Wiesenbach (MP 094), see figures 2 and 3. Due to the deposited mining waste containing arsenic ore the arsenic concentrations of Aue Dump 371/II are higher than in Aue Dump 371/ I. The uranium concentrations in the pore waters of Aue Dump 371/I are reaching three times higher values than the pore waters of Aue Dump 371/II. Contaminated pore water was analysed in the ground water in the narrow strips of faulted rock beneath both dumps reaching Zwickauer Mulde River.

Investigation of stable isotopes

The mean residence time of water percolating through the dump is about 2 to 5 months. The ^3H contents in the ground water of the first aquifer yield a mean residence time of about 10 to 20 years. The pore waters of the dump refer to a mixing system of precipitation and an older component. The results of the ^{34}S contents show no significant acidification potential of the dump material. Dump waters is draining into the nearby rivers by interflow as indicated by stable isotopes, partly. The results of the isotopic and hydro chemical investigations are summarised in figures 2 and 3.

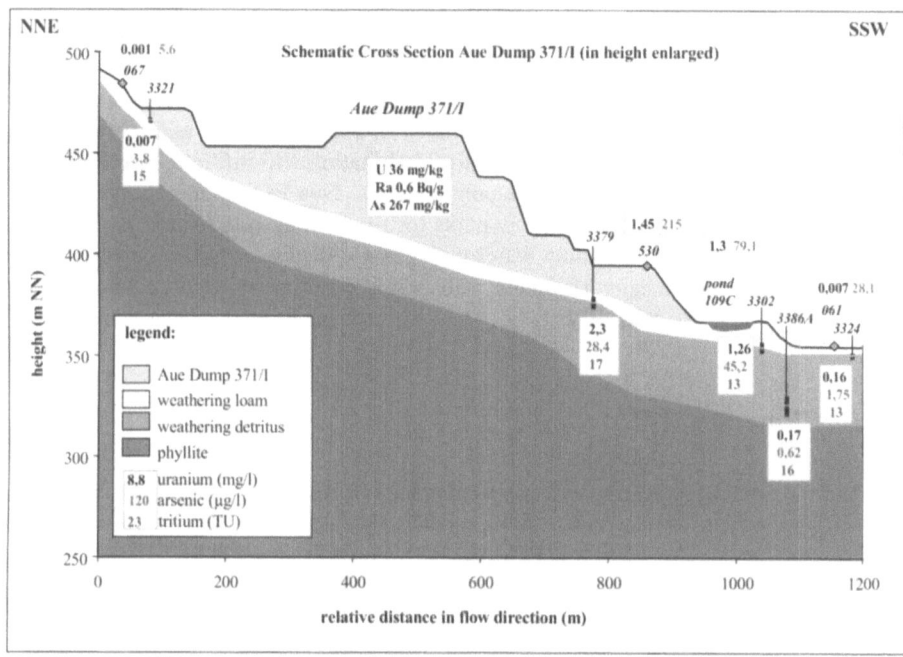

Fig. 2. Schematic cross section of Aue Dump 371/I due to investigation results.

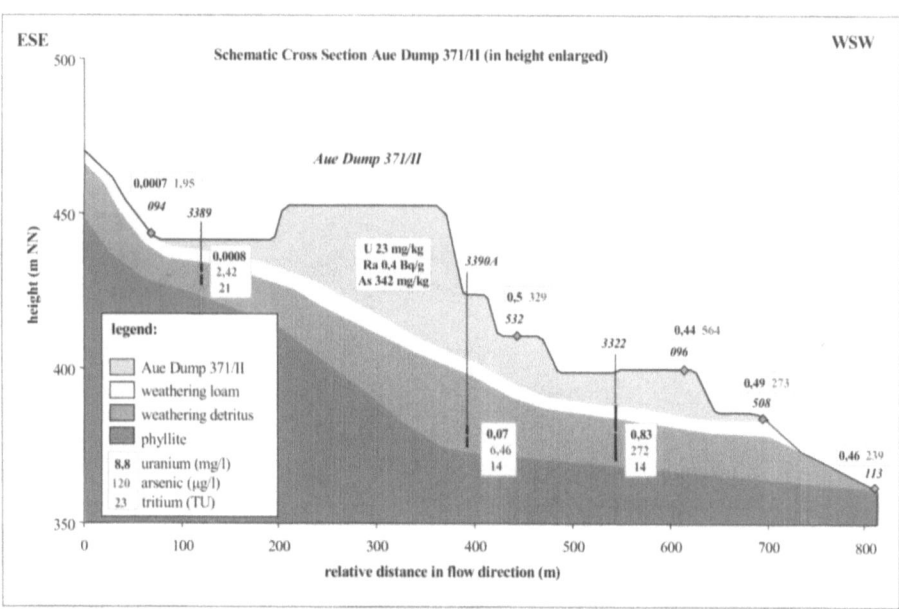

Fig. 3. Schematic cross section of Aue Dump 371/II due to investigation results.

Case studies: Uranium tailings

Conclusions

The impact of Aue Mine Dump 371 on the aquatic environment is determined by the evolution of the contaminants dissolved in the dump water and the amount of water released from the dump per year. Precipitation will infiltrate into the mostly uncovered dump body and eluate contaminants. Due to the heterogeneous structure of the dump, preferential flow paths for percolating waters are assumed to be responsible for the dominating water with low residence times, indicated by variations in $\delta^{18}O$ and recent ^3H values. Subsurface contaminant transport is bound to the weathering zone, mainly.

References

Meyer J, Jenk U, Schuppan W, Knappik R (1998) Hydrogeochemical Aspects of the Aue pit Flooding. In: Merkel B, Helling C (eds.): Uranium-Mining and Hydrogeology, GeoCongress II, S. 124-129, Verlag Sven von Loga, Köln.

Distribution of heavy elements and radionuclides on small uranium mining dumps after decades of soil development

Arndt Weiske, Joachim Rotsche

Dresden University of Technology, Institute of General Ecology and Environmental Protection, Pienner Straße 19, D-01737 Tharandt. Germany. Tel: 49 351 463 31393. Fax: 49 351 463 31399. E-mail: weiske@forst.tu-dresden.de

Abstract. Distribution of uranium and other trace elements was investigated in the surroundings of four approximately 50-year-old small uranium mining dumps. The vertical distribution of uranium was similar in soil profiles both above and downstream of the dump. There was no evidence of enrichment of uranium resulting from a local influence of the dump. Isotope ratios (Pb, U, Th) obtained with quadrupole ICP-MS, showed a sufficient precision level for a screening analysis and monitoring of element translocation into the surrounding of dumps.

Introduction

Beside the large scale devastations from the former uranium mining activities, there are a number of small dumps due to the very first Soviet uranium mining period in the Erzgebirge Mountains (WISMUT 1999). They represent quite small sites, but adjacent surroundings of forest and farm land could present risks from uranium and radionuclide contamination through migration beyond the dump boundaries. This prompted a study investigating the distribution of uranium and other elements in order to establish the extent of possible contamination and hence evaluate the extent of the possible risks for land use. Four dump sites were chosen in the vicinity of Breitenbrunn. One of them was investigated by means of a transect. Two approaches were followed in investigating the migration of elements from the dump: (i)evaluation of gradient distribution from dump related elements along an arbitrary transect line; (ii)determination of the vertical distribution at points on the transect.

Further information about uranium decay equilibrium and screening for daughter elements was made in the course of standard trace element analysis with quadrupole ICP-MS. Hence, the findings about elements and isotope ratios are presented and discussed in this paper.

Methods

Four small dumps from the uranium mining of the period between 1948 and 1955 in the vicinity Breitenbrunn in the western Erzgebirge Mountains were chosen as described in Table 1.

Table 1. Sampling Sites

Site 1	Margarethe mine dump, a forest stand on medium slope
Site 2	Dump Antonshöhe west, a forest stand on medium slope
Site 3	Dump Antonshöhe east, urban and livestock meadow on medium slope
Site 4	Dump Junskoje, forest stand on steep slopes, dump base near the valley floor

A transect line was set cutting across the dump at site 1. Samples were collected along the transect from 5 sampling points above, and 12 downstream the dump. Sampling was according to the soil horizon boundaries (AG Boden 1994), which were found in the Podsolic brown earth. An additional sampling point was set outside the transect at the lower end of the dump on a ditch influenced by seepage water discharge.

Sampling procedures at sites 2, 3 and 4 were similar. Sampling points were chosen on the top of dump plateau, the slope (bare soils), the base, and surroundings. They were horizontally replicated three times. A assortment of samples (dump slope, dump base, dump surroundings <1m distance) was subjected to the screening for parameters of the uranium decay series.

The soil samples were prepared according to standard procedures (Dudel et al. 1997). An agate ball mill (FRITSCH) was used for milling, and a microwave oven (CEM Model MDS 200, USA) for digestion. All samples were diluted with 2% nitric acid in dilution ratios of 10:1, and sometimes 100:1 for analysis. All chemicals were used in suprapure (Merck) quality, and de-ionised water (18.0 MΩ cm^{-1}) from a cardridge device (Barnsted). Analytics for trace element were performed with a quadrupole ICP-MS, a PQ2+ from VG Elemental (Dudel et al. 1997). Common isotopes of trace elements were counted with standard dwell time of 10.24 milliseconds, while the dwell time for ^{234}U and ^{230}Th was prolonged to 102,4 milliseconds. An external calibrations was used for the total element concentration, and for the isotopic ratios, the drift corrected count integrals were used.

Results and discussion

Fig. 1 shows the behaviour of dump related elements in an organic layer at Site 1. Increased element levels on the part of the transect below the dump was not clearly observed as the direct influence of the dumping. Generally, there was no difference between the mean value in the upward and downward zone of the transect observed. However, there was higher variation in the latter. The variation is greatly reduced in the upward zone due to the geological background. In the downstream zone a higher variation was observed. Thus, a hypothesis of linear decrease of element enrichment with a rising distance from the dump is rejected. Instead, it can be assumed that impact from the dump into the downstream zone with flushing precipitation etc. takes place with a high spatial heterogeneity.

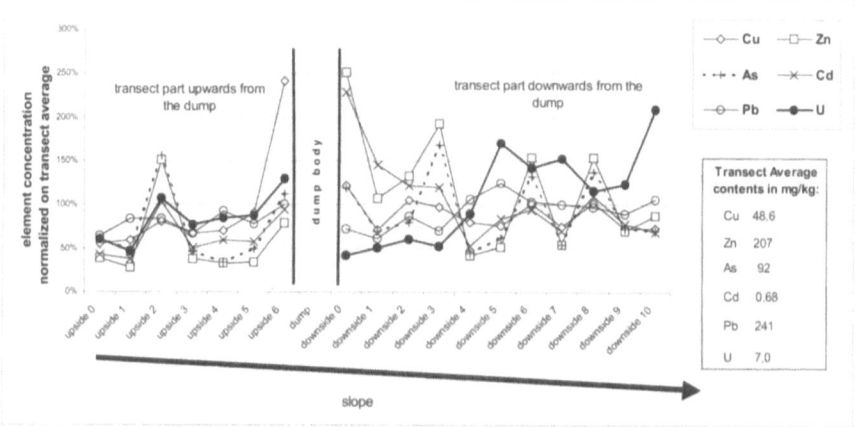

Fig. 1. Relative element concentrations in Oh layer along the site #1 transect.

The vertical distribution of elements was analysed in order to find an increasing effect of element output from the dump in the downstream part of the transect, according to a gravity gradient. Fig. 2 shows the vertical distribution of elements in the transect. Data transformation was done according to Klinger & Fiedler (1995). Element data of single soil profiles had been normalised to a profile mean value, and then, a transect mean value was made, one for upward, and one for the downward part of the transect.

Lead, uranium and arsenic are element type with topsoil enrichment (Alloway 1999). An increased topsoil enrichment in the downward part of the transect was not evident (Fig. 2a-c). The vertical distributions of lead, uranium and arsenic, show the same trend in both parts of the transect. The enrichment in topsoil of lead (Fig. 2a) was found to lie on the lower boundary reported by Colbourn &Thornton (1978) for mining contaminated soils. Uranium (Fig. 2b) showes a similar behaviour like lead. It can be stated, that the organic layer is the most effective sink for mobile uranium in the soil. A significant deviating behaviour was observed for arsenic (Fig. 2c). While lead and uranium show maximum concentration in organic layer, it was in Ah horizon for arsenic. Such a difference was not described by Alloway (1999). It should be a subject of further investigation to focus on root influence and chemical speciation. Samples from all sites (dump slope, base, and surroundings) had been chosen for a screening of parameters of the uranium decay

Case studies: Uranium tailings

series. Aim was, to test the possibility to win such information under the condition of standard trace element analysis.

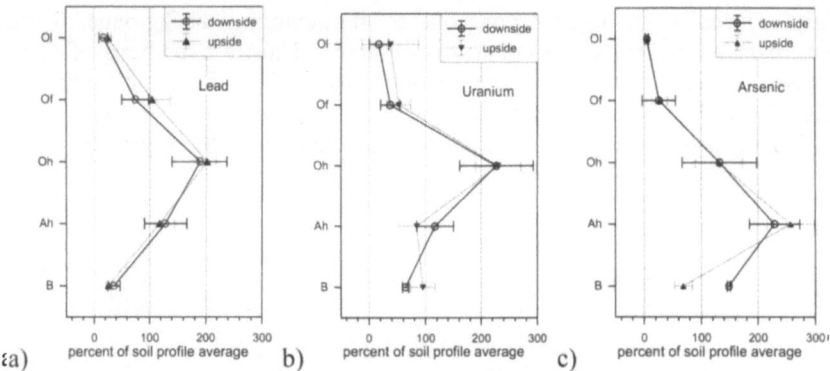

Fig. 2a-c. Vertical distributions of lead, uranium and arsenic in the transect at site 1

The ^{206}Pb isotope is the stable decay product of ^{238}U. Fig. 3 shows the content of the ^{206}Pb isotope in selected samples from the sites. For comparison a pure tailing sample (Lengenfeld) is added. Bracke & Satir (1995) reported lead contaminants in Schlema, and Horn et al. (1995) worked with tracers in the Königstein Uranium mine. Both groups used high precision data, obtained from TIMS. They reported a strong enrichment of ^{206}Pb in lead, which originated from uranium ore.

Similar data can be obtained with ICP-MS, as an extension of standard trace element analysis. Precision level is lower than TIMS, but it's sufficient for a screening analysis to mark lead from uranium ore (Fig. 3). The highest value for ^{206}Pb was observed in the reference sample from the Lengenfeld tailing, with a known enrichment of uranium ore residues. From Fig. 3, it can be concluded, that use of ICP-MS is giving a good finger-print information. From all analysed samples, only yC samples showed lead with an increased level of ^{206}Pb. These were materials from a non-covered dump slope and dump plateau in Antonshöhe east. This Pb shows the direct presence of ore residues in the dump.

All other samples, especially organic layers including dump bases, represent a "normal crustal enrichment" (Horn et al. 1995). This means that no high amounts of lead had been transported from the dump to the surroundings. Fig. 4 shows the ^{234}U/^{230}Th ratios for the same samples as discussed for the lead Isotopes. Similarly, a limited level of precision for these data must be stated, because a validation is still in progress. The standard method for determination of ^{230}Th is also TIMS, or High resolution ICP-MS (ThermoFinigan 2001).

The highest values of the ^{234}U/^{230}Th ratio were observed in humic topsoil layers at the dump bases. The lack of equilibrium as shown in original uranium has its origin in a selective accumulation of uranium, without of decay product, such as ^{230}Th. This is due to the higher mobility of uranium, leached and transported to the dump base, while ^{230}Th is remaining in the dump.

Case studies: Uranium tailings

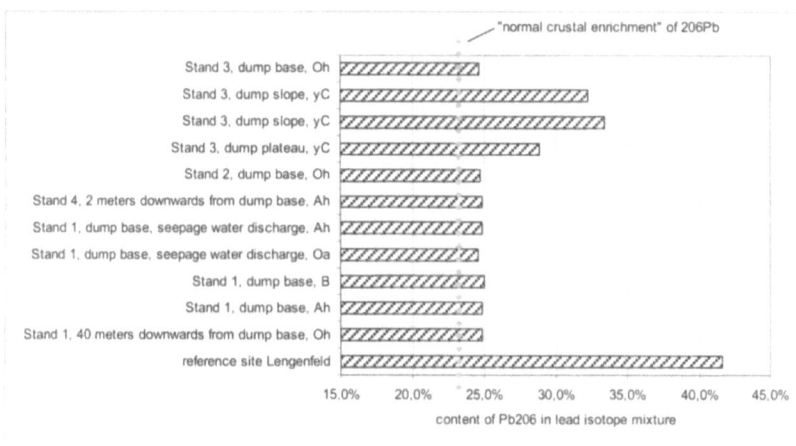

Fig. 3. Distribution of ^{206}Pb in the lead isotope composition in selected samples of all four Sites. Additionally a sample from the Lengenfeld tailing as a reference

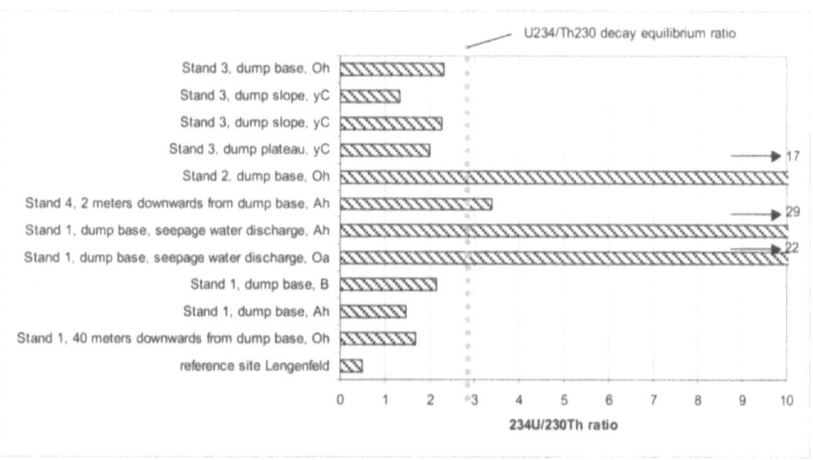

Fig. 4. ^{234}U/^{230}Th ratio for selected samples of all four Sites. Additionally a sample from the Lengenfeld tailing as a reference

The lead at the dump base shows "normal crustal enrichment" (Horn et al. 1995). It can't have its origin in the dump slope, where the ^{206}Pb is strongly enriched (Fig. 3). In conclusion, at the dump slope ^{230}Th and lead have a decreased mobility, while transportation of uranium to the dump base takes place.

The reference sample from the Lengenfeld tailing shows the highest similarity to the yC-Samples from the dump slope:

- ^{206}Pb is enriched because of the presence of uranium ore minerals

Case studies: Uranium tailings

- the $^{234}U/^{230}Th$ ratio is shifted to a uranium depletion. The reason is selective leaching in the dump, and the industrial uranium extraction process in the tailing material.

Conclusions

A severe contamination in the surroundings of small uranium mining dumps was not observed, but a accumulation at the dump base was observed.

An extension of standard analysis parameters in quadrupole ICP-MS is possible in order to get information on uranium decay equilibrium in samples from dumps and surroundings. Information on ^{206}Pb and ^{230}Th was found to be available at screening quality. It is proposed to use these easy available $^{234}U/^{230}Th$ and $^{206}Pb/^{208}Pb$ ratios to use for characterization of dumps and their surroundings.

Acknowledgements

Much thanks to Mrs. cand. geol. M. Seidel for allowing to use some of her samples from Lengenfeld.

References

AG Boden (1994) Bodenkundliche Kartieranleitung, Hannover.

Alloway BJ ed. (1999) Schwermetalle in Böden. Berlin Heidelberg: Springer

Wismut (1999) Chronik der Wismut. WISMUT GmbH Chemnitz (on CD)

Colbourn P, Thornton I (1978) J Soil Sci 29: 513-526

Bracke G, Satir M (1995) Lead isotope ratios as fingerprints of contamitants at the uranium mining site of Schlema-Alberoda/Saxony. .- In Uranium-Mining and Hydrogeology.- (Eds. Merkel B, Hurst S, Löhnert E, Struckmeyer, W) 1-3, Sven v. Loga, Köln

Dudel EG, Brackhage C, Dabrunz M, Passek U, Rotsche J, Stolz L (1997): Freisetzung radioaktiver Elemente aus offenen und abgedeckten Bergbau-halden: Einflüsse auf eine land- und forstwirtschaftliche Nutzung. Abschlußbericht, TU Dresden, Tharandt

Horn P, Hölzl S, Nindel K (1995) Uranogenic and thorogenic lead isotopes (206Pb, 207Pb, 208Pb) as tracers for mixing of waters in the Königstein Uranium-Mine, Germany.- In Uranium-Mining and Hydrogeology.- (Eds. Merkel B, Hurst S, Löhnert E, Struckmeyer, W) 281-289, Sven v. Loga, Köln

Klinger T, Fiedler HJ (1996): Zur Spurenelement-Vertikalverteilung in Waldbodenprofilen Ostdeutschlands. Chemie der Erde – Geochemistry 56: 65-78

SÄCHSISCHES LANDESAMT FÜR UMWELT UND GEOLOGIE (2000) Bodenatlas des Freistaates Sachsen. Teil 3 – Bodenmessprogramm - Bodenmessnetz Raster 4 km x 4 km (electronic record)

ThermoFinigan (2001): Neptune – A User instructing software. Thermo Finnigan MAT GmbH 2001

Natural attenuation in a wetland under unfavorable conditions - Uranium tailing Schneckenstein / Germany

Anja Landgraf, Britta Planer-Friedrich, Broder Merkel

University of Mining and Technology Freiberg, Department of Geology, Gustav-Zeuner-Str.12, 09599 Freiberg / Germany, a.landgraf@web.de

Abstract. The uranium tailing at Schneckenstein / Germany is one of the tailings which was abandoned in the late 50's respectively early 60's. 45 years after the closure of the site, the downstream environment is still receiving considerably amounts of uranium and other contaminants. While background uranium concentrations for surface water and ground water in the region are around 1 µg/l, the tailing seepage shows uranium concentrations between 400 and 1500 µg/l with a yearly uranium load of 70 to 100 kg. Due to drainpipes and dam leakage a small wetland developed at the dam foot. This small wetland was investigated with respect to its removal capacity for metal(loid)s, especially uranium, and arsenic.

Introduction

Historical development

Historical mining activities within the Schneckenstein area, in the southwestern part of Saxony / Germany (Fig. 1), date back to the year 1506, when tin, arsenic and pyrite were mined. From 1934 to 1946 the Sachsenerz AG Freiberg mined and processed tin within the Falkensteiner tunnel. From 1938 on processing residues were deposited within a first small tailing, that was still used when Wismut SDAG took over the processing plant in 1947 and established their Wismut object No. 32 (VEB Wolfram-Zinnerz Pechtelsgrün 1963).

Wismut processed Uranium ores from several small uranium mines in the vicinity (e.g. Gottesberg) but as well from the entire state of Saxony and Thuringia. Uranium contents in these ores were considerably low (200-1200 gU/t, average

700 gU/t). Processing was done by radiometric, gravitational and chemical leaching (both acid and basic), enriching the uranium concentration to about 4 mg U/L.
The capacity of the milling site was 150 kt per year. With an uranium yield of
80% 1200 kt ore were processed during the 11 years of production (Gottschalk
1997, Kutschke 1998).

Fig. 1. Location of the Schneckenstein area in the southwestern part of Saxony / Germany

The sandy to silty residues resulting from the chemical leaching of the Uranium
ores required larger deposit sites than the first small tailing that was already used
by Sachsenerz AG Freiberg for tin processing. In May 1948 the dam for a first settling pond (IAA II) was built, in 1951 a new dam for a second settling pond followed (IAA I) (Fig. 2). The volume of both tailings comprises 700,000 m^3 (Rat
des Bezirkes Karl-Marx-Stadt 1965).

Just a few years after the mine was shut down in 1957 and handed over to VEB
Wolfram-Zinnerz Pechtelsgrün, severe erosion problems occurred leading to instability of the northern dam site. Sediments from the tailings were spilled to the
Bodabach, a small creek with a yield of about 5 L/s, and deposited downstream in
numerous small basins and lakes towards the village of Tannenbergsthal.

In the early sixties the tailing's surface was covered by a 1 to 5 m thick heap
material layer, surface water was diverted by hanging ditches, the northern dam's
inclination was reduced from 1:1 to 1:2.5 and eight existing drainage pipes were
lengthened (Fig. 3). Their drainage water merged with the water from the eastern
hanging ditch and finally discharged to the Bodabach creek. Reforestation with
spruce and birch followed, and lupin was planted for soil amelioration (VEB
Bergsicherung Schneeberg 1965). No further rehabilitation work or monitoring
was done until nowadays.

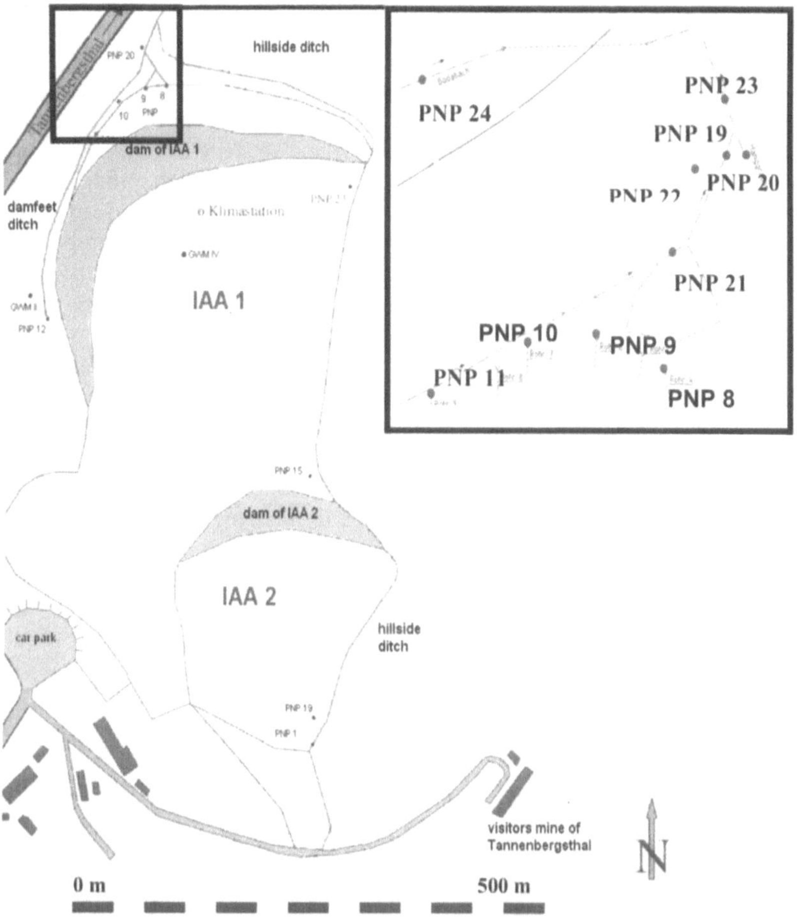

Fig. 2. Uranium milling and mining site Schneckenstein: tailing IAA 2 built in 1948, tailing IAA 1 built in 1951; zoom to the wetland at the northern tailing foot with sampling points

Fig. 3. Reshaping the northern tailing dam inclination from 1:1 to 1:2.5 and lengthening of the eight existing drainage systems in the late 1960s; sampling points as indicated in Fig. 2

Case studies: Uranium tailings

Tailing hydrology - wetland

According to Merkel et al. (1998a) the vertical seepage rate through the tailing is roughly 0.1 m per year based on Tritium profiles taken from four drillings in both tailings. Assuming an effective pore volume of 0.2, a seepage rate of 20 mm per year is resulting. Taken into account an area of 0.06 km^2 for both tailings, a discharge of about 0.04 L/s leaking from the tailing can be assumed. However, the tailing leakage is mixing with ground water (proximately 1 L/s) flowing beneath the tailing in the weathered zone of the underlying granite. By means of the old drainage systems this ground water is drained partly at the dam foot, however, flowing as well in the shallow groundwater of the Bodabach valley.

Nowadays due to continuing erosion and hang sliding on the tailing dam some of the old drainage pipes are totally or partially damaged. Only five tubes of the former drainage system can still be identified. In front of PNP 9 a small puddle has formed, maybe the discharge of a further former drainage system. Most of the tubes show only seasonal discharge, PNP 10, with 10-15 L/s the only one with continuous flow, is likely to be the discharge of a former spring which was tapped by the Wismut before constructing the tailing site IAA I (Rat des Bezirkes Karl-Marx-Stadt 1965). The water from the remaining tubes, as well as diffusive seepage from the tailing dam discharges to a wetland of about 30 m x 60 m, before merging with the eastern hanging ditch and the Bodabach (Fig. 4).

Fig. 4. Old drainage systems discharging to the Schneckenstein wetland (PNP 11 left and PNP 10 left)

It is well known by now that wetlands can serve as natural water treatment systems for mine waters, reducing contaminant potential by precipitation, sorption, ion exchange, microbial degradation, etc. In the study presented the remediation effect of the wetland at the Schneckenstein tailing dam site for natural attenuation was investigated. The external conditions differ significantly from other wetlands, known to be efficient in contaminant removal. The Schneckenstein wetland has formed on a rough gravel layer, the relics of the dam reconstruction measures. Mud and sediment, as potential sinks for contaminants by means of coprecipitation, sorption or ion exchange, form only a small layer of several centimeter thickness at the most. Due to high average annual precipitation of about 900-1000 mm/a, low average annual temperatures of about 5°C combined with a short growing season, bioactivity is low and the organic upper layer is also reduced in thickness (1-3 cm at the most). Additionally since the downward gradient of the wetland is relatively steep and the discharge is pretty high (10-15 L/s), retention times are low.

Measurements were conducted on discharge and concentrations at different control points upstream and downstream of the wetland. Onsite parameters, major cation and anions, N- and Fe-species, DOC, TOC, metal(loid)s (ICP-MS, AAS), were analyzed. Radium, and radon were determined by means of the emanation method. Additionally some soil and moss samples were taken and analyzed by means of solid phase AAS. Special emphasis was put on arsenic and uranium.

Results and Discussion

Fig. 5 shows arsenic concentrations in the aquatic phase at the control points in the vicinity of the dam foot. No significant differences between the three sampling campaigns can be seen.

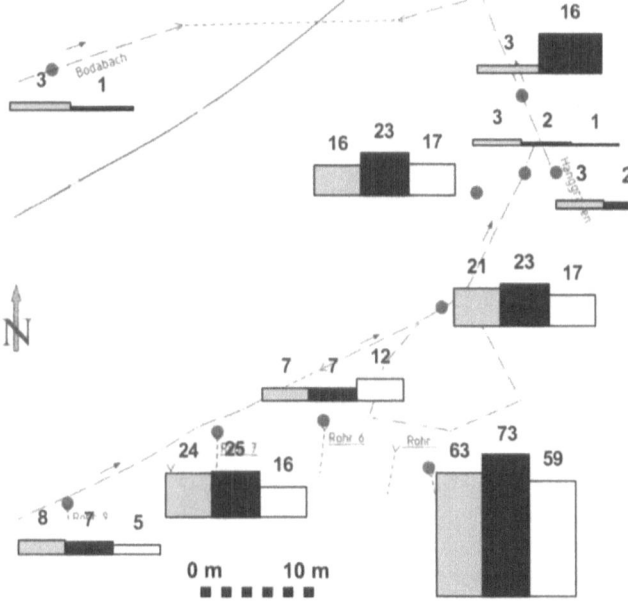

Fig. 5. Arsenic concentrations in µg/L (left column: sampling 08/2001, middle: 10/2001, right 05/2002); sampling points as indicated in Fig. 2

Sampling point PNP_8 has elevated concentrations with 59 to 73 µg/l. However, the major part of the yearly load of 7 to 8 kg of arsenic is linked with the drainage of the tapped spring (PNP_10) with relatively low concentrations but higher discharge (table 1).

Table 1. Wetland budget for arsenic calculated from two sampling events

Arsenic		10/2001			05 / 2002		
		Q [L/s]	As [µg/L]	As [kg/a]	Q [L/s]	As [µg/L]	As [kg/a]
input	PNP 11	1.6	7.41	0.37	3.0	5.26	0.50
	PNP 10	10.0	25	7.88	12.0	16	6.05
	PNP 9	1.1	6.87	0.24	1.1	11.5	0.40
	PNP 8	1.4	72.6	3.21	2.6	58.8	4.82
	PNP 22	0.3	2.25	0.02	0.3	1.88	0.01
total input				**11.72**			**11.91**
output	PNP 19	11.9	22.7	**8.52**	14.0	16.9	**7.46**
difference				**3.20**			**4.33**

On contrary to arsenic the uranium load of the tapped spring contributes only about 50% to the total load (table 2). The highest concentration for uranium was found for the sampling point PNP_9 with about 1000 mg/l (Fig. 6).

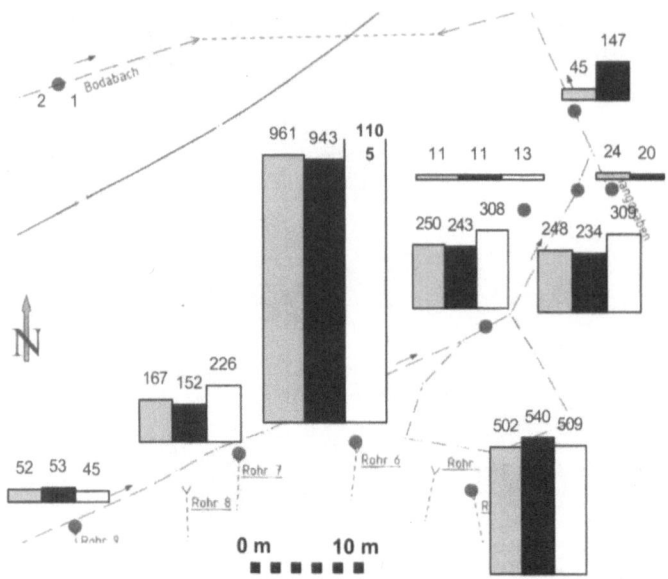

Fig. 6. Uranium concentrations in µg/L (left column: sampling 08/2001, middle: 10/2001, right 05/2002); sampling points as indicated in Fig. 2

Case studies: Uranium tailings

According to Merkel et al (1998a) uranium concentrations range from 800 to 1500 µg/l at this sampling point correlated with C_{org}-concentrations between 2 and 15 mg/l, which can be taken as an indicator for complexation of uranium by humic matter (Merkel et al. 1998b). PNP_9 is located in the deepest part of the tailing which could be the reason for the high concentrations.

Table 2. Wetland budget for uranium calculated from two sampling events

Uranium		10 / 2001			05 / 2002		
		Q [L/s]	U [µg/L]	U [kg/a]	Q [L/s]	U [µg/L]	U [kg/a]
Input	PNP 11	1.6	53.4	2.69	3.0	44.6	4.22
	PNP 10	10.0	152	47.93	12.0	226	85.53
	PNP 9	1.1	943	32.71	1.1	1105	38.33
	PNP 8	1.4	540	23.84	2.6	509	41.73
	PNP 22	0.3	11.1	0.09	0.3	12.8	0.10
total input				**107.3**			**169.9**
output	PNP 19	11.9	234	**87.82**	14.0	309	**136.42**
difference				**19.5**			**33.5**

The budgets of both arsenic and uranium show some evidence that the wetland has a small but significant impact on the arsenic and uranium load, removing about 20% of the annual load. This is confirmed by the increased arsenic concentrations found in moss with 32 to 70 mg/kg respectively arsenic in soil samples with 740 to 2100 mg/kg.

Conclusions

The downstream environment of the Scheckenstein uranium tailing is receiving a considerable load of uranium and as well of arsenic though the milling and dump site got abandoned 40 years ago. A small wetland developing over time at the dam foot is diminishing the uranium and arsenic load by roughly 20% despite its small size and unfavorable climatic boundary conditions. Thus it is likely that a constructed wetland could improve the environmental situation.

References

Gottschalk S (1997) Hydrogeologische Untersuchungen am Uran-Tailing Schneckenstein, unveröffentlichte Diplomarbeit, Institut für Geologie, TU Bergakademie Freiberg.
Kutschke S (1998) Weiterführende hydrogeologische Untersuchungen an der Industriellen Absetzanlage Schneckenstein, unveröffentlichte Diplomarbeit, Institut für Geologie, TU Bergakademie Freiberg.

Merkel B, Preußer R, Namoun T, Gottschalk S, Kutschke S (1998a) Natural Leaching of Uranium from the Schneckenstein Uranium Mine Tailing. In: Merkel B. & Hellig C. (Eds) (1998): Uranium Mining and Hydrogeology II. Proc. of the Intern. Conference and Workshop, Freiberg, Germany, Verlag Sven von Loga, Köln

Merkel B, Peter HJ, VOLKE P (1998b) Investigation of Uranium Sorption on Humic Acids. In: MERKEL B. & HELLING C. (Eds) (1998) Uranium Mining and Hydrogeology II. Proc. of the Intern. Conference and Workshop, Freiberg, Germany, Verlag Sven von Loga, Köln

Rat des Bezirkes Karl-Marx-Stadt (1965) Abteilung für Wismutangelegenheiten; Sicherung des Schlammabsetzbeckens in Schneckenstein; Brief an den VEB Bergsicherung Schneeberg; Karl-Marx-Stadt.

VEB Bergsicherung Schneeberg (1965) Sicherung des Schlammabsetzbeckens Schneckenstein, unveröffentlichtes Gutachten, Schneeberg

VEB Wolfram-Zinnerz Pechtelsgrün (1963) Protokoll über die Befahrung der Klärteiche Objekt 32 bei Tannenbergsthal i.V. mit anschließender Besprechung am 24.05.1963, Pechtelsgrün

Remediation of uranium milling site Lengenfeld/ Vogtland, needs and objectives

Michael Schöpe, Roland Hähne, Bernd Tunger, Ralf Löser

Abstract. The tailings pond Lengenfeld was used to deposit the residues of the uranium ore processing. An investigation program was realised to clarify environmental impact caused by release of radioactive and non radioactive pollutants via water pathway. The assessment was done both for the present and for the long-term situation. The results were used to give recommendations for the remediation of the tailings pond.

History

In 1934 a processing plant in Lengenfeld was set up because of the good location conditions particularly the availability of enough water in the stream Plohnbach. The plant was used for processing of tungsten ore from the mine of Pechtelsgrün near Lengenfeld. The sandy and silty processing residues were deposited as waste rock damps in the surroundings of the processing plant. In addition the residues were used also as building material.

In 1947 the purpose of the plant was changed. Now the SDAG (Soviet German share company) WISMUT used the plant for processing uranium ore from various mining regions in Saxony und Thuringia. The tailings were deposited on waste rock dumps in the first time. Later a tailings pond was established in a northern side valley of the Plohnbach. A dam was built to block off the valley and the stream in the valley was laid out side of the dumping area. During the work of the plant a further dam was built around the pond and the main dam was expanded. Also collecting trenches for seepage water and a drainage system were constructed. In 1961 the discharge of the residues of the Uranium ore processing was stopped with the end of the processing activities. Altogether an amount of about 890.000 m³ tailings with an average thickness of 6 to 8 m was deposited on an area of 14 ha. The tailings pond has no explicit base barrier.

From 1968 until 1989 the tailings pond was used by the baryte and fluorite industry. In the seventies the main dam was strengthened once again to extend the capacity of the tailing pond. The residues of baryte and fluorite processing cover the uranium tailings with a layer thickness of about 8 m. In some areas the thickness of the baryte and fluorite cover less than 1 meter was found, so the cover of the tailings is partially relative low.

The western dam and the main dam of the tailings pond mainly consist of sandy residues of the ore processing. After a dam break in 1954 the dam was stabilised with granite blocks. More than 100000 m³ tailings probably flowed out by the dam break. Thereby the downstream area particularly two small lakes in the Plohnbach valley and a march area near the tailings pond was contaminated by tailings of the uranium ore processing. Also the western dam was broken in a local area in 1985. However the environmental impact was relatively less at this second incident.

Composition of the tailings of the uranium ore processing

In the nineties an investigation programme was realised. In this connection a lot of samples of tailings were analysed on radioactive and not radioactive content. The table 1 shows a summary of the results of this investigation.

Table 1. Content of the tailings

Nuclide	Unit	Mean	Min	Max
Ra-226	[Bq/g]	5	1.3	20
U	[mg/kg]	183	8	385
As	[mg/kg]	442	10	800
Pb	[mg/kg]	124	30	400
Cu	[mg/kg]	405	20	800
Ni	[mg/kg]	160	10	400
Mn	[mg/kg]	1244	300	2500

In the result of the investigation it can be noticed that with regard to the environment influencing the radioactive components and arsenic play the essential roll.

Environmental impact

The tailings of uranium ore processing are covered with the sandy residues of baryte and fluorite processing with a thickness of about 1 to 8 m. Therefore a significant influence of the environment caused by the uranium tailings by air pathway is excluded. Therefore the consideration to the environmental impact is concentrating on the water pathway. To assess the present situation it is necessary to quantify the amount and the quality of seepage water from the tailings pond as well as the quality of ground and surface water in the down stream area. For this the surface water and the ground water was considered.

Surface Water

In order to estimate the influence of the contaminated seepage water to the water quality downstream a water balance for the concerning area was generated. The data base were results of the investigations of the surface water. To investigate the surface water 11 measuring points (MP1 – MP11) were determined. The measuring points were selected considering

- the relevant streaming points of the natural catchment area upstream of the tailings pond,
- all essential components of seepage water of the tailings pond,
- the surface water downstream of the tailings pond (see figure 1).

The selected measuring points enable the calculation of a complete balance of chemical water components for the location. However the water samples only allow conclusions for a limited time period, conditions for a longer time period cannot be characterised by this way. The investigation of the surface water samples gives the possibility to answer essential questions for urgency of remediation and the necessary remediation measures.

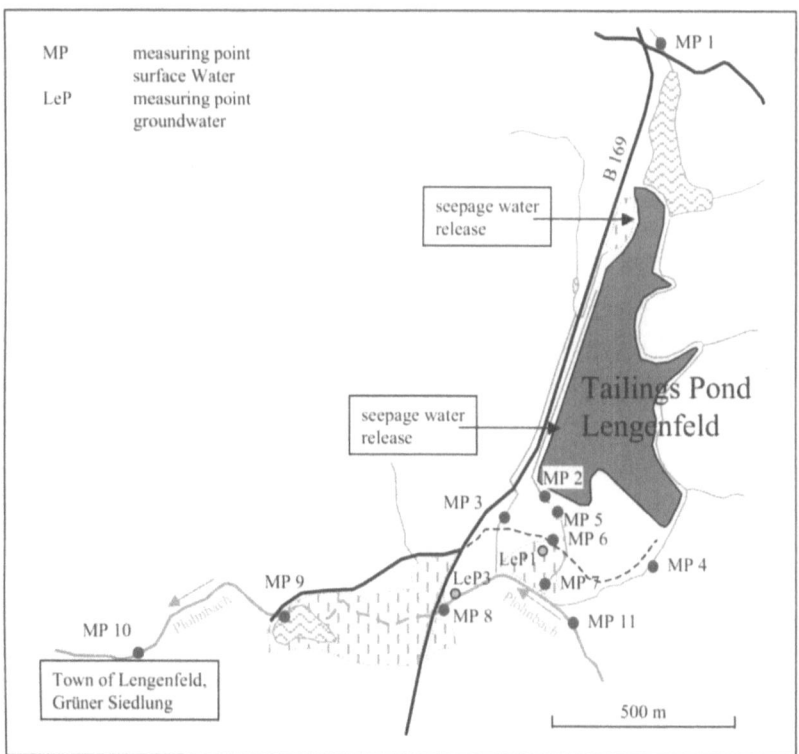

Fig. 1. Position of measuring points for surface and ground water

Case studies: Uranium tailings

For a better explanation of the associations between various surface water streams and the measuring points at the location figure 2 shows a scheme with streams and the positions of the measuring points.

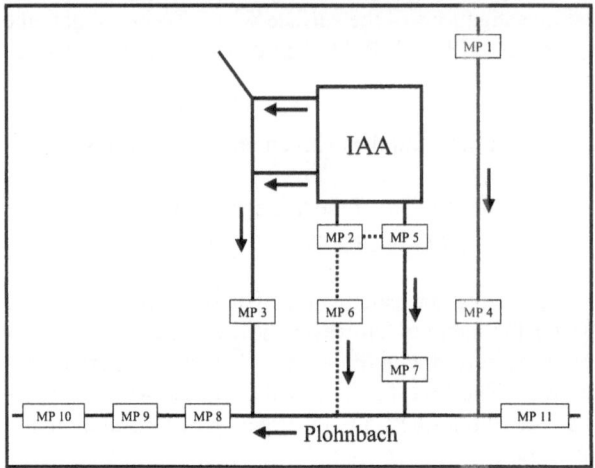

Fig. 2. Scheme of the locations of measuring points

Results of in situ measurement

While taking samples for analytical investigation the in situ parameters of sampled water and the flow rate were measured (see table 2).

Table 2. Results of in-situ measurement

Point No.	pH-value	Conductivity [µS/cm]	O₂-Concentration [mg/l]	Temperature [°C]	Flow rate [l/s]
MP 1	7.42	296	6.3	16.4	4.5
MP 2	7.06	1490	6.7	14.8	1.0
MP 3	7.39	456	6.0	14.8	2.5
MP 4	7.43	318	5.9	16.6	5.5
MP 5	7.09	1530	6.3	14.6	1.0
MP 6	7.20	1056	3.0	19.2	0.5
MP 7	7.17	1436	5.7	17.9	1.5
MP 8	7.55	288	5.8	18.5	135
MP 9	7.35	336	3.9	21.6	-
MP 10	7.33	325	5.3	18.6	145
MP 11	7.34	256	5.7	18.5	120

The waters of MP 2, MP 5 and MP 7 are very similar in respect of the electric conductivity. This water are clearly influenced by the tailings pond and can be declared as seepage water.

Near MP 6 the seepage water of the tailings pond flows diffuse through a swampy area. The concentration of the pollutants degreases. The water temperature and the low content of oxygen is an indication of a low flow rate. On the other hand, the water near MP 7 flows in a ditch to the stream Plohnbach without a significant dilution.

The electric conductivity of the Plohnbach water increases from the uninfluenced surface water before the tailings pond (MP11: 256 µS/cm) to the receptor (MP10: 325 µS/cm) downstream. This shows the influence of the leakage water of the tailings pond with higher electric conductivity to the surface water of the stream Plohnbach.

Results of analytical investigations

The 11 samples of surface water were investigated for the parameters uranium, arsenic, sulphate and hydrocarbonate. Additional radionuclides were analysed to determine the nuclide vectors of six samples.

Table 3. Results of analytical investigations

Sample No.	As [µg/l]	U [mg/l]	Ra-226 [mBq/l]	Pb-210 [mBq/l]	Po-210 [mBq/l]	SO$_4$ [mg/l]	HCO$_3$ [mg/l]
MP 1	0.35	0.001	0.3	0.1	0.1	56	27.5
MP 2	282	0.470	90	10	21	542	296
MP 3	5.5	0.020	5.9	1.0	2.4	81.5	88.5
MP 4	1.9	0.002	0.6	0.1	0.2	53.1	48.8
MP 5	1500	0.480	105	15	89	569	312
MP 6	13.3	0.250	166	25	22	341	241
MP 7	41.2	0.440	82	15	34	477	293
MP 8	1.9	0.006	17	10	17	61.3	48.8
MP 9	3	0.005	24	10	13	65.9	67.1
MP 10	3.5	0.011	3.2	0.6	1.3	71.2	58
MP 11	1.7	0.001	0.3	0.1	0.1	54.7	45.8

The concentration of arsenic of MP 2 and MP 5 is extremely high. A remarkable fact is the evident reduction of arsenic to MP 6 and MP 7, which are located in a low distance from MP 2 and MP 5, on the other hand differences in the specific activities are not so clear.

To show the consistence of the analytical results the changes in uranium and sulphate concentration of the Plohnbach were checked between the water before (MP 11) and behind (MP 8) the tailings pond. The influence of the load of seepage water (MP 2, MP 3, MP 5) and of the stream from northern catchment area (MP 4) to the Plohnbach was calculated.

Table 4. Comparison of concentration change for uranium and sulphate

Measuring Point	Flow rate L/s	Concentration		Load	
		U [mg/l]	SO₄ [mg/l]	U [kg/a]	SO₄ [t/a]
Measured data					
Inflow MP 11	120	0.001	54.7	4	207
Outflow MP 8	130	0.006	61.3	25	251
Calculated data for Outflow on basis of the investigation results					
Weighted mean MP 2 to MP 5 and MP 11	130	0.009	62.9	37	258

The comparison of the measured and calculated data for the Outflow shows that the sulphate concentrations are nearly the same, but the calculated uranium concentrations is higher than the measured concentration. Because of the aerobe condition in the swamp area no reduction of sulphate content occurs, but the measurements show the sorption of uranium especially by organic substances. Obviously about 1/3 of the uranium content in seepage water is absorbed in the swamp area between the main dam of the tailings pond and the stream Plohnbach. So the swap area acts as a "wetland", witch improves the water quality. The subject of current investigations is the "wetland" function in order to determine the main processes, the effectiveness and the long time effects.

Nuclide Vector

For estimation of the radiation dose it is necessary to investigate the nuclide vector. Therefore the activity concentration of typical water of the location were analysed. The results of the investigation are shown in the following table 5.

Table 5. Nuclide vectors

Nuclide	U-238 [mBq/l]	U-234 [mBq/l]	Th-230 [mBq/l]	Ra-226 [mBq/l]	Pb-210 [mBq/l]	U-235 [mBq/l]
Inflow (MP 11)	17	24	0,50	0,39	0,07	0,78
Seepage water (MP 5)	5183	4875	118	111	16	240
Outflow (MP 10)	110	195	51	15	7	5

Groundwater

The analytical results of the groundwater monitoring were used to assess the groundwater pathway. The measuring point Lep1/91 represents the groundwater run off directly influenced by seepage water of the tailings pond and the tailings which are here caused by the dam break. The measuring point Lep3/94 represents the quality of remote groundwater downstream of the tailings pond. The following average analytic results are available:

Table 6. Quality of groundwater

Measuring Point	U [μg/l]	Ra-226 [mBq/l]	As [μg/l]
LeP1/91	1165	35	2
LeP3/94	38	66	4

The observation of migration process, witch is necessary to prevent environmental damages, is given by the current monitoring system around the tailings pond.

Environmental Assessment

Principle of assessment

The environmental assessment has to be done is valid for an existing situation. Therefore the assessment criterion in regard to radiation protection is the guide value of 1 mSv/a for the effective dose.

In regard to non radioactive pollutants Arsenic is the relevant species for environmental assessment. In the aspect of the water quality an arsenic concentration limit of 10 μg/l is used as a value of reference (German regulation for drinking water - TVO).

Present situation

At this time there isn't any relevant use of surface or groundwater in the area between the tailings pond and MP 10 (see figure 1). Therefore the inhabitants of the "Grüne Siedlung Lengenfeld", witch is located downstream near MP 10, will be considered as potential user. At present use of ground or surface water as drinking water can be excluded, because the existing water supply system.

The private use of Plohnbach water for irrigation in gardens is used to assess the pathway of exposure. Besides both radiological exposure and exposure in regard to conventional pollutants are investigated.

Assessment of radiation protection

To estimate the effective dose it was supposed that the concerned persons cover with their own harvest about 50 % of the annual need of fruit and vegetable. For the assessment the concentration of radio nuclides in outflow water (MP 10) is used (see table 4). The results of dose calculation is shown in the following table 7.

Table 7. Dose valuation for the actual situation

Path of exposition	Dose [mSv/a]	
	Adult	Child
Water of Plohnbach in gardens	0,04	0,03

The calculated dose results are clearly less than 1 mSv/a. Therefore it can be noticed that in the current situation there is no unacceptable radiation exposition for the population arising from the radio nuclides release of the tailings pond by water pathway.

Assessment of non radioactive pollutants

As a possible pathway of exposure will be also considered the private water consumption of Plohnbach water in the gardens of "Grüne Siedlung Lengenfeld". At the potential point of water in take MP 10 a concentration of arsenic of about 3,5 µg/l was found. This is less than a half of the arsenic limit of the German regulation for drinking water (TVO). That means, the water of stream Plohnbach at MP 10 could be used as drinking water with regard to arsenic. Therefore the water may be used in gardens without an unacceptable exposition for the health of the population.

Long-term situation

Assessment of radiation protection

To assess the long-term situation a period of at least 200 years have to be considered. The development of the contaminated swamp area near the tailings pond in the future, especially as a residential area, is not probable. A use of the area of the tailings pond including the near surroundings e.g. for relaxation or for horticulture cannot be excluded. According to a sufficient conservative assessment it was assumed, that a water in take for use in gardens occurs in the swamp area, before the surface water reaches the Plohnbach.

Furthermore a use of groundwater in the area downstream the tailings pond as drinking water can be not excluded. But a water in take in the area immediately close to the main dam (LeP1/91) is not probable. Therefore the quality of groundwater of the measuring point LeP3/94 was used to assess these exposure pathway. The calculated dose are represented in the following table for both exposure paths.

The result is, for people which are exposed to both exposure pathways simultaneously a exceeding of the limit of ≥ 1 mSv/a cannot be excluded for long-term situation. The dose mainly results from the directly use of seepage water close to the main dam of the tailings pond.

Table 8. Dose valuation for long-term situation

Path of exposition	Dose [mSv/a]	
	Adult	Child
seepage water use for ir-regation in gardens	0.74	0.47
Groundwater of LeP3/94, use as drinking water	0.08	0.12

Assessment of non radioactive pollutants

The average concentration of arsenic in seepage water is about 100 µg/l, observed over a period of several years. That doesn't correspond to the limit of the German regulation of drinking water (TVO) of 10 µg/l. On the other hand the groundwater don't show any exceeding of TVO limits, neither close to the tailings pond nor in a larger distance downstream.

A potential relevant exposure of non-radioactive substances in the long-term situation is given exclusively by use of seepage water of the tailings pond.

Conclusions and recommendations

The subject of the environmental assessment of the tailings pond in Lengenfeld is the present and the long-term situation. The results for the present situation can be used to determine the urgent needs for remediation. The results of the long-term situation primary serve for the specification of the necessary remediation measures.

The goal of the remediation of the tailings pond with regard to environmental impact is to reach an acceptable situation, witch can be kept over a long time period without additional extensive measures for maintaining this condition.

As a result of the investigations no unacceptable environmental impact caused by radioactive and non radioactiv release in the currrent situation. Therefore there is no urgent need for remediation actions.

Supposing a private use of seepage and ground water in the area close to the tailings pond in the long-term case, a potential effective dose results, witch exceeds the recommended limit of 1 mS/a. There is an additional potential risk for the receptors because of the relative high concentration of arsenic in the seepage water, if the water directly will be used as drinking water. Therefore there are needs for remediation actions for the tailings pond Lengenfeld in the long term.

The essential aim of remediation is the avoidance the use of seepage water around the western and southern area of the tailing pond. So the following recommendations for remediation were given:

- By forming the surface contour of the main dam the diffuse and hidden outflow of seepage water to the surface must be removed. A central drainage system to collect the seepage water should be built to avoid an unregulated in take of contaminated water and to allow a controlled discharge into the Plohnbach.

- In connection with the restoration of the western dam a drainage for the seepage water should be installed. This drainage must be connected with the central drainage system near the main dam.

- The remediation of the tailings pond must be done considering the requirements of landscape design, environmental conservation and land use.

- It is recommended cover the tailings pond with natural soil material and initiate a dense vegetation cover to increase the water consumption of vegetation and to decrease the water amount infiltrating into the tailings. Thereby the amount of pollution release can be reduced for a long term, without that any expensive measures are necessary for the maintenance of the condition after the remediation.

- The monitoring of ground and surface water quality should be continued to recognise changes in concentrations, water amounts and loads of pollutants in the seepage water and the ground and surface water downstream.

Hydrogeochemical characterisation of surface water, sorption of metal(loid)s on sediments and exchange processes within the wetland Lengenfeld / Germany

Manja Seidel, Silke Mannigel, Britta Planer-Friedrich, Broder Merkel

University of Mining and Technology Freiberg, Department of Geology, Gustav-Zeuner-Str.12, 09599 Freiberg / Germany, mseidel@student.tu-freiberg.de

Abstract. Lengenfeld in Saxony, Germany, was the first uranium milling and processing site of the former Wismut SAG. In 1954 a dam collapse happened and roughly 50.000 m^3 of tailings were spilled into the low lands of the Plohnbach river forming several wetland areas. Both sediments as well as surface and seepage water were investigated downstream of the tailing dam with special respect to uranium and arsenic. It was shown that arsenic speciation (As(III), As(V), Mono- and Dimethylarsenic) is highly dependent from seasonal variation. Arsenic is strongly bound to ironhydroxides and thus removed effectively from the water. However, for uranium no removal happens to occur in the Lengenfeld wetland since a major part of the uranium in the sediment is readily exchangeable.

Introduction

After the end of the Second World War uranium ore deposits within the Erzgebirge and Vogtland areas received more and more economic importance. Mining and processing was operated by the Wismut SAG. In 1947 Wismut opened their first uranium processing plant in Lengenfeld, a small town in the Eastern Vogtland/ Germany (Fig. 1).

Ores were delivered from mines in the Vogtland area as well as from the Erzgebirge and in the late 1950s also from Thuringia (Ronneburg). The ores, which showed exclusively low uranium concentrations (0.03-0.5%, Schall 1995), were processed mechanically at first and later by alkaline or acid leaching. During the first period of mechanical processing heaps were deposited along a small river named "Plohnbach". During the later stages of leaching a settling pond was built up. The tailings deposited vary considerably in grain size and chemistry due to the testing of different processing methods. Of the total 1.5 Mio m³ residues from the uranium ore processing, 890 000 m³ of tailing material were deposited (Wismut 1992).

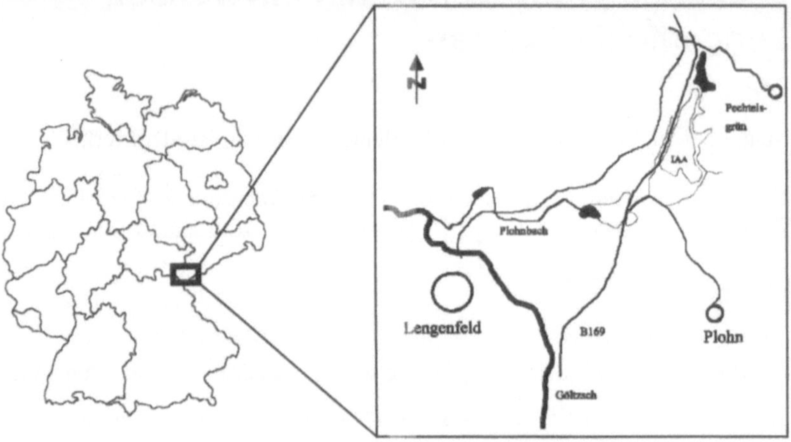

Fig. 1. Location of the tailing site (IAA) northeast of the town Lengenfeld, Eastern Vogtland / Germany

In 1954, the main dam of the uranium tailing collapsed during a flooding event. Tailing material (roughly 50.000 m³) swept into the wide Plohnbach river valley. The fine tailings, deposited up to four kilometres south of the tailing to the former lake "Lenkteich", as a low permeability layer, dammed both river and tailing water and created an extensive wetland area within the Plohnbach valley.

After the uranium processing was abandoned at Lengenfeld and shifted to the big tailing and milling sites Crossen and Seeligenstedt, the "Fluorite and Baryte GmbH" took over in 1961 and covered the tailings with 5-8 m of residues from fluorite and barite processing.

Current investigations of the wetland and the nearby surrounding were focused on water and sediment samples, taking into account existing data from examinations by the "Radiation Protection Agency" (BfS) and the engineer enterprise "C&E Chemnitz".

Hydrogeology

Background

The study area is located at the windward side of the Ore Mountains (Erzgebirge). Mean annual precipitation is 813.5 mm, with an evaporation of 504.4 mm 62% (meteorological station Rodewisch, 5 km southeast of Lengenfeld tailing). Mean surface runoff is 240 mm/a (Schall 1995).

The tailing itself is located in a N-S striking side valley of the river "Plohn-bach", which drains the study area, finally discharging to the northwest into the river Göltzsch. Small creeks discharging from the west and east towards the tailing site are diverted by hanging ditches around the tailing, discharging to the Plohnbach, too.

Precipitation infiltrates into the tailing or is collected in depressions around the settling pond (small wetlands). Infiltration of water through the dams, which are built up from weathered granite, is only know from the northern part. This amount is however negligible (C&E 1997). General groundwater flow direction excluding the tailing site itself is from the east to the west. (Schall 1995).

Leakage from the main dam created a wetland of about 100 x 200 m at the southern tailing dam foot. South of the old processing plant the Plohnbach river is damned, creating a small lake with an adjacent wetland of about 200 x 500 m.

Methodology

Surface water samples were taken from the leakages beneath the tailing dam, the tailing side drainages, the Plohnbach river, and several locations within the larger wetland itself (sampling points 1-21, Fig. 2). Groundwater was not investigated during this study.

The first water samples were taken in July 2001. Onsite parameters, major cation and anions, N- and Fe-species, DOC, TOC, metal(loid)s (ICP-MS, AAS), were considered. Further sampling campaigns in October 2001, respectively March, Arpil 2002 focused on the area around the southern tailing dam foot and its drainage to the Plohnbach. Special attention was paid to uranium and arsenic, the major contaminants.

Arsenic species were determined according to the arsenic speciation technique from Le et al. (2000). It is based on on site separation of As (V), As (III), dimethy-larsenic (DMA) and monomethylarsenic (MMA) on ion exchangers. 2mL one-way plastic syringes are filled with exchanger material (DOWEX 50WX8, and 1X8, SERVA) and pre-conditioned with 6 ml of 1 N HCl and 6 ml methanol (50 %). Onsite 10 mL of water sample are passed through the exchangers. While DMA is sorbed on the cation exchanger and As(V) and MMA on the anion ex-changer, As(III) as zero valent complex passes through both exchangers and is collected in solution. In the laboratory arsenic eluted from the cation-exchange cartridge with 1MHCl is quantified for DMA. The anion-exchange cartridge is first eluted with 60mMacetic acid for the determination of MMA and then with 1MHCl to elute As(V).

Results

Contaminant potential decreases significantly from the leakages beneath the tail-ing dam in the east to Plohnbach river and Lenkteich in the west (Fig. 2). The highest concentrations for arsenic (171 µg/l) and uranium (228 µg/l) during

July2001 were found right at the dam's foot. Within the small lake and adjacent wetland arsenic concentrations are already below detection limit (3 µg/l), and also uranium decreased significantly. The major part of this effect is probably owed to dilution by the Plohnbach river.

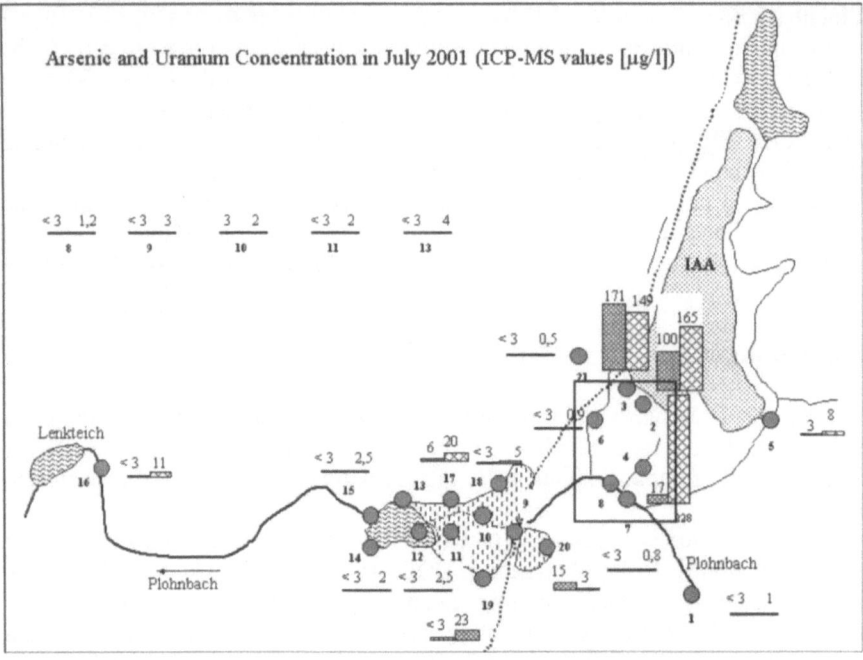

Fig. 2. Sampling points 1- 21 for hydrogeochemical screening in July 2001; arsenic and uranium concentrations (in µg/L) show a significant decrease from the leakages at the tailing dam to Plohnbach river and Lenkteich, rectangle = Fig. 3

Focusing however only on the small wetland right at the dam foot (Fig. 3) it becomes obvious, that at least arsenic concentrations already decrease within the wetland itself, pointing out the efficiency of the wetland for arsenic removal (probably by precipitation and sorption, see chapter Sediments – Results). For uranium concentrations stay more or less the same within the wetland, concentration reduction is only achieved by dilution after discharge of the tailing seepage water to the Plohnbach river.

Arsenic speciation was conducted at selected sites in October 2001 and April 2002 (Fig. 4a). Speciation distribution showed a significant difference between the two sampling times. Remarkable is the generally low content of As(V), normally the predominant species in surface water. This is correlated with considerably reducing conditions and high bioactivity within the wetland. This fact becomes also apparent for sampling point 20 (Fig. 4a). This sampling point is located in a small wetland with almost stagnant flow conditions and shows an almost negligible percentage of inorganic arsenic. For the other three points As(III) predominates in autumn (October sampling), followed by monomethylarsenic, the organic arsenic compound with one CH_3-group. With the beginning of the growing season (April sampling) the percentage of the organic species DMA with two CH_3-groups increases significantly, becoming the predominant arsenic species. Researches on this phenomenon are ongoing with an increased number of sampling sites.

Comparing the results from current investigation on uranium with investigations from the engineering office C & E (1999) (Fig. 4b), the results for the sampling points 3 & 20 are comparable. It is remarkable however that, for the sampling points 2 & 4 concentrations decreased significantly from 440 to 480 µg/l uranium in 1999 to 310 to 137 µg/l uranium in 2002.

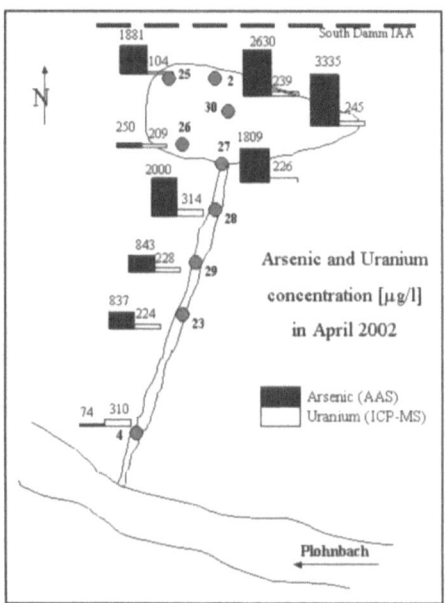

Fig. 3. Sampling points 2, 4, 23, 25-30 at the foot of the southern tailing dam; arsenic shows a significant decrease within the wetland, while uranium is almost unchanged

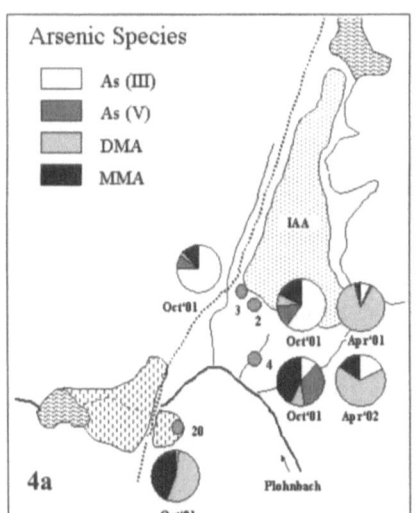

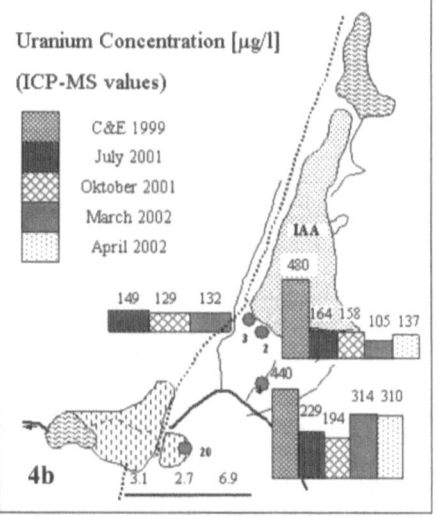

Fig. 4a. Development of arsenic species distribution from October 2001 to April 2002; **4b.** Development of uranium concentrations from 1999 to 2002 (sampling points 2, 3, 4, 20)

Case studies: Uranium tailings

Sediments

Methodology

During a first screening, sediments, mud cores and borestick samples were taken at 26 locations (S 1-26 in Fig. 5). Arsenic (with solid phase AAS), organic matter (TOC) and radioactivity using gamma spectrometry were determined. Based on these results and data from the "Radiation Protection Agency" (BfS), the "Wismut AG" and the engineering office "C&E, Chemnitz", regions of interests were defined and sites for six drillings were chosen (D 1-6 in Fig. 5). Analysis for the drilling samples included gamma spectrometry, grain size analysis and sequential extractions using microwave technique (Yu 1999). Four fractions were extracted: (1) exchangeable metals (1M NH_4Oac); (2) carbonate bound metals (1M NaOAc-HOAc); (3) metals bound to Fe/Mn-oxides (1M $NH_2OH.HCl$ + 0.1M HNO_3); (4) metals bound to organic matter (30% H_2O_2). The residual fraction was calculated as difference of the total concentration (aqua regia extraction) and the sum of fraction 1 to 4.

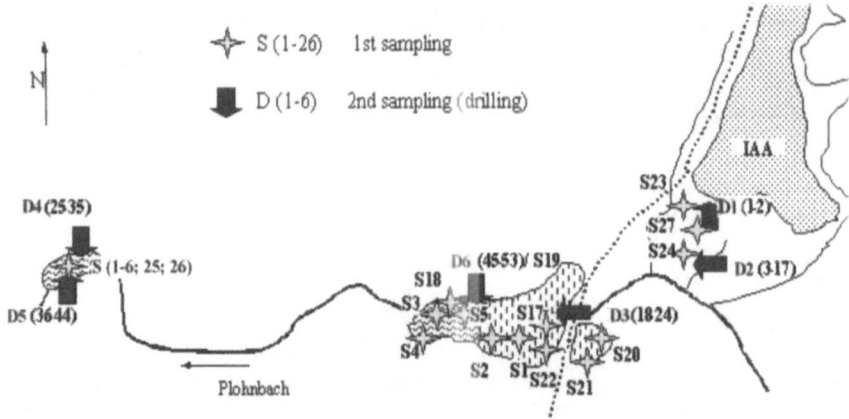

Fig. 5. Location of the screening sites (S1-26) and the drilling sites (D1-6)

Results

Fig. 6 shows uranium and arsenic concentrations obtained by aqua regia extractions (assumed to be total concentrations) from sediment samples of the six drillings D 1-6. Uranium concentrations vary from 3 to 1200 mg/kg. Increased concentrations were found at the Lenkteich area as well as the large wetland adjacent to the former processing plant. It is remarkable that uranium is enriched mostly in surface horizons up to a depth of about 1.50m (on average approx. 360 mg/kg).

Samples taken at or adjacent to the southern tailing dam as well as from the wetland located southwest of the tailing, showed much lower concentrations (average approx. 40 mg/kg with single peaks of 120 to 170 mg/kg). Gamma spectrometry confirms those general trends.

Arsenic concentrations reflect the general tendencies of uranium. Average concentrations in horizons close to the surface in the Lenkteich area are about 460 mg/kg. Contrary to uranium also samples from the area south of the tailing dam show increased arsenic concentrations of 200 to 300 mg/kg. These results correspond well with the results from hydrogeological investigations, that arsenic concentrations in the aqueous phase already decrease within the wetland itself, pointing out the efficiency of the wetland for arsenic removal by sorption, while uranium concentrations stay more or less the same within the wetland.

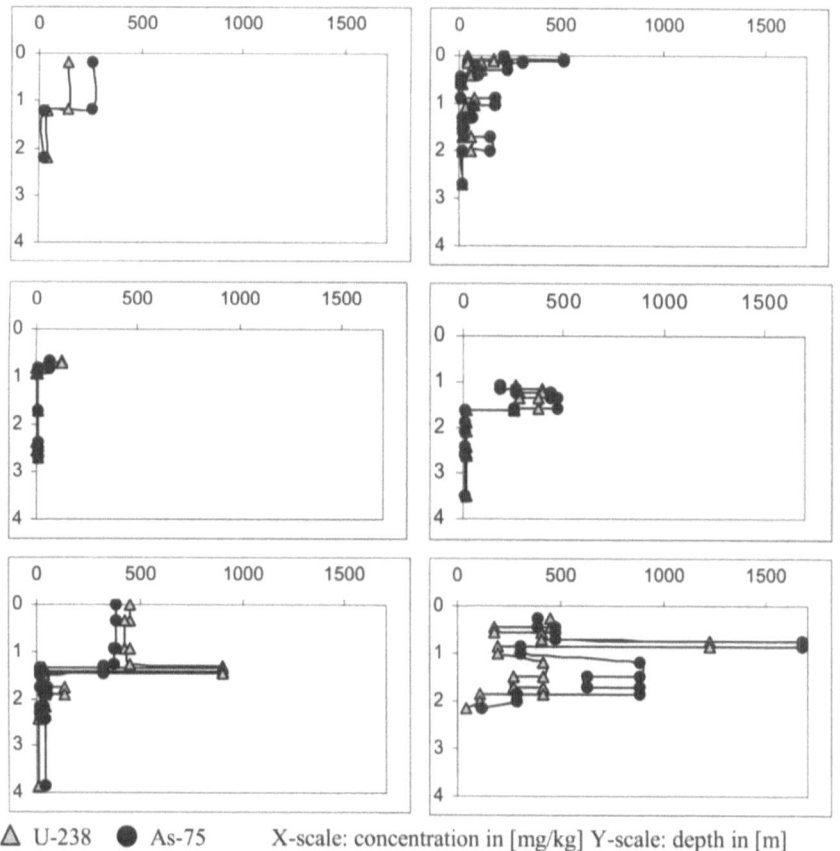

△ U-238   ● As-75   X-scale: concentration in [mg/kg] Y-scale: depth in [m]

Fig. 6. Depth depending arsenic and uranium concentrations in the drillings D1-6 (aqua regia extractions, analyzed by ICP-MS; from the upper left to the lower right D1 and D2, D3 and D6, D4 and D5, locations according to Fig. 5)

Case studies: Uranium tailings

The results of sequential extraction are shown exemplary for drilling No. D5 (Fig.7). Nine horizons were distinguished within this drilling situated in the Lenkteich area (Table 1). The total drilling depth was 2.15 m. Gamma spectrometry as well as ICP-MS analyses for uranium show the highest activities, respectively concentrations for the upper horizons. Older analysis from Wismut (1992) confirm these findings. Maximum values are found for horizon No. 39, defined as a silty-clayey material. This horizon is supposed to be part of the tailing material distributed to the Lenkteich during the dam failure in 1954.

Table 1. Drilling No. 5 (γ-spectrometry done in the laboratory on 10 g of sample corrected for background values, eU = equivalent uranium from radium)

Number of sample	Drilling depth in [m]	γ Spectr. Total Counts [Bq]	γ Spectr. U [Bq]	ICP-MS U-238 [mg/kg]	Activity [Bq/kg] (calc. from conc.)	Soil textural classes [<2mm]	depth of bore hole measurement [m]	eU in ppm (Wismut 1992)
36	0.27-0.45	1.23	0.18	**447**	5503	Slu		
37	0.45-0.56	0.17	0.12	175	2158	Ss		
38	0.56-0.73	**2.73**	**0.51**	405	4987	Us		
39	**0.73-0.85**	7.70	1.37	1222	15032	Lu	**0.00-1.00**	1140
40	0.85-1.18	1.11	0.27	194	2391	Su2		
41	1.18-1.85	**2.48**	**0.60**	420	5160	Slu	1.00-2.50	210-290
42	1.48-1.70	2.18	0.29	276	3394	Su2		
43	1.85-2.00	1.63	0.13	114	1398	Ut3		
44	2.00-2.15	no sample	no sample	44	540	Su3	2.50 -	10-40

Concerning the results of sequential extraction (Fig. 7) a correlation between total uranium concentration and percentage distribution of the single extraction fractions was not found. Sample No. 39 with the highest uranium concentrations shows uranium bound mainly to oxides and the residual fraction with only 8% of easily extractable uranium (F1), 2% bound to carbonates (F2), 42% bound to Fe-/Mn-Oxides (F3), 8% to organic matter (F4), as well as 40% in the residual fraction (F5). Compared to this is e.g. sample No. 44 with the lowest uranium concentrations and a more or less even distribution of uranium bound to the different fractions (F1: 25%; F2: 10%; F3: 20%; F4: 15%; F5: 30%).

Arsenic distribution shows little general trends so far, however an inverse correlation with uranium is apparent (Fig.7). Within sample 39 the percentage of arsenic increases from fraction 1 to 4, while the residual fraction shows the lowest concentrations contrarily to uranium.

Remarkable in general are the high percentages of easily exchangeable uranium (8-25 %, respectively 90-250 mg/kg), as well as uranium bound to organic matter (9-25%, respectively 7.5-107 mg/kg) (Fig. 8a). Arsenic on contrary is rather bound to iron oxides and the residual fraction, so in general harder to remobilize than uranium (Fig. 8b). This observation confirms again the tendency for uranium to be transported and remobilized from the wetland at the tailing dam foot further downstream along the Plohnbach river, while the retention capacity for arsenic is higher, decreasing further distribution.

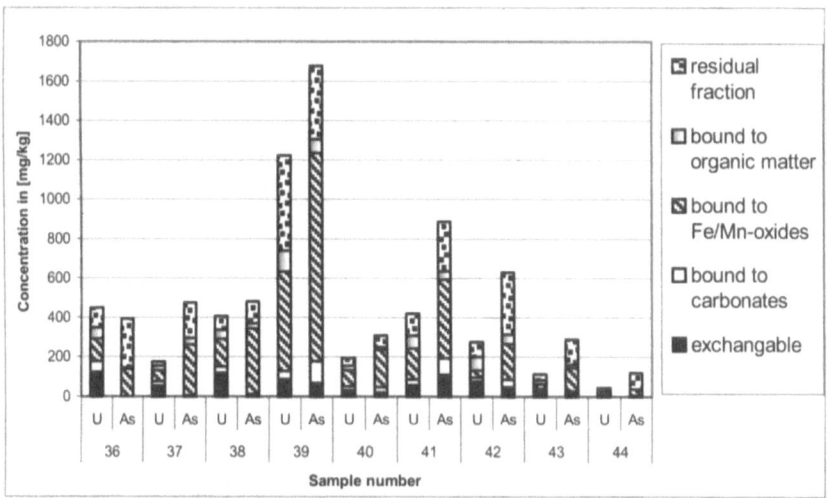

Fig. 7. Uranium and arsenic concentrations obtained by sequential extraction for drilling No. D5 (analyzed by ICP-MS), fractions in relation to total concentrations

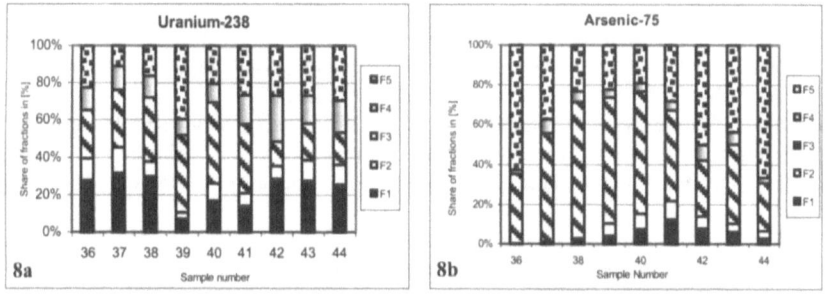

Fig. 8. Uranium (8a) and arsenic (8b) concentrations obtained by sequential extraction for drilling No. D5 (analyzed by ICP-MS); fractions summed up to 100%

Conclusions

First results of hydrogeochemical investigations on natural attenuation processes within the wetland at the Lengenfeld tailing foot show that removal capacities for arsenic are high, lowering concentrations from about 3 mg/L at the tailing foot to 0.07 mg/L before discharging to the Plohnbach river. High arsenic concentrations in the sediments, mostly bound to iron oxides and residual fraction, confirm the importance of sorption within the wetland. For uranium no concentration decrease within the wetland was found, sorption seems to play a minor role only. Dilution by the Plohnbach river however is sufficient to reduce uranium concentrations from 230 µg/L at the tailing dam to less than 20 µg/L, mostly 1-2 µg/L, all over the adjacent wetland and further downstream. Results from sequential extractions show that uranium in general is easy to mobilize, mostly bound to easily exchangeable fractions and organic matter.

Rehabilitation measures, that just started in April 2002, including excavation of parts of the most severely contaminated sediments beneath the tailing, may alter the hydraulic conditions in a way that remobilisation from sediments and further transporting of contaminants via the Plohnbach can not be excluded, especially for uranium.

References

Consulting und Engineering GmbH (C&E) (1997) Bilanzierung des Wasserhaushaltes für die IAA Lengenfeld im Vogtland. Bericht für die GVV mbH Bergwerk Lengenfeld

Consulting und Engineering GmbH (C&E) (1999) Umweltbewertung des Wasserpfades der IAA Lengenfeld/ Vogtland. Bericht für die GVV mbH Nachsorgebetrieb Lengenfeld

Le X.C., Yalcin S., Ma M. (2000) Speciation of submicrogram per liter levels of arsenic in water: on-site species separation integrated with sample collection. Environ.Sci.Technol. Vol.34, No.11, p.2342-2347; American Chemical Society

Schall I. (1995) Hydrogeologische und hydrochemische Untersuchungen im Bereich der ehemaligen Uranerzaufbereitungsanlage Lengenfeld (Vogtland). Geol. Diplomarbeit, Rheinisch-Westfälische Technische Hochschule Aachen

Wismut AG Chemnitz (1992) Abschlußbericht über die Untersuchungen zur Ermittlung des Radioaktivitäts- und Schadstoffinventars auf der Verdachtsfläche 9 (Lengenfeld/Vogtland). Pilotprojekt im Rahmen des Altlastenkatasters Südregion (TP 2 – Phase 2)

Yu HH (1999) Cadmium Chromium and Lead in environmental Samples: Speciation and determination. Dissertation of New Jersey Inst. Of Technology. Department of Chemical Engineering, Chemistry and Environmental Science

Environmental Risk Management at Uranium Tailings Ponds in Mailuu-Suu, Kyrgyzstan

U.G. Aleshyn[1], I. A. Torgoev[1], G. Shmidt[2]

[1]SIC "GEOPRIBOR", Bishkek, Kyrgyzstan
[2]Eco-Institut, Darmstadt, Germany

Abstract. The complex of natural and geotechnical factors, disaster scenarios and conditions of reliable rehabilitation of uranium tailings in Mayluu Suu is considered. The forecast of consequences of destruction of the most dangerous tailing #3 is made. This destruction can take place by means of a landslide. The radioactive contamination of water from this landslide is calculated.

Technical maintenance and rehabilitation of abandoned objects of abandoned mines (AM) – one of the major directions on providing longtime ecological safety of the territories several times bigger than natural risks in those areas
Three major activities are necessary to start the procedure of rehabilitation:

- investigation of the territory T_1, where the site is located (near field)
- investigation of the territory T_2, which is adjacent to the object (far field)
- Investigation of the engineering construction, e.g. territories $T_1 \cap T_2$ (Fig.1).

The results should be presented in the form of maps displaying geological, hydro geological, seismic tectonic and meteorological conditions. For this purpose it is necessary to determine the level of usability of engineering construction solutions and natural conditions in terms of risks.

In such complex natural and geotechnical system (tailing of radioactive waste rock (RWR)), pollution of the environment can be the result in the long term. The factors, which are initiating the whole sequence, can be presented not only by natural phenomena but also by geotechnical reasons and anthropogenic pressure. An example is given in the following scheme:

elimination of the flora on the slopes of the tailing's bowl → development of exogenetic geological processes → creation of additional weight on the tailing surface by products of slopes denudation plus landslides deposits → loose of stability of the tailing during high humidity period of time → destruction and movement of the tailing in the form of the landslide towards the river-bed → closing of the riverbed by means of the landslide, accumulation of water → break of the landslide-dam with radioactive materials → creation of the mudflow with destruction of buildings along the river-bed and radioactive waste of large territories → contamination of large agricultural territories where the water is used for irrigation. Such sequence of negative events can take place in Mayly Suu town on the territory of the former hydro steel producing plant # 3 (Fig.2). The conditions of radioactive safety on the territory Mayluu Suu were investigated earlier (Aleshin and others, 2000).

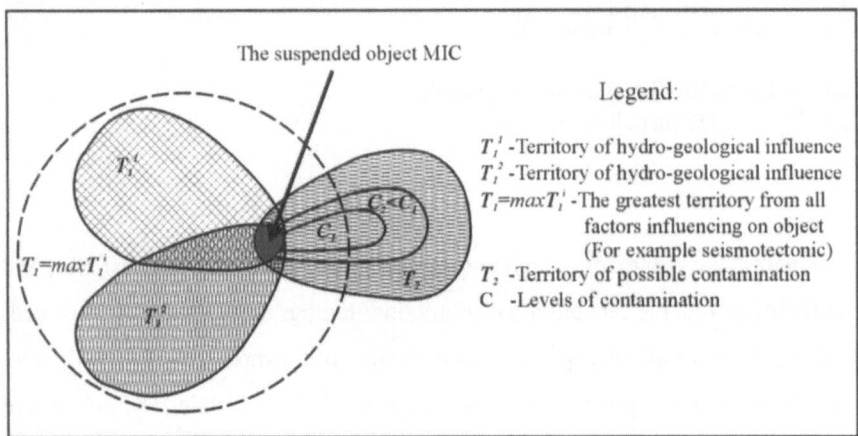

Fig. 1. Scheme of interrelations in natural geotechnical systems

Overall scheme of the safety analysis of radioactive wastes burials can be based on the logical sequence of events for risks analysis (Slepcov, 1997; Kochkin, 1998). This scheme is described in details in our previously published article (Aleshyn, Torgoev, 2001)

$$\{NTC/AC\} \rightarrow \{P_{nf}\} \rightarrow \{M_{nf}\} \rightarrow \{D\} \rightarrow \{P_{ff}\} \rightarrow \{M_{ff}\} \rightarrow \{C\} \rightarrow \{L\}, \quad (1)$$

Where the arrows mean logical expressions "if..., then...".

Systematic consideration of the factors are influencing on burial's safety is showing that besides natural and geotechnical factors (NTC) should be taken into consideration "acting conditions" (AC) of anthropogenic character. Common list of such factors starting from critical processes in close Pnf and Pff far fields contain all reason – consequence sequences of the longtime storage of radioactive wastes in geological environment; M_{nf} and M_{ff} movement of the substance in close and far fields, D – destruction of engineering constructions and barriers, C – deposit of radio nuclides in the environment, L – fixed (valuable) losses. These losses are easy to detect by numeric marks. All sequences of events lead to the one consequence – deposit of radio nuclides in the environment.

Case studies: Uranium tailings

Fig. 2. Territory with high ecological stress.

List of actions for the minimization of risks during storage includes:

- A_1 - anthropogenic activity which leads to the actuality of the natural sources and NTC conditions of critical processes. This activity was previously done when the area and technology of burring was chosen (e,g, choose of area close to tectonically active fracture; in watered gorge, by the landslide and so on);

- A_2 - activity which can eliminate negative critical process or stabilize (reduce) intensity of it's development (for example, reduce the pressure to the pioneer dumb or stress in the direction of movement along slipping surface and so on);

- A_3 – possible activity on the creation of anti processes are not allowing critical processes reach "abnormal movement of substance" (for example, creation of drain, reducing of landslides on the upper slopes);

- A_4 - possible activity which is reducing the possibility of destruction during abnormal movements of natural substance (seismic enforcement of constructions; hydro insulation of the bed; build of mudflow dumb, prisms and so on);

- A_5 - activity on the reduce of possibility of the development of critical processes in the far field (training of population on the living in contaminated area; creation of alarm system on the destroy of tailing; creation of drinking water deposits and so on);

- A_6 - possible activity on the creation of anti processes and devices are not allowing to critical processes reach "abnormal movement of substance" in the far field (for example, lowering the level of the ground waters which are connected to filter waters of the tailings; enforce river benches by gabions where it is possible mudflow during the tailing destroy; build of dumps in the river – bed for capturing of tailing's materials in close zone and so on);

Case studies: Uranium tailings

- A_7 - activities against deposit of radio nuclides in the environment, including human bodies in the far zone (activities on the cleaning and dis - contamination of the bottom deposits; use of medicine is taking out of human bodies radio nuclides; limited use of the dangerous territory for agricultural purposes; cultivation of plants less absorbing radio nuclides and so on);
- A_8 - possible activities are reducing negative impact to the environment with fixed destruction and use of methods from the $\{A_i\}_{i=2}^7$ (mostly deal with relocation of citizens from the dangerous area and reduce of agricultural use of the territory).

All activities including planning of the engineering work, monitoring, research are related to different parts of the negative events sequence and they are part of the tailings rehabilitation program on certain territory.

Obviously the more requirements to the rehabilitation of the tailings the lower risk of contamination by radio nuclides of the environment.

The D events – destruction of the tailings can be considered in terms of several scenarios. In terms of tailing safety scenarios we consider possible developments in the burial system (jointly with mining landscape) for a long time which is determined by insulation requirements. The amount of such scenarios in real geological environment can be as much as possible. They should be summarizing into 4 categories: catastrophic, emergency, uncertain and favorable.

In catastrophic scenarios eject of radio nuclides behind engineering barriers and local area of the geological environment is predicted in the result of immediate destroy of the burial system. In this situation it will be radioactive contamination of the environment: $P(M_{ff})$, $P(C) \rightarrow 1$

The problem is to determine indicators of the affected zone. The scenarios of this group can follow from the following factors:

- earthquake more than 9 scores or with intensity that affects the stability of the tailing;
- landslide, mudflow or water stream with erosive opening of the tailing;
- moving of the tectonic blocks along the fractures in the zone of tailing;
- immediate decrease of waste rocks hardness, tailing materials due to intensive watering of the rocks by meteorological factors, and in the far perspective – change of the climate.

In the emergency scenarios it is thought that entrance of the radio nuclides from the waste into environment is transferred to the far zone mostly by water streams. Risk of the contamination of the environment $P(C/(M_{ff}, P_{ff}, D))$ is high enough ~ 0,1...0,5.

Such type of scenarios is due to the following factors:

- fast relocation of the radio nuclides till major water stream (river) on the relatively short distances;
- high dissolvent of the radio nuclides in the underground waters;
- partial destruction of the tailing and relocation of the buried materials on the small distances due to loss of the stability of the engineering construction.

Case studies: Uranium tailings

Major ecological risk during the destruction of the tailing #3 is connected to the deposit of radioactive materials into the riverbed of Mayly Suu. The major danger for the citizens in this case is radioactive radiation and to ingest radio nuclides by drinking the water. There is also a potential danger of contaminating agricultural areas of the neighboring country.

Taking into consideration the possibility of destruction by different natural processes (landslide, earthquake) the most dangerous tailing #3 and it's location below tailing #18, the process of emitting waste into the river can be developed by 2 scenarios:

(1) slow slip of the tailing's materials to the river and leaching by water. The level of the water reminds the same. During flooding in 1958 –59 all buildings for living were destroyed and gabions were installed and in our days they require some repair. (the biggest flooding was observed in April, 1958 after failure on tailing # 7);

(2) immediate slip of the tailings to the river affecting the territory of "Kyrgyzelectroizolit" plant and the road. Following events will create the danger of flooding and radioactive contamination of the large territory within central and lower part of Mayluu Suu town. It is not possible in the short period of time to build temporary constructions in the riverbed.

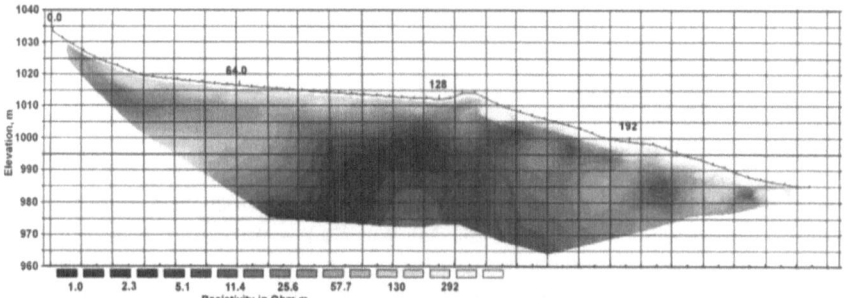

Fig. 3. Geoelectrical tomography of the tailing #3 in Mayly Suu (profile through the dumb)

Tailing #3 is located in the zone of transit and/or deposit of the landslide mass during the landslide activation in the dome part of the Central anticlinale, on the edge of chalk deposit hill surface with high slope, which limited the bowl of the tailing from the Eastern side with heights difference up to 100 m.

The loss of stability of the tailing #3 is based on 2 factors:

- constant and increasing weight pressure on the surface due to denudation deposit of small dispersion rocks from the high slope surface of the main anticlinale, and also due to small landslides from the dome of the anticlinale;
- by the increase of the tailings water content due to the fact that bedding waters are appearing by the foothills of the main anticlinale, mostly in the first part of the year (Fig. 3).

The most important results of the catastrophic phase modeling of the tailing #3 destroy by the landslide are the following:

Case studies: Uranium tailings

- in any situation of destruction by a landslide, about 32 – 48 % of tailing's materials will be transported into the riverbed of Mayly Suu. The tailings will back-up the river to a height of 10 – 15 m and a width of about 50-60 m
- 10 000 sq.m. of the "Kyrgyzelectroizolit" plant will be in the affected area for the most possible scenarios of the landslide. The total activity of the radio nuclides will about $4 \cdot 10^{13}$ up to $8 * 10^{13}$ Bk;
- the fastest speed of the landslide will be 36 m/s, catastrophe replacement phase will take from 14 to 17 sec; the landslide will reach the plant within 10 sec;
- The volume of the water which will be accumulated by the landslide dam is about 120 000 m^3 ; the average flow of Mayly Suu in the spring time is about 50 m^3/s in spring time, thus the reservoir will be filled within 40 min. In the extreme situation with a flow of 100 m^3/s it will be filled within 20 min; In the flooding zone will be a road on the left side, partially the territory of the "Kyrgyzelectroizolit" plant including electrical department and electric station (Fig. 4);
- The tailing #18 with a volume 3 000 m^3 is located in the transit zone of landslide mass and it will be destroyed
- total activity of the radio nuclides in the landslide will be not less than 0,7 … $1,0*10^{14}$ Bk, and an α-activity of $4 \cdot 10^{13}$ Bк. The mass equivalent for it is 500 … 600 tones of uranium. Safe concentration of uranium in the water is assumed to be $4*10^{-6}$ g/l.

The conclusion is the following: after the landslide dam will be destroyed in 40 - 60 min the concentration of uranium in the river Mayly Suu will exceed this safety norm hundreds or thousands fold.

Risk methodology which is applied to the management of abandoned objects on regions with ecological stress like the Mayluu Suu is providing systematic approach in the formulation and solving of tasks, also it is allowing to determine the logic of tailings rehabilitation projects management and the following stages should be implemented:

- to define boundaries of negative impact which should be eliminated in terms of rehabilitation project for the conserved object in multi-dimensional rage of parameters: territory T, time interval τ^*, levels of environment contamination C^*, social and financial losses L^*, the first two can be expressed in the form of limitations, and by means of probability scale $Pa(C)^*$, $P_a^*(L/c)$, $P_a(L)$;
- based on research and engineering investigations conducting investigations of dangerous processes (P), relocation of the material (M), destruction of geological environment and engineering constructions in far and near zones based on the scheme of sequential sequences with establishing links between events with their probability;
- determination of decision making management processes based on loss probability P(L): determine possible strategies on rehabilitation $\{A_i\}$, separate versions with known probabilities $P^{Ai}(\cdot)$ of the negative processes development with further calculation and getting the values $P^{A(7)}(L/c)$, $P_t^{A8}(L)$.

Case studies: Uranium tailings

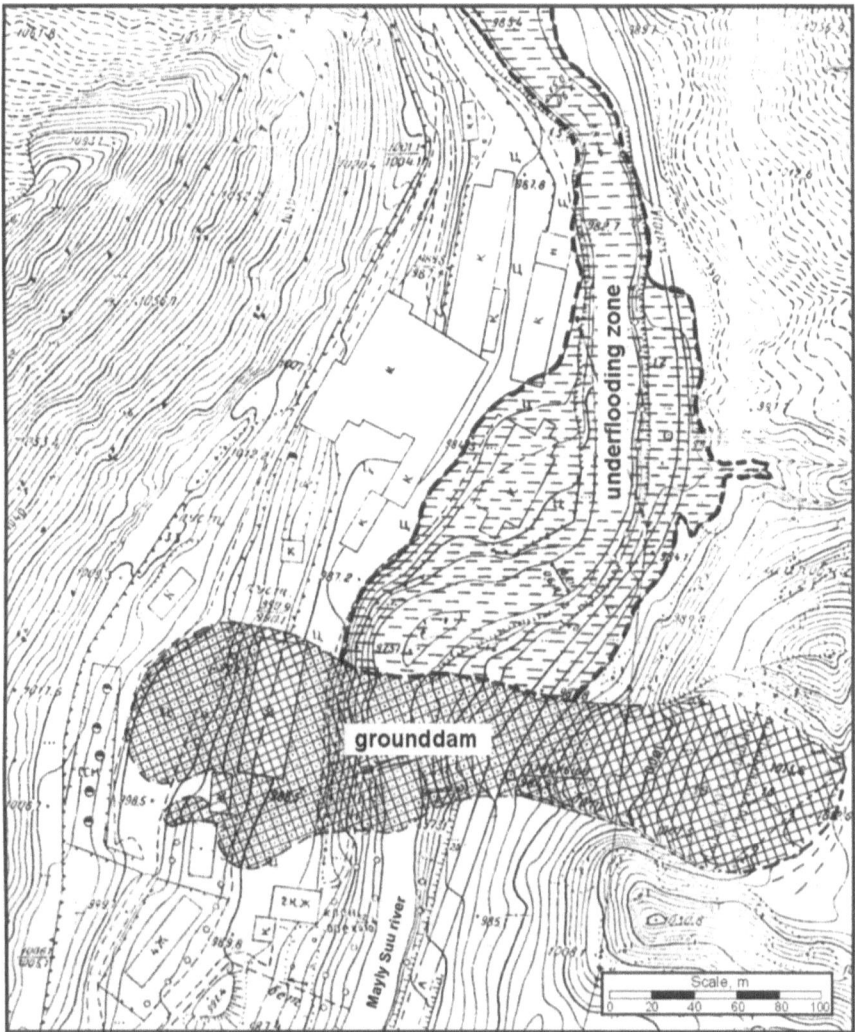

Fig. 4. Map of a zone of a sliding defeat and under flooding at destruction tailing # 3

This logical scheme leads to the absolutely transparent evaluation of the project's quality :

$$P^{A(7)}(L/C) \le Pa^*(L/C) \text{ by } \tau \ge \tau^*, T \ge T^*,$$

it also gives reasons for necessity and queue of the rehabilitated objects.

Creation of the most complete list of natural geotechnical events and relevant conditions (NTC/AC) is important stage of the management methodology. It is enough if it is allowing to evaluate certain territory $(T_1 \cup T_2)$ with conserved objects and it is counting all natural and geotechnical conditions which have negative impact on the safety of the further object's maintenance. Such list is important due to loss of the one important condition is increasing environment contamination risk. On the other side, not taking into consideration such approach will lead to the evaluation of conditions and conducting rehabilitation activities which do not relate to the safety commitments.

References

Aleshin U.G. and others (2000) Radiation ecology of Mayluu Suu /"Geopribor", Bishkek "ILIM", 2000, 96 p. (Rus)

Slepcov B.G. (1997) Some issues of the "risk management" in the engineering geology practice/ Geo ecology 1997#7,p.p. 29 – 40 (Rus)

Kochkin B.T. (1998) Risk management problems selecting place for the burial of radioactive waste/ Geo ecology 1998 #5 p.p. 37 – 50 (Rus)

Aleshin U.G. , Torgoev I.A. (2001) Ecology risk management on the territories of conserved objects in mining industry / Problems of the geo mechanics and geo technical utilization of mining territories – Bishkek "ILIM" , 2001, p.p. 436 – 456 (Rus)

The nature and extent of uranium contamination from tailings dams in the Witwatersrand gold mining area (South Africa)

Frank Winde, Abraham Barend de Villiers

Potchefstroom University (South Africa)

Abstract. A major problem in the field of environmental contamination by uranium in South Africa is the lack of a synoptic overview of the nature and extent of the problem. At 14 selected sites within all seven goldfields of the Witwatersrand basin, mechanisms and the extent of tailings-related U-contamination were investigated. These sites are spread along the so-called Golden Arc stretching over a distance of approximately 300 km. Results suggest that U-pollution by slimes dams in South Africa is widespread and of extraordinary spatial dimension.

Introduction

Due to the nature of the ore from which the gold and uranium has been extracted in the South African gold fields, the tailings dams ("slimes dams" in the SA context) contain a substantial amount of U and other heavy metals. U-concentrations in some auriferous ore-bodies of the Witwatersrand basin vary from ten to several hundred ppm U_3O_8 (CfG 1998). Gold mining in the Witwatersrand basin started already in 1886. But it was only in connection with the Manhattan project after WWII that from 1952 onwards U was produced by some mines as a by-product of gold (Ford 1993). That means, that all tailings deposited before 1952 contain U in the original ore-content since the radionuclid was never extracted. By the end of 1995 a total of 144,481 tons of U had been produced in South Africa for the world market (CfG, 1998). Due to declining world demand in the 1980's, U-production had been ceased at many South African gold mines and U-production dropped from a maximum of some 6000t/a to currently less than 1000t/a (Venter 2001). As a result of this decline the amount of U that is dumped on slimes dams, has steadily increased. With concentrations far above the natural background U-bearing tailings deposits are a potential source of contamination, mainly facilitated by the following mechanisms and pathways:

a) Particle-bound transport of U, including erosion of sidewalls of slimes dams (triggered by highly intensive rain events); major and minor breaches in side-walls (dam failures) and aeolian transport of dust from dry and unprotected slimes dams.

b) Solute transport of U, including seepage from slimes dams that mobilises heavy metals from tailings particles and pollutes groundwater and adjacent streams, non-recyclable and highly polluted process-water seeping from evaporation pans into underlying aquifers; point-discharge of U-polluted groundwater from underground mine workings into nearby streams; spills of process-water (pipeline bursts etc.) into storm-water drainage systems or directly into the environment.

c) Gaseous pathway; as a result of the radioactive decay of U, the radioactive gas radon, known to cause lung cancer, emanates from slimes dams into the atmosphere. This is of concern in cases, where tailings were used as building material for houses, which tend to accumulate the gas inside.

This paper focuses on aspects of particle-bound and solute transport of U and associated stream pollution, which are often neglected in relevant studies of contaminant-dispersion. The major aim is to identify important pathways for U-transport from slimes into surrounding areas and to estimate the magnitude of the associated contamination. This will include the following:
- estimating the total amount (mass) of deposited tailings;
- determining U-concentrations and –mass in slimes dams;
- characterising the geographical distribution of slimes deposits in terms of: surface area covered by the slimes dams, adjacent areas with particularly sensitive land uses (settlements, agriculture, etc.), the vicinity of streams as major pathways of contaminant-transport and preferred sinks for off-site U-re-concentration.

Methodology and study area

Based on the evaluation of available data from literature and other sources, 14 study sites within all of the seven gold fields of the Witwatersrand basin, were selected. The fieldwork at the sites comprised documentation of erosion (where applicable), on-site determination of water quality parameters (pH, EC, Eh, T, DO) in seepage, groundwater and streams, as well as sampling of relevant water and sediments for lab analyses of U and other heavy metals. U was analysed as U_{nat} by IC-OES (after dissolving solids in *aqua regio*) and laser-phosphorescence (water).

Results

Mass and U-content of tailings

Since the beginning of gold mining in the Witwatersrand basin in 1886 a total of about 6 million tons of tailings has been produced (Wymer 2001). Average U_3O_8-concentrations in tailings of 46 active mines in the Witwatersrand basin range from 10ppm (Evander goldfield) to 343ppm (Far West Rand) (Wymer 1999). The mass-weighted average U_3O_8-concentration for all tailings is 104ppm (=88ppm U_{nat}). Considering that the compilation by Wymer (1999) does not include old depositions with uncertain ownership and currently produced tailings, from which U was not extracted, a somewhat higher average of about 100ppm U_{nat} can be reasonably assumed. Although not classified as U-tailings gold mining deposits in SA, in fact, contain about the same or even higher U-concentrations than many "genuine" U-tailings in Germany, the USA and Namibia, where mainly low-grade U-ore was mined (Winde 1998, 2000; Robinson 1995, Kehrberg 2000). Compared to an estimated global total mass of 0.5 billion t U-tailings in 18 countries (Waggitt, 1994), South Africa has over ten times more U-bearing tailings. With some 6 billion t of tailings produced by gold mines since 1886, approximately 600,000t of U_{nat} in slimes dams are currently exposed to the biosphere (Winde 2001a, 2001b).

Geographical distribution

It is estimates that the area covered by slimes dams in the Witwatersrand is in the order of 400km² (CfG 1998). The geographical distribution of gold mines within the Witwatersrand basin is displayed in Fig. 1.

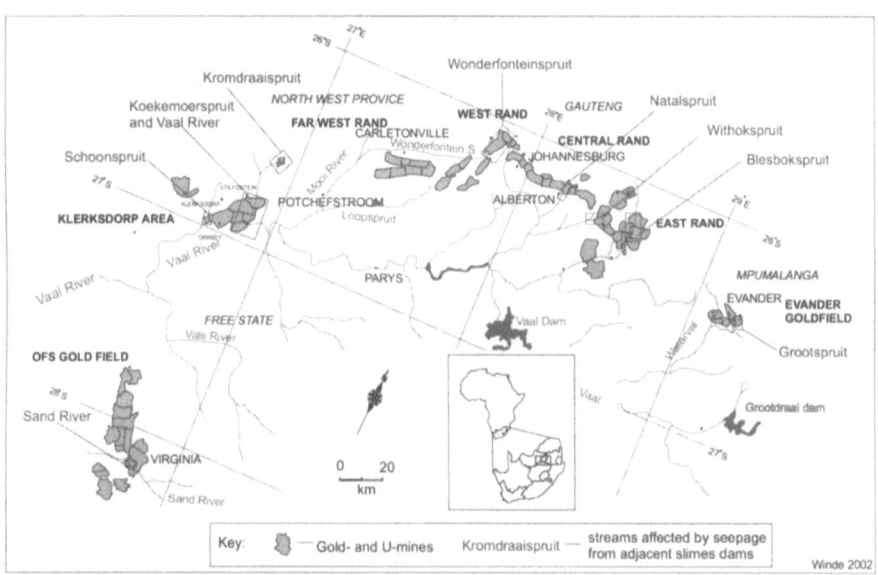

Fig. 1. Gold mining areas and slimes dams affected streams in the Witwatersrand basin

Apart from the enormous size of the covered area it is often the specific location of slimes dams in relation to adjacent land use or natural conditions that worsens the effects of associated environmental pollution.

One of the major problems experienced with the location of these slimes dams is the fact that a fairly large percentage of all the gold mines are located in the

Case studies: Uranium tailings

headwater regions of streams, thereby affecting the most pristine part of streams and polluting water courses from their very source. Mining related stream pollution in headwater regions is particularly high because of the low potential of these small catchments to dilute seepage from tailings deposits and discharged mining waters. For a tributary of the upper Natalspruit in central Johannesburg it was found that approximately 20% of the headwater-basins is covered by mining residues, resulting in about a quarter of the stream flow consisting of highly contaminated seepage that diffusely enters the stream. This situation is likely to be worse in dry winter times, when seepage is not diluted by rainwater. Since such stream water is often used by dwellers of informal settlements without appropriate treatment it directly affects human health. This is of particular concern as those settlements are preferably located next to mines, which are perceived as potential employers. Since virtually all the gold mines in the Witwatersrand basin are located within the drainage basin of the Vaal River. As an important source for potable water for many areas in the Gauteng province, as the most densely populated region in South Africa, the river finally receives a large proportion of mining contaminants via groundwater and streams.

Mechanisms and extent of U contamination from slimes dams

Erosion of slimes dams

The erosion occurring on the sidewalls of slimes dams is to a large extent dependent on the condition of the sidewalls themselves. In the case of old slimes dams that were not rehabilitated most of the sidewalls are extremely eroded and is the source of large volumes of pollutants (Fig. 2).

Fig. 2. Eroded slimes dam near Dominionville (Photo: Winde 9/2001)

In some cases where rehabilitated sidewalls are neglected, clear signs of degradation are visible. These solid particles originating from dams can reach the environment through suspension in an overland flow. Such a redistribution of the eroded material is usually restricted to distances fairly close to the dam itself. This is due to the fact that this geomorphic process is restricted to fairly low gradients on flat surfaces (Weissenstein et al. 2000). A more effective way in which the particles can be transported is obviously through stream flow. The distance over which such particles can be transported is determined by the flow mechanics operative in the individual stream. The transportation of slimes particles in streams had been observed in several cases. Slimes were found as far as 2.5km downstream from an old slimes dam located at the old Machavie gold mine (Kromdraai Spruit).

Breaches occurring in sidewalls of slimes dams

Breaches frequently occur in the sidewalls of slimes dams. Material from such breaches is normally confinable inside the surrounding berms. However if the berms are not properly maintained or the breach occurring is a major one, spills can occur into the surrounding environment. Minor spills are fairly frequent. The amount and distance over which these spills are distributed is normally small. Major breaches do however occur from time to time. These spills can spread over large areas, into streams and even settlements.

Dust pollution.

The spread of solid particles by aeolian action was not investigated in this study. It is however known that dust originating from slimes dams can travel fairly large distances. During one such windstorm near Stilfontein fine dust was observed more than 4km from the dam.

Solute transfer of pollutants

One of the results from this study re-affirms the fact that water moving through slimes dams contains solute matter obtained through various chemical reactions taking place between the slimes and the water passing through it. The movement of seepage from slimes dams is enhanced by elevated phreatic surfaces formed in the dams that are well above ground water levels. As a result seepage migrates into the groundwater as well as into adjacent streams throughout the year. The formation of artificial wetlands on the sides of some slimes dams attests to the fact that although the winter rainfall is well below the evaporation figures a constant supply of seepage is obtained from these dams.

The low winter rainfall also affects the quality of water from slimes dams entering the environment. In wet summer-periods rainwater can help to dilute the seepage from the dams. During winter this dilution effectively comes to an end. This results in increased concentrations of pollutants in water that eventually reaches streams and rivers or the underground water.

Apart from diffuse water pollution dissolved transported contaminants also form secondary accumulations in adjacent areas, by re-concentrating in sediments

Case studies: Uranium tailings

and salt crusts along the aqueous pathway. Due to possible re-mobilisation mechanisms these accumulations act as secondary sources of contamination. While seepage-contaminated groundwater next to the Koekemoer Spruit contained 0.5ppm dissolved U re-concentration in the floodplain soils reached values varying between 20 and 30ppm U. Sediments in the adjacent stream channel displayed between 40-54ppm. The highest re-concentration was found in sulphate crusts on topsoil of the floodplain containing some 1200ppm U (Winde 2001b). In the case of the Wonderfontein Spruit re-concentration of U taking place in a farmers dam downstream from the pollution sources is an example of such processes. The U concentration in sediment samples obtained from the dam showed concentrations of U as high as 490ppm (Wade et al. 2000). According to the Nuclear Act of 1993 any material containing radionuclids of uranium and thorium decay chains with a specific activity of >200Bq/kg is deemed to be radioactive material. Converted to parts per million this equals a concentration of 16ppm. At many of the study sites it is therefore imperative that provisions of the Act are imposed.

It was frequently found that water emanating from old slimes dams exhibited low pH values, indicating acid mine drainage that accelerates leaching of U and other heavy metals from tailings particles. Seepage from a site in the Klerksorp area (Boetrand) measured as low as pH 2.56. Groundwater with a pH of 3.31 (EC of 6.29mS/cm) has been measured in the catchment of the upper Natalspruit, with a pH value of 3.21 and an EC of 1.82mS/cm in the stream itself, where it left the headwater basins. It is to be expected that these acid conditions, which favour solute speciations of metals, allow for fairly long distances of downstream transport and a high bioavalability of contaminants. In those cases re-concentration of U and other heavy metals may occur far away from the primary source of contamination.

Summary and conclusions

With more than ten times the global total-mass of U-bearing tailings deposited in the Witwatersrand basin alone, mining related U-contamination in South Africa constitutes an environmental problem of an extraordinary dimension. It is aggravated by the fact, that a high proportion of the 400km² area that was covered by tailings deposits during the last 100 years, is located in densely populated regions. Dust plumes from slimes dams, erosion and breaches of dam walls, and solute transport of leached pollutants expose nearby dwellers to toxic contaminants and radiation. Due to the associated re-concentrations of U and other toxic heavy metals in the environment, often occurring far away from the source of pollution and in much higher concentrations, solute transportation of leached contaminants along the aqueous pathway is of particular concern. This is problematic in headwater regions where highly contaminated streamwater frequently is used for human consumption.

Recommendations

Compared to the impacts of mine-water discharges on the quality of receiving streams and technical possibilities for reducing erosion and breaches of tailings dams the diffuse contamination of adjacent water courses by tailings seepage and the associated transport of dissolved U received only little attention and is not yet completely understood (Funke 1990, Bain et al. 1994, Coetzee 1995, Everrett and Quibell 1995, Hearne and Bush 1996,Kempster et al. 1996, IWQS 1999, Wade et al. 2000, Coetzee et al. 2001). A focus of future research therefore should be on the migratory behaviour of U in the environment. Understanding mechanisms and controlling factors of transport will allow for identifying preferred sinks and for preventing secondary accumulations within the biosphere (Winde, in press).

Simultaneously, an inventory of all tailings deposits in SA should be compiled containing comprehensive information about their pollution potential and the sensitivity of surrounding environments. Based on this, high-risk areas can be identified and rehabilitation programmes prioritised. Such an inventory should be GIS-based and should include all types of deposited mining residues with attached information to:

- location, age, ownership/responsible authority;
- type of land use and ownership of surrounding areas;
- covered area, height, volume, mass;
- general state (active or decommissioned, erosion state, etc.);
- content of radionuclids and chemo-toxically constituents and
- the pollution potential of the deposits (determined by a currently developed index taking acid generation potential, metal mobility and re-concentration potential of heavy metals into account; Winde and de Villiers, in prep.).

Acknowledgement

This study is part of a comparative research project conducted in Germany, Southern Africa and Australia which was funded by the German Academy of Natural Scientists Leopoldina (project number BMBF-LPD 9801-17).

References

Bain CAR, Schoonbee HJ, de Wet LPD, Hancke JJ (1994): Investigations into the concentration ratios of selected radionuclids in aquatic ecosystems affected by mine drainage effluents with reference to the study of potential pathways to man. WRC report no. 313/1/94. Pretoria.

CfG (Council for Geoscience) (1998): The Mineral Resources of South Africa. Handbook 16. CTP Bookprinters, Cape Town, 740pp.

Coetzee H (1995): Radioactivity and the leakage of radioactive waste associated with Wit-watersrand gold and uranium mining. In: Merkel, B.; Hurst, S.; Löhnert, E. P. & W. Struckmeier [Eds.] (1995): Uranium-Mining an Hydrology. Proceedings of the Internationals Conference and Workshop in Freiberg, Germany, October 1995. GeoCongress, 1. Köln. 34-39.

Coetzee H, Wade P, Winde F (2001): Understanding environmental geophysical anomalies – an interdisciplinary case study from the West Rand. Extended abstract: The South African Geophysical Association, Biennial Technical Meeting and Exhibition, 9-12 October 2001, Drakensberg (South Africa), 5pp.

Everrett MJ, Quibell G (1995): The impact of diffuse load reductions from the mines in the Orkney, Stilfontein area on the water quality in the Vaal River. Final Report. Pretoria.

Ford MA (1993): Uranium in South Africa. Journal of The South African Institute of Mining and Metallurgy, volume 93, no.2, February 1993, pp 37-58

Funke JW (1990): The water requirements and pollution potential of South African gold and uranium mines. Pretoria. 172pp.

Hearne CL, Bush RA (1996): Investigation into the impact of diffuse seepage from gold mines in the Klerksdorp region on water quality in the Vaal River, Phase 1. Report No. CED/011/96 for the Klerksdorp Mine Managers Association (KMMA). Johannesburg. Unpublished.

IWQS (Institute for Water Quality Studies) (1999): Report on the radioactivity monitoring programme in the Mooi River (Wonderfonteinspruit) catchment. Report No: N/C200/00/RPQ/2399.

Kehrberg S (2000): Geohydrologist at the Rössing Uranium Mine, personal communication, Swakopmund, Namibia.

Kempster PL, van Vliet HR, Looser U, Parker I, Silberbauer MJ, du Toit P (1996): Overview of radioactivity in water sources: uranium, radium and thorium. Final report. IWQS-No: N/0000/00/RPQ/0196. Pretoria.

Robinson WP (1995): Ground water contamination at uranium mill sites in the United States reclaimed by the Department of energy (DOE) Uranium Mill Tailings Remedial Action Project (UMTRAP): a review of seven selected sites. In: Merkel B, Hurst S, Löhnert EP, Struckmeier W [Eds.]: Uranium-Mining and Hydrology. Proceedings of the International Conference and Workshop in Freiberg, Germany, October 1995. GeoCongress, Köln.

Venter I (2001): South African mining sector facing radiation challenge. Mining Weekly, vol 7 no 43 (November 9-15 2001), 2-3.

Wade PW, Woodbourne S, Morris WM, Vos P and Jarvis NV (2000) Tier 1 risk assessment of radionuclids in selected sediments of the Mooi River. WRC-Project No K5/1095.

Waggitt P (1994): A review of worldwide practise for disposal of uranium mill tailings. Technical memorandum 48. Canberra, pp44

Weißenstein K, de Villiers AB, Frühauf M., Sinkala T, Coetzee H, Freyer K, KönigW and Wartheman G (2000): Mapping and evaluation of pollution in Mine Environments in Southern Africa using GIS and EMS. Environmental Information for Planning, Politics and the Public. Metropolis-Verlag, Marburg. Vol 1: (103-113).

Wendel G (1998): Radioactivity in mines and mine water – sources and mechanisms. The Journal of The South African Institute of Mining and Metallurgy. March/April, pp 87-92.

Winde F (1998): Untersuchungen zum wassergebundenen Stofftransfer im Bereich industrieller Absetzanlagen des Uranbergbaus. In: Frühauf M; Hardenbicker U (ed.): Beiträge zur 3. Tagung Geographischer Umweltforschung in Mitteldeutschland. 6-7/11/1997, 11-31, Halle (Saale).

Winde F (2000): Gelöster Stofftransfer und fluviale Prozeßdynamik in Vorflutern des ost-thüringischen Uranbergbaugebiets. Jenaer Geogr. Schr. 9, 111-127.

Winde F (2001a): Fluvial processes and uranium transport – pitfalls from the Wismut region (Germany) and the Klerksdorp gold field (South Africa). 4th International Conference of the Society of South African Geographers., 2-5 July 2001 in Goudino Spa (South Africa), Conference proceedings (CD), 5pp.

Winde F (2001b): Slimes dams as source of uranium contamination of streams – the Koekemoer Spruit (Klerksdorp gold field) as a case study. Chamber of Mines of South Africa (ed.): Conf. on Environ. Responsible Mining in Southern Africa, 25-28/9/2001, Muldersdrift (South Africa), vol. 1, 2c1-2c10 pp.

Winde F (in press): U-contamination of fluvial systems – mechanisms and processes. Part I-III: Proceed. of the 4th Conf. of the IGU-Study Group on Water Sustainability. 4-7 July 2001, Zaragoza, Spain, 13pp, 18pp, 21pp.

Winde F, de Villiers AB (in prep.): Assessing the pollution potential of slimes dams.

Wymer D (1999): Compilation of volumes and uranium concentration of milled ore and tailings of South African goldmines. Chamber of Mines, Johannesburg, South Africa.

Wymer D (2001): The impact of gold mining on radioactivity in water and foodstuffs. Chamber of Mines of South Africa (2001): Conf. on Environ. Responsible Mining in Southern Africa, Muldersdrift/Johannesburg, 25-28/9/2001, vol 1, 2C-19 – 2C-30 pp.

Study courses at the Technical University Freiberg
Akademiestr. 6   09599 Freiberg   Germany
http:/www.tu-freiberg.de   dekan.F3@fggb.tu-freiberg.de

Mineralogy

1st – 4th term (undergraduate studies)
+ Mathematics, Physics, Chemistry
* Geology
* Mineralogy
+ Field trips, mapping exercises
* oral examination after 4th term

5th – 9th term (graduate studies)
* Mineralogy
+ Geoscientific methods
* Choice of one out of three modules:
    – Geochemistry/Environmental Geochemistry
    – Economic Geology/Petrology
    – Technical and Applied Mineralogy
+ Field training and excursions, Laboratory training courses
* Undergraduate thesis in the 7th and 8th term (200 hours)
* Masters thesis in the 9th term

Admission
* „Abitur"/Baccalaureate (A-level) or
+ Specialized college degree or
+ equivalent access entitlement

Duration/Degree: 9 semesters/Master of Science in Mineralogy (Dipl.-Min.)
Start: winter term

Job opportunities
Assessment and remediation of environmental damage; Planning and control of disposal sites; Analysis and safety control; Product development; Protection of historical monuments;

Product development and waste disposal in materials research; Environmental consultancy; Exploration, extraction and treatment of ores and other mineral resources; Geoscientific research institutes; Regional and federal geological surveys

Experience on application of continuous drain trench during the remediation of tailings ponds in Hungary

Jenő Bánik, József Csicsák, Zsolt Berta

MECSEKÉRC Environment Corp. H-7633, Pécs Esztergár L. str. 19.

Abstract. In the course of the four-decade-lasting uranium ore mining in Hungary two tailings ponds of altogether 165 ha area were constructed, in which approx. 20.3 million tons of solid material and about 32 million m^3 of technological solutions were dumped. The ore production was finished in 1997, now remediation takes place. A decontamination system was built treat the underground water pollution developed in the surroundings of tailing ponds. The free water was removed from the tailings ponds, reshaping and covering of tailing ponds and, previously to them, the stabilization of fine-grain tailing surface are on their way. In the course of remediation activities, horizontal drains were used for removing of contaminated groundwater for the stabilization of a previously damaged dam section and for the acceleration of the consolidation of fine tailing surface. The paper presents the experiences of application, with special regard to the results achieved during the application for the acceleration of the consolidation of fine tailings.

Overview of operation and abandonment of the uranium industry in Hungary

The uranium ore mining in Hungary was based on the deposit discovered in the southern part of the country, in Baranya County, in the vicinity of the city of Pécs, in the Permian sedimentary sandstone formations of the Western Mecsek Mts. Five mining plants were constructed in the course of industrial scale production since 1957, on a continuous mining plot of 65 km^2 area.

Initially, the oxidized ore produced from the Permian sedimentary sandstone was transported into the Soviet Union, after radiometric sorting. At the beginning of the sixties an ore mill plant was built, together with the tailing pond. The ore was processed with acid, producing annually about 500 t of uranium as concentrate.

At the middle of the eighties the predominant part of mine production got to deeper and deeper levels, increasing the specific cost of production gradually. This situation combined with the declining economic position of Hungary and the unfavorable word market uranium prices lead uneconomic conditions, thus the Government made a decision on the abandonment and closure of uranium ore mining at the end of year 1997.

Description of tailing ponds

In the course of uranium ore mining, two separated tailing ponds were constructed for the disposal of tailings, and approx. 20.3 million tons of solid material and approx. 32 million m^3 of technological solution was disposed there between 1962 and 1997. The predominant part of solid tailings, 15.8 million tons, got onto the Tailing Pond #I. The amount of solid tailings stored on the Tailing Pond #II is 4.5 million tons. The average uranium content of tailings is 67.8 g/t. The two tailing ponds are located at 2.2 km to the SE from the Mill, along the stream Pécsi-víz, their total area is 165 ha. Tailing Pond #I is a so-called lowland-type pond with ring dike, it was put in operation in August 1962, while Tailing Pond #II is a valley-type one, constructed with dams on its three sides on the southward ascending slope of the Görcsöny Ridge. It was in operation since 1979.

The average grain size distribution of ground ore can be characterized as follows:

>0.3	mm	-	8%
0.3-0.2	mm	-	13%
0.2-0.1	mm	-	30%
0.1-0.06	mm	-	17%
<0.06	mm	-	32%

The tailings were built up in the following way:

The tailings got on the ponds were separated with hydrocyclones into two fractions of different grain size composition. The coarse material was used to heighten the banks, while the finer grain settled in the inner area of the tailing pond.

Thus the dam bodies mainly consist of sand (50-70%), partially of silt (10-30%) and a minor amount of slime (10-20%). Moving toward the water area, these proportions gradually changed into the opposite composition, and the silty slime under the water can contain even several percent of material of grain size below 2 μm.

The direct tailings discharge means that the space inside the dam crest heightened with hydrocyclones was filled with tailings released directly through hoses from the ring duct. The uniform material distribution could be achieved through the appropriate adjustment of control valves and periodical moving of the hoses. In this case the coarser grains deposited close to the dam crest and the finer ones inside, too, but the coarser grain material got anyway farther from the dam crest than in the hydrocyclone operation mode. With this method the dam crest was able to be widened, increasing the safety of the pond.

Outlining the remediation concept of tailing ponds

The tailing ponds were constructed without bottom sealing, therefore considerable part of the technological solutions discharged with the tailings has passed into the groundwater, thus considerable inorganic salt pollution (Mg-Na-SO$_4$-Cl type) developed in the surroundings, however, the migration of radionuclides is very low, fortunately.

Considering that the tailing ponds are located above aquifers, between two potable water reserves playing important role in the water supply of the city of Pécs and partially of the neighboring settlements, in the course of remediation the preventing of further pollution is an important task.

Accordingly, the tailing ponds remediation had to follow the statements of a previously completed feasibility study:

- Reconstruction and building of toe drains and seepage water ditches providing the dam stability and belt ditches providing the external water discharge, for direct damage decreasing.
- Preventing the further spreading of inorganic salt pollution reached the groundwater, reclamation of already escaped contaminants. For this purpose an active water pumping system and a desalination plant were built in the vicinity of the tailing ponds.
- The natural flow of rainfall from the surface of tailing ponds towards the belt ditches has to be provided through the development of an appropriate shape, in addition to the on-site remediation of tailing ponds. For this purpose hollow cuts should be constructed in the surface of tailing ponds through the dam body.
- For the long-term stability a cover layer structure has to be developed, which appropriately provides the radon barrier (<0.74 Bq/m^2/s), decreases the precipitation water infiltration as far as possible, and gives protection against biointrusion.

Numerous studies and plans were completed during the preparation of remediation activities, involving several companies having international remediation experiences. Among others, the consideration of stabilizing options of fine tailings surfaces and its practical feasibility study were performed within the scope of a PHARE Multi-country Project, managed by C&E and WISMUT GmbH. According to the international practice, a 1.5-1.8 m thick, composite structure cover layer was planned to provide sufficient external protection, and the WISMUT GmbH and the Golder Hungary Ltd. took part in its calculation regarding infiltration, using numeric modeling.

Case studies: Uranium tailings

Application of drain trenches in the course of remediation

Till now, drain trenches were used in three fields in the course of remediation of tailing ponds.

A water remediation system was built to remove the polluted groundwater developed around the tailing ponds. Within its scope an altogether almost 3000 m long drain trench wall consisting of three separate sections were constructed in 6-9 m depth. In the case of walls below 6 m a pre-cutting was needed, in which the equipment moved and the drain pipe could be installed into the appropriate depth. In the drain trench constructed with continuous trenching technology single-walled, ribbed, ring-perforated soft polyethylene (SPE) drain pipe of 160 mm diameter wrapped with geotextile of 170 g/m^2 was laid, together with 4 m thick gravel filter layer. The pollution removing system provides the removal of approx. 700,000 m^3 of contaminated water annually through 15 pumping wells.

An approximately 300 m long section of the northern dam of Tailing Pond #II soaked through during the operation because of the inadequately working drainage system and therefore the dam slipped. Thus strengthening of the slipped section and reconstruction of drainage system was required. This problem was solved through two drain trenches installed into the sandy dam body in two levels, initially having drained nearly 60 m^3/d water from the affected dam section.

The drain trench construction used for surface stabilization on the tailing pond

Morphological plan of Tailing Pond #I

According to the statements of the Feasibility Study concerning the long-term stability, the remediation plan of Tailing Pond #I of lowland type with ring dike contained the construction of two water discharging valleys into the originally concave surface through the southern dam, with a hollow. Accordingly, two hollows should be constructed on the tailings pond partially cutting into the current surface, reaching the slime core on some places as well.

According to the remediation plans pre-load had to be applied in the line of the hollow, on a well-defined area to ensure the machine working and to stabilize the slime. The area of fine tailings was covered with geotextile and geogrid prior to the starting of the pre-load (Fig.1), then vertical drains were installed in a 5×5 m network, down to 6 m depth.

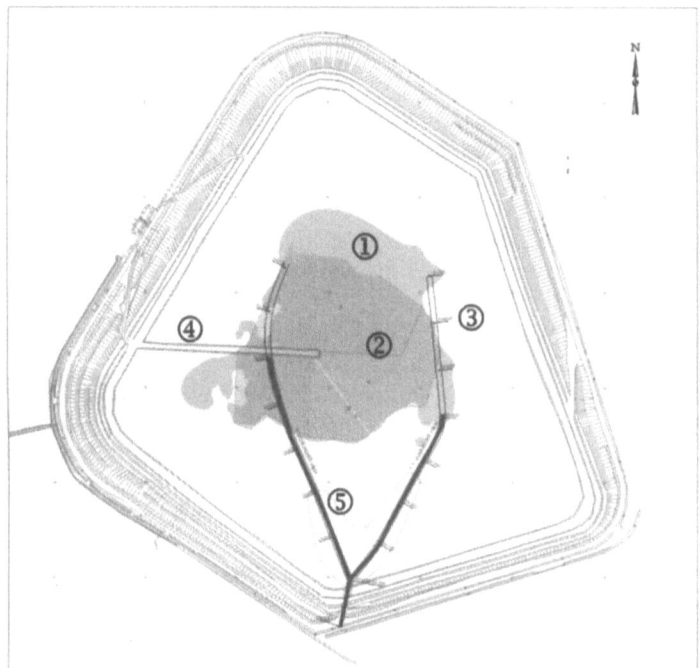

Fig. 1. Map of Tailings Pond No. I.   Legend: 1.: Fine Tailings; 2.: Covered area with geogrid and geotextile; 3: Planned water discharge valleys; 4: Rock-dam; 5.: Drainage Trench.

The "idea" of using drain trenches

Considering the strict schedule of the accomplishment given for the remediation of tailing ponds, we assessed before the starting of the pre-load, that the scheduling of hollow construction (pre-load, pre-cutting) is needed for the further acceleration of dewatering process.

During the preparation activities, active dewatering experiments were performed previously on two areas on the transitional zone of Tailing Pond #I, giving favorable results.

In connection with the experimental water pumping we've come to the conclusion that as a result of using the two types of technology applied in the course of tailing pond construction (hydrocyclones and direct tailings discharge) there are thin (~0.5-15 cm) sand layers interbedding in the transitional zone and partially reaching into the slime core, therefore the gravitational dewatering is possible.

Thus the idea has arisen on the basis of the favorable experiences on the fields of drain construction application discussed above, that in the line of Tailing Pond #I hollow crossing the slime core, too, the dewatering of the area could be accelerated and therefore the consolidation as well. Thus the work of vehicles moving on the area becomes safer, and the hollow construction can be accelerated.

In connection with this, the draining of the transitional zone was expected, which will have an effect on the acceleration of dewatering of the slime core as well, according to our plans. It was an important part of the plan that the installation should be done to increase further the degree of draining utilizing the water conducting capability of the existing rock dam or "mole" serving for operational purposes, which has good water conducting capability, according to the previous experiences. For this purpose the line of the western branch was continued to the rock dam.

Our aim was not the entire dewatering of the tailing pond or the slime core, just the acceleration of consolidation in the line of hollow.

Considering that our tailing ponds are located between two drinking water resources playing important role in the water supply of the city of Pécs, the authorities also welcomed the solution, through which lesser contaminated water will enter into the aquifers under the tailing ponds, as a result of the load developing in the course of covering.

Constructing the drain trench

Constructing the drain trench planned in the hollow line of the Tailing Pond #I – considering that the trace line crossed the slime core as well – the adequate safety, and accordingly the bottom pressure less than 20 kN/m^2 was a requirement.

The work was completed by the Békés Drén Ltd. with an automotive, laser-controlled equipment of 31 t weight using continuous trenching technology of 4-6 m depth, which had to be modified slightly to meet the technological requirements (structure lightening, chain track widening).

Constructing the drain trench single-walled, ribbed, ring-perforated soft polyethylene (SPE) drain pipe of 160 mm diameter wrapped with geotextile of 170 g/m^2 and washed, sorted pebbly sand filter material of 0.5-4 mm grain size were installed. During the pre-calculations the forecasted, stabilized water volume was determined to be about 40 m^3/d, after the entire installation.

In the course of preparatory works the planned trace line was surveyed with test boreholes, and with shear vane test on the slime core. Providing the appropriate downgrade a pre-cutting was required on an approx. 150 m long section of the dam, affecting only the area of sand dam. In this section non-perforated pipe was installed, the trench above the drain pipe was refilled with the own material of the dam. The drain outlet was connected into a shaft of the ring pipe of the existing drainage system. The drain was constructed with even declination of 1‰, except the first approx. 50 m section of the outlet, where the descent is steeper. The western branch is 587 m, while the eastern one is 440 m long.

During the construction, on the slime core the draining equipment moved on an approx. 1 m thick banking.

Summary of results

The water volume drained by the completed drain trench was measured together with the water volume flowing from the existing drainage system. The previous total daily water volume of 150-160 m^3 increased sometimes above 250 m^3/d at the time of completing the drain trench. The stabilized yield is 200-210 m^3/d (Fig.2).

The effect on the different sections of the completed drain trench was controlled on water monitoring points on the tailing pond and with shear vane tests.

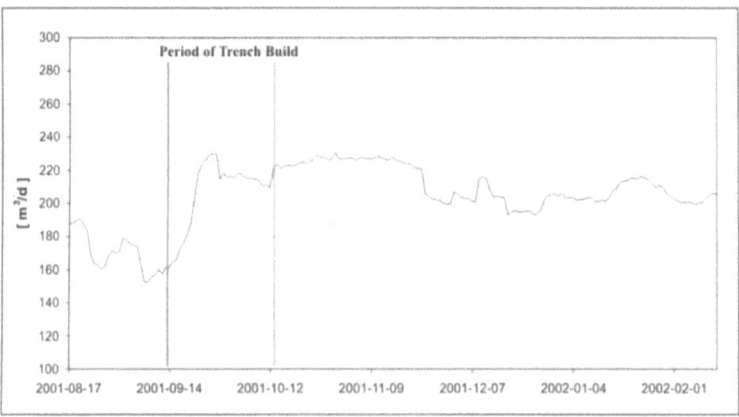

Fig. 2. Yield change of seepage water in consequence of build drainage trench.

Process of dewatering of the tailing ponds

Inside the tailing ponds water level monitoring wells are used to control the dewatering process. After the free water was removed, during the approx. 1 year long period before the installation, the water levels inside the tailing ponds barely changed (Fig. 3).

After the free water was removed, considerable water level lowering could be detected only in the sandy parts close to the dams. It resulted in slow water level lowering in the transitional zone as well. No lowering trend could be detected inside the slime core on the water level measurement data lines of more than one year term, except variations under the effect of precipitation. However, after the drain trench was constructed, almost one meter of water level lowering could be detected within half year, in the vicinity of the trace line. The succeeding in the utilization of features of the existing rock dam also can be verified, since the water level lowering could be detected in its surrounding as well.

The dewatering process was followed by the acceleration of consolidation processes, which is verified with the data of performed shear vane tests. It is clearly visible on the logs prior to the drain laying and on the ones made about half year after the laying that the shear strength values increased in the line of drain. Considering these results the technical intervention can be regarded as successful.

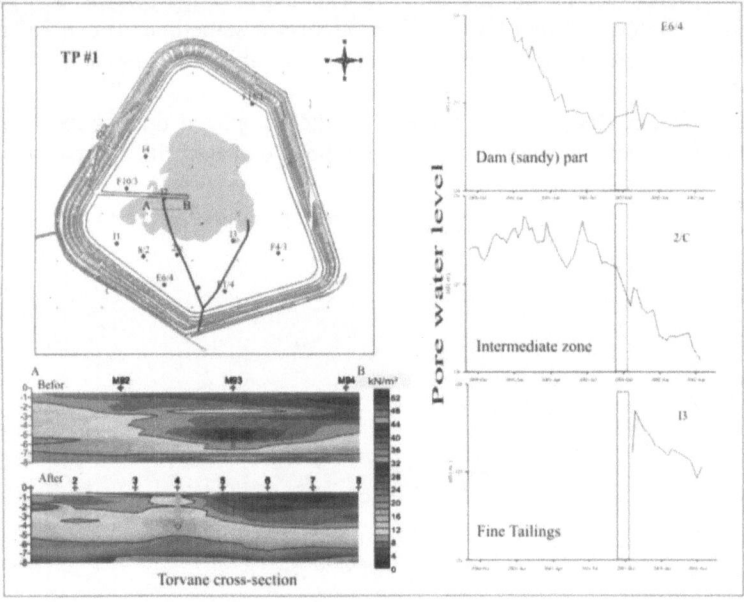

Fig. 3. Impact of the drainage trench on the pore water level and the shear strength values of the fine tailings.

Summary

The strict schedule of the remediation of tailing ponds in Hungary made it necessary to plan and complete such work phases – in addition to the application of conventionally used techniques (geotextiles, vertical drains) – that can accelerate consolidation (e.g. pre-loading). On the Tailing Pond #I the continuous drain trench technology was applied to accelerate consolidation, utilizing the good water permeable sandy interbeddings occurring in the transitional zone and partially in the slime core as well. The drain constructed directly to stabilize the bottom line of the hollow verified, that with this technology considerable excess water can be removed from the tailing ponds and it was verified by the fact, that the water level inside the slime core lowered by 1 m in the vicinity of the line of drain, thus the tailings consolidation accelerated, which process can be verified with other measurements as well.

Methodology to assess the radiological impact of a repository for uranium mill tailings after remediation (short-term impact)

Anne-Christine Servant

Institut de Radioprotection et de Sûreté Nucléaire
BP 17 92265 Fontenay-aux-Roses Cedex France

Abstract. A radiological impact study consists in assessing the various individual radiological exposures taking into account all the possible exposure pathways, for population groups representative of local population. It is based on the calculation of the total added effective dose indicative of the fact that the repository radiological impact is added to the relatively high natural background radiological level of areas where uranium-mining sites are located. Population groups that receive the maximum impact from the site are considered as reference groups that are monitored to ensure that the site impact remains acceptable.

Introduction

The options adopted for the remediation of a repository for uranium mill tailings must be such that the current radiological and chemical impact can be reduced to levels that are as low as reasonably achievable using the best available techniques at acceptable cost.

The objective of the assessment of the radiological impact of a repository for uranium mill tailings after remediation is to demonstrate the ability of the repository to ensure protection for people, taking into account the current legislation.

Respect of the effective dose limit of 1 mSv.year^{-1} required by the European Directive 96/29/Euratom whereas former French regulation limit was 5 mSv.year^{-1} leads to the re-assessment of the radiological impact of the repositories based on more realistic scenarios.

Presentation of the nearby environment of the repository

Local meteorological conditions

It is necessary to know precisely the climatic processes that govern the atmospheric dispersion of dust and radon. Exceptional events (e.g. rainy events) also have to be compiled so as to study their influence on water management.

Geology - hydrogeology

A detailed description of the local geology has to be provided as well as relevant data linked with hydrogeology. Any information that could help the understanding of underground water movements has to be mentioned. If a local modeling has been realized it should be presented.

Hydrographical system

Detailed descriptions of the various watercourses that cross the site have to be presented. The flow (particularly floods and low water) of the watercourses, as well as drained surfaces, catchment areas, fittings and tributaries have to be mentioned.

Agricultural environment

Local cultivations, breeding, and activities linked with food production have to be described. Soil occupation is important to describe as well as the origin of irrigation water. Fishing and aquaculture activities should be described if relevant.

Water uses

Domestic and agricultural uses of water have to be clearly known and described. Possible uses for aquaculture, leisure or tourism should also be described if relevant.

Description of the human environment

Demographic data have to be supplied for the area subjected to the impact of radionuclides. These data should deal not only with the settlement but also with working populations and other populations (schoolchildren, tourists...).

Case studies: Uranium tailings

Site description

Description of the wastes

All the potential sources of radioactivity have to be developed. For each radioactivity source, quantities, mass activities, physical and chemical stabilization treatments, mineralogy, and geochemistry have to be described.

Repository structures- remediation options

The repository structures as well as the remediation options have to be described and their durability should be discussed.

Water management on the site

Water management objectives are organization of a selective collect of non-contaminated water, and erosion phenomena prevention. The various systems used as well as their efficiency and durability have to be described.

Various workings on the site

A precise list of the various workings on the site (open pits, underground mines, barren rocks dumps) that may interact with the repository has to be developed.

Development of the site conceptual model and exposure pathways

Conceptual model

The conceptual model is a discursive and graphical representation of the situation. It represents a synthesis of the knowledge acquired concerning the site and its surroundings, and concerning the environment vulnerability.

It helps to understand and to explain the situation in terms of contamination sources and real as well as potential populations exposures pathways.

Fig.1 shows an example of representation using compartments.

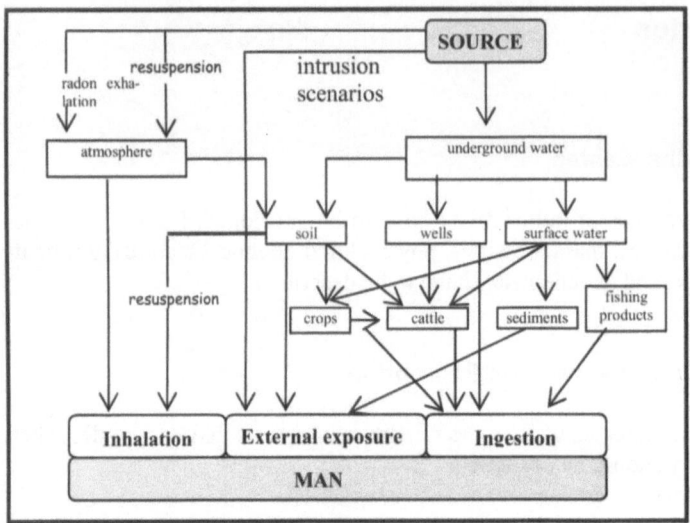

Fig. 1. Example of conceptual model

Exposure pathways

External exposure

External exposure comes from the particles deposited and from the radiations coming from the site, radiation coming directly from the site being generally highly predominant. It can be calculated using measured average dose rate.

Internal exposure via inhalation

Internal exposure due to inhalation of short-lived daughters of radon 222 and radon 220 is a major pathway. It is calculated from the data of volume Potential Alpha Energy concentrations of short-lived daughters of radon 222 and radon 220 in the atmosphere. Dust coming from the site can also contribute to the internal exposure via inhalation, calculated using the volume activity due to long-lived alpha emitters in the dust in suspension in the atmosphere.

Internal exposure via ingestion

Ingestion of contaminated foodstuffs is an internal exposure source. Use of contaminated water in the purpose of irrigation or animals watering can be a major contamination source. Drinking water contamination has to be taken into account, whether it comes from a well or from the distribution system.

Case studies: Uranium tailings

Development of population groups and scenarios

Location of population groups

The objective is to determinate the geographic location of the inhabited area, residents of which receive the highest impact from the studied mining area. Giving the generally large quantity of inhabited areas around the site, it can be necessary to use a rational method.

Composition of population groups

The higher the impact is, the finest the assessment should be, and so the smallest the population groups could be.

People in a population group can be adults or children, but in one group, people should belong to the same age group. There is no methodology to determine a priori, i.e. before measurements in the environment and dose calculation, that one specific age group should be adopted.

Scenario definition

Trying to be realistic, in accordance with article 45 of the European directive 96/29/Euratom, must lead to determine scenarios corresponding to local real lifestyles.

In order to perform a realistic assessment, timetables have to be as close as possible to life habits in the local environment of the site. Data concerning food consumption have to be given and should be accompanied by as many statistic data as possible. Origin of the food produces has to be known finely especially when they are cultivated locally.

While determining the diets associated with each population group, it is necessary to take into account the age group, the professional activity, the residence place and the self-sufficiency level.

Measurements in the environment

Radiological impact assessment just after remediation can be totally based on the monitoring network installed by the operator or by the Authorities in the nearby environment. This network also enables to check the evolution of the source term.

The measurements have to be realised at least on a yearly basis, and evidently depends on the exposure pathways that have been identified and on the local lifestyles.

Location of measurements and analysis

Outside measurements

These measurements concern the mean external exposure dose rate, the Potential Alpha Energy (PAE) of radon 222 and of radon 220, and long-lived alpha emitters in suspended dust.

Necessary measurement places on the site are all the places where contaminant products are stored, and a part of the site where there is no storage of contaminated products.

In the nearby environment measurement stations have to be installed in all the places inhabited by the population groups, and all the places where the population groups work and have spare-time activities.

Measurements inside the buildings

These measurements concern the Potential Alpha Energy (PAE) of radon 222 and of radon 220.

Necessary measurement places on the site are the existing buildings that are used for other professional activities.

In the nearby environment measurement stations have to be installed in some houses of the population groups, in some buildings where those people work (farms, schools...) and in some leisure buildings (gymnasium....).

Water

Sampling for analysis has to be done at water catchment for agricultural use, at consumers' taps, and if relevant in the swimming and fishing areas.

Food chain

Radionuclides that have to be analyzed are at least: 238 U, 230 Th, 226 Ra, 210 Po and 210 Pb.

Sampling has to be done in the gardens, in the fields and in the greenhouses, among products bought locally, and if relevant in gathering areas.

Presentation of results

Results have to be presented with:
- the location of the measurement station or of the sampling or of the cultivation place if it is a transformed product,
- a description of the measurement station (distance to the mining area, location on the wind rose, local topography, presence or absence of a shield between

the site and the measurement station, location with regard to the hydrological downstream of the site)
- the sampling period
- the climatic parameters that can influence the result,
- all the parameters that are used to transform the result of measurements in usable concentrations or dose rate (for example water content, sampled air volume).

The objectives are to get at least, during the whole studied year, an average data for the external exposure dose rate every quarter, an average data for APE of radon 222 and of radon 220 and for the activity of long life alpha emitters in the suspended dust each month. Concerning drinking water, one concentration value for each radionuclide each month, during the studied year, and one concentration value for each radionuclide and each type of food locally produced.

Initial radiological level

"Natural" radioactivity comes from cosmic and telluric radiation.

If there is detailed information on the radiological state of the area before the beginning of mining operations it can be used as initial level. However generally environmental data are not precise enough and cannot be used. Some measurement stations should then be installed in the natural surroundings, in order to get some relevant data without any influence of the site.

Those stations have to be installed in a geological context similar to the geological context of the site, upwind and upstream the mining site, on drinking water distribution systems different from those used by the studied population groups.

The objective should be to get, for each measurement station in the site nearby environment a corresponding natural surroundings station, which is a station located in a similar geological environment and a similar topographic situation.

Site impact calculation

Impact assessment is based on the determination of the total added effective dose resulting from the exposition of the various identified population groups. It corresponds to the difference between the current effective dose and the effective dose estimated for the initial situation.

Dose coefficients

Dose coefficients enable to connect incorporated quantities (through ingestion or inhalation) and effective doses. They depend on the age group. Dose coefficients that have to be used are issued from the European Directive 96/29 Euratom.

Sensitivity analysis - uncertainty

Generally a number of parameter that is imperfectly known plays a part in the dose assessment. Therefore there is necessarily uncertainty in the dose calculation.

In the impact study there has to be an estimation of the uncertainty on the dose results. Although a complete uncertainty analysis is not necessary, there must be a concise uncertainty analysis based on a sensitivity analysis of the parameters that highly influence the dose results and on those that are the most uncertain. It is a necessary process in order to be able to assess the highest possible dose received by population groups.

Comparison with regulatory limits

In a situation where all the industrial activities are normally operating, the members of the population groups should not be exposed to an added effective dose exceeding the annual limit laid down by law. French current added effective dose limit is 5 mSv.year^{-1}. But as this limit is to be changed into 1 mSv.year^{-1} due to the European directive 96/29 Euratom, it is of course the 1 mSv.year^{-1} limit that has to be taken into account.

Conclusion

The radiological impact assessment of a repository for uranium mill tailings after remediation allows to demonstrate the ability of the repository to ensure protection for people, taking into account the current legislation.

In addition, through this impact study, reference groups, which are population groups that receive the maximum impact from the site, can be defined. Reference groups will be monitored afterwards to ensure that the site impact on man remains acceptable.

Moreover the impact study helps to define a monitoring network in order to ensure that the impact on the environment remains acceptable.

It is essential to remind that the impact on man has to be re-assessed if population groups with different lifestyles are noticed afterward (for example after the construction of a new residence near the site, or a new camping site…), or if the monitoring of the environment shows some changes in the measured parameters values.

A radiation dose estimate for visitors of the South Alligator River valley, Australia, from remnants of uranium mining and milling activities

Andreas Bollhöfer, Bruce Ryan, Kirrilly Pfitzner, Paul Martin, Michelle Iles

Environmental Research Institute of the Supervising Scientist (*eriss*), Locked Bag 2, Jabiru, NT 0886, Australia

Abstract. In this paper we estimate the annual radiation dose received from remnants of uranium mining and milling activities by members of three groups visiting the South Alligator River valley: local Aboriginal people, Parks Australia North Rangers, and tourists. We take into account three possible pathways of radiation exposure: external gamma radiation, inhalation and ingestion of radioactive elements, and include information on occupancy times and traditional Aboriginal diet.

Introduction

The South Alligator River valley in Kakadu National Park in the Northern Territory, Australia, is a famous destination for tourists. It is also of important cultural significance to local Aboriginals, the Jawoyn people. In the 1950s and 60s, however, it was mainly renowned for its mineral resources and uranium mining activities.

The upper South Alligator River valley's uranium deposits are characterized by a high uranium content and some have concentrations up to 2.5% U_3O_8. During peak activity, at least 16 different uranium ore bodies were known in the area with 13 of them eventually being mined (Waggitt 2001). Milling of the material extracted from some of the uranium ore bodies occurred at the South Alligator Mill between 1957-64. The mill was located a few hundred meters east of the Rockhole Mine Creek, where tailings were deposited on the ground with very little containment within the area of the South Alligator River catchment. Mining in the upper South Alligator River valley ceased in 1964, the minesites and camps were abandoned and no substantial effort to rehabilitate or clean up the area was made.

In 1984 exposed tailings were identified by *eriss* staff. In 1985-86 most of the tailings were removed by *Pacific Goldmines* to extract gold. Hazard reduction works were carried out in 1991-92 during which the South Alligator mill and village were decommissioned and buried. A survey in October 1991 revealed that tailings material had become exposed in an erosion gully next to Gunlom Road, a road frequently used by visitors of the area. Gamma dose rates of up to 5-6 μGy per hour were measured. The tailings were covered with rocks to prevent further erosion. A routine inspection in 1999 revealed that some of the tailings had become exposed again and were spread towards the west along Gunlom road by graders involved in roadworks.

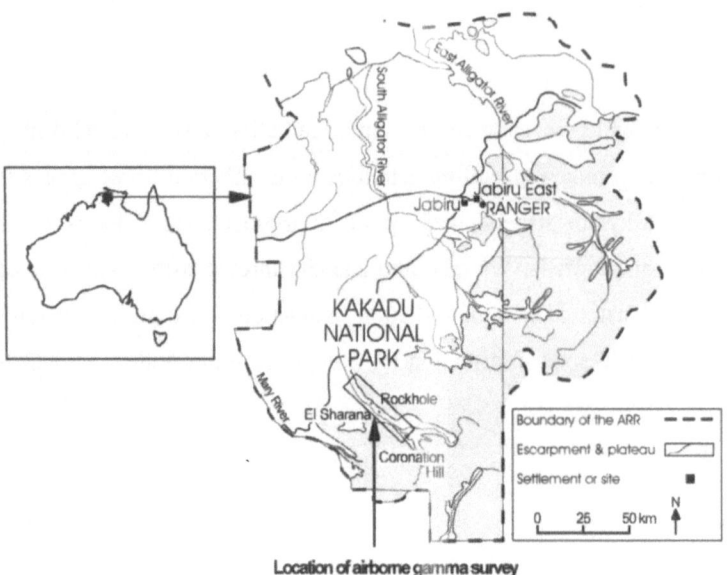

Location of airborne gamma survey

Fig. 1. Map of the Alligtor Rivers Region encompassing Kakadu National Park and the location of the airborne gamma survey.

A ground-based gamma survey was conducted in 1999 at the Rockhole mill tailings site (Tims et al. 2000) followed by an airborne gamma survey in 2000 (Pfitzner and Martin 2000) (figure 1). Data from the airborne gamma survey were evaluated for the Rockhole tailings site and the remaining portion of the South Alligator River valley (Pfitzner et al. 2001a,b) to quantify the magnitude and extent of radiological contamination from the mining activities in the 1950s and 1960s.

This paper deals with radiation levels at various disused and abandoned mining and milling locations in the South Alligator River valley. The aim is to give an estimate of the radiation dose received by local Aboriginals, Park Rangers and tourists visiting the area. It takes account of external gamma doses received, the dose received via the inhalation of dust and radon progeny, and the ingestion of radionuclides in bush foods and water. Information from Traditional Landowners and Park Rangers on site occupation times and food consumption has been incorporated into the dose assessments.

Methods

Airborne survey and external gamma dose rate

Comparison of the overall results of the airborne survey with the results from various groundtruthing exercises showed that the field based measurements follow the general trend of the airborne survey i.e. areas of the highest airborne count rates (Rockhole tailings, El Sherana mine, Palette mine) exhibited the highest activities during the groundtruthing. It proved very useful in highlighting areas of comparatively high radiation levels that were later identified during the field-based measurements. However, the direct comparison of field data with individual image pixels proved unsuccessful (Pfitzner et al. 2001a,b). Reasons for this include:

- Areas of higher activities appear in small pockets at most of the sites (Pfitzner et al. 2001b; Waggitt 1996). The nominal line spacing of the airborne survey was 50 m, with individual readings integrated over an average distance of 50 m along each line. The values obtained were interpolated to yield a 12m x 12m image grid, but the true resolution of the image is based upon the line spacing.
- Inaccuracy of the groundtruthing GPS, which was at times up to 5-9 m or higher and inaccuracies due to low signal of the GPS on board the plane.
- Topography of the area. This sometimes deviated significantly from the assumed 2π (i.e. flat plane) geometry of the source surface.
- Variations and differences in soil moisture, during both the airborne and field surveys.

During the groundtruthing, emphasis was given to identify "hotspots" and therefore averaging the field-based readings overestimates the true soil activity and can not be used for a comparison with the average values from an airborne survey. Hence a different approach was chosen to calculate a gamma dose rate using the counts of the airborne survey.

In situations such as that of the South Alligator River minesites where there are small, well-defined areas with very high radiation fields (i.e. more than an order of magnitude greater than "background") surrounded by areas of relatively evenly-distributed radiation fields, the maximum airborne signal received is largely influenced by, and often dominated by, the small, high-activity areas. If we assume that we have identified these areas during our groundtruthing survey and plot the maximum groundtruthing dose rates versus the maximum airborne count rate in that region forcing the linear fit through zero, we may expect a reasonable linear relationship with the slope being the calibration factor (gamma dose rate per air-

Case studies: Uranium tailings

borne counts per second). This factor has been estimated to be 0.0022 ± 0.0009 µSv hr^{-1} per counts s^{-1} (95% confidence). To estimate a dose rate above background a background dose rate of approximately 0.1 µGy per hour (Tims et al. 2000) was subtracted for each individual pixel of the airborne survey. The average external gamma dose rates for a particular site as a whole was then calculated from the airborne signal, averaged over all pixels (>350 counts per second) representing the site. This is more likely to represent a true average of the external gamma dose rate. For details see Bollhöfer et al. (2002).

Inhalation of radionuclides in dust particles

In 2000 a study along Gunlom road was conducted to estimate airborne dust concentration and consequent radiation exposure via the inhalation of long lived alpha activity (LLAA) in dust. Samples were collected on filters operating at a flow rate of 2 litres per minute and airborne alpha activities have been measured with a *Daybreak* gross alpha counter.

As we are dealing with dust stirred up by vehicles we assume an activity median aerodynamic diameter (AMAD) of 5 µm similar to what is assumed for the inhalation of particulates by workers on mines (ICRP68 1994). Therefore we used the effective dose coefficients given in ICRP68 for the calculation of the total DCF, assuming an AMAD of 5 µm. In addition, five scrape samples were collected from the surface of Gunlom Road and measured for their radionuclide activities. Using the average $^{238}U/^{226}Ra$ activity ratio of those samples of 0.6 ± 0.3, we calculated the total DCF as an activity weighted average assuming that no Rn is retained in the dust particles (Zapantis 2000). A DCF of 5.5 µSv per α-decay was calculated for the inhalation of Gunlom Rd dust. The dose received from the inhalation of dust while driving along Gunlom road was subsequently calculated via:

$$D_{LLAA} = DCF \cdot LLAA \cdot BR \cdot IT \tag{1}$$

DCF: 5.5 µSv per α-decay
LLAA: long-lived airborne α-activity concentration
BR: breathing rate: 1.2 m^3/h
IT: inhalation time [h].

Inhalation of radon progeny

In 1991, a radon survey was conducted in the upper South Alligator River valley (for details of the Rn station see Akber et al. 1991). Average Rn concentrations have been estimated at the Rockhole mill site, Rockhole mill run off, El Sherana weighbridge, battery and burial sites. From the Rn activity concentration [Bq/m^3]

we were able to make an estimate of the airborne Rn progeny potential alpha energy concentration [$\mu J/ m^3$] and calculated the dose received using:

$$D_{RDP} = DCF_{RDP} \cdot PAEC \cdot IT \tag{2}$$
$$PAEC = f\alpha \cdot E\alpha \cdot N\alpha \tag{3}$$

DCF_{RDP}: dose conversion factor, 1.1 $\mu Sv/(\mu Jh/m^3)$ (ICRP65 1993)

$PAEC$:   Potential Alpha Energy Concentration [$\mu J/m^3$]

IT: inhalation time

$E\alpha$:   PAEC per Bq of ^{222}Rn in equilibrium with its short-lived progeny, 5.6·10^{-3} $\mu J/Bq$

$f\alpha$:   average equilibrium factor: 0.2 – 0.5

$N\alpha$:   radon activity per m^3 of sampled air.

Ingestion of radionuclides

Bush foods

The assumed annual consumption of bush foods for Aboriginal people of the South Alligator River valley is shown in Table 1. This group of Aboriginal people would theoretically be the group most at risk in the study area because of their semi traditional lifestyle.

Table 1. *Diet of the critical group in the South Alligator River valley, shop bought food not included (Ryan in prep.).*

Animal	Flesh	Organs eaten	Weight kg/yr
Wallaby/Kangaroo	7	7	110
Buffalo	7	By older people	50
Cow	7	By older people	50
Pig	7		30
Fish	7	7	60
Goose	7	7	25
Vegetarian	7		25
Reptile	7	7	30
Mussel	7		2

Koperski and Bywater (1985) estimated annual bushfood consumption totals at 450 kg per Aboriginal inhabitant in the region of the Ranger Uranium Mine (RUM) whereas Johnston (1987) estimated 290 kg. In ICRP Publication 23 (1975) the per caput estimate of food supplies for Reference Man from Oceania is 677 kg/yr. Shop bought food has been estimated to be approximately 30% for Aboriginals living around the RUM (Johnston 1987). In the case of the South Alligator River valley area, there are access problems caused by the distance of the group from a major shopping centre, and lack of transport. Consequently, the shop bought food component has been estimated to be 25% of total food makeup. An access time average for Aboriginal people has been put at 9 months a year, due to flooding of the area in the wet season. If the shop bought food estimation and the access estimation were applied to the 677 kg/yr intake of the ICRP's reference man this would then give an annual bushfood intake of 380 kg/yr in agreement with table 1.

With ingestion dose coefficients given in ICRP72 (1996), the respective concentration factors (Martin et al. 1998; Martin 2000) and the estimated annual bushfood consumption (table 1) total annual committed doses received via the ingestion of various bush foods for adults were approximated.

Water

Surface water samples were taken in 1989 from the South Alligator River, which is the main drinking water source of local Aboriginal people, and along the Rockhole Mine Creek. Radionuclide concentrations were measured by gamma and alpha spectrometry. In addition, during the 2001 ground survey, water from permanent waterholes in the mine pits of Scinto V and Koolpin mines, a permanent spring at Scinto V and from Stag Creek close to El Sherana were collected and U concentrations determined by ICPMS. Effective doses were calculated using ingestion dose coefficients from ICRP 72 (1996).

Mussels

In 2000 *eriss* staff and Jawoyn Aboriginal people collected mussel samples from the South Alligator River in the vicinity of the Rockhole Mine Creek confluence and the exposed tailings area. Gamma spectroscopy was used to analyse samples for ^{226}Ra and ^{210}Pb, alpha spectroscopy was used to determine ^{230}Th and ^{210}Po (Iles et al. in prep.). Subsamples were taken and due to its higher sensitivity ICP-MS was used to determine uranium concentrations. Effective doses were calculated using the measured radionuclide concentrations, ingestion dose coefficient from ICRP 72 (1996) and assuming an annual consumption of 2 kg of mussels.

Results

External gamma radiation, inhalation of radon progeny and dust

In order to calculate the dose received in a year from external gamma radiation and the inhalation of radon progeny it is crucial to know the occupation times at the different sites. This can be difficult without the knowledge and input from locals and Traditional Owners. Table 2 was formulated after seeking advice from Traditional Owners and Parks Australia North (PAN) staff.

Case studies: Uranium tailings

Table 2. *Occupancy times (bold) for the different locations in the South Alligator River valley and calculated external gamma (γ-dose) and Rn progeny (RDP) doses received per year.*

Location	Aboriginal occupancy time hr/yr		PAN Staff occupancy time hr/yr		Park visitor occupancy time hr/yr	
	γ-dose [μSv]	RDP [μSv]	γ-dose [μSv]	RDP [μSv]	γ-dose [μSv]	RDP [μSv]
Rock Hole Mine	**4**		**6**		**0**	
	5.5	0.9#	8.2	1.4#	0	0
Tailings and mill site	**30**		**4**		**1**	
	54	7	7.2	0.9	1.8	0.23
El Sherana Camp	**4**		**30**		**0**	
	0	0.1	0	0.6	0	0
El Sherana weighbridge	**1**		**4**		**1**	
	1.2	0.14	4.6	0.6	1.2	0.14
El Sherana burial site	**1**		**2**		**1**	
	0.9	0.02	1.7	0.04	0.9	0.02
El Sherana pit	**0**		**1**		**0**	
	0	0	2.5&	0.23	0	0
Scinto V	**0**		**2**		**1**	
	0	0	2.9	0.23#	1.4	0.23
Scinto VI	**1**		**2**		**1**	
	1.3	0.23#	2.7	0.23#	1.3	0.23#
Saddle Ridge	**2**		**4**		**0**	
	2.3	0.46#	4.7	0.92#	0	0
Palette and skull	**2**		**4**		**0**	
	3.1	0.46#	6.3	0.92#	0	0
Koolpin	**1**		**2**		**0**	
	0.9	0.23#	1.9	0.46#	0	0
Coronation Hill	**4**		**6**		**0**	
	5.3	0.92#	7.9	1.4#	0	0
Total	74.5	10.5	50.6	8.5	6.6	0.85
Total annual doses [μSv]	**85**		**59**		**7**	

assuming a dose rate of 0.23 μSv per hour from RDP as estimated for the tailings area
& using 2.5 μSv per hour measured during the ground survey in the pit

The maximum dose rate for the inhalation of dust while driving along Gunlom Road amounts to 0.4 μSv per hour. Assuming that dust is stirred up and inhaled for 5 minutes each time driving across the area 0.04 μSv per trip will be received. Tourists may drive along this road a couple of times during their visit, PAN Rangers and local Aboriginal people possibly on a daily basis. Assuming the tailings site had to be crossed twice a day during the dry season (240 working days) the total inhalation time would amount to 40 hrs. The maximum committed effective dose that PAN Rangers and local Aboriginal people may receive from the inhalation of dust therefore amounts to about 20 μSv per year.

Consumption of food

Buffalo concentration factors (CFs) were used as an analogue for wallaby/kangaroo and cow as there are no CF's available for these animals. The CFs for fish come from a combination of fish species reported by Martin (2000) these include fork tailed catfish (*Arius leptaspis*), archer fish (*Toxotes chatareus*), barramundi (*Lates calcarifer*), eel-tailed catfish (*Plotosidae*), freshwater mullet (*Liza alata*), long tom (*Scleropages jardini*) and Tarpon (*Megalops cyprinoides*). Turtle flesh CFs were used for the reptile CFs as these are higher than goanna and crocodile CFs (Martin 2000) again giving a more conservative dose estimate.

Furthermore, we assume that animals such as wallaby, buffalo, cow and pig spend approximately 1% of the time grazing on contaminated areas and 99% on uncontaminated areas. Using the results from the airborne survey an average uranium concentration of the contaminated soil in the vicinity of the Gunlom tailings can be estimated to 2800 Bq per kg soil the average background soil uranium activity concentrations is ~45 Bq per kg. Assuming radioactive equilibrium of all progeny in the soil and with concentration factors reported by Martin et al. (1998) and Martin (2000) average flesh radionuclide concentrations can be calculated. With the ingestion dose coefficients given in ICRP72 (1996) and the estimated annual consumption (table 1) total annual doses received via the ingestion of various bush foods for adults were approximated (table 3). From the total annual dose received, however only about 160 μSv can be attributed to enhanced radionuclide flesh concentrations due to grazing on tailings and contaminated soils.

Table 3. Total activity consumed annually [Bq] by adult Aboriginal people and annual dose received via the consumption. Approximately 100 μSv are background.

Food item	^{238}U	^{230}Th	^{226}Ra	^{210}Pb	^{210}Po	dose [μSv]
Wallaby/Kangaroo**	1.2	0.2	2.1	1.3	18.4	24
Buffalo	0.5	0.1	0.9	0.6	8.3	11
Cow**	0.5	0.1	0.9	0.6	8.3	11
Pig	0.4	0.4	0.9	0.5	106.6	129
Fish	0.9	0.3	17.1	2.5	13.0	22
Goose	0.1	0.1	1.7	0.6	9.3	12
Fruit/Yams	0.6	-	3.2	-	0.5	~2
Reptile#	0.8	0.3	5.4	4.3	36.0	48
Total dose [μSv]	**0**	**0**	**8**	**8**	**240**	**257**

** *Buffalo values were used as default, #Turtle flesh CF used*

The committed effective dose above background from the consumption of 2 kg of 7 year old mussels from a downstream tailings site in the South Alligator River has been calculated on the basis of ^{238}U, ^{230}Th, ^{226}Ra, ^{210}Pb and ^{210}Po, the background dose was calculated from the average radionuclide activities of mussels from a control site (S4) in the South Alligator River (Iles et al in prep; Bollhöfer et al. 2002). The committed effective dose due to mining and milling residues above background amounted to ~50 µSv, more than 85% can be attributed to ^{226}Ra and ^{210}Pb. An additional 3 µSv are received, assuming 1095 litres of South Alligator River water being drunk by Park Rangers and adult Aboriginals in 9 month (i.e. 4 litres per day). Therefore, the total dose received from the consumption of food by Aboriginal people amounts to ~210 µSv per year and is negligible for Park Rangers and tourists, respectively.

Table 4 summarizes our results taking into account different pathways of radiation exposure.

Table 4. Annual doses in µSv received by the three population groups via the different pathways.

	Local Aboriginals	PAN staff	Park visitors
External gamma	74	51	7
Inhalation of dust	17	17	Neg
Rn progeny	11	8	1
Food consumption	213	3	Neg
Total	**~300**	**~100**	**~10**

Acknowledgements

The authors wish to thank Bessi and Beryl Smith, and Mary River PAN District Supervisor Greg Ryan, for their advice on occupation times at the different sites and on food consumption. We also wish to thank Dr. Chris Humphrey for assistance with mussel collection and ageing, Peter Waggitt for collection of airborne dust samples along Gunlom Road, Therese Fox and Matthew Noske for sample preparation. Parks Australia North and *eriss* jointly funded the acquisition of the airborne gamma survey.

References

Akber R, Pfitzner J, Whittlestone S (1991) A mobile station for radon measurements. In: Internal Report 33. Proceeding: Workshop on environmental radiochemistry and radionuclide measurement, Supervising Scientist, Canberra. Unpublished paper.

Bollhöfer A, Ryan B, Pfitzner K, Martin P (2002) A radiation dose estimate for visitors of the South Alligator River valley from remnants of uranium mining and milling activities. Internal Report 386, Supervising Scientist, Canberra. Unpublished paper.

ICRP 23 (1975) ICRP Publication 23, Reference Man: Anatomical, Physiological and Metabolic Characteristics,

ICRP 65 (1993) ICRP Publication 65, Protection against Radon-222 at home and at work.

ICRP 68 (1994) ICRP Publication 68, Dose coefficients for Intakes of Radionuclides by Workers.

ICRP 72 (1996) ICRP Publication 72, Age-dependent doses to members of the public from the intake of radionuclides: part 5. Compilation of Ingestion and Inhalation dose coefficients.

Iles M, Bollhöfer A, Paul M, Ryan B (in prep.) Investigation of the effects of exposed tailings on freshwater mussels (*Velesunio angasi*) in the South Alligator River, NT: Data summary. Internal Report, Supervising Scientist Canberra. Unpublished paper.

Johnston A (1987) Radiation exposure of members of the public resulting from operation of the Ranger uranium mine. Technical Memorandum 20, Supervising Scientist for the Alligator Rivers Region, AGPS, Canberra.

Koperski J, Bywater J (1985) Radionuclide Analysis of Bush Food. Radiation Protection in Australia, Vol. 3(2): 80-84.

Martin P (2000) Radiological impact assessment of uranium mining and milling. PhD thesis. Queensland University of Technology, Brisbane.

Martin P, Hancock GJ, Johnston A, Murray A (1998) Natural-series radionuclides in traditional North Australian Aboriginal foods. J. Environ. Radioactivity 40: 37–58.

Pfitzner K, Martin P (2000) Airborne gamma survey of the upper South Alligator River valley: first report. Internal Report 353, Supervising Scientist, Canberra. Unpublished paper.

Pfitzner K, Martin P, Ryan B (2001a) Airborne gamma survey of the upper South Alligator River valley: second report. Internal Report 377, Supervising Scientist, Canberra. Unpublished paper.

Pfitzner K, Ryan B, Bollhöfer A, Martin P (2001b) Airborne gamma survey of the upper South Alligator River valley: third report. Internal Report 383, Supervising Scientist, Canberra. Unpublished paper.

Ryan B (in prep.) Ranger Uranium Mine: A Radiological Assessment, Masters Thesis, Monash University, Melbourne.

Tims S, Ryan B, Waggitt P (2000) Gamma radiation survey of exposed tailings in the area around Rockhole mine. Internal Report 332, Supervising Scientist, Canberra. Unpublished paper.

Waggitt PW (2001) Remediation of Abandoned Uranium Mines in the Gunlom Trust Area, Northern Australia. In: *Proceedings of 8th International Conference on Environmental Management ICEM'01*. Bruges, Belgium, 30 September – 4 October 2001. Pub. American Society of Mechanical Engineers.

Zapantis A (2000) Derivation of the dose conversion factor for the inhalation of uranium ore dust considering the effects of radon loss. 24th Annual Conference of the Australasian Radiological Protection Society Inc. Margaret River, WA, 24–27 August, 1999.

Infiltration to Groundwater at High Altitude

Adrian Brown

Adrian Brown Consultants, 1875 Lawrence Street, Denver, Colorado, USA

Abstract. Infiltration (or recharge) is the ultimate source of groundwater, and controls mine inflow, groundwater supply, and flow and solute transport behavior in large-scale groundwater systems. However, infiltration is rarely directly measurable, and is difficult to quantify. An 18 square kilometer area at 3,000 meters above sea level in the Rocky Mountain Cordillera of North America has been dewatered for the last century by a system of drainage tunnels, which has provided a rare opportunity to directly measure infiltration. Infiltration to deep groundwater was found to average 175 mm per year to the district, or 39% of the total precipitation of approximately 445 mm per year. These infiltration rates are significantly higher than the normal range of 5% to 10% of total precipitation that are identified at lower altitudes in this region, due to the effects of altitude, temperature, storm intensity, and thin soil cover in the district.

Setting

The setting of the study is the central Rocky Mountain Cordillera, in central Colorado, USA. At 3,000 meters elevation on the western flank of Pike's Peak near the town of Cripple Creek is located one of the world's great gold ore bodies, the Cripple Creek Mining District. Approximately 40 million ounces of gold has been recovered from this district since 1890.

The geology of the Cripple Creek Mining District is characterized by a large diatreme plug that intruded the Pikes Peak Granite and metamorphic gneiss and schist during the late Oligocene. Initial emplacement of the diatreme was followed by several episodes of volcanism resulting in a brecciated rock mass with intrusive igneous rocks (e.g., numerous phonolite and lamphrophyre dikes and sills). The dikes generally trend northwest through the diatreme complex. The void spaces within the volcanic breccia were mineralized by hydrothermal solutions associated with the rising magma body (Pontius, 1997). Post-Laramide erosion has exposed a portion of the diatreme at the surface where it now out outcrops over approximately 18 square kilometers.

The occurrence and movement of groundwater within the Mining District is controlled primarily by the diatreme complex and surrounding Precambrian country rocks. The relatively permeable diatreme complex receives a significant quantity of water from local precipitation, which averages 445 mm per year. Groundwater is stored in the fractured volcanic breccia within the diatreme where flow is restricted by the surrounding less permeable Precambrian strata. Although the volcanic breccia is well cemented and relatively impervious, groundwater movement within the diatreme complex occurs primarily through secondary permeability features such as faults, fractures, joints, contacts between the dikes and breccia, and likely, the old underground mine drifts and shafts.

Regional Drainage

Gold and silver tellurides were first discovered in the Cripple Creek Mining District in 1891. During the early 1890's mining of the ore body was conducted near the ground surface in shallow shafts and drifts. Starting in 1895, mining occurred below the water table and mine dewatering became necessary. Early accounts by Lindgren and Ransome (1906) indicate that the mean water table elevation was at 2,926 meters above mean sea level (mAMSL) and depth to water ranged from 10 to 300 meters below ground surface prior to dewatering.

As mining advanced to greater subsurface depths, several drainage tunnels were driven through the Mining District. Some of the major drainage tunnels within the Mining District include the Standard Tunnel (completed in 1896), the Moffatt Tunnel (1903), the El Paso Tunnel (1903), the Roosevelt Tunnel (1910), and the Carlton Tunnel (1941). Historical accounts suggest that these tunnels were relatively dry when driven through the low permeability Precambrian country rocks that surround the diatreme complex. Significant quantities of water were encountered when the tunnels were driven into the diatreme breccia. Maximum flow rates generally occurred on initial break-through of the tunnels into the diatreme, and into mine workings. The peak flows were significant: 500 liters per second (L/s) in the Roosevelt Tunnel, 1,100 L/s in the Standard Tunnel, and 1,600 L/s in the Carlton Tunnel (Lindgren and Ransome, 1906; Vivian, 1941). The deepest drainage tunnels within the District, the Roosevelt and Carlton Tunnels, were driven at approximate elevations of 2,470 mAMSL and 2,100 mAMSL, respectively

The Roosevelt and Carlton Tunnels had the biggest dewatering impacts on the hydrology of the diatreme. For example, from 1917 to 1918, the discharge from the Roosevelt Tunnel dropped from 500 L/s to 250 L/s, while water levels were lowered 200 meters across the Mining District (Vivian, 1941).

The deeper Carlton Tunnel produced a large flow of water after being driven through the New Market Fault, which is located below the Ajax Mine shaft. This effectively drained 200 meters of standing water in the Ajax Mine in approximately 2 weeks (Vivian, 1941b). Flow within the tunnel decreased steadily, and by 1949 was ineffective in draining some portions of the Mining District. Two drainage tunnels, the Vindicator and Cresson Laterals, were subsequently driven through the Portland Independence structural zone to promote more widespread drainage. The two laterals increased flow in the Carlton to approximately 370 L/s by 1955, but flow gradually diminished to approximately 125 L/s by 1959. Since 1959, flow has generally decreased slightly, stabilizing at approximately 100 L/s.

The history of all flows from dewatering tunnels and flows pumped from mines in the district is presented in Figure 1. This shows the increase in inflow resulting from the installation of the Standard and El Paso Tunnels, the Roosevelt Tunnel, and the Carlton Tunnel. Each tunnel system is approximately 300 meters deeper than the prior tunnel system, resulting in drainage of additional storage water from the rockmass. By about 1960, all the stored water in the rockmass had been removed, resulting in an approximate steady-state regional flow of 100 L/s. This outflow is equal to the infiltration to the diatreme surface.

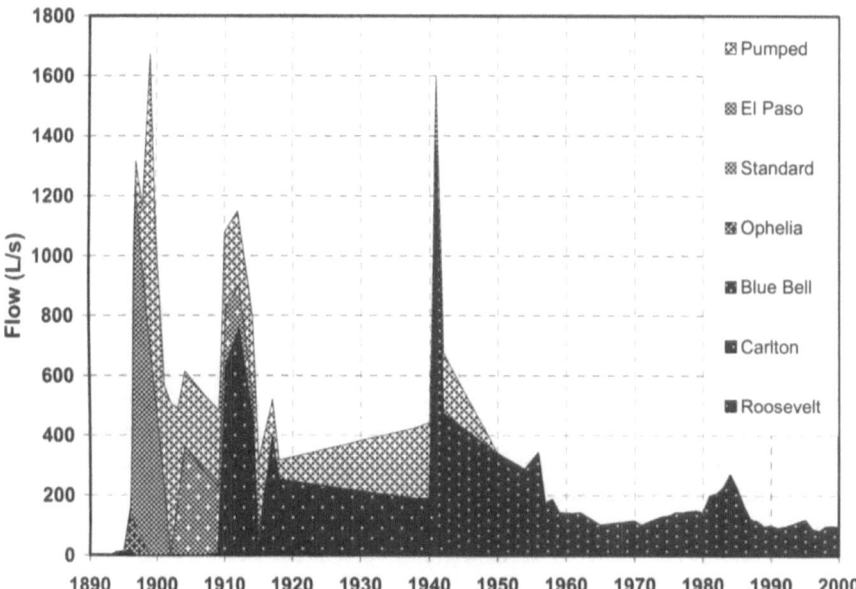

Fig. 1. Diatremal drainage flow from the Cripple Creek Mines 1890-2000.

Infiltration

The rate of deep groundwater infiltration to the diatremal area can be determined by dividing the steady-state flow rate into the area that is being drained:
- Infiltration flow to diatreme   = 100 L/s
- Ground surface diatreme area   = 18 km^2
- Infiltration rate to diatreme   = 175 mm/yr

The actual infiltration to the diatreme is probably slightly smaller than this value; some of the water that is flowing from the Carlton Tunnel is being drawn from the immediately adjacent granitic rockmass. Monitoring well information indicates that the zone of capture from the dewatering of the diatreme extends only a very short distance into the granite, and this has been ignored in this evaluation.

The infiltration rate of 175 mm/year can be compared to the average annual precipitation of 445 mm/year (based on records at two adjacent weather stations, Cripple Creek and Victor). The ratio of infiltration to precipitation is computed to be:

Infiltration rate to diatreme	= 175 mm/yr
Total precipitation to diatreme	= 445 mm/yr
Percentage of precipitation infiltrating	= 39%

This is a very high infiltration rate; typical infiltration rates on similar terrain at lower elevations in Colorado (around 2,000 mAMSL) are 5% to 10% of incident precipitation (see for example Musgrave 1955). The high infiltration rate appears to be the result of the following factors:

1. Low temperatures. The average annual temperature in the mining district is 5.7°C. Five months have average temperatures below or at freezing. This limits the amount of plant growth, and correspondingly limits evapotranspiration. Vegetation on the diatreme is limited, with sparse conifer forest and open dry grassland predominating.

2. Permeable soils and rocks. The soils that cover the area are sandy, and are underlain by relatively permeable rocks. This combination increases shallow infiltration; surface runoff in the area is rare. It also reduces retention time near the surface, further limiting opportunities for evaporation or evapotranspiration.

3. Montane weather. Weather patterns in the mountains of Colorado are highly variable. Most precipitation occurs in the district as summer thunderstorms, and 39% of the total precipitation occurs in storms delivering in excess of 10 mm per day. This precipitation pattern encourages infiltration due to surface flooding.

Even taking these factors into account, the infiltration rate is high, and suggests that infiltration evaluations in modeling and other processes may be understated, at least in elevated areas.

Other Parameters

The mine flow information collected over the years of observation allow estimates of key hydraulic parameters for the diatreme rockmass, including hydraulic conductivity and drainable porosity.

Hydraulic Conductivity

The hydraulic conductivity of the overall diatreme rockmass has been computed using the responses to the installation of the drainage galleries during the mining campaigns. Based on these data, the bulk hydraulic conductivity of the diatremal rock is computed to be 4×10^{-7} meters per second (m/s).

Drainable Porosity

The drainage of the rock prior to reaching steady state allows a computation of the drainable porosity of the rock in the diatreme. The total stored water removed by the mine dewatering is computed by subtracting the infiltrating water (assumed constant) from the removed water for the period 1890 to 2000:

Total volume dewatered	$= 1.24 \times 10^9 \, m^3$
Total infiltration during dewatering	$= 0.34 \times 10^9 \, m^3$
Total volume removed from storage	$= 0.90 \times 10^9 \, m^3$

The total rock volume from which the water was removed is computed by multiplying the area of the diatreme by the total groundwater level reduction:

Area of diatreme	$= 1.81 \times 10^6 \, m^2$
Groundwater level reduction	$= 800 \, m$
Volume of rock dewatered	$= 14.5 \times 10^9 \, m^3$

The drainable porosity of the rock is computed by dividing the storage water by the volume of rock from which it was obtained:

Total volume removed from storage	$= 0.90 \times 10^9 \, m^3$
Volume of rock dewatered	$= 14.5 \times 10^9 \, m^3$
Drainable porosity of rock	$= 6.2\%$

Computed Infiltration

In many situations it is not possible to directly measure infiltration to the deep groundwater system. In such cases, computational methods are used to evaluate the infiltration that will occur. These methods in general rely on water balance analyses, applying precipitation to the ground surfaced, and partitioning it into runoff, evapotranspiration, and infiltration. A widely used method of performing this complex calculation is the U.S. Environmental Protection Agency's Hydrologic Evaluation of Landfill Performance (HELP) Model (Peyton and Schroeder 1994). This model was used to compute the expected infiltration to the Cripple Creek Diatreme, using the following parameters:

1. Daily precipitation data for 10 years for the area, duplicated to provide a synthetic 20 year record (Earthinfo 1997).
2. Daily temperature data for 10 years for the area, duplicated to provide a synthetic 20 year record (Earthinfo 1997)

3. Monthly solar insolation information for Colorado Springs, a nearby city where this parameter is measured).
4. Monthly evapotranspiration data synthetically generated (Miller et al. 1973).
5. A typical geological section for the diatremal area (Pontius 1997).
6. Geohydrology parameters obtained from site testing, and from typical values for site materials (Peyton and Schroeder 1994).

Using this information, a one-dimensional flow model was created and run in the HELP Model. The results are indicated in Table 1.

Table 1. HELP model results for infiltration to Cripple Creek Diatreme.

Run	20A	20B	20C	20D	20E
Depth: 0m-0.15m	Loam	Loam	Co Sand	Gravel	Gravel
Depth: 0.15m-1m	Loam	Loam	F Sand	F Sand	Co.Sand
Depth: 1m-3m	Rock	Rock	Rock	Rock	F. Sand
Depth: 3m-6m	Rock	Rock	Rock	Rock	Rock
Depth: 6m-15m	Rock	Rock	Rock	Rock	Rock
Leaf Index	2	1	1	1	1
Evaporation Depth (m)	0.5	0.4	0.4	0.4	0.4
Precipitation (mm/y)	445	445	445	445	445
Runoff (mm/y)	2	2	2	2	2
Evapotranspiration (mm/y)	425	409	392	382	353
Storage (mm/y)	0	0	1	1	7
Infiltration (mm/y)	17	34	50	60	83
Infiltration/Precipitation	4%	8%	11%	14%	19%

The HELP analyses performed cover a wide range of possible soil types, and evapotranspiration options, including very high permeability soils and very low evapotranspirative conditions. The infiltration rates computed with this model range from 17 mm/year (4% of precipitation) to 83 mm/year (19% of precipitation). These values are significantly less than the measured infiltration of 175 mm/year. Accordingly, it appears that at least in this case, which is one of the few where actual infiltration is known, an accepted tool that is widely used for computing infiltration greatly underestimates infiltration.

Conclusion

Infiltration to the deep circulating groundwater system has been measured in a high altitude setting in Colorado, using steady state flow to a drainage system beneath an 18 square kilometer area. The infiltration rate was found to be 175 mm/year, approximately 39% of the total precipitation to the area of 445 mm/year.

This measured infiltration rate is significantly higher than normally expected infiltration rates, which are 5% to 10% of incident precipitation. It is also significantly higher than the range of infiltration rates computed using standard infiltration models. Based on this rare opportunity to accurately measure infiltration, it would appear that estimation of infiltration, at least in high altitude settings, is not reliable using accepted proportions of precipitation, and accepted computational methods.

References

EarthInfo (1997). NCDC Summary of the Day, West1 1997. CD-ROM Databases. Boulder, Colorado, USA.

Lindgren W, Ransome FL (1906) Geology and Gold Deposits of the Cripple Creek District, Colorado. United States Geological Service Professional Paper No. 54.

Miller JF, Frederick RH, Tracy RJ (1973) Precipitation Frequency Atlas of the Western United States, Volume III-Colorado. NOAA Atlas 2

Musgrave GW (1955) How Much of the Rain Enters the Soil? USDA Water Yearbook of Agriculture, Washington, D.C. pp 151-159.

Vivian CH (1941) Carlton Drainage Tunnel to Give Cripple Creek New Life. Colorado Bureau of Mines.

Pontius JA (1997). Field Guide-Gold Deposits of the Cripple Creek Mining District, Colorado, USA. Pikes Peak Mining Company

Peyton RL, Schroeder P (1994) The Hydrologic Evaluation of Landfill Performance (HELP) Model. EPA/600/R-94/168A.

IMWA – Objectives and Aims

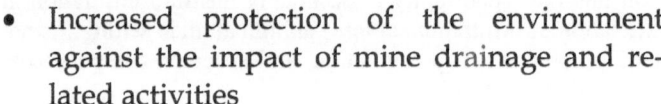

Objectives

- Increased protection of the environment against the impact of mine drainage and related activities
- Improved utilisation of mine waters
- Improved technology and economy of mine drainage operations
- Improved exploitation of mineral deposits consistent with the desirable standards of safety against water hazards

Aims

- To promote the development of science and technology concerned with water in mining
- To promote, develop and coordinate co-operation among persons and organisations of different countries, engaged in scientific or engineering work in the field of mine water problems and related sciences
- To encourage and facilitate research, education and training related to all scientific and practical problems of water in mining and allied operations
- To promote implementation of improved technology to mine drainage practice
- To promote exchange of scientific and engineering information
- To encourage the exchange of visits of persons between the countries, either as individuals or teams
- To organise international meetings and to promote regional and national conferences and symposia with international participation
- To promote publication and distribution of papers and discussions relating to the objectives of IMWA

Reliability of rainfall intensity prediction method for mine dewatering design in tropical region

Rudy Sayoga Gautama

Department of Mining Engineering
Institut Teknologi Bandung
Jl. Ganesha 10, Bandung 40132, Indonesia
Tel. +62-22-250 2239; Fax. +62-22-250 4209
E-mail : r_sayoga@mining.itb.ac.id

Abstract. Convective rainfall type which is very common in most mines in tropical region is characterized by its high intensity in short duration. For mine dewatering design purposes it is necessary to analyze the extreme rainfall intensity for a certain return period. Extreme value type I probability distribution is the most common used method for storm rainfall analysis. In this paper, the 2-year return period rainfall intensities are calculated for various data series and time periods. The results are compared to the observed rainfall data to see whether the prediction is considerably in line with the actual rainfall.

Background

Rainfall is derived from atmospheric water. The condensation is generated by dynamic or adiabatic cooling due to the vertical transport of air masses. According to the conditions that generate vertical air motion precipitation may be classified into three categories, e.g. convective, orographic and cyclonic. Convective rainfall is typical of the tropics and characterized by high intensity rainfall in a short duration. The heated air expands with a resultant reduction in weight, increasing quantities of water vapor are taken up, the warm moisture-laden air becomes unstable and vertical currents are developed. Dynamic or adiabatic cooling takes place, causing condensation and precipitation. Orographical precipitation results from the mechanical lifting of moist horizontal air currents over natural barriers such as mountain ranges. Both rainfall types commonly occur in tropical regions like Indonesia.

With a rainfall rate of 2000 mm to 5000 mm per annum, rainwater is the main water problem in almost all open pit mines in tropical regions like Indonesia. The tropic climate of Indonesia is influenced by the monsoon which divides the yearly season into dry and rainy seasons. At most mine sites the rainfall type is convective and in some mines orographic.

In a mining area the high intensity of convective rainfall will significantly reduce the effective working hours of mining operation. Certain rainfall intensity will cause muddy conditions, particularly in mining fronts, and the slippery haul roads. Due to operational and safety reasons the operation in such conditions is usually delayed. Data of November 2001 in Bukit Asam Coal Mining Area shows that total delay time due to rainfall and slippery haul road condition is 184.75 hours and it is 25.7% of the scheduled monthly working time.

To minimize the impact of rainwater on the mining operation it is necessary to define the appropriate design parameter for mine dewatering facilities. The main parameter is a predictive rainfall intensity with a certain level of probability of occurrence which accounts the hydrologic risk of dewatering facilities.

Storm Rainfall Analysis

Rainfall is considered as a random event and it varies geographically, temporally and seasonally. In designing the mine dewatering facilities it is necessary to understand the rainfall characteristics of the mining area. The most important characteristic is the extreme rainfall intensity and its probability of occurrence.

The method of storm rainfall analysis has been described in Chow et al (1988), Viessman Jr et al (1977) and Gautama (1997). Using historical rainfall data design rainfall intensity for a certain return period could be determined by applying extreme value probability distribution analysis (Chow et al, 1988; Kite, 1977). The data analyzed are assumed to be independent identically distributed, and the storm rainfall system is considered to be stochastic, space-dependent and time-independent.

The extreme value type I (EVI) probability distribution function (Chow et al, 1988) is

$$P(x<x_T) = F(x) = \exp(-\exp[-(x-u)/\alpha]) \tag{1}$$

$$\alpha = \sqrt{6}. \, s \, /\pi \tag{2}$$

If u = mode of distribution or point of maximum probability density,

$$u = \bar{x} - 0.5772 \, \alpha \tag{3}$$

A reduced variate y can be defined as

$$y = (x-u)/\alpha \tag{4}$$

If reduced variate of the return period T is :

$$y_T = -\ln[\ln\{T/(T-1)\}] \tag{5}$$

for the EVI distribution, x_T is related to y_T by Eq. (4), or

$$x_T = u + \alpha\, y_T \tag{6}$$

Case study : Bukit Asam Coal Mining Area

Bukit Asam Coal Mining Area is one of the primary coal mining areas in Indonesia. It is located at the southern part of Sumatra Island in a transition zone between lowland in the west and Bukit Barisan mountainous range in the east. The area characterized by a hilly morphology with an elevation range between 50 to 300 m above sea level.

Rainfall Characteristics

The climate is influenced by a monsoon system. The annual rainfall for the last 12 years varies from 1577.5 mm/year to 3341.3 mm/year with an average of 2732.4 mm/year (see Fig. 1).The extremely low annual rainfall in 1997 is believed as the impact of El Nino phenomena.

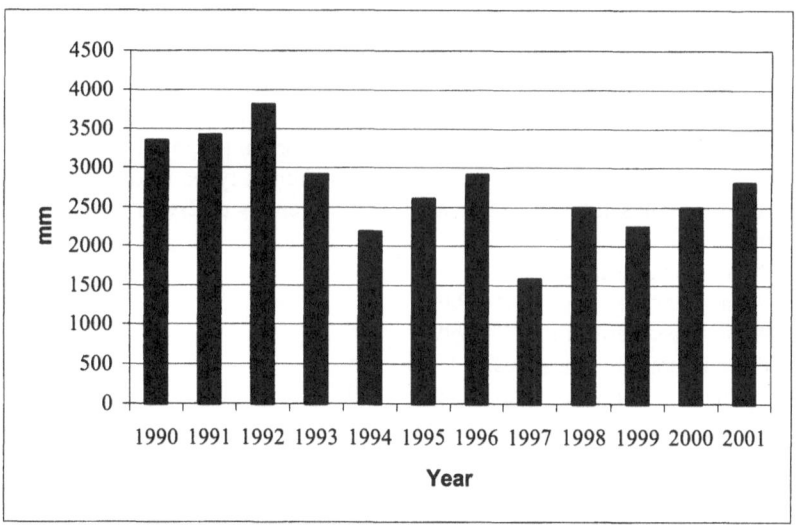

Fig. 1. Annual rainfall of Bukit Asam Coal Mining Area from 1990 - 2001

The monthly rainfall distribution which shows the maxima, minima and average monthly rainfall is given in Fig. 2. Considering the average values May to September are considered as dry months with rainfall less than 150 mm/month. The maxima values show that the monthly rainfall rate during rainy season, October to April, may be more than 500 mm.

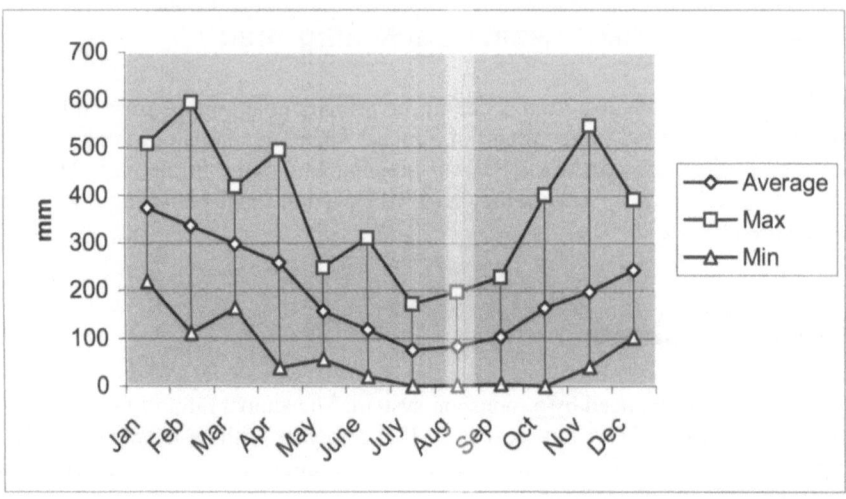

Fig. 2. Monthly Rainfall Distribution of Bukit Asam Coal Mining Area (data 1990-2001)

Rainfall Intensity Analysis

Using data derived from automatic rainfall recorder since 1990 storm rainfall for Bukit Asam Coal Mining Area has been analyzed. For the purpose to study the reliability of rainfall intensity prediction method, different types of data series and range have been chosen to determine the rainfall intensity of 2-year return period. Those are :

- Annual data series of 1990-1994
- Annual data series of 1995-1999
- Annual data series of 1990-1999
- Partial duration data series of 1990-1994
- Partial duration data series of 1995-1999
- Partial duration data series of 1990-1999

In hydrological analysis it is suggested to use annual data series to determine an extreme event with a defined recurrence period. But the problem faced by most of mines in Indonesia is the availability of long term rainfall data because the mines are developed in remote and undeveloped areas where usually no rainfall station available in the nearby area. In some mines the rainfall station was first installed during the construction phase.

Three types of annual data series will be compared to each other as well as to partial duration data series. The predefined base value for partial duration data series is the lowest annual maximum value of the records. The data analyzed is 60-minute-duration rainfall recorded at weather station in Bukit Asam Coal Mining Area.

The calculated 60-minute 2 year return period rainfall using Eqs. (2), (3), (4), (5) and (6) for various data series are shown in Table 1 as well as the respective sample size (n), mean ($\bar{x}$) and standard deviation (s) of the sample.

Table 1. 60-minute Duration Storm Rainfall Analysis of Bukit Asam Coal Mining Area

Data Series	Sample size, n	Mean, $\bar{x}$	Standard deviation s	2 year return period rainfall, x_T
Annual series of 1990-1994	5	71.82	24.90	67.73
Annual series of 1995-1999	5	56.66	5.19	55.81
Annual series of 1990-1999	10	64.24	18.75	61.16
Partial duration series of 1990-1994	18	61.76	15.08	59.28
Partial duration series of 1995-1999	13	54.79	4.29	53.54
Partial duration series of 1990-1999	34	58.23	11.85	56.29

Reliability Analysis and Discussion

The reliability of the results of frequency analysis depends on how well the assumed probabilistic model applies to a given set of hydrologic data.

Annual data series

The calculated 2-year return period rainfall of annual data series of 1990-1994 is 67.73 mm/hr which is the highest compared to the results of other annual data series and of partial duration data series. It is related to the high rainfall years as well as annual maximum values which can be seen from $\bar{x}$ and s of the sample (see Table 1). The exceedence of the 2-year return period rainfall intensities of annual data series for various time periods is given in Table 2.

A 2-year return period rainfall intensity means that the event has the probability to be equaled or exceeded once in two years or twice in four years. It may be noted that the 2-year return period rainfall intensity of 1990-1994 data series is over predicted during time period of 1995-2001 since it is never being equaled during that period. On the other hand the 2-year return period rainfall intensity of

1995-1999 annual data series is less than the predicted model because it has been exceeded twice by the annual maxima in 2 years (2000-2001). The best fit to the predicted model is the calculated rainfall intensity of annual data series of 1990-1999.

The exceedences by annual maxima are between 0.11 % and 0.58% of the total rain days of the respective time periods. The total exceedences are in general twice as much as the exceedence by annual maxima (see Table 2).

Partial duration data series

As in annual data series, the calculated rainfall intensity of partial duration data series of 1990-1994 is higher than the other partial duration data series because of higher rainfall intensities measured during this period as it is shown by the $\bar{x}$ value in Table 1. The calculated 2-year return period rainfall intensity of 1990-1994 partial duration data series is exceeded three times by annual maxima during 1995-2001 (see Table 2). It means that the frequency of occurrence of observed extreme rainfalls during that period is approximately as predicted by the model. The result of 1995-1999 and 1990-1999 partial duration series are under predicted since they have been exceeded 5 times in 7 years (1995-2001) and twice in 2 years (2000-2001) respectively.

The exceedences by annual maxima lie between 0.24% and 0.58% of the total rain days of the respective time periods. The total exceedences are between 0.36% and 1.46% of the total rain days (see Table 2).

Annual vs. Partial Duration Series

As it is shown in Fig. 3, the annual data series give higher results than partial duration data series. Comparing the results between annual and partial duration series of 1990-1994 during the time period of 1995-2001, it can be seen that the 2-year return period rainfall calculated from partial duration data series has the better prediction than that of annual data series. But for the 2-year return period rainfall of 1995-1999 data series, both series show a similarity in the results during the period of 2000-2001.

In term of exeedence by the annual maxima both annual and partial duration series are in general showing a similarity. But the total exceedences of the results derived from partial duration series are higher than those derived from annual series.

Table 2. Number of exceedence of the 2-year return period rainfall intensity

Remarks	X, T=2 yr	1990-1994 am	1990-1994 a	1995-1999 am	1995-1999 a	1990-1999 am	1990-1999 a	1995-2001 am	1995-2001 a	2000-2001 am	2000-2001 a
Total raindays (days)		933		838		1771		1181		343	
Annual series											
Period 1990-1994	67.73	2	3	0	0	2	3	0	0	0	0
% exceedence of total raindays		0.21	0.32	0	0	0.11	0.17	0	0	0	0
Period 1995-1999	55.81			3	6			5	10	2	4
% exceedence of total raindays				0.36	0.72			0.42	0.85	0.58	1.17
Period 1990-1999	61.16					3	5			1	2
% exceedence of total raindays						0.17	0.28			0.29	0.58
Partial duration series											
Period 1990-1994	59.28	3	6	2	3	5	9	3	5	1	2
% exceedence of total raindays		0.32	0.64	0.24	0.36	0.28	0.51	0.25	0.42	0.29	0.58
Period 1995-1999	53.54			3	7			5	12	2	5
% exceedence of total raindays				0.36	0.84			0.42	1.02	0.58	1.46
Period 1990-1999	56.29					7	15			2	4
% exceedence of total raindays						0.40	0.85			0.58	1.17

Notes : am = exeedence by annual maxima; a = total exeedence by extreme rainfalls

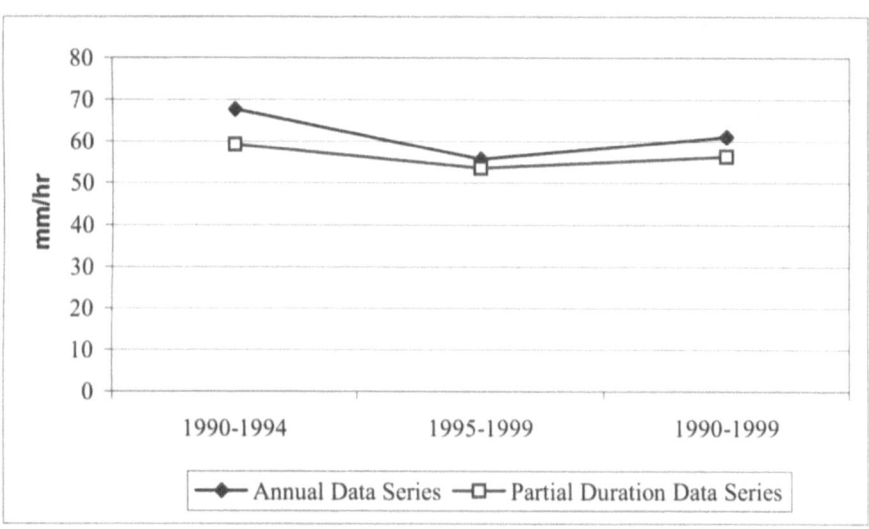

Fig. 3. Comparison of 2-year return period rainfall intensity for various data time period and data series.

Design parameter of mine dewatering facilities

The calculated 2-year return period rainfall intensity may be used as design parameter of mine dewatering facilities such as drainage ditches and pit collecting sump. As shown in Table 2, for the best prediction the use of annual data series with a sufficient record length is suggested. Big difference in magnitude between data series of 1990–1994 and 1995–1999 gives a significant difference in the results. It seems that minimum ten year data is required to make a good prediction so it can consider the climate variations.

If sufficient data is not available, for instance the data time period is less than 5 years, partial duration data series may be used. It is suggested, if possible, that the base value selected is the lowest annual maxima of the respective time period. Before using the calculated rainfall intensity for the design of mine dewatering facilities, it is advised to study the regional climate characteristics for a better prediction of the probability of exceedence in the future.

References

Chow V.T, Maidment D.R. and Mays L.W.(1988) Applied Hydrology, McGraw-Hill Book Company

Gautama R.S. (1997) Storm Rainfall Analysis : An Important Factor in Designing Mine Dewatering Facilities in Tropical Region, Mine Water and The Environment, Proceedings of 6th International Mine Water Association Congress, Bled, Slovenia, 235-242

Harto S (1993) Analisis Hidrologi, Penerbit PT Gramedia Pustaka Utama, Jakarta

Kite G.W (1977) Frequency and Risk Analysis in Hydrology, Water Resources Publication, Fort Collins, Colorado

Viessman Jr W, Knapp J.W, Lewis G.L & Harbaugh T.E (1977) Introduction to Hydrology, Dun-Donnelley Publisher, New York

Chances and risks of the geothermal use of mine water in Schlema / Saxony

Roland Hähne[1], Rolf Stoll[1], Frank Kabus[2,] Frank Pelz[3]

[1]C&E Consulting und Engineering GmbH, Jagdschänkenstr. 52, D-09117 Chemnitz
[2]Geothermie Neubrandenburg GmbH, Lindenstr. 39, D-17008 Neubrandenburg
[3]WISMUT GmbH, Jagdschänkenstr. 29, D-09117 Chemnitz

Abstract. The flooded mine Schlema-Alberoda, Saxony with a mined cavity of 35 million m³ and a water temperature of about 27 °C represents an important geothermal reservoir and is situated directly under the village of Schlema. Independent from the local conditions four principal possibilities how to use the warm water are applicable:

- Withdrawal of warm water from the flooded mine
- Heat extraction in the flooded mine
- Use of overflowing water prior to water treatment
- Use of treated water

Before heat mining can be realised it has to be evaluated how heat extraction changes the hydrochemical situation and the concentrations of contaminants in water respectively consequences for the treatment plant, which is going to purify the out flowing mine water.

Introduction

The flooded mine Schlema-Alberoda, Saxony with a cavity of 35 million m³ and a water temperature of about 27 °C represents an important geothermal storage and is situated directly under the village of Schlema. Recently the flooding process is in the final stage.

Due to the very deep development of the former uranium mine to a depth of about 1800 m and caused by the thermal-kinetic convection in the flooded cavities relative high temperatures are expected in the mine water permanently.

During the flooding and the period directly after the flooding high concentrations of typical contaminants like Arsenic, Uranium, and Radium will occur in the water. Therefore a treatment of 700 to 1000 m³/h of out flowing water discharging to the Zwickauer Mulde River with a temperature of 26°C is in planning.

The water filled mine cavities are spread over an area from Niederschlema to Oberschlema / Schneeberg. After flooding being completed the water surface will reach a level between 10 and 60 m below the ground surface of the village of Schlema.

Independent on the local specific conditions actually four possibilities of geothermal use of the flooding water are applicable (Fig. 1):

- Withdrawal of warm water from the flooded mine
- Heat extraction in the flooded mine
- Use of overflowing water prior to water treatment
- Use of treated water

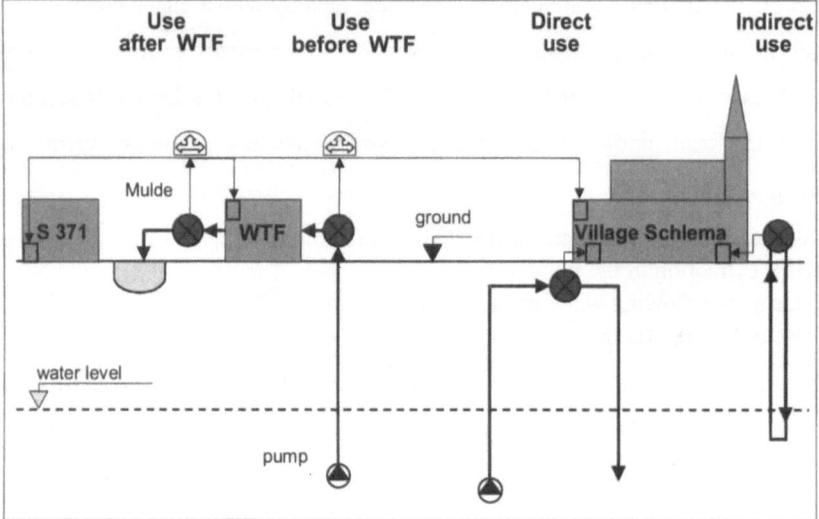

Fig. 1. Alterternative techniques of geothermal use of flooding water in the region Schlema-Alberoda (WTF- Water treatment facility)

Options of geothermal use

Direct Use in the Flooded Mine

One technique could be pumping of mine water via old shafts or boreholes to the subsurface or to the Markus-Semmler-Adit level directly towards potential consumers. In that case heat is extracted from the water by means of heat exchangers and the water is pumped back to the mine cavities. This may be an efficient form from the energetically and economically viewpoint but negative convection and hydrodynamic processes cannot be excluded. Therefore it is necessary to proof by field tests that this direct use of the geothermal potential will not lead to an additional mobilization of contaminants in the mine and in thus to a negative impact on the functionality of the water treatment plant.

Indirect Use of flooding water

The indirect use of flooding water by a heat probes installed in mining cavities via short boreholes is efficient especially if this can be realized close to the location of the consumer. Such probes consist of a tube with circulating cool water inside in a secondary circuit and warm mine water outside. By heat exchange a heat extraction of about 60 W/m is realistically. By means of a heat pump the warm water will be cooled and pumped back to the circulation tube. However, local convection processes with a negative impact at the water treatment plant can not be excluded neither.

Use of overflowing water before the water treatment

The heat use from the out flowing mine water with a temperature of about 30 °C prior to the water treatment plant is highly efficient from the energetically viewpoint. A possible negative impact on the water treatment technology depends on the degree of cooling of the mine water.

Use of treated water

Without any disadvantageous impact on the water treatment technology is the heat use of water discharge from the treatment plant. The treated water has a uniform temperature of about 26 °C. With an amount of water of 700 m³/h and cooling the treated water to 10 °C a heat power of 19 MW will be available. In this case costs are depending on the distance of the potential consumer from the water treatment plant.

Conclusions

Investigations showed that in case of direct or indirect use of mine water from the mine cavities via boreholes or old shafts the geothermal use is probably not without problems. The probability of an induction or an increase of convection streams with mobilization of contaminants from the mine is relatively high especially for parts of the mine without heat use which are not or not efficently connected with the thermal convection system in the flooded cavities. However, the volume of isolated parts of the flooded cavities as well as the amount of potential contaminants which can be mobilized are unknown. The hydraulic flow paths are unknown and therefore technical measures to prevent additional convection are not practicable. Thus without clarification the possible impact on the functionality of the water treatment plant the heat extraction from isolated areas of the flooded cavities is not recommended.

Finally the geothermal use of water both before and behind the water treatment plant is possible and is depending on the heat demand, the distance of the user to the plant, the input and output temperatures, and the efficiency of the heat exchanger.

When the water treatment plant will be closed down in 20 to 30 years after concentration of contaminants in the water have decreased to a certain level a further use of the renewable geothermal potential by direct or indirect methods is possible.

Thus the region around Schlema nowadays and in future has a great challenge to produce energy by use of natural heat from mine water very efficiently and will be very attractive for consumers and investors. It can be an example that an abandoned mining region may develop to a center of sustainable energy production.

Regional Heat Flow: An Aid to Geothermal Exploration in Chattisgarh, India

S.K. Sharma

24 National Road, Dehradun 248001, India, E-mail: SKS105@rediffmail.com

Abstract. A well known anomalous mineralization of Uranium up to 4400 ppm in the Abujhman Basin in Central India and the data on regional heat flow studies depicting a heat flow pattern ranging from 60 to 105 mW/m^2 has given a lead to locate a potential geothermal field in Proterozoic crystalline basement rocks of the Chattisgarh Basin, North of Abujhman Basin. The Proterozoic rocks comprise coarse granite/gneiss which is found in juxtaposition with Lower Gondwana sandstone by a ENE-WSW trending fault. On the basis of the above measurements and the geology of the region, five shallow wells have been drilled in the Tattapani area in Chattisgarh to exploit geothermal resources for various purposes including the generation of electricity.

Introduction

Geothermal reservoirs located in the earth's interior constitute an important energy source currently being sought after for meeting energy demands. Various mega features associated with geothermal energy resources that can be observed near the surface of the earth are spreading ridges, transform faults, subduction zones etc., which form a vast network that divides our plant into distinct lithospheric regions. The plate boundaries comprising to fractured zones, seismicity, large number of volcanoes and radio elemental distribution of the basement rocks are some of the typical source regions for the geothermal energy. Surface manifestations such as geysers, hot springs etc. can be seen near these features which can be investigated for delineating geothermal sources.

In a geothermal system, water transports the heat through convection from deep crustal levels i.e., from a heat source at depth to the heat sink, usually the surface of the earth. The system consists mainly of the three elements – heat source, a reservoir and a fluid. The heat source could be magnetic intrusion or radioactive decay. The reservoir is permeable rock with high temperature from which the circulating fluids absorb the heat. The fluid in many geothermal systems is water. Among these three elements, the prime requisite is a heat source at deeper levels. The other two elements under favorable conditions can be created.

Though, geophysical investigation of such structures relevant to geothermal sources is important but for our present studies, the investigated Uranium mineralization has played a significant role in the identification and delineation of Tattapani geothermal areas in Chattisgarh Basin.

The Study Area

The study area is located near the border of the Chattisgarh and Bihar states and falls close to the regional Balarampur fault zone and spread over an area of about 400 m x 600 m and perhaps, the ENE-WSW trending Narmada-Son Lineament (NSL) zone is one of the most significant lineament in the Indian shield. Tattapani fault is a major fault in the region forming a boundary between Archean formation towards the south and the Gondwana formation towards the north (Fig. 1) in the western part of the study area.

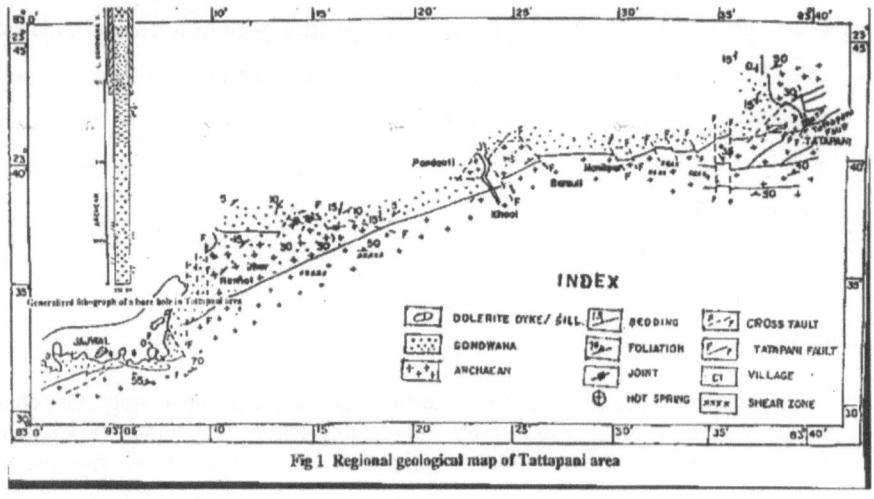

Fig 1 Regional geological map of Tattapani area

Fig. 1. Regional geological map of Tattapani area

Geology of the Area

Detailed geological studies have been carried out by Geological Survey of India (G.S.I.) (Thussu et al. 1987). The surface geology presents different rock types, which include quartzite, schist, gneisses and phyllite of Proterozoic Archean age.

These are generally folded and cross-folded. The Archean rocks show foliation in the direction N 70^0E –S 70^0 W and N 80^0W -S80^0E with dips of 75^0 to 85^0 towards north. Granites, granodionites, pegmatites, amphibolites and quartz veins represent the intrusive rocks of the region. The phyllites are mainly carbonaceous and graphites are associated with Pyrite mineralization. The granites and granodiorites occur as plutons while the amphibolites occur as dykes and sills and the pegmatite as intrusive. Quartz veins are associated with magnetic mineralization. Table 1 presents the stratigraphic sequence of the formations.

Gondwana sequence of formations in this area, lying unconformable over the Proterozoics, consists of conglomerates, sand stones, shale and minor coal beds. These show low dips (15^0) with a general strike direction of N 80^0 W – S 80^0 E.

Fig. 1 above presents the geological and structural map together with the generalized lithograph of a bore hole in the Tattapani area (Thussu et al. 1987). In general, as can be seen from the figure, the major geological formations are separated by ENE-WSW regional faults, which in turn are affected by a number of cross faults striking in NNW-SSE and NE-SW direction. Several faults are also observed towards south of hot spring near Balarampur and Lurgutta and are located within the Proterozoic sequence in the area (Ravishanker et al. 1987).

Table 1. Stratigraphic sequence of Tattapani area, Chattisgarh Basin (Thussu et al. 1987)

Age	Formation	Rock Type
Recent		soil/river alluvium
Quaternary		hydrothermal altered clay
--- Unconformity ---		
Lower Tertiary To Upper Cretaceous	Deccan Trap	dolerite sills and dykes intruding Gondwana sediments (near Jajawal)
Mesozoic to Lower Permian	Gondwana Post Barakar Super Group (Mahadeva)	brown to reddish brown possibly sandstone showing current bedding and ripple marks and subordinate conglomerate, no coal seams
Permian	Barakar Formation	Sandstone and shale with grit bands, gray shale with plant fossils and coal seams.
	Talchir Formation	green splintery shale sandstone with minor grit bands and conglomerates
------------------------------------- Unconformity ---		
Proterozoic		phyllite, graphitic at some places, quartzite, grey gneiss, biotite schist, actinolite, remolite schist kyanite-sillimanite schist, garnet gneiss, hornblende gneiss, granulite, amphibolite, augengneiss, diorite, granulite and pink gneiss.

Uranium Mineralization

The discovery of uranium occurrence at Bogan, lead to investigations in detail and a number of uranium/thorium anomalies were located all along the northeastern margin in the Gundul sandstones spread between Bogan and Gundul in south of Tattapani area. Table 2 shows radiometric assay results of upper and lower horizons of all the escarpments indicating average U/Th ratio from 3.5 to 21.57 in the lower zone and from 0.17 to 0.23 in the upper zone (Chaturvedi et al. 1998).

Table 2. Radiometric Assay Results of Gundul Sandstone* (NE-Abujhmar)

Escarpment	Horizon		%eU3O8	%U3O8	%ThO2	U/Th
Bogan	Upper	n=4	0.046	0.016	0.090	0.172
	Lower	n=11	0.050	0.054	0.003	17.386
Maram	Upper	n=5	0.049	0.023	0.084	0.274
	Lower	n=11	0.066	0.067	0.003	21.572
Kachor	Upper	n=3	0.059	0.024	0.107	0.217
	Lower	n=3	0.006	0.007	0.002	3.500
Gundul	Upper	n=4	0.027	0.012	0.051	0.227
	Lower	n=11	0.010	0.016	0.003	5.473
Mangdon	Upper	n=3	0.021	0.007	0.040	0.169
	Lower	n=5	0.010	0.013	0.002	6.278

*After Chaturvedi et al. 1998

Discussions

Provenance for the Abujhmar sediments particularly of northeastern part, is Bengpal Granite-Gneisses, Bailadila Iron Formation and Nandgaon rhyolites and sandstones. All these basement rocks have anomalous concentration or uranium (2 to 120 ppm) and thorium (4 to 880 ppm) which might have been mobilized and concentrated in the sediments of Abujhmar Basin (Cahaturvedi 1998).

Major lineaments and related faults have been the disposition of rocks in and around Abujhmar Basin. From the study of basin structures, it appears that some of the basement lineaments/faults, reactivated after the deposition of sediments and the emplacement of basic lava (Deccan Trap) took place along some of these major faults/lineaments and covered major parts of Abujhmar Chattisgarh basin.

Several episodes of basaltic emplacements are known. The extensive basaltic magmatism must have generated geothermal gradient within the basin and possibly remobilized uranium from within the sediments to suitable locales in Chattisgarh Basin. With analyses for U and Th in representative rocks, it has been possible to look at a plot of heat flow versus heat generation for nearly a dozen places, spread around the Tattapani area. A good number of thermal conductivity measurements have been affected by Verma et al. (1970) and Panda (1985) published the heat flow map of India in 1985, part of which has been used for the present studies to make plot (Fig 2).

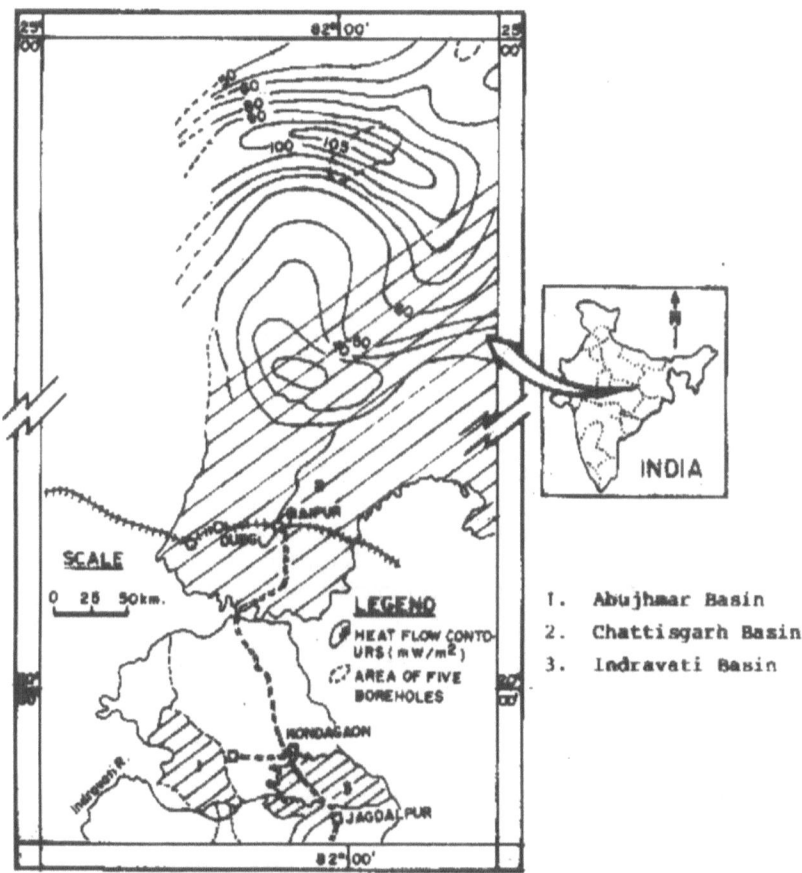

Fig.2 Heat flow map of part of Chattisgarh Basin

Fig. 2. Heat flow map of part of Chattiagarh Basin

Conclusion

A statistically significant correlation between heat flow and uranium concentrations was found, implying that variations in surface heat flow are primarily to be ascribed to variations in the heat generated by radioactivity in the crustal column. Thus it is envisaged that the uranium derived from granite-gneiss probably migrated upward towards north - east along certain fractures and yielded an heat flow source for hot springs in and around Chattisgarh Basin which from Abujhmar basin, of the order of around 100 mW/m^2 (Fig. 2). This formed the basis of geothermal exploration in Tattapani area, Chattisgarh basin in 1988 and as many 5 shallow bore holes have been drilled down to a depth of 350m in an area of 2 km^2 which yielded a cumulative geofluid at 60 L/min with a surface temperature of 105^{0}C. This hot geo fluid at this temperature is capable of generating 300 Kwe electricity through ORC based binary power plant besides various direct heat applications.

References

Chaturvedi S N, Saha A, Saxena V.P. (1998) Uranium potential for sandstone type deposits around NE part of the Proterozoic Abujhman Basin, Bastar District, Madhya Pradesh, India. Proc. Nat. Symp : Recent Researchers in Sedimentary Basin : 139-149

Panda P K (1985) Geothermal maps of India and their significance in resources assessments. Petroleum Asia Journal: 203

Ravisankar, Thussu J.L, Prasad J M (1987) Geothermal studies at Tattapani hot sprint area, Surguja District, Central India. Geothermics: 61-76

Thussu J L, Prasad J M, Saxena R K, Prakash G, Muthuraman K (1987) Geothermal energy potential of Tattapani hot spring belt, Surguja, M.P. Rec. of GSI: 115, 30-35.

Verma, R K, Hamza V M, Panda P K (1970) Further study of the correlation of heat flow with age of basement rocks. Tectonophysics: 301

Rare Earth Elements (REEs) as Natural Tracers in Mine Waters

Christian Wolkersdorfer

TU Bergakademie Freiberg, Lehrstuhl für Hydrogeologie, Gustav-Zeuner-Str. 12, 09596 Freiberg/Sachsen; e-mail: c.wolke@web.de

Abstract. Rare earth elements (REE) in mine waters from two European mines are analysed and discussed: Straßberg/Harz (Germany) and Georgi-Unter-bau/Tyrol (Austria). From the REE patterns differences in the mine and surface waters can be seen, which are attributed to the mineralogy. Furthermore, the REEs' qualities as natural tracers are described and the results interpreted in conjunction with the findings of artificial tracer tests conducted at the two sites. Accordingly, REEs can be used as natural tracers in mines to study stratification and hydrodynamic conditions.

Introduction

Trace elements are widely used to characterize the geochemical evolution or petrologic composition of rocks, sediments or archaeological artefacts (Petrascheck 1956; Taylor and McLennan 1988; Seim and Tischendorf 1990; Knacke-Loy et al. 1992). Though, trace elements and in some cases REEs (Rare Earth Elements) are regularly analysed during hydrogeochemical investigations of mine waters (e.g. Wolkersdorfer 1996; Sanden 1997; Davis et al. 1998), systematic investigations of trace elements or REEs in mine waters of different mining sites have rarely been conducted. Important case studies of REE pattern distributions and their behaviour in mine waters were given recently by Worrall and Pearson (2001) in Great Britain and Protano and Riccobono (2002) in Italy. Contrary to a lack of REE studies for mine waters, the use of REEs as tracers in ground and surface waters has been reported and their complexes studied more frequently (e.g. Davis et al. 1998; Johannesson et al. 1994, 1996a, 1996b, 2000; Johannesson and Lyons 1995)

In this article, the results of REE investigations in a German fluorspar mine and an Austrian silver-barite mine will be reported. Both cases were supplemented by mine water tracer tests to examine the hydrodynamic conditions in the flooded mine (Wolkersdorfer et al. 2002; Wolkersdorfer and Hasche 2001).

Case Studies

Straßberg/Harz Fluorspar Mine (Germany)

In 1991 the Straßberg fluorspar mine was closed due to economic and environmental reasons (Kuyumcu and Hartwig 1998). Situated in the Mid Harz Fault
Zone of the eastern Harz Mountains approximately 30 km south of Quedlinburg
and 6 km west of Harzgerode, the Straßberg mine was the most important producer of fluorite in the former GDR (Mohr 1978). Besides fluorite, the hydrothermal polymetallic mineralisation of the vein structures comprises approximately
eight ore and less than a dozen secondary minerals (Kuschka and Franzke 1974;
Junker et al. 1991).

At the beginning of mining silver, copper, and lead were the targets of the miners. From the 18th century until 1990, mining focused on fluorspar, which was
mainly found in the deeper parts of the mine (Bartels and Lorenz 1993). Between
1950 and 1970, the VEB Harzer Spatgrube joined the three most important deposits of the Straßberg mining district by driving two deep adits on the 5th and 7th
level (from north to south: Brachmannsberg pit: No 539 Shaft, Straßberg pit: Fluor
Shaft and Glasebach pit: Glasebach Shaft).

After installing a 3-adit-system in 1998 to control the mine drainage, the mine's
owner suggested to conduct a tracer test within the flooded part of the mine. The
aim of the tracer test was to investigate the hydrodynamic conditions within the
mine, the pathways of the water between the three pits and the potential of a mine
water stratification in the three pits. Therefore, a multi-tracer test with 15 μm diameter microspheres and club moss spores (*Lycopodium clavatum*) was carried out.
Accompanying, field parameters, major ions and trace elements were analysed.

From the results of the tracer test it became clear that all parts of the mine are
hydraulically well connected. Generally, the flow direction during the tracer test
was from north to south, thus explaining the similar chemical composition of the
mine water in the Fluor and Glasebach shafts. Finally, it could be shown that under the current flow regime, with the 3-adit-system working, no stratification will
be achieved.

Georgi-Unterbau Silver-Barite mine (Austria)

Some kilometres south of Brixlegg in Tyrol/Austria, the historical silver-barite
mines of the famous Kogel mining area can be found. One of these underground
pits is the Georgi-Unterbau with its flooded underground works. Since medieval
times silver-bearing fahlore was mined and between 1947 and 1968 barite was
mined.

In 1887 the miners started to drive the Georgi-Unterbau, which opened up rich
fahlore and barite resources. Starting in 1900, a 100 m deep two compartment
blind shaft was sunk, connecting the 20, 40, 70 and 100 m main levels and the 10,
75 and 80 m sublevels (Pirkl 1961; Mutschlechner 1984; Krischker 1990; Hanne-

berg and Schuster 1994) as well as the 14-Nothelfer pit and the Barbara pit with each other. It was partially to fully flooded since the 1950's (Hießleitner 1951; Schmidegg 1953) and pumped out again in 1984 and 1988 for ore prospecting (Krischker 1990). Since 1990 the blind shaft has been flooded and therefore, stationary hydraulic conditions exist.

All parts of the mine are within the Devonian Schwaz Dolostone (*Schwazer Dolomit*) of the Northern Tyrol Greywacke area (*Nordtiroler Grauwackenzone*). Typically, the Schwaz Dolostone is a very hard, light white to light grey dolostone, being highly brecciated and fissured in the area investigated. The dolostone hosts silver and mercury bearing fahlores (there under Schwazite) as well as barite, the mineralisation being bound, but not restricted, to the breccia zones (Pirkl 1961, Wöbking 1982). Grundmann and Martinek (1994) and Schnorrer (1994, 1996) described 20 ore minerals being characteristic for the Schwaz Dolostone and a total of over 132 minerals for the Schwaz-Brixlegg area, there under REE minerals. According to Arlt and Diamond (1998), who investigated fahlores of the Schwaz-Brixlegg mining area, the fahlores of the Georgi-Unterbau comprise of 41 % Cu; 0.5 % Ag; 2 % Fe; 5 % Zn; 0.7 % Hg; 0.02 % Mn; 16 % Sb; 9 % As; and 26 % S. Unfortunately, no data on REE concentrations in the rocks of the Greywacke and their interpretation can be found in the literature.

A first physico-chemical investigation of the flooded blind shaft was conducted in December 2000. It became clear that the mine water is stratified, clearly showing two water bodies separated from each other at the 40 m level. In addition to the temperature and salinity measurements water samples were taken to be analysed on site and in the laboratory. Further investigations, including two tracer tests with 15 μm microspheres and uranine, were conducted in August 2001 and February 2002.

Sampling Procedures

Samples in the Straßberg mine were collected between May and July 2000 on a weekly basis in the three shafts at depths of 1—2 meters below the surface and the Uhlenbach brook. Only one sample was taken at the Siptenfelde brook. The Georgi-Unterbau was sampled in December 2000, May 2001, August 2001, and February 2002 with a down whole sampler in 10, 30, 55 and 90 m depths of the blind shaft.

In all samples, the temperature, salinity, redox potential, and pH value were measured directly after sampling with a multi parameter instrument Ultrameter 6P (Myron, Carlsbad CA). Fe^{2+} and total iron were measured with a Hach DR/890 and Hach DR/2500 (Hach, Loveland CO) and acidity as well as alkalinity with a Hach Digital Titrator. All samples were filtered through 0.45 μm cellulose acetate filters (Sartorius, Göttingen) using a 1000 mL Nalgene Bottle Top Filter (Nalge Nunc, Rochester NY). Two aliquots were used for analyses, one without acidification (500 mL), the other one with acidification (50 mL) and were stored in a cool place between sampling and analysing.

Main ions were analysed at TU Bergakademie Freiberg with ion chromatography and trace elements with an ICP-MS at TU Dresden/Tharandt. Fluoride concentrations were measured electrochemically with a WTW F500 (Wissenschaftlich-Technische Werkstätten, Weilheim) and SE20/EB reference electrode (Sensortechnik Meinsberg, Meinsberg).

Results and discussion

Depending on the pH of the water, the samples contained REE means between 0.2 and 17 µg L^{-1}, with the higher concentrations at low pH values (Table 1). From the 51 samples reported here, the Straßberg/Harz Flour Shaft contained the highest REE concentrations, whereas the Georgi-Unterbau only contained between 0.2 and 1 µg L^{-1} of rare earth elements. All REE concentrations are normalized against the Post-Archean Australian Shales PAAS Standard (Taylor and McLennan 1988), as no standard has been defined for mine waters yet.

Water samples of the Georgi-Unterbau, because of their higher pH values, regularly have Tm, Yb, and Lu concentrations below the detection limit. Taylor and McLennan (1988) also have shown that waters with high pH values, such as sea waters, have Tm and Lu concentrations below the detection limits.

Figs. 1 and 2 illustrate the REE patterns of the Straßberg/Harz mine and the Georgi-Unterbau, respectively. For comparison reasons, in Fig. 1 also the normalized mean of REEs in the Georgi-Unterbau is reported. Fig. 2 gives details of the three sampling campaigns in different depths of the flooded blind shaft at the Georgi-Unterbau mine. All samples show characteristic enrichments in REEs with intermediate masses (MREEs) compared to REEs with low (LREEs) or high (HREEs) masses; an effect observed by many authors before, but still not understood in detail (see Protano and Riccobono 2002 and Johannesson et al. 1996a, b). Only the samples from Georgi-Unterbau, the Glasebach Shaft and the Uhlenbach brook show significant Eu anomalies. These are missing in the No 539 and Flour Shaft patterns. As seen in Fig. 1, based on the relative abundances, three groups of water types can be distinguished at the Straßberg mining site: mine water in the

Table 1. Number of samples taken and sum of rare earth elements as well as pH at each location.

Sampling point	number of samples	Σ REE, µg L^{-1}	pH, 1
Fluor Shaft	7	16.96	6.45
Glasebach Shaft	5	3.22	6.84
No 539 Shaft	11	2.64	6.62
Siptenfelde brook	1	1.87	–
Uhlenbach brook	5	0.93	7.18
Georgi-Unterbau 10 m	3	0.99	7.75
Georgi-Unterbau 30 m	3	0.17	7.65
Georgi-Unterbau 55 m	3	0.15	7.69
Georgi-Unterbau 90 m	3	0.25	7.77

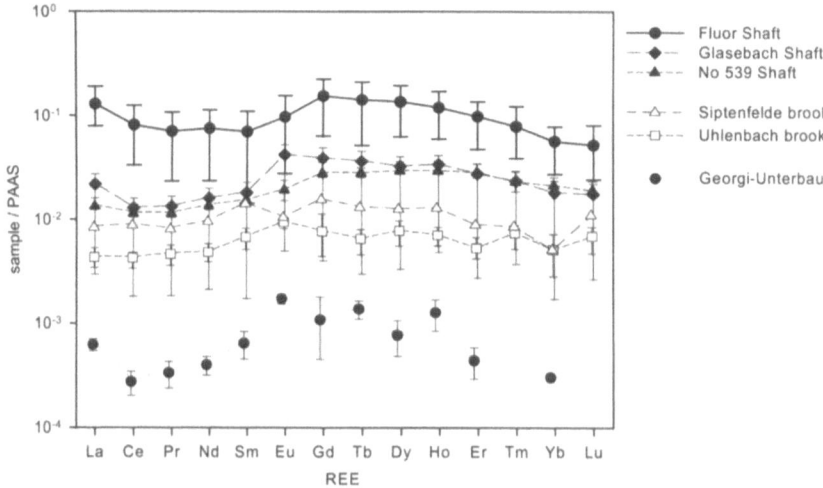

Fig. 1. Rare earth element (REE) patterns for the Straßberg/Harz and Georgi-Unterbau mine and surface waters normalized to Post Archean Australian Shales (PAAS). Filled symbols: mine waters; open symbols: surface waters. Error bars for 25 % and 75 % percentiles.

Flour Shaft; mine water in the Glasebach and No 539 shaft; and surface waters in the Siptenfelde and Uhlenbach brook. Between the surface waters and the mine waters there are 1.5 magnitudes of difference, which might be attributed to the different contact times with the different rock types. As the Siptenfelde brook is the sewage water discharge of the settlement Siptenfelde with a population existing mainly of elder people, the small Gd anomaly might be deduced from anthropogenetic effects as observed elsewhere (Bau and Dulski 1996). Such an effect can't be observed in the mine waters and the Uhlenbach brook, which dewaters a forest only. Furthermore, the mine water of the Glasebach Shaft is more closely related to the water samples of the No 539 than to the Flour Shaft. This is interesting in so far, as from the relative arrangement of the shafts, the No 539 and Flour Shaft waters should have been more similar. Nevertheless, the tracer test with microspheres and *lycopodium* spores proved this chemical result to be correct.

In the case of the Georgi-Unterbau, the REE patterns show a higher spatial and temporal variability than the Straßberg ones. This seems to be unusual, as the geological situation for the Georgi-Unterbau mine is less complicated than for the Straßberg mine. On the other hand, the mineralisation – even if most of the ore has been mined – in the Northern Tyrol Greywacke is exceptionally varied as described in the chapter "Case Studies".

First of all, the difference of 0.5 magnitudes in the relative abundances between the winter and spring/summer sampling campaigns is remarkable. It is assumed, that this difference is due to seasonal variations and therefore is currently investigated in more detail. All MREEs are enriched, and the HREEs are mostly under the detection limits. Moreover, a strong positive Eu anomaly and a negative Nd anomaly can be observed. However, the three elements with the biggest variability (Pr, Nd, and Eu) don't have systematic, depth dependent REE patterns. Yet, it is not clear if the differences are due to the analytic precision or temporal variations and changing stratification in the flooded shaft.

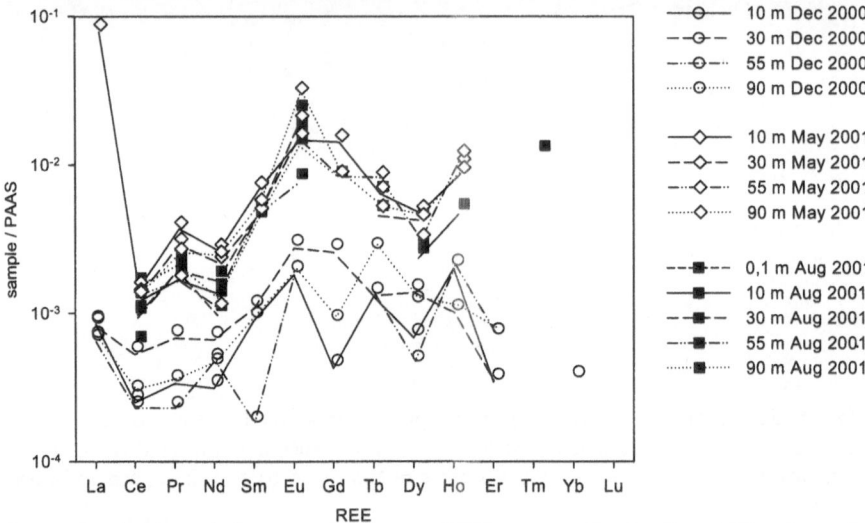

Fig. 2. Depths depended rare earth element (REE) patterns for the Georgi-Unterbau mine waters normalized to Post Archean Australian Shales (PAAS). Data of three sampling periods in December 2000, May, and August 2001, respectively.

Conclusions

Mine and surface waters from the Straßberg/Harz and Georgi-Unterbau/Tyrol mines with distinct mineralisations show significant differences in REE concentrations and PAAS normalized REE patterns. As a general rule, mine waters with low pH values tend to be enriched in REEs, whereas higher pH values result in lower REE concentrations. Contrary to the Straßberg mine waters, the Brixlegg ones show a distinct relative enrichment in MREEs, which might again be due to the mineral assemblage of the deposits. Consequently, the REE concentrations seem to be controlled by both, physico-chemical parameters of the water and the local geology.

In the Straßberg case, the mine water of the No 539 Shaft and the Glasebach Shaft have similar PAAS normalized REE patterns. Thereby, a good connection between the Glasebach pit and the Brachmannsberg pit must be assumed and has been approved by an artificial tracer test. As the Flour Shaft is between the two other shafts, and also hydraulically connected to them, it can be concluded, that the hydraulic connection of different shafts in a single mine does not necessarily result in an overall mixing of the water pool even if no stratification can be observed in the shafts.

On the other hand, the PAAS normalized REE patterns in the Georgi-Unterbau blind shaft clearly prove, that REEs can be used to study stratification in flooded underground workings. Especially Pr, Nd, Eu, and Dy seem to be good tracers in well buffered carbonated waters, as their variation exceeds those of the other REEs. However, the temporal and depth dependent variations show, that not all factors controlling REE mobilization seem to be understood yet.

Acknowledgements

Thanks to Hans-Joachim Kahmann and Klaus Heinrich of the Straßberg/Harz and Robert Stibich of the Georgi-Unterbau mines, who made this study possible and who gave every help that was needed. Andrea Hasche, Katy Unger, and Thomas Wackwitz conducted the field and laboratory studies, interpreted the results in their diploma theses and improved an earlier version of this paper. Financial support was given by the DFG Graduate School 272 and the EU project PIRAMID (contract no. EVK1-CT-1999-00021).

References

Arlt T, Diamond L W. 1998. Composition of tetrahedrite-tennantite and "schwazite" in the Schwaz silver mines, North Tyrol, Austria. Mineralog Magaz; 62(6):801—820.

Bartels C, Lorenz E. 1993. Die Grube Glasebach – ein Denkmal des Erz- und Fluoritbergbaus im Ostharz. Der Anschnitt; 45(4):144—158.

Bau M, Dulski P. 1996. Anthropogenic origin of positive gadolinium anomalies in river waters. Earth Planet Scien Letters; 143(1—4):245—255.

Davis A, Anderson J, Byrns C. 1998. Methods to differentiate between groundwater solute sources. Proceedings 15th Annual National Meeting – American Society for Surface Mining and Reclamation; 15:54—61.

Grundmann G, Martinek K-P. 1994. Erzminerale und Gangarten des Bergbaugebietes Schwaz-Brixlegg. Mineralien Magazin, Lapis; 19(7—8):28—40.

Hanneberg A, Schuster H. 1994. Geschichte des Bergbaus in Schwaz und Brixlegg. Mineralien Magazin, Lapis; 19(7—8):13—21.

Hießleitner G. 1951. Geologischer Gutachtensbericht über den Schwerspatbergbau Großkogl bei Brixlegg. Unveröffentlichtes Gutachten; Archiv Montanwerke Brixlegg.

Johannesson K H, Lyons W B. 1995. Rare-earth element geochemistry of Colour Lake, an acidic freshwater lake on Axel Heiberg Island, Northwest Territories, Canada. Chemical Geology; 119(1—4):209—223.

Johannesson K H, Lyons W B, Fee J H, Gaudette H E, McArthur J M. 1994. Geochemical processes affecting the acidic groundwaters of Lake Gilmore, Yilgarn Block, Western Australia; a preliminary study using neodymium, samarium, and dysprosium. J Hydrol; 154(1—4):271—289.

Johannesson K H, Stetzenbach K J, Hodge V F, Lyons W B. 1996a. Rare earth element complexation behavior in circumneutral pH groundwaters; assessing the role of carbonate and phosphate ions. Earth and Planetary Science Letters; 139(1—2):305—320.

Johannesson K H, Lyons W B, Yelken M A, Gaudette H E, Stetzenbach K J. 1996b. Geochemistry of the rare-earth elements in hypersaline and dilute acidic natural terrestrial waters; complexation behavior and middle rare-earth element enrichments. Chemical Geology; 133(1—4):125—144.

Johannesson K H, Zhou X, Guo C, Stetzenbach K J, Hodge V F. 2000. Origin of rare earth element signatures in groundwaters of circumneutral pH from southern Nevada and eastern California, USA. Chemical Geology; 164(3—4):239—257.

Junker R, Krause H, Schumann K, Siemroth J. 1991. Sekundärminerale des Neudorf-Straßberger-Gangzuges im Unterharz. Der Aufschluss; 42(2):95—100.

Knacke-Loy O, Satir M, Pernicka E. 1992. Provenance studies on Troian Bronze Age pottery; mineralogical and geochemical investigations. Berichte Dt Mineralog Ges; 1992(1):148.

Krischker A G. 1990. Die Baryt-Fahlerz-Lagerstätte St. Gertraudi/Brixlegg. Innsbruck: Unveröff. Dipl.-Arb. Univ. Innsbruck: 206.

Kuschka E, Franzke H J. 1974. Zur Kenntnis der Hydrothermalite des Harzes. Z geol Wiss; 2(12):1417—1436.

Kuyumcu M, Hartwig H-J. 1998. Aufgaben der GVV – Gesellschaft zur Verwahrung und Verwertung von stillgelegten Bergwerksbetrieben mbH. Bergbau; 49(1):18—22.

Mohr K. 1978. Geologie und Minerallagerstätten des Harzes. Stuttgart: Schweizerbart: 388.

Mutschlechner G. 1984. Erzbergbau und Bergwesen im Berggericht Rattenberg. Innsbruck: Eigenverlag Gemeinden Alpbach, Brixlegg, Rattenberg und Reith im Alpbachtal: 163.

Petrascheck W E J. 1956. Recherche des gites mineraux par examen geochimique des eaux de sources et de fissures en Autriche. Revue de l'industrie minerale, mines; num. spec. 1R:304—309.

Pirkl H. 1961. Geologie des Trias-Streifens und des Schwazer Dolomits südlich des Inn zwischen Schwaz und Wörgl (Tirol). Jb Geol B-A; 104(1):1—150.

Protano G, Riccobono F. 2002. High contents of rare earth elements (REEs) in stream waters of a Cu-Pb-Zn mining area. Env Poll; 117(3):499—514.

Sanden P, Karlsson S, Duker A, Ledin A, Lundman L. 1997. Variations in hydrochemistry, trace metal concentration and transport during a rain storm event in a small catchment. J Geochem Expl; 58(2—3):145—155.

Schmidegg O. 1953. Die Erzlagerstätten am Reiter Kopf und am Reiter Kogel. Schlern-Schriften; 101:17—25.

Schnorrer G. 1994. Die Sekundärminerale des Bergbaugebietes Schwaz-Brixlegg in Tirol. Mineralien Magazin, Lapis; 19(7—8):41—68.

Schnorrer G. 1996. Neue Sekundärminerale des alten Bergreviers Schwaz-Brixlegg in Tirol. Lapis; 21(12):16—19.

Seim R, Tischendorf G. 1990. Grundlagen der Geochemie. Leipzig: VEB Deutscher Verlag für Grundstoffindustrie: 632.

Taylor S R, McLennan S M. 1988. The Significance of the Rare Earths in Geochemistry and Cosmochemistry. In: Gschneider K A J, Eyring L. Handbook on the Physics and Chemistry of Rare Earths, vol. 11, chapt. 79. Amsterdam: Elsevier: 485—578.

Wöbking H. 1982. Untertägige Eigenpotentialmessungen in den Bergbauen Großkogel bei Brixlegg und Falkenstein bei Schwaz. Berg- und Hüttenm Mh; 127(12):476—483.

Wolkersdorfer Ch. 1996. Hydrogeochemische Verhältnisse im Flutungswasser eines Uranbergwerks – Die Lagerstätte Niederschlema/Alberoda. Clausthaler Geowiss Diss; 50:1—216.

Wolkersdorfer C, Feldtner N, Trebušak I. 2002. Mine Water Tracing - A Tool for Assessing Flow Paths in Flooded Underground Mines. Mine Water and the Environment; 21(1):7-14.

Wolkersdorfer Ch, Hasche A. 2001. Tracer Test in the abandoned Fluorspar Mine Straßberg/Harz Mountains, Germany. Wissensch Mitt; 16:57—67.

Worrall F, Pearson D G. 2001. Water-rock interaction in an acidic mine discharge as indicated by rare earth element patterns. Geochim Cosmochim Acta; 65(18):3027—3040.

Infiltration water dating from tritium measurements in mining dumps: methodic specifics and case study

Jens Mibus[1], Peter Szymczak[2], Detlef Hebert[3]

[1] Forschungszentrum Rossendorf e.V., Institute of Radiochemistry, P.O. Box 51 01 19, D-01314 Dresden / Germany
[2] G.E.O.S. Freiberg Ingenieurgesellschaft mbH, P.O. Box 1162, D-09581 Freiberg / Germany
[3] Freiberg University of Mining and Technology, Institute of Applied Physics, Bernhard-von-Cotta-Straße 4, D-09596 Freiberg / Germany

Abstract. The utilization of lumped parameter models for the determination of groundwater residence times is restricted to steady state hydraulics. However, for mining objects such as pits and dumps a transient behaviour is more typical even after closure. Nevertheless information on the mean residence time is needed for risk assessment. Therefore, a combination of the linear and the piston flow model is proposed to take the transient aspect into consideration. The application to a mining dump is discussed.

Introduction

Mine waste dumps and industrial tailings often emit toxic substances such as sulphuric acid, heavy metals, or radionuclides through seepage discharge. An important parameter to characterize these transfer processes is the mean residence time τ_m of the infiltration water usually determined from tritium measurements. Lumped parameter models are used among others for the evaluation of concentrations of environmental tracers in groundwater (Maloszewski and Zuber 1982, Richter and Szymczak 1994). Investigations in abandoned mining areas require an application of these aquifer models to mining dumps. However, some specifics are to be considered to prevent misinterpretations.

Technological depositions with a thick unsaturated zone exhibit a strong heterogeneity of flow paths. This spatial variability is ignored by lumped parameter models. Hence stochastic approaches are used for modeling (van Genuchten 1991).

On the other hand there are sources of systematic errors. Mining dumps exhibit a transient hydraulic behaviour during the deposition of waste material, where lumped parameter models are not applicable. In praxis information is needed on the turnover time of infiltration water for risk assessment and to design remediation measures. An attempt is made to give a reasonable approximation for the problem using an adequate combination of lumped-parameter models.

Hydrogeologic problem

In recent mining periods mine wastes usually were dumped cone-shaped with a total volume of several million cubic meters. In coarse grained, high permeable material large portions of the dump are flown through by infiltrating water unsaturatedly. Water saturation is reached in deeper parts of the dump as shown schematically in Fig. 1. The mean residence time in the unsaturated zone is much higher than that in the saturated zone. The latter works as hydraulic short circuit, where all age components are mixed.

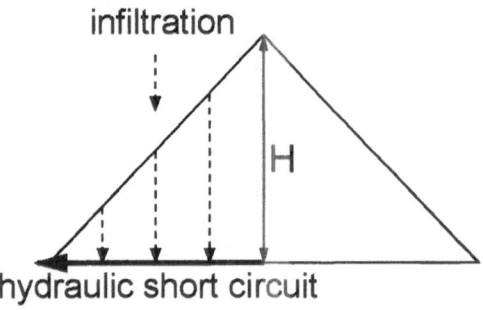

Fig. 1. Geometric set up and simplified flow scheme in a cone-shaped dump

The geometrical set-up and the simplified flow scheme of the dump is sufficiently approximated by the linear model. The linear model represents an unconfined aquifer with linear increasing thickness H. The weighting function f(t) is characterized by an equipartition of all age components from $\tau = 0$ to $\tau = \tau_{max}$ (Maloszewski and Zuber, 1982) as shown in Fig. 2a.

The model applicability explicitly assumes steady state hydraulics, i.e., volume and flux have to be constant. In the course of the deposition of the dump the reservoir size (volume and thickness) increase more or less continuously. The infiltration water is partly or even completely consumed to build up a static reservoir of adhesive water in the initial dry pores. The water percolation begins not until the field capacity is reached. This deceleration depends on deposit rates, recharge rates and capillary capacity and increases the portion of older components in the effluent waters compared to model assumptions. The rigorous application of the linear model results in an systematic overestimation of the mean residence time. They even may exceed the life span of the dump, which is not plausible.

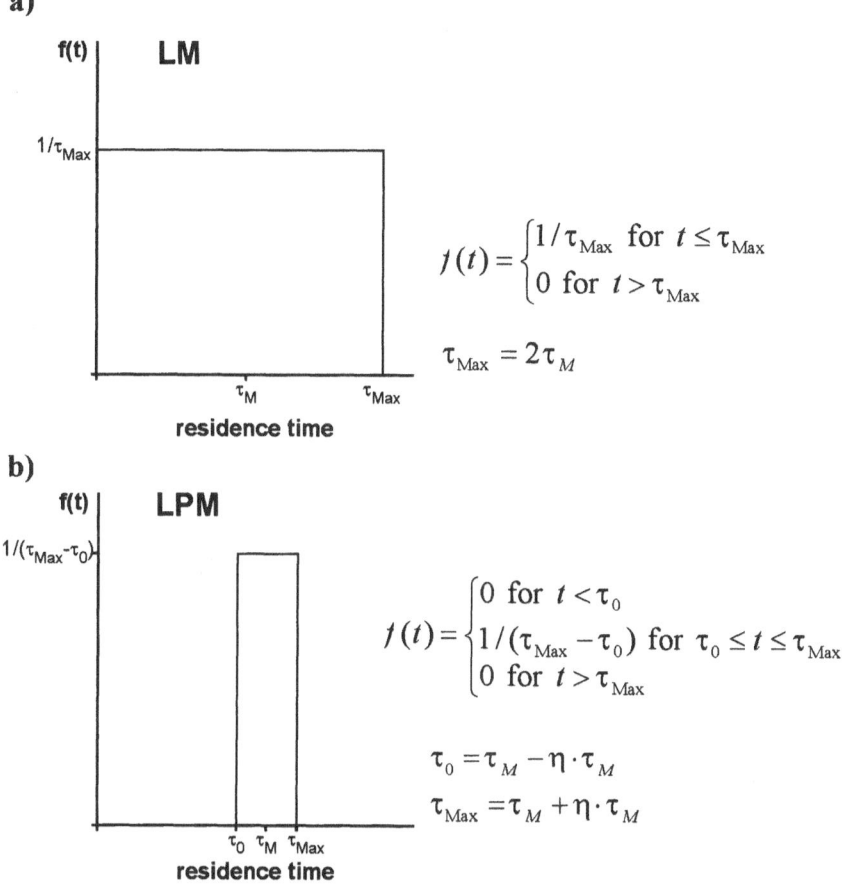

a)

$$f(t) = \begin{cases} 1/\tau_{\text{Max}} & \text{for } t \leq \tau_{\text{Max}} \\ 0 & \text{for } t > \tau_{\text{Max}} \end{cases}$$

$$\tau_{\text{Max}} = 2\tau_M$$

b)

$$f(t) = \begin{cases} 0 & \text{for } t < \tau_0 \\ 1/(\tau_{\text{Max}} - \tau_0) & \text{for } \tau_0 \leq t \leq \tau_{\text{Max}} \\ 0 & \text{for } t > \tau_{\text{Max}} \end{cases}$$

$$\tau_0 = \tau_M - \eta \cdot \tau_M$$

$$\tau_{\text{Max}} = \tau_M + \eta \cdot \tau_M$$

Fig. 2. Weighting function of the linear model LM (a) and the combined linear and piston-flow model LPM (b). Parameters are explained in the text.

Enhanced model

Two extreme cases are to be considered at first.
1. During deposition the volume increase is large enough, to completely accumulate the infiltration water. For this case applies:

$$\theta_{FC} \frac{dV}{dt} \geq v_R \cdot A$$

where θ_{FC} is the volumetric water content at field capacity, v_R the infiltration or recharge rate, V the volume and A the base area of the dump. In this case no flow occurs. The tritium concentration depletes only by radioactive decay.

2. The volume of the dump does not change (e.g. after termination of mine waste dumping). Over long time periods the dump is stationary percolated by infiltration water. The storage volume is not used.

Case 1 requires very high deposition rates of waste material or very low infiltration rates which are not to be expected in reality. It can rather be assumed that a certain portion of the infiltration water is stored and an excess portion starts to percolate (hereinafter referred to as *excess infiltration*). Both basic processes occur simultaneously and may be described by a parallel connection of two lumped parameter models (Fig. 3).

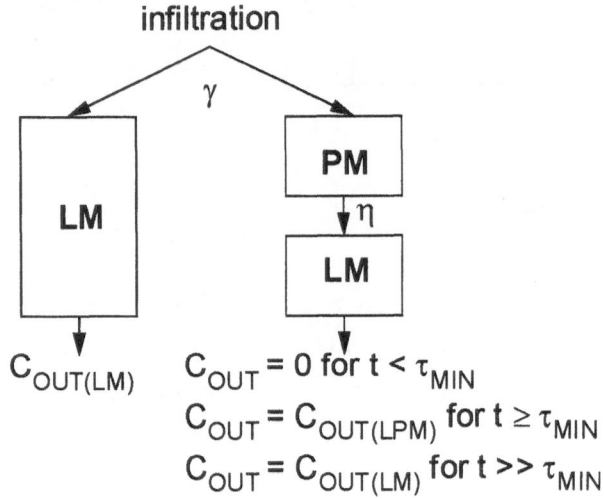

Fig. 3. Scheme of the combination of lumped-parameter models. Parameters are explained in the text.

Box 1 represents the reservoir percolated from the begin of the deposition of the dump. This process is approximated by a linear model (LM). The sole model parameter is the mean residence time τ_m.

Box 2 represents the storage set-up during the deposition of the dump. After termination of the deposition this reservoir is percolated analogue to box 1. This deceleration due to storage set-up results in a time shift of the weighting function by a dead time τ_0. The behavior of this reservoir may be described by a series connection of a linear and a piston-flow model (LPM). Model parameters are the mean residence time τ_m and the percentage of the LM at the LPM (η) where dead time τ_0 and maximal residence τ_M time are derived from (Fig. 2b).

The overall model is formulated as a mixing equation:

$$C_{OUT} = \gamma \, C_{OUT,LM} + (1 - \gamma) \, C_{OUT,LPM}$$

C_{OUT}:   output-function of the overall model
$C_{OUT, LM}$:   output-function of the instantaneously percolated reservoir (LM)
$C_{OUT, LPM}$:   output-function of the reservoir with retarded outflow (LPM)
γ: weighting coefficient

The output-function C_{OUT} is equivalent to the tritium concentration in the seepage (in *tritium units* T.U., 1 T.U. = 0.118 Bq/kg H_2O; Moser and Rauert, 1980). The weighting coefficient γ is estimated from the volume fraction of the percolating excess infiltration at the total infiltration volume.

$$\gamma = \frac{V_{excess\ infiltration}}{V_{total\ infiltration}}$$

With proceeding time the percentage η of the LM within box 2 (LPM) moves. Immediately after termination of deposition only the piston-flow model works. The parameter η increases with increasing exchange of the stored water to reach 100 % at stationary flow. Then the overall process may be described by a LM.

The neglect of the diffusive exchange of water between both reservoirs is a possible source of error. This causes an equalization of both residence times to a certain degree. A consideration within the lumped-parameter models is not possible here.

Application

Hydrological and geochemical investigations were carried out at a mining dump (former shaft "Koenen II" near Nienstedt) of Kupferschiefer mining in the southeastern Harz foreland (Dunger, 1999; Mibus, 2001). Tritium concentrations in the seepage were analysed by TRICAR Laboratory at the Freiberg University of Mining and Technology over a time period of about two years (07/1995 to 01/1998). The dump was built up from 1964 to 1990. It consists of mainly carbonatic dead rock, has a total volume of V = 5 million m^3, a height of H = 100 m, and a base area of A = 11 hectares.

Application of the LM to the tritium measurements results in a mean residence time of $\tau_m = 20 \pm 5$ a. For the LM applies $\tau_{max} = 2\,\tau_m$. The life span of the dump, however, was only 32 a. So the tritium-age is systematically to high.

Assuming a typical value for field capacity of a sandy gravel of $\theta_{FC} = 5$ % (Busch and others, 1990) and an initial water content of the hard rock of 1 % a static reservoir of about 200,000 m^3 in the whole dump has to be set up. According to model balances from Dunger (1999) the mean infiltration rate is 115 mm/a. Over the time period of t = 32 a total water volume of about 400,000 m^3 infiltrated. Thus about one half of the infiltration is retarded in the storage volume ($\gamma = 0.5$).

The combined model is applied to measured values using the computer code MULTIS (Richter and Szymczak, 1994). The estimation of the parameter η (percentage of the LM the LPM within box 2) is difficult. Since the termination of deposition in 1990 portions of the stored old water are washed out, the solution for $\eta = 5\ \%$ is proposed here.

The application of the combined model results in two mean residence times.

mean residence time of the LPM: 22 a
mean residence time of the LM: 16 a
mean quadratic deviation: 1.75 T.U.

The mean residence time of the LPM corresponds to retarded outflow of the storage reservoir. The mean residence time of the LM in contrast represents the unretarded outflow.

Discussion

These values of residence times derived with the combined model exhibit a better consistency with the life span of the dump compared to the application of the sole LM. Potentially the difference of both residence times is even higher due to neglecting the exchange between the reservoirs.

The continued wash out of stored old water causes further decrease of the PM and thus the mean residence time within the LPM. The finally stationary percolation of both reservoirs, each described by one LM, does not necessarily result in a unique mean residence time. It is rather to be assumed that the percolation of the excess infiltration follows preferential flow paths and thus is bound to coarse grained and disturbed zones. The storage in contrast occurs in fine grained regions. This classification is in concordance with the conception on the heterogeneity of flow in the vadose zone. Thus the weighting function of the seepage always contains a fast component and a temporarily stored component which is not only determined by the thickness of the aeration zone.

Conclusions

The proposed extension of the modeling approach for the determination of residence time of seepage in mining dumps facilitates the simplified consideration of the transient process of deposition. The implementation provides results, consistent with the physical life span of the dump and permits a reasonable interpretation. The mean residence time of the water in the dump has consequences for the prediction of pollutant discharge (such as salts, metals, or radio nuclides) in the future (Mibus 2001).

References

Busch K F, Luckner L, Tiemer K (1990) Geohydraulik. Gebrüder Borntraeger, Berlin Stuttgart

Dunger C (1999) Sickerwasser- und Schadstoffbewegung aus ausgewählten Bergehalden Sachsens und Sachsen-Anhalts. Ph.D. thesis. Freiberg University of Mining and Technology, Freiberg

Maloszewski P, Zuber A (1982) Determining the turnover time of groundwater systems with the aid of environmental tracers 1. Models and their applicability. J. Hydrol. 57: 207-231

Mibus J (2001) Geochemische Prozesse in Halden des Kupferschieferbergbaus im südöstlichen Harzvorland. Ph.D. thesis. Freiberg University of Mining and Technology, Freiberg

Moser H, Rauert W (1980) Isotopenmethoden in der Hydrologie. Gebrüder Borntraeger, Berlin Stuttgart

Richter J, Szymczak, P (1994) MULTIS - Ein Computerprogramm zur Auswertung isotopenhydrogeologischer Daten auf der Grundlage konzeptioneller Boxmodelle. Freiberg University of Mining and Technology, Freiberg

van Genuchten M Th (1991) Recent progress in modeling water flow and chemical transport in the unsaturated zone. In: Kienitz G, Milly P C D, van Genuchten M Th, Rosbjerg D, Shuttleworth W J (eds.) Hydrological interactions between atmosphere, soil and vegetation. IAHS Publication No. 204, 169-183

IMWA – What for and for what?

The International Mine Water Association (IMWA) is a multidisciplinary expert organisation which links up experts of various branches of natural sciences and technology.

IMWA was founded 1979 in Granada/Spain as a result of interest shown in the increasing problems associated with water in the mining industry at the "First International Mine Drainage Symposium". Since then we have organised a congress every three years. IMWA also holds annual symposia and workshops on topical subjects which have included: Hydrogeology of Coal Basins; Mine Water and Pumping; Acid Mine Water; First African Mine Water Symposium; Water Resources at Risk; Engineering in Karst; Mine Water and Environment.

Membership of IMWA is open to everyone with an interest in mine water and we also distribute to members a quarterly journal of "Mine Water and the Environment" for which relevant papers are invited for publication.

Concerning mining, mine water, and the environment, there are still many open issues. If this is true on the theoretical level, it is even more true on the performance level. Precisely due to these facts an international debate and exchange of expertise and of ideas is worth its effort.

Dear professional and expert, do not hesitate to take an active part at this debate. With a membership IMWA is offering you the opportunity to do so. You are most warmly invited to contribute and to profit from it as much as you can.

IMWA –Meetings

Budapest 1982 – Granada 1985 – Nottingham 1986 – Kattowice 1987 – Melbourne 1988 – Lisbon 1990 – Ljubljana·Pörtschach 1991 – Chililabombwe 1993 – Nottingham 1994 – Denver 1995 – Portoroz 1996 – Bled 1997 – Johannesburg 1998 – Seville 1999 – Kattowice 2000 – Belo Horizonte 2001 – Freiberg 2002 – South Africa 2003 – Ireland 2004 – Italy 2005

Gamma-natural radiation field in Mesozoic rocks and in Uranium-bearing Phosphate Mineralization of the Ionian zone, Albania.

A. Serjani , K. Onuzi, F. Dafa, A. Papuciu, E. Bedini

Geological Research Institute, Tirana, ALBANIA

Abstract. Three radiolithostratigraphical levels were distinguished in the Ionian zone. The most important are Middle Jurassic uranium-bearing phosphorites. Some theoretical studies and calculations of intensities of gamma-radiation of mineral body were performed and correlations between intensity of gamma-radiation field and useful compounds of phosphate ores discovered. Physical-geological and mathematical models for determining the coefficients "Ko", "F", radioactive equilibrium "C", and emanation "a" were conducted.

Introduction

During a scintillometer Survey of Ionian zone in scale 1:50 000 (Nasi et al., 1970), measurements of natural gamma radiation intensity of carbonate-chert rocks were carried out. Characteristics of the distinctions of the values of gamma field in different sections were studied in details (Dafa, Serjani, 1987), compiling the radio-lithological-stratigraphical column of the Ionian zone. Three radio-lithostratigraphical levels were distinguished:
- Middle Jurassic uranium-bearing phosphorites, $I\gamma$= 30-300 $\mu R/h$
- Phosphate horizon of Coniacian, $I\gamma$= 18-50 $\mu R/h$
- Bituminous schists in Triassic-Jurassic boundary, $I\gamma$= 16-24 $\mu R/h$

The lower level of anomaly of the Triassic-Jurassic age is presented by transition pack build up from bituminous dolomites and thin bituminous beds. The second level of gamma radiation is linked with disconformities in sedimentation and uranium-bearing phosphorites formed during Middle Jurassic.

The Upper level shows the widest level of regional character. It belongs to the phosphate-carbonate-siliceous horizon of Coniacian widespread almost all over Ionian zone and was studied in details during prospecting works for phosphorite ores. Although the content of phosphate beds in this horizon sometimes is up to

30% P_2O_5, the uranium content is very low. Thus in this case phosphate beds contain the normal uranium amounts as the rest of phosphorites.

Uranium-bearing phosphorite mineralization is distributed in Ionian zone in Albania and in Greece (Stayropodis & Basjakos 1981). This mineralization is found in disconformities of sedimentation typically between massive dolomite limestones of Lower-Middle Liass at the bottom, and carbonate siliceous packs of Middle or Upper Jurassic on top.

Theoretical studies were performed and intensities of gamma-radiation for mineral bodies with non uniformity of distribution of radioactive matter were calculated. A correlation between intensity of gamma-radiation field and useful compounds of the phosphate ores were found. Some physical-geological and mathematical models for the determination of the coefficients "K_o", "F", of radioactive equilibrium "C" and emanation of "α" were conducted as well (Dafa 1989). Radiometrically the system is a simple: uranium is in equilibrium with radium with absolute absence of thorium and potassium.

Short Geological data on uranium-bearing phosphorites

In 1967 for the first time in Albania the uranium-bearing phosphate mineralization (Nasi et al., 1967) were discovered. Until 1990 many geological and geophysical prospection works and studies have been carried out.   Some outcrops and two small deposits were found and prospected in Kurveleshi and Çika anticline belts. Almost in all outcrops and especially in Noraj (24 Maj) (fig. 1) and Bogazi (fig. 2) deposits were studied. Phosphate uranium-bearing ores of the Middle Jurassic are widespread almost all over surface of the Ionian zone in Albania and Greece.

Phosphate mineralization of the Middle Jurassic is found only in sections with disconformities. At the bottom of the disconformities in sedimentation always massive limestones of the Lower Liass "Pantocrator limestones" were found. The gap in sedimentation is not always co-associated with phosphate mineralization. In Ionian zone two small deposits and some outcrops of Middle Jurassic uranium-bearing phosphorites were prospected. These occurrences are widespread mainly to the northwestern Greece and to the Southwestern Albania. On Albanian territory Noraj ("24 May") an uranium-bearing phosphorite deposits (where uranium-bearing phosphate bed is up to 4 m thick), and Bogazi deposit of uranium-bearing phosphate veins, pockets, nets, and dissemination were detected. Many occurrences in Kurveleshi and Çika anticline belts were prospected as well. In most cases prospecting of above mentioned deposits and occurrences was done by drillings and galleries.

Uranium-bearing mineralization occurs in two morphological forms:
1.   Bedded mineralization (stratiform phosphorites)
2.   Mineralization of different morphological types (veins, pockets, nets) filling fractures and cracks of massive limestones of Lower-Middle Liassic.

Phosphate bed contains up to 20-25 % P_2O_5 and 0.01-0.005 % U (intensity of gamma-radiation varies from 80 up to 200 Mkr/h). Bedded mineralization occurs rarely and phosphate beds are usually very thin (0.2 up to 1-2 m). It lies some meters above the contact of disconformities within bedded limestones of Dogger, and is intercalated with thin chert layers. There are found a lot of macrofauna type *possedonia* in the phosphate bed.

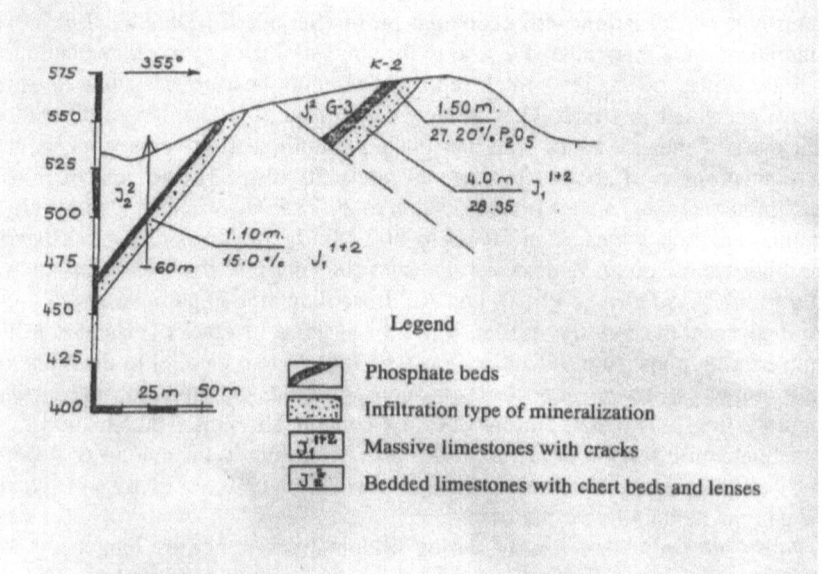

Fig. 1. Geological section of Noraj (24 Maj) uranium bearing deposit (after Husi R. and Serjani A., 1980)

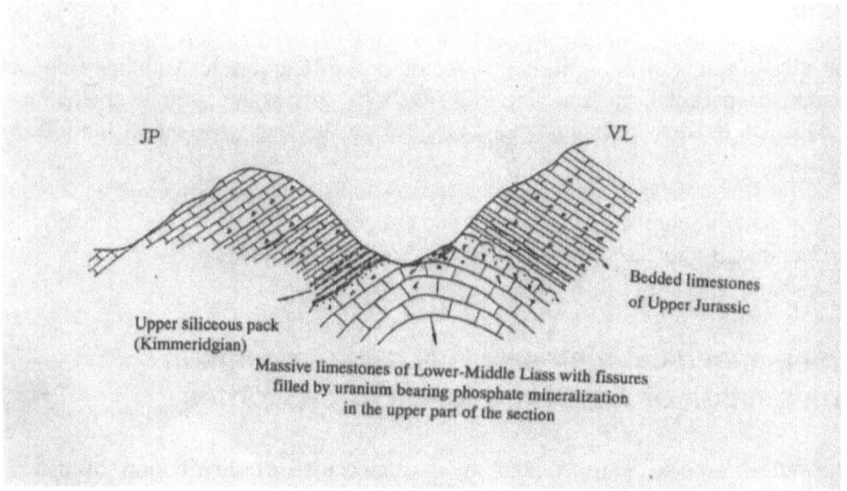

Fig. 2. Geological section of Bogazi anticline (after Nasi et. al. 1969)

IMWA: Mine water: underground and surface mines

The second morphological type of mineralization (infiltration) is more wide-spread in the Ionian zone and especially in Kurveleshi and Çika anticline belts. This type is characterized by high content of P_2O_5 and U (2-3 % up to 24 % P_2O_5, whereas the intensity of gamma-radiation varies from 30 up to 3000 Mkr/h, but the mineralization is spread into massive limestones and it is difficult to do exploitation and beneficiation with economic profit (Serjani A., Dafa F., 1993). This mineralization is more intensive next to the contact of the gap in sedimentation.

Radiometric anomalies, which represent uranium-bearing phosphate ores, are often interrupted in strike. They possess sizes from 50-100 m up to 600-800 m. Thickness of mineral zones with phosphate uranium-bearing veins, pockets, nets, disseminations is of about 10-20 m and rarely 30-40 m. The content of mineral ores (mineral zones) varies from 3-5 % up to 20-25 P_2O_5, whereas the intensity of gamma-radiation varies from 30 up to 300 Mkr/h. In Jurassic uranium-bearing phosphorites of Ionian Zone were analyzed showing rare earth elements rates of 0.04 - 0.07 %, as well as Pb, Zn and Ag. Rose diagrams of gamma-natural radiation and fissures show the existence of two systems of cracks correlated to uranium-bearing phosphate mineralization. The first system parallel to the plane disconformities is more important for concentration of mineralization. (Dafa, Papuçiu, 1984). The uranium ore does not contain Thorium. The situation of radioactive equilibrium is stable up to 0.8-1.2. According to the studies of different physical-geological models coefficients of correlation between P_2O_5 and U content and gamma radiation were calculated.

Phosphate beds were formed during Middle Jurassic age by biogenetic sedimentaion. The mineralization of vein type was formed epigenetic during infiltration of phosphate solution into massive limestones of the Lower-Middle Liassic. The phosphorites were formed in the photic zone on submarine mountains whereas black shales in basins a few hundred meters deep (Chiotis, Vechios, 1992).

Geological and geophysical studies bear witness about good perspective of second type of mineralization mainly in central part of Kurveleshi Anticline Belt and in southern part of Çika Anticline Belt. The most interesting geological structures in Albanian territory for study and geological prospecting we consider the following areas:

o Çiflig-Bogaz-Dishat-Janjar-Sajatha to the southern direction in Greece.
o Northern direction of Sotira overturned anticline.
o Deep levels and eastern flank of Noraj ("24 Maj") deposit.

Some analytical-mathematical achievements in prospection of uranium-bearing phosphorites

Theoretical aspects were studied about regularities of distribution of natural gamma-radiation field of uranium bearing phosphate bed and especially of infiltration types of mineralization which is more widespread. Correlations between intensity of gamma-radiation field and useful compounds of the phosphate ores were discovered by Dafa (1989).

Considerable resources of phosphate bearing beds and especially bodies, veins, disseminations, nets of the mineralization of infiltration type are available. Sizes of ore bodies and the content of useful components in two deposits and in many outcrops were estimated.

Application of geophysical methods such as radiometric testing and documentation, quantitative interpretation of gamma-lodging diagrams made it possible to assess the quantity of ores and the quantity of chemical analysis necessary for the evaluation of deposits respectively. Fig (2) displays the distribution of mineralization in combination with measurements of natural gamma-radiation of fissure systems in massive limestones. Geological-radiometrical survey on scale 1:25 000 and 1:10 000 accompanied by follow up survey at scale 1:2 000 up to 1:500 was done to provide the necessary information on the distribution of natural gamma-radiation.

Theoretical solutions and mathematical models for uranium-bearing mineralization with very irregular distribution convinced us to use radiometry for quantitative evaluation and run physical-geological models. According to the studies of different physical-geological models coefficients for evaluation of mineralization and reserves of ores were determined:

"K_o" = 52 MR/h which represent the intensity of gamma-natural radiation for content of 0.01 % U in equilibrium with Ra.

"F" = 37 MR/h which represent the intensity of gamma-natural radiation, corresponding to the content of 4 % P_2O_5.

Conclusions

1) Three lithostratigraphical markers in Mesozoic formations of the Ionian Zone were distinguishes by means of radiometric methods:
- Bituminous schists in contact between dolomites of Upper Triassic and massive dolomitic limestones of Lower Jurassic.
- Uranium-bearing phosphate mineralization of the Middle Jurassic.
- Upper Cretaceous phosphate horizon.

2) Radiometric anomalies helped to explore uranium-bearing mineralization of Middle Jurassic age in two types:
- Bedded mineralization
- Infiltration type of mineralization which is widespread in Ionian Zone.

3) Uranium is in equilibrium with radium with absolute absence of Thorium and Potassium.

4) Rose diagrams of gamma-natural radiation and fissures show the existence of two systems of cracks correlated to uranium-bearing phosphate mineralization. The first system parallel to the plane disconformities is more important for concentration of mineralization than the other one.

5) Two deposits and many outcrops of uranium-bearing phosphorites ores were prospected. Geological and geophysical studies bear witness about good perspective of this mineralization in Central part of Kurveleshi anticline Belt and in Southern Part of Çika anticline Belt. The reevaluation of the uranium-bearing phosphate deposits in Ionian Tectonic Zone is recommended

References

Chiotis E., Vechios, P. (1992) Paleogeographic conditions of phosphorites and black shales deposits in Epirus; Contribution to the exploration for phosphorites and hydrocarbons. Abstracts of the 6-th Congress of the Geol. Soc. Of Greece. Athens.

Dafa F., Papuçiu A. (1984) Rreth rozave te çarjeve dhe marredhenieve midis perberesve ne mineralizimin fosfatik jurasik. Bul.Shk. Gjeol. Nr.4, Tirane

Dafa F., Serjani A. (1987) Mbi radiolitostratigrafine e depozitimeve mesozoike ne Zonen Jonike. Bul.Shk. Gjeol. Nr.3, Tirane.

Dafa F., Serjani A. (1994) Application of Geophysical methods in the searth of phosphorites in Albania. presentation to the 7-th Congress of the Geol. Soc. Of. Greece. Thessaloniki.

Dafa F. (1999) Studim i perhapjes se fushes se radioaktivitetit natyror ne formacionet e zones Ionike dhe vleresimi i uraniumit ne fosforitet jurasike. Disertacion, Tirane.

Langora Ll., Nasi V., Zeqja K. (1976) Te dhena gjeologo-mineralogjike mbi shfaqjet fosfatike jurasike te zones Jonike. permb. Stud. Nr.1 Tirane.

Mechairos, J. et al. (1979) Decouverte d'importants depost de phosphorites en Epir (Greece) C.R.Acad. Sc. Paris, 288, Serie D.

Serjani A. (1988) Epokat fosforogjenike dhe depozitimet fosfatmbajtese te vendit tone. Bul.Shk.Geol. Nr.3, Tirane.

Serjani A. (1988) Te dhena litostratigrafike per depozitimet Mesozoike ne zonen Ionike. Nafta dhe gazi, Nr.1, Fier.

Serjani A. (1990) Chemical and Mineralogical composition of phosphorites of the Kurveleshi anticline Belt. Mineral. Wealth, 67, Athens.

Serjani A., Dafa F. (1993) Geological-Geophysical prospection of the phosphorites of Albania. Contribution to the 2-nd Congress of the Hellenic Geophysical Union. Florina, Greece.

Stayropodis, I.D., Basjakos, I.E. (1981) The Carbon scintilometer Survey of Epirus, Ionian islands and areas of west "Sterea Hellas" Mineral Wealth, 10, Athens.

Xhomo A., Kondo A., Papa A., Balluku I., Kanani J., Alikaj N., Nasi V. (1968) Transgresioni i Titonian-Neokomianit ne zonen Jonike. Permb. Stud. Nr.9-10, Tirane.

Gases released by water: impact on underground mines and on the surface

Giuseppe Sammarco[1], Onofrio Sammarco[2]

[1] Richard Ginori 1735 S.p.A., Sesto Fiorentino, FI, Italy
[2] Bureau of Mines, Dept of Earth and Environment, Tuscan Region, Via Goldoni 6, 58100 Grosseto, Italy; e-mail disminer@gol.grosseto.it

Abstract. There are many areas in the world subject to CO_2 and, subordinately, H_2S emissions that may be locally intensified by mining activities. Since these gases are highly soluble in water and are almost associated with it, this paper analyzes the mechanism of their release from the water and the factors that affect it. Above all it takes into account the effects of mining activities on this mechanism as well as the consequences of gas release on the mining work. It also analyzes all the other factors that concur with the foregoing to determine the spatial-temporal distribution of the gases in the atmosphere. By citing some specific cases it highlights the importance of promptly predicting and thereby preventing hazardous situations also through indirect detection of concentrations of dissolved gas.

Introduction

Carbon-dioxide discharges are concentrated in the world along zones of seismicity that are also characterized by active or recent volcanic activity, such as the circum-Pacific belt, from southern South America to New Zealand and the alpine belt of Southern Europe and Asia Minor. The carbon-dioxide discharges generally seem to be derived from the metamorphism of carbonates and from mantle sources (Barnes et al. 1978). The high presence of carbonatic rocks and the high frequency of metamorphic events in the case of the first origin and the poor solubility of CO_2 in magma, in the case of the second origin, and the poor reactivity and water solubility of CO_2 in both cases explain the high quantities of this gas that are discharged into the atmosphere where, because of its high specific weight after discharge it tends to accumulate. In mines excavated in the above areas there may be gas emissions comprising mainly CO_2, and then H_2S. In general terms H_2S should be considered a thousand times more dangerous than CO_2. These gases reach the mine either with the water that flows into it or alone after having been released from the water in which they were dissolved or had somehow reached. The fact that these gases are associated with water, a finding also confirmed by a census of gas emissions in the Peri-Tyrrhenian area of Central Italy (Duchi and Minissale 1995) that revealed the presence of water in 107 out of 109 emissions (Sammarco 2000 b), can be explained as follows: In general, the tectonic processes would generate not only metamorphism and hence, if carbonatic rocks are involved, CO_2, but also fractures which can permit both the flow of gases to the surface as well as water circulation and the formation of aquifers even in the same areas where the CO_2 had accumulated. Carbon dioxide and hydrogen sulphide are much more water-soluble than the other gases that are frequently found in natural emissions (Table 1).

Table 1. Solubilities of gases in water at various temperatures when the partial pressure of the gas is 1 atm. (101, 3 kPa).

Gas	Solubility (g/l)			
	10°C	20°C	30°C	40°C
CH_4	0.0308	0.0251	0.0208	0.0179
CO_2	2.3501	1.7208	1.3203	1.0025
H_2S	5.2311	3.9737	3.1349	2.5547
N_2	0.0235	0.0195	0.0168	0.0148
O_2	0.0543	0.0443	0.0373	0.0330

For the reasons given above, this paper refers primarily to CO_2 and H_2S gases which, since they are highly soluble in water are often released from it and are markedly affected by its dynamics. The paper primarily analyses the factors which by governing the mechanisms of gas release from water affect in the end the gas concentrations in the atmosphere.

The deductions presented here apply to the places in the above mentioned areas that are subject to CO_2 discharges and hence accumulations of that gas in the water and atmosphere especially if there are working mines in those places. In the areas where these workings develop they may cause marked increases in water and mainly gas permeability that can continue even for decades after the mine is abandoned, promoting the accumulation of those fluids in the areas.

The mechanism of gas release from water: influences on and as consequence of mining.

Bunsen coefficient values at various temperatures:
$$\alpha = v_o/Vp \tag{1}$$
where v_o is the S.T.P. volume of gas which at partial pressure p dissolves in the volume of water, V, and Henry's law
$$p = HC, \tag{2}$$

where H is the Henry coefficient and C is the concentration of gas dissolved in water at a given temperature and partial pressure p, make it possible to express the solubility of a gas in water as a function of temperature and partial pressure (Glassston 1963).

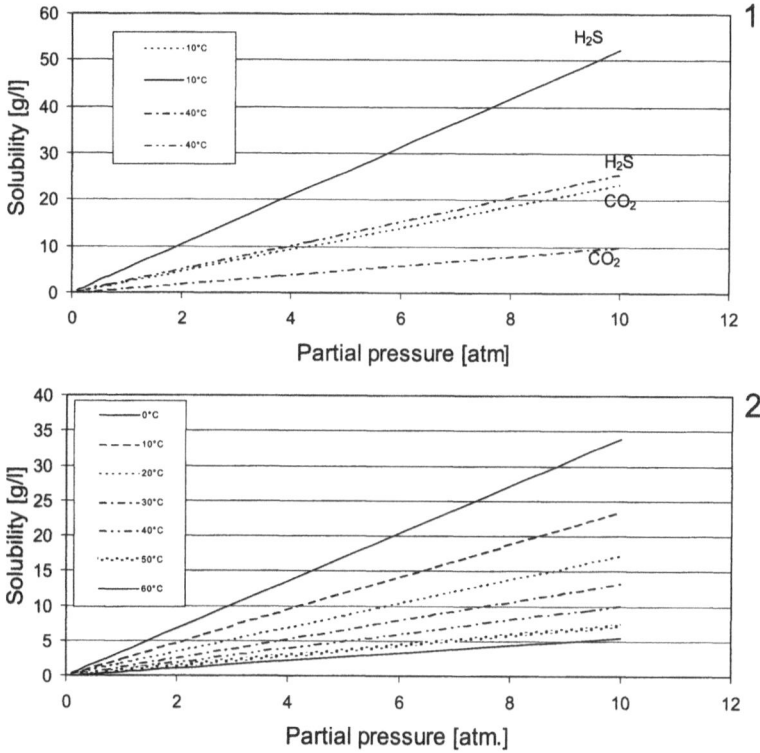

Fig. 1. Solubility of the CO_2 and H_2S gases in water as a function of the partial pressure at 10 and 40°C, (1). Solubility of CO_2 in water as a function of the partial pressure at various temperatures (2).

Figure 1. shows the approximate solubility of CO_2 and H_2S as a function of the partial pressure and for two temperature values having considered the Henry's law valid for partial pressures exceeding 1 atmosphere (101.3 kPa), too, and for concentrations of non-diluted gases and solubility independent of the system's total pressure. Using the same approximations, figure 1.2 shows the solubility of CO_2 alone, as a function of the partial pressure and for different temperature values (Perry and Green 1998; Stumm and Morgan 1996). It is obvious that before reaching saturation conditions, the more soluble the gas is in water, the bigger is the quantity of gas which can be dissolved.

Once saturation is reached, as shown specifically for CO_2 and H_2S in figure 1.1, at constant temperature and at a given partial pressure decrease, as well as at constant pressure and at a given temperature increase, the more soluble the gas is in water, the higher will be the quantity of gas that is released. At a constant pressure and at a given temperature increase, the lower the temperature at which the increase begins, the greater the amount of the gas that is released (Fig. 1.2).

Both figures clearly show that significant amounts of both H_2S and CO_2 are released as the temperature rises and, above all, partial pressure decreases.

There are various circumstances under which we can have water with dissolved gases. Just as there are many situations and modes with which the waters can be subject to pressure decreases and hence find themselves in conditions that will release gas.

In the case of surface water bodies, pressure decreases occur either in the backwater water when it is agitated or when its level drops, or in the water that is discharged from it. In this case the more the water is mixed where it flows the more is mixed and broken down where it falls or where there is great turbulence the more the pressure decreases. In these circumstances if the water contains dissolved gases they will be released according to ways which depend on those with which the pressure is reduced.

Pressure is reduced in subterranean water bodies when they are intercepted and exposed locally to the atmosphere, such as during mine excavations, or in environments with pressures lower than that of the water body at the point of interception, such as when drilling a bore-hole. As to the interceptions that occur when excavating underground mines, the following two alternatives can be cited.

1. Interception with a gallery of an unconfined or confined aquifer in formations with primary permeability.

2. Interception with mining work at several levels, of a water body in a karst cavity isolated from or somehow connected to superficial water bodies (Fig. 2).

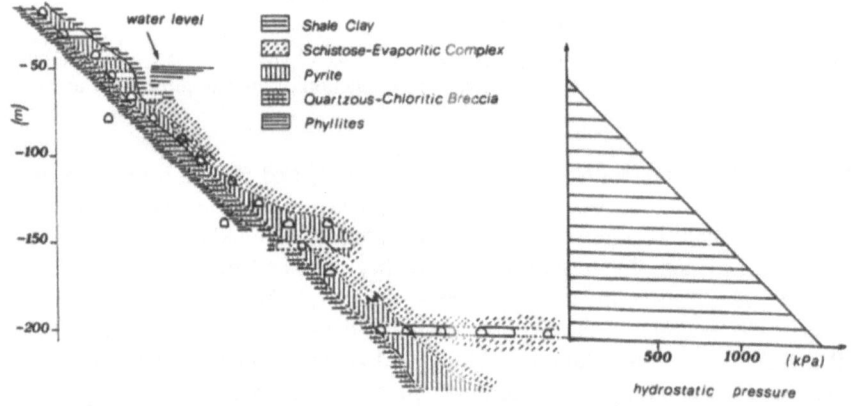

Fig. 2. Hydrostatic pressure in a water body, on the right, intercepted at various altitudes by an underground mining, on the left.

In the first case as the water flows towards the gallery its pressure decreases with the pressure head and with a gradient that increase with the decreasing permeability of the formation and the decreasing distance from the gallery. If the water were to contain dissolved gases, because of these pressure decreases once it is saturated it would release gas which would flow together with the water in the gallery and by itself into the dry part of the formation, if it is an unconfined aquifer. In any case, the greater release of gas should occur when the water flows into the gallery because it is at the outlet where the greatest pressure reduction occurs. The gas that was released and flowed into the dry part of the formation would then enter the gallery with flow rate depending on the permeability of the formation and difference between the pressure of the gas in the formation and the pressure of the air in the gallery.

In the second case (Fig.2), as soon as the excavations have come close to the edge of the flooded cavity and compromise its integrity to the point of fracturing it and thus due to the hydrostatic pressure permit the water to flow into the mine, the water will immediately enter it. In this circumstance the water pressure in the cavity will start to decrease very slowly at a distance from the fracture that is little greater than its width (Nebbia et al. 1960), and decrease faster and faster as it gets closer to it, to finally reach atmospheric pressure at the outlet. If the water in the cavity were to contain dissolved gases, they would be released as soon as the pressure starts to decrease, and after up to the outlet in the same manner as which the pressure is decreasing. The gases released in this way could remain in part in the cavity and go back into solution or rise to the surface of the water and spread above it; in part they could be carried by the water that flows into the mine. In addition to the gases already released, the gases that are released from the water at the outlet and along its underground path will also enter the mine.

It is evident that as the hydrostatic pressure increases with depth, for a given amount of discharged water, the more the gallery is deep the more these gas releases will be intense. In the case of very high pressures, which could occur besides that owing to high hydrostatic heads also for high temperatures, these gaseous emissions could become explosive due to the rapid expansion of the gases. As examples we can mention instantaneous irruptions of water and gas and, for antonomasia, phreatic explosions for which, however, the energy is provided mainly by the expansion of steam formed as a result of the pressure decrease.

Influences on the released gas emissions into atmosphere

The actions of the factors that can affect the immission into and diffusion in the atmosphere of released gases depend on specific circumstances. Gas released by water can reach the surface atmosphere or the atmosphere in an underground mine directly or after having traversed permeable formations and/or cavities. Only in the latter case the flow rate of the gas that flows into atmosphere could depend significantly on atmospheric pressure: if the gas released by the water should enter a cavity before going into the atmosphere, its flow rate when it finally does enter the atmosphere depends on the atmosphere pressure, as long as its pressure, upstream from the outlet of the cavity, is lower than twice the atmospheric pressure (Sammarco 1984, 2000). In this case the gas flow rate depends, in particular, on the atmospheric pressure, speed with which it varies, trend with which it has varied and size of the cavity (Sammarco 2000 b).

Actually, there could be increases in atmospheric pressure that would invert the flow between the cavity and the atmosphere causing air to enter the cavity, thus when the flow of gas would be restored, the gas would enter the atmosphere diluted with the air which would then contribute to reducing gas concentrations outside the cavity.

It is obvious that in the case of partially flooded cavities, increases and decreases in the water level will correspond, respectively, to decreases and increases in atmospheric pressure, even if the firsts will occur and influence these gas discharges with ways that could be quite different from those of the seconds.

There is no doubt that the manner in which this gas will appear on the surface or underground will also depend on the formations permeability to both water and gas. And the permeability changes with the geomorphology either owing to natural events or as a consequence of the mining work itself.

Gas concentrations in the atmosphere: effects of the combined action of the various influencing factors

The factors reviewed are those that control the distribution and intensity of the emissions of gases released from water that reach the atmosphere either directly or via the formations into which they flow after release. High intensity emissions promote high concentrations of gas that may occur due to reduced dilutions which, in turn, may also be caused by equally reduced diffusion of the gas in atmosphere. On the surface gas diffusion depends on the morphology of the area around the discharge and meteorological factors, mainly winds; in underground mines it depends on the geometry of the mine network, microclimatic conditions and mainly ventilation. In both situations the specific weight of the gas or, in the case of mixtures, the specific weights of the various components also contribute to gas diffusion.

Due to the combined action of these factors the distribution of gas concentrations in the atmosphere can vary greatly from case to case. Sometimes they also vary greatly over time and in an unpredictable manner due to the significant and at times unpredictable changes that characterize some of these factors.

One example of the excessive variation of the gas concentrations in atmosphere is the case of two gas discharges into the atmosphere from bore-holes that connect the abandoned underground mine *Abbadia S. Salvatore*, where sulphydric waters flow too, with the surface. The gas, that also contains CO_2 (~96%) and H_2S (~0,20%), released by these waters enters the mine cavity and the surrounding natural cavities, and reaches the surface via two bore-holes and the 7 meter high vent pipe put on each. In this case the concentrations in atmosphere of the gases discharged from the two vent pipes depend on the piezometric situation around the mine, the water level in it, the atmospheric pressure, because the gas enters, after having been released from the water, the mentioned cavities prior to reaching the outside atmosphere, the evolution of those factors and primarily the winds.

Systematic readings of gas concentrations were taken over a period of 500 days at 12 points one meter from the soil, around each of the emissions, and situated on 4 orthogonal directions at distances of 1.50, 5.00 and 10.00 meters from each. Table 2 refers to the concentration of H_2S in the atmosphere around one of the two bore-holes from which the gas exits at a mean flow rate of 150 l/s. The table shows the measuring points, the number of measurements taken at each point and the minimum, mean, maximum and standard deviation values of the concentrations found. The concentration distributions were anything but stationary. The high variability of the H_2S concentrations at each point is due mainly to the just as high variability of the winds that have a strong impact on the diffusion of the gas.

Table 2. Abbadia S. Salvatore Mine. H_2S concentrations in atmosphere round Ermeta borehole: number of the look-outs in each point and minimum, maximum, average and standard deviation values of the H_2S concentrations which have been observed in each of these points.

Location	Number	Minimum (ppm)	Maximum (ppm)	Average (ppm)	Standard Deviation (ppm)
Dir. 1 at 1.5 m	83	0.00	60.00	6.01	15.90
at 5.0 m	83	0.00	43.00	6.59	12.32
at 10.0 m	83	0.00	22.00	4.46	7.31
Dir. 2 at 1.5 m	83	0.00	59.00	4.48	12.07
at 5.0 m	83	0.00	58.00	8.72	14.06
at 10.0 m	83	0.00	31.00	3.13	6.91
Dir. 3 at 1.5 m	83	0.00	60.00	6.10	16.27
at 5.0 m	83	0.00	53.00	4.69	10.78
at 10.0 m	83	0.00	25.00	1.55	4.58
Dir. 4 at 1.5 m	83	0.00	20.00	0.24	2.20
at 5.0 m	83	0.00	20.00	0.36	2.44
at 10.0 m	83	0.00	20.00	0.54	2.90

A example of distribution of gas concentrations very different from those cited and that could be considered almost stationary given the slight variations in time that characterize it can be found in the distribution of the CO_2 concentrations 0.20 meters above a waterway, with a flow rate of approximately 300 l/s, that flowing along a gallery collects the drainage water of the Niccioleta mine. The readings were taken starting from the point where the water flows onto the surface. Only after the mine was abandoned the CO_2 was recognized in the water flowing at a rate exceeding 20 l/s. It probably comes from discharges of waters containing CO_2 or CO_2 discharges alone that probably developed following cave-ins in submerged parts of the mine. Figure 3 shows trends of the CO_2 concentration along the waterway obtained with monthly readings: the variations over time of those concentrations were very low in relation to the interval of time during which they occurred. The same figure 3 shows that in the section of the waterway that is between 8 and 30 meters from the outlet, where a small waterfall and high slopes occur, the concentration gradients are much higher than in other sections. The low variability over time of the trend of the concentration of CO_2 in atmosphere along the waterway depends on the very low ventilation around the surface of the water where the readings were taken.

Air velocity certainly has a determining influence on the distribution of gas concentrations. Therefore, on the surface where the winds are irregular in direction and intensity the concentration distributions will differ from one direction to another and will vary over time even in the same direction. In active underground mines, whereas, where air motion is guided by the same galleries and its speed is on an average constant except, for some transients (Sammarco 1971), the distribution of the gas concentrations will be essentially stationary if they are caused by gas discharges that do not change in position and in intensity, or that are controlled by adjusting the ventilation.

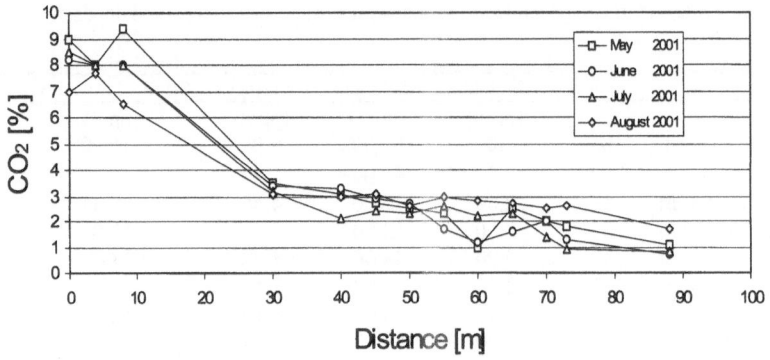

Fig. 3. Niccioleta mine. CO_2 released from the drainage: limited variability during the time of the CO_2 concentration in the air over the water course.

Hazards and forecasts

When gas is released by water the consequent risks depend on specific circumstances. Here are a few examples.

If CO_2 were released with other gases, or if after release it is in an atmosphere with other toxic or anyhow harmful gases, in addition to a decrease in the oxygen concentration in the air, and the effects of each gas, there could be synergic effects as well, and unfortunately there is little knowledge about them.

In the case that the CO_2 develops from water at a high temperature, in addition to favouring the development of the gas that temperature would intensify evaporation and the steam could come to be in air in not negligible concentrations in order to reduce, together with the CO_2, the oxygen concentration. In figure 4 that shows the volume concentration of steam as a function of the temperature at different values of relative humidity, it is possible to see that at temperatures above 37°C and relative humidity of 1, the steam concentrations exceed 6%. Hence the reductions in the oxygen concentrations are greater than 1%. It is obvious that since it is dealing with CO_2 and not an explosive gas, these reductions increase rather than diminish the risk.

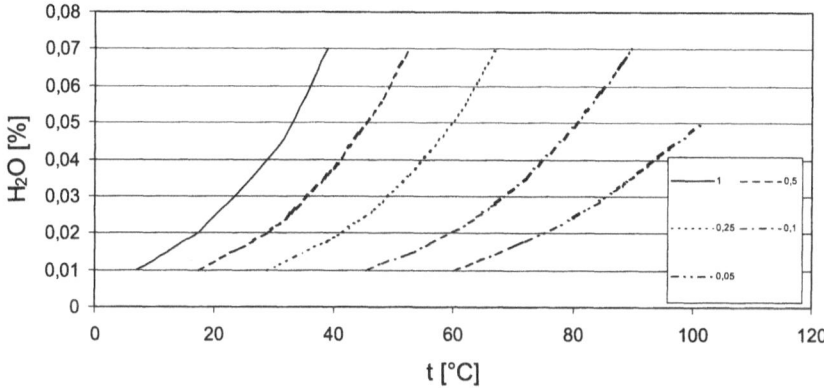

Fig. 4. Steam concentration in volume as a function of the temperature at various relative humidity values.

Another not uncommon case is that of accumulations of water subject to level changes and which contain CO_2 or which this gas can be discharged into. In these circumstances the increase of the water mass and that of the hydrostatic pressure in the same mass increase the backwater's capacity to accumulate gas. If the water level of the backwater were to drop and/or the water flow into the valley the gas would be discharged into the atmosphere with considerable risks. The occurrence at the Selvena mine during a flood of its deepest parts is highly significant.

As the piezometric surfaces around the mine dropped, high concentrations of CO_2 appeared in a tunnel above the drainage tunnel (Sammarco 2000 b).

The primary preventive measure is to ascertain whether there are situations that can generate risks. In the specific case it is first necessary to ascertain whether there is gas in the water that can develop. To facilitate and accelerate this check it would be advantageous to be able to determine the concentrations of dissolved gas indirectly by reading other quantities immediately readable and that could give a reliable indication of the concentration. We took over 30 mineral waters containing CO_2 in solution, and correlated the concentration of CO_2 dissolved in each with the concentrations of ions in solution and with the values of the electric con-

ductivity and of the pH. We found that high CO_2 concentrations in solution correspond to just as high conductivity values. For CO_2 concentrations exceeding 600
mg/l we found conductivity values, K, above 850µ S/cm (fig.5) and pH values below 6.7.

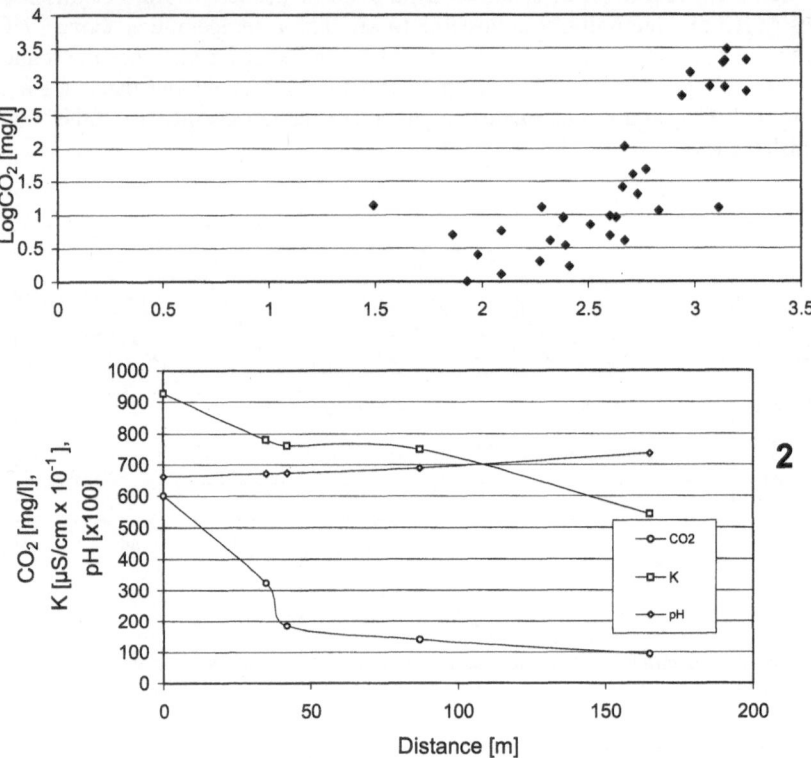

Fig. 5. Correlation between concentration of CO_2 dissolved and water conductivity.

Confirmations of these correlations are shown in Figure 5 that refers to the degassing of a thermal water raised to the surface by CO_2 gas. Degassing is obtained
by making the water, which presents concentrations of free CO_2 of 602 mg/l at the
outlet, run over 165 meters where small waterfalls are produced in the initial section. Figure 6 shows how in what manner decreasing the CO_2 in solution, decrease
the calcium and magnesium concentrations, and, in particular, significantly decreases the conductivity and slightly increases the pH.

The pH and mainly electric conductivity monitoring can, with the appropriate
precautions, be effectively utilized in determining whether or not there are excessive concentrations of CO_2 in solution or if a water containing this gas has been
sufficiently degassed.

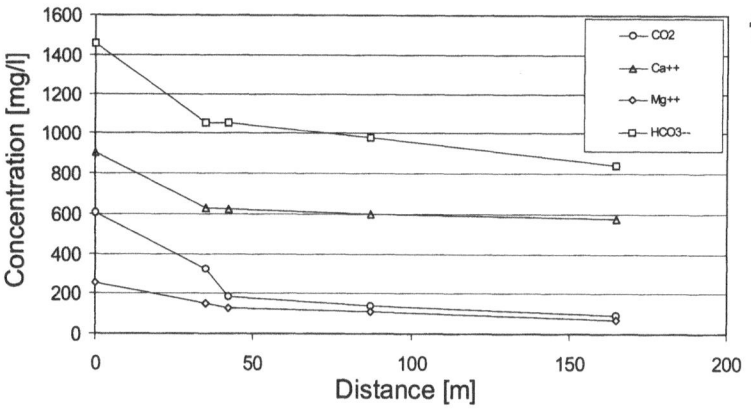

Fig. 6. Effects of the degassing of a thermal water: by decreasing the CO_2 dissolved in water, the Ca^{++} and Mg^{++} concentrations decrease, 1, pH increases and the conductivity decreases, 2.

Conclusions

The gases covered by this paper are carbon dioxide and hydrogen sulphide. They are frequently found in geothermal fluids, and due to their high solubility in water as opposed to other gases, they are released from water with an as much high intensity when the factors that control their dissolution vary.

To predict and prevent excessive concentrations of gas in the atmosphere in the anything but rare cases, in which the gas reaches the atmosphere after having been released from water, we must take into account all the possible dynamics of the water as well as their repercussions on the release mechanism. We must never underestimate the possibility of a water body to accumulate gas when it extends and to return it when it diminishes. Nor should we underestimate its capacity to compress and expand the gas that was released and flowed into any cavities, when the water level respectively rises or lowers. All these factors must be considered along with the others that, at the end, affect gas concentrations in the atmosphere in order to combat those that contribute to increasing gas concentrations and favour those that tend to reduce them.

It is evident that in order to be able to control gas concentrations in the air it is important to be aware of them in water. With specific reference to carbon dioxide, concentrations of this gas in water, with appropriate precautions, can be readily checked by testing the electric conductivity of the same water.

References

Barnes I, Irwin W P, White D E (1978) Global distribution of carbon dioxide discharges and major zones of seismicity. Open-file Rep 78-39: 53-63

Duchi V, Minissale A (1995) Distribuzione delle manifestazioni gassose nel settore peritirrenico tosco-laziale e loro interazione con gli acquiferi superficiali. Boll Soc Geol It 114: 337-351 (in Italian)

Enciclopedie des gaz – Gas encyclopaedia (1976). Elsevier Amsterdam – New York

Glasstone S (1963) Trattato di chimica-fisica. Manfredi Milano: 870 (in Italian)

Nebbia G, Ippolito G, Russo Spena A, Viparelli M (1960) Idraulica Treves Napoli: 512 (in Italian)

Perry H R, Green W D (1998) Perry's chemical engineers' handbook. McGraw-Hill NY

Sammarco O (1971) Le perturbazioni della ventilazione nelle miniere. L'industria mineraria II XXII: 467-476 549-559 (in Italian)

Sammarco o (1984) L'influenza della pressione atmosferica sulle venute di gas in miniera. L'industria mineraria III V: 5-19 (in Italian)

Sammarco O (2000) Water and gas emissions: dynamics and mutual influences. Proc 7th International IMWA Congress, September, Ustron Poland: 645-659

Sammarco G, Sammarco O (2000) Gas da acque termo-minerali: modalità di liberazione, rischi e cautele. Atti Problemi di geoingegneria: le risorse termo-minerali, Ottobre, Piacenza Italy: 45-58 (in Italian)

Stumm W, Morgan J J (1996) Aquatic chemistry. Wiley & Sons NY: 1022

Some Experiences with Modelling Water Quality of Post-Mining-Lakes in East Germany

Florian Werner, Mike Müller, Felix Bilek

Dresdner Grundwasserforschungszentrum e.V., Meraner Str.10, 01219 Dresden

Abstract. Predictive modelling of water quality is an important tool to aid decision making in the process of transforming the East German lignite mining districts into post mining landscapes. A major concern is to determine weather water quality goals are sustainable or not. Initial contents of reactive minerals in the overburden material must be known to estimate the source of water-soluble matter that can enter the hydrosphere during mine-flooding. The interaction of surface and ground water as well as the strong influence of erosion demand the use of complex and spatially resolving models.

Introduction

Almost a decade has passed since the framework for the transformation of the industrial mining areas in East Germany into post-mining-landscapes has been set up. Few lakes have reached their final water levels yet, most are still in the process of being flooded. Water quality problems arise from a legal point of view where flooded lakes discharge highly saline or acidic water into rivers. If water quality goals for the large lakes are not reached diverting management strategies can be favoured. If a water quality goal were indeed sustainable, but the establishment of stable post mining conditions just takes longer than expected, water treatment by adding neutralizing substances seems feasible. If the regional hydrologic flow system creates mass fluxes that are too large or long lasting to counteract with anthropogenic action, water treatment of the lakes itself does not seem feasible. Up to now biologic or passive systems are limited in their capability of handling high mass fluxes. To decide on these issues predictive modelling is needed. The problem with predictive modelling of these complex systems seems to be the gathering of representative parameters as well as the development of adequate simulation codes. The source of dissolvable matter in the re-wetted aquifers, the release of matter from eroded bank sediments and the fate of the dissolved substances in the stratified lakes and the downstream aquifers must be evaluated. The main pollutants are inorganic and resulting from acidic mine drainage. The open pit strip mines exploited tertiary lignite seams that were covered with 50 to 100 m of quaternary and tertiary sediments. In the Lausitz district hydraulic conductivity of the overburden is quite high with an approximate mean of $1 \cdot 10^{-4}$ m/s. Natural groundwater levels were to a large extent close to the ground surface. These levels will in most cases not be reached in the post mining landscape as they are controlled to protect meanwhile congested and rural areas from being swamped. The control of groundwater levels is achieved by setting lake water levels accordingly (and thus increasing groundwater seepage rates). In the urban areas of Senftenberg and Hoyerswerda additional drainage ditches become necessary to control groundwater levels. For the city of Hoyerswerda large horizontal filter wells were installed. In the Lausitz area 10 out of over 30 lakes that will be established will have a water volume of more then $9 * 10^7$ m^3.

Description of models used

The modelling of water quality focuses on mass balances for the vicinity of the flooded mines itself (Werner et al. 2001). The water balance of these lakes is in almost all cases strongly dependent on groundwater in- and ex-filtration. 3 D groundwater flow models are used to identify these fluxes. To assign mass fluxes to these water fluxes, monitoring wells were positioned around the future lakes. These mass balances were lumped on virtual geochemical reactors. The spatial resolution of these reactors was stepwise extended from 0 D to 2 D. At each balance point a geochemical reaction calculation was employed. The exchange of O_2 and CO_2 between the exfiltrated groundwaters and the atmosphere as well as the precipitation of metals like iron and aluminium are believed to be key processes in the formation of the water quality and are modelled using existing speciation codes.

The mass balance has to include the ground and surface water fluxes as well as the mass fluxes that are driven by erosion and consequent leaching of (overburden) soils in the lakes. The latter was identified to be of major importance. Biological primary production within these lakes is important to consider as it (i) directly acts on the acid-base system and (ii) supplies an electron donor that can be used in redox reactions.

A number of limitations and unknowns were encountered while trying to predict the future water quality of post mining lakes. These led to the improvement of the conceptual and consequently the numeric models.

Characterizing dumps in their ability to act as pollutant soure

An active mine in the Niederlausitz mining district was investigated. The sulphides pyrite and markasite were identified as potentially acid releasing compounds the sulphides. Potential acid buffering compounds were defined to be carbonates. Theses compounds were analyzed in all stratigraphic units of the overburden. A detailed description is given in Berger (2000). Samples were taken from fresh cut slopes in the open pit. Vertical averaged samples were produced from the stratigraphic units. Sulphides were analyzed with the evolved gas method and regularly checked with the chromium reducing sulphur method. Carbonates were analyzed by measuring the carbon dioxide release of acidified samples. The findings were backed with mineralogical methods.

Borehole data from the mining prospecting were used to create a three dimensional block model of the undisturbed strata. Post mining geological conditions were derived from pre mining conditions and effects of typical technology applied in the mine by means of a mixing key. 3 to 4 excavators carry the sediment from different stratigraphic units on a conveyor belt bridge which dumps the sediments using four shoots. The stratigraphic units and the fractions thereof that are reached by the single excavators, vary with their spatial distribution. The reach of an excavator is called a "cut". It is defined by a certain height interval. Which statigraphic unit will at a given coordinate fit into which cut is a simple geometric problem. The first cut is the uppermost portion of the overburden. On the investigated site it is completely moved by a separate conveyer belt lane to cover the dump. The second cut is i.e. distributed as follows: 70% into 4th shoot, 20% into 3rd shoot and 10 % into 2nd shoot. Each cut is distributed in an individual way into the shoots which reflect horizontal levels of the dump. This mixing key was applied to the block model.

The typical exposition times that are found in the mine were mapped. Hand drillings in slopes of known exposition times were compared to oxidation experiments in moisture cells. As a result soil specific oxidation rate functions were determined. A software tool was developed that performed the mixing of the sediments and the calculation of the reacted fractions of the sulphides as a function of soil type and exposition time. Initial total sulphur in the overburden ranges between about 0.1 and 3 wt. %. Maximum sulphide concentrations reach up to almost 1.8 wt.%. The highest sulphide concentrations were found in the sediments adjoining the coal seams. The dominant grain size in these units is silt and fine sand. The units that are comprised of medium to coarse sands contain sulphide concentrations of less than 0.5 wt.%.

The total carbon concentration is positively correlated with the total sulphur concentrations. Inorganic carbon is only found in the quaternary units. The tertiary sediments and thus the conveyer bridge dump itself is free of carbonates.

A cross section cut through the model of the dump is shown in Fig. 1. The three parameters initial sulphide concentration, final sulphide concentration and reacted sulphide concentration are shown (in this order from top to bottom). The model clearly reflects the finding that most of the oxidation is taking place on the top of the uncovered dump. The map showing the surface of the dump gives a picture of the horizontal changes in the sulphide concentrations of the top layer of the dump. As each stratigraphic unit was given a homogenous sulfide concentration, this reflects the changes in the thickness of the units. The prediction of sulphide oxidation is limited to the temporal frame that was investigated in the mine. It was assumed that this process looses its significance once the dump is covered and recultivated.

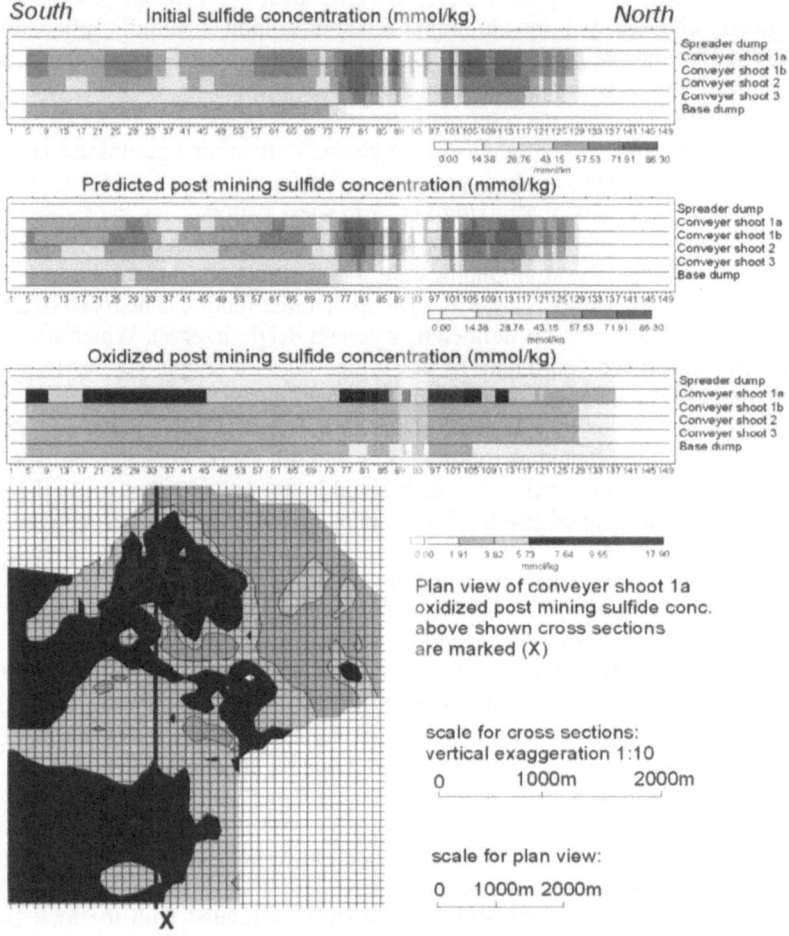

Fig. 1. Spatial distribution of the sulphide concentration in an overburden dump. Concentrations are calculated on the basis of sediment mixing and exposition time dependent reactions.

The created database serves as a tool to evaluate the effects of groundwater rise in the post-mining landscape. The block model of the dump serves as a basis for a predictive reactive transport calculation. The release of oxidation products into the moving groundwater was simulated using a dual porosity model.

Erosion driven mass fluxes

Mass fluxes that are driven by erosion depend on the total amount of bank sediments that is eroded and the ability of these sediments to release compounds into the surface water.

Estimating the amount of eroded sediments

Given a certain lake level the banks of a lake tend to form a characteristic profile. Wave erosion is of maximum intensity in early times and reclines after a wave breaking profile has established (Wagner 1996). This profile is a typical beach formation with a shallow inclination. Waves loose their energy before they reach the steeper slopes. In case of a rising lake level wave erosion is not reclining with time. The degree of erosion depends on the speed the water level is raising. Interruptions in the flooding process will increase the total amount of eroded sediments. An estimation method published by Wagner (1996) is used to calculate the eroded amounts. Parameters needed for this method are
1. Initial geometry of the slopes
2. Characteristic wave length and amplitudes
3. Prevailing wind speed and direction
4. Soil physical parameters (grain size, porosity)
Besides wave driven erosion storm driven erosion is considered. The banks of post mining lakes are often prone to erosion as vegetation is sparse. This is especially true for sediments containing acidified oxidation products. Abel (2000) measured erosion rates of up to 95 kg/ m^2 yr for a slope of a developing post mining lake. Attempts to model these rates on the basis of the universal soil loss equation failed because of the extensive rill and gully formation. Storm driven erosion still is of less importance compared to wind driven erosion if the water level is moving. For stagnant water levels and the times the lake has reached its final level, storm driven erosion gains importance.

Estimating the ability of sediments to release compounds

Bank sediments are mapped using soil pH as parameter to be determined in the field. The mapping is used as a basis to distinguish characteristic units that are subsequently sampled. Base titrations with soil-water mixtures showed that the release of acidity from the samples is often strongly dependant on the base used. Addition of NaOH seemed to overestimate the acidity release as it affected the cation exchange composition in a non natural way. To derive parameters that can be used in the balance model the ion release from the sediments was split up in two mechanisms that can be identified with defined lab methods:
- Release of ions stored in pore water or soluble secondary minerals
- Release of ions stored bound to cation exchange sites
These mechanisms yield a different behaviour concerning the ion release into water of changing composition. This is believed to be of importance, as the lake water is changing its composition in the course of flooding. As the ion release due to

erosion is used in predictive modelling lab tests have to be universal and can not be adapted to the prevailing (at this time unknown) conditions.

Ions bound to pore water and easily soluble secondary minerals do not seem to react significantly on the composition of the water used for extraction as long as solubility barriers are not reached. The latter was not observed under field conditions. In contrast ions bound to exchange sites react strongly on changing composition of the water used for extraction. The fraction of ions bound by either mechanism is not fixed and can vary according to our observations in the order of magnitudes. This means that each sample has to be investigated to identify if the effect of water quality dependant ion release is significant or not. If it is significant a reaction modelling of a water-cation-exchange-system is performed each time step erosion is considered.

Coupling of existing modells

To simulate water flow and transport in ground and surface water, the release of ions from dump aquifers, erosion driven ion input into surface water, and water quality in lakes the simulation package MODGLUE (Müller et al. 2002) was developed. This package combines existing models with specifically developed sub models.

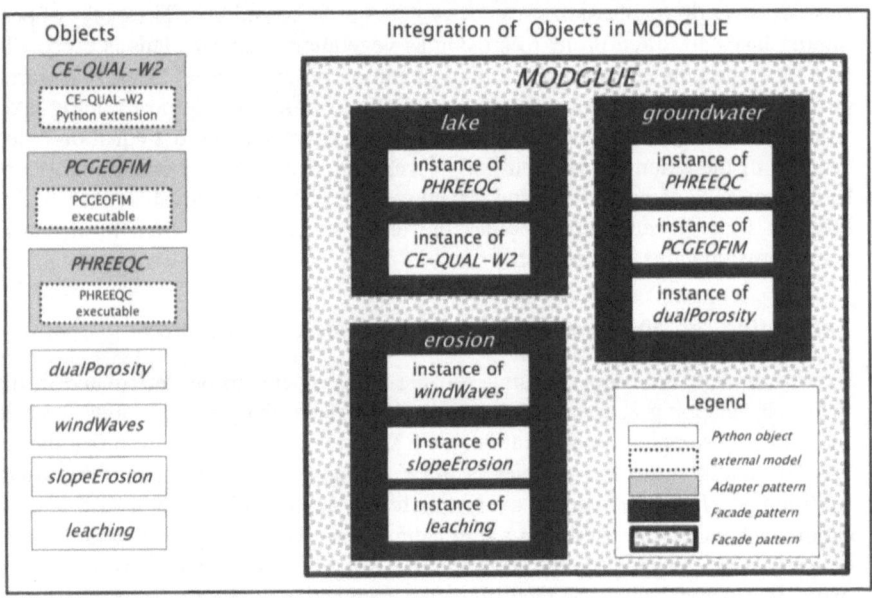

Fig. 2. Structure of objects linked in MODGLUE

The existing models that were used are:

- PCGEOFIM (Sames and Boy, 1999) a Three-dimensional finite volume groundwater flow and transport model

- CE-QUAL-W2 (Cole and Buchak, 1995) a Two-dimensional, laterally averaged, hydrodynamic and water quality model
- PHREEQC2.0 (Parkhurst and Appelo, 1999) a computer program for speciation, batch reactions, one dimensional transport, and inverse geochemical calculations

The lake model CE-QUAL-W2 had to be modified to allow the interaction of its quality model with inorganic processes and the subsequent changes of alkalinity in the lake water. Limitation of primary production by the availability of dissolved inorganic carbon was included.

The PCGEOFIM model was modified to include the dual porosity model that is used to simulate the release of oxidation products from dump aquifers. PHREEQC2.0 already offers a basic interpreter to implement user defined algorithms which was heavily used to describe all redox reactions on the basis of balancing electron donors and acceptors rather then as redox equilibria.

All models were modified to allow for a data exchange at predefined time intervals. The source codes of these models were written in Fortran (PCGEOFIM and CE-QUAL-W2) and C (PHREEQC2.0). New models (Erosion) as well as the parts concerning the coupling and data exchange itself were written in the object oriented language Python (van Rossum 2000)

Conclusions

This study describes experiences that have been gained with modelling water quality in post mining environments. It reveals shortcomings that have to be dealt with. The described investigation of an active mine to gather important geochemical information can not be conducted after the dumps are constructed. Thus this information is not available for most parts of the post mining landscape. Detailed information on erosion is often missing as well. The models that were used were improved in the course of gathering these experiences. Finally highly complex tools were created and presented in a condensed manner. The work with these tools is often limited by the availability of parameters.

The focus of this study is on the process of improving the conceptual model by the insight that is gained in field and laboratory observations. Site specific characteristics and questions that are highly dependent on the specific needs of the decision making process do not allow to recommend single models for exclusive use.

Acknowledgements

The presented experiences in modelling were made possible by funding of the German Department for Education and Research (BMBF) and the Lausitz and Middle German Mining Administration Agency (LMBV mbH).

References

Abel A, Michael A, Zartel A, Werner F (2000) Impact of erosion-transported overburden dump materials on water quality in Lake Cospuden evolved from a former open cast lignite mine south of Leipzig, Germany. Environmental Geology 39 (6): 683-688

Berger W (2000) Composition an changes in composition by redox reactions in sediments of the Lausitz Mining district (in German). Doctoral-Thesis University of Cottbus. Proceedings des Dresdener Grundwasserforschungszentrums e.V. 14: 177p

Cole T M, Buchak E M, (1995) CE-QUAL-W2: A Two-Dimensional, Laterally Averaged, Hydro-dynamic and Water Quality Model Version 2.0, US Army Corps of Engineers, Waterways Experiment Station.

Müller M, Werner F, Luckner L. (2001) A Tool for Concunjunctive Groundwater and Surface Water Quality Modeling in Post Mining Areas, 3rd International Conference on Water Resources and Environment Research, 22-25 July 2002, Dresden. accepted for publication. 5p

Parkhurst D L, Appelo C, (1999) Users's Guide to PHREEQC (Version 2)–a Computer Program for Speciation, Reaction-Path, 1D-Transport, and Inverse Geochemical Calculations, U.S. Geological Survey.

Sames D, Boy S, (1999) PCGEOFIM, Programsystem for Computation of GEOFIltration and GeoMigration.

Wagner (2000) Slopes affected by waves (in German) Dresdner Wasserbauliche Mitteilungen, TU Dresden, Insitut für Wasserbau und Hydromechanik, Wasserbaukolloquium 96, Heft 9: 177-198

Werner F, Bilek F, Luckner L (2001) Impact of regional groundwater flow on the water quality of an old post-mining lake, Ecological Engeneering 17: 133-142

van Rossum, G (2000) Python Reference Manual. iUniverse.com.

Natural processes caused by soil resaturation in the Patnow open-pit cone of depression

Krzysztof Polak, Mariusz Czop, Jerzy Klich, Jacek Motyka'

University of Mining and Metallurgy, 30-059 Krakow, al. Mickiewicza 30, Poland

Abstract. Patnow Open-Pit is one of the largest brown-coal post-mining workings in Poland. Natural flooding processes started in March 2001. Groundwater discharge to the open-pit is aided by water taken from a dewatering system based on a barrier of wells. This article presents the first stage of research and includes hydrogeological conditions in Pątnow area, piezometric water level data and chemical analysis of groundwater inflows and water supplied to the Patnow workings. It also describes some geotechnical problems connected with open-pit flooding.

Introduction

The process of exhausting brown coal resources in opencast mines forces the liquidation of mine industrial plant and the abandoning of open-pits. This abandoning process will take place in some large open-pits in Poland in the near future. The liquidation of mine industrial infrastructure and equipment is a short process and in addition it is more expensive.

In most cases the excavation of brown coal in Poland is located in areas where the natural groundwater level is close to terrain surface. The dewatering of orogen allowing excavation of beds is typical at polish mines and has often run for many tens of years. The water recharge is large in the majority of open-pit mines – up to a few cubic meters per one tone of lignite. The open-pit dewatering process results in a cone of depression extending a few kilometres.

Taking into account the hyrogeological conditions in polish brown coal mines, the open-pits are supposed to be reclaimed by their flooding by water. The liquidation of mine's dewatering system will result in mine flooding by water inflow from aquifers. The experiences from other countries show that the restoration process is extremely complicated.

Recovery water level rate in open-pits can have an influence on the chemistry of water in the artificial lake. The chemical composition of lake's water will determine the direction of reclamation processes especially in the restoration of water biology.

The recovery of piezometric water level within the cone of depression could bring the intensification of geotechnical phenomena in open-pit slopes (Dmitruk et al. 1998, Hajdo et al. 2000.). This process could result in tunneling, landslides and finally could change the boundary of artificial lake.

Patnow open-pit is one of the oldest in Konin region. The open-pit excavation started in 1957. The average output of Patnow reached 2.9 million ton per year and 17 million m^3 of overburden per year. Average depth of the open-pit reached 59 meter below the terrain surface. The last part of Patnow's bed was excavated until mid-2001, but the flooding process started in March/April 2001. The slopes of working were profiled with a gradient from 1:7 to 1:12. (University of Mining and Metallurgy supported research; project no. 11.11.100.270).

Geological and hydrogeological conditions

Patnow brown coal beds are located in Łódzka basin, Polish Lowland. Its geological profile consist of:
- Cretaceous sandy-clay marls, upper Jurassic calcareous sandstones,
- Tertiary sands under coal seat, brown coal bed and Pliocene clays,
- Quaternary sands, gravels and boulder clays

The geological conditions determine the following aquifers in the vicinity of Patnow open-pit:
- Cretaceous aquifer,
- Miocene aquifer,
- Quaternary aquifer.

Cretaceous and Miocene aquifers are continuous and regular. They both form the common Tertiary-Cretaceous aquifer with good hydraulic connections. Quaternary aquifer is generally non-continuous. In this aquifer water occurs locally in non-continuous aquifers formed in sandy lens located within boulder clays and shallow sands.

In the nearest neighbourhood of open-pit are located:
- in the South – Goslawskie Lake, Patnowskie Lake
- in the East – Wasowskie Lake, Mikorzynskie Lake, Slesinskie Lake.

In the natural conditions, lakes drained all aquifers trough the Quaternary erosion channel. As the dewatering process occurred, the direction of underground water flow was altered – with inflow to open-pit coming from the lakes (Figure 1).

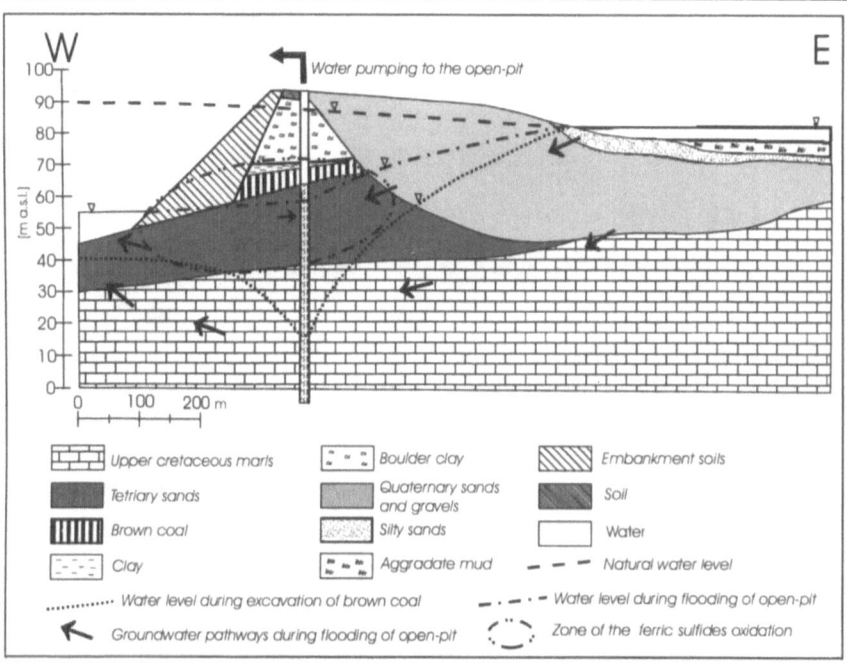

Fig. 1. Hydrogeological scheme in the East part of Patnow open-pit

Open-pit dewatering system

Patnow open-pit dewatering system consisted of:
- Dewatering galleries with shaft and dug wells; they have been closed by working face,
- Pumping plant located on the bottom of working,
- Barrier of 35 wells that was constructed to lower groundwater level during excavation in neighbourhood of safety pillar at the Mikorzynskie Lake.

The discharge from the dewatering system for the period 1986 - 1999 is presented in Figure 2.

Landslides

As the excavation of the beds located near to the safety pillar had been stopped and the slope had been profiled with overburden, the well barrier activity was finished and the recovery of water level was initiated. Only the pump station continued it's dewatering operation for the open-pit. The total recovery of water level reached 15 m at the well barrier line. The first local landslide processes occurred about one year after the dewatering in wells had been stopped.

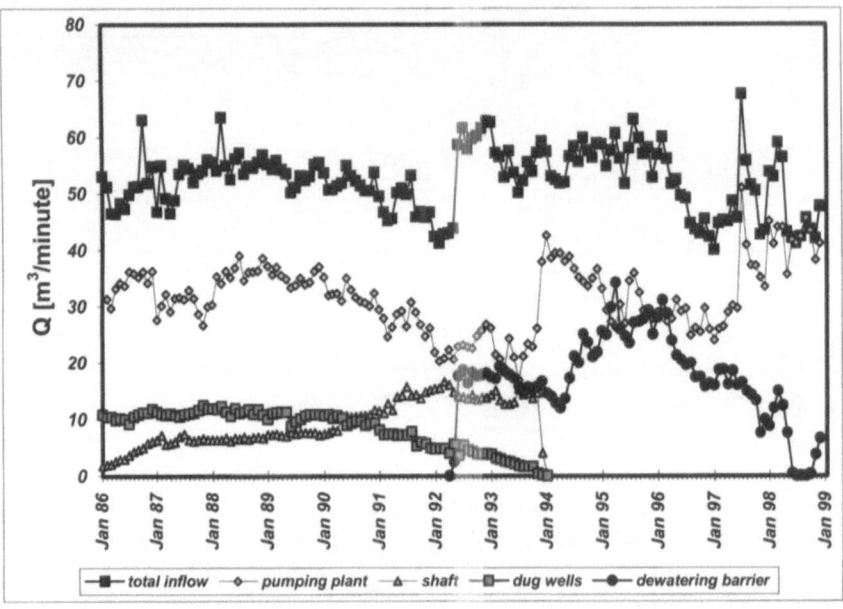

Fig. 2. Total and partial inflow to the dewatering system of Patnow open-pit.

The local landslides and tunneling have followed the increase of water pressure in Quaternary aquifer and the occurrence of water thrust for dumping ground of low permeability (boulder clays) in the northern part of open-pit. The yield of water leakage from exhibited aquifer reached ca. 1 m³/minute at the beginning. After some time it stabilized at the level from a few to over ten L/minute. The area of slope damage reached ca. 6 ha.

The damage to slope not profiled by overburden has followed the increase of water level. This slope was steep and located in Tertiary sands. The tunneling from scarp of open-pit has occurred there. The total outflow of water from the aquifer reached ca. 2.5 m³/minute and the area of surface damage reached ca. 7 ha. The ground floating with outflow was stopped after the construction of an embankment with overflow spillway.

Open-pit flooding

At the beginning of April 2001 natural flooding process commenced. The increase of ground water level is monitored. Selected wells of the barrier restarted pumping in June 2001. Water is now pumped directly to a reservoir which helps to decrease the hydraulic gradient in the area of slope. Furthermore it helps to decrease the amount of water flowing through the resaturated zone. This is important due to the fact that water flowing through the resaturated zone leaches weathering minerals and this process could worsen water quality in reservoir. So, the reservoir is supplied with unpolluted water.

Figure 3 presents the increase of water volume in Patnow reservoir. It has been changing linearly so far. It confirms the fixed groundwater inflow to the open-pit of 27,5 m^3/min. If inflow from well barrier is 12 m^3/min, then the groundwater inflow through the bottom reaches ca.15.5 m^3/min.

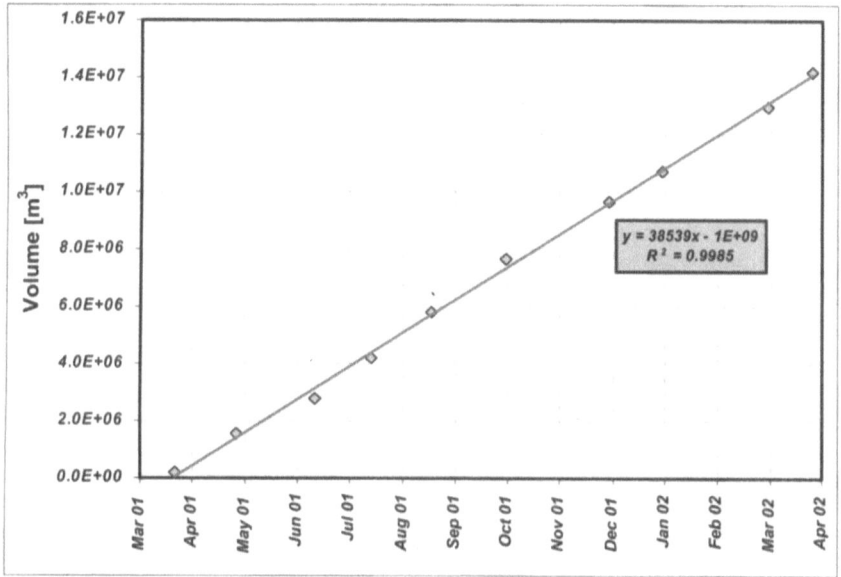

Fig. 3. The total volume of water changes in Patnow reservoir

Water quality

The water level drawdown induced by the dewatering of open-pit leads to the artificial expansion of unsaturated zone. The availability of oxygen from air initiates the weathering process of less-stable minerals. The oxidation of ferric sulphides (pyrite and marcasite-FeS_2) process is one of the most dangerous processes to the environment. The oxygen and water initiate the decomposition of mentioned minerals and it results in a change of sulphur redox form from S(-II) to S(+VI). Sulphate ions and hydrogen cations are formed. The process of ferric sulphide oxidation consists of a few connected chemical reactions. (Singer, Stumm 1970). The summary stoichiometric reaction is shown below.

$$3FeS_2 + 8O_2 + 8H_2O \longrightarrow 3Fe^{2+} + 6SO_4^{2-} + 16H^+$$

The process of pyrite oxidation and the process of environment acidification are connected closely. They are both associated with the excavation of sulphide metallic ores and hard and brown coal. Ferric sulphides are located within coal and metal deposits because they were formed under reducing conditions, which support pyrite and marcasite formation. There are many key studies showing how the process of ferric sulphide oxidation determines the chemical composition of groundwater within post-mining areas. (Fernandez-Rubio 1986, Smith et al. 1994; King et al. 1995; Kropka 1995, Rogoż 1996, Banks et al. 1997, Deutsch 1999, Motyka et al. 1999, Adamczyk A.F. et al. 2000, Iribar et al. 2000, Razowska 2000, Czop et al. 2001, Hidalgo 2001, Marszałek and Wąsik 2001, Steven et al. 2001). It is well known, the oxidation of pyrite from hard coal and waste rock causes pollution and has a negative influence on surface water and groundwater and causes environment aciditation (Anderson and Youngstrom 1976, Davis and Boegly 1981, Yu 1996, Foos 1997, Schüring et al. 1997, Geldenhuis and Bell 1998, Deutsch 1999, Szczepańska and Twardowska 1999, Brake et al. 2001, ab Kim et al. 2002).

In the area surrounding Patnow open-pit the conditions supporting ferric sulphide oxidation processes have been created. It is because the coal deposit consists of total sulphur in 1,5 - 1,8% and it is connected with the presence of ferric sulphide (Osika et al. 1970). In addition there is the significant amount of organic matter within Miocene sands under the bottom of brown coal deposit. There is ferric sulphide in organic matter as well. There are pyrite weathering processes occurring within all open-pit cone of depressions in Konin region.

As the brown coal deposit was exhausted the open-pit was abandoned and flooding process started. The open-pit is recharged with water coming from an easterly direction. Water from Quaternary aquifer flows to the Miocene sands and inflows into the open-pit. The well barrier takes over the role of discharge and it is directed into the reservoir. The water has low electrolytic conductivity on average ca. 505 μS/cm. In the water there are: calcium – 82 mg/L, magnesium – ca. 18 mg/L and small amounts of sulphide ions, on average 40,7 mg/L.

The resaturation of the zone where the sulphide mineral oxidation processes took place has followed the groundwater recovery. The groundwater stream dissolves the products of sulphide mineral decomposition. It results in the increase of calcium, magnesium and sulphates concentration and the decrease of pH value. In long term above mentioned process could result in the aciditation of reservoir water and in the worsening of its quality due to the concentration of sulphides, magnesium, iron, aluminium and microelements. This phenomena is observed in abandoned open-pits in East Germany (prior DDR; Luckner 1997, Hurst et al. 2002).

The Table 1 presents the chosen physical and chemical indicators of water in the open-pit. They confirm that the quality of reservoir water is much worse than the quality of water pumped by wells barrier. The biggest differences are observed in the sulphide ion concentration. The changes of water quality are observed in the reservoir as well. For example the samples taken in June and November 2001 show the doubling of sulphides average concentration (Table 1).

Table 1. The average of chosen physical and chemical indicators of water in wells barrier and Patnow reservoir.

Sample location		pH	γ [μS/cm]	Ca [mg/L]	Mg [mg/L]	SO$_4$ [mg/L]
Wells barrier		8.41	505.11	81.86	17.69	40.71
Reservoir	Average	8.20	750.86	117.74	27.08	174.73
in open-pit	June 2001	8.44	679.33	114.05	24.01	125.37
	November 2001	8.01	804.50	120.50	29.38	211.75

The process of weathering minerals dissolving will go on and follow the groundwater recovery in the unsaturated zone. It might cause the further worsening of water quality in Patnow reservoir and reduce ability to restore water biology.

Leakage has the main influence on the chemistry of reservoir water. The scale of changes to the quality of groundwater might be estimated on the basis of the quality of leakage above the water level in reservoir. Some analysed leakage had pH value of 3-4. Its sulphide concentration reached 2000 mg/L. A high concentration of iron, aluminium, silica, strontium and manganese was observed as well. The chemical composition of leakage is probably the result of sulphuric acid buffering through the decomposition of clayey minerals from boulder clay.

Conclusions

Natural processes could have a negative influence on the environment in the vicinity of flooding open-pits. The rate of water level recovery in the open-pit can have an influence on the chemistry of water in the artificial lake. Hydrostatic pressure increase in aquifers could change conditions of open-pit slopes stability. Flooding processes aided by dewatering systems in Patnow open-pit reduces the negative effects of processes connected with groundwater level recovery. The pumping of unpolluted groundwater reduces the amount of water migration trough the weathered soils. It also lowers hydrostatic waters pressure on scarp of the artificial lake. An increase in pumping rate could help reduce the flooding time, which is important from a foreshore abrasion point of view.

References

Anderson W C, Youngstrom M P (1976) Coal pile leachate – quantity and quality characteristics. J. Environ. Eng., Div. ASCE 102:1239 – 1253.

Banks D, Younger P L, Arnesen R-T, Iversen E R, Banks S B, 1997 - Mine-water chemistry: the good, the bad and the ugly. Environmental Geology 32 (3): 157-174.

Brake S S, Connors K A, Romberger S B, (2001 b) A river runs through it: impact of acid mine drainage on the geochemistry of West Little Sugar Creek Pre- and post-reclamation at the Green Valley coal mine, Indiana, USA. Environmental Geology 40 (4-5): 1471 - 1481.

Brake S S, Dannelly H K, Connors K A, (2001 a) Controls on the nature and distribution of an alga in coal mine-waste environments and its potential impact on water quality. Environmental Geology 40 (4-5): 458 - 469.

Czop M, Motyka J, Szuwarzyński M (2001) Sulphates in groundwater inflowing to the Zn–Pb Trzebionka mine (south Poland). In: Proc. of the Conf. Modern Problems of Hydrogeology, Vol. X/1: 291-299 (in Polish)

Davis E C, Boegly W J (1981) A review of water quality issues associated with coal storage. J. Environ. Qual. 10:127 – 133.

Deutsch W J (1999) Groundwater geochemistry, fundamentals and applications to contamination, Lewis Publishers, New York, 226pp.

Dmitruk S, Hawrysz M, Batog A (1998) Slopes stability in phase of liquidation open –pit mine, In: Proc. of the Conf. XXI School of stratomechanics, p. 61-70 (in Polish)

Fernandez-Rubio R, Fernandez Lorca S, Esteban Arlegui J (1986) Abandono de minas. Impacto hidrologico. Inst. Geol. y Minero de Espana E.T.S. de Inguenieros de Minas, Madrid, 267 pp.

Foos A (1997) Geochemical modeling of coal mine drainage, Summit County, Ohio. Environmental Geology 31 (3/4): 205 - 210.

Geldenhuis S, Bell F G (1998) Acid mine drainage at a coal mine in the eastern Transvaal, South Africa. Environmental Geology 34 (2/3): 234 - 242.

Hajdo S, Klich, Polak K (2000) – Hydrogeological conditioning of slopes stability during the process of liquidation based on Patnow open-pit example. In: Proc. of the Conf. Reclamation and derelict land management in land degraded by mining activity (in Polish)

Hidalgo MC, Benavente J, (2001) Controls on groundwater chemistry in the Linares lead-copper abandoned mines (Spain). In: Seiler K.-P., Wohnlich S. (eds.) - New Approaches Characterizing Groundwater Flow: 1199-1202.

Hurst S, Schneider P, Meinrath G (2002) Remediating 700 years of Mining in Saxony: A Heritage from Ore Mining. Mine Water and the Environment 21 (1): 3-6.

Iribar V, Izco F, Tames P, Antiguedad I, da Silva A, (2000) Water contamination and remedial measures at the Troya abandoned Pb-Zn mine (The Basque Country, Northern Spain). Environmental Geology 39 (7): 800-806.

Kim J J, Kim S J, Tazaki K (2002) Mineralogical characterization of microbial ferrihydrite and schwertmannite, and non-biogenic Al-sulfate precipitates from acid mine drainage in the Donghae mine area, Korea. Environmental Geology 42:19–31.

King T V V (ed.) (1995) Environmental considerations of active and abandoned mine lands. Lessons from Summitville. Colorado. U.S. Geological Survey Bulletin 2220, 40pp.

Kropka J (1995) Zinc and lead in the water from Zn-Pb mines of Bytom region (Silesia, Poland). In: Proc. of the Conf. Modern Problems of Hydrogeology, Vol. VII/2: 87-92 (in Polish).

Luckner L (1997) Bedeutung der Fremdwasserflutung für die Wiedernutzbarmachung der vom Braunkohlentagebau beanspruchten Flächen. Glückauf, v 133, 5.

Marszałek H, Wąsik M (2001) Hydrogeochemical anomaly in waters of pyrite deposit area in Wiesciszowice (Western Sudetes Mts., SW Poland). In: Seiler K.-P., Wohnlich S. (eds.) - New Approaches Characterizing Groundwater Flow: 1031-1034.

Osika R (ed.) (1970) Geology and mineral resources of Poland, Geological Publishers. Warsaw, 878 pp.

Razowska L (2000) Hydrogeochemical changes in Czestochowa region (south Poland) induced by iron mine flooding. Biul. PIG (Polish Geological Institute) 390: 35-96 (in Polish).

Rogoż M (1996) Impact of the mine cessation on water environment. In: Archives of mining sciences. 41 (1): 105-130 (in Polish).

Schüring J, Kölling M, Schulz H D (1997) The potential formation of acid mine drainage in pyrite-bearing hard-coal tailings under water-saturated conditions: an experimental approach. Environmental Geology 31 (1/2): 59 - 65.

Singer P C, Stumm W (1970) Acid mine drainage: the rate - determining step. Science, v. 167: 1121 - 1123.

Smith K S, Plumlee G S, Ficklin W H (1994) Predicting Water Contamination from Metal Mines and Mining Wastes. U.S. Geol. Survey, Denver CO, Third Int. Conf. on the Abatement of Acidic Drainage, Workshop 2, April 24, 112 pp.

Steven N M, Badenhorst F P, Harris C, Spath A (2001) An integrated hydrogeological and water chemistry study at the Navachab Gold Mine, Namibia. In: Seiler K-P, Wohnlich S (eds.) - New Approaches Characterizing Groundwater Flow: 1297-1301.

Szczepanska J, Twardowska I (1999) Distribution and environmental impact of coal-mining wastes in Upper Silesia, Poland. Environmental Geology 38 (3): 249-258.

Yu J-Y (1996) Pollution of Osheepcheon Creek by abandoned coal mine drainage in Dogyae area, eastern part of Samcheok coal field, Kagwon-Do, Korea. Environmental Geology 27 (4): 286 - 299.

Operational and treatment performance of an unique Reducing and Alkalinity Producing System (RAPS) for acidic leachate remediation in Lancashire, UK.

Adam Jarvis, Adrian England

IMC Consulting Engineers, PO Box 18, Common Road, Huthwaite, Sutton-in-Ashfield, Nottinghamshire, UK. NG17 2NS.

Abstract. A Reducing and Alkalinity Producing System (RAPS) has been constructed to treat the acidic tip leachate from a former coal mine at Deerplay, Lancashire, UK. The tip leachate has a mean pH of 3.5, and is strongly net-acidic (up to 380 mg/L as $CaCO_3$). In addition it contains elevated concentrations of iron ($\leq$ 60 mg/L) and aluminium ($\leq$ 30 mg/L). The flow-rate varies with weather conditions, but has been as high as 6.5 L/second.

The RAPS has been engineered such that the leachate flows down through the compost, but then upwards through the limestone (and then into an aerobic wetland). The system has been operating since September 2001. The net-acidic water is converted to a net-alkaline water during its passage through the RAPS. pH increases from 3.5 to approximately 7.0. Whilst initially high, the ammonia content of the RAPS effluent decreased to baseline levels over the course of 6 weeks.

By assessing the weathering rate of limestone, from the effluent water quality, it is calculated that the system should operate for 20 years without the need for media replacement. Other operational issues associated with the RAPS are discussed, and plans for future monitoring and research are outlined.

Introduction

The Deerplay mine water treatment scheme is one of over a dozen such projects now completed by the UK Coal Authority. The Coal Authority commissioned IMC Consulting Engineers to undertake the design and construction of the scheme. The treatment system is located approximately 400 m above sea level on a steep, south-facing slope. There were in fact two discharges to be addressed at the site:

1. A ferruginous, and net-alkaline discharge with a flow-rate of approximately 20 L/second, arising from deep coal mine workings and
2. An acidic, iron and aluminium-rich discharge of tip leachate, with a flow-rate in the range 1 – 3 L/second.

Originally the net-alkaline mine water discharged into the River Calder catchment, to the north, but the water is now intercepted by pumping, and the treated water is discharged to the River Irwell, which drains south. The treatment of the net-alkaline discharge comprises aeration, settlement, and tertiary treatment using an aerobic wetland. This system is discussed in detail by Barnes (2000).

The net-acidic tip leachate has always drained to the River Irwell, but is now mixed with the treated net-alkaline mine water (though not for the purposes of treatment), prior to discharge to the river. It is the treatment of this acidic water that is the focus of this paper.

The acidic waters arise from the spoil heap of the former mine at the site. The discharge emanates from a buried pipe, which is thought to be fractured at one or more points along its length, allowing the ingress of waters percolating through the pyritic tip material. The objective of treatment was to neutralise acidity and remove metal contaminants (predominantly iron and aluminium). However, the steep nature of the site precluded the use of a compost wetland for neutralisation. Anoxic limestone drains (ALDs) were inappropriate due to the elevated iron and aluminium concentrations (see below), and active chemical treatment was to be avoided, due to cost and the isolated situation of the treatment scheme. Kepler and McCleary (1994) first proposed the use of RAPS (albeit under a slightly different name – Successive Alkalinity and Reducing Systems) for exactly the type of situation encountered at Deerplay. Such a system was therefore designed, and this paper reports the operational and treatment performance of this unique system since its commissioning in September 2001. This is only the third RAPS to be built in the UK, the others being at Pelenna, south Wales, and Bowden Close, County Durham (see Younger et al., 2002 for details). However, the system at Deerplay is unique due to its configuration (see below).

Discharge water quality

Table 1 illustrates the quality of the tip leachate at Deerplay. All analyses were undertaken by a nationally approved laboratory. The validity of measurements such as pH and alkalinity, which may change during transit and storage, has been confirmed with on-site determinations.

The data illustrate that the discharge is strongly net-acidic, with mean pH and acidity concentration of 3.3 and 165 mg/L as $CaCO_3$ respectively. Mean total iron and aluminium concentrations are 21.5 mg/L and 13.6 mg/L respectively.

Table 1. Water quality of the Deerplay tip leachate discharge, for the period September 2001 to April 2002.

Determinand	Mean	Range	n^a
Flow-rate (L/s)	1.30	0.18 – 6.45	30
pH	3.3	2.9 – 4.0	82
Conductivity ($\mu S/cm$)	1588	62 – 2280	39
Alkalinity (mg/L as $CaCO_3$)	0	0	82
Acidity (mg/L as $CaCO_3$)	165	0 – 384	82
Chloride (mg/L)	8	0 – 53	39
Sulphate (mg/L)	863	325 – 1360	39
Iron (total) (mg/L)	21.5	1.5 – 61.0	81
Iron (ferrous) (mg/L)	8.4	1.2 – 57.5	82
Aluminium (mg/L)	13.6	3.6 – 31.5	81
Manganese (mg/L)	4.25	2.02 – 6.70	81
Ammoniacal nitrogen (mg/L)	0.9	0.2 – 3.8	39

[a]total number of measurements made

System configuration

To date RAPS have been configured such that the compost overlies the limestone (e.g. Demchak et al., 2001; Kepler and McCleary, 1994). However, engineering constraints meant that this was not possible at Deerplay. Therefore at Deerplay water flows downwards through the compost, and then upwards through the limestone bed i.e. the compost bed and limestone bed are laid out side by side. This has the advantage of ensuring that there is no possible opportunity for short-circuiting through the limestone. Water exits the compost bed via dendritic network of pipes (overlain by a 300 mm layer of limestone to prevent blockage), and then flows up through the limestone bed. Effluent water level (and therefore water level over the compost) is controlled by an adjustable section of pipe at the exit of the limestone bed.

The key objective of a RAPS is to generate alkalinity, and thus neutralise acidity. For this to happen there must be sufficient alkalinity generation potential within the system, and sufficient residence time to allow the relevant chemical and microbiological reactions to occur. Hedin et al. (1994a) established that 14 hours was the optimum time for alkalinity generation within a limestone bed. Therefore the limestone bed for the Deerplay RAPS contains 220 m^3 of carboniferous limestone. On the basis that the limestone has a porosity of 50%, and at a design flow-rate of 2 L/s, the calculated residence time is therefore 15.3 hours.

In terms of compost content and configuration, design criteria are less specific. Demchak et al. (2001) suggest a minimum compost depth of 500 – 600 mm to in-duce the reducing conditions necessary to prevent subsequent armouring of the limestone by aluminium and ferric iron. Younger et al. (2002) reiterate this rec-ommendation. The compost at Deerplay is in fact approximately 2 m deep, to en-sure effective reduction. This may in fact prove to be a problem in the future, if permeability begins to decrease (as is likely), and therefore hydraulic head through the compost increases. However, removal of a layer of compost would be a simple enough task if this transpires to be the case.

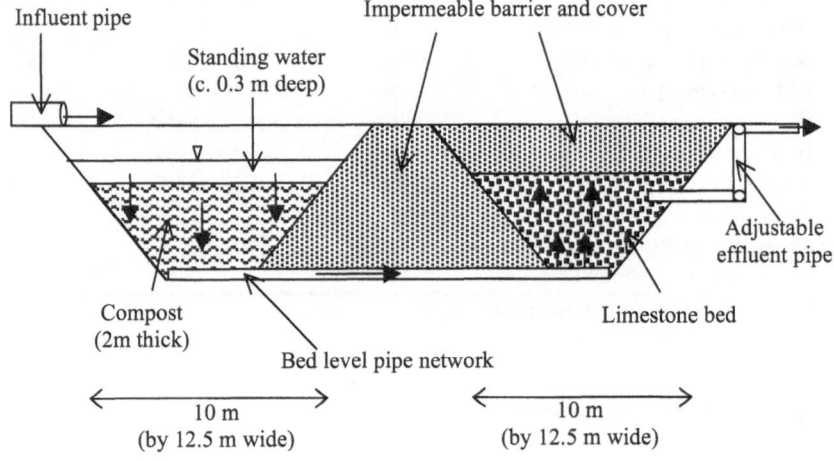

Fig. 1. Conceptual cross-section of the Deerplay RAPS (not to scale)

There is a depth of at least 300 mm of water overlying the compost, and a total freeboard of approximately 1 m. An overflow pipe is built into the compost bed in case water levels exceed the maximum water level. In addition sampling ports were installed at 3 depths during construction, in both the compost and limestone layers. The overall layout and dimensions of the system are illustrated in Fig. 1.

Treatment and operational performance

The Deerplay RAPS has been operating since September 2001. Samples have been collected from the influent and effluent channels on at least a weekly basis. For some key variables such as pH and acidity samples were collected on a daily basis for the first 3 months of operation.

Mean pH increases from 3.3 to 6.7 across the system. Alkalinity, which is con-sistently zero in the influent waters, is at least 100 mg/L as $CaCO_3$ in the effluent waters, as illustrated in Fig. 2.

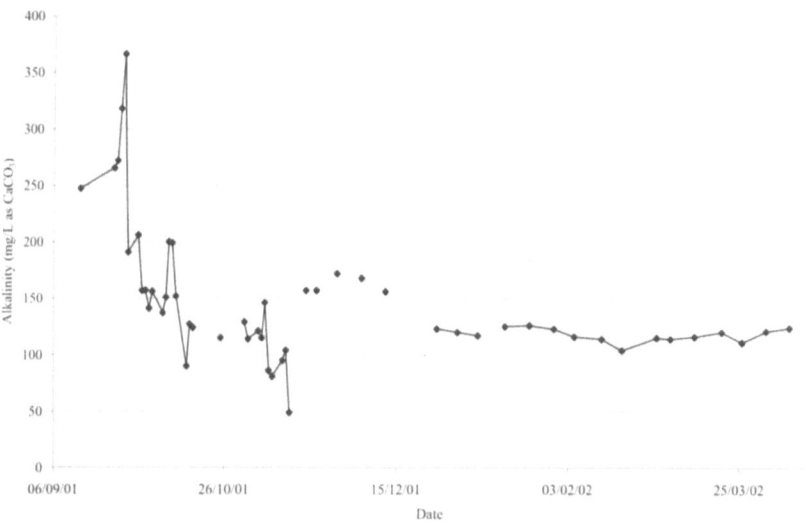

Fig. 2. Deerplay RAPS effluent alkalinity for the period September 2001 to April 2002 (Influent alkalinity is always zero)

Fig. 2. demonstrates that there is something of a 'honeymoon' period in terms of alkalinity generation, which lasts for approximately 1 month. This pattern has been observed in other passive treatment systems, and is probably due here to initially rapid dissolution of limestone dust. However, since the end of December 2001 alkalinity generation has levelled off, and is now consistently in the region of 120 mg/L as $CaCO_3$. Since the mean influent acidity is 165 mg/L as $CaCO_3$, the net generation of alkalinity is 285 mg/L as $CaCO_3$. This is at the top end of the range of alkalinity production considered capable by passive limestone systems (Younger et al., 2002).

Influent iron, aluminium and sulphate concentrations are highly variable, as indicated by the ranges of values shown in Table 1. Increases in concentrations of these variables are correlated with increases in acidity concentration, as shown in Fig. 3. This is the pattern that would be expected. However, these trends are not reflected in effluent concentrations i.e. high influent acidity concentrations are not necessarily reflected by high effluent concentrations of Fe, Al and SO_4. This suggests that the system effectively buffers the effects of changing contaminant concentrations. However, fluctuating flow-rates may have an influence in this regard. It is difficult to assess what influence such variation has at this stage, because influent and effluent measurements of flow-rate have only been sporadic.

Mean effluent iron (total) concentration is 15.5 mg/L (cf. influent concentration of 21.5 mg/L). Thus, over the operating period of the system there is not much reduction. A closer inspection of the data reveals that the RAPS goes through phases of being both a net sink for, and a net source of, iron. The reason for this is not clear at present.

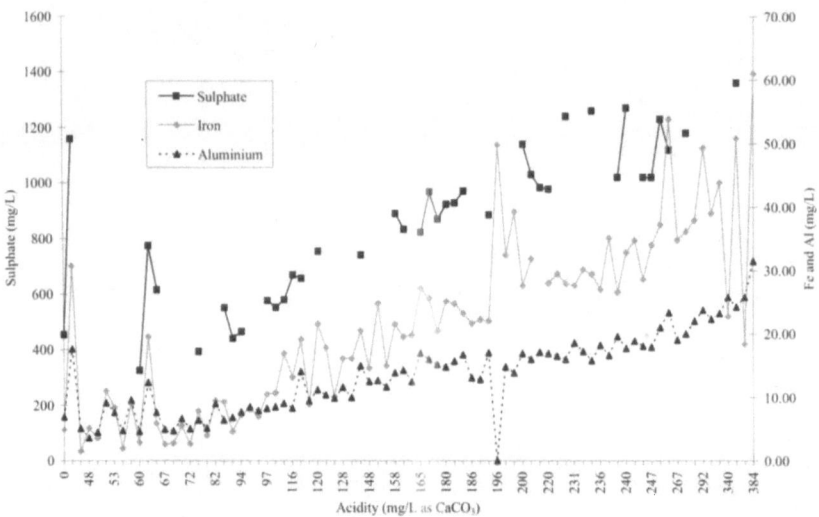

Fig. 3. Relationship between increasing acidity concentration and increases in concentration of Fe, Al and SO_4, for tip leachate at Deerplay.

In contrast aluminium is consistently removed by the RAPS. Mean effluent concentration is 0.51 mg/L (range 0.01 – 3.16 mg/L), significantly lower than influent concentration (see Table 1). It is not clear at this stage where the aluminium is removed. It seems most likely that it is removed as a hydroxide floc on the surface of the compost. However, it is also conceivable that it is armouring the limestone. This seems unlikely given the depth of compost available to induce reducing conditions, but will be a cause for concern if it proves to be the case. Newcastle University (UK) has recently begun collecting water from the sampling ports installed through the system. The results of this work should reveal the fate of metal contaminants in the RAPS.

Reductions in sulphate concentrations in compost based systems are indicative of the activity of Sulphate Reducing Bacteria (SRBs), which in turn is a sign that reducing conditions pertain in the compost. For the first 3 months of operation of the RAPS at Deerplay there was no indication that sulphate concentrations were decreasing. However, the absolute change in sulphate concentration would not need to be great to remove all of the iron as ferrous monosulphide, since sulphate concentrations are typically an order of magnitude higher than iron concentrations. Nevertheless, in recent months there has been a measurable decrease in sulphate concentration, perhaps suggesting that sulphate reduction is becoming a significant process. Again, continuing sampling will reveal whether this is the case.

Understandable concerns have often been expressed by regulatory agencies about the potential increase in organic matter arising from compost based systems. Measurements of NH_3-N at the Deerplay RAPS demonstrate that such increases are short-lived, confirming findings at other compost-based passive treatment systems (e.g. Jarvis, 2000). Fig. 4. illustrates this point, showing that NH₃-N levels returned to almost baseline levels within a month of commissioning.

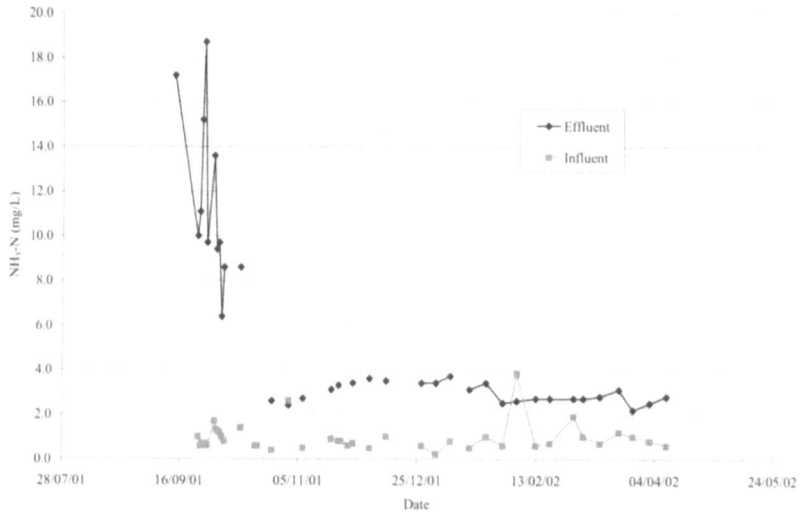

Fig. 4. Reductions in NH₃-N concentration in effluent from the Deerplay RAPS over time.

Longevity of treatment system

The details above illustrate that the RAPS at Deerplay is highly effective at generating alkalinity, as it was designed to do. The RAPS occupies a footprint of some 250 m². This is a fraction of the size of a compost wetland that would be required: using the design formula of Hedin et al. (1994b) an area of approximately 4000 m² would be needed.

The most significant issue relating to RAPS is perhaps the longevity of such systems, and in particular how quickly the compost and limestone media will become exhausted or clogged.

Younger et al. (2002) (pp. 104-112) outline a method for assessing weathering rates in mined systems that can equally be applied to the dissolution of limestone in a RAPS. Using these methods, and knowing the rate of alkalinity generation in the RAPS at Deerplay, it has been possible to calculate that the limestone (assumed to be 80% calcite) in the RAPS at Deerplay will be completely exhausted in 21 years. Of course the limestone is likely to require replacement prior to complete exhaustion, but this time period is nevertheless encouraging.

It is far more difficult to make such predictions about the longevity of the compost substrate, both in terms of the physical reduction in permeability that will inevitably occur, and the exhaustion of carbon substrates, essential for the survival of the SRB populations that create reducing conditions.

More intensive sampling of the system by Newcastle University will certainly help to answer some of these questions. In addition, IMC is an industrial partner in a recently awarded research programme led by Bangor and Newcastle Universities. This research project will investigate the very issues of changes in carbon and sulphur cycling in compost based systems, and will hopefully answer some of the key questions relating to the longevity of RAPS.

Acknowledgements

We would like to thank the UK Coal Authority, and in particular Keith Parker, for allowing us permission to publish these data. We would also like to thank our colleagues at IMC, especially Alistair Byfield, for their assistance with this paper. ECS Engineering services Ltd and TES Bretby undertook the sample collection and laboratory analysis respectively, and we are grateful for their assistance.

References

Barnes, T.M. (2000) Treatment of the gravity minewater discharge at Deerplay Mine, Burnley, UK. Prceedings of the 7th International Mine Water Association Congress, Ustron, Poland, 11-15 september 2000. pp. 344-351.

Demchak, J., Morrow, T., Skousen, J. (2001) Treatment of acid mine drainage by four vertical flow wetlands in Pennsylvania. Geochemistry: Exploration, Environment, Analysis 1:1

Hedin, R.S., Watzlaf, G.R., Nairn, R.W. (1994a) Passive treatment of acid mine drainage with limestone. Journal of Environmental Quality 23:1338-1345.

Hedin, R.S., Nairn, R.W., Kleinmann, R.L.P. (1994b) Passive treatment of polluted coal mine drainage. Bureau of Mines Information Circular 9389. United States Department of Interior, Washington DC. 35 pp.

Jarvis, A.P. (2000) Design, construction and performance of passive systems for the treatment of mine and spoil heap drainage. Unpublished PhD thesis, Department of Civil Engineering, University of Newcastle, UK, 231 pp.

Kepler, D.A. and McCleary, E.C. (1994) Successive Alkalinity Producing Systems (SAPS) for the treatment of acidic mine drainage. Proceedings of the International land reclamation and Mine drainage Conference and the 3rd International Conference on the Abatement of Acidic Drainage. (Pittsburgh, PA; April 1994). Volume 1: Mine Drainage, pp. 195-204.

Younger, P.L., Banwart, S.A., Hedin, R.S. (2002) Mine Water: Hydrology, Pollution, Remediation. Kluwer academic Publishers, Dordrecht. 442 pp.

Mine Water Recovery in the Coal Mining District of Aachen - Impacts and Measures to Control Potential Risks

K.-H. Heitfeld[1], M. Heitfeld[1], P. Rosner[1], H. Sahl[2] & K. Schetelig[1]

[1] Ingenieurbüro Heitfeld-Schetelig GmbH, Preusweg 74, D - 52074 Aachen

[2] EBV Aktiengesellschaft, Roermonder Straße 63, D - 52134 Herzogenrath

Abstract. After 800 years of mining activities, coal mining came to an end in the district of Aachen in 1992. Dewatering measures were terminated at the end of 1993/ beginning of 1994. Since then, mine water in the coal mining districts of Aachen and South-Limburg has been allowed to recover in a controlled, step-by-step-manner in order to ultimately achieve natural groundwater conditions. During mine water recovery potential impacts to the ground surface are anticipated and acted upon; it is assumed that within several years following completion of mine water recovery all major impacts of mining activities in the area will have ceased.

Introduction

The coal mining district of Aachen forms a transition zone between the carboniferous deposits of the Ruhr area to the east and the coal deposits in the Netherlands and Belgium to the west. For several hundred years transnational mining of coal deposits was conducted along the German/Dutch border. Coal mines located in the Dutch South-Limburg district were abandoned in 1975.

When the EBV Aktiengesellschaft shut down the coal mine Emil Mayrisch in Alsdorf (Rheinland) in 1992, coal mining activities in the district of Aachen ended after 800 years. This shut down also terminated the transnational mining tradition in the mining districts of Aachen and South-Limburg.

The coal district of Aachen and South-Limburg is located north of the city of Aachen and includes an area of approximately 400 km^2 (Fig. 1).

The small river Wurm runs from south to north through the central part of the coal mining district of Aachen and South-Limburg. In this area the coal deposits are covered only by a thin layer consisting of unconsolidated sediments of quaternary and tertiary age; coal deposits are exposed at ground surface in the Wurm valley.

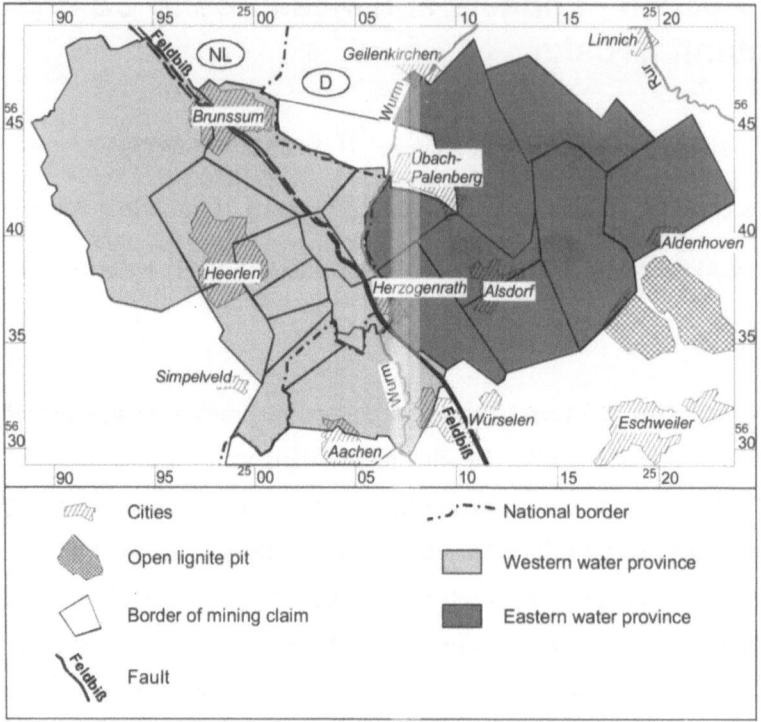

Fig. 1. Location plan of the mining fields of the coal mining district of Aachen and South-Limburg

In the Wurm area coal mining was abandoned in 1969. Following closure of the Dutch mines located to the northwest in 1975, coal mining activities continued only in the eastern part of the district of Aachen, east of the Feldbiß fault.

When the withdrawal of the technical equipment in the mine fields east of the Feldbiß fault was terminated at the beginning of 1994, the remaining mine dewatering works were shut down. The final draw-down cone resulting from dewatering activities covered an area of approximately 400 km².

Regional Geological Setting

The coal mining district of Aachen and South-Limburg is located in the transition zone between the Eifel and the Lower Rhine Basin. The carboniferous deposits are exposed along the Wurm valley north of Aachen; in most other areas these deposits are overlain by up to several hundred meters of unconsolidated deposits.

The coal bearing formations of the Aachen district generally consist of sandy, silty and clayey deposits of Namur C to Westfal B age. The carboniferous deposits run from SW to NE and dip generally at a shallow angle of less than 20 degrees to the south and southeast. At distances of several hundred meters to several kilometres anticlines, faults and up-throw faults occur. In addition to fold tectonics there are also NW-SE running cross-faults which subdivide deposits into large blocks. The main cross-faults - Feldbiß, Sandgewand and Diagonal-Sprung - form a step-fault system. The fault displacement is between 100 and 500 meters. In the Aachen area variscian folded bedrock is overlain by tertiary sediments. Towards the end of the Oligocene the sedimentation of lignite deposits started; the deposits are up to 250 m thick. The most recent deposits of tertiary age consist of 200 to 300 m of gravel/sand and clay. During the quaternary age the district of Aachen and South-Limburg was covered by terrace deposits of the Maas and Rhine rivers.

Hydrogeological Conditions

Groundwater Conditions in the Carboniferous Bedrock

During operation of the dewatering system east and west of the Feldbiß fault two distinct and hydraulically separate water provinces were established (Fig. 2).

In the eastern water province the 865 m-niveau (-735 m below sea level) was the lowest point of the central dewatering system. Mine water isolated in the Maria mine field discharged towards the central dewatering system of the eastern groundwater province via a connecting tunnel at the -440 m level.

The western water province consisted of the German Gouley-Laurweg mine field and the Dutch mine fields in the northwest and west. After shut-down of the central dewatering system, groundwater in the Dutch mine fields was pumped only from shaft Beerenbosch II, located in the southeast of the South-Limburg district. The groundwater level in the field Gouley-Laurweg was maintained at -167 m (below sea level) by pumping groundwater from the Von-Goerschen-shaft.

During operation of the coal mines a hydraulic connection existed between the mine fields Domaniale and Gouley-Laurweg; this allowed groundwater seepage from the German mine field into the Dutch mine field Domaniale.

Groundwater Conditions in the Overlying Sediments

In the mining district of Aachen and South-Limburg groundwater is extracted from the tertiary sediments in connection with dewatering measures for open pit lignite mining. The open pit Inden is located Southeast of the coal mining district of Aachen (Fig. 2); in the year 2000 a total of 92 million m^3 groundwater were extracted in order to keep the open pit dry. Since 1975, shallow groundwater conditions have also been impacted by dewatering measures in the open pit Hambach, located approximately 10 km further to the northeast. Groundwater extraction from the Hambach open pit amounted to approximately 330 million m³ in 2000. In addition, several drinking water production facilities extract groundwater from various aquifers in the overlying sediments.

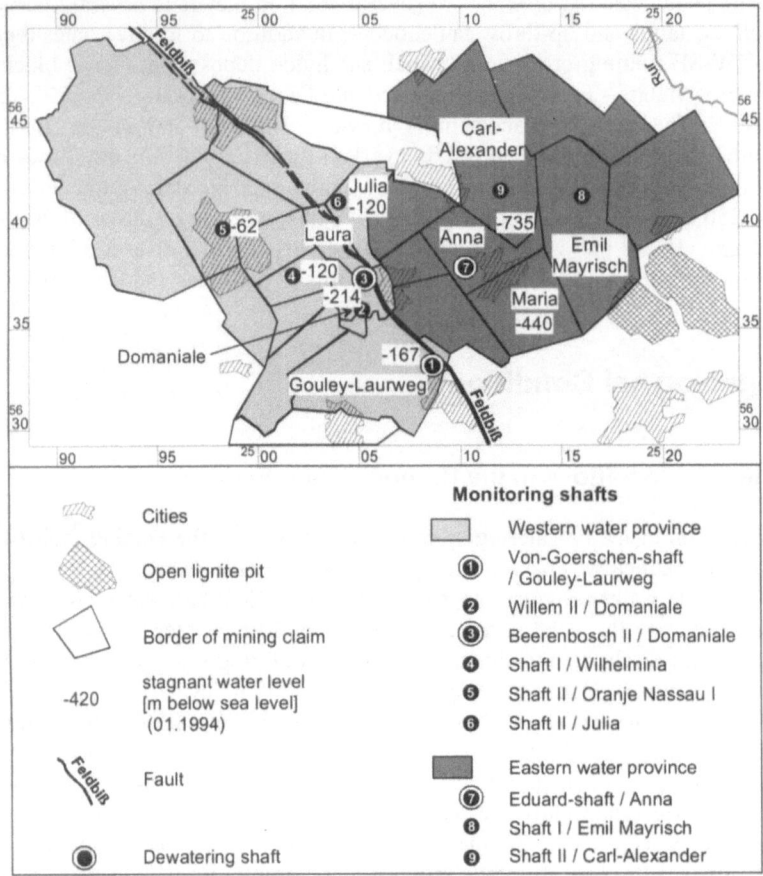

Fig. 2. Stagnant water levels in the coal mining district of Aachen and South-Limburg

Risk Assessment of Mine Water Recovery and Conceptual Design of a Controlled Step-by-Step Recovery

In 1990, EBV Aktiengesellschaft submitted a final operations plan to the mining authority in Aachen; the plan included the recovery of mine water in the coal district of Aachen and South-Limburg to the natural groundwater level.

Problems associated with the recovery scheme consisted mainly of the water level difference of approximately 550 m in the coal mines east and west of the Feldbiß fault. If water levels at adjacent mining structures (distances of only 18 m in some places) significantly differ, the remaining rock might suddenly collapse and inflowing water might gush into the eastern mine field. In particular, the sudden decrease of the water level west of the Feldbiß fault might cause the destruction of shaft fillings in this area. Detailed rock-mechanical and hydraulic investigations showed that a controlled recovery of mine water levels would inhibit extensive groundwater inflow as well as preventing any risk to the ground surface.

In addition, the impact of the groundwater recovery process on potential risks caused by former mining shafts had to be assessed; former methods used to secure abandoned mine shafts do not meet present-day safety standards. As approved by the mining authority, the dewatering system was shut down in the shafts west of the Feldbiß fault in January of 1994. The dewatering shaft located in the Wurm valley was kept operational in order to be able to actively control and regulate the mine water recovery process. At the same time the dewatering system for the eastern mine field was shut down at the end of 1993; any existing deep shafts in this area were secured according to the guidelines to ensure long-term stability.

The EBV Aktiengesellschaft conducted an intensive measuring program in order to monitor the mine water recovery process, including measurements of mine water levels and groundwater levels in monitoring wells installed in the overlying sediments, hydrochemical investigations and geodetic measurements. Pump tests of several months duration are conducted regularly in the former dewatering shafts in order to further assess hydraulic connections between different mine fields and to balance water inflow. Test results are particularly important as a basis for hydraulic calculations with respect to predicting future mine water recovery (further flooding of adjacent mine structures) and hydrochemical conditions of mine water at the time of reaching water levels present in the receiving water bodies.

Results of the Mine Water Recovery to Date

Water Provinces West of the Feldbiß Fault

Results of mine water recovery in the shafts west of the Feldbiß fault are summarized in fig. 3. During the initial phase since 1994 those mine fields in the south where the dewatering levels had been lowest (Gouley-Laurweg and Domaniale) were flooded first. During this period, water levels initially recovered at a rate of a maximum of 1.3 m/d and subsequently reduced to a rate of 0.3 m/d. After app. 16 months the stagnant mine water level in the northern Dutch district was reached, resulting in an overall uniform water level in all of the western province of -62 m (below sea level). Above -62 m, mine water in the Dutch mine fields rose considerably slower at a rate of 0.01 to 0.03 m/d due to the larger area which had to be flooded. During this phase of the mine water recovery a relatively high mine water level was established in the German mine field Gouley-Laurweg.

Mine water recovery in the western water province was regulated in such a way that the hydraulic gradient is directed from south to north. As a result, highly mineralised mine water from the north was pushed back.

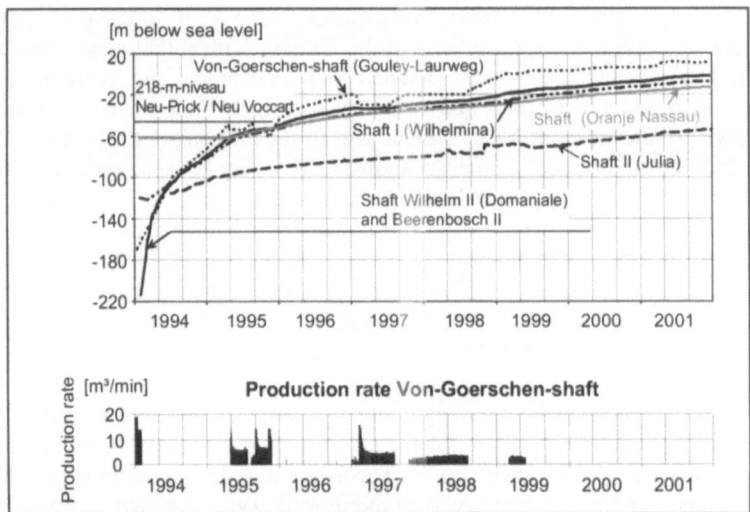

Fig. 3. Mine water recovery in the coal district of Aachen and South-Limburg west of the Feldbiß fault (monitoring period 1994 to 2001)

Mine water recovery is also monitored by regular hydrochemical investigations. Results of hydrochemical investigations conducted in the shafts west of the Feldbiß fault are summarized in fig. 4.

Mine water present in the mine fields Gouley-Laurweg and Domaniale are of the sodium-hydrogencarbonate-sulfate-type. The relatively high sodium content (> 500 mg/l) indicates the strong impact of highly mineralised deep water. The high hydrogen-carbonate-content (> 1,000 mg/l) also indicates the impact of infiltration groundwater originating in the overlying sediments.

During operation of the dewatering system the mineral content of the mine water in the Von-Goeschen-shaft was relatively low, with electric conductivity values of app. 1,500 to 2,000 µS/cm. In contrast, the electric conductivity of mine water in the northern Dutch mine field Domaniale (as determined in the shaft Beerenbosch II) was app. 9,000 µS/cm with iron concentrations of 10 to 24 mg/l.

During the initial mine water recovery phase in the mine field Gouley-Laurweg the total mineral contents slowly increased continuously to app. 2,800 µS/cm (12.1996).

Generally, groundwater chemistry in the mine fields Gouley-Laurweg and Domaniale gradually adjust. This is basically a consequence of highly mineralised deep groundwater and mine water from the Dutch mine fields being pushed back due to an overall increase of mine water levels. Further recovery of mine water levels will lead to further reduction of total mineral contents.

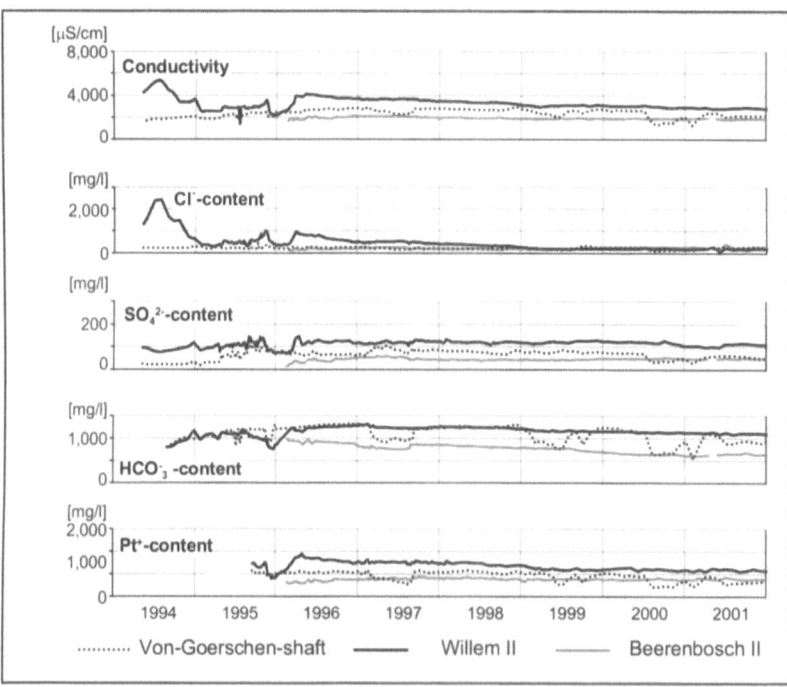

Fig. 4. Mine water chemistry in the coal district of Aachen and South-Limburg west of the Feldbiß fault (monitoring period 1994 to 2001)

Water Provinces East of the Feldbiß Fault

After the lower parts of the mines east of the Feldbiß fault were flooded the rate of mine water recovery was very uniform (Fig. 5). The groundwater level recovered at a rate of 0.20 m/d up to the level of accumulated mine water in mine field Maria of approximately -440 m (below sea level). Only when flooding the major mining floors the rate of mine water recovery decreased significantly. When exceeding the level of accumulated mine water in mine field Maria the mine water recovery rate in all of the eastern water province decreased significantly due to the larger volumes of mine cavities to be flooded. The average mine water recovery rate was 0.1 m/d.

   Mine water in shaft 1, Emil Mayrisch is of the sodium-chloride-type, and is supplemented primarily by inflow of deep groundwater. Low concentrations of hydrogen-carbonate, sulphate and alkaline earths indicate that there is little impact of infiltrating surface water.

   Electric conductivity values in shaft 1, Emil Mayrisch remained relatively constant at between 6,060 and 6,815 µS/cm until March, 1996, when mine water recovery levels reached -710 m (below sea level).

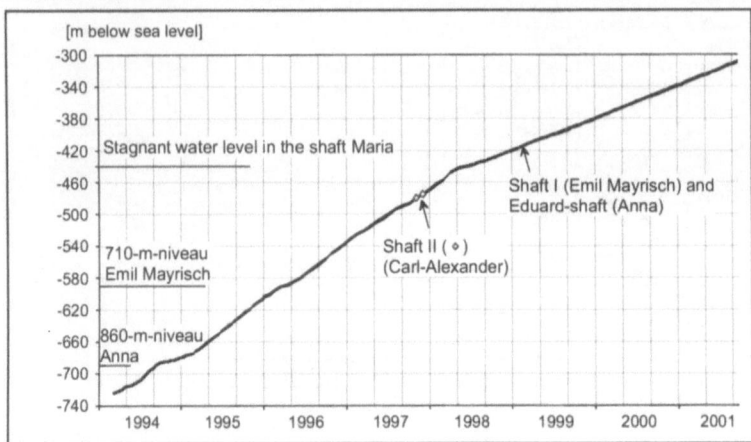

Fig. 5. Mine water recovery in the coal district of Aachen east of the Feldbiß fault (period 1994 to 2001)

Above the -710 m (below sea level), electric conductivity levels increased slightly to between 7,500 and 8,000 µS/cm (see Fig. 6). This indicated a considerable decrease in electric conductivity values compared to those measured during the operation of the mine dewatering system, which is due to the fact that highly mineralised groundwater present along deep fault zones is forced back.

Contrary to mine water present in shaft I, Emil-Mayrisch, groundwater accumulated in the Eduard-shaft, Anna, is of the sodium-chloride-hydrogen-carbonate-type. Significant hydrogen-carbonate concentrations (approximately 700 to 800 mg/l) indicate a considerable impact by infiltrating surface water. The electric conductivity of 2,800 µS/cm in the mine water present of the Eduard-shaft is considerably lower than that of the mine water at Emil Mayrisch.

Impacts on Groundwater in the Overlying Sediments

Monitoring of groundwater conditions in the overlying sediments started rather early in connection with the mine water recovery process. This data is available for the assessment of potential impacts of mine water recovery on overlying sediments groundwater quality. Based on a comparison between 1998 groundwater monitoring results and those obtained during long term monitoring it can be concluded that the mine water level recovery east of the Feldbiß fault does not impact groundwater conditions in the overlying sediments.

Impacts of Mine Water Recovery on Ground Surface Conditions

Flooding of Unconsolidated Fills in Old Mine Shafts

There are app. 850 former shafts in the coal mining district of Aachen; 90 % of these are old mine shafts resulting from ground owner operated mining activities which were sunk prior to the beginning of the 19th century. As part of the controlled mine water recovery available data on former mine shafts was reviewed and assessed in terms of the potential risk posed by each individual shaft. Criteria for risk assessment include location and size of shaft, kind of fills, preciseness of location, depth of shaft, surrounding land use, and thickness and type of overlying sediments. The flooding of unconsolidated fills of old mine shafts during mine water recovery poses another potential risk. During flooding the fill column is subject to buoyancy and cohesion is reduced. Buoyant material might collapse into open cavities and result in collapsing of the mine shaft. Predictions of projected mine water recovery levels and associated potential impacts on the overlying sediments are essential for an overall assessment. These documentations and evaluations have been compiled for the coal mining district of Aachen; they will have to be adjusted accordingly to incorporate future monitoring results.

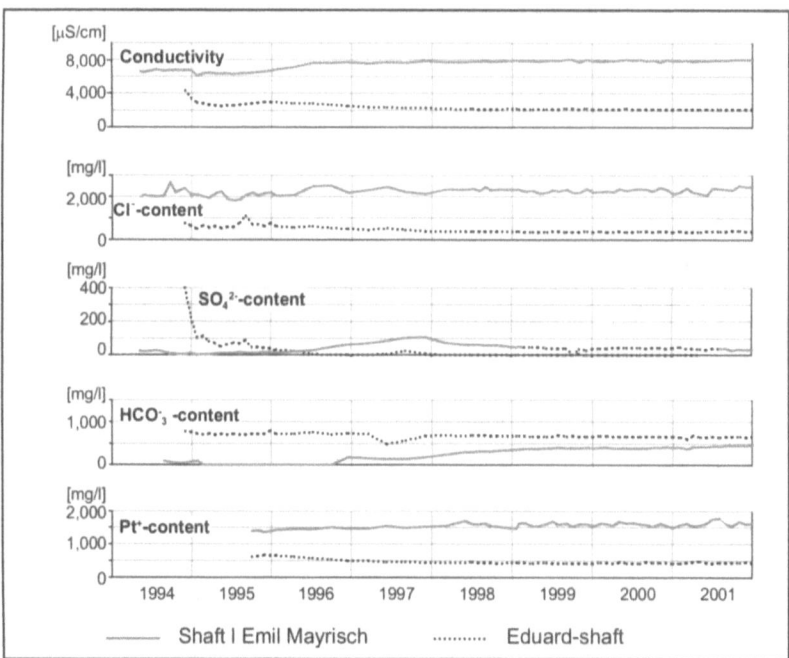

Fig. 6. Mine water chemistry in the coal district of Aachen east of the Feldbiß fault (monitoring period 1994 to 2001)

Terrain Heaving

To date there is only little information available on the impacts of mine water recovery in mining areas on ground surface conditions.

PÖTTGENS (1990) evaluates conditions in the coal mining district of South-Limburg. Geodetic monitoring results obtained between 1975 and 1985 in connection with mine water level recovery were evaluated. According to PÖTTGENS, the amount of terrain heaving is estimated to amount to approximately 2 to 3 % of the ground subsidence caused by mining activities.

In agreement with the responsible mining authority a geodetic monitoring program was developed to accompany the mine water level recovery. The objective of this monitoring program was to document the spatial distribution of terrain heaving as basis for the assessment of potential future damage at the ground surface level. Monitoring data to date has been compiled in iso-maps showing lines of corresponding amounts of terrain movement. Data evaluation indicates that movements of app. 10 mm/a have occurred in the area west of the Feldbiß fault since shut-down of the dewatering system. Terrain heaving generally occurred in large, extended areas; differences in heaving rates are anticipated only along tectonic fault lines or the edges of mine fields.

In the area east of the Feldbiß fault the groundwater level was lowered by a total of about 800 m as part of the overall dewatering scheme. It is anticipated that during the mine water level recovery terrain heaving of a total of 0.25 m will occur.

Outlook

Experience gained in the coal mining district of Aachen to date following shut-down of the mine dewatering systems indicate that potential risks associated with the recovery of mine water levels to the original ground water levels can be controlled even in very old former mining areas. The controlled mine water recovery process has to be managed based on a detailed analysis of mining related and hydrogeological conditions.

Literature

BERGAMT DÜREN (29.06.1993): Zulassung zum Abschlußbetriebsplan der EBV Aktiengesellschaft vom 22.10.1990.- 10 S.; Düren.

INGENIEURBÜRO HEITFELD-SCHETELIG GMBH (Januar 1997): Gesamtbericht über die Ergebnisse der Untersuchungen im Zusammenhang mit der Einstellung der Wasserhaltung durch die Eschweiler Bergwerks-Verein AG.- 190 S.; Aachen.

PÖTTGENS, J.J.E. (1990): Bodenhebungen durch ansteigendes Grubenwasser.- Unveröffentlichter Vortrag; 11 S., 7 Abb.; Heerlen.

Flooding and Post-flooding Scenarios – Dynamics and Geochemistry

Harald Kalka, Horst Märten, Rudolf Münze

UIT GmbH Dresden, Zum Windkanal 21, D-01109 Dresden, Germany

Abstract. The geochemical and geohydraulic modeling of flooding processes is the scientific background for forecast and optimization of remediation. Basic principles are deduced from the dynamical description of flooding and post-flooding scenarios in different types of uranium mines. The time evolution is treated as a transition from dominant *oxidation* processes during flooding to long-term *reduction* processes influenced by reactive materials (organics, scrap iron). The spatial-time structure of the model depends on local conditions. The constructed models include advection, kinetics as well as equilibrium processes.

Box model (compartment model)

For the description of flooding processes in mining sites there is *no* standard software available. The well-known code families for reactive transport (Domenico and Schwartz 1998) are no adequate tools to describe the processes within a network of mine cavities because of the complicated – and in most cases unknown – combinations of flows through cavities, porous media, and/or fracture networks. Therefore a new approach for an *efficient* simulation of flooding dynamics has been developed and applied in several cases.

This paper reports some basic ideas and experiences obtained from modeling of flooding scenarios in uranium mines at different locations.

Main principles

To solve the problem numerically with adequate resolution, the mine site is decomposed in several boxes or compartments (up to about 10^3 boxes). The boxes are coupled by hydraulic flows (*internal* couplings to other boxes and *external* coupling to the outside). The structure of the box-system depends on the local conditions (cf. application cases in Figs. 4 - 6).

The main principles of the flooding models can be summarized as follows:
- Spatial-time structure: box model (synonym: compartment model)
- Dynamical description with high time resolution: $\Delta t \approx 1$ h
- Strict mass and charge conservation (local and global)
- Modular system including PHREEQC (Parkhurst and Appelo 1999) as a "subroutine"
- Combination of advection, kinetics and chemical equilibrium.

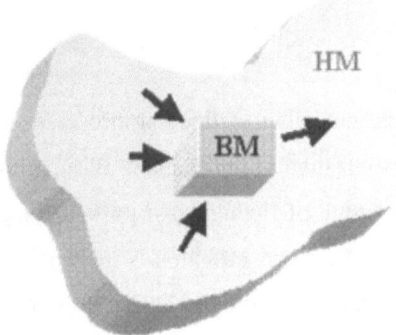

The box model (BM) represents a 3D distribution of cavity volumes in the mine. As shown in Fig.1, the BM is embedded in a large area hydro-geological model (HM) for ground-water flow simulations. The HM generates the time-dependent external flows as input for the box model.

Fig. 1. Box model (BM) of the mine embedded in a large-area hydrogeological model (HM).

An advantage of the box model is that any box can be equipped with (i) proper mechanisms of mass transformation (geochemistry) and (ii) technological devices like pumps, valves, pipelines, and reactive materials.

Basic equations

The model is based on a "transport equation" which describes the mass change within one box i due to advection and mass transformations (sink/source term):

$$\frac{dm_i}{dt} = \left(\frac{dm_i}{dt}\right)_{adv} + \left(\frac{dm_i}{dt}\right)_{src} \tag{1}$$

This equation holds for any chemical element or species with mass $m = cV$ (c – concentration, V – water volume) where $c(t)$ and $V(t)$ are time-dependent quantities. The advection term describes the hydraulic processes (without reactions) and is given by

$$\left(\frac{dm_i}{dt}\right)_{adv} = \sum_j \left(Q_{j\to i}c_j - Q_{i\to j}c_i\right) + Q_i^{in}c^{in} - Q_i^{out}c_i \tag{2}$$

where $Q_{i\to j}(t)$ denotes the flow rate (dV/dt) from box i to box j. $Q_i^{in}(t)$ and $Q_i^{out}(t)$ represent the external flow *from* and *to* the outside region, respectively. Whereas $Q_{i\to j}(t)$ are calculated from the hydraulic conditions at time t (using Darcy's Law), the quantities $Q_i^{in}(t)$ and $Q_i^{out}(t)$ are input values (obtained from the hydrogeological model).

The source term in Eq.(1) summarizes all mass transformations between solid and aqueous phases discussed below in more detail.

Linkage between geohydraulics and geochemistry

During the flooding process, the hydraulic conditions (flow pattern, water volume in each box, water level which bisects the mine into a saturated and an unsaturated zone) changes from time step to time step. In this way, the geochemistry which do strongly rely on the hydraulic conditions becomes also time-dependent.

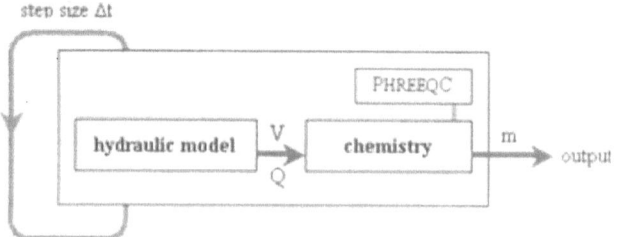

Fig. 2. Two-step-algorithm for geohydraulics and geochemistry.

From a numerical point of view, the flooding dynamics is simulated within a *two-step algorithm*: In each time step of size Δt (≈ 1 h) the hydraulic model calculates the internal flows $Q(t)$ and the water volume $V(t)$, which are the input quantities for the geochemical model (see Fig.2).

Geochemistry

The model predicts the composition $c = c(t)$ of mine water (10 - 20 elements, 100 - 200 species) and its changes in time.

Main processes

Using the geochemical code PHREEQC as a "subroutine" the flooding model includes the following standard processes:
- Mixing of water flows
- Speciation and complexation
- Precipitation and co-precipitation / dissolution
- Oxidation-reduction reactions

Additionally, in-situ water treatment (neutralization, incorporation of reactive materials) can be simulated in a given box, i.e. at any position in the mining site.

To calculate the chemical transformations between different phases (pore fluids, secondary phases, mine water and underground water) PHREEQC is called in the order of 10^6 times during a program run. The software codes have been optimized concerning computing time and accuracy (mass balance uncertainty < 1 %).

From oxidation to reduction processes

As shown in Fig.3 flooding is treated as a transition stage were *oxidation* processes gradually cease, and concurrently in the flooded mine, *reduction* processes influenced by reactive materials (wood, scrap iron etc.) become important.

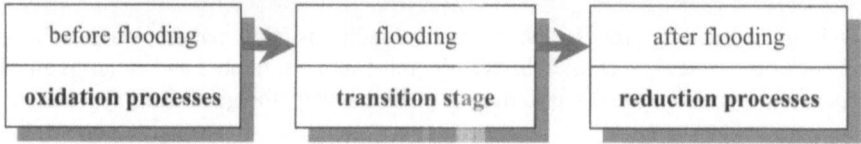

before flooding	flooding	after flooding
oxidation processes	transition stage	reduction processes

Fig. 3. Flooding as a transition stage between oxidation and reduction processes.

In this way, for large times ($t \rightarrow \infty$) the mine water in the upper levels of the flooded mine (near the land surface) does asymptotically approach a composition close to the water composition before mining.

In the model, the "switch-off" of the oxidation processes and the "switch-on" of the reduction processes automatically occur when the water level increases and the boxes are filled during flooding:

- Box empty – only oxidation processes (dissolution of primary minerals)
- Box partly filled – dissolution of secondary minerals
- Box full – reduction processes and dissolution of backfill materials; precipitation of hydroxides, carbonates and sulfides

Each of these processes is described by an individual source term in Eq.(1).

Source terms – the origin of mass and contaminants

To include chemical reactions and kinetic processes the source terms in Eq.(1) should be specified. Typical terms are given by (abbreviated notation):

$$\left(\frac{dm_i}{dt} \right)_{src} = rF_i \qquad \text{and} \quad r = r_0 f(pH) \tag{3}$$

$$\left(\frac{dm_i}{dt} \right)_{src} = m_i^{sekm} \, \delta(t - t_i) \tag{4}$$

$$\left(\frac{dm_i}{dt} \right)_{src} = Q_i^{seep} c_i^{seep} \tag{5}$$

Eq.(3) describes the dissolution kinetics of a reactive material or primary mineral (r – specific reaction rate, F – reactive surface in box i). Here the reaction rate depends on pH value. Eq.(4) is used for the dissolution of secondary minerals at time t_i when box i is flooded (m_i^{sekm} – total mass of a secondary mineral in box i).

An alternative description of oxidation processes (pyrite oxidation etc.) is given by Eq.(5) where mass is "produced" by the seepage water inflow rate Q_i^{seep} entering box i with the concentration c_i^{seep}. Here, the seepage water composition c_i^{seep} can either be calculated within a so-called geochemical infiltration model (Kalka et al. 1998) or directly obtained from monitoring data sampled *before* flooding.

The advantage of the "effective kinetics" in Eq.(5) is that the seepage water (taken from monitoring data) already contains all information about the pyrite oxidation and other related reactions inside the rock matrix (dissolution of heavy metals and radionuclides, neutralization by calcite and dolomite minerals). Thus, using the composition of the seepage water as input, there is no data-need for the specific rates r and reactive surface F (which are unknown in most cases).

In summary, Eq.(3) can be used for the corrosion of backfill materials as well as for the dissolution of reactive materials like wood and scrap iron. The latter are responsible for the reduction processes under *saturated* conditions. On the other hand, Eq. (5) is a good candidate for an efficient description of the "pyrite oxidation" and all relating processes in the *unsaturated* zone.

Post-flooding processes

When flooding is completed the mine system is still in a physical and chemical non-equilibrium.

The origin of the *physical* instability is as follows: The mass production (due to oxidation processes and dissolution of secondary minerals) depends on the cavity volumes (and on the former mine activities) which are locally distributed. This pattern causes a layer structure in the flooding water where water of high density (salinity) might occur above low-density water. Hence, the instable state induces a density-driven *vertical convection*, which has been included in the box model. Calculations for the Ronneburg mine have shown that in view of these "density forces" effects arising from the geothermal gradient are negligible.

Due-to the presence of organic matter and scrap iron the flooding water is in a *chemical* non-equilibrium. To achieve equilibrium reduction processes take place (sulfate reduction). The description of such processes is difficult and lies beyond the "state of the art". Nevertheless first attempts are made by using kinetics of zero order in form of Eq.(3) with the following rates:

- Dissolution of "CH_2O" $r_0 = 1.1 \cdot 10^{-10}$ mol/m^2/s
- Dissolution of metallic Fe $r_0 = 3.6 \cdot 10^{-7}$ mol/m^2/s
- Backfill corrosion (CaO) $r_0 = 2.0 \cdot 10^{-6}$ mol/m^2/s

Each rate is strongly pH-dependent, $r = r0 \cdot f(pH)$, where in the case of organic matter $f(pH)$ has its maximum at pH = 7. Here, the backfill corrosion is included because of its influence on pH value of the flood water. In the calculations, quasi-equilibrium is assumed at each time step t for calling PhreeqC. As time elapses the sulfate reduction leads to the precipitation of the U(IV)-phase UO2(a) and sulfides (FeS, Covellite, Millerite, Orpiment, ZnS(a) etc.). One open question is the "incubation period" for the microbiological processes. This parameter can only be deduced from observations in mines flooded a long time ago.

Examples for model application

The application of the box model to flooding simuluations requires an adjustment to the local conditions at each uranium mine in two respects:

(i)   Spatial discretization of model space in relation to geohydraulics
(ii)   Specification of the inner box structure in relation to geochemistry

Ronneburg uranium mine complex (Thuringia)

The Ronneburg mining site is the most extensive *uranium mine complex* of WISMUT. Between 1952 and 1990, approximately 125 Mio. m³ uranium ore were recovered from underground and open-pit mines. Within the 164 square kilometer area (63 mi.²) of the Ronneburg Revier, 65 km² were developed for mining purposes. 113 000 t U were recovered in total. 2 926 km of mine works were built underground. The open pit mine had a depth of 230 m. Roughly 60 % of the site has been contaminated by mining activities. This site is unique worldwide, particularly due to its location in a densely populated area.

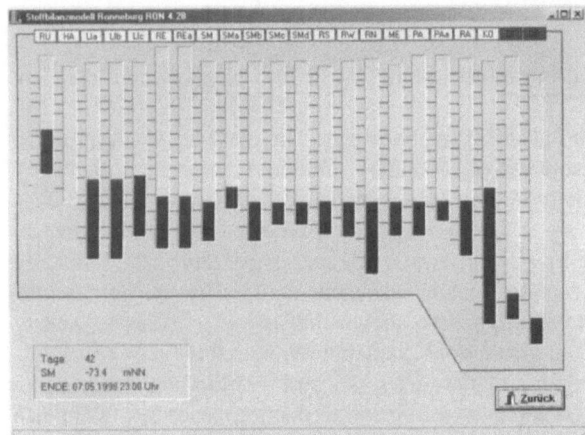
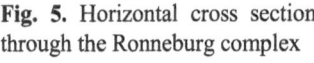

Fig. 4. Vertical cross section (online-graphic of the Ronneburg model)

Model space: The model considers the various types of dewatered cavity volumes for each mine level and each mining field (= 22 × 22 boxes) as well as the hydrogeological connections (vertical and horizontal) between the boxes. Devices like pumps and pipelines are also considered.

Fig. 5. Horizontal cross section through the Ronneburg complex

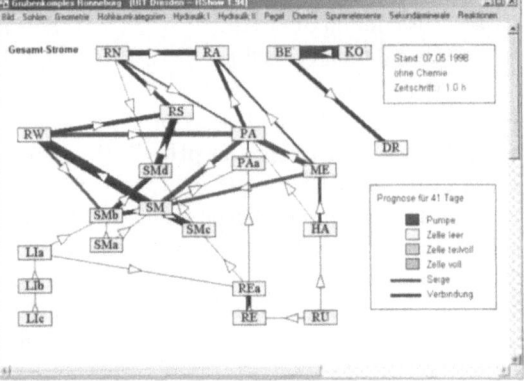

Königstein in-situ leaching mine (Saxony)

The sedimentary uranium deposit Königstein is located about 30 km SE of Dresden. Mining started in 1967. By 1984 leaching had replaced conventional mining entirely. Since 1991 the remediation of the under-ground mine works had been prepared. Mine flooding started in January 2001.

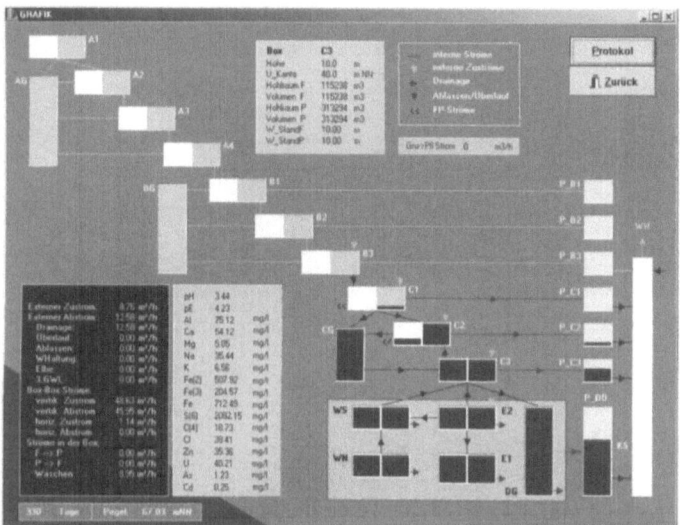

Fig. 6. Online-graphic of the Königstein model software

A "snapshot" of the flooding process is shown in Fig. 6. The model space consists of 28 boxes. The special feature of the Königstein mine is the following: Due to the in-situ leaching technology most of the contaminants are stored in the pore water of the sandstone. Therefore, each box had to be decomposed into two subspaces: (i) "pore volume" P (containing the highly contaminant pore water) and (ii) "cavity volume" F (with connections to other boxes). Both subspaces are coupled by a density driven mass exchange to simulate the washing-out process in the sandstone. Furthermore, "re-flooding"-scenarios have also been considered, i.e. intermediate decline of water level.

Schwartzwalder mine (Colorado)

From 1953 to the end of the nineties (with a standby from 1989 to 1995) about $1.9 \cdot 10^6$ tons ore with an average uranium grade of 0,47 % had been recovered from Schwartzwalder mine located near Golden, Colorado. The mine is a vein-type uranium deposit (the largest known in U.S.) located in Precambrian metamorphic rock of amphibolite grade. It has been developed from the steep canyon of Ralston Creek down to a level in 2 100 ft depth.

The box model similar to the Ronneburg application case was used to simulate different flooding scenarios in connection with proposed technical reclamation measures (bulkheads, hydraulic connections). Based on the calculations an optimum reclamation concept has been developed. As shown in Fig.7, the flooding processes in the mine were combined with the total water balance at the site including the optional water treatment. Because of the spatial structure of the mine works a 2-column/ 21-level model space had been chosen, i.e. 42 boxes.

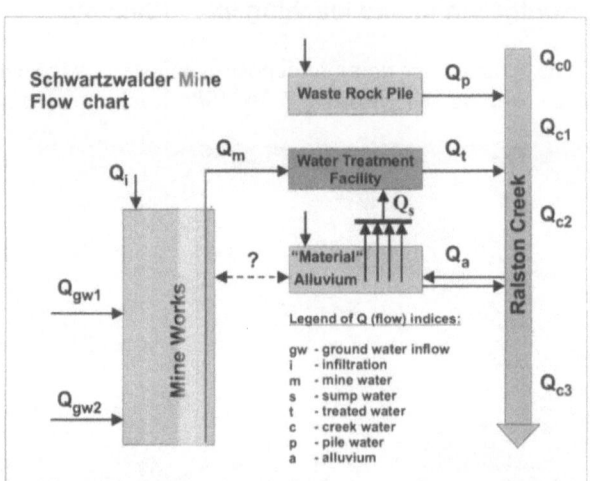

Fig. 7. Flow chart for combining flooding dynamics with total water balance

Conclusions

The dynamic box model combining the hydrogeological and geochemical description in a consistent manner represents a novel approach for the simulation of flooding scenarios. Various remediation measures can be considered explicitly.

A model complexity up to box numbers in the order of 10^3 are feasible, provided that reliable input data for the characterization of mine structure, hydrogeology and geochemistry are available. In general, the model has to be carefully calibrated on the basis of existing monitoring data, thus, reproducing the previous and present water balance and water quality data.

The dynamic 3D-compartment model is a powerful tool for the forecast of water quality in the flooded mine regions. It is the basis for:

- Evaluation of water quality development during mine flooding
- Development of water management strategies (in situ / ex situ)
- Permitting procedures governed by authorities

Acknowledgment. The excellent long-term cooperation with WISMUT experts is gratefully acknowledged.

References

Domenico P A, Schwartz F W (1998) Physical and Chemical Hydrogeology. John Wiley & Sons, New York

Kalka H, Märten H, Hagen J, Münze R (1998) Proc. Int. Conf. On Uranium Mining and Hydrogeology II, Freiberg, Germany, September 1998, Sven von Loga, Köln: 470-479

Parkhurst D.L., Appelo C.A.J. (1999) User's guide to PHREEQC (version 2). Water-Resources Investigation Report 99-4259, Denver Colorado

Groundwater contamination in the area of Zn-Pb ore mines situated in the southern part of the Olkusz-Zawiercie Triassic aquifer (Poland)

Jacek Motyka[1], Andrzej J. Witkowski[2]

[1] Faculty of Mining, University of Mining and Metallurgy, Mickiewicza Av. 30, 30-059 Cracow, Poland, e-mail: motyka@uci.agh.edu.pl
[2] Department of Hydrogeology and Engineering Geology, University of Silesia, Bedzinska Str. 60, 41-200 Sosnowiec, Poland, e-mail: awitkows@us.edu.pl

Abstract. In areas of Zn-Pb ores mines significant contamination of groundwater have been noticed. The analysis of the spatial distribution of such pollution indicators as SO_4, Cl, NO_3, Fe and Zn indicates different exogenic and endogenic origin. Processes of oxidation of sulphide minerals taking place in the rocks are the most significant factors influencing concentrations of SO_4, Fe and Zn in mine water in Olkusz area. The inflow of pollution from the surface of the ground is an additional factor considerably affecting the chemical composition of groundwater (concentration of NO_3 and Cl and locally SO_4 and Fe).

Introduction

The Olkusz-Zawiercie karst-fissured Triassic aquifer (Fig.1) is one of the major sources of potable water for the Upper Silesian agglomeration. Unfortunately this generally very vulnerable aquifer is subjected to high negative human impact related to mining, industry, urbanisation and agriculture. Particularly bad situation is observed in southern part of this aquifer in area of Zn-Pb ore mines where a progressive deterioration of groundwater quality has been noticed. Before intensive mining activity TDS of groundwater in carbonate Triassic rocks in Olkusz region varied between 250 and 500 mg/dm^3 and average concentrations of SO_4 was 37 mg/dm^3, Zn - 0,52 mg/dm^3, Pb - 0,06 mg/dm^3 (Adamczyk and Wilk 1976). Unfortunately currently groundwater quality has been significantly changed in this area. High concentrations of SO_4 (over 5 000 mg/dm^3), Cl (locally up to 600 mg/dm^3), Fe (up to 400 mg/dm^3), Zn (up to 50 mg/dm^3) and also NO_3 (up to 60 mg/dm^3) in areas of Zn-Pb ores mines have been noticed. Based on the analysis of the spatial distribution of such pollution indicators as SO_4, Cl, NO_3, Fe and Zn major sources of groundwater contamination have been defined in this paper.

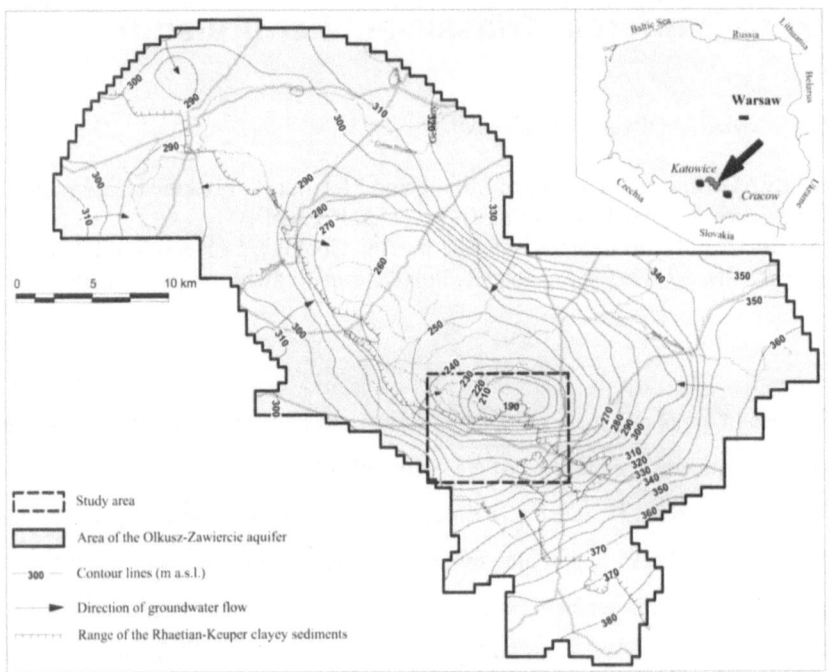

Fig. 1. Hydrogeological sketch of the Olkusz-Zawiercie Triassic aquifer (according to Witkowski at al. 2001)

Groundwater system

The Olkusz region belongs to the Silesian-Cracow Monocline built up of the Triassic and Jurassic formations discordantly overlying folded and faulted Palaeozoic basement. There are Quaternary, Jurassic, Triassic and Palaeozoic aquifers in the hydrogeological profile of the Olkusz area.

Most important and resourceful is the Triassic karst-fissured carbonate aquifer (dolomites and limestones with marl interbeddings). Generally two water-bearing horizons can be differentiated within this aquifer: the Muschelkalk horizon and the Roethian one. These two horizons are often considered jointly as one aquifer (Różkowski ed. 1990). Considered Triassic aquifer is partly covered by clayey Rhaetian-Keuper sediments (Fig.1). Hydraulic structure of fissured and karstified Triassic rocks consists of three types of spaces: pores, fissures and caverns. Limestones represent fissured-cavernous type of the aquifer while dolomites represent porous-fissured-cavernous type (Motyka 1998). Fissures and karstic channels are favourable pathways of groundwater flow while the pore space is the main water reservoir. It resulted in a vertical and horizontal differentiation of rock permeability. The thickness of this aquifer ranges from a several to about 150m.

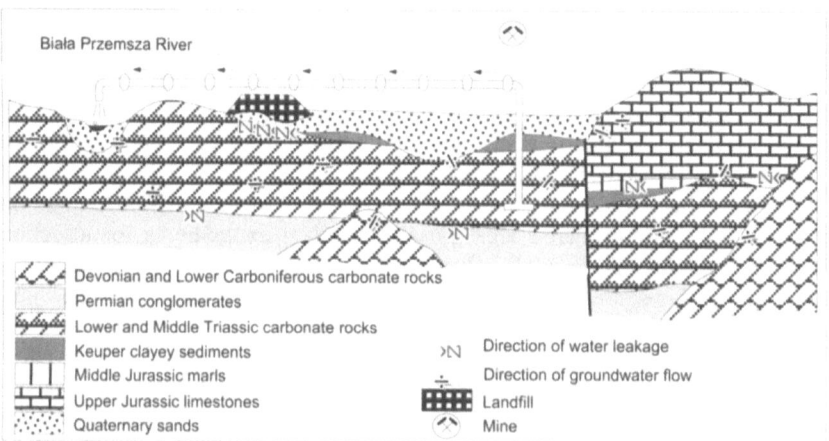

Fig. 2. Groundwater circulation scheme

Triassic aquifer is recharged directly in outcrop areas or indirectly through permeable Quaternary, Jurassic or Palaeozoic sediments. Other sources of recharge include water downward leakage from the shallow aquifers through poorly permeable Upper Triassic, Lower and Middle Jurassic sediments as well as water seepage from rivers, mainly from The Biala Przemsza river (Fig.2). Important, from the point of view of amount and quality of recharged water, hydraulic contacts in Triassic aquifer are of the erosive, sedimentation-transgressive and tectonic types (Fig.2) (Motyka 1988).

In natural conditions, Triassic aquifer in the Olkusz region was drained by springs and rivers. Currently this aquifer is intensively drained by three Zn-Pb ores mines ("Bolesław" – abandoned at the end of 1996, "Olkusz" and "Pomorzany") and numerous groundwater intakes.

Mining drainage in this region has lasted for more than 400 years. Significant increase of this drainage has been observed from 1975 after starting of exploitation by the "Pomorzany" mine – the third and biggest mine in this region. Average total amount of water pumped out by three mines is about 330 m^3 per minute. This intensive long lasting drainage has caused significant changes in hydrodynamic conditions in the whole region. Lowering of groundwater table up to over 120 m resulted in changes of flow directions, increase of hydraulic gradients and creation of the extensive regional cone of depression covering an area of about 470 km^2 (Fig.1). In this way a new extended unsaturated zone has been originated. Thickness of this zone exceeds 80 m in the area of the "Olkusz" and "Pomorzany" mines.

Potential sources of groundwater contamination

In the study area a negative impact of exogenic and endogenic factors on ground-water quality is observed. Mentioned extended unsaturated zone is an important endogenic (Motyka and Witkowski 1999; Adamczyk et al. 2000).

Quantity of the potential and real load of contaminants originated in this zone mainly depends on the amount and type of sulphides, intensity and length of time of oxidation processes occurring there. Because of very irregular occurrence of sulphides reliable estimation of the amount and nature of the potential load of contaminants originated in this zone is difficult.

A lot of real and potential sources of groundwater contamination situated on the surface of the considered area are an important exogenic factor affecting groundwater quality. There are 9 industrial and 7 municipal landfills (4 unauthor-ised), 2 tailing dams, areas without sewerage network, leaky septic tanks, 14 pet-rol stations, 4 treatment plants, 2 animal husbandry.

The most important sources of pollution deteriorating quality of groundwater in the Triassic carbonate formations in the area of Olkusz are:
- previous discharge area of lignosulfonates from the Paper and Cellulose Mill in Klucze which were deposited in the sands of Desert of Błędów. Approximately about 450 000 ton of lignosulfonates were deposited in that period from 1930 till 1979.
- tailing dams from which leachates are enriched in sulphates and heavy metals.
- industrial landfill where rich in metals waste from the rolldown furnaces and acid waste from the sulphuric acid plant were deposited until quite recently
- municipal landfill (without any protective measures) located in the abandoned quarry in the area of the "Bolesław" mine
- uncontrolled leaks of sewers from a leaky septic tanks

Spatial distribution of the major groundwater pollution indicators

Groundwater quality assessment in the area of considered mines have been based mainly on results of sampling of natural inflows to Zn-Pb ores mines done in the period 1996-2001. Moreover, data from existing observation wells situated in the area of "Boleslaw" mine have been taken into account. Because of the peculiarity of the examined area and overlapping of different factors causing water contami-nation there was performed an analysis on spatial variability of chosen pollution indicators only. The following components were taken into account: sulphates, iron and zinc – as indicators of both anthropogenic and geogenic contamination, and nitrates and chlorides as indicators of anthropogenic contamination.

Analysis of sulphates concentrations in groundwater in the Olkusz area have been already presented by authors (Motyka and Witkowski 1999). Highest sul-phates concentrations exceeding 5000 mg/dm^3 have been noticed in the central

part of "Pomorzany" mine. Concentrations of sulphates in analysed groundwater are a result of overlapping impact of a few factors:
- weathering of metal sulphides and simultaneous buffering processes of acid solution originated as a result of this weathering,
- effluents from landfills, concentrated in the area of "Bolesław" mine,
- ascension of groundwater from the bedrock through the Permian conglomerates, where gypsum has been noticed (in some parts of "Pomorzany" mine).

It can be deducted from the spatial distribution of sulphate concentrations (Motyka and Witkowski 1999) that the processes of oxidation of sulphide minerals taking place in the rocks are the most significant factors influencing these concentrations.

The distribution of iron and zinc concentrations in waters of Triassic aquifer in the region of the analysed Zn-Pb ore mines is conditioned by the influence of exogenic and endogenic factors. Pure waters with unchanged chemical composition or changed insignificantly by anthropogenic factors (agriculture, municipal waste, urban sewage) flow from north-west, east and south-east. This is why iron concentration in water inflows to workings in "Olkusz" mine and eastern part of "Pomorzany" mine generally does not exceed the detection limit, i.e. 0,01 mg/dm^3. (Fig.3). This part of Triassic aquifer is also characterised by zinc concentration below 1 mg/dm^3 (Fig. 4).

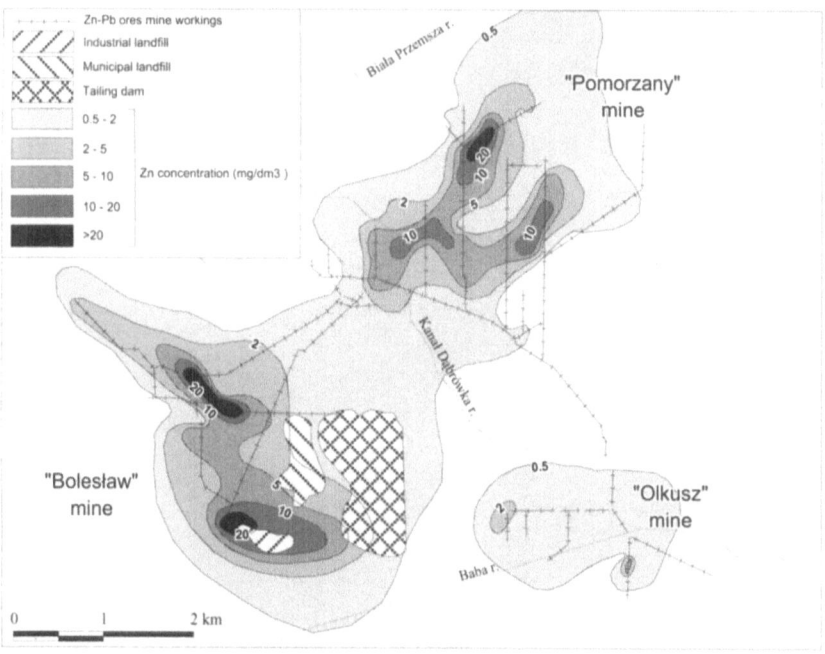

Fig. 3. Spatial distribution of iron in groundwater of the Triassic aquifer

Due to lowering of the primary groundwater table and deepening of the unsaturated zone, and oxidation processes of metal sulphides (mainly pyrite and marcasite) in mine water increasing concentrations of SO_4 as well as Fe and Zn have been observed.

Oxidation processes of sulphate metals are most responsible for high iron and zinc concentrations, exceeding 1 mg/dm^3 Fe (locally 40 mg/dm^3) (Fig.3) and 2 mg/dm^3 Zn (locally 90 mg/dm^3) (Fig.4), in the central part of "Pomorzany" mine.

Increased iron concentrations in water inflows in the north-western part of "Pomorzany" mine are caused by the influence of lignosulfonates, which flow from a place several kilometres away, where they were poured.

The reasons for increased iron and zinc concentrations in water inflows in "Bolesław" mine and in surrounding groundwater are more complicated. This is caused not only by weathering processes of sulphate metals but various contamination sources as well. The most important are industrial and municipal landfills and tailing dam. Effluents from this waste disposals have an affect on a chemical composition of mine water inflows in southern and south-eastern parts of "Bolesław" mine. The highest concentrations of Fe and Zn exceeding 20 mg/dm^3 have been observed there. (Fig.3 ,4).

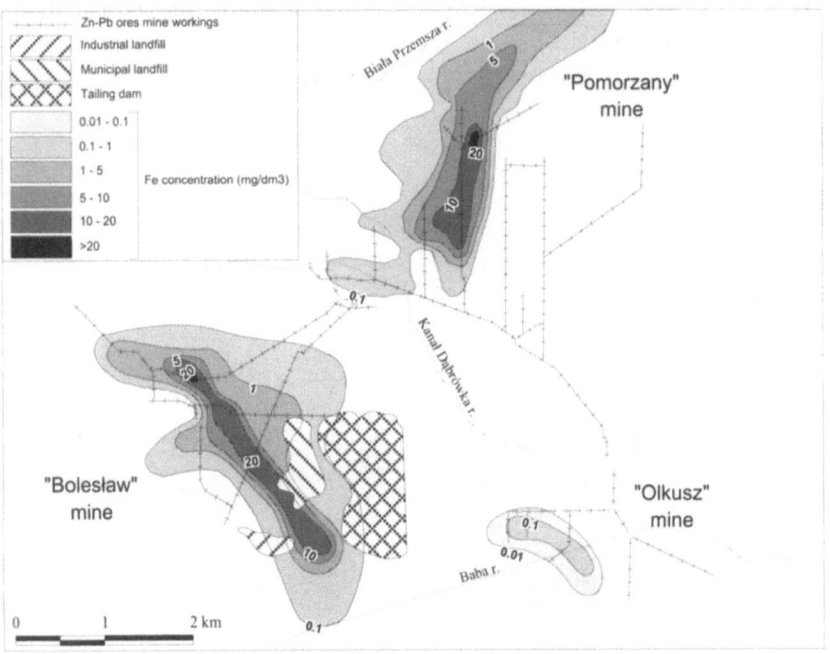

Fig. 4. Spatial distribution of zinc in groundwater of the Triassic aquifer

Nitrates in examined groundwater occur in south-western, south-eastern and north-eastern parts of the analysed area (Fig.5). The distribution of increased concentrations is of mosaic type and is connected with two major factors, i.e. occurrence on the surface such contamination sources as communal landfills, animal farms, built-up areas with no sewage network and natural vulnerability of Triassic aquifer to contamination. The highest concentrations (about 60 mg/dm³) are reported in the southern part of "Bolesław" mine. They are connected with the uncovered character of the aquifer and obvious influence of the mentioned municipal landfill, as well as flowing, mainly from the west and south, groundwater contaminated with municipal sewage. Negative impact of this landfill can be also proved by high chloride concentrations observed in its neighbourhood, up to 600 mg/l in examined groundwater (Adamczyk et al. 2000).

Visibly increased concentrations of nitrates were reported also in the region of "Olkusz" mine (Fig.5). This contamination results from both negative impact of built-up areas with no sewage network and contaminated water flowing from the east, from Jurassic deposits.

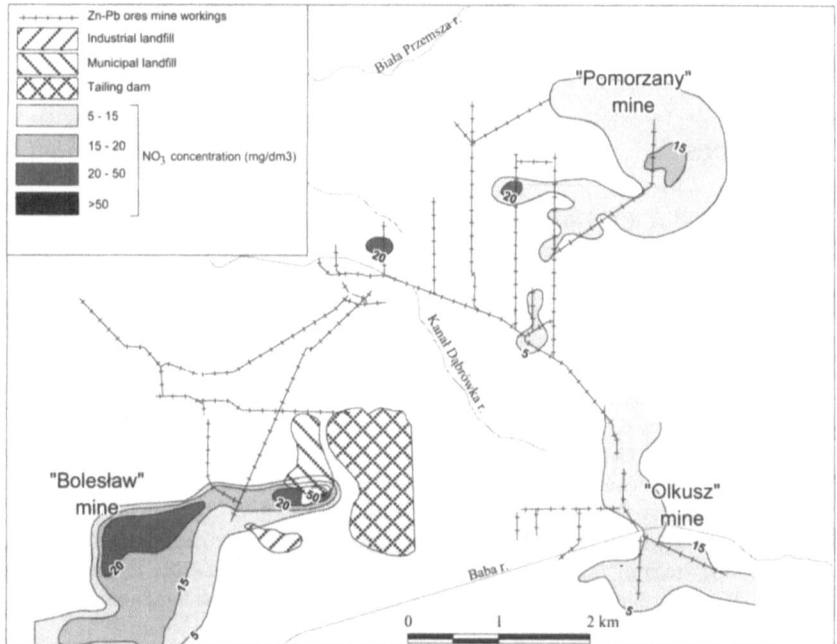

Fig. 5 Spatial distribution of nitrates in groundwater of the Triassic aquifer

Observed increased concentrations of nitrates in the eastern part of "Pomorzany" mine, because of insulating Keuper deposits occurrence, are connected with the flow of contaminated water through hydrogeological windows from Jurassic deposits. Increased nitrates concentrations in the remaining parts of the mine are due to the water flow from Quaternary and Jurassic deposits in the zones of local erosive or anthropogenic (cavings, boreholes) contacts.

Conclusions

Observed deterioration of groundwater quality in considered area is a result of overlapping negative impact of endogenic and exogenic factors. Analysis of the spatial distribution of SO_4, Fe, Zn concentrations in mine water shows that the processes of oxidation of sulphides minerals taking place in the rocks are the most significant factors influencing these concentrations. Comparison of mentioned constituent concentrations to concentrations of NO_3 points on additional significant impact of local human sources of pollution (tailings dams, municipal and industrial landfill, injected lignosulfonates) on groundwater quality. Negative impact of the exogenic factors on groundwater is visible not only in very vulnerable, uncovered parts of the Triassic aquifer in the area of "Bolesław"mine but also in area of "Pomorzany" mine in parts of local hydraulic contacts between Quaternary, Jurassic and Triassic aquifers.

References

Adamczyk A.F, Wilk Z (1976) Differentiation of the chemical composition of mine water of ores mines in the Olkusz area (in Polish). Spraw. z Pos. Kom. Nauk. PAN, Oddz. W Krakowie, T. XX/2: 408-410

Adamczyk Z, Motyka J, Witkowski A (2000) Impact of Zn-Pb ore mining on groundwater quality in the Olkusz region. Mine Water and the Environment (A.Różkowski, M.Rogoż ed.). Proceed. of 7th IMWA Congress, Katowice: 27-37

Motyka J (1988). Triassic carbonate sediments of Olkusz-Zawiercie ore-bearing district as an aquifer (in Polish), Scientific bulletins of Stanisław Staszic Academy of Mining and Metal., No. 1157, Geology, Bull. 36, Kraków: 1-109

Motyka J (1998) A conceptual model of hydraulic networks in carbonate rocks, illustrated by examples from Poland. Hydrogeology Journal 6: 469-482

Motyka J, Witkowski A.J (1999) Sulphates in groundwater of the karst-fractured Triassic aquifers in areas of intensive mining drainage (the Olkusz and the Bytom regions). Mine, Water & Environment (R.F.Rubio ed.), Vol.I, IMWA International Congress, Sevilla: 189-195

Różkowski A. ed (1990) Fissure-karst groundwater basins of the Cracow-Silesian monocline and problems of their protection. Monograph CPBP 04.10, Vol 57 (in Polish), Wydawnictwo SGGW-AR, Warszawa: 1-123

Witkowski et al. (2001) Groundwater vulnerability and quality of karst-fissured aquifers in areas of intensive drainage (case study: Olkusz-Zawiercie and Chrzanow aquifers). Project KBN 9 T12 B01116. University of Silesia. Unpublished report (in Polish): 1-167

Arsenic contamination of natural origin in stream sediments of Lake Mulargia Reservoir, Southern Sardinia

S.Fadda[1], M.Fiori[1], S.M.Grillo[2], C. Matzuzzi[1], P. Valera[2]

[1]Centro Studi Geominerari e Mineralurgici del CNR - P.za d'Armi, (Ca) Italy
[2]Dipartimento di Geoingegneria e Tecnologie Ambientali -P.za d'Armi, (Ca) Italy

Abstract. High As contents have been detected in the stream sediments of Lake Mulargia and in the tributaries draining the basin which carry mineral matter mostly containing Fe, As and Sb sulphides. Systematic sampling has shown the presence of significant amounts of arsenic originating from natural sources alone, as it is supplied by the surrounding mineralised rocks. The seasonal water level fluctuations peculiar of this lake cause oxidation and remobilization of most metals. Arsenic is readily released in the acid waters, but the constantly high-pH of the main body of water causes adsorption and co-precipitation of this element on the abundant newly formed Fe-hydroxides. These trap the As, as the negligible arsenic contents in the lake waters clearly demonstrate.

Introduction

Lake Mulargia reservoir was completed in 1963. It has a maximum capacity of about 350 Mcu.m, and a roughly triangular shape with a maximum surface area of about 124km^2. It forms part of the Flumendosa Reservoir system which comprises six interconnected artificial lakes of different size (Fig.1). The entire system supplies almost all of its water to the distribution network located chiefly in the Campidano plain, from the L.Alto Flumendosa to Simbirizzi area (Botti et al.2001). Water levels in the lake are strongly affected by the warm and semiarid climate of the island of Sardinia, characterised by two short rainy periods and a long dry season lasting at least six months, during which extensive lake bottom areas gradually become exposed all along the lake edges.

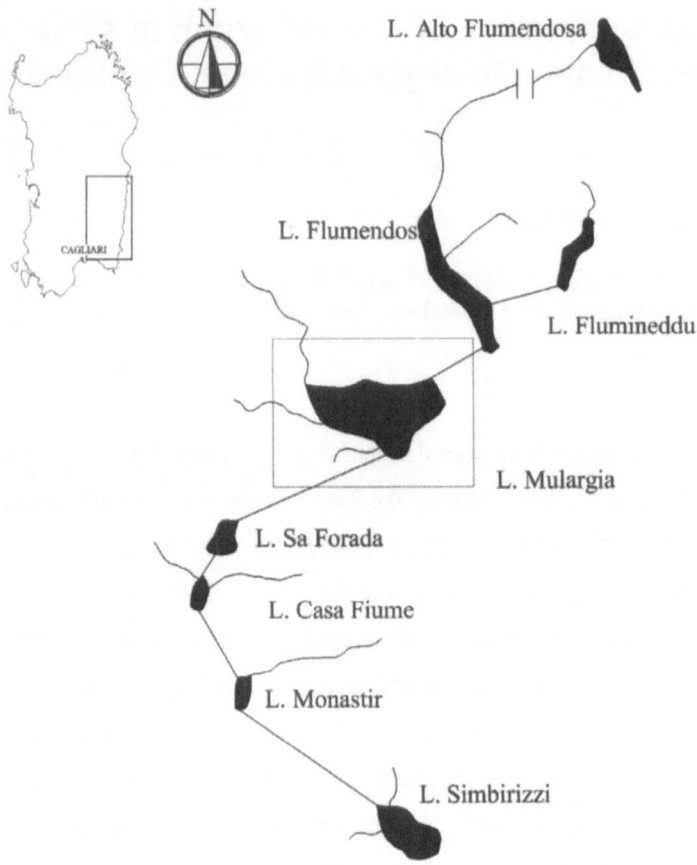

Fig. 1. Sketch maps of the Flumendosa-Campidano reservoir system.

The lake system is over dimensioned because multi annual cycles overlap seasonal wet and dry spells with alternating wet and longer drought periods. As a consequence, water levels in these lakes are prone to wide variations and the bottom sediments are cyclically exposed. As the water stored in the reservoir is used for agricultural and municipal purposes the consequent As release in solution may constitute a public health hazard.

During the exceptional drought of the late eighties (Botti et al.2001, Fig.2), samples were collected in the deepest part of the lake, at its edges and in the now submerged drainage framework which has been reconstructed with the aid of old maps of the area before the artificial basin was inundated. The analytical data from more than 200 samples have been statistically treated and processed with a GIS to create maps showing the As distributions and to highlight those areas of arsenic dispersion and accumulation. Observation of the maps reveals that the terrains containing As-bearing minerals are probably the neighbouring outcropping rocks where the pyrite-arsenopyrite association is frequently observed in the ore bodies containing prevalently Fe, As and Sb sulphides.

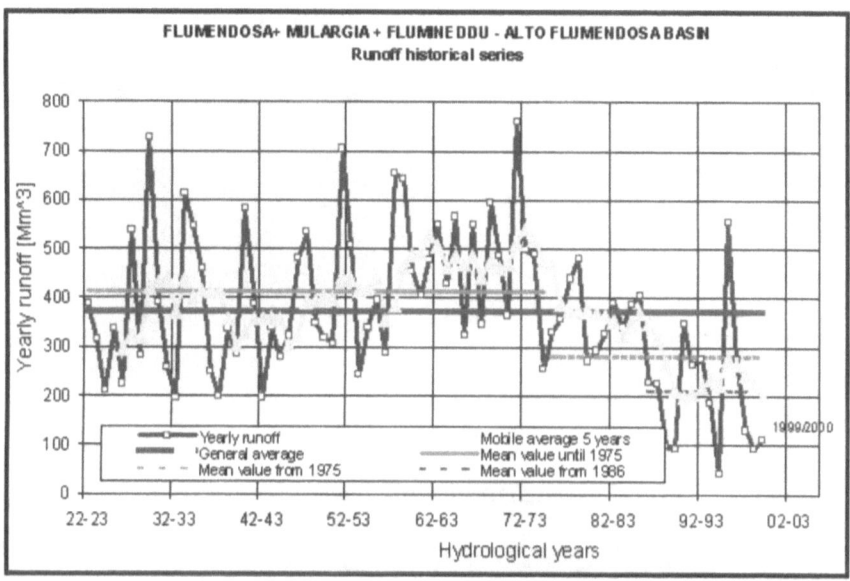

Fig. 2. Runoff historical series starting from 1922. The driest six-years in the last decade.

Geological context

The geology of the area around Lake Mulargia includes a Palaeozoic crystalline basement overlain by younger sediments and volcanic rocks (Fig.3). The lower to middle Palaeozoic basement (Carmignani 1982) comprises predominantly clastic meta sediments, mostly shales and meta sandstones with marble lenses, and acid meta volcanic rocks, "porphyroids". These are covered by lacustrine sediments of Permian age, mainly conglomerates to siltstones, overlain by porphyritic rocks. The series crops out along the northern and eastern sides of the lake (Marcello 1961). On the northern side rests the transgressive Jurassic series, including a quartzose conglomerate passing to sandstone and clay and finally to carbonate beds. Again on the northern flank of the lake is a new continental phase, represented by Eocene conglomerates. The Miocene transgression is represented by platform limestones and marls (western side), while the last continental phase (Pliocene to present) is represented by alluvium (northern side) and flood basalts (eastern side).

The terrains in the drainage basin include complexes hosting heavy-metal bearing mineralizations, mostly base metal sulphides accompanied by hydrothermal alteration These sulphides may be the principal arsenic bearing minerals in the source rocks.

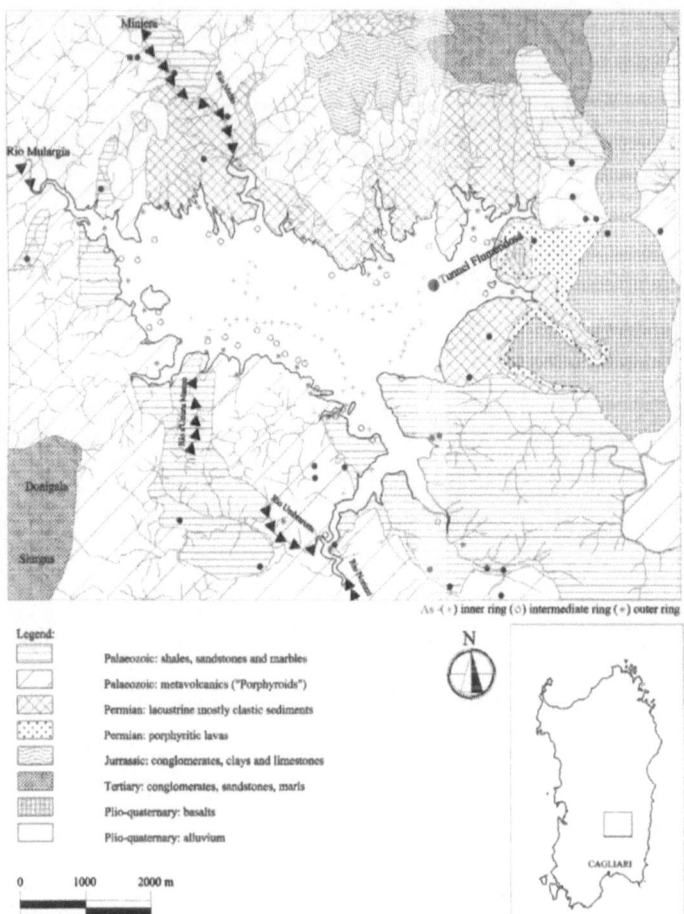

Fig. 3. Geological sketch map of terrains surrounding the lake Mulargia and schematich maps showing the sampling patterns with the three rings of samples collected. Crosses: "bottom" samples (first or inner ring); circles: "middle lake" samples (second or intermediate ring); asterisks: "near shore" sample (third or outer ring). Dots represent the sampling sites in the rocks surrounding the lake.

Within the Palaeozoic rocks, the most important supply source is the disused Genna Ureu mine, where Fe, As, Sb and W minerals occur with minor Cu, Au, Ag, Bi, Ni and Co contents. This mine and the surrounding meta volcanic rocks ("porphyroids") are situated in the Rio Mulargia drainage basin, which enters the lake at the western corner. Other small manifestations consist of pyrite lenses and disseminations, occurring all around the lake's perimeter, in particular in the Rio Norizzi basin, with a drainage area of about $15km^2$ and drained by minor tributaries flowing into the southern corner of the lake. Besides the mine pits, trenches, galleries, small mining explorations, unexplored mineralized outcrops and numerous mine waste dumps, another anthropogenic factor comes into play: Lake Mulargia receives water from Lake Flumendosa through a tunnel (Fadda 1996), situated near the eastern end on the northern side. The Flumendosa reservoir drains a much wider basin where several mixed-sulphides ore bodies occur, including a disused mine within the lake itself (Fig.3).

Sampling, analysis and data treatment

A total of 140 points were sampled and more than 200 samples collected. The bottom sediments were collected in 1990 in the deepest part of the lake and in 1991 in the outer parts. Three "rings" were sampled: the first in the central bottom area, following the tributaries draining into the lake based on old maps indicating the submerged drainage pattern. The second concerned the intermediate zone and the third the stream sediments along the lake's edge at its maximum level. Inter-stream, exclusively lacustrine sediments were also collected (Fig.3). In addition to the lake sediments, samples of rock were taken from the main rock facies around the lake in order to ascertain the influence of the surrounding terrains on the sediments' metal content.

Analyses were performed on the upper part (20cm) of all the samples weighing on average around 300 g, dried then split into two. One half was split again and this quarter was sieved to -2mm and prepared for As determination. Chemical analysis were carried out with ICP-MS after solubilization by mixed acid attack performed in a fully automatic microwave oven.

Chemical determinations for As in the sediments were treated using statistical methods so as to detect the existence of any homogeneous populations and then determine their statistical parameters. The results concerning the kind of the studied populations were first verified through the Henry's line test, then the population parameters calculated (Table 1).

Table 1. Main statistical parameters for As. Anomaly thesholds are given, for a normal population by: means plus 1,2 and 3 standard deviations. (As limit thresholds of intervention for stream sediments fixed by regulation of Canada: 33 ppm and Holland: 55 ppm).

| Element | Mean value | Standard Deviation | | Anomaly thresholds | |
	m(ppm)	s(ppm)	possible	probable	sure
As-inner	19.00	9.20	28.20	37.40	46.60
As-intermediate	19.30	1.23	20.53	21.76	22.99
As-outer	73.70	1.23	74.93	76.16	77.39
As-population	67.36	1.30	68.66	69.96	71.26

Discussion and conclusions

The geogenic sources of arsenic can be traced to the geological formations of the drainage basin which have a high natural geochemical background of heavy metals and where arsenic bearing ores occur, thus producing sites with considerable contamination potential. Samples of the Palaeozoic rocks cropping out around the lake yielded an arithmetic mean of about 250 ppm for As which is simply indicative and over-estimated as it is affected by the ore mineral disseminations occurring quite near the lake shore. The highest As values along with the major anomalies were in fact observed near the shore and appeared to be related to sulphide occurrences in the vicinity of the lake. The high values observed near the stream inlets in the north-western corner (Rio Mulargia, Rio Melas) are related with research and mining explorations about 1km upstream in the W-Sb-As-Au bearing ore body of the disused Genna Ureu mine. As values as high as 100 ppm have been detected in the southern corner of the lake at the inlet of the Rio Norizzi where the Rio Umbrarutta and Rio S'Utturu Mannu also drain into the lake (Fiori et al.2001).

In Italy there is no legislation regulating metal concentrations in stream sediments so the limits established for arsenic by Canada (Canadian Council of Ministers of Environment, CCME 1995) and Holland (data from Italian Ministry of the Environment) regulations have been considered here. The background level for an element, obtained from classic statistical techniques, is the mean value plus the coefficient of variation for the normal population and should represent the upper limit for a value that is certainly not anomalous. Thus a comparison of this value with risk parameters should give an idea of the environmental situation in a given area. The As background level of 68 ppm (Table 1) in the near-shore samples (third or outer ring) is higher than the Canadian and Dutch limits. However, far higher As concentrations have been detected locally, for instance 129 and 192 ppm in the Norizzi basin (S.M.Grillo et al. 2001), and these can be regarded at risk level. Fig.4 shows As anomaly maps constructed using the statistical parameters given in Table 1. The peculiar characteristics of arsenic concentrations can be clearly observed, the highest values, along with the major anomalies, occurring in the outer sampling ring which comprises both the stream inlets and sulphide occurrences fairly near to the edge of the lake at its maximum level. The maps also clearly show that, regarding bottom area and middle lake sampling, large portions of the lake yielded As contents consistently lower than the risk level.

Arsenic minerals in the mineralization and disseminations are represented by mixed sulphides and the pyrite-arsenopyrite (FeS_2-As(S)) association is frequently observed. These may be considered to be the chief As-carrier minerals in the source rocks while the nature of the As carrier phases in the derived residual soils and stream sediments is under study. When subjected to weathering these sulphides can release soluble arsenic into natural waters

$$(FeS_2\text{-}As(S)) + 7/2\ O_2 + H_2O = Fe^{2+}\ (aq) + SO_4^{2-} + As\ (aq)$$

where arsenic forms no single cations but reacts readily to inorganic acids. During transportation the dissolved predominant arsenic species are in the higher (+V) oxidation states and are very effectively adsorbed onto ferric oxide-hydroxide goethite ($FeOOH+H_2AsO_4^-$), manganese oxides or hydroxides as well as onto clay minerals and organic matter. Its behaviour appears to be mainly controlled by these sorption processes even if a slow release may be possible and can be related to pH increase in the aqueous phase.

It is quite likely that at least a part of the original readily weatherable sulphides still persist in the stream sediments. The grains deposited in proximal sediments may undergo oxidation during the dry season; the smaller grains are totally destroyed and the largest ones reduced in size. Floodwaters carry away the rest of the sulphide grains along with As oxyanions which in turn are adsorbed or co-precipitated onto Fe hydroxide as pH reaches a sufficiently high value settling in the inner sediments. Furthermore, during recurrent drought, when precipitation is not sufficient to completely fill the lake and water deliveries empty it almost entirely, most of the bottom sediments undergo alternate phases of dryness and are reworked by the incipient river flows after the first heavy rains. The supply of new material includes anthropogenic organic matter contained in the largely domestic wastewater discharged by several villages in the main valleys. Another contribution to organic matter derives from the agricultural and pastoral activity in the vicinity of the lake, in the tributary valleys and in those parts of the lake bottom regularly exposed during the dry season. It is possible, provided that reducing conditions are attained in the deepest part of the thick sediments, that sulphides form from the Fe-hydroxide flocks and the metal ions adsorbed thereon. In other words, once deposited the arsenic may undergo reduction by organic carbon in the deeper parts of the bottom sediments with dissolution of arseniferous iron-oxydroxide, formation of new sulphides and soluble arsenic mainly as neutral $H_3As O_3$ complex which in turn may undergo exchange processes (Sahu 2001)

$$FeOOH + H_2AsO_4^- + CH_2O = Fe^{2+} + H_3AsO_3 + HCO_3^-$$

Though the cyclic filling-emptying of the lake favours the mobilization of arsenic and other elements into acid solutions, the constantly high pH (always above 8) of the main body of water in the lake prevents the elements remaining in solution (Fiori 1999). Arsenic trapping can be correlated with the rapid precipitation of abundant iron oxydroxides and hydroxides-sulphates for which this metalloid has a strong sorptive affinity as is well known in water treatment plants. This leads to the immobilization of most arsenic in sediments settling near the shore lines and the inlets of the tributaries as is also indicated by the decreasing concentration of this contaminant towards the lake bottom. However, small quantities of As may also be released locally under anoxic conditions in the deepest parts of the lake. Apart from the obvious dilution effect, the cyclic mobilisation-reprecipitation processes are certainly effective in keeping the soluble As sufficiently low as is clearly demonstrated by its negligible contents in the waters of this lake.

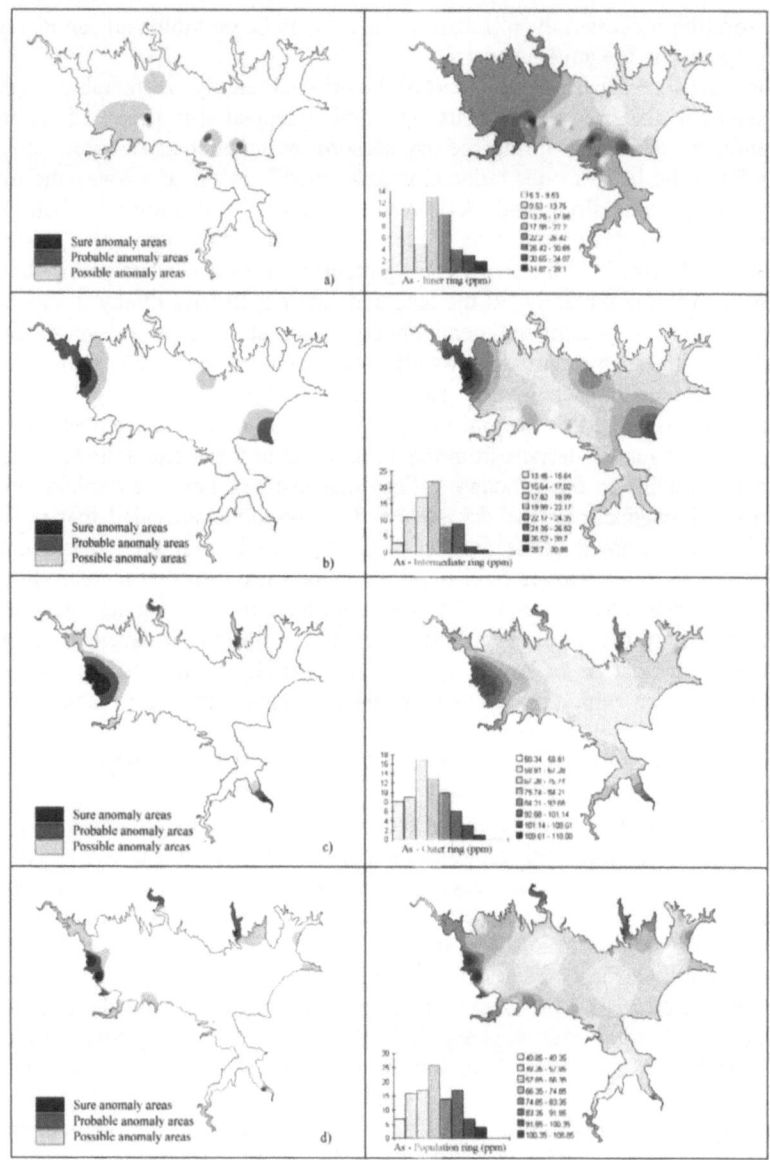

Fig. 4. On the left distribution of As according to the anomaly thresholds given in table 1. On the right distribution of As according to the frequency classes of his histograms relative to the three samples collection. (a - inner ring; b - intermediate ring; c - outer ring; d - total population).

Acknowledgements

This study was supported by the Centro Studi Geominerari e Mineralurgici del CNR, Facoltà di Ingegneria, Piazza d'Armi, 09123 Cagliari, Italy

References

Botti P., Dessena M.A., Fiori M., Grillo S.M., Marcello A., Matzuzzi C., Pretti S., Vacca S. (2001). "The Medio Flumendosa Reservoir". Field Trip Guide of the "Tenth International Symposium on Water-Rock Interaction": 1-20

Carmignani L., Cocozza T., Ghezzo C., Pertusati P.C. & Ricci C.A. (1982) "Lineamenti del basamento Sardo. Guida alla Geologia del Paleozoico Sardo". Soc. Geol. It., 11-23

Fadda S., Fiori M., Grillo S.M., Manca M.G., Marcello A., Matzuzzi C., Pretti S. (1996) "Influenza della vecchia attività mineraria nel bacino del Lago Flumendosa e sui contenuti di metallo nei sedimenti del Lago Mulargia" Atti del Congresso Internazionale per il Centenario dell'Associazione Mineraria Sarda 1896-1996, 137-149

Fiori M., Grillo S.M., Manca M.G., Marcello A., Matzuzzi C., Pretti S. (1999) "Spatial distribution patterns of heavy metals in lake bottom sediments: an example from the Mulargia reservoir (Italy)" Proceedings of the Mine, Water &Environment, International Congress, 157-162

Fiori M.,Grillo S.M., Marcello A., Pinna D., Pretti S. (2001) "Heavy metal pollution of natural origin in the Rio Norizzi basin (Southwestern Sardinia, Italy). The As aspect". Proceedings of the Tenth International Symposium on Water-Rock Interaction: 1091-1094

Marcello A. (1961) "Il bacino Permico del Rio Mulargia (medio Flumendosa) e le sue successive modificazioni geologiche".Ph-th, 1-39

Sahu S.J., Roy S., Jana J., Nath B., Bhattachrya R., Chatterjee D., Dey Dalal S.S. (2001) "Water chemistry and sediment-water interaction responsible for mobilization of arsenic in groundwater in Bengal delta plain ". Proceedings of Tenth International Symposium on Water-Rock Interaction: 1139-1142.

Arsenic Rich Waste Rock Disposal under Subaqueous and Anoxic Conditions

Michael J. Rinker[1], R.V. Nicholson[1,2], Don Lush[1] and Nan H. Lee[3]

[1]Beak International Inc., 14 Abacus Rd. Brampton, ON Canada. L6T 5B7
[2]University of Waterloo, Waterloo, ON Canada N2L 3G1
[3]Cogema Resources, 45th Street West, Saskatoon, SK Canada S7K 3X5

Abstract. Waste rock from the Cigar Lake project in Northern Saskatchewan, Canada, was evaluated for in-pit disposal under subaqueous and anoxic conditions. Column tests were conducted to assess arsenic release rates and sequential leach tests were conducted to form the basis of a mass balance approach for predicting arsenic concentrations over time. The arsenic concentration in pore water was a function of the pore water residence time and this suggests a new paradigm for interpretation of column tests under subaqueous conditions. The relatively 'steady state' concentration of arsenic in pore water correlated well with the 'leachable' fraction of arsenic in the solids. The relationship provided a tool to assess the risk of arsenic leaching in wastes using solids assays.

Introduction

The Cigar Lake Project is currently the world's largest undeveloped uranium proect with an estimated reserve sufficient to sustain mining for up to 40 years. Over the life of the Cigar Lake Project, approximately 1.32 million cubic metres of waste rock will require environmentally appropriate management. One effective way to manage this waste rock is to place it in an environment similar to that from which it was extracted. That is, surrounded by water and isolated from atmospheric conditions, such as at the bottom of a flooded open pit. Arsenic was the primary constituent of concern in Cigar Lake waste rock. Radionuclides, including uranium, were at relatively low concentrations and comparable to background levels.

Waste rock studies were conducted in 2000 to estimate pore water concentrations in waste rock under anaerobic and subaqueous conditions and to quantify the leachable mass fractions of arsenic from solids with a wide range of values for arsenic content (mass balance approach).

Column studies were conducted to refine estimates of chemical constituent concentrations in pore water associated with waste rock and to assess the relative effect of cement, that may be incidentally added to the waste rock during mining on leaching behaviour of the waste rock. The pore water concentration will be used to estimate constituent loadings to groundwater moving through the waste rock and eventually discharging to the environment (surface water bodies) down gradient to the open pit. Sequential leach tests were used to compare the leaching efficiencies of selected solutions to extract "leachable" constituents, in order to estimate the leachable fraction that could contribute to the pore water concentrations when waste rock is submerged in water. The tests were designed to provide some insight into the geochemical mechanism controlling arsenic release and to develop a kinetic and mass balance approach to predict constituent loadings to groundwater passing through the waste rock in the post closure period.

Methods

Tests were conducted on altered and unaltered graphitic metapelite drill core samples (1 year old) and fresh drill core samples (1 month old) from Cigar Lake. The samples had a range of arsenic concentrations that bracketed the expected values for waste rock that will be disposed in the Sue C pit.

Samples of waste rock from a three year-old stockpile at Cigar Lake were also tested to compare and contrast the results from weathered samples in the stockpile to those with the relatively fresh drill core samples. Waste rock samples were obtained from the stockpile by coring eight holes (114 mm diameter) using a sonic drill.

Column tests were conducted on Cigar Lake waste rock to refine estimates of arsenic concentration in pore water and to assess the relative effect of cement contained in the waste rock on leaching constituents from the waste rock. The conditions for column tests conducted for this study included the following:

- 20 kg of waste rock per column;
- column filled with water and sealed to minimize ingress of oxygen;
- no water column over the waste rock;
- porewater volume of 5 L; and
- sample volumes of 150 mL were removed bi-weekly for chemical analysis, resulting in a pore water residence time of approximately 1.5 years for each column.

The results from these column tests were carried out in 2000 compared to results from previous waste rock investigations that were conducted in 1994 (CLMC, 1995) to evaluate water quality associated with submerged waste rock, and to determine representative concentrations of arsenic and other constituents in pore water that would be used for the assessment of waste rock disposal options. The test conditions for these previous tests included the following:

- 1.5 to 4 kg of waste rock per column;
- columns filled with water and sealed to prevent ingress of oxygen;
- each column included a volume of water overlying the rock that was also sampled in some cases; and
- 1 litre of water was removed (and replaced) weekly for analysis resulting in pore water residence times that ranged from a few days to 1.5 weeks.

Sequential leaching tests were conducted on waste rock samples. Sequential leaching refers to a method of repeated washing of the same solid sample with a fresh leach solution for a selected number of cycles and solution types. All sequential leach tests had water to solids ratio of 20:1 (1 L of water to 50 g of solids). Several leach solutions were tested initially to assess the most effective condition to leach arsenic. The leaching solutions that were investigated included de-ionized water, de-ionized water with pH adjusted to 5 with acetic acid, de-ionized water with pH adjusted to 5 with phosphoric acid, de-ionized water with pH adjusted to 10 with NaOH, and de-ionized water with hydroxylamine-hydrochloride (HAH; 0.01 M) with a pH of 4.5.

Waste rock samples were placed in a plastic vessel and solution was added. The vessel was agitated for a specified period of time (typically 24 hr and 48 hr intervals) and sample aliquots were removed from the leaching solutions and submitted for chemical analysis. A fresh leach solution was added to the vessel and the procedure was repeated. The total leachable amount of any constituent was interpreted as the sum of the constituent leached in all sequential steps performed on the sample.

Samples evaluated in this study include drill core (1 month to 1 year old) and samples collected from a three year old stockpile produced during test mining and development.

Results

Column Tests

Solids

The elemental composition of selected constituents in waste rock in the column tests is provided in Table 1. The results show that the waste rock samples tested in 1994 are very similar to samples tested for this study. Arsenic values found in the waste rock are relatively high compared to the average crustal abundance of 1 to 13 mg/kg (Taylor and McLennan, 1985), and is the primary constituent of concern for Cigar Lake waste rock. The sulphide-sulphur values were greater than

2% and suggest that there is a relatively high risk to acidic drainage and metal leaching if the waste rock is disposed of on land indefinitely.

Table 1. Concentration of selected constituents in waste rock used for column tests.

Constituent	1994 Column Test	2000 Column Test
Sulphide (%)	2.5	2.1
Arsenic (mg/kg)	460	444
Uranium (mg/kg)	14.5	18
Nickel (mg/kg)	280	181

Pore Water Concentrations

Two column tests were conducted for this study carried out in 2000. Both columns contained Cigar Lake waste rock, and one column contained rock with cement added (0.75 weight percent). The concentrations of sulphate and arsenic and pH in pore water as a function of time are shown in Fig. 1. An attempt was made to place waste rock of similar arsenic concentration and mineralogy in each column but it is possible that there was variation between the columns.

In general, the results show that marginally higher pH values were observed in pore water associated with waste rock that contained cement (mean pH value 7.7) relative to values observed in pore water associated with waste rock that did not contain cement (mean pH value 7.2).

Sulphate concentrations exhibited increasing trends over the first two weeks, followed by marginally decreasing values thereafter. The maximum sulphate concentrations were 640 mg/L for the test with cement, and 850 mg/L for the test without cement. Arsenic concentrations increased to values of 40 to 55 mg/L over the first 6 to 15 weeks of testing followed by more slowly rising values thereafter. The mean, relatively steady-state, arsenic concentrations in pore water were 53 mg/L for the test with cement and 65 mg/L for the test that did not contain cement.

Selected results for the 1994 column test results are provided in Fig. 1. In general, the results show that sulphate was rinsed from the waste rock over the first few weeks of testing, and relatively low steady-state concentrations were observed thereafter. In contrast, arsenic concentrations exhibited initial increases with relatively steady-state values over time. The maximum observed arsenic concentration in leachate was 4 mg/L. The pH values were relatively constant for the duration of the tests with a mean value of 7.5.

The trends observed for sulphate and the maximum values observed for arsenic in the tests conducted in 2000 are significantly different than those observed in the 1994 tests. Because the waste rock samples were similar in both the 1994 and the current tests, it is likely that the experimental methodology accounted for the difference. The differences in experimental methods and limiting assumptions are discussed below.

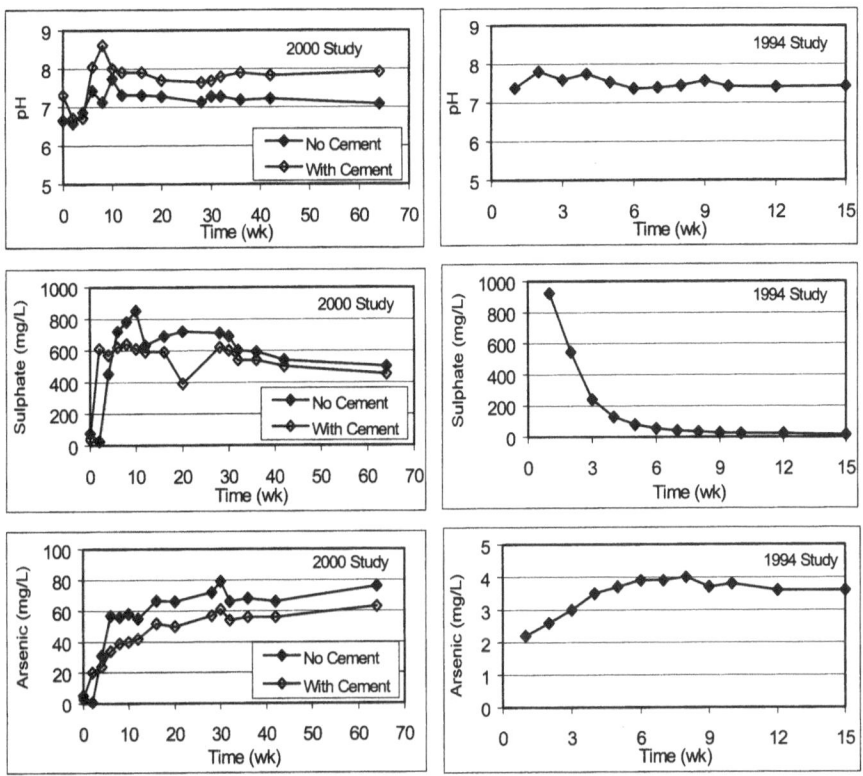

Fig. 1. Selected results for column tests conducted in 2000 (this study) and in 1994 (previous study).

The trends observed for sulphate and the maximum values observed for arsenic in the tests conducted in 2000 are significantly different than those observed in the 1994 tests. Because the waste rock samples were similar in both the 1994 and the current tests, it is likely that the experimental methodology accounted for the difference. The differences in experimental methods and limiting assumptions are discussed below.

Sequential Leaching of Drill Core Material

Preliminary leach tests were conducted to refine the methodology used to determine the leachable inventory of arsenic in waste rock samples. The solution pH effects were evaluated by pH adjustment to values of 5 and 10 with acetic acid and sodium hydroxide respectively. Reductive dissolution was investigated using hydroxylamine hydrochloric acid (HAH), and sorption was investigated using phosphoric acid at pH 5 (~300 mg/L PO_4). Leaching with de-ionized water was also assessed.

The solution with a pH of 10 and the weak phosphoric acid solution were the most effective at leaching arsenic. There was no indication that dissolving ferric iron solids with HAH affected the quantity of arsenic leached from the solids. However, the response of arsenic leaching in other solutions was consistent with sorption of arsenic onto solids such as clays. This was further supported by the lower or depressed arsenic concentrations observed in the acetic acid (pH 5) solutions, and the higher concentrations associated with the pH 10 solutions. Because of the indication that sorption appeared to be important, a solution containing phosphate was considered to be adequate to quantitatively extract the total leachable arsenic.

For the purpose of determining the leachable inventory of arsenic in the Cigar Lake waste rock, the rock samples were subjected to 2 rinses with de-ionized water (to determine the water leachable fraction) followed by 2 rinses with phosphoric acid (to determine the total leachable fraction) with solution:solids ratios of 20:1.

Leaching Drill Core

Six leach tests were conducted on Cigar Lake core samples. The results of the leach tests are presented in Table 2 that presents the equivalent mass of arsenic leached per mass (%) of arsenic in the solids. These results show that the percentage of arsenic leached by water from the solids ranged from about 9% to about 17% of the arsenic inventory for core samples. The additional phosphate solution resulted in the cumulative quantity of leachable arsenic that approached 30%.

Leaching of Stockpile C Samples

Sequential leach tests were also conducted on samples collected from Stockpile C at the Cigar Lake mine site. These tests were conducted to determine the soluble arsenic inventory in material that had been stored on surface for several years and exposed to oxidation. A total of 15 samples from eight drill holes were tested.

The leach test results are presented as the equivalent mass of arsenic leached per mass (%) of solids in Table 2. These results account for the total quantity of arsenic leached from the solids after four sequential extractions. The quantity of arsenic leached with water ranged from 1% to 24% of the arsenic inventory in the samples. The additional phosphate solution resulted in a cumulative quantity of leachable arsenic that was identical to that observed for the drill core material (28%).

Table 2. Water and total leachable fractions of arsenic in Cigar Lake waste rock..

Sample	Water Leachable		Total Leachable	
	Range (%)	Mean (%)	Range (%)	Mean (%)
Core Samples (n=6)	9 to 17	12.4	27 to 29	28
Stockpile Samples (n=15)	1 to 24	7.4	12 to 67	28

Discussion

Pore Water Concentration

The arsenic concentrations in pore water that were observed in the 1994 tests were significantly different from those observed in the column test in this study. This difference primarily reflects the relatively short pore water resident time (1.5 weeks) for the 1994 tests compared to longer resident times in this study (1.5 years). The short resident time is acceptable for constituents that dissolve to "equilibrium" or that leach readily in each new pore volume (e.g., sulphate). However, it is not necessarily applicable for arsenic because insufficient time between sampling events prevented arsenic from dissolving to "equilibrium" concentrations which may be expected if the water remained in contact with the waste rock for several decades in a typical disposal scenario, for example.

The results of this study suggest that the submerged waste rock tests may require residence times of several months to exhibit concentrations that would be representative of those observed after disposal. If the residence times in the previous experiments were longer, pore water concentrations would have been higher than those observed. The concentrations that could occur with longer residence times can be calculated by summing the total arsenic released from the waste rock (collected as sampled during the test) and then dividing by the volume of resident pore water.

The results for all tests conducted over the past 8 years on Cigar Lake waste rock under saturated conditions are summarized in Table 3 showing the solids arsenic content, the maximum observed leachate arsenic concentrations, and recalculated arsenic concentrations that could occur if the residence times were longer. The solids arsenic content of samples ranged from 20 to more than 708 $\mu g/g$, and maximum observed arsenic concentrations ranged from 0.1 to 70 mg/L.

The observed or recalculated concentrations of arsenic in pore water were compared to the amount of arsenic in the solids (i.e., waste rock arsenic content). This correlation between solids content and pore water concentrations is shown in Fig. 2. Testing of samples with a range of arsenic concentrations provided a relationship between solids chemistry and pore water chemistry and a means to predict pore water concentrations for submerged waste rock that will be disposed at the Sue C open pit. The waste rock to will be hauled and disposed of in the Sue C pit has estimated arsenic content of 175 µg/g. The source term concentration (initial pore water concentration) for the Cigar Lake waste rock was predicted to be 17.5 mg/L using this relationship. Table 3. Summary of test programs conducted on Cigar Lake Project waste rock under saturated conditions.

Table 3. Summary of test programs conducted on Cigar Lake Project waste rock under saturated conditions.

Test Program	Solids As Content (µg/g)	Maximum Observed Leachate As Conc. (mg/L)	Recalculated Pore Water As Conc. (mg/L)
CW-93-7B[1]	24	0.1	1.1
CW-93-8B[1]	20	0.5	5.2
CW-93-9B[1]	708	11.0	60
CW-93-10B[1]	686	11.4	75
CW-94-13S1[2]	460	4.0	38
CW-94-13S2[2]	460	3.5	35
CW-94-13S3[2]	460	7.2	45
CW-94-13S4[2]	460	9.1	55
Bulk Leach Test[3]	51	4.0	7
CW-00-14C1[4]	444	55	55[5]
CW-00-14C2[4]	444	70	70[5]

[1] 1993 test program conducted on graphitic metapelite from Cigar Lake (CLMC, 1995).
[2] 1994 test program conducted on graphitic metapelite from Cigar Lake (CLMC, 1995).
[3] Bulk leach test conducted on a blend of rock types from Cigar Lake (CLMC, 1995).
[4] Tests conducted for this study.
[5] Pore water concentrations were not calculated for these tests because the solids dissolved to near equilibrium concentrations during the course of testing.

Influence Of Waste Exposure Time Prior To Submergence

During this and previous studies, the degree of weathering or exposure to the atmosphere and moisture was considered to influence the concentration of arsenic in the pore water. Previous hypothesis suggested that longer weathering periods prior to testing resulted in higher concentrations in pore water. This study, however, exhibited results that suggested that more weathered samples resulted in lower concentrations of arsenic in pore water as shown in Table 2. This trend is evident in the distribution of sequential leach results on the fresh core and weathered stockpile samples as shown in Fig.3 reflecting the average "water leachable" values shown in Table 2.

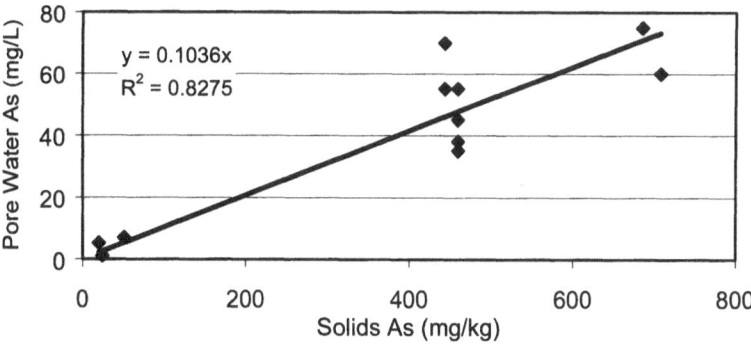

Fig. 2. Recalculated or observed arsenic concentrations in pore water as a function of arsenic content in the solids

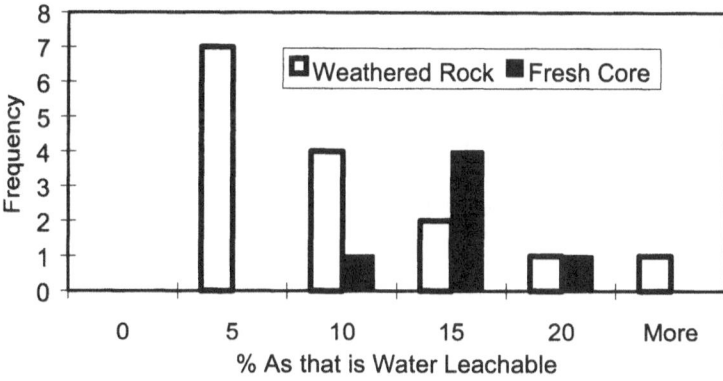

Fig. 3. Distribution of sequential leach test results.

Leachable Inventory and Implications for Management

The behaviour of arsenic during sequential leaching suggests that there is a readily soluble fraction, as well as a fraction that is more strongly sorbed on the solids. While leachable sulphate represented a small fraction of the total sulphur inventory (less than 1%), water leachable arsenic represented 7.4% of the total arsenic, on average, in the stockpile samples and 12.4% of the total arsenic, on average, in the core samples. The contrast in behaviour between arsenic and sulphur·in the rock suggests that the controls on reactions that transform the primary solids in the rock to soluble forms are significantly different for these two elements. This may imply that the kinetics of the reactions are significantly different, or that the arsenic transformation process is not completely controlled by oxidation processes.

A significant amount of the transformation reaction for arsenic occurs rapidly after the rock is disturbed, and the amount of leachable arsenic appears to decrease with time. The primary arsenic must transform to a more leachable form when disturbed and cannot be in a leachable form *in situ*; there is no evidence of high levels of arsenic in local groundwater and arsenic would likely have leached from the rock over the geologic history of the deposit. It is well known that arsenic occurs in several mineral phases in the rock associated with uranium mineralization in Saskatchewan. The different rates of transformation would be consistent with the depletion of a very reactive phase or phases, and slower reaction rates associated with the arsenic in the other remaining phases. For example, the rate of arsenopyrite (FeAsS) oxidation is known to be similar to that of pyrite (FeS_2) and could not explain the rapid transformation that resulted in the high proportions of leachable arsenic observed in this study. The rate of oxidation of the gersdorffite ([Ni,Co]AsS), however, is not known but this finely disseminated infilling mineral may exhibit very high rates of oxidation after disturbance and could be responsible for the high values for leachable arsenic.

The rapid release of arsenic initially upon excavation and exposure together with the subsequent slower rate or restabilization of arsenic after months of further exposure suggest that waste rock can be stored on surface for several years without substantially increasing the amount of leachable arsenic in the rock. There may no appreciable benefit from immediate submergence of the waste rock, and the results of this test program suggests that waste rock can be stored on surface during mine operation and hauled in campaigns without significant negative geochemical consequences.

References

Cigar Lake Mining Corporation (1995) The Cigar Lake Project Environmental Impact Statement: Mine Rock Management. Supporting Document No. 1.

Taylor, S.R. McLennan S.M. 1985 The Continental Crust: Its Composition and Evolution. Blackwell Scientific Publ., Oxford, England, 312 pp.

Mobilisation of trace elements from an epithermal high sulphidation mineralization in the sediments of Sa Forada artificial lake. South-Sardinia, Italy

M.Fiori[1], S.M.Grillo[2], C. Matzuzzi[1]

[1]Centro Studi Geominerari e Mineralurgici del CNR - P.za d'Armi, (Ca) Italy
[2]Dipartimento di Geoingegneria e Tecnologie Ambientali -P.za d'Armi, (Ca) Italy

Abstract. The Sa Forada man-made lake is an intermediate reservoir of the Flumendosa water supply system that carries a silty clayey material which has already formed a fairly thick deposit on the lake floor. Because water flowing into the lake drains the ore bodies, one may expect dissolution and subsequent transport of several trace elements (mostly Cu, As, Sb, Au) into the lake with potentially environmentally hazardous concentrations in the water and sediments.

Introduction

The Furtei mine is a typical example of a volcanic-hosted high-sulphidation epithermal gold deposit. Mining commenced in 1997 for confirmed reserves totalling 2,150,000 metric tons containing 2.82 g/t Au. The research area includes a small man-made lake, called Sa Forada, located at the Furtei Mine in the southern part of Sardinia. It receives water from the Flumendosa-Mulargia lakes and supplies first a power-plant, then the Campidano plain, mostly for agricultural purposes. The volume of water draining into the lake is extremely modest, but carries with it material coming from an intensely mineralized area currently under exploitation. Consequently the elements contained therein may have a non-negligible influence on the characteristics of the water discharged from the reservoir.

Sardinia has a semiarid climate, and the shallow soils in this region are less than 50 cm deep the natural vegetation consisting of Mediterranean scrub, dominated by plants of the genus "cystus".

This study concerns the bottom sediments of the lake sampled prior to commencement of mine workings. The metal content distribution in the sediments, which corresponds to their supply source in natural conditions and before mining operations began, allow to derive hypotheses as to their mobilisation and transport.

Study Area

The study area is characterised by a volcano-sedimentary sequence of Oligo-Miocene age, overlaying marine and fluvial-lacustrine sediments of late Eocene-Early Miocene age, which in turn rest on a Paleozoic basement, composed of meta-sediments, metapelite and quarzite. The outcropping volcano-sedimentary sequence consists of andesitic sub-volcanic domes and of volcaniclastic products such as fall, surge and ash-flow deposits. These outcrops are locally and partially overlain by Miocene and Quaternary fluvial-lacustrine and marine sediments.

Most of the volcano-sedimentary complex in the area have undergone hydro-thermal alteration and mineralisation. Four types of alteration have been recognized: propylitisation, argillisation, strong argillisation and silicification. Hydro-thermal alteration was overprinted by supergenic, secondary argillisation, mainly after oxidation of the sulphide phases. Pyritisation is widespread in all types of alterations considered with major concentrations occurring in the innermost areas of intense alteration, where it practically grades into the ore bodies.

The main mineralised bodies, already explored and mined in their upper oxidized portion, are the Santu Miali, Is Concas and Sa Perrima group. The paragenetic study revealed typical high-sulphidation assemblages, in order of abundance: pyrite, enargite-luzonite, tetrahedrite, tennantite, chalcopyrite, covellite, digenite, galena, sphalerite, arsenopirite, wurzite, tellurides, native Au, and native Te (Fiori et al. 2001).

The drainage basin and sampling

The natural drainage basin of Sa Forada covers an area of a few square kilometres, delimited by a roughly semicircular watershed culminating in the hills of Santu Miali, Coronas Arrubias and Sa Perrima (Fig.1).

Near the outcrops, deep oxidation related with the abundant pyrite transforms these bodies into an assemblage of Fe hydroxides, quartz, sulphates (gypsum, jarosite, barite) and minor arsenates and other secondary minerals. Gold easily survives as native Au, and remnants of sulphides and sulphosalts are also commonly present.

The mineralised zones drained by this basin include part of the Santu Miali hillside but not the two main ore bodies, and an outcrop forming the ridge of Sa Perrima. At Coronas Arrubias pyrite prevails, but minor gold values have been detected.

The bottom sediments of lake Sa Forada were sampled when it was emptied for dam maintenance works. The samples were collected following the pre-existing drainage pattern, both along the talweg lines and in the inter-talweg surfaces. Sediment thickness ranged from a few centimetres near the shoreline, to a few metres near the dam. Given the small size of the lake, 44 samples were collected.

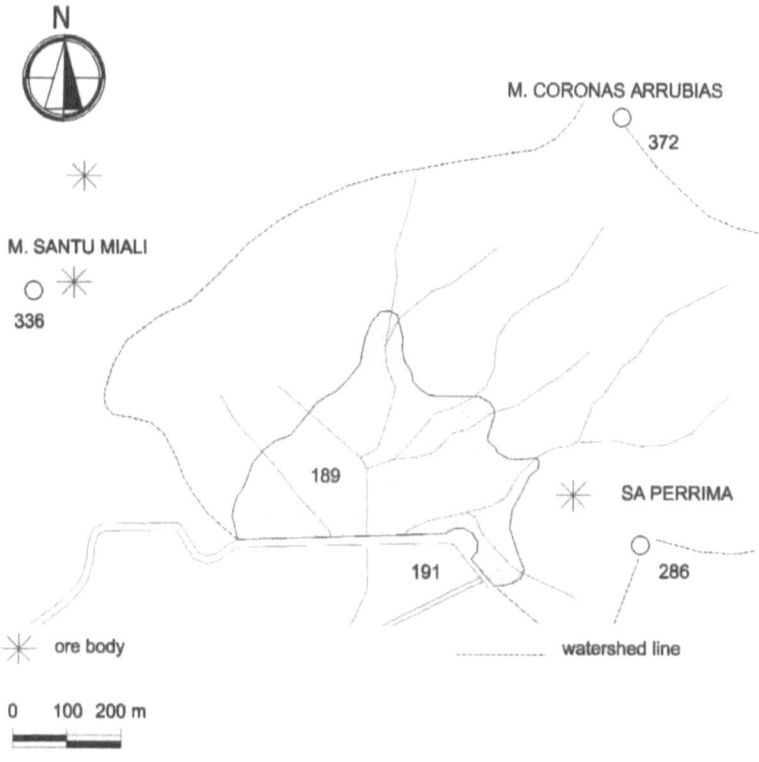

Fig. 1. Schematic map of the lake Sa Forada area, with drainage pattern.

Sample preparation and analysis

The samples, averaging some 300 g in weight, were first dried then split into two batches, one of which was stored for further controls where necessary. The remaining lot was further split into two and one half was sieved to -2mm and prepared for analysis. On the other half particle size analysis was performed on most samples. The following analyses were then conducted:
- chemical determinations using AAS and ICP-MS for the trace elements typically contained in the surrounding rocks and ore deposits;
- diffractometric analysis on all samples, to determine the distribution of both the most common and some minor minerals characterising the local sources of clasts.

The choice of the elements to be investigated stems from the present knowledge of the ore mineral assemblages in the known ore bodies.

This assemblage includes Au as well as the base metals and other elements commonly contained in high-sulphidation epithermal occurrences, plus Sn and W, often been observed in local ore samples.

Data treatment

From the chemical analyses, kriging was performed on the analytical values of minor elements and the results then mapped.

The data for trace elements in the lake sediments were treated using statistic methods and their populations were defined. For most of the elements examined, the statistical analysis of the results showed that generally at least one main population exists. Given the comparatively small number of samples, the data were arithmetic processed.

The statistical parameters given in Table 1 concern, for each element, the statistical population as derived from total sample distributions.

The mean population values have been compared with the mean world concentrations for siltitic-pelitic rocks and sediments reported by Wedepohl (1978). This comparison showed that Zn, Cd, Sb, As, Te, Pb and Au have mean concentrations significantly higher than those reported by Wedepohl, Cu content is higher and W slightly higher than the "normal" mean whereas Co, Bi and Sn appear fairly "normal".

Table 1. Main statistical parameters for the studied elements. Anomaly thesholds are given, for a normal populaton by: means plus 1,2 and 3 standard deviations.

Element	Mean value	Standard Deviation		Anomaly thresholds	
	m(ppm)	s(ppm)	possible	probable	sure
Zn	199.22	54.52	253.74	308.26	362.78
Hg	0.19	0.09	0.28	0.37	0.46
Cu	46.93	11.1	58.03	69.13	80.23
As	25.84	9.24	35.08	44.32	53.56
Sb	3.99	1.18	5.17	6.35	7.53
Te	0.25	0.09	0.34	0.43	0.52
Pb	52.54	17.12	69.66	86.78	103.9
Au	0.024	0.012	0.036	0.048	0.060

Discussion and conclusion

Our results indicate widespread and elevated concentrations of nearly all the elements examined. The distribution pattern shown in Figs.2, 3, compiled using Arcview geographic information software, shows that despite the small number of samples examined, a correlation does exist between the sources of mineral supply and the relative samples of stream sediments.

All the examined metals display regular distributions, with unimodal histograms and few clearly anomalous values. On the other hand, for most of these elements mean concentrations are significantly higher than those reported for similar sediments in other parts of the world, or are comparable with the highest known values. These observations are consistent with the small size of the drainage basin, its lithological homogeneity, and the abundance of several ore minerals in the basin. The mode of transport of these metals is governed by their chemico-physical properties. Gold, a typical "clastic" element, reaches the basin mostly after acid leaching of the sulphides. The Te in the paragenesis of metallic ore bodies occurs as tellurides, Te-rich tetrahedrite and in the native form. Even for high oxidising potential these stable mineral phases tend to release the Te, especially in arid climates, in highly oxidising and acid leaching environments (Wedepohl 1978), Hence, in the basin concerned, tellurium and its mineral phases may have travelled in the form of both clasts and ions.

Tin does not occur in the paragenesis in its most common form of cassiterite but as stannite which was detected by means of electron microprobe. The presence of tungsten is more complicated in that no mineral phases containing this element have been recognised among the mineral paragenesis.

As far as arsenic is concerned, the main As-carrier are sulphosalts and the strongest anomalies have been observed at the lake edge. This finding is consistent with the sudden precipitation of the ions due to adsorption by the clay fractions and iron oxides and hydroxides and rapid settlement of the clasts. The fact that the highest concentrations of As, Te, Sb, Hg, contained in enargite e tennantite mineralogical phases, were found in the eastern portion of the lake strongly suggest that the minerals reach the basin within clasts. The high content of mobile elements in the analyzed material can be explained, at least for Cd and Pb, by the fact that their ability to be adsorbed onto colloidal particles increases with pH. The same considerations hold for Cu and especially for Zn which are readily mobilised and have been observed to accumulate in the innermost parts of the basin.

Given the small size of the basin and the semiarid climate it can be reasonably assumed that the leaching solutions are strongly acid, though only minor amounts are involved. On the other hand, the Sa Forada reservoir continuously receives waters from the much larger lakes of the Flumendosa system, characterised by a far higher pH, about 8 (Fadda et al. 1996). Thus the seasonal acid waters carrying metallic ions blend with the slightly alkaline lake water. The resulting increase in pH leads to a much greater amount of soluble elements being bound on the colloidal particles and on the clay fraction and Fe hydroxides.

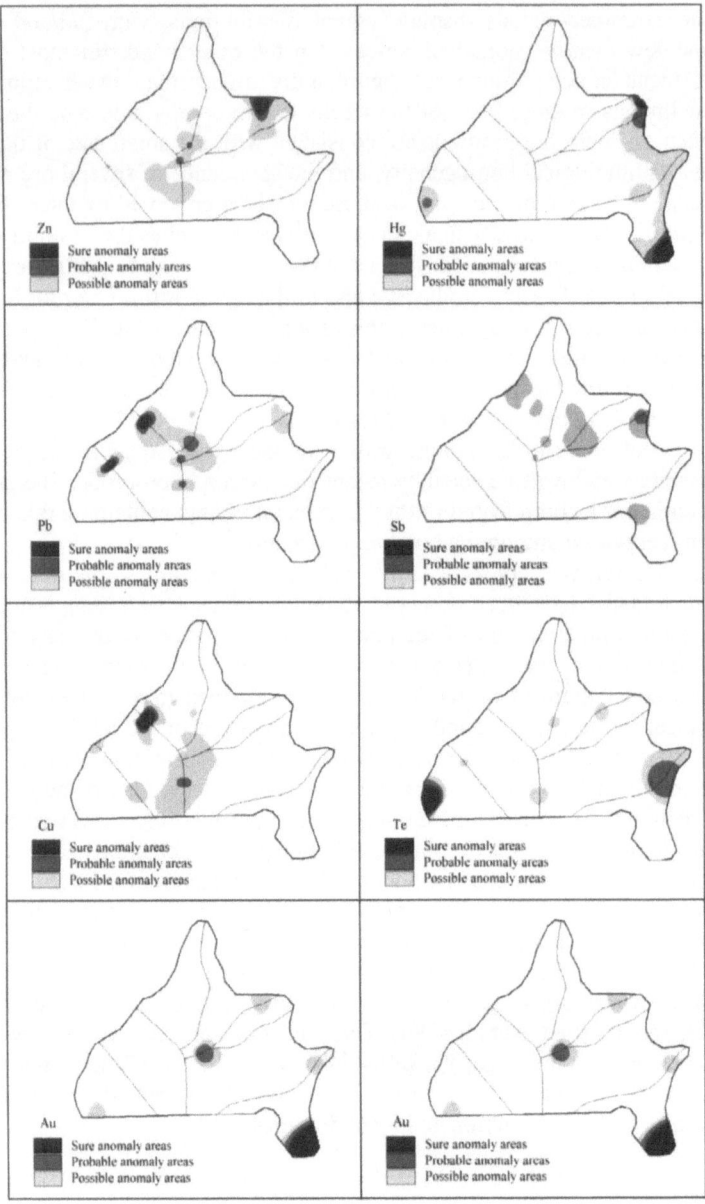

Fig. 2. Distribution of the study elements according to the anomaly thresholds given in table 1.

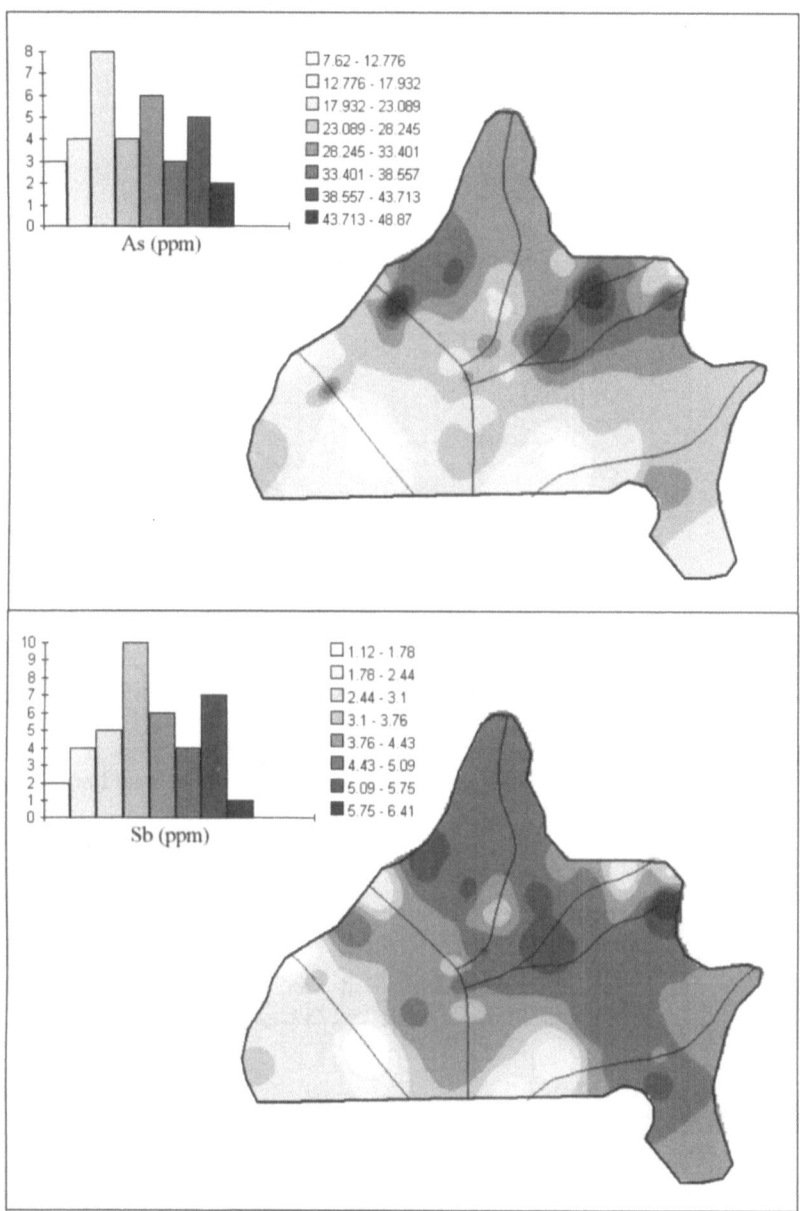

Fig. 3. Distribution of As and Sb according to the frequency classes of his histograms.

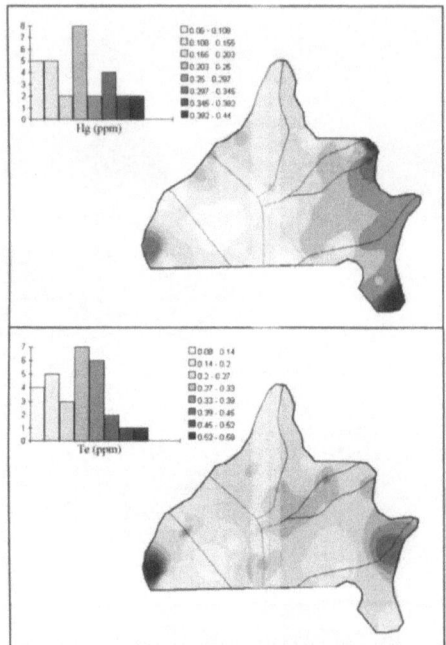

Fig. 4. Distribution of Hg and Te according to the frequency classes of his histograms.

Moreover, the anomalous values are generally spatially related to the metal sources. The above findings suggest the lake sediments act as a trap both for mineral clasts and metallic ions.

Acknowledgements

This study was supported by the Centro Studi Geominerari e Mineralurgici del CNR, Facoltà di Ingegneria, Piazza d'Armi, 09123 Cagliari, Italy

References

Fadda S., Fiori M., Grillo S.M.,Manca M.G., .Marcello A., Matzuzzi C.,Pretti S. (1996). Influenza della vecchia attività mineraria nel bacino del Lago Flumendosa e sui contenuti di metallo nei sedimenti del Lago Mulargia" Atti Congr. Intern. A.M.S., 137-149
Fiori M., Grillo S.M. (2001). The Furtei high sulphidation epithermal gold deposit (Sardinia, Italy): mineral assemblage and its evolution. Proceeding of the International Congress "Mineral Deposits at the Beginning of the 21st Century, 739-742
Wedepohl H.K. (1978) Abundance in common sediments and sedimentary rocks. Wedepohl H.K.(ed) Handbook of geochemistry. Springer.

Impact of ore deposits and anthropogenic activities on local hydrochemistry at "Silberberg", S-Germany

Christine Vornehm[1], Steffen Bender[1], Gerhard Lehrberger[2]

[1]Department for Geo- and Environmental Sciences, Ludwig-Maximilians-University Munich, Luisenstr. 37, D-80333 Munich
[2]Lehrstuhl für Allgemeine, Angewandte und Ingenieur-Geologie, Technische Universität München, Arcissstr. 21, D-80290 München

Abstract. Massive sulfide ores have been mined at the mountain „Silberberg" in Bavaria, South Germany. Metal- and sulfur-rich tailing dumps were deposited at the surface. Due to road construction the tailing dump material has been moved, which result in a remobilization of anions and cations. Four water types can be classified: mine water, shallow ground water and a mixture of both types. One well with deeper ground water does not fit in this scheme. Experimental results point to an impact of ore oxidation processes or an influx of high mineralized water.

Introduction

The aim of the investigation at the former sulfide mine at "Silberberg" mountain in the Bavarian Forest (South Germany) (Fig. 1) is to investigate the drainage system and hydro chemical processes. Possible impacts for the hydrochemistry are 1) the ore deposit and 2) anthropogenic activities (road construction and traffic, tourism, forestry). The study, which is part of the project "LOWRGREP" (land use optimization in hard rock areas with regards to ground water resources protection), is funded by the European Community. Main task of this project is focused on the impact on ground water beside of anthropogenic activities and the challenges for water management in hard rock areas.

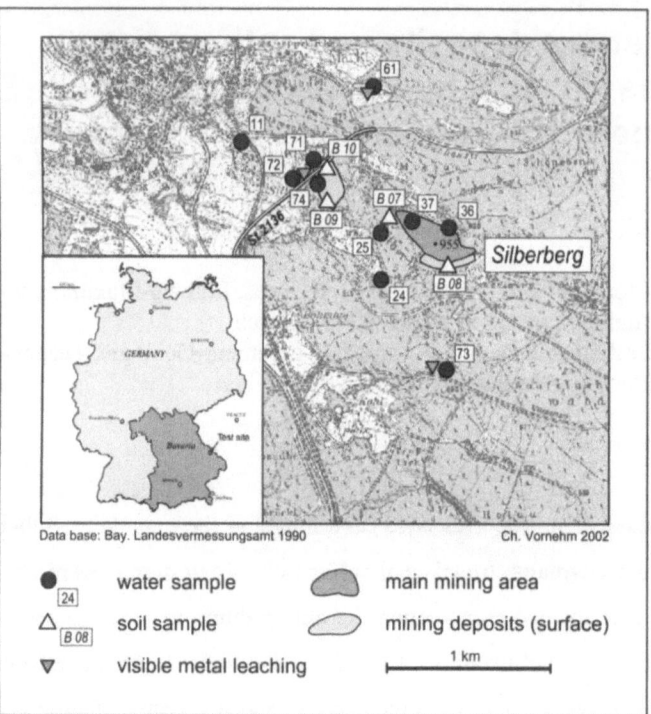

Fig. 1. Investigations in the "Silberberg" area

The ore deposit

General overview

The ore deposit is situated in Cordierit-Sillimanit-paragneisses, which are part of the moldanubic monotonous group (BayGLA 1998). These metamorphic units are part of the Bavarian section of the Bohemian Massif (BayGLA 1996). Up to now, the genesis of the ore deposit is not totally identified. The ore of the "Silberberg" is part of a massive zone of deposits that had been built during Palaeozoic or Prae-cambrian age. It originated in a submarine sedimentary basin, where tectonical activity in a sheer zone favored the inflow of iron-, zinc- and sulfur-bearing hydro-thermal fluids. Chemical reactions with surrounding sea water and regional metamorphic processes finally generated the iron- and zinc-sulfides. High contents of barium in ores and feldspars found in the region support the theory of their hydrothermal origin (Obermüller 1993, Linhardt 1985).

The ores predominantly consist of sphalerite (ZnS), galena (PbS), chalcopyrite (CuFeS$_2$), pyrite (FeS$_2$) and phyrrhotite (FeS). Generally, these minerals are associated by Ag, As, Au, Cd and Ni. (Alloway 1996). Inside the mine "stalactites" of mostly iron hydroxide (62 % Fe$_2$O$_3$) were formed by precipitation out of iron-rich acidic drop-water (pH 2.6). RFA-analysis show that these iron oxides contain further amounts of sulfur, fluorine, small amounts (< 1 %) of SiO$_2$, Al$_2$O$_3$, MgO, K$_2$O and the trace elements Zn (66 ppm), V (49 ppm), Ni (27 ppm), Cr (25 ppm), U (15 ppm) and Cu (12 ppm) (Vornehm et al 2002).

Mining activities

Since the Middle Ages massive sulfide ore has been excavated in open pit and underground mining (Blendinger and Wolf 1971). Main purpose was the production of sulfates and concentrated sulfuric acid for the chemical industry and iron oxide powder ("Polierrot") for polishing in the glass industry. During the mining activities until 1962, large quantities of ore and rocks have been excavated. On the surface sulfur- and metal-rich dumps were deposited near adits and shafts. Furthermore, around the former roasting and leaching plants, tailings were deposited.

After the ending of mining activities, the residual dumps, that have been stored in terraces around the mine, were covered with soil material, which was the base for forest vegetation. In the end of the 1990s parts of these tailing dumps were removed due to road construction, stored next to the other tailing dump material and protected with plastic canvas covers. With beginning of the road constructing activities, heavy metal concentration in surface runoff and seepage water increased significantly. The leaching of the metals can be observed at a dam of a new constructed road and in a nearby pond (Fig. 1).

Characterization of the covering material

Several samples have been taken in July 2001 from the tailing dumps (B 09 and B10 in Fig. 1), at the former roasting sites (B 08) and from soil in uninfluenced parts of the region ("base value"). Geochemistry was investigated by RFA-analysis and leaching behavior was determined by batch experiments.

Geochemistry

Compared to the local chemical situation of the soil, all tailings are characterized by higher contents of Fe$_2$O$_3$, S and F$^-$ and lower contents of SiO$_2$ and Al$_2$O$_3$ (Fig. 2). The highest contents of Fe$_2$O$_3$ and S are found in the tailings of the former roasting plant (B08), because the iron-sulfate had not yet been leached like in the tailing dumps (B09, B10).

Regarding trace elements, the concentration of Zn is a characteristic marker for the tailing material. It derives from sphalerite (ZnS), the most important ore of the "Silberberg" mine. Whereas the "base value" ranges between 30 and 60 ppm (Fig. 3), the amount in the tailings is clearly higher (800 and 10,000 ppm). Copper, lead and barium can be used as characteristic elements, too. Arsenic has been found only in the material of the former roasting plant (B08) with a quantity of 53 ppm, in the other tailing material it is below detection limit (20 ppm).

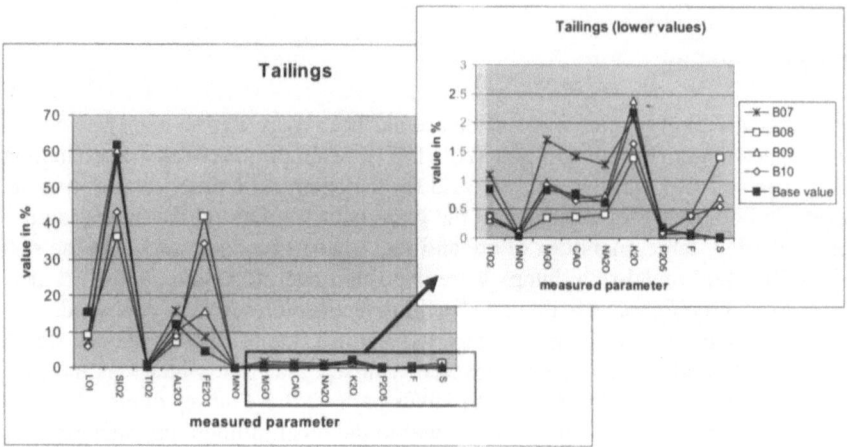

Fig. 2. RFA-analysis of the tailing material, main components

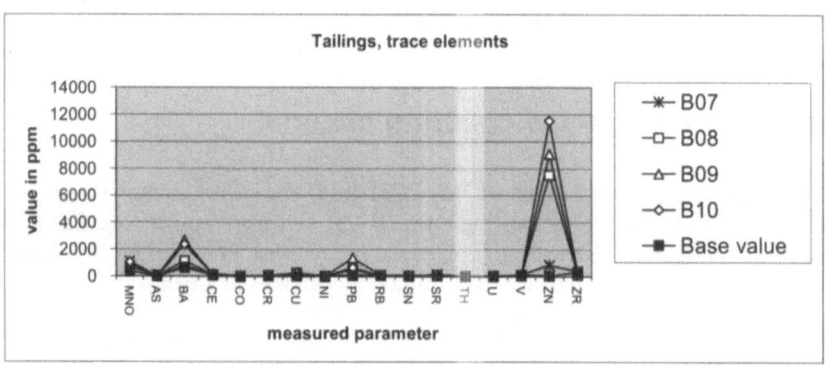

Fig. 3. RFA-analysis of the tailing material, trace elements (Vornehm et al. 2002)

Increased contents of manganese seems to be bound to manganese bearing garnets, which are often associated with phyrrhotite and galena-bearing rocks (Obermüller 1993), two typical ores of this deposit. A maximum of 1093 ppm MnO was analysed at soil sample B07 at the foot of "Silberberg" mountain. The tailing dump material has values between 900 and 1060 ppm MnO (B09, B10) whereas the base values range between 200 and 800 ppm.

The geochemistry of the soil sample B 07, taken at the foot of the "Silberberg" mountain, follows the general trend of the local "base value" with only slightly higher contents of the characteristic "Silberberg"-metals Zn and Fe_2O_3 (Fig. 2 and 3).

Leaching behavior

To determine the leaching capacity with water, selected samples of uninfluenced soil ("base value) and tailing dump material were leached according to DIN (1984). The eluat of the tailing dumps (B08, B09, B10) has a specific electric conductivity over 2000 µS/cm and pH ranges between 3.4 and 3.8. In comparison the eluat of sample B07, taken at the foot of "Silberberg" mountain, has a conductivity of 39 µS/cm and a pH of 4.2, lying in "base value" range (EC 30 – 130 µS/cm, pH 2.9 – 4.5). Figure 4 and 5 depict the leached amounts of selected cations and anions. Sample B06 represents the "base value" soil (EC 92, pH 3.4).

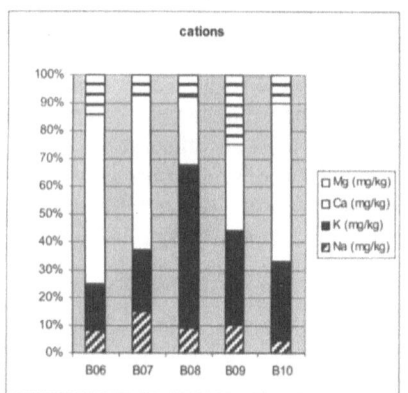

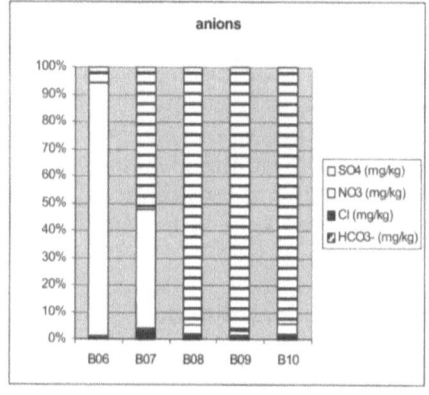

Fig. 4. cations of leaching experiment   **Fig. 5.** anions of leaching experiment

The results of the leaching experiment show that calcium is the dominating alkaline and earth-alkaline ion in most samples of soil and tailing damp material. In sample B08 (tailing dump material), potassium is predominant (60 %). Compared to the local "base value" potassium is increased in all samples from tailing dump material (>20 %). Therefore it can be used as characteristic for the mining area. Both calcium and potassium can be leached as well from feldspars and clay minerals. Compared to potassium, magnesium can not be used as a marker, because no significant difference between tailing dump material and the "base value" can be monitored. Concerning sodium, no significant difference can be seen between the samples, too (Fig. 4).

Using SO_4^{2-}, NO_3^- and Cl^- simplifies the classification of the samples. In tailing dump material sulfate is the most important anion (over 90 %), whereas the local soil is dominated by nitrate (around 80 %). The soil sample B07 from the foot of the "Silberberg" mountain is characterized by a domination of sulfate (almost 70 %) and additionally increased amounts of nitrate (30 %). Bicarbonate does not leach from any of the soils, neither from tailing dump material nor from local un-influenced soil, which points to the poor buffering capacity of the rocks (Fig. 5).

Concerning the most important metals for this region, iron was leached in very small amounts (below 0.3 mg/kg), whereas for zinc, concentrations between 0.4 and 11.4 mg/kg (tailing dump material) and 0.4 mg/kg (former roasting places) were found. Amounts of dissolved Zn up to 0.4 mg/kg have been observed in local soil as well.

Water conditions

Hydro geological overview

To describe the hydro chemical situation at the "Silberberg" several springs and one well were sampled. The springs, which are located in gneisses are mostly cap-tured in the weathering zone (3–6 m). Some wells in the region further show in-fluence of higher mineralized water originated in deeper circulating water. Their hydro chemical composition is hard to describe because local wells seldom exceed depths over 40 m. But investigations and models in similar regions point to the possible mixing of the deeper ground water from the fractured aquifer with the shallow ground water in the porous weathering zone (Bender 2000).

Hydro chemical situation

Three different types of water can be found in the surroundings of the "Silber-berg": mine water, shallow ground water and mixtures of both types (Vornehm et al 2002). The origin of the water from a well at the foot of "Silberberg" mountain is not cleared yet: it could be either influenced by the ore deposit or by deeper cy-cling ground water (fourth water type). The equivalent concentrations of the most important dissolved species are depicted in the vertical diagram (Fig. 6).

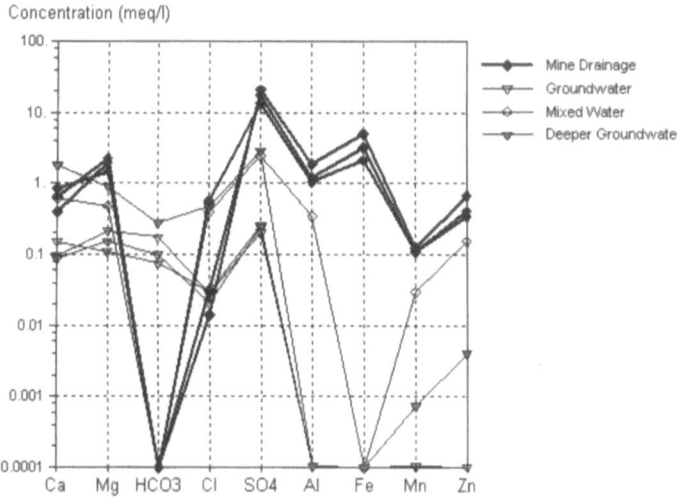

Fig. 6. Classification of the waters, using a vertical diagram (Vornehm et al. 2002)

The mine water is characterized by low pH-values (2.4 – 3.0) and high specific electric conductivities (EC) between 800 and 1900 µS/cm. It can be classified as Fe-SO$_4$-type. Compared to the other local water the amounts of Al, Zn, Mn and Cu are increased.

Shallow ground water is low mineralized (EC: 50 – 60 µS/cm) with pH-values about 5.5 to 6.0. The alkaline earth-HCO$_3$-water type shows regionally higher contents of Na$^+$ and SO$_4$$^{2-}$. The amounts of heavy metals are very low (Vornehm et al. 2002).

The mixed water found in the drainage system of the mine is acidic (pH-values about 4) with a mineralization below 300 µS/cm. Due to the alkaline earth metals that are added with the inflow of ground water, this water is of a CaMg-SO$_4$-type. Whereas nearly no dissolved iron can be detected, the following dissolved metal were found: Al, Zn, Mn, Cu, Co, Ni, Cd, Pb, As and Cr.

Deeper cycling ground waters have pH-values like the shallow ground water (5.5 – 6.0) and a mineralization of about 350 µS/cm. The CaMg-SO$_4$-water type shows increased HCO$_3$$^-$ concentrations (up to 20 mg/L) in comparison to the shallow ground water (1–12 mg/L).

Hydro chemical processes

Besides leaching of the ores and surrounding rocks by infiltrating rainwater, oxidation is the most important hydro chemical process. Main oxygen sources for the reaction are atmospheric oxygen and saturated oxygen in rainwater and seepage water. The oxidation releases sulfates such as melantherite, which characterize the waters in the vicinity of the ore deposit. Due to the oxidation processes of the ores, heavy metals (such as Fe^{2+}, Zn^{2+}, Cu^{2+} and so on) and high amounts of H^+ are released. All mine waters show pH-values between 2.5 and 3.0. However pH-values under 2.5 do not appear, not even in very concentrated drop water inside the mine, which can be due to buffering by products of silicate-weathering. Considering the very high amounts of sulfates in the mine water (around 1000 mg/L) it can be supposed that bacteria catalyze the process of ore oxidation (Nordstrom 2000).

The increased pH-values of the shallow ground water (up to 6.0) are the result of buffering reactions (Vornehm et al. 2002), but the sulfate domination still points to the ore oxidation.

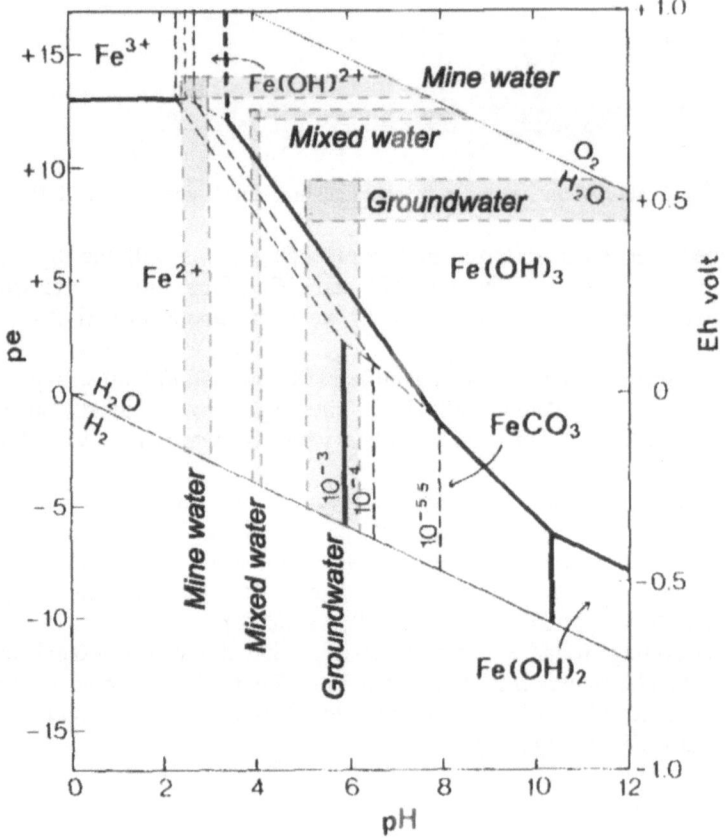

Fig. 7. pH-Eh-diagram for iron species (Appelo and Postma 1999) with the different water types

As a result of redox-processes and buffering reactions, the pH-Eh-conditions are changing along the flowpathes. Fig. 7 shows the pH-Eh-diagram for the iron species according to the observed water types. Regarding to mine water, different combinations of species are possible (Fe^{2+}, Fe^{3+}, $Fe(OH)_3$ and $[FeOH]^{2+}$). On site measurements of iron species (Fe^{2+}) showed that in the mine water (sample 37), 98 % of the total iron is present as Fe^{3+}. In water outlet no. 72, still 80 % of iron is dissolved as Fe^{3+}. Due to the abundant supply of oxygen and protons inside the mine, the ferrous iron from the oxidation of the ores is converted to dissolved ferric iron according to the following equation:

$$Fe^{2+} + \tfrac{1}{4} O_2 + H^+ = Fe^{3+} + \tfrac{1}{2} H_2O$$

In shallow ground water the existence of ferrous iron is still possible, but no dissolved iron can be detected in the water. One reason for this fact is shown by the leaching experiments, where no iron can be leached from the solid phase.

A further source for iron decrease in the water can be observed at the mine water outlet. Due to a super saturation regarding Fe_2O_3, an iron-oxide-phase is precipitating (Vornehm et al. 2002).

Conclusion

Four different water types are typical for the "Silberberg" area: 1) acidic mine waters, 2) shallow ground water, 3) mixed waters of mine water and shallow ground water and 4) deeper cycling ground water. However the genesis of the last water is not totally identified.

Leaching experiments show that calcium is the predominant earth alkaline and alkaline cation. Some samples show increased values of potassium (tailing dumps) and magnesia. The leaching behavior can be identified more clearly by the anions. Samples influenced by the ore deposit contain sulfate as predominant anion, whereas in other samples nitrate is dominating. Iron in soil and tailing dump material seems to be strongly fixed.

References

Appelo C A J, Postma D (1999) Geochemistry, groundwater and pollution.

Bayerisches Geologisches Landesamt (BayGLA) (1996) Erläuterungen zur Geologischen Karte von Bayern 1:500.000.

Bayererisches Geologisches Landesamt (BayGLA) (1998) Geologische Karte 1:25.000, Blatt 6944 Bodenmais.

Bayerisches Landesvermessungsamt (1990) Topographische Karte 1:25.000, Blatt 6844 Lam und 6944 Bodenmais.

Bender S (2000) Klassifikation und genetische Entwicklung der Grundwässer im Kristallin der Oberpfalz/Bayern. Münchner Geologische Hefte, B 10.

Blendinger H, Wolf H (1971) Die Magnetkieslagerstätte Silberberg bei Bodenmais und weitere Erzvorkommen im Hinteren Bayerischen Wald. Der Aufschluß, Sonderbd. 21: 108–139.

DIN (1984) DIN 38414, Teil 4. Deutsche Einheitsverfahren zur Wasser-, Abwasser- und Schlammuntersuchung, Bestimmung der Eluierbarkeit mit Wasser (S4).

Linhardt E (1985) Petrographische Untersuchungen am Silberberg bei Bodenmais im Hinblick auf den Stoffbestand der Rahmengesteine; Diplomarbeit am Institut für Mineralogie und Petrographie der Ludwig-Maximilians-Universität München.

Nordstrom D K (2000) Advances in the Hydrogeochemistry and Microbiology of Acid Mine Waters. Intern Geol Rev 42: 499–515.

Obermüller Th (1993) Der Silberberg bei Bodenmais/Bayerischer Wald. Der Aufschluß 44: 201-224.

Vornehm Ch, Bender S, Lehrberger G, Wohnlich S (2002) Hydrochemical and Drainage Processes in vicinity of a former sulphide mine in South Germany. GeoProc2002, in press.

Quantitative analysis of supply from anthropogenic sources to mine workings of closed zinc-lead ore mines in the Bytom Trough (southern Poland)

Janusz Kropka

Faculty of Earth Sciences, University of Silesia, Będzińska Str. 60, 41-200 Sosnowiec, Poland

Abstract. The central pumping station Bolko in Bytom enables continuous dewatering of mined-out workings in lead-zinc ore mines in the Triassic Bytom Trough. Water of anthropogenic origin is of great importance in the balance of water flowing into workings and pumping stations: leakage from water supply networks and sewerage systems and seepage from rivers and collector trenches. The average total amount of supplying ore workings from analysed anthropogenic sources of water and sewer in the years 1999-2000 was 12.2 m^3/min, equivalent to a recharge rate of 341 mm/y. Its share in total water inflow was 40%.

Introduction

The Bytom Trough constitutes a complicated, multi-layer (Quaternary, Triassic, Carboniferous) aquifer system, which is the location of a system of two-level mining exploitation. The upper one, in the Middle Triassic beds, was for the exploitation of zinc-lead ore deposits, finished in 1989; while the lower one if for active exploitation of hard coal deposits in the layer of productive Carboniferous. Difficult mining conditions result from the fact that areas of the mentioned coal mines were or are located beneath a densely developed building setting belonging to the urban area of Bytom and Piekary Śląskie cities. The cities have an approximate population of 200,000. It is necessary to drain ore workings in closed mines since there are active coalmines situated beneath. Otherwise, exploiting coal deposits in the layer of productive Carboniferous would be endangered by groundwater inflow. Atmospheric precipitation, 801 and 915 mm in the years 1999 and 2000 respectively, together with waters of anthropogenic origin constitute the main sup-

plying source. Anthropogenic sources comprise water that seeps from surface watercourses during the breakdown of water mains and urban water supply and
sewerage systems. It is extremely complicated to specify clearly anthropogenic
components in the balance of water flowing into ore workings.

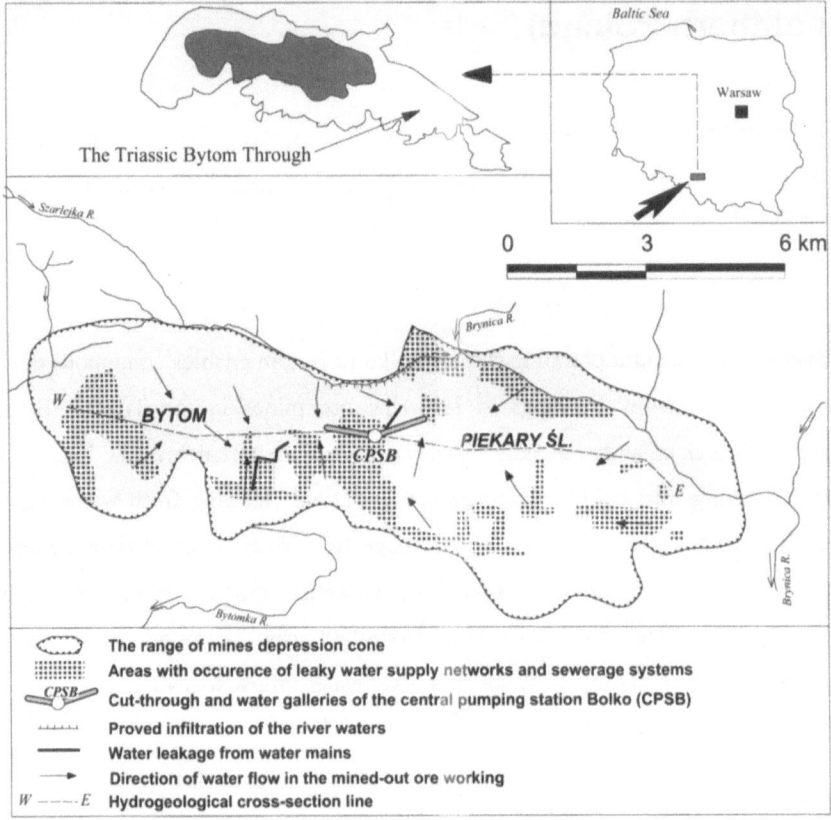

Fig. 1. Hydrogeological sketch of the mined-out workings in lead-zinc ore mines in the Triassic Bytom Trough.

Geology, mining operation and hydrogeology of the study area

The study area is located in the southern part of Poland, as indicated in Fig. 1. The
Bytom region of ore mining is situated in the western part of Silesian–Cracow
area where zinc-lead ore deposits occur (Wilk et al. 1990). The Triassic Bytom
Trough is one of several fold-block alpine structures of lower strata within the
border of Silesian-Cracow monocline. The northern, northern-eastern and southern
borders are tectonic-erosive, while the western is hydrodynamic (Fig.1). In the

Triassic profile, with total thickness from several to maximum ca. 230-250 metres, terrigene deposits of Lower and Middle Bunter Sandstone and carbonate deposits of Roethian and Muschelkalk occur. The Triassic cover consists mainly of Quaternary deposits, greatly differentiated as far as thickness and lithological development are concerned (Fig.2).

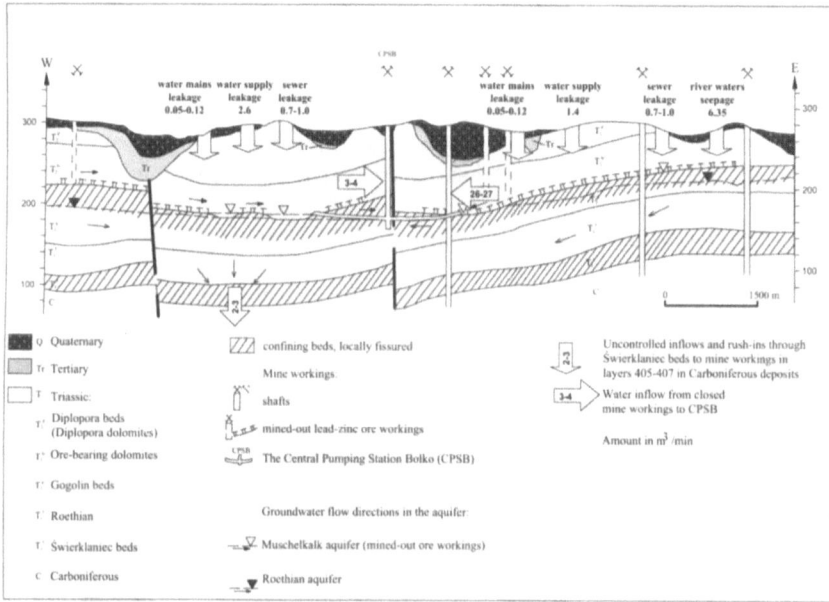

Fig. 2. Hydrogeological cross-section.

The first documented historical reports on zinc and lead ores mining activity within the discussed area date back early 12th century (Majorczyk 1985). Industrial mining exploitation, started in the late 19th century, reached the depth of ca. 110 metres under the surface level, and caused intensive transformations in water relationships. In the mid-seventies of the 20th century there were five active zinc and lead ore mines. Zinc and lead ore deposits within the discussed area are related to the so-called ore bearing Dolomite series of the Middle Triassic. The subject of exploitation was both oxidized ore deposits (galmei) and sulphide ores (zinc blende and galenite) (Baranowski 1980; Gałkiewicz 1980). Thickness of exploited deposits ranged from 2.0 to 6.5 m, 3.0-4.0 m on average. The main exploitation levels were situated at the depth (from surface) from 90-100 in the western and central and 64-96 m in eastern part of mining area (Fig.2). Mines not working at present are connected with each other by means of underground workings. The area of mine workings in the last few years of exploitation (1983-1989) was ca. 39.0-40.0 km². Mine workings were situated within the borders of a mining area of 61.4 km². Total range of draining influence of ore mines on groundwaters in Quaternary and Muschelkalk deposits is ca. 57.8 km² (Fig.1).

Exploitation of hard coal deposits is performed in productive Carboniferous, within the influence of zinc-lead ore exploitation. The period of intensified hard coal exploitation started in the seventies of the 19th century. Ten hard coal mines were still working 11 years ago, when zinc-lead ore mining was definitely finished. Currently, according to the state for 1 Jan. 2002, five coalmines are still active. Generally, below ore workings, the present deposits of Orzesze Beds (Westphalian A) and Ruda, Saddle and Poruba Beds (Namurian C, B and A accordingly), with thickness from 1.2-9.0 metres, there were, and still are, exploited. Intensive, mainly roof caving exploitation of hard coal, within the level of 180-930 deep (under the area surface), leads to serious disturbances and deformations in Carboniferous rock mass, which are then transferred to carbonate Triassic rocks lying above, and finally to the surface. Surface deformations play a role in the breakdowns of urban water supply and sewerage systems on areas of dense building settings, as well as the loss of tightness in watercourse beds and basin resulting in no water runoff.

Basically, three independent aquifers were distinguished in the natural hydrogeological profile of the Triassic multi-aquifer formation (Kropka 1996). The main Muschelkalk and Roethian aquifers, consisting of dolomites and limestone, are karst-fissured aquifers. The first of them is of great significance for inflow to workings of liquidated zinc-lead ore mines (Fig.1, 2). Muschelkalk aquifer comprises a series of dolomitic limestone, ore bearing limestone, and minor marly dolomites. This series is highly porous and cavernous, partially brecciated, providing advantageous conditions for receiving, storing and transmitting water. This layer is isolated from the lower one by vitriolic clays and marly-limestone clays, and from the upper one by part of Gogolin strata. In natural conditions, the water table in Muschelkalk aquifer occurs at the ordinate +265 m above the sea level, while thickness ranged from several up to maximum ca. 50-55 metres. Recharge conditions in Muschelkalk aquifer within the Triassic Bytom Trough are greatly differentiated. Long lasting ore exploitation in the central part of the Bytom Trough caused lowering of the water table up to the level of ore workings, i.e. to the ordinate 170-240 m.a.s.l. (metres above sea level). At present, the Triassic rock mass above ore workings level is drained (Fig.2).

Abandoning mining exploitation in 1989 and final liquidation of ore mines in the Bytom Trough (years 1990-1991) did not stop drainage of ore workings. Ceasing drainage and, in consequence, flooding ore workings and Triassic rock mass would cause a real water inflow danger for active hard coal mines. This is why the system of central drainage in liquidated mines was introduced, which is designed to operate until hard coal deposits lying below are completely exploited. The main role in the centralised system of draining abandoned ore workings is played by the Bolko shaft, 129.3 m deep, with a newly built central pumping chamber, with pumping capacity ca. 36.0 m³/min, and hollowed drainage workings directed to the west and east (Fig.2). Central pumping station at Bolko shaft (CPSB) started to work in 1988, and in the period from March 1989 to August 1990, by means of the two mentioned workings, it took over waters coming from liquidated pumping stations in the five ore mines. In the period from August 1990 to December 2000 water flowing into CPSB was ca. 30.0 m³/min on average. The main factors that differentiate the share of water inflow from the western and eastern parts to CPSB are following:

- cover, predominantly occurring in the western part, from several to ca. 70 metres thick, consisting of clay and loamy Tertiary and Quaternary deposits, while in the eastern part the dominant role is played by areas of advantageous supply; and
- entering a part of water from ore workings through Bunter Sandstone beds to the workings in coal mines in 1991, at the western side of CPSB.

Anthropogenic sources supplying mine workings

The factors that decide the amount of anthropogenic water supplying inflow to mine workings in abandoned ore mines is water and sewer infiltration from the following sources:

Rivers and collector trenches

The problems connected with water seepage from surface watercourses to mine workings in ore mines were presented, among others, by Plotnikov and Roginets (1989), Sawicki (2000) and Wilk et al. (1990). Rivers and collector trenches flowing through mining areas belonging to ore mines created, until 1989, the most serious water inflow danger to mining works in the Bytom Trough. The reasons for subsidence and damage to tightness of river beds, resulting in water infiltration and as a consequence an increase in the amount of water flowing to workings, was mining exploitation of hard coal deposits in a direct neighbourhood of watercourses. Among other factors, seepage from the Brynica and Szarlejka rivers intensified in the 1970s, caused temporary increase in waterflow to two ore mines, from 1.6-2.7 m^3/min to 10.7 m^3/min, and from 7.0 m^3/min to 13.0 m^3/min, respectively (Kropka, in press). River waters, highly polluted, flow in regulated and partly self-tightened river and stream beds. However, three series of hydrometric measures performed in 1999-2000, proved that water escaped in two sections of the mentioned rivers in total amount of ca. 6.35 m^3/min (Fig.1, 2).

Water mains

Large diameter water main pipelines (several hundred mm dia.) transect the Triassic Bytom Trough from north to south. Intensive deformation of the rock mass has caused two sections of pipeline to be particularly vulnerable to cracking and leakage (Fig.1). It is estimated that, as a result of several breakdowns of the mentioned pipelines, ca. 0.1 up to ca. 0.25 m^3/min of water infiltrated to the rock mass and then to ore workings.

Leaky water supply networks in Bytom and Piekary Śląskie cities

Leakage from water supply is a major source of urban recharge (Lerner 1997). Often cited average values of groundwater recharge rate from this source in the suburbs of Lima (Peru), Tokyo (Japan) and Birmingham (England) are 360, 440 and 180 mm/y respectively (Foster et al. 1997; Lerner 1997). According to Lerner (1997) loss rates of 20-25% are considered normal in the UK. The cities of Bytom and Piekary Śląskie take water for consumption purposes from outer sources. Leakage from water supply is intensified by old, highly exploited network with many worn-out connections, which are easily affected by changes of pressure in water mains and activity of underground mining. In the early 1990s, loss resulting from infiltration within the area of discussed cities amounted to 35-40%. The amount of loss is the difference between the amount of water pumped into the district and the amount of water bought by single users of the water supply network. In the period 1996-2000, water supply agencies allocated significant financial resources to rehabilitating infrastructure of urban water supply network. Thanks to the effectiveness of these actions the number of breakdowns has fallen and infiltration has decreased. Infiltration rates are presented in Table 1.

Leaky sewerage systems (urban and industrial waste water)

Although sewer leakage occurs, there are almost no estimations of quantities, and no proved methods of identification and quantification. Some publications suggest that an average leakage rate should be no lower than 5% (Lerner 1997). Much higher standards are possible in areas where the provision of waste water disposal is performed by sanitation without sewerage network (e.g. septic tanks) and ductile-iron pressurized sewage systems (Foster et al. 1999). Over 99% of all households in the analysed districts of Bytom city are connected to a complete sewer system to remove waste water for further treatment in water treatment stations. The districts in Piekary Śląskie, within the borders of mining activity are covered with sewerage system in ca. 70%. The leakage of sewers is not well documented within the discussed area. The analyses made so far suggest that an average leakage rate ranges from 7-12% in areas with sewerage system and 30-50% in areas where provision of waste water disposal is performed by sanitation without sewerage network (Table 1).

Discussion and conclusions

The average total water inflow from closed mine workings to CPSB in the years 1999-2000 amounted to ca. 30.0 m³/min. Total closed workings recharge in the western part of the system remains on the level of 5.0-7.0 m³/min, out of which 3.0-4.0 m³/min flows into the pumping station, while the remaining 2.0-

3.0 m³/min infiltrates to shallow mine workings in hard coal mines. Inflow of water from the eastern part to the pumping station was 26.0-27.0 m³/min (Fig.2).

The average total amount of ore workings recharge in the western region (within two districts of the Bytom city) from analysed anthropogenic sources of water and sewer in 1999-2000 was 3.4-3.7 m³/min. Its share in total recharge was 53-68%. The average amount of analogous recharge from anthropogenic sources in the eastern region (mainly in four districts of the Piekary Śląskie city) in the years 1999-2000 was 8.5-8.9 m³/min. Its share in total water recharge was 33%.

Table 1. Major anthropogenic sources of mine-out ore workings recharge (m³/min/mm).

Mined-out ore workings	Year	Water mains leakage	Water supply leakage	Sewer leakage	River waters seepage	Total recharge
The western part	1999	0.05-0.12	2.7	0.7-0.9	-	3.4-3.7
		-	337	87-112	-	(424-449)
	2000	0.05-0.12	2.4	0.7-1.1	-	3.2-3.6
		-	299	87-137	-	(387-436)
The eastern part	1999	0.05-0.12	1.5	0.6-0.9	6.35	8.5-8.9
		-	172	69-103	-	(241-275)
	2000	0.05-0.12	1.3	0.7-1.1	6.35	8.4-8.9
		-	149	80-126	-	(229-275)
The average total amount in the years 1999-2000		0.1	4.0	1.7	6.35	12.2
		-	239	102	-	(341)

Annual summaric maintenance costs of CPSB in the analysed period ranged from 8.8 to 9.3 million zloty. It must be emphasized that 22-23% of these costs results from expenditure on electric energy. It is extremely complicated to specify clearly anthropogenic components in the balance of water flowing into CPSB and it requires further examination. The author realises that generalisations and estimations made during calculations were necessary. Results indicate that for average meteorological-hydrogeological conditions, the contribution of water from anthropogenic origin as inflow to closed ore workings is about 39-42%, i.e. from ca. 11.6 to ca. 12.6 m³/min. Waters of anthropogenic origin leads to deterioration of water quality and an increase in the cost of maintaining pumping stations. This increase is best illustrated in a one-year period from 1 Sept. 1998 to 31 Aug. 1999, in which the unit cost of electric energy of pumping 1 m³ water out to the surface was 0.1295 złoty. Sealing the whole system by means of, first of all limiting water seepage from anthropogenic sources, and simultaneously limiting water inflow by 20-30% as compared to present amount, will contribute to savings in expenditure (on electric energy alone) by the sum of 408,000—613,000 zloty per year.

References

Baranowski H (1980) The Bytom Depository Region. Hydrogeological conditions of the zinc-lead ore deposits of the Silesia-Cracow region (A.Różkowski, Z.Wilk ed.) (in Polish). Pr. Inst. Geol. (series not numbered): 113-139

Eiswirth M, Hotzl H (1997) The Impact of leaking sewers on urban groundwater. Groundwater in the Urban Environment. Vol.1: Problems, Processes and Management (J.Chilton et al. ed.). Proceed. of the XXVII IAH Congress, Nottingham. Balkema, Rotterdam: 399-404

Foster S, Morris B, Lawrence A, Chilton J (1999) Groundwater impacts and issues in developing cities-An introductory review. Groundwater in the Urban Environment. Selected City Profiles (J.Chilton ed.). International Contributions of Hydrogeology Vol. 21. Balkema, Rotterdam: 3-16

Gałkiewicz T (1980) Zinc and Lead Ore Deposits. Hydrogeological conditions of the zinc-lead ore deposits of the Silesia-Cracow region (A.Różkowski, Z.Wilk ed.) (in Polish). Pr. Inst. Geol. (series not numbered): 53-56

Kropka J 1995 The influence of the coal mining activity on groundwater in the Triassic carbonate series (Bytom). Ecological impact of underground coal mining and activities of associated industries. XIII International Congress on Carboniferous-Permian. Guide to Excursion I1. Polish Geological Institute: 9-12.

Kropka J (1996) Circulation routes, resources and groundwater management in Triassic Bytom reservoir in conditions of active mining activity (in Polish). Prz.Geol., 8: 845-849

Kropka J. in press – Hydrogeology in Bytom Region of Zinc-Lead Ore Deposits Exploitation. Hydrogeology of Polish natural resources and mining regions (Z.Wilk ed.)

Kropka J, Respondek J (2000) Hydrogeological and mining problems of the central dewatering system in mined-out ore workings in The Bytom Trough (southern Poland) (in Polish). Prz. Geol., 48: 727-735

Lerner D.N (1997) Too much or too little: Recharge in urban areas. Groundwater in the Urban Environment. Vol.1: Problems, Processes and Management (J.Chilton et al. eds.). Proceed. of the XXVII IAH Congress, Nottingham. Balkema, Rotterdam: 41-47

Plotnikov N.I, Roginets I.I (1989) Hydrogeology of Ore Deposits. Balkema, Rotterdam: 1-290.

Sawicki J (2000) The changes of natural infiltration of precipitation into aquifers as the result of a deep mining groundwater drainage (in Polish). Ofic. Wyd. Pol. Wrocł., Wrocław: 1-174.

Wilk Z, Adamczyk A.F, Nałęcki T ed. (1990) Influence of mining activities on water environment in Poland (in Polish). Monograph CPBP 04.10, Vol 27, Wyd. SGGW-AR, Warszawa: 1-220.

Influence of chemical reactions on the flow system and contaminant transport in a former salt mine

St. Wilhelm[1], J. Poppei[1] , G. Mayer[1], R. Schwarz[1], G. Klubertanz[1], P. Siegel[1], B. Förster[2]

[1]Colenco Power Engineering AG, Groundwater Protection and Waste Disposal, Mellingerstrasse 207, CH-5405 Baden, Switzerland
[2]GSF-Forschungszentrum für Gesundheit und Umwelt, Forschungsbergwerk Asse, Am Wald 2, D-38319 Remlingen, Germany

Abstract. We have investigated the influence of large-scale chemical reactions like gas production and host rock dissolution on driving forces for fluid flow and contaminant transport in an underground waste repository located in a former salt mine. Numerical simulations were conducted to study the influence particularly of these two chemical effects. The tool of choice was the newly developed KAFKA, a code considering multi-complex physical and chemical processes of two-phase flow and nuclide transport. The sensitivity of the chemical reactions was illustrated with a series of simulation scenarios. The paper describes the modelling concept, the processes involved and – for exemplifying scenarios – the influence of gas production and host rock dissolution on the potential release of nuclides into the biosphere.

Introduction

For post-operational phase scenarios of underground radioactive waste disposals in ancient salt mines it is assumed that formation waters of different origin may penetrate the repository. This would cause mobilization of the radionuclides and, potentially, release from the repository into the biosphere.

Safety analysis and long-term prognoses rely on models. These models have to reflect the fluid flow and transport processes with consideration of various physical and chemical processes such as convergence, advection, diffusion, dispersion, dissolution/precipitation or sorption of contaminants, as well as radioactive decay. If some of these processes are coupled, the models become increasingly complex.

The main focus of this paper is on the effect of two large-scale chemical reactions: gas production and host rock dissolution. These processes have a major impact on fluid flow by altering physical properties and the geometry of the domain.

Additional complexity results from the fact that old mines are generally geometrically very complex. It is, therefore, necessary to simplify the geometry in order to achieve a reasonably simple (in terms of geometry) model (Fig.1).

Fig. 1. Example of an actual, complex mine building (left picture) and the inferred "skeleton model"

In the following section we discuss the basic concepts of our newly developed numerical tool KAFKA = Kompartimentmodell für die Ausbreitung und die Fluiddynamik in einer konvergierenden Untertageanlage für Abfälle = Compartment model for fluid flow and contaminant transport in converging underground waste repositories (Colenco 2001). The intention is not to demonstrate the mathematical foundations of KAFKA, but to give an overview of the processes integrated.

Modelling the flow and transport system

The following processes and couplings are taken into account:
- Two-phase flow
- Diffusion and dispersion
- Sorption and limited solubility of radionuclides
- Radioactive decay
- Time-dependent gas generation due to corrosion and microbial degradation

- Dependency of pore volume (porosity) on convergence and creeping
- Dependency of flow and transport processes on varying permeability and cross-sections
- Dependency of pore volume on dissolution and precipitation
- Dependency of pressure on the chemical composition of the fluid.

The numerical solution is based on a finite volume approximation. For each finite volume, the equation of mass balance is solved for all constituents (e.g. nuclides, water, gas, salts subjected to dissolution). The processes described are (1) the volume change due to convergence and dissolution, and (2) the mass changes due to transport and chemical reactions. These changes impose alterations in dependent parameters, e.g. fluid pressure and density, which drive flow and transport.

Two-phase flow is described by the Brooks-Corey model and a generalised Darcy's law is used. The account of volume changes caused by convergence is based on the formulations in PSE 1985.

The discretisation of the domain into finite volumes implies a geometric abstraction of the mine's structure. In our concept we summarise the complex geometry of the excavated structures into four principal structural elements. These are (1) caverns, i.e. backfilled or open excavations with a high hydraulic conductivity, (2) tunnels, i.e. man-made horizontal hydraulic connections with a distinct hydraulic conductivity and which may serve as hydraulic barriers, (3) shafts, i.e. man-made vertical hydraulic connections with a distinct hydraulic conductivity, and (4) teeters, i.e. vertical connections with a hydraulic conductivity caused by stress conditions in the rock. Those structural elements are attributed specific properties with respect to hydraulic and chemical behaviour. Caverns (type 1) are, for example, treated as perfect mixing tanks while tunnels (type 2) portray conventional two-phase flow behaviour. The result is a "skeleton model" of the underground mine structures (Fig.1). Obviously, expert knowledge is essential in collapsing sets of actual mine structures into schematic model elements.

Chemical processes

The behaviour of the repository with respect to the release of radionuclides is influenced by gas generation and by the dissolution/precipitation of salts in contact with aqueous solutions.

Of the two large-scale chemical processes, the build-up of the gas pressure in the mine is obviously a direct result of chemical processes such as anaerobic corrosion of metal inventory and microbial degradation of organic matter. Based on the type of waste and the chemical environment, the effective gas generation rates as a function of time can be calculated with a separate chemical simulation model. A deduced parameterization of the temporally dependent, effective gas generation rates is then used as input for KAFKA.

In a similar manner, the dissolution and precipitation of salt (here referred to as fluid-salt interactions) in contact with aqueous solutions has been investigated in detail with an independent geochemical code, i.e. The Geochemist's Workbench

(Bethke 1998). In the system considered here, a NaCl-saturated solution is assumed to enter the mine and react with Mg-bearing salts such as carnallitite. This process results in a) a loss of carnallitite volume, b) the generation of additional solution by liberating crystal-bound water and c) the formation of a loose conglomerate of secondary minerals. Thus the fluid-salt interactions are expected to have an influence on the convergence and flow processes, as well. In KAFKA, the Mg content of the solution is used as a measure of the system's reactivity.

Model description

We chose a simple geometric model to demonstrate possible influences of gas formation and carnallitite dissolution (Fig.2). A cavern (Cav1) containing the nuclide inventory and the gas source is connected to a second cavern (Cav2) containing carnallitite. Cavern Cav2 is connected to a vertical shaft with a backfilled tunnel (Tun).

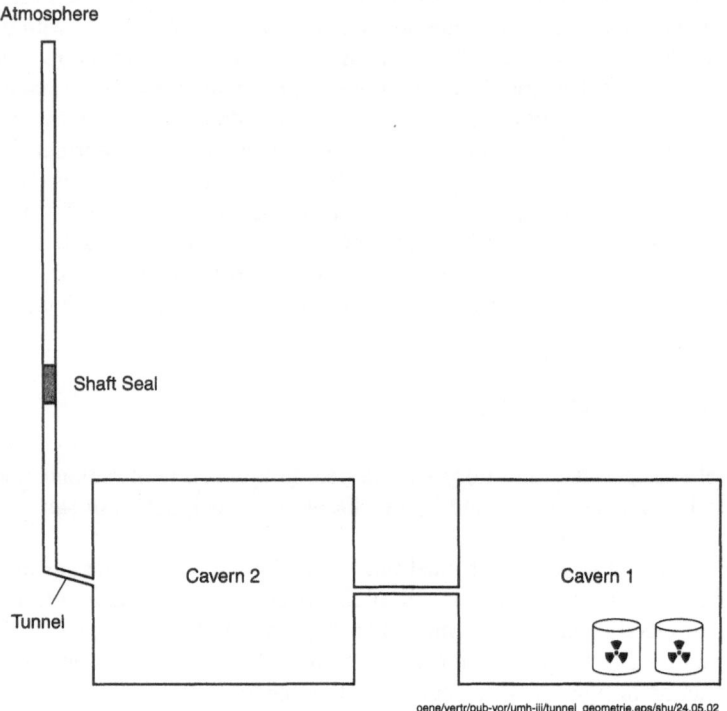

oene/vertr/pub-vor/umh-iii/tunnel_geometrie.eps/shu/24.05.02

Fig. 2. Geometric configuration of the model

The caverns, the horizontal connectors between them and the tunnel are subject to convergence, whereas the shaft itself is assumed to be stable. Part of the shaft is plugged with 50 m of filling and closing materials with a permeability of 10^{-13} m^2. The tunnel (Tun) is filled with sorbing backfill materials and has a permeability of 10^{-13} m^2. The connector between the caverns is given a permeability of 10^{-11} m^2.

The diffusion is described by an overall pore diffusion coefficient of 10^{-10} m^2/s. The dispersion length generally applied is 10 m.

The waste materials in Cav1 are mobilized instantaneously. The gas production follows an empirical exponential relationship described by the gas production rate and the final total gas volume:

$$V_g(t) = V_{gk}\left(1 - e^{-\lambda_{gk}t}\right)$$
(1)

with

V_{gk} ...   total gas volume at infinity (m^3)

λ_{gk} ...   production rate (1/s).

NaCl-solute enters at a constant rate through the top of the shaft and infiltrates into all structural elements of the model.

A simplified decay chain (N+2) and iodine (^{129}I) are used to demonstrate the influence on radionuclide concentrations under the consideration of solubility, sorption and decay. The inventory is assumed to contain the following nuclides, each with a general initial activity of 10^{10}Bq: Am^{242}, Cm^{246}, Pu^{238}, Pu^{242}, U^{234}, U^{238}, Ra^{226} and I^{129}. Both radioactive decay as well as ingrowths are taken into account.

The solubility limits and the K$_D$-values of the elements (the latter are required for the backfill materials in the tunnel) are listed in Table 1.

Table 1. Solubility limits and retardation coefficients used for the simulations

	Solubility limit (mol/m^3)	K$_D$ (m^3/kg)
Am	$1.0 \cdot 10^{-3}$	0.7
Cm	$1.0 \cdot 10^{-3}$	0.7
Pu	$1.0 \cdot 10^{-3}$	0.7
U	$1.0 \cdot 10^{-3}$	$1.4 \cdot 10^{-2}$
I	$1.0 \cdot 10^{40}$	0
Th	$1.0 \cdot 10^{-3}$	0.7
Ra	$1.0 \cdot 10^{-4}$	$2.1 \cdot 10^{-3}$

The parameters outlined in this section are used for the computational cases introduced in the following sections.

Since the transport effects of sorption, diffusion, dispersion and precipitation/dissolution are not the focus of this paper, their influence on the nuclide concentration is not discussed.

Influence of chemical reactions

The sensitivity of chemical reactions was investigated with a series of simulation scenarios.

1. Influence of fluid-salt interaction only, i.e. no gas generation:
-   "reference": no interaction to represent the reference case
-   "half carn": interaction is limited to half of the total amount of carnallitite mass in Cav2
-   "total carn": unlimited interaction with the total amount of carnallitite mass, i.e. $3 \cdot 10^8 kg$.

The dissolution of carnallitite results not only in an increased pore volume (Fig.3), but also in an increase of solute out flux from the shaft into the atmosphere. The latter is due to the increase in the amount of fluid available and an increase in the volume exposed to converge (Fig.4).

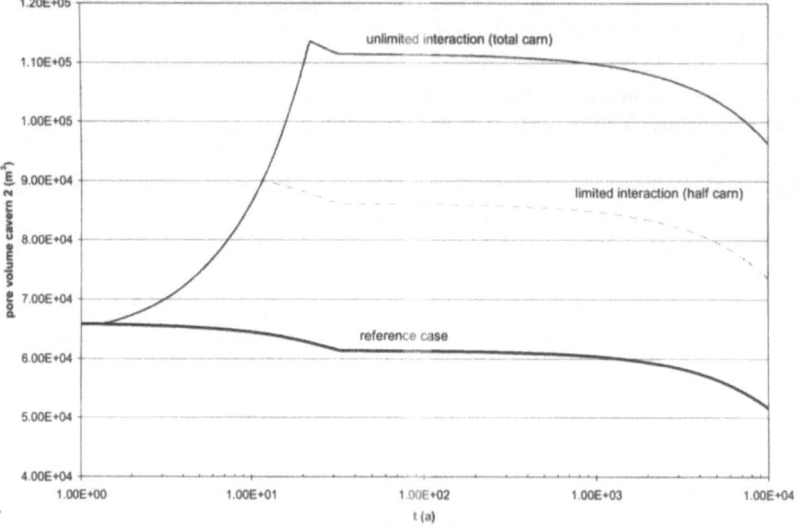

Fig. 3. Influence of carnallitite dissolution on the pore volume of the carnallitite bearing cavern Cav2

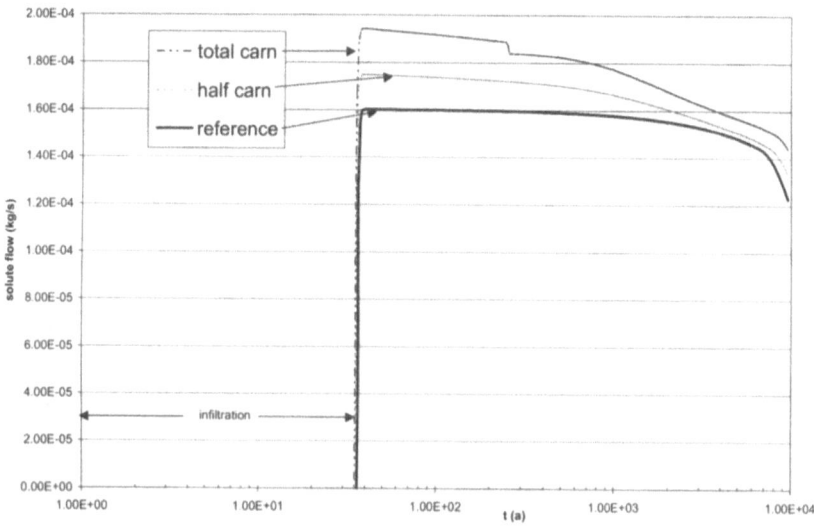

Fig. 4. Resulting solute outflux at the top of the shaft

The nuclide release into the atmosphere at the top of the shaft is element-specific as it is dependent on sorption and solubility limits (despite diffusion and dispersion). Already after about 10,000 years, the release rate for the non-sorbing iodine nuclides reaches a plateau, i.e. an equilibrium. The release rate for uranium, however, shows a marked lag (Fig.5). The release rates shown in Fig.5 are related to the initial nuclide concentrations in the inventory.

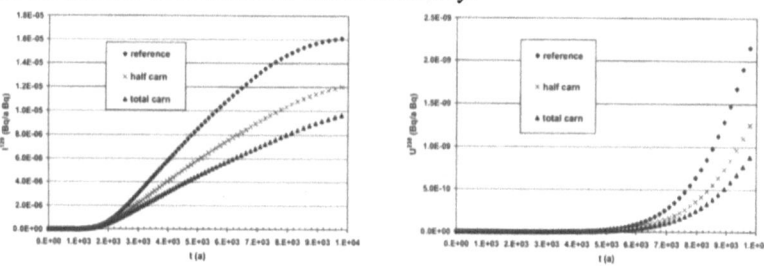

Fig. 5. Resulting annual nuclide release rate at the top of the shaft (related to the initial inventory) for iodine and uranium

It may be concluded that the fluid-salt interaction generally has an effect of retardation on the release of nuclides into the atmosphere.
2) Influence of the gas production only, i.e. no fluid-salt interaction:
- "reference": no gas production to represent the reference case
- "slow gas": reduced gas production rate with $\lambda_{gk}=10^{-11}$ 1/s and $V_{gk}=5\cdot10^{5}$ m^{3}
- "fast gas": increased gas production rate with $\lambda_{gk}=10^{-10}$ 1/s and $V_{gk}=5\cdot10^{5}$ m^{3}

-    "less gas": less gas volume produced with $V_{gk}=2.5 \cdot 10^5$ m^3 and $\lambda_{gk}=10^{-11}$ 1/s.

The simulation results show that once the mine structures have filled with NaCl-solute, gas production enhances the flow of solute (now transporting radionuclides) back out into the atmosphere (Fig.6). A slow gas production rate prolongs the solute out flux and, thereby, the release of radionuclides. Therefore, depending on sorption conditions, especially late time releases are boosted by slow rates. After 10,000 years, the release rate of U^{238} is increased by a factor of 2.4 and that of I^{129} by a factor of 1.2 when compared to the reference case without gas production (Fig.7).

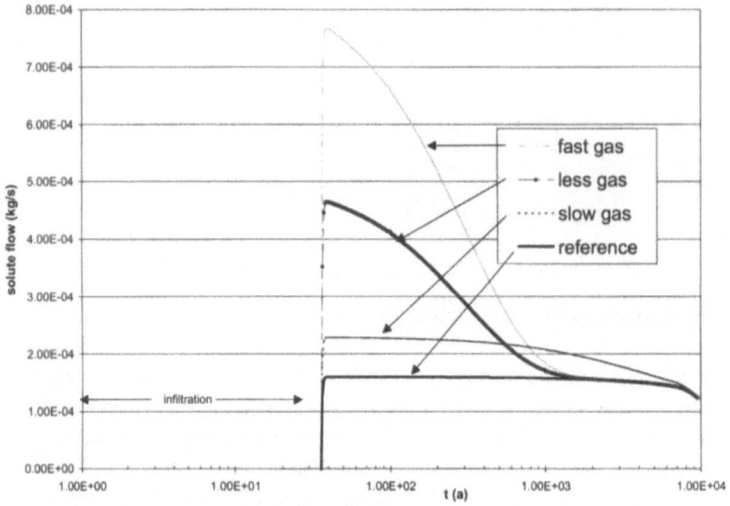

Fig. 6. Influence of gas production on solute flow through the top of the shaft

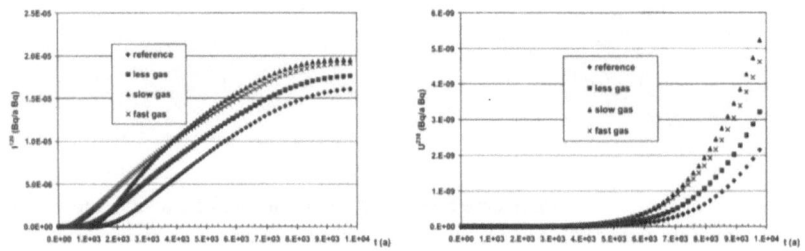

Fig. 7. Resulting annual nuclide release rate at the top of the shaft (related to the initial inventory) for iodine and uranium

3) A "realistic" scenario, i.e. with both gas generation and fluid-salt interaction:

"with gas and half carn": interaction is limited to half of the total amount of carnallitite mass and fast gas production rate with full volume of gas produced.

Fig.8 illustrates the results as exemplified by two nuclides. The non-sorbing iodine is found to be released at a rate reduced by a factor of 0.9 (after 10,000 years) as compared to the reference case, whereas the release rate of U^{238} nuclides is increased by a factor of 1.2. At sorbing conditions (late time release) the reduction by carnallitite interaction cannot be superimposed due to gas induced release.

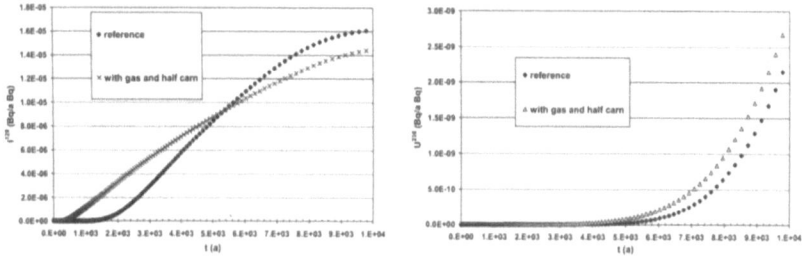

Fig. 8. Comparison of results from realistic scenarios (high gas production rate, small carnallitite amount) with reference case

Conclusions / outlook

The behaviour of the repository with respect to the release of radionuclides is influenced by gas generation and by the dissolution/precipitation of salts in contact with aqueous solutions.

The sensitivity of chemical reactions was investigated with a series of simulation scenarios.

It may be concluded that the fluid-salt interaction generally has an effect of retardation on the release of nuclides into the atmosphere. However, the effect of potential dissolution of flow barriers (containing carnallitite) is not investigated here. This would possibly result in an enhancement of outflux.

Once the mine structures have filled with NaCl-solute, the gas production enhances the flow of solute back out into the atmosphere. A slow gas production rate prolongs the solute out flux and, thereby, the release of radionuclides.

The simulation tool KAFKA has proved to provide an adequate reflection of the complexity of physical and chemical processes involved. Further simulations will consider the higher complexity of geometric conditions (e.g., Fig.1) and the treatment of the uncertainty of parameters in a probabilistic manner.

References

Bethke CM (1998) The Geochemist's Workbench®, Version 3.0. Univ. of Illinois, Urbana, 184 p.
KAFKA (2001) User's guide, Colenco internal report
PSE (1985) Projekt Sicherheitsstudien Entsorgung (PSE): Abschlussbericht, Fachband 15: Einzeluntersuchungen zur Radionuklidfreisetzung aus einem Modellsalzstock, Berlin

Environmental control made in fifty three opencast mines of the Basque country: attention to acoustic and water contamination.

Avilés González Cristina[1], Sánchez Conejo Alfonso[2]

[1,2] Mining engineering, metallurgic and material's science Department. University of The Basque Country. Colina de Beurko s/n 48901 Barakaldo, Vizcaya. Spain. +34639539956   iipavgoc@lg.ehu.es

Abstract. In the Basque country fifty three opencast mines exists which partly impacts the environment in a high grade; the Department of Mining engineering, metallurgic and material's science of the University of the Basque country signed a collaboration agreement with the autonomic government in order to control the exploitation and restoration works, and the possible damages to the environment. In this paper we expose the methodology and conclusions of the study of acoustic and water contamination, proposing also the properly reclamation actions.

Introduction

In the Basque country fifty three opencast mines exists, which causes negative environmental impacts that must be controlled and reduced in order to obtain an effective environmental managing system. The quarries of the Basque country exploit different raw materials, such as limestone, sandstone and subvolcanic rocks used as aggregates, and also ornamental rocks. The exploitation methods are differents; drilling and blasting method is the usual exploitation method in the quarries of aggregates, causing several negative impacts on the villages in the vicinity.

The project here exposed is the result of an agreement between the autonomic government of the Basque country and the department of mining engineering, metallurgic and material's science of the University of the Basque Country. The main objectives of the project are:

1. Measure and control of vibrations and air blast.
2. Measuring, control and minimization of atmospheric contamination.
3. Measuring, control and minimization of acoustic contamination.
4. Analysis and control of the impact caused over the surface water and groundwater.

5. Control of the exploitation and restoration works.

In this paper we have focused our attention in the exposition of the measurement, control and minimization of the acoustic and water contamination.

Measurement, control and minimization of acoustic contamination.

Noise, as a contaminant issue of every mining activity (independently from the air blast), is caused from two main sources: treatment plants for raw materials and movable equipments. To identify the main sound sources, all the processes associated to the mining cycle are analyzed: hole perforation, blasting, loading, transport, treatment and raw materials storing, which may cause high acoustic levels.

To typify and quantify the sound levels, digital sound devices are used: portable sound devices measuring the pressure sound, make the balance by the type A curve according to the norm UNE 20.493/92 and calculate the continuous equivalent sound level (Leq); they also register the peak instant values originated in a measurement.

Measurement proceedings

Different measurement points has been established to value the potential sound incidence on the villages. The main factor to select these points was the potential existence of inconveniences in the population; therefore the measurements have been made in the vicinity of the buildings nearer to the quarries, or where acoustic impacts may be higher. Also, it is selected to measure in the buildings exterior to minimize the inconveniences over the population and to reduce their awareness.

In reference to the measurement techniques and without specific sound laws in the Basque country the recommendations of the norm ISO 1996 are followed. Therefore the measurements are made in the exterior of the buildings with sound device placed at 1.5 m height and separated a minimum of 3.5 m from elements which can modify the signal. The microphone is placed at 1 m from the operator and orientated directly to the sound source.

In each measurement point all climatic informations (temperature, humidity, wind directions), sound sources, predominant sources, activity level of each source, and registered time are gathered.

We selected a temporary sampling due to the high number of sound sources and to their accidental nature; in this sampling the sound levels present during a certain period of time is measured (Leq and the peak-fast value as the norm ISO 1996 recommends), repeating these measurements in a systematic and punctual way over time.

Therefore in a statistical study following systematic and punctual measurements, the evolution of the sound level is analyzed, determining the most common value and his extreme values, approaching these in high proportion to the former, as much as high number of measurements is made.

The measurements of the Leq,60 (Leq during 1 minute) in each register point, is made every 5 minutes during a period of 20 minutes, so obtaining 5 measures; this actions is repeated every 2 hours covering from 7.00 h to 22.00 h. With this procedure and through an strict measure of the Leq,60 and peak-fast values, and the predominant sound sources and their characterization, we can model the uncertain function we are looking for: sound in function of time. An example of a register obtained in a measurement point can be seen in figure number 1.

Sono Mañaria P4-Zallobenta/Oct-2001

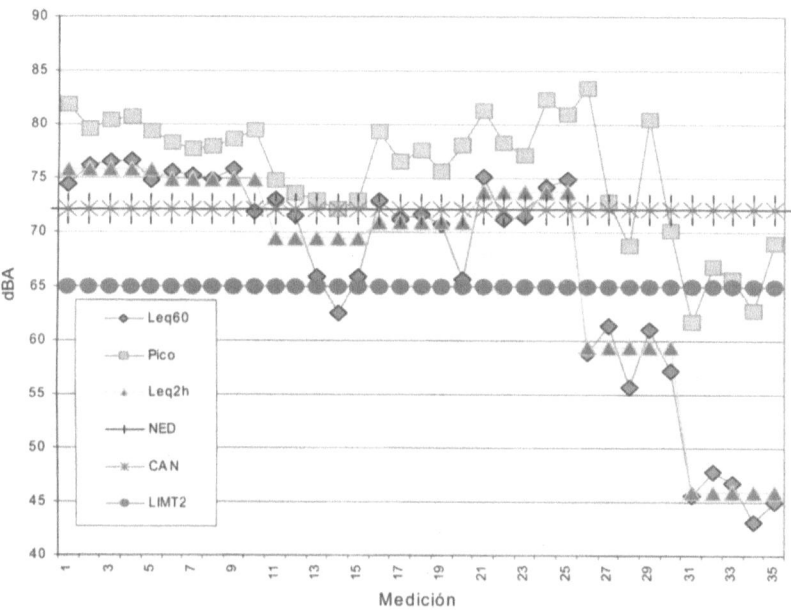

Fig. 1. Register obtained in a measurement point.

However, medium values are needed to allow a quick interpretation of the results, to compare the calculated value with the prevention criterion. Nowadays the most used criterion is NED or diurnal equivalent level. (Diurnal Leq media measured between 7.00 h to 22.00 h.)

Along the day it has been confirmed that the quarries inactivity periods are between 13.00 h. to 14.00 h. and between 19.00 h. to 8.00 h.; during this periods the activity of noise sources not caused by the quarries are constant, so it can be obtained the value of medium Leq without the quarries activity and using the theory of the addition of sound sources we can estimate the noise originated by the quarries.

Prevention criterion of damages

As a prevention criterion of damages and due to the fact that measures are made in the exterior of the buildings, we adopt the values proposed by the Public works and Urban Development Ministry, in 1993. (table 1)

It has been considered that the receivers are type II and NED values are applicated, so there's no activity in the quarries between 22.00 to 7.00 h.

Table 1. NED: Diurnal equivalent level (Diurnal Leq media measured between 7.00 h. and 22.00 h.). NEN: Nocturnal equivalent level (Nocturnal Leq media measured between 22.00 h. and 7.00 h.)

RECEIVER TYPE		ALLOWED INMISION LEVELS in dB A	
		NED	NEN
I	Sanitary, academic and cultural areas; natural reserves.	60	50
II	Housing, hotels and sports areas.	65	55
III	Offices, public services and commercial centers.	70	60
IV	Industries and traveling stations.	75	75

Sound level estimation.

Previously to obtain NED, according to the theory of the composition of sound levels, as is recommended by the ISO 1996/1 norm and from the measurements of Leq60, we must estimate Leq2h by the following equation:

$$Leq_{2h} = 10\log\left[\frac{1}{T}\sum_{i=1}^{N} t_i 10^{\frac{Leq60i}{10}}\right]$$

(1)

The quantity of NED is obtained from the Leq2h values, applying the theory of the composition of sound levels, by the equation number 2:

$$NED = 10\log\left[\frac{1}{7}\sum_{i=1}^{N} 10^{\frac{Leq2hi}{10}}\right]$$

(2)

Expressing NED for no uniforms lapses of times, we obtain the equation number 3 which defines NED in a more general way.

$$NED = 10\log\left[\frac{1}{15}\sum_{i=1}^{N} t_i 10^{\frac{Leqti}{10}}\right]$$

(3)

In this equation ti is the hours interval where is valid the value of Leqti (Usually 2 hours) and Leqti is the valid value for an interval of ti in hours. Usually ti=2hours obtained from 5 measures of Leq60 each 5 minutes in the programmed ti interval.

Equation number 2 is deduced from the equation number 3, supposing seven periods of 2 hours and underestimating the latest hour from 21.00 h. to 22.00 h. where we have confirmed a low activity level of all the sound sources. It has been confirmed that NED value and also CAN value from equation number 2, is 0.5 dB (A) higher than NED value obtained from the equation number 3 and, therefore, more restrictive.

A first estimation of the sound level originated by the quarries (which has been named CAN), is obtained applying the theory of the addition of sound sources and using the values of Leq corresponding to inactivity periods, resulting in the following equation:

$$CAN = 10\log\left[10^{\frac{NED}{10}} - 10^{\frac{Leq}{10}}\right]$$

(4)

It must be said that the great variation of sources, as well as the diversity of generated noise types (continuous, fluctuating, transitories and impacts) and their uncertain character, make it very difficult to characterize the emission sound levels.

Proposed actions to minimize the acoustic impacts

From the beginning of the project it has been confirmed that in certain moments of the acoustic sampling of some quarries, the levels established in the prevention criterion of damages has been repeatedly exceeded. In these cases it has been designed recommendations to minimize the acoustic impacts.

The usual methods to minimize the acoustic impact are: reduce the causes, isolate the sources, and absorb or attenuate the noise between source and receiver.

Reduce the causes means that all the equipment of the quarries, fixed and movable must be adequate to the new standards to reduce noise; in Spain this standard is regulated by The Royal Order 1215/1997 and The Royal Order 1389/1997 but in most of the quarries the equipment, due to be previous to this orders, are not completely adequate to them. It must be remembered that the mining equipment causes high costs, and the quarries are industrial activities of limited economic resources, so the purchase cannot be done immediately.

Therefore, the used methods to reduce noise are isolating the sources and absorb the noise. In this way the use of acoustic barriers made with the materials of the quarries (waste dump and stock piles) is, in some cases, an economic and efficient solution. Another proposed action includes the covering of the treatment installations with isolating materials. The use of films of rubber to reduce the noise caused by the impact of the materials over metallic surfaces is also an appropriate solution.

Also it has been confirmed very interesting and convenient, to propose new methods of discharging materials in order to reduce the discharge height. Of course, is always necessary to carry out the adequate maintenance in the mobile and fixed equipment to avoid vibrations, looseness and frictions. Also with a correct maintenance, the tonal noise can be reduced resulting in a less annoying noise.

Analysis and control of the impact on surface and underground water

All the processes involved in these quarries potentially affects the water management; the landscape of the Basque country is very rough with high mountains and deep valleys and the quarries are located in areas of high inclination, which affects the natural process of sedimentation and erosion. Furthermore, long time periods occur between the exploitation of an area and the re-vegetation operations, thus erosion is likely to happen and surface runoff is affected.

The raw materials of the quarries are limestones and sandstones with a high permeability. The quarries of limestone are frequently affected by karst processes, with exokarstic forms. One of the limestone quarries also has the peculiarity of being exploited and located in an area which suffered in the past underground mining activity, so there exists a series of abandoned galleries through which underground water seeps and circulates, creating a complex underground circulation.

Due to the fact that the quarries are located near the villages and exploits permeable levels, there are a lot of wells for drinking water in the proximity, so special care was needed to maintain the quality of these waters. One of the opencast mines has tailing dams and water storages which also needs a specific water quality monitoring plan.

In order to carry out the follow-up of the water quality and to control the possible influence of flows from the exploitations, two sampling points in each quarry have been defined, far enough from each other to contrast the obtained results. The first point is placed in an area of the stream not affected by the works, and the second point is located in a place where the water comes from the area affected by the quarry.

It must be said that the treatment of the raw materials does not need the addition of any chemical product, so no chemical contamination was expected. The parameters investigated are the level of suspended solids and hydrocarbons, both parameters usually affected by the activity of the quarries. Suspended solids come from the possible flow from the water runoff and from the eventual flow of fine materials from the sediment dams; the hydrocarbons stem from the possible incorrect maintenance of the mine equipments.

An effective system has been developed for sampling, which warrants that the final results from the analysis have optimum liability and representative value.

As prevention criteria the *Spanish Law for Water* in its annex to the Hydraulic Public Dominion as it appears in table 2 was established.

Table 2. Limit values of suspended solids and hydrocarbons in water, according to the Spanish Law for Water.

LIMIT VALUES (mg/l)			
Parameter	Table 1	Table 2	Table 3
Suspended solids	300	150	80
Hydrocarbons	40	25	20

The periodicity of this water control is every two months. Only in few occasions and related with extreme meteorological conditions the limit values have been exceeded.

In order to complete the evaluation of a potential water contamination, we begin this year to control also the pH and have introduced biotic controls from which we expect to obtain additional results to evaluate and contrast the previous analysis.

Finally, for a better environmental water management of the quarries, a global water management plan, which includes all the mining companies, is needed to work with the water management in a congruent and responsible way.

IMWA – Membership Application

Name, Surname, Title ....

Date, Country of Birth ....

Citizenship ....

Address (Business)

Street ....

City and State ....

Telephone No. .... Fax No. ....

E-Mail ....

Address (Residence)

Street ....

City and State ....

Telephone No. ....

Membership Class (please check one)

❑ Individual ❑ Student ❑ Corporate

Date of Application .... Signature ....

Annual Membership fees

Individual:US $40.00   (US $42.00 with credit card)
Student: US $20.00   (US $22.00 with credit card)
Corporate: ask for corporate membership fees

Payment enclosed for calendar year 20 ....

Please send application and checks payable to IMWA to

Adrian Brown – Adrian Brown Consultants, Inc.
1875 Lawrence Street, Suite 500
Denver, CO 80202, USA; E-mail: treasurer@imwa.info
Fax: +1/303/698-9241

Credit Card Payment (Visa, MasterCard only)

amount due ....

valid thru   /   cardholder's name ....

Investigations into seepage water acidification of a tailing dam of the tin ore processing

Werner Klemm, Yvonne Krellmann

Freiberg University of Mining and Technology, Institute of Mineralogy, Brennhausgasse 14, D-09596 Freiberg

Abstract. The changes in the chemical composition of water and solid phases in the zone of seepage water and ground water of a tailing dam of a tin ore processing plant are described. To clarify the causes of a steep decrease of the pH value and in connection with this a partly considerable increase in concentration for several elements, geochemical and mineralogical investigations of waters and tailings were carried out in the dam area. Mineralogical investigations of the tailings from the slope area clearly showed symptoms of dissolution and corrosion. The causes of the current state and several possibilities for the regeneration of an almost neutral pH in seepage water, connect with a reduction of the element discharge, will be shown and explained.

Task

Tailing dams of the ore processing usually contain elevated contents of sulphide minerals. Their oxidation often takes place in a relatively short time and produces compounds, some of them with a very good water solubility. Depending on the ore paragenesis present, these tailings dams have an almost unlimited reservoir of several heavy metals and arsenic. Therefore, tailings dams' seepage waters are polluted by these elements over long time periods.

Long-term monitoring of two tin ore processing tailing dams of comparable content, which are located in Ehrenfriedersdorf, unexpectedly showed differing developments. The younger and larger dam 2 (1969—1990), which is located on a gentle slope, emanated water of an almost unchanged quality over a period of ten years. Contrary to this, water from the older dam 1 (1942—1969), which is situated in a valley cut, without obvious reasons and in a very short time showed a steep decrease of the pH value from 6.8 to 4 together with an increase in several elements' concentrations (Fe, Al, Mn, Zn, As). This phenomenon could not be observed at the main and the eastern infiltration ditch of the tailing dam 2, which otherwise contains the same material.

The determination of the reason(s) for the change in quality of the seepage water of Tailing Dam 1 was of fundamental importance in view of the anticipated "remedy" for water cleaning and its necessary time span as well as for the possible influence of the ageing process of tailing dams. Possible ageing effects as causes for the acidification and the increased discharge of As and heavy metals would then have to be expected in future for a larger number of younger tailing dams.

Mineralogical and geochemical investigations of the tailing material

The tailing material mainly consists of quartz and micas (ca. 96%). Feldspars, fluorite, tourmaline, and chlorite, as well as remains of sulphides (arsenopyrite, pyrite, löllingite) are present in smaller amounts (Fig.1).

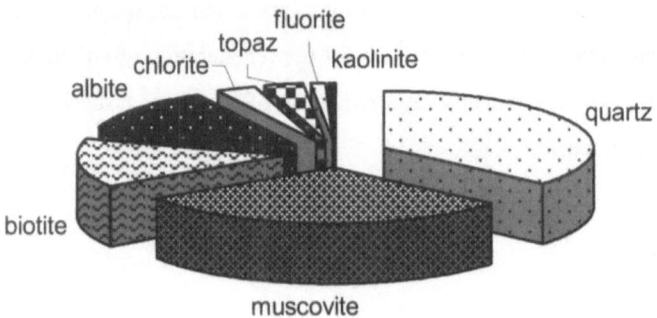

Fig. 1. Mineralogical phase composition of tailing of Tailing Dam 1.

The sand and silt fraction is mainly composed of the stable minerals quartz and mica, whereas the prevailing components of the clay fraction are weathering products like clay minerals (especially kaolinite) and iron hydroxides.

The grain size distribution of the tailing material can be found in the following Fig. 2.

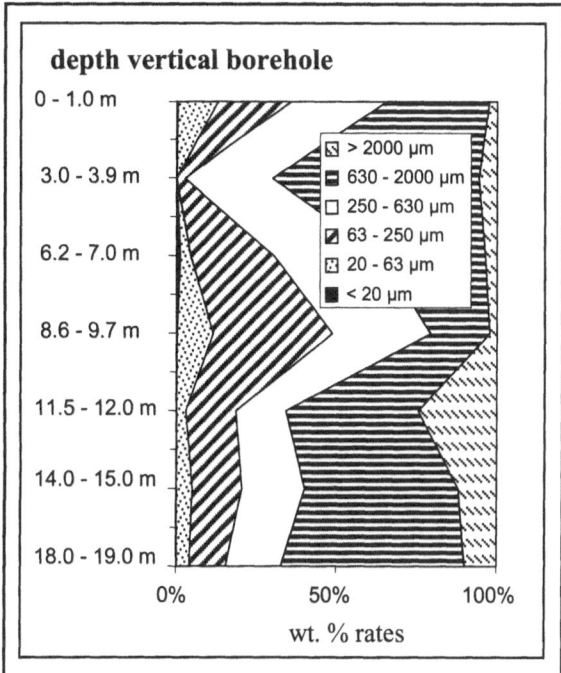

Fig. 2. Grain size profile of vertical borehole (from Krellmann 2001).

The weathering processes are mainly controlled by the oxidation of the sulphides and Fe^{2+} (which also produces H^+ ions), as well as the hydrolysis of the silicates in acidic conditions. The weathering of the feldspars is already in an advanced state. Remains of the feldspars (mainly albite) are left as light coloured crusts on most of the dark micas. Intense oxidation of Fe^{2+} to Fe^{3+} and its subsequent precipitation by hydrolysis preferably occurs in the outer zone of the dam and on the groundwater level, well marked by the strong red-brown colouring of these zones.

The material discharge by the seepage water is decisively influenced by the bonding capacity of the elements, but also by the pH value and the redox potential. Investigations of a depth profile of gauge well $4A_{new}$ were carried out by sequential extraction technique according to Salomon and Förster (1984). Because of the heterogeneous material distribution the measurement results showed large variations for the elements under consideration, comparable to the overall composition. The oxidation of the dominating sulphide arsenopyrite has advanced considerably. In a depth of 4—17 m iron and arsenic are present in sulphide bond only in a very small amount (< 1%). Of prime importance is the water soluble portion, which is presented in Table 1 together with the *aqua regia* soluble content (DIN 38414).

Table 1. H_2O and *aqua regia* soluble contents of As, Al, and heavy metals in material from Tailing Dam 1 (from Krellmann 2001).

	Al	Fe	As	Co	Mn	Ni	Zn
H_2O-soluble	8 – 159	3.8 – 196	14 – 80	0.02 – 1.8	1.9 – 17	0.17 – 5.6	0.49 – 36
aqua regia soluble	21,000 – 2,800	29,000 – 63,000	820 – 13,700	5.6 – 14.6	374 – 988	11.7 – 51.6	180 – 1,809

The pH value and several element concentrations were measured at the captation zone for the seepage waters (measuring point 33) between May 1993 and March 2000. Between August 1993 and April 1996 the pH value, which until then had only been slightly varying, dropped from 6.8 to 4, a phenomenon that was accompanied by an increase in concentration of main components and trace elements, like Fe^{2+}, Al, Mn, As, and Zn (Fig. 3). Nickel and cobalt have only been measured from March 1996 onwards, hence an increase in their concentration can only be assumed.

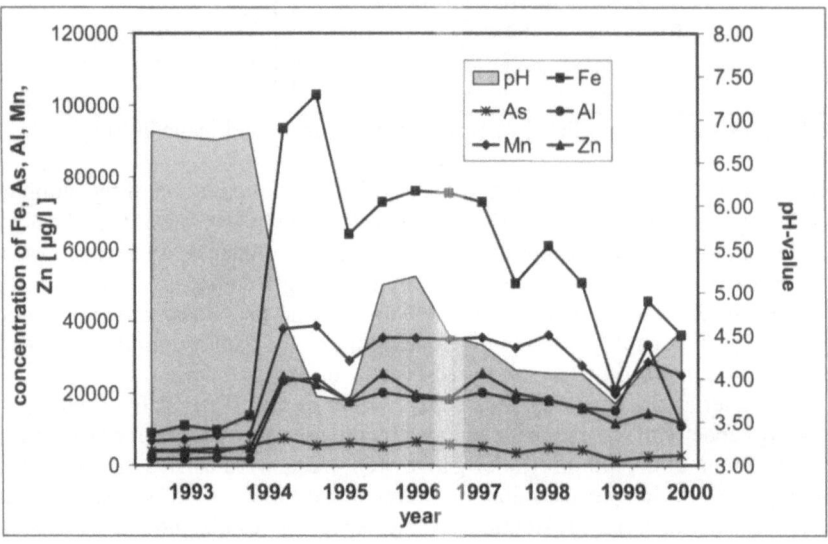

Fig. 3. Time dependent concentration trend of pH, arsenic, aluminum and heavy elements (Mn, Zn, Fe; from Krellmann 2001).

Geochemistry of the groundwater in the tailing dam

Gauge wells in the tailing dam enabled measurements of the groundwater level and samples to be taken (Fig.5). Geochemical analyses showed a clear increase of the overall mineralisation (Ca^{2+}, SO_4^{2-}) as well as of the concentrations of Fe, Al, Mn, As, Ni, Co, Zn occurring alongside a decrease of the pH value from 9 to 6.3 in the direction of flow of the groundwater (Fig.4). This tendency only changed in the area of the dam slope between the gauge well ($4A_{new}$) and the seepage water captation zone (measuring point 33). Together with the decrease of the pH value from ca. 6 to 4 the concentrations of the elements diminished irregularly. Whilst the values of Fe and As showed a sharp decrease, the changes for Al, Mn, Ni, and Zn were only moderate. Reasons for this could be a higher flow velocity of the seepage water because of the coarser grain fraction in this area and the lateral in-flux of seepage waters, which may dissolve the dam water because of its lower pollution. Also the lower compression of the dam slope offers much better path-ways for the influx of air, thus intensifying the oxidation processes.

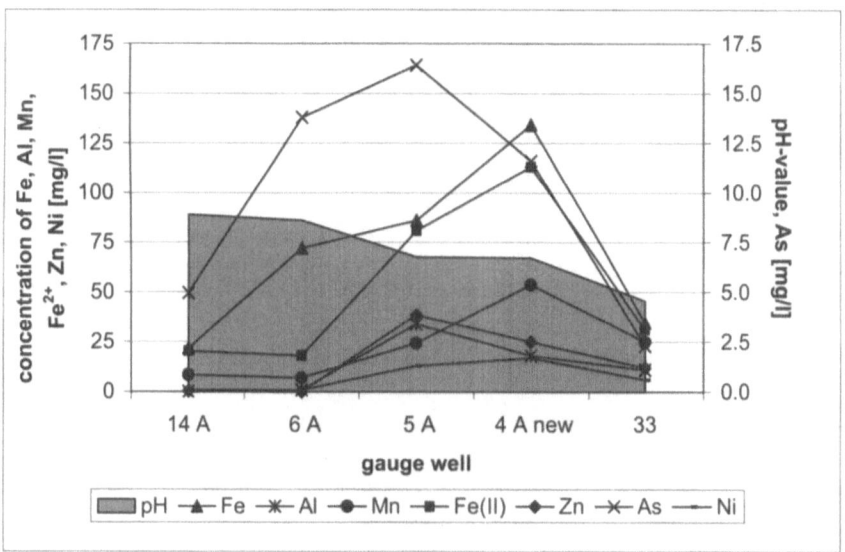

Fig. 4. Variation of pH value and trace elements concentration in the ground water of the Tailing Dam 1.

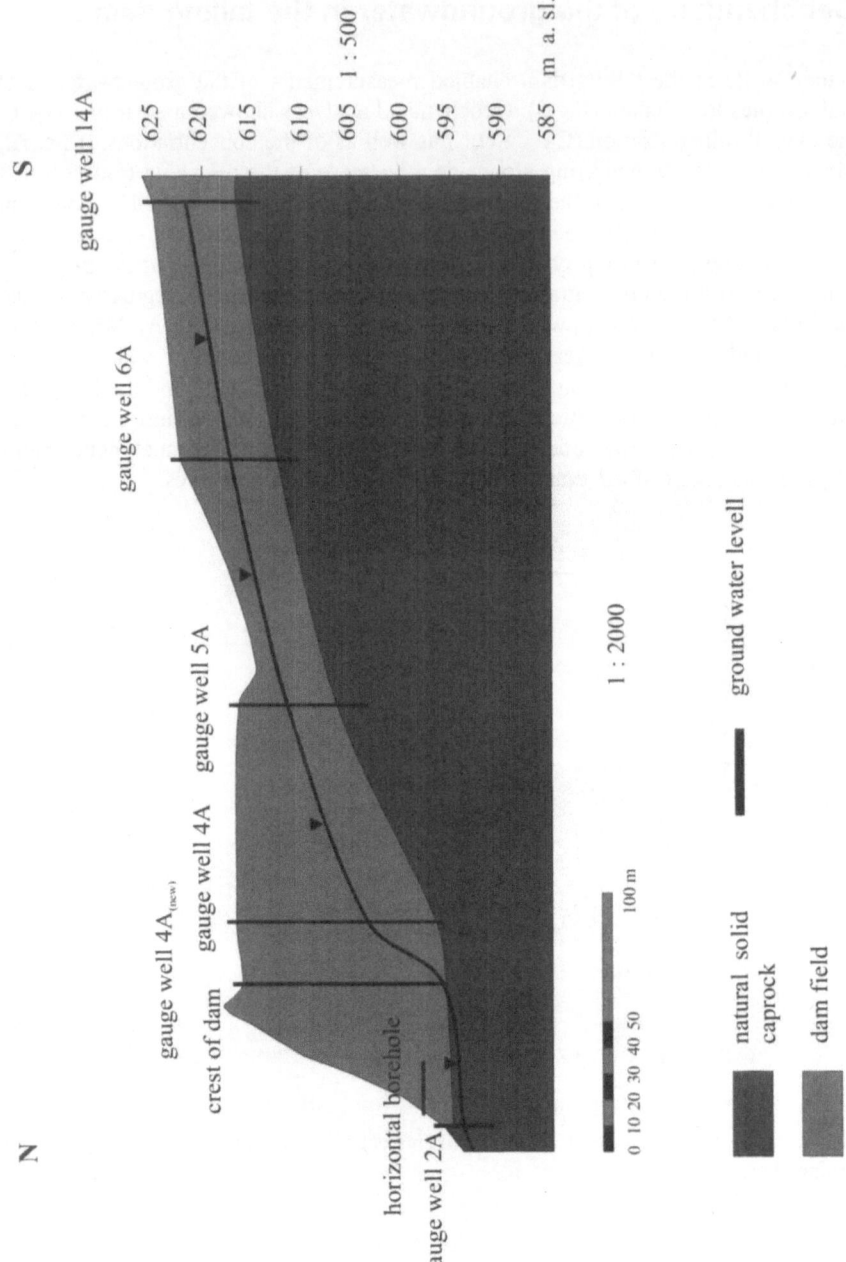

Fig. 5. Profile of Tailing Dam 1 (from Krellmann 2001).

Possible causes for the decrease of the pH value

Several possible causes for the decrease of the pH value and the change of the water quality, which is closely linked to this, can be named:

- Intensification of the sulphide oxidation inside the dam
- Depletion of the silicate buffer of the tailing material
- Geochemical reactions with increased "acid production"
- Changes in the hydrological regime

Of these, the first one, namely the intensification of the sulphide oxidation, appears to be hardly probable as the main reason for the acidification of the seepage water being emanated, because the sulphide concentrations are already very low. Contrary to this, the oxidation of Fe^{2+} to Fe^{3+} and its subsequent hydrolysis to iron hydroxides offers a reaction that is in accordance with the change of the pH value towards stronger acidity as well as with the considerable decrease of the concentration of iron (111—140 to 34—78 mg/l). The content of As is reduced by adsorption at the flocculating iron hydroxides parallel with the diminution of the Fe concentration. The elements that are not affected by this process (Mn, Al, Ni, and Co) show a much slighter decrease in composition, which can be explained as a "dilution" by laterally inflowing waters.

If this is to be seen as the dominating process for the diminution of the pH value, changes in the buffering capacity of the area under consideration must have taken place. However, laboratory experiments with tailing material showed that in batch experiments with seepage water (pH 4) after ca. 20 hours an almost constant pH value of 6 is reached (Fig.6).

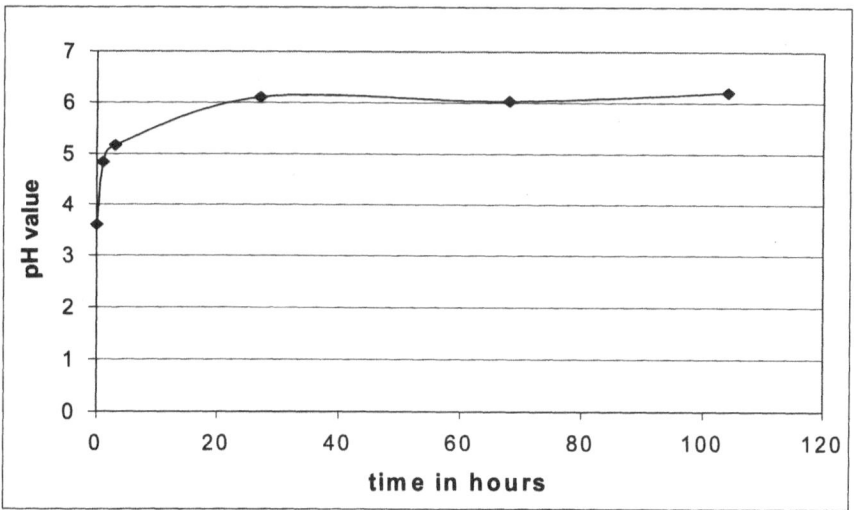

Fig. 6. Buffer effect of the tailings in shacking experiment with seepage water from the Tailing Dam1.

Besides evidence of a sufficient buffering capacity of the tailing material the experiment provided some indication as to the necessary reaction time. If this is undershot, decrease of the pH value of the emanated seepage water is the result.

In the years 1993—1996 a ditch was dug at the dam base during clean-up operations of the seepage water captation zone. As a result of these works the seepage water emanated at a level 1.5 m lower than before. The pH buffering was subsequently limited because the loss of the water tailback in the dam base had reduced the dwell time of the water and the originally existing reaction time accordingly.

Conclusions for steps to be taken

The certain exclusion of an ageing effect as the cause for the decrease of the pH value and the increase of the discharge of As and heavy metals also precludes similar reactions for younger tailing dams. The deterioration of the seepage water quality in the dam under investigation is caused by changes in the hydrological regime, which provokes the decrease of the pH value towards higher acidity and alongside this the increase of the element discharge. From these findings the following steps for the reduction of the water pollution with As and heavy metals can be proposed:

- Reduction of the water throughput through the dam by branching off of accruing groundwater before its entry into the dam body with suitable drainage systems and sealing of the tailing dam surface against meteoric water
- Increase of the dwell time of the seepage water in the dam body in order to improve the pH value buffering by the restoration of the original water outlet level by building in packing material at the dam base

References

DIN 38414-7 Schlamm und Sedimente (Gruppe S) – Aufschluss mit Königswasser zur nachfolgenden Bestimmung des säurelöslichen Anteils von Metallen (S 7)

Krellmann, Y. (2001) Geochemische und mineralogische Untersuchungen an Spülhalden der Zinnerzaufbereitung zur Verringerung des Arsenaustrages durch Sickerwässer. Diplomarbeit, TU Bergakademie Freiberg

Salomons, W., Förstner, U. (1984) Metals in the Hydrocycle. Springer-Verlag, Berlin, Heidelberg: 23 – 50

Index